"**Monumental analysis of the history and politics of oil . . . Engagingly written and a landmark of research.**"
—*Newsweek*

"If y⟶ ⟶ant to know what really makes the world go round, Yergin's colorful hist⟶ ⟶ the petroleum industry is indispensible."
— *⟶e*

"D⟶ ⟶es to become the standard text on the history of oil."
— ⟶slie H. Gelb, front page, *The New York Times Book Review*

"Th⟶ is no doubt about Yergin's basic thesis: Oil is power, big power . . . Yergin ⟶tly has a lot to tell us."
— ⟶eodore C. Sorensen, front page, *The Washington Post Book World*

"A ⟶mpelling history . . . that clarifies the contemporary world situation."
— ⟶s *Angeles Times*

"Ye⟶n has not written the history of oil but the history of the world from the poi⟶ ⟶f view of oil. And he has written it very well, with an eye for the relevant and ⟶en amusing detail. . . . He marveled at his discoveries and, thanks to his grea⟶ ⟶terary gifts, he is able to make us marvel as well. . . . Yergin is finally as mu⟶ ⟶ psychologist as he is a geologist and a historian—one who knows that oil is s⟶ ewhere, deep down, in everybody's emotions under two other names: we⟶ ⟶ and power."
— ⟶bert Mabro, front page, *Chicago Tribune Book World*

"Im⟶ ⟶ssive mastery . . . Daniel Yergin is as well equipped as anyone to build the ⟶ ⟶ge between oil and world diplomacy. . . . He attempts nothing less than a rewi⟶ ng of world history, to bring oil out of the garage into the cabinet-rooms."
— ⟶nthony Sampson, author of *The Seven Sisters*, *The Spectator*

"Mo⟶ than a gripping tale of international politics, *The Prize* chronicles oil's role ⟶ ⟶haping the twentieth century's 'Hydrocarbon Society' of expressways, subu⟶ ⟶s—and pollution—as well as 'Hydrocarbon Man,' who shows little inclinatio⟶ to give up the conveniences of automobiles, suburban homes and other oil-b⟶ ⟶d essentials of life."
—*Atlanta Constitution*

"Dazzling. . . . a masterful study of how oil has dominated and shaped world events in the twentieth century."
—Jeremy Campbell, *London Evening Standard*

DANIEL YERGIN

THE

PRIZE

THE EPIC QUEST FOR OIL, MONEY & POWER

FREE PRESS
New York   London   Toronto   Sydney

*f*P
FREE PRESS
A Division of Simon & Schuster, Inc.
1230 Avenue of the Americas
New York, New York 10020

This Free Press trade paperback edition December 2009

FREE PRESS and colophon are trademarks of Simon & Schuster, Inc.

For information about special discounts for bulk purchases,
please contact Simon & Schuster Special Sales at:
1-800-456-6798 or business@simonandschuster.com.

Designed by Irving Perkins Associates, Inc.

Manufactured in the United States of America

7   9   10   8   6

The Library of Congress has cataloged the Simon & Schuster edition as follows:
Yergin, Daniel.
The prize: the epic quest for oil, money, and power / Daniel Yergin.
p.   cm.
Includes bibliographical references and index.
1. Petroleum industry and trade—Political aspects—History—20th century.
2. Petroleum industry and trade—Military aspects—History—20th century.
3. World War, 1914–1918—Causes.   4. World War, 1939–1945—Causes.
5. World politics—20th century.
I. Title.
HD9560.6.Y47   1990
338.2'782'0904—dc20   90-47575
CIP

ISBN-13: 978-1-4391-1012-6
ISBN-10:   1-4391-1012-3

To Angela, Alexander, and Rebecca

Contents

List of Maps

Prologue

WINSTON CHURCHILL CHANGED his mind almost overnight. Until the summer of 1911, the young Churchill, Home Secretary, was one of the leaders of the "economists," the members of the British Cabinet critical of the increased military spending that was being promoted by some to keep ahead in the Anglo-German naval race. That competition had become the most rancorous element in the growing antagonism between the two nations. But Churchill argued emphatically that war with Germany was not inevitable, that Germany's intentions were not necessarily aggressive. The money would be better spent, he insisted, on domestic social programs than on extra battleships.

Then on July 1, 1911, Kaiser Wilhelm sent a German naval vessel, the *Panther*, steaming into the harbor at Agadir, on the Atlantic coast of Morocco. His aim was to check French influence in Africa and carve out a position for Germany. While the *Panther* was only a gunboat and Agadir was a port city of only secondary importance, the arrival of the ship ignited a severe international crisis. The buildup of the German Army was already causing unease among its European neighbors; now Germany, in its drive for its "place in the sun," seemed to be directly challenging France and Britain's global positions. For several weeks, war fear gripped Europe. By the end of July, however, the tension had eased—as Churchill declared, "the bully is climbing down." But the crisis had transformed Churchill's outlook. Contrary to his earlier assessment of German intentions, he was now convinced that Germany sought hegemony and would exert its military muscle to gain it. War, he now concluded, was virtually inevitable, only a matter of time.

Appointed First Lord of the Admiralty immediately after Agadir, Churchill vowed to do everything he could to prepare Britain militarily for the inescapable day of reckoning. His charge was to ensure that the Royal Navy, the symbol and very embodiment of Britain's imperial power, was ready to meet the German challenge on the high seas. One of the most important and contentious questions he faced was seemingly technical in nature, but would in fact have vast implications for the twentieth century. The issue was whether to convert the British Navy to oil for its power source, in place of coal, which was the traditional fuel. Many thought that such a conversion was pure folly, for it meant that the Navy could no longer rely on safe, secure Welsh coal, but rather would have to depend on distant and insecure oil supplies from Persia, as Iran was then known. "To commit the Navy irrevocably to oil was indeed 'to take arms against a sea of troubles,' " said Churchill. But the strategic benefits—greater speed and more efficient use of manpower—were so obvious to him that he did not dally. He decided that Britain would have to base its "naval supremacy upon oil" and, thereupon, committed himself, with all his driving energy and enthusiasm, to achieving that objective.

There was no choice—in Churchill's words, "Mastery itself was the prize of the venture."[1]

With that, Churchill, on the eve of World War I, had captured a fundamental truth, and one applicable not only to the conflagration that followed, but to the many decades ahead. For oil has meant mastery through the years since. And that quest for mastery is what this book is about.

At the beginning of the 1990s—almost eighty years after Churchill made the commitment to petroleum, after two World Wars and a long Cold War, and in what was supposed to be the beginning of a new, more peaceful era—oil once again became the focus of global conflict. On August 2, 1990, yet another of the century's dictators, Saddam Hussein of Iraq, invaded the neighboring country of Kuwait. His goal was not only conquest of a sovereign state, but also the capture of its riches. The prize was enormous. If successful, Iraq would have become the world's leading oil power, and it would have dominated both the Arab world and the Persian Gulf, where the bulk of the planet's oil reserves is concentrated. Its new strength and wealth and control of oil would have forced the rest of the world to pay court to the ambitions of Saddam Hussein. The result would have been a dramatic shift in the international balance of power. In short, mastery itself was once more the prize.

Over the previous several years, it had become almost fashionable to say that oil was no longer "important." Indeed, in the spring of 1990, just a few months before the Iraqi invasion, the senior officers of America's Central Command, which would be the linchpin of the U.S. mobilization, found themselves lectured to the effect that oil had lost its strategic significance. But the invasion of Kuwait stripped away the illusion. Oil was still central to security, prosperity, and the very nature of civilization. This remains true in the twenty-first century.

Though the modern history of oil begins in the latter half of the nineteenth century, it was the twentieth century that was completely transformed by the advent of petroleum. The role of oil—and anxiety about its supply—is a primary consideration of the era of globalization that characterizes the first decades of the twenty-first century.

Three great themes underlie the story of oil. The first is the rise and development of capitalism and modern business. Oil is the world's biggest and most pervasive business, the greatest of the great industries that arose in the last decades of the nineteenth century. Standard Oil, which thoroughly dominated the American petroleum industry by the end of that century, was among the world's very first and largest multinational enterprises. The expansion of the business thereafter—encompassing everything from wildcat drillers, smooth-talking promoters, and domineering entrepreneurs to highly trained scientists and engineers, great corporate bureaucracies, and state-owned companies—embodies the evolution of business, of corporate strategy, of technological change and market development, and indeed of both national and international economies. Throughout the history of oil, deals have been done and momentous decisions have been made—among men, companies, and nations—sometimes with great calculation and sometimes almost by accident. No other business so starkly and extremely defines the meaning of risk and reward—and the profound impact of chance and fate.

As we look forward, it is clear that mastery will certainly come as much from a computer chip as from a barrel of oil. Yet the petroleum industry continues to have enormous impact. Of the top ten companies in the Fortune 500 global ranking in 2008, six are oil companies. Until some alternative source of energy is found in sufficient scale, oil will still have far-reaching effects on the global economy; major price movements can fuel economic growth or, contrarily, drive inflation and help kick-start recessions. Today, oil is the only commodity whose doings and controversies are to be found regularly not only on the business page but also on the front page. And, as in the past, it is a massive generator of wealth—for individuals, companies, and entire nations. In the words of one tycoon, "Oil *is* almost like money."[2]

The second theme is that of oil as a commodity intimately intertwined with national strategies and global politics and power. The battlefields of World War I established the importance of petroleum as an element of national power when the internal combustion machine overtook the horse and the coal-powered locomotive. Petroleum was central to the course and outcome of World War II in both the Far East and Europe. The Japanese attacked Pearl Harbor to protect their flank as they grabbed for the petroleum resources of the East Indies. Among Hitler's most important strategic objectives in the invasion of the Soviet Union was the capture of the oil fields in the Caucasus. But America's predominance in oil proved decisive, and by the end of the war German and Japanese fuel tanks were empty. In the Cold War years, the battle for control of oil between international companies and developing countries was a major part of the great drama of decolonization and emergent nationalism. The Suez Crisis of 1956, which truly marked the end of the road for the old European imperial powers, was as much about oil as about anything else. "Oil power" loomed very large in the 1970s, catapulting states heretofore peripheral to international politics into positions of great wealth and influence, and creating a deep crisis of confidence in the industrial nations that had based their economic growth upon oil. Oil was at the heart of the first post–Cold War crisis—Iraq's 1990 invasion of Kuwait. And oil figured much in the reconfiguration of international relations that came with the dramatic petroleum price increase, 2004–2008, the return of resource politics, and the new importance of China and India in the world market.

Yet oil has also proved that it can be fool's gold. The Shah of Iran was granted his most fervent wish, oil wealth, and it destroyed him. Oil built up Mexico's economy, only to undermine it. The Soviet Union—the world's second-largest exporter—squandered its enormous oil earnings in the 1970s and 1980s in a military buildup and a series of useless and, in some cases, disastrous international adventures. And the United States, once the world's largest producer and still its largest consumer, must import between 55 and 60 percent of its oil supply, weakening its overall strategic position and adding greatly to an already burdensome trade deficit—a precarious position for a great power.

With the end of the Cold War, a new world order took shape. Economic competition, regional struggles, and ethnic religious rivalries replaced traditional ideology as the focus of international—and national—conflict, aided and abetted by the proliferation of modern weaponry. A new kind of ideology— religious extremism and jihadism—came to the fore. Yet oil remained the strategic commodity, critical to national strategies and international politics.

A third theme in the history of oil illuminates how ours has become a "Hydrocarbon Society" and we, in the language of anthropologists, "Hydrocarbon Man." In its first decades, the oil business provided an industrializing world with a product called by the made-up name of "kerosene" and known as the "new light," which pushed back the night and extended the working day. At the end of the nineteenth century, John D. Rockefeller had become the richest man in the United States, mostly from the sale of kerosene. Gasoline was then only an almost useless by-product, which sometimes managed to be sold for as much as two cents a gallon, and, when it could not be sold at all, was run out into rivers at night. But just as the invention of the incandescent light bulb seemed to signal the obsolescence of the oil industry, a new era opened with the development of the internal combustion engine powered by gasoline. The oil industry had a new market, and a new civilization was born.

In the twentieth century, oil, supplemented by natural gas, toppled King Coal from his throne as the power source for the industrial world. Oil also became the basis of the great postwar suburbanization movement that transformed both the contemporary landscape and our modern way of life. In the twenty-first century, we are so dependent on oil, and oil is so embedded in our daily doings, that we hardly stop to comprehend its pervasive significance. It is oil that makes possible where we live, how we live, how we commute to work, how we travel— even where we conduct our courtships. It is the lifeblood of suburban communities. Oil (and natural gas) are the essential components in the fertilizer on which world agriculture depends; oil makes it possible to transport food to the totally non-self-sufficient megacities of the world. Oil also provides the plastics and chemicals that are the bricks and mortar of contemporary civilization, a civilization that would collapse if the world's oil wells suddenly went dry.

For most of the twentieth century, growing reliance on petroleum was almost universally celebrated as a good, a symbol of human progress. But no longer in the twenty-first century. With the rise of the environmental movement, the basic tenets of industrial society are being challenged; and the oil industry in all its dimensions is at the top of the list to be scrutinized, criticized, and opposed. Efforts are mounting around the world to curtail the combustion of all fossil fuels—oil, coal, and natural gas—because of the resultant smog and air pollution, acid rain, and ozone depletion, and because of the specter of climate change. The last has now become a central focus of national policies and international negotiation. Oil, so central a feature of the world as we know it, is now accused of fueling environmental degradation; and the oil industry, proud of its technological prowess and its contribution to shaping the modern world, finds itself on the defensive, charged with being a threat to present and future generations. This puts a new imperative on technological innovations to mitigate the environmental challenges.

Yet Hydrocarbon Man shows little inclination to give up his cars, his suburban home, and what he takes to be not only the conveniences but the essentials of his way of life. The peoples of the developing world give no indication that they want to deny themselves the gains of an oil-powered economy. Any notion of scaling back the world's consumption of oil will be influenced by the population growth ahead—with more and more of the world's people demanding the "right" to the benefits that come from consumption. Total world oil consumption grew almost 30 percent between 1990 and 2008—from 67 million to 86 million barrels per day. Oil demand in India more than doubled and in China, more than tripled. Thus, the stage has been set for a great balancing between, on the one hand, environmental protection and reduction of carbon and, on the other, economic growth, the benefits of Hydrocarbon Society, and energy security. Today, this is evident in the restarting of the race between the internal combustion engine and the electric car, a competition that was supposedly decided at the beginning of the twentieth century.

These, then, are the three themes that animate the story that unfolds in these pages. The canvas is global. The story is a chronicle of epic events that have touched all our lives. It concerns itself both with the powerful, impersonal forces of economics and technology and with the strategies and cunning of businessmen and politicians. Populating its pages are the tycoons and entrepreneurs—Rockefeller, of course, but also Henri Deterding, Calouste Gulbenkian, J. Paul Getty, Armand Hammer, T. Boone Pickens, and many others. Yet no less important to the story are the likes of Churchill, Adolf Hitler, Joseph Stalin, Ibn Saud, Mohammed Mossadegh, Dwight Eisenhower, Anthony Eden, Henry Kissinger, George H. W. Bush and his son George W. Bush, and Saddam Hussein.

Yet for all its conflict and complexity, there has often been a "oneness" to the story of oil, a contemporary feel even to events that happened long ago and, simultaneously, profound echoes of the past in recent and current events. At one and the same time, this is a story of individual people, of powerful economic forces, of technological change, of political struggles, of international conflict and, indeed, of epic change. It is the author's hope that this exploration of the economic, social, political, and strategic consequences of our world's reliance on oil will illuminate the past, enable us better to understand the present, and help to anticipate the future.

PART I

THE
FOUNDERS

CHAPTER I

Oil on the Brain: The Beginning

THERE WAS THE MATTER of the missing $526.08.

A professor's salary in the 1850s was hardly generous, and in the quest for extra income, Benjamin Silliman, Jr., the son of a great American chemist and himself a distinguished professor of chemistry at Yale University, had taken on an outside research project for a fee totaling $526.08. He had been retained in 1854 by a group of promoters and businessmen, but, though he had completed the project, the promised fee was not forthcoming. Silliman, his ire rising, wanted to know where the money was. His anger was aimed at the leaders of the investor group, in particular, at George Bissell, a New York lawyer, and James Townsend, president of a bank in New Haven. Townsend, for his part, had sought to keep a low profile, as he feared it would look most inappropriate to his depositors if they learned he was involved in so speculative a venture.

For what Bissell, Townsend, and the other members of the group had in mind was nothing less than hubris, a grandiose vision for the future of a substance that was known as "rock oil"—so called to distinguish it from vegetable oils and animal fats. Rock oil, they knew, bubbled up in springs or seeped into salt wells in the area around Oil Creek, in the isolated wooded hills of northwestern Pennsylvania. There, in the back of beyond, a few barrels of this dark, smelly substance were gathered by primitive means—either by skimming it off the surface of springs and creeks or by wringing out rags or blankets that had been soaked in the oily waters. The bulk of this tiny supply was used to make medicine.

The group thought that the rock oil could be exploited in far larger quantities and processed into a fluid that could be burned as an illuminant in lamps. This new illuminant, they were sure, would be highly competitive with the "coal-oils" that were winning markets in the 1850s. In short, they believed that, if they could obtain it in sufficient quantities, they could bring to market the inexpensive, high-quality illuminant that mid-nineteenth-century man so desperately needed. They were convinced that they could light up the towns and farms of North America and Europe. Almost as important, they could use rock oil to lubricate the moving parts of the dawning mechanical age. And, like all entrepreneurs who became persuaded by their own dreams, they were further convinced that by doing all of this they would grow very rich indeed. Many scoffed at them. Yet, persevering, they would succeed in laying the basis for an entirely new era in the history of mankind—the age of oil.

To "Assuage Our Woes"

The venture had its origins in a series of accidental glimpses—and in the determination of one man, George Bissell, who, more than anybody else, was responsible for the creation of the oil industry. With his long, towering face and broad forehead, Bissell conveyed an impression of intellectual force. But he was also shrewd and open to business opportunity, as experience had forced him to be. Self-supporting from the age of twelve, Bissell had worked his way through Dartmouth College by teaching and writing articles. For a time after graduation, he was a professor of Latin and Greek, then went to Washington, D.C., to work as a journalist. He finally ended up in New Orleans, where he became principal of a high school and then superintendent of public schools. In his spare time, he studied to become a lawyer and taught himself several more languages. Altogether, he became fluent in French, Spanish, and Portuguese and could read and write Hebrew, Sanskrit, ancient and modern Greek, Latin and German. Ill health forced him to head back north in 1853, and passing through western Pennsylvania on his way home, he saw something of the primitive oil-gathering industry with its skimmings and oil-soaked rags. Soon after, while visiting his mother in Hanover, New Hampshire, he dropped in on his alma mater, Dartmouth College, where in a professor's office he spied a bottle containing a sample of this same Pennsylvania rock oil. It had been brought there a few weeks earlier by another Dartmouth graduate, a physician practicing as a country doctor in western Pennsylvania.

Bissell knew that amounts of rock oil were being used as patent and folk medicines to relieve everything from headaches, toothaches, and deafness to stomach upsets, worms, rheumatism, and dropsy—and to heal wounds on the backs of horses and mules. It was called "Seneca Oil" after the local Indians and in honor of their chief, Red Jacket, who had supposedly imparted its healing secrets to the white man. One purveyor of Seneca Oil advertised its "wonderful curative powers" in a poem:

> The Healthful balm, from Nature's secret spring,
> The bloom of health, and life, to man will bring;
> As from her depths the magic liquid flows,
> To calm our sufferings, and assuage our woes.

Bissell knew that the viscous black liquid was flammable. Seeing the rock oil sample at Dartmouth, he conceived, in a flash, that it could be used not as a

4

medicine but as an illuminant—and that it might well assuage the woes of his pocketbook. He could put the specter of poverty behind him and become rich from promoting it. That intuition would become his guiding principle and his faith, both of which would be sorely tested during the next six years, as disappointment consistently overwhelmed hope.[1]

The Disappearing Professor

But could the rock oil really be used as an illuminant? Bissell aroused the interest of other investors, and in late 1854 the group engaged Yale's Professor Silliman to analyze the properties of the oil both as an illuminant and lubricant. Perhaps even more important, they wanted Silliman to put his distinguished imprimatur on the project so they could sell stock and raise the capital to carry on. They could not have chosen a better man for their purposes. Heavyset and vigorous, with a "good, jolly face," Silliman carried one of the greatest and most respected names in nineteenth-century science. The son of the founder of American chemistry, he himself was one of the most distinguished scientists of his time, as well as the author of the leading textbooks in physics and chemistry. Yale was the scientific capital of mid-nineteenth-century America, and the Sillimans, father and son, were at the center of it.

But Silliman was less interested in the abstract than in the decidedly practical, which drew him to the world of business. Moreover, while reputation and pure science were grand, Silliman was perennially in need of supplementary income. Academic salaries were low and he had a growing family; so he habitually took on outside consulting jobs, making geological and chemical evaluations for a variety of clients. His taste for the practical would also carry him into direct participation in speculative business ventures, the success of which, he explained, would give him "plenty of sea room . . . for science." A brother-in-law was more skeptical. Benjamin Silliman, Jr., he said, "is on the constant go in behalf of one thing or another, and alas for Science."

When Silliman undertook his analysis of rock oil, he gave his new clients good reason to think they would get the report they wanted. "I can promise you," he declared early in his research, "that the result will meet your expectations of the value of this material." Three months later, nearing the end of his research, he was even more enthusiastic, reporting "unexpected success in the use of the distillate product of *Rock Oil* as an illuminator." The investors waited eagerly for the final report. But then came the big hitch. They owed Silliman the $526.08 (the equivalent of about $5,000 today), and he had insisted that they deposit $100 as a down payment into his account in New York City. Silliman's bill was much higher than they had expected. They had not made the deposit, and the professor was upset and angry. After all, he had not taken on the project merely out of intellectual curiosity. He needed the money, badly, and he wanted it soon. He made it very clear that he would withhold the study until he was paid. Indeed, to drive home his complaint, he secretly handed over the report to a friend for safe-keeping until satisfactory arrangements were made, and took himself off on a tour of the South, where he could not easily be reached.

The investors grew desperate. The final report was absolutely essential if they were to attract additional capital. They scrounged around, trying to find the money, but with no success. Finally, one of Bissell's partners, though complaining that "these are the hardest times I ever heard of," put up the money on his own security. The report, dated April 16, 1855, was released to the investors and hurried to the printers. Though still appalled by Silliman's fee, the investors, in fact, got more than their money's worth. Silliman's study, as one historian put it, was nothing less than "a turning point in the establishment of the petroleum business." Silliman banished any doubts about the potential new uses for rock oil. He reported to his clients that it could be brought to various levels of boiling and thus distilled into several fractions, all composed of carbon and hydrogen. One of these fractions was a very high-quality illuminating oil. "Gentlemen," Silliman wrote to his clients, "it appears to me that there is much ground for encouragement in the belief that your Company have in their possession a raw material from which, by simple and not expensive processes, they may manufacture very valuable products." And, satisfied with the business relationship as it had finally been resolved, he held himself fully available to take on further projects.

Armed with Silliman's report, which proved a most persuasive advertisement for the enterprise, the group had no trouble raising the necessary funds from other investors. Silliman himself took two hundred shares, adding further to the respectability of the enterprise, which became known as the Pennsylvania Rock Oil Company. But it took another year and a half of difficulties before the investors were ready to take the next hazardous step.

They now knew, as a result of Silliman's study, that an acceptable illuminating fluid could be extracted from rock oil. But was there enough rock oil available? Some said that it was only the "drippings" from underground coal seams. Certainly, a business could not be built from skimming oil stains off the surfaces of creeks or from wringing out oil-soaked rags. The critical issue, and what their enterprise was all about, was proving that there was a sufficient and obtainable supply of rock oil to make for a substantial paying proposition.[2]

Price and Innovation

The hopes pinned on the still mysterious properties of oil arose from pure necessity. Burgeoning populations and the spreading economic development of the industrial revolution had increased the demand for artificial illumination beyond the simple wick dipped into some animal grease or vegetable fat, which was the best that most could afford over the ages, if they could afford anything at all. For those who had money, oil from the sperm whale had for hundreds of years set the standard for high-quality illumination; but even as demand was growing, the whale schools of the Atlantic had been decimated, and whaling ships were forced to sail farther and farther afield, around Cape Horn and into the distant reaches of the Pacific. For the whalers, it was the golden age, as prices were rising, but it was not the golden age for their consumers, who did not want to pay $2.50 a gallon—a price that seemed sure to go even higher. Cheaper lighting fluids had been developed. Alas, all of them were inferior. The most popular was camphene, a derivative of turpentine, which produced a good light but had the

6

unfortunate drawback of being highly flammable, compounded by an even more unattractive tendency to explode in people's houses. There was also "town gas," distilled from coal, which was piped into street lamps and into the homes of an increasing number of middle- and upper-class families in urban areas. But "town gas" was expensive, and there was a sharply growing need for a reliable, relatively cheap illuminant. There was that second need as well—lubrication. The advances in mechanical production had led to such machines as power looms and the steam printing press, which created too much friction for such common lubricants as lard.

Entrepreneurial innovation had already begun to respond to these needs in the late 1840s and early 1850s, with the extraction of illuminating and lubricating oils from coal and other hydrocarbons. A lively cast of characters, both in Britain and in North America, carried the search forward, defining the market and developing the refining technology on which the oil industry would later be based. A court-martialed British admiral, Thomas Cochrane—who, it was said, provided the model for Lord Byron's *Don Juan*—became obsessed with the potential of asphalt, sought to promote it, and, along the way, acquired ownership of a huge tar pit in Trinidad. Cochrane collaborated for a time with a Canadian, Dr. Abraham Gesner. As a young man, Gesner had attempted to start a business exporting horses to the West Indies, but, after being shipwrecked twice, gave it up and went off to Guy's Hospital in London to study medicine. Returning to Canada, he changed careers yet again and became provincial geologist for New Brunswick. He developed a process for extracting an oil from asphalt or similar substances and refining it into a quality illuminating oil. He called this oil "kerosene"—from *Keros* and *elaion*, the Greek words, respectively, for "wax" and "oil," altering the *elaion* to *ene*, so that his product would sound more like the familiar camphene. In 1854 he applied for a United States patent for the manufacture of "a new liquid hydrocarbon, which I denominate Kerosene, and which may be used for illuminating or other purposes."

Gesner helped establish a kerosene works in New York City that by 1859 was producing five thousand gallons a day. A similar establishment was at work in Boston. The Scottish chemist James Young had pioneered a parallel refining industry in Britain, based on cannel coal, and one also developed in France, using shale rock. By 1859, an estimated thirty-four companies in the United States were producing $5 million a year worth of kerosene or "coal-oils," as the product was generically known. The growth of this coal-oil business, wrote the editor of a trade journal, was proof of "the impetuous energy with which the American mind takes up any branch of industry that promises to pay well." A small fraction of the kerosene was extracted from Pennsylvania rock oil that was gathered by the traditional methods and that would, from time to time, turn up at the refineries in New York.[3]

Oil was hardly unfamiliar to mankind. In various parts of the Middle East, a semisolid oozy substance called bitumen seeped to the surface through cracks and fissures, and such seepages had been tapped far back into antiquity—in Mesopotamia, back to 3000 B.C. The most famous source was at Hit, on the Euphrates, not far from Babylon (and the site of modern Baghdad). In the first century B.C., the Greek historian Diodor wrote enthusiastically about the ancient

7

bitumen industry: "Whereas many incredible miracles occur in the Babylonian country, there is none such as the great quantity of asphalt found there." Some of these seepages, along with escaping petroleum gases, burned continuously, providing the basis for fire worship in the Middle East.

Bitumen was a traded commodity in the ancient Middle East. It was used as a building mortar. It bound the walls of both Jericho and Babylon. Noah's ark and Moses' basket were probably caulked, in the manner of the time, with bitumen to make them waterproof. It was also used for road making and, in a limited and generally unsatisfactory way, for lighting. And bitumen served as a medicine. The description by the Roman naturalist Pliny in the first century A.D. of its pharmaceutical value was similar to that current in the United States during the 1850s. It checked bleeding, Pliny said, healed wounds, treated cataracts, provided a liniment for gout, cured aching teeth, soothed a chronic cough, relieved shortness of breath, stopped diarrhea, drew together severed muscles, and relieved both rheumatism and fever. It was also "useful for straightening out eyelashes which inconvenience the eyes."

There was yet another use for oil; the product of the seepages, set aflame, found an extensive and sometimes decisive role in warfare. In the *Iliad*, Homer recorded that "the Trojans cast upon the swift ship unwearied fire, and over her forthwith streamed a flame that might not be quenched." When the Persian King Cyrus was preparing to take Babylon, he was warned of the danger of street fighting. He responded by talking of setting fires, and declared, "We also have plenty of pitch and tow, which will quickly spread the flames everywhere, so that those upon the house-tops must either quickly leave their posts or quickly be consumed." From the seventh century onward, the Byzantines had made use of *oleum incendiarum*—Greek fire. It was a mixture of petroleum and lime that, touched with moisture, would catch fire; the recipe was a closely guarded state secret. The Byzantines heaved it on attacking ships, shot it on the tips of arrows, and hurled it in primitive grenades. For centuries, it was considered a more terrible weapon than gunpowder.[4]

So the use of petroleum had a long and varied history in the Middle East. Yet, in a great mystery, knowledge of its application was lost to the West for many centuries, perhaps because the known major sources of bitumen, and the knowledge of its uses, lay beyond the boundaries of the Roman empire, and there was no direct transition of that knowledge to the West. Even so, in many parts of Europe—Bavaria, Sicily, the Po Valley, Alsace, Hannover, and Galicia, to name a few—oil seepages were observed and commented upon from the Middle Ages onward. And refining technology was transmitted to Europe via the Arabs. But, for the most part, petroleum was put to use only as the all-purpose medicinal remedy, fortified by learned disquisitions on its healing properties by monks and early doctors. But, well before George Bissell's entrepreneurial vision and Benjamin Silliman's report, a small oil industry had developed in Eastern Europe—first in Galicia (which was variously part of Poland, Austria, and Russia) and then in Rumania. Peasants dug shafts by hand to obtain crude oil, from which kerosene was refined. A pharmacist from Lvov, with the help of a plumber, invented a cheap lamp suited to burning kerosene. By 1854, kerosene was a staple of commerce in Vienna. By 1859, Galicia had a thriving kerosene oil business, with over 150 villages involved in oil mining, led by such families as Backenroth-Bronicki. Altogether, European crude production in 1859 had been estimated at thirty-six thousand barrels, primarily from Galicia and Rumania. What the Eastern European industry lacked, more than anything else, was the technology for drilling.

In the 1850s, the spread of kerosene in the United States faced two significant barriers: There was as yet no substantial source of supply, and there was no cheap lamp well-suited to burning what kerosene was available. The lamps that did exist tended to become smoky, and the burning kerosene gave off an acrid smell. Then a kerosene sales agent in New York learned that a lamp with a glass chimney was being produced in Vienna to burn Galician kerosene. Based upon the design of the pharmacist and the plumber in Lvov, the lamp overcame the problems of the smoke and the smell. The New York salesman started to import the lamp, which quickly found a market. Though its design was subsequently improved many times over, that Vienna lamp became the basis of the kerosene lamp trade in the United States and was later re-exported around the world.[5]

Thus by the time that Bissell was launching his venture, a cheaper quality illuminating oil—kerosene—had already been introduced into some homes. The techniques required for refining petroleum into kerosene had already been commercialized with coal-oils. And an inexpensive lamp had been developed that could satisfactorily burn kerosene. In essence, what Bissell and his fellow investors in the Pennsylvania Rock Oil Company were trying to do was discover a new source for the raw material that went into an existing, established process. It all came down to price. If they could find rock oil—petroleum—in sufficient abundance, it could be sold cheaply, capturing the illuminating oils market from products that were either far more expensive or far less satisfactory.

Digging for oil would not do it. But perhaps there was an alternative. Salt "boring," or drilling, had been developed more than fifteen hundred years earlier in China, with wells going down as deep as three thousand feet. Around 1830, the Chinese method was imported into Europe and copied. That, in turn, may have stimulated the drilling of salt wells in the United States. George Bissell was still struggling to put his venture together when, on a hot day in New York in 1856, he took refuge from the burning sun under the awning of a druggist's shop on Broadway. There in the window, he caught sight of an advertisement for a rock oil medicine that showed several drilling derricks—of the kind used to bore for salt. The rock oil for the patent medicine was obtained as a byproduct of drilling for salt. With that coincidental glimpse by Bissell—following on his earlier ones in western Pennsylvania and at Dartmouth College—the last piece fell into place in his mind. Could not that technique of drilling be applied to the recovery of oil? If the answer was yes, here at last was the means for achieving his fortune.

The essential insight of Bissell—and then of his fellow investors in the Pennsylvania Rock Oil Company—was to adapt the salt-boring technique directly to oil. Instead of digging for rock oil, they would drill for it. They were not alone; others in the United States and Ontario, Canada, were experimenting with the same idea. But Bissell and his group were ready to move. They had Professor Silliman's report, and because of the report they had the capital. Still, they were not taken very seriously. When the banker James Townsend discussed their idea of drilling, many in New Haven derided it: "Oh Townsend, oil coming out of the ground, pumping oil out of the earth as you pump water? Nonsense! You're crazy." But the investors were intent on going ahead. They were convinced of the need and the opportunity. But to whom would they now entrust this lunatic project?[6]

The "Colonel"

Their candidate was one Edwin L. Drake, who was chosen mainly by coincidence. He certainly brought no outstanding or obvious qualifications to the task. He was a jack-of-all-trades and a sometime railroad conductor, who had been laid up by bad health and was living with his daughter in the old Tontine Hotel in New Haven. By chance, James Townsend, the New Haven banker, lived in the same hotel. It was the sort of hotel where men gathered to exchange news and shoot the breeze, a perfect setting for the thirty-eight-year-old Drake, who was friendly, jovial, and loquacious, and had nothing else to do. So he would pass the evenings entertaining his companions with stories drawn from his varied life. He had a vivid imagination, and his stories tended to be dramatic, exaggerated tales, in all of which Drake himself played a central, heroic role. He and Townsend talked frequently about the rock oil venture. Townsend even persuaded Drake to buy some stock in the company. Townsend then recruited Drake himself to the scheme. He was out of work and thus available, and since he was on leave as a conductor, he had a railroad pass and could travel for free, which was most helpful to the financially pinched speculative venture. He had another attribute that would be of great value: He could be very tenacious.

Dispatching Drake to Pennsylvania, Townsend gave him what turned out to be a valuable send-off. Concerned about the frontier conditions and the need to impress the "backwoodsmen," the banker sent ahead several letters addressed to "Colonel" E. L. Drake. Thus was "Colonel" Drake invented, though a "colonel" he certainly was not. The stratagem worked. For a warm and hospitable welcome was received by "Colonel" E. L. Drake, when, in December of 1857, he arrived, after an exhausting journey through a sea of mud, on the back of the twice-weekly mail wagon, in the tiny, impoverished village of Titusville, population 125, tucked into the hills of northwestern Pennsylvania. Titusville was a lumber town, whose inhabitants were deeply in debt to the local lumber company's store. It was generally expected that the village would die when the surrounding hills had all been logged and that the site would then be reclaimed by the wild.

Drake's first job was simply to perfect the title to the prospective oil land, which was on a farm. This he quickly accomplished. He returned to New Haven, intent on the much more daunting next step, drilling for oil. "I had made up my mind," he later said, that oil "could be obtained in large quantities by Boreing as for Salt Water. I also determined that I should be the one to do it. But I found that no one with whom I conversed upon the subject agreed with me, all maintaining that oil was the drippings of an extensive Coal field or bed."

But Drake was not to be dissuaded or diverted. He was back in Titusville in the spring of 1858 to commence work. The investors had established a new company, the Seneca Oil Company, with Drake as its general agent. He set up operations about two miles down Oil Creek from Titusville, on a farm that contained an oil spring, from which three to six gallons of oil a day were collected by the traditional methods. After several months back in Titusville, he wrote Townsend, "I shall not try to dig by hand any more, as I am satisfied that boring is the cheapest." But he begged the New Haven banker to send additional funds immediately. "Money we must have if we are to make anything. . . . Please let me know at once. Money is very scarce here." After some delay, Townsend managed to send a thousand dollars, and with it Drake tried to hire the "salt borers"—or drillers—that he needed if he were to proceed. But salt drillers had a reputation for extreme partiality to whiskey and frequent drunkenness, and he wanted to be very careful whom he hired. So he would tie compensation to successful completion at the rate of one dollar per foot drilled. The first couple of drillers he engaged simply disappeared or begged off. In truth, though they dared not tell Drake so to his face, they thought he was insane. Drake knew only that he had nothing to show for his first year in Titusville, and the bleak winter was at hand. So he devoted himself to erecting the steam engine that would power the drill bit, while the investors back in New Haven fretted and waited.

Finally, in the spring of 1859, Drake found his driller, a blacksmith named William A. Smith—"Uncle Billy" Smith—who came with his two sons. Smith knew something about what needed to be done, for he made the tools for the salt water drillers, and the little team now proceeded to build the derrick and assemble the necessary equipment. They assumed they would have to go several hundred feet into the earth. The work was slow, and the investors in New Haven were becoming more and more restive at the lack of progress. Still, Drake stuck to his plan. He would not give up. Eventually, Townsend was the only one of the promoters who still believed in the project, and, when the venture ran out of money, he began paying the bills out of his own pocket. In despair, he at last sent Drake a money order as a final remittance and instructed him to pay his bills, close up the operation, and return to New Haven. That was toward the end of August 1859.

Drake had not yet received the letter when, on Saturday afternoon, August 27, 1859, at sixty-nine feet, the drill dropped into a crevice and then slid another six inches. Work was called off for the rest of the weekend. The next day, Sunday, Uncle Billy came out to see the well. He peered down into the pipe. He saw a dark fluid floating on top of the water. He used a tin rain spout to draw up a sample. As he examined the heavy liquid, he was overcome by excitement. On Monday, when Drake arrived, he found Uncle Billy and his boys standing guard over tubs, washbasins, and barrels, all of which were filled with oil. Drake attached a common hand pump and began to do exactly what the scoffers had ridiculed—pump up the liquid. That same day he received the money order from Townsend and the command to close up shop. A week earlier, with the last of the funds in hand, he would have done so. But not anymore. Drake's singlemindedness had paid off. Just in time. He had hit oil. Farmers along Oil Creek rushed into Titusville shouting, "The Yankee has struck oil." The news spread like wildfire and started a mad rush to acquire sites and drill for oil. The population of tiny Titusville multiplied overnight, and land prices shot up instantaneously.

Success with the drill did not, however, guarantee financial success. It meant new problems. What were Drake and Uncle Billy to do with the flow of oil? They got hold of every whiskey barrel they could scrounge in the area, and when all the barrels were filled, they built and filled several wooden vats. Unfortunately, one night the flame from a lantern ignited the petroleum gases, causing the entire storage area to explode and go up in fierce flames. Meanwhile, other wells were drilled in the neighborhood, and more rock oil became available. Supply far outran demand, and the price plummeted. With the advent of drilling, there was no shortage of rock oil. The only shortage now was of whiskey barrels, and they soon cost almost twice as much as the oil inside them.[7]

"The Light of the Age"

It did not take long for Pennsylvania rock oil to find its way to market refined as kerosene. Its virtues were immediately clear. "As an illuminator the oil is without a figure: It is the light of the age," wrote the author of America's very first handbook on oil, less than a year after Drake's discovery. "Those that have not seen it burn, may rest assured its light is no moonshine; but something nearer the clear, strong, brilliant light of day, to which darkness is no party . . . rock oil emits a dainty light; the brightest and yet the cheapest in the world; a light fit for Kings and Royalists and not unsuitable for Republicans and Democrats."

George Bissell, the original promoter, was among those who had wasted no time in getting to Titusville. He spent hundreds of thousands of dollars frantically leasing and buying farms in the vicinity of Oil Creek. "We find here an unparalleled excitement," he wrote to his wife. "The whole population are crazy almost . . . I never saw such excitement. The whole western country are thronging here and fabulous prices are offered for lands in the vicinity where there is a prospect of getting oil." It had taken Bissell six years to get to this point, and the ups and downs of his journey gave him reason to reflect. "I am quite well, but very much worn down. We have had a hard time of it, very. Our prospects are most brilliant that's certain. . . . We ought to make an immense fortune."

Bissell did indeed become very wealthy. And, among his many philanthropies, he donated the money for a gymnasium to Dartmouth, where first he had seen the bottle of rock oil that inspired his vision. He insisted that the gym be equipped with six bowling alleys "in remembrance of disciplinary troubles into which he had fallen as an undergraduate because of his indulgence in this sinful sport." It was said of Bissell in his later years "that his name and fame is a 'household word' among oil men from end to end of the continent." James Townsend, the banker who had taken the greatest financial risk, was denied the credit he thought he deserved. "The whole plan was suggested by me, and my suggestions were carried out," he later wrote bitterly. "The raising of the money and sending it out was done by me. I do not say it egotistically, but only as a matter of truth, that if I had not done what I did in favor of developing Petroleum it would not have been developed at that time." Yet he added, "the suffering and anxiety I experienced I would not repeat for a fortune."

As for Drake, things did not go well at all. He became an oil buyer, then a partner in a Wall Street firm specializing in oil shares. He was improvident, not a good businessman, indeed a gambler of sorts when it came to commerce. By 1866, he had lost all his money, then became a semi-invalid, racked with pain, living in poverty. "If you have any of the milk of human kindness left in your bosom for me or my family, send me some money," he wrote to one friend. "I am in want of it sadly and am sick." Finally, in 1873, the state of Pennsylvania granted him a small lifetime pension for his service, bringing him some measure of relief in his final years from his financial difficulties, if not his physical pain.

Toward the end of his life, Drake sought to stake out his place in history. "I claim that I did invent the driving Pipe and drive it and without that they could not bore on bottom lands when the earth is full of water. And I claim to have bored the first well that ever was bored for Petroleum in America and can show the well." He was emphatic. "If I had not done it, it would have not been done to this day."[8]

The First Boom

Indeed, all the other elements—refining, experience with kerosene, and the right kind of lamp—were in place when Drake proved, through drilling, the final requirement for a new industry, the availability of supply. And with that, man was suddenly given the ability to push back the night. Yet that was only the beginning. For Drake's discovery would, in due course, bequeath mobility and power to the world's population, play a central role in the rise and fall of nations and empires, and become a major element in the transformation of human society. But all that, of course, was still to come.

What followed immediately was like a gold rush. The flats in the narrow valley of Oil Creek were quickly leased, and by November of 1860, fifteen months after Drake's discovery, about seventy-five wells were producing, with many more dry holes scarring the earth. Titusville "is now the rendezvous of strangers eager for speculation," a writer had already observed by 1860. "They barter prices in claims and shares; buy and sell sites, and report the depth, show, or yield of wells, etc. etc. Those who leave today tell others of the well they saw yielding 50 barrels of pure oil a day. . . . The story sends more back tomorrow. . . . Never was a hive of bees in time of swarming more astir, or making a greater buzz."

Down at the bottom of Oil Creek, where it flowed into the Allegheny River, a small town called Cornplanter, named after a Seneca Indian chief, was renamed Oil City and became the major center, along with Titusville, for the area now known as the Oil Regions. Refineries to turn the crude into kerosene were cheap to build, and by 1860, at least fifteen were operating in the Oil Regions, with another five in Pittsburgh. A coal-oil refiner visited the oil fields in 1860 to see the competition for himself. "If this business succeeds," he said, "mine is ruined." He was right; by the end of 1860, the coal-oil refiners either were out of business or had moved quickly to turn themselves into crude-oil refiners.

Yet all the wells thus far were modest producers and had to be pumped. That changed in April 1861, when drillers struck the first flowing well, which gushed at the astonishing rate of three thousand barrels per day. When the oil from that well shot into the air, something ignited the escaping gases, setting off a great explosion and creating a wall of fire that killed nineteen people and blazed on for three days. Though temporarily lost in the thunderous news of the week before—that the South had fired on Fort Sumter, the opening shots of the Civil War—the explosion announced to the world that ample supplies for the new industry would be available.

Production in western Pennsylvania rose rapidly—from about 450,000 barrels in 1860 to 3 million barrels in 1862. The market could not develop quickly enough to match the swelling volume of oil. Prices, which had been $10 a barrel in January 1861, fell to 50 cents by June and, by the end of 1861, were down to 10 cents. Many producers were ruined. But those cheap prices gave Pennsylvania oil a quick and decisive victory in the marketplace, swiftly capturing consumers and driving out coal-oils and other illuminants. Demand soon caught up with available supply, however, and by the end of 1862 prices rose to $4 a barrel and then, by September 1863, to as high as $7.25 a barrel. Despite the wild fluctuation of prices, the stories of instant wealth continued to draw the throngs to the Oil Regions. In less than two years one memorable well generated $15,000 of profit for every dollar invested.[9]

The Civil War hardly disrupted the frantic boom in the Oil Regions; on the contrary, it actually gave a major stimulus to the development of the business. For the war cut off the shipment of turpentine from the South, creating an acute shortage of camphene, the cheap illuminating oil derived from turpentine. Kerosene made from Pennsylvania oil quickly filled the gap, developing markets in the North much more quickly than might otherwise have been the case. The war had an even more significant impact. When the South seceded, the North no longer benefited from the foreign revenues from cotton, one of America's major exports. The rapid growth of oil exports to Europe helped compensate for that loss and provided a significant new source of foreign earnings.

The end of the war, with all its turbulence and dislocations, released thousands and thousands of veterans who poured into the Oil Regions to start their lives again and seek their fortunes in a new speculative boom that was fueled by the incentive of prices, which rose as high as $13.75 a barrel. The effects of the frenzy were felt up and down the East Coast, as hundreds of new oil companies were floated. Office space for those new companies ran short in the financial district of New York, and shares were sold so rapidly that one new company disposed of its entire issue in just four hours. A British banker was amazed by the "hundreds of thousands of provident working men, who prefer the profits of petroleum to the small rates of interest afforded by savings banks." Washington, D.C., was no more immune to the craze than New York. Congressman James Garfield, who became a substantial investor in oil lands—and, later, President of the United States—reported to an oil-lease salesman that he had discussed oil with a number of other members of Congress, "who are in the business, for you must know the fever has assailed Congress in no mild form."[10]

Nothing revealed the feverish pitch of speculation better than the strange story of the town of Pithole, on Pithole Creek, some fifteen miles from Titusville. A first well was struck in the dense forest land there in January 1865; by June, there were four flowing wells, producing two thousand barrels per day— one third of the total output of the Oil Regions—and people fought their way in on the roads already clogged with the barrel-laden wagons. "The whole place," said one visitor, "smells like a corps of soldiers when they have the diarrhoea." The land speculation seemed to know no bounds. One farm that had been virtually worthless a few months earlier was sold for $1.3 million in July 1865, and then resold for two million dollars in September. In that same month, production around Pithole Creek reached six thousand barrels per day—two-thirds of all the production in the Oil Regions. And, by that same September, what had once been an unidentifiable spot in the wilderness had become a town of fifteen thousand people. The *New York Herald* reported that the principal businesses of Pithole were "liquor and leases"; and *The Nation* added, "It is safe to assert that there is more vile liquor drunk in this town than in any of its size in the world." Yet Pithole was already on the road to respectability, with two banks, two telegraph offices, a newspaper, a waterworks, a fire company, scores of boarding houses and businesses, more than fifty hotels—at least three of which were up to elegant metropolitan standards—and a post office that handled more than five thousand letters a day.

But then, a couple of months later, the oil production abruptly gave out—just as quickly as it had begun. To the people of Pithole, this was a calamity, like a biblical plague, and by January 1866, only a year from the first discovery, thousands had fled the town for new hopes and opportunities. The town that had sprung up overnight from the wilderness was totally deserted. Fires ravaged the buildings, and the wooden skeletons that were left were torn down to be used for building again elsewhere or burned as kindling by the farmers in the surrounding hills. Pithole returned to silence and to the wilderness. A parcel of land in Pithole that sold for $2 million in 1865 was auctioned for $4.37 in 1878.

Even as Pithole died, the speculative boom was exploding elsewhere and engulfing neighboring areas. Production in the Oil Regions jumped to 3.6 million barrels in 1866. The enthusiasm for oil seemed to know no limits, and it became not only a source of illumination and lubrication, but also part of the popular culture. Americans danced to the "American Petroleum Polka" and the "Oil Fever Gallop," and they sang such songs as "Famous Oil Firms" and "Oil on the Brain."

> There's various kind of oil afloat, Cod-liver, Castor, Sweet;
> Which tend to make a sick man *well,* and set him on his *feet.*
> But our's a curious *feat* performs: We just a *well* obtain,
> And set the people crazy with "Oil on the brain."
>
> There's neighbor Smith, a poor young man, Who couldn't raise a dime;
> Had clothes which boasted many rents. And took his "nip" on time.
> But now he's clad in dandy style, Sports diamonds, kids, and cane;
> And his success was owing to "Oil on the brain."[11]

Boom and Bust

The race to find the oil was swiftly followed by another race to produce it as quickly and in as much volume as possible. The drive for "flush production" often damaged the reservoirs, leading to premature exhaustion of gas pressure, and thus far less recovery than would otherwise have been the case. Yet there were several reasons why this became the standard practice. One was the lack of geological knowledge. Another was the large and quick rewards that were to be attained. A third was the nature of leasing terms, which put a premium on producing as quickly as possible.

But, most important in shaping the legal context of American oil production, and the very structure of the industry from the earliest days, was the "rule of capture," a doctrine based on English common law. If a game animal or bird from one estate migrated to another, the owner of the latter estate was perfectly within his rights to kill the game on his land. Similarly, owners of land had the right to draw out whatever wealth lay beneath it; for, as one English judge had ruled, no one could be sure of what was actually going on "through these hidden veins of the earth."

As applied to oil production, the rule of capture meant that the various surface owners atop a common pool could take all the oil they could get, even if they disproportionately drained the pool or reduced the output of nearby wells and neighboring producers. Inevitably, therefore, the owners of adjacent wells were in heated competition to produce as much as they could as swiftly as possible, to avoid having the pool drained by another. The impetus to rapid production contributed to the instability of both production and prices. Oil was not the same as game birds, and the rule of capture led to considerable waste and damage, to the detriment of ultimate production from a given pool. But there was another side to the rule's effects. It created room for many more people to enter the industry and to master the required skills than would have been the case under more restrictive rules. And, by building up production more quickly, it also helped to make possible a wider market.[12]

Fueled by the rule of capture—and the race for riches—the wild drive to produce created in the Oil Regions a chaotic scene of heaving populations, of shacks and quick-built wooden buildings, of hotels with four or five or six straw mattresses crowded into a single room, of derricks and storage tanks, with everyone energized by hope and rumor and the acrid scent of oil. And, everywhere, there was one inescapable factor—the perennial mud. "Oil Creek mud attained a fame in the earlier and subsequent years, that will ever be fresh in the memory of those who saw and were compelled to wade through it," two writers observed at the time. "Mud, deep, and indescribably disgusting, covered all the main and by-roads in wet weather, while the streets of the towns composing the chief shipping points, had the appearance of liquid lakes or lanes of mud."

There were some who looked at all the boom and hustle, and at the "sharpers" who came for the quick dollar, and remembered the quiet Pennsylvania hills and villages before oil burst on the scene. They asked what had happened and marveled that human nature could be so transformed—and debased—by the specter of riches. "The oil and land excitement in this section has already become a sort of epidemic," wrote a local editor in 1865. "It embraces all classes and ages and conditions of men. They neither talk, nor look, nor act as they did six months ago. Land, leases, contracts, refusals, deeds, agreements, interests, and all that sort of talk is all they can comprehend.

Strange faces meet us at every turn, and half our inhabitants can be more readily found in New York or Philadelphia than at home. . . . The court is at a standstill; the bar is demoralized; the social circle is broken; the sanctuary is forsaken; and all our habits, and notions, and associations for half a century are turned topsy-turvy in the headlong rush for riches. Some poor men become rich; some rich men become richer; some poor men and some rich men lose all they invest. So we go."

The editor had a final thought. "The big bubble will burst sooner or later."[13]

The bubble did burst—the inevitable reaction to the speculation and frantic overproduction. Depression engulfed the industry in 1866 and 1867; the price of oil dropped as low as $2.40 a barrel. Yet, while many stopped drilling, some did not, and new fields were opened up beyond Oil Creek. Moreover, innovation and organization were being imposed upon the industry.

From the first discoveries, teamsters, lashing their horses, had clogged the roads of the Oil Regions with their loads of barrels. They were more than just a physical bottleneck. Holding a monopoly position, they charged exorbitant rates; it cost more to move a barrel over a few miles of muddy road to a railway stop than to transport it by rail from western Pennsylvania all the way to New York. The teamsters' stranglehold on transportation led to an ingenious effort to develop an alternative—transportation by pipeline. Between 1863 and 1865, despite much scoffing and public ridicule, wooden pipelines proved that they could carry oil much more efficiently and cheaply. The teamsters, seeing their position challenged, responded with threats, armed attacks, arson, and sabotage. But it was too late. By 1866, pipelines were hooked up to most of the wells in the Oil Regions, feeding into a larger pipeline gathering system that connected with the railroads.

The refiners needed to acquire oil and that, too, was chaotic. Purchasing of oil had first been done on a hit-or-miss basis by buyers on horseback, riding from well to well. But, as the industry grew, a more orderly marketing system emerged. Informal oil exchanges, where buyers and sellers could meet and agree on prices, developed in a hotel in Titusville and at a curbside exchange, near the railway tracks, in Oil City. Beginning in the early 1870s, more formal oil exchanges emerged in Titusville, in Oil City, elsewhere in the Oil Regions, and in New York. Oil was bought and sold on three bases. "Spot" sales called for immediate delivery and payment. A "regular" sale required the transaction to be completed within ten days. And the sale of "futures" established that a certain quantity would be sold at a certain price within a specified time in the future. The futures prices were the focus for speculation, and oil became "the favorite speculative commodity of the time." The buyer was bound either to take the oil and pay the contracted price—or to pay or receive the difference between the contracted price and the "regular" price at the time of settlement. Thus, buyers could make a handsome profit—or suffer a devastating loss—without even taking possession of the oil.

By the time the Titusville Oil Exchange opened in 1871, oil was already on its way to becoming a very big business, one that would transform the everyday lives of millions. Altogether, the decade of the 1860s had been one of dizzy advance from Drake's lunatic experiment. Here was truly the lasting proof of "the impetuous energy with which the American mind takes up any branch of industry that promises to pay well." George Bissell's intuition and Edwin Drake's discovery and the perseverance of both these men had opened a turbulent era—a time of ingenuity and innovation, of deals and frauds, of fortunes made, fortunes lost, fortunes never made, of grueling hard work and bitter disappointments, and of astonishing growth.[14]

And what might be expected of oil's future? There were those who looked at what had happened so quickly in western Pennsylvania and saw much greater opportunities ahead. They envisioned the industry on a scale that few in the Oil Regions could begin to imagine, and yet at the same time they were also repelled and disgusted by the chaos and disorder, the fluctuations and the frenzy. They had their own very strong ideas about how the oil business ought to be organized and proceed. And they were already at work, according to their own plans.

CHAPTER 2

"Our Plan":
John D. Rockefeller
and the Combination of
American Oil

A CURIOUS AUCTION took place one February day in 1865 in Cleveland, Ohio, then a bustling city that had profited from both the Civil War and the oil boom and now stood to prosper from the great era of America's industrial expansion. The two senior partners in one of the city's most successful oil refineries had fallen into yet another of their chronic disputes over the speed of expansion. Maurice Clark, the more cautious partner, threatened dissolution. This time, the other partner, John D. Rockefeller, surprised him by accepting. The two men subsequently agreed that a private auction should be held between the two of them, the highest bidder to get the company; and they decided to hold the auction immediately, right there in the office.

The bidding began at $500, but climbed quickly. Maurice Clark was soon at $72,000. Rockefeller calmly went to $72,500. Clark threw up his hands. "I'll go no higher, John," he said. "The business is yours." Rockefeller offered to write out a check on the spot; Clark told him, no, he could settle at his convenience. On a handshake they parted.

"I ever point to that day," Rockefeller said a half century later, "as the beginning of the success I have made in my life."

That handshake also signaled the beginning of the modern oil industry, which brought order out of the chaos of the wild Pennsylvania boom. The order would take the form of Standard Oil, which, as it sought total dominance and mastery over the world oil trade, grew into a complex global enterprise that carried cheap illumination, the "new light," to the farthest corners of the earth. The company operated according to the merciless methods and unbridled lust of late-nineteenth-century capitalism; yet it also opened a new era, for it developed into one of the world's first and biggest multinational corporations.[1]

"Methodical to an Extreme"

The mastermind of Standard Oil was the young man who won that auction in Cleveland in 1865. Even then, at the age of twenty-six, John D. Rockefeller already made a forbidding impression. Tall and thin, he struck others as solitary, taciturn, remote, and ascetic. His unbending quietness—combined with the cold, piercing blue eyes set in an angular face with a sharp chin—made people uneasy and fearful. Somehow, they felt, he could look right through them.

Rockefeller was the single most important figure in shaping the oil industry. The same might arguably be said for his place in the history of America's industrial development and the rise of the modern corporation. Admired by some as a genius of management and organization, he also came to rank as the most hated and reviled American businessman—in part because he was so ruthless and in part because he was so successful. His lasting legacy would be strongly felt, in terms of his profound influence on the petroleum industry and on capitalism itself, as well as the continuing impact of his vast philanthropy—and in terms of the darker images and shadows he would cast permanently into the mind of the public.

Rockefeller was born in 1839 in rural New York State, and lived almost a full century, until 1937. His father, William Rockefeller, traded in lumber and salt and then, moving the family to Ohio, turned himself into "Dr. William Rockefeller," who sold herbal remedies and patent medicines. The father was often away on long absences from the family; the reason, some have suggested, was that he maintained another wife and family in Canada.

The son's character was already set at a young age—pious, single-minded, persistent, thorough, attentive to detail, with both a gift and a fascination for numbers, especially numbers that involved money. At the age of seven, he launched his first successful venture—selling turkeys. His father sought to teach him and his brothers mercantile skills early. "I trade with the boys," the father was reported to have boasted, "and skin 'em and I just beat 'em every time I can. I want to make 'em sharp." Mathematics was the young Rockefeller's best subject in high school. The school stressed mental arithmetic—the ability to do calculations quickly in one's head—and he excelled at it.

Intent on achieving "something big," Rockefeller went to work at age sixteen in Cleveland for a produce-shipping firm. In 1859, he formed his own partnership with Maurice Clark to trade produce. The firm prospered from demand generated both by the Civil War and by the opening of the West. Maurice Clark would later testily recall that Rockefeller was "methodical to an extreme." As the firm grew, Rockefeller stuck to his habit of holding "intimate conversations" with himself, counseling himself, repeating homilies, warning himself to beware of pitfalls, moral as well as practical. The firm dealt in Ohio wheat, Michigan salt, and Illinois pork. Within a couple of years of Colonel Drake's discovery, Clark and Rockefeller were also dealing in, and making money from, Pennsylvania oil.

Oil and the stories of instant wealth had already captured the imagination of entrepreneurial men in Cleveland, when, in 1863, a new railroad link placed Cleveland in a position to compete in the business. Refinery after refinery sprang into existence along the railway tracks into Cleveland. Many of the refineries were desperately undercapitalized, but this was never true of the one owned by Rockefeller and Clark. At the beginning, Rockefeller thought that refining would merely be a sideline to the produce business, but within a year, as the refinery became quite profitable, he became convinced otherwise. Now, in 1865, with the auction and Clark out of the way, Rockefeller, already a moderately wealthy young man, was the master of his own business, which was the largest of Cleveland's thirty refineries.[2]

The Great Game

Rockefeller won this, his first victory in refining, at a perfect time. For the end of the Civil War in that same year, 1865, inaugurated in the United States an era of massive economic expansion and rapid development, of fiery speculation and fierce competition, and of combination and monopoly. Large-scale enterprises rose in conjunction with technological advances in industries as diverse as steel, meat packing, and communication. Heavy immigration and the opening of the West made for rapidly growing markets. Indeed, in the last three and a half decades of the nineteenth century, as at no other time in American history, the business of America was truly business, and it was to this magnet that the energies, ambitions, and brains of young men were irresistibly drawn. They were caught up in what Rockefeller called "the Great Game"—the struggle to accomplish and build, and the drive to make money, both for its own sake and as a register of achievement. That game, played with new inventions and new techniques of organization, turned an agrarian republic, so recently torn by a bloody civil war, into the world's greatest industrial power.

As the oil boom progressed, Rockefeller, throwing himself wholeheartedly into the Great Game, continued to pour both profits and borrowed money into his refinery. He built a second one. He needed new markets for his growing capacity, and in 1866 organized another firm in New York to manage both the Atlantic Coast trade and the export of kerosene. He put his brother William in charge. In that year, his sales exceeded two million dollars.

Yet, while the markets for kerosene and lubricants had grown, they were not growing fast enough to match the growth in refinery capacity. Too many companies were competing for the same customers. It didn't take much in terms of capital or skills to set oneself up as a refiner. As Rockefeller later recalled, "All sorts of people went into it: the butcher, the baker, and the candlestick-maker began to refine oil." In fact, Rockefeller and his associates became quite concerned when they learned that a German baker they liked had foolishly traded his bakery for a low-quality refinery. They bought him out in order to get him back to baking.

Rockefeller devoted himself to strengthening his business—by expanding facilities and striving to maintain and improve quality, and yet always controlling costs. He took the first steps toward integration, the process of bringing supply and distribution functions inside the organization, in order both to insulate the overall operation from the volatility of the market and to improve its competitive position. Rockefeller's firm acquired its own tracts of land on which grew the white oak timber to make its own barrels; it also bought its own tank cars, and its own warehouses in New York, and its own boats on the Hudson River. At the beginning, Rockefeller also established another principle, which he religiously stuck to thereafter—to build up and maintain a strong cash position. Already, before the end of the 1860s, he had built up sufficient financial resources so that his company would not have to depend upon the bankers, financiers, and speculators on whom the railways and other major industries had come to rely. The cash not only insulated the company from the violent busts and depressions that would drive competitors to the wall, but also enabled it to take advantage of such downturns.

One of Rockefeller's great talents could already be discerned; he had a vision of where his company and the overall industry were going, and yet at the same time he persisted in commanding the critical daily details of its operations. "As I began my business life as a bookkeeper," he later said, "I learned to have great respect for figures and facts, no matter how small they were." Rockefeller immersed himself in all details and aspects of the business, even the unpleasant ones, and literally so. He kept an old suit that he would wear whenever he went out to the Oil Regions to tramp around in the muddy fields, buying oil. The result of his single-minded enterprise was that, by the latter part of the 1860s, Rockefeller owned what was probably the largest refinery in the world.[3]

In 1867, Rockefeller was joined by a young man, Henry Flagler, whose influence in the creation of Standard Oil was almost as great as Rockefeller's. Going to work at age fourteen as a clerk in a general store, Flagler had succeeded, by his mid-twenties, in making a small fortune distilling whiskey in Ohio. He had sold out in 1858 because of moral scruples about alcohol—if not his own, then at least those of his parson father. He then threw himself into salt manufacturing in Michigan. But, in circumstances of chaotic competition and over-supply, he went broke. It was a sobering experience for a man to whom making money had, initially, come so easily.

Still, Flagler was an eternally buoyant man, determined to rebound, though now matured by his hard-won lessons. His bankruptcy left him with a deep-seated belief in the value of "cooperation" among producers and a no less deep-seated aversion to what he later called "unbridled competition." Cooperation and combination, he had concluded, were necessary to minimize the risks in the uncertain world of capitalism. He had also learned another lesson; as he later said, "Keep your head above water and bet on the growth of your country." Flagler was ready and eager to wager on post-Civil War America.

Flagler was to become the closest colleague Rockefeller ever had, and one of his closest friends. His relationship with the remote Rockefeller was to lead Flagler to another adage: "A friendship founded on business is better than a business founded on friendship." Energetic and striving, Flagler was well matched to the dour, careful Rockefeller, who was delighted to acquire a partner so "full of vim and push." To a critic, however, Flagler looked somewhat different—"a bold, unscrupulous self-seeker [who] made no bones about conscience. He did whatever was necessary to success." Many years later, after having made one great fortune with Rockefeller, Flagler set off on a second conquest, the development of the state of Florida. He would build the railways down the east coast of Florida, all the way to the Keys, in order to open up what he called the "American Riviera," and was to found both Miami and West Palm Beach.

But that was well into the future. Now, in these building years, Rockefeller and Flagler worked in close harness. They sat in the same office, with their desks back to back, passing drafts of letters to customers and suppliers back and forth to each other until the missives said exactly what they wanted to say. Their friendship was the business, which they were constantly and obsessively discussing—in the office, during lunch at the Union Club, or as they walked between the office and their nearby homes. "On those walks," Rockefeller said, "when we were away from the office interruptions, we did our thinking, talking, and planning together."

Flagler devised and ran the transportation arrangements, which would prove central to the success of Standard Oil. For they gave the company a decisive power against all competitors, and it was on this base that the company's position and formidable prowess were built. Without Flagler's expertise and aggressiveness in this realm, there might well have been no Standard Oil as the world came to know it.

The size, efficiency, and economies of scale of Rockefeller's organization enabled it to extract rebates—discounts—on railway freight rates, which lowered its transportation costs below what competitors paid, providing it with a potent advantage in terms of pricing and profit. These rebates would later be a source of great controversy. Many charged that Standard forced the rebates to enable it to undercut competitors unfairly. But so intense was the competition among railroads for freight that rebates and discounts of one kind or another became common practice across the nation, especially for anyone who could guarantee large, regular shipments. Flagler, with the strength of the Standard Oil organization behind him, was very good at driving the best deal possible.

Standard, however, did not stop with rebates. It also used its prowess to win "drawbacks." A competing shipper might pay a dollar a barrel to send his oil by rail to New York. The railroad would turn around and pay twenty-five cents of that dollar back, not to the shipper, but to the shipper's rival, Standard Oil! That, of course, gave Standard, which was already paying a lower price on its own oil, an additional enormous financial advantage against its competitors. For what this practice really meant was that its competitors were, unknowingly, subsidizing Standard Oil. Few of its other business practices did as much to rouse public antipathy toward Standard Oil as these drawbacks—when eventually they became known.[4]

"Now Try Our Plan"

While the market for oil was growing at an extraordinary rate, the amount of oil seeking markets was growing even more rapidly, resulting in wild price fluctuations and frequent collapses. Toward the end of the 1860s, as overproduction caused prices to plummet again, the new industry went into a depression. The reason was simple—too many wells and too much oil. The refiners were hit no less than the producers. Between 1865 and 1870, the retail price of kerosene fell by more than half. It was estimated that refining capacity was three times greater than the market's needs.

The costs of overcapacity were obvious to Rockefeller, and it was in these circumstances, with most refiners losing money, that he launched his effort to consolidate the industry in his own grasp. He and Flagler wanted to bring in more capital, but without jeopardizing control. The technique they used was to turn their partnership into a joint stock company. On January 10, 1870, five men, led by Rockefeller and Flagler, established the Standard Oil Company. The name was chosen to indicate a "standard quality of product" on which the consumer could depend. At the time, kerosene of widely varying quality was sold. If the kerosene contained too much flammable gasoline or naphtha, as sometimes happened, the purchaser's attempt to light it could be his last act on this earth. Rockefeller held a quarter of the stock in the new company, which, at that time, already controlled a tenth of the American refining industry. But that was only the beginning. Many years later, Rockefeller would look back on the early days and muse: "Who would ever have thought it would grow to such a size?"

Newly constituted, armed with more capital, Standard used its strength to pursue even more vigorously the railroad rebates that gave it further advantage against its competition. But overall business conditions continued to deteriorate, and by 1871 the refining industry was in a complete panic. Profit margins were disappearing altogether, and most refiners were losing money. Even Rockefeller, though head of the strongest company, was worried. By this time, he was a leading business figure in Cleveland, and a pillar of the Euclid Avenue Baptist Church. He had married Laura Celestia Spelman in 1864. In her high school graduation essay, "I Can Paddle My Own Canoe," she had written, "The independence of *woman* in thought, deed, or will is one of the problems of the age." While giving up her dream of paddling her own canoe upon marrying Rockefeller, she became his closest confidante, even reviewing his important business letters. Once in their bedroom, he had earnestly promised her that if he ever had fears about business, he would tell her first. Now, in 1872, in the midst of the refinery depression, he was sufficiently concerned to feel that he had to reassure her. "You know," he said, "we are independently rich outside of investments in oil."

It was at this anxious time that Rockefeller conceived his bold vision of consolidating nearly all oil refining into one giant combination. "It was desirable to do something to save the business," he later said. An actual combination would do what a mere pool or association could not: eliminate excess capacity, suppress wild fluctuations of price—and, indeed, save the business. That was what Rockefeller and his colleagues meant when they talked of "our plan." But the plan was Rockefeller's, and he guided its execution. "The idea was mine," he said much later. "The idea was persisted in, too, in spite of the opposition of some who became faint-hearted at the magnitude of the undertaking, as it constantly assumed larger proportions."

Standard Oil geared up for the campaign; it increased its capitalization to facilitate takeovers. But events were moving in another direction as well. In February 1872, a local railway official in Pennsylvania became confused and abruptly put up rates, suddenly doubling the cost of carrying crude from the Oil Regions to New York. Word leaked out that the increase was the doing of an unknown entity called the South Improvement Company. What was this mysterious company? Who was behind it? The independent producers and refiners in the Oil Regions were aroused and alarmed.[5]

The South Improvement Company was the embodiment of another scheme for stabilization of the oil industry and would become the symbol of the effort to achieve monopoly control. Rockefeller's name was to be ever more associated with it, but though he was one of the principal implementers of the scheme, the idea actually belonged to the railroads, which were trying to find a way out of bitter rate wars. Under the scheme, railroads and refiners would band together in cartels and divide markets. The refiners would not only get rebates on their shipments, but also receive those drawbacks—rebates from the full rates paid by nonmember refiners. "Of all the devices for the extinction of competition," one of Rockefeller's biographers has written, "this was the cruelest and most deadly yet conceived by any group of American industrialists."

Though still cloaked in mystery, the South Improvement Company enraged the Oil Regions. A Pittsburgh newspaper warned that it would create "but one buyer of oil in the whole oil region," while the Titusville paper said it was nothing less than a threat to "dry up Titusville." At the end of February, three thousand angry men trooped with banners into the Titusville Opera House to denounce the South Improvement Company. Thus was launched what became known as the Oil War. The railroads, Rockefeller, the other refiners—these were the enemy. Producers marched from town to town to denounce "the Monster" and "the Forty Thieves." And now, united against monopoly, they launched a boycott of the refiners and the railroads that was so effective that the Standard refineries in Cleveland, which normally employed up to twelve hundred men, had only enough crude to occupy seventy. But Rockefeller had absolutely no doubts about what he was doing. "It is easy to write newspaper articles but we have other business," he told his wife during the Oil War. "We will do right and not be nervous or troubled by what the papers say." At another point in the battle, in a letter to his wife he set down one of his lasting principles: "It is not the business of the public to change our private contracts."

By April 1872, however, both the railroads and the refiners, including Rockefeller, had decided that it was time to disown and scuttle the South Improvement Company. The Oil War was over, apparently won by the producers. Later, Rockefeller would say that he had always expected the South Improvement Company to fail, but went along for his own purposes. "When it failed, we would be in a position to say, 'Now try our plan.' " But Rockefeller had not even waited for the South Improvement Company to fail. By the spring of 1872, he had already won control over most of Cleveland's refining and some of the most important refiners in New York City—making him the master of the largest refinery group in the world. He was ready to take on the entire oil industry.

The 1870s were to be marked by ever-rising production. Producers repeatedly tried to restrict production, but to no avail. Storage tanks overflowed, covering the land with black scum. The gluts became so large and prices fell so low that crude oil was run out into streams and onto farms because there was nowhere else to put it. At one point, the price dropped to forty-eight cents a barrel—three cents less a barrel than housewives in the Oil Regions were paying for drinking water. The recurrent efforts to organize shutdown movements always failed. New territories were continually being opened by the drill, which undermined any stability in the industry. Moreover, there were far, far too many producers to organize any meaningful restraints. Estimates of producing firms in the Oil Regions in the last quarter of the nineteenth century ranged as high as sixteen thousand. Many of the producers were speculators, others were farmers, and many of them, whatever their backgrounds, were highly individualistic and unlikely to take "a long view" and think of the common good, even if a workable plan had presented itself. Rockefeller, with his passion for order, looked with revulsion at the chaos and scramble among the producers. "The Oil Regions," he later said with acid disdain, "was a mining camp." His target was the refiners.[6]

"War or Peace"

The objective of Rockefeller's audacious and daring battle plan was, in his words, to end "that cut-throat policy of making no profits" and "make the oil business safe and profitable"—under his control. Rockefeller was both strategist and supreme commander, directing his lieutenants to move with stealth and speed and with expert execution. It was no surprise that his brother William categorized relations with other refiners in terms of "war or peace."

Standard began, in each area, by attempting to buy out the leading refiners, the dominant firms. Rockefeller and his associates would approach their targets with deference, politeness, and flattery. They would demonstrate how profitable Standard Oil was compared with other refiners, many of which were struggling through hard times. Rockefeller himself would use all his own considerable talent for persuasion in the pursuit of a friendly acquisition. If all that failed, Standard would bring a tough competitor to heel by making him "feel sick" or, as Rockefeller put it, by giving him "a good sweating." Standard would cut prices in that particular market, forcing the competitor to operate at a loss. At one point, Standard orchestrated a "barrel famine" to put pressure on recalcitrant refiners. In another battle, seeking to bring an adversary to heel, Henry Flagler instructed: "If you think the perspiration don't roll off freely enough, pile the blankets on him. I would rather lose a great deal of money than to yield a pint to him at this time."

The Standard men, moving in great secrecy, operated through firms that appeared to be independent to the outside world, but had in fact become part of the Standard Group. Many refiners never knew that their local competitors, which were cutting prices and putting other pressures on them, were actually part of Rockefeller's growing empire. Through all the phases of the campaign, the Standard men communicated in code—Standard Oil itself was "Morose." Rockefeller never wavered in his defense of the secrecy of his operations. "It is all too true!" he once said. "But I wonder what General of the Allies ever sends out a brass band in advance with orders to notify the enemy that on a certain day he will begin an attack?"

By 1879, the war was virtually over. Standard Oil was triumphant. It con-

26

trolled 90 percent of America's refining capacity. It also controlled the pipelines and gathering system of the Oil Regions and dominated transportation. Rockefeller was unemotional in victory. He bore no grudge. Indeed, some of the conquered were brought into the inner councils of Standard's management to become devoted allies in subsequent stages of the campaign. But even as Standard Oil reached its commanding position at the end of the 1870s, unexpected challenges appeared.[7]

New Threats

At the very end of the 1870s, just when Rockefeller thought he had everything virtually tied up, Pennsylvania producers made one last effort to break out of Standard's suffocating embrace with a daring experiment—the world's first attempt at a long-distance pipeline. There was no precedent for the project, named the Tidewater Pipeline, and no guarantee at all that it was technically possible. The oil would travel eastward 110 miles from the Oil Regions to a connection with the Pennsylvania and Reading Railroad. Its construction was carried out with both deception and dispatch. Fake surveys were even taken to throw Standard off as to its route. Many doubted right up to the last moment that the pipeline would work. Yet, by May of 1879, oil was flowing in the pipeline. It was a major technological achievement, comparable to the Brooklyn Bridge four years later. It also introduced a new stage in the history of oil. The pipeline would become a major competitor with the railroad for long-distance transportation.

The clear success of Tidewater, and the revolution it implied in transportation, not only caught Standard by surprise, but also meant that its control of the industry was suddenly again in jeopardy. The producers had an alternative to Standard Oil. The company sprang into action, building in short order four long-distance pipelines from the Oil Regions to Cleveland, New York, Philadelphia, and Buffalo. Within two years, Standard was a minority stockholder in Tidewater itself and had an arrangement to pool shipments with the new pipeline company to manage competition, though Tidewater did retain some independence of operation. The refining consolidation completed, these pipeline developments marked the next major stage of Standard's integration of the oil industry. Very simply, with the partial exception of the Tidewater, Standard controlled almost every inch of pipeline into and out of the Oil Regions.[8]

There remained only one way to hold this giant in check, and that was through the political system and the courts. At the end of the 1870s, producers from the Oil Regions launched a series of legal assaults in Pennsylvania against discriminatory rates. They denounced "the overweening control of the oil business by the Standard Oil Company," castigated it as an "Autocrat" and as "this gang of thieves," and sought the indictment of its principal officers for criminal conspiracy. Meanwhile, legislative hearings in New York State on railroads focused on Standard Oil's rebate system. The investigations and legal proceedings in the two states together marked the first public revelation of the activities of Standard Oil, its reach and extent, and its manipulation of rebates and drawbacks. A Penn-

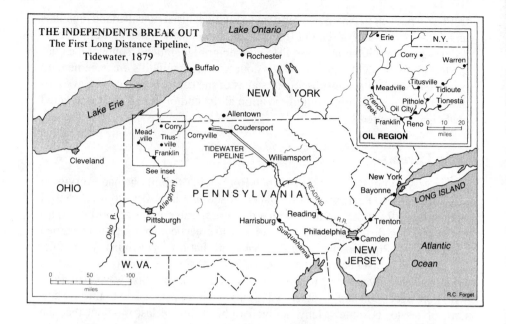

THE INDEPENDENTS BREAK OUT
The First Long Distance Pipeline,
Tidewater, 1879

Lake Ontario

Rochester

Buffalo

NEW YORK

Allentown

Coudersport

Corryville

TIDEWATER
PIPELINE

Williamsport

Cleveland

See inset

OHIO

PENNSYLVANIA

New York

Bayonne

LONG ISLAND

Pittsburgh

Harrisburg

Reading

Trenton

Philadelphia

Camden

NEW
JERSEY

Atlantic

Ocean

W. VA.

Lake Erie

Mead-
ville
Titus-
ville
Franklin
Corry

0   50   100
miles

Erie   N.Y.

Corry

Warren

Meadville   Titusville   Tidioute

Pithole   Tionesta

Oil City

Franklin   Reno

OIL REGION

0   10   20
miles

R.C. Forget

sylvania grand jury indicted Rockefeller, Flagler, and several associates for conspiracy to create a monopoly and injure competitors. A vigorous effort was
made to extradite Rockefeller to Pennsylvania. He was alarmed enough to exact
a promise from the Governor of New York not to approve any extradition order,
and the attempt eventually failed.

Still, the cumulative effect on public opinion of the varying exposes was
devastating for the company—and lasting. The veil had been lifted, and the public was outraged by what it saw. The charges against Standard were brought together for the first time by Henry Demarest Lloyd, in a series of editorials for the
Chicago Tribune, and then in an article entitled "The Story of a Great Monopoly," which was published in the *Atlantic Monthly* in 1881. So great was the attention and interest that the issue went through seven printings. Lloyd declared
that the Standard Oil Company had done everything to the Pennsylvania State
Legislature except refine it. Yet the article had little immediate impact on Standard's business. Lloyd's was the first major exposé of Standard Oil, but it was to
be far from the last. The mysterious figure of John D. Rockefeller could no
longer maintain his invisibility. In the Oil Regions, mothers would warn their
children, "Rockefeller will get you if you don't mind."[9]

The Trust

While the courts and public opinion had to be kept at bay, an ingenious internal
order and control was created in the vast empire that Rockefeller had conquered.
To begin with, there was no clear legal basis for the association of these various
refineries around the country. Thus, in an affidavit, Rockefeller could later say,
with a straight face and without perjuring himself, that Standard Oil itself did
not own or control a host of companies that it manifestly did control. One exec-

utive from the group could explain to a committee of the New York State Legislature that relations among 90 percent or so of the refineries in the country were "pleasant" and that they just happened to work together "in harmony." And another could assure the same committee that his own firm had no connection to Standard Oil and that his only personal relationship was as "a clamorer for dividends." That was the real clue to the organization. It was the *stockholders* of Standard Oil, not Standard Oil itself, who owned shares in the other firms. At that time, corporations themselves could not own stock in other corporations. The shares were held in "trust," not for the Standard Oil Company of Ohio, but on behalf of the stockholders of that corporation.

The legal concept of the "trust" was refined and formalized in the Standard Oil Trust Agreement, which was signed on January 2, 1882. It was a response to the judicial and political attacks of the late 1870s and early 1880s. There was a more personal reason, as well. Rockefeller and his partners had begun to think about mortality and inheritance, and they had concluded that the death of one of them would likely lead, under the existing system, to confusion, controversy over values, litigation, and bitterness. A trust would get the ownership organized and clarified, with little left to future debate.

In preparing the trust, "every foot of pipeline was measured, every particle of brickwork was estimated." A board of trustees was set up, and in the hands of those trustees was placed the stock of all the entities controlled by Standard Oil. Shares in turn were issued in the trust; out of the 700,000 total shares, Rockefeller held 191,700 and Flagler, next, had 60,000. The trustees held the shares in the individual companies on behalf of the forty-one shareowners of the Standard Oil Trust, and were charged with "general supervision" of the fourteen wholly owned and twenty-six partly owned companies. Their responsibilities included the selection of directors and officers—among whom they might include themselves. It was the first great "trust," and it was perfectly legal. But this was also why the "trust," formerly a device for protecting widows and orphans, became a term of derogation and hatred. Meanwhile, separate Standard Oil organizations were set up in each state to control the entities in those states. The trust agreement made possible the establishment of a central office to coordinate and rationalize the activities of the various operating entities—a task made more urgent by the growing scale of the business. And the trust gave Rockefeller and his associates "the shield of legality and the administrative flexibility they needed to operate effectively what had become virtually global properties."

That took care of the legal form. But what of the practical problem of managing the new entity? How to integrate into the new trust so many independent entrepreneurs and so many enterprises producing so many products—kerosene and fuel oil, plus some three hundred by-products? What evolved was a system of management and coordination by committee. There was a Domestic Trade Committee, an Export Trade Committee, a Manufacturing Committee, a Staves and Heading Committee, a Pipe Line Committee, a Case Committee, a Lubricating Committee, and later a Production Committee. Daily reports flowed into the committees from around the country. On top of it all was the Executive Committee, composed of the top managers, which set the overall policies and directions. The Executive Committee did not issue orders so much as requests,

29

suggestions and recommendations. But no one doubted its authority or control. The relationship between headquarters and the field was suggested by a comment Rockefeller made in a letter: "You gentlemen on the ground can judge better than we about the matter, but let us not drift into arrangements where we cannot control the policy."[10]

A basic strategy that had governed Standard in the 1870s became even more explicit in the 1880s—to be the low-cost producer. This required efficiency in operations, mastery of costs, a drive for scale and volume, constant attention to technology, and a ceaseless striving for ever-larger markets. Refining operations were consolidated in the quest for efficiency; by the middle 1880s, just three Standard refineries—in Cleveland, Philadelphia, and Bayonne, New Jersey—produced upward of a quarter of the world's total supply of kerosene. The focus on costs, sometimes calculated to the third decimal place, never wavered. "It has always been my rule in business to make everything count," Rockefeller once said. Using its superior communications, the company took advantage of the arbitrage and played the spreads among prices in the Oil Regions, Cleveland, New York, and Philadelphia, as well as in Antwerp and elsewhere in Europe. The company also used an extraordinary system of corporate intelligence and espionage to keep track of market conditions and competitors. It maintained a card catalog of practically every buyer of oil in the country, showing where virtually every barrel shipped by independent dealers went—and where every grocer, from Maine to California, obtained his kerosene.

A central theme underlay Rockefeller's management; he *believed* in oil, and his faith never wavered. Any drop in the price of crude was not a reason for anxiety, but an opportunity to buy. "Hope if crude oil goes down again . . . our Executive Committee will not allow any amount of statistics or information . . . to prevent their buying," he instructed in 1884. "We must try and not lose our nerve when the market gets to the bottom as some people almost always do." Shortly after, he added, *"We will surely make a great mistake if we do not buy."*

The senior management included Rockefeller, his brother William, Henry Flagler, and two others who altogether controlled four-sevenths of the stock. But it also extended to perhaps a dozen others as well, virtually all of them willful, assertive individuals who had been successful entrepreneurs—and, originally, competitors of Rockefeller. "It is not always the easiest of tasks to induce strong, forceful men to agree," Rockefeller later said. The only way such a grouping could work was by consensus. Choices and decisions were debated and argued, but action was taken only when, as Rockefeller insisted, the problems had been turned around and around, the various contingencies anticipated, and, finally, agreement formed about the right direction. "It is always, I presume, a question in every business just how fast it is wise to go, and we went pretty rapidly on those days, building and expanding in all directions," Rockefeller recalled. "We were being confronted with fresh emergencies constantly. . . . How often we discussed those trying questions! Some of us wanted to jump at once into big expenditures, and others to keep to more moderate ones. It was usually a compromise but one at a time we took these matters up and settled them, never going as fast as the most progressive ones wished, nor quite so carefully as the conservatives desired." He added that they "always made the vote unanimous in the end."

The senior managers were frequently to be found shuttling back and forth on the day and night trains between Cleveland and New York and Pittsburgh and Buffalo and Baltimore and Philadelphia. In 1885, the trust itself moved into new headquarters, a nine-story office building at 26 Broadway, in lower Manhattan, which soon became a landmark of sorts. From there the entire enterprise was directed, starting with the Executive Committee, its membership being whoever was in town that day. The senior executives lunched together daily in a private dining room at the top of the building. Over the meal, vital information was exchanged, ideas examined, and consensus built. And under Rockefeller's leadership, these former competitors built a company whose activities and scale were unprecedented—a new type of organization, and one that had evolved with astonishing rapidity. The men around the lunch table at 26 Broadway were an unusually talented group. "These men are smarter than I am a great deal," William Vanderbilt of the New York Central Railroad told the New York State Legislature. "They are very enterprising and smart men. I never came into contact with any class of men so smart and able as they are in their business."[11]

"The Wise Old Owl"

But the smartest was certainly John D. Rockefeller. At the time the trust was formed, he was in his early forties, already one of the half-dozen richest men in America. He was the guiding force of the company, single-minded in his devotion to its growth and the cause of combination, scathing in his disdain for the "waste" of unbridled competition—and with no shortage of self-righteousness about his purpose. He was also strangely, and deliberately, inaccessible. Later in life, he recited a little rhyme from memory:

> A wise old owl lived in an oak,
> The more he saw the less he spoke,
> The less he spoke, the more he heard,
> Why aren't we all like that old bird?

He had resolved from the beginning of his business career to "expose as little surface as possible." He was analytical and suspicious, and he kept his distance from people. His remoteness and icy, penetrating stare were unnerving. On one occasion, Rockefeller met in Pittsburgh with a group of refiners. After the meeting, several of the refiners went off to dinner. The talk centered on the taciturn, ungregarious, menacing man from Cleveland. "I wonder how old he is," a refiner said. Various other refiners offered their guesses. "I've been watching him," one finally said. "He lets everybody else talk, while he sits back and says nothing. But he seems to remember everything, and when he does begin he puts everything in its proper place . . . I guess he's 140 years old—for he must have been 100 years old when he was born."

Many years later, one who worked for Rockefeller described him as "the most unemotional man I have ever known." Yet, of course, there was a man behind the mask. The 1870s and 1880s were years when "our plan" reached its fruition. But those years of consolidation and integration, of unexpected political and press attacks, were also years of great strain and tension. "All the fortune that I have made has not served to compensate me for the anxiety of that period," Rockefeller once said. His wife, too, would remember that time as "days of worry," and he himself would recall that he seldom got "an unbroken night's sleep."

He sought relaxation and relief in different ways. Late in the day, during business meetings, he would lie down on a couch, tell his colleagues to continue, and participate in the discussions while stretched out on his back. He kept a primitive muscle extender in his office. He had a special love for horses, fast horses, and he would take them out for a carriage ride at the end of the day. An hour's fast driving—"trot, pace, gallop, everything"—followed by a rest and dinner would rejuvenate him. "I was able to take up the evening's mail and get ten letters off." [12]

In Cleveland, outside of business, his life centered on his Baptist church. He was superintendent of the Sunday school, where he left an indelible impression on one of the students, a friend of his children. Many years later, she recalled: "I can see Mr. Rockefeller yet as he led the exercises in Sunday School, his long sharp nose, and long sharp pointed chin pointed out over the childish audience, his pale blue eyes never changing in expression. He spoke with such deliberation always that he seemed to drawl, yet that he really enjoyed his position no one could doubt. Take away his piety and you remove his greatest avocation."

Rockefeller loved his Forest Hill estate, outside Cleveland, and devoted himself to its details—the building of a fireplace, constructed of special red-glazed bricks; the planting of trees; the cutting of new roads through the woods. He continued his hobby on a grander scale when he moved to his vast new estate in the Pocantico hills, north of New York City. There he directed the landscaping, constructed views, and worked at laying out new roads himself with stakes and flags, sometimes until he was exhausted. His passion for landscaping drew on the same talents for organization and conceptualization that had made him so formidable in business.

Yet even while becoming the richest man in America, he maintained a curious frugality. He insisted, to the distress of his family, on wearing the same old suits until finally they became so shiny that they had to be replaced. One of his favorite dishes remained bread and milk. Once, in Cleveland, he invited a prominent local businessman and his wife to stay at his Forest Hill estate for the summer. The couple spent a pleasant six weeks. They were, however, surprised afterward to receive a bill of six hundred dollars from Rockefeller for board.

He was not without a sense of humor, even of playfulness, though he displayed it only in the most restricted circles. "Have been in the dentist's chair," he once reported to his colleague Henry Flagler. "Think would have preferred to write you, or even read your letters, but could not help myself!" He would entertain his own family at dinner by singing, or by putting a cracker on his nose and then catching it in his mouth, or even by balancing a plate on his nose. He loved to sit with his children and their friends on the front porch and play a game called "Buzz." You began to count and every time you came to a number with a seven in it, you were supposed to say "Buzz" instead; otherwise, you were out. Somehow, Rockefeller, despite his gift for mathematics, just could never get beyond 71. The children always found this hilarious.

Rockefeller had begun making small donations to his church as soon as he started earning money. As time went on, the donations swelled, and he devoted increasing efforts to giving away a significant part of the wealth he had accumulated. He applied to philanthropy the same kind of methodical investigation and careful consideration that he brought to business; eventually, his donations would extend through the sciences, medicine, and education. In the nineteenth century, however, much of his philanthropy was oriented to the Baptist church, whose most powerful layman he had become.

At the end of the 1880s, he committed himself to the creation of a great Baptist university, and, in that cause, he provided the endowment, as well as the organizational focus, for the establishment of the University of Chicago. He continued to be by far its largest donor. Though he paid keen attention to its development, he did not interfere in its academic workings, save to insist that it stay within its budget. He refused to allow any buildings to be named after him so long as he was alive, and visited the university only twice in its first ten years. The initial visit was in 1896, on its fifth anniversary. "I believe in the work," he told a university convocation. "It is the best investment I ever made in my life. . . . The good Lord gave me the money, and how could I withhold it from Chicago?" He listened as a group of students serenaded him:

> John D. Rockefeller, wonderful man is he
> Gives all his spare change to the U. of C.

By 1910, the "spare change" that Rockefeller had given to the university added up to $35 million, compared to $7 million from all other sources. And, altogether, to all his causes, he was to give away some $550 million.

He carried over his habits of business to his private life. These were the decades of the Gilded Age, when the "robber barons" made immense fortunes and created extravagant and riotous lifestyles. His New York townhouse and his Pocantico estate were opulent indeed, but Rockefeller and his family somehow stood apart from the garishness, ostentation, and vulgarity of the age. He and his wife sought to inculcate their own values of probity into their children and so avoid having them ruined by inherited riches. Thus, the children would have only one tricycle among them so that they might learn to share. In New York City, young John D. Rockefeller, Jr., would be made to walk to and from school even as other children of the rich were carried back and forth in rigs, accompanied by grooms, and he earned pocket money working on his father's estates for the same wages as the laborers.

In 1888, Rockefeller packed himself off, with his family and two Baptist ministers, to Europe for three months. Though he did not know French, he would scrutinize each item on every bill. "Poulets!" he would exclaim. "What are poulets?" he asked his son John Junior. Told that they were chickens, he would go on, reading the next item, asking what it was. "Father," John Junior would later recall, "was never willing to pay a bill which he did not know to be correct in all its items. Such care in small things might seem penurious to some people, yet to him it was the working out of a life principle." [13]

A Marvel to the Eye

The company Rockefeller founded and guided to unparalleled prosperity continued to expand during the 1880s and into the 1890s. Scientific research was incorporated into the business. Great attention was devoted both to the quality of the product and to the neatness and cleanliness of the operations, from refinery to the local distributor. The growth of the marketing system—down to the final consumer—was an imperative of the business. The company needed markets to match its huge capacity, which forced it to seek aggressively "the utmost market in all lands," as Rockefeller put it. "We needed volume." And it surely and steadily moved to ever-higher volumes. For the growth in the use of oil, largely in the form of kerosene, was stupendous.

Oil and the kerosene lamp were changing American life—and the clock by which Americans lived. Whether living in the towns and cities of the East or the farms of the Midwest, consumers usually bought their kerosene either from their grocer or from their druggist, both of whom were supplied by a wholesaler, most of whom, in turn, were supplied by Standard Oil. As early as 1864, a New York chemist described the impact of this new illuminating oil. "Kerosene has, in one sense, increased the length of life among the agricultural population," he wrote. "Those who, on account of the dearness or inefficiency of whale oil, were accustomed to go to bed soon after the sunset and spend almost half their time in sleep, now occupy a portion of the night in reading and other amusements; and this is more particularly true of the winter seasons."

Practical advice on the use of kerosene—showing its quick and widening acceptance—was provided in 1869 by the author of *Uncle Tom's Cabin*, Harriet Beecher Stowe, who assisted her sister with a book entitled *American Woman's Home or Principles of Domestic Science*. "Good kerosene gives a light which leaves little to be desired," they wrote, as they advised their readers what type of lamps to buy. But they warned against poor quality and impure oils, which were responsible for "those terrible explosions." In the mid-1870s, five to six thousand deaths a year were attributed to such accidents. Regulation was spotty and slow in coming, which is why Rockefeller insisted on consistency and quality control, and why he had chosen the name Standard. [14]

In larger urban areas, kerosene still faced competition from manufactured or "town" gas, now extracted from coal or naphtha, another fraction of crude oil. But kerosene still had a considerable cost advantage. According to one publication, in New York, in 1885, kerosene could supply a family's needs for about ten dollars a year, while "it was not uncommon for the gas bill of the more well-to-do householders to run that much per month." In rural life, there was no such competition. "A look at the stock of a good, lively country store at the time of the Philadelphia Centennial in 1876 would have been enough to convert any citizen to a belief in progress," a student of the country store has written. "Lamps and lamp chimneys, and the whole class of merchandise known as 'kerosene goods' would seem to be a marvel to eyes that had strained to see at night by means of a lighted rag, soaked in beef tallow and draped over the edge of a dish."

Kerosene was by far the most important product coming out of refineries, but not the only one. The other products included naphtha; gasoline, used as a solvent or turned into a gas for illuminating individual buildings; fuel oil; and lubricants for the moving parts in train engines and railway cars, agricultural implements, cotton spindles, and later bicycles. Other products were petroleum jelly, trademarked as "Vaseline" and made into a base for pharmaceutical products, and paraffin, which was used not only for candle making and food preservation, but also for "paraffin chewing gum," which was "highly recommended for constant use in ladies in sewing circles."

In its effort to reach the consumer, Standard Oil moved to gain control over the marketing side of the business. By the mid-1880s, its control of marketing must have been almost equivalent to its control of refining—in the 80 percent range. And its tactics in acquiring that huge market share were just as ruthless. Its salesmen would "make a fist" and seek to intimidate both rivals and errant retailers who dared to carry competing products. Standard pushed a series of innovations to make its marketing more efficient and lower costs. Much effort was made to do away with the bulky, leaky, awkward, and expensive barrel. One innovation was the railway tank car, which eliminated the need to pile barrels into boxcars. Standard also replaced barrels on the streets of America with horse-drawn tank cars, which could disburse to a retailer anything from a pint to five gallons of kerosene. Wooden barrels—though they were to continue to define the measurement of oil—were eventually reserved only for the hinterlands, from which it was assumed they would not return.[15]

"Buy All We Can Get"

But Standard had stayed out of one critical part of the business—the production of oil. It was too risky, too volatile, too speculative. Who knew when any particular well might go dry? Better to let the producers carry that risk and stick to what could be rationally organized and managed—refining, transportation, and marketing. As one of the members of the Executive Committee wrote Rockefeller in 1885, "Our business is that of manufacturers, and it is in my judgment, an unfortunate thing for any manufacturer or merchant to allow his mind to have the care and friction which attends speculative ventures."

But a sense of precariousness underlay Standard's great globe-girdling system. There was always the fear that the oil would run out. This gift that came from the earth might disappear with the suddenness with which it had appeared. Flush production quickly exhausted the capability of wells to produce. Insofar as American oil production was concerned, Pennsylvania was the entire game, the only game; and perhaps what had happened in different areas of the state might be the fate of the entire Oil Regions. The rise and fall of Pithole was a stark warning of what could come. And who knew when? Could the industry survive even another decade? And, without crude, what value would there be to all the hardware and all the capital investment—the refineries, the pipelines, the tanks, the ships, the marketing systems? Various experts cautioned that the Oil Regions would soon be depleted. In 1885, the State Geologist of Pennsylvania warned that "the amazing exhibition of oil" was only "a temporary and vanishing phenomenon—one which young men will live to see come to its natural end."

That same year, John Archbold, a top executive of Standard, was told by one of the company's specialists that decline in American production was almost inevitable and that the chances of finding another large field "are at least one hundred to one against it." These warnings were sufficiently persuasive to Archbold that he sold some of his shares in Standard Oil at seventy-five to eighty cents on the dollar. Around the same time, Archbold was also told about signs of oil in Oklahoma. "Are you crazy?" he replied. "Why, I'll drink every gallon produced west of the Mississippi!"

But, just at that moment, the industry was about to break out of Pennsylvania—and with dramatic suddenness. The scene was northwestern Ohio, where flammable gas springs in the vicinity of Findlay had been known since the earliest settlements. In the mid-1880s, oil was discovered there, igniting a great boom in the region, which straddled the border with Indiana and became known as the Lima-Indiana fields. The newly discovered fields were so prolific that, by 1890, they accounted for a third of United States oil production![16]

Rockefeller was poised to make his last great strategic decision—to go directly into oil production. No less than his colleagues, he had great antipathy for oil producers. Yes, they were speculators, they were unreliable, they behaved like greedy miners in a gold rush. Yet here, in Lima, was an opportunity for Standard to gain control of its raw materials on a very large scale, to apply its rational management to the production of oil, to balance supplies and inventories against its market needs. In short, Standard would be able to insulate itself to a considerable degree against the fluctuations and volatility of the oil market—and against the disorder of the "mining camp." And that was the direction in which Rockefeller very definitely wanted Standard to go.

The signs of depletion in Pennsylvania were a warning that it was time to be bold, and Lima offered the indisputable evidence that the oil industry had a future beyond Pennsylvania. But there were two major obstacles. One was the quality of the petroleum. It had very different properties from that of Pennsylvania, including a most unappealing sulfuric odor, like rotten eggs. Some called the Lima crude "skunk juice." There was no known way to remove the odor, and until such a way was found, the Ohio oil had only a very limited market.

The second obstacle was located at 26 Broadway—the obstinacy of Rockefeller's more cautious colleagues. They thought the risk much too great. As a starting point, Rockefeller argued that the company should buy up all the oil it could and store it in tanks all over the region. The oil was flowing in such huge volumes out of the Ohio ground that the price dropped from forty cents a barrel in 1886 to fifteen cents a barrel in 1887. But many of Rockefeller's colleagues strongly opposed the policy of buying oil for which there was not yet any good use. "Our conservative brethren on the Board," as Rockefeller called them, "held up their hands in holy terror and desperately fought a few of us." Eventually, however, Rockefeller prevailed, and Standard Oil put more than 40 million barrels of Lima oil in storage. Then, in 1888 and 1889, Herman Frasch, a German chemist employed by Standard, figured out that, if the crude oil were refined in the presence of copper oxide, the sulfur could be removed, eliminating the problem of the rotten-eggs smell and thus making Lima oil an acceptable source of kerosene. Rockefeller's Lima gamble proved to be well worth it; after Frasch's breakthrough, the price of Lima oil immediately doubled from the fifteen cents a barrel at which Standard had acquired it to thirty cents, and continued to climb.

Rockefeller pushed the company toward the final step of buying up a large number of producing properties. The most rowdy, disorderly participants of the new industry were the producers—both in the way they managed their fields and in their business relationships. Here was a chance to impose a more orderly, more stable structure. His colleagues were, as before, reluctant, even opposed. Rockefeller was insistent. He carried the day. Of the leases available for purchase he simply ordered "Buy all we can get." By 1891, though virtually absent from production a few years earlier, Standard was itself responsible for a quarter of America's total output of crude oil.[17]

Standard committed itself to building the world's largest refinery at a place called Whiting, amidst desolate sand dunes on the shore of Lake Michigan in Indiana, to process the Lima crude. There, as everywhere, Standard's cult of secrecy—which would ultimately help undermine the entire organization—was at work. It was completely obvious that Standard was building a refinery. Still, a reporter from the *Chicago Tribune* found it impossible to get any information out of a Mr. Marshall, the close-mouthed manager of the construction project. "As to what was being done at Whiting he was entirely ignorant," the reporter wrote. "They might be erecting a $5 million dollar oil refinery or they might be putting up a pork packing establishment. He didn't think it was a pork packing establishment, but he wasn't sure."

Then there was the matter of the price itself. For many years, prices had reflected the often-feverish trading in oil certificates on the various oil exchanges in the Regions and New York. Through the 1880s, the Joseph Seep Agency, the buying arm of Standard Oil, bought oil on the open market like everyone else, by acquiring "certificates" on these exchanges. When the Seep Agency did buy directly at the wellhead, it averaged the day's highest and lowest prices from the exchanges. Increasingly, however, Seep bought directly from producers, and the independent refiners followed suit. Transactions on the exchanges fell steadily over the early 1890s.

Finally, in January 1895 Joseph Seep closed down the era of the oil exchanges with a historic "Notice to Oil Producers." He announced that "dealing" on the exchanges was "no longer a reliable indication of the value of the product." From then on, he declared, in all purchases "the price paid will be as high as the markets of the world will justify, but will not necessarily be the price bid on the exchange for certificate oil." He added, "Daily quotations will be furnished you from this office." As either purchaser or owner of between 85 and 90 percent of the oil in Pennsylvania and Lima-Indiana, Seep and Standard Oil now effectively determined the purchase price for American crude oil, though always bound by supply and demand. Said one of Rockefeller's colleagues: "We have before us daily the best information obtainable from all the world's markets. And we make from that the best possible consensus of prices, and that is our basis for arriving at the current price."[18]

The Upbuilder

In every dimension, the scale of Standard's operations was awesome, overwhelming competitors. Yet it was not a complete monopoly, not even in refining. Somewhere around 15 to 20 percent of oil was sold by competitors, and the directors of Standard were willing to live with that. Control of upwards of 85 percent of the market was sufficient for Standard to maintain the stability it cherished. Reflecting upon his landscaping and tree growing, Rockefeller observed in old age, "In nursery stock, as in other things, the advantage of doing things on a large scale reveals itself." Standard Oil could certainly be numbered at the top of the list of "other things." Rockefeller created the vertically integrated petroleum company. Many years later, one of Rockefeller's successors at Standard Oil of Ohio, who had, as a young lawyer, worked with him, mused on one of Rockefeller's great achievements. "He instinctively realized that orderliness would only proceed from a centralized control of large aggregations of plant and capital, with the one aim of an orderly flow of products from the producer to the consumer. That orderly, economical, efficient flow was what we now, many years later, call 'vertical integration.' " He added, "I do not know whether Mr. Rockefeller ever used the word 'integration.' I only know he conceived the idea."

Some commentators were puzzled by Rockefeller's accomplishments. The United States government's authoritative *Mineral Resources* declared in 1882: "There seems to be little doubt that the company has done a great work, and that through its instrumentality oil refining has been reduced to a business, and transportation has been greatly simplified; but as to how much evil has been mixed with this good, it is not practicable to make a definite statement."

For others—Standard's competitors and a good part of the public—the judgment was incontestable and completely negative. To many producers and independent refiners Standard Oil was the Octopus, out to grasp all competitors, "body and soul." And to those throughout the oil industry who suffered from Rockefeller's machinations—from the ceaseless commercial pressures and the "good sweatings," from the duplicity and secret arrangements—he was a bloodless monster, who hypocritically invoked the Lord as he methodically set about destroying people's livelihoods and even their lives in his pursuit of money and mastery.

Some of Rockefeller's colleagues were grieved by the drumbeat of criticism. "We have met with a success unparalleled in commercial history, our name is known all over the world, and our public character is not one to be envied," one wrote to Rockefeller in 1887. "We are quoted as the representative of all that is evil, hard hearted, oppressive, cruel (we think unjustly). . . . This is not pleasant to write, for I had longed for an honored position in commercial life." [19]

Rockefeller himself was not so troubled. He was, he thought, only operating in the spirit of capitalism. He even sought to enlist Protestant evangelists and Social Gospel clergy in the defense of Standard Oil. Mostly he ignored the criticism; he remained confident and absolutely convinced that Standard Oil was an instrument for human betterment, replacing chaos and volatility with stability, making possible a major advance in society, and delivering the gift of the "new light" to the world of darkness. It had provided the capital and organization and technology and had taken the big risks required to create and service a global market. "Give the poor man his cheap light, gentlemen," Rockefeller would tell his colleagues in the Executive Committee. As far as he was concerned, Standard Oil's success was a bold step into the future. "The day of combination is here to stay," Rockefeller said after he had stepped aside from active management of the company. "Individualism has gone, never to return." Standard Oil, he added, was one of the greatest, perhaps even the greatest, of "upbuilders we ever had in this country."

Mark Twain and Charles Dudley Warner, in their novel, *The Gilded Age*, grasped the character of the decades after the Civil War—a time of "the manufacture of giant schemes, of speculations of all sorts . . . [and of] inflamed desire for sudden wealth." Rockefeller was in some ways the true embodiment of his age. Standard Oil was a merciless competitor that would "cut to kill," and he became the wealthiest of all. Yet, whereas many of the other robber barons amassed their wealth by speculation, stock and financial manipulation, and outright fraud—cheating their stockholders—Rockefeller built his fortune by taking on a youthful, wild, unpredictable, and unreliable industry, and relentlessly transforming it according to his own logic into a highly organized, far-flung business that satisfied the basic hunger for light around the world.[20]

"Our plan" was to succeed even beyond Rockefeller's boldest visions, but it would ultimately fail. In the United States, public opinion and the political process would revolt against combination and monopoly, and what came to be seen as unacceptable arrogance and immoral business behavior. At the same time, new individuals and new companies—operating beyond Rockefeller's reach in the United States and in faraway places like Baku, Sumatra, Burma, and later Persia—would rise up to prove themselves hardy and persistent competitors. And some would do more than survive; they would flourish.

CHAPTER 3

Competitive Commerce

THOUGH THE REST OF THE WORLD was waiting for the "new light" from America, it had been no easy thing to get the first shipment of oil off to Europe. Sailors were terrified about the possibility of explosions and fires that might result from carrying kerosene as a cargo. Finally, in 1861, a Philadelphia shipper obtained a crew by getting the potential recruits drunk and virtually shanghaiing them aboard the sailing ship. That cargo made its way safely to London. The door to global trade was opened, and American oil quickly won markets throughout the world. People everywhere would begin to enjoy the benefits of kerosene. So, virtually from the very beginning, petroleum was an international business. The American oil industry could not have grown to the size it did and become what it was without its foreign markets. In Europe, the rapid increase in the demand for American oil products was stimulated by industrialization, economic growth, and urbanization, and by a shortage of fats and oils that had afflicted Continental Europe for more than a generation. The development of the various markets was speeded by United States consuls in Europe, who were eager to push this new "Yankee invention," as one put it, and who, in some instances, purchased oil out of their own pockets to distribute to potential customers.

Consider what the global demand meant. The substance for the popular form of lighting worldwide was provided not merely by one country, but, for the most part, by one state, Pennsylvania. Never again would any single region have such a grasp on supply of the raw material. Almost overnight, the export business became immensely important to the new American oil industry and to the national economy. In the 1870s and 1880s, kerosene exports accounted for over half of total American oil output. Kerosene was the fourth-largest U.S. export in value; the first among manufactured goods. And Europe was by far the largest market.

By the end of the 1870s, not only was one state dominant, but so was one company—Standard Oil. Eventually, at least 90 percent of the exported kerosene passed through Standard's hands. Standard was satisfied with a system in which its role ended in an American port. It was confident in its overwhelming position and was prepared to conquer the planet from its American base. John D. Rockefeller would, indeed, be able to impose "our plan" on the entire world. At the same time, the company took enormous pride in its product. Petroleum, said Standard Oil's chief foreign representative, has "forced its way into more nooks and corners of civilized and uncivilized countries than any other product in business history emanating from a single source."

There was, of course, a danger—the potential of foreign competition. But the men at 26 Broadway discounted that possibility. The only way such competition could arise was on the basis of some new source of cheap and abundant crude. The Pennsylvania Geological Report of 1874 proudly commented on how thoroughly the state's oil dominated the markets of the world. It mentioned in passing that there was a question whether "the drill in other countries . . . would find oil." But this was only an issue "that some day may interest us." The authors of the report were so sure of America's dominant role that they saw no purpose in further pursuing the question at the time. Yet they were already in error.[1]

"The Walnut Money"

Among the most promising markets for the "new light" was the vast Russian empire, which was beginning to industrialize, and for which artificial light had a special importance. The capital city, St. Petersburg, was so far north that, in the winter, it had barely six hours of daylight. As early as 1862, American kerosene reached Russia, and in St. Petersburg, it quickly won wide acceptance, with kerosene lamps swiftly replacing the tallow on which the populace had almost entirely depended. The United States consul at St. Petersburg reported happily in December 1863 that it was "safe to calculate upon a large annual increase of the demand from the United States for several years to come." But his calculations could not take into account future developments in a distant and inaccessible part of the empire, which would not only foreclose the Russian market to American oil but would also spell the undoing of Rockefeller's global plans.

For many centuries, oil seepages had been noted on the arid Aspheron Peninsula, an outgrowth of the Caucasus Mountains projecting into the landlocked Caspian Sea. In the thirteenth century, Marco Polo reported hearing of a spring around Baku that produced oil, which, though "not good to use with food," was "good to burn" and useful for cleaning the mange of camels. Baku was the territory of the "eternal pillars of fire" worshiped by the Zoroastrians. Those pillars were, more prosaically, the result of flammable gas, associated with petroleum deposits, escaping from the fissures in porous limestone.

Baku was part of an independent duchy that was annexed to the Russian empire only in the early years of the nineteenth century. By then, a primitive oil industry had already begun to develop, and by 1829 there were eighty-two hand-dug pits. But output was tiny. The development of the industry was severely restricted both by the region's backwardness and its remoteness and by the corrupt, heavy-handed, and incompetent Czarist administration, which ran the minuscule oil industry as a state monopoly. Finally, at the beginning of the 1870s, the Russian government abolished the monopoly system and opened the area to competitive private enterprise. The result was an explosion of entrepreneurship. The days of hand-dug oil pits were over. The first wells were drilled in 1871–72; and by 1873, more than twenty small refineries were at work.

Shortly after, a chemist named Robert Nobel arrived in Baku. He was the eldest son of Immanuel Nobel, a clever Swedish inventor who had emigrated in 1837 to Russia, where the military establishment excitedly took up his invention of the underwater mine. Immanuel built up a considerable industrial company, only to have it fail when the Russian government made one of its periodic swings from domestic to foreign procurement. One son, Ludwig, built upon the ruins of his father's business a new company, a great armaments concern; he also developed the "Nobel wheel," which was uniquely suited to the wretched Russian roads. Another son, Alfred, gifted in both chemistry and finance, and picking up on a suggestion from his tutor in St. Petersburg about nitroglycerine, created a worldwide dynamite empire, which he ran from Paris. But the eldest son, Robert, had no such good fortune; he was unsuccessful in a variety of businesses, and finally returned to St. Petersburg to work grudgingly for Ludwig.

Ludwig obtained a huge contract to manufacture rifles for the Russian government. He needed wood for the rifle stocks, and in the quest for a domestic supply, he dispatched Robert south to the Caucasus to search for Russian walnut. In March 1873, Robert's journey took him to Baku. Though a great polyglot trading emporium between East and West, Baku was still very much a part of Asia with the minarets and the old mosque of the Persian shahs, and with its population of Tatars, Persians, and Armenians. But the recent oil development had begun to bring great change; and Robert, immediately on his arrival in Baku, was caught up in the fever. Without consulting his brother—after all, he was the eldest and, therefore, held certain prerogatives—Robert took the twenty-five thousand rubles that Ludwig had entrusted to him for buying wood—the "walnut money"—and instead bought a small refinery. The Nobels were in the oil business.[2]

The Rise of Russian Oil

Robert quickly set about modernizing and making more efficient the refinery he had bought with Ludwig's money. With additional funds from his brother, he established himself as the most competent refiner in Baku. In October 1876, the first shipment of Nobel's illuminating oil arrived in St. Petersburg. In that same year, Ludwig came to Baku, to see for himself. Skilled in dealing with the imperial system, Ludwig won the blessing of the Grand Duke, brother of the Czar and the viceroy of the Caucasus. But Ludwig Nobel was also a great industrial leader, capable of conceiving a plan on the scale of Rockefeller. He set about analyzing every phase of the oil business; he learned everything he could about the American oil experience; he harnessed science, innovation, and business planning to efficiency and profitability; and he gave the entire venture his personal leadership and attention. In a very few years, Russian oil was to take on and even surpass American oil, at least for a time; and this Swede, Ludwig Nobel, would become "the Oil King of Baku."

Long-distance transit was a critical problem. The oil was shipped in wooden barrels from Baku over an inefficient and lengthy route—carried by boat six hundred miles north on the Caspian Sea to Astrakhan, then transferred to barges for the long journey up the Volga River, eventually reaching one or another rail line to which it was transferred for further shipment. Handling costs were enormous. Even the barrels were costly. No local wood was available in sufficient quantity, and wood was brought from a distant part of the empire or imported from America, or secondhand American barrels were bought in Western Europe. Ludwig conceived a solution to the barrel problem that would have far-reaching implications. It was to ship the oil in "bulk"—that is, in large tanks built into the ships.

The idea had great merit, but in practice it faced considerable ballast and safety problems. The captain of a ship that had been wrecked while carrying oil in bulk explained: "The difficulty was that the oil seemed to move quicker than water, and in rough weather, when the vessel was pitched forward, the oil would rush down and force the vessel into the waves." Ludwig figured out how to solve the ballast problem and commissioned the first successful bulk tanker, the *Zoroaster*, which was put into service in 1878 on the Caspian. By the middle 1880s, Ludwig's conception had also proved itself on the Atlantic, launching a major revolution in oil transport. Meanwhile Ludwig was constantly pushing his Baku refinery to be among the most scientifically advanced in the world. His was the first company anywhere in the world to have a permanent staff position for a professional petroleum geologist.

The great, highly integrated oil combine built by Ludwig soon dominated the Russian oil trade. The evidence of the Nobel Brothers Petroleum Producing Company could be found throughout the empire: wells, pipelines, refineries, tankers, barges, storage depots, its own railroad, a retail distribution network— and a multinational workforce that was treated better than virtually any other working group in Russia, and whose members proudly called themselves "Nobelites." The rapid development of Ludwig Nobel's oil empire in the first ten years of its existence has been described as "one of the greatest triumphs of business enterprise in the entire nineteenth century."[3]

Russian crude production, which was less than six hundred thousand barrels in 1874, reached 10.8 million a decade later, equivalent to almost a third of American production. By the early 1880s almost two hundred refineries were at work in the new industrial suburb of Baku that was, appropriately enough, known as Black Town. They emitted so dense a cloud of dark, smelly oil smoke that life in Black Town was compared by one visitor to "confinement in a chimney-pot." This was the expanding industry that the Nobels dominated. Their company was producing half of all Russian kerosene, and triumphantly telling its stockholders that "American kerosene has now been completely forced out of the Russian market."

But the company suffered from discord among the Nobel brothers themselves. Robert resented Ludwig's intrusion into his preserve, and eventually went back to Sweden. Ludwig was a builder, constantly seeking to expand, which meant that Nobel Brothers was continuously hungry for new capital. Alfred, well remembering how their father had failed through overexpansion and overcommitment, was much more cautious. "The main point of criticism," Alfred scolded Ludwig, "is that you build first and then look around for the wherewithal." He advised Ludwig to speculate with company shares on the stock market as a way to generate additional capital. In reply, Ludwig told Alfred to "give up market speculation as a bad occupation and leave it to those who are not suited for really useful work." Despite their disagreements, Alfred provided crucial assistance both in the form of his own money and in his help in arranging loans elsewhere, including a major borrowing from the Crédit Lyonnais. That transaction set a significant precedent in that it may have been the first loan for which future petroleum production was used as collateral.

While Nobel Brothers dominated distribution of oil within the Russian Empire, beyond those borders Russian oil was hardly a factor. Geography locked the oil into the empire. For example, to reach a Baltic port meant "2,000 miles, intermittent water and rail transportation through western Russia." To make matters worse, severe winter weather precluded the shipment of kerosene on the Caspian between October and March, with the result that many refiners simply shut down for half the year. Even parts of the empire were inaccessible; in the city of Tiflis (now Tbilisi), it was cheaper to import kerosene from America, 8,000 miles away, than from Baku, 341 miles to the east.

There were also limits to the market within the Russian empire; illumination was far from a necessity for the vast peasantry and not something they could afford in any event. Ever-growing production forced the producers of Baku to look hungrily beyond the borders of the empire. Seeking an alternative to the northern route dominated by Nobel, two other producers—Bunge and Palashkovsky—won government approval to begin building a railroad that would go west from Baku over the Caucasus to Batum, a port on the Black Sea that had been incorporated into Russia in 1877 as the result of a war with Turkey. But in the midst of construction, the price of oil dropped, and Bunge and Palashkovsky ran out of money. They were in desperate straits.

Their rescue came from the French branch of a family that, among the wars and governments and industries it had bankrolled, had also already financed many of Europe's new railroads. They owned a refinery at Fiume, on the Adriatic, and were interested in acquiring lower-priced Russian crude for it. They loaned the money to complete the railroad that Bunge and Palashkovsky had begun, acquiring in exchange a package of mortgages on Russian oil facilities. They also arranged guaranteed shipments of Russian oil to Europe at attractive prices. They were the Rothschilds.

This was a time of fervent anti-Semitism in Russia. An 1882 Imperial Decree had forbidden Jews to own or rent any more land within the empire; and, after all, the Rothschilds were the most famous Jews in the world. But in their case, the decree did not seem to matter. Russian oil was a project of the Paris Rothschilds. That meant, in particular, of Baron Alphonse—who had organized France's reparations payments after its defeat by Prussia in 1871, was considered one of the best-informed men in all of Europe, and was said to own the best pair of mustaches on the Continent—and of his younger brother, Baron Edmond, who sponsored Jewish settlement in Palestine. The Rothschild loan allowed the railroad from Baku to be completed in 1883, turning Batum almost overnight into one of the world's most important oil ports. In 1886, the Rothschilds formed the Caspian and Black Sea Petroleum Company, known ever after by its Russian initials—"Bnito." They built up their storage and marketing facilities in Batum; the Nobel Brothers quickly followed suit. The Baku-Batum railroad opened a door to the West for Russian oil; it also initiated a fierce, thirty-year struggle for the oil markets of the world.[4]

The Challenge to Standard Oil

With the arrival of the Rothschilds on the scene, the Nobels were suddenly faced with a major competitor, soon to become the second-largest Russian oil group. Though these two competitive groups discussed amalgamation, they could find no common ground beyond expressions of friendly intent, and their rivalry remained intense. There were others whose intentions were decidedly hostile. Standard Oil could not afford to ignore the Russian oil industry. Russian kerosene was now competing with American illuminating oils in many countries in Europe. In response, Standard Oil stepped up its intelligence-gathering effort about foreign markets and the new competitors. Reports began to flow into 26 Broadway from all over the world, including some from American consuls who were also on the Standard payroll. The intelligence was disturbing. No longer could Standard complacently count on its overwhelming dominance.

Standard Oil's management figured that the Czarist government would never allow it to buy out Ludwig Nobel altogether. But it could try instead to acquire a substantial number of Nobel shares, and retain the invaluable Ludwig in the management—just as it had retained the best of the competitors it had bought out in the United States. In 1885, W. H. Libby, Standard's top business-diplomat and ambassador-at-large, opened talks with the Nobels in St. Petersburg. Ludwig Nobel was not interested. Instead he concentrated on strengthening his own marketing network and building up his sales—in Europe. He had no choice. The spectacular increase of Russian oil production forced Nobel, and the other Russian oil men, to seek new markets beyond the empire. Baku was characterized by a series of astonishing oil "fountains" or gushers, with such names as "Kormilitza" (the Wet Nurse) and Golden Bazaar and Devil's Bazaar. One called "Droozba" (Friendship) gushed for five months at the rate of forty-three thousand barrels per day, most of it wasted. By 1886, there were eleven fountains, then a host of new ones in a newly opened field. Altogether Russian oil production rose tenfold between 1879 and 1888, reaching 23 million barrels, which was equivalent to more than four-fifths of American production. As the flood of oil rapidly rose in the 1880s, it needed to find its way to markets.

Faced with the aggressive Nobel's new sales campaign in Europe, and deeply alarmed by the growing production from Baku, Standard concluded that it would have to take actions beyond mere discussions. In November of 1885, it dropped its prices in Europe—just as it would when attacking a competitor in the United States. Its local agents started rumor campaigns in various European countries about the quality and safety of Russian kerosenes. They also resorted to sabotage and bribery. Despite the ferocity of the Standard assault, Nobel and the Rothschilds fought back fiercely and successfully, and Standard's executives watched with dismay as the region of what they ominously labeled "Russian competition" broadened across the map.[5]

At 26 Broadway in New York City, some members of Standard's Executive Committee had been pushing for Standard to set up its own marketing companies in foreign countries, rather than sell to independent local merchants, so that it could compete more aggressively. Moreover, the development of bulk shipment in tankers brought new economies of scale to the business. John D. Rockefeller himself, exasperated with the slowness of decision, even wrote a chiding poem to the Executive Committee in 1885:

> We are neither old nor sleepy and must "Be up and
> doing, with a heart for any fate;
> Still achieving, still pursuing, learn to labor and
> to wait."

In 1888, the Rothschilds took a new step in the competition; they established their own importing and distributing companies in Britain. Nobel Brothers did likewise. Finally galvanized into action, Standard set up its first foreign "affiliate," the Anglo-American Oil Company, just twenty-four days after the official organization of the Rothschilds' new enterprise in Britain. It also established new affiliates on the Continent—joint ventures in which it shared ownership with leading local distributors. Standard Oil had become a true multinational enterprise.

Still its competitors could not be stayed. The Rothschilds lent money to smaller Russian producers, in turn tying up rights to their production at advantageous prices. The Baku-Batum railroad suffered from a great bottleneck; the seventy-eight-mile stretch over the three-thousand-foot peak was so difficult that only half a dozen cars could be hauled over at any given time. In 1889, the Nobel Brothers completed a forty-two-mile pipeline through the mountain. What made all the difference was the use of four hundred tons of Alfred's dynamite. In this new era of what Libby, Standard's roving ambassador, called "competitive commerce," America's share of the world export trade in illuminating oil fell from 78 percent in 1888 to 71 percent in 1891, while the Russian share rose from 22 percent to 29 percent.

The prolific Baku fields continued to throw up new petroleum fountains and ever more oil. But there had been one dramatic change in the Russian oil industry. While Ludwig Nobel's patience and determination did not abate in the face of the never-ending obstacles, physically he was worn out. In 1888 at the age of fifty-seven, the Oil King of Baku died of a heart attack while vacationing on the French Riviera.

Some of the European newspapers confused the Nobel brothers and instead reported the death of Alfred. Reading his own premature obituaries, Alfred was distressed to find himself condemned as a munitions maker, the "dynamite

46

king," a merchant of death who had made a huge fortune by finding new ways to maim and kill. He brooded over these obituaries and their condemnations, and eventually rewrote his will, leaving his money for the establishment of the prizes that would perpetuate his name in a way that would seem to honor the best in human endeavor.[6]

The Son of the Shell Merchant

Still, there was the Russian kerosene, flowing out of Batum in ever-increasing quantities, in search of markets. The Nobels, at least, had a firm grip on the internal Russian market. But for the others, especially for the Rothschilds, the problem of "disposal" was growing with each passing year. Somehow, the Rothschilds had to find their way around Standard Oil and into the world market. They looked with special interest to the East, to Asia, where they saw hundreds of millions of potential customers for the "new light." But how to get the oil to them?

The Rothschilds in Paris knew a shipping broker in London named Fred Lane, who watched out for their oil interests there, and they shared their problem with him. Though always a backstage figure, Lane was to be one of the important oil pioneers. He was a big, burly man of great intelligence and with a talent for making friends and mediating interests. He was willing to back up his friendships and business alliances, which were usually one and the same, with his own capital. A "go-between *par excellence*," he was eventually to be known as "Shady Lane," not because he was crooked, for he was not, but because he sometimes appeared to be representing so many different interested parties simultaneously in a transaction that it was hard to know for whom he was really working.

Lane was truly expert in shipping; and now he had a solution to offer the Rothschilds. For he, in turn, knew a certain merchant of rising prominence, Marcus Samuel. He put the Rothschilds in touch with Samuel. The result would be an audacious scheme that might not only solve the problem of Russian oil, but also take the form of a veritable worldwide coup that, if successful, would loosen the iron grip of Rockefeller and Standard Oil on the kerosene trade of the world.

By the end of the 1880s, Marcus Samuel had already gained some prominence in the City of London. It was no mean achievement for a Jew—and a Jew not from one of the old Sephardic families, but from the East End of London, a descendant of immigrants who had come to Britain in 1750 from Holland and Bavaria. Samuel had the same name as his father, Marcus Samuel, most unusual for a professing Jew. The elder Marcus Samuel had begun his own business career trading on the East London docks, buying curios from returning sailors. In the census of 1851, he was listed as a "shell merchant"; among his most popular products were the little knickknack boxes covered with seashells, known as a "Gift from Brighton," which were sold to girls and young ladies at English seaside resorts in the mid-Victorian years. By the 1860s, the elder Marcus had accumulated some wealth and, in addition to seashells, was importing everything from ostrich feathers and partridge canes to bags of pepper and slabs of tin. He

47

was also exporting an expanding list of manufactures, including the first mechanical looms sent to Japan. In addition, in what was to prove of great importance to his son, the elder Samuel had built up a network of trusted relationships with some of the great British trading houses—run mainly by expatriate Scots—in Calcutta, Singapore, Bangkok, Manila, Hong Kong, and other parts of the Far East.

The younger Marcus was born in 1853. And in 1869, at age sixteen, after some schooling in Brussels and Paris, he went to work on his father's ledgers. At that very moment in America, John Rockefeller, fourteen years older than Samuel, was about to begin his decade-long campaign to consolidate the oil industry. Throughout the entire world, new technology was radically transforming trading and international commerce. In 1869, the Suez Canal was opened, knocking four thousand miles off the journey to the Far East. Steamships were taking over from sail. In 1870, the direct telegraph cable from England to Bombay was completed, and shortly after, Japan, China, Singapore, and Australia were all brought into the telegraph network. For the first time, the world was knitted together by global communications through the telegraph wire. Swift information now eliminated the months of waiting and suspense. Shipping was no longer a speculative venture, and explicit deals could be made in advance. These were all tools that the younger Marcus Samuel would use to build his wealth.

After the death of his father, Marcus, in partnership with his brother Samuel Samuel, developed a considerable trading operation. For several years, Samuel Samuel was resident in Japan, and the brothers had two firms—M. Samuel & Co. in London and Samuel Samuel & Co. in Yokohama, later removed to Kobe. The brothers played an important role in the industrialization of Japan, and before he was thirty, Marcus had made his first fortune out of the trade with Japan. The two brothers went on to do business throughout the Far East, in cooperation with those trading houses with which their father had first forged the relationship. At the time, Marcus and Samuel Samuel were the only British Jews prominent in the trade with the Orient.

Marcus Samuel was always the trader, the idea man, and Samuel Samuel, two years younger, always the loyal adherent and sidekick. Marcus was the more complicated, and as the years went by, his considerable charm gave way to a remoteness that almost seemed to be a mask. Short and stout, with heavy eyebrows, he was totally unprepossessing in appearance. But he was capable of bold vision, and he was adventurous, ingenious, quick to act, and single-minded when he chose to be. He talked in a very soft voice, sometimes hardly audible, making people strain to hear him and perhaps making himself all the more persuasive. He also instilled trust in people, so much so that for two decades, he depended for his credit not on bankers but on those Scottish merchants in the Far East. Marcus had more on his agenda than simply accumulating wealth for its own sake. He had a craving for position. As an outsider, as a Jew born in the East End of London, he would put his considerable energies into seeking and winning acceptance for the name Samuel at the highest levels of British society.

Samuel Samuel, in contrast to his brother, was warm-hearted, generous, gregarious, and in addition always late. He had a fondness for silly riddles, some of which he cherished for half a century or more. Let a guest come to lunch on a sunny day and he would be told by Samuel, "It's a lovely day for the race." What race? "The human race," Samuel would reply triumphantly.

Marcus did not believe in overhead; indeed he profoundly disbelieved in it. He operated out of a small office in Houndsditch in the East End, behind which was his warehouse, crammed to the ceiling with Japanese vases, imported furniture and silks, seashells and feathers, and every other kind of knickknack and curio. The perishable commodities were disposed of immediately on arrival. His operating staff was lean, another way of saying he had virtually no staff at all. He had little capital, depending instead on the credit extended to him by the Far Eastern trading houses. He also used the trading houses as his foreign agents, saving more on organization and administration. And to charter ships, he used the shipping brokerage firm of Lane and Macandrew, whose senior partner, Fred Lane, could frequently be found in the cramped offices, off a narrow alley, that belonged to M. Samuel and Company.[7]

The Coup of 1892

Marcus Samuel's entire business experience had conditioned him to be swift in grasping an opportunity, and here with the Rothschilds was an astonishing one. He moved quickly to lay the groundwork with Lane. The two men made a prospecting trip to the Caucasus in 1890. It was there that Samuel observed a primitive bulk tanker and saw in a flash that bulk tankers—the ship as a floating bottle, like modern tankers—would be much more efficient. Samuel then traveled out to Japan, and back through the Far East, seeking to persuade the Scottish trading houses with which he customarily did business to sign on with his new venture. Without them, he could not go ahead. He needed more than their cooperation; they would also have to finance the enterprise. And they all agreed to join his scheme.

Altogether, Marcus Samuel carried out a study of the opportunity and the requirements of success with a meticulous care that was uncharacteristic of the normally fast-moving trader. But he knew how large were the risks—and the stakes. He recognized that there was no point in trying to break into the market unless he and his partners could undersell Standard Oil—or at least avoid being undersold by Standard Oil. In order to assure that result, the campaign would have to be waged in all markets simultaneously; otherwise, Standard Oil would slash prices in markets where the Samuel group was competing and subsidize the price cuts by raising prices where they were not present. And, finally, speed and—to the greatest extent possible—secrecy were essential. He knew he was girding for a war with a merciless opponent.

But exactly how was Samuel to fight this war? He could tote up a long and daunting list of requirements. He needed tankers, so that the kerosene could be shipped in tanks, rather than cases. The savings on space and weight, and the gain in volume, would greatly reduce shipping costs per gallon. Like Rockefeller with the railroads, Samuel understood the absolute need to master transportation costs. The type of tanker then in operation simply would not do. Samuel needed a new, larger, technologically more advanced type of tanker, and he commissioned the design and construction of such ships. He needed guaran-

49

teed supplies of kerosene from Batum, in sufficient volume and priced to reflect the savings gained by not having to tin the kerosene. He needed access to the Suez Canal, which would cut the voyage by four thousand miles, pulling costs down further and increasing his competitive advantage against Standard, whose oil traveled to the Far East on sailing ships around the Cape of Good Hope. But the Suez Canal was closed to tankers on grounds of safety; indeed, Standard's tankers had already been refused entrance. But that did not deter Samuel. He would batter down the door. Samuel also required large storage tanks in all of the major Asian ports. He needed tank cars or tank wagons to carry the kerosene into the hinterlands. Finally, he and his partners in this venture, the trading houses, would have to establish inland depots where the bulk shipments of kerosene could be broken down and put into receptacles for the local wholesale and retail trade. And this demanding enterprise, involving detailed long-distance organization and coordination of markets, engineering, and politics, had to be kept as secret as possible!

Samuel found it difficult to work out the actual deal with the Rothschilds and Bnito. The Rothschilds were of two minds: They were never quite sure whether they wanted to compete with Standard or reach an accommodation. To M. Aron, the Rothschilds' chief oil man, Standard was always "cette puissante compagnie" ("this powerful company")—not to be trifled with. But finally, in 1891, after long negotiations and in the face of falling prices, Samuel won his contract with the Rothschilds, which gave him the exclusive rights for nine years, until 1900, to sell Bnito's kerosene east of Suez. That contract was what he wanted, he had always been sure he would get it, and he had been proceeding at full speed on the other fronts.

The tankers that he had already ordered represented a significant technological advance. In order to further reduce costs, his tankers would be capable of being steam-cleaned and then filled for the return trip with goods from the Orient, including food that would by definition have to be untainted by the taste of oil. The tankers also had to meet the safety requirements of the Suez Canal Company. Fear of explosions, fully justified by the early experience with tankers, made safety a major concern. Unlike the tankers that Standard used between the East Coast of the United States and Europe, Samuel's were to be designed with a host of new safety features, such as tanks that allowed for expansion and contraction of kerosene at different temperatures, thus minimizing the risk of fire and explosion.

Opposition quickly arose to allowing Samuel's tankers into the Suez Canal. Already, by the summer of 1891, the press was darkly reporting rumors of a "powerful group of financiers and merchants" under "Hebrew influence" who were trying to take tankers through the Suez Canal. Then, one of the most eminent firms of solicitors in the City of London, Russell and Arnholz, launched a strong lobbying campaign against granting permission to Samuel, including a lengthy correspondence with the Foreign Secretary himself. The solicitors were very concerned, ever so concerned, about safety in the canal. What might happen to ships, what might happen on hot days, what might happen during sandstorms? There were so many things to worry about, one hardly knew where to begin. They refused to reveal who their client was, even when the Foreign Secretary inquired what *British* interest they were representing. But there was hardly any question that the client was Standard Oil. Soon, Russell and Arnholz was hastily alerting the British government to a new danger: If British merchants were permitted to put tankers into the canal, Russian shipping concerns would surely also win the same right. And if the Russian naval officers and seamen, who would undoubtedly man these vessels, got into the canal, they were very likely to undertake all kinds of mischief, including seeking "to block the navigation of the Canal" and "destroy all the shipping in it."

But Samuel had powerful allies both in the Rothschild family, whose English branch had financed Benjamin Disraeli's purchase of the Suez Canal shares in 1875, and in the influential French Banque Worms. Moreover, the Foreign Secretary saw the passage of British tankers through the canal as very much in Britain's interest, and he was not going to let a firm of solicitors, however eloquent, sway him. Lloyds of London rated Samuel's new tanker design safe.[8]

Meanwhile, M. Samuel & Co. had already embarked upon a campaign to build storage tanks throughout Asia to receive the oil. The Samuel brothers sent out their nephews, Mark and Joseph Abrahams, to find the sites and supervise the construction of the tanks, and to work with the trading houses to set up the distribution systems. Joseph had India and Mark the Far East. Mark was paid five pounds a week and was further rewarded by constant long-range interference, carping, criticism, and insults from his uncles. They hammered at him both about keeping costs down and about speeding up work—two quite contrary objectives. They showed no sympathy for him in his lengthy negotiating and haggling with an endless series of consular officials, harbormasters, merchants, and Asian potentates. When Mark purchased his own secondhand rickshaw to keep costs down, he could not win his uncles' approval. And to make matters even more difficult, as if he did not have enough to do, they also hounded him to keep busy, on the side, selling coal they were trying to export from Japan. Yet, through it all, Mark was buying the sites and building storage tanks throughout the Far East, including a new site on Freshwater Island, off Singapore, and thus outside the jurisdiction of an obstructionist harbormaster.

On January 5, 1892, despite all the objections of the eminent solicitors from the City of London, the Suez Canal gave its official approval to passage for tankers built according to M. Samuel's new design. "The new scheme is one of singular boldness and great magnitude," the *Economist* commented four days later. "Whether it is true, as its opponents insinuate, that it is purely of Hebrew inspiration, we are not concerned to inquire; nor does it appear why such a circumstance should count against it. . . . If simplicity is an element of success, the scheme certainly seems full of promise. For instead of sending out cargoes of oil in cases costly to make, expensive to handle, easy to be damaged, and always prone to leak, the promoters intend to ship the commodity in tank-steamers *via* the Suez Canal, and to discharge it wherever the demand is greatest into reservoirs, from which it can be readily supplied to consumers."

Mark had already made progress in the Far East. He acquired an excellent site in Hong Kong, and he hurried to buy a site in Shanghai before the Chinese New Year since "it can be got cheaper because the Chinese have to pay all their debts contracted during the past year & they are requiring money." Having traveled constantly back and forth among the other ports of the Far East, he finally returned to Singapore in March 1892 to find yet another scolding letter from his uncles, insisting on haste and greater haste. The clock was ticking. One never knew when or how Standard Oil would launch a counterstrike.

The first tanker was nearing completion at West Hartlepool. It was called the *Murex*—named for a type of seashell, as were all of Samuel's subsequent tankers. It was a memorial to the elder Marcus, the shell merchant. On July 22, 1892, the *Murex* sailed from West Hartlepool for Batum, where it filled its tanks with Bnito's kerosene. On August 23, it passed through the Suez Canal, headed for the East. It discharged part of its cargo at Freshwater Island, Singapore; then, its load sufficiently lightened to allow it to pass over a difficult sand bar, it sailed on to Mark's new installation in Bangkok. The coup was launched.

Taken by surprise by the swiftness with which Samuel had moved, Standard's shocked representatives rushed to the Far East to assess the dangers. The implications were enormous, for, as the *Economist* noted, "If the sanguine anticipations of the promoters are realized, the Eastern case-oil trade must needs become obsolete." Standard Oil's agents were too late; Samuel's kerosene was everywhere. Thus, Standard could not cut prices in one market and subsidize them by raising prices elsewhere.

The coup was indeed brilliant and the execution superb—with one exception. For Samuel and the Far Eastern trading houses had committed a small oversight, and yet one that almost destroyed their venture. They had assumed that they would deliver the kerosene in bulk to various localities, and that the eager customers would line up with their own receptacles to be filled. The customers were expected to use old Standard Oil tin cans. But they did not. Throughout the Far East, Standard's blue oil tins had become a prized mainstay of the local economies, used to construct everything from roofing to birdcages to opium cups, hibachis, tea strainers, and egg beaters. They were not about to give up such a valuable product. The whole scheme was now threatened—not by the machinations of 26 Broadway or by the politics of the Suez Canal, but by the habits and predilections of the peoples of Asia. A local crisis was created in each port, as the kerosene went unsold, and despairing telegrams began to flow into Houndsditch.

In the quickness and ingenuity of his response to the crisis, Marcus proved his entrepreneurial genius. He sent out a chartered ship, filled with tinplate, to the Far East, and simply instructed his partners in Asia to begin manufacturing tin receptacles for the kerosene. No matter that no one knew how to do so; no matter that no one had the facilities. Marcus persuaded them they could do it. "How do *you* stick on the wire handles?" the agent in Singapore wrote to Samuel's representative in Japan. Instructions were sent. "What color do you suggest?" cabled the agent in Shanghai. Mark gave the answer—"Red!"

All the trading houses in the Far East quickly established local factories to make the tin containers, and throughout Asia, Samuel's bright and shiny red receptacles, fresh from the factory, were soon competing with Standard's blue ones, battered and chipped after the long voyage halfway around the world. Perhaps some customers were buying Samuel's kerosene more for the useful red can than for its contents. In any case, red roofs and red birdcages—as well as

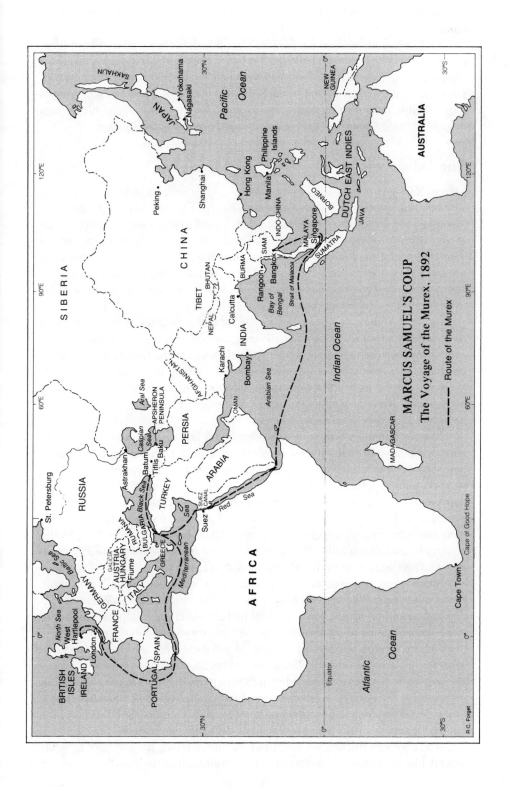

MARCUS SAMUEL'S COUP
The Voyage of the Murex, 1892

––––– Route of the Murex

R.C. Forget

red opium cups, hibachis, tea strainers, and egg beaters—began to replace the blue.

And so the day was saved. Samuel's coup had worked, and in record time. By the end of 1893, Samuel had launched ten more ships, all of them named for seashells—the *Conch*, the *Clam*, the *Elax*, the *Cowrie*, and so on. By the end of 1895, sixty-nine tanker passages had been made through the Suez Canal, all but four in ships owned or chartered by Samuel. By 1902, of all the oil to pass though the Suez Canal, 90 percent belonged to Samuel and his group.[9]

The Alderman

Marcus Samuel was not only on the edge of a great success in business, he was also beginning to achieve some station in British life. In 1891, in the midst of planning for his global coup, he had taken time off to stand for and win election as an alderman of the City of London. Though it was largely honorific, he savored the post. But then in 1893, the year after the coup, all—both business and social—seemed for naught. Samuel became seriously ill; his physician diagnosed cancer and gave him no more than six months to live. The prediction was to prove slightly off the mark—by some thirty-four years. Still, the threat of imminent death did motivate Samuel to put his business affairs into a somewhat more orderly form. The result was the creation of a new entity, the Tank Syndicate, composed of the Samuel brothers, Fred Lane, and the trading houses of the Far East. They shared all profits and losses on a global basis; such an arrangement was necessary if they were to be able to fight Standard Oil in whatever market it chose and absorb the resulting losses. The Tank Syndicate grew quickly and became increasingly successful.

Marcus Samuel's fortune was accumulating rapidly, not only from oil and tankers, but also from the longer-standing trade links with the Far East, principally Japan. The Samuel brothers made money as the principal provisioners of weapons and supplies to Japan during its 1894–95 war with China. And so it happened that within a very few years of the *Murex*'s first passage through the Suez Canal, Marcus Samuel, a Jew from the East End, had become a very rich man, one who went riding every morning in Hyde Park, who owned a splendid country house in Kent called the Mote, with its own five-hundred-acre deer park, and who had one son at Eton and another already entered.

Samuel had, however, one serious fault as a businessman. Unlike his rival, Rockefeller, he lacked talent for organization and administration. Where Rockefeller had an instinct for order, Samuel had an addiction to improvisation. For him, organization was an afterthought; he ran everything out of his hat, which made his continuing success all the more astonishing. He was operating, among other things, a large steamship line as part of his oil enterprise, and yet he had no one in his office with any knowledge or experience of actually managing such an organization. He simply depended upon Fred Lane. The day-to-day operations of the fleet were run out of a small room in Houndsditch that contained nothing but a table, two chairs, a small wall map of the world, and two clerks.

And compare Rockefeller's owl-like unfathomability, his masklike face, his quiet deliberation, his drawing out of judgment and consensus from the gentlemen in Room 1400, to the violent quarrels—the combat, anger, and recriminations—by which Marcus and Samuel arrived at decisions. Sometimes a clerk would be summoned to bring a piece of information to Samuel's office and while he waited, as one employee would recall, "the two brothers would always go to the window, their backs to the room, huddled together close, their arms around each other's shoulders, heads bent, talking in low voices, until suddenly they would burst apart in yet another dispute, Mr. Sam with loud and furious cries, Mr. Marcus speaking softly, but both calling each other fool, idiot, imbecile, until suddenly, for no apparent reason, they were in agreement again. There would be a quick, decisive exchange of final views. Then Mr. Marcus would say: 'Sam, speak to him on the telephone,' and would stand at his brother's shoulder while the telephoning took place." And that was how their deals were done.[10]

"This Struggle to the Death"

The rapid rise of Russian production, the towering position of Standard Oil, the struggle for established and new markets at a time of increasing supplies—all were factors in what became known as the Oil Wars. In the 1890s, there was a continuing struggle involving four rivals—Standard, the Rothschilds, the Nobels, and the other Russian producers. At one moment, they would be battling fiercely for markets, cutting prices, trying to undersell one another; at the next, they would be courting one another, trying to make an arrangement to apportion the world's markets among themselves; at still the next, they would be exploring mergers and acquisitions. On many occasions, they would be doing all three at the same time, in an atmosphere of great suspicion and mistrust, no matter how great the cordiality at any given moment. And, at each juncture, there was the Standard Oil Trust, that remarkable organism that was always ready to absorb generously its fiercest rivals—or, as Standard executives put it, "assimilate" them.

In 1892 and 1893, the Nobels, Rothschilds, and Standard came close to bringing virtually all oil production into one system, dividing the world among them. "In my opinion," noted M. Aron, who represented the Rothschilds' interests in the negotiations, "the crisis has reached its end, for everybody, in America and Russia, is exhausted by this struggle to the death that has gone on so long." Baron Alphonse, the head of the French Rothschilds, was himself keen to get matters settled; but, mortally afraid of publicity, he resisted an invitation that Standard was pressing on him to come to New York. Finally, Libby of Standard Oil assured Baron Alphonse that, with so many foreigners visiting America on account of the Chicago World's Fair, the arrival of the Rothschild group would not be much noted. Reassured, the Baron made it to New York and to 26 Broadway. After the meeting, a Standard Oil executive reported to Rockefeller that the Baron was very courteous and remarkably fluent in English, adding that the Rothschilds would "immediately begin the steps toward control in Russia, and are quite confident of their ability to accomplish it." But the Baron had also gently but firmly insisted that Standard Oil bring the American independents into the contract. With great effort, slowed not only by rivalries but by a cholera epidemic that gripped Baku, the Rothschilds, joined by the Nobels, did succeed in getting all the Russian producers to agree to form a common front, as a prelude to a grand negotiation with Standard. But despite its 85 to 90 percent control of American oil, Standard could not deliver the critical missing element, the independent American refiners and producers, to the grand scheme, and the proposed agreement collapsed.

In response, in the autumn of 1894, Standard launched another worldwide price-cutting campaign. The Rothschilds regarded Samuel as a tool with which to improve their bargaining position with Standard, and were very tough in their interpretation of their contract with him. Understandably, Samuel complained bitterly and loudly—loudly enough for Standard Oil to hear. Suspecting that the dissatisfied Samuel could be the weak link in the Rothschilds' position, Standard opened negotiations with him. It presented a proposal much like those it had made to competitors in America who had left the fray and joined the fraternity, save that the offer to Samuel was on a far grander scale. He would be bought out for a great deal of money, his enterprise would become part of Standard Oil, and he would become a director of Standard, though free to pursue his civic interests. Altogether, it was a very attractive offer. But Samuel rejected it. He wanted to keep the independent identity of his enterprise and his fleet, flying the flag of M. Samuel and Company, and he wanted it all to remain British. For it was British success on British terms on which he was intent, not integration into an American entity.

Standard Oil immediately returned again to the Russian producers, and on March 14, 1895, it signed the long-sought grand alliance with the Rothschilds and the Nobels "on behalf of the petroleum industry of the U.S." and "on behalf of the petroleum industry of Russia." The Americans were to get 75 percent of the world export sales, the Russians 25 percent. But the agreement never came into force. The specific reason would seem to have been the opposition of the Russian government. Once again, the would-be grand alliance had collapsed. Standard responded with new price-cutting campaigns.

If Standard Oil could not regain control over the world oil market and its international competitors through a grand alliance with the Russian producers, there was an alternative, a way to beat the Russians at their own game. A significant part of the Russian advantage came from the fact that Batum was 11,500 miles from Singapore, compared to Philadelphia's 15,000 miles. But Standard could turn the tables if it could acquire access to crude much closer to the Asian market, or, indeed, in Asia itself. Thus, Standard's attention turned to Sumatra, in the Dutch East Indies, from which the steaming time to Singapore, across the Strait of Malacca, could be measured in hours. And its eyes fell, in particular, on a Dutch company that, after years of struggle, had successfully carved out a profitable business from the jungles of Sumatra. This company was now beginning to make a sizeable impact on markets throughout Asia with its own brand, Crown Oil, and in so doing, it was opening up the world's third major producing province. It was called Royal Dutch.[11]

Royal Dutch

Seepages had been commented upon in the Dutch East Indies for hundreds of years, and small amounts of "earth oil" had been used for relief of "stiffness in the limbs" and other traditional medicinal purposes. By 1865, no fewer than fifty-two oil seepages had been identified through the archipelago. But there matters languished, while American kerosene went on to capture the world.

One day in 1880, Aeilko Jans Zijlker, a manager of the East Sumatra Tobacco Company, happened to be visiting a plantation in the marshy coastal strip of Sumatra. The youngest son of a Groningen farming family, Zijlker had come out to the lonely life of the East Indies two decades earlier, after a failed love affair. Now, while he was traipsing around the plantation, a powerful storm came up, and he took refuge for the night in a darkened, unused tobacco shed. With him was a *mandur*, or native overseer, who lit a torch. Its bright flame caught the drenched Zijlker's attention. He thought the fire must be the product of an unusually resinous wood. How had the *mandur* acquired the torch? Zijlker asked. The *mandur* replied that the torch had been daubed over with a kind of mineral wax. For longer than anyone could remember, the locals had been skimming this wax from the surface of small ponds, and then putting it to many uses, including caulking boats.

The next morning, Zijlker had the *mandur* take him to one of the ponds. He recognized the smell; imported kerosene had been introduced a few years earlier into the islands. The Dutchman collected a little of the muddy substance and sent it off to Batavia for analysis. The results enthused Zijlker, for the sample yielded between 59 and 62 percent kerosene. Zijlker made up his mind to develop the resource and threw himself wholeheartedly into the venture. His new obsession would demand his every ounce of devotion over the next decade.

His first step was to win a concession from the local Sultan of Langkat. The concession, which became known as Telaga Said, was in northeast Sumatra, six miles of jungle away from the Balaban River, which emptied into the Straits of Malacca. It was not until 1885 that the first successful well was drilled. The drilling technology itself was backward and ill-suited to the terrain, and progress continued to be very slow over the next few years. Zijlker was continually strapped for cash. But he finally gained prestigious sponsorship at home, in the Netherlands, from the former head of the central bank of the East Indies and the former governor general. Moreover, as a result of the efforts of these powerful sponsors, the Dutch king himself, William III, was willing to grant the use of the title "Royal" in the name of this speculative enterprise, a license normally reserved for established, proven companies. That imprimatur was to have lasting value. The Royal Dutch company was launched in 1890, and the first flotation of its stock was oversubscribed four and a half times.

Zijlker was triumphant. Ahead, he could see vindication of the labors of ten years. "What won't bend must break," he wrote in a letter. "Throughout the entire exploration, my motto was: whoever is not with me is against me, and I shall treat him accordingly. I know well enough that this motto earns me enemies, but I know also that had I not acted as I did, I should never have accomplished the business." Those words might well have stood as the epitaph of Aeilko Jans

Zijlker. For, returning to the Far East in the autumn of 1890, a few months after the launching of the company, he stopped at Singapore, and there he died suddenly, his vision still unrealized. His grave was marked with an inconspicuous monument.

The leadership of the enterprise in the inhospitable, swampy jungles of Sumatra passed to Jean Baptiste August Kessler. Born in 1853, Kessler had established himself in a successful trading career in the Dutch East Indies. He ran into serious business reverses that sent him back to Holland, broken and in poor health. Royal Dutch offered him a chance to begin again, and he took it. Kessler was a born leader, with an iron will, and with the ability to concentrate all his own energy and that of those around him on a single objective.

When he arrived at the drilling site in 1891, he found the entire enterprise in chaos, with everything, from the equipment shipped from Europe and America to the local finances, in total disarray. "I do not feel very cheerful about this business," he wrote to his wife. "An enormous amount of money has been lost by precipitate action." The working conditions were awful. After days of nonstop rain, the men sometimes labored in water up to their waists. The site ran out of rice, and a team of eighty Chinese workmen had to wade and swim to a village fifteen miles away to bring back a few sacks. There were also the inevitable pressures from Holland to speed things up, to stick to schedules, to keep the investors happy. Somehow, working both day and night, often racked with fever, the obsessed Kessler forced the pace.

In 1892, a six-mile pipeline linking the wells in the jungle to the refinery on the Balaban River was completed. On February 28, the entire crew gathered to wait nervously for the oil to arrive at the refinery. They had calculated how long it would take, and now, watches in hand, they counted the minutes. The moment came, and it went, but there was no oil. Depression settled over the anxious onlookers. Kessler, fearing that defeat was at hand, turned away. But then suddenly they all froze. A "roar as of a mighty storm" announced the arrival of the oil, and it quickly poured "with incredible driving force" into the first still of the Royal Dutch refinery. The crowd burst into cheers, the Dutch flag was raised, and Kessler and the crew toasted the future prosperity of Royal Dutch.

The company was now in business. By April of 1892—while Marcus Samuel was preparing to send his first cargo through the Suez Canal—Kessler himself had delivered to market the first few cases of kerosene, christened Crown Oil. Still, prosperity was hardly at hand. Royal Dutch's financial resources were quickly strained by the continuing requirements, and its very existence was threatened by its inability to raise working capital. Kessler left for Holland and Malaysia in the frantic search for new funds. Though the company was selling twenty thousand cases of kerosene a month, it was still losing money.

Kessler managed to secure the capital. He returned to Telaga Said in 1893, where he found the entire operation in a deplorable state. "Half-heartedness, ignorance, indifference, dilapidation, disorder, and vexation are everywhere apparent," he reported. "And it is in these circumstances that we have to expand the enterprise if we wish to make ends meet." He pushed the operation as hard as he could, summing up the danger in a few pithy words: "To stagnate means to liquidate."

All sorts of obstacles had to be overcome, including the arrival of almost three hundred marauding pirates from another part of Sumatra, who temporarily cut communications between the drilling site and the refinery and then set fire to some of the outbuildings with, ironically, the traditional oil torches that had first caught the eye of Zijlker more than a decade earlier. Yet, no matter what the difficulty, Kessler kept pushing. "If things go wrong," he wrote his wife, "my job and my name are gone and perhaps my sacrifices and my extraordinary exertions will be repaid with censure into the bargain. Heaven preserve me from all that misery."

Kessler persevered and succeeded. Within two years, he had increased production sixfold, and Royal Dutch had finally become profitable. It was even able to pay a dividend. Yet being a producer was not enough; if Royal Dutch were to survive, it needed to establish its own marketing organization throughout the Far East, independent of middlemen. Royal Dutch also began to use tankers and to build its own storage tanks near its markets. The immediate danger was that Samuel's Tank Syndicate would move too swiftly ahead and gain a hammerlock on the business. But, in a timely piece of protectionist intervention, the Dutch government excluded the Tank Syndicate from the ports of the East Indies, telling its own producers that the Tank Syndicate thus "need not be for the time being an object of terror" to the local industry.

Royal Dutch's business was growing at an astonishing pace; between 1895 and 1897, its production increased fivefold. Yet neither Kessler nor the company wanted to crow too loudly about its success. Kessler warned at one point that, until Royal Dutch could obtain additional concessions, "we must pretend to be poor." For, he explained, he did not want to draw other European and American interests to the East Indies, or to Royal Dutch. His principal worry was, of course, Standard Oil, which if too aroused, would wield its potent weapon— price cutting—and push Royal Dutch to the wall.[12]

"Dutch Obstacles"

But Royal Dutch could hardly remain invisible to its competitors. Its rapid growth, along with that of other producers in Asia, created a new distress for Standard Oil, matching that already created by the Russian producers. Standard Oil investigated all possible options. Early on, it began negotiating for a concession in Sumatra, but quickly gave that up in the wake of a native revolt. It searched for production opportunities in every corner of the Pacific, from China and Sakhalin to California.

In 1897, Standard dispatched two representatives to Asia to assess what could be done in the face of the Royal Dutch threat. In the East Indies, they met Royal Dutch's local manager and visited the company's installations; they called on Dutch government officials; they gathered intelligence from homesick American drillers. The representatives warned 26 Broadway against a "promiscuous search through such an enormous expanse" of steaming jungle. Much better, they told New York, to buy existing production and establish a partnership with an authentic Dutch enterprise—not only because "the ways of the Dutch Colonial Government are past finding out," but also because "you will always find it difficult to keep enough Americans here, of good business ability to make the management." Standard's objective, they insisted, should be to "assimilate" the successful companies. And that meant, above all, Royal Dutch.

To the Dutch, Standard Oil may have looked like a terrifying competitor. But Standard, for its part, had no lack of respect for the intrepid Dutch company. Standard's agents were impressed by everything from Kessler's leadership to Royal Dutch's favorable economics to its new marketing system. "In the whole history of the oil business," they reported, "there has never been anything more phenomenal than the success and rapid growth of the R. D. Co." When the Standard Oil men said good-bye to the Royal Dutch managers in Sumatra, there was something almost wistful in their farewell. "Would not it be a pity that two such big concerns as you and we own should not go together," one said.

To complicate matters further, it soon became apparent that Samuel's syndicate was also hungrily eyeing Royal Dutch. In late 1896 and early 1897, intense discussions were taking place between the two groups. But their objectives were quite different. Royal Dutch was looking for a joint marketing arrangement in Asia. Marcus and Samuel Samuel wanted more; they wanted to buy out Royal Dutch. Much was said of mutual interest, but that was about it. After one visit to the Dutch directors in The Hague, a visit characterized mostly by silence and stone coldness, Sam wrote back to Marcus: "A Dutchman sits and says nothing till he gets what he wants but of course in this case he won't." There was no progress. Yet, despite their competition, Marcus and Kessler maintained a friendly relationship. "We are still open to negotiate with you, if you think there is a possibility of coming to business," Marcus wrote cordially to Kessler in April 1897. "We feel quite certain that in the long run terms must be arranged between us, or ruinous competition to both will take place."

Standard Oil knew such discussions were going on, and could not be confident that they would not eventually lead to some kind of powerful combination arrayed against the company. One executive warned, "Every day makes the situation more serious and dangerous to handle. If we don't get control of the situation soon, the Russians, Rothschilds, or some other party may." Standard had already tried and failed to acquire Ludwig Nobel's and Marcus Samuel's companies. Now, in the summer of 1897, W. H. Libby, Standard Oil's chief foreign representative, presented Kessler and Royal Dutch with a formal proposal. The capital of Royal Dutch would be quadrupled, with Standard Oil taking all the additional shares. Standard Oil, Libby stressed, had no intention at all of getting Royal Dutch "into its power." Its objectives, he assured Kessler, were modest; it was "only seeking a favorable capital investment." Kessler could hardly believe Libby or the sincerity of his pledge. On Kessler's strong recommendation, Royal Dutch's board rejected the offer.

Standard Oil, disappointed, began discussions about acquiring another concession in the Dutch East Indies, but both Dutch government officials and Royal Dutch successfully intervened. "Dutch obstacles are about the most difficult in the World for Americans to remove," a Standard Oil official declared, "for

Americans are always in a hurry and Dutchmen never." Still, Royal Dutch did not feel secure. Its directors and management knew how Standard Oil had operated in America—buying up shares in offending competitors quietly, and then putting them out of action. To forestall such a stratagem, the directors of Royal Dutch created a special class of preference stock, the holders of which controlled the board. To make acquisition even more difficult, admission to this exclusive rank was by invitation only. One of Standard's agents unhappily reported that Royal Dutch would never merge with the American company. It was not merely a "sentimental barrier" on the part of the Dutch that blocked the way, he said; there was a practical matter, as well. The managers of Royal Dutch greatly enjoyed receiving 15 percent of the company's profits.[13]

CHAPTER 4

The New Century

THE "OLD HOUSE" was what some independent producers called Standard Oil among themselves. It rose up as a vast and imposing structure, casting its shadow in all directions, dominating every inch of the oil landscape in the United States. While foreign competitors were challenging the "Old House" abroad, there was a certain resignation throughout the United States; it seemed inevitable that Standard would end up owning or controlling everything. Yet developments in the 1890s and the first decade of the new century would pose threats to the preeminence of the Old House. The markets on which the oil industry was based were about to shift drastically. At almost exactly the same time, the producing map of the United States would also change dramatically, and significant new American competitors would emerge to challenge Standard's dominance. Not only was the world becoming too large even for Standard Oil, so was the United States.[1]

Markets Lost and Gained

At the end of the nineteenth century, demand for artificial light was met mostly by kerosene, gas, and candles, where it was met by anything at all. The gas was derived by local utilities from coal or oil or by direct production and transport of natural gas. All three of those sources—kerosene, gas and candles—had the same serious problems; they produced soot, dirt, and heat; they consumed oxygen; and there was always the danger of fire. For that last reason, many buildings, including Gore Hall, the library of Harvard College, were not illuminated at all.

The dominance of kerosene, gas, and candles would not last. The polymath inventor Thomas Alva Edison—among whose major innovations were the mimeograph, the stock ticker, the phonograph, storage batteries, and motion pictures—had turned to the problem of electric illumination in 1877. Within two years, he had developed the heat-resistant incandescent light bulb. For him, invention was not a hobby, it was a business. "We have got to keep working up things of commercial value—that is what this laboratory is for," he once wrote. "We can't be like the old German professor who as long as he can get his black bread and beer is content to spend his whole life studying the fuzz on a bee!" Edison immediately applied himself to the question of commercializing his invention, and in the process, created the electric generation industry. He even worked very carefully to price electricity so that it would be highly competitive—at exactly the equivalent of the town gas price of $2.25 per thousand cubic feet. He built a demonstration project in Lower Manhattan, whose territory just happened to include Wall Street. In 1882, standing in the office of his banker, J. P. Morgan, Edison threw a switch, starting the generating plant and opening the door not only on a new industry but on an innovation that would transform the world. Electricity offered superior light, it needed no attention from its user, and it was hardly resistible where available. By 1885, 250,000 light bulbs were in use; by 1902, 18 million. The "new light" was now derived from electricity, not kerosene. The natural gas industry had to shift its markets to heating and cooking, while the United States market for kerosene, the staple of the oil industry, leveled out and was increasingly restricted to rural America.

The new technology of electricity was quickly transferred to Europe as well. An electric light system was installed in the Holborn Viaduct Station in London in 1882. So swiftly and so thoroughly did electricity—and the electrical industries—penetrate Berlin that the city was called *Elektropolis*. The development of electricity in London was more haphazard and disorganized. In the early twentieth century, London was served by sixty-five different electric utilities. "Londoners who could afford electricity toasted bread in the morning with one kind, lit their offices with another, visited associates in nearby office buildings using still another variety, and walked home along streets that were illuminated by yet another kind."

To those who had access to it, electricity was a great boon. But its rapid development was deeply threatening to the oil industry, and, in particular, to the Old House. What kind of future could Standard Oil—with its massive investment in production, refineries, pipelines, storage facilities, and distribution—look toward if it were to lose its major market, illumination?[2]

Yet just as one market was about to slip away, another was opening—that of the "horseless carriage," otherwise known as the automobile. Some of those vehicles were powered by the internal combustion engine, which harnessed a channeled explosion of gasoline for propulsion. It was a noisy, noxious, and none too reliable means of transportation, but vehicles powered by internal combustion gained credibility in Europe after a Paris–Bordeaux–Paris race in 1895, in which the remarkable speed of fifteen miles per hour was achieved. The next year, the first auto track race was held in Narragansett, Rhode Island. It was so slow and so boring that there was first heard the cry, "Get a horse!"

Nevertheless, in the United States, as well as in Europe, the horseless carriage quickly captured the minds of entrepreneurial inventors. One such person was the chief engineer of the Edison Illuminating Company in Detroit, who quit his job so that he could design, manufacture, and sell a gasoline-powered vehicle that he named after himself—the Ford. Henry Ford's first car was sold to one man, who in turn sold it to another, one A. W. Hall, who told Ford that he had caught "the Horseless Carriage fever." Hall would deserve a special place in the hearts of all future motorists as the first recorded purchaser of a used car.

By 1905 the gasoline-powered car had defeated its competitors for automotive locomotion—steam and electricity—and had established total suzerainty. Still, there were doubts about the ruggedness and reliability of the car. Those questions were laid to rest, once and for all, by the San Francisco earthquake of 1906. Two hundred private cars were pressed into service for rescue and relief, fueled by fifteen thousand gallons of gasoline donated by Standard Oil. "I was skeptical about the automobile previous to the disaster," said the acting chief of the San Francisco fire department, who commanded three cars for round-the-clock work, "but now give it my hearty endorsement." That same year a leading journalist wrote that the automobile "is no longer a theme for jokers, and rarely do we hear the derisive expression, 'Get a horse!' " Even more than that, the car had become a status symbol. "The automobile is the idol of the modern age," said another writer. "The man who owns a motorcar gets for himself, besides the joys of touring, the adulation of the walking crowd, and . . . is a god to the women." The growth of the automobile industry was phenomenal. Registrations in the United States rose from 8,000 in 1900 to 902,000 in 1912. In a decade, the automobile went from a novelty to a familiar practicality, changing the face and mores of modern society. And it was all based on oil.

Heretofore, gasoline had been an insignificant part of the output of the refining process, with some small value for solvents and as a fuel for stoves, but with little other use. In 1892, an oil man had congratulated himself for managing to sell gasoline for as much as two cents a gallon. That changed with the motorcar, which turned gasoline into an increasingly valuable product. In addition to gasoline, a second major new market for petroleum was developing with the growth in use of fuel oil in the boilers of factories, trains, and ships. Yet even as the worrying question of future markets for oil was swiftly being resolved, a new question was asked with increasing pessimism: How were these exploding markets going to be supplied? Pennsylvania was clearly in decline. The Lima field in Ohio and Indiana was inadequate. Were new oil reserves to be found? And where? And who would control them?[3]

Breakouts

Standard's hold on the oil industry had begun to erode even before the end of the nineteenth century. Some producers and suppliers were at last able to escape the trust's vise of gathering systems and pipelines and refineries to win some measure of real independence. In the early 1890s, a group of independent oil men in Pennsylvania, teaming up with refiners, organized the Producers' and Refiners' Oil Company. Recognizing that they had no real chance against the Old House if they could not find a way to get their petroleum out of the Oil Regions and to the seaboard at competitive cost, they undertook to construct their own pipeline.

The construction workers were forced to brave armed attacks from railway men, as well as steam, hot water, and hot coals poured over them from locomotives. Such may have been the "gloved hand" of Standard Oil at work. Nevertheless, the pipeline got built.

In 1895, these various independent interests formed the Pure Oil Company to organize marketing overseas and on the East Coast. Pure Oil was set up as a trust, with the trustees designated "champions of independence." Standard Oil, as was its wont, persistently tried to buy out and gain control of Pure's constituent parts; but, despite some close calls, it failed to do so; and within a few years, Pure turned itself into a fully integrated company, with significant export markets. While Pure was small compared to the mammoth Standard Oil, the independent producers and refiners had at last realized their dream: They had successfully challenged Standard Oil and had managed to insulate themselves from it. And Standard Oil, though certainly through no choice of its own, was now forced to accustom itself to the distasteful reality of significant and lasting domestic competition.[4]

But Pure was entirely based in Pennsylvania. The conventional wisdom remained that oil was a phenomenon of the Eastern United States, and pessimism continued to be the order of the day when it came to new supplies. Yet new oil fields were being discovered farther west across the continent—in Colorado and Kansas.

There was another land even farther west, across the Rockies—California. Asphalt seepages and tarpits had signaled to some the possible presence of oil. A heavily promoted boomlet had developed north of Los Angeles in the 1860s. The distinguished Yale professor Benjamin Silliman, Jr., who had provided the imprimatur for George Bissell's and Colonel Drake's venture in the 1850s, and who was always interested in extra work, took on a job as a consultant to various of the California oil promotions. He did not hold back in his enthusiasm. The value of one ranch "is its almost fabulous wealth in the best of oil," he wrote, and of another, "the amount of oil capable of being produced here is almost without limit." Silliman's research, however, was not exactly overwhelming. While he had visited some of the areas on which he had passed judgment, others he had seen only from a horse-drawn stagecoach while traveling to Los Angeles, and one he had not seen at all. The reason that his tests showed such a high kerosene potential was that the sample he analyzed had been salted with a first-rate refined Pennsylvania kerosene taken from the shelves of a general store in Southern California. The Los Angeles boom fizzled by the end of the 1860s, severely tarnishing the prospects for California. Professor Silliman's reputation was hurt even more. Indeed, so great was the humiliation and disgrace that Silliman, heretofore one of the preeminent figures in American science, was forced to resign his professorship of chemistry at Yale.

Yet, only a decade or so later, Silliman was to be vindicated. Modest production began in the regions that he had praised—in Ventura County and at the northern end of the San Fernando Valley, north of Los Angeles, which was then a town of all of eight thousand. At one point, there was widespread fear that cheap foreign oil would flow in, aided by a removal of the tariff on imported oil, and so stifle the local California industry. But as the result of adroit political maneuvering, the tariff on foreign oil was not reduced, but indeed was actually doubled. In the early 1890s, the first large find, the Los Angeles field, was discovered, followed by additional major finds in California's San Joaquin Valley. The growth of California production was dramatic—from 470,000 barrels in 1893 to 24 million barrels in 1903—and, for most of the next dozen years, California was to lead the nation in oil production. By 1910, its output would reach 73 million barrels, more than that of any foreign nation, and 22 percent of total world production.

The dominant producer in California was Union Oil (now Unocal), the only major American corporation outside of Standard Oil to have maintained a continuous independent existence since 1890 as a major integrated oil company. Union and the other smaller California companies were kindly disposed toward professional geologists, which contrasted sharply with the attitude in other parts of the country. Indeed, the profession of oil geology in the United States first established itself in California. Between 1900 and 1911, forty geologists and geological engineers were employed by California companies, which was probably more than were employed in the rest of the United States combined, or for that matter, in any other part of the world. Though Union Oil itself eluded its grasp, Standard quickly developed a hammerlock on much of the petroleum marketing and distribution in the West. In 1907, operating as Standard Oil of California, it began to move directly into production. Though California had by the turn of the century emerged as a major oil province, it was far from the rest of the nation, isolated, and its external markets were in Asia and not east of the Rockies where most of the citizens of the United States happened to live. California might as well have been another country from a business point of view. The answer to the growing oil thirst of the rest of the United States would have to be found elsewhere.[5]

Patillo Higgins's Dream

Patillo Higgins, a one-armed mechanic and lumber merchant, and a self-educated man, was possessed by an idea. He was convinced that oil would be found beneath a hill that rose above the flat coastal plain near the little town of Beaumont, in southeast Texas, some nineteen miles inland from Port Arthur on the Sabine Lake, which connected to the Gulf of Mexico. The idea first occurred to him when he took his Baptist Sunday school class for an outing on the hill. He came across a half dozen little springs, with gas bubbling up into them. He poked a cane into the ground in the area and lit the gases that escaped. The children were thoroughly amused; Higgins was puzzled and intrigued. The hill, over which wild bulls roamed, was called Spindletop, after, it was said, a local tree that grew like an inverted cone. Higgins called it the Big Hill, and he simply could not get it out of his mind. Later he said it was something about the small rocks that he lifted out of the springs that told him it was an oil field. He never could quite say what it was about the rocks. But it was something.

Absolutely sure there was oil in the Big Hill, Higgins ordered a book on geology and read it eagerly. In 1892, he organized the Gladys City Oil, Gas, and Manufacturing Company, named for one of the little girls in his Sunday school class. The company had a most imposing letterhead—a sketch of two dozen oil tanks, the smoking chimneys of a dozen plants, and several brick buildings—but the company's efforts came to nothing. Additional tries by Higgins were equally unsuccessful.

Minor oil production was just beginning elsewhere in Texas. The civic leaders of a little town called Corsicana had concluded that their fervent hopes of promoting commercial development would be frustrated by lack of water. They organized a water company, which began drilling in 1893. To their initial chagrin, they found oil. The chagrin quickly turned to excitement, much drilling followed, and the Texas oil industry was born. In Corsicana a new, more efficient method, rotary drilling, was borrowed from water-well contractors and applied to the search for oil. But Corsicana was still small stuff; by 1900, its production would reach only 2,300 barrels per day. Meanwhile, in Beaumont, Patillo Higgins refused to give up his dream and continued to promote the oil potential of Spindletop. Various geologists descended from the train in Beaumont, reviewed the prospect, and pronounced Higgins's notion nonsense. A member of the Texas Geological Society went even further and published an article in 1898, warning the public against investing in Higgins's dream. Higgins would not relent; he siphoned gas from the hill into a couple of five-gallon kerosene tins and burned it in a lamp at home. His fellow townsmen said that he was hallucinating and might be mad. But Higgins would not give up.

In a last act of desperation, he placed an advertisement in a magazine, seeking someone else to drill. There was only one reply—from a Captain Anthony F. Lucas. Born on the Dalmatian coast of the Austro-Hungarian empire and educated as an engineer, Lucas had joined the Austrian Navy and then emigrated to the United States. He had had considerable experience prospecting the geological structures known as salt domes in search of both salt and sulfur. And Big Hill was a salt dome.

Lucas and Higgins made a deal, and the captain commenced drilling in 1899. His first efforts failed. More professional geologists ridiculed the concept. They told him that he was wasting his time and money. There was no chance that a salt dome could mean oil. Captain Lucas could not convince them otherwise. He was discouraged by the professionals' rejection of what he called his "visions," and his confidence was shaken. He ran out of money, and he needed new funds if he was to continue. He won a hearing from Standard Oil, but was turned away empty-handed.

With nowhere else to go, Lucas went to Pittsburgh to see Guffey and Galey, the country's most successful firm of wildcatters. They were his last hope. In the 1890s, James Guffey and John Galey had developed that first major oil field in the midcontinent, in Kansas, which they sold to Standard Oil. Galey was the true wildcatter, the explorer. "Petroleum had John Galey bewitched," a business associate would later say. In turn, Galey had an amazing ability to find oil. Though he diligently studied and applied the geological theories of the day, some of his contemporaries thought he could literally smell oil. Quiet and low-key, he was unstoppable and indefatigable on the hunt. Indeed, the search for the treasure counted for him far more than the treasure itself. As he once said, the only geologist who could tell with certainty whether oil would be found was "Dr. Drill."

James Guffey was more flamboyant. He had once been chairman of the Democratic party, dressed like Buffalo Bill, and even had long white hair flowing out from underneath his broad-brimmed black hat. "An example of the generally accepted type of an American," a British visitor said. A contemporary American oil publication saw Guffey somewhat differently. "Dash and push had characterized his operations from the very first and he had not then, nor now, reached the point in life when he was content to travel by freight train if there was an express or flyer to be had." Guffey was the promoter and deal-maker. In this case, he drove a hard bargain with Lucas; in exchange for the financial backing of Guffey and Galey, Captain Lucas could retain only an eighth of the deal. As for Higgins, Guffey was sorry, but he would get nothing from Guffey and Galey. If Lucas felt sentimental and was so inclined, he could split his share with Higgins.

John Galey went to Beaumont and surveyed the area. As the drilling site, he chose a spot next to the little springs with bubbling gas that Patillo Higgins had found. He drove a stake into the ground to mark the spot. With Captain Lucas out of town at that moment hiring drillers, Galey turned to Mrs. Lucas and said, "Tell that Captain of yours to start that first well right here. And tell him that I know he is going to hit the biggest oil well this side of Baku."[6]

Drilling began in the autumn of 1900, using the techniques of rotary drilling that had been pioneered in Corsicana. The townspeople in Beaumont had pretty much decided that Lucas and his crew were, like Patillo Higgins, plain crazy and hardly worthy of attention. Just about the only people who came around to see what was happening were boys out shooting rabbits. The drillers fought their way through the hundreds of feet of sand that had frustrated all previous efforts. At about 880 feet, oil showed. Captain Lucas excitedly asked the lead driller, Al Hamill, how much of a well it might be. Easily fifty barrels per day, Hamill replied, thinking of the Corsicana wells he knew that might get up to twenty-two barrels per day.

The drillers took Christmas off and resumed their exhausting work on New Year's Day, 1901. On January 10, the memorable happened: Mud began to bubble with great force from the well. In a matter of seconds, six tons of drill pipe catapulted out of the ground and up through the derrick, knocking off the top, and breaking at the joints as the pipe shot further upward. Then the world was silent again. The drillers, who had scattered for their lives and were not sure what they had seen, or even if they had actually seen it, sneaked back to the derrick to find a terrible mess, with debris and mud, six inches deep, all over the derrick floor. As they started to clean the mess away, mud began to erupt again from the well, first with the sound of a cannon shot and then with a continuing and deafening roar. Gas started to flow out; and then oil, green and heavy, shot up with ever-increasing force, sending rocks hundreds of feet into the air. It pushed up in an ever-more-powerful stream, twice the height of the derrick itself, before cresting and falling back to the earth.

Captain Lucas was in town when he heard the news. He raced to the hill in his buckboard, pushing his horse at a dead run. As he got to the hill, he fell out of the buckboard and rolled onto the ground. He stood up, fighting to catch his breath, and ran to the derrick. "Al! Al! What is it?" he shouted through the din.

"Oil, Captain!" replied Hamill. "Oil, every drop of it."

"Thank God," said Lucas, "thank God."

Lucas 1 on Spindletop, as the well became known, was flowing not at fifty barrels per day, but at as much as seventy-five thousand barrels per day. The roar could be heard clearly in Beaumont; some people thought it was the end of the world. It was something never seen before anywhere—except in the "oil fountains" of Baku. The phenomenon came to be called a gusher in the United States. The news flashed across the nation and was soon on its way around the globe. The Texas oil boom was on.

What followed was riotous. The mad scramble for leases began immediately, with some plots traded again and again for ever more astounding prices. A woman garbage collector was thrilled to get $35,000 for her pig pasture. But, soon, land that had only two years before sold for less than $10 an acre now went for as much as $900,000 an acre. Much land was sold and resold on the basis of small, error-ridden maps, and with actual titles totally unclear. The town swelled with sightseers, fortune seekers, deal-makers, and oil field workers; each train disgorged new hordes drawn by the dream of instant wealth embodied in the dark gusher. One Sunday alone, excursion trains dropped off at Beaumont some 15,000 people, who tramped through the mud and slime and oil just to see this new wonder of the world. Upward of 16,000 people were said to be living in tents on the hill. Beaumont's own population ballooned in a matter of months from 10,000 to 50,000.

Tents, lean-tos, shacks, saloons, gambling houses, whorehouses—all sprang up in Beaumont to serve the various needs of the lusting population. According to one estimate, Beaumont drank half of all the whiskey consumed in Texas in those early months. Fighting was a favorite pastime. There were two or three murders a night, sometimes more. Once sixteen bodies were dredged out of a local river, their throats slit, the victims of a night's mayhem. One of the most popular entertainments in the saloons was betting on how long it would take a rattlesnake to eat a bird that was put into its cage. Even more popular were the prostitutes who swarmed into Beaumont, and the names of some of Beaumont's madams—Hazel Hoke, Myrtle Bellvue, and Jessie George—became legendary. At the barbershops, folks stood in line an hour to pay a quarter for the privilege of bathing in a filthy tub. People did not want to waste time when there was oil business to be done, so spaces near the head of the long line at the outdoor conveniences went for as much as one dollar. Some people made forty or fifty dollars a day, standing in line and selling their spaces to those who didn't have time to wait.

There were, of course, many more losers than winners, and there were endless frauds to make sure that money changed hands quickly. The stock salesmen, with shares of dubious value at best, were so numerous and so busy that Spindletop became known to some as "Swindletop." A fortune-teller named Madame la Monte did a brisk business telling her customers where new gushers could be found. Even better was the "boy with the X-ray eyes," who could see through the earth and find oil. Thousands of shares were bought in the company promoting the talented lad.

Within months, there were 214 wells jammed in on the hill, owned by at least a hundred different companies, including one called the Young Ladies Oil Company. Some of these companies were drilling on postage-stamp-size sites, just large enough for one derrick. As the Spindletop wells continued to flow, a glut of oil developed very quickly. By midsummer of 1901, oil went for as little as three cents a barrel. By comparison, a cup of water cost five cents, providing testament of a sort to the initial prolificacy of Patillo Higgins's Big Hill.[7]

The Deal of the Century

No one needed markets for his oil more than James Guffey, who was the major producer at Spindletop. But he had no intention of being swallowed up by Standard Oil, so he wanted other customers. He soon found a very large one. For among those most electrified by the news from Spindletop was the alderman of the City of London, next in line to be Lord Mayor, Sir Marcus Samuel. He had recently rechristened his rapidly growing company Shell Transport and Trading—again, like the names of his tankers, in honor of his father's early commerce in seashells. Now, Samuel and his company saw the oil flowing from the Texas plain as a way to diversify away from Shell's dependence on Russian production and to obtain oil that could be exported directly to Europe. Texas production would strengthen Samuel's hand against all competitors. Another factor also riveted Marcus Samuel: The Texas crude, while a poor source for illuminant, was well suited for use as fuel oil for ships. One of his consuming passions was the conversion of coal-burning vessels to oil—his oil. He proudly announced in 1901 that his company "may clearly claim to be the pioneers of ocean consumption of liquid fuel."

So, when the news from Spindletop reached London, it immediately set off frantic and comical efforts by Shell, first to find out where Beaumont was—it could not to be found in the office atlas at all—and then to make contact with Guffey. The Shell people had never before heard of Guffey, and he took some tracking down. Guffey allowed that, for his part, he had never heard of Shell, which rankled and offended London, and resulted in further cables and letters pointing out that Shell was a company "of great magnitude," the second-largest oil company in the world, and "Standard Oil Co.'s most dangerous opponent." Meanwhile, intelligence that Standard Oil's tankers were regularly picking up cargoes of Spindletop oil from Port Arthur only increased Shell's anxiety to move swiftly. Samuel dispatched his brother-in-law to the New World—to New York, then to Pittsburgh, then to Beaumont—to seek a contract with the unknown Guffey. The negotiations were hastily pursued. Shell made no independent geological evaluation; it did not even bother to hire an American lawyer to review the eventual contract. At one point, the brother-in-law had to scurry around to buy a wall map of the world to explain to Guffey Shell's activities elsewhere in the world. After his tour and discussions with Guffey, the brother-in-law felt confident in reassuring Samuel, back in London, on a crucial point—

that "there is no likelihood of failure of supplies." The only thing to worry about was overproduction.

By June of 1901, only half a year after the gusher had burst out at Spindletop, the two companies had completed their negotiations and signed a contract. For the next twenty years, they agreed, Shell would take at least half of Guffey's production at a guaranteed twenty-five cents a barrel—a minimum of almost 15 million barrels. It could take more if it desired. To each side, this appeared to be the deal of the new century. Marcus Samuel ordered four new tankers to be built swiftly to implement what he regarded as another great coup—the new Texas trade.

Spindletop was to remake the oil industry, and with its huge volumes move the locus of American production away from Pennsylvania and Appalachia and toward the Southwest. Spindletop also helped open up one of the main markets of the twentieth century and the one Marcus Samuel was championing—fuel oil. This, however, was more by default rather than design; the Texas oil was of such poor quality that it could not be made into kerosene by existing processes. So it went, primarily, not for lighting, but for heat and power and locomotion. A host of industries in Texas converted almost immediately from coal to oil. The Santa Fe Railroad went from just one oil-fired locomotive in 1901 to 227 in 1905. Steamship companies, as well, rushed to switch from coal to oil. These conversions, the result of Spindletop, pointed to a major shift in industrial society.

Spindletop also became the training ground for the oil industry of the Southwest. Farm boys and city boys and ranch hands all learned the tricks of the trade there. A new language was even born on the hill, for it was at Spindletop that a "well borer" first became known as a "driller," a skilled helper as a "roughneck," and a semiskilled helper as a "roustabout." A cash-short "shoe-stringer" would "poor boy" a well by splitting his interest with his crew, the landowner, his supply house, his boardinghouse owner, his favorite saloon keeper and, if need be, his most cherished madam, as well.

The boom at Spindletop, with all its madness and frenzy and honky-tonk, was to be repeated many times over in the Southwest in the course of the next few years, beginning with other salt domes along the Gulf Coast of Texas and Louisiana. But the Gulf Coast was about to meet its match in Oklahoma. A string of Oklahoma oil discoveries, beginning in 1901, culminated in the great Glenn Pool, near Tulsa, in 1905. More strikes followed in Louisiana. Meanwhile, North Texas ranchers who were trying to drill for water instead encountered oil, setting off another boom. Still, Oklahoma, not Texas, became the dominant producer in the area, with over half of the region's total production in 1906; only in 1928 did Texas recapture the number-one rank, a position it would continue to hold in the United States until the present day.[8]

Gulf: Not Saying "By Your Leave"

James Guffey, the promoter who had backed Lucas, became a national symbol of instant wealth—on his way, it was said, to being another Rockefeller. That was the appearance, at least. Guffey himself may have even believed it for a while. After all, he had made the largest oil deal in the world, to last twenty years, with Marcus Samuel of Shell. But, by the middle of 1902, within a year and a half of the Spindletop strike, Guffey and his company were in real trouble. The underground pressure gave out at Spindletop because of overproduction, and especially because of all those derricks on postage-stamp-sized plots, and production on the Big Hill plummeted. But the problems of Guffey Petroleum were also of its own making; James Guffey's skills were those of the promoter, not the manager. As a manager, he was about as poor as the quality of his oil.

This situation greatly distressed the Pittsburgh bankers who had put up the original capital to back Guffey and Captain Lucas—Andrew W. and Richard Mellon. Their father, Judge Thomas Mellon, had handed over the family bank to Andrew when he was only twenty-six; and he and his brother had built Mellon and Sons into one of the nation's great banks, central to America's nineteenth-century industrial development. The two brothers had a special feeling of affection and respect for John Galey, Guffey's partner. Galey's father and their own, Judge Mellon, had both come over as small boys from Ireland to the United States on the same boat. They knew John Galey was a great finder of oil, even if they worried about his financial carelessness. In 1900, Galey's partner, Guffey, had managed to convince the Mellons to put up the three hundred thousand dollars for the wildcat at Spindletop, plus several million dollars more to get Spindletop into production. Now, in 1902, only a few months later, with the pressure and flow having given out at Spindletop, the Mellons feared that Guffey would lose not only their money, but also that of the other investors they had brought in on the deal.

They thought they had a solution in the person of their nephew, William C. Mellon, only a decade or so younger than the two banker brothers. One could count on William. At age nineteen, he had heard about an oil strike in a town near Pittsburgh called Economy. The smell of oil, and the excitement of the business, captured him; and he threw himself into it. In the next few years he scrambled all over Appalachia, looking for oil and finding it. He once brought in a thousand-barrel-a-day well in a church graveyard. The church did handsomely out of it.

William knew he was caught up in a fever. "For a great many" of the oil men, he was to recall, "the oil business was more like an epic card game, in which the excitement was worth more than great stacks of chips. . . . None of us was disposed to stop, take his money out of the wells, and go home. Each well, whether successful or unsuccessful, provided the stimulus to drill another." But his uncle Andrew had instilled in him the lesson that such was not the way to run a serious business. Rather, the aim should be to integrate—to control every stage of operations. "The real way to make a business out of petroleum," said Andrew, was "to develop it from end to end; to get the raw material out of the ground, refine it, manufacture it, distribute it." Any other way, and one was at the mercy of Standard Oil.

William acted on his uncle's advice. Despite opposition from Standard Oil and the Pennsylvania Railroad, he built up an integrated oil company, which produced in western Pennsylvania, refined at both ends of the state, transported by its own pipeline, and sold from Philadelphia to Europe. By 1893, the Mellons' company was shipping an estimated 10 percent of total United States exports, and it had a million barrels in storage. Then Standard Oil offered to buy the Mellons out. They were not sentimental; they built businesses and then sold them and went on to something else, and this was the time to cash out on their oil company. The Mellons made a considerable amount of money from the sale. William went into the streetcar business, thinking he was through with oil forever. Now seven years later, and only twenty-seven, William discovered that he was wrong. At the behest of his uncles, he went down to Spindletop to inspect the family's investment. He reported back that they would never get their money out so long as Guffey was in charge.

As they had seven years earlier, the Mellons offered the new enterprise to Standard Oil. But Standard said no because of the legal assaults that Texas kept launching against the company and, in particular, against John D. Rockefeller. "We're out," a Standard director explained. "After the way Mr. Rockefeller has been treated by the state of Texas, he'll never put another dime in Texas."

After that, said a disappointed William Mellon, there was only one solution to "just about as bad a situation as I had ever seen," and that was "good management, hard work, and crude oil." The first obstacle was James Guffey, whom William Mellon regarded as an incompetent blowhard. Mellon took over the management control of the intertwined Guffey Petroleum and Gulf Refining companies, both founded in 1901. Of course, Guffey was deeply resentful; after all, the press had pronounced him the greatest oil man in the United States. Sometimes William Mellon found that he had to be quite arbitrary and harsh with the greatest oil man in the United States.

"The main problem," Mellon said, "was to translate crude petroleum into money." Something had to be done about Guffey Petroleum's contract with Shell, which committed the American company to sell half of its production to Shell for twenty-five cents a barrel for the next twenty years. That contract had been drafted when production seemed unlimited, even unstoppable, when the company needed markets, and when oil was selling for ten or even three cents a barrel—a fine profit by any calculation. Though the contract was to run twenty years, the world had changed a great deal in less than two. In the latter part of 1902 and into 1903, as a result of the plunge in production at Spindletop, oil was selling for thirty-five cents or more a barrel. So, in order to meet the contract, Guffey Petroleum would have to buy oil from third parties and then sell it at a loss to Shell. Guffey may still have thought this was the deal of the century; Mellon certainly did not. He thought it was a rotten deal, and knew that he had to get out of it quickly.

But Marcus Samuel was counting heavily on the contract. Thus, the bad news from Texas—that Guffey's oil supplies had failed—was a great shock. Whatever the pain for Guffey, Shell had every reason to want to keep to the letter of the contract, or if not, to be generously compensated for its cancellation. Samuel ordered that the four new tankers that had been built to transport Texas oil be converted to carry Texas cattle to the East End of London, making the best out of a bad situation. But this was only meant to be a temporary expedient until the oil shipments could be resumed. He prepared to sue; but the outcome of a court battle, an American legal expert warned him, was not at all certain, as the contract had been so poorly and incompetently drawn in the first place.

Andrew Mellon himself came to London to pursue the matter, and traveled down to Kent to talk with Samuel at his estate, the Mote. Mellon "greatly admired the Park," Samuel wrote in his diary of August 18, 1903. The next day, Samuel added to his diary, "Went to London by the 9:27 train upon important business. . . . Had very busy day in negotiating with Mr. Mellon to try to avoid legal proceedings with Guffey Co. but did not succeed in reaching a *modus vivendi* and subsequently consulted solicitors." Andrew Mellon was courteous, charming, mild in manner, but persistent and absolutely firm. By the beginning of September, the two sides did reach a *modus vivendi*, a new agreement. The deal of the century—so critical to Marcus Samuel's vision—was replaced by a contract that guaranteed Shell practically nothing in the way of oil. Guffey Petroleum—and the Mellons—were completely off the hook.[9]

Meanwhile, William Mellon was pursuing a strategy that would be central to the oil industry for the entire twentieth century—to tie together all the disparate activities of the industry and build a coherent, integrated oil company. His strategy was intentionally different from that of Standard Oil. Mellon observed that Standard exerted its power and protected and enhanced its position because it was practically the sole buyer of crude oil and because of its control of transportation. "Standard made the price," said Mellon, and practically every producer was dependent on the company. While producers could and did do well out of the arrangement, they nevertheless were "at the mercy of this company." Mellon worried that, eventually, as more fields were discovered and developed in Texas, Standard would extend its pipeline system into the state, and the Mellons' operation would inevitably become drawn into Standard's production system. That was not what he was after; his ambitions were larger than merely to be an appendage of Standard. Echoing his uncle's lesson, Mellon concluded "that the way to compete was to develop an integrated business which would first of all produce oil. Production, I saw, *had* to be the foundation of such a business. That was clearly the only way for a company which proposed to operate without saying 'by your leave' to anybody." And the Mellons had no intention of saying "by your leave" to anyone, least of all to Standard Oil.

One of the biggest problems facing Mellon was the fact that the capacity of the company's new refinery at Port Arthur was about equal to that of the production of the entire state of Texas. Moreover, it was dependent upon poor-quality oil that could give out at any time. But then, in 1905, with the discovery of the Glenn Pool in Oklahoma, better-quality oil was available. Here was the way out of the problem—oil of "marketable Pennsylvania quality in Texas quantities." But the company would have to move fast. Standard Oil was busy extending its pipeline network from Independence, Kansas. "Unless we could hitch onto that Oklahoma field," Mellon warned his uncles, their whole enterprise might fail. In order to speed the forced-pace construction of a 450-mile pipeline from Port Arthur to Tulsa, Mellon put four crews to work, one starting south from Tulsa; one starting north from Port Arthur; and two starting in the middle and working toward each end. It was a race against time—and against Standard Oil. By October of 1907, oil from Glenn Pool was flowing through the pipeline into the Port

Arthur refinery, and the Mellons were firmly established as major players in the oil industry.

The construction of the pipeline had been matched by corporate reconstruction. The Mellons would not pour money into the existing ramshackle setup. William Mellon engineered a reorganization of Guffey Petroleum and Gulf Refining that resulted in the Gulf Oil Corporation. It was now resolutely a Mellon company. Andrew Mellon became president; Richard B. Mellon, treasurer; and William, vice-president. Guffey was pushed completely aside. "They throwed me out," he bitterly complained later on.

And what became of the pioneers of Spindletop? "Owing to the fact that Mr. Guffey and the Mellon group had a lot of money and I had not," Captain Anthony Lucas subsequently said, "I accepted their offer and sold my interest to them for a satisfactory sum." He set himself up in Washington, D.C., as a consulting engineer and geologist. Three years after his discovery at Spindletop, he returned to Beaumont and surveyed the derrick-covered but now depleted hill, which had been so rapidly overproduced. After traipsing all over the oil field, he was moved to an epitaph. "The cow was milked too hard," he said. "Moreover, she was not milked intelligently."

As for Patillo Higgins, he started a lawsuit against Captain Lucas, who, lacking in sufficient sentiment, had cut him out. He also founded the Higgins Oil Company, but sold out to his partners. He tried to launch an integrated oil company, the Higgins Standard Oil Company, but that venture failed because the public had become wary of any more stock offerings bearing the imprint of "Swindletop." Still, it seems that Higgins made a sizeable amount of money along the way, and thirty-two citizens of Beaumont once signed a public letter declaring that he deserved "the whole honor of discovering and developing" Spindletop. He had not been so crazy after all.

Neither James Guffey nor John Galey was able to hold on to his money. "Difficult times came upon both men as they aged, and a comeback became less and less attainable," wrote Galey's nephew. "They had muffed numerous opportunities to attain great wealth because, perhaps, of not playing the trump card at the right time. Such opportunity rarely comes. Spindletop was the great venture of Guffey and Galey as a partnership. Thereafter they struggled with trifling drilling projects here and there, largely financed through their waning prestige as the greatest oil finders of the first half-century of petroleum in this hemisphere."

Guffey, the promoter, spent the last decades of his long life—he lived to the age of ninety-one—deeply in debt. His residence in a mansion on Fifth Avenue, in Pittsburgh, was maintained until his death through the courtesy of his creditors. Galey, the oil finder, was paid only a "dribble" of the $366,000 that Guffey owed him as a result of their Spindletop deal. Toward the end of his life, Galey toured parts of Kansas, sniffing out deals, in the company of Al Hamill, who had been the driller at Spindletop. One day, a heavy snow came up, and they could not get about. So the two men decided to call it quits and head home. Then Galey made a painful admission. He had never been so poor in his entire life as he was right then. Could Hamill cash a check signed by Mrs. Galey? Instead, Hamill paid Galey's hotel bill and put him onto the train home through the snow. That was the last try at an oil deal by John Galey, the man who could smell oil; he died soon after.

As for William Mellon, he served for many years as president and chairman of Gulf Oil, as it became one of the major oil companies of the world. In 1949, shortly before his death, he remarked, "The Gulf Corporation has grown so big I have lost track of it."[10]

Sun: "To Know What to Do with It"

Among the thousands and thousands who descended from the train in Beaumont, Texas, on the news of Captain Lucas's discovery was one Robert Pew, who arrived just six days after the gusher at Spindletop, on the instruction of his uncle J. N. Pew. Robert Pew quickly saw the opportunity afforded not only by the oil but by the good transportation prospects available via the Gulf of Mexico. He did not, however, like the weather or the town or the people or the boom, or much of anything else about Texas, and he became ill and left. He was replaced by his brother J. Edgar Pew, who arrived packing a revolver, which both his brother and uncle had insisted he would need for personal protection in the brawling atmosphere of Beaumont.

The Pews may have been strangers to Beaumont, but not to oil; they had already been in the hydrocarbon business for a quarter century. In 1876 in western Pennsylvania, J. N. Pew and a partner had begun to collect natural gas, then regarded as a waste product, and to sell it—first as an oil field fuel. In 1883, they became the first group to supply a major city—Pittsburgh—with natural gas as a substitute for manufactured town gas. They built up a substantial business. But Standard Oil had turned its attention to gas, forming the Natural Gas Trust in 1886, and eventually J. N. Pew followed the same track as the Mellons with their first venture in oil in the 1890s; he sold his gas business to Standard.

Pew had also begun to produce oil from the Lima field in 1886. Searching the heavens for a body to name his new company after, he finally decided on the sun because of its prominence above all other bodies in the sky. The Sun Oil Company did not achieve similar prominence in the industry during the next decade and a half, but it did manage to carve out a respectable oil business in the shadow of Standard Oil.

Upon arriving in Beaumont in 1901, J. Edgar Pew acquired leases for the Sun Oil Company; but he and his family knew from previous experience that production was not enough. "You could buy millions of barrels of oil at five cents a barrel," J. Edgar was later to say, "but the point was to know what to do with it." So Sun also acquired storage facilities in the region. At the same time, it built a refinery at Marcus Hook, outside Philadelphia, to receive the Texas crude shipped by boat, and set about developing long-term markets. As Spindletop's decline became evident, the company expanded elsewhere in Texas, acquiring production and establishing its own major pipeline system in the region. By 1904, Sun was one of the handful of companies preeminent in the Gulf Coast oil trade.[11]

"Buckskin Joe" and Texaco

One more major oil company was to be born out of the maelstrom at Spindletop. It was the handiwork of Joseph Cullinan, who was among the foremost pioneers of Texas oil development. In 1895, Cullinan had left a promising career in Standard's pipeline arm to form his own oil equipment company in Pennsylvania. He had acquired the nickname "Buckskin Joe," because his aggressive, abrasive personality and his drive to get a job done reminded those who worked for him of the rough leather used for oil field gloves and shoes.

In 1897, Cullinan was invited to make a quick visit to Corsicana, Texas, to advise the town fathers on further oil development. Instead of merely advising, he settled in, and became the dominant oil figure in Corsicana. Within a day of Captain Lucas's gusher at Spindletop, Cullinan was on the spot in Beaumont to inspect the scene. He knew instantly that this was something wholly different and on a much greater scale than Corsicana. His first step in Beaumont was to create the Texas Fuel Company, for crude oil purchasing and marketing. Cullinan's equipment expertise came in handy; his Texas Fuel Company had an advantage over would-be competitors because Cullinan had already built storage facilities just twenty miles away.

Soon Cullinan also gained control of valuable leases that a syndicate of former politicians had accumulated on Spindletop itself. The syndicate was led by James Hogg, the three-hundred-pound ex-governor and progressive champion of Texas. The former governor was also a tough businessman: "Hogg's my name," he once explained, "and hog's my nature." Hogg's group had acquired its key lease position from James Guffey, who, whatever his failings as a manager, had the sound political instincts appropriate to a former chairman of the Democratic party. For, Guffey later explained, the sale of such obviously valuable leases was the price of political insurance. "Northern men were not well respected in Texas in those days," he said. "Governor Hogg was a power down there and I wanted him on my side because I was going to spend a lot of money." Hogg had a more specific virtue as well; he was the great opponent in Texas of Standard Oil. While governor, he had even tried to extradite Rockefeller from New York to stand trial, and Hogg's participation provided some protection against Standard's familiar tactics when confronted with a new adversary.

For the capital he needed to develop his leases, Cullinan turned to Lewis H. Lapham, a New Yorker who owned U.S. Leather, the centerpiece of the leather trust, and John W. Gates, a flamboyant Chicago financier known as "Bet-a-Million" Gates because of his willingness to make a wager on anything. To his Texas partners, who worried about the predominance of "foreign" capital, Cullinan reassuringly declared, "The Tammany crowd will find their match in the Southerners." His prediction would prove true—up to a point.

Cullinan, with his wide experience and natural talent for leadership, quickly emerged as the foremost oil man in Beaumont. When a flaming inferno swept through Spindletop in September 1902, he commanded the efforts to control the fire; and this he did, virtually nonstop, for a week, until the fire was out and he collapsed with exhaustion. His eyes seared by the gas fumes, he even lost his sight for a few days; but, confined to bed with bandages around his eyes, he continued to hold conferences and provide direction. Among those working for Cullinan were Walter B. Sharp, who had drilled Patillo Higgins's first unsuccessful attempt on Spindletop in 1893 and was now a premier driller, and another expert driller named Howard Hughes, Sr. In the spring of 1902, Cullinan established the Texas Company in order to consolidate his various operations and better enable him to exert his personal and autocratic control.

Unlike James Guffey, Cullinan knew how to manage an oil company, and unlike Guffey-Gulf, the Texas Company was profitable from the beginning. In its first year of business, it sold its oil at an average price of sixty-five cents a barrel. Since Cullinan had put the oil into storage during the time of flush production, at an average price of twelve cents a barrel, the company did very well. The Mellons, trying to sort out their Guffey problems, almost consummated a merger of Gulf with Cullinan's Texas Company. But the smaller oil producers, raising the specter of a new oil trust, managed to turn the proposed deal into the hottest issue in the Texas legislature; the chief lobbyists for each side even ended up having a very public fist fight in a hotel lobby in Austin. Finally, the Texas legislature came out against the merger, and that killed its chances.

Cullinan then turned his full attention to expanding the Texas Company. It built its own pipeline from the Glenn Pool in Oklahoma down to Port Arthur in Texas. It registered the name Texaco as a trademark in 1906, and came up with the green "T" superimposed on the red star as its symbol. It began manufacturing gasoline, and by 1907, only six years old, it was able to exhibit a full range of some forty different products at the Dallas State Fair. By 1913, its gasoline production had overtaken illuminating oils as its most important product. Early on, Cullinan had predicted "that the time will come—perhaps in no distant day—when we will want our general office in Houston instead of Beaumont, as . . . Houston seems to me to be the coming center of the oil business for the Southwest." Soon after, braving the oppressively steamy heat of Houston's summer, he moved the office to that city, though significant parts of the business were also run from New York.

Buckskin Joe's autocratic style of management began to grate on his investors and led to the first of the clashes between Texas and New York that would shape the company. One of the senior executives wrote Lapham to complain that Cullinan "thinks he knows everything and must butt into everything. . . . He looks upon us here in New York as the tail of the dog, and a very small tail at that." When the major stockholders tried to rope Cullinan in, he rebelled and launched a proxy fight to try to regain control. The transplanted Pennsylvanian sought to turn the battle into a sectional struggle, Texas versus the East. In his statement to stockholders, he proclaimed that the company's "original management, its corporate attitude and activities were branded with the name *Texas* and Texas ideals," and that its "headquarters and governing authorities should be kept and maintained in Texas." But, of course, that was not what the fight was all about. The real issue was Cullinan's one-man rule.

New York had the votes, and Cullinan was badly defeated in the proxy fight. He tried to be philosophical. "It was a good boarding-house brawl," Buckskin Joe wrote to an old associate from Pennsylvania, "and some furniture was broken but our side was whipped fair and I'll be looking for another job soon." He did and went on to new successes in oil. But thereafter he stuck to exploration and producing, and left refining and marketing to others.[12]

"How Can We Control It?"

The development of the new oil fields of the Gulf Coast and the midcontinent undermined the seemingly impregnable position of Standard Oil. These new sources of oil, combined with the rapidly emerging markets for fuel oil and gasoline, opened the doors to a host of new competitors that, as William Mellon had put it, did not have to say "by your leave" to Standard or anyone else. To be sure, Standard's sales had continued to grow in absolute terms. Its sales of gasoline, reflecting the new age, more than tripled between 1900 and 1911 and, indeed, by 1911, for the first time exceeded those of kerosene. And Standard Oil was attuned to the further technological changes that were at hand. When the Wright Brothers' airplane first flew into the air at Kitty Hawk, North Carolina, in 1903, its engine burned gasoline and used lubricants that had been brought to the beach in wooden barrels and blue tin cans by salesmen from Standard Oil. But, in terms of overall market shares in oil products in the United States, Standard's position of overwhelming dominance was receding. Its control of refining capacity declined from over 90 percent in 1880 to only 60 to 65 percent by 1911.

As a result of the explosion of production on the Gulf Coast, the Old House also saw its control over crude oil production in the United States—and its ability to "establish" prices—slipping away. At the same time, development of crude sources abroad was reducing its power in the international marketplace. Of course, Standard's position seemed impregnable to those on the outside, but that was not how it was seen from inside the Old House. "Look at things now— Russia and Texas," Standard director H. H. Rogers lamented to a visitor. "There seems to be no end of the oil they have there. How can we control it? It looks as if something had the Standard Oil Company by the neck." It was, he added ominously, "something bigger than we are."[13]

CHAPTER 5

The Dragon Slain

THE OLD HOUSE was under siege. Its commercial competitors, both in the United States and around the world, could not be overcome. Moreover, a political and judicial war was being waged throughout the United States against Standard and its ruthless business practices. It was not a new challenge; Rockefeller and his associates had been criticized and vilified from the inception of the Standard Oil Trust. Standard Oil executives never really understood such criticism. They thought it was cheap demagoguery, uninformed jealousy, and special pleading. They were sure that, in its relentless pursuit of its own interests and enrichment, Standard Oil was not only checking the scourge of "unbridled competition," but was also truly, as Rockefeller himself said, perhaps the greatest of "upbuilders" that the nation had ever known.

To the public at large, however, that was not at all how things looked. Standard's critics saw a powerful, devious, cruel, entrenched, all-pervasive, and yet mysterious enterprise. It was accountable to no one except a handful of arrogant directors, and it mercilessly tried to destroy all who stood in its way. This view was part of the prospect of the age. The growth of Standard Oil had not occurred in a vacuum. It was a product of the swift industrialization of the American economy in the last few decades of the nineteenth century, which within a remarkably short time had transformed a decentralized and competitive economy of many small industrial firms into one dominated by huge industrial combinations called trusts, each one sitting astride an industry, many with interlocking investors and directors. This rapid change was deeply alarming to many Americans. As the nineteenth century gave way to the twentieth, they looked to government to restore competition, control the abuses, and tame the economic and political power of the trusts, those vast and fearsome dragons that roamed so freely across the country. And the fiercest and most feared of all the dragons was Standard Oil.

The Holding Company

The renewed legal assaults against Standard began from the states, with anti-monopoly suits brought by Ohio and Texas. In Kansas, the governor pushed a scheme to build a state-owned oil refinery, which would compete with Standard's and would be staffed by penitentiary inmates. At least seven other states, plus the territory of Oklahoma, launched legal actions of one kind or another. But Standard was slow to apprehend the full extent of the popular opposition to its business practices. "I think this anti-Trust fever is a craze," one senior executive wrote Rockefeller in 1888, "which we should meet in a very dignified way & *parry every question* with answers which while perfectly truthful are evasive of *bottom facts*." The company continued to keep everything as secret as it could. When Rockefeller testified in one of the Ohio suits, he was so unforthcoming that a New York newspaper headlined, "John D. Rockefeller Imitates a Clam."

Moving to marshal all necessary resources to the battle at hand, Standard hired the best and most expensive legal talent. It also sought to influence the political process, perfecting the art of the timely political contribution. "Our friends do feel that we have not received fair treatment from the Republican Party," wrote Rockefeller when forwarding a contribution to the party in Ohio, "but we expect better things in the future." But Standard Oil did not stop with contributions. It put the Republican Senator from Ohio on a legal retainership—his fee in 1900 alone was $44,500. And it considerately made loans to a powerful Senator from Texas, then known as the "foremost Democratic leader in America," who needed money to pay for a six-thousand-acre ranch he had purchased outside Dallas. It used an advertising agency that, in the course of purchasing advertising space in newspapers, also planted news articles friendly to Standard Oil. It set up or took over what were called "blind tigers"—companies that looked to the outside world like totally independent distributors, but of course were not. In 1901, for instance, a company named Republic Oil was established to market in Missouri. Its advertisements bore such headlines as "No Trust" and "No Monopoly" and "Absolutely Independent." But it secretly reported to 75 New Street in New York, which just happened to be the back door of 26 Broadway.

While some of the states achieved temporary victories against Standard, none ultimately succeeded in their attacks. In one instance, after the Standard Oil companies were expelled from Texas and their properties put into receivership, the receivers convened a meeting in the Driskill Hotel in Austin to sell off all the properties. And sell them off they did—to agents of Standard Oil.[1]

Still, the legal assaults forced further changes in Standard's organization. In 1892, in response to a court decision in Ohio, the trust was dissolved and the shares were transferred to twenty companies. But control remained with the same owners. The companies were grouped together as the "Standard Oil Interests." Under this new arrangement, the Executive Committee at 26 Broadway gave way to an informal meeting of presidents of the various constituent firms that constituted the Standard Oil Interests. Letters were no longer addressed to the Executive Committee, but rather simply to the "gentlemen upstairs."

But the "gentlemen" were not happy with the reorganization of the "Standard Oil Interests." Further protection was necessary in response to continuing pressures and in order to put the company on a firmer legal foundation. They found the solution to their problems in New Jersey. That state had revised its laws to permit the establishment of holding companies—corporations that could own stock in other corporations. It was a decisive break with traditional business law in the United States. New Jersey also sought to make its business environment hospitable to this new form of combination. Thus, in 1899, the owners of the Standard Oil Interests established Standard Oil of New Jersey as the holding company for their entire operation. Its capitalization was increased from $10 million to $110 million, and it held stock in forty-one other companies, which controlled yet other companies, which in turn controlled still other companies.

During this time, a momentous change of another kind also took place within Standard Oil. John Rockefeller had already amassed vast wealth, he was tired, and he began to plan for retirement. Though he was only in his mid-fifties, the constant strain of business, and of the attacks, was taking its toll. After 1890, his complaints of digestive problems and fatigue had become more frequent. He said that he was being crucified. He took to keeping a revolver by his bed at night. In 1893, he came down with a stress-related disease, alopecia, which not only caused him a good deal of physical distress, but also robbed him of all his hair—which, afterward, he sought to remedy variously with a skullcap or a wig. His formerly spare form now gave way to corpulence. His plans to step aside were temporarily postponed by a series of crises—the Panic of 1893 and the ensuing depression, and the growing vigor of competition both at home and abroad. Still, Rockefeller began to distance himself, and finally, by 1897, he had—not yet sixty years of age—stepped aside, turning administrative leadership over to one of the other directors, John D. Archbold.[2]

The Successor: The Oil Enthusiast

There had been little question but that John Archbold would be the successor. More than any other of the senior Standard executives, he was expert in all phases of the business. He had been one of the most powerful figures in the American oil industry during the preceding two decades; for the next two decades, he would be *the* most powerful. His was a long career.

Short, and younger-looking than his age, Archbold was determined and indefatigable, always keen to "go ahead," and totally consumed by the demands and rightness of his cause. As a boy, during the Presidential campaign of 1860, he had sold badges bearing the likenesses of the candidates. His brother took the better sales district; John far outsold him. At age fifteen, with the blessing of his Methodist minister ("God is willing that he should go"), Archbold boarded a train by himself in Salem, Ohio, to seek not his salvation but rather his fortune in

Titusville and in oil. He started off as a shipping clerk, his salary so meager that he slept on a bed under the office counter. He became an oil broker—always in motion, caught up then, as for the rest of his life, in what was known as "oil enthusiasm." Such enthusiasm was badly needed in the helter-skelter of the Oil Regions. "His daily round then was a hard job," an associate was to recall of the young oil broker. "There was always a foot or more of oil-soaked mud in the main streets of Titusville, and around the wells along the Creek it was just as bad, sometimes up to a man's thigh, but John Archbold cared nothing for it. He would wade through it, lilting a song if there was oil to be bought or bargained for."

Archbold had no diversions other than work. He learned to use humor to defuse a tense situation, which became most valuable during the subsequent controversies and strife. Much later, when asked if Standard Oil had looked out only for its own interests, he dryly replied, "We were not always entirely philanthropic." He also learned how to keep events, no matter how troubling, in perspective. He figured out how to make himself, and prove himself, very useful to others—particularly to John D. Rockefeller. He had caught Rockefeller's eye early, in 1871, when on registering at a hotel in Titusville, Rockefeller had seen a signature above his own. It was that of the young broker and refiner, who had signed in as "John D. Archbold, $4 a barrel." Rockefeller was taken by such self-confident advertising—at a time when oil could not fetch anywhere near such a price—and made special note.

An activist, Archbold became secretary of the Titusville Oil Exchange. During the affair of the South Improvement Company and the Oil War of 1872, when Rockefeller and the railroads tried to monopolize control over the output of oil, he emerged as one of the leaders of the Oil Regions, denouncing Rockefeller in most scathing terms. Yet Rockefeller recognized someone who grasped the fundamentals of the Oil Regions, a man totally dedicated to the business, who could be aggressive and ruthless, and yet was flexible and adaptable. That last certainly proved to be the case in 1875, when Rockefeller invited him into the combine. Archbold swiftly accepted. His first task was to acquire secretly all the refineries along Oil Creek. He took up the charge with absolute determination. In a period of a few months, he bought or leased twenty-seven refining properties—and worked himself into a serious physical collapse.

Archbold rose quickly toward the top of Standard Oil. "He would make up his mind with one flash of his dark snapping eyes, and then was smiling again," recalled one of his colleagues. But he still had to clear one major obstacle with Rockefeller—his "unfortunate failing," as it was called. He liked alcohol too much, and Rockefeller insisted that he sign the temperance pledge—and stick to it. He did what Rockefeller wanted. And now, just fifty, yet already a veteran of more than three decades in the oil industry, Archbold brought vigor and experience to his new post as the number-one man in Standard Oil. Rockefeller, while remaining in touch with 26 Broadway, from then on devoted himself to his estates, his philanthropy, his golf, and the management of his money, which was ever increasing. Between 1893 and 1901, Standard Oil paid out more than $250 million in dividends, of which by far the greater part went to a half dozen men—and fully one-quarter of the total to Rockefeller. Such was the cash mountain that Standard Oil threw forth that one financial writer described the company as "really a bank of the most gigantic character—a bank within an industry, financing this industry against all competitors."

Meanwhile, Rockefeller, relieved of day-to-day responsibility, regained his health under his new regimen. In 1909, his doctor predicted that he would live to be a hundred because he followed three simple rules: "First, he avoids all worry. Second, he takes plenty of exercise in the open air. Third, he gets up from the table a little hungry." Rockefeller kept abreast of developments in the company, but he did not actively involve himself in its management. Nor would Archbold have allowed that.

Archbold did visit Rockefeller on Saturday mornings to discuss the business with its largest stockholder. And Rockefeller retained the title of president, which proved to be a major error of judgment. In adherence to Standard's policy of complete secrecy, no effort was ever made to make his retirement known, and so Rockefeller would still be held personally responsible for whatever Standard Oil did. Thus, insofar as the public was concerned, Rockefeller continued to be synonymous with Standard Oil. He was the lightning rod for all the criticism, all the rage, all the attacks. Why did he retain the president's title? His colleagues may have thought that his name was needed to hold the empire together—the factor of awe. Perhaps it was out of due respect for his stock holdings. But shortly after the turn of the century, one of the other senior directors, H. H. Rogers, privately offered quite another reason: "We told him he had to keep it. These cases against us were pending in the courts; and we told him that if any of us had to go to jail, he would have to go with us!"[3]

"The Red Hot Event"

The assault on Standard Oil gained force at the end of the nineteenth century. A powerful new spirit of reform—progressivism—was gaining ascendancy in America. Its principal aims were political reform, consumer protection, social justice, better working conditions—and the control and regulation of big business. The last had emerged as an urgent issue as a great merger wave swept across America, with rapid growth in the number of trusts. The Standard Oil Trust, the nation's first, had been established in 1882. But the movement toward combination really gathered speed in the 1890s. According to one count, 82 trusts with a total capitalization of $1.2 billion had been formed before 1898. An additional 234 trusts were organized, with a total capitalization in excess of $6 billion, between 1898 and 1904. Some saw the trust—or monopoly—as capitalism's ultimate achievement. For others it was a perversion of the system that threatened not only farmers and laborers, but also the middle classes and entrepreneurial businessmen, who feared that they would be economically disenfranchised. The trust issue was characterized in 1899 as "the great moral, social and political battle that now confronts the whole Union." Trusts were one of the most important issues in the Presidential campaign of 1900, and shortly after his victory, President William McKinley told his secretary, "The trust question has got to be taken up in earnest and soon."

One of the first to take it up, Henry Demarest Lloyd, had continued his scathing attacks on Standard Oil in book form, in *Wealth Against Common-wealth*, in 1894. In his wake, a group of fearless journalists set about to investigate and publicize the evils and ills of society. These writers, who set the progressive agenda, were to become known as "muckrakers," and they were at the center of the progressive movement. For, as one historian has observed, "The fundamental critical achievement of American Progressivism was the business of exposure." At the top of the agenda was the exposure of business.

The magazine that touched off the whole muckraking campaign was *McClure's*. It was one of the country's leading periodicals, with a circulation in the hundreds of thousands. Its publisher was the temperamental, expansive, and imaginative Samuel McClure. He was also an idiosyncratic man; on one trip to Paris and London, he collected a thousand neckties. He had already collected a talented group of writers and editors back in New York, and they were eagerly looking for a large theme. "The great feature is Trusts," McClure wrote to one of them in 1899. "That will be the red-hot event. And the magazine that puts the various phases of the subject that people want to be informed about will be bound to have a good circulation."

The editors of the magazine decided to focus on one specific trust to illustrate the process of combination. But which one? They debated the Sugar Trust and talked about the Beef Trust, but discarded both. One of the writers then suggested the discovery of oil in California. No, replied the managing editor, a woman named Ida Tarbell. "We have got to find a new plan of attacking it," she said. "Something that will show clearly not only the magnitude of the industries and commercial developments, and the changes they have brought in various parts of the country, but something which will make clear the great principles by which industrial leaders are combining and controlling these resources."[4]

Rockefeller's "Lady Friend"

By this time, Ida Minerva Tarbell had already established herself as America's first great woman journalist. She was a tall woman, six feet, with a grave, quiet authority about her. After graduating from Allegheny College, she had gone off to Paris to write a biography of Madame Roland, a leader of the French Revolution who ended up on the guillotine. Tarbell devoted herself to career and never married, though later in life she was to become a celebrant of family life and an opponent of women's suffrage. At the beginning of the twentieth century, she was in her mid-forties, and already well known as the author of popular, but carefully crafted, biographies of Napoleon and Lincoln. Her manner and her appearance made her seem older than her age. "Her life largely consisted of holding people off," recalled another woman who was the literary editor at *McClure's*. "She seemed to the naked eye to have no coquetteries at all." With the issue of trusts firmly on the table at *McClure's*, Tarbell considered undertaking her own investigation. The obvious target was the Mother of Trusts; she decided to take it on. Making a pilgrimage with McClure to a mud bath at an ancient spa in Italy, she won his approval. So Ida Tarbell began the research that would eventually topple Standard Oil.

Life is not without its ironies, and the book that emerged from Tarbell's research would stand as the final revenge of the Oil Regions against their conquerors. For Ida Tarbell had grown up in the boom-and-bust communities of the Oil Regions. Her father, Frank Tarbell, had gone into business as a tank maker just months after Drake's discovery, and in the 1860s had done rather well, setting himself up, for a time, in the great boom town of Pithole. When the field there suddenly gave out and the bustling little metropolis went to ruin, he paid six hundred dollars for the town's leading hotel, which had just been built for sixty thousand dollars. He tore it down, piled up wagons with the French windows, the fine doors and woodwork, the lumber, and the iron brackets, and carried them all off to Titusville, ten miles distant, where he used them to build a handsome new house for his family. In that remnant and reminder of one of the most extreme of all the booms-and-busts, Ida Tarbell spent her adolescence. (Later, she considered writing the story of Pithole—"nothing so dramatic as Pithole in oil history," she said.)

Frank Tarbell allied himself with the independent oil producers in 1872 in the Oil War against the South Improvement Company; and thereafter, like so many in the Oil Regions, his working life was to be dominated by the struggles against the advance of Standard Oil and the pain that went with it. Later, Ida Tarbell's brother William was to become one of the senior officers of the independent Pure Oil Company and set up its German marketing operation. From both her father and brother, Tarbell imbibed the precariousness of the business—it was like "playing cards," as her brother William put it. "Often I wish I was in some other business and if I ever hit it rich," he wrote her in 1896, "you bet I'll put most of it into something safe." She remembered the agonies and financial difficulties her father had endured—the mortgaged house, the sense of failure, the apparent helplessness against the Octopus, the bitterness and divisions between those who did and those who did not come to terms with Standard Oil.

"Don't do it, Ida," her now-elderly father implored her when he learned that she was investigating Standard Oil for *McClure's*. "They will ruin the magazine."

One evening, at a dinner party given by Alexander Graham Bell in Washington, the vice-president of a Rockefeller-aligned bank took Tarbell aside; he seemed to be politely threatening exactly what her father had warned her about when he raised a question about the condition of *McClure's* finances.

"Well, I'm sorry," Ida Tarbell replied sharply, "but of course that makes no difference to me."[5]

She would not be stayed. An indefatigable and exhaustive researcher, she also became a sleuth, absorbed and obsessed by her case, convinced that she was on to a great story. Her research assistant, whom she sent traipsing down the back streets of Cleveland to search out those who had reason to remember, wrote her, "I tell you this John D. Rockefeller is the strangest, most silent, most mysterious, and most interesting figure in America. The people of this country know nothing about him. A brilliant character study of him would make a tremendous trump card for *McClure's*." Tarbell intended to play that card.

But how was she going to gain access directly to Standard? Help came from an unexpected direction. After John Archbold, H. H. Rogers was the most senior and powerful director of Standard Oil, as well as a prominent speculator in his own right. He was responsible for Standard's pipeline and natural gas interests. But Rogers's own interests did not end with business. In one of the great services to American letters, he had, a decade earlier, taken control of Mark Twain's tangled and bankrupt finances, put them right, and thereafter managed and invested the famous author's money so that Twain could, as Rogers instructed him, "stop walking the floors." Rogers once explained, "It rests me to experiment with the affairs of a friend when I am tired of my own." Rogers loved Twain's books, and would read them aloud to his wife and children. The two men became very close friends; Twain played billiards on a table Rogers had given him. But, when it came to his own business, Rogers was a very tough man, with little sentimentality. It was he, after all, who had once made the classic statement to a commission investigating Standard Oil, "We are not in business for our health, but are out for the dollars." In *Who's Who*, he listed himself simply as "Capitalist"; others called him "Hell Hound Rogers" because of his speculative forays into Wall Street. He thought that Rockefeller disapproved of him because he was, in his own words, "a born gambler." And, indeed, with the stock market closed on the weekends, Rogers, itching for some action, would invariably start up a poker game.

It was at Twain's urging that Rogers took over the financing of the education of the blind and deaf Helen Keller, enabling her to go to Radcliffe. Twain himself was ever grateful to Rogers, once describing him not only as "the best friend I ever had," but also, "the best man I have known." Ironically, Twain, a sometime publisher, had been offered the opportunity to publish Henry Demarest Lloyd's attack on Standard Oil, *Wealth Against Commonwealth*. "I wanted to say," he wrote to his wife, "the only man I care for in the world; the only man I would give a *damn* for; the only man who is lavishing his sweat and blood to save me & mine from starvation and shame, is a Standard Oil fiend. . . . But I didn't say that. I said I didn't want *any* book; I wanted to get out of the publishing business."

Twain came and went as he pleased from Rogers's office at 26 Broadway, and sometimes lunched with the "gentlemen upstairs" in their private dining room. One day Rogers mentioned that he had learned that *McClure's* was preparing a history of Standard Oil. He asked Twain to find out what kind of history. Twain was also a friend of McClure's, and he inquired of the publisher. One thing led to another, and Twain ended up arranging for Ida Tarbell to meet Rogers. She now had her connection.

Her meeting with Rogers took place in January 1902. She was apprehensive about encountering the powerful Standard Oil tycoon face-to-face. But Rogers greeted her warmly. He was, she immediately decided, "by all odds the handsomest and most distinguished figure in Wall Street." They swiftly established a special rapport, for it emerged that, when Tarbell was a small girl, Rogers had lived not only in the same town in the Oil Regions, running a little refinery, but on a hillside just below the Tarbell family. He told her how he had lived in a rented house—at a time when to live in a rented house was a "confession of failure in business"—in order to be able to have more money to buy stock in Standard Oil. He said that he well remembered Tarbell's father and the sign for "Tarbell's Tank Shops." He said that he had never been happier than in those early days. He may have been sincere—or a good psychologist who had done his homework. He succeeded in charming Ida Tarbell; years later she was still to call him fondly "as fine a pirate as ever flew his flag in Wall Street."

Over the next two years, she met regularly with Rogers. She would be ushered in one door and out another; company policy forbade visitors to encounter one another. She was sometimes even granted the use of a desk at 26 Broadway. She would bring case histories to Rogers, and he would provide documents, figures, justifications, explanations, interpretations. Rogers was surprisingly candid with Tarbell. One winter day, for instance, she boldly asked him in what way did Standard "manipulate legislation."

"Oh, of course, we look after it!" he replied. "They come in here and ask us to contribute to their campaign funds. And we do it—that is, as individuals. . . . We put our hands in our pockets and give them some good sums for campaign purposes and then when a bill comes up that is against our interests we go to the manager and say: 'There's such and such a bill up. We don't like it and we want you to take care of our interests.' That's the way everybody does."

Why was he so forthcoming? Some suggested that it was a form of revenge against Rockefeller, with whom he had fallen out. He himself offered a more pragmatic explanation. Tarbell's work, he told her, "will be taken as a final expression on the Standard Oil Company," and, since she was going to write it in any event, he wanted to do everything he could to have the company's case "made right." Rogers even arranged for her to see Henry Flagler, by then already deeply immersed in his own grand development of Florida. To Tarbell's irritation, all Flagler would say—piously—was that "we were prospered," apparently by the Lord. Rogers had broadly hinted that he might be able to deliver an interview with Rockefeller himself, but it did not eventuate. Rogers never said why.

Tarbell's overall goal, she told a colleague, was "a narrative history of the Standard Oil Company." It was not "intended to be controversial, but a straightforward narrative, as picturesque and dramatic as I can make it, of the great monopoly." Rogers—proud of his accomplishments and of his company—was under the same impression.[6]

Whatever Tarbell's original intent, her series—which began appearing in *McClure's* in November 1902—proved to be a bombshell. Month after month, she spun the story of machination and manipulation, of rebates and brutal competition, of the single-minded Standard and its constant war on the injured independents. The articles became the talk of the nation and opened doors to new informants. After the first few months, Tarbell returned to Titusville to see her family. "It is very interesting to note now that the thing is well underway, and I have not been kidnapped or sued for libel as some of my friends prophesied," she said, "people are willing to talk freely to me." Even Rogers continued to receive her cordially as the articles were coming out, despite all. But then she published an installment that revealed how Standard's intelligence network operated, putting intense pressure on even the smallest of the independent retailers. Rogers was furious. He broke off their relationship and refused to see her again. She remained totally unrepentant about what she had written. More than anything else, she later said, the "unraveling of this espionage charge . . . turned my stomach against the Standard." For "there was a littleness about it that seemed utterly contemptible compared with the immense genius and ability that had gone in to the organization. Nothing about the Standard had ever given me quite the feeling that that did." And that feeling, more than anything else, gave the acid edge to her labors and to her exposé.

Altogether, Tarbell's series ran for twenty-four successive months, and was then published in November 1904 in book form as. *The History of Standard Oil Company*, complete with sixty-four appendices. It was a work of great clarity and force, a considerable accomplishment—under the handicap of limited access—in its mastery of the complex history of the company. But beneath its controlled surface coursed a raging anger and a powerful condemnation of Rockefeller and the cutthroat practices of the Trust. In Tarbell's narrative, Rockefeller, despite his much-professed devotion to Christian ethics, emerged as an amoral predator. "Mr. Rockefeller," she wrote, "has systematically played with loaded dice, and it is doubtful if there has been a time since 1872 when he has run a race with a competitor and started fair."

The publication of the book was a major event. One journal described it as "the most remarkable book of its kind ever written in this country." Samuel McClure told Tarbell, "You are today the most generally famous woman in America. . . . People universally speak of you with such reverence that I am getting sort of afraid of you." Later, from Europe, he reported that even in the Continental newspapers "your work is constantly mentioned." As late as the 1950s, the historians of Standard Oil of New Jersey, hardly friendly to Tarbell's book, were to declare that it "probably has been more widely purchased and its contents more widely disseminated throughout the general public than any other single work on American economic and business history." Arguably, it was the single most influential book on business ever published in the United States. "I never had an animus against their size and wealth, never objected to their corporate form," Tarbell explained. "I was willing that they should combine and grow as big and rich as they could, but only by legitimate means. But they had never played fair, and that ruined their greatness for me."

Ida Tarbell was not yet quite done with her story. She followed up in 1905 with a final attack, a furious personal portrait of Rockefeller. "She found him," her biographer has written, "guilty of baldness, bumps and being the son of a snake oil dealer." Indeed, she took his physical appearance, including his illness-induced baldness, as a sign of moral decrepitude. Perhaps it was the ultimate revenge of a true daughter of the Oil Regions. For, as she was finishing that last article, her father, one of the independent oil men who had fought Rockefeller and been vanquished, lay dying in Titusville. As soon as she completed the manuscript, she rushed off to his deathbed.

And what of Rockefeller's reaction? As the articles were coming out, an old neighbor, dropping in to visit the oil tycoon, brought up the subject of what he called Rockefeller's "lady friend"—Ida Tarbell.

"I tell you," Rockefeller replied, "things have changed since you and I were boys. The world is full of socialists and anarchists. Whenever a man succeeds remarkably in any particular line of business, they jump on him and cry him down."

Afterward, the neighbor described Rockefeller's attitude as that "of a game fighter who expects to be whacked on the head once in a while. He is not the least disturbed by any blows he may receive. He maintains that Standard has done more good than harm." On other occasions, Rockefeller was overheard to use a pet name for his "lady friend"—"Miss Tar Barrel."[7]

The Trust-Buster

Tarbell was by no means a socialist. If there was a program to her attack on Standard Oil, it was an appeal for a countervailing force to corporate power. To Theodore Roosevelt, who had become President in 1901 upon the assassination of William McKinley, the countervailing force could be only one—government.

Theodore Roosevelt embodied the progressive movement. The youngest man ever to enter the White House up to that time, he was forever bursting with energy and enthusiasm. He was described as "a steamroller in trousers" and as "the meteor of the age." A journalist wrote that, after visiting him, "you go home and wring the personality out of your clothes." With equal passion, Roosevelt embraced reform causes of all sorts—from the mediation of the Russo-Japanese War to the promotion of simplified spelling. For the former he received the Nobel Peace Prize in 1906. As to the latter, in the same year, he sought to have the Government Printing Office adapt three hundred simplified spellings of familiar words—for instance, "dropt" for "dropped." The Supreme Court refused to accept such simplifications in legal documents, but Roosevelt steadfastly kept to them in his own private letters.

It was he who first used the term "muckraker" to describe the journalists of the progressive movement. He meant it derisively, for he felt that their attacks against politicians and corporations were too negative and too focused on the "vile and debasing." He feared that their writings would fuel the flames of revolution and push people toward socialism or anarchism. Still, he soon took their agenda as his own—including the regulation of railroads and the horrendous meat-packing industry, and the protection of food and drugs. At the center of his program stood the control of corporate power—which would earn him the sobriquet of "Trust-buster." Roosevelt was not opposed to trusts per se. Indeed, he saw combinations as the logical, inevitable feature of economic progress. He once said that combination could be turned back by legislation no more easily than the spring floods on the Mississippi. But, said the President, "we can regulate and control them by levees"—that is, by regulation and public scrutiny. Such reform was essential in his view to short-circuit radicalism and revolution and preserve the American system. Roosevelt distinguished between "good" trusts and "bad" trusts. Only the latter deserved to be atomized. And on that cause he would not be stayed. Altogether, his administration launched at least forty-five antitrust actions.

The Mother of Trusts was to have center stage in the ensuing battles. Standard Oil was one of Roosevelt's most useful targets; it became the favorite dragon of this irrepressible knight—there was no better opponent against which to joust. Still, Roosevelt sought the support of big business in his 1904 campaign, and the executives of Standard Oil tried to reach out to him. When a friendly congressman, who was also chairman of a Standard subsidiary, informed Archbold that Roosevelt thought Standard Oil was antagonistic toward him, Archbold replied that, on the contrary, "I have always been an admirer of President Roosevelt and have read every book he ever wrote, and have them, in the best bindings, in my library."

The congressman had a bright idea. A presidential author might certainly be subject to flattery, especially one who had proved as prolific as Roosevelt. He would apprise Roosevelt of Archbold's admiration, and use that gambit to arrange a meeting. "The 'book business' fetched down the game at the very first shot," the congressman wrote triumphantly to Archbold. But he added a word of warning: "You had better read, at least, the titles of those volumes to refresh your memory before you come over." Flattery may have gotten Archbold in the front door, but not much further. "Darkest Abyssinia," he angrily said a few years later, "never saw anything like the course of treatment we received at the hands of the administration following Mr. Roosevelt's election in 1904."

Before election day, the Democrats had made a major issue of big business contributions to the Republican campaign, including one hundred thousand dollars from Archbold and H. H. Rogers. Roosevelt ordered the hundred thousand dollars returned, and thereupon, in a burst of publicity, promised every American what became his slogan, a "square deal." Whether the money was ever actually returned was another question. Attorney General Philander Knox told Roosevelt's successor, William Howard Taft, that, when he had walked into Roosevelt's office one day in October 1904, he had heard the President dictating a letter directing the return of the money to the Standard Oil Company.

"Why, Mr. President, the money has been spent," Knox said. "They cannot pay it back—they haven't got it."

"Well," replied Roosevelt, "the letter will look well on the record, anyhow."

Immediately after Roosevelt's election in 1904, his administration launched an investigation of Standard Oil and the petroleum industry. The result was a searing critique of the trust's control of transportation, amplified by a personal denunciation of the company by Roosevelt himself. The pressure was so obviously building against Standard that Archbold and H. H. Rogers hurried to Washington in March 1906 to see Roosevelt and ask him not to proceed with legal action against the company. "We told him that we had been investigated and investigated, reported on and reported on," Archbold wrote to fellow director Henry Flagler after the meeting with Roosevelt, "but that we could stand it as long as the others could. He listened patiently to all that we had to say and I think was fairly impressed. . . . It can hardly have failed to do good with the President."[8]

The Suit

Archbold was deluding his colleagues—and himself. For, in November of 1906, the moment arrived: In the Federal Circuit Court in St. Louis, the Roosevelt Administration brought suit against Standard Oil, charging it under the Sherman Antitrust Act of 1890 with conspiring to restrain trade. As the suit progressed, Roosevelt fanned the flames of public outrage. "Every measure for honesty in business that has been passed in the last six years has been opposed by these men," the President publicly declared. Privately, he told his attorney general that Standard's directors were "the biggest criminals in the country." The War Department announced that it would not buy oil products from the combine. Not to be outdone, the Democrats' perennial Presidential candidate, William Jennings Bryan, declared that the best thing that could happen to the country would be to put Rockefeller in jail.

Standard Oil realized that it was in a battle for survival. The tables were turned, and now the government was subjecting the company to a "good sweating." As one executive wrote to Rockefeller: "The Administration has started out on a deliberate campaign to destroy the Company and everybody connected with it, and to use every resource at its disposal to accomplish that end." In its defense, Standard marshaled grand legal talent, some of the most distinguished names in American jurisprudence. The government's case was led by a corporate lawyer named Frank Kellogg, who, two decades later, would become Secretary of State. Over a course of more than two years, 444 witnesses gave testimony, and 1,371 exhibitions were introduced. The full record was to cover 14,495 pages bound in twenty-one volumes. The Chief Justice of the Supreme Court later described the transcript as "inordinately voluminous . . . containing a vast amount of conflicting testimony relating to innumerable, complex, and varied business transactions, extending over a period of nearly forty years."

Meanwhile, other suits and cases were also proceeding against Standard. Occasionally, Archbold tried to make light of the judicial and administrative onslaught. "For nearly forty-four years of my short life," he told a large banquet audience, "I have been engaged in somewhat strenuous effort to restrain trade and commerce in petroleum and its products, in the United States, the District of Columbia, and in foreign countries. I make this confession, friends, as a confidential matter to you, and in the strong conviction and belief that you will not give me away to the Bureau of Corporations." But, despite the bantering, he and his colleagues were deeply apprehensive. "The Federal authorities are doing their utmost against us," he wrote privately in 1907. "The President names the judges, who are also the jury, who try these corporation cases . . . I do not suppose they can eat us although they may succeed in inciting a mob to do damage. We shall do our very utmost to protect our shareholders. Further than this it is impossible for me or anyone to say."

In another case, in that same year, a federal judge with the memorable name of Kenesaw Mountain Landis—who would later become the first commissioner of baseball—levied a huge fine against Standard Oil for violating the law by accepting rebates. He also denounced the "studied insolence" of Standard's lawyers and regretted "the inadequacy of the punishment." Rockefeller was playing golf with friends in Cleveland when a messenger boy appeared with the judge's decision. Rockefeller tore open the envelope, read the contents, and put it in his pocket. He then broke the silence by saying, "Well, gentlemen, shall we proceed?" One of those present could not contain himself. How large was the judgment, he asked?

"The maximum penalty, I believe—twenty-nine million dollars," replied Rockefeller. Then, as an afterthought, he added, "Judge Landis will be dead a long time before this fine is paid." With that single outburst, he resumed his golf, seemingly unperturbed, and went on to play one of the best games of his life. Indeed, Landis's judgment was eventually overturned.[9]

But then in 1909, in the main antitrust suit, the Federal court found in favor of the government and ordered the dissolution of Standard Oil. Theodore Roosevelt, now out of office and on his way back from a big-game-hunting trip in Africa, heard the news while on the White Nile. He was exultant. The decision, he said, "was one of the most signal triumphs for decency which has been won in our country." For its part, Standard Oil wasted no time in appealing to the Supreme Court. Twice the case had to be reheard by the Supreme Court, owing to the deaths of two justices. Both industry and the financial community waited nervously for the outcome. Finally, in May of 1911, at the end of a particularly tedious afternoon, a mumbling Chief Justice Edward White said, "I have also to announce the opinion of the Court in No. 398, the United States against the Standard Oil Company." The stuffy, somnolent, oppressively hot courtroom suddenly came to life, straining to hear. Senators and congressmen rushed over to the chamber. For the next forty-nine minutes Chief Justice White spoke, but often so inaudibly that the justice to his immediate left had to lean over several times and suggest he raise his voice so that his momentous words could actually be heard. The Chief Justice introduced a new principle—that the judicial evaluation of restraint of trade under the Sherman Act should be based upon the "rule of reason." That is, "restraint" would be subject to penalty only if it was unreasonable and worked against the public interest. And, in this case, it obviously did. "No disinterested mind," the Chief Justice declared, "can survey the period in question [since 1870] without being irresistibly drawn to the conclusion that the very genius for commercial development and organization . . . soon begat an intent and purpose to exclude others . . . from their right to trade and thus accomplish the mastery which was the end in view." The justices upheld the Federal court decision. Standard Oil would be dissolved.

At 26 Broadway, the directors had gloomily gathered in the office of William Rockefeller to await the verdict. Little was said as the minutes went by. Archbold, his face taut, bent over the ticker, scanning for some word. When the news came, everybody was shocked. No one had been prepared for the devastating extent of the Supreme Court's decision; Standard was given six months to dissolve itself. "Our plan" was to be shattered by judicial fiat. There was dead silence. Archbold started to whistle a little tune, just as he had done many years earlier when, as a boy, he had waded through the deep mud of Titusville to buy and bargain for oil. Now, he walked over to the mantel. "Well, gentlemen," he said after a moment's further consideration, "life's just one damn thing after another." Then he began to whistle again.[10]

The Dissolution

In the aftermath of the decision, the directors of Standard faced an immediate and momentous question. It was one thing for a court to order a dissolution. But how exactly was this vast, interconnected empire to be broken up? The scale was simply enormous. The company transported more than four-fifths of all oil produced in Pennsylvania, Ohio, and Indiana. It refined more than three-fourths of all United States crude oil; it owned more than half of all tank cars; it marketed more than four-fifths of all domestic kerosene and was responsible for more than four-fifths of all kerosene exported; it sold to the railroads more than nine-tenths of all their lubricating oils. It also sold a vast array of by-products—including 300 million candles of seven hundred different types. It even deployed its own navy—seventy-eight steamers and nineteen sailing vessels. How was all this to be dismembered? There was only silence from 26 Broadway and the rumors were many. Finally, in late July of 1911, the company announced its plans for dismantling itself.

Standard Oil was divided into several separate entities. The largest of them was the former holding company, Standard Oil of New Jersey, with almost half of the total net value; it eventually became Exxon—and never lost its lead. Next largest, with 9 percent of net value, was Standard Oil of New York, which ultimately became Mobil. There was Standard Oil (California), which eventually became Chevron; Standard Oil of Ohio, which became Sohio and then the American arm of BP; Standard Oil of Indiana, which became Amoco; Continental Oil, which became Conoco; and Atlantic, which became part of ARCO and then eventually of Sun. "We even had to send out some office boys to head these companies," one Standard official sourly commented. These new entities, though separated and with no overlapping boards of management, nonetheless generally respected one another's markets and carried on their old commercial relationships. Each had rapidly growing demand in its own territory, and competition among them was slow to develop. That lassitude was reinforced by one legal oversight in the breakup. Apparently, no one at 26 Broadway had given any thought to the ownership of trademarks and brand names. So all the new companies started out selling under the same old brand names—Polarine, Perfection Oil, Red Crown gasoline. That fact greatly limited the ability of one company to encroach on another's territory.

Public opinion and the American political system had forced competition back into the transportation, refining, and marketing of oil. But, if the dragon was dead, the rewards of dismemberment were to prove considerable. The world had been changing too fast for Standard Oil; its system of controls had become too rigid—especially for the men in the field. With dissolution, they would have the opportunity to run their own shows. "The young fellows were given the chance for which they had been chafing," recalled the man who was to become head of Standard of Indiana. For executives of the various successor companies, it was a great liberation no longer to have to petition 26 Broadway for approval of every capital expenditure over five thousand dollars—or any hospital donation over fifty dollars.[11]

The Liberation of Technology

Among the other consequences of the dissolution was the unexpected liberation of technological innovation from the rigid and controlling grip of 26 Broadway. Standard of Indiana, in particular, moved quickly with a breakthrough in refining to help support the still-infant auto industry at a critical moment, and thus to preserve what would become oil's most important market in the United States.

With existing refining know-how, the highest yield of natural gasoline that could be wrung out of a barrel of crude naturally was 15 to 18 percent of the total refined product, or, at most, 20 percent. That did not matter when gasoline was virtually a waste product, an explosive and flammable fraction for which there was hardly any market. But the situation had changed quickly with the rapid growth in the number of gasoline-powered motorcars. It was becoming evident to some in the oil industry that the supply of gasoline would soon become very strained.

Among those who saw the problem most clearly was William Burton, the head of manufacturing for Standard of Indiana. He was a Johns Hopkins Ph.D. in chemistry, one of the very few scientists working in American industry. He had joined Standard in 1889 to work on the problem of getting the "skunk juice" smell out of Lima crude. In 1909, two years before the dissolution decree, Burton, anticipating the coming gasoline shortage, had directed his small research team, staffed by other Johns Hopkins Ph.D.s, to tackle the problem of increasing gasoline output. He also made a critical decision: He began his research without authorization by 26 Broadway and even without the knowledge of the Indiana subsidiary's directors in Chicago. The lab, he told his scientists, was to try every conceivable idea. The aim was to "crack"—or break down—the larger hydrocarbon molecules of less desirable products into smaller molecules that could provide auto fuel.

The blind alleys were many. But, finally, the researchers experimented with "thermal cracking"—putting a relatively low-value product, gas oil, simultaneously under high pressure and high temperatures—up to 650 degrees and beyond. It had never been done before. The scientists were cautious, and rightly so, for danger was ever present. There was precious little knowledge about how oil behaved under such conditions. Practical refinery men were frightened. As the experiments progressed, the scientists had to clamber around the burning-hot still, caulking leaks, at considerable personal risk, because the regular boiler men refused to do the job. But Burton's idea worked; the gas oil yielded up a "synthetic gasoline" product, which more than doubled the share of usable gasoline from a barrel of crude—up to 45 percent. "The discovery of this thermal cracking process was destined to be one of the great inventions of modern times," wrote a student of the industry. "As a result the petroleum industry was the first big industry to be revolutionized by chemistry."

Discovery was one thing; there was still the question of commercializing the innovation. Burton had applied to Standard Oil headquarters in New York City for a million dollars to build a hundred stills for thermal cracking. But 26 Broadway had turned him down flat, without even an explanation. New York thought the whole idea was foolhardy. Privately, one director said: "Burton wants to blow the whole state of Indiana into Lake Michigan." Immediately after dissolution, however, the directors of the now-independent Standard of Indiana, who had much more direct contact with and personal confidence in Burton, gave him the green light—although one director joked, "You'll ruin us."

The go-ahead came just in time. With the extraordinary growth of the automobile fleet, the world was already on the edge of a gasoline famine. In 1910, gasoline sales had exceeded kerosene for the first time, and demand was galloping ahead. The Gasoline Age was at hand, but the developing shortage of the fuel was a great threat to the nascent auto industry. The price of gasoline rose from nine and a half cents in October 1911 to seventeen cents in January 1913. In London and Paris, motorists were paying fifty cents a gallon, and in other parts of Europe, up to a dollar.

But, by early 1913, within a year of Standard Oil's dissolution, the first of Burton's stills was in operation, and Indiana announced the availability of a new product—"motor spirits"—gasoline made from thermal cracking. Looking back, Burton recalled: "We took some awful risks, and we were awfully lucky not to have any smash-ups early in the game." His thermal cracking process introduced flexibility into refinery output, something the industry had never had before. The refiner's output was no longer arbitrarily bound by the atmospheric distillation temperatures of the different components of crude oil. Now he could manipulate the molecules and increase the output of more desirable products. Moreover, cracked gasoline actually had a much better antiknock value than natural gasoline, which meant more power and allowed for higher-compression engines.

The success of the process created a dilemma for Standard of Indiana. A great internal debate raged over whether or not to license its patents. Some said it would simply strengthen competitors. But in 1914, Standard of Indiana did begin to license the process to companies outside its own markets, on the premise that the resulting revenues would be "all velvet." The velvet proved substantial, as the royalties flowed from fourteen companies between 1914 and 1919. Indiana licensed the process to all companies on the same terms. But one company kept trying to cut a better deal—Standard of New Jersey. The former parent thought it deserved sweeter terms, and that it could force them out of Indiana. But Standard of Indiana would not budge. Finally, in 1915, Jersey capitulated and became a licensee on Indiana's terms. For many years after, it was said that the most galling thing the president of Jersey Standard had to do each month was to sign the fat royalty checks—made payable to Standard of Indiana.[12]

The Winners

A new era had quickly come into existence in the oil industry, around the turn of the century. It was born of several coincidences: the rapid rise of the automobile; the discovery of the new oil provinces in Texas, Oklahoma, California, and Kansas; new competitors; and technological advances in refining. Added to all these, of course, were the far-reaching implications of the break-up of Standard Oil and the resulting restructuring of the industry.

Just before the dissolution, one of John D. Rockefeller's advisers had thought that Rockefeller should sell some of his Standard Oil shares, as the price he assumed was at its top and would fall with the breakup. Rockefeller refused; he knew better. The stock shares in the successor companies were distributed pro rata to the shareholders of Standard Oil of New Jersey. But if the dragon had been dismembered, its parts would soon be worth more than the whole. Within a year of the dissolution of Standard Oil, the value of the shares of the successor companies had mostly doubled; in the case of Indiana, they tripled. Nobody came out of this better or richer than the man who owned a quarter of all the shares, John D. Rockefeller. After the break-up, because of the increase in the price of the various shares, his personal worth rose to $900 million (equivalent to $9 billion today).

In 1912, Theodore Roosevelt, four years out of office, was making a new run at the White House, and Standard Oil was once again his target. "The price of stock has gone up over one hundred percent, so that Mr. Rockefeller and his associates have actually seen their fortunes doubled," he thundered during the campaign. "No wonder that Wall Street's prayer now is: 'Oh Merciful Providence, give us another dissolution.' "[13]

C H A P T E R 6

The Oil Wars:
The Rise of Royal Dutch,
the Fall of Imperial Russia

IN THE AUTUMN of 1896, a youngish man, already tempered by life in the Far East and with a minor reputation in oil circles, passed through Singapore on his way from Britain to an isolated, virtually unknown stretch of jungle called Kutei, on the east coast of Borneo. His movements were quickly noted and as quickly reported to New York by a Standard Oil agent in Singapore: "A Mr. Abrahams, said to be a nephew of M. Samuel's, of the . . . Samuel's Syndicate, has arrived from London and immediately departed for Kutei where it is rumored that the Samuels people have large oil concessions. As Mr. Abrahams is the gentleman who started the Russian tank oil business at Singapore and Penang, erecting and building the plant at both places, his visit to Kutei might mean something." Indeed, it did. For Mark Abrahams had been dispatched by his uncles to develop the oil concession that Samuel's oil combine desperately needed to maintain its position—and perhaps even to assure its survival.

In this undertaking Marcus Samuel was driven by an imperative of the oil business. Those in oil are always in quest for balance. An investment in one part of the business forces them to make new investments in another part, to protect the viability of the existing investment. Producers need markets if their oil is to have value. As Marcus Samuel once said, "The mere production of oil is almost its least value and its least interesting state. Markets have to be found." Refiners, meanwhile, need both supply and markets; a refinery that goes unused is little more than scrap metal and used pipe. And those who run a marketing system need oil to pass through it; otherwise they, too, have nothing but financial losses. The intensity of those needs varies at different times, but the underlying imperative is a constant of the industry.

And, by the late 1890s, Marcus Samuel, with his huge investment in tankers and storage facilities, very definitely needed a secure supply of oil. As a trader, as a merchant, he was too vulnerable. The contract for the Rothschilds' Russian oil would run out in October 1900. Could he count on a renewal? At best, his relations with the Rothschilds were rocky, and the banking family could always turn around and make a deal with Standard Oil. Beyond that, it was dangerous to be dependent upon Russian oil alone. Arbitrary changes in transport rates within Russia kept the economics in continual confusion, complained Samuel, making the Russian oil trade a hand-to-mouth business, and "placing those engaged in the Russian trade at a great disadvantage with their powerful American competitors." There were other dangers, as well: The growing volumes of oil from the Dutch East Indies, with shorter routes and lower freight rates, threatened his ability to remain competitive in the Far East; and at any moment Standard Oil could marshal its resources to launch an all-out war, aimed at destroying Shell. Samuel knew, quite simply, that he needed his own production, his own crude, to protect his markets and his investments—indeed, in order to assure the survival of his enterprise. And, in the words of his biographer, "He went all but berserk in his search for oil."[1]

The Jungle

In 1895, through the efforts of an elderly, obsessed Dutch mining engineer, who had spent all his adult life in the jungles of the East Indies, Samuel was able to obtain rights to a concession in the region of Kutei in east Borneo. The concession stretched for more than fifty miles along the coast, and reached inland into the jungle. It was to this overgrown, desolate destination that Mark Abrahams was dispatched to be the man on the spot. Abrahams had no experience at all in drilling for and refining oil; rather, he had organized the construction of storage tanks in the Far East, but that hardly prepared him for the new and much more difficult enterprise on which he was now embarking.

The irrelevance of Mark Abrahams's skills was mirrored on a larger scale in the case of Marcus Samuel himself. The very way he did business—the antipathy to organization and to systematic analysis and planning, the lack of sound administration and competent functionaries—made the job in the Borneo jungle far more difficult. Ships were always arriving at the wrong time, bearing the wrong equipment, without even a manifest of the cargo. Loads were dumped on the beach, forcing the workers to stop everything else in order to try to gather and organize and make sense out of what had been dropped; all sorts of equipment ended up being left to rust in the tall grass.

Even without the haphazard, disjointed management from London, the job would have been extremely difficult. Borneo was far more isolated from the outside world than even Sumatra; the nearest depot from which any supplies or equipment could be obtained was a thousand miles away, in Singapore. The only communication to Singapore was via the odd ships that might pass by every week or two. The workforces, isolated from one another in different parts of the concession, were in constant battle with the jungle. A four-mile path they arduously cut through the jungle to a place called Black Spot, where there were oil seepages, was overgrown again within a few weeks. The project had to depend on imported Chinese coolies for laborers; the local headhunters were not exactly eager for steady work. Disease and fever constantly attacked everyone working on the sites. Frequently, when Abrahams was sitting up at night to write reports home, he himself was half-delirious with fever. The death rate among all the workers—the Chinese, the European managers, and the Canadian drillers—was high. Some died on shipboard, even before arrival. Every piece of wood with which they tried to build anything, be it a house or a bridge or a pier, soon rotted. Their constant companion was the "hot, steaming, rotting, destructive, tropical rain."

Once again, the Samuels in London and Mark Abrahams in Borneo renewed the stormy, explosive, abusive correspondence that they had exchanged in the days of constructing the storage tanks in the Far East. Poor Mark Abrahams—whatever he did, no matter how hard and daunting his work, was not good enough for his uncles. His uncles could not begin to understand the reality of the jungle. When Marcus Samuel complained that the houses built for the Europeans were lavish "villa residences" that looked like "a couple of pleasure resorts," Abrahams replied angrily that "Your 'Villa Residences' " were so makeshift that "the least gale of wind, or heavy rain, takes away the whole of the roof. The houses in which we lived on first arrival were only fit to accommodate pigs."

Yet, despite all, the first oil was struck in February 1897; the first gusher in April 1898. Much additional effort, however, would be required to go from discovery to commercial production. Moreover, the chemical characteristics of the Borneo crude were such as to yield little kerosene. It could, however, be used, unrefined, as a fuel oil. This quality of the heavier Borneo oil became the foundation for a vision to which Samuel thereafter zealously clung—what he called the "tremendous role which petroleum can play in its most rational form, that of fuel." Here, on the eve of the twentieth century, he looked ahead to prophesy, and rightly so, that oil's great future would be not as a source of illumination, but as a source of power. And Marcus Samuel was to become the most vociferous proponent of the conversion of shipping from coal to oil.

That historic development had actually begun in a small way in the 1870s, when *ostatki*, as the waste residue from kerosene refining was called in Russia, was first successfully used to fuel ships on the Caspian Sea. Pure necessity drove this innovation: Russia had to import coal from England, a very expensive proposition, and wood was scarce in many areas of the empire. Subsequently, the new Trans-Siberian Railway began to use oil fuel, supplied by Samuel's syndicate through Vladivostok, rather than coal or wood. Moreover, the Russian government encouraged oil's use as a fuel in the 1890s to speed overall economic development. In Britain, railways did in some cases switch from coal to oil—to reduce smoke in urban areas or for special safety reasons, such as when carrying members of the Royal Family. But, for the most part, coal held on to its massive market share; indeed, it was the basis for the vast development of heavy industry in North America and Europe. It also fueled the world's commercial and naval fleets. And Samuel met the greatest resistance to his vision in the market about which he cared the most—the Royal Navy. He was to pound on its door for more than a decade, to little avail.[2]

Shell Emerges

Still, Marcus Samuel found consolations. While painful progress was being wrought in Borneo, he was making progress on his own road to acceptance and status. He became a justice of the peace in Kent, and in London, a master of the Spectacle Makers' Company, one of the most venerable of all the ancient guilds. He also received a knighthood after one of his tugs, said to be the most powerful such vessel in the world, dislodged a British warship that had gone aground at the entrance to the Suez Canal. In 1897, Samuel took a major step in the organization of his business. It was a defensive move. He wanted to ensure the loyalty of the various trading houses that formed the Tank Syndicate in the Far East. To that end, he made all of them shareholders of a new company that incorporated the whole of his oil interests and tanker fleets, as well as the storage installations belonging to the various trading houses. It was called the Shell Transport and Trading Company.

Meanwhile, Samuel was ballyhooing the Borneo enterprise far beyond what was justified either by the immediate commercial prospects or by the reality of the painfully difficult and frustratingly slow work in the jungle. But in order to advance his contract renegotiations with the Rothschilds, he had to make it seem that he would soon have alternative supplies from his own fields in Kutei in Borneo. The stratagem worked. The Rothschilds were persuaded, and they renewed the contract to supply Shell with Russian oil—on terms, it should be added, more favorable to Shell than previously. Yet, while Shell's position now appeared stronger, its fortunes were, in fact, precariously balanced. For Marcus Samuel was boldly riding on the crest of a rising market, and like any wave, it would eventually break.

The end of the nineteenth century was marked by a worldwide boom in oil. Demand was growing rapidly, supplies were tight, and prices were rising. The Boer War in South Africa, which started in 1899, pushed prices up further. But in the autumn of 1900, the price of oil began to crumble. A disastrous harvest led to famine and an economic depression in the Russian empire. Domestic demand for oil fell away, and the Russian refiners now began producing as much kerosene as they could for export, which caused a glut on the world market. Prices collapsed. In China, one of Shell's most promising markets, the Boxer Rebellion erupted against foreigners, disrupting the country and the entire Chinese economy. Not only was there no longer an active market, but Shell facilities were pillaged.

These and other adverse developments all converged on the vulnerable Samuel. When prices dropped, Shell's tanks were full of expensive oil. Shell had continued to expand its shipping fleet, and now freight rates also plummeted. To make matters worse, Borneo was falling far short of expectations. Production was developing slowly. The poorly designed refinery was proving a disaster. Fires, explosions, technical malfunctions, and accidents continually interrupted its operations and killed workers. Despite the bad news, Samuel maintained his dignity and composure and, as is required of the entrepreneur in times of trouble, his front. He was still to be found almost every morning on his favorite horse, Duke, riding through Hyde Park. Another British oil man, who would from time to time encounter Samuel on horseback, observed with some acuity that Samuel rode his horse much as he rode his vast business, always looking as though he were about to fall off, but never quite doing so.[3]

Royal Dutch in Trouble

Meanwhile, in Sumatra, the competing Royal Dutch had continued its dramatic increases in production and further stepped up its investment in tankers and storage facilities. A celebration of its coming eminence was planned for New Year's Eve, December 31, 1897, at the company's refinery site on Sumatra. The evening was highlighted by fireworks and a holiday reception for the new tanker, *Sultan of Langkat*, welcomed by the Sultan himself. But the festivities were marred by a rumor circulating through the night—that a considerable amount of water had been found in the oil tanks, indicating that there might be something wrong with the wells. The rumor could not be stamped out.

The rumor was true—Royal Dutch's wells were beginning to produce not oil, but salt water. Its prolific field was in decline. By July 1898, the word was out, and panic gripped the oil section of the Amsterdam stock exchange. The value of Royal Dutch's shares plummeted. Standard Oil missed the chance to pick up Royal Dutch on the cheap. So did Marcus Samuel, much to his later regret.

Royal Dutch desperately tried to find new production. No fewer than 110 times did it drill for oil in Sumatra, and no fewer than 110 times did it fail to find new oil. But the company would not give up. Eighty miles or so to the north of its existing concession in Sumatra, it sought a new drilling site at a seepage in the little principality of Perlak, a frontier territory still troubled by a native rebellion. The local ruler, who made his money in the pepper trade, was most eager to augment his revenues with oil money. An expedition to Perlak was led by Hugo Loudon, a young engineer who had already demonstrated a depth of technical and administrative competence, backed up by experience that stretched from land reclamation in Hungary to railway construction in the Transvaal. He also happened to be the son of a former governor general of the East Indies and had unusually effective diplomatic skills. Those talents were particularly requisite in Perlak, where Loudon successfully advanced Royal Dutch interests not only with the Rajah of Perlak, but also with the leaders of a local rebellion, who had declared a holy war against the Rajah.

Loudon included several professional geologists in his group, and drilling started on December 22, 1899. The expertise of the geologists made a difference, for only six days later, the crew struck oil. Now, just in time for the new century, Royal Dutch was back in business, and once again in a very big way. It quickly called upon geological talent to find and develop oil elsewhere in the Indies. And with those substantial new supplies of high-quality oil, Royal Dutch was ready to invade the budding gasoline markets of Europe.[4]

"A Pushing Fellow"

In November 1900, Jean Baptiste August Kessler, the man who, more than any other, was responsible for the survival of Royal Dutch, cabled to The Hague from the Far East that he was "in a very nervous condition." Worn out by the strains of the business, he set off for the Netherlands and home. He got only as far as Naples, where, in December 1900, he suffered a heart attack and died. The next day a driving young man named Henri Deterding, age thirty-four, was installed as "interim manager." The "interim" lasted a very long time; for the next three and a half decades Deterding would dominate the world of oil.

Henri Wilhelm August Deterding was born in Amsterdam in 1866, the son of a sea captain who died when the boy was six. The family funds went to support the education of Henri's older brothers, while Henri was left to feel the full weight of ever-deepening genteel poverty. At school, he stood out for his special talent—like Rockefeller, he was very good at doing quick mathematical computations in his head. On leaving school, instead of going to sea and becoming a captain like his father, as he had intended, Deterding went into the more prosaic world of banking in Amsterdam, where he soon mastered accounting and finance. For a hobby, he took up the study of balance sheets of companies, trying to figure out who was doing well and who was not, and why, and what kind of strategies the various companies might be pursuing. Thus began the development of what his business associates would later call his "lynx-eye for balance sheets and figures." Much later, his inspirational advice to young men starting out was, "You will go a long way in business if you train yourself to be able to appraise figures almost as rapidly and as shrewdly as a good judge of character can sum up his fellow-men."

When Deterding's promotion in the bank did not proceed with the speed that he thought was his due, he did what many young Dutchmen of the time would do—he shipped out to the East Indies to seek opportunity. He went to work for the Nederlandsche Handel-Maatschappij, the Netherlands Trading Society, a famous old banking concern. Managing its office first at Medan and then at Penang on the west coast of Malaya, he learned how to make money. "By generally sniftering round wherever business could be done," he was later to say, "and without this flair for sniftering, no man starting from the bottom can make money on a large scale—I discovered fresh avenues whereby additional financial grist rolled into the bank's till." Deterding earned quite appreciable sums for the bank by exploiting the differences among various cities in the Far East in exchange and interest rates.

"Sniftering around" also led to oil, where, on his first venture, he made more money for his bank. When Royal Dutch suffered its severe shortage of working capital in the early 1890s, it was to Deterding that Kessler, spurned everywhere else, had finally turned. The two men had known each other since their boyhoods in Amsterdam. Deterding figured out an ingenious solution: He agreed to lend the necessary working capital, using the kerosene stored in inventory as collateral. Royal Dutch survived, and the Netherlands Trading Society found a new way to make money. Kessler was grateful and impressed.

Not long after, when Kessler decided that he had to set up Royal Dutch's own trading organization through the Far East, he wrote to Deterding to ask for suggestions as to who might run it. Kessler knew exactly the sort of person he must have—"a first-rate businessman, a pushing fellow, with seasoned experience and a good eye for business." Who fit that bill better than Kessler's correspondent, Henri Deterding himself? In 1895, Kessler offered the job to Deterding, who, frustrated with the life of a banker, accepted. He immediately began to aggressively build up its marketing system through the Far East. His aims were to bring Royal Dutch to parity with its competitors, and to insulate it from those competitors. His grand ambition was to become, as he was later to say, "an international oil man."

Henri Deterding was short and dynamic, with very wide open eyes that had a startling effect on people. When he laughed, all his teeth showed. Hardy and vigorous, he believed fervently in exercise, both for its own sake and as the way to work out business problems. In Europe in later years, even when he was on the "shady side" of sixty, he would, every morning before going to work, in winter as well as summer, first go swimming, then spend forty-five minutes horseback riding. He made a powerful and compelling impression on everyone with whom he came in contact. He had what was described as an "irresistible magnetism" and an "almost aggressive charm," both of which he used to persuade others to join in his causes and campaigns. But unlike Marcus Samuel, he was not motivated by a quest for status, for position. The Dutch historian F. C. Gerretson, chronicler of Royal Dutch, and for many years private secretary to Deterding, summed up his real purpose: "Now Deterding was not aiming at something exalted and wonderful: to serve the public interest, to create a new economic order, to build up a mighty commercial concern. His purpose was that of any merchant, great or small, something extremely matter-of-fact: to make money." Whatever else Deterding became, he was always "a merchant in heart and soul."

In time, Deterding would jokingly begin to call himself a "Higher Simpleton." He certainly did not mean it as a term of self-derision, but as a guide to his working theory—to reduce each problem to its simplest terms, to its essential elements. "Simplicity rules everything worth while, and whenever I have been up against a business proposition which, after taking thought, I could not reduce to simplicity, I have realized that it was hopelessly wrong and I have let it alone."

One "simple" idea dominated his mind during his early years with Royal Dutch—the need for amalgamation among the new oil companies. He saw it as the only way to protect Royal Dutch against Standard Oil. *"Eendracht maakt macht"*—unity gives power. So ran the old Dutch proverb that he took as a touchstone. He also sought cooperation as a way to bring stability to the industry. Like Rockefeller, he was repelled by the wild fluctuations in price. Unlike Rockefeller and Standard Oil, he did not want to use price cutting as a competitive tool; rather, he wanted to work out price-setting arrangements and peace treaties among the warring companies. That was better even for the consumer in the long run, he would argue, because more stable and predictable returns would encourage more capital investment and greater efficiency. But, with this one simple idea of amalgamation went another, though hardly one he trumpeted—that in any amalgamation Royal Dutch would eventually have to occupy first place. Still, Deterding's intentions were not regarded as altogether benign by others. To the Nobels, he later appeared not as a paragon of conciliation, but as nothing less than "a terrible sort of being whose mission was to slaughter everybody and pick up the carcass."[5]

The First Step Toward Combination

Together, Shell and Royal Dutch controlled over half of the Russian and Far Eastern oil exports. The "ruinous competition" between the two would provide the starting point from which Deterding was to embark on a momentous negotiation to achieve amalgamation with his great rival, Marcus Samuel. The character of this global enterprise would be determined by the long struggle between two men—each a businessman of great talent and daring, each of daunting ego, but one, Marcus Samuel, more subject to flattery and sentiment and more interested in position; the other, Henri Deterding, driven more than anything else by his quest for raw power and for money itself. On the fundamental question—which of them would lead any new combination—the two men were wholly at odds. Marcus Samuel had no doubt who should be the leader—he himself, because of Shell's visible preeminence and its far-flung activities. But Deterding had absolutely no intention of being, as he said, anybody's second fiddle.

These two were not going to get anywhere negotiating directly with each other. They badly needed a middleman, and who better than that middleman par excellence when it came to oil, the shipping broker Fred Lane? After all, "Shady" Lane was the London representative of the Rothschilds' oil interests; he was Samuel's friend, consultant, confidant—and trusted coconspirator in the great oil coup of a decade earlier. He had just met Deterding, but they had instantly hit it off, and they were bound to become very close friends as well. Lane began by negotiating a truce in a price war in the Far East between Royal Dutch and Shell and by putting an end to what he called the damaging "battledore and shuttlecock game of accusation" between Samuel and Deterding. His efforts helped to create the right mood for discussions to begin. From the outset, however, there was a major difference of purpose. Samuel wanted a simple marketing arrangement between the two firms. Deterding wanted out-and-out "joint management." Lane had to advise Deterding that, while "in the long run joint management was inevitable," for the moment, Samuel's opposition was "insuperable." Matters became even more complicated when, in the middle of October 1901, Marcus Samuel sailed to New York to visit none other than the gentlemen at 26 Broadway, for the apparent purpose of negotiating an alliance with Standard Oil. "There is here Sir Marcus Samuel," John Archbold wrote to Rockefeller. "This company represents by all means the most important distributing Agency for Refined Oil throughout the World, outside of our own interests. He is here undoubtedly to take up with us the question of some sort of an alliance, preferable on his part of the sale to us of a large interest in their Company." But, despite extensive talks, the two sides could not agree on how much Shell was worth; Standard was skeptical of the value that Samuel set on his operations. Yet Samuel was not an entrepreneur for nothing. When he returned to London, he gave the impression of impending triumph, displaying great talent in stirring up enthusiasm about Shell, a company that was, in fact, in deep trouble.[6]

The "British Dutch"—and Asiatic

While Samuel was in New York, Lane had been diligently trying to sketch out the basis for a negotiation between Royal Dutch and Shell. But the major question remained unanswered: Was there simply to be a dividing up of the market, or was there to be an out-and-out combination? It was on November 4, 1901, that Lane went to see Samuel for what was to prove a decisive discussion. Lane hammered at one point. A simple marketing arrangement would be meaningless if more and more oil kept coming into the market, destroying prices. Production had to be controlled as well. That, in turn, made the conclusion clear: "There is no solution except the absolute amalgamation of the businesses." Once Samuel too had come to this conclusion, he became graciousness personified and "cordially" declared himself won over. There would have to be a new organization, which would have the ability to limit production. From that fateful meeting dated the first steps that led eventually toward the establishment of the Royal Dutch/Shell Group.

Deterding was in a rush to get the deal completed; he was afraid Standard Oil would beat him to Shell. His fears were justified. Two days before Christmas 1901, Standard Oil, despite its earlier reluctance, finally made an offer for Shell, and it was huge—$40 million was a great deal of money in 1901 (on the order of $500 million today). Samuel's family urged him to accept. Samuel himself went down for the holidays to the Mote, his estate in Kent, to struggle with the choices. He faced one of the most agonizing decisions of his life: to accept a fantastically large sum, acquire almost unimaginable wealth, and become one of the most important personages in the Standard Oil empire, or to take his chances with Deterding and Royal Dutch. There was enormous reason to pause and waver. But then, right after Christmas, Samuel's meditations were abruptly broken by an urgent telegram from Lane summoning him back to London. Deterding had given in on the key points, Lane told him. Samuel signed a hurriedly drafted agreement with Royal Dutch on the afternoon of December 27, 1901. It was hand-carried on the night boat to Deterding. That same evening, Samuel sent a telegram to New York rejecting Standard's offer and breaking off negotiations.

What Samuel wanted was equality. Standard could be very generous in terms of money, but it was insisting, as it always insisted, upon control, which would thus pass from a British to an American entity, and that, no matter how much the money, Marcus Samuel could not countenance. He was too much of a patriot. Still, he and Deterding did not yet have a thorough agreement, only the barest outline. With his usual singleness of purpose, Deterding succeeded in getting the other major producers in the Netherlands East Indies to bind themselves together in a new combination, with Royal Dutch in the driver's seat. Deterding now had half of what he wanted—effective control and management of the oil output of the Dutch East Indies. But what kind of sales combination was it to be with Shell? Deterding had talked about the "joint management" of Samuel and Deterding. But once Standard Oil was off the stage, Shell's position was weakened, and Deterding began to focus on another of his very simple ideas, one that was inordinately appealing to him. There should be only one man in charge, and that man should be Henri Deterding.

106

Deterding delivered an ultimatum. Accept his scheme for the organization, which limited Shell's and Samuel's control over the management, he told Samuel, or he would not even bother to cross the Channel for any more negotiations. "Neither of us can afford to waste time," said the Dutchman. He got his way. Samuel would be the chairman of the new company, but Deterding would be its manager and chief executive, with responsibility for the day-to-day direction of affairs. Deterding could ask for no more. Soon after, the two key documents were signed. One set up the Committee of Netherlands Indian Producers and the other a new company called "The Shell Transport Royal Dutch Petroleum Company"—soon to be known as the "British Dutch." Thus was conceived the company that would emerge as a true global rival to Standard Oil.

Then, a third party, the Rothschilds, decided that in spite of their distaste for Samuel and Shell, they could not afford to be left out. If the Rothschilds wanted in, Deterding argued to a dubious Samuel, bring them in at all costs. "Delay dangerous," he said. "If this chance has slipped this time, we shall never get it again. Once we are combined with the Rothschilds, everybody knows that we hold the future, but we cannot do without their name." Samuel was finally persuaded.

In June 1902, a chastened Samuel signed a new overarching agreement with Deterding and the Rothschilds. "British Dutch" would disappear into a new, larger combination, the Asiatic Petroleum Company. The results of the business, Samuel now promised his stockholders, would be much improved because the "whole organization" would no longer be based exclusively upon marketing Russian oil, with all its insecurity and risks. "It is a matter of sincere congratulation to all concerned," he ringingly concluded, "that the war which we have been engaged in with our Dutch friends has now ended, not only in peace, but in an offensive and defensive alliance."[7]

Deterding Triumphant

The British Dutch and now Asiatic companies represented the first major steps toward amalgamation. But this initial agreement still had to be turned into a working contract. Meanwhile, Shell's financial and market position was continuing to deteriorate to the point of peril, and Deterding threatened to withdraw from the entire arrangement. Samuel had to face the possibility that everything would fail.

Such failure could not have been more ignominious, for on September 29, 1902, Samuel, senior alderman, was due to be elected Lord Mayor of London. At the end of August, he asked Deterding to come to the Mote. The Dutchman was very impressed by the English country house; he had never seen one before, and he determined that he would own one, too. Samuel was frank about his current troubles. Deterding understood Shell's weakness, but he also knew that the Dutch "flag" would not be sufficient for the global enterprise he had in mind; he needed a more powerful "flag"—the Union Jack. Thus he reassured Samuel that he would seek to restore Shell's fortunes through the medium of the new Asiatic Petroleum Company.

In order to manage the new company, Deterding took up residence in London (though since 1897 he had been using a London cable address—

"Celibacy"). And from Asiatic's offices in London, Deterding controlled and balanced the combined resources of Royal-Dutch and Shell, a substantial part of the Rothschilds' Russian oil exports, and the output of the independent producers in the Netherlands East Indies. He now began buying and selling oil on a vast scale, with great skill and success. Through his chairmanship of the Committee of Netherlands Indian Producers, he began to restrict production there and to work a quota system.

While Deterding was furiously focusing his energies on the nascent Asiatic, Marcus Samuel was firmly fixed on something else that had nothing to do with the oil business—his official installation as Lord Mayor of London on November 10, 1902. It was surely to be the grandest day of his life, for he was to attain the highest honor to which a London merchant could aspire—and all the more important to Marcus Samuel, a Jew from the East End and the son of a seashell merchant. When the great day came, he had the procession of carriages, which bore him and his family and various dignitaries, include in its route the Jewish quarter, Portsoken Ward, his birthplace. The day culminated at the Guildhall in a grand banquet, filled with notables, honoring Marcus Samuel. Among the guests was Deterding, who distanced himself from the event, as though watching some quaint native ritual. "I certainly should not think it worth a white tie to attend a second time," he derisively wrote to one of his colleagues. "The Lord Mayor's show was very fine, according to the view here, but in Dutch eyes it was more like the ceremonial parade of [a] circus."

Samuel was thereafter caught up in ceremonial duties, reception after reception, speech after speech. Almost a month passed before he turned his attention back to the oil business. Even then he was to be continually involved with the business of being Lord Mayor, with its many duties and the official trips, and all the visiting dignitaries. One of his responsibilities was to interview personally every lunatic who was to be certified insane at the Mansion House, and some were to think he spent more time with the lunatics than he did with the oil men. Samuel enjoyed the ritual and position of Lord Mayor greatly, but the strain also took its toll on him. During his year as Lord Mayor, he had to cope with ill health and constant headaches, and in the midst of everything else, he had to have all his teeth removed.

There were pains of another kind, too. On the last Saturday in December 1902, Samuel took the early morning train up from the Mote, in Kent, to attend the funeral for the Archbishop of Canterbury, lunch with the sheriffs of the City, and then attend a play. On Sunday, he viewed weapons presented by Lord Kitchener from the Boer War; on Monday morning, he presided in the City, and only then, at last, turned to pressing private business—a letter waiting for him from Fred Lane. It was nothing less than devastating. Samuel's old friend and partner was resigning from the board of Shell. It was not just the press of activities consequent on his having become deputy managing director of Asiatic. Lane launched into a bitter critique of the way Marcus Samuel ran his company. "You are, and have always been, too much occupied to be at the head of such a business," he wrote. "There seems only one idea: sink capital, create a great bluster, and trust to providence. Such a happy-go-lucky frame of mind in business I have never seen before. . . . Business like this cannot be conducted by an occasional glance in one's spare time, or by some brilliant *coup* from time to time. It is steady, treadmill work." Unless "some very radical change is made," Lane prophesied, "the bubble will burst" and then nothing "will be sufficient to save the company." Samuel met with Lane; they talked; they corresponded further. Trading blame and accusations, they became angrier and angrier. The breach could not be healed. So Lane left the board; on each side there was to be a lasting and bitter sense of betrayal.

Meanwhile, Asiatic was still being constructed; the final deal was not yet done, and that engendered continuing disputes over control and policy—and power. The historian of Royal Dutch wrote that Deterding only wanted everybody to act "rightly and fairly." Samuel's biographer had a different view; Deterding was so intent to get his way that he was driven into "a state of unreasonable rage and unreasoning venom" that was "close to dementia." Sure that he held the winning hand, Deterding was unwilling to compromise. At one point, he declared, "I am feeling entirely fit and able to withstand ten Lord Mayors."

Finally, by May of 1903, ten contracts had been agreed to that established Asiatic, which was a third owned by each of the parties. The new company would regulate production in the East Indies, carry out sales in the Far East, and also control the sale of East Indies gasoline and kerosene in Europe. The greatest achievement of all, Deterding triumphantly assured his own board, was that Royal Dutch emerged paramount in every part of the agreement. Perhaps most important, the managing director of Asiatic would also be the managing director of Royal Dutch—Henri Deterding. Samuel insisted that the term of the managing director be limited to three years. Deterding dug in his heels. "Twenty-one years and not a day less," he declared, which was another way of saying the appointment would be permanent. He won on that, too.

The first meeting of the Asiatic board took place in July 1903, with Marcus Samuel in the chairman's seat. Deterding, speaking without notes, seemed to know where every ship was at that moment, its destination, its cargo—and the prices awaiting it in each port. Marcus Samuel was much impressed.[8]

"The Group"—Samuel Surrenders

Deterding threw himself with irrepressible energy into the new enterprise. When the chairman of the Royal Dutch board warned him that he was pushing himself too hard, Deterding replied, "It so happens that in the oil business one has to seize one's opportunities quickly; otherwise, they escape." He was not a gambler, but a calculated risk taker, and his method was working. In short order, Royal Dutch assimilated most of the independent producers in the East Indies, where the oil was particularly suited to the manufacture of gasoline. Automobiles were starting to become familiar sights on the roads of Britain and the Continent; and, under Deterding's bullwhip, Asiatic won an important share of the growing European gasoline market.

As things were going ever better for Royal Dutch, they were getting progressively worse for Shell. Not only had the Texas supplies from Spindletop given out, but the British Admiralty remained committed to coal and refused to take seriously Samuel's vision for fuel oil for the Royal Navy. Thus, the large market that Samuel fervently hoped for—the Navy—simply was not there. Then, too, Royal Dutch discovered Borneo crudes suitable for fuel oil, shattering Samuel's hope to have a monopoly on its production. Standard's price wars took a continuing toll. And there was also the animus of Fred Lane, who had turned bitterly on Shell and used his position as deputy managing director of Asiatic to settle his own scores. Deterding, wearing two hats, certainly did what he could to advance the position of Royal Dutch against that of the dilapidated Shell. Limping along, with collapse in the air, Shell was barely able to pay 5 percent dividends, while Royal Dutch's were at the rates of 65 percent, 50 percent, and then in 1905, an immensely satisfying 73 percent.

What was left for Shell to do? The clock was running out for Marcus Samuel. In the winter of 1906, his most talented employee, a young man named Robert Waley Cohen, told him the bad news—a consolidated marketing company was insufficient. The only way that Shell could survive was to amalgamate completely with Royal Dutch on the best terms he could get. The idea devastated Samuel. After all, he had almost single-handedly created a great global oil company. But there seemed hardly any choice. Facing up to what had now become inevitable, he raised with Deterding the desirability of amalgamation. Deterding agreed. Yes, it was desirable. But on what basis? Fifty-fifty, replied Samuel, as in their original British Dutch agreement. Absolutely not, said Deterding. He was blunt. The days of the "British Dutch" were past; the relative position of the two companies had changed dramatically. The ratio would have to be sixty for Royal Dutch and forty for Shell. "The property and interests of Shell would henceforth be managed by a foreigner!" Samuel responded. He would never be able to justify it to his stockholders.

There they left the matter for several months, but when the position of Shell showed no improvement, Samuel was forced to bring up the issue again with Deterding. "I should be prepared," said Samuel, "to leave the management to Royal Dutch, if you, Deterding, could give me some absolute guarantee that it would be in the interest of the Royal Dutch to manage the Shell properly."

Deterding offered only one guarantee. Royal Dutch would buy a quarter of the shares of Shell, and thus would, as a shareholder, have Shell's best interests at heart. Samuel asked for time to think it over. Deterding refused. "I am at present in a generous mood. I have made you this offer, but if you leave this room without accepting it, the offer is off." Samuel saw no obvious alternative. He accepted. His struggle with Deterding had gone on for a half decade. But, finally, it was over. Deterding had won.

The union was cemented in 1907, and out of it emerged the Royal Dutch/Shell Group. The first joint marketing company, four years earlier, had been called the "British Dutch"—the order of names reflecting the seniority. But now "Royal Dutch" came first. The change in order was deliberate; Deterding was, after all, the victor. Over the years, the new combine was sometimes simply known as "the Group." All the oil production and refining assets were lodged in a Dutch company, Bataafsche Petroleum Maatschappij; and all the transport and storage in an English company, Anglo-Saxon Petroleum Company. Both Royal Dutch and Shell became holding companies, with Royal Dutch holding 60 percent of the stock in the operating subsidiaries and Shell 40 percent. There was no

Royal Dutch/Shell board and, indeed, no legal entity called Royal Dutch/Shell. The "Committee of Managing Directors" had no specific legal status; rather, it was composed of active members of the boards of the two holding companies. Royal Dutch did buy a quarter of the shares of Shell, the bond of good faith that Samuel had demanded, but over the years it disposed of all save one last symbolic share.

Deterding established his working office in London, which became the financial and commercial center of Royal Dutch/Shell; he also acquired a country estate in Norfolk, where he took up the life that he had envied, that of an English country squire. The technical side of the business, production and refining, was based in The Hague. As events transpired, the former corporate distinctions faded; it did not matter in which part of the business profits were made, as they were all split on the same sixty-forty basis.

Indeed, all parts of the business were run by the same people, of whom three were key. Deterding was the first, of course. The second was Hugo Loudon, the Dutch engineer who had rescued Royal Dutch with new discoveries in Sumatra when its initial wells gave out. The third was young Robert Waley Cohen. Of an old Anglo-Jewish family, Waley Cohen had graduated from Cambridge University with a degree in chemistry, went to work for Marcus Samuel in 1901, and then moved as Shell's man into Asiatic. After the amalgamation, he played a major role in bonding the parts together. Deterding concentrated on the business side of the business, constantly traveling and negotiating; Loudon focused on the technical. Waley Cohen was Deterding's de facto commercial deputy, making decisions in Deterding's absences, picking up and concluding one set of negotiations when Deterding moved on to the next, and bucking Deterding up at those times that the Dutchman began to have misgivings and second thoughts.

Defeated by Deterding and forced by necessity to give up his control, Samuel initially regarded himself as a failure. There was no glory for him in the amalgamation. "I am a disappointed man," he told the newspaper reporters. Immediately after the merger, Samuel treated himself to a 650-ton yacht to assuage his hurt and took himself to sea. But the humiliation quickly healed. The two tycoons made an effort to get along with each other. Deterding consulted Samuel, made him much richer, and after his death was to speak of him as "our chairman." In turn, it did not take Samuel long to see what Deterding could accomplish; already, by 1908, he was telling Shell stockholders that Henri Deterding was "nothing less than a genius." Even if he did not rule, Samuel presided for over a decade as chairman of Shell Transport and Trading and was actively involved in a wide range of the Group's business. He grew even more wealthy, became an engaged philanthropist, continued to be celebrated or caricatured in the newspapers as events warranted, and went on promoting the use of his beloved fuel oil for shipping. During his years as chairman, he maintained an amicable relationship with Deterding. But there was never any question about the nature of that relationship. Deterding was the boss.[9]

"To America!"

The completion of the amalgamation in 1907 meant that the world oil market was now dominated by the original giant, Standard Oil, and a growing giant, the Royal Dutch/Shell Group. "If the Standard had tried three years ago to wipe us out, they'd have succeeded," Deterding said in 1910, but proudly added, "Now things are different." The competition between the two, however, remained fierce and bitter, and that same year he made a pilgrimage to 26 Broadway, seeking conciliation. What he encountered instead was an offer to buy Royal Dutch/Shell for $100 million. "I am sorry to have to place on record that my visit to this city . . . has been so useless," was his acid response. He felt humiliated for, he said, the issues of cooperation "are at present not considered as being worthy of discussion with the manager and chairman of the various Companies who, next to your Company, are doing the largest oil trade in the world."

Standard Oil replied to Deterding's rejection with a new price-cutting campaign, opening another phase in the oil wars. As if that was not enough, it also established a Dutch subsidiary to seek oil concessions in southern Sumatra. The Group no longer had any choice; it had to counterattack, and that meant one thing: "To America!" It became the slogan for the policy of Royal Dutch/Shell between 1910 and 1914. If the Group was not active in America, it would always be vulnerable to Standard's price cutting, for Standard could sell surplus gasoline at cut-rate prices in Europe, as it had sold surplus kerosene, while maintaining higher domestic American prices and thus also its profits. That position gave Standard a staying power that the Group did not have; it could use its American profits to subsidize losses resulting from marketing wars in Europe and Asia.

Deterding moved in two directions. The first was on the West Coast, where in 1912 he set up a marketing operation for Sumatra gasoline and then the following year went directly into oil production in California. The second direction took the Group toward the mid-continent. Keen to get in on the Oklahoma boom, Deterding dispatched a new special agent to the United States to organize the whole thing quickly. The agent was the man who had organized Shell's original network of storage tanks in the Far East in the early 1890s, and its Borneo foray in the late 1890s—none other than Mark Abrahams, Marcus Samuel's nephew, now fresh from launching an oil exploration company for the Group in Egypt.

Heading for Oklahoma was hardly like going to Borneo, but still Abrahams did not quite know what to expect when he set out from New York for Tulsa in July 1912. So he had his little party carry its own typewriter, in case there were no typewriters in Tulsa, and he stashed $2,500 in a money belt, in case there were no reputable banks in the little boom town that was already proclaiming itself "the Oil Capital of the World." Once ensconced in Tulsa, he proceeded to acquire a number of small oil companies and incorporated them into a new company, Roxana Petroleum. Deterding had now achieved his larger goal, which might have been called defensive expansion. He was on Standard's home ground. When Mark Abrahams, his task completed, returned to London, Deterding sent a jubilant letter to Hugo Loudon: "At last we *are* in America!" [10]

Russia in Turmoil

Galling as it was for Samuel to have lost control to Deterding in the amalgamation of Shell and Royal Dutch, events soon proved the move a wise one, given Shell's dependence on Russian oil. Russia's industrial economy had gone through stupendous growth under the favorable policies of Count Sergei Witte, the powerful finance minister from 1892 to 1903. Trained as a mathematician, Witte had risen from a position as a lowly railroad administrator to become the master of the Russian economy by sheer ability—a most unusual means of ascent in the Czarist empire. As Finance Minister, Witte oversaw the rapid, large-scale industrialization of Russia and of the oil industry in particular, fueled by a vast infusion of foreign capital. Conservative critics attacked his program; the Minister of War complained of "too hurried development" in the oil region, especially by "foreign capitalists, foreign capital, and Jews." But Witte stuck to his development strategy.

Witte was truly an exception, a man of great talents in a government populated by people of little ability. The entire system was rotten with corruption, prejudice, and incompetence. The font of ineptitude was the Czar himself. Nicholas II was highly vulnerable to flattery, a dangerous characteristic in an autocrat, and he and his court descended into mysticism and unreality, immersing themselves in cults and surrounding themselves, as Witte said, with "imported mediums and home-bred 'idiots' passing as saints." The Czar could "not relinquish his 'Byzantine' habits," said Witte prophetically. "But inasmuch as he does not possess the talents of either Metternich or Talleyrand, he usually lands in a mud puddle—or in a pool of blood." Witte could only pray that God should deliver "us from the tangle of cowardice, blindness, craftiness, and stupidity."

Nicholas II was contemptuous of all the non-Russian minorities in his multinational empire and sanctioned the repression that, in turn, made them into rebels. By the early 1900s, the whole empire was in turmoil. In 1903, the Minister of Interior was forced to admit to Witte that the reign of Nicholas II was already a colossal failure. With a few inconsequential exceptions, the minister declared, the empire's entire population was alienated and dissatisfied. The Caucasus—home of the Russian oil industry—was one of the worst-run parts of the ill-run empire. Living and working conditions in the area were deplorable. Most workers were in Baku without their families, and in Batum, the working day was often fourteen hours, with two hours of compulsory overtime.

Baku became the "revolutionary hotbed on the Caspian." Hidden away deep in the heart of its Tatar quarter was a large cellar that stretched under several buildings. Here was the home of "Nina"—the name given to the secret, large printing operation into which the mats of Vladimir Ilyich Lenin's revolutionary paper, *Iskra*, were smuggled, from Europe via Persia, to be printed for circulation within the country. To the continued befuddlement of the Czarist police, "Nina" became the source of a massive flow of revolutionary materials. The oil industry was the unknowing accomplice; its national distribution system provided a perfect vehicle for clandestinely distributing propaganda throughout the country. Baku and the oil industry also provided the training ground for a host of eventual Bolshevik leaders, including a future Soviet President, Mikhail Kalinin, and a future marshal of the Soviet Union, Klementi Voroshilov. The alumni included a still more important figure, a young Georgian, a former seminarian and son of a shoemaker. His name was Joseph Djugashvili, though he operated in the underground under the name "Koba"—Turkish for "Indomitable." Only later did he begin to call himself Joseph Stalin.

In 1901 and 1902, Stalin became the chief socialist organizer in Batum, masterminding strikes and demonstrations against the local oil industry, including a prolonged strike against the Rothschilds' interests. Stalin was among the many arrested after the strikes, the first of his eight arrests. He repeatedly escaped from exile, only to find himself landed again and again back in a Czarist prison. In 1903, the oil workers of Baku went out on strike, setting off a wave of labor strife across Russia, culminating in the first general strike in the empire. The country was in disarray, and the government in crisis. No wonder Marcus Samuel, the Rothschilds, and others worried about their dependence on Russia as their source of oil supply.[11]

The Czarist regime needed a diversion, and, as so many others have done before and since, it sought its diversion in a foreign adventure, hoping to unite the nation and restore the prestige of its rulers. And, like many others, it chose the wrong opponent—in this case, Japan. Competition for control over Manchuria and Korea, particularly the Yalu Valley, had made war with Japan a distinct possibility ever since 1901. The Czar, who had been wounded in an assassination attempt on a trip to Japan a decade earlier, had no respect for the Japanese; even in official documents he called them "monkeys." St. Petersburg turned aside every effort by the Japanese to work out some sort of accommodation. Count Witte had sought to head off conflict; his removal from the Finance Ministry in 1903 convinced the Japanese that war was inevitable. That suited the Czar and his circle. "Russia's internal situation" required something drastic, said the Minister of Interior. "We need a little victorious war to stem the tide of revolution." It was obvious that war was only a matter of time.

The Russo-Japanese War began in January 1904 with Japan's successful surprise attack against the Russian fleet at Port Arthur. Thereafter, the Russian forces lurched from one military disaster to the next, culminating in the burial at sea of the entire Russian fleet at the Battle of Tsushima. The war did not stem the tide of revolution, but rather hastened it. In December 1904, the Baku oil workers went out on strike again, and won their first collective labor agreement. A few days after the strike ended, revolutionaries put out a proclamation, "Workers of the Caucasus, the hour of revenge has struck." Its author was Stalin. The next day, in St. Petersburg, police fired on a group of workers marching on the Winter Palace to submit a petition to their Czar. This was Bloody Sunday, the beginning of the Revolution of 1905—what Lenin called the Great Rehearsal.

When the news reached Baku, the oil workers again went out on strike. Government officials, fearful of revolution, provided arms to the Moslem Tatars, who rose up to massacre and mutilate Christian Armenians, including the leaders of the oil industry. A legend arose afterward about one of the wealthiest Armenian oil men, one Adamoff. A crack shot, he stationed himself on the balcony of his house, and with the aid of his son, held off a siege for three days, until finally he was killed, the house set fire, and his forty dependents either burned to death or dismembered.

Strikes and open rebellion spread again throughout the empire in September and October of 1905. In the Caucasus, it was race and ethnic conflict, and not socialism, that drove events. Tatars rose up once more in an attack on the oil industry throughout Baku and its environs, intent on killing every Armenian they could find, setting fire to buildings where Armenians had taken refuge, pillaging every piece of property on which they could lay their hands. "The flames from the burning derricks and oil wells leaped up into the awful pall of smoke which hung over the inferno," one survivor was to write. "I realized for the first time in my life all that can possibly be meant by the words 'Hell let loose.' Men crawled or dashed out of the flames only to be shot down by the Tatars . . . I thought the scene might well be compared with the last days of Pompeii. It was made worse than anything that could have taken place at Pompeii by the ping of rifle and revolver bullets, the terrific thunder of exploding oil tanks, the fierce yells of the murderers, and the dying screams of their victims." The smoke was so thick that at two in the afternoon, the sun could not be seen. Then, as if to provide proof that the last days were truly at hand, a terrifying earthquake shook the entire region.

The news from Baku had a profound effect on the outside world. Here, for the first time, a violent upheaval had interrupted the flow of oil, threatening to make a vast investment worthless. Standard Oil wasted no time in taking advantage of the disarray in Russia; it moved quickly and successfully to regain the markets for American kerosene in the Far East that had formerly been lost to Russian oil. As for the Russian industry itself, the tally was dismaying: Two-thirds of all the oil wells had been destroyed and exports had collapsed.

By the end of 1905, the revolution was spent. The Russo-Japanese War was also over, its conclusion mediated at the behest of the belligerents by President Theodore Roosevelt at Portsmouth, New Hampshire. In October 1905, the Czar granted, albeit completely against his will and grain, a constitutional government, which included a Parliament, the Duma. Though the revolution was over, the oil region remained in turmoil. The oil workers of Baku elected Bolshevik deputies to the Duma; Nobel's chief in Batum was murdered in the street. In 1907, strikes swept through Baku, again threatening to become a general strike, while the Czar stupidly undermined the constitution that might ultimately have preserved him and his dynasty. Also in 1907, the Bolsheviks sent Stalin back to Baku, where he directed, organized, and as he said, fomented "unlimited distrust of the oil industrialists" among the workers. Those years in Baku were one of the few times that Stalin actually involved himself in the day-to-day struggles of the working class. In 1910, he was arrested in the midst of preparations for another general strike, imprisoned, and exiled to the desolate north of Russia. It was in Baku that he had honed the revolutionary and conspiratorial skills—and the ambition and cynicism—that would help make his future.[12]

Return to Russia

It was not only the political upheavals and racial and labor tensions that were undermining the Russian petroleum industry. Russia's great advantage had been large-scale production at comparatively cheap cost. But chaotic and sloppy drilling and production had led to deterioration in production capacity and irreversible damage in the fields around Baku, hastening exhaustion. All this pushed operating costs up sharply. Political instability discouraged the large new investment that was required. Meanwhile, the Russian government unwisely raised internal transport tariffs to help satisfy the ravenous appetites of its treasury. The result was to increase further the price of Russian oil products on the world market, making them even less competitive. Its price advantage had turned into a disadvantage. Increasingly, Russian oil was a residual, to be bought when other petroleum was not available.

Important changes in the overall structure of the European oil industry were occurring, as well. A major new source of oil was emerging in Europe itself—Rumania, where a minuscule supply had long been eked out of hand-dug pits on the slopes of the Carpathian Mountains. In the 1890s, investment by Hungarian and Austrian banks, combined with modern technology, began to push up the country's production dramatically. But the situation was really transformed at the beginning of the twentieth century by the entry into Rumania of Standard Oil, the Deutsche Bank, and Royal Dutch. These three groups ended up controlling much of the Rumanian industry, and their impact was enormous. Rumanian output grew sevenfold in the first decade of the twentieth century. Deutsche Bank, with its new Rumanian production, joined the Nobels and Rothschilds in 1906 to form the European Petroleum Union—the EPU. Over the next two years, the EPU negotiated specific market division agreements with Standard Oil's distributors throughout Europe, giving the EPU 20 to 25 percent of various markets, with the rest going to a satisfied Standard Oil. A similar market share agreement was worked out for Britain.

Though the haphazardly produced Baku supply was in decline, new Russian fields were being opened up at about this same time. Their development was aided by improved technology and production methods and by speculative fever for oil on the London Stock Exchange, which provided capital. One field was at Maikop, fifty miles east of the Black Sea coast. Another was Grozny, in Georgia, northwest of Baku. But even with new production the Rothschilds had wearied of their Russian oil venture. They wanted out. The anti-Semitism and anti-foreign sentiment in Russia had deeply disturbed them, as had the growing political instability; they knew firsthand of the strikes, the arson, the murders, the revolution. But the immediate commercial reasons for selling out were no less compelling. Profits were now low or nonexistent. All of the Rothschilds' oil assets had depended upon Russian production; they did not have international geographical balance. Why not instead find security with a concern that was globally diversified?

In 1911, the Rothschilds began negotiating with Royal Dutch/Shell over the sale of their entire Russian oil organization. The deal was not easily made. The ever-present Fred Lane represented the Rothschilds in the transaction. "I can assure you to get Deterding to do something is not an easy task," "Shady" Lane wrote to the worried head of the Rothschilds' oil interests. "His habit is to allow things to remain as open as possible and he sits like an owl upon it thinking it over to ascertain whether he has done badly or not quite so good as he imagined, or whether he cannot do something better, so that one never knows where one is until things are definitely 'signed.' " Still, by 1912, the deal was done. The Group paid the Rothschilds in the form of stock, in both Royal Dutch and Shell—making them among the largest shareholders in each. That way, the Rothschilds transformed their uncertain and insecure Russian assets into substantial holdings in a rapidly growing, diversified international company with outstanding prospects.

At the turn of the century, a frantic Marcus Samuel had done everything in his power to cut Shell's dependence on uncertain Russian supplies. Now, a decade later, Deterding had engineered Royal Dutch/Shell's reentry into Russia in a very big way. As a result of the transaction, the Group acquired the largest Russian producing, refining, and distributing operation after Nobel. When asked by a Nobel representative why he would want to come into Russia, Deterding answered bluntly that "his intention was to make money." Overnight, the Group became a major economic force in Russia, controlling at one point, it was estimated, at least a fifth of the entire Russian production. The acquisition of the Rothschilds' interests, in turn, gave the Group a globally balanced portfolio of production—53 percent from the East Indies, 17 percent from Rumania, and 29 percent from Russia. Obviously, there was significant risk going into Russia. But the advantages from integrating this additional output into its worldwide system were immediate. As to the risks, time would tell.

Overall, the Russian oil industry, particularly around Baku, continued to decline in the decade before the First World War. Its technology was stagnating and falling behind that of the West. Its time of greatness, when it was the dynamic element in the world market, had passed. Between 1904 and 1913, Russia's share of world petroleum exports dropped from 31 to 9 percent. Yet those who had, in one way or another, participated in the Russian oil industry during its heyday could look back with nostalgia. For the Nobels, the Rothschilds, and Marcus Samuel, it had been a source of enormous wealth and considerable power. But nostalgia could take many forms, and it belonged not only to the oil men but also to their adversaries. "Three years of revolutionary work among the workers of the oil industry tempered me as a practical fighter and as one of the local practical leaders," Stalin was to say in the 1920s, on the eve of his accession to the Bolshevik throne. "I first discovered what it meant to lead large masses of workers. There in Baku I received, thus, my second baptism in revolutionary combat. There I became a journeyman for the revolution." [13]

Though the revolutionary upheaval that began in 1905 set in motion developments that would turn Baku into a commercial backwater in the world oil market for two decades, it would remain the most important source of oil on Europe's immediate periphery. For that reason, revolution notwithstanding, Baku would become one of the great and decisive prizes in the global conflicts that were still ahead.

CHAPTER 7

"Beer and Skittles" in Persia

A DAPPER GENTLEMAN from Persia, Antoine Kitabgi, carrying the title of general, arrived in Paris toward the end of 1900. Variously said to be of Armenian or Georgian origin, Kitabgi had held several positions in the Persian government, including director general of the customs service. He was, said a British diplomat, "well versed in Western matters—being able to draw up a concession and initiate commercial movements." Those were the skills appropriate to his mission. For although the ostensible reason for his visit was the opening of a Persian Exhibition in Paris, Kitabgi's main purpose was something else: He was a salesman—his aim was to find an investor in Europe willing to assume a petroleum concession in Persia. Kitabgi was serving not only his own ends—he certainly expected suitable compensation—but also those of the Persian government, which had important political and economic interests at stake. While the finances of the government of Persia were always muddled, one thing about them was certainly obvious: The government was desperately short of money. The reason? The answer was provided by the Prime Minister—"the Shah's prodigality."

What was to flow from General Kitabgi's efforts would prove to be a business transaction of historic proportions. Though its fate would hang by a thread for years, the deal would initiate the era of oil in the Middle East, eventually propelling that region to the center of international political and economic contention. And Persia itself—or Iran, as it would be known from 1935 onward—would emerge into a prominence on the world stage that it had not enjoyed since the days of the ancient Persian and Parthian empires.[1]

"A Capitalist of the Highest Order"

In Paris, Kitabgi sought the aid of a retired British diplomat, who, after some consideration, reported back: "Concerning the oil, I have spoken to a capitalist of the highest order, who declares himself disposed to examine the affair." The capitalist in question was one William Knox D'Arcy. Born in Devon, England, in 1849, D'Arcy had emigrated to Australia, where he became a solicitor in a small town. He also developed an unquenchable passion for horse racing. By nature, D'Arcy was always willing to take a chance, and he took a flyer and organized a syndicate to get an old gold mine back into operation. The mine turned out still to be very rich in gold, and in due course D'Arcy returned to England and the life of an extremely wealthy man. After the death of his first wife, he married a prominent actress, Nina Boucicault, who entertained lavishly; Enrico Caruso even came to sing at their dinner parties. In addition to his house in London, D'Arcy maintained two estates in the country and had the only private box at the Epsom racing track aside from the royal box. He was an investor, a speculator, a putter-together of syndicates, not a manager, and he was looking for a new investment. The prospect of petroleum in Persia attracted him, he was again willing to take a chance, and, in so doing, he would become the founder of the oil industry of the Middle East.

Oil seepages had been noted for centuries in Persia, where the oozings were used for such purposes as the caulking of boats and the binding of bricks. In 1872, and again in 1889, Baron Julius de Reuter, founder of the Reuters news agency, had obtained Persian concessions that provided, among other things, for the development of oil. But both concessions generated great protest within Persia and considerable opposition from Imperial Russia, as well as much waste in haphazard and unsuccessful efforts to find oil. Both ended by being terminated. In the 1890s, a French geologist began to publish reports, based upon his extensive research in Persia, that pointed to considerable oil potential. His work was known to various parties, including General Kitabgi, who, eager to ensnare D'Arcy, promised the millionaire nothing less than "the presence of a source of riches incalculable as to extension." How could one not be interested? But first the concession had to be won.

On March 25, 1901, D'Arcy's own representative left Paris, arriving in Tehran, via Baku, on April 16. The negotiations in the Persian capital proceeded slowly and intermittently, and D'Arcy's man passed his time buying rugs and embroidery. The inveterate intermediary, Antoine Kitabgi, was busier. According to the British minister to Persia, Sir Arthur Hardinge, Kitabgi "secured in a very thorough manner the support of all the Shah's principal Ministers and courtiers, not even forgetting the personal servant who brings His Majesty his pipe and morning coffee."[2]

Russia Versus Britain

Persia could claim a national identity stretching back to the ancient empire of Cyrus the Great and Darius I, which by the fifth century B.C. stretched from India all the way into what today are modern Greece and Libya. Later the Parthian empire emerged out of the region now known as Iran and became the redoubtable eastern rival of the Roman empire. Persia itself was a great crossroads for trade and conquest between Asia and the West. Wave after wave of armies and entire peoples passed through and, in some cases, settled there. Alexander the Great swept in from the West; Genghis Khan and the Mongols, from the East. At the end of the eighteenth century, an avaricious dynasty known as the Qajars succeeded in winning control over a country that had fragmented into the principalities of contending warlords and tribal confederacies. The Qajar shahs ruled uneasily for a century and a half. In the nineteenth century, a country habituated to invasion found itself subject to a new form of foreign pressure—the diplomatic and commercial competition between Russia and Britain for dominance over Persia, which inevitably became a preoccupation of the Qajar shahs, as they sought to play the two great powers off against each other.

The rivalry between Britain and Russia turned Persia into a major issue in Great Power diplomacy. Lord Curzon, Viceroy of India, described Persia as one of "the pieces on a chessboard upon which is being played out a game for the domination of the world." Beginning in the 1860s, Russia had embarked on a relentless drive of expansion and annexation in Central Asia. The Russians were looking beyond Central Asia, as well, toward controlling neighboring countries and acquiring a warm-water port. To Britain, Russia's expansion was a direct threat to India and the routes thereto. Any resources put into bolstering Persia against the Russian advance, a British diplomat had said in 1871, were "a sort of premium on the Insurance of India." Russia was on the move throughout the region; in 1885 it launched an attack on neighboring Afghanistan, which came very close to precipitating war between Russia and Britain.

Russia renewed its pressure on Persia around the turn of the century. In the face of this new push, the British sought ways to keep Persia intact, to serve as a buffer between Russia and India. The two great powers wrangled for influence over Persia through concessions and loans and other tools of economic diplomacy. But, as the new century opened, the British position was precarious, for Persia was in clear danger of falling under Russian sway. Russia was seeking to establish a naval presence in the Persian Gulf, while Persia's economy was already to a considerable degree integrated into that of Russia. The Shah, Muzaffar al-Din, was "merely an elderly child," in the words of Hardinge, the British minister, and "the Persian monarchy itself was an old, long-mismanaged estate, ready to be knocked down at once to whatever Foreign Power bid highest, or threatened most loudly its degenerate and defenceless rulers." Hardinge feared that the foreign power would most likely be Russia, for the "Shah and his ministers were in a state of complete vassalage to Russia, owing to their own reckless extravagance and folly." The Russians were not much concerned about the economics of the relationship; as one Russian official put it, "What interest do we have in trading with seven or eight million lazy ragamuffins?" Rather, the Russians wanted to assert their political dominion over Persia and exclude the other Great Powers. To Hardinge, an "all-important" objective of British policy was to resist so "detestable" an incursion.

Here was where D'Arcy and his oil scheme could help. A British oil concession would assist in righting the balance against Russia. And thus Britain gave

its support to the venture. When the Russian minister found out about the negotiations over D'Arcy's concession, he angrily sought to block them. He did succeed in slowing the pace. But then D'Arcy's man in Tehran threw another five
thousand pounds onto the table, since, he reported back to D'Arcy, "the Shah
wanted some ready money and stood out for some on signing the concession."
That extra money did the trick, and on May 28, 1901, Shah Muzaffar al-Din
signed the historic agreement. It provided him with twenty thousand pounds in
cash, with another twenty thousand pounds' worth of shares, as well as 16 percent of "annual net profits"—however that term was to be defined. (And the definition was to prove very contentious.) In turn, D'Arcy received a concession
good for sixty years, covering three-quarters of the country.

From the beginning, D'Arcy had deliberately excluded from his proposed
concession the five northern provinces, closest to Russia, in order to "give no
umbrage to Russia." But the rivalry between Britain and Russia was hardly finished. The Russians now sought to build a pipeline from Baku to the Persian
Gulf that would not only expand their kerosene exports into the Indian market
and Asia but also, and more important, project Russia's strategic influence and
power in Persia, throughout the Gulf region, and onto the shores of the Indian
Ocean. The British argued hard against the project, both in Tehran and St. Petersburg. Hardinge, the minister in Tehran, warned that the "preposterous" concession for a pipeline, even if it was never built, would "afford an excuse for
covering Southern Persia with surveyors, engineers and protecting detachments
of Cossacks, preparing a veiled military occupation." The British opposition
succeeded; the pipeline was not built.[3]

D'Arcy's negotiator in Tehran was exuberant over the deal he had made. Not
only would the scheme benefit D'Arcy, but it would also "have far-reaching
effects, both commercially and politically for Great Britain and cannot fail to
largely increase her influence in Persia." The Foreign Office, though refusing to
assume any direct responsibility, was certainly willing to give political support
to D'Arcy's efforts. But Hardinge, the man on the spot, was more skeptical. He
knew Persia—its political system, its people, the geographical and logistical
nightmares, and the decidedly unpromising history of recent concessions in the
country. He suggested caution: "The soil of Persia, whether it contains oil or not,
has been strewn of late years with the wrecks of so many hopeful schemes of
commercial and political regeneration that it would be rash to attempt to predict
the future of this latest venture."

What then drew D'Arcy to such a risky enterprise—to "wild-catting on a
colossal scale in a distant unsettled land," in the words of one historian? The
answer, of course, was the irresistible lure of immense wealth, the chance to
become another Rockefeller. Moreover, D'Arcy had gambled before, on the
Australian gold mine, and with tremendous success. Yet no doubt, if D'Arcy had
been able to predict accurately what lay ahead, he would have held back from
this new venture. It was a vast gamble, on a much grander scale than his Australian mine, with many more players than he had reckoned on, and a complex
political and social dimension that had been wholly absent in Australia. In short,
it was not a reasonable business proposition. Even the estimate for expenditures

was to be grossly understated. At the outset, D'Arcy had been advised that it would cost ten thousand pounds to drill two wells. Within four years, he was to be out of pocket in excess of two hundred thousand pounds.[4]

The First Go

D'Arcy had no organization, no company, only a secretary to handle his business correspondence. To put together and run the operations on the ground in Persia, he hired George Reynolds, a graduate of the Royal Indian Engineering College with previous drilling experience in Sumatra. The first site chosen for exploration was at Chiah Surkh, an inaccessible plateau in the mountains of northwestern Persia, near what would later become the Iran-Iraq border, closer to Baghdad than Tehran, and three hundred miles from the Persian Gulf. The terrain was hostile, the entire country altogether had barely eight hundred miles of road, and large parts of the region were ruled by warring tribes that hardly recognized Tehran's authority—let alone any concession it might grant. Persian Army commanders would rent out their soldiers as gardeners or workmen to local landowners and pocket the wages for themselves.

The population was abysmally lacking in technical skills, and indeed, the hostility of the terrain was more than matched by the hostility of the culture toward Western ideas, technology, and presence. In his memoirs, Hardinge discussed in some detail the dominant Shia sect with its religious zeal, its resistance to political authority, and its fierce antagonism toward all from the outside world, be they Christians or Sunni Moslems. "The hatred of the Shiahs for the first four Caliphs was, and is still, so strong that some of the more enthusiastic members of the sect have, from time to time, sought to hasten their own entrance into Paradise by defiling the tombs of these usurpers and especially that of Omar, the chief object of their hatred at Mecca. It could only be restrained by the doctrine of 'Ketman' or pious dissimulation . . . which renders it lawful for a good Moslem to appear to dissemble or even lie, for a really pious purpose." He then went on to apologize for giving so much attention to the clash between Shia and Sunni and to the influence of Shia faith on the political system of Persia: "I have touched on this question at perhaps unnecessary length, but it played—and I think still continues to do so—an important part in Persian politics and thought." And indeed it would continue to do so.

The task ahead was daunting. Each piece of equipment had to be shipped to Basra on the Persian Gulf, transshipped three hundred miles up the Tigris to Baghdad, then carried by man and mule over the Mesopotamian plain and through the mountains. Once the pieces had arrived, Reynolds and his motley crew of Poles, Canadians, and Azeris from Baku struggled to put the machinery together and somehow get it to work. To the Azeris, even the introduction of the lowly wheelbarrow was startling, a major innovation.

D'Arcy himself worried from London that things were not moving fast enough. "Delay serious," he telegraphed George Reynolds in April 1902. "Pray expedite." But delay was the order of the day; actual drilling only commenced a half year later, at the end of 1902. The equipment kept breaking down, the insects were incessant, supply of food and parts was a constant problem, and the general working conditions were ruinous. The "infernal heat" in the workers' quarters got up to 120 degrees.

Then there were the problems of politics. The work camp had to maintain a separate "Mohamedan Kitchen" because of the frequent appearance of various local dignitaries who all seemed, said Reynolds, "very keen on receiving a substantial present from us, especially in the shape of some shares of our Company." On top of everything else, Reynolds had to be a diplomat of the first order to deal with the petty feuds and open warfare between the various tribes. And the small band in the drilling camp had to be constantly alert to the threat from the Shia faithful. "The Mullahs in the North are exciting the population as much as they can against the foreigners," Reynolds's deputy warned D'Arcy. "The real fight is now between the Shah and the Mullahs for the control of the public affairs."[5]

"Every Purse Has Its Limits"

Even in such uncompromising circumstances, the work proceeded, and in October 1903, eleven months after drilling had begun, there were the first shows of oil. But D'Arcy quickly discovered that he had gotten himself into something far more difficult and much more expensive than he had imagined: a financial struggle that would threaten the venture at each step. "Every purse has its limit," he anxiously wrote in 1903, "and I can see the limits of my own." As expenditures continued to mount, he realized he could not go it alone. He needed to be bailed out. Otherwise, the concession would be lost.

D'Arcy applied to the British Admiralty for a loan. The idea for the loan was not his own, but rather had been inspired by one Thomas Boverton Redwood, "the *éminence grise* of British oil policy before the First World War," and a man who had a profound influence on the course of international oil developments in the first two decades of this century. Immaculately dressed, with an orchid in his buttonhole, Redwood was often mistaken for a handsome leading actor of the day, a mix-up in which he took obvious pleasure. Redwood's achievements in petroleum were wide ranging. A chemist by training, he patented what later proved to be a valuable process of distillation; in 1896, he published *A Treatise on Petroleum*, which, several times revised, remained the standard work for the next two decades. Already, at the turn of the century, he was Britain's premier oil expert; his consulting firm was used by almost every British oil company, including D'Arcy's venture. Redwood also became the leading outside adviser on petroleum to the British government. He saw the advantages to the Royal Navy of burning fuel oil, rather than coal; and, strongly suspicious of both Standard Oil and Shell, he wanted to see oil reserves developed by British companies from sources under British control.

Redwood was a member of the Admiralty's Fuel Oil Committee. To say that he was familiar with D'Arcy's concession and its difficulties would be an understatement, for he advised D'Arcy at every step, and it was surely he who brought D'Arcy's plight to the attention of the Fuel Oil Committee, whose chairman in turn encouraged D'Arcy to put in for the loan. In his letter of application, D'Arcy outlined the financial pressures he faced; he had spent £160,000 on exploration to date, with at least another £120,000 to be expected. Advised that the loan would be approved, D'Arcy was told to expect in return to give the Admiralty a contract for fuel oil. Both the Admiralty and the Foreign Office supported the proposal. But the Chancellor of the Exchequer, Austen Chamberlain, thought there was no chance that the House of Commons would approve any such loan. He turned it down.

D'Arcy was desperate. "It is all I can do to keep the bank quiet and something *must* be done," he wrote after the loan was refused. By the end of 1903 he was overdrawn £177,000 at Lloyds Bank and was forced to put up some shares in his Australian gold mine syndicate as collateral. But in mid-January 1904, the second well at Chiah Surkh turned into a producer. "Glorious news from Persia," a jubilant D'Arcy declared, adding an altogether sincere personal comment— "the greatest relief to me." But discovery or no discovery, tens of thousands pounds more, perhaps hundreds of thousands more, would be required to carry on the job, and D'Arcy no longer had access to such resources.

In his search for new investors, D'Arcy tried to secure a loan from Joseph Lyons and Company—to no avail. He dallied for a few months with Standard Oil, but without result. He went to Cannes to see Baron Alphonse de Rothschild, but the Rothschilds decided that they had enough to do with their new links to Shell and Royal Dutch in Asiatic Petroleum. Then, to make matters worse, the flow at Chiah Surkh shrank to a trickle, and Boverton Redwood had the unhappy task of telling his client that the wells would never repay their cost and that they should be closed down—and the entire exploration effort shifted to the southwest of Persia. By April 1904, D'Arcy's overdraft had increased further, and Lloyds Bank was demanding the concession itself as security. Less than three years after its inception, the Persian venture was on the verge of collapse.[6]

The "Syndicate of Patriots"

But there were those in the British government who were alarmed that D'Arcy might be forced to sell out to foreign interests or lose the concession altogether. What concerned them were matters of grand strategy and high politics and Britain's relative position among the Great Powers. For the Foreign Office, the main issues were Russian expansionism and the security of India. In May 1903, the Secretary, Lord Lansdowne, had risen in the House of Lords to make a historic statement: The British government would "regard the establishment of a naval base or of a fortified port in the Persian Gulf by any other power as a very grave menace to British interests, and we should certainly resist it with all the means at our disposal." This declaration, said a delighted Lord Curzon, Viceroy of India, was "our Monroe Doctrine in the Middle East." For the Admiralty, the issue was more specific: the possibility of obtaining a source of secure supplies of fuel oil for the British fleet. The battleships, the heart of the Royal Navy, were committed to coal for their fuel. Oil was being used, however, to propel smaller ships. Even that reliance aroused fear about whether there were sufficient quantities of oil in the world on which to base a significant element of British strength. Many doubted it. Those in the Admiralty who did favor oil over coal for propulsion still saw it only as an adjunct, at least until a large, secure supply of petroleum could be identified. Persia might provide that source, and thus D'Arcy's venture deserved support.

The Treasury's rejection of D'Arcy's loan application seemed terribly shortsighted to the Foreign Office, and Lord Lansdowne immediately expressed concern that "there is danger of whole petroleum concession in Persia falling thus under Russian control." Hardinge, the minister in Tehran, concurred, warning that the Russians might well gain control of the concession and then use it to expand their reach, with dire political consequences. He argued that British majority control in the concession should be maintained at all costs.

The Russians were not the only worry. D'Arcy's visit to Cannes to see the Rothschilds, with the threat that the concession might pass under French control, galvanized the Admiralty back into action. The chairman of the Fuel Oil Committee hurriedly wrote D'Arcy to ask that, before entering into any deals with foreign interests, he allow the Admiralty the opportunity to arrange for its acquisition by a British syndicate. So the Admiralty had assumed the role of matchmaker, and none too soon. Lord Strathcona, an eighty-four-year-old self-made millionaire with impeccable "imperial" credentials, was asked to become head of a "syndicate of patriots." After he was assured that the venture was in the interests of the Royal Navy—and in addition, that he would have to invest no more than fifty thousand pounds of his own money—Strathcona agreed, not because of its commercial possibilities, as he later recalled, "but really from an imperial point of view."

Now the Admiralty had a figurehead. But with whom was it to make the match? The answer was a firm called Burmah Oil. An offspring of the network of trading houses in the Far East, Burmah had been founded by Scottish merchants in 1886, with headquarters in Glasgow. It had transformed primitive oil gathering by Burmese villagers into a commercial industry with a refinery in Rangoon and markets in India. By 1904, it also had a tentative agreement to provide fuel oil to the Admiralty, for Burma was regarded as a secure source owing to its annexation into India in 1885. But the Scottish directors of Burmah Oil worried that supply in Burma would prove limited and that successful development in Persia would flood the Indian market with abundant new sources of cheap kerosene. Thus, they were willing to listen to the Admiralty's overtures.

The oil consultant Boverton Redwood acted as intermediary. He was an adviser to Burmah, as well as to D'Arcy, and he told Burmah's directors that Persia could prove rich in oil and that a marriage between the two companies made eminent sense. The Admiralty, meanwhile, insisted that the Persian concession "should remain in British hands and especially from the point of view of supplies for the navy of the future." But the cautious Scottish merchants, for their part, did not talk grandly and abstractly, nor would they be rushed. They had very practical questions—most important, could Persia be considered under British protection? The Foreign Office, prompted by the Admiralty, reassured them on this point. The impatient D'Arcy, in an attempt to speed up the negotiations, invited Burmah's vice-chairman to watch the Epsom Derby from his private box, near the winning post. The rich food and drink so upset the vice-chairman's liver that he was sick four times in the next few weeks, and he never again accepted an invitation from D'Arcy to see the races.

Meanwhile, the Admiralty increased its pressure on Burmah Oil to save D'Arcy, and Burmah Oil in turn obviously needed the Admiralty, both for the fuel oil contracts, which were being negotiated in detail at exactly the same time, and to help protect its markets in India. Finally, in 1905, almost exactly four years to the day after the concession had been initialed by the Shah in Tehran, the match was consummated between D'Arcy and Burmah in London. Their agreement established the so-called Concession Syndicate; D'Arcy's operation became a subsidiary, and D'Arcy himself a director of the new enterprise. In effect, Burmah became a very special kind of investor, for it provided new capital as well as the management and expertise to carry on. Given the bleak history of previous concessions in Persia and his own lack of luck to date, D'Arcy may well have had no alternative. The important point was that his venture had been saved. At least exploration could now go forward, and D'Arcy still had a chance to get his money out of the deal. The matchmakers, too, were satisfied. As the historian of Burmah Oil put it, D'Arcy's needs "coincided exactly with those of the Foreign Office, anxious about the route to India, and of the Admiralty, seeking reliable fuel oil supplies." Henceforth profit and politics would be inextricably linked in Persia.[7]

To the Fire Temple: Masjid-i-Suleiman

The establishment of the Concession Syndicate was followed by the shift of exploration to southwestern Persia. Under the direction of George Reynolds, the wells were plugged at Chiah Surkh, the camp was closed, and the equipment— some forty tons worth—was dismantled, carried back to Baghdad, shipped down the Tigris back to Basra, and then transshipped to the Iranian port of Mohammerah. Eventually it would be shipped by river, wagon, and mules (as many as nine hundred) to new sites, where there were also indications of oil. Drilling first commenced at Shardin.

But there was another potential site at a place called Maidan-i-Naftan, "the Plain of Oil." The specific spot, Masjid-i-Suleiman, was named for a nearby fire temple. Reynolds had first made his way to that roadless spot somewhat circuitously. In late November of 1903, he had been marooned in Kuwait, trying to arrange passage back to England, feeling altogether dispirited about D'Arcy's venture in Persia and its financial problems, and just about ready to pack it all in. But in Kuwait he encountered a British official, Louis Dane. Dane was traveling around the Persian Gulf with Lord Curzon, who was making a grand tour of the region to celebrate the Lansdowne Declaration and to underline British interests in the Gulf. Dane himself was compiling a gazetteer of the Gulf and surrounding lands, and he had come across several references to Maidani-Naftan in both old and recent accounts of travelers. The accounts reminded him of Baku.

At Dane's strong urging—"it seems a thousand pities to turn up what may be an immense national benefit"—and with the support of Lord Curzon, Reynolds set off for Maidan-i-Naftan. He had reached the desolate region in February 1904, and had reported back that the rocks were saturated with oil. Now, two years later, in 1906, he returned to Masjid-i-Suleiman and found even more extensive indications of oil. When Boverton Redwood saw Reynolds's report, he was exultant. It contained, he announced, the most important and promising information to date.

The operation at Masjid-i-Suleiman would prove immensely difficult and trying—not "all beer and skittles," as Reynolds sarcastically informed the Burmah managers in Glasgow. Work was delayed by sickness caused by contaminated drinking water, which, said Reynolds, was "best described as water with dung in suspension." He added, "The materials afforded for food here are rather trying for any digestion, so that teeth natural or false, are essential if a man is to retain his health." That point was well taken. When a British military officer later assigned to the concession developed a toothache, he had to survive days of agony—the pain in no way assuaged by the knowledge that the nearest dentist was fifteen hundred miles away in Karachi. At least when it came to sex, the workers could find relief closer to home, a mere 150 miles away in Basra, at what was, by coincidence, euphemistically called the "dentist."[8]

George Reynolds was the man who held the whole thing together. Already around fifty when he first arrived in Persia in September of 1901, he would proceed to carry out an unusually difficult enterprise under endlessly trying circumstances. He was at one and the same time engineer, geologist, manager, field representative, diplomat, linguist, and anthropologist. In addition, he had a most valuable knack for jerry-rigging machinery when parts broke or were simply missing. He was taciturn, tough, and tenacious. It was his determination and obstinate commitment that kept the project going when there was every reason—from illness, to extorting tribesmen, to mechanical frustration, to searing heat and unforgiving winds, to endless disappointment—to waver. Arnold Wilson, the lieutenant of the British guards at the site, described Reynolds as "dignified in negotiation, quick in action, and completely single-minded in his determination to find oil." In short, said Wilson, Reynolds was "solid British oak."

Reynolds could also be a stern taskmaster. He ordered his men to behave like "reasonable beings," not "drunken beasts," and made sure they understood that Persian women were definitely off limits. But the true bane of his existence was not the desert, nor even the local tribesmen. Rather, it was the new investor, Burmah Oil, which he constantly feared would lose its will. The managers in Glasgow seemed unable to comprehend the immense difficulties of the circumstances under which Reynolds worked and could not resist second-guessing him, questioning and impugning his judgment. Reynolds responded with searing and impolitic sarcasm that enveloped the weekly reports he sent back to Scotland. "You really amuse me," he wrote to his contact in Glasgow in 1907, "by instructing me how to run a contumacious Parsee and an alcoholic driller, both suffering from swelled heads." The dislike was mutual. "The type machine would not reproduce the words I would like to say about the man," this Glasgow manager once said.[9]

Revolution in Tehran

The physical rigors and isolation—and conflicts with the management back in Glasgow—were by no means the only obstacles to success. The Shah's government was in an advanced state of decay, and the foreigners' concessions were a

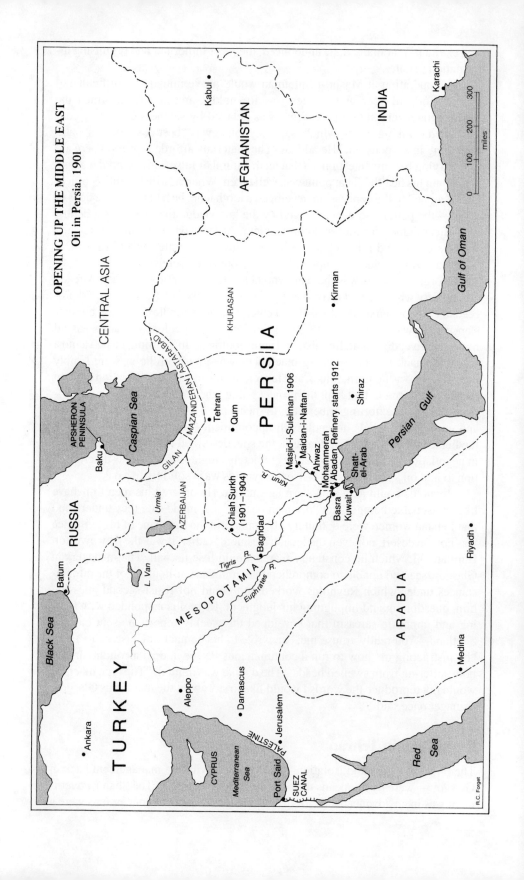

OPENING UP THE MIDDLE EAST
Oil in Persia, 1901

R. C. Forget

major political sore. The conservative religious opponents of the Shah's regime took the lead in attacking despotism. They joined forces with merchants and groups seeking liberal reforms. In July 1906, the government attempted to arrest a prominent preacher, who had blamed the people's misery on "the great luxury of Monarchs, some clerics, and the foreigners." Riots ensued in Tehran as many thousands of Persians, fired up by the mullahs, took to the streets. The bazaars closed; a general strike gripped the capital; and a large crowd, estimated at around fourteen thousand people, mostly from the bazaars, sought refuge in the garden of the British legation. The result was the end of the Shah's regime, a new constitution, and the establishment of a Majlis or Parliament, which put an investigation of the concession at the top of its agenda. But the new political system proved unstable, and its authority was very weak outside the capital.

Even more troublesome was the matter of local rulers. The new drilling site was in the winter grazing pasture of the Bakhtiari, the most powerful tribal confederacy in Persia, and one over which Tehran had very little say. The Bakhtiaris were nomads, driving flocks of sheep and goats and living in open goat-hair tents. In 1905, Reynolds made an arrangement with some of the Bakhtiaris under which, in exchange for a high fee and promise of a share in profits, they agreed to provide "guards" for the concession. However, among the main things to be guarded against were the Bakhtiaris themselves, and the agreement fell apart because of the constant family feuds and tribal tensions, as well as what seemed the Bakhtiaris' inveterate tendency toward extortion. Reynolds described one of the Bakhtiari leaders as "a man as full of intrigue as the egg of a nightingale is pregnant with music." D'Arcy, continually informed of the problems, could only complain, "Of course Baksheesh is at the root of it all."

The increasing tempo of harassment and threats from local tribes led to new fear for the safety of the enterprise and its works. D'Arcy asked the Foreign Office for protection, and a guard force was eventually dispatched. This was done, the Foreign Office grandly said, because of "the importance attached by His Majesty's Government to the maintenance of British enterprise in South West Persia." But it was not much on which to lean—a total of two British officers and twenty Indian cavalry. Meanwhile, the clash between Britain and Russia eased; in 1907, as part of the Anglo-Russian Convention, the two countries sought to put their differences to rest by agreeing to partition Persia into spheres of influence. Both sides had good reasons. Russia had been weakened by its devastating defeat in the Russo-Japanese War and the turmoil of the Revolution of 1905, and St. Petersburg now saw great merit in reaching an accord with London. For their part, the British, in addition to their long-standing fear of "spontaneous infiltration" of Russian influence toward India, were now beginning to worry more about German penetration into the Middle East. Under the 1907 convention, northern Persia was to be under Russian sway, the southeast under British, and the middle a neutral zone. But that middle area happened to be the location of the new drilling sites. The immediate impact of the explicit division of the country into spheres was, as the new British minister in Tehran observed, to give "a great impetus" to the "already existing anti-foreign sentiment." The partition of Persia was also one of the steps that led to the formation of the Triple Entente of

Britain, Russia, and France that, seven years later, would be at war with Germany and the Austro-Hungarian and Turkish empires.[10]

Racing the Clock

The drilling site, Masjid-i-Suleiman, would be "the last throw of the concessionary dice." It also presented Reynolds and his crew with the greatest logistical problems yet. The first difficulty was that there was no road. One had to be carved out of the desert in the face of all sorts of hazards, including a torrential rain that wiped out most of a half-year's effort. Finally, the road was completed, the equipment moved in, and in January of 1908, drilling began at this last site.

But time was fast running out for the Concession Syndicate. Burmah Oil was most unhappy with the slow progress and the large outflow of money. Its vice-chairman suggested that "the whole thing" might "go smash." All of this put Burmah increasingly at odds with D'Arcy, who was totally committed to the project and who, in turn, was impatient with the Scottish caution. In April 1908, the Burmah board told D'Arcy in no uncertain terms that the money was exhausted and that, unless he himself put up half of the additional funds required, work would stop.

"Of course, I cannot find £20,000 or anything," D'Arcy plaintively complained, "and what to do I know not." But he shrewdly concluded that Burmah was too committed to back out. The Burmah directors set an April 30 deadline for D'Arcy's reply; he simply ignored it, letting the day come and go with no response. He was playing for delay—to give Reynolds in Persia more time. Relations between Burmah and D'Arcy sank to a new low.

With no word from D'Arcy, Burmah acted on its own. It sent off from Glasgow, on May 14, 1908, a letter to Reynolds, saying that the project was over, or nearly so, and that he should be prepared to pack up. The letter instructed Reynolds to carry the two wells at Masjid-i-Suleiman down to no more than sixteen hundred feet. If no oil was found by that depth, Reynolds was ordered to "abandon operations, close down, and bring as much of the plant as is possible down to Mohammerah" and from there, ship the equipment on to Burma. The end of the Concession Syndicate seemed very near. So much for that dream of "riches incalculable" that had been dangled before D'Arcy years earlier. A cable was sent ahead to Reynolds, alerting him to be prepared for an important directive that was being dispatched by post. But, such being the mails in that part of the world, the letter itself was not received in Persia for several weeks. That delay was just what the headstrong Reynolds badly needed.

For even as the letter was making its way to Persia, excitement at the drilling site began to mount. A smell of natural gas could now be detected from one of the wells. Then a drill bit came unscrewed and was lost in the hole; several days were spent fishing for it in temperatures that reached 110 degrees in the shade. The drilling was now going through the hardest rock yet. Vaporous gas, in the powerful sunlight, could clearly be seen rising from the hole. On the night of May 25, 1908, the temperatures were so hot that Arnold Wilson, the British lieutenant of the Indian cavalry guards, went to sleep on the ground outside his tent.

Shortly after 4:00 A.M. on the twenty-sixth, he was awakened by shouting. He rushed to the site. A gusher of petroleum, rising perhaps fifty feet above the top of the drilling rig, was smothering the drillers. The accompanying gas was threatening to suffocate the workers.

Oil had, at last, been struck in Persia. It was just two days short of seven years since the Shah had signed the concession agreement. Lieutenant Wilson's may well have been the first report to get back to England. At least according to legend, he sent it in code: "See Psalm 104 verse 15 third sentence." At that place, the Bible reads: "that he may bring out of the earth oil to make a cheerful countenance." Unofficial word reached D'Arcy at a dinner party. He was delighted, but determined to keep his enthusiasm in check. "I am telling no one about it until I get the news confirmed," he insisted. Confirmation came very shortly, and a few days later, while the first well continued to gush, oil was struck in the second well. Three weeks or so after that, Reynolds received the letter of May 14 from Burmah Oil, ordering him to begin winding down operations. It was a striking echo of the letter half a century earlier that had told Colonel Drake to quit his operations at Titusville, which arrived just as he struck oil. In this case, by the time Reynolds received the letter, he had already sent a cable back to Glasgow, sarcastically saying: "The instructions you say you are sending me may be modified by the fact that oil has been struck, so on receipt of them I can hardly act on them." The letter itself proved every prejudice Reynolds had about Burmah's management in Glasgow and gave him a good deal of bitter satisfaction.

Reynolds remained in Persia as chief engineer for a couple of years after the strike at Masjid-i-Suleiman. Yet, despite the discovery, his conflicts with Burmah continued to worsen. D'Arcy tried to protect him, telling Burmah's directors that Reynolds was "a man who will never by a stupid action imperil the Concession." But such support could not save him in the face of the hostility that had built up toward him in Glasgow, and in January 1911, he was unceremoniously fired. In his own memoirs, Arnold Wilson offered an epitaph for Reynolds's service: "He was able to endure heat and cold, disappointment and success, and to get the best out of every Persian, Indian, and European with whom he came in contact, except his Scottish employers, whose short-sighted parsimony had so nearly wrecked a great enterprise. . . . The service rendered by G. B. Reynolds to the British empire and to British industry and to Persia was never recognized. The men whom he saved from the consequences of their own blindness became very rich, and were honoured in their generation." In firing Reynolds, the directors of Burmah Oil did manage some grudging praise for him, and they gave him a thousand pounds as a token for his troubles.[11]

The "Big Company": Anglo-Persian

On April 19, 1909, the Glasgow branch of the Bank of Scotland was mobbed by fevered investors. Never before had the premises seen such a scene. "Oil on the brain" had suddenly gripped the dour Scottish industrial city. The public stood ten deep at the counter, clutching application forms. At times during the day, it was altogether impossible even to enter the building. The newly incorporated Anglo-Persian Oil Company was going public, and this was the day for the public offering of its stock.

For some months, it had been clear that a very rich source of oil had been found in Persia. All involved were agreed that a new corporate structure now had to be devised to work the concession. But the actual shaping was attended by the inevitable and endless wranglings of lawyers. Moreover, the British Admiralty took exception to the draft prospectus's "making public" its encouragement of Burmah's stake in Persia. "As the Admiralty is our prospective good customer, we cannot afford to stamp on their corns," admitted the vice-chairman of Burmah, and the prospectus was toned down. Objections also arose from an unexpected source, Mrs. D'Arcy. With a flair for the theatrical that befitted the one-time actress, she remonstrated with her husband about the omission of his name from the company's title. Though he refused to make an issue of it, Mrs. D'Arcy persisted. "This, I think, a great mistake as far and wide his name is associated with this Persian business," she wrote to D'Arcy's lawyer. "I am making a last bid for fame to you."

Her bid failed. Still, while Burmah Oil had taken the majority of the ordinary shares, D'Arcy came out well in the end. He was compensated for the exploration expenses that had so sorely tested his pocket, and he received shares worth a market value of £895,000 (£30 million or $55 million today). Yet D'Arcy could see the venture slipping further from his grasp. "I feel like signing away a child," he lamented on the day he came to final agreement with Burmah Oil. True, the links of paternity were not quite broken. D'Arcy became a director of the new company, and he pledged his continuing interest—"I am just as keen as ever." But the influence of this "capitalist of the highest order," and, as his wife had feared, his very name, faded away even before William Knox D'Arcy's death in 1917. It was small consolation that Anglo-Persian kept the name "D'Arcy" merely for an exploration subsidiary.

A major new source of oil had been proved, loosely at least under British protection. Anglo-Persian itself very quickly emerged as a significant company. By the end of 1910, it already employed 2,500 people. But still, the organization of its operations in Persia was a complex and problematic business, made even more Byzantine by the clash of corporate and political authorities. Arnold Wilson, by then acting consul in the region, became the de facto adviser on local affairs to the company, which he found to be a continually trying experience. "I have spent a fortnight upon Oil Company business, mediating between Englishmen who cannot always say what they mean and Persians who do not always mean what they say. The English idea of an agreement is a document in English which will stand attack by lawyers in a Court of Justice: the Persian idea is a declaration of general intentions on both sides, with a substantial sum in cash, annually or in a lump sum."

An oil field at least ten miles square was soon proved in the area, creating a new problem—how to get the crude oil out and then get it refined. A 138-mile pipeline, crossing two ranges of hills and a desert plain—its route initially marked out by sticks and calico flags—was built in a year and a half. Six thousand mules were enrolled in the effort. The site chosen for a refinery was

Abadan, a long, narrow island of mud flats and palm trees in the Shatt-al-Arab, the extended estuary of the Tigris, Euphrates, and Karun rivers. The laborers were mainly Indians from Burmah's Rangoon refinery, and the construction was badly done. On its first test, in July 1912, the refinery immediately broke down. Thereafter, it operated far below capacity. The quality of its products was also poor; the kerosene had a yellowish tinge and filmed up lamps. "It has been," an exasperated director of Burmah said in September 1913, "one chapter of misfortunes after another since the Refinery first tried to start."

In October of 1912, Anglo-Persian took a significant step to assure itself of markets by making an arrangement with Asiatic, the trading arm of Royal Dutch/Shell. Beyond local markets, Anglo-Persian would sell crude and all its gasoline and kerosene through Asiatic, but reserved the rights to its fuel oil, on which it was preparing to base its strategy for future growth. At this stage, Anglo-Persian simply could not afford the costs of challenging the established giants to a marketing war. Shell, for its part, wanted to contain any new threats; as Robert Waley Cohen wrote to his colleagues in The Hague, "the situation of these people, apparently with very large supplies, made them rather a serious menace in the East."

But the menace was mitigated by the fact that Anglo-Persian soon found itself in deep financial trouble. Once again, the very survival of the Persian venture was in doubt. By the end of 1912, the company had exhausted its working capital. John Cargill, the chairman of Burmah Oil, was blunt. "What a hell of a mess Persian things are in," he wrote. "It's all very well to say 'don't worry,' but my name and business reputation are too closely associated with the Anglo-Persian Oil Company to admit of my not being terribly anxious and worried over the present horrible state of affairs."

Millions of pounds were needed for development, but there was no obvious way to get new capital. Yet, without an infusion of funds, the effort in Persia would grind to a stop, or the whole enterprise might simply be swallowed by Royal Dutch/Shell. A few years earlier, Burmah had saved the day. Now a new savior would have to be found.[12]

CHAPTER 8

The Fateful Plunge

IN JULY OF 1903, during one of his many moments of despair, William Knox D'Arcy, disappointed and worn down by the slow and expensive progress of his oil venture in Persia, had taken himself off for a cure at the spa at Marienbad, in Bohemia. His spirits were lifted there, however, not only by the treatment but also by an acquaintance he made—that of Admiral John Fisher, then the Second Sea Lord of the Royal Navy and already long known as the "oil maniac." That chance meeting would eventually lead to the transformation of D'Arcy's venture and would push oil to the center of national strategies.

Admiral Fisher had been a regular visitor to Marienbad since recovering at the spa, many years earlier, from a case of chronic dysentery. But on this particular visit, Fisher too had arrived a disappointed man. Shortly before, the first test of fuel oil in a British battleship had taken place aboard HMS *Hannibal*. The ship had steamed out of Portsmouth Harbor, burning good Welsh coal, with a trail of white smoke. At a signal, it switched to oil. Moments later, the ship was completely enveloped in a dense black cloud. A faulty burner had turned the test into a disaster. It was a bitter defeat for the two leading proponents of oil fuel for the Navy, both of whom were in attendance—Admiral Fisher and Marcus Samuel of Shell. Shortly after, a dejected Fisher had set off for Marienbad, where by coincidence he met D'Arcy.

The two men immediately discovered that they shared an enthusiasm for oil, and D'Arcy hurriedly sent for maps and papers concerning the Persian venture to show Fisher. In turn, Fisher was cheered and enormously impressed by what he was told by D'Arcy, whom he called the "gold-mine millionaire." D'Arcy, Fisher wrote, "has just bought the south half of Persia for OIL. . . . He thinks it's going to be a great thing: I am thinking of going to Persia instead of Portsmouth, as he tells me he wants someone to manage it for him!" D'Arcy understood Fisher to have promised some kind of help. Though help would come—first, behind the scenes, and then in a very significant public way—it would never be anywhere near as swift as D'Arcy would have wished.[1]

"The God-father of Oil"

John Arbuthnot Fisher, who would be memorialized by Marcus Samuel as "the God-father of oil," became First Sea Lord in 1904. For the next six years, "Jacky" Fisher would dominate the Royal Navy as no other man had ever done. Born in Ceylon of an impoverished planter family, Fisher went to sea in 1854, at age thirteen, as a naval cadet on a sailing ship. He had advantages of neither birth nor rank, but rather advanced by sheer intelligence, tenacity, and force of will. To one contemporary, he was "a mixture of Machiavelli and a child." Overwhelming all with whom he came in contact, he was a "tornado of energy, enthusiasm, and persuasive power." Once, after being subjected to some forceful argument by Fisher, King Edward VII himself told the admiral, "I wish you would stop shaking your fist in my face."

Aside from family, dancing, and religion (including a prodigious recall of biblical quotations), Fisher had only one consuming passion—the Royal Navy. He dedicated himself fully to modernizing it, furiously seeking to shake it free of its ingrained habits, its complacency, its cobwebbed traditions. He pursued his goals with unswerving determination. An officer who served under him said, " 'Jacky' was never satisfied with anything but 'Full Speed!' " A self-proclaimed zealot in his causes, he was the Royal Navy's greatest proponent of technological change. His "Golden rule" was never "to allow ourselves to be out 'classed.' " First achieving some reputation in the Navy as an expert on torpedoes, he went on to champion the submarine, the destroyer, Kelvin's compass, advances in firepower, eventually naval aviation—and, all along, petroleum. "Oil fuel," he wrote as early as 1901, "will absolutely revolutionize naval strategy. It's a case of 'Wake up England!' " He wanted to convert the fleet from coal propulsion to oil. The benefits would be faster speed and greater efficiency and maneuverability. But he was in a minority; the other admirals felt more secure depending on Welsh coal, and insisted on continuing to do so.

While First Sea Lord, Fisher maintained his interest in the project to which D'Arcy had introduced him at Marienbad. Intent on seeing oil fields developed under British control, he provided much of the impetus for the Admiralty's support of the Persian concession and then for the pressure on the Burmah Oil Company to come to D'Arcy's rescue. His principal objective was always the same—to bring the Royal Navy into the industrial age and to have it prepared when war came. Earlier than most, he was convinced that Britain's enemy would be the formidable industrial rival that had arisen on the Continent—imperial Germany. And he would push both the Royal Navy and the British government toward oil, for he was no less convinced that oil fuel would be a critical element in the inevitable conflict ahead.[2]

"Made in Germany"

Though the specific subjects of direct dispute between Germany and Britain were surprisingly few, many factors contributed to the growing enmity between them at the turn of the century—including the marked insecurity of the Kaiser, a grandson of Queen Victoria, toward his uncle, Britain's King Edward VII. But no other single factor counted for so much as the burgeoning naval race between Britain and Germany—the competition for size and technological advance of their two fleets. It dominated relations between the two nations; within each, it captured the attention of the press, shaped public attitudes and discussion, fed the rising nationalistic passions, and fueled the deepest anxieties. It was the focus of their antagonism. "So far as contemporary opinion was concerned," one historian has written, "it was the naval question above everything else which exacerbated Anglo-German relations."

By the late 1890s, the German government had inaugurated its full-scale attempt at *Weltpolitik*—the drive for global political, strategic, and economic prominence, for recognition of Germany as a world power, and for what was referred to in Berlin as "world political freedom." The heavy-handed, occasionally crude, and blatantly aggressive way in which the "new" Germany sought to assert itself on the world stage only disconcerted and increased the alarm of other powers. Even one of the Kaiser's own chancellors was to criticize the nation's "strident, pushing, elbowing, overbearing spirit." It was a manner that seemed to reflect and to be made worse by the character of Kaiser Wilhelm himself. He was a temperamental, erratic, prejudiced, petulant, and mercurial monarch. One prominent German despaired of the Kaiser's ever becoming wiser with age.

To many Germans, living in the heyday of post-Bismarckian empire, a single obstacle, above all others, seemed to stand in the way of their dream of world power—British supremacy on the high seas. Germany's aim was, in the words of one of its admirals, to break "England's world domination so as to lay free the necessary colonial possessions for the central European states who need to expand." That meant, first of all, building a Navy to rival Britain's. As the Kaiser himself declared, "Only when we can hold out our mailed fist against his face will the British lion draw back." The Germans launched their naval challenge in 1897. Though they fully expected that achievement of their goal would take considerably more than a decade, they were counting on the British to tire eventually of the cost of the rivalry. The actual effect on the British would be quite the opposite: The challenge alarmed and galvanized them to their own strenuous efforts. For naval supremacy was central to England's conception of its world role and to the security of the British empire. The new menace from Germany was even more alarming when measured against the pressures and problems that Britain was experiencing as it struggled to cope with imperial responsibilities and burdens larger than its capabilities to manage, man, and pay for. Industrial leadership was slipping away from it—to the United States and, worse, to Germany. In 1896, an admonitory work entitled *Made in Germany* became a best seller in England. Britain, moaned a Cabinet minister, was "the weary Titan."[3]

Admiral Fisher had no doubt that it was Germany and only Germany that was the future enemy. He feared that it would strike out of the blue, probably on a long holiday weekend—so, over the years, his aides were always kept on special duty and thus missed many such holiday weekends. Pushed by Fisher, the British government responded to the German challenge with the modernization of its fleet and an expanded construction program. By 1904, the naval race was on in full—fueled on both sides by "a runaway technological revolution" in the size and speed of battleships, in the range of and accuracy of their firepower, and in the development of new weapons like the torpedo and submarine.

In both countries, the race took place against a backdrop of social and labor unrest, of domestic conflicts, of financial and budgetary constraints. Britain underwent a classic guns-or-butter debate. The ruling Liberal party was torn between the "navalists," who supported a "big Navy" policy and an expanded Admiralty construction budget, and the "economists," who wanted to contain naval expenditures and instead put more money into the social and welfare programs they thought necessary to maintain domestic peace. The ensuing debate was very bitter. "Is Britain going to surrender her maritime supremacy to provide old-age pensions?" the *Daily Express* declaimed. From 1908 on, the "economists" in Prime Minister Herbert Asquith's Liberal Cabinet were led by David Lloyd George, the Welsh solicitor who was Chancellor of the Exchequer, and for a time, by Winston Spencer-Churchill, who had dropped the Spencer while still at school, so he would not have to wait and be "the last of all" in line. Now, in British politics, he was "the young man in a hurry."[4]

Enter Churchill

Winston Churchill was the nephew of the Duke of Marlborough and son of the brilliant but erratic Lord Randolph Churchill and his beautiful American wife, Jennie Jerome. He had entered Parliament as a Conservative in 1901, at age twenty-six. Three years later, he bolted from the Tory party over the question of free trade and crossed over to the Liberals. His political conversion did not impede his progress. He was soon President of the Board of Trade and, by 1910, Home Secretary. He lived for politics and grand strategy. On the day of his marriage, even as he stood in the vestry in the moments before the ceremony, he talked and gossiped of politics. He threw himself into the leadership of the "economists" campaign. Battling against Fisher's expanded naval program, he and Lloyd George championed an Anglo-German naval agreement as a way to reduce the Navy's budget and so free money for social reform. For all this Churchill was much criticized. But he would not budge. Belief in the inevitability of war between Britain and Germany, he declared, was "all nonsense."

But in July 1911, the German gunboat *Panther* sailed into the Moroccan port of Agadir—in that clumsy ploy meant to assert Germany's insistence on its place in the African sun. The *Panther* episode backfired, consolidating anti-German feeling both in Britain and on the Continent, especially in France. Churchill's views were instantaneously transformed. From that moment on, he had no doubt: Expansionism was the German goal, and the growth of the German fleet served no purpose save to threaten Britain—a threat that had to be countered. Germany, he now concluded, meant to make war. Britain, thus, had to marshal its resources to maintain its supremacy; and Churchill, though still

Home Secretary, began to express intense interest in the strength of the Royal Navy and to question whether it was really ready for a bolt out of the blue. He was outraged that senior officials chose to go on shooting holidays in Scotland at the height of the Agadir crisis. At the end of September 1911, the crisis ended, and thereupon, Churchill himself went off to Scotland to stay with Prime Minister Asquith. On the way back from a game of golf, the Prime Minister quite abruptly asked him if he would like to become First Lord of the Admiralty, the top civilian post for the Royal Navy.

"Indeed I would," Churchill replied.[5]

Now the Admiralty would have, as its civilian head, a man who could channel his enormous energy, vision, concentration, and powers of exposition to the task of assuring Britain's victory in the naval race. "The whole fortunes of our race and Empire," Churchill said, "the whole treasure accumulated during so many centuries of sacrifice and achievement, would perish and be swept utterly away if our naval supremacy were to be impaired." His guiding precept during those three years before the outbreak of the First World War was clear: "I intended to prepare for an attack by Germany as if it might come the next day."

His ally in that campaign would be Admiral Fisher, who, almost twice his age, had just retired from the Navy. Fisher had been entranced by Churchill ever since their initial meeting at Biarritz in 1907. So close were they that Fisher may well have been the first to be told of Churchill's impending marriage. Despite a falling out over his earlier criticism of the Navy's budget, Churchill, on becoming First Lord, immediately sent for the old admiral and, after spending three days with Fisher at a country house at Reigate, won him back. Thereafter, it would be said that Fisher had become Churchill's "dry nurse." He certainly emerged as the dominating unofficial adviser. Churchill regarded Fisher as the source for a decade of "all the most important steps taken to enlarge, improve, or modernize the Navy," and he found the admiral, who bombarded him endlessly with memos, to be "a veritable volcano of knowledge and of inspiration." Fisher offered tuition on the widest variety of subjects.

One of the most significant lessons to be learned concerned petroleum—which, Fisher argued, would prove integral to the strategy of supremacy. He set out to make sure that Churchill was properly educated about the virtues of oil over coal for His Majesty's Navy. Alarmed by reports that the Germans were building oil-powered ocean liners, Fisher felt a new urgency to shove the Royal Navy "over the precipice" of oil, and as rapidly as he could. To speed Churchill's education, the admiral conspired with Marcus Samuel of Shell. More than a decade earlier, those two men had come to an instant meeting of minds on oil's potential role; their relationship was cemented when Samuel confidentially informed Fisher that a German shipping line had made a ten-year contract for oil—with part of the supply secretly destined for experimentation by the German Navy. "How right you have been & how right you are now!" Samuel wrote Fisher at the end of November 1911. "The development of the internal Combustion engine is the greatest the world has ever seen for so surely as I write these lines it will supersede steam and that too with almost tragic rapidity . . . I am heartsick as I know you are at the machinations of the permanent officials at the

Admiralty & it will require a strong & very able man to put right the injury they have inflicted so far.

"If Winston Churchill *is* that man I will help him heart and soul."[6]

Speed!

Shortly thereafter, Fisher arranged for Marcus Samuel to meet Churchill in order to make the case for oil. But Churchill was not all that impressed with the chairman of Shell Transport and Trading. In a follow-up note to Churchill, Fisher first apologized for Samuel: "He is not as good at exposition but he began as a pedlar selling 'sea' shells! (Hence the name of his Company) and now he has six million sterling of his own private money. 'He's a good teapot though he may be a bad pourer'!" Fisher then explained that he had promoted the meeting with Samuel to convince Churchill that oil was available in volumes sufficient to make a confident commitment to it for the propulsion of the Royal Navy. He lectured Churchill on oil's advantages over coal: "Remember oil like coal don't deteriorate and you can accumulate vast stores of it in submerged tanks so as to be free from destruction by fire or bombardment or incendiaries and *east of Suez oil is cheaper than coal!*" Fisher added that Samuel had invited him to join the Shell board but that he had declined: "I'm a pauper and I am deuced glad of it! but if I wanted to be rich I would go in for oil! When a cargo steamer can save 78 percent in fuel and gain 30 percent in cargo space by the adoption of internal combustion propulsion and practically get rid of stokers and engineers—it is obvious what a prodigious change is at our doors with oil!" The admiral was scornful of the delays in converting to oil and warned Churchill of the dangerous consequences. "Your old women will have a nice time of it when the new American Battleships are at sea burning oil alone and a German Motor Battleship is cocking a snook at our 'Tortoises'!"[7]

When Churchill arrived at the Admiralty, the Navy had already built or was building fifty-six destroyers solely dependent on oil and seventy-four submarines that could only be driven by oil. Some oil was also sprayed in the coal furnaces of all ships. But the most important part of the fleet—the battleships, the capital ships that were the very backbone of the Navy—burned coal. What both Churchill and the Navy wanted was to create a new breed of battleships, with yet bigger guns and stronger armor but also with the greater speed necessary to draw ahead and circle around the head of the enemy's line. "Sea fighting is pure common sense," Fisher reminded Churchill. "The first of all necessities is SPEED, so as to be able to fight—*When* you like, *Where* you like, and *How* you like." The British battleships of the day could get up to twenty-one knots. But, as Churchill observed, "much greater speed" would introduce "a new element into naval war." In a study conducted at Churchill's behest, the War College estimated that with twenty-five knots, a new "Fast Division" could get the better of the emerging new German fleet. In short, the Royal Navy wanted an extra four knots—and there seemed no way to get it without oil.

Churchill's education was complete. Oil allowed not only higher speeds, he recognized, but also greater rapidity in getting up to speed. Oil offered further advantages in the operation and manning of the fleet. It allowed a greater radius of action. It permitted refueling at sea (at least on calm seas), without occupying a quarter of the ship's manpower in the effort, as was the case with coal. Moreover, it greatly reduced the stress, time, exhaustion, and discomfort that went with coaling and cut the required number of stokers by more than half. Oil's advantage in terms of operations, as well as speed, could count the most at the most critical time—in battle. "As a coal ship used up her coal," Churchill later wrote, "increasingly large numbers of men had to be taken, if necessary from the guns, to shovel the coal from remote and inconvenient bunkers to bunkers nearer to the furnaces or to the furnaces themselves, thus weakening the fighting efficiency of the ship perhaps at the most critical moment in the battle. . . . The use of oil made it possible in every type of vessel to have more gun-power and more speed for less size or less cost."

The three naval programs of 1912, 1913, and 1914 constituted the greatest addition—in terms of sheer power and cost—in the history of the Royal Navy up to that time. All the ships of those three programs were based on oil—not a coal-burning ship among them. (Some of the battleships were originally to be coal burning, but were switched to oil.) The key decision was taken in April of 1912, with the inclusion in the naval budget of a Fast Division, the *Queen Elizabeth* class—composed of five oil-fired battleships. With this "fateful plunge," Churchill wrote, "the supreme ships of the Navy, on which our life depended, were fed by oil and could only be fed by oil."

That commitment, however, raised a very serious problem—where was the oil to be found, would there be enough, and would it be a militarily and politically secure supply? Churchill's great gamble was to push for conversion to oil *before* the supply problem had been solved. He eloquently summarized the issue: "To build any large additional number of oil-burning ships meant basing our naval supremacy upon oil. But oil was not found in appreciable quantities in our islands. If we required it we must carry it by sea in peace or war from distant countries. We had, on the other hand, the finest supply of the best steam coal in the world, safe in our mines under our own land. To commit the Navy irrevocably to oil was indeed 'to take arms against a sea of troubles.' "Yet, if the difficulties and risks could be surmounted, "we should be able to raise the whole power and efficiency of the Navy to a definitely higher level; better ships, better crews, higher economies, more intense forms of war power"—in a word, "mastery itself was the prize of the venture."[8]

The Admiral Cracks the Nut

Churchill established a committee to study the issues raised by converting from coal to oil, including pricing, availability, and security of supply. The committee in turn recommended the establishment of a royal commission to investigate these matters more thoroughly. Churchill's obvious choice to head such a commission was the retired Admiral Fisher. There was only one obstacle—Jacky Fisher himself. The volcanic admiral was once again furious with Churchill, this time because he disapproved of some promotions Churchill had made. "You have betrayed the Navy," Fisher wrote to Churchill from Naples in April 1912. "This must be the last communication with you in any matter at all."

It required a good deal of cajolery, the blandishment of a Mediterranean cruise on an Admiralty yacht with Churchill and Prime Minister Asquith in attendance, and a most forceful letter to win over the irascible admiral. "My dear Fisher," Churchill wrote:

> We are too good friends (I hope) and the matters we are concerned with are too serious (I'm sure) for anything but plain language.
> This liquid fuel problem has got to be solved, & the natural, inherent, unavoidable difficulties are such that they require the drive & enthusiasm of a big man. I want you for this, *viz*, to crack the nut. No one else can do it so well. Perhaps no one else can do it at all. I will put you in a position where you can crack the nut, if indeed it is crackable. But this means that you will have to give life & strength, & I don't know what I have to give in exchange or in return. You have got to find the oil; to show how it can be stored cheaply: how it can be purchased regularly & cheaply in peace, and with absolute certainty in war. Then by all means develop its applicn in the best possible way to existing & prospective ships. . . .
> When you have solved the riddle, you will find a vy hushed attentive audience. But the riddle will not be solved unless you are willing—for the glory of God—to expend yourself upon its toils.

Churchill could not have done any better by way of flattery. Without undue modesty, Fisher wrote to his wife, "I really have to admit that they are right when they all unanimously say to me that no one else can do it." He accepted the post, and shortly after—so as to avoid conflict of interest—sold the shares he held in Shell at a prospective loss.[9]

A distinguished group was assembled to sit for the Royal Commission on Fuel and Engines, including the ever-present oil expert Sir Thomas Boverton Redwood, with the orchid in his buttonhole. Fisher threw himself into the job, working, he said, as hard as he had ever worked. His urgency increased when he learned that the German Navy was going forward with oil propulsion. *"They have killed 15 men in experiments with oil engines and we have not killed one! And a d——d fool of an English politician told me the other day that he thinks this creditable to us."*

The commission issued the first part of its report in November 1912 and two further sections in 1913. It stressed both the "overwhelming advantages in favour of oil fuel" over coal and oil's vital importance to the Royal Navy. It maintained that sufficient supplies existed throughout the world, although it did call for much-expanded storage facilities because, as Fisher put it, "Oil don't grow in England." At last, Marcus Samuel's dream of an oil-fueled British Navy looked to become a reality. But one question remained: who would reap the profits? The likely choices were only two: the powerful and entrenched Royal Dutch/Shell Group, and the much smaller and still-struggling Anglo-Persian Oil Company.[10]

The Shell Menace

Though Anglo-Persian's creation was the result of the combined efforts of William Knox D'Arcy, George Reynolds, and Burmah Oil, Charles Greenway was the man who really fashioned the company. It was as manager of a Scottish trading house in Bombay that he had first begun to deal in oil. The Scottish merchants associated with Burmah Oil asked him to assist in the beginning stages of Anglo-Persian, and within a year he was its managing director. He dominated the company for the next two decades. When he started, he was virtually a one-man band; by the time of his retirement, he presided over an integrated oil company, actively engaged throughout the world. Later in life, he became known as "Champagne Charlie" and was caricatured as "Old Spats and Monocle." Though "decorous, even fastidious" in manner, Greenway was tenacious and always ready for a brawl. He was also unbending and obstinate in pursuing his central objectives: to build Anglo-Persian into a major force in world oil; to make it the national champion of Great Britain; to resist the unwelcome and suffocating embrace of Royal Dutch/Shell—and to ensure his own unquestioned control of the new concern. He would do whatever was necessary to achieve his goals, including the pursuit of a ceaseless vendetta against Royal Dutch/Shell, which became both a useful tactic and a personal obsession.

Britain's "fateful plunge" inevitably spurred even fiercer rivalry between Royal Dutch/Shell and Anglo-Persian. In that battle, Anglo-Persian was at a definite disadvantage; it once more found itself under intense financial pressure. As far as Greenway was concerned, time was growing short, and he was forced to pursue several goals at once: obtain the capital to develop the Persian resources, build up the oil company, develop secure markets, and—despite its marketing agreement with Royal Dutch/Shell—avoid being absorbed by that company. In Anglo-Persian's weak financial position, there was only one obvious alternative to Shell, and that was the British Admiralty. Greenway offered the Admiralty a twenty-year fuel contract and campaigned hard for a special relationship that would rescue the company from its financial straits.

Greenway's recurrent theme, both in testifying before Fisher's commission and throughout Whitehall, was that, without government aid, Anglo-Persian would disappear into Shell. If that happened, Greenway warned, Shell would be in a monopoly position and would extract monopoly prices from a hapless Royal Navy. He stressed Samuel's "Jewishness" and Deterding's "Dutchness." Shell, he said, was controlled by Royal Dutch, and the Dutch government was susceptible to German pressure. Control by Shell, he told Fisher's commission, would eventually place Anglo-Persian "under the control of the German Government itself."

There was, Greenway altruistically allowed, a price to be paid by him and his colleagues for being so concerned about Britain's national interest. But, he confided, he and his associates, all patriotic Englishmen, were willing—indeed, more than willing—to sacrifice the economic advantage that would accrue from affiliating with Shell and instead keep the company independent. All they asked in return was just some small consideration from the British government—just a guarantee or contract "that will at any rate give us a moderate return on our capital." He emphasized repeatedly that Anglo-Persian was a natural adjunct to British strategy and policy and was a significant national asset—and that all the company's directors saw it just that way.[11]

Greenway's message was well received. Immediately after his testimony to the royal commission, Fisher detained him for some time outside in Pall Mall, to talk privately. Something had to be done at once, Fisher insisted. Greenway was greatly pleased, for, despite Fisher's friendship with Marcus Samuel, the admiral was completely clear on exactly what it was that needed to be done. "We must do our d——st to get control of the Anglo-Persian Company," he wrote, "and to keep it for all times as an absolutely 'all-British' Company."

Greenway's arguments won support elsewhere as well. The Foreign Office, concerned as it was with Britain's position in the Persian Gulf, generally found the case convincing. The priority for the Foreign Office was that the Anglo-Persian concession, "embracing as it did the entire oil fields of Persia . . . should not pass under the control of a foreign syndicate." Britain's political predominance in the Persian Gulf "is largely the result of our commercial predominance." At the same time, the Foreign Office was persuaded by the more specific needs of the Royal Navy. "Evidently," Sir Edward Grey, the Foreign Secretary, commented, "what we must do is to secure under British control a sufficient oil field for the British Navy." Though sometimes irritated and made suspicious by Greenway's harping upon the "Shell menace" and the much-touted patriotism of Anglo-Persian Oil, the Foreign Office stuck to that position. "It is clear that diplomatic assistance alone will be useless in preserving the independence of the APOC," the Foreign Office warned the Admiralty at the end of 1912. "It is pecuniary assistance in some form that they require."[12]

Aid for Anglo-Persian

That pecuniary assistance would have to involve the Admiralty. Initially, the Admiralty was not at all interested in developing such a special relationship with Anglo-Persian; it feared becoming involved in a business "subject to much speculative risk." But three decisive factors changed the Admiralty's outlook. First, there were growing doubts about the availability and reliability of petroleum supplies from sources other than Persia. Second, the price of fuel oil was increasing dramatically, doubling between January and July of 1913 alone, in response to rising maritime demand around the world—a critical consideration as the construction of oil-fired battleships had begun even while the protracted political battle over the Navy's budget continued to rage.

The third factor was Churchill, who was pushing decisions and forcing senior Navy officers to analyze the availability, needs, and logistics of oil in both peace and war. In June 1913, Churchill presented the Cabinet with a key memorandum on "Oil Fuel Supply for His Majesty's Navy," which called for long-term contracts to assure adequate supplies at secure prices. A governing principle was "keeping alive independent competitive sources of supplies," thus frustrating "the formation of a universal oil monopoly" and safeguarding "the Admiralty from becoming dependent on any single combination." The Cabinet agreed in principle, as Prime Minister Asquith wrote to King George V, that the government should "acquire a controlling interest in trustworthy sources of supply." But exactly how? Greenway then met with members of the Cabinet, and in the course of their discussions the long-sought-after answer to that question began to emerge: namely, the arresting idea that the government itself become a shareholder in Anglo-Persian as a way to legitimize its financial support.[13]

On July 17, 1913, Churchill, in a statement to Parliament that the *Times* of London described as an authoritative presentation on the national interest in oil, took the idea one step further. "If we cannot get oil," he warned, "we cannot get corn, we cannot get cotton and we cannot get a thousand and one commodities necessary for the preservation of the economic energies of Great Britain." In order to assure dependable supplies at reasonable prices—because the "open market is becoming an open mockery"—the Admiralty should become "the owners or, at any rate, the controllers at the source" of a substantial part of the oil it required. It would begin by building up reserves, then develop the ability to deal in the market. The Admiralty should also be able "to retort, refine . . . or distil crude oil"—disposing of surplus as need be. There was no reason to "shrink from making this further extension of the vast and various business of the Admiralty." Churchill added, "On no one quality, on no one process, on no one country, on no one route and on no one field must we be dependent. Safety and certainty in oil lie in variety and variety alone."

Though there was no specific commitment to Anglo-Persian, the Cabinet decided to send a commission to Persia to investigate whether Anglo-Persian could actually deliver on any of its promises. The new refinery at Abadan was experiencing enormous problems. One of the directors of Burmah Oil had described it as nothing more than a "scrap heap." Even the fuel oil it produced—confidently named "Admiralty"—had flunked the Admiralty's own qualifying test. But, on the eve of the commission's arrival, the company hastily introduced cosmetic improvements, orchestrated by a new refinery manager hurriedly rushed in from Rangoon. The ploy worked. "It seems to be a thoroughly sound concession, which may be developed to a gigantic extent with a large expenditure of capital," Admiral Edmond Slade, former Director of Naval Intelligence and head of the commission, privately informed Churchill. "It would put us into a perfectly safe position as regards the supply of oil for naval purposes *if we had the control of the company* and at a very reasonable cost." In his official and influential report at the end of January 1914, Slade added that it would be "a national disaster if the concession were allowed to pass into foreign hands." Slade even managed to find some kind words to say about the operation of the Abadan refinery.[14]

A Victory for Oil

Admiral Slade's report was heaven-sent for Anglo-Persian. The company's financial situation was steadily deteriorating and indeed was nothing less than desperate. Now, however, Slade had blessed the operation and, on the all-important issue, pronounced it a secure source for the Royal Navy; the way was open to bring matters to a conclusion. On May 20, 1914, less than four months after Slade's report, the deal was wrapped up with the signing of an agreement between the company and the British government. But there was still one last obstacle; the Treasury insisted that any appropriation required Parliamentary approval, and that test had yet to be passed.

On June 17, 1914, Churchill rose in the House of Commons to introduce a historic measure. The bill he proposed had two essential elements: First, the government would invest £2.2 million in Anglo-Persian, acquiring in turn 51 percent of the stock; second, it would place two directors on the company's board. They would have a veto on matters involving Admiralty fuel contracts and major political matters, but not on commercial activities. Another contract was drawn up separately, so it could be kept secret; it provided the Admiralty with a twenty-year contract for fuel oil. The terms were very attractive, and in addition, the Royal Navy would get a rebate from the company's profits.

The debate in the House was highly charged. Charles Greenway sat in the official box with senior Treasury officials in case Churchill needed any special information. Also present in the Commons was the member from Wandsworth, one Samuel Samuel, who, working for many years by the side of his brother, Marcus Samuel, had helped to create Shell—and who, that day, became increasingly fidgety and aggravated as Churchill spoke.[15]

"This afternoon we have to deal, not with the policy of building oil-driven ships or of using oil as an ancillary fuel in coal-driven ships," Churchill began, "but with the consequence of that policy." The oil consumer, he declaimed, had freedom of choice neither in regard to fuels nor in regard to sources of supply. "Look out upon the wide expanse of the oil regions of the world. Two gigantic corporations—one in either hemisphere—stand out predominantly. In the New World there is the Standard Oil. . . . In the Old World the great combination of the Shell and the Royal Dutch, with all their subsidiary and ancillary branches, has practically covered the whole ground, even reached out into the New World." Churchill proceeded to argue that the Admiralty, along with all private consumers, had been subjected to "a long steady squeeze by the oil trusts all over the world."

Early in the debate, Samuel Samuel popped up three times to object to Churchill's characterizations of Royal Dutch/Shell. He was ruled out of order. "He had better hear the case for the prosecution," Churchill acidly said after the third interruption, "before he offers an argument for the defense." Samuel resumed his seat but not his composure.

"For many years," Churchill went on, "it has been the policy of the Foreign Office, the Admiralty, and the Indian Government to preserve the independent British oil interests of the Persian oil-field, to help that field to develop as well as we could and, above all, to prevent it being swallowed up by the Shell or by any foreign or cosmopolitan companies." Since the government was going to give such a boost to Anglo-Persian, it was but reasonable, he added, that it share in the rewards. And "over the whole of these enormous regions we obtain the power to regulate developments according to naval and national interest." Declaring that "all the criticisms" of such a plan "so far, have flowed from one fountain," Churchill then launched an attack on that fountain—Royal Dutch/Shell and Marcus Samuel—though adding, "I do not wish to make any attack upon the Shell or the Royal Dutch Company."

"Not the least!" Samuel Samuel called out from the back bench.

Churchill's oratory was full of sarcasm. Were the bill to fail, he said, Anglo-Persian would become part of Shell. "We have no quarrel with the 'Shell.' We have always found them courteous, considerate, ready to oblige, anxious to serve the Admiralty, and to promote the interests of the British Navy and the British Empire—at a price. The only difficulty has been price." With the leverage of Persian oil "at our disposal, we do not think we shall be treated with less courtesy, or less consideration, or shall we find these gentlemen less obliging, less public spirited, or less patriotic than before. On the contrary, if that slight difference of opinion which has hitherto existed about prices—I am obliged to return to that vicious and sordid matter of prices—were removed, our relations would be better; they would become . . . the sweeter, because no longer leavened with the sense of injustice."

Samuel finally had his chance, later in the debate, to reply. "I do protest most strongly on behalf of one of the greatest British commercial industrial companies, that the attacks that have been made are wholly unjustifiable." He catalogued Shell's services to the Navy and its championing of oil-powered propulsion. He asked the government to make public the prices that Shell had charged, which had been kept secret, and which, he said, would prove that the company had never gouged the Admiralty.

"The attack we have heard had nothing on earth to do with the question before the Committee," said another M.P., Watson Rutherford. Criticizing Churchill for raising the specter of monopoly and for "Jew-baiting," he declared that the rising prices of fuel oil had resulted not from "the machinations of some trust or ring" but from the fact that an international market for fuel oil—as opposed to those for gasoline, kerosene, and lubricants—had only arisen in the "last two or three years, in consequence of these new uses which have been found for this oil. . . . There is a world shortage," he continued, "of an article which the world has only lately begun to see is required for certain special purposes. That is the reason why prices have gone up, and not because evilly-disposed gentlemen of the Hebraic persuasion—I mean cosmopolitan gentlemen—have put their heads together in order to try and force prices up."

Churchill's proposal for government ownership of a private company was indeed unprecedented, save for Disraeli's purchase of shares in the Suez Canal a half century earlier—a step also taken on strategic grounds. Some M.P.s, representing their local interests, argued for the development of oil from Scottish shale and liquids from Welsh coal (many years later known as synthetic fuels). Both, they said, would provide more reliable supplies. Yet, despite the strong criticism inside Parliament and out, the oil bill passed by an overwhelming vote—254 to 18. The margin was so large that it surprised even Greenway. After the vote, he asked Churchill, "How did you manage to carry the House with you so successfully?"

"It was," Churchill replied, the "attack on monopolies and trusts that did it."[16]

But his assault on foreigners and "cosmopolitans" also helped. Moreover, Churchill had been more than a little cynical in his presentation. For there was no evidence that Shell had ever served the Admiralty poorly. Indeed, years before, Marcus Samuel had actually asked the government to place a director on the board of Shell. And while Churchill had taken a dislike to Marcus Samuel, who had been Lord Mayor of London, he had developed a most favorable opinion of Deterding, who was, after all, the foreigner.

Here, in the matter of Deterding, Churchill was following Admiral Fisher's lead. Fisher wrote to Churchill that Deterding, "is Napoleon and Cromwell rolled into one. *He is the greatest man I ever met* . . . Napoleonic in his audacity: Cromwellian in his thoroughness! . . . Placate him, don't threaten him! Make a contract with him for his fleet of 64 oil tankers in case of war. Don't abuse the Shell Company. . . . [Deterding] has a son at Rugby or Eton and has bought a big property in Norfolk and [is] building a castle! Bind him to the land of his adoption!" Churchill did exactly that. Despite the new agreement, Anglo-Persian was not to be the sole supplier to the Admiralty, and in the spring of 1914, he took over personally in negotiating with Deterding on Shell's fuel oil contract with the Navy. Deterding was responsive to Churchill's attention. "I have just received a most patriotic letter from Deterding," Fisher wrote to Churchill on July 31, 1914, "to say he means you shan't want for oil or tankers in case of war— *Good Old Deterding!* How these Dutchmen do hate the Germans! Knight him when you get the chance."[17]

Deterding was a practical man and understood the rationale for the Anglo-Persian arrangement. Still, there were those perplexed by the government's purchase. The Viceroy of India, Lord Hardinge, had served two years in Tehran, leaving him with a lasting suspicion of all things Persian. His view, and that of his senior officials in India, was that it was altogether unwise to become dependent upon a most insecure foreign source of oil when Britain was blessed with an abundance of secure coal. As the Secretary of State for India declared, "It is rather as though the owners of the *premier cru* vineyards in the Gironde went about preaching the virtues of Scotch whisky as a beverage."

The critics had a point. Why the troubles of Scotch whisky when one produced a fine wine? Quite simply, the decision was driven by the technological imperatives of the Anglo-German naval race. Even as the Germans sought equality, the British Navy was committed to maintaining naval supremacy, and oil offered a vital edge in terms of speed and flexibility. The deal assured the British government a large supply of oil. It provided Anglo-Persian with a much-needed infusion of new capital and a secure market. It spoke directly to the need for survival of Anglo-Persian, and indirectly, to that of the empire. Thus, by the summer of 1914, the British Navy was fully committed to oil and the British government had assumed the role of Anglo-Persian's majority stockholder. Oil, for the first time, but certainly not the last, had become an instrument of national policy, a strategic commodity second to none.

As First Lord of the Admiralty, Churchill would often say that his goal was to have the Navy ready, as though war might erupt the very next day. Yet during the weeks leading up to the June 17, 1914, Parliamentary debate, Europe had seemed more at peace, and war farther away, than had been the case for several years. No major issue riled the passions of the Great Powers. Indeed, British naval units were making courtesy visits to German ports at the end of June. Later, many would look back on those spring and early summer days of 1914 with nostalgia, as the dusk of an era, the end of childhood, a time of unusual, even unnatural calm. It would not last. One June 28, 1914, eleven days after Parliament approved Churchill's bill, Archduke Franz Ferdinand of Austria was assassinated at Sarajevo. It was not until August 10, 1914, that the Anglo-Persian Oil Convention would receive its Royal Assent. By then, the world had changed. Russia mobilized on July 30. On August 1, Germany declared war on Russia and mobilized its armies. At 11:00 P.M. on August 4, after Germany had ignored a final British ultimatum against violating Belgium's neutrality, Churchill flashed a message to all of His Majesty's ships: "COMMENCE HOSTILITIES AGAINST GERMANY." The First World War had begun.[18]

PART II

THE
GLOBAL
STRUGGLE

CHAPTER 9

The Blood of Victory: World War I

IT WAS SUPPOSED to be a short war, over in a few weeks or, at most, a few months. Instead, it sank into stalemate and dragged on and on. All the mechanical ingenuity of the late nineteenth and early twentieth centuries was drafted into the conflict. And, when it was over, people groped to understand why it had occurred and what it had been about. Many reasons were proffered—from blunder, arrogance, and stupidity to the accumulated tensions of international rivalries and industrial society. The reasons also encompassed the secular religion of nationalism; the sclerosis of the Austro-Hungarian, Russian, and Turkish empires; the collapse of the traditional balance of power; and the ambitions and insecurities of the recently risen German Reich.

The Great War would prove a disaster for the victors as well as the vanquished. An estimated 13 million people died, and many millions more were wounded and displaced. It was also a catastrophe for the political systems of much of Europe, and for the economies of all concerned. Such was the dismal effect of the First World War that a new upheaval would breed in its aftermath. Indeed, so terrible was the cataclysm that one of the great twentieth-century historians of international relations would look back from his old age, a half century later, and recall the war as "the well-spring of our discontents."

It was a war that was fought between men and machines. And these machines were powered by oil—just as Admiral Fisher and Winston Churchill had foreseen, but to a much greater extent than even they or any other leader had expected. For, in the course of the First World War, oil and the internal combustion engine changed every dimension of warfare, even the very meaning of mobility on land and sea and in the air. In the preceding decades, land warfare had depended on inflexible railway systems that could carry troops and supplies to a railhead, as had occurred in the Franco-Prussian War of 1870–71. From the railhead onward, the troops' movement had been circumscribed by physical endurance, muscular capabilities, and the legs of man and beast. How much could be carried, how far and how fast—all that would change with the introduction of the internal combustion engine.

The extent of this transformation far outpaced anything conceived by strategists. Horses were still the basis of planning at the outbreak of the war—one horse for every three soldiers. Moreover, the reliance on horses greatly complicated the problems of supply, for each horse required ten times as much food as each man. At the beginning of the war, at the First Battle of the Marne, one German general cursed that he did not have a single horse that was not too exhausted to drag itself forward across the battlefield. By the end of the war, whole nations would lie exhausted; for the oil-powered engine, while simplifying the problems of mobility and supply, also multiplied the devastation.

Yet at first, insofar as the land war was concerned, it hardly seemed likely that oil would be of great significance. Boasting superiority in iron and coal and a better rail transport system, the German General Staff, with its methodical plans, assumed that the campaign in the West would be swift and decisive. During the initial month of hostilities, the German armies did press forward pretty much according to plan. By early September 1914, one battle line stretched 125 miles, from northeast of Paris to Verdun, where it joined another battle line that stretched to the Alps—the two lines altogether encompassing two million fighting men. The right flank of the German Army was just forty miles from Paris, headed directly for the City of Light. At this critical moment, the internal combustion engine would prove its strategic importance—in a totally unexpected way.[1]

The Taxi Armada

The French government, along with one hundred thousand civilians, had already evacuated Paris. The fall of the capital seemed imminent, and it looked as if France might soon be suing for peace, perhaps from Bordeaux. General Joseph Césaire Joffre, the Commander in Chief of the French Army, considered ordering his troops to drop back to the south and east of Paris, leaving the city mostly unguarded. But the military governor of Paris, General Joseph Gallieni, had other ideas. Aerial reconnaissance convinced him that an opportunity existed to hit the German lines and stop the advance. He tried to convince the British Army to assist him, but to no avail. They would not take him seriously. The old general, with his shaggy mustache and wearing black-buttoned boots, yellow leggings, and an ill-fitting uniform, hardly looked the image of the spit-and-polish officer. "No British officer would be seen speaking to such a comedian," said one eminent British commander. But in an emotional angry nighttime phone call on September 4—what Gallieni later called his *"coups de téléphone"*—he finally persuaded General Joffre to launch a counterattack.

On September 6, 1914, through forests and fields of ripe grain and under scorching heat, the French went on the offensive, scoring some early successes. But then the Germans brought up more troops. The French now found themselves in a truly precarious position. Their own desperately needed reinforcements were in the immediate environs of Paris, but there seemed no way to get them to the front. They certainly could not move by railway; the French system was effectively disrupted. If they marched on foot, they would never arrive in time. And many more men were needed than could be moved by the paltry number of military vehicles available. What else could be done?

General Gallieni would not give up. He seemed to be everywhere in Paris, in his baggy, untidy uniform, organizing and rallying his forces. Despite his shoddy appearance, Gallieni was no comedian. He was a military genius and a master of improvisation, and in the face of bleak necessity, he was the first to grasp the possibilities of yoking motor transport and the internal combustion engine to the exigencies of warfare.

Already, a few days earlier, he had ordered the formation of a unique transport squad, to be held in reserve in case the city had to be evacuated. It was composed of a number of Paris taxicabs. But now, on September 6, it became obvious to Gallieni that the existing taxi reserve was much too small and that all available taxis would have to be transformed at once into a troop transport system. At 8:00 P.M., sitting in his headquarters at a lycée on the boulevard des Invalides, Gallieni had his inspiration: He decided that an armada of taxis would have to be organized to move thousands of troops to the front.

Gallieni ordered that every one of the three thousand available taxis be sought out and commandeered. Policemen and soldiers immediately began to stop cabs, demanded that they disgorge their paying passengers on the spot, and directed them to drive to the Invalides.

"How will we be paid?" one driver asked the lieutenant who had flagged him down. "By the meter or on a flat rate?"

"By the meter," the lieutenant said.

"All right, let's go," replied the driver, making sure to put down his flag before starting off.

By ten in the evening, within two hours of Gallieni's order, scores of taxis were already converging at the esplanade des Invalides. A first group set out in the dark for Tremblay-les-Gonesse, a small town to the northwest of Paris. The following morning a second army of taxis gathered at the Invalides. They took off in a great convoy, up the Champs-Elysées, along the rue Royale and the rue Lafayette, then left the city for another staging point to the east, at Gagny. During the day of September 7, while the taxis regrouped at their gathering spots, the fighting—and with it, the war—hung in a critical balance. "Today destiny will deliver a great decision," Helmuth von Moltke, the German Commander-in-Chief, wrote to his wife. "What torrents of blood have flowed!"

Once night had fallen, each taxi was crammed with soldiers—under the personal watch of General Gallieni, who noted, with a mixture of amusement and understatement, "Well, at least it's not commonplace." Then the overloaded vehicles, their meter flags down, began to set off in convoys of twenty-five to fifty toward the battlefield—"this forerunner of the future motorized column," as one historian later wrote, driving as only Parisian taxicab drivers can, speeding and passing and repassing each other, their headlamps darting points of light along the dark roads.

Thousands and thousands of troops were rushed to the critical point on the front by Gallieni's taxicabs. They made the difference. The French line was strengthened, and the troops fought all along it with new vigor beginning with the dawn on September 8. On September 9, the Germans fell back and began to retreat. "Things are going badly, the battles east of Paris will not be decided in our favor," Moltke wrote to his wife as the German armies reeled. "Our campaign is a cruel disillusion. . . . The war which began with such good hopes will in the end go against us."

The taxicab drivers, hungry and tired after two days with no sleep, returned to Paris, where they were besieged by the curious and were paid their fares. They had helped save Paris. They had also demonstrated, under General Gallieni's improvisational tutelage, what motorized transport would mean in the future. Later, a grateful city rechristened the broad roadway that traverses the esplanade des Invalides as the avenue du Maréchal Gallieni.[2]

Internal Combustion at War

The French counterattack of September 6–8, 1914, combined with a concurrent British assault, was of decisive importance—the turning point in the First Battle of the Marne, and the end of the much-planned German offensive. It also decisively changed the character of battle and ended any chance that it would be a short war. When the Germans halted their retreat, the opposing forces dug trenches on both sides and settled in for what was to prove a long, bloody, senseless war of attrition—the static war of defense. Indeed, for more than two years, the lines on the western front were to move no more than ten miles in either direction. The widespread use of the machine gun, combined with trenches and barbed-wire entanglements, gave primacy to the defense and thus guaranteed the stalemate. "I don't know what is to be done," said a frustrated Lord Kitchener, the British War Secretary. "This isn't war."

The only obvious way to break the stalemate of trench warfare was with some kind of mechanical innovation that would enable troops to move across the battlefield with greater protection than their own skin and uniforms. As the military historian Basil Liddell Hart expressed it, what was needed was "a specific antidote for a specific disease." The first military man "who diagnosed the disease and conceived the antidote" was a British colonel, Ernest Swinton, a writer of popular war fiction who, as a result of his earlier work on the official British history of the Russo-Japanese War, had already foreseen the potential impact of the machine gun. Later, he paid close attention to various military experiments with the agricultural tractor, which had recently been developed in the United States. When dispatched early in the war to France, to be an official "eyewitness" at general headquarters, he put two and two together and came up with the idea for the antidote—an armored vehicle that was powered by the internal combustion engine and moved on traction, impervious to machine gun bullets and barbed wire.

Yet what was needed was not necessarily wanted. Entrenched opponents in the high command of the British Army did not take the idea seriously and did everything they could to squelch it. Indeed, it might well have died altogether had it not been taken up and championed by Winston Churchill. The First Lord of the Admiralty appreciated military innovation and was outraged at the failure of the Army and the War Office to begin developing such vehicles. "The present war has revolutionized all military theories about the field of fire," he told the Prime Minister in January 1915. And, in the face of the Army's resistance, Churchill doled out Navy funds for the continuing research needed to develop the new vehicle. Reflecting the Navy's temporary sponsorship, the new machine was known as the "land cruiser" or the "landship." Churchill called it the "caterpillar." To maintain secrecy, it needed a code name while it was being tested and transported, and various names—among others, the "cistern" and the "reservoir"—were considered. But finally it became known by another of its code names—the "tank."

The tank was first used, prematurely, in 1916 at the Battle of the Somme. It played a more important role in November 1917, at Cambrai. But it had its most decisive impact on August 8, 1918, at the Battle of Amiens, when a swarm of 456 tanks broke through the German line, resulting in what General Erich Ludendorff, who was deputy to Supreme Commander Paul von Hindenburg, later called the "black day of the German Army in the history of the war." The "primacy of the defense" was over. When the German High Command declared in October 1918 that victory was no longer possible, the first reason it gave was the introduction of the tank.

Another reason was the extent to which the car and truck (the lorry, as the British call it) had succeeded in mechanizing transport. While the Germans had held the advantage when it came to railway transport, the Allies were to gain the upper hand insofar as cars and trucks were concerned. The British Expeditionary Force that went to France in August 1914 had just 827 motor cars—747 of them requisitioned—and a mere 15 motorcycles. By the last months of the war, British Army vehicles included 56,000 trucks, 23,000 motorcars, and 34,000 motorcycles and motor bicycles. In addition, the United States, which entered the war in April 1917, brought another 50,000 gasoline-driven vehicles to France. All these vehicles provided the mobility to move troops and supplies swiftly from one point to another as the need arose—a capability that proved critical in many battles. It was rightly said after the war that the victory of the Allies over Germany was in some ways the victory of the truck over the locomotive.[3]

The War in the Air and at Sea

The internal combustion engine had an even more dramatic impact in a new arena for war—the air. The Wright brothers had made their first flight at Kitty Hawk in 1903. But until the Italians made use of airplanes in fighting against the Turks at Tripoli in 1911–12, the conventional attitude of the military toward the airplane had been summed up by the French General Ferdinand Foch, who dismissed aviation as "good sport, but for the Army the aeroplane is worthless." At the outbreak of the war in 1914, the "trade," as the British military called the aviation industry, barely numbered a thousand people, and by January 1915, five months later, the British industry had managed to build just 250 planes—sixty of them experimental.

Even so, the airplane had been immediately pressed into military service, and the potential of its impact had quickly become apparent. "Since war broke out," a British aviation writer observed in early 1915, "the aeroplane has done such surprising things that even the least imaginative begin to realize that it affords a vital adjunct to naval and military operations, and possibly even a vehicle for ordinary use when war ceases." The development of air power required the quick build-up of an industrial infrastructure; the automobile industry provided a major part of the base, especially for the engines. As the war stretched on, aviation developed swiftly, driven by rapid-fire innovation. By July 1915, every machine that had been in the air at the outbreak of the war, less than a year earlier, had become obsolete.

The first significant use of aviation in the war had been for reconnaissance and observation. Air combat initially involved pilots shooting at each other with rifles and handguns. Then machine guns were fitted on scouting planes, and new mechanisms were developed to synchronize their firing with the rotation of the propellers, so that the pilot would not accidentally shoot his own propellers. Thus, the fighter plane was born. By 1916, planes were flying in formation, and tactics of aerial combat had been developed. Tactical bombing—in conjunction with infantry combat—was introduced, and it was used by the British both against the Turks, with devastating effect, and also to stop the onrush when the Germans broke through the British front in March 1918. The Germans took the lead in strategic bombing, launching assaults directly against England, with zeppelins and then with bombers, and so violating the insularity of the British Isles in what became "the first Battle of Britain." The British replied in the closing months of the war with air attacks on targets inside Germany.

The war constantly pushed the pace of innovation. By the last months of the struggle, the speed of the most advanced aircraft had more than doubled, to over 120 miles per hour, and they operated with a ceiling of nearly 27,000 feet. The overall production numbers told the same story of rapid development. In the course of the war, Britain produced 55,000 planes; France, 68,000; Italy, 20,000; and Germany, 48,000. In its year and a half in the war, the United States produced 15,000 planes. Such proved to be the utility of what had, before the war, been dismissed as merely "good sport." What the Chief of the British Air Staff said of the Royal Air Force could well be applied to military aviation in general: "The necessities of war created it in a night."

By contrast, the prewar naval race, which had so aggravated relations between Britain and Germany, produced a stalemate. At the outbreak of the war, Britain's Grand Fleet was superior to Germany's High Seas Fleet. In the Battle of the Falkland Islands in December 1914, the Royal Navy defeated a German squadron, and by that victory deprived Germany of access to the trading centers of the world. Yet, despite the central role that the naval rivalry had played in leading the two countries to war, the Grand Fleet and the High Seas Fleet met only once in major engagement—at the Battle of Jutland on May 31, 1916. The outcome of that legendary encounter has been debated ever since. The German fleet was victorious in a tactical sense, succeeding as it did in escaping from a trap. But, strategically, the British won, for they were able to dominate the North

Sea for the rest of the war and keep the German fleet penned up in its home bases.

Events thus proved Churchill and Fisher generally right in forcing the conversion of the Royal Navy to oil, for it did give the British fleet an overall advantage—greater range, greater speed, and faster refueling. The German High Seas Fleet was primarily coal burning; it had no stations outside Germany at which to resupply, and thus its range and flexibility were more limited. In truth, its reliance on coal made its very name, the High Seas Fleet, a misnomer. But then Germany had never been in the position that Britain was—able to make a calculated bet on its ability to maintain access to petroleum during war.[4]

Anglo-Persian Versus Shell

Britain's acquisition of shares in Anglo-Persian had been made for exactly that purpose of ensuring oil supplies. But war had come even before the purchase could be completed, let alone the relationship between government and company sorted out. Moreover, the enterprise in Persia was still only of minute significance, accounting in 1914 for just less than one percent of total world oil output. But as production grew, its strategic value would be enormous, and the British commitments, both to oil fuel and to the company, had to be protected. Yet it was not at all evident that this could actually be done. Ironically, less than a month after the war began, it was Churchill himself, the champion of oil and of the Anglo-Persian acquisition, who despaired of Britain's ability to defend the Persian oil fields and refinery. "There is little likelihood of any troops being available for this purpose," he said on September 1. "We shall have to buy our oil from elsewhere."

The forces of the Ottoman Empire were the chief threat. Immediately after Turkey's entry into the war on Germany's side in the autumn of 1914, its troops were threatening the Abadan refinery site in Persia. They were repulsed by British soldiers, who went on to capture Basra—a city of critical importance, as it guarded the strategic approaches from the West toward the Persian oil. Control of Basra also secured the safety of the local rulers friendly to British interests, including the Amir of Kuwait. The British wanted to extend their defensive line further to the northwest, if possible to Baghdad itself. Again, one of the major considerations was to secure oil fields, as well as to counteract German subversion in Persia. At the same time, the oil potential of Mesopotamia (in what is in present-day Iraq) was beginning to loom larger in British military and political planning. In 1917, after a degrading defeat at the hands of the Turks, the British did finally succeed in capturing Baghdad.

Oil production in Persia itself was little disturbed during the war, except early in 1915, when local tribesmen, riled up by German agents and the Turks, damaged the pipeline from the oil fields to Abadan. Five months elapsed before the oil was flowing satisfactorily again. Despite problems in the quality of Abadan's refined products and wartime shortages of equipment, a great industrial enterprise was taking root in Persia, driven by military demand. Oil production in Persia grew more than tenfold between 1912 and 1918—from 1600 barrels per day to 18,000. By late 1916, Anglo-Persian was meeting a fifth of the British Navy's entire oil needs. The company, which had often been about to go broke in its first decade and a half of existence, started to make quite substantial profits.

Anglo-Persian's character was also changing, as its managing director Charles Greenway pursued a clear and determined strategy to transform Anglo-Persian from exclusively a crude producer into an integrated oil company—"to build up," in his words, "an absolutely self-contained organization" that would sell products to "wherever there may be a profitable outlet for them without the intervention of any third parties." In the midst of the world war, Greenway was positioning the company for postwar competition. His most important step was the purchase from the British government of one of the largest petroleum distribution networks in the United Kingdom, a company called British Petroleum. Despite its name, it had belonged to the Deutsche Bank, which used it as the outlet in the United Kingdom for its Rumanian oil; after the outbreak of the war, the British government had taken over the German-controlled company. Now, with its acquisition of British Petroleum, Anglo-Persian acquired not only a major marketing system, but also what would subsequently prove a most useful name. Anglo-Persian also developed its own tanker fleet. The very base of Anglo-Persian was changed by these transactions. Up until 1916–17, over 80 percent of its fixed assets were in Persia; in the very next fiscal year, only half were in Persia, with the rest in the tankers and the distribution system. It had indeed become an integrated company.

But Greenway had a second objective as well, which he pursued no less passionately—to turn Anglo-Persian into the oil champion of the British empire. He often reiterated his aim to make Anglo-Persian the nucleus of an "All-British Company . . . free from foreign taint of any kind"—an obvious reference to Royal Dutch/Shell. Greenway revived the "Shell menace," attacking "the schemes of Sir Marcus and his associates for securing a worldwide monopoly of the oil trade." Again and again, Greenway and his supporters charged Royal Dutch/Shell with disloyalty to British interests, with "making large profits out of the sale of Oil Products to Germany" and with having become "a serious National menace."[5]

These charges were both unfair and untrue. The merchant Deterding, who had himself naturalized and spent the war years in London, strongly identified his own interests and those of his company with the Allies. As for Marcus Samuel, he was, simply, a fierce British patriot, and he paid the price. One of his two sons, who had run a settlement house for poor boys in the East End of London before the war, was killed in France leading his platoon into action. Samuel and his wife published posthumously a small volume of the young man's poems as a memorial. Of his two sons-in-law, one was also killed in action, while the other died after the war from the effects of trench warfare.

Samuel himself masterminded an audacious scheme that proved of critical importance to the entire British war effort. Toluol, an essential ingredient for the explosive TNT, was generally extracted from coal. In 1903, a chemist from Cambridge University had discovered that toluol could also be extracted in significant amounts from Shell's Borneo crude. Samuel tried to win the Admiralty's interest, but the Admiralty greeted his report with great skepticism and rejected his offer of supplies. Eleven years later, at the beginning of the war, the offer was again made, and again rejected. Even when presented with evidence of German TNT almost certainly derived from the Borneo crude, the Navy was not interested. But the picture changed rapidly. By the end of 1914, the coal-based production of toluol was inadequate, and Britain was perilously close to running out of explosives. It needed toluol from oil, but there were no facilities to make it. The toluol-extracting factory that might have been built in Britain by Shell had instead been built in Rotterdam, in the neutral Netherlands, by the Dutch arm of the group. It was clear, moreover, that German companies were using the Rotterdam factory's output to make TNT.

Samuel and his colleagues conceived a daring plan, which was swiftly put into effect. In the middle of the night at the end of January 1915, the plant in Rotterdam was disassembled, part by part, each piece numbered and camouflaged, and then carried to the docks and loaded onto a Dutch freighter, which slipped out into the darkness to rendezvous at sea with British destroyers. A cover story was leaked to German agents that such an evacuation was to take place—but that it would occur a day later than the actual event. That following night, whether by coincidence or not, a similar Dutch freighter was torpedoed by the Germans at the mouth of Rotterdam's harbor. The parts of the toluol plant, meanwhile, were transported to Britain, and were re-erected in Somerset within weeks. That plant, along with a second one that Shell subsequently built, provided 80 percent of the British military's TNT. It was for this achievement, in large part, that Samuel was awarded a peerage after the war.

Despite Greenway's continuing assaults on the patriotism of Royal Dutch/Shell, the company became integral to the Allies' war effort; in effect, Shell acted as the quartermaster general for oil, acquiring and organizing supplies around the world for the British forces and the entire war effort and ensuring the delivery of the required products from Borneo, Sumatra, and the United States to the railheads and airfields in France.

Shell, thus, was central to Britain's prosecution of the war. Government officials, concerned about alienating Shell just when it was needed most, began to react negatively to the continuing attacks on the Group by Greenway and his supporters. Indeed, Greenway so overplayed his hand that he eventually turned many in the government against Anglo-Persian. They suspected Greenway's arrogation of the patriot's mantle and questioned his strategy of trying to build an integrated company with interests beyond Persia. There was much discussion and debate in Whitehall, as officials tried to figure out exactly what should be the government's objective for this company, in which it had just acquired a 51 percent stake. Was it only, as a skeptical Treasury official said, "to secure navy supplies" and no more? Or was it to help create an integrated state-owned oil company, a national champion, and then to assist that company in expanding its commercial interests worldwide? Some sought to tie the commercial ambitions of the company to Britain's postwar needs, looking to a time when "the Nation would secure an independent position in oil as it now holds in coal." But Arthur Balfour, Churchill's successor as First Lord of the Admiralty, wondered in August 1916 about the competence of government "to be responsible for the policy of a huge combine dealing with a prime necessity of modern life." Various forms of government-sanctioned mergers were also debated, including schemes for making British interests, rather than Dutch, predominant within the Royal Dutch/Shell Group. These proposals came to naught during the war. Much more urgent and pressing matters were at hand.[6]

"A Dearth of Petrol"

As late as 1915, the supply of oil to feed the engines of war raised little sense of anxiety in Britain. But that changed at the beginning of 1916. A "dearth of petrol" was reported by the *Times* of London in January 1916. And the following May, the *Times* called "for a sharp definition of where motoring for business ends," adding that " 'joy-riding' may have to go altogether" in the face of "the demands of the war services."

The reasons for the emerging oil crisis were twofold. One was the growing shortage of shipping tonnage—owing to the German submarine campaign—which constricted supplies of oil, along with all other raw materials and food, to the British Isles. The internal combustion engine had provided Germany with its only clear advantage at sea—the diesel-driven submarine. And Germany responded to the British economic blockade of Germany and Britain's overall superiority on the seas by instituting deadly submarine warfare, aimed at choking off supplies to the British Isles as well as to France. The other reason for the crisis was the rapidly growing demand for oil—to meet wartime needs both on the battlefield and on the home front. Fearing shortage, the government instituted a system of rationing. The relief was only temporary.

Pressure on supplies returned at the beginning of 1917 when Germany unleashed its unrestricted submarine campaign against Allied shipping. Ultimately that campaign proved to be a blunder of immense proportions, for it led the United States to forsake its neutrality and declare war against Germany. Still, the effects of the submarine attacks were large and quickly felt. Tonnage lost in the first half of 1917 was twice that lost in the comparable period in 1916. Between May and September, Standard Oil of New Jersey lost six tankers, including the brand new *John D. Archbold*. Among the many tankers that Shell lost during the war was the *Murex*, which had been the first vessel dispatched by Marcus Samuel through the Suez Canal in 1892 to carry out his great coup. The Admiralty's policy was to maintain stocks equivalent to six months of consumption, but, by the end of May 1917, they were less than half that level, and already, the shortfall in oil supplies was constraining the mobility of the Royal Navy. So serious had the situation become that it was even suggested that the Royal Navy stop building oil-driven ships and go back to coal![7]

The grave shortages of 1917 gave a strong push to official efforts in Britain to develop a coherent national petroleum policy. A variety of committees and offices, including a Petroleum Executive, were established to coordinate oil policy—both to contribute to better prosecution of the war and to try to enhance Britain's oil position in the postwar years. Similarly, the French government established a Comité Général du Pétrole, modeled on Britain's Petroleum Executive and headed by a Senator, Henry G. Bérenger, to respond to the growing crisis. But it was recognized in both countries that the only real solution to the crisis was to be found in the United States. Shipping—tankers—held the key to the supply situation.

What have been described as "desperate" telegrams were dispatched from London to America, declaring that the Royal Navy would be immobilized, putting the "fleet out of action," unless the United States government made more tonnage available. "The Germans are succeeding," the American ambassador in London despairingly wrote in July 1917. "They have lately sunk so many fuel oil ships, that this country may very soon be in a perilous condition—even the Grand Fleet may not have enough fuel. . . . It is a very grave danger." By the autumn of 1917, Britain was exceedingly short of supplies. "Oil is probably more important at this moment than anything else," Walter Long, the Secretary of State for the Colonies, warned the House of Commons in October. "You may have men, munitions, and money, but if you do not have oil, which is today the greatest motive power that you use, all your other advantages would be of comparatively little value." In that same month, pleasure driving in Britain was summarily and completely banned.

France's oil position was also degenerating rapidly in the face of Germany's unrestricted submarine campaign. In December 1917, Senator Bérenger warned Prime Minister Georges Clemenceau that the country would run out of oil by March 1918—just when the next spring offensive was set to begin. Supplies were so low that France could sustain no more than three days of heavy German attacks, such as those experienced at Verdun, where massive convoys of trucks had been needed to rush reserves to the front and hold off the German assault. On December 15, 1917, Clemenceau urgently appealed to President Wilson that an additional hundred thousand tons of tanker capacity be made immediately available. Declaring that gasoline was "as vital as blood in the coming battles," he told Wilson that "a failure in the supply of gasoline would cause the immediate paralysis of our armies." He ominously added that a shortage might even "compel us to a peace unfavorable to the Allies." Wilson responded quickly, and the necessary tonnage was swiftly made available.

But more than ad hoc solutions were needed. The oil crisis was already forcing the United States and its European Allies into much tighter integration of supply activities. An Inter-Allied Petroleum Conference was established in February 1918 to pool, coordinate, and control all oil supplies and tanker shipping. Its members were the United States, Britain, France, and Italy. It proved effective at distributing the available supplies among the Allied nations and their military forces. By the very nature of their domination of the international oil trade, however, Standard Oil of New Jersey and Royal Dutch/Shell really made the system work—though they continually argued about who was making the larger contribution. That joint system—along with the introduction of convoys as an antidote to the German U-boats—solved the Allies' oil supply problems for the rest of the war.[8]

The Energy Czar

The Inter-Allied Petroleum Conference was also created in response to domestic American energy problems. Clearly, American oil had become an essential element in the conduct of the European war. In 1914, the United States had produced 266 million barrels—65 percent of total world output. By 1917, output had risen to 335 million barrels—67 percent of world output. Exports accounted for a quarter of total U.S. production, with the bulk going to Europe. Now that access to Russian oil had been closed off by war and revolution, the New World had become the oil granary for the Old; altogether, the United States was to satisfy 80 percent of the Allies' wartime requirement for petroleum.

Nevertheless, America's entry into the war greatly complicated the American oil picture. For there needed to be adequate supplies for many purposes— the American military, the Allies' forces, the American war industries, and normal civilian use. How to assure sufficient supplies, efficient distribution, and appropriate allocation? This became the charge of the Fuel Administration, established by President Wilson in August 1917 as part of the overall economic mobilization. All of the belligerent states faced a parallel challenge—to harness the industrial economies that had emerged over the preceding half century to the requirements of modern warfare. In each country, the needs of mobilization expanded the role of the state in the economy and created new alliances between government and private business. The United States and the American oil industry were no exception.

The head of the Oil Division in the Fuel Administration was a California petroleum engineer named Mark Requa, who became America's first energy czar. His main job was to forge a new and unprecedented working relationship between the government and the oil industry. The Oil Division worked in close liaison with the National Petroleum War Service Committee, whose members were the leaders of major companies, and whose chairman was Alfred Bedford, president of Standard Oil of New Jersey. It was this committee that organized the supply of American oil for the war in Europe. It placed the major orders from the various Allied governments with American refiners and played a central role in arranging the shipping. In essence, it was the agency on the American side that pooled the oil supplies for Europe. This new pattern of close cooperation between business and government stood in marked contrast to the battle between government and Standard Oil just a decade earlier. Trust busting seemed far away, as the industry was now pushed to run itself as a single body, under the leadership of the once-hated Standard Oil of New Jersey.[9]

In 1917, the surging demand for American oil began to hit the limit of available supplies. The gap was being closed only by using up inventories and by importing more oil from Mexico. On top of that, the bitterly cold winter of 1917–18 and the overall pace of industrial activity combined to create a shortage of coal in the United States—so severe that local officials commandeered coal trains passing through their jurisdictions, and policemen had to stand guard over industrial coal piles to prevent pilfering. Orphanages and asylums ran out of fuel, and inmates died of frostbite. Even the wealthy were complaining of empty coal bins and chattering teeth. In January 1918, the Fuel Administration ordered almost all industrial plants east of the Mississippi to close for a week in order to free fuel for hundreds of ships filled with war materials for Europe that were immobilized in East Coast harbors for want of coal. Thereafter, the factories were ordered to remain closed on Mondays to conserve coal. "Bedlam broke loose," observed Colonel Edward House, Woodrow Wilson's political confidant. "I have never seen such a storm of protest."

The coal shortage stimulated a sharp increase in the demand for oil, and oil prices rose accordingly. By early 1918, average crude prices were double what they had been at the beginning of 1914. Refiners were offering bonuses and premiums in order to obtain supplies, while producers were withholding supplies on the expectation of still higher prices. This situation greatly alarmed the government. On May 17, 1918, Requa, the energy czar, warned the industry that there was "no justification" for "any further advance in the price of crude oil" and called for "voluntary" price controls on the part of the oil industry. Standard Oil of New Jersey was agreeable to Requa's call for such price restraint. Not so the independent producers. But without "voluntary" controls, Requa bluntly told a group of producers in Tulsa, there would be direct government controls. Moreover, he reminded them, it was the government that helped producers obtain steel and other drilling supplies (the oil industry took a twelfth of the country's output of iron and steel), and it was the government that provided draft exemptions for oil field workers. These arguments were persuasive. In August 1918, maximum prices were set in each producing region and prices leveled off for the remainder of the war.

Still, demand continued to outstrip supply, not only because of the war but also because of the phenomenal growth in the number of automobiles in the United States. The number of cars in use had almost doubled between 1916 and 1918. Petroleum shortages seemed imminent, which could threaten the war effort in Europe and restrict essential activities in the United States. An "appeal"—not a mandatory order—for "Gasolineless Sundays" was made. The only exemptions were for freight, doctors, police, emergency vehicles, and hearses. Inevitably, the call aroused suspicions and complaints, but it was for the most part faithfully observed, even in the White House. "I suppose," declared President Wilson, "I must walk to church." [10]

The Man with the Sledgehammer

Despite periodic alarms and critical moments of shortages of supply, the Allies never suffered from a protracted oil crunch. Germany did, as the Allied blockade succeeded in choking off supplies to Germany from overseas. That left only one source available to them—Rumania. And while Rumania's output on a worldwide scale was comparatively small, it was the largest European producer, excluding Russia. Germany was heavily dependent on it. The activities of the Deutsche Bank and other German firms had already, before the war, tied a significant part of the Rumanian oil industry to the German economy. For the first two years of the war, Rumania remained neutral, waiting to see which side was likely to win. But finally, in August 1916, in the wake of Russian success on the eastern front, Rumania declared war against Austria-Hungary, thus bringing it almost immediately into a state of war with Germany as well.

Victory in this Eastern theater was essential for Germany. "As I now saw quite clearly, we should not have been able to exist, much less to carry on the war, without Rumania's corn and oil," said General Erich Ludendorff, who was the true mastermind of Germany's war effort. German and Austrian troops advanced on Rumania in September of 1916, but the Rumanians dug in and managed to hold on to the mountain passes, which protected the Wallachian Plain, where the oil production was concentrated. In mid-October the Germans and Austrians captured a vast amount of petroleum products, including a large cache of gasoline belonging to the Allies, held in storage at a Rumanian oil port on the Black Sea. There had been a plan to destroy all the facilities and oil supplies, but in the confusion of battle it had never been executed. And now the great prize itself—the Rumanian oil fields and refineries—seemed almost within Germany's grasp.

Could it be denied to the Germans? On October 31, 1916, the subject was urgently discussed in London by the British Cabinet War Committee. "No efforts should be spared to ensure, in case of necessity, the destruction of the supplies of grain and oil, as well as of the oil wells," the committee concluded. But the Rumanian government was reluctant to consider destroying its national treasure, especially while there was still some hope on the battlefield. That hope faded by November 17, when the Germans succeeded in breaking through the Rumanian resistance in the mountain passes and began pouring down through the mountains and across the Wallachian Plain.

The British government took matters into its own hands and recruited Colonel John Norton-Griffiths, M.P., to organize the destruction of the Rumanian oil industry. A larger-than-life figure, Norton-Griffiths was one of the great engineering contractors of the British empire. He had undertaken construction projects in almost every corner of the world—railways in Angola and Chile and Australia, harbors in Canada, aqueducts in Baku, sewage systems in Battersea and Manchester. On the eve of World War I, he was in the midst of promoting a plan for a new subway for Chicago. Handsome, physically imposing, and with the strength and endurance of a prizefighter, Norton-Griffiths was a charming swashbuckler and persuasive showman. Men invested in his projects, women were attracted to him. He was considered "one of the most dashing men of the Edwardian era." He was also a man of fiery temperament, rebellious nature, and uncontrollable rages. He lacked discipline and perseverance, and some of his projects were spectacular financial flops. But he did achieve prominence as a Parliamentary back-bencher, variously known as "Hell-fire Jack," "the Monkey Man" (for having eaten a monkey while in Africa) and—since he was a thoroughgoing imperialist—by the sobriquet he treasured most, "Empire Jack."

Norton-Griffiths's first great engineering feat during World War I was to adapt techniques he had previously developed for the Manchester sewers to the challenge of tunneling beneath German lines and trenches, where underground mines were then placed and detonated. His methods were proved at Ypres. But he had alienated many commanders as he careened about Flanders in his two-ton Rolls-Royce, which was permanently supplied with crates of champagne, and he was recalled from the front. Still, there was no one better suited for the Rumanian mission. On November 18, 1916, the day after the Germans broke through the Rumanian lines, "Empire Jack" arrived in Bucharest, via Russia, accompanied only by his manservant. As the Germans continued their advance, the Rumanian government, under Allied pressure, finally agreed to the policy of destruction.

The destruction teams now swung into action, with "Empire Jack" at the forefront. The first fields went up in flames on November 26 and 27. The teams followed the same general procedure at each site. Explosives were placed in refineries. Then, petroleum products in storage were allowed to flow into the refineries, creating lakes several inches, or even feet, deep. Equipment was brought in and dumped into the pools of oil. And then, with matches and burning straw, the entire facility was set afire. Those who challenged Norton-Griffiths or stood in his way were overwhelmed by the sheer force of his personality. If that proved insufficient, he would deliver a powerful kick or pull out his revolver and shout, "I don't speak your blasted language."

Apparatus in the fields was smashed; derricks were dynamited; wells were plugged with stone, spikes, mud, broken chains, drillbits, and whatever else was handy; pipelines were crippled; and huge oil storage tanks were set ablaze, exploding with great roars. At some installations, "Empire Jack" insisted on setting the blaze himself. In one engine house, after lighting the flammable gases, he was blown out by the blast, with his hair afire. That didn't stop him. Again and again, Norton-Griffiths took the lead in swinging a huge hammer to wreck derricks and pipes, leaving an indelible memory in Rumania of "the man with the sledgehammer."

The oil valleys were ablaze, with red flames rising high into a sky completely filled with a dense, black, asphyxiating smoke that blotted out the sun. Yet beyond the valleys could be heard the sound of the big guns, growing closer all the time. The last field to be set afire was Ploesti itself. The work was completed just in time. For, on December 5, only a few hours after the facilities went up in flames, the Germans entered the town of Ploesti. Norton-Griffiths barely escaped by car, just ahead of the German cavalry. "To lay waste the land" had been his mission, he said, but as a builder, the destruction sickened him, and though awarded military honors for his efforts, he was uncharacteristically loath to talk about this exploit in later years.

After the war, General Ludendorff admitted that Norton-Griffiths's efforts "did materially reduce the oil supplies of our army and the home country." The German general grudgingly added, "We must attribute our shortages in part to him." Altogether, some seventy refineries and an estimated eight hundred thousand tons of crude oil and petroleum products had been destroyed in Rumania under Norton-Griffiths's tutelage. It took five months before the Germans could begin to get the fields back into production, and for all of 1917, production was only a third of what it had been in 1914. The Germans applied themselves methodically to undoing Norton-Griffiths's work, and by 1918, they had pushed production back up to 80 percent of the 1914 level. The Rumanian oil was sorely needed. The Germans might well have not been able to continue the war without it. As a historian of Britain's Imperial Defense Committee later observed, Germany's timely capture of the Rumanian oil industry, along with the Rumanian grain, "made just the difference between shortage and collapse" for the German side. But only for a time.[11]

Baku

Even as the Germans were getting the Rumanian fields back into operation, General Ludendorff set his sights on a greater prize, which might help meet the enormous and rising need for oil and so turn the tide of battle in Germany's favor. It was Baku, on the shores of the Caspian Sea. The collapse of the Czarist regime in early 1917, the rise of the Bolsheviks later in the year, and the fragmentation of the Russian empire—all held out some hope for the Germans that they might be able to get their hands on oil supplies from Baku. They began to seek access to Baku petroleum in March 1918 with the Treaty of Brest-Litovsk, which ended hostilities between Germany and revolutionary Russia. However, the Turks, the ally of Germany and Austria, had already begun to advance toward Baku. Fearing that success by their ally would lead to the wanton destruction of the oil fields, the Germans promised the Bolsheviks that they would try to restrain the Turks in exchange for oil. "Of course, we agreed," said Lenin. Joseph Stalin, who by then had emerged as one of the leading Bolsheviks, telegraphed the Bolshevik Baku Commune, which controlled the city, ordering it to comply with this "request." But the local Bolsheviks were in no mood to go along. "Neither in victory nor in defeat will we give the German plunderers one drop of oil produced by our labor," they replied.

The Turks, in their quest for the Baku prize, spurned Berlin's entreaties and continued their advance toward the oil region. By the end of July, they were laying siege to the city, and by early August had captured some of the producing fields. The Armenian and Russian residents of Baku had long been imploring the British for help. Finally, in mid-August 1918, the British intervened with a small force that made its way through Persia. The troops were charged with saving Baku and keeping the oil from the enemy. If need be, they were (in the words of the War Office) to follow the Rumanian plan and "destroy the Baku pumping plant, pipeline and oil reservoirs."

The British stayed in Baku only a month, but that was enough to deny Baku oil to the Germans at the critical moment. It was, Ludendorff was to say, "a serious blow for us." Then the British withdrew and the Turks captured the city. In the maelstrom, the local Moslems, abetted by the Turks, once again—as in the revolutionary days of 1905—began to pillage and destroy, in the process killing every Armenian they could find, even those lying in hospital. Meanwhile, Bolshevik commissars from the Baku Commune were captured by revolutionary rivals. Twenty-six of them were taken to a desolate spot in the desert, 140 miles east of the Caspian Sea, and there executed. One of the few to escape was a young Armenian named Anastas Mikoyan, who eventually got to Moscow to tell Lenin what had happened. But, by the time the Turks took Baku, it was too late to do the Germans and their oil supply any good.[12]

Floating to Victory

The denial of Baku at that juncture was, in fact, a decisive blow for Germany. The pressure on its oil supplies was growing ever more acute. By the desperate month of October 1918, the picture was grim. The German Army had all but exhausted its reserves, and the German High Command was anticipating a grave petroleum crisis in the coming winter and spring. In October, it was estimated in Berlin that the battle at sea could be continued for only six to eight months. The war industries that operated on oil would run out of supplies within two months; the entire stock of industrial lubricants would be exhausted within six months. Limited land operations could be carried out with supplies on a strictly rationed basis. But air and mechanized land warfare would cease absolutely within two months.

The validity of these estimates was never tested, for within a month, an exhausted Germany surrendered. The armistice was signed at five in the morning, November 11, 1918, in Marshal Foch's railway car in the Forest of Compiègne. Six hours later, it went into effect. The war was over.

In London, some ten days after the Armistice, the British government hosted a dinner for the Inter-Allied Petroleum Conference at Lancaster House, with the distinguished Lord Curzon as chairman. He had once been the Foreign Office's great Persian expert; he had been Viceroy of India, in which capacity he had supported D'Arcy's oil venture in Persia on strategic grounds. He had been a member of the War Cabinet, and was shortly to become Foreign Secretary. Now he rose to tell the assembled guests that "one of the most astonishing things" he had seen in France and Flanders during the war "was the tremendous army of motor lorries." Then he resoundingly declared, "The Allied cause had floated to victory upon a wave of oil."

Senator Bérenger, the director of France's Comité Général du Pétrole, was even more eloquent. Speaking in French, he said that oil—"the blood of the earth"—was "the blood of victory . . . Germany had boasted too much of its superiority in iron and coal, but it had not taken sufficient account of our superiority of oil." Bérenger also had a prophecy to make. Continuing in French, he said, "As oil had been the blood of war, so it would be the blood of the peace. At this hour, at the beginning of the peace, our civilian populations, our industries, our commerce, our farmers are all calling for more oil, always more oil, for more gasoline, always more gasoline." Then he broke into English to drive home his point—"More oil, ever more oil!" [13]

CHAPTER 10

Opening the Door on the Middle East: The Turkish Petroleum Company

SOME TEN DAYS after Curzon and Bérenger had raised their glasses in toast to the "blood of victory," French Premier Georges Clemenceau came to London to pay a visit to British Prime Minister David Lloyd George. The guns had already been silent for three weeks, and the issues of the postwar world could not be postponed. The questions were momentous and inescapable—how to make the peace and how to reorganize a world in shambles. Oil was now inextricably linked to postwar politics. And this topic was very much on the minds of Clemenceau and Lloyd George as they drove through the cheering crowds in the streets of London. Britain wanted to assert its influence over what was loosely known as Mesopotamia, the provinces of the now defunct Turkish Ottoman Empire that would later be known as Iraq. The area was thought to be highly prospective of oil. But France had a claim to one part of the region—Mosul, northwest of Baghdad.

What specifically did Britain want? That was the question Clemenceau asked when the two men finally reached the French embassy.

Would France give up its claim to Mosul, Lloyd George responded, in exchange for British recognition of French control over neighboring Syria?

France would, Clemenceau replied—so long as it received a share of the oil production from Mosul.

To this Lloyd George assented.

Neither Prime Minister bothered to inform his respective foreign minister. Indeed, their casual verbal agreement was not a settlement at all; rather, it was the beginning of the great postwar struggle for new oil sources in the Middle East and throughout the world. It would pit the French against the English, but it would also draw in the Americans. No longer would the competition for new oil lands be primarily restricted to a battle among risk-taking entrepreneurs and aggressive businessmen. The Great War had made abundantly clear that petroleum had become an essential element in the strategy of nations; and the politicians and bureaucrats, though they had hardly been absent before, would now rush headlong into the center of the struggle, drawn into the competition by a common perception—that the postwar world would require ever-greater quantities of oil for economic prosperity and national power.[1]

The struggle would focus on that one particular region—Mesopotamia. In the decade before the war, Mesopotamia had already been the object of intricate diplomatic and commercial competition for oil concessions, stimulated by favorable reports of its petroleum potential. The wrangling had been encouraged by a dilapidated Turkish empire that was chronically in financial arrears and eager to find new ways to generate revenues. One player in the prewar years was a German group, led by the Deutsche Bank, which aimed to project German influence and ambitions into the Middle East. Arrayed against it was a rival group, sponsored by William Knox D'Arcy and eventually merged into the Anglo-Persian Oil Company. It was championed by the British government as a counterweight to Germany.

Then, in 1912, the British government was alarmed to discover a new player on the scene. It was called the Turkish Petroleum Company, and it turned out that the Deutsche Bank had transferred its claims for a concession to this entity. The Deutsche Bank and Royal Dutch/Shell each held a quarter of the new company. The largest share, half of the total equity, was held by the Turkish National Bank, which, ironically, happened to be a British-controlled bank set up in Turkey to advance British economic and political interests. But there was one additional player, a man who would be admired by some as the "Talleyrand of oil diplomacy" and scorned by others—an Armenian millionaire named Calouste Gulbenkian. It was Gulbenkian who had put the entire Turkish Petroleum Company deal together. Upon closer examination, it turned out that he was the silent owner of 30 percent of the Turkish National Bank, which made him a 15 percent owner of the Turkish Petroleum Company.[2]

Mr. Five Percent

Calouste Gulbenkian was the second generation of his family in the oil business. He was the son of a wealthy Armenian oil man and banker, who had built his fortune as an importer of Russian kerosene into the Ottoman Empire, and who had been rewarded by the Sultan with the governorship of a Black Sea port. The family actually lived in Constantinople, and there occurred Calouste's first recorded financial transaction. Given, at age seven, a Turkish silver piece, the boy took it off to the bazaar, not to buy a sticky candy as might have been expected, but to exchange it for an antique coin. (Later in life he would create one of the world's great collections of gold coins, and he took special pleasure in acquiring J. P. Morgan's superb collection of Greek gold coins.) Unpopular as a schoolboy—throughout his life there was never to be any great love lost between him and the rest of humanity—the young Calouste often spent his after-school hours in the bazaar, listening to deals being made, sometimes making small ones himself, imbibing the arts of Oriental negotiation.

He was sent off to secondary school in Marseilles, to perfect his French, and then to King's College, London, where he studied mining engineering and wrote a thesis on the technology of the new petroleum industry. He graduated in 1887, at the age of nineteen, with a first-class degree in engineering. A professor at King's suggested that the obviously talented young Armenian student go off to France for graduate studies in physics, but his father overruled the idea. Such a notion, he said, was "academic nonsense." Instead, his father sent Calouste to Baku, from which the family's fortunes had, in large part, derived. The young man was fascinated by the oil industry that he was seeing for the first time. He was also drenched by a gusher, but the oil being "fine and consistent," he did not find the experience unpleasant. Though pledging to return, he never bothered to visit oil country again.

Gulbenkian wrote a series of highly regarded articles on Russian oil, which appeared in a leading French magazine in 1889, and he turned the articles into a prestigiously published book in 1891—making himself a world oil expert by the time he was twenty-one. Almost immediately after, two officials of the Turkish Sultan asked him to investigate oil possibilities in Mesopotamia. He did not visit that area—he never did—but he put together a competent report based on the writings of others, as well as on talks with German railway engineers. The region, he said, had very great petroleum potential. The Turkish officials were persuaded. So was he. Thus began Calouste Gulbenkian's lifelong devotion to Mesopotamian oil, to which he would apply himself with extraordinary dedication and tenacity over six decades.

In Constantinople, Gulbenkian tried several commercial ventures, including selling carpets, none of them particularly successful. But he did master the arts of the bazaar—trading and dealing, intrigue, baksheesh, and the acquisition of information that could be put to advantageous use. He also developed his lifelong passion for hard work, his capacity for vision, and his great skills as a negotiator. Whenever he could, he would control a situation. But when he could not, he would follow an old Arab proverb that he liked to quote, "The hand you dare not bite, kiss it." In those early business years in Constantinople, he also cultivated his patience and perseverance, which some said were his greatest assets. He was not prone to budge. "It would have been easier," someone later said, "to squeeze granite than Mr. Gulbenkian."

Gulbenkian possessed one other quality. He was totally and completely untrusting. "I have never known anybody so suspicious," said Sir Kenneth Clark, the art critic and director of the National Gallery in London who helped Gulbenkian in later years on his art collection. "I've never met anybody who went to such extremes. He always had people spying for him." He would have two or three different experts appraise a piece of art before he bought it. Indeed, as he got older, Gulbenkian became obsessed with bettering a grandfather who had lived to the age of 106 and, to that end, employed two different sets of doctors so he could check one against the other.

Perhaps such suspiciousness was a necessary survival mechanism for an Armenian living precariously between opportunity and persecution in the last years of the Ottoman Empire. It was in 1896, during one of the periodic government-sanctioned Turkish massacres of Armenians, that Gulbenkian fled by ship to Egypt. He made himself invaluable to two powerful Armenians—an oil million-aire from Baku and Nubar Pasha, who helped rule Egypt. Those connections opened the doors of both oil and international finance to him, and he was able to set himself up in London as a sales representative for Baku oil.

Once in London, Gulbenkian met and allied himself with the Samuel brothers and with Henri Deterding. His son Nubar later wrote that Gulbenkian "and Deterding were very close for over twenty years. One never knows . . . whether in the end it was Deterding who used my father or my father who used Deterding. Whichever way round it was, their association was very fruitful to them both as individuals and to the Royal Dutch/Shell Group as a whole." To Shell, Gulbenkian brought deals, especially acquisitions, and arranged financing.

One of the very earliest transactions he offered was the Persian concession that eventually went to D'Arcy. He and Deterding had looked at the original prospectus for the concession, promoted in Paris by the Armenian Kitabgi, but rejected it because, Gulbenkian later said, it was "a very wild cat, and it looked so speculative that we thought it was a business for a gambler." Thereafter rue-fully watching the growth of Anglo-Persian, he framed a motto—"Never give up an oil concession"—that would be a guiding principle for the rest of his life. He would apply it, first and foremost, and with relentless tenacity through many tribulations, next door to Persia, in Mesopotamia. In 1907, he persuaded the Samuels to open a Constantinople office under his charge. Anti-Armenian senti-ment had waned for the time being, and he was altogether busy. In addition to pursuing many other business interests, he was financial adviser to the Turkish government itself, and to its Paris and London embassies, and was a major stockholder in the Turkish National Bank. It was from this base that he brought the rival British and German interests, and then Royal Dutch/Shell as well, into the entity called the Turkish Petroleum Company—a task, he said, requiring great delicacy, and "not, in any way, a pleasant one."[3]

From 1912 onward, once the Turkish Petroleum Company had come into existence, the British government directed its efforts toward trying to force the company to amalgamate with D'Arcy's Anglo-Persian syndicate and jointly pursue a concession. Finally, the British and German governments were able to agree on a unification strategy, and to force its execution. According to the "For-eign Office Agreement" of March 19, 1914, British interests were to predomi-nate in the combined group. The Anglo-Persian Group held 50 percent interest in the new consortium, while the Deutsche Bank and Shell each had their 25 per-cent. There was still Gulbenkian to contend with. Under the agreement, the Anglo-Persian Group and Shell each gave up the "beneficiary interests" of 2.5 percent of the total shares to the Armenian. That meant that he could not vote the shares, but he would enjoy all the financial benefits of such a shareholding. And so Mr. Five Percent was born, and that was how he was known ever after.

Thus, a decade of rivalry and squabbling was brought to an end. But the sig-natories had taken upon themselves a very significant obligation, one that would haunt many people down through the decades. They had all agreed to the "self-denying clause": None would be involved in oil production anywhere in the Ottoman Empire—save jointly "through the Turkish Petroleum Company." The only areas to which the self-denying clause did not apply were Egypt, Kuwait, and "transferred territories" on the Turco-Persian border. That clause would establish the foundation for oil development in the Middle East—and for titanic struggles—for many years thereafter.[4]

"A First-Class War Aim"

In a diplomatic note on June 28, 1914, the Grand Vizier promised that the Mesopotamian concession would be formally granted to the now-reconstituted Turkish Petroleum Company. Unfortunately, that was the very day that the Austrian Archduke Franz Ferdinand was assassinated in Sarajevo, triggering the First World War. The timing would leave a major question unanswered: Had the concession actually been granted, or had only a nonbinding promise of a concession been made? On the answer would hang much argument. But for the time being, the war put an abrupt end to Anglo-German cooperation in Mesopotamia and, apparently, interred the Turkish Petroleum Company as well.

But the oil potential of Mesopotamia was not forgotten. In late 1915 and early 1916, a British official and a Frenchman hammered out an understanding for the postwar order in Mesopotamia. Known by their names as the Sykes-Picot Agreement, it rather casually assigned Mosul in northeastern Mesopotamia, one of the most promising potential oil regions, to a future French sphere of influence. This "surrender" of Mosul immediately outraged many officials in the British government, and strenuous effort was thereafter directed toward undermining it. The issue became more urgent in 1917 when British forces captured Baghdad. For four centuries, Mesopotamia had been part of the Ottoman Empire. That empire, which had once stretched from the Balkans to the Persian Gulf, was now over, a casualty of war. A host of independent and semi-independent nations, many of them rather arbitrarily drawn on the map, would eventually take its place in the Middle East. But, at the moment, in Mesopotamia, Britain had the controlling hand.

It was the wartime petroleum shortage of 1917 and 1918 that really drove home the necessity of oil to British interests and pushed Mesopotamia back to center stage. Prospects for oil development within the empire were bleak, which made supplies from the Middle East of paramount importance. Sir Maurice Hankey, the extremely powerful secretary of the War Cabinet, wrote to Foreign Secretary Arthur Balfour that, "oil in the next war will occupy the place of coal in the present war, or at least a parallel place to coal. The only big potential supply that we can get under British control is the Persian and Mesopotamian supply." Therefore, Hankey said, "control over these oil supplies becomes a first-class British war aim."

But the newly born "public diplomacy" had to be considered. In early 1918, to counter the powerful appeal of Bolshevism, Woodrow Wilson had come out with his idealistic Fourteen Points and a resounding call for the self-determination of nations and peoples after the war. His own Secretary of State, Robert Lansing, was appalled by the President's broadside. The call for self-determination,

Lansing was sure, would result in many deaths around the world. "A man, who is a leader of public thought, should beware of intemperate or undigested declarations," he said. "He is responsible for the consequences."

But the British government, though no less appalled by what it considered Wilson's high-minded vagueness, had to take the President's popular appeal into account in formulating its postwar objectives. Foreign Secretary Balfour worried that explicitly pronouncing Mesopotamia a war aim would seem too old-fashionably imperialistic. Instead, in August 1918, he told the Prime Ministers of the Dominions that Britain must be the "guiding spirit" in Mesopotamia, as it would provide the one natural resource the British empire lacked. "I do not care under what system we keep the oil," he said, "but I am quite clear it is all-important for us that this oil should be available." To help make sure this would happen, British forces, already elsewhere in Mesopotamia, captured Mosul after the armistice was signed with Turkey.[5]

Clemenceau and His Grocer

The entire experience of wartime, beginning with the armada of taxis that saved Paris in the first weeks of the war, had convinced the French no less than the British that access to oil was now a matter of great strategic concern. Before World War I, Georges Clemenceau was supposed to have said, "When I want some oil, I'll find it at my grocer's." During the war he changed his mind, and at war's end, he sought to obtain oil for France, not from his grocer's, but—like the British—from the Middle East. On December 1, 1918, Clemenceau, following his drive with Lloyd George through the cheering throngs of London, apparently surrendered France's claim to Mosul. But in turn, Clemenceau won not only British support for a French mandate over Syria, but also a guarantee that France would receive a share of any oil found in British-controlled Mosul.

The exchange in London between the two Prime Ministers, in fact, settled nothing. Rather, it initiated a protracted series of stormy negotiations, filled with acrimony and mutual recriminations, between their respective governments. Indeed, in the spring of 1919, during the Paris Peace Conference, at a meeting of the Big Three dealing with Syria and oil, Clemenceau and Lloyd George rancorously disagreed as to what they had "agreed" on in London and repeatedly accused each other of bad faith. The discussion turned into a "first-class dog-fight," which, save for the on-site peacemaking of Woodrow Wilson, might have become an actual fistfight.

The matter remained unresolved and a major bone of contention until, finally, the Allied Supreme Council met—though with the United States no longer participating—in April 1920, to settle their many outstanding differences, including oil and the Middle East. Lloyd George and France's new premier, Alexandre Millerand, hammered out the compromise San Remo Agreement: France would get 25 percent of the oil from Mesopotamia, which itself would become a British mandate under the League of Nations. The vehicle for oil development remained the Turkish Petroleum Company; and the French acquired what had been the German share in it, which had been seized by the British during the war. In turn, the French gave up their territorial claim to

Mosul. Britain, for its part, made absolutely clear that any private company developing the Mesopotamian oil fields would very definitely be under its control. There was only one remaining question: Was there, in fact, any oil in Mesopotamia? No one knew.[6]

The French were looking at another way to enhance their oil position—by creating a state company, their own national champion. Rejecting a proposal for partnership with Royal Dutch/Shell from Henri Deterding, Raymond Poincaré, who became premier in 1922, insisted that this new company be "entirely French" in terms of control. To that end, he turned in 1923 to an industrial magnate, Colonel Ernest Mercier. Mercier was well-qualified for the task. A *Polytechnicien* and a war hero who had been wounded trying to help protect the Rumanian oil fields from the advancing Germans, he was also a technocrat devoted to modernizing the French economy. He had already put together a modern electric industry in France. Now he would try to do the same for oil. The new company was to be called the Compagnie Française des Pétroles, CFP, for short, and it was to be the "instrument" of "liberation" for France. While the French government appointed two directors and approved all others, the company was to be private.

Mercier's assignment was made more difficult by the reluctance of French companies and banks to invest in the new firm. They had none of the speculative, even feverish enthusiasm for new oil ventures that gripped British and American investors, even though this one would be underwritten by the state. Mesopotamia looked like very high risk—"so full of international difficulties," Mercier was later to say. "None of the initial investors begged for the favor of being admitted into the CFP." Nevertheless, Mercier did eventually succeed in finding sufficient investors—ninety banks and companies—so that the Compagnie Française des Pétroles could be launched by 1924. This new firm took up the French shares of the Turkish Petroleum Company.

But the French government remained unsatisfied that its objectives and interests were sufficiently safeguarded. In 1928, a special Parliamentary commission reported on the future organization of the domestic oil market, the largest in Europe after Britain's. It opposed both a "free market" and state monopoly. Instead, it called for a hybrid—a quota system, under which the state allocated market shares to various private refining companies in order to assure diversity of supply and guarantee the viability of French refining companies. In addition, tariffs and various other legal protections would be established to protect the French refiners against foreign competition. Legislation of March 1928 outlined the main objectives of a new "constitution" for French oil: to curtail the "Anglo-Saxon oil trusts," to build a domestic refining industry, to bring order to the market, and to develop the French share of Mesopotamian oil. To ensure that CFP would actively embody French interests under the new system, the state acquired a direct 25 percent ownership and increased the number of government directors, while the share of foreign ownership fell sharply. CFP was ready, in the words of a French deputy, to become "the industrial arm of government action." And the French government had now positioned itself as a major contender in the struggle to obtain the oil riches of the Middle East.[7]

Amalgamation?

For the British government, the sailing was not so smooth. It continued its efforts, started during the war, to upset the sixty-forty Dutch-British split and bring Royal Dutch/Shell under British control, by having British rather than Dutch shareholders predominate. To Marcus Samuel, such a result would be of great sentimental importance, and thus very appealing. But Henri Deterding was not much interested in sentiment; his interest was only business. British protection and sponsorship could count for much more than Dutch in a postwar world convulsed by revolution, diplomatic competition, and nationalist movements. But there was a further prize, or bait, for Shell to agree to surrender its Dutch predominance: Mesopotamian oil and the Turkish Petroleum Company. By passing under British control, Shell could guarantee its title to Mesopotamian oil.

From the viewpoint of the British government, bringing Shell under British control would greatly enhance Britain's worldwide oil position. But the British government wanted to name at least one director and approve others on the board of the restructured Shell, much as in its arrangement with Anglo-Persian. Deterding simply would not countenance that. British predominance was one thing; British government interference in the business was another. Deterding would not risk giving up any commercial control. He also began to see disadvantages in too close an association with the British government, particularly in terms of obtaining acreage in North and South America. Royal Dutch/Shell was the target of persistent attacks in America, where it was mistakenly thought that the Group was an arm of the British government. The criticism was vigorous enough to make Deterding very reluctant to pass under explicit British control.

Yet, despite all the delays, disappointments, and loss of patience, Deterding and Shell continued to be mightily interested in amalgamating with Anglo-Persian. They saw great merit in gaining control of Anglo-Persian before it could become a fearsome direct competitor. Amalgamation would strengthen Shell in its worldwide competition with Standard of New Jersey and the other American companies. It would end the preferential relationship of Anglo-Persian as fuel oil vendor to the key British market, the Royal Navy. Deterding was also repelled by what he saw as waste and duplication in the way the industry was functioning. "The world," he would soon write to the president of Standard Oil, was "suffering from over-production, over-refining, over-transporting, and—last but not least—over-retailing."

Anglo-Persian had already had to face difficulties because of government ownership. Many countries, said a Foreign Office official, assumed that "every action of the company" resulted from "direct Government inspiration," hampering both company and government. Latin American countries, in response to American prodding, banned concessions to government-controlled oil companies, which meant, specifically, Anglo-Persian. Its link to the British government could prove especially dangerous on Anglo-Persian's home soil, Persia. The company was already seen as standing much too close to the British government in the eyes of Reza Shah, the one-time military commander who had made himself ruler of the country. How secure would the company—and Britain's position—be with the new Shah? Anglo-Persian's entire position in the country was highly vulnerable; as one British official observed, the "whole revenue is at present derived from an area of a few square miles in Persia. Any interruption, either from natural causes or through hostile action, of the output of this small field would be disastrous."

A merger with Shell, some British government officials were convinced, would diversify Anglo-Persian's interests and thus reduce the risk. And, in the process, the government would obtain its long-desired control over Shell. And Shell was still willing—at least up to a point. "The whole question of control," said Robert Waley Cohen of Shell in 1923, was "very largely nonsense. It is a matter of sentiment, but if by transferring control to the Hottentots we could increase our security and our dividends, I don't believe any of us would hesitate for long."

To be sure, there was no shortage of opposition to amalgamation, beginning on political grounds. Public hostility to "oil trusts" was not much less in Britain than in the United States. But the strongest opposition came from the Admiralty, which continued to be antagonistic to Shell. The Navy's original rationale still remained; the government, as one official commented, "did not go into the Anglo-Persian Company to make money but to form an independent Company for national reasons." The Admiralty had also become deeply attached to its right to obtain fuel oil from Anglo-Persian at a substantial discount from the going market price, especially as the Navy's budgets were under constant threat of cutback. And, of course, Anglo-Persian itself vehemently opposed the amalgamation. Charles Greenway had not fought so hard to turn the enterprise into an integrated oil company in order for it to become merely an addendum to the hated Shell.[8]

Reenter Churchill

How, against such entrenched opposition, was Shell to effect its takeover of Anglo-Persian? Robert Waley Cohen had a brain wave. In the course of a carefully orchestrated dinner, he approached Winston Churchill with a most interesting proposition. Would the former M.P. and distinguished former Cabinet member consider taking up a project on Shell's behalf? The assignment? To lobby for an amalgamation of both Anglo-Persian and Burmah Oil with Shell, whereby Shell might end up purchasing the government's shares in Anglo-Persian. Burmah was also supportive of such a combination. Churchill would really be working for Britain, Cohen stressed, for if his effort was successful, it would secure British control over a worldwide oil system.

The offer could not have been better timed. For in the summer of 1923, Churchill, the "champion of oil," was out of a job. He had been defeated in his Parliamentary constituency at Dundee East, had just purchased a new country estate, Chartwell, and was writing at a furious pace in order to make ends meet. "We shall not starve," he promised his wife. After his discussions with Churchill, Cohen said, "Winston at once saw the picture complete." Still, Churchill said he had to think about it. He did not want to damage the political career to which he was totally devoted. Moreover, he needed to earn a living, and he would have to put aside the fourth volume of his work on the Great War—*The World Crisis*. So, of course, there would have to be a fee.

Yes, of course.

After brief consideration, Churchill accepted the offer. But about the fee? Churchill wanted ten thousand pounds if the deal did not go through, and fifty thousand pounds if it did.

Cohen was taken aback by the magnitude of Churchill's terms, but it was decided that the sum could be split between Shell and Burmah. As the chairman of Burmah remarked, "We couldn't very well haggle or bargain" with Churchill. Burmah's officers worried about how to pay the money, since if the recipient of such a large fee was not disclosed on the books, the auditors would not approve. Finally it was decided to set up a secret account.

Thus, Churchill went to work for Burmah and, more so, for Shell, the very same company that—while First Lord of the Admiralty, a decade earlier, engaged in his battle to bring the Navy into the oil age—he had so roundly castigated. Shell's voraciousness, he had then insisted to the House of Commons, was the central reason for the government to buy shares in Anglo-Persian and guarantee its independence. Now he was prepared to undo all that, to persuade the government to sell those same shares in the cause of what he now saw as larger political and strategic interests. Shell would pick up those shares, thus shifting the balance within the Royal Dutch/Shell Group from Dutch to British predominance.

Churchill wasted no time. In August 1923, he called on the Prime Minister, Stanley Baldwin, who, Churchill wrote to his wife, was "thoroughly in favor of the Oil settlement on the lines proposed. Indeed he might have been Waley Cohen from the way he talked. I am sure it will come off. The only thing I am puzzled about is my own affair. . . . It is a question of how to arrange it so as to leave no ground of criticism." Prime Minister Baldwin was certainly persuaded that the British government should quit the oil business. He even had a definite figure in mind for the purchase of the government's shares. "Twenty million pounds would be a very good price," he told Churchill. It was almost ten times what the government had paid less than ten years earlier, an excellent return on a speculative investment.

But before anything further could be done, there was an outside intervention. Baldwin called a snap general election at the end of 1923, and Churchill, the job not yet done, resigned his commission, returned the initial fee, and charged back into his natural and beloved fray, politics. A minority Conservative government came back into power, but quickly fell, and was replaced by Britain's first Labour government, which resolutely rejected both the amalgamation and the selling off of the government stake. In the autumn of 1924, the Conservatives came back into power, but they, too, were now opposed to selling the government stake. "His Majesty's Government," the Undersecretary of the Treasury wrote to Charles Greenway, chairman of Anglo-Persian, "have no intention of departing from the policy of retaining these shares." The minister responsible for the Treasury was the new Chancellor of the Exchequer, none other than the newest convert to Conservatism, Winston Churchill.[9]

Oil Shortage and the Open Door

The Middle East was not to be the preserve of European oil interests alone. The American companies were embarking on a campaign to develop new oil supplies worldwide, which would inevitably thrust them into the Middle East. A fear of imminent depletion of oil resources—indeed, a virtual obsession—gripped the American oil industry and many in government at the end of World War I and well into the early 1920s. The wartime experience—"Gasolineless Sundays" and the part played by oil in battle—gave a tangibility to the fear. When, in 1919, a retiring official wrote him that lack of foreign oil supplies constituted the most serious international problem facing the United States, President Wilson sadly agreed: "There seemed to be no method by which we could assure ourselves of the necessary supply at home and abroad." The anticipated rapid depletion of American oil resources was gauged against the rise in demand: American consumption had increased 90 percent between 1911 and 1918 and was expected to grow even faster after the war. America's love affair with the automobile was becoming ever more intense. The increase in the number of registered motor vehicles in the United States between 1914 and 1920 was astonishing—a jump from 1.8 to 9.2 million. The fear of shortage was such that one Senator called on the U.S. Navy to reconvert from oil back to coal.

The leaders of engineering and scientific geology shared the fear. The director of the United States Bureau of Mines predicted in 1919 that "within the next two to five years the oil fields of this country will reach their maximum production, and from that time on we will face an ever-increasing decline." George Otis Smith, the director of the United States Geological Survey, warned of a possible "gasoline famine." What to do? The answer, he said, was to go overseas; the government should "give moral support to every effort of American business to expand its circle of activity in oil production throughout the world." He warned that the known American oil reserves would be exhausted in exactly nine years and three months.

At the same time, there was much discussion about the potential of the shale oil locked up in the mountains of Colorado, Utah, and Nevada. It was predicted in 1919 that "within a year petroleum will probably be distilled from these shales in competition with that obtained from wells." *National Geographic* excitedly declared that "no man who owns a motor-car will fail to rejoice" because shale oil would provide the "supplies of gasoline which can meet any demand that even his children's children for generations to come may make of them. The horseless vehicle's threatened dethronement has been definitely averted." Alas for the proponents of shale oil, the costs of development were woefully underestimated. In Britain, where similar shortfalls were anticipated, Anglo-Persian was doing research on extracting liquid fuels from coal, and the British government had given over two acres in Dorset to the cultivation of Jerusalem artichokes in the hope that this plant could produce alcohol in commercial quantities to be used as automobile fuel.

Large price increases gave powerful support to the expectation of shortage. Between 1918 and 1920, the price of crude in the United States jumped 50 percent, from two to three dollars a barrel. Moreover, the winter of 1919–20 saw an actual shortfall in fuel oil supplies. The United States, it was generally thought, would soon have to become a significant importer of oil. And that raised the specter of international competition and a clash with Britain. Both the United States oil industry and the American government firmly believed that Britain was pressing its own aggressive policy to preempt the rest of the world's oil resources before the Americans could move. Thus, Washington rallied quickly to support the oil companies in their quest for foreign supplies. The principle invoked was that of the "Open Door"—equal access for American capital and business.

The British reacted to this campaign with varying mixtures of skepticism, injury, outrage, and implacability. They noted that the United States produced two-thirds of the world's crude oil. "I don't expect that you or any other oil man in America really believes that your supplies are going to be exhausted in the next 20 or 30 years," John Cadman, director of the Petroleum Executive, wrote incredulously to an American friend. But the fears of both shortage and competition pushed American companies to seek out new supplies on a worldwide basis, either by exploration or by purchase of existing production. And the shift in strategy would be supported by technological improvement—in tankers, pipelines, and drilling—that helped to overcome the physical difficulties and distances that before the war would have been forbidding obstacles to global exploration or production.[10]

American eyes fastened on the Middle East, particularly Mesopotamia, under British mandate. But the door was manifestly not open there. When two Standard Oil of New York geologists slipped into the territory, the British civil commissioner handed them over to the chief of police of Baghdad.

The news of the San Remo Agreement of 1920, the understanding between the British and French over the division of any possible Mesopotamian oil, stunned Washington and the oil industry. The accord was thunderously denounced in the American press as old-fashioned imperialism; it was regarded as all the more obnoxious because it seemed to violate the principle of equal rights among the victorious Allies. Jersey was deeply worried. It feared a double alliance—one between the British and the French, and one between Shell and Anglo-Persian—that would shut it out of production and markets around the globe. The company protested vigorously to the State Department, which no less vigorously denounced the agreement as a violation of the cherished principles of the Open Door. Congress passed the Mineral Leasing Act of 1920, which denied access to drilling rights on public lands to foreign interests whose governments denied similar access to Americans. It was aimed, specifically, at the Dutch in the East Indies and the British in Mesopotamia.

Cynical observers were struck by the degree to which the Wilson Administration, the embodiment of progressivism, now gave support in its final phase to the oil companies—particularly to Jersey, the most prominent heir to the dragon that had been slain by the Supreme Court just a decade earlier. The British ambassador to Washington marveled at how the rapprochement between the Wilson Administration and the Standard Oil interest "completely reversed the prewar relationships under which it was nothing less than courting disaster for any member of the administration to incur the suspicion of an affiliation with the oil interests." The specter of oil shortage and the suspicion of British treachery did much to firm this new alliance. So did the wartime experience of business-government collaboration; Standard Oil of New Jersey, alone, had supplied a quarter of all the oil used by the Allies. There were other reasons as well for the turnabout. Progressivism and reform had spent their force. And the American businessman was again to be seen, as in the 1880s and 1890s, as a hero, and government would be his supporter, not his adversary.

The new Republican Administration of Warren Harding, which came into office in 1921, was an outright champion of business, and it proved even stronger than its predecessor in defending American oil interests, from Mexico to the Dutch East Indies—and including Mesopotamia. Tension between the United States and Britain mounted. But then something strange happened. The British became conciliatory and signaled a new openness to American participation in Mesopotamia. Why? For one thing, they recognized that there was ambiguity about the legal status of the Turkish Petroleum Company. Had it won a concession in 1914—or only the promise of a concession? In addition, the British had many other economic and strategic considerations on the agenda with the United States, and it wanted American cooperation. London was also concerned about anti-British sentiment in the United States, which was at a high point. There was even talk in Congress of retaliating with an embargo on the shipment of American oil to Britain. Moreover, failure to allow American participation in the Mesopotamian development would only be a permanent irritant—or worse—in Anglo-American relations. By contrast, direct American involvement could be a real plus: the British were anxious to see the region's petroleum resources developed as rapidly as possible in order to provide revenues to the new British-backed government that was emerging in Mesopotamia, thus reducing pressure on the British Treasury. American capital and technology would certainly speed the process. Finally, Shell at least believed that American participation would strengthen the hand of the companies in any political difficulties that might arise in that unstable part of the world. Calouste Gulbenkian added his voice, advising the Permanent Undersecretary of the Foreign Office that it would be better to have the Americans "inside" than to have them "outside," competing—and challenging the concession. The Permanent Undersecretary was persuaded, and he very firmly instructed Anglo-Persian and Royal Dutch/Shell that it was in the British national interest to include the Americans—and as soon as possible. Afterward, he wrote Gulbenkian to say that the Armenian had been "instrumental in bringing in American participation."[11]

"The Boss": Walter Teagle

But which American companies was the United States government to support? Would it not appear more than a little unseemly to exert so much diplomatic energy solely on behalf of a single company, Jersey Standard? Various influential people, including Commerce Secretary Herbert Hoover, suggested that a syndicate of American companies be formed to operate in Mesopotamia. Hoover, in particular, knew the oil business and its risks well; he had been active in it before the war, and in fact had sold some Peruvian oil properties to Walter Teagle of Jersey, who had described the future President in his notes at the time as "a queer looking fellow—seersucker suit & white tennis shoes." Now, at a meeting in Washington in May of 1921, Hoover, as Commerce Secretary, and Secretary of State Charles Evans Hughes explained frankly to a group of oil men that the United States could not swing the door open on behalf of one company alone, but could do so for a representative group. For its part, Jersey recognized that it could never count on sustained government support if it went the course alone, and so Teagle put together a consortium of several leading companies. Only recently this new group would have been attacked by the government on grounds of restraint of trade; now it was supported as a national champion in promoting the Open Door and access to foreign oil.

Following the establishment of this American group, the State Department backed away from the inevitable clash with European oil interests. While monitoring developments closely, it would stand apart from the actual negotiations. Walter Teagle, a businessman and not a politician or diplomat, would speak for the American syndicate, and in July 1922, he sailed for London to begin negotiations on American participation in developing whatever petroleum resources were to be found in Mesopotamia. He could have had no idea how lengthy and difficult the course would be.[12]

On one side stood Teagle, representing not only Standard Oil, but also the entire consortium of American companies. Arrayed against him were Henri Deterding, Charles Greenway, and for the French, Colonel Ernest Mercier of CFP. But hovering close to the table was Calouste Gulbenkian. All of Teagle's opponents were partners in the Turkish Petroleum Company, which controlled the Mesopotamian concession—or at least presumed it did.

Gulbenkian, more than anybody else, would prove to be Teagle's chief antagonist in the unfolding drama. The contrast between the two men seemed enormous in almost every respect. Short and unprepossessing, Gulbenkian was suspicious and uncommunicative. Teagle loomed over almost everybody; he was six foot three, and a man of considerable girth—sometimes getting up as high as three hundred pounds when he was losing one of his battles with his almost unquenchable passion for chocolate. He appeared direct and forthright, the very embodiment of the friendly American. Whereas Gulbenkian was a lone operator, Teagle was the head of the world's largest oil company, by far the largest of the successor companies to the Standard Oil Trust. Known as "the Boss," he singularly dominated Standard Oil of New Jersey, and was one of the most prominent and familiar figures throughout the oil business. Gulbenkian preferred anonymity.

Yet there were strange similarities between the two men. Teagle, too, was born to oil. As Gulbenkian was second-generation oil business, so was Teagle, on his father's side. On his mother's side, he was actually third generation; his maternal grandfather was Maurice Clark, the partner whom John D. Rockefeller had bought out at the critical "auction" held in Cleveland in 1865. Teagle's father, originally from Wiltshire in England, was one of the most successful independent refiners in Cleveland, and for years he had resisted the onslaughts of the

Standard Oil Trust. He hated Standard Oil, and had been one of the heroic bat-
tlers against it depicted in the pages of Ida Tarbell's history of the trust.

Both Gulbenkian and Teagle had been outstanding students of petroleum
technology. At Cornell University, Teagle seemed to be manager or organizer of
almost every student activity. He wrote his undergraduate thesis on the desulfu-
rization of crude oil and scored an unheard-of perfect one hundred in industrial
chemistry. Like Gulbenkian, he was encouraged by his professor to study for an
advanced degree, and his father responded as sharply as Gulbenkian's had—in
Teagle's case, with a terse telegram, "Come home at once." Back in Cleveland,
Teagle went to work firing a still in the family refinery, at nineteen cents an hour.
Then his father sent him out on the road. Teagle proved himself to be a formidable, aggressive, and persuasive salesman. But he was again summoned home to
help sell out the family business to the enemy his father had so long resisted—
Standard Oil. His father could not carry the strain any longer. Better to be bought
out than to struggle on. Moreover, Standard Oil had spotted the talented young
Teagle, and it wanted not only the owner's business, but also the owner's son.

The family business was now reconstituted as Republic Oil, and the young
Teagle was made its boss. His skills soon became apparent—a mastery of the
whole range of the oil business; a prodigious memory for technical, commercial,
and administrative details; unflagging energy; a capacity to reason through a
problem and find a solution; and beneath the outward charm, a relentlessly demanding and dominating personality. The years on the road had taught him what
Gulbenkian had learned in the bazaar—always go for the best deal possible. "He
haggled over everything," a colleague from the Republic Oil days remembered.
"He'd trade and trade and trade. If it was company money, he'd think he was
paying too much for a five cent cigar and try to get it for four."

Teagle rose rapidly and by 1908, he was head of Standard Oil's Foreign Export Committee. He had a keener understanding than Standard's other senior executives of the new dynamics of the international marketplace. He also
developed a better understanding of Henri Deterding and promoted conciliation
with Royal Dutch/Shell. Once, in order to settle a particularly acrimonious competitive situation in the Far East, Teagle spent two days shooting grouse with
Deterding in Scotland—they were both excellent wing shots—two days playing
poker, and then worked out the matter. Yet their mutual respect, even what could
be described as friendship, could not overcome the fundamental suspicion that
governed their relationship. Far too much was at stake. Bluntly stated, each man
totally distrusted the other. Deterding, Teagle once said, "frequently changes his
mind and usually forgets to tell you that such is the case." Teagle never ceased to
see Royal Dutch/Shell as the most dangerous and deadliest of competitors.

In 1909, Teagle became a director of Standard Oil, taking the chair of the
powerful H. H. Rogers, who had, among other things, been Ida Tarbell's inside
source. Teagle was only thirty-one. One newspaper predicted that he had been
picked to fill the "John D. shoes" and reported that—in contrast to Rogers, who
had been Mark Twain's admirer and patron—Teagle's favorite authors were
Dun and Bradstreet. Teagle believed that a kind of managerial paralysis had set
in at Standard Oil, primarily as a result of the antitrust suit and other legal assaults. One of the costs, Teagle thought, was the failure by the company to adjust

to the new global competition and to develop its own crude production from foreign sources.

In 1917, at the age of thirty-nine, Teagle became president of Standard Oil of New Jersey. He was a new style leader. He was not a substantial stockholder, in contrast to the previous generation; he was a professional manager, and his arrival reflected a change in American business and the nature of the corporation. He would later completely restructure Standard's operations. Yet he also represented a continuity with the company's past—after all, he was the grandson of Rockefeller's original partner—and he made sure that the continuity was clear to others. On becoming president, he installed Rockefeller's old rolltop desk—number 44—in his own office and set about reinvigorating the moribund company. He had observed firsthand the cost of excessive secrecy as measured in the public antipathy to the old Standard Oil, and put much effort into better public relations. He created a new in-house magazine, *The Lamp*, and made himself its de facto editor. He instituted an "open door" for the press. He was available, yes, friendly and hearty with reporters, and apparently candid and forthright. But what he said was also carefully controlled and calibrated. Still, it was a striking difference from the old regime.

With the end of World War I, Teagle saw that the company faced a major problem—crude supply. His efforts to push the company into crude production had continually been blocked by the traditional opposition to such a "risky" activity, as reflected in the comment of one veteran director, who said, "We're not going to drill dry holes all over the world. We're a marketing company." Now, Teagle feared that oil shortages would become chronic in the postwar world. He believed that Standard Oil was at a great disadvantage, as its crude production was only equivalent to 16 percent of its refinery output. Meanwhile, his old rival Deterding was pursuing a global strategy of building up diversified sources of crude around the world. Teagle knew of the efforts of the British government to merge Shell and Anglo-Persian. He fully expected an ever-harsher global competitive environment, and he feared that Standard Oil of New Jersey was not ready for it. To meet the challenge, Teagle overrode his opponents and pushed the company into domestic acquisitions, as well as into a new commitment to foreign oil production. In 1920, at the fiftieth-anniversary celebration for Standard Oil, he bluntly enunciated his strategy: "The present policy of the Standard Oil Company is to be interested in every producing area no matter in what country it is situated." And wherever in the world there looked to be the possibility of oil, Standard Oil of New Jersey intended to be there.[13]

That was why, in the summer of 1922, Teagle was in London, facing the partners in the Turkish Petroleum Company. The discussions were fruitless and, after a month, Teagle returned home. The negotiations were continued by correspondence. By December 1922, the frustrated Americans were seriously thinking of walking away completely. It was no easy matter to divide up Mesopotamia, or Iraq, as the British mandate was now called, at such a crowded table.

The participants argued over who would get what share of Iraqi oil. They debated whether they would maintain the self-denying exclusion from the earlier agreement and thus not participate in oil production in most of the rest of the former Ottoman Empire except through the Turkish Petroleum Company. Then there was the acrimonious matter of revenues, which proved to be the most contentious issue of all. Teagle and Greenway of Anglo-Persian wanted the oil sold to the participating stockholders at cost, without any profits on it. That way, they would preclude a battle with Iraq over definition of profits and just pay it a royalty, and the American companies would avoid additional British taxes. But this proposal did not please Iraq, which wanted a direct share of earnings. Nor did it sit at all well with Calouste Gulbenkian, who was most interested in receiving his dividends in money—not oil.

To make matters more problematic, the new, much-shrunken nation-state of Turkey was challenging the border with Iraq and was trying to undermine the legal basis of the Turkish Petroleum Company—all of which highlighted the risk that the oil companies would be running in that part of the world. To blunt these risks, the British government, taking advantage of its League of Nations mandate over the region, put pressure on Iraq to grant a new concession, but without swift result. For the British government had a most uneasy relationship with the regime it had recently established in Iraq. The two parties could not even agree on what the word "mandate" meant.[14]

Faisal of Iraq

During the war, London had encouraged Hussein, the Sharif of Mecca, to take the lead in raising an Arab revolt against Turkey. This he did, beginning in 1916, aided by a few Englishmen, of whom the most famous was T. E. Lawrence— Lawrence of Arabia. In exchange, Hussein and his sons were to be installed as the rulers of the various, predominantly Arab, constituents of the Turkish empire. Faisal, the third son of Hussein, was generally considered the most able. Lawrence, enchanted at meeting Faisal during the war, described him as "an absolute ripper" and the perfect person to command the revolt in the field. After the war, Faisal cut a romantic figure at the Versailles Conference, even capturing the imagination of the dry American Secretary of State, Robert Lansing, who wrote that Faisal's "voice seemed to breathe the perfume of frankincense and to suggest the presence of richly colored divans, green turbans, and the glitter of gold and jewels."

The British put Faisal on the throne of the newly created nation of Syria, one of the independent states carved out of the extinct Turkish empire. But a few months later, when control of Syria passed to France under the postwar understandings, Faisal was abruptly deposed and turned out of Damascus. He showed up at a railway station in Palestine, where, after a ceremonial welcome by the British, he sat on his luggage, waiting for his connection.

But his career as a king was not yet over. The British needed a monarch for Iraq, another new state, this one to be formed out of three former provinces of the Turkish empire. Political stability in the area was required not only by the prospect for oil, but also for the defense of the Persian Gulf and for the new imperial air route from Britain to India, Singapore, and Australia. The British did not want to rule the region directly; that would cost too much. Rather what Churchill, then the head of the Colonial Office, wanted was an Arab government, with a constitutional monarch, that would be "supported" by Britain under League of Nations mandate. It would be cheaper. So Churchill chose the out-of-work Faisal as his candidate. Summoned from exile, he was crowned King of Iraq in Baghdad in August 1921. Faisal's brother Abdullah—originally destined for the Iraqi throne—was instead installed as king "of the vacant lot which the British christened the Amirate of Transjordan."

Faisal's task was enormous; he had not inherited a well-defined nation, but rather a collection of diverse groups—Shia Arabs and Sunni Arabs, Jews and Kurds and Yazidis—a territory with a few important cities, most of the countryside under the control of local sheikhs, and with little common political or cultural history, but with a rising Arab nationalism. The minority Sunni Arabs held political power, while the Shia Arabs were by far the most numerous. To complicate things further, the Jews were the largest single group among inhabitants of Baghdad, followed by Arabs and Turks. To this religious and ethnic mosaic, Britain sought to import constitutionalism and a responsible parliament. Faisal depended upon Britain to support his new kingdom, but his position would be gravely impaired if he were seen as being too beholden to London. The British government had to cope not only with Arab nationalism in Iraq but also with the oil men, who were clamoring for some word on the status of the Iraqi concession. Britain was all for oil development, hoping that the potential oil revenues would help finance the new Iraqi government and further reduce its own financial burdens.

But oil exploration and development in Iraq could not begin without a new, sounder concession granted by the government. For one thing, Washington consistently refused to recognize the validity of the 1914 grant to the Turkish Petroleum Company. Allen Dulles, the chief of the Division of Near Eastern Affairs in the State Department, carefully monitored the long negotiations for the U.S. government. In 1924, he told Teagle that the United States government believed that the Turkish Petroleum Company's claim to a concession was "invalid." As Dulles had explained on another occasion, "The information we have is sufficient to knock the case of the Turkish Petroleum Company into a cocked hat." Yet the various Iraqi cabinets, fearful of nationalistic sentiments and domestic criticism—which sometimes expressed itself in the form of assassination—were most reluctant to take responsibility for signing over a revised concession to the foreigners. Negotiations between the Turkish Petroleum Company and the Iraqi government were, thus, slow, difficult, and invariably bitter. But at last, on March 14, 1925, a new concession agreement was signed. It satisfied the American government; it gave the illusion of holding open the Open Door. But that last, Gulbenkian later noted, was mere "eyewash."[15]

The Architect

Everything seemed settled at last, even the boundary with Turkey, except for one stumbling block—Calouste Gulbenkian and his 5 percent. Throughout the negotiations, Gulbenkian had remained a strange, solitary figure. He went to great lengths to avoid meetings, but scrutinized every word of memoranda, and replied with a torrent of telegrams. Isolation also marked his personal connections. "Oil friendships are very slippery," he once said. That certainly proved true of his formerly close business relationship with Deterding, which ruptured in the middle 1920s. "We worked most harmoniously for over twenty years," Gulbenkian later explained, "but, as it has very often been the case in the oil business, personal jealousies, divergencies of opinions separated us." Others said that their quarrel was the result of a struggle for the affections of a White Russian lady, Lydia Pavlova, former wife of a Czarist general. For a time the two men collaborated on that lady, as on oil. Once, when Deterding found that he could not come up with the three hundred thousand dollars he owed Cartier's for the emeralds he had impulsively bought for her, Gulbenkian arranged a bridging credit until Deterding's next draw from Royal Dutch/Shell. But, in due course, Lydia Pavlova became the second Mrs. Deterding, and the outcome led to bad blood between the two men. Deterding and Gulbenkian also had a nasty dispute over the profits from a Venezuelan oil company that Gulbenkian had brought to the Royal Dutch/Shell Group. Deeper questions of ego were at stake, as well. At least, that was the view of Nubar Gulbenkian, who had the unique vantage point of having been personal assistant both to his father and to Deterding—leaving the latter position only when the two men angrily severed their relationship. As Nubar explained, Deterding came to resent the "persnickety interference" of Gulbenkian, while Gulbenkian could not stand the "overbearing grandeur of Deterding."

With or without Deterding, Gulbenkian continued to be involved in manifold business activities, including an effort to secure an exclusive concession for the marketing of Soviet caviar. He had left his wife installed among his art treasures—his "children," as he called them—in the mansion he had built on avenue d'Iena in Paris. He himself alternated between suites at the Ritz in Paris or, in London, at the Ritz or the Carlton Hotel, attended by a succession of mistresses, at least one of whom at all times, on the basis of "medical advice," had to be eighteen years or younger in order to rejuvenate his sexual vigor. He could be seen once or twice a day, taking his constitutional in the Bois de Boulogne or in Hyde Park, his limousine trailing behind him. The rest of the time he sought to keep out of sight, devoting himself to his worldwide business interests, keeping in constant contact by a stream of telephone calls and telegrams.

The companies in the American consortium, particularly Standard, remained committed to developing new oil sources around the world. Iraq loomed very large in their plans. But Gulbenkian stood in the way, and he would not budge. Of overwhelming importance to him was his 5 percent of the Turkish Petroleum Company—to be paid in cash, which the Americans opposed. His break with Deterding only strengthened his obstinacy, taxing ever more Deterding's and Teagle's—and everybody else's—patience. Teagle was once driven to say that Gulbenkian was "most difficult in a difficult situation." Gulbenkian was convinced, in his own words, that "the oil groups headed by the American had only one aim, that is, by hook or by crook to wipe out" his rights. But he was absolutely confident in his position. The Armenian wanted money, not crude oil. "How would you like it," he asked a newspaper reporter, "if you had a small interest in an oil company and it was proposed that your dividends be paid in a few gallons of oil?"

186

Teagle finally decided that he would have to see Gulbenkian in person. He arranged that they should lunch together at the Carlton Hotel in London. After working his way through many courses, Teagle got to the point. He adopted what he thought would be an appealing line in discussing the royalty that Gulbenkian demanded. "Surely, Mr. Gulbenkian, you're too good an oil merchant not to know that the property won't stand any such rate as that."

Gulbenkian's face went red, and he furiously banged the table. "Young man! Young man!" he shouted. "Don't you ever call me an oil merchant! I'm not an oil merchant and I'll have you distinctly understand that!"

Teagle was taken aback. "Well, Mr. Gulbenkian," he began again, "I apologize if I have offended you. I don't know what to call you or how to classify you if you aren't an oil merchant."

"I'll tell you how I classify myself," the Armenian replied hotly. "I classify myself as a business architect. I design this company and that company. I designed this Turkish Petroleum Company and I made a room for Deterding and I made a room for the French and I made a room for you." His fury was unabated. "Now, the three of you are trying to throw me out on my ass."[16]

Toward the Red Line

Meanwhile, it was yet to be determined if oil was going to be found in commercial quantities in Iraq. Only in 1925 did a joint geological expedition—representing Anglo-Persian, Royal Dutch, and the American companies—arrive in Iraq. Even as the stalemate with Gulbenkian continued, the geologists carried out their exploration with rising excitement. One of the Americans reported back to New York that he knew of no other region in the world where the promise of drilling was greater.

Gulbenkian still refused to give any ground. But then why should he? It had been almost thirty-five years since he had written his original report on Mesopotamia and its oil for the Sultan. Almost fifteen years had passed since he had put together the Turkish Petroleum Company. He had paid the expenses out of his own pocket to keep that ramshackle scheme going during the First World War. He had waited patiently for so long; what did a little more delay matter? He was already a fabulously rich man. And he knew that any geological success in Iraq would only strengthen his position by putting pressure on Teagle and the other Americans to come quickly to some agreement.

The response to the flow of news from the geologists proved Gulbenkian right. A settlement, Teagle recognized, was now imperative. Drilling began in April 1927, which meant that delay was no longer possible on the business front. The stalled negotiations started to move again at the same time, as Teagle reluctantly began to give ground to Gulbenkian. Finally, an agreement was in sight.

It was none too soon. One of the drilling sites was at Baba Gurgur, about six miles northwest of Kirkuk, in what was primarily the Kurdish region. There, for thousands of years, two dozen holes in the ground had been venting natural gas, which was always alight. They were thought to be the "burning fiery furnace" into which Nebuchadnezzar, King of Babylon, had cast the Jews. It was there, too, that the local inhabitants—so Plutarch had written—had set afire a street sprinkled with oil seepages to impress Alexander the Great. And it was there, at 3:00 A.M. on October 15, 1927, from a well known as Baba Gurgur Number 1—in which the drill bit had barely passed fifteen hundred feet—that a great roar was heard, reverberating across the desert. It was followed by a powerful gusher that reached fifty feet above the derrick, carrying in it rocks from the bottom of the hole. The countryside was drenched with oil, the hollows filled with poisonous gas. Whole villages in the area were threatened, and the town of Kirkuk itself was in danger. Some seven hundred tribesmen were quickly recruited to build dikes and walls to try to contain the flood of oil. Finally, after eight and a half days, the well was brought under control. It had flowed, until capped, at ninety-five thousand barrels per day.[17]

The leading question had been answered. There *were* petroleum resources in Iraq—potentially so bountiful that they were, after all, well worth all the wrangling. Now a final settlement became urgent. The negotiations had to be completed. At last on July 31, 1928, nine months after the initial discovery—almost six years to the day since Teagle had first sailed to London to nail down an agreement—the full contract was signed. Royal Dutch/Shell, Anglo-Persian, and the French would each receive 23.75 percent of the oil, as would the Near East Development Company, which was created at this time to hold the interests of the American companies. As to the main sticking point, Gulbenkian would receive his 5 percent interest in oil, but he could immediately sell the petroleum to the French at market prices, thus automatically transmuting crude oil into his desired and beloved cash.

There remained the question of the critical "self-denying" clause, by which all the participants agreed to work jointly together—and only jointly—in the region. As Gulbenkian later told it, at one of the final meetings he called for a large map of the Middle East, then took a thick red pencil and drew a line along the boundaries of the now-defunct Turkish empire. "That was the old Ottoman Empire which I knew in 1914," he said. "And I ought to know. I was born in it, lived in it, and served in it." Gulbenkian may have, however, been adding his own personal embellishment to what had already been decided. For, several months earlier, the British, using Foreign Office maps, and the French, with maps from the Quai d'Orsay, had already fixed the same boundaries. Whoever the author of the boundaries, this far-reaching oil settlement was thereafter called "The Red Line Agreement."

Within the red line were eventually to be found all the major oil-producing fields of the Middle East, save for those of Persia and Kuwait. The partners bound themselves not to engage in any oil operations within that vast territory except in cooperation with the other members of the Turkish Petroleum Company. So the self-denying clause of the 1914 Foreign Office Agreement was reborn fourteen years later as the Red Line Agreement. It set the framework for future Middle Eastern oil development. It also became the focus for decades of bitter conflict.

Many years later, when it was said that Gulbenkian had defeated him on the deal with the Turkish Petroleum Company, Walter Teagle looked back. Remembering those arduous and time-consuming negotiations, he said, "It was a damn bad move! Should have gone in by ourselves three years earlier."

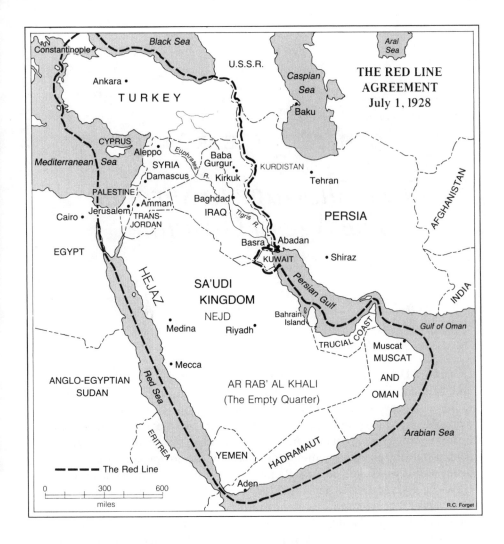

THE RED LINE
AGREEMENT
July 1, 1928

— — — The Red Line

0   300   600
miles

It was certainly a great victory for Gulbenkian—the culmination of thirty-seven years of concentration, and a testament to his perseverance and tenacity. It was the deal for which he had waited his entire adult life. It would be worth tens of millions of dollars to him. To mark the grand event, he chartered a boat that summer and set off on a Mediterranean cruise with his daughter Rita. Off the coast of Morocco, he caught sight of a type of ship he had never seen before. It looked very strange to him, with its funnel jutting up at the extreme stern of the long hull. He asked what it was.

An oil tanker, Rita told him.

He was fifty-nine years old, he had just made one of the greatest oil deals of the century, he was the Talleyrand of oil, and he had never before seen an oil tanker.[18]

CHAPTER 11

From Shortage to Surplus: The Age of Gasoline

IN 1919, a U.S. Army captain, Dwight D. Eisenhower, depressed by the tedium and the scrimping that seemed likely to be the chronic condition of peacetime military life, gave some thought to leaving the Army to take up a job offer in Indianapolis from an Army buddy. But then he heard that the Army wanted an officer to join a cross-country motor caravan that was being organized to demonstrate the potential of motor transportation and to dramatize the need for better highways. He volunteered, if only to relieve his boredom and to arrange a cheap family vacation in the West. "A coast-to-coast convoy," he later said, "was, under the circumstances of the time, a genuine adventure." He was to remember the trip as "Through Darkest America with Truck and Tank."

The journey started on July 7, 1919, with the dedication of Zero Milestone, just south of the White House lawn. Then the caravan took off. It included forty-two trucks; five staff, observation, and reconnaissance passenger cars; and a complement of motorcycles, ambulances, tank trucks, mobile field kitchens, mobile repair shops, and Signal Corps searchlight trucks. The vehicles were in the hands of drivers whose language, as well as driving skills, suggested, at least to Eisenhower, that they were more familiar with teams of horses than the internal combustion engine. The first three days, the convoy managed five and two thirds miles an hour—"not quite so good," said Eisenhower, "as even the slowest troop train." It never got much better. The record of the trip was a log of broken axles, broken fan belts, broken spark plugs, and broken brakes. As for the roads, they varied, "from average to nonexistent," said Eisenhower. "In some places, the heavy trucks broke through the surface of the road and we had to tow them out one by one, with the caterpillar tractor. Some days when we had counted on sixty or seventy or a hundred miles, we would do three or four."

Having left Washington on July 7, the caravan did not arrive in San Francisco until September 6, where the drivers were greeted with a parade, followed by a speech by the Governor of California, who compared them to the "Immortal Forty-Niners." Eisenhower was looking ahead. "The old convoy," he recalled "had started me thinking about good, two lane highways." Eventually, three and a half decades later, he would, as President of the United States, champion a vast system of interstate highways. But in 1919, Eisenhower's snail-paced mission "Through Darkest America" signified the dawn of a new era—the motorization of the American people.[1]

"A Century of Travel"

"This is a century of travel," Henri Deterding had written in 1916 to one of the senior Shell executives in the United States, "and the restlessness which has been created by the war will make the desire for travel still greater." His prediction was quickly borne out in the post–World War I years, with consequences that transformed not only the oil industry, but indeed, the American and then the global way of life.

The transformation occurred with astonishing rapidity. In 1916, the year of Deterding's prophecy, some 3.4 million autos were registered in the United States. But, through the 1920s, with peacetime prosperity at hand, the cars were rolling off the assembly lines in ever more staggering numbers. By the end of the decade, the number of registered cars in America had zoomed to 23.1 million. Each of those cars was being driven farther and farther each year—from an average of 4,500 miles per car in 1919 to an average of 7,500 in 1929. And each of those cars was powered by gasoline.

The face of America was changed by the vast invasion of automobiles. In *Only Yesterday*, Frederick Lewis Allen presented a portrait of the new visage of the 1920s. "Villages which had once prospered because they were 'on the railroad' languished with economic anemia; villages on Route 61 bloomed with garages, filling stations, hot-dog stands, chicken-dinner restaurants, tearooms, tourists' rests, camping sites and affluence. The interurban trolley perished. . . . Railroad after railroad gave up its branch lines. . . . In thousands of towns, at the beginning of the decade a single traffic officer at the junction of Main Street and Central Street had been sufficient for the control of traffic. By the end of the decade, what a difference!—red and green lights, blinkers, one-way streets, boulevard stops, stringent and yet more stringent parking ordinances—and still a shining flow of traffic that backed up for blocks along Main Street every Saturday and Sunday afternoon . . . the age of steam was yielding to the gasoline age."

The impact of the automobile revolution was far greater on the United States than anywhere else. By 1929, 78 percent of the world's autos were in America. In that year, there were five people for each motor vehicle in the United States, compared to 30 people per vehicle in England and 33 in France, 102 people per vehicle in Germany, 702 in Japan, and 6,130 people per vehicle in the Soviet Union. America was, indubitably, the leading land of gasoline. The change in the basic orientation of the oil industry was no less dramatic. In 1919, total United States oil demand was 1.03 million barrels a day; by 1929, it had risen to 2.58 million barrels, a two-and-a-half-times increase. Oil's share of total energy consumption had over the same period risen from 10 to 25 percent. By far the biggest growth was registered by gasoline—more than a fourfold increase. Gasoline and fuel oil together accounted for fully 85 percent of total oil consumption in 1929. As for kerosene, its production and consumption were negligible by comparison. The "new light" had given way to the "new fuel."[2]

"The Magic of Gasoline"

The transformation of America into an automotive culture was accompanied by a truly momentous development: the emergence and proliferation of a temple dedicated to the new fuel and the new way of life—the drive-in gasoline station. Before the 1920s, most gasoline was sold by storekeepers, who kept the motor fuel in cans or other containers under the counter or out in back of the store. The product carried no brand name, and the motorist could not be sure if he was getting gasoline or a product that had been adulterated with cheaper naphtha or kerosene. Moreover, such a system of distribution was cumbersome and slow. In the infancy of the auto age, some retailers experimented with gasoline wagons that delivered fuel from house to house. That idea never really caught on, partly because of the frequency with which the wagons tended to explode.

There had to be a better way, and there was—the drive-in station. The signal honor of building the first drive-in station is attributed to several different pioneers, but according to the *National Petroleum News*, the distinction belonged to the Automobile Gasoline Company in St. Louis in 1907. The oil trade publication reported in a small story tucked on an inside page, under the headline "Station for Autoists," that "a new way of reaching the auto gasoline trade direct is being tried with reported success in St. Louis by the Auto Gasoline Co." The oil man who brought the innovation to the editor's attention had chortled and said, "Now get a good laugh out of this dump." While the editor never personally saw the first station, he did visit Automobile Gasoline's second station in St. Louis, and it was, in his view, truly a dump. A small tin shack housed a couple of barrels of motor oil. Outside, two old hot-water tanks were set on high brackets, with lengths of garden hose from each so as to drain the gasoline by gravity into automobile tanks. The whole operation was set on a muddy corner lot. That was pretty much what all the early stations looked like—small, cramped, dirty, ramshackle structures, equipped with one or two tanks, and barely accessible to the street via a narrow, unpaved path.

The real growth and development of the gas station did not come until the 1920s. In 1920, certainly no more than 100,000 establishments sold gasoline; fully half of them were grocery stores, general stores, and hardware stores. Few of those stores were selling gasoline a decade later. In 1929, the estimated number of retail establishments selling gasoline had grown to 300,000. Almost all of them were gas stations or garages. The number of drive-in gasoline stations, specifically, had grown from perhaps 12,000 in 1921 to about 143,000 in 1929.

The stations were everywhere—big city street corners, main streets in small towns, country crossroads. East of the Rockies, such facilities were called "filling stations"; west of the Rockies, they were known as "service stations." And their future was heralded when, in 1921, a celebrated super station opened in

Fort Worth, Texas, with eight pumps and three different approaches from the street. But California, and specifically Los Angeles, was the true incubator of the modern service station, a standard structure with huge signs, restroom facilities, canopies, landscaped grounds, and paved entrances. The standardized "cracker box" gasoline stations, pioneered by Shell, proliferated at an astonishing rate across the nation, and by the end of the 1920s, they were making money not only from gasoline sales but also from what were called "TBA"—tires, batteries, and accessories. Standard of Indiana was turning stations into grander emporiums that sold a whole range of petroleum products in addition to gasoline, from motor oil to furniture polishes to oil for sewing machines and vacuum cleaners. A new type of pump quickly became the order of the day all across the country, one in which the gasoline was forced into a glass bowl atop the pump, where it could be seen by the customer, reassuring him as to the purity of the product, before it flowed through the hose and into the fuel tank of his car.

And, as the service stations spread and competition heated up, they hoisted aloft the signs and symbols of the new age: Texaco's star, Shell's scallop shell, Sun's radiant diamond, Union's "76," Phillips's "66" (suggested not only by the highway, but by Heinz's "57 Varieties"), Socony's flying horse, Gulfs orange disc, Standard of Indiana's red crown, Sinclair's brontosaurus, and Jersey Standard's patriotic red, white, and blue. Competition forced the oil companies to develop trademarks to assure national brand identification. They became the icons of a secular religion, providing drivers with a feeling of familiarity, confidence, and security—and of belonging—as they rolled along the ever-lengthening ribbons of roads that crossed and crisscrossed America.

Gas stations were also the source for what one expert described as "uniquely American contributions to the development and growth of cartography"—the oil company road map. The first road map specifically directed toward the automobile was probably the one that appeared in the *Chicago Times Herald* in 1895 for a fifty-four-mile race that the newspaper was sponsoring. But it was only in 1914, when Gulf was opening its first gasoline station in Pittsburgh, that a local advertising man suggested handing out free maps of the region at the facility. The idea caught on rapidly as Americans took to the road in the 1920s, and the maps soon became staples.

Customers were courted with many other amenities and attractions. By 1920, Shell of California was providing free uniforms to attendants and paying for up to three launderings a week. It prohibited the attendants from reading magazines and newspapers while on duty, and its manual forbade the accepting of tips: "Air and water service is a gratuity which you are expected to render the public, showing no distinction as to whether the individual is a Shell customer or not." By 1927, the "service station salesmen," as they were called, were expected to ask the customer, "Can I check the tires for you?" They were also forbidden to allow "personal opinions and prejudices" to get in the way of service: "Salesmen should be careful in their attendance upon Oriental and Latin classes of customers and refrain from using broken English in conversation with them."

Advertising and publicity helped create the major regional and national brands. And it was Bruce Barton, an advertising man, who sought to carry the sale of gasoline to its most uplifting heights. Barton spoke with immense authority. He had already assured himself immortality of sorts with *The Man Nobody Knows*, the nation's number-one best seller in both 1925 and 1926, which proved that Jesus was not only "the most popular dinner guest in Jerusalem," but also "the founder of modern business," and "the greatest advertiser of his day." Now, in 1928, Barton called upon oil men to reflect upon "the magic of gasoline." He urged them to "stand for an hour beside one of your filling stations. Talk to the people who come in to buy gas. Discover for yourself what magic a dollar's worth of gasoline a week has worked in their lives.

"My friends, it is the juice of the fountain of eternal youth that you are selling. It is health. It is comfort. It is success. And you have sold merely a bad smelling liquid at so many cents per gallon. You have never lifted it out of the category of a hated expense. . . . You must put yourself in the place of the man and woman in whose lives your gasoline has worked miracles."

The miracle was that of mobility; people could go where they wanted whenever they wanted. This was an uplifting message for men in the oil business, who worried about margins, volumes, inventories, market share, and greasy uniforms. If not quite a religion, the sale of gasoline at retail outlets had become, by the end of the decade, a big and very competitive business.[3]

The Tempest in the Teapot

Because the price of gasoline now affected the lives and fortunes of so many Americans, it became axiomatic by the 1920s that gasoline prices, whenever they went up, would become a source of rancor, a subject to be reported by the press, discussed by governors and senators and even presidents, and investigated by various branches of the U.S. government. In 1923, after a price run-up, the populist Senator from Wisconsin, Robert ("Fighting Bob") La Follette, conducted highly charged hearings on gasoline prices. He and his subcommittee warned that, "if a few great oil companies" were permitted to continue "to manipulate oil prices for the next few years, as they have been doing since January 1920, the people of this country must be prepared, before long, to pay at least $1 a gallon for gasoline." His warning would lose much of its punch as surplus mounted and gasoline prices plummeted. In April 1927, retail gasoline prices fell to thirteen cents a gallon in San Francisco, and to ten and a half cents in Los Angeles, a far cry from La Follette's dire prediction.

But, if La Follette was off-base about the dynamics of gasoline prices, he was directly on target in another drama, of which his investigation into gasoline prices was only a sideshow. For he led the initial crusade in the Senate that uncovered one of the most famous and bizarre scandals in the nation's history—Teapot Dome.

Teapot Dome in Wyoming, named for the shape of a geological structure, was one of three oil fields (the other two were in California) that had been set aside as "naval oil reserves" by the Taft and Wilson Administrations as one result of the pre–World War I debate about converting the U.S. Navy from coal to oil. The argument had been similar to the simultaneous one in Britain, which had so engaged Winston Churchill, Admiral Fisher, and Marcus Samuel. While recognizing the superiority of oil over coal and, of course, acknowledging the preeminent position of the United States when it came to production, the Americans, like the British, had worried greatly about the possibility of what one American naval officer called a "failure of supply . . . menacing the mobility of the fleet and the safety of the nation." What would happen if oil gave out at the critical moment? Yet the advantages of petroleum were irresistible, and the decision was made to convert the United States fleet, with the key year being 1911, the same year as in Britain. The next year, in order to alleviate the supply anxieties, Washington began to establish the naval petroleum reserves in areas of potential production. They were to constitute "a supply laid up for some unexpected emergency," which could be brought into production in time of war or crisis. But there had been a long battle in Washington over the establishment of these reserves and about whether private interests would be able to lease them for partial exploitation. That debate was, in turn, part of a continuing public policy battle in twentieth-century America between those championing the development of resources on public lands by private interests, and those advocating the conservation and protection of those resources under the stewardship of the Federal government.

When Warren G. Harding, chosen as the Republican candidate because among other reasons he "looked like a President," won the White House in 1920, he sought, like any good politician, to appeal to both sides in the resource debate, celebrating "that harmony of relationship between conservation and development." But, in selecting Senator Albert B. Fall from New Mexico to be Secretary of the Interior, Harding could hardly disguise his choice of development over conservation. Fall was a successful and politically powerful rancher, lawyer, and miner—"the frontiersman, the rough and ready, two-fisted fighter," said one magazine, "who looks like an old-time Texas sheriff and is said to have handled a gun in his younger days with all the speed and accuracy of a Zane Grey hero." Fall's "belief in the unrestrained disposition of the public lands was as typically Western as his black, broad-brimmed Stetson hat and his love of fine horses." Those on the other side of the debate saw Fall differently. He was described as a member of the "exploitation gang" by one leading conservationist. "It would have been possible to pick a worse man for Secretary of the Interior," the conservationist added, "but not altogether easy."

Fall succeeded in wresting control of the naval oil reserves away from the Navy Department and placing it in the Interior Department. The next step would be to lease the reserves to private companies. His activities had not gone unnoticed. In the spring of 1922, just before the leases were signed, Walter Teagle of Standard Oil unexpectedly appeared in the office of advertising man Albert Lasker, who had directed Harding's campaign publicity and was, at the time, head of the United States Shipping Board. "I understand," Teagle told Lasker, "the Interior Department is about to close a contract to lease Teapot Dome, and all through the industry it smells. I'm not interested in Teapot Dome. It has *no* interest whatsoever for Standard Oil of New Jersey, but I *do* feel that you should tell the President that it *smells*."

With some reluctance, Lasker went to see the President and repeated Teagle's message. Harding paced up and down behind his desk. "This isn't the first time that this rumor has come to me," he said, "but if Albert Fall isn't an honest man, I'm not fit to be President of the United States." Both propositions were soon tested to their limit.[4]

Fall leased Teapot Dome to Harry Sinclair in an exceedingly sweet deal that assured Sinclair Oil a guaranteed market—the U.S. government. He also leased a more bountiful California reserve, Elk Hill, to Edward Doheny. Both were among the best-known of American oil men. They were entrepreneurs, "new men" who had risen up on their own abilities to create major enterprises outside the old Standard Oil inheritance. Doheny was something of a legend. He had begun his career as a prospector. Laid up when he broke both legs falling down a mine shaft, he had put the time to good use by studying to become a lawyer. He was also said to have fought off a mountain lion with a knife. By the 1920s Doheny had amassed a vast fortune, and his company, Pan American, was actually a larger crude oil producer than any of the Standard Oil successor companies. Doheny himself scrupulously made a point to patronize and befriend politicians of both parties.

So did Harry Sinclair, the son of a small-town druggist in Kansas, who trained to be a druggist himself. But, at age twenty, he lost the family drugstore in a speculation. Broke, he tried to make a living selling lumber for drilling rigs, and then took to buying and selling small oil properties in southeast Kansas and the Osage Indian territory of Oklahoma. Luring investors, he began to build up a host of tiny oil companies, one per lease. He was a masterful trader and a forceful, assertive businessman, with unbridled self-confidence, who would defer to no one, least of all his investors. Said one of his colleagues, "Where he sat, there was the head of the table." He simply insisted on getting his way. He put all his chips on the Glenn Pool in Oklahoma, and made a fortune from it. He went into the newly discovered Oklahoma oil fields, awash in oil because of flush production and not yet connected to pipelines, and bought all the oil he could get at ten cents a barrel. He then threw up steel storage tanks, waited for the pipelines to be completed, and sold the oil for $1.20 a barrel.

By World War I, Sinclair was the largest independent oil producer in the midcontinent. But having to sell to the large, established, integrated companies, and pay heed to them, galled him no end. He raised $50 million and in 1916 swiftly put together his own integrated oil company, which was soon among the ten largest in the country. The absolute monarch of his company, Sinclair was ready to fight for business almost anywhere in the country. He got into the habit of thinking that when he wanted to do something, nothing should stand in his way. And one thing he had wanted was Teapot Dome.

The Interior Department signed its contracts with Doheny and Sinclair in April 1922 amid swirling rumors, as one conservationist said, "about Mr. Fall being quite friendly with large interests of an olcaginous nature." Senator La Follette began to investigate. He discovered that naval officers who had opposed the shift of the reserves from the Navy Department to Interior and their subsequent leasing had been transferred to distant and inaccessible stations. His suspicions were further aroused. But they remained only suspicions a year later when, in March 1923, Fall resigned as Secretary of the Interior, still very much a solid and respected, though increasingly controversial, public figure.

By this point, the Harding Administration was sinking into a deep mire of scandal and wrongdoing. Harding himself was struggling to cope with accusations that he maintained a full-time mistress. "I have no trouble with my enemies," the sad President said as his private railroad car rolled across the Kansas plain, "I can take care of them. It is my . . . friends that are giving me my trouble." Shortly after, in San Francisco, he suddenly died—a doctor said of "an embolism," but a newspaper editor countered that it was "an illness that was part terror, part shame, and part utter confusion!" He was succeeded by his Vice-President, Calvin Coolidge.

Meanwhile, the Senate's Public Lands Committee had taken up the matter of Teapot Dome. There were still no hard facts, and some were saying that the whole thing was no more than "a tempest in a teapot." But then items of considerable interest began to emerge. Fall had undertaken extensive and expensive renovations on his New Mexico ranch about the time of the leasing of Teapot Dome. He had also bought a neighboring ranch partly with hundred dollar bills he lifted out of a small tin box. How had he suddenly become flush with money? Pushed on the question of the sudden improvement in his finances, Fall said he had received a one-hundred-thousand-dollar loan from Ned McLean, the publisher of the *Washington Post*. Interviewed in Palm Beach—sinus trouble having supposedly kept him from traveling—McLean admitted the loan, but then said Fall had, a few days later, returned his checks uncashed. More embarrassing revelations came out. Sinclair's secretary testified that Sinclair had once told him he should give Fall twenty-five or thirty thousand dollars if he ever asked for it. And Fall did ask. Sinclair himself, who had suddenly departed for Europe on very short notice, hastily left Paris for Versailles in order to dodge reporters.

Then came the real bombshell. On January 24, 1924, Edward Doheny told the Senate committee that he had provided the one hundred thousand dollars to Fall, which his son had personally carried in cash "in a little black bag" to Fall's office. No, it was not a bribe, definitely not, Doheny insisted, just a loan to an old friend; they had prospected together for gold decades earlier. He even produced a mutilated note supposedly signed by Fall, though the signature had been ripped off. Doheny explained that his wife held the signature portion, so as not to embarrass Fall with demand for an inconvenient repayment, should Doheny happen to die. It was friendship compounded with thoughtfulness.

Fall himself said he was too sick to testify, which reminded some people of an incident only a few years earlier. The very partisan Fall was one of the two Senators who went to the White House in 1920 to investigate whether Woodrow Wilson was really suffering from a stroke or had, as rumored, actually lost his mind. "Mr. President, we all have been praying for you," Fall earnestly declared on that day in 1920. "Which way, Senator?" the feeble Wilson replied. Now people said that Fall's illness ought to be investigated. Reputations were being ruined left and right as the bizarre story continued to unfold. Investigators learned that telegrams in an old Justice Department code had passed between McLean, the *Washington Post* publisher, from Palm Beach, and various persons in Washington, D.C. An ex-train robber from Oklahoma appeared to testify before the Senate committee. Harry Sinclair, on trial for contempt of the Senate for refusing to answer questions, hired the Burns Detective Agency to shadow members of the jury, which could not exactly be considered in the best tradition of Anglo-

Saxon jurisprudence. By 1924, said *The New Republic*, all of Washington was "wading shoulder-deep in oil. . . . The newspaper correspondents write of nothing else. In the hotels, on the streets, at the dinner tables, the sole subject of discussion is oil. Congress has abandoned all other business."

The 1924 Presidential election was at hand, and Calvin Coolidge was intent on winning the White House in his own right. His main interest in oil at that point was to stay as far away as possible from the subject and avoid any taint from the Teapot Dome scandal. Coming to Coolidge's defense, a Republican Congressman proclaimed that Coolidge's only connection to Teapot Dome was that he had been sworn in by the light of an oil lamp. Even that was too close for comfort. The Democrats intended to play the scandal as a potent election issue. But they underestimated the political skills of Calvin Coolidge. They also overlooked their own vulnerability—Doheny was, after all, a Democrat who had provided lucrative employment to at least four former members of Woodrow Wilson's Cabinet. He had also paid $150,000 in legal fees to William McAdoo, Woodrow Wilson's son-in-law and the front-runner for the 1924 Democratic nomination. McAdoo lost that place when the fees became public knowledge, and the Democratic nomination went instead to John W. Davis. It even turned out that Doheny had discussed an oil "proposition" in Montana with the Democratic Senator who just happened to be heading the Senate's investigation of Teapot Dome.

As the public clamor over Teapot Dome mounted, Coolidge counterattacked: He fired Harding's underlings, denounced wrongdoing, and appointed twin special prosecutors—one Democratic and one Republican. Thereafter he effectively distanced himself from the scandal and, in the 1924 Presidential campaign, did everything he could to live up to the title of "Silent Cal." His strategy was to neutralize issues by ignoring them—a campaign of silence. And on nothing else was he so silent as on the subject of oil. The strategy worked. Amazingly enough, the great Teapot Dome scandal never became an issue at all in the campaign, and Coolidge won handily.

The scandal itself dragged on through the rest of the decade. In 1928, it was discovered that Sinclair had channeled several hundred thousand dollars more to Fall through a bogus company, the Continental Trading Company, which meant that Fall had received at least $409,000 for his services to his two old friends. Finally, in 1931, the corrupt and greedy Fall went to jail, the first Cabinet officer convicted and imprisoned for a felony committed while in office. Sinclair was sentenced to prison for six and a half months for contempt both of court and of the Senate. On his way to jail, he stopped to attend a board meeting of the Sinclair Consolidated Oil Corporation, where the other directors formally tendered him "a public vote of confidence." Doheny was judged innocent and never went to jail, leading one Senator to complain, "You can't convict a million dollars in the United States."[5]

The Colonel and the Liberty Bonds

The scandal had even wider repercussions when further investigations revealed that the bogus company, Continental Trading, was really a mechanism by which a group of prominent oil men had received kickbacks in the form of government Liberty Bonds on purchases of oil made by their own companies. Harry Sinclair had used part of his kickback as payoff money to Fall, passing on the bonds. He had also given some of the bonds to the Republican National Committee. The nation was shocked to learn that among those receiving Liberty Bond kickbacks was one of America's most distinguished, successful, and forceful oil men, Colonel Robert Stewart, chairman of Standard of Indiana.

A broad-faced, bulky man, Stewart had ridden with Teddy Roosevelt's Rough Riders. Unlike the heads of many of the other major oil companies he had never had a day of practical oil field experience. He had first gone to work for Standard of Indiana as an attorney, and he had ridden his legal skills to the top of the company. That was not so surprising; after all, the legal challenges before and after dissolution had dominated and redefined the oil industry, and since 1907 Stewart had been at the center of every single major case involving Standard of Indiana. Autocratic, commanding, and combative, he infused the company with an aggressiveness that made it into the nation's number-one marketer of gasoline during the 1920s. "Colonel Bob," as he was called, was among the most respected and admired leaders not only of the oil industry but of all of American business. Who could believe that someone so upstanding would stoop to besmear himself in the slush of Teapot Dome? Yet, after years of evading questions about his involvement with Continental Trading and the Liberty Bonds, Stewart finally admitted receiving about $760,000 in bonds.

As Stewart became ever more deeply embroiled in the Teapot Dome controversy, the largest stockholder in Standard of Indiana, who had until then hardly interfered in the company's management, urged Stewart "to remove any just ground for criticism." Stewart would not cooperate. Finally, in 1928, the stockholder decided he had given Stewart every chance and concluded he would have to go. The stockholder was known as "Junior"; he was the only son of John D. Rockefeller.

John D. Rockefeller, Jr., was a short, shy, serious, and reclusive man. He worshiped his father and had wholeheartedly imbibed his lessons about thrift. As a student at Brown University, the younger Rockefeller had surprised his college classmates by hemming his own dish towels. But, more than anything else, he had been rigorously and repeatedly schooled by his mother in "duty" and "responsibility" and concerned himself with probity. He found his own life's vocation, independent of his father, in the systematic giving away of a significant part of the family fortune, though much would still, of course, be left over. He also involved himself in a wide variety of civic and social causes, once going so far as to chair an official investigation into prostitution, on behalf of the city of New York.

The younger Rockefeller even established a dialogue with Ida Tarbell, his father's "lady friend" and muckraking nemesis. He had met her at a conference in 1919 and had gone out of his way to be extremely polite, even chivalrous to her. A few years later, he asked Tarbell to review a series of interviews with his father that were to be the basis of a book he was planning. To facilitate matters, he himself delivered the materials to Tarbell's apartment in Gramercy Park in Manhattan. After studying the interviews, Tarbell told him that the elder Rockefeller's comments were self-serving and sidestepped all the charges made against him. "Junior" was persuaded. "Miss Tarbell has just read the biography manuscript and her suggestions are most valuable," Rockefeller wrote to a colleague. "It seems clear that we should abandon any thought to the publication of the material in anything like its present incomplete and decidedly unbalanced form."

That was in 1924. Now, four years later, the younger Rockefeller was no less aroused by the specter of wrongdoing in Standard of Indiana than Ida Tarbell had been by the wrongdoing in the old Trust. By profession, he was a philanthropist, not an oil man, and he had made a habit of staying away from the business of the successor companies. To much of the country, the father remained a great villain; now the son broke into the public scene in quite a different guise—as a reformer. And he was intent on carrying the mantle of reform to the heart of Standard Oil of Indiana. He told a Senate committee that, in the affair of Colonel Stewart, nothing less than the "basic integrity" of the company and indeed of the whole industry was at stake. But the Rockefeller interests directly controlled only 15 percent of the stock in the company. When Stewart refused to resign voluntarily, Rockefeller launched a proxy fight to oust him. The colonel counterattacked vigorously. "If the Rockefellers want to fight," he declared, "I'll show them how to fight." He had a strong business record; in the last ten years of his leadership, the company's net assets had quadrupled. And now, for good measure, he declared an extra dividend and a stock split to boot. Some saw the bitter struggle as a battle between East and West for the control of the oil industry; others said the Rockefellers wanted to reassert their control over the entire industry. But the Rockefeller forces were not clamorous for dividends; they wanted victory, and they organized and campaigned hard. And, in March of 1929, they won, with 60 percent of the stockholders' votes. Stewart was out.

John D. Rockefeller, Jr., had intervened directly, and in a highly visible way, in the affairs of one of the successor companies of his father's Standard Oil Trust. He had done so not for mere profits but in the name of decency and high standards, and to safeguard the oil industry from new attacks from government and the public—and to protect the Rockefeller name. He was much berated for his efforts. "If you look up the record of your father in the early days of the old Standard Oil Company," one angry supporter of Stewart wrote to Rockefeller, "you will find it pretty well smeared with black spots ten times worse than the charges you lay at the door of Col. Stewart. . . . There is not enough soap in the world to wash the hands of the elder Rockefeller from the taint of fifty years ago. Only people with clean hands should undertake to blacken the character of other and better men."

A college professor disagreed. "No endowment of a college nor support of a piece of research," he wrote, "could have done more it seems to me to educate the public toward honest business." American capitalism, and the oil industry, could never again be as rapacious as it once had been; now the future of the industry and of business was at stake, not the fortunes of a few men. And the oil industry had its public image to consider. But if the younger Rockefeller's hands were clean, the entire "Teapot Dome" scandal—from Fall, Doheny, and Sinclair to Stewart—had picked up where the Standard Oil Trust had left off in ingraining in the public mind a nefarious image of the power and corruption of "oil money."[6]

Geophysics and Luck

There were many in America, at the beginning of the automotive age, who worried that supplies of the "new fuel" were about to give out. The years 1917–20 had been generally disappointing in terms of new discoveries. Leading geologists prophesied gloomily that the limits on U.S. production were near. Post–World War I pressure on supplies reinforced the expectation of shortage among refiners as well. Some refineries could run at only 50 percent of their capacity because crude oil was in short supply, and local retailers around the country kept running out of kerosene and gasoline. Indeed, shortage was so much the dominant view in the industry that Walter Teagle of Standard Oil of New Jersey once remarked that pessimism over crude supplies had become a chronic malady in the oil business.

But the wheel had already begun to turn. The search for new sources of supply was nothing short of frantic, fueled by the expectation of shortage itself and reinforced by the powerfully alluring incentive of rising prices. Oklahoma crude, which had been $1.20 a barrel in 1916, rose to $3.36 by 1920 as refiners, who had run short, bid up the price; and a record number of oil wells were drilled.

The technology for finding oil was also about to improve. Up to 1920, geology, as it applied to the oil industry, had meant what was known as "surface geology," the mapping and identification of likely prospects on the basis of the visible landscape. But, by 1920, surface geology had gone almost as far as it could. Many of the visible prospects had been identified. Explorers had to find a way to "see" underground, in order to figure out whether the subsurface structures were the kind that might trap oil. The emerging science of geophysics provided that new way of "seeing."

Many of the geophysical innovations were adapted from technology that had been drafted into use during World War I. One was the torsion balance, an instrument that measured changes in gravity from point to point on the surface, thus providing some sense of the subsurface structure. Developed by a Hungarian physicist before the war, it was used by the Germans during the First World War, in trying to get the Rumanian fields back into production. Another innovation was the magnetometer, which measured changes in the vertical components of the earth's magnetic field, giving further hint of what lay beneath the surface.

The seismograph also joined the arsenal of oil exploration, proving to be the most powerful new weapon of all. The seismograph had originally been developed in the mid-nineteenth century to record and analyze earthquakes. The Germans put it to work during the war to locate enemy artillery emplacements. That led directly to its use in the oil industry in Eastern Europe. What was called refraction seismology was introduced into the U.S. oil industry about 1923–24, initially by a German company. Dynamite charges were set off, and the resulting energy waves, refracted through underground structures, were picked up by listening ears—"geophones"—on the surface, which helped to identify underground salt domes, where oil might be found. The reflection seismograph, introduced about the same time and soon to supplant the refraction technique, recorded the waves that bounced off rock interfaces underground, which allowed the shapes and depths of all kinds of underground structures to be plotted. Thus a whole new world was opened up to exploration, irrespective of surface signs. Though many of the major fields were still discovered through surface geology in the 1920s, geophysics became more and more important, even in the fields initially identified by the more traditional methods. Oil men had indeed found the way to "see" underground.

They also found new ways to see aboveground. During the Great War, aerial surveillance had been used by the combatants in Europe for troop spotting. The technique was quickly adopted by the oil industry, making possible a broad view of surface geology that simply was not available to someone on the ground. As early as 1919, Union Oil hired two former lieutenants, who had done aerial work in France for the American Expeditionary Force, to photograph sections of the California landscape. Another important innovation was the analysis of microscopic fossils—micropaleontology—brought up from various drilling depths. This technique provided further clues to the type and relative ages of sediments thousands of feet underground. And at the same time, major improvements were being made in the technology of drilling itself, which permitted more rapid, more informative, and deeper drilling, thus expanding potential. The deepest wells in 1918 reached six thousand feet; by 1930, they were ten thousand feet. One final factor played an important role, one never easily analyzed but seemingly always present in the oil industry—luck. Certainly, luck was at work in the 1920s. How else to explain the fact that so much of America's oil was discovered during that decade?

One of the most significant of those discoveries was made at Signal Hill, which rises up some 365 feet behind Long Beach, just south of Los Angeles. From its peak, the local Indians had once signaled to their brethren on Catalina Island. Later still, the hill loomed large in the eager eyes of real estate developers. In June 1921, it was in the process of being subdivided into residential lots when a Shell exploratory well, Alamitos Number 1, blew in. The discovery created a stampede. Many of the lots, though already sold to prospective homeowners, were not yet built upon, and money flew all over the hill as oil companies, promoters, and amateurs scrambled to get leases. The parcels were so small and the forest of tall wooden derricks so thick that the legs of many of them actually interlaced. So keen were the would-be drillers that some property owners were able to get a 50 percent royalty. The next-of-kin of persons buried in the Sunnyside Cemetery on Willow Street would eventually receive royalty checks for oil drawn out from beneath family grave plots. True believers thought they could get rich buying a one five hundred thousandth share of a one-sixth interest in an oil well that had not yet even been drilled. Signal Hill was to prove so prolific that, almost unbelievably, some of those buyers actually made money on their investments.

Signal Hill was only the most dramatic of a large number of substantial discoveries in and around Los Angeles, which made California the nation's number-one producing state in 1923, and the source that year of fully one quarter of the world's entire output of oil. Even so, fears of shortage were still very much in the air. "The supply of crude petroleum in this country is being rapidly depleted," the Federal Trade Commission warned in 1923, in a study of the oil industry. But in that same year, American crude oil production exceeded domestic demand for the first time in a decade.[7]

The Tycoon

Henry Doherty was an anomaly in the oil business. With his oversized glasses and Van Dyke beard, he looked more like a stage version of "the professor" than a businessman of substance. But he was among the great entrepreneurs of the 1920s, controlling a host of companies, including Cities Service. One writer called him "the nearest approach" on Wall Street to Ned the Newsboy of the Horatio Alger stories. The description was apt; Doherty had begun his working life at age nine, selling newspapers on the streets of Columbus, Ohio. He dropped out of school when he was twelve. "I had not been in school more than ten days before I grew to hate school worse than Satan," he once explained. But with hard work, pluck, subsequent enrollment in night school, and training in engineering, he rose to be a director of no fewer than 150 companies. His empire was composed of gas and electric utilities serving various metropolitan areas, thus the name "Cities Service." When one of his companies, drilling for gas in Kansas, struck oil, Doherty quickly also became an oil man. More than a bit eccentric, he was a prolific writer of success epigrams: "Never give orders—give instructions. . . . Make a game out of your work. . . . The greatest dividend in human life is happiness." His favorite form of relaxation was driving a car through New York City traffic; fresh air was a great enthusiasm, and health an obsession.

A tough, resourceful businessman, Doherty gave no ground to his opponents. He was also an independent thinker, who enjoyed his role as the intellectual gadfly of the oil industry. He was no less tenacious and aggressive in campaigning for his ideas than for his deals, and he was convinced that the very way in which the industry operated out in the field was a threat to its future and needed to be changed. He was insistent, tiresomely so, on one theme: The "rule of capture" had to be eliminated.

The "rule of capture" had continued to govern the industry's operations since its early days in western Pennsylvania, and it had repeatedly been sanctioned by the courts, based upon the English common law regarding migratory wild beasts and game. To some property owners who complained to one court that their oil was being drawn off by their neighbors, the justices had scant solace to offer: "Only go and do likewise." Because of the rule, every operator everywhere in the United States put down his wells and produced as rapidly as he could, draining not only the oil under his own property but also that under his neighbor's property, before his neighbor drained his own. This doctrine fueled the riots of flush production and the wildly fluctuating prices that came with every new discovery.

Doherty believed that the multiplication of wells and rapid production resulting from the rule of capture exhausted the underground pressure in a field more quickly than need be the case. The consequence? Much oil that might otherwise be produced would be left underground, unrecoverable, because there was not enough pressure from gas—and also from water, it was later understood—to provide the "lift," or push, to get the oil to the surface. Recognizing how important oil had been in World War I, Doherty feared what it would mean for the United States in another war if wanton—and what he called "extremely crude and ridiculous"—production practices were to prevent vast stores of oil from ever being recovered.

Doherty had a solution to the problem. Fields should be "unitized." That is, they should be tapped as single units, with the output apportioned to the various owners. In that way, oil could be recovered at the controlled rates judged most sound by current engineering knowledge, thus maintaining the underground pressure. When Doherty, and subsequently many others, talked about "conservation," they meant such measured production practices, which would aim to ensure the largest ultimate recoverable resource, and not reduced or more efficient consumption. But how was Doherty's "conservation" to be accomplished? It was here that Doherty shocked most others in the industry. The Federal government, he argued, would have to take the lead, or at least sanction industry cooperation. And there would have to be public enforcement of technologically superior production practices.

For much of the 1920s, Doherty's views were shared only by a small minority of oil men, and he was widely attacked, and indeed, savagely abused. Some critics said that he got his facts out of the *World Almanac*. Many in the industry disputed his assessment of production technology, and regarded his call for Federal involvement as a betrayal of the industry. The larger companies were willing to talk about voluntary cooperation and self-regulation to manage production, but no more than that. Many independents did not want to have anything at all to do with unitizing fields and controlling production, whether voluntarily or not. They wanted their chance to get rich too.

Doherty fought back. He filibustered at meetings and conferences. He wrote endless letters. He was a "diabolical needler" of other oil men. He sought every occasion to push his views. Three times he tried to get the board of the industry's American Petroleum Institute to consider his proposals, and three times he was turned away. Barred at one API meeting from presenting his ideas, Doherty hired his own hall to address whoever would listen. Others started to call him "that crazy man." In turn, he declared that an "oil man is a barbarian with a suit on." But he did, after all, have a friend who was interested in his ideas—President Calvin Coolidge. In August 1924, Doherty wrote a long letter to the President: "If the public some day in the near future awakens to the fact that we have become a bankrupt nation as far as oil is concerned, and that it is then too late to protect our supply by conservation measures, I am sure they will blame both the men of the oil industry and the men who held public offices at the time conservation measures should have been adopted. A deficiency of oil is not only a serious war handicap to us but is an invitation to others to declare war against us."[8]

Once Coolidge had won the 1924 election and put the Teapot Dome scandal safely behind him, he could turn to oil. Responding to Henry Doherty's arguments, he established the Federal Oil Conservation Board to investigate conditions in the oil industry. Echoing his friend Doherty, the frugal President explained that the wasteful production methods were nothing less than a threat to the industrial, military, and overall security position of the United States. "The supremacy of nations may be determined by the possession of available petroleum and its products," Coolidge declared.

The Federal Oil Conservation Board stimulated further research on the physical properties of oil production, which, in turn, lent increasing support to Doherty's views. While the American Petroleum Institute was declaring that waste in the industry was "negligible," the new board argued that natural gas was "more than a commodity of smaller commercial value associated with oil," and in fact provided the underground pressure that pushed the oil to the surface. To dissipate gas through helter-skelter production was to lose that essential pressure, and thus to leave large amounts of petroleum unrecovered underground.

As the research results mounted, some of the better-informed began to swing over to Doherty's side. William Farish, the president of Humble, the Jersey affiliate that was the largest producer in Texas, had scorned Doherty's ideas in 1925. By 1928, he was thanking Doherty for making the industry see the virtues of "better production methods." Farish became a strong advocate of unitization—operating fields as single units. In the changing circumstances of the second half of the decade, he decided, the emphasis had to be on low-cost production. Unitization was one of the best ways to achieve that because fewer wells were needed, and more reliance could be placed on natural underground pressure as opposed to pumping.

Henry Doherty was technically far in advance of his brethren in comprehending how oil came to the surface, and how flush production damaged reserves. But he grossly underestimated the possibilities for finding new sources of oil. He insisted, in his 1924 letter to Coolidge, that a great shortage was at hand. Others did not hesitate to disagree with Doherty's bleak assessment of America's oil prospects. A bitter opponent of government involvement in the industry, J. Howard Pew of Sun Oil, sarcastically commented in 1925 that the nitrates in the soil would disappear, timber reserves be depleted, and the rivers of the world change their course before petroleum reserves were exhausted. "My father was one of the pioneers in the oil industry," Pew declared. "Periodically ever since I was a small boy, there has been an agitation predicting an oil shortage, and always in the succeeding years the production has been greater than ever before."[9]

The Rising Tide

It was Pew, not Doherty, who would prove to be the more accurate prophet on that score. The spring of 1926 saw the first of the major discoveries in what became known as the Greater Seminole field in the state of Oklahoma. The frenzy that ensued marked one of the most rapid developments of an oil field that the world had ever seen. It was a breakneck drilling competition, wanton and wasteful, again driven by the rule of capture. The traditional chaos and confusion of boomtowns reigned—the streets clogged with equipment, workers, gamblers, hucksters, and drunks; hastily built wooden structures; the stifling odor of escaping gas and the acrid smell of burning oil from wells and pits. Prices broke under the impact of the new finds. But still the oil flowed from that single field, reaching 527,000 barrels per day on July 30, 1927, a mere sixteen months after the first major discoveries. Other major discoveries followed in Oklahoma. Texas was about to catch up. A series of major discoveries in the late 1920s, including the huge Yates field, established the Permian Basin, a vast, sun-scorched, dusty, and desolate region of West Texas and New Mexico, as one of the great concentrations of oil in the world.

Another factor was at work to swell the tide. Technology was not only contributing to higher production, it was also altering the requirements of consumption. The spread of techniques for cracking, which increased the amount of gasoline that could be extracted from each barrel by changing the molecules, reduced the need for crude. One barrel of oil that was cracked could produce as much gasoline as two barrels of crude that went uncracked. It was then discovered that cracked gasoline was actually preferable to "straight run" gasoline because it had much superior antiknock properties. Thus, though the demand for gasoline increased, the demand for crude oil did not grow at the same rate, adding to the rising surplus.

By the end of the decade, the gloomy predictions of the early 1920s had been washed away by the flood of oil that seemed to flow unendingly out of the earth. American consumers simply could not absorb all the oil that was being produced, and more and more of it poured out of the ground, only to flow into a growing army of storage tanks around the country. But oil men were still driven to produce to the maximum. The effects were devastating. Flush production— "too many straws in a tub"—damaged reservoirs, reducing the ultimate recoverable resource. And the huge oversupply of crude totally disrupted the market and rational planning, thus creating sudden price collapses.[10]

Yet, ironically, as discovery followed discovery, adding further to the unprecedented glut, opinion in the industry began to shift toward Henry Doherty's remedy for shortage—conservation and production control. The reason was no longer to forestall an imminent shortage, since the mounting proof of the opposite was now all too evident. Rather, it was to prevent the ruinous floods of flush production that so violently shook the pricing structure.

But who would control production? Was it to be done voluntarily or under the government's aegis? By the Federal government or by the states? Even within individual companies there were sharp debates. A major split developed within Jersey Standard, with Teagle in favor of voluntary control, while Farish, the head of the Humble subsidiary, concluded that the government had to be involved. "The industry is powerless to help itself," Farish wrote to Teagle in 1927. "We must have government help, permission to do things we cannot do today, and perhaps government prohibition of those things (such as waste of gas) that we are doing today." When Teagle suggested that "practical men" from the industry should develop a program of voluntary self-regulation, Farish replied sharply, "There is no one in the industry today who has sense enough or knows enough about it to work out this plan." He added, "I have come to the conclusion that there are more individual fools in the petroleum industry than in any other business."

The smaller independent producers were opposed to any form of government regulation. "No state corporation commission will tell me how to run my business," an independent oil man, Tom Slick, thundered to a group of cheering producers in Oklahoma. Dissatisfied with the API, the small producers formed their own organization, the Independent Petroleum Association of America, and launched a campaign for a quite different form of government intervention, a tariff on imported oil. The main objective was to exclude the Venezuelan oil that the majors were importing. The independents tried to get an oil tariff added to the Smoot-Hawley Act in 1930, but although this infamous piece of legislation raised the tariff rates for just about everything else, it did not do so for oil. Representatives from the East Coast, and influential groups like the American Automobile Association, did not want higher fuel oil and gasoline prices and opposed the tariff. Moreover, the independent oil men alienated potential supporters by inept and altogether unsubtle lobbying. In the words of one of their backers in the Senate, they were "rather foolish in the writing of telegrams and letters." Meanwhile, the question of production controls remained unresolved and bitterly debated, and the tide of oil continued to rise.[11]

Emerging Competition

The oil industry had confronted chronic imbalances of supply and demand since its very first days in the hills of western Pennsylvania, and it had responded with a drive toward consolidation and integration to assure and regulate supplies, gain access to markets, stabilize prices, and protect and expand profits. Consolidation had meant the acquisition of competitors and complementary companies. Integration meant the yoking together of some or all segments of the industry, upstream and downstream, from exploration and production at the wellhead to refining and retail sales. The great Standard Oil Trust had skillfully managed to integrate in both directions, only to be attacked and dissolved by the Supreme Court. But in the uncertain supply and demand climate of the 1920s, the same old strategies among the once-cozy Standard Oil successor companies, as well as among other companies, reemerged, turning them into vigorous competitors. There was also a new dimension in the competition. Oil companies were becoming marketers, for the first time selling automotive fuel at retail, directly to motorists, at the brand-name stations that were springing up all across the American landscape. Oil wars were not only being fought for supply and markets in foreign lands, but were also erupting in an equally fierce struggle for markets on the main streets of America. And, in its efforts to court consumers, as well as in its inherent propensity toward consolidation and integration, the American oil industry began to take on its modern and familiar outline.

The 1911 dissolution had left Standard Oil of New Jersey a huge refining company with virtually no oil of its own, making it highly dependent on other companies and thus vulnerable to the whims of suppliers and of the marketplace. As part of his central strategic objective of expanding Standard Oil of New Jersey's secure crude sources, Walter Teagle sought domestic as well as foreign supplies. As early as 1919, Jersey had bought just over half of Humble Oil, a leading Texas producer that badly needed capital. Humble quickly put Jersey's money to good use; by 1921 it was the largest producer in the state of

Texas, substantially contributing to Teagle's goal of assuring access to crude. Standard of Indiana, which had also begun as a refiner, moved aggressively to assure itself its own crude supply, from both the Southwest and Wyoming, and thus protect the investment in its refining system. It also purchased Pan American Petroleum, which was one of the leading American companies in Mexico. Meanwhile, major crude producers were going downstream to assure themselves of markets. The Ohio Oil Company, later Marathon, had been the largest of Standard Oil's producing companies before the 1911 dissolution. Now it began moving into refining and marketing through acquisition, and did so just in time. Between 1926 and 1930, the company's production almost doubled; it eventually controlled, among other things, half of the immensely prolific Yates field in Texas. And it needed direct access to markets.

The Phillips Petroleum Company was created by Frank Phillips, an exbarber and ex–bond salesman who had developed considerable flair in putting together oil deals. Perhaps because he was also a banker, he was particularly adept at overcoming skepticism among investors and thus, at raising money in New York, Chicago, and other major cities. Put off by the boom or bust of oil, he was about to desert the business to start a network of banks through the Midwest, when the entry of the United States into World War I pushed up oil prices and drew him back into petroleum. By the mid-1920s, Phillips and his brother had built the company into one of the major independents, in the same league as Gulf and the Texas Company.

In November 1927, to accommodate a growing surplus of oil, Phillips opened its first refinery in the Texas Panhandle, and in the same month, its first service station in Wichita, Kansas. To start things off in Wichita, company officials planned to offer any purchaser a coupon for ten free gallons of gas. But they had to get Frank Phillips's permission first. "Sure, go ahead," Phillips replied. "It isn't worth as much as water anyway. Give 'em all you want to." The company moved into refining and marketing at an even more dizzying pace than its growth as a crude producer. By 1930, within three years of opening its first station, Phillips had either built or acquired 6,750 retail outlets in twelve states.

Competitive pressures pushed other companies to follow suit and break through the wholesale wall into the retail trade by acquiring their own gas stations and additional marketing facilities. They had built refineries to handle the new crude supplies; now they had to be sure they would have markets and direct outlets to consumers. Between 1926 and 1928, Gulf expanded its retail operations rapidly into the North Central states. Two of the most aggressive firms, the Texas Company and Shell, were both marketing in all forty-eight states by the end of the 1920s. Moreover, established retailers had to expand into new areas to try to protect profitability as new competitors invaded their established territories.[12]

These invasions finished the work of the Supreme Court. A kind of shadow Standard Oil Trust had persisted for a decade after the 1911 dissolution. The various successor companies to the Trust had remained tied together by contracts, habits, personal relationships, old loyalties, and common interests, as well as by shared dominant stockholders. Given the historical associations of these companies and the common effort of World War I, in which they all worked amicably together, that was not surprising. Each of the successor refining companies—such as Jersey, Standard Oil of New York and of Indiana, and Atlantic—had been based in a specific geographical region. And for a decade or so they respected one another's borders, more or less.

But in the 1920s, they began invading one another's territories and challenging one another's businesses. Atlantic Refining entered the established markets of both Standard of New Jersey and of New York—in the words of its 1924 annual report, "as a matter of protection, rather than of desire." Jersey and other Eastern successor companies got into a bitter and highly publicized price war with several of the Midwestern successors, including Standard of Indiana. When this happened, no less critical an authority than Ida Minerva Tarbell wrote with astonishment: "It certainly looks very much as if the Standard Oil Company might be crumbling—crumbling within; as if something had happened to it which the great dissolution suit had not been able to bring about. The parent company making a price for oil and its strong young relative of the West refusing to follow is something that has not happened in forty years." To those, she said, "who have watched the course of this extraordinary concern from its rise," this new development "is almost unbelievable."

Though many politicians continued to attack the "Standard Oil Group," the concept of total control was increasingly obsolete by the mid-1920s. Rather, the successor companies were turning themselves into large, fully integrated companies that, along with several so-called "independents," like the Texas Company and Gulf, were coming to dominate the industry. Instead of one giant, there were many very big companies. A 1927 study by the Federal Trade Commission found that "the separated Standard companies" controlled 45 percent of the output of refined products, compared to 80 percent control of refined products by the Standard Oil Company two decades earlier. The cozy relationship among the successor Standard Oil companies had dissipated. "There is no longer unity of control of these companies through community of interest," the FTC study found. On the critical and never-ending question of control of prices, the FTC was skeptical that the Standard Oil companies were in a position to manipulate prices in any lasting way: "The price movements for the longer periods are substantially controlled by supply and demand conditions. . . . No recent evidence was found of any understanding, agreement or manipulation among large oil companies to raise or depress prices of refined products."[13]

"Those Sunkist Sons of Bitches"

The breakup of the Standard Oil Trust into a multitude of newly aggressive companies greatly intensified competition in the game. Adding to the heat was the appearance of many new companies based variously on crude oil discoveries or the expansion of gasoline refining and marketing. These developments, combined with the thrust toward integration, spurred a powerful wave of mergers. Rockefeller's impulse toward acquisition and consolidation lived on, in an effort not to exert total control—that was no longer possible—but to protect and improve competitive position. Standard of New York, for instance, bought a major

California producer and refiner, and later merged with the Vacuum Oil Company to form Socony-Vacuum and develop the brand name Mobil. Standard of California acquired one of the other major California producers.

Shell grew rapidly in these years, in part through an aggressive campaign of acquisition. But it continued to abide by a policy of involving American investors as well, reflecting a dictum that Deterding had laid down in 1916. "It is, of course, always galling (apart from political considerations) in any country to see an enterprise doing well without local people being interested," he had written. "It is contrary to human nature, however well a concern like that may be directed, or however much it may have the interest of the people at heart, not to anticipate there will be a kind of jealous feeling against such a company." But even the cynical Deterding, the merchant at heart, found himself put off by some aspects of the merger-and-acquisition business in the United States. What particularly aroused him were the doings of American investment bankers. "Of all the grasping individuals I have ever met," he wrote to the president of one of Shell's American subsidiaries, "the American bankers . . . absolutely take the cake."

No less notable were the mergers that *almost* happened. In 1924, Shell came close to buying a production company called Belridge, well-situated on a prolific field of the same name near Bakersfield, California. The price was to be $8 million, but Shell decided it was too high and passed on the deal. Fifty-five years later, in 1979, Shell finally got around to buying Belridge—for $3.6 billion. In the early 1920s, Shell also found itself embroiled in exactly the "kind of jealous feeling" that Deterding had cautioned against. Through an acquisition, Shell came to own a quarter of Union Oil of California, and gaining full control would have made the company very strong indeed in the United States. But the California stockholders of Union Oil rose up in righteous indignation, invoking patriotism against "parties foreign to California and entirely unknown to us." They managed to embroil the United States Senate, the Federal Trade Commission, and various Cabinet officers, warning one and all that the deal was "viciously inimical to the interests" of the United States. They eventually forced Shell to sell off its holdings in Union, though Shell's disappointment was somewhat mitigated by the fact that it made a 50 percent return on what had turned out to be a two-year investment.

The Texas Company and Phillips came close to merging. So did Gulf and Standard Oil of Indiana. And, between 1929 and 1933, Standard Oil of New Jersey and Standard of California devoted much managerial time to negotiating the terms of a merger. To keep the conversations secret and off "the wires," Walter Teagle traveled to one rendezvous at Lake Tahoe in a private railroad car under an assumed name. But talks ultimately collapsed, partly because of the tough negotiating stance of Standard of California's president, Kenneth Kingsbury, and his associates—"King Rex" and "those sunkist sons of bitches," as they were known to the Jersey people. Personalities aside, a more important reason for the failure of the merger was Jersey's accounting system, which—to Walter Teagle's great anger and chagrin—could not satisfactorily establish either Jersey's book value or its true profitability.[14]

One thing did unite virtually the whole industry: Though scientific understanding of oil production had advanced by the end of the 1920s, opposition to direct regulation by the federal government was overwhelming. The tycoon Henry Doherty, outraged that the bulk of the oil industry denounced his incessant calls for regulation, predicted, "The oil industry is in for a long period of trouble . . . I do not know how long it will take, but I will stake the last shred of my reputation that the day will come when every oil man will wish we had sought Federal legislation." But Doherty was sick of the debate; his own health had broken from the strain of the long battle. He decided that he had suffered enough abuse, and from then on, he would try to leave it to others. "If a man has ever gotten a dirtier deal from an Industry than I have gotten from the Oil Industry, I would certainly like to meet him," he wrote in 1929. "I often wish to God I had never gone into the oil business and more often I wish that I had never tried to bring about reforms in the oil business."

No one paid much attention to his prophecy of future difficulties. For, as the decade ended, the new corporate giants were preoccupied with sorting out their competitive positions, and the prospects both for stabilization and for adjustment in the supply-demand balance looked reasonable without government intervention. But then everything fell apart. The fevered stock market took an unprecedented plunge in October 1929, heralding the Great Depression, which would mean unemployment, poverty, and hardship throughout the nation—and an end to the growth in demand for oil. And then, in the autumn of 1930, just as the nation was coming to the reluctant conclusion that the stock market collapse was no mere "correction" but rather portended a general economic disaster, a throw of the dice led to the discovery of the largest oil field ever found in the forty-eight states—the Black Giant—one that, by itself, could have met a very substantial part of the entire American demand. And, with that, Henry Doherty would turn out to have been right on the mark.[15]

CHAPTER 12

"The Fight for New Production"

THE EQUATION—oil equals power—had already been proven on the battle-fields of World War I, and from that conflict emerged a new era in relations between oil companies and nation-states. These relations were, of course, fueled by the volatile dynamics of supply and demand: who had the oil, who wanted it, and how much was it worth. Yet now more than the economics of the market-place had to be factored into the equation. If oil was power, it was also a symbol of sovereignty. That inevitably meant a collision between the objectives of oil companies and the interests of nation-states, a clash that was to become a lasting characteristic of international politics.

Mexico's Golden Lane

In the early years of the twentieth century, exploration in the Western Hemisphere outside the United States, was centered, above all, in Mexico. The two dominating companies were Pan American Petroleum, led by Edward L. Doheny, who later became embroiled in Teapot Dome, and Mexican Eagle, led by the Englishman Sir Weetman Pearson, later to be Lord Cowdray. Doheny, already a successful oil man in California, had first gone to Mexico in 1900 to scout oil territories at the invitation of the head of the Mexican State Railways, who, owing to the scarcity of fuel wood, was anxious to see oil developed somewhere along his line.

Pearson's interests were much more far-reaching; he was one of the greatest of the great nineteenth-century engineer contractors. Talented and highly innovative technologically, he was also a daring entrepreneur. He seemed born to engineering, for he was gifted in mathematics and was slow, steady, meticulous, and persevering in character. The round and unprepossessing Pearson also had the gifts of the natural commander. He turned down places at both Cambridge and Oxford in favor of the family engineering business, based in Yorkshire. His early years of hard, dirty work left him with a lifelong preoccupation with scrubbed hands and clean fingernails. It was all part of his never-ending attention to the details of work.

The "Pearson touch"—his knack for success on a grand scale—was much admired. But he had few illusions about how it worked. To his daughter, he wrote: "Dame Fortune is very elusive; the only way is to sketch a fortune which you think you can realize and then go for it baldheaded." To his son, he added, "Do not hesitate for one second to be in opposition to your colleagues or in overriding their decisions. *No business can be a permanent success unless its head is an autocrat*—of course the more disguised by the silken glove the better." He proved his own adages again and again. He was responsible for several of the engineering marvels of the late nineteenth century, including Blackwall Tunnel under the River Thames; the four tunnels under the East River in New York, built for the Pennsylvania Railroad; and Dover Harbor. Eventually, the empire he established would include everything from the *Financial Times*, the *Economist*, and Penguin Books to the investment bank of Lazard's in London, as well as an oil-service company. But Mexico would provide the basis for the greater part of his fortune.

The lure of the "Pearson touch" was such that President Porfirio Díaz, Mexico's dictator, had invited him to Mexico to undertake the first of several major projects—the Grand Canal that drained Mexico City, to be followed by the Vera Cruz Harbor and the Tehuantepec Railway that connected the Atlantic and the Pacific. From the moment he arrived in Mexico to begin doing business, Pearson worked hard to ingratiate himself with the Mexicans, and in particular with Díaz and those around him, with everything from favors and presents, including fancy European objects d'art, to one hundred thousand pounds to found a hospital bearing his name. He seemed always willing to make concessions to Mexican sensitivities in ways that Americans would not. His English connections also impressed the Mexicans; in Parliament, where Pearson sat for several years, he was known as the "Member for Mexico." But Pearson also owed his position in Mexico to Díaz's cold political calculation. "Poor Mexico," the dictator was supposed to have once remarked, "so far from God and so close to the United States." Díaz and the politicians around him could not permit Americans to dominate their economy completely; thus, Díaz had good reason to invite a world-famous engineer from a distant country to undertake major engineering projects and then to give him every opportunity to expand his activities in Mexico.

In 1901, on a trip to Mexico, Pearson missed a rail connection in the Texas border town of Laredo. Forced to spend the night, he discovered that the town was, as he put it, "wild with the oil craze" that had spread across the state from the Spindletop discovery three months earlier. Recalling the report of an employee about seepages in Mexico, he examined every oil prospectus he could find on short notice in Laredo, and then cabled his manager to "move sharply" to acquire prospective oil lands. "And be sure that we are dealing with principals," he ordered. Oil, he reasoned, would be a good fuel for his new Tehuantepec Railway. All of this was accomplished in the course of a nine-hour stopover. Pearson's Mexican oil venture was launched. He expanded his area of exploration to include Tabasco, and hired none other than Captain Anthony Lucas, who had brought in the well at Spindletop, to assist in Mexico.

Large expenditures and intense commitment followed. Yet, after almost a full decade, Pearson's Mexican Eagle had little to show in terms of production. "I entered lightly on this enterprise," a chastened and depressed Pearson wrote to his son in 1908, "not realizing its many problems, but only feeling that oil meant a fortune and that hard work and application would bring satisfactory results." To his wife, he was even more plaintive. "I cannot help but think what a craven adventurer I am compared to men of old," he wrote her. "I am slothful and horribly afraid of two things—first that my pride in my judgment and administration should be scattered to the winds and secondly that I should have to begin life again. These fears make me a coward at times. I know that if my oil venture had to fizzle out entirely that there is enough left for me to live quietly. . . . Yet until it is a proved success I continue nervous & sometimes despondent."

Finally, in 1909, acknowledging that his own knowledge of the oil business was "superficial," he fired the English consulting geologists whom he had been using, the famous Sir Thomas Boverton Redwood and his firm, and instead hired Americans formerly associated with the U.S. Geological Survey. They proved their mettle, for in 1910 Pearson, now known as Lord Cowdray, made major strikes, beginning with the fabulous Potrero del Llano 4, which flowed at 110,000 barrels per day and was considered the biggest oil well in the world. These discoveries ignited a boom in Mexico; they also, virtually overnight, made Mexican Eagle one of the world's leading oil companies. Production was centered along the "Golden Lane," not far from Tampico, along which seventy- to one-hundred-thousand-barrel wells were soon not uncommon.

Mexico quickly became a major force in the world oil market. The quality of its crudes was such that they were mainly refined into fuel oil, which competed directly with coal for industrial, railway, and shipping markets. By 1913, Mexican oil was even being used on Russian railroads. During World War I, Mexico became a critical source for the United States, and by 1920, it was meeting 20 percent of domestic American demand. By 1921, Mexico had, with rapidity, achieved an astonishing position: It was the second-largest oil producer in the world, with an annual output of 193 million barrels.[1]

Yet, by then, the political environment of Mexico had changed dramatically. In 1911, the eighty-one-year-old President Díaz—distracted, some were to say, by a toothache that had gone septic—was overthrown, inaugurating the Mexican Revolution. The subsequent and continuing violence drastically reduced foreigners' taste for investing in the country. E. J. Sadler, the head of Jersey's Mexican operations, was captured by bandits as he carried the company's payroll, beaten savagely, and left for dead. Somehow, he survived and made his way back to the camp. But, thereafter, he never carried more than twenty-five dollars in cash, always wore a cheap gold watch that could be surrendered to assailants, and had a visceral aversion to becoming more involved in Mexico. Oil camps belonging to Mexican Eagle were overrun and held for a time by rebels, and some of its employees were killed. In October 1918, the last month of World

War I, Cowdray was approached by Calouste Gulbenkian, on behalf of Henri Deterding. Royal Dutch/Shell, said Gulbenkian, would like to purchase a substantial part of the stock in Mexican Eagle and take over its management, and thus "leave Lord Cowdray with a perfect peace of mind."

Two decades of oil development in Mexico had made Cowdray not only weary but also wary of further risk. The Englishman had had enough. He had no desire, he explained to a British government official, to "carry indefinitely, and single-handed, the financial burden of this huge business." Cowdray quickly accepted the proposal that Gulbenkian had put before him; and—if not with perfect peace of mind, at least with some relief and with a great addition to his wealth—he stepped aside. He had gotten his timing just right, for Mexican Eagle hardly proved to be one of Shell's best acquisitions. Almost immediately after the sale, salt water started to intrude into the big producing wells that Shell had purchased from him. That salt water was very bad news—it meant the beginning of a decline in oil output. The same process was soon observed by other petroleum companies. The problem could have been conquered with more capital, better technology, and new exploration. But, in the midst of the revolutionary turmoil, the foreign companies were loath to step up their investment. Indeed, their days in Mexico were on the wane. For, as it turned out, more far-reaching in its impact on oil company activities than the lawlessness and physical danger of the revolution itself was the fierce struggle that developed between Mexican nationalists and revolutionaries, on one side, and foreign investors, on the other.[2]

The emerging conflict in Mexico would establish an essential and lasting line of battle between governments and oil companies that would soon become familiar around the world. In Mexico, the issue came down to two things: the stability of agreements and the question of sovereignty and ownership. To whom did the benefits of oil belong? The Mexicans wanted to reassert a dormant principle. Until 1884, resources in the country beneath the ground, in the "subsoil," had belonged first to the crown and then to the nation. The regime of Porfirio Díaz had altered that legal tradition, giving over ownership of subsoil resources to the farmers and ranchers and the other surface landowners, who, in turn, welcomed foreign capital, which eventually controlled 90 percent of all oil properties. One of the major objectives of the revolution had been the restoration of the principle of national ownership of those resources. That was achieved and enshrined in Article 27 of the Constitution of 1917. It became the center of the battle. Mexico had recaptured the oil but could not develop or market it without foreign capital, while investors had little desire to bear the risk and expense of development without secure contracts and the prospect of profits.

In addition to the nationalization of the subsoil, various other actions by successive Mexican regimes—regulations and tax hikes—fueled continuing conflicts with oil companies. Some of the oil companies, led by Edward Doheny, succeeded in whipping up strong sentiment in Washington for military intervention to protect "vital" American-owned oil reserves in Mexico. The battle was made even more complicated by the efforts of Mexico to raise revenues to pay off foreign loans on which it had defaulted. Leading American bankers were keen to see Mexico make good on its debts, for which it needed oil revenues.

And thus they took Mexico's side against the American oil companies and strongly opposed the companies' call for intervention and punitive sanctions.

The oil confrontation made relations between Mexico and the United States continually turbulent. Washington habitually withheld diplomatic recognition from the changing Mexican regimes, and more than once, the two countries seemed close to war. To the Americans, important interests and rights, including those of private property, were being attacked, and contracts and bargains were being broken. When Washington looked south toward Mexico, it saw instability, insecurity, banditry, anarchy, a dangerous threat to the flow of a strategic resource, and welshing on contracts. But when Mexico looked toward Washington and American oil companies, it saw foreign exploitation, humiliation, the violation of sovereignty, and the enormous weight, pressure, and power of "Yankee imperialism." The oil companies, for their part, felt increasingly vulnerable and endangered, which led to reduced investment and a rapid retreat in terms of activity and personnel. The effects quickly registered on output, which plummeted, and Mexico soon ceased being a world oil power.[3]

General Gómez's Venezuelan "Hacienda"

The expectations for world oil demand, the fear of shortage, the new role of oil in national power that had been proved by the war, and of course, the profits to be made—all these fueled what Royal Dutch/Shell called "the fight for new production" in its 1920 annual report. It declared, "We must not be outstripped in this struggle to obtain new territory . . . our geologists are everywhere where any chance of success exists." Venezuela was at the top of the list, and not only for Royal Dutch/Shell. The change in the political environment in Mexico was stimulating a wholesale migration by oil men to Venezuela. There, centuries before, early Spanish explorers had observed how the Indians used the oil seepages to caulk and repair their canoes. Now Venezuela offered, in contrast to Mexico, a friendly political climate. It was the handiwork of General Juan Vicente Gómez, the cruel, cunning, and avaricious dictator who, for twenty-seven years, ruled Venezuela for his personal enrichment.

Venezuela itself was an underpopulated, impoverished, agricultural nation. Ever since the country's liberation from Spain in 1829, local caudillos had governed the various regions. Of the 184 members of the legislature in the mid-1890s, at least 112 managed to claim the rank of general. Seizing power in 1908, Gómez set about centralizing power and turning the country into a personal fiefdom, his own private hacienda. Barely literate, he ruled through his cronies and family—by one count, he fathered ninety-seven illegitimate children. He installed his brother as his vice-president, a post the brother held until he was murdered by Gómez's son. Before World War I, Gómez affected a Teddy Roosevelt big game hunter garb. During the war, he was pro-German and dressed up in imitation of the Kaiser. Woodrow Wilson called him a "scoundrel," a mild epithet for a man who kept his tight grip on the country through terror and brutality. The British minister to Caracas was more blunt; he described Gómez as "an 'Absolute Monarch' in the most medieval sense of the word." Whatever the state of his literacy, Gómez knew what he wanted, which, in addition to absolute political power, was vast wealth. His poor country needed revenues if it was to develop economically, and if he was to become rich. The two objectives blended as one. Revenues meant foreign capital. Oil was Gómez's opportunity; but he shrewdly recognized that, in order to lure foreign investors, he would have to guarantee a stable political and fiscal environment.[4]

By 1913, Royal Dutch/Shell was already at work in Venezuela, in the environs of Lake Maracaibo, and minor commercial production began in 1914. In 1919, with the postwar surge of interest in Venezuela, Jersey Standard sent its scouts to look over the country, among them a geologist who decided to skip the Maracaibo Basin altogether. "Anyone who stays there a few weeks," he explained, "is almost certain to become infected with malaria or liver and intestinal disorders which are likely to become chronic." He recommended against investing in Venezuela. But a Jersey manager who had come along on the trip disagreed. To him, what counted in Venezuela, much more than malaria and the liver and intestinal disorders, was the commitment of Royal Dutch/Shell. "The fact that they have spent millions there leads us to suspect that there is considerable oil in this country," he reported. Failure to develop production in Latin America could endanger Standard Oil's ability to remain predominant in supplying Latin America.

Getting a concession on General Gómez's "hacienda" was not, however, as easy as it looked. Standard Oil's representative managed to arrange to see the general himself, rather than going through the normal host of intermediaries. The general seemed encouraging, and Standard, with some confidence, put in a bid. But, on that exact day, the same concession was also bid on by one Julio Méndez, who happened to be Gómez's son-in-law and who coincidentally won the concession—and immediately sold it to another company. Eventually, Jersey did acquire a good deal of acreage, some from other American companies and some from Julio Méndez, including 4,200 acres under Lake Maracaibo. That last was considered a big joke. A Jersey official suggested that the company also buy a boat so that, if the 4,200 underwater acres proved valueless as an oil concession, the company could go into the fishing business.

Even on dry land, the search for oil was difficult and hazardous in Venezuela. There were virtually no roads passable by auto, and very few even by ox cart. The geologists traveled by canoe or mule. The country had never even been accurately mapped; it was discovered that rivers indicated on maps did not exist or, if they did, were tributaries of entirely different systems from those depicted. Disease seemed to hit inescapably almost everyone who came into the country. "Mosquitoes were the worst and largest of any place I ever saw," recalled one of the American geologists. The geologists also had to cope with another insect that specialized in laying its eggs under human skin. Medical care was inaccessible, primitive, or nonexistent. In addition to everything else, the geologists and the drillers who followed had to contend with hostile Indian tribes. One Jersey driller was killed by an arrow as he sat on the porch of a mess hall; thereafter, all jungle growth within arrow's range was ordered cut back. As late as 1929, Shell protected the cabins of its tractors with several layers of a special cloth, dense enough to stop Indian arrows.

Gómez's desire to draw in foreign capital led his government to seek the

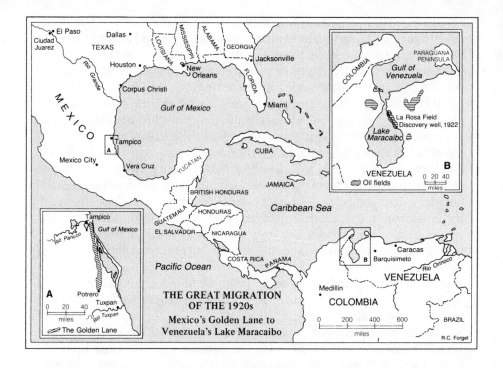

THE GREAT MIGRATION
OF THE 1920s
Mexico's Golden Lane to
Venezuela's Lake Maracaibo

help of both the American minister in Caracas and American companies in drafting what became the Petroleum Law. It set the terms for concessions, taxes, and royalties, and, at least once a concession was granted, Venezuela under Gómez provided political predictability and administrative and fiscal stability—in sharp contrast to Mexico. Yet, even as late as 1922, the year of the new Petroleum Law, there was some question whether there would be any significant oil development at all. The exploration results had been interesting but no more, while the capital and effort required were proving quite large. In 1922, some American geologists, who had already spent four years mapping the country for Shell, offered a gloomy assessment of oil prospects for Venezuela and the entire South American continent. What they saw there was "a mirage." Ten cents spent increasing production in the United States, they said, would "be more productive of profits than a dollar spent in the tropics." They even went so far as to argue that shale oil in the United States could be produced more cheaply than oil from Venezuela and elsewhere in Latin America.

Their judgment was premature. In December of that same year, Shell's Barroso well, in the La Rosa field in the Maracaibo Basin, blew out with an uncontrolled flow that was estimated at one hundred thousand barrels per day. The La Rosa field, which had not appeared particularly promising at first, had been selected and staked out by the local Shell manager, George Reynolds. He was the very same George Reynolds—the "solid British oak"—who had resolutely guided Anglo-Persian's project in Persia to its first discoveries a decade and a half earlier in the face of enormous obstacles, and then had been let go with nothing except a minuscule bonus. A decade and a half earlier, his persistence had opened up the Middle East to oil production. Now he had done the same for Venezuela.[5]

The La Rosa strike confirmed that Venezuela could be a world-class producer. The discovery inaugurated a great oil frenzy. Over a hundred groups, mostly American, but some British, were soon active in the country. They extended from the largest companies down to independent oil men like William F. Buckley, who obtained a concession to build an oil port. The oil rush provided enormous opportunity for General Gómez to enrich himself. His family and cronies, the *Gomecistas*, would, infallibly, obtain choice concessions from the government and then resell them, at considerable profit, to the various foreign companies, passing on kickbacks to the general himself. Later, to formalize such matters, the general and his friends set up a paper outfit called Compañía-Venezolana de Petróleo, but more familiarly known as "General Gómez's company." Gómez and his *Gomecistas* developed the playing-off of the various foreign suitors into a fine art. The companies had no choice—no choice, that is, if they wanted to participate in what would become the great Venezuelan oil boom of the 1920s.

Development proceeded at breakneck speed. In 1921, Venezuela produced just 1.4 million barrels. By 1929, it was producing 137 million barrels, thus making it second only to the United States in total output. That year, oil provided 76 percent of Venezuela's export earnings and half the government's revenues. The country had already become Royal Dutch/Shell's largest single source of production, and, by 1932, Venezuela was also Britain's largest single supplier, followed by Persia and then the United States. Venezuela had, within less than a decade, emphatically become an oil country. And it had won the competition for foreign capital. Investment on a large scale was required for exploration and development in Venezuela, and thus, despite the many players, the scene was really dominated by a few companies. In the 1920s, most of the production was directly or indirectly accounted for by just three—Royal Dutch/Shell, Gulf, and Pan American. The last was Edward Doheny's company, which still remained one of the dominant producers in Mexico. In 1925, Pan American was purchased by Standard of Indiana.

This scale of foreign investment in Venezuelan oil would probably not have occurred had not Gómez provided a relatively hospitable political environment. But how long would the stability last? A representative of Lago, Standard of Indiana's subsidiary, told a U.S. State Department official in 1928: "President Gómez could not live forever, and there was always the danger that a new Government, perhaps with more radical tendencies, might seek to confiscate the oil properties and follow some of the policies which have been adopted in Mexico." Thus, for safety's sake, Lago built its huge export refinery for Venezuelan oil not in Venezuela, but on Aruba, a Dutch island just off the coast. Shell did likewise on another Dutch island, Curaçao.

Unlike Shell and the other companies, Jersey had had no exploration success to speak of in Venezuela, despite large expenditures. In New York, the executive responsible for Venezuela was known as the "nonproducing production director." Finally, in 1928, applying new technology to a concession discarded by another company, Jersey made its first significant strike. The development of underwater drilling technology opened up the rich deposits under Lake Maracaibo, ultimately raising a vast amount of oil from beneath the lakebed. No one joked anymore about Jersey's going into the fishing business.

In 1932, at the bottom of the Great Depression, Standard of Indiana became worried that a proposed new American tariff on imported oil—$1.05 per barrel of gasoline, 21 cents on crude and fuel oil—would effectively shut Venezuelan oil out of the United States. Indiana did not have a foreign marketing system to which the oil could be diverted. It was also apprehensive about additional capital requirements in the midst of the Depression, as well as the possibility that its Mexican assets would be nationalized. Adding it all up, the risks looked large— too large from Indiana's point of view—and it sold off to Jersey its foreign operations, including its large Venezuelan position. Jersey paid, in part, with stock, and thus, for a time, Indiana became the largest single stockholder in Standard of New Jersey.[6]

Duel with the Bolsheviks

But it was in the Eastern Hemisphere, not the Western, that the collision between petroleum and politics was most dramatic. Before the war, Russian oil had been one of the most important elements in the global market. But now that oil was in the hands of the new communist government of the Soviet Union. How would it choose to play the game, and by whose rules?

Royal Dutch/Shell had the most at stake, owing to its purchase, just before World War I, of the Rothschilds' large oil interests in Russia. After the Bolshevik Revolution, many parties were busy trying to acquire Russian oil fields on the cheap. Gulbenkian was said to be picking up properties from emigré Russians at "bargain basement" prices. Never one to pass up a deal, he was also buying the art treasures that the cash-hungry emigrés brought out with them in their luggage.

Unlike the Rothschilds, the Nobel family had held tightly to its interests in Russian oil. But during the Revolution, the Nobels fled the country, one group disguised as peasants, another escaping by sled and foot across the border into Finland. After three-quarters of a century, their dynasty in Russia was over. They eventually made their way to Paris, where they holed up in the Hotel Meurice and tried to figure what could be salvaged of their oil empire—and how.

The answer was a fire sale. The Nobels offered Deterding their entire Russian oil operation. The country was still in chaos and civil war, and the outcome was not at all certain. Deterding grasped at once what was being offered: the opportunity to become master of Russian oil. There was just one catch—the assumption that the Bolsheviks would be defeated. He took the lead in forming a syndicate, along with Anglo-Persian and Lord Cowdray's interests, to negotiate with the Nobels. He was convinced that the Bolshevik regime could not last. "The Bolsheviks will be cleared, not only out of the Caucasus," he wrote to Gulbenkian in 1920, "but out of the whole of Russia in about six months." Still, as an insurance policy, Deterding sought a guarantee of political support from the British Foreign Office. When the Foreign Office refused, he insisted that the

Nobels retain a minority share, or that, at best, the Group buy an option until "the establishment of some settled form of Government." The Nobels wanted out, completely, and in the face of Deterding's implacability, the negotiations broke down.

But another potential suitor was waiting in the wings, and one that was, frankly, much more attractive to the Nobels, not only because of its resources but also because of its nationality, which promised to bring with it the political support of the American government. It was Standard Oil of New Jersey. Here, in these far different and more threatening times, was the opportunity to realize, at last, that long-sought alliance of American and Russian oil that the Nobels had originally tried to forge in the 1890s.[7]

Jersey, in turn, was interested. Walter Teagle and his colleagues remembered all too well the impact Russian oil had once had on the old Standard Oil Trust, frustrating its efforts to create a universal petroleum order. They knew that the Mediterranean markets could be supplied more cheaply by Russian oil than with oil exported from the United States. Russian exports had ceased during World War I, but if production were sufficiently restored and new technology applied, it could once again drive American petroleum from European markets. Better that Standard Oil have a say over Russian oil than have it in the hands of a competitor. "It seems to me that there is no other alternative but for us to accept the risk and make the investment at this time," commented Teagle. "If we do not do it now, I think we will be debarred from ever exercising any considerable influence in the Russian producing situation."

Jersey and the Nobels started intense negotiations—despite the strong possibility that the Nobels were trying to sell properties they might no longer own. That risk became more real in April 1920, when the Bolsheviks recaptured Baku and promptly nationalized the oil fields. The British engineers who worked in Baku were thrown into prison, while some of the "Nobelites" were to be put on trial as spies. Yet, so attractive was the deal if the Bolsheviks failed, and so strong the conviction that they would, that Jersey and the Nobels continued their discussions. In July 1920, less than three months after the nationalization, the deal was consummated. Standard Oil bought controlling rights to half of the Nobel oil interests in Russia at what was definitely a bargain basement price—$6.5 million down, with a commitment of up to another $7.5 million. In exchange, Standard gained control over at least one-third of Russian oil output, 40 percent of refining, and 60 percent of the internal Russian market. But, notwithstanding what the Western oil men wanted to believe, the risk was very high indeed—and all too evident. What if the new Bolshevik regime did, after all, survive? Having already nationalized the oil fields, it might operate them itself or put them on the international auction block.

In the duel between capitalists and communists that followed, the Bolsheviks were represented by the skilled and resourceful Commissar for Foreign Trade, Leonid Krasin. Tall, with chiseled features and a sharply pointed beard, he was urbane and persuasive, and seemingly reasonable, not at all the bloodthirsty fanatic that Westerners expected. He also had an eye for the ladies. He "looks, every inch of him" swooned an Englishwoman, "the highly bred and highly trained human being, a veritable aristocrat of intellect and bearing." Like no other of his comrades, Krasin understood the capitalists, for he had been one himself. Before the war, he had served as the altogether respectable manager of the Baku Electric Company and then as the Russian representative of the big German combine Siemens. At the same time, however, Krasin was also, secretly, the chief technocrat and, in Lenin's own words, "finance minister" of the Bolshevik Revolution. "I am a man who has no shadow," he liked to say. During the war, in his official role in the Czarist state, he had been one of the main architects of the Russian war economy, putting strains on his relations with his fellow revolutionaries. One dispute with his Bolshevik comrades so distressed him that he gave up meat and subsisted on a diet of mare's milk. But the Bolsheviks needed him and his managerial skills—he was the only big businessman in the Bolshevik hierarchy—and he emerged from the Revolution with a double brief, commissar not only for foreign trade but also for transport. From these posts he would cast a considerable shadow.

As Standard was wrapping up its negotiations with the Nobels, Krasin arrived in London to discuss trade relations on behalf of the Bolshevik government. On May 31, 1920, he went to 10 Downing Street at the invitation of Prime Minister David Lloyd George—a historic moment, in that it was the first time an emissary of the Soviet Union had been received by the head of government of a great Western power. His appearance stirred intense curiosity among the British, combined with repugnance. Lord Curzon, the Foreign Secretary, staring into the fireplace, hands firmly clasped behind his back, refused to shake hands with Krasin until Lloyd George sternly rebuked him: "Curzon! Be a gentleman!"

For many months thereafter, the Anglo-Soviet discussions proceeded, though with great difficulty. Lenin himself passed a secret message to Krasin in London: "That swine Lloyd George has no scruples or shame in the way he deceives; don't believe a word he says and gull him three times as much." While the negotiations dragged on, Krasin proved himself singularly adept at stimulating the appetites of British businessmen eager for trade. But his was the weaker hand. For the Soviet Union was a country headed toward economic disaster, beset as it was by woeful industrial underproduction, inflation, severe lack of capital, and a widespread food shortage that was turning into famine. It desperately needed foreign capital to develop, produce, and sell its natural resources. Toward that end, in November 1920, Moscow announced a new policy of offering concessions to foreign investors.

Then, in March 1921, Lenin went further. He announced what became known as the New Economic Policy, which provided for a much-expanded domestic market system, a return of private enterprise, and a broadening of the Soviet commitment to foreign trade and selling concessions. It was not that Lenin had experienced a change of heart; he was only responding to immediate and dire necessity. "We cannot by our own strength restore our shattered economy without equipment and technical assistance from abroad," he declared. To gain that assistance, he was ready to give extensive concessions "to the most powerful imperialist syndicates." Characteristically, his first two examples referred to oil—"a quarter of Baku, a quarter of Grozny." Petroleum could once again, as in Czarist times, become the most lucrative export commodity. A Bolshevik newspaper called it "liquid gold."

Lenin's rapprochement with the West met strong opposition from other comrades, including the ever-suspicious Stalin. The businessmen coming to the Soviet Union, Stalin warned, would include "the best spies of the world bourgeoisie," and wider contacts would lead to dangerous revelations of Russia's weaknesses. Nevertheless, a week after Lenin's announcement of the New Economic Policy, and in affirmation of it, Krasin signed the Anglo-Soviet Trade Agreement in London. He then proceeded with great dexterity to approach the various companies, dangling offers of new concessions for oil—at the same time using rumors and hints to play one company off against another.

Deterding decided that he was not disappointed to have lost the Nobel deal. Not at all. He, like the Nobels, was convinced that the entrance of Standard Oil into Russia brought a strong measure of insurance to all foreign investors, including Royal Dutch/Shell, with its ownership of the former Rothschild properties. "We have already several good seats and a very great part of the food on the Russian table," he lectured Gulbenkian. "Dining is very much better in company with other people who have also got a very big interest in the dinner." But Deterding had no intention of accepting passively the Bolsheviks' efforts to sell off what he regarded as his properties and so exclude him from the table. Nor, for that matter, did Walter Teagle.[8]

In Search of a United Front

In 1922, Jersey, Royal Dutch/Shell, and the Nobels began to fashion what became known as the Front Uni. Their aim was to create a common bloc against the Soviet threat to their Russian oil properties and their trade. Ultimately a dozen other companies joined it. All the members pledged to fight the Soviet Union together and not allow themselves to be picked off individually. They agreed to seek compensation for nationalized property and to refrain from dealing with the Russians independently of one another. To be sure, there as elsewhere, "the brethren of oil merchants" hardly trusted one another, let alone the Soviets. Thus, despite the mutual pledges and promises, the Front Uni stood on very unsteady legs from its birth. And the crafty Leonid Krasin, who well understood capitalists and their competitive instincts, continued to play the companies off against one another with a master's skill.

Meanwhile, in many markets around the world, the companies were feeling the rising pressure of competition from cheap Russian oil. The Soviet oil industry, virtually dormant from 1920 to 1923, thereafter revived quickly, helped by imports of large amounts of Western technology, and the USSR soon reentered the world market as an exporter. Within Jersey, the senior executives faced a dilemma. Should they, whatever their property claims, start buying cheaper Russian oil, or should they continue to stand back on both moral and business grounds? Teagle now regretted the investment in the Nobel enterprise. "I am," he said, "convinced that instead of sitting up with a sick child of this character and nursing it along for years, we could have taken the same amount of money and invested it elsewhere in the oil business in such a way that the investment would have immediately become productive."

Heinrich Riedemann, the head of Standard's German operations, saw things somewhat differently. Private companies, he concluded, would not easily be able to defend their rights against confiscation and nationalization. "The participation of a government in industrial and business enterprise as in Russia is new and unheard of in the history of business," he said. "None of us like the thought of helping Soviet ideas," he added. "But if others should be willing to come in, what would then have been the use if we had kept aloof?" Indeed, other Western groups were already knocking at the door, some quietly and some with great fanfare, seeking concessions across the whole of the Soviet Union—all the way from Baku, in the Caucasus, to the island of Sakhalin, off the coast of Siberia. The properties in the Caucasus were those already claimed by Jersey, Shell, and others. To make matters worse, the Soviets were selling oil from those properties as if it were their own.

There was one possible way to outmaneuver the Soviets: Jersey and Shell could form a joint organization to buy Russian oil. Teagle did not like the idea at all. "I know I am old fashioned in feeling this way," he said, "but somehow or other the idea of trying to be on friendly terms with the man who burglarizes your house or steals your property has never appealed to me as the soundest course that could be pursued toward him." Yet, once other American companies began to purchase more Russian oil and used it to compete directly with Jersey, opposition within the company to doing business with the Russians gave way. A joint Jersey-Shell buying organization was finally established in November 1924, and the two companies began exploring the options for doing business with the Soviets. Privately, Teagle was bitter about how the whole matter had been handled. It was the classic business problem of not enough time, of the day never being long enough for long-term thinking. "As I look back over what we have done during the past six or eight months, I am rather impressed with the fact that a matter so important as this Russian purchase situation should have been handled by us without really giving the subject the consideration which its importance justified," he wrote to Riedemann. "It is certainly to be regretted that we have so many things to do and our business day is so fully occupied that somehow or other we seem to make mistakes which could have been avoided if we had really spent the time necessary to think the matter through to a logical conclusion."

Cooperation with Royal Dutch/Shell was one thing, but cooperation with the Soviets remained repugnant to Teagle. In a letter to Riedemann he wrote, "Affording the Soviet a market for petroleum not only is actually becoming a receiver of stolen goods, but operates to encourage the thief to persist in his evil courses by making theft readily profitable." Riedemann tried to calm the agitated chief executive. "Man is a strange being," he replied, on Christmas Eve 1925, "and in spite of all disappointments he still starts every year with new hopes. So let us do the same."

A Jersey-Shell agreement for a joint purchase arrangement with the Soviets soon appeared imminent. It even provided that 5 percent of the purchase price would be set aside to compensate former owners. Both Teagle and Deterding remained personally skeptical of the whole undertaking. And, when the proposed agreement fell apart in early 1927, Deterding was almost gleeful. "I am so glad that nothing came of these Soviet deals," he wrote to Teagle. "I feel that everybody will regret at some time that he had anything to do with these robbers, whose only aim is the destruction of all civilization and the re-establishment of brute force."

Sentiment of a sort had apparently entered into Deterding's business calculations. After his marriage to the White Russian émigré Lydia Pavlova, he seemed to become more staunchly and outspokenly anti-communist. Deterding even telegraphed John D. Rockefeller, Jr., beseeching him to block the various Standard successor companies from buying Russian oil. The Dutchman reported that he had "begged" Rockefeller, "for humanity's sake," that "all decent people" should abstain "from helping the Soviets to get hard cash." The regime, he told Rockefeller, was "anti-Christ." Surely, he added, Rockefeller did not want his companies "to have bloodstained profits. . . . The Soviet murderous system will be soon at an end if your Companies will not support it."[9]

Price War

In spite of Deterding's entreaties, two of the other Standard Oil successor companies, Standard Oil of New York and Vacuum, were going their own way in dealing with the Soviets. Standard of New York built a kerosene plant for the Russians at Batum, which it leased back. Both companies now contracted to buy large amounts of Russian kerosene, particularly for India and other Asian markets. Socony needed Russian oil to supply its markets in India. Shell had other sources from which to substitute in India; Socony did not.

Deterding went into an awful rage. He denounced Socony's president, C. F. Meyer, as "a man who has neither honor nor intelligence"; and in 1927, in retaliation for what he saw as Socony's treachery, he launched a brutal price war in India, which he soon carried to other markets around the world. Socony counterattacked, cutting prices in still other markets. Deterding also orchestrated a press campaign against Standard of New York for buying "communist" oil. Since the distinctions among the successor companies to the Standard Oil Trust were not very clear—not only to the public, but also in the mind of Deterding, who was too suspicious to believe in them anyway—Standard Oil of New Jersey was dragged into the fray. To the intense discomfort of Walter Teagle, Jersey was also accused of buying "communist" oil. That had been Deterding's intention all along. "We are facing now such enormous events that the Jersey Cy. should settle these things herself," Deterding wrote menacingly to a Jersey executive. He left no doubt that he expected Jersey to bring Socony sharply to heel. Jersey "after all is the biggest American oil Cy. and anyhow is the one with the biggest future," he said, "and the minor New York Cy. should-be made to understand that she is the servant of the Jersey Cy., and not her boss." As Deterding had intended, Jersey was driven to criticize publicly the other two companies on their Russian purchases. Teagle was able to find some solace in the affair. "The thinking people in Europe, for the first time," he said, "now realize that there is a real and genuine difference between one Standard Oil Company and another."

Jersey's executives suspected Deterding had withdrawn from the joint purchase agreement with the Soviets because of pressure from the British government. But, with the price war in full swing, a senior British official assured the

Americans that such was not the case. "Sir Henri Deterding always got himself into hot water because of his tactlessness," wrote the official. When the Russians had questioned the compensation arrangement in the proposed joint purchase venture of Shell and Jersey, "Sir Henri completely lost his head and told them that he would prevent any one from buying Soviet oil. . . . This was an utterly stupid and at the same time thoroughly characteristic thing for him to do. . . . It was obvious that other purchasers could not be kept away and that it is partly rage at his own futility which made Sir Henri come out with his blast on the Standard Oil of New York and the statement of his determination to undersell."

But at a dinner party in The Hague, at the home of one of his Dutch directors, Deterding offered his own version of events. "After years of comparative peace," he declared, "we found ourselves suddenly attacked in Burma, where the Standard Oil Company of New York began to import Soviet kerosene. Considering our best defence in this instance to be a good offence, I immediately accepted the challenge and ever since we have been thrusting and parrying in our efforts to find weak spots in the other's armour. I believe that, for the moment at least, the position of the Royal Dutch with regards to Soviet oil is once again clearly understood."

The two errant American companies, Vacuum and Standard of New York, would not concur. Deterding's aim, the president of Vacuum was convinced, was really to cut his company off from inexpensive oil for its export systems, which happened to compete directly with Royal Dutch/Shell, and then to win "a monopoly supply of Russian petroleum available for export." And when Jersey accused the two companies of betraying American principles, Vacuum's president observed that American businessmen and farmers were busily selling cotton and other products to Russia. "Is it more unrighteous," he asked, "to buy from Russia than to sell to it?" That would be a long-persisting question.

By the late 1920s, the major companies were weary of the whole matter of Russian oil. The effort either to regain their properties or to recoup their investment had become a lost cause. Moreover, the gusher at Baba Gurgur in Iraq turned their attention to the new sources of supply in the Middle East. The Jersey board decided to adopt a neutral stance—neither to seek a contract with the Soviets nor to participate in a boycott. Riedemann summed up the matter in the autumn of 1927. "Personally," he said, "I have buried Russia."

If so, it would prove to be a lively corpse, as growing volumes of Soviet oil entered a sated world market. The vicious and zealous price war that Deterding had instigated, in India and elsewhere, was aimed at this Russian petroleum, but it was to have far larger consequences for all who played in the game of international oil.[10]

CHAPTER 13

The Flood

HIS NAME WAS Columbus Joiner, though afterward he would be known as Dad Joiner, because he was the "daddy" of what happened. He was seventy years old in 1930 and walked bent forward from the waist, as though looking for something on the sidewalk—the result of rheumatic fever. He was virtually a caricature of the classic down-on-his-luck, woebegone but always optimistic, silver-tongued and ever-persuasive wildcat promoter. He had a silky-smooth complexion, quite unusual for a man his age, which he attributed to eating carrots. His formal schooling had totaled just seven weeks, but he had been tutored at home on the family farm in Alabama, taught to read with only the Bible as his text, and had learned to write by copying out the Book of Genesis. He had absorbed the language of the King James version, and he knew how to spin an enticing web out of promised wealth. When the need arose, he could also write lush and tender love letters to widows, whom he had first learned about when their names appeared in the newspaper obituaries of their well-to-do husbands. His interest, to be sure, was not in their lonely hearts; it was in their purses.

Joiner was only one of the many bit players in the big oil promotion of the 1920s. Oil stocks and oil deals were madly enticing in the fevered speculative climate of the decade, and they were virtually irresistible to anyone with a taste for gambling. "How would you like to get started by investing $100 in an oil company and have it later worth over $50,000?" one promoter asked the graduating Yale class of 1923. "Your chance to ride with us and be with us as we accumulate a momentum of success is now." Some promoters did their selling face to face, even taking would-be investors on tours of oil fields, where they were filled with "cold lunch and hot air." Others, finding promotion by mail more convenient, would send out letters filled with all sorts of wild pitches and promises, and in return would receive mailbags filled with cash, money orders, and checks—no questions asked. One promoter, Dr. Frederick Cook, who claimed, among other things, to have beaten Admiral Peary to the North Pole, was mailing up to three hundred thousand letters a month, from which he made, in one year, some two million dollars—before he was apprehended by Federal authorities. In sheer audacity, few could have matched the General Lee Development Company. Two promoters discovered a certain Robert A. Lee, a descendant of General Robert E. Lee, and prevailed upon him to tell investors around the country, "I would rather lead you and a thousand others to financial independence than to have won Fredericksburg or Chancellorsville."

By comparison, Dad Joiner was strictly a small-time operator. But he did have the saving grace that he actually wanted to wildcat for oil rather than simply separate the gullible from their dollars. He worked out of Dallas, hanging around with the other promoters in the lobby of the Adolphus Hotel, a baroque landmark built by the Busch beer family of St. Louis. Eventually Joiner's eyes fell on East Texas, a drought-ridden, dirt-poor region of rolling hills, piney woods, and sandy soil that had never come out of the agricultural depression at the end of the First World War. Neither of the two main towns in the area, Overton and Henderson, could claim even one paved road. To the careworn inhabitants of that area, Dad Joiner held out a great and hopeful vision—that, under their scarred and barren soil, lay an ocean of oil, a "treasure trove all the kings of earth might covet."

Most of the geologists who knew anything of his East Texas scheme scoffed at Joiner; and when they were not scoffing, they were laughing. There was no oil in East Texas. But Joiner was convinced there was—or had allowed himself to become convinced—by "Doc Lloyd," a mysterious, self-educated, self-proclaimed geologist, who weighed over three hundred pounds and favored sombreros and riding boots. Some said Doc Lloyd was also a veterinarian; others said that he was a pharmacist, and that he had run "Dr. Alonzo Durham's Great Medicine Show" across the country, selling patent medicines concocted from oil. Lloyd was not his real name. The reason he had changed it became clear later, when his photograph appeared in newspapers across the nation. Thereupon, it was said, quite a number of women, some with children in tow, boarded the train for East Texas from points around the country, seeking to catch up with their missing husband.

Doc Lloyd had provided Dad Joiner with a description of the geology of the East Texas region. To say it was misleading would be an understatement; it was totally incorrect, fabricated. Lloyd was what was called a "trendologist"; he drew up a map of the major oil fields of the United States, showing trend lines from all of them intersecting in East Texas. But Doc Lloyd did one memorable thing; he told Joiner exactly where to drill, when almost everybody thought the idea was completely ridiculous.

Joiner mailed out a prospectus, which included Doc Lloyd's fictitious description of the geology of East Texas, to the sucker list he kept. And somehow he scraped together the money to begin drilling on the farm of one Daisy Bradford in Rusk County. To keep going, he had to call on every ounce of his considerable persuasiveness, especially with women. "Every woman has a certain place on her neck, and when I touch it they automatically start writing me a check," the old wildcatter once said. "I may be the only man on earth who knows just how to locate that spot." Then he grinned. "Of course, the checks are not always good." He was boasting. Such money as he could raise barely dribbled in.

To the oil industry at large, of course, Dad Joiner was virtually invisible, just another of the thousands of threadbare promoters, each with an idea, the promise of riches, and the gift of gab. For three years, beginning in 1927, while the industry's leaders carried on their furious debate about shortage and glut and regulation, Joiner—the poorest of the poor boys—and his motley crew were drilling amid the dense pine trees of East Texas with rusted, third-hand equipment, constantly tormented by breakdowns and accidents, always short of even the barest cash. He paid his workers partly in "royalty rights" to various bits of acreage. When he had no cash at all, they would return to their farms or do odd jobs, but eventually they would drift back. Dad Joiner issued so many "certificates" against the possibilities of discovery, sold at a deep discount, that they became a local currency in the area. A geologist from Texaco came by and uttered the by-then-time-honored taunt, "I'll drink every barrel of oil you get out of that hole." But despite the constant discouragement, Joiner and his little band of workers and supporters managed to keep the faith.

The power of faith would soon prove itself. Dad Joiner's luck began to turn in early September 1930, when a well, Daisy Bradford Number 3, tested positively. "It's not an oil well yet," Joiner protested to some who were watching, but he did not protest all that vigorously. Word circulated. On the road to the well, a shantytown sprang up overnight where the hopeful could gather to wait. It was called Joinerville in homage to the would-be prophet. Thousands of people arrived to become part of the vigil. Expectation, as if indeed of a religious event, a promised miracle, hung in the air. Something would happen, people were sure, and they wanted to be there to see it. During those early Depression days, hamburgers normally sold for sixteen or seventeen cents, but in Joinerville, they cost a quarter each. That was only a faint harbinger of what was to come.

A month later, at eight in the evening on October 3, 1930, a gurgling could suddenly be heard from the well. The man in charge of the drilling spun around to the assembled crowd and shouted, "Put out the fires! Put out your cigarettes! *Quick!*" The earth trembled. A column of oil and water shot high above the derrick. And the crowd went mad. Men looked up into the sky, shouting and cheering as the oil sprayed down on them. It *was* a miracle. Dad Joiner *was* a prophet. A crewman became so excited that he pulled a pistol from his pocket and started firing into the oil spray in the sky. Three men quickly jumped him and wrestled the gun away. One spark could have ignited the volatile escaping gas, causing the well to explode, killing everybody on the spot.[1]

The Black Giant

"Joiner's Wildcat a Gusher," was the headline the next morning in the *Henderson Daily News*. But the first reaction among the industry leaders to news of Dad Joiner's bonanza was either skepticism or outright disbelief. That changed to wonderment and frenzy when, over the next three months, two other wells in the area, also drilled as wildcats, blew in. Ultimately the East Texas reservoir proved to be forty-five miles long, and five to ten miles wide, 140,000 acres altogether. The field became known as the Black Giant. Nothing to compare with it had ever before been discovered in America. And the boom that followed made all the others—in Pennsylvania, at Spindletop, elsewhere in Texas, at Cushing, at Greater Seminole and Oklahoma City, at Signal Hill in California—look like dress rehearsals. In early 1931, as the rest of the country was gripped by the gloom of the Great Depression, East Texas was exuberant, and it was going crazy. People poured in from everywhere; they crowded together in tent cities and shantytowns, and this fundamentalist region of stern values and abstinence suddenly became home to a host of honky-tonk districts that catered to every kind of vice. By the end of April 1931, six months after Dad Joiner's Daisy Bradford Number 3 came in, the area was producing 340,000 barrels per day, and a new well was being spudded every hour.

In the wake of such a sudden, vast new supply, the inevitable happened: Prices fell and then fell more. They had been as high as $1.85 in Texas in 1926. In 1930, they averaged about a dollar a barrel. By the end of May 1931, the price was as low as fifteen cents a barrel, and some oil was being sold at six cents a barrel. There was even oil, deeply distressed, that went for two cents. Still the orgy of drilling continued. By the first week of June 1931, one thousand wells had been completed, and East Texas was producing five hundred thousand barrels per day.

People with a quick eye for profit rushed in hastily to build dozens and dozens of cheap, pint-sized refineries called "teakettles" that produced a volatile "Eastex gasoline." Small filling stations, in turn, sprang up to sell "Eastex" at discount prices. With so much supply, everyone had to battle for markets, and stations selling "Eastex" were driven to offer, with each fill-up, such premiums as a crate of tomatoes or a free chicken dinner.

Alas, Dad Joiner could not completely give himself over to exultation. To be sure, the discovery at Daisy Bradford Number 3 and the subsequent development of the Black Giant were glorious vindication. But he had been very cavalier in his promotion, to put it generously. He had sold more "interests" than there were interests to sell. Some leases had been sold several times over, and at least one lease eleven separate times. Legally, he was very vulnerable, and he knew it.

A local newspaper sprang to the defense of the man who had caused East Texas to be reborn. "Is he," the editor wrote of the prophet of East Texas, "to be the second Moses to be led to the Promised Land, permitted to gaze upon its 'milk and honey,' and then denied the privilege of entering by a crowd of slick lawyers who sat back in palatial offices cooling their heels and waiting while old 'Dad' worked in the slime, muck and mire of slush-pits and sweated blood over his antiquated rig, down in the pines . . . ? " It did indeed look as if Joiner could lose it all. Of his five thousand acres of leases, he had clear title to just two.

Salvation was to appear, however, in the person of a stout man who wore a straw boater and a string tie. His name was Haroldson Lafayette Hunt, always called "Boy" by Joiner, but more generally known as H. L. "Boy" was a failed cotton farmer who had already demonstrated two formidable and not unrelated talents: one for gambling and the other, like Rockefeller and Deterding, for swift and complex mental arithmetic. A decade earlier, he had opened a gambling hall in the oil boom town of El Dorado, Arkansas. When the Ku Klux Klan threatened to burn it down, Hunt had prudently switched to oil, and had done rather well in both Arkansas and Louisiana. He was at this time, for all practical purposes, already maintaining two wives, each with a growing family. Hearing rumors about Joiner's well before it blew in, "Boy" had shown up to observe progress and had befriended the beset old wildcatter.

Hunt stepped forward when the tidal wave of woes fell upon Joiner after his first discovery, but before other wells had begun to indicate the true magnitude of the oil field. Staked by the owner of a men's clothing store back in El Dorado, Hunt holed up with Joiner for an interminable negotiating session in Room 1553 at the Baker Hotel and hammered away at him, trying to make a deal. Unbeknownst to Joiner, Hunt was being fed secret reports on the progress of the Deep Rock well, three-quarters of a mile from Joiner's discovery. During the session, he got the all-important word that a second major discovery was at hand, proving not only that Joiner's well was no fluke but also that the field could be very large. Hunt did not share the news with Joiner, and indeed kept suggesting that Deep Rock could well be dry. After more than thirty-six nonstop hours in Room 1553, Dad Joiner succumbed. At some point between midnight and 2:00 A.M. on Thanksgiving, November 27, 1930, he signed all his rights over to "Boy." Hunt ordered up a plate of cheese and crackers to celebrate their deal.

Hunt proceeded to settle the jungle of claims against Joiner and quickly became the largest independent in East Texas. His deal with Joiner gave him what he afterward called his "flying start." He went on to make an immense fortune. Later, he achieved notoriety as a patron of right wing causes, a promoter of health foods, and an inveterate enemy of white flour and white sugar.

Altogether, Hunt paid Dad Joiner $1.33 million—$30,000 up front, the rest out of production. When Joiner later learned that Hunt had given $20,000 to the head driller on the Deep Rock well, who had secretly provided advance word to Hunt's scouts on the oil show, he angrily brought suit, charging fraud. Hunt was emphatic that he had not tricked the old man. "We had traded," he declared. Joiner suddenly thought better of his suit and withdrew the claim. He spent the money he received from Hunt on new wildcats, searching for another Black Giant, the next East Texas—as well as romancing his "secretary" and other young women. Dad Joiner was almost eighty-seven when he died, and he was wildcatting virtually to the end. But he never hit again. At his death, his net worth amounted to not much more than his car and his house.[2]

Anarchy in the Oil Field

The flood of East Texas crude soon brought prices down all over the country, and a continuation of the price collapse could have spelled ruin for even the largest producers. There was some expectation that, as with other major discoveries, the underground reservoir pressure would eventually fall because of rapid production, causing output to decline, and prices would return to "normal." But East Texas, in its magnitude, was something wholly unique. Who knew when its output might begin to decline? And who would still be in business when that day

231

came? The rush of competitive production in East Texas—and elsewhere—meant "competitive suicide" for the entire oil industry.

It was imperative to devise some system to control production and stabilize prices. And that meant bringing the East Texas field into harness, in the face of violent opposition from the local producers and royalty owners, as well as from smaller refiners, who liked cheap crude. The situation was further complicated by the fragmented nature of ownership in East Texas, and by the large share of production coming from the independents. Owing to the slow start of the majors, small producers owned or controlled a good part of the East Texas field, and they were the ones most likely to produce at breakneck speeds. For the independents, any surrender of their freedom of action was "a deadly threat" to their relative advantage over the hated big companies.

In the feud between the majors and the independents, the agent of order was to be, despite its name, the Texas Railroad Commission, which had originally been created in 1891 by Governor Jim Hogg to assert populist control over the railroads. By the beginning of the 1930s, it had become a favorite dumping ground for political patronage. It also lacked technical competence. Yet it was the agency that had been given some mandate over oil, though its authority was critically circumscribed. Its counterpart in Oklahoma, the Commerce Commission, had been empowered since 1915 to regulate oil production to match the market's demand, a power that was very explicitly denied to the Texas Railroad Commission. The Texas commission was permitted to regulate so as to prevent "physical waste" in oil production. But it was specifically forbidden by legislation, under the influence of the independents, from controlling production to stop "economic waste." That meant it was denied the right to market proration—that is, it could not cut back everybody's production to bring total output down to a level sufficient to satisfy demand.

Nevertheless, the Railroad Commission set out to do exactly that. To do so, however, it had to operate under a disguise, which was that of preventing physical waste. The commission charged that flush production would lead to potential oil output being lost forever. Specifically, it said that, if prices were too low, the numerous wells that brought up only a few barrels per day—so-called stripper wells—would not be able to produce economically, and therefore would be shut down. That counted as "physical waste." But the Federal courts repeatedly checkmated the commission's efforts to prorate cutbacks on that rationale. At one point, the commission itself was held in contempt of court. All its efforts were continually overwhelmed by the ever-rising production flowing out of East Texas.

With prices falling far below the cost of production and no cure in sight, fear and demoralization gripped the entire American petroleum industry—as Frederick Godber, a Shell director from London, discovered when he came to the United States in the late spring of 1931. Part of the purpose of Godber's mission was to be sure that the economies and cuts in Shell's American operations that had been demanded by its European headquarters were actually being implemented. Godber laid down the line in the United States: Offices were too elaborate; company cars were too many and of too high a grade. With satisfaction, he could report back to Deterding and the other directors, "Enormous economies are taking place."

In his meetings with the senior management of many of the major American companies, Godber encountered unmitigated gloom. The chairman of Standard of Indiana, Godber reported, "is very depressed, almost panicky, and quite clearly in a very nervous state of mind." Godber saw Walter Teagle of Standard Oil of New Jersey. "Even New Jersey as a unit has no very definite policy," he observed. Teagle, he added, "is very pessimistic, feels there is nothing to do but sit back and rely on lowering prices, feels there is no co-operation from most of the other companies and that this will not be possible except after they have all made large losses." In sum, Godber reported, "much of the industry's troubles are due to known causes over which individuals have little or no control and which cannot be remedied until laws are passed in the various producing states permitting enforcement of laws preventing waste and excessive drilling. . . . These laws might perhaps have been pushed before, but there is much prejudice to overcome, particularly in Texas."

Meanwhile, production was still going up in Texas and continuing to surge in neighboring Oklahoma. At the beginning of August 1931, while Federal judges were considering the constitutionality of Oklahoma's prorationing laws, the governor, "Alfalfa Bill" Murray, proclaimed a state of emergency, declared martial law, and ordered the state militia to take control of the major oil fields. He would keep them closed, he announced, until the oil "price hits one dollar."

"A dollar a barrel" became the rallying cry throughout the oil states.

By August 1931, both East Texas and the petroleum market as a whole were in total anarchy. Production in East Texas was now over a million barrels per day, equivalent to almost half of the entire American demand, and crude prices had plunged to thirteen cents a barrel. Oil from Texas was even underselling Russian petroleum in Europe. Prices at the wellhead in Texas and elsewhere in the United States were so far below production costs, which averaged around eighty cents a barrel, that they portended ruin for most oil producers in Texas and around the country. The day after producers in East Texas called for a voluntary shutdown to help boost prices, output actually went up still further. Violence was in the air; there was talk that dynamite was being brought in to blow up wells and pipelines. The Texas economy, and perhaps law and order, were on the verge of collapse.[3]

For some time, Texas Governor Ross Sterling, a founder and former chairman of Humble Oil, had been vacillating about what to do. But now he had no choice; he had to act. He, in effect, declared war on East Texas. On August 17, 1931, he announced that East Texas was in a "state of insurrection" and "open rebellion," and sent in several thousand National Guardsmen and the Texas Rangers, who showed up on horses, as the recent rains had made the roads impassable to motor vehicles. They set up their base on what was to be dubbed "Proration Hill" and, operating from horseback, shut down production within a matter of days. An eerie quiet settled in over East Texas as work in the oil fields ceased. Even the chickens, which had happily feasted on the millions of insects drawn each day by the continuous gas flares, were forced to "return to the prosaic ante-petroleum practice of scratching for worms." Ancillary activities were also brought to a halt. The commanding general of the National Guard banned the wearing of "beach pajamas," a garb favored by the busy prostitutes, and their business went into a sharp slump as well.

The oil shutdown actually worked; prices in the field rose from thirteen cents a barrel. The Texas Railroad Commission continued to issue prorationing orders, which were now enforced by state troopers. By April 1932, prices were almost back up to the magic dollar—ninety-eight cents. In the course of 1932, the Railroad Commission issued nineteen separate prorationing orders for East Texas, and each was declared invalid by the judiciary. Still, the market held firm and the stronger prices finally persuaded many independents, and the politicians who were responsive to them, of the value of the across-the-board allocation of cutbacks—prorationing. In November, Governor Sterling finally decided to give the commission the specific power it needed to counter "economic waste." He called a special session of the state legislature and rammed through a bill allowing market prorationing. The new law was facilitated by a better understanding of the dynamics in the East Texas reservoir. The production pressure came not from gas, as had often been the experience elsewhere up until then, but from water—"water drive." Rapid, chaotic production would damage the water drive and prematurely stunt overall output.

With the passage of the new law, market prorationing went into effect in Texas. Yet, despite the new powers of the Texas Railroad Commission to control output, the spring of 1933 looked to be as bad as, or even worse than, the summer of 1931. The commission had set the quota for East Texas far too high, wretchedly high, twice what was suggested by new engineering knowledge about "bottom hole" pressure. In addition, hundreds of thousands of barrels of oil were being illegally produced above the allowable quotas. This excess became known as "hot oil," a term that was first coined in the East Texas field. It was said that, one chilly night, a state militiaman was talking to an operator suspected of producing above the allowable limit. The militiaman was obviously shivering, and the thoughtful operator suggested that he lean against a tank containing some of the suspect oil. "It's hot enough," said the operator, "to keep you warm."

It was also hot enough to keep the oil industry in turmoil. The "hot oil" was being smuggled out of Texas and across borders into other states. The same was happening in Oklahoma, where prorationing was also supposed to be in effect. Altogether, between the high allowables and the hot oil, production in East Texas was again going completely out of control. The Texas Company slashed its posted price from seventy-five cents a barrel to ten cents a barrel. So vast was the flood, and so glutted the market, that some "hot oil runners" were having trouble finding markets at even two cents a barrel. To stanch the flow, several pipelines were mysteriously dynamited.

A demoralized William Farish, the president of Humble, wrote to Walter Teagle that only the shock and pain of very low prices could convince the independents that their long-term interests lay in control of production and unitization. Perhaps the point had been reached, added Farish, at which "the law of the tooth and claw" was the only recourse left to bring some order. At ten cents a barrel—and the even lower spot prices—that point had already been reached. But the oil industry also realized that it desperately needed outside help; state governments were not enough. Emergency help would have to come from another direction, from Washington. Some Texas producers urgently petitioned for Federal supervision of the Texas industry for the duration of the emergency. The alternative, they said, was not only the bankruptcy of independents but nothing less than the complete collapse of the oil industry as a whole.

And now, just in time, there was a new administration in Washington, Franklin Roosevelt's New Deal. It was activist, ready to wage war on the Depression, committed to reviving the economy, and wholeheartedly prepared to intervene everywhere. The Federal government was keenly attentive to what was happening in Texas. Oil prices were too low, and it was willing to do whatever seemed necessary to rescue them.[4]

The Reformer

Roosevelt was inaugurated on March 4, 1933. To the politically sensitive position of Secretary of the Interior, a post still tainted by memories of Albert B. Fall and the Teapot Dome scandal, he appointed Harold L. Ickes. Described at first meeting by another member of Roosevelt's Cabinet as "a plump, blond, bespectacled gentleman," Ickes was a Chicago lawyer who had been a leading figure in progressive Republican and Progressive party politics for many years. He had managed Theodore Roosevelt's Chicago campaign in 1912, and in 1932, was chairman of the Western Committee of the National Progressive League for Franklin Roosevelt. As his reward for helping Roosevelt win the Presidency, he set his heart on becoming Secretary of the Interior. He mobilized leading progressives to campaign for him and won the job. Roosevelt later explained that he liked the cut of Ickes's jib. He also wanted a progressive Republican with Western credentials. What he got in Ickes was a man of powerful liberal convictions, strong passions, stinging polemics, pervasive suspicions, oversensitivity to any slight (real or imagined), immense self-righteousness, great dedication to duty, and a deeply ingrained moral conscience.

Ickes had been raised in a poor household by a stern Calvinist mother. As a boy, he was not even permitted to whistle on Sunday—a ban that was only lifted when he brought evidence to his mother of a minister observed whistling on a Sunday. Ickes was so good a student in high school that, when the Latin teacher fell ill, he took over and taught the lessons himself. As high-school class president, he also first practiced what he would later refine into an art: the impulsive resignation on grounds of high principle, only to have the resignation rejected by the powers that be. His high-school class would not accept his resignation. Decades later, neither would Franklin Roosevelt. To one of the several resignations Ickes was to proffer to Roosevelt, the President simply replied: "You are needed. . . . Resignation *not* accepted!"

As a young attorney filled with "restless reforming energy," Ickes joined a host of campaigns in Chicago—against corruption and monopoly and social injustice; in favor of civil rights, women's trade unions, and the ten-hour day. At one point, he even became secretary of the Straphanger's League, the better to campaign for public transport. He also developed into an effective political manager—though always, it seemed, of reformers running against great odds, which led him to joke about his own "uncanny ability to pick losers!" But, finally, in 1932 he had picked his winner—Franklin Roosevelt. And as Roosevelt's Interior Secretary, though always professing a selfless devotion to principle and duty, Ickes relished the accumulation of power, and very much intended to be a "strong man" who could say "no." In addition to being Secretary of the Interior, he readily assumed the post of Oil Administrator, and also held the key New Deal position of Public Works Administrator.

Ickes threw himself into the intricate administrative details of all three positions. "With the spotty record of Interior always in mind," he later wrote, "I slaved away over endless mountains of documents, contracts, and letters, refusing to sign anything that I had not personally read, lest one day it should rise to haunt me in the steam of another Teapot." The "oil-besmeared Albert B. Fall," as Ickes described him, had finally gone to prison in 1931, but he never seemed to be far from Ickes's thoughts. After Harry Sinclair, one of Fall's two paymasters, visited him in 1933, Ickes wrote in his diary, "I kept wondering whether the ghost of Albert B. Fall, carrying a little black satchel, might not emerge from one of the gloomy corners of this office." The legacy of Teapot Dome made Ickes acutely fearful of corruption and consistently mistrustful of the oil industry. He was intent on restoring the morale and the reputation of the Interior Department. And to ensure that new financial scandals and fraud did not occur, he even set up his own internal investigating unit, which used wiretaps as a standard operating technique.

His own tenure in office, however, was almost immediately threatened by scandal of a different sort. Ickes had long been a partner in a terribly unhappy marriage, and soon after he was appointed secretary, he became romantically involved with a much younger woman. He found jobs in the Interior Department both for the lady and for her "fiancé"—in Washington for the woman and, quite conveniently, in the Midwest for the fiancé. It did not take long before anonymous letters threatening revelation of the affair began to appear, some of which found their way to various newspapers. The White House eventually became involved, at least to some degree, and Ickes's own investigative branch established that the author of the letters was—not exactly a surprise—the fiancé. The romance faded away by 1934. The next year Ickes's wife was killed in a car accident. Three years later, Ickes married a woman who was four decades younger than he was, and oddly connected to him. She was the younger sister of the wife of his stepson, who himself had recently committed suicide. Ickes "asked" Roosevelt's permission before the marriage. The President waved aside the age difference. There had been a similar gap in age between his own parents, Roosevelt said.[5]

Bombarded from the moment he took office with a variety of opinions regarding the oil business, Ickes learned quickly and firsthand how "thorny" the petroleum situation was. On May 1, 1933, he wrote Roosevelt about the impending "utter demoralization" of the oil industry. Admitting that he could not sort out the fierce debate raging among majors and independents over the price collapse, overproduction, and waste, he declared: "But we do know that oil has been selling at ten cents a barrel in the East Texas field. We do know that this situation can not continue much longer without disastrous results to the oil industry and to the country."

The industry itself, as well as the elected representatives from the oil states, were crying out for action from Washington. Even a large segment of the independents were, in the words of the president of the Independent Petroleum Association of America, supporting legislation to "place unprecedented authority in the hands of the Secretary of the Interior." But, while most were in agreement that action was needed, they were far from agreeing as to what specifically should be done.

On May 5, 1933, Ickes was handed a telegram as he was going into a Cabinet meeting; prices in East Texas had fallen to as low as four cents. Later that same day, he received another telegram, this one from the Governor of Texas, saying that "the situation is beyond the control of the state authorities." Three days later, Ickes warned that "the oil business has about broken down and . . . to continue to do nothing," he added, will "result in the utter collapse of the industry," with a huge loss in terms of the nation's oil reserves. Harold Ickes and the New Deal were ready and willing to step in and do something.

The crisis of the oil industry was addressed initially under the aegis of the National Industrial Recovery Act and the National Recovery Administration that it spawned, the system of business-government cooperation that was meant to stimulate economic recovery, reduce competition, strengthen the position of labor and, in the process, more or less gloss over antitrust laws. But oil would be handled differently from most other industries in that control was eventually lodged not in the NRA, but in the Interior Department, where Harold Ickes held undisputed sway.

Rooted in the progressive, trust-busting tradition of Ida Tarbell and Theodore Roosevelt, Ickes had spent much of his career campaigning against "interests." He was not exactly an advocate or even a friend of business, and he was grimly amused by how the once-proud businessmen, now shell-shocked by the Great Depression, were seeking help from the Federal government. "So many of these great and mighty" from the business world, he observed after attending a dinner at the United States Chamber of Commerce, "were crawling to Washington on their hands and knees these days to beg the Government to run their businesses for them."

Neither politics nor experience nor temperament had made Ickes sympathetic to the business of oil, but he was to come to its rescue and champion its future. In his view, the stakes were very high indeed. "There is no doubt about our absolute and complete dependence upon oil," he said. "We have passed from the stone age, to bronze, to iron, to the industrial age, and now to an age of oil. Without oil, American civilization as we know it could not exist."[6]

The Government Acts

Ickes started with prices. As he saw it, the price of oil, like that of other commodities, was too low. Prices of all such raw materials needed to be shored up, in order to help restore purchasing power to the economy. Oil men, like producers of other raw materials, could not continue to sell their products below cost. Ten cents a barrel would contribute to the prolongation of the Depression. For prices to be raised, production had to be controlled, and to bring production under control, Ickes began with an all-out campaign against the "hot oiler," who was, he said, equipped with "sly animal cunning." The leaks of hot oil, though composed of thousands of different streams, were adding up to a mighty flow—estimates in 1933 ran as high as half a million barrels per day. This bootleg oil was secretly siphoned off from pipelines, hidden in camouflaged tanks that were covered with weeds, moved about both in an intricate network of secret pipelines and by trucks, and then smuggled across state borders at night. At every clandestine step, the way was smoothed by graft and payoffs. It all added up to a big and a profitable business. Matters were made even worse by the fact that any firming of prices only provided an incentive to produce more hot oil, which flooded the market, and thus brought the price down again.

Hot oil was the great weakness of prorationing as it had so far developed, undermining all efforts to stabilize price. In order to sustain prorationing, there had to be a way to put teeth into the system, to police it, to plug the huge leak. The problem could not be solved solely at the level of Texas, Oklahoma, and the other states. The Federal government had to take on the role of policeman. But on what basis? The answer was to be found in the power of the Federal government to regulate interstate commerce. Legislation, hurriedly passed in 1933, gave the President the explicit power to ban and interdict "hot oil"—petroleum produced in excess of state-mandated levels—from entering into interstate commerce. Roosevelt himself was appalled by what he called "the wretched conditions" in the oil industry; and on July 14, 1933, he signed an executive order that would, Ickes wrote in his diary, "stop the carrying into interstate or foreign commerce of any petroleum, or the products thereof, produced in violation of the law of the state of their origin." Ickes added, "Under the Executive Order, I am given broad powers not only to issue regulations, but to enforce these regulations."

Ickes immediately dispatched Federal investigators into the East Texas field to examine refinery records, test oil gauges, inspect tanks, even to dig up pipelines to measure the accuracy of sworn records. Subsequently, Federal "certificates of clearance" were required to move any oil out of East Texas. Ickes acted with a vengeance to promote the arrest and prosecution of those who were called the "hot oil boys." To one impatient congressman, Ickes pledged, "I have been moving heaven and earth in the matter." Federal officials carried the brunt of this whole effort to interdict hot oil, as the state of Texas was by now so broke that it could not even afford to send in additional Texas Rangers.

The Oil Code, established under the National Industrial Recovery Act, gave Ickes an extraordinary additional power—to set monthly quotas for each state. A few years earlier, such government intervention would have been enough to ignite a rebellion among oil men everywhere; now it was welcomed and greeted with relief by many in the battered industry. Ickes was in charge, and he gloried in it. On September 2, 1933, aiming to reduce the country's oil production by three hundred thousand barrels per day, Ickes sent out telegrams to the governors of the oil-producing states telling them what each state's quotas—levels of production—would be. It was a historic act, a fundamental alteration in the way the industry operated. The days of flush production were over. With prorationing, the rule of capture was overturned. What had made sense for deer and game birds on estates in medieval England would no longer do when the very structure of the American oil industry seemed likely to be washed away by the otherwise uncontrollable fury of the flood of unrestrained production.

The restoration and stabilization of prices could have been sought in another way—by the government's actually fixing prices. There was strong support for Federal price fixing from some in the industry who had been battered by the price collapse. "If you do not give us price regulations," said a representative of Standard of California in 1933, "you can make codes from now to doomsday and you will get nowhere." But there was also much opposition. Some feared that if the government began setting prices, it would then regard the oil industry as a public utility and start regulating profits as well. Ickes himself showed great eagerness, for a time, to take on the task of setting oil prices, which was enough to stir further apprehension. In fact, fixing a price could well backfire, by providing a big incentive to overproduction. Price setting, when compared to the regulation of production, also looked to be more difficult, more complex, more public, and surely much more contentious. Regulation of production was definitely the preferred method. And despite efforts to get that task assigned directly to Washington, the job was kept at the state level, where it would be less controversial, closer to the actual world of oil production, and much less visible.

The new system of Federal-state partnership was gaining substantial headway by the end of 1934. "We seem to be making good progress in matter of Hot Oil in E. Texas," one of Roosevelt's aides informed the President in December. But then, in the very next month, January 1935, the Supreme Court abruptly dealt the new system a potential death blow. It overturned the subsection of the National Industrial Recovery Act under which hot oil was prohibited, setting off a new crisis. Without control over hot oil, the whole system would collapse. To keep contraband oil—produced in excess of approved levels—out of interstate commerce, new legislation was speedily drafted and passed. The law became known as the Connally Hot Oil Act in honor of its champion, Senator Tom Connally of Texas. Then, in June 1935, the Supreme Court delivered an even worse blow. It declared much of the National Industrial Recovery Act unconstitutional. The specific case had nothing to do with oil; rather it involved "sick chickens" that had been sold by a poultry jobber in New York City in violation of an NRA code. Still, the voiding of the NIRA, among many other things, stripped Ickes of his power to set mandatory quotas for the states.

Yet the aftereffects were not anywhere near as devastating as would have been the case even a year or two earlier. By this time, a framework for the regulation of the oil industry had been set in place and a consensus established; both survived the demise of the NIRA. The system still involved Federal-state cooperation. As it now worked, the Connally Hot Oil Act provided sufficient police powers to curtail contraband oil. In addition, the Federal government, specifically the Bureau of Mines, prepared demand estimates for the coming period; then it "assigned" to each state a suggested share of that demand—a sort of informal, voluntary "quota." With the end of the NIRA, the states were not required to accept that level; indeed, to show its independence, the Texas Railroad

Commission, which had by this time become more professional and technically competent, occasionally exceeded the Texas "quota" by a little bit. But, essentially, the states adopted the Federal estimates as their own, and hewed to them, even though they were no longer mandatory.

A state could, of course, have greatly exceeded its quota. But, in so doing, it would have run the risk of reprisal from the Federal government and other states, and would have faced the peril of encouraging other states to overproduce as well, resulting in yet another glut and another price collapse. Thus, each state essentially accepted and acted on a federally suggested quota, then proceeded to prorate its output to fill its own share of the projected demand. The memory of ten cents a barrel was still strong both among oil producers and among the state governments that depended on oil revenues. And after all, major discoveries could be made again. As one legal expert wrote in the 1930s, "One must be somewhat of a prophet to feel that the experiences in the East Texas Field will never be repeated."[7]

The role of the states was further formalized in 1935 with the establishment of the Interstate Oil Compact. The development of what the chairman of the Texas Railroad Commission called "this treaty" among the oil-producing states occasioned a major fight between Oklahoma and Texas. Oklahoma wanted to establish something akin to a cartel, which would have both the explicit authority to allocate the Bureau of Mines estimates of American oil demand to each of the oil states and the legal power to enforce adherence to the quotas. Texas was resolutely opposed to such a cartel. It did not want to surrender its sovereignty. Texas won, and the Interstate Oil Compact became much less than the cartel that some had hoped. Still, it provided a forum for states to exchange information and plans, to standardize legislation, and to coordinate prorationing and conservation in production.

There was, however, one further building block without which the system could not have worked—a tariff to check the flow of foreign oil. Otherwise, cheap imports would have simply flooded into the American market, negating any restraints on domestic production and creating a second stream of "hot oil," outside the regulatory system. Despite the failure to add an oil duty to the 1930 Smoot-Hawley Act, agitation for such a tariff had continued to mount. In 1931, the main importing companies had agreed to reduce their imports "voluntarily," to ward off attacks from independents, who preferred to blame low prices on the majors and foreign oil, rather than on their own rather wanton habits of production. But the voluntary restrictions on imports failed, as might have been expected.

By 1932, the distress in the industry and in the oil-producing states was sufficient to get a tariff passed through Congress and signed into law. A duty of twenty-one cents a barrel was imposed on crude and fuel oil, and $1.05 on gasoline. The tariff picked up support for another reason: It was a good source of government revenues in the midst of the Depression. The tariff was put in place just in time to provide the barrier to the inflow of foreign oil, and such a barrier was required if the new prorationing system was to work. The duties—supported by a "voluntary agreement" about the volume of imports in 1933 between

Ickes and the main importing companies—did their job. In the late 1920s and early 1930s, imports had been equivalent to between 9 and 12 percent of domestic demand. (Of course, tariff proponents rarely noted that the United States remained a net oil exporter, and that American oil exports were as much as two times the volume of imports.) After the passage of the tariff, oil imports fell to a level equivalent to only 5 percent of domestic demand.

The country hardest hit was Venezuela; it had been supplying over half of U.S. crude imports, with 55 percent of its total oil production going to the United States in the form of crude and oil products. That country's industry, which had boomed during the 1920s, went into a severe contraction; ships filled with expatriate oil men and their families sailed for home. Meanwhile, the companies operating in Venezuela hurried to reorient their exports toward the European market, and Venezuela overtook the United States as Europe's largest supplier. By the mid-1930s, Venezuela had regained its previous high production level. But, for the domestic American oil industry, the tariff provided the protective dike behind which the rest of the regulatory system could subsequently be put into place.[8]

Stability

If some system of regulation appeared logical, even inevitable, the circumstances under which it emerged were nasty and disordered; the debates surrounding it, bitter and accusatory; and the entire process, rancorous and desperate. Moreover, it emerged in an incremental way, piece by painful piece, responding to the unfolding of events. It took East Texas and ten cents a barrel to shock the industry and producing states into moving in this direction. The process was facilitated by the major advances in petroleum engineering and in the understanding of the dynamics of oil production that had started in the mid-1920s. But it also required the Great Depression and the New Deal to make it happen. It was devised by an unlikely alliance of Texas and Oklahoma oil men, patronage politicians in Austin and Oklahoma City, and Ickes and other New Deal liberals in Washington. Though innately suspicious of one another, they nevertheless worked together to bring stability to an industry particularly prone to boom and bust because of the nature of oil discovery and the traditional way of exploiting newly discovered reserves. The terrors of 1933 had been banished. "There is," the chairman of the Texas Railroad Commission wrote proudly to Roosevelt in 1937, "a complete cooperation and coordination at the present time between the Federal Government and the Oil producing states in this common effort to conserve this natural resource." He was only modestly exaggerating.

Despite its haphazard growth, the regulatory system as it finally evolved did indeed possess a powerful underlying logic. It rewrote the book on production and even, to some degree, on what constituted "ownership" of oil reserves. It brought a whole new approach to production, technically as well as legally and economically. And it established a new direction for the American oil industry. Many years later, others operating on an even larger scale would seize on it as a compelling model.

Two working assumptions were central to the system. One was that the demand for oil would not be particularly responsive to price movements: That is, oil at ten cents a barrel would not mean a far greater demand than oil at a dollar a barrel. Demand could be taken as a given, and at least in the Depression, many found that a reasonable thing to think. The second assumption was that each state had its "natural" share of the market. If those shares changed dramatically, the overall system could be threatened. That was exactly what occurred in the late 1930s, when significant discoveries in Illinois made that state the nation's fourth-largest oil producer. Illinois did not belong to the Interstate Oil Compact. It was a new producer, it wanted into the market, and it went its own way to carve out its own share. Texas and Oklahoma cut back substantially on their production to make way for the Illinois crude. They did not do this happily. There were recriminations and calls for the entire system to be scrapped. Texas announced that it might abandon prorationing altogether and go it alone. Yet the system withstood even the onslaught of new petroleum from Illinois.

Prices themselves were not fixed by the government under the system. Its advocates, whether in Austin or Washington, were insistent on this point. Still, setting production levels to match market demand did establish a level of crude output that could be marketed at a stable price. From 1934 through 1940, the average price of oil in the United States varied between $1.00 and $1.18 per barrel. The magical "dollar a barrel" rallying cry had been realized. The system worked. The flood was stayed. And, in the process, both the management of petroleum surplus and the relationship between oil companies and the government had been forever changed.[9]

CHAPTER 14

"Friends"—and Enemies

MALCOLM AND HILLCART was a real estate agency in the town of Fort William, on the west coast of Scotland, seventy-five miles north of Glasgow. It dealt in the rental of estates for hunting and fishing; and in anticipation of the summer season of 1928, it had prepared the "particulars" for a property called Achnacarry Castle, which was located a dozen or so miles away, in Inverness-shire. Like real estate agents around the world, Malcolm and Hillcart spared no adjectives. "The Castle is beautifully situated on the banks of the River Arkaig and is one of the most interesting historical spots in the Scottish Highlands," it said. "The surrounding scenery is probably unsurpassed in Scotland." The shooting—over fifty thousand acres—and fishing were excellent; one might anticipate as many as 90 stags, 160 brace of grouse, and 2,000 fish. The main house itself—built at the beginning of the nineteenth century "in the Scottish baronial style," though modernized with electricity, hot water, and central heating—had nine bedrooms, plus additional rooms with beds, and another four bedrooms were attached to the garage. For all of August 1928, the estate was available for three thousand pounds; though, to be sure, the renter had to bring his own servants, except for the head housemaid, who would be provided.

What better place could there be to spend some time in a relaxed setting with old friends? Henri Deterding took the property for the month. One old friend who joined him was Walter Teagle, the head of Standard Oil of New Jersey. There was nothing surprising in that; after all, the two men had made it a point to go hunting together on occasion over the years. But this time, the list of old friends was longer; there was Heinrich Riedemann, Jersey's chief man in Germany; Sir John Cadman of Anglo-Persian; William Mellon of Gulf; and Colonel Robert Stewart of Standard of Indiana. Their holiday entourage included secretaries, typists, and advisers, who were housed in a specially secured cottage seven miles away.

Though great effort had gone into keeping the get-together secret, word leaked out and the London press galloped northward, only to be told that the oil men had gathered merely for some grouse shooting and fishing. But then why the great effort at secrecy? "The grouse were to have no warning," the *Daily Express* hypothesized. No further word could be extracted, not even from the butler, about what the oil men might be saying as they traipsed across the moors or sat together in the evening, talking over drinks. Uninterested in the sports and bored with the conversation, Deterding's two teenage nieces—"hellions," in Teagle's phrase—poured molasses into Riedemann's bed and tied his pajamas in knots. The stiff German was furious. As for the shooting, it was lousy, Teagle later said. But that did not matter, for it was not grouse the oil men were after. They were searching for a solution to the dilemmas of overproduction and overcapacity in their troubled industry. More than trying to bring another truce to the oil wars, they were after a formal treaty for Europe and Asia—one that would bring order, divide markets, stabilize the industry, and defend profitability. Achnacarry was a peace conference.[1]

This was a year before the 1929 stock market crash and the beginnings of the Great Depression—and two years before Dad Joiner's discovery in East Texas. But already, oil was surging from the United States, Venezuela, Rumania, and the Soviet Union, flooding the world market, weakening prices and threatening "ruinous competition." The flow of Russian oil, in particular, had carried the oil men directly to Achnacarry. The vicious price war that Deterding had launched against Standard Oil of New York in retaliation for its purchases of Russian oil had spread to many markets around the world. The battle had gotten out of hand and turned into bitter global warfare, prices were collapsing, and none of the oil companies could feel secure in any market.

Achnacarry reflected the temper of the times. Industrial rationalization, efficiency, and the elimination of duplication were the values and objectives of the day in Europe and the United States, celebrated by both businessmen and government officials, as well as by economists and publicists. Mergers, collaboration, cartels, marketing agreements, and associations were the various instruments for achieving those goals, and they constituted the pattern of international business in the 1920s and, even more so, in the 1930s, with the coming of the Depression. Profits would be preserved and costs controlled through the "efficiencies" of collaboration. As in the days of John D. Rockefeller and Henry Flagler, "unbridled competition" was the danger that had to be fought off. But it was no longer possible to seek to eliminate commercial rivalry through total control, a universal monopoly. No one firm was powerful enough to "sweat" the others into submission. Nor would political realities allow it. And so a concordat, rather than conquest, was now the objective of the oil men of Achnacarry.[2]

The Hand of the British Government

The meeting at Achnacarry was not only the work of the oil companies. Behind the scenes, hidden from most observers, the British government was prodding and pushing the companies toward collaboration in the pursuit of its own economic and political goals.

At the juncture of these various interests stood Sir John Cadman, the successor to Charles Greenway as chairman of Anglo-Persian. By 1928, Cadman was arriving at the peak of his influence, operating on the same plane as Deterding and Teagle, but also with unparalleled credibility in the eyes of the British government. Growing up in a family of mining engineers, Cadman had begun his own career as a coal mine manager. (He won awards for saving miners during underground disasters.) In time, he became professor of mining at Birmingham University, where he had shocked the academic establishment by introducing a new course in "petroleum engineering," so novel that an academic opponent denounced it as "flagrantly advertised" and "a blind alley" with "a freak title." By the outbreak of World War I, Cadman was one of the outstanding experts on oil technology. During the war, as head of the Petroleum Executive, he demonstrated considerable skills both at politics and in managing people. In 1921, he became technical adviser to Anglo-Persian; six years later, the government's candidate, he became its chairman.

By that time, petroleum production was growing throughout the world, and the total output of Cadman's own company, from Persia as well as Iraq, was poised to increase fourfold. "It was essential that new markets should be found," Cadman said flatly. Anglo-Persian had two choices: Fight its way into those new markets, with the consequent large investment and unavoidable competition, or set up joint ventures with established companies and so divide the markets with them.

Cadman chose the second path, making an arrangement to pool markets and facilities in India with Shell as well as with Burmah, which happened to be Anglo-Persian's second-largest shareholder after the British government. The next target was Africa, where Anglo-Persian and Royal Dutch/Shell proposed to form an "alliance," under which they would split markets fifty-fifty. But, in order to proceed with this new venture, Anglo-Persian sought in early 1928 the permission of its majority stockholder, the British government. And the government was not at all sure that it should approve. The Admiralty expressed its customary fear that Anglo-Persian might be absorbed by Shell, which would go against the most basic tenets of government policy. The Foreign Office and the Treasury were anxious about alienating the United States. They worried that such a combination might lead the United States—"in the present state of irritability of American public opinion"—to charge that the two companies were making " 'war' on American interests represented by the Standard Oil Company." That indictment could easily be extended to the British government, owing to its majority ownership in Anglo-Persian, and that would have most unfortunate political consequences. Moreover, the resulting tensions might lead—so the reasoning went—to pressure on the government to sell its holdings in Anglo-Persian, which would be a catastrophe for the Royal Navy and not very good at all for the Exchequer, which was most attached to its attractive dividends.

Once again Winston Churchill, now Chancellor of the Exchequer, was to play a pivotal role. At first, he had many doubts about the proposed African combination. "The moment when Sir Henri Deterding is at 'war' with the Standard," he said, "seems a singularly inopportune one for the British government to be drawn into the quarrel." But as Churchill reflected further on the matter, he came to the conclusion that combination was the best policy. It was also the cheapest. "The alternative to the proposed working arrangement was for the Anglo-Persian Oil Company to fight for the market in Africa," he told the Committee on Imperial Defense. That would require a great deal more money, and would mean that he—on behalf of His Majesty's government, the largest shareholder—would have to approach Parliament for it. He had done so once before, in 1914, when he had convinced the government to buy shares in Anglo-Persian; and he did not want to go through a similar episode again, especially when it was likely to prove much more controversial. The direct interests of the British government in oil matters were best kept out of sight.

The government, thus, gave its firm support to Cadman's efforts to form his African "alliance" with Shell. Its overall position was laid out in a joint memorandum from the Treasury and the Admiralty in February 1928. "Such a policy will in the long run be more in the interest of the consumer than cutthroat competition." The arrangement could have additional benefits, the memorandum went on to say; it might promote "similar alliances elsewhere"—especially with Standard Oil of New Jersey.

That last provision, aside from approval of the African arrangement itself, was the most important thing to come out of the government deliberations. The government had given Anglo-Persian a mandate to talk with Standard Oil and to seek out similar market arrangements with the Americans in order "to allay their jealousies and show that we are not out to quarrel." For, unhindered by the American tradition of antitrust, the British government was partial to combination. As one British official wrote at the time, "Our experience has broadly been that the amalgamation of oil interests has not resulted in the consumer suffering."

Cadman, now representing government policy as well as Anglo-Persian, pursued a concordat with the American companies. Once the African deal with Shell was done, he wrote to Teagle of Jersey to propose "a small 'clearing-house' for matters of the very highest policy" for their respective companies, plus Royal Dutch/Shell. These various developments were among the major influences leading to Deterding's invitation to Teagle, Cadman, and the others to join him, in August of 1928, for a little shooting and fishing at Achnacarry Castle in the Scottish Highlands.[3]

"The Problem of the Oil Industry"

The two weeks of discussion that ensued on the banks of the River Arkaig resulted in a seventeen-page document, agreed to but not signed, that was called the "Pool Association." It became better known as the Achnacarry or "As-Is"

Agreement. The document summarized the "problem of the oil industry"—overproduction, the effect of which "has been destructive rather than constructive competition, resulting in much higher operating costs. . . . Recognizing this, economies must be effected, waste must be eliminated, the expensive duplication of facilities curtailed."

But the heart of the document was the "As-Is" understanding: each company was allocated a quota in various markets—a percentage share of the total sales, based upon its share in 1928. A company could only increase its actual volumes insofar as the total demand grew, but it would always keep to the same percentage share. Beyond this, the companies would seek to drive down costs, agreeing to share facilities and to be cautious in building new refineries and other facilities. In order to increase efficiency, markets would be supplied from the nearest geographical source. That would mean extra profits, since the sales price would still be based upon the traditional formula—American Gulf Coast price plus the going freight rate from that coast to the market—even if the oil was coming from a closer location. That provision was central, for it established a uniform selling price, and adherents to the "As-Is" Agreement did not have to worry about price competition—and price wars—from other adherents.

A few months later, the industry leaders agreed to control production as well. Participants in the Achnacarry system could increase their output above the volumes indicated by their market quotas, but only so long as they sold this extra production to other pool members. In order to implement the agreement, an "Association," managed by one representative from each company, was set up to carry out the necessary statistical analysis of demand and transportation and to allocate the actual quotas.

An important participant in the European oil trade was, however, noticeable by its absence from the agreement—the Soviet Union. Clearly, the Soviets had to be brought into the "As-Is" system if it was to have any chance of success. For, by 1928, a Soviet company, Russian Oil Products, was the fourth-largest importer into the United Kingdom. The Soviets had regained prewar production levels, and oil had become the Soviet Union's largest single source of hard currency earnings. Remarkably enough, considering Deterding's and Teagle's distaste for doing business with the Soviet Union, the major companies reached an understanding with the Russians in February 1929, which gave the Soviet Union a guaranteed share of the British market. With Russia apparently roped in, at least in part, there was only one major exception to this amicable division of the world's oil markets. But it was a very large one: The agreement explicitly excluded the domestic U.S. market, in order to avoid violating American antitrust laws.[4]

The Achnacarry Agreement, crafted in the isolated beauty of the Scottish Highlands, harked back to the turn of the century, when Rockefeller and Archbold, Deterding and Samuel, the Nobels and the Rothschilds all strenuously sought a grand concord in the world oil market, but failed in the attempt. This time around, the oil companies were no more successful in implementing their new agreement than they had been in keeping their meeting at Achnacarry secret in the first place. While the companies involved in the "As-Is" Agreement were by far the dominant firms, there were enough "fringe" players who did not belong and who did not hesitate to nibble away at the market share of the major companies. In fact, nonmembers of the "As-Is" Agreement found it to their liking. They could price just a bit beneath the large companies and win market share. Even if the members did respond with tough price competition, forcing the nibblers out of one market, these smaller companies could move on to another.

In particular, it was critical to win control over American oil exports, which amounted to about a third of all oil consumed outside the United States. And immediately upon Teagle's return from Achnacarry, a number of American companies, seventeen in all, combined to form the Export Petroleum Association, which would jointly manage their oil exports and allocate quotas among them. They were acting under an American law called the Webb-Pomerene Act of 1918, which allowed U.S. companies to do abroad what the antitrust laws did not permit them to do at home—come together in a combination—so long as the combination's activities took place exclusively outside the United States. But the association's negotiations with the "European Group" fell apart over the question of how to allocate output between the American and European companies. Moreover, the association never attained the critical mass—at the most, it controlled only 45 percent of American exports—while seventeen companies were simply too many to come to satisfactory agreement on prices and quotas. The failure of this attempt to cartelize U.S. oil exports further undermined the determined efforts at Achnacarry.

All over the world, there were too many producers and too much production outside the "As-Is" framework. "The figures we had before us," J. B. Kessler, a Royal Dutch/Shell director, wrote to Teagle, "showed that, of the potential world production, a large part is controlled by companies which are not controlled either by you or us or any of the few other large oil companies. From this followed that the present balance in the world's oil production cannot possibly be maintained by you and we only." It did not take long for the accuracy of Kessler's prediction to be confirmed. Discoveries and output in the United States were building up to the great crescendo of East Texas. Oil was also coming onto the world market from other sources, such as Rumania. In the surge of uncontrollable production, the Achnacarry Agreement was washed away. And the oil companies once again began attacking one another's markets.[5]

Discord Within "Private Walls"

The Big Three—Jersey, Shell, and Anglo-Persian—tried to reformulate an alliance in 1930, but this time in a less grandiose manner. They revised the "As-Is" understanding in the form of a new Memorandum for European Markets. Instead of seeking a global arrangement, their operating companies in the various European markets would try to make "local arrangements," dividing market shares, with "Outsiders." Yet once again, the system proved rather ineffective in the face of the still-rising volumes of American, Russian, and Rumanian oil. The Soviets, in particular, never hesitated to cut prices when they saw the opportunity to earn more revenues. Normal commercial considerations did not apply for them; the Russian trading organizations were charged by the Kremlin with earning as much foreign currency as possible, in whatever way they could, in order to pay for the machinery that Russia needed for its industrialization drive. Despite the continuing efforts, the companies found it impossible to fashion an "orderly" and durable marketing arrangement with the Russians.

By 1931, Jersey, for one, had grown disenchanted with unworkable global alliances. "In view of the collapse of the Export Association, our as-is arrangement with the Royal Dutch should be abrogated," E. J. Sadler, Jersey's head of production, told his colleagues. "Jersey at present makes a great sacrifice in protecting other companies in uneconomical movements or situations." He advocated that Jersey junk the whole effort at cooperation and instead make war on the Shell group. "Now is the best time to fight the Royal Dutch, as they are most vulnerable in the Far East. . . . In this region we have never made any profit, and a price war would cost us almost nothing." In a grim meeting in March 1932, Deterding and other senior executives of Royal Dutch/Shell made the bleakness of the worldwide situation abundantly clear to M. Weill, who oversaw the Rothschilds' large interest in Royal Dutch/Shell. Sales volumes had plummeted, Weill reported afterward to Baron Rothschild. "Prices are bad everywhere, and except for a few rare places, money is not being made."

In November 1932, Sir John Cadman addressed the American Petroleum Institute. He devoted his speech to extolling the merits of "cooperation"— "strictly, of course, within the laws of every country." His remarks clearly belied the widespread view that the "As-Is" arrangements were a secret conspiracy, unknown to the larger world. After all, here was John Cadman, chairman of Anglo-Persian, standing up before the entire membership of the American Petroleum Institute to declare that "the principle of 'As-Is' " has "become the keystone of cooperation in international petroleum trading outside the United States."

Cadman went on to warn the delegates, "It is still raining outside," and with catastrophe looming at the very depths of the Depression, the companies could not give up the effort to seek shelter from the storm and stabilize the industry. They came up with a new version of the "As-Is" understanding: the Heads of Agreement for Distribution, of December 1932, which "should be used as a guide to representatives in the field for drawing up rules for local cartels or local Agreements." The initial adherents to the Heads included Royal Dutch/Shell, Jersey, Anglo-Persian, Socony, Gulf, Atlantic, Texas, and Sinclair. The new arrangement was managed by two "As-Is" committees, one in New York, oriented to supply, and one in London, oriented to distribution. A central "As-Is" secretariat was established in London to carry out the statistical and coordinating tasks under the agreements. Internal "As-Is" departments were also established in at least some of the companies. But there were many points of friction in the new arrangement, including chronic cheating and the problem of what to do about "virginal markets," that is, markets in which participants had not previously traded but were now seeking to enter.[6]

As the Great Depression progressed, so did the problems in the oil industry, and the companies tried yet again to improve the "As-Is" system, this time devising the 1934 Draft Memorandum of Principles, which provided for looser cooperation agreements. So severe was the Depression's pinch that the new memorandum called for "economy in competitive expenditure." Money was to be saved, and competitive differences among companies reduced, by cutting back on advertising budgets. The number of road signs and billboards was to be reduced; newspaper advertising "to be restrained within reasonable limits"; and "premiums to racing drivers" to be cut or eliminated. The little promotional gifts so dear to motorists, such as cigarette lighters, pens, and calendars, were to be sharply limited or done away with altogether. Nothing was left unscrutinized; even the number and type of signs in gas stations were "to be standardized to reduce unnecessary expenditure."

Such agreements, whatever their scale or actual effectiveness, were inevitably controversial, arousing passionate and pervasive criticism on one side and self-righteous defense on the other. Many looked at them and found only proof of a giant conspiracy directed against the interests of consumers. There was great apprehension about international cartels in any industry and, especially, any show of camaraderie among the giants of oil. Yet these arrangements, outside the United States, did not run counter to the laws of various countries. On the contrary, both the temper of the time and the pressure of government policies, in addition to the business environment, pushed toward some form of collaboration and cartelization.

Within the walls of each company that was a party to the agreements, the senior management would refer to the management of other companies as "friends": "Friends in London say . . ." or "Friends not yet made up mind." But it was not friendship that was at work, not a "brotherhood of oil." Rather it was a sense of desperation in a depressed world economy and in the face of stagnant demand that really brought the oil companies together. They were aggressive competitors, and they never forgot that. The efforts at collaboration were paralleled by pervasive distrust, wariness, and deep-seated rivalry. Even as they were talking cooperation, they were plotting new attacks. Just a few months after Achnacarry, Shell entered the East Coast market in the United States and proceeded to expand its business very rapidly. Jersey was infuriated; one of its executives denounced the Shell move as "warranted only by ambition." Then, in 1936, Henri Deterding learned that Jersey was discussing the sale of its entire Mexican operation to William Davis, an independent oil man with interests both in the United States and in Europe. Davis was one of those "fringe" players who made enforcement of the "As-Is" agreements so difficult. "We are engaged together in resistance to Davis' activities," Deterding angrily wrote to Jersey, "and surely it is hopelessly inconsistent with sound warfare to sell the enemy a complete set of much needed munitions with which more effectively to conduct the war with us." In fact, all during the 1930s, as Jersey discussed cooperation with Shell, it was continually giving serious consideration to merging its foreign business with Socony so that it could better take on Shell.

Moreover, conflict constantly erupted over implementing what was agreed to, or even agreeing as to what had been agreed to. The 1934 Draft Memorandum of Principles contained the provision that outside auditors would review the various adherents' trade numbers. A senior executive of Standard-Vacuum, the joint venture of Jersey and Socony in Asia, was outraged. He and his colleagues "are unanimously opposed to this," he said in December 1934. "Not only is the thought of having outside auditors go over our books objectionable for obvious reasons, but it seems to us that the As-Is would be on a very weak foundation if the parties concerned cannot trust each other to the extent of giving correct information on the volume of trade." He added, "We should by all means keep the operation of the As-Is within the private walls of those interested." But things did not go easily even inside those private walls. In December 1934, Shell director Frederick Godber was in the Far East, where he reported, "Figures trade clearly show Texas Co. unnecessarily aggressive and will end year with larger share of trade than that to which they are entitled." He added that the other companies would have to apply "sterner measures" against the errant Texas Company. Agreements notwithstanding, the impulse to compete could not be completely checked.

And how successful was the allocation process itself? The results for the United Kingdom, for instance, showed considerable unevenness. Shell and Anglo-Persian formed a combined marketing system in Britain, Shell-Mex/BP. The ratio of sales between that group and Jersey's affiliate, with some notable exceptions, remained relatively constant. But the two groups' combined share of the *total* market fluctuated considerably as oil entered Britain from a variety of sources.

As unstable as they were, the "As-Is" agreements became much more effective with the Draft Memorandum, from 1934 onward. Three factors contributed to its comparative success. In the United States, Federal and state authorities, led by Harold Ickes, finally brought production under control. In the Soviet Union, the quickening pace of industrialization stimulated domestic oil demand, constricting the amount of oil available for export. And the large companies finally succeeded in getting some control imposed on Rumanian output. Even so, there were not very many years of respite. In early 1938, Jersey gave verbal notice of termination of the "As-Is" agreements. And, in large part, any surviving "As-Is" activities came to an end in September 1939, with the outbreak of World War II.[7]

Nationalism

The "As-Is" agreements did not take place in a vacuum. They were meant to defend not only against the glut of oil and then the Depression, but also against the emergence of powerful political forces in Europe and elsewhere. "Throughout the European continent, government policies confronted those of private foreign oil companies—and the scope of the confrontation was unprecedented," one historian has written. "It is little wonder that in a defensive manner they discussed among themselves means of coping with the abnormal conditions of trade."

During the 1930s, the forms of political pressure on the oil companies were many. Governments imposed import quotas, set prices, and placed restrictions on foreign exchange. They forced companies to blend alcohol made from surplus crops into motor fuel, and to use other petroleum substitutes. They levied a multitude of new taxes and intervened to control the direction of the export and import trade in oil to match bilateral trade agreements and larger political links. They blocked profit remittances, forcing investment in domestic facilities that lacked economic justification, and insisted that companies maintain extra inventories. As a result of the Depression, autarchy and bilateralism were the order of the day in the 1930s, with consequent pressure to circumscribe the major oil companies. For, warned the President of the Board of Trade in London, there was now a "general tendency in all foreign countries to-day to force or encourage the establishment in their own lands of national companies in the place of non-national subsidiaries."

It became standard practice for European governments to compel foreign companies to participate in national cartels and to divide the market between foreign and local companies. In country after country, the government insisted that the foreign companies help build up local refining capacity. The French government, operating under the 1928 legislation, allocated specific shares in the market to each company. In France, said a Jersey executive, "failure to cooperate with a national commercial endeavor—whatever the sacrifice in dollars or in principles—invariably provoked retaliatory legislation which was more costly to private interests than the original government proposals." In Nazi Germany, regulations and manipulation of all kinds were mounting, as that government geared up for war. Overall, in the second half of the 1930s, with the worst years of the Depression behind, the most important objective for the major oil companies became to insulate and protect themselves against government intervention. "We are now confronted with nationalistic policies in almost all foreign countries, as well as decidedly socialistic tendencies in many," said Orville Harden, a vice-president of Jersey, in 1935. "These problems are between the Government, on the one hand, and industry as a unit, on the other. They are becoming continually more serious, and a very large part of management's time is devoted to efforts at their solution."

That same year, an observer of the oil industry, noting the intensification of political and economic nationalism in Europe, summed it all up very simply: Operations in the oil business in Europe, he said, "are 90 percent political and 10 percent oil." The same seemed to be true in the rest of the world.[8]

The Shah's New Terms

At the very bottom of the Depression, Shah Reza Pahlavi of Persia became infuriated at discovering that, as an observer put it, "oil is not gold in these days." The Shah's country had become an oil state; petroleum royalties from Anglo-Persian provided two-thirds of its export earnings and a substantial part of government revenues. But, with the Depression, the royalties from Anglo-Persian had plummeted to the lowest level since 1917. Appalled and outraged, the Shah blamed the company, and he decided to take matters into his own hands. At a Cabinet meeting on November 16, 1932, to the surprise of his ministers, he abruptly announced that he was unilaterally canceling Anglo-Persian's concession. It was the thunderbolt that no one had really believed that the Shah would dare deliver. His action threatened the very existence of Anglo-Persian.

The Shah's announcement, though unexpected, was the culmination of four years of negotiation and tension between Persia and the Anglo-Persian Oil Company. In 1928, John Cadman had observed "that concessionaries can regard their future as safeguarded against the rising tide of economic nationalism in proportion to the extent in which the national interests and their own approach identity." But Cadman had found it most difficult to create that identity. Indeed, the Persians charged that William Knox D'Arcy's 1901 concession violated national sovereignty; they also wanted more money out of the concession, much more. In 1929, Cadman thought he had worked out a deal with the Shah's Minister of Court, Abdul Husayn Timurtash, whereby the Persian government would not only have obtained much higher payments; it would also have acquired 25 percent of the company's shares along, possibly, with board representation and a share of the company's overall global profits. But the proposed deal could never be closed. There were recriminations and accusations on both sides. The discussions continued, but every time an agreement seemed to be at hand, the Persians would submit new amendments and revisions and would ask for still more.

The prime reason for the inability to come to final agreement lay in the character of the highly personal autocracy in Persia and the huge man who stood at the top. Reza Khan had used his command of the Cossack Brigade to make himself the unchallenged and unchallengeable leader of the country. He was a tough, domineering, brutal, and blunt man, who, said a British minister to Tehran, "does not waste time in exchanging the delicately phrased but perfectly futile compliments so dear to the Persian heart." Reza Khan had become War Minister in 1921 and then Prime Minister in 1923. He dallied with the notion of making himself President, but then decided otherwise, and instead in 1925 crowned himself Reza Shah Pahlavi, founder of the new Pahlavi dynasty. Thereafter he set out to modernize his country, but in an erratic and chaotic fashion. The Shah's greatest fault, said Timurtash, was "his suspicion of everybody and everyone. There was really nobody in the whole country whom His Majesty trusted and this was very much resented by those who had always stood faithfully by him."

The Shah was contemptuous of his subjects; he told one visitor that the Persian people were "bigoted and ignorant." He was also bent on consolidating the fractious country and centralizing control in his own hands, which meant reducing all other competing centers of power. He began with the clerics, the mullahs who led the traditionalists and religious fundamentalists strongly opposed to his efforts to create a modern, secular nation. In their eyes, the Shah was guilty of many sins; he had, after all, abolished the compulsory wearing of the veil by women. He was spending money on public health services and was widening educational opportunities. But he would not be stayed. He once even personally beat an ayatollah who had questioned the appropriateness of the dress of female members of his family as they entered a shrine. The mullahs as a group were beaten down into a sullen but still rebellious submission. "It has often been said," observed a visitor, "that the Shah's greatest achievement is his victory over the Mullahs."

In the Shah's opinion, Anglo-Persian was like the mullahs—an independent power center, and he was intent on reducing its power and influence as well. But he also relied on its royalty payments to fulfill his ambitions. With the drastic fall in oil revenues for Persia, the local press and politicians, responding to the dictates of the Shah, intensified their attacks on the company, criticizing and challenging everything from the validity of D'Arcy's original concession to the fact that there was refrigeration of food at the refinery site at Abadan, which was considered irreligious.

Moreover, the Shah had become angry with Anglo-Persian's majority stockholder, the British government, about other issues. He was trying to assert Persian sovereignty over Bahrain, while Britain insisted on maintaining its protectorate over the island sheikhdom. And he was furious at the British for their diplomatic recognition of Iraq, which he regarded as an invention of British imperialism. The management of Anglo-Persian could endlessly repeat that the company operated as a commercial entity, independent of the government, but no Persian would ever believe such an assertion. Hearing such statements, they "could only imagine that duplicity was being compounded with complicity."

Matters came to a head in November 1932, with the Shah's unilateral cancellation of the Anglo-Persian concession. His action was a direct challenge to the British government, whose military security had been tied by Churchill in 1914 to Persian oil. Britain could not passively accept the Shah's action. But what to do? The issue was sent to the League of Nations. With consent of all concerned, the League put the matter into abeyance, to give the contending parties time to try to work out a new arrangement. Five months later, in April 1933, Cadman himself went to Tehran to try to salvage the situation. After meeting with the Shah, he noted, "There is no doubt that H.M. [His Majesty] is after money." By the third week of April, negotiations were again deadlocked; and Cadman, frustrated and exasperated, returned to the Palace for yet another discussion with the Shah. To emphasize that a breakdown was at hand, that his own patience was at an end, and that he was ready to depart, Cadman instructed his pilot to carry out a trial flight and taxi the plane in such a way that it would be visible from the windows of the Shah's palace during the course of the meeting.

The point was not lost on Reza Shah, who now retreated. Persian demands were moderated. By the end of April 1933, a new agreement was finally forged. The concession area was reduced by three-quarters. Persia was guaranteed a fixed royalty of four shillings per ton, which protected it against fluctuations in oil prices. At the same time, it would receive 20 percent of the company's worldwide profits that were actually distributed to shareholders above a certain minimum sum. In addition, a minimum annual payment of £750,000, irrespective of other developments, was guaranteed. The royalties for 1931 and 1932 were to be recalculated on the new basis, and the "Persianization" of the workforce was to be accelerated. Meanwhile, the duration of the concession was extended from 1961 to 1993. "I felt that we had been pretty well plucked," noted Cadman afterward. But the essential position of Anglo-Persian had been preserved.[9]

The Mexican Battle

Of all the nationalistic challenges to the oil companies, the greatest came in the Western Hemisphere. There, in one of the world's most important petroleum-producing countries, the companies were caught up in a bitter battle against the full force of fervent nationalism, which challenged the very legitimacy of their activities. The setting was Mexico, and the focal point of dispute was paragraph 4 of Article 27 of the Mexican Constitution of 1917, the clause that declared that underground resources—the "subsoil," as it was called—belonged not to those who owned the property above, but to the Mexican state.

To the companies, of course, that was dangerous dogma. In the years immediately after the adoption of the 1917 constitution, they fought hard against implementation of Article 27, invoking support of the American and British governments along the way. They maintained that the property rights that they had acquired before the revolution, and in which they had invested so heavily, could not retroactively be seized by the state. Mexico insisted that it had owned the subsoil all along, and what the companies possessed did not constitute company property, but only concessions, granted by the fiat of the state. The result was a standoff—in effect, an agreement to disagree.

But the Mexican government did not want to push too far in the late 1920s. It needed the companies to develop and market petroleum. It also, more generally, sought foreign investment to promote the country's "reconstruction;" and driving out the oil companies would hardly have been good advertising. Thus, the Mexican government devised a loose, face-saving formula that kept the companies working but preserved its claim to ownership of the subsoil. This modus vivendi was hardly easy; it was punctuated by periods of sharp acrimony and bitter rhetoric. The tension rose so high in 1927 that a rupture between the Mexican and American governments appeared imminent, with the possibility of another U.S. military intervention, as had occurred when Woodrow Wilson dispatched troops to Mexico during the revolution. The risk seemed real enough to President Plutarco Elías Calles that he ordered General Lázaro Cárdenas, the military commander in the oil zone, to prepare to set the oil fields on fire in the event of a U.S. invasion.

Yet, from 1927 onward, a greater stability and a calming emerged in relations between the oil companies and the Mexican government, and between the two governments themselves. But, by the mid-1930s, that new détente was coming undone. One reason was the economic condition of the industry. Mexico was losing its ability to compete in the world oil market, particularly against Venezuela, because of higher production costs, increasing taxation, and the exhaustion of existing fields. Venezuelan oil was even being landed in Mexico for refining at Tampico because it was cheaper than Mexican oil! The largest foreign oil company was Cowdray's old Mexican Eagle, now partly owned and more or less completely managed by Royal Dutch/Shell. This group was responsible for about 65 percent of Mexico's total production. American companies produced another 30 percent, led by Standard Oil of New Jersey, Sinclair, Cities Service, and Gulf. Rather than risk new investments in the face of the unsettled conditions in the country, most of the companies were simply trying to maintain what they had. As a result, the country's production fell dramatically. In the early 1920s, Mexico had been the world's second-largest producer. A decade later, production had fallen from 499,000 barrels per day to 104,000 barrels per day, a drop of 80 percent. That was a sore disappointment to a Mexican government that had been counting on a buoyant oil industry to deliver higher revenues. It blamed the foreign companies exclusively, rather than acknowledging the effects of a depressed international market and of domestic conditions that were decidedly inhospitable to foreign investment.[10]

The political environment was also changing in Mexico. Revolutionary fervor and nationalism were surging again, and syndicalist trade unions were rapidly growing in membership and power. These changes were personified in the figure of General Lázaro Cárdenas, the former War Minister who became President at the end of 1934. A man of striking appearance, he had, said the British minister, "the long, mask-like face and inscrutable, obsidian eyes of the Indian." The son of an herbalist, Cárdenas had been able to attend school only to the age of eleven, though he remained for the rest of his life a voracious reader of everything from poetry to geography, but especially of history, and more especially the history of the French Revolution and of Mexico. At the age of eighteen, having already worked as a tax collector, a printer's devil, and a jailkeeper, he enlisted in the Mexican Revolution. Recognized for his valor, his self-contained modesty, and his leadership, he was a general by the age of twenty-five and became a protégé of Plutarco Calles, the *jefe máximo,* the "maximum chief" of the revolution. In the 1920s, while other of the new military leaders moved to the right, Cárdenas stayed on the left. As Governor of his home state of Michoacán, he devoted much energy to promoting education and to breaking up large estates in order to give the lands to the Indians. He was sober and puritanical in his own way of life, a supporter of prohibition and an opponent of casinos.

When Cárdenas was elected president, he packed his old mentor General Calles into exile and demonstrated that he was his own man and not a puppet. Adept at playing one group off against another and asserting his own supremacy, he went on to create the political system that would dominate Mexico until the end of the 1980s. Oil and nationalism would prove central to that system. Cárdenas was, in fact, the most radical of any of the Mexican Presidents. "His leftist inclinations make him the bugbear of capitalism," the British minister said of him in 1938, "but all things considered it is to be regretted that there are not more men of his calibre in Mexican life." Cárdenas aggressively pushed land reform, education, and an expensive program of public works. Labor unions became far more powerful during his presidency. He publicly identified himself with the masses and incessantly toured the country, often arriving unannounced to listen to the complaints of the peasants.

To Cárdenas, a fervent nationalist as well as a political radical, the foreign oil industry in Mexico was a painful and sore presence. As military commander in the oil region in the late 1920s, he had developed a considerable dislike for the foreign companies. He resented what he saw as their arrogant attitude and the way that they treated Mexico as "conquered territory"—at least so he was to write in his diary in 1938. And once he assumed the presidency, a shift to radicalism was inevitable. In early 1935, a few months after Cárdenas's inauguration, one of Cowdray's lieutenants in Mexican Eagle complained that "politically the country is quite Red." The oil companies had known how to do business in pre-Cárdenas Mexico, a world of blackmail, bribery, and payoffs, but they were ill-equipped to handle the new realities.

Mexican Eagle itself was caught in a crossfire between its local management, trying to adjust to the new spirit of radicalism in the country, and Royal Dutch/Shell, which had overall managerial control, though only a minority of the stock. Henri Deterding, said the resident manager, "was incapable of conceiving of Mexico as anything but a Colonial Government to which you simply dictated orders." He tried to "disillusion" Deterding. Not only did he fail, but Deterding, in turn, accused him of being "half a Bolshevik." The manager could only fulminate. "The sooner," he said, "that these big international companies learn that in the world of to-day, if they want the oil they have got to pay the price demanded, however unreasonable, the better it will be for them and their shareholders."

Standard Oil of New Jersey was also in no mood to accommodate to the new political realities. Everette DeGolyer, the prominent American geologist who just before World War I had made the huge discovery that was the basis for the "Golden Lane" and the growth of the Mexican oil industry, had maintained his Mexican contacts. Now he was worried by the implacable stance of the American companies. He privately urged Eugene Holman, head of Jersey's production department, to "work out a partnership deal with the Mexican Government which would satisfy their national aspirations and leave the Jersey in a position where it could ultimately retire its capital and, meanwhile, earn a reasonable return upon it." Holman adamantly rejected the idea. "The matter was so important as a precedent in other areas," he told DeGolyer, "that the company would prefer to lose everything that it had in Mexico rather than acquiesce in a partnership which might be regarded as a partial expropriation."

Pressure continued to build against the foreign companies. Indeed, developments in Mexico were but the sharpest expression of a growing confrontation throughout much of Latin America between foreign oil companies and rising nationalism. In 1937, the shaky new military government of Bolivia, anxious to win public support, accused Standard Oil's local subsidiary of tax fraud and confiscated its properties. The action won great applause in Bolivia and attracted much attention throughout Latin America. Meanwhile in Mexico, by 1937 wages had supplanted the chronic debates over taxes, royalties, and the legal status of the oil concessions as the number-one point of contention. The oil workers' union went out on strike in May 1937, with the other unions planning to undertake a general strike in support. Cárdenas was spending much of his time away from Mexico City—in the Yucatán, supervising land distribution to the Indians, and in the small port of Acapulco, where he was overseeing the development of a hotel and bathing beach. But now, with wholesale turmoil threatening, he intervened; the industry could not be closed down, nor a general strike tolerated. Instead, the President set up a commission to review the companies' books and activities.[11]

There was little basis for dialogue. Professor Jesús Silva Herzog, the key member of the review commission, described company officials as "men without respect who were unaccustomed to speaking the truth." The dislike was mutual. To the British ambassador, Silva Herzog was "a notorious but sincere communist." Silva Herzog's commission declared that the oil companies had been making lucrative profits while raping the Mexican economy and had contributed nothing to the country's broader economic development. It not only recommended much higher wages, pegged at a total annual bill of 26 million pesos, but also called for a host of new benefits: a forty-hour week, up to six weeks' vacation, retirement pensions equivalent to 85 percent of wages at age fifty. The commission also said that all foreign technicians should be replaced by Mexicans within two years.

The companies retorted that the commission had woefully misinterpreted their books and misrepresented their profitability. The total average combined profit of all the companies during the years 1935–37, they claimed, was no more than 23 million pesos, compared to the 26 million pesos in additional wages that were now being demanded. The companies also said that if they were forced to comply with the commission's recommendations, they would have to close down. They were, of course, gambling that the government would not take that chance; they believed that Mexico lacked the personnel, the skills, the transportation facilities, the markets, and the access to capital that would be required in the event of a government takeover.

The companies appealed the commission's recommendations. The government not only confirmed them but also added retroactive penalties. In anticipation of what might happen next, Mexican Eagle evacuated the wives and children of employees. As the charges and countercharges flew, the stakes grew ever higher. The companies feared the establishment of a precedent and model that could threaten their activities around the world. From the beginning, Cárdenas had intended to extend government control over the oil industry. But now, his own personal prestige and power were increasingly engaged. He could not afford to be seen as retreating before the foreign companies; nor could he allow himself to be outflanked by the militant unions on his left. He had to remain in command of an explosive situation. But he was also pulled along by events and circumstances. At one point, he complained to a friend that he was "in the hands of advisers and officials who never tell him the whole truth and rarely give full effect to his instructions." He added that "it was only when he went into things himself that he could ever get at the facts."

Though Mexican Eagle, a British company, was by far the largest producer, much of the agitation against the oil companies was based upon the strong anti–United States sentiment that seemed to unite the country. "The one respect in which I have found Mexicans of all classes completely unanimous," observed an English diplomat, "is their conviction that it is a fixed principle of American policy to prevent the economic development and political consolidation of their country." Yet, ironically, the diplomatic support that the American companies had previously counted on was now a thing of the past. The Roosevelt Administration had adopted a "Good Neighbor" policy toward Latin America, and the New Deal viewed the Mexican government's stance with some empathy. From a foreign policy point of view, Washington was keen to avoid alienating Mexico at a time when concern for hemispheric defense and fear of an impending war were both beginning to mount. Thus, there was little pressure from the North to counterbalance the unions' radical demands.

The crisis deepened when the Mexican Supreme Court upheld the judgment against the companies. The companies, in turn, upped their wage offers twice, but still not high enough for the union leadership or the Mexican government. On March 8, 1938, Cárdenas met privately with oil company representatives. The result was a further stalemate on the wage issue. Later that same night, Cárdenas, by himself, made up his mind to expropriate—if need be. On March 16, the oil companies were officially declared to be "in rebellion." Even so, Cárdenas continued to negotiate; the two sides were getting closer. The companies finally accepted the 26 million peso wage hike. But they would not budge in their opposition to transferring management decision making and administrative control to the unions.

On the night of March 18, 1938, Cárdenas met with his Cabinet. He told them that he intended to take over the oil industry. It was better to destroy the oil fields, he said, than let them be an obstacle to national development. At 9:45 P.M., he signed the expropriation order, and then broadcast the momentous news to the nation from the Yellow Room in the presidential palace. His words were greeted with a six-hour parade through Mexico City. The ensuing struggle was to be fierce and drawn out. For Mexico, what had occurred was a great symbolic and passionate act of resistance to foreign control, which would be central to the spirit of nationalism that tied the country together. To the companies, the expropriation was absolutely illegitimate, a violation of clear agreements and formal commitments, a denial of what they had created by risking their capital and energies.[12]

The expropriated companies joined in a united front and tried to negotiate—not about compensation, in which they had no confidence, but to get their properties back. Their efforts were to no avail. But beyond the specifics of Mexico, there was a much graver concern. If the expropriation was seen as succeeding, said one Shell director, "a precedent is established throughout the world, particularly in Latin America, which would jeopardize the whole structure of international trade and the security of foreign investment." Therefore, it was imperative for the companies to respond as vigorously as possible, and they indeed sought to organize embargoes against Mexican oil around the world, charging that such exports were stolen goods. The company that had the most to lose was Mexican Eagle; in addition to being controlled by the Royal Dutch/Shell Group, its stockholders were largely British. The British government took a very strong stand against Mexico. It insisted to the Mexicans that the properties be returned. Instead of replying, Mexico severed diplomatic relations.

A similar break with the United States was only barely averted in the immediate aftermath of the expropriation. Over the next couple of years, Washington tried to exert pressure, primarily economic, on Mexico, but these efforts were half-hearted. In fact, the American companies felt that they were far from getting the support that was appropriate. In the era of Roosevelt's Good Neighbor Policy, and in light of the New Deal's criticism of "economic royalists" and, specifically in the late 1930s, of the oil industry, the American government could hardly act harshly against Mexico or oppose the sovereign right of expropriation so long as what Roosevelt called "fair compensation" was offered. Moreover, with the rapidly deteriorating international situation at the center of his concerns, Roosevelt did not want to further aggravate relations with Mexico, or any other country in the hemisphere, with consequences that could benefit the Axis powers. Cárdenas had judged the balance of world politics correctly.

Washington could already see the unsettling effects from the British-led embargo and the efforts to close off traditional markets to Mexico. Nazi Germany became Mexico's number-one petroleum customer (and at discount prices or on barter terms), with Fascist Italy next. Japan became a major customer as well. Japanese companies were also exploring for oil in Mexico and were discussing the construction of a pipeline from the oil fields across the country to the Pacific. In the view of the Roosevelt Administration, additional American pressure would, perversely, only enable the Axis powers to strengthen their footholds in Mexico.

The much stiffer British position toward Mexico was also driven more by strategic than by commercial considerations. But the strategic issues were seen through a different lens. As outlined by the Oil Board and the Committee of Imperial Defense in May of 1938, Britain's problem was this: Just eight countries accounted for 94 percent of world oil production. The neutrality legislation enacted by the Congress and isolationism in the United States could conceivably foreclose American petroleum to Britain in a crisis. Russian exports had fallen to low levels, and might cease altogether in a war. "The Dutch East Indies, Roumania and Iraq, because of their geographical situation, are regarded as doubtful sources of supply in certain eventualities," said the Oil Board. That left Iran, Venezuela, and Mexico. Yet only a few years earlier, in the clash with Reza Shah, Anglo-Persian had almost lost its treasured Iranian concession.

All of this meant that, in a military crisis, production in the Latin American countries would be essential to Britain, not only "because of their size of production, but because they are favourably placed from a sea transport point of view." Therefore, every effort had to be made "to ensure that the Mexican policy is not followed by other Latin-American countries." London was particularly worried about Venezuela, which was supplying upward of 40 percent of Britain's total petroleum needs. The strategic issues—"defense requirements" and access to oil in wartime—were, the Foreign Office reiterated, "the paramount consideration" driving its entire policy. While the United States was Mexico's neighbor and had many important interests at stake, when it came to oil, Mexico was far more important to Britain than to the United States.[13]

"As Dead as Julius Caesar"

After the outbreak of war in Europe in September 1939, the interests of the expropriated American oil companies and of the United States government diverged even more sharply. As far as the Roosevelt Administration was concerned, national security was much more important than restitution for Standard Oil of New Jersey and the other American companies. Washington did not want Nazi submarines refueling in Mexican ports, nor German "geologists" and "oil technicians" wandering over northern Mexico, near the U.S. border, or in the south, in the direction of the Panama Canal. Indeed, the United States was now busy trying to tie Mexico into a hemispheric defense system. Therefore, it was important to get the oil issue out of the way as quickly as possible. Moreover, in the event of American entry into the war, the U.S. government wanted access to Mexican oil supplies, as had been the case during World War I, and it cared increasingly little about who actually owned those supplies. The expropriation was the major obstacle to cooperation with Mexico, U.S. Ambassador

Josephus Daniels told Roosevelt in 1941, and there was no sense trying to restore and defend a status "as dead as Julius Caesar."

The strategic considerations were such that, by the autumn of 1941, shortly before Pearl Harbor, Washington decided to push for a settlement. The crux of the matter by now was certainly not restoration; it was compensation—but how much? Widely differing estimates of the value of the companies' assets in Mexico were broached—from a Mexican figure of $7 million to a company figure of $408 million. The most critical aspect was the value of the subsoil reserves. A joint U.S.–Mexican commission, appointed by the two governments, was charged with developing a compensation scheme; it found a novel and creative solution. It simply came up with the judgment that 90 percent of all the subsoil reserves owned by the companies had already been produced by the time of expropriation! Under this clever formulation, there was no point in arguing further about who actually owned the subsoil or the value of the reserves, since most of the oil was supposedly gone in any event. On that basis, the commission proposed a compensation settlement of about $30 million, spread out over several years.

The companies reacted with bitter outrage at the compensation figure. They argued that they had gone ahead, in the 1920s, to seek out foreign oil supplies partly at the behest of a U.S. government seriously concerned about future security of supply, and now they were being abandoned and betrayed by that same government. But Secretary of State Cordell Hull finally made it clear that, while the companies were under absolutely no obligation to accept the award, neither should they expect any further assistance or support from Washington. The Administration's position was, to put the matter bluntly, take it or leave it, and in October 1943, a year and a half after the valuation had been proposed, the American companies took it.

There was also, however, a price for Mexico. A national oil company, Petróleos Mexicanos, was established, owning almost the entire oil industry in Mexico. But the oil business was no longer export-oriented; its focus shifted to the domestic market and to producing cheap oil as the predominant fuel for Mexico's own economic development. Mexican exports became a small factor in the world's markets. Moreover, the industry was also hamstrung by shortage of capital and lack of access to technology and skills. The insistence on that large wage hike—the "magic figure" of 26 million pesos—had been the *casus belli* of the expropriation of the oil fields. But inescapably, nationalism had to make some concessions to economic reality. In the aftermath of the expropriation, not only was the promised wage hike indefinitely postponed, wages were, in fact, cut.

Britain was in no hurry to effect its own settlement, or even to restore diplomatic relations with Mexico. It still feared that compromise with Mexico would, in the words of Alexander Cadogan, the Permanent Undersecretary of the Foreign Office, "put ideas" into the "heads" of Iran and Venezuela. "Of course, when the war is over that question will assume an entirely different aspect." As it turned out, Mexican Eagle and Shell did not settle with Mexico until 1947, two years after the war's end. In this instance, patience paid off; even considering the fact that Mexican Eagle was the largest foreign company that had operated in Mexico, it won, proportionately, a far better deal than the Americans—$130 million, to be exact.

Mexican Eagle knew, at least, that it had the British government firmly behind it. The American companies, in contrast, believed that they had been grievously wronged not only by Mexico, but also by their own government. On one thing, however, both the British and American companies could agree; the Mexican expropriation was the biggest trauma that the industry had experienced in many years—since the Bolshevik Revolution, perhaps even since the 1911 dissolution of the Standard Oil Trust. For Mexico, the settlements with the foreign companies confirmed the rightness of its course. The 1938 nationalization was seen as one of the greatest triumphs of the revolution. Mexico was the complete master of its oil industry, and Petróleos Mexicanos—Pemex—would emerge as one of the first and most important of the state-owned oil companies in the world. Mexico had, indeed, established a model for the future.[14]

CHAPTER 15

The Arabian Concessions: The World That Frank Holmes Made

AMONG THE MILLIONS whose lives were displaced and diverted by the First World War was one Major Frank Holmes. But then, he had learned to be a wanderer long before the war. Born on a farm in New Zealand in 1874, he had first gone off to work in a South African gold mine and then, for the next two decades, specializing in gold and tin, had followed the itinerant life of a mining engineer all around the world—from Australia and Malaya to Mexico, Uruguay, Russia, and Nigeria. Holmes was robust and sturdy in stature, and assertive and headstrong in manner. A competitor once described him as "a man of considerable personal charm, with a bluff, breezy, blustering, buccaneering way about him." During World War I, he became a quartermaster in the British Army, and it was while on a beef-buying expedition to Addis Ababa, in Ethiopia, in 1918, that he first heard from an Arab trader about oil seepages on the Arabian coast of the Persian Gulf. As a mining engineer, Holmes found his interest piqued. Later, while stationed in Basra, in the area that would become Iraq, he took a further interest in what could be learned about the activities of the Anglo-Persian Oil Company on the Persian side of the border, and in what he was told about indications of petroleum along the Arabian coast.

After the war, Holmes helped set up a company, the Eastern and General Syndicate, to develop business opportunities in the Middle East. In 1920, he established the syndicate's first venture, a drugstore in Aden. But Holmes's heart was not in drugstores; it was in what had become his consuming passion and obsession—oil. He was convinced that the Arabian coast would be a fabulous source of petroleum, and he pursued his dream with unswerving stamina. A promoter par excellence, with a gift for making people believe in him, he traveled up and down the Arabian side of the Gulf, from one impoverished ruler to the next, spinning his vision, promising them wealth where they saw only poverty, seeking always to put another concession into his kit.[1]

Holmes undertook his campaign under the watchful, skeptical, and suspicious eyes of various British officials in the area, who were charged with overseeing the foreign relations of the local potentates and with protecting His Majesty's interests in the region. They saw Holmes as an unscrupulous troublemaker with a "capacity for mischief," who was trying, in pursuit of a quick profit, to undermine British influence in the area. To one official, Holmes was nothing more than "a rover in the world of oil." Perhaps the most damning comment of all came from another, who simply declared that Holmes was not "a particularly satisfactory individual." But that was not the judgment of the Arabs along the coast. To them, Major Holmes would become something altogether different—"Abu Naft," the "Father of Oil."

Leaving behind the drugstore in Aden, Holmes set up headquarters for his oil campaign on the small island of Bahrain, just off the coast of Arabia. He had been attracted to Bahrain by the reports of oil seepages. The Sheikh had no interest in oil but was, however, keenly interested in fresh water, which was in short supply. Holmes drilled for water, struck it, and made a nice profit on it. More important, the grateful ruler, as had been promised, in exchange rewarded him in 1925 with an oil concession.[2]

Holmes had already sewn up other oil rights. In 1923, he won an option to a concession in al-Hasa, in what was to become the eastern part of the kingdom of Saudi Arabia, and in the next year, to the Neutral Zone between Saudi Arabia and Kuwait, which was jointly controlled by the two countries. He also tried, unsuccessfully, to get one in Kuwait proper. And, as if all that were not enough to keep him busy, he was commuting to Baghdad from Bahrain, trying to put together a rival bid in Iraq against the Turkish Petroleum Company, thus assuring the further enmity of various governments and companies.

Holmes's activities alarmed, in particular, the Anglo-Persian Oil Company, which did not want anyone else operating within its "sphere of influence," causing trouble that could interfere with its operations in Persia. The company, to be sure, was convinced that there was no oil to be found in Arabia. In the words of John Cadman, the geological reports "leave little room for optimism," and one of the company's directors had declared in 1926 that Saudi Arabia appeared "devoid of all prospects" for oil. (Albania, the director had added, was the promising oil play.)[3]

In an effort to promote their prospects, in the face of such skepticism, Holmes and his Eastern and General Syndicate retained a prominent Swiss geologist to investigate eastern Arabia. But that effort backfired when the professor, whose expertise in Alpine geology singularly ill-prepared him for the desert, produced a report that declared that the region did "*not present any decided promise for drilling on oil*" and that exploration there "would have to be classified as pure gamble." Word of the damning report leaked out into the London financial community, making it even more difficult for the syndicate to raise the money it needed to support Holmes in his concession hunt and further drilling.

By 1926, the syndicate was in deep financial trouble. Holmes was continually obliged to fork out money for travel expenses, gifts and gratuities, and entertainment. So bleak was the syndicate's financial predicament that it was driven to try to sell all of its concessions to Anglo-Persian, but the company said no. After all, there was no oil in Arabia. And Holmes met a decidedly frigid reception when he tried to obtain capital in the City of London. Despite his persistence and salesmanship, he could not get anywhere. "Holmes was the worst nuisance in London," one English businessman recalled. "People ran when they saw him coming."[4]

Bahrain and the New York Sheikhs

With no likelihood of success in Britain, Holmes sailed for New York to work with an American named Thomas Ward—and hoping to find better luck with what Holmes called "the really big New York Sheikhs." But he met only further rejection. An executive from Standard Oil of New Jersey told him that Bahrain was too far away and too small to be of interest—after all, it was no larger on the map than his pencil point. Other companies were not interested because their attention was focused on their efforts to become part of the Turkish Petroleum Company.

But, at last, one American company did show a flicker of interest in Bahrain—Gulf Oil. It was committed to developing production on a diversified basis around the world as a hedge against general shortage or declines in specific producing areas. The company had almost been wiped out in its early years, at the turn of the century, when production gave out at Spindletop. Holmes provided Gulf with rock samples and a "greasy substance," along with a report of oil traces in his water wells in Bahrain. All this was sufficiently intriguing that, in November 1927, Gulf took overall rights claimed by Eastern and General to Arabian concessions, and agreed to work with Holmes's group to try to secure a concession in Kuwait. But a problem quickly emerged with the option. In 1928, Gulf became part of the American group in the Turkish Petroleum Company and thus a signatory to the Red Line Agreement, which precluded any one of the companies from operating independently in any area within the confines of the lines specified on the map. That clearly ruled out Saudi Arabia, as well as Bahrain. The companies had to act in unison or not at all. Notwithstanding Gulf's implorings, the TPC board was not prepared to take up Holmes's entire Arabian package. So, while Gulf could pursue Kuwait because it was outside the Red Line, it had to surrender its interest in Bahrain.

Gulf executives brought the Bahrain concession to the attention of Standard of California, which, like Gulf, was aggressively committed to developing foreign oil supplies, but which, despite very large expenditures, did not have one drop of foreign oil to show for its efforts. So Standard of California, known as Socal, took up Gulf's option to Bahrain. Unlike Gulf, Socal was not part of the Turkish Petroleum Company, and thus was not bound by the Red Line restriction. Socal set up a Canadian subsidiary, the Bahrain Petroleum Company, to hold the concession.[5]

And then both Socal, in Bahrain, and Gulf, in Kuwait, ran smack into a wall: the implacable opposition of the British government to the entry of American companies into the area. Before World War I, in an effort to ward off German

265

penetration in the Gulf region, Britain had made agreements with the local sheikhs, including those of Kuwait and Bahrain, that oil development should be entrusted only to British concerns, and that the British government would be in charge of their foreign relations. Thus, London insisted on a "British nationality clause" in any concession agreement, whether it be in Bahrain or Kuwait. Such a clause required that oil development be carried out by "British interests" and exclude American. The requirement meant that neither Gulf nor Socal could develop their concessions.

A rather nasty series of negotiations followed between Socal and Gulf on one side, backed up by the United States government, and the British government on the other. To the American companies, the "nationality clause" seemed nothing less than a cunningly constructed barrier to keep them out of the various sheikhdoms along the Gulf. Yet, in truth, the British government felt harried, beleaguered, and very much on the defensive before the greater American power, as it struggled to maintain positions it regarded as crucial to the empire.[6]

But in 1929, the British government reconsidered its position; the entry of American capital, it decided, would in all probability encourage more rapid and widespread development of oil in areas Britain controlled, which would be to the benefit both of local rulers, who always needed money and might otherwise ask Britain for further subsidies, and of the Royal Navy, which needed reliable oil supplies. Moreover, the diplomatic pressure from the United States was intensifying. The British government was willing to back off, at least insofar as Bahrain was concerned. So it made a deal with Socal. The American company could exercise the Bahrain option, though only under certain conditions that would guarantee Britain's position and political primacy. For instance, all communications from the company to the Amir were to pass through the political agent, the local representative of the British government.

The Bahrain Petroleum Company began drilling a little over a year later, in October 1931. And, on May 31, 1932, it hit oil. Petroleum had been discovered on the Arab side of the Gulf. Though only modest in production, the Bahrain discovery was a momentous event, with far wider implications. The established companies were quite shaken by the news. Over the course of a decade, Major Holmes, with his obsession about oil, had become a figure of condescension and ridicule. But now his instincts, and his vision, had been vindicated, at least to some small degree. Was he to be proved right on a much grander scale? After all, the tiny island of Bahrain was only twenty miles away from the mainland of the Arabian Peninsula where, to all outward appearances, the geology was exactly the same.[7]

Ibn Saud

In the early 1930s, Britain's political agent in Kuwait spoke of the ruler of neighboring Saudi Arabia as "the astute Bin Saud, who always takes the 'long view.' " In fact, Ibn Saud did not have the luxury of taking a very long view during those years. He had a pressing problem; he needed money for his treasury, and quickly. That was what led him to think about oil. To be sure, he was very skeptical about the prospects for oil in his country. And he was not at all happy about what the development of an oil industry would mean for his kingdom, in the unlikely event that petroleum was discovered. Foreign capital and technicians could disturb, perhaps even disrupt, traditional values and relationships. Granting a concession to *explore* for oil was, however, quite another thing, so long as it was reciprocated with the appropriate financial considerations.

Abdul Aziz bin Abdul Rahman bin Faisal al Saud was just over fifty years old. He had an imposing physical presence; at six feet, three inches tall, with a barrel chest, he towered over most of his countrymen. The impression he had made during a visit to Basra more than a decade earlier still held. "Though he is more massively built than the typical nomad sheikh, he has the characteristics of the well-bred Arab, the strongly marked aquiline profile, full-fleshed nostrils, prominent lips and long, narrow chin, accentuated by a pointed beard," observed a British official in Basra at that time. "He combines with his qualities as a soldier that grasp of statecraft which is yet more highly prized by the tribesmen." Ibn Saud, also known as 'Abd al-'Aziz, certainly applied his talents both for war and for statecraft to a most remarkable achievement in nation building, the creation of modern Saudi Arabia. And his later amassing of immense wealth was no less remarkable for a ruler who, in his early days, could carry his entire national treasury in the saddlebags of a camel.[8]

The Saudi dynasty had been established by Muhammad bin Saud, the emir of the town of Dariya in the Nejd, the plateau in central Arabia, in the early 1700s. There he took up the cause of a spiritual leader, Muhammad bin Abdul Wahab, who espoused a stern puritanical version of Islam that would become the religious mortar for the dynasty and its state. The Saudi family, allied with the Wahabis, began the rapid program of conquest that within half a century carried them to domination of much of the Arabian Peninsula. But the expansion of the Saudi realm alarmed the Ottoman Turks, who mobilized against them and defeated them in 1818. Muhammad's great grandson, Abdullah, was taken to Constantinople, where he was beheaded. In due course, Abdullah's son Turki reestablished the Saudi kingdom, this time centered in Riyadh, but this first Saudi restoration fell apart because of the struggle for power between two of Turki's grandsons. For a time, a third grandson, Abdul Rahman, was the nominal governor of Riyadh, under the sway of a hated rival family, the Al-Rashid. But in 1891 Abdul Rahman fled into exile, along with his household, including his son Abdul Aziz—the future Ibn Saud—who made part of the journey in a bag hung from a camel saddle. Abdul Rahman and his household wandered for two years, spending some months with a nomadic tribe deep in the desert. Eventually, they were invited by the Sabah family, which ruled Kuwait, to take up residence in that small city-state on the Persian Gulf.

For his part, Abdul Rahman had two goals: to reestablish the Saudi dynasty as master of Arabia, and to make universal the Wahabi branch of Sunni Islam. His son, Ibn Saud, would be the instrument to both ends. Mubarak, the Amir of Kuwait, took the young Saudi prince under his wing and gave him an expert education in *realpolitik* and the judicious making of foreign policy. Mubarak helped teach him, as Ibn Saud later put it, how to "consider our advantage and our harm." The boy was also given a stern religious education, prescribed a Spartan life, and learned at an early age the arts of war and of survival in the desert. He was soon given an opportunity to apply those arts when the Turks incited the Rashids, the traditional enemy of the Saudis, to attack Kuwait, which was then under British protection. As a diversionary measure, the Amir of Kuwait dispatched Ibn Saud, then just twenty years old, to try to retake Riyadh from the Rashids. He led a small band across the desert, only to have his first assault repulsed. On the second attempt, combining stealth with force, Ibn Saud entered the city by night and, at dawn, slew the Rashids' governor. In January 1902, his father proclaimed him, at age twenty-one, Governor of Nejd and Imam of the Wahabis. He had begun the second al-Saud restoration.

Over the next several years, in one military campaign after another, Ibn Saud established himself as the acknowledged ruler of central Arabia. Around that time, he also made himself the leader of the Ikhwan, or Brotherhood, a new movement of intensely religious warriors, whose rapid spread in Arabia provided Ibn Saud with a pool of devoted soldiers. During 1913–14, he brought eastern Arabia, including the large, inhabited al-Hasa oasis, under his control. And since the population there was mainly Shia Moslems—whereas the Saudis were Sunni, and not merely Sunni but of the strict Wahabi sect—he paid special attention to the administration and schools of al-Hasa, regularizing the Shias' status and preventing their harassment. Despite the tenets of Wahabism, Ibn Saud was an astute politician who knew that it was in his political interest not to encroach too far upon the sensitivities of the Shias. "We have thirty thousand of the Shi'ah, who live in peace and security" in the Hasa, he once said. "No one ever molests them. All we ask of them is not to be too demonstrative in public on their fete-days."

The last territories critical to the Saudi empire were added in the years immediately after World War I. Ibn Saud captured northwest Arabia. Then, in 1922, the British High Commissioner, exasperated at disputes involving Ibn Saud and the Amir of Kuwait, took a red pencil and himself fixed the boundaries between them. He also delineated two "neutral zones" along Ibn Saud's borders, one shared with Kuwait, the other, with Iraq—called "neutral" in both cases because the Bedouin would be able to pass back and forth to graze their flocks and because they would be jointly administered. By December 1925, Ibn Saud's troops, the ferocious Ikhwan, had captured the Hejaz, the holy land of Islam, on the western side of the peninsula, bordering on the Red Sea. There lay the port of Jidda and the two holy cities of Mecca and Medina. In January 1926, in the Great Mosque of Mecca, after congregational prayers, Ibn Saud was proclaimed King of the Hejaz, making the Saudi dynasty the keeper of the holy places of the world's Moslems. And so, at the age of forty-five, Ibn Saud was the master of Arabia. In the course of a quarter of a century of skillful war making and astute politics, he had reestablished Saudi ascendancy over nine-tenths of the Arabian Peninsula. The restoration was virtually complete.[9]

But then the warriors of his expansion, the Ikhwan, began to criticize Ibn Saud for backsliding from Wahabism. They declared that the instruments of modernity that were starting to find their way into the kingdom—the telephone, the telegraph, radio, the motorcar—were all tools of the devil, and they rabidly criticized Ibn Saud for having any truck at all with the infidel British and other foreigners. Increasingly insubordinate, in 1927 they rose up in rebellion against him. But he defeated them, and by 1930 had destroyed the Ikhwan movement. Ibn Saud's control of Arabia was now secured. Thereafter, the scales would tilt away from conquest and toward caution, to safeguard and enhance the nation that he had constructed over a thirty-year period. To commemorate the consolidation, the name of the realm was changed in 1932 from the "Kingdom of the Hejaz and Nejd and Its Dependencies" to the name by which it is known today— Saudi Arabia.[10]

But just at the moment when Ibn Saud's efforts appeared to have won the crown of success, yet another threat appeared. Simply put, Ibn Saud was fast running out of money. With the onset of the Great Depression, the flow of pilgrims to Mecca—all Moslems able to do so were expected to try to make at least one such pilgrimage in the course of their lives—slowed to a trickle, and they were the major source of the King's revenues. The kingdom's finances fell into desperate straits; bills went unpaid; salaries of civil servants were six or eight months in arrears. Ibn Saud's ability to dispense tribal subsidies constituted one of the most important glues bonding a disparate kingdom and unrest developed throughout his realm. To make matters worse, the King had just embarked on an expensive and varied development program, ranging from the establishment of a domestic radio network, to tie the country together, to the reconstruction of the water supply system for Jidda. Where could alternative sources of money be found? Ibn Saud tried to collect taxes a year in advance. He dispatched his son Faisal to Europe to look for aid or investment, but he was unsuccessful. As his financial problems continued to grow, the King did not know where to turn for help.[11]

The Sorcerer's Apprentice

Perhaps valuable resources were hidden under the soil of his realm. Such was the idea suggested to King Ibn Saud by a companion during an automobile ride, probably in the autumn of 1930. The companion was an Englishman, a former official in the Indian Civil Service, who had set himself up as a merchant in Jidda and had, just a few months earlier, converted to Islam under the tutelage of Ibn Saud. The King had personally given him his Islamic name, Abdullah. But his real name was Harry St. John Bridger Philby, known as Jack to his English friends, and now, perhaps, remembered best as the eccentric father of one of the most notorious double agents of the twentieth century, Harold "Kim" Philby, who became the head of anti-Soviet counterespionage in British intelligence, while actively spying for the Soviets. He might well have taken lessons from his father on how to play multiple roles. Indeed, many years later, on reading Kim Philby's own account of his years as a double agent, the retired court interpreter for Ibn Saud could only marvel that Kim was "a true replica of his father."

The father, Jack Philby, was a relentless contrarian, an inveterate rebel against authority and convention. In Jidda, he once paraded his pet baboons in public to demonstrate that he could do without the human company of the small European community. Raised in Ceylon and a graduate of Trinity College, Cambridge, Philby began his career in the Indian Civil Service. He was a member of the British political mission in Baghdad and Basra during World War I, which provided his introduction to the Arab world. A gifted linguist, he took the opportunity to study Arabic, which led him to take a deep interest in the genealogy of Arab tribes and potentates. That, in turn, developed into a lifelong fascination with perhaps the most powerful of all the potentates at that time, Ibn Saud, whom he first met while on a mission to Riyadh in 1917. That meeting, which included thirty-four hours of personal interviews with Ibn Saud, set the course of the rest of Philby's life.

In 1925, angered by British policy in the Middle East, Philby quit the Indian Civil Service, of which he was still a member, though he was then serving in Transjordan. He returned to Saudi Arabia to establish a trading company in Jidda. He also renewed his friendship with Ibn Saud, and over time became an informal adviser to the King, traveling and hunting in his party, even joining in the evening deliberations of the King's Privy Council. Ibn Saud took a special interest in Philby. On the eve of his conversion to Islam in 1930, Philby would recall, the King told him "how nice it would be for me when I became a Muslim and could have four wives." But first he was required to go through the painful process of adult circumcision. Some said that Philby had no particularly strong religious convictions, and that he became a Moslem in order to facilitate his business dealings and his ease of movement through the country. The conversion did enable him to pursue one of his obsessions, for which he became deservedly famous—as explorer, mapper, and chronicler of Arabia. Over the years, his arduous journeys took in great parts of the peninsula, from a lonely expedition through the Rub al-Khali, the Empty Quarter in the southeast of Arabia, to a search for the ancient Jewish communities in northwestern Arabia. In recognition of his efforts, he was eventually to receive the Founder's Medal of the Royal Geographical Society.

Philby would don the appropriate bowler hat on his trips back to England or put on the white jacket required at dinner in the colonies of the empire, and even in Arabia, he took five o'clock tea and fanatically kept up with the cricket scores from Lord's. Yet, despite all that, he remained at odds with Britain and British policy, which he saw as "traditional western dominance in the eastern world." In contrast, he was to recall proudly, "I was surely one of the first of the champions of eastern emancipation from all foreign controls." Certainly the British found Philby most troublesome. "Since he retired from Govt. service 5 years ago, Mr. Philby has lost no opportunity of attacking & misrepresenting the Govt. & and its policy in the Middle East," commented one British official. "His methods have been as unscrupulous as they have been violent. He is a public nuisance, & it is largely due to him & his intrigues that Ibn Saud—over whom he unfortunately exercises some influence—has given us so much trouble during the last few years." Another official denounced Philby as "an arch-humbug." [12]

Whatever the extent of his influence, Philby knew very well Ibn Saud's severe financial problems and the threat they posed to the kingdom. During that automobile ride in the autumn of 1930, Philby, noting that Ibn Saud was particularly despondent, said, as cheerfully as he could, that the King and his government were like folk asleep on buried treasure. Philby was convinced that great mineral wealth lay beneath the desert. But its development required prospecting, explained Philby, and that meant foreign expertise and foreign capital.

1

"Colonel" Edwin Drake, in top hat, stands in front of the first oil well near Titusville, Pennsylvania, in 1859. The title of "Colonel" had been invented to impress the local backwoodsmen, who thought Drake was crazy for trying to drill for oil.

George Bissell, the father of the oil industry, figured that a business could be made out of selling "rock oil." An advertisement for patent medicine gave him the idea of drilling for it.

2

The fever that swept America in the wake of the first oil boom was captured in the popular music of the day.

3

The Shoe and Leather Petroleum Company, foreground, at Oil Creek, Pennsylvania, in 1865.

4

John D. Rockefeller, repelled by the chaos of the oil business, created the great Standard Oil Trust, which soon dominated the industry and made him the richest man in America.

5

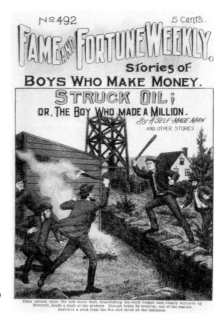

For boys with pluck, the oil business was a sure way to fame and fortune.

6

Arthur Bates, an oil dealer in Geneva, Ohio, delivered kerosene—the "new light"—door-to-door from his horse-drawn tank wagon.

8

Ida Tarbell, muckraker and America's first great woman journalist, fearlessly exposed Standard Oil. Her main target, John D. Rockefeller, called her "Miss Tar Barrel."

Tarbell followed up her attack on Standard with a scathing personal sketch of Rockefeller himself in 1905.

10

9

Standard director H. H. ("Hell Hound") Rogers became Tarbell's inside source as the result of an introduction from Mark Twain, whom Rogers had saved from bankruptcy.

The trust was busted. The headlines the day after the Supreme Court decision of May 15, 1911.

11

It was said that John Galey, the greatest wildcatter of the 19th century, could smell oil. He chose the drilling site in Texas that led to the creation of the Gulf Oil Company, vastly enriching the fortunes of the Mellon family of Pittsburgh.

13

On January 10, 1901, Captain Anthony Lucas's well blew in at Spindletop, the dramatic beginning of the oil industry in Texas.

12

14

The boom town of Beaumont, near Spindletop, whose prostitutes were arrested and then displayed on the balcony of the Crosby House. Each woman's fine was announced, and the man who paid it could keep her for twenty-four hours.

The French Rothschilds, led by Baron Alphonse (said to own the finest moustache in Europe) pioneered the development of the vast oil resources of Russia in competition with members of another famous family, the Nobels.

The young Joseph Stalin fomented strikes and insurrection among the oil workers of Baku in the early 1900s. Forty years later, as leader of the Soviet Union, he would fiercely defend the same oil fields from the invading Germans.

The Baku oil fields, set ablaze during the Revolution of 1905. Political and social upheaval eventually drove the Rothschilds and the Nobels out of Russia.

18

Marcus Samuel, the London merchant who conceived and executed the great coup of 1892 that broke Standard Oil's grip on the worldwide kerosene market.

The discovery of oil in the jungle at Telaga Said beginning in 1885, on the island of Sumatra in the Dutch East Indies, made Royal Dutch a major competitor in world markets.

Henri Deterding, the "pushing fellow" who forged the merger of Royal Dutch with Shell and for the next quarter century was the world's most powerful oil man.

19

A shell box of the type that Marcus Samuel's father sold at Victorian seaside resorts. In honor of his father, Samuel named his new venture "Shell."

20

Henry Ford sitting in the first car he built in 1896. The automobile created a new market for oil just as the industry was losing its main market—for kerosene—to Thomas Edison's innovations in electricity.

22

23

Automobile races kindled wild enthusiasm for the new invention. The winner of this race, in 1905, made the 4400-mile trip from New York City to Portland, Oregon, in exactly 44 days.

By 1909, when auto lovers went motoring, no one any longer shouted "Get a horse!"

24

A "capitalist of the highest order," English financier William Knox D'Arcy acquired the huge Persian concession in 1901.

Nasir al-Din Shah granted several oil right concessions before his successor, Muzaffar al-Din, signed the D'Arcy Concession.

After five years of drudgery and disappointment, the engineer George Reynolds (left)—"solid English oak"—finally discovered oil in Persia in 1908, opening up oil development in the Middle East.

25

26

27

First Lord of the Admiralty Winston Churchill in 1911 with Admiral Jacky Fisher (right), the "oil maniac" who persuaded Churchill to convert the Royal Navy from coal to oil in preparation for a war with Germany.

28

In September 1914, as the huge German army closed in on Paris, the French commandeered the city's taxicabs to rush additional troops to the front. Oil had gone to war.

29

The rapid mechanization of the battlefield in World War I, including the tank and airplane, brought a new mobility to war and made oil an essential strategic commodity.

30

31

32

In 1919, Captain Dwight Eisenhower led an Army expedition through "darkest America" to dramatize the need for roads suitable for the new automotive age. Sometimes the convoy could do no more than six miles an hour.

A motorist fills up at a service station in Fargo, North Dakota, while gasoline is delivered to the station in a dependable horse-drawn wagon.

33

America's love affair with the automobile began in earnest in the 1920s when gasoline was abundant—and cheap.

34

35

The 1920s saw the birth of the secular temple of modern American civilization—the drive-in gasoline station. This is opening day at Phillips's first gasoline station in Wichita, Kansas, 1927.

Oil companies promoted brand names and trademarks in the 1920s to differentiate their products and win customer loyalty.

36   37   38

Calouste Gulbenkian, the wily and tenacious "Mr. Five Percent," became immensely rich by defending his interests in Iraqi oil. He always kept at least one mistress under the age of 18, even in his 80s, because his doctor insisted it was necessary to maintain his vigor.

39

John Cadman, chairman of Anglo-Iranian (later British Petroleum), at the airport in Tehran in 1933. He had just rescued his company's concession in Iran, which had been summarily nationalized by Shah Reza Pahlavi.

40

Standard Oil of New Jersey (later Exxon) emerged from the breakup of the Standard Oil Trust as America's biggest and most powerful oil company. It was dominated by "The Boss," Walter Teagle, grandson of Rockefeller's original partner.

41

42

The geologist Everette Lee DeGolyer, sitting on a porch near Tampico after his discovery in 1910 of what became Mexico's Golden Lane. By 1921, Mexico was the world's second largest oil producer.

The great "oil hunt" after World War I led to the Los Barroso gusher in 1922 and the Venezuelan oil boom.

43

44

Venezuela's Lake Maracaibo became one of the world's great oil sources, even though drilling was considered so risky that oil men joked about going into the fishing business.

45

As boom followed boom in Oklahoma in the late 1920s, mud was the only thing that could slow down the movement of men and oil.

Dad Joiner, promoter and wildcatter, shakes hands with the corpulent Doc Lloyd just after discovering the huge East Texas oil field on October 3, 1930. H. L. Hunt (smoking a cigar) rescued Joiner from financial difficulties—and in the process acquired his immensely valuable leases.

46

King Ibn Saud, founder of modern Saudi Arabia. When he began his campaign to restore his dynasty, he could carry his entire treasury in the saddlebag of a camel.

47

"Jack" Philby—father of spy Kim Philby—persuaded Ibn Saud in 1930 to open the doors of his country to oil exploration. Initially, the King would have preferred to find water.

48

49

Japanese noodle vendor Kokichi Mikimoto's success in developing cultured pearls in the late 1920s devastated the pearl-diving industry of Kuwait, forcing the tiny country to look to oil exploration for an alternative source of revenues.

50

President Lazaro Cardenas, announcing the expropriation of the foreign petroleum companies in Mexico on March 18, 1938—and fueling the bitter conflict between the producing countries and the international oil industry.

"Oh, Philby," the King replied, "if anyone would offer me a million pounds, I would give him all the concessions he wanted."

Philby admonished the King that no one would give a million pounds or anything like that without some preliminary exploration. The King was much more interested in exploring for water than oil. In that case, Philby had a name to suggest—Charles Crane, an American plumbing tycoon and philanthropist, who had a special interest in the Arab world, and who would, said Philby, "give one of his eyes for the pleasure of shaking hands with Your Majesty." Crane was sponsoring development projects in neighboring Yemen, and Philby knew he was then in Cairo. Why not invite him to Saudi Arabia, said Philby.

Ibn Saud extended the invitation, and on February 25, 1931, Crane arrived in Jidda, where the King received him with considerable pomp and lavish banquets. He was entertained by a mesmerizing sword dance performed by several hundred of the King's own bodyguards. As gifts, the King gave Crane bundles of rugs, daggers, and swords, as well as two pedigreed Arabian horses. The two men discussed the parched and stony desert and the possibility that there might be underground rivers beneath the Nejd. Crane recounted how he had obtained dates from Egypt and personally introduced their cultivation in the California desert, in a town called Indio, successfully irrigating them with artesian wells. Now, out of his new friendship for Ibn Saud, he made available, at his own cost, an American mining engineer named Karl Twitchell, who was then working on one of Crane's projects in Yemen, to investigate the water potential of the kingdom. After making an arduous 1,500-mile journey to assay the likelihood of artesian water beneath the Arabian desert, Twitchell showed up in Jidda in April 1931 with bad news: there were no prospects of artesian wells.[13]

A year later, in March 1932, as the gap between revenues and expenditures in his kingdom continued to widen, the King had a visit in Riyadh from an especially astute observer of his problems, Sheikh Ahmad, the Amir of Kuwait. The Sheikh had made the trip by motorcar over three hundred miles of desert sand and gravel, which had left him with an abiding lesson: All cars traveling that route should carry at least five passengers as "only five persons could get cars out of sand."

The two rulers pledged eternal loyalty to each other. When Sheikh Ahmad described Ibn Saud as his "elder brother," the King broke into tears, and he in turn declared, "Just as the Al Saud and the Al Sabah standards had flown side by side in every victory or defeat, during the last three hundred years, so he prayed and believed it would continue to do so in the future."

Sheikh Ahmad was struck by Ibn Saud's apparent ill-health and the general strain he was showing. Sharing his impressions with the British political agent on his return to Kuwait, the Sheikh said, "Gone were the days when he was the hardest man in his kingdom and led every raid and foray." Sheikh Ahmad begged the King "to go slow in the matter of expenditure," for otherwise "he would most assuredly 'crash.' " In particular, the Sheikh "talked 'straight' on the matter of the object of obvious waste that he saw all round him"—motorcars. Yes, for the King, a few good luxury cars were necessary. Beyond that, however, Sheikh Ahmad urged Ibn Saud to reduce the number of his cars by three-quarters "and standardize by concentrating on Fords and Chevrolets." That advice having been delivered, Sheikh Ahmad drove back across the desert in a gift presented to him by Ibn Saud—a large, eight-cylinder Cadillac limousine from the King's own fleet.

The two men had also discussed oil exploration. The King admitted that he had permitted some preliminary examination to be carried out, but added that "he was not anxious in the least to grant concessions to foreigners." Yet, given his financial difficulties, *did* he have a choice? Twitchell, the American engineer, had, in fact, reported some promising oil prospects in the al-Hasa, in the eastern part of the country. Then, on May 31, 1932, Standard Oil of California made its oil discovery on Bahrain. That abruptly and significantly increased the attractiveness of al-Hasa—and made Ibn Saud, on consideration, less averse to foreign investment in his kingdom. Twitchell, though insisting to Ibn Saud that he was only an engineer and not a promoter, nevertheless agreed at the King's behest to try to locate interest and capital in the United States.[14]

The Negotiation

Months before its Bahrain discovery, Standard Oil of California had already begun to inquire about a concession in al-Hasa. Now, contacted by Twitchell, Socal was not only delighted and immediately receptive, but also retained him as one of its negotiators. Twitchell returned to Saudi Arabia in February 1933 in the company of Lloyd Hamilton, a lawyer for Socal, to initiate their negotiations with Ibn Saud's minister of finance, Abdullah Suleiman. They were up against a cunning and masterly opponent. Suleiman was the brother of the King's private secretary. A Nadji by birth—most of the other senior administrators were Syrians, Egyptians and Libyans—he had, as a young man, been an assistant to an Arab merchant in Bombay, where he had learned much about trading and business. The King had nicknamed him "my support." In fact, this "frail little man of uncertain age" was the most powerful person in Ibn Saud's inner circle, with responsibility not only for finance, but also for defense and the Pilgrimage. He was "the ultimate eminence grise, always self-effacing and keeping himself in the wings," said Ibn Saud's interpreter, but with "power and influence so monumental that I often thought of him as the uncrowned King of Arabia."

Suleiman was certainly the most important man in the kingdom outside the royal family. He carried an enormous workload that was based upon an accounting system for public finances that he had invented and that only he could understand. He was inveterately secretive and peremptory, and kept as much control of affairs for himself as he possibly could, making sure that no potential rivals encroached on his terrain. Though he could generally act on his own authority, he took care, on the matter of oil, to send long messages to the King. In his negotiations with Socal about a concession in al-Hasa, Suleiman knew exactly what he wanted—a very large sum of money, and as soon as possible. Whether oil was there could be left to later.

Twitchell and Hamilton were not, however, the only contestants for access to the Hasa. The Iraq (formerly Turkish) Petroleum Company dispatched Stephen Longrigg as its representative. Longrigg, formerly a British official in Iraq, was also in effect representing the interests of Anglo-Persian, which, because of its participation in the IPC and the Red Line Agreement, could not go it alone. "So the stage is set," the British minister, Andrew Ryan, reported to London in March 1933, "the dramatis personae being an avid Abdullah Suleyman, who thinks of oil in Hasa as already a marketable commodity; Twitchell and Hamilton featuring Standard Oil of California; Longrigg . . . representing the Iraq Petroleum Company." But in his cast of characters Ryan had left out the most important player of all—the King. And he also made a major misjudgment in writing that Harry St. John Bridger Philby would be among "the minor possible personages." Philby was no mere bit player.[15]

At the time of the Bahrain strike in May 1932, Socal had sought out Philby in order, in the words of a Socal director, to "get in touch with His Majesty Ibn Saud." Philby dallied with Socal. But he knew that competition among the various oil companies would get a better deal for his friend the King, so he also made contact with the Iraq Petroleum Company through its dominant member, Anglo-Persian, alerting them to Socal's interest in al-Hasa. "I am not in any way committed to serve the interests of the said concern," he wrote to a senior Anglo-Persian geologist, "but I am generally disposed to help anyone practically interested in such matters and capable of being useful to the Government to get a move on." Ultimately, Philby signed on as an adviser to Socal. But he kept that arrangement secret. At the same time, he kept up his contacts with the IPC—so successfully that its representative, Longrigg, regarded him as a confidant. In fact, Philby's primary loyalty was, and would remain, to the King.

Philby took pleasure from his new association with Socal; helping an American company to succeed in Arabia would be yet another way to tweak the lion's tail and frustrate British interests in the area. The arrangement with Socal also brought him great personal relief. Though he was pursuing many projects for his trading company, he shared the problem of the rest of the kingdom: He was not being paid. He needed money urgently—among other things, in order to pay the Cambridge University fees for his son, Kim. For his services, Socal agreed to give Jack Philby one thousand dollars a month for six months, plus bonuses both upon the signing of a concession contract and on the discovery of oil. Thus, Kim Philby was able, after all, to pursue his studies at Cambridge, where he took the first steps on the way to becoming a Soviet spy.

As the negotiations dragged on, the Saudis left no doubt that their prime objective was a large up-front payment. "It is no good my holding out to you hopes that you can secure the concession without a substantial quid pro quo," Philby wrote to Socal. "The main point is that Ibn Saud's Government owes a good deal of money, and has had to default on its payments to its creditors. Its only hope of being able to pay them now depends on the mortgaging of its potential resources."[16]

The positions of the two Western groups were strikingly at odds. While Socal was very interested in obtaining the concession, the Iraq Petroleum Company, with Anglo-Persian behind it, was in quite a different frame of mind. Longrigg had confided to Philby "that they did not need any more oil, as they already had more in prospect than they knew what to do with. At the same time they were vitally interested to keep out all competitors." Thus, for IPC, its efforts were more prophylactic than prospective. In addition, since IPC—Anglo-Persian really—continued to be skeptical about the oil potential in al-Hasa, it was in no mood to make any large commitment in Saudi Arabia. Longrigg's brief, as he explained to the British minister, was certainly "not to buy a pig in a poke by paying big money at this stage for the right of extracting problematic oil."

Though others were growing frustrated by the pace of negotiations, Philby, who delighted in being a mystery man, was glorying in his multiple roles—working as a paid agent for Socal, acting as an adviser to the Saudis, coaching IPC and serving as Longrigg's confidant, and casually dropping in conversation with the various oil men what the King had said to him on their most recent auto ride up to Mecca. And it was not only oil that occupied Philby; he was also busily engaged in obtaining a monopoly on the importation of motor vehicles for the Saudi government and for the pilgrim-transport company, as well as in setting up the country's wireless system.[17]

Despite its eagerness, Socal was offering only about a fifth of what the Saudis were asking. In early April 1933, one of Socal's executives wrote Philby about "the unfortunate impasse to which our negotiations have come. . . . The country is practically unknown as to oil possibilities; and it would be the height of folly for an oil company to pay out large sums of money before having had a look at the geology of the area." Socal did not have to worry too much about IPC and Anglo-Persian. For their part, they were unwilling to come up with anything but a small fraction of what Socal was offering to pay. Finally Philby advised Longrigg, "You might just as well pack up; the Americans are far and away higher than that." And Longrigg did exactly that, departing abruptly and leaving the field open to Socal. Meanwhile, Philby was coaxing Socal and Suleiman toward what he called "this detente," which meant a much higher offer from Socal.

By May 1933, the final draft of the concession agreement between Socal and Saudi Arabia was ready for the King's "pleasure." After some pro forma discussion in the Privy Council, Ibn Saud told Abdullah Suleiman, "Put your trust in God, and sign." The agreement provided for a £35,000 ($175,000) payment in gold up front—£30,000 being a loan, and £5,000 as the first year's royalty paid in advance. After eighteen months, a second loan of £20,000 ($100,000) would be made. The total loan was to be repaid only out of any oil royalties the government was due. In addition, the company would make another loan of £100,000 ($500,000) in gold on the discovery of oil. The concession was good for sixty years, and covered about 360,000 square miles. On May 29, 1933, the agreement was signed. Ibn Saud had won the very substantial cash payments that he wanted. The King and his finance minister had also insisted on terms that provided a great incentive to Socal to move as expeditiously as it possibly could.[18]

The only remaining problem was how to obtain that much gold. Because America had just gone off the gold standard, Socal's efforts to dispatch the gold directly from the United States were turned down by Assistant Secretary of the Treasury Dean Acheson. But finally, the Guaranty Trust's London office, acting on behalf of Socal, obtained thirty-five thousand gold sovereigns from the Royal Mint, and they were transported to Saudi Arabia in seven boxes on a ship belonging to the P&O line. Care had been taken that all the coins bore the likeness of a male English monarch, and not Queen Victoria, which, it was feared, would have devalued them in the male-dominated society of Saudi Arabia.

The gaining of the concession by an American company would inevitably begin to change the web of political interests in the region. When Philby told Sir Andrew Ryan, the British minister, that Socal had won the concession, he was "thunderstruck, and his face darkened with anger and disappointment," which delighted Philby no end.* Britain's loss would indeed be America's gain, though Washington was slow to realize it. Despite recurrent protestations from Socal, the Roosevelt Administration refused to establish diplomatic representation, wearily repeating that there was no need for it. It was not until 1939 that the U.S. minister to Egypt was also accredited to Saudi Arabia, and only in 1942 did the United States establish a permanent, one-man legation in Saudi Arabia.

Anglo-Persian and the Iraq Petroleum Company soon realized that they had blundered by being too timid and too miserly. The members of the IPC recriminated among themselves but determined not to make the same mistake again. In 1936, the group obtained a concession for the Hejaz, the western part of Saudi Arabia, extending from Transjordan all the way down to Yemen. The terms were much higher than those agreed to three years earlier by Socal. The only drawback was the IPC never found oil in its concession.[19]

Kuwait

Saudi Arabia was not the only country on the Arabian Peninsula in which oil interest was mounting. Intermittent negotiations for a concession in neighboring Kuwait had been dragging on for a decade. The exploration effort in Bahrain had upset the Amir of Kuwait, Sheikh Ahmad. "It was a stab to my heart," he told Major Holmes in 1931, "when I observed the oil work at Bahrein and nothing here." The jovial and thickset Ahmad, who had become Amir of Kuwait in 1921, prided himself on his modernity; by the mid-1930s, he wore dress slacks and patent leather shoes beneath his robes. He was also an enthusiast of the British Navy, and the walls of his drawing room were decorated with photographs of British officers and war sloops. But he seemed to be engaged in a balancing act, arising out of Kuwait's precarious position. As a senior British diplomat explained, the Sheikh had "set out to run a rather dangerous policy" of trying "to play His Majesty's Government, the Iraqi government and King Ibn Saud off against each other."

Such balancing had always been the central problem for Kuwait, as a small state trying to assure its independence and freedom of action midst larger powers. It had long played a commercial role owing to its location near the head of the Persian Gulf and along the trade and pilgrimage route between Basra and

---

* Philby was to do well out of his work for Socal. But being able to pay for the education of his son Kim at Cambridge did not necessarily guarantee the outcome he had hoped for. Kim had done miserably, getting thirds—almost flunking—on his final exams. Jack Philby wanted his son to go into the Civil Service. But two of Kim's tutors refused to support his application because of the young man's already obvious communist proclivities. Jack Philby was outraged. Kim was entitled to his "leanings towards communism" and should not be victimized for "views honestly held," the father wrote to one of the tutors in 1934. "The only serious question is whether Kim definitely intended to be disloyal to the government while in its service." He doubted that. Jack Philby's wife was more pragmatic. "I do hope," she wrote of their son, "he gets a job to get him off this bloody communism."

275

Mecca. Its emergence as an independent principality dates from the middle of the eighteenth century, when nomadic tribes from the interior of the Arabian Peninsula settled there and, in 1756, selected a sheikh from the Al Sabah family as the ruler. By the nineteenth century, it had become the emporium for commerce in the upper Gulf. Though paying some tribute to the Ottoman Empire, it successfully resisted direct application of Turkish authority. At the end of the nineteenth century, Britain wanted to blunt German penetration, as represented in the Berlin–Baghdad railway, and Kuwait wanted to assure its independence of the Ottomans. As a result, Britain assumed responsibilities for Kuwait's foreign affairs and later established a protectorate over the emirate.

Now Sheikh Ahmad was being courted both by Anglo-Persian and Gulf. Having acquired Major Holmes's purported and controversial option, Gulf was working through Holmes and his Eastern and General Syndicate (which the Foreign Office had taken to calling Gulf's "jackal"). Anglo-Persian was still skeptical about prospects for oil in Kuwait. Moreover, if exploration did happen to prove successful, it would only add more oil to a world market already laboring under great surplus. And there was always the fear among Anglo-Persian executives that in Iran, the site of their most valuable concession, the Shah would "revive the accusation that they are frittering away their energies elsewhere than in Persia." Then why was Anglo-Persian pursuing a concession in Kuwait? The reason was that it could not take the chance of simply standing aside in Kuwait, if only to block someone else from getting a concession. Anglo-Persian's prime interest was defensive—to prevent another company from moving in on what it called its "flank," threatening to undermine its position and influence in Persia and Iraq. The risk was too great. Kuwait was, as Sir John Cadman continued to insist, within Anglo-Persian's "sphere of influence."[20]

Financial need also fueled Sheikh Ahmad's interest in courting concessionaires. Like all the other sheikhdoms down the coast of the Persian Gulf, Kuwait was suffering severe economic hardship. The local pearling trade had been Kuwait's number-one industry and principal source of foreign earnings. Whether or not he knew the name, Sheikh Ahmad had good reason to be intensely annoyed with a Japanese noodle vendor from Miye prefecture, one Kokichi Mikimoto, who had become obsessed with oysters and pearls and had devoted many difficult years to developing the technique for cultivating pearls artificially. Eventually, Mikimoto's efforts paid off, and by 1930, large volumes of Japanese cultured pearls were beginning to appear on the world's jewelry markets, practically destroying the demand for the natural pearls that divers brought up from the waters off Kuwait and elsewhere in the Persian Gulf. Kuwait's economy was devastated; export earnings plummeted, merchants went bankrupt, boats were laid up onshore, and divers returned to the desert. Ahmad and his principality needed a new source of revenues; the welcome prospect of oil had appeared just in time.

The little country faced a number of other economic difficulties. The Great Depression had more generally crippled the economies of Kuwait and the other sheikhdoms. So bad had conditions become that slaveowners along the Arab coast were selling off their African slaves at a loss, to avoid the maintenance costs. Moreover, Sheikh Ahmad was angry with Britain for failing to give him what he deemed adequate support in various controversies with neighboring Saudi Arabia and Iraq. The Sheikh believed that the entry of an American oil company into Kuwait would bring American political interest, which he could use to bolster his position against Britain, as well as against regional rivals. Yet with all that said, the Sheikh knew that he dare not alienate Britain. He still depended primarily on it for Kuwait's political and military security against all his neighbors—Saudi Arabia; Iraq, which was challenging his rights; and Persia, which did not even acknowledge Kuwait's existence and legitimacy. Kuwait was a very small state; it was Britain's imperium that reigned over the Gulf, and the Sheikh recognized the practical value of the Royal Navy.[21]

For its part, the British government wanted to do everything it could to maintain its influence and position in the region, and that meant trying to ensure that any concession went to an English company. But how to do it? Though the British Nationality Clause had been put aside in the case of Bahrain, London continued to insist upon it for Kuwait, which would have effectively barred Gulfs participation with the Eastern and General Syndicate by restricting development only to a British-controlled firm. Gulf protested the exclusionary policy to the U.S. State Department, which in turn pressed the issue with the British at the end of 1931.

The British Admiralty vigorously insisted on retaining the nationality clause, not only on the familiar strategic and military oil supply grounds, but also because of the supposed difficulty that Britain would face in guaranteeing "the protection of American citizens in the hinterland of Kuwait." That might even result in "American warships intervening in Gulf affairs in order to provide the protection" that Britain "might not be able to offer." But the central fear was, as one official put it, of Britain's "losing influence and position to another, richer nation, in an arena critical to its imperial interests." After further reflection, however, key departments of the British government—the Foreign Office, the Colonial Office, and the Petroleum Department—were all prepared to jettison the nationality clause. "The last thing we want," said one Foreign Office official, "is an oil war" with the United States. Indeed, American capital could contribute to political stability and economic development in the area, which was in British interests. In April 1932, the British government put the nationality clause aside. At that juncture, there seemed no great cost, and no real reason not to do so. After all, Anglo-Persian seemed hardly interested in exploring for petroleum in Kuwait. Sir John Cadman, the chairman of Anglo-Persian, told the Foreign Office that any oil found in Kuwait would not "be of interest to the Anglo-Persian Oil Company." He added, "The Americans are welcome to what they can find there!"[22]

Gulf and the United States government were pleased by the Cabinet's decision to eliminate the nationality clause. But no one was more jubilant than Major Holmes. He attributed the "wonderful victory," at least in good part, to an individual he decided was the most popular man in England, the American ambassador Andrew Mellon—the former U.S. Treasury Secretary and scion of the family that controlled Gulf Oil. Having arrived at his new post in 1932, at the age of seventy-seven, Andrew Mellon was more than comfortable in London. He enjoyed the fact that he could get a legal drink (Prohibition was still in force in the United States); he had been married in England; he habitually dressed in English-cut suits. And he knew how to do business in Britain. Almost exactly three decades earlier, he had gone to England to try to persuade Marcus Samuel that Shell should let the nascent Gulf Oil Company out of its supply contract, which had gone sour when the underground pressure at Spindletop dwindled away. With his quiet and persistent charm, Mellon had succeeded.

In 1932, however, he was under a cloud. There were several instances during his tenure as Secretary of the Treasury in which, it was said, companies belonging to the huge Mellon empire had gotten special treatment or support. Such reports had already led to an effort in Congress to impeach him as Treasury Secretary when Hoover abruptly appointed him to the Court of St. James's. Some described his ready acceptance as a form of voluntary and prudent self-exile.

Mellon was not merely the family patriarch and uncle of Gulf's chairman, William Mellon; he was also the man who had bankrolled Gulf and pushed it into becoming an integrated oil company. He continued to regard Gulf as a Mellon family company and took a very personal interest in it. He had already intervened to facilitate State Department help for Gulf in its quest to open the door to Kuwait. When he sailed for London as ambassador, which could put him in the middle of the Kuwait fight, the Undersecretary of State had squeamishly tried to establish fair ground rules: "It would always be very easy to lean over backwards too far for the sake of preventing criticism," he cabled the American embassy in London. "In all that we do, therefore we must accord the Gulf Oil Company no more or no less, but precisely the same, assistance that we should accord to any other bona fide American company under similar circumstances." But it would be mighty hard to maintain such a distinction. Even in the State Department, Gulf was described as a "Mellon interest"; the British would refer, interchangeably, to Gulf and the "Mellon oil group." And Andrew Mellon himself never gave much indication that he recognized the distinction. He referred to Gulf as "my company" (hardly without foundation, as Mellon interests owned most of the stock) and he would act on that premise.[23]

While jettisoning the nationality clause for Kuwait, London had nevertheless announced that it would insist upon reviewing all bids and recommending to the Amir which one he ought to accept. That should not have been too complicated a matter, as Cadman had just flatly announced that Anglo-Persian was not interested. But then, in May 1932, Socal made its discovery in Bahrain, transforming the situation and perspective along the entire Arabian coast. Anglo-Persian quite abruptly changed its mind. Cadman hurriedly wrote to the Foreign Office to disavow his recent statement of no interest; for now Anglo-Persian had suddenly decided that it very much wanted to bid for a concession in Kuwait. No one was more pleased with Anglo-Persian's change of heart than the Sheikh himself, who stated with eloquent simplicity a fundamental maxim of business. "Yes, I have now two bidders," he said, "and from the point of view of a seller that is all to the good."

The next move lay with the British government; it was up to the Petroleum Department, in particular, to review not only Gulf's offer, but the new bid from Anglo-Persian, and to afford an "opinion" to the Amir. But as the review of the two bids dragged on in London, Holmes and Gulf—and the U.S. government—

became suspicious, believing that the delay was a ruse that would lead to a recommendation in favor of Anglo-Persian's offer. The American embassy kept after the matter, though the State Department did not wish to appear to be acting "merely for the personal benefit of Mr. Mellon." But, by the autumn of 1932, when no recommendation seemed to be forthcoming, Mellon lost patience and decided to forget about decorum and pursue the matter directly with the Foreign Office. After all, this was business. Perhaps his sense of urgency increased as it became evident that the unpopular Herbert Hoover was soon to be turned out of the White House, thus bringing Mellon's own stint as ambassador to a quick termination. "The fact that the American Ambassador has so keen a personal interest in securing the concession for his own group and that his term of office is now coming to an end," observed a senior Foreign Office official, "may also afford some explanation of the repeated and insistent representations." Indeed, so vigorous was Mellon's pursuit that one State Department official recommended that the Secretary of State suggest to Mellon "that he go easy on this question."

The Petroleum Department finally disgorged its analysis of the two bids, which the British political agent in Kuwait transmitted to the Amir in January 1933. But it resolved nothing; all it did do was open a new, more acrimonious stage of competition between Anglo-Persian and Gulf, involving an acid exchange of accusations and threats. But Anglo-Persian felt its hand weakening. Its position in Persia, the company's real treasure, was in jeopardy, owing to the Shah's unilateral repudiation of its concession there in November 1932.[24]

There was, indeed, an alternative to a bidding war—cooperation. Each company was impressed by the other's strong determination, and by the powerful forces behind it. While Anglo-Persian saw America's wealth and its potentially great political influence, Gulf saw entrenched British power in the region. John Cadman raised the possibility of amalgamation with Ambassador Mellon, but with no clear answer. Shortly after Mellon vacated his post and went back to the United States, Cadman was distressed to learn that the word in oil circles in America was that "Andy Mellon had returned determined to keep his hands on Kuwait."

At the end of March 1933, Cadman left London on his way to Persia to negotiate with the Shah over the canceled concession. He stopped in Kuwait, fully prepared to discuss the details of a concession with the Amir. Learning of Cadman's imminent arrival, Major Holmes not only arranged to see the Sheikh Ahmad a few hours before Cadman's appointment but also extracted a promise that he would be given the opportunity to top whatever offer Cadman might put on the table. In his own meeting at Dasman Palace, Cadman tried to get the Sheikh to agree that an "All British Company" would better serve his purposes. The Sheikh replied that "it was a matter of indifference" to him "what nationalities were involved so long as the payments stipulated in the Agreement were made." Cadman then put his own bid, all prepared, on the table and produced a gold pen, which he gave to the Amir in order to sign the agreement. He told the Sheikh that he would double the offer "if the Sheikh was prepared to sign the Agreement forthwith." But, added Cadman, "he was unable to leave his higher offer open." Unfortunately, the Sheikh could express only his most earnest regrets. He had promised Holmes that the Gulf group would be given an opportunity to better any offer Cadman might present, and he obviously could not go back on his word.[25]

Cadman was surprised and upset. Now he was absolutely convinced that an agreement had to be made with Gulf; the Sheikh's "two buyers" had to be reduced to one, at all costs. Otherwise, the Sheikh could continue to play one group off against the other, pushing up the price. Moreover, the only way that Anglo-Persian could absolutely guarantee that it did not lose out in a bidding match was by establishing a joint venture with Gulf. Strenuous discussions followed between the two companies, and by December 1933, they came to final terms, establishing a new fifty-fifty joint venture. It was called the Kuwait Oil Company. Yet still fearful of the expansionist power of the American companies, the Foreign Office insisted that actual operations on the ground of the Kuwait Oil Company had to be "in British hands." The result was a further agreement in March 1934, between the British government and the Kuwait Oil Company that assured, despite Gulf's 50 percent, British dominance over development within the country.

The actual negotiations to obtain a concession for the new Kuwait Oil Company from Sheikh Ahmad were entrusted to two men, the venerable Frank Holmes for Gulf, and the much younger Archibald Chisholm for Anglo-Persian. When the two crossed over through the customs post from Iraq into Kuwait, they were met by a letter from the political resident, genially offering his "welcome to the heavenly twins." Indeed, the competition between the two companies appeared to have run its course. One Sunday morning, not long after their arrival in Kuwait, Holmes and Chisholm found themselves sitting beside each other at the tiny church service run by an American mission. The lesson that day was from Beatitudes, and when the words "blessed are the pure of heart" were spoken, Holmes jabbed Chisholm in the ribs. "At last," whispered the redoubtable major, "you and I are pure in heart about each other."

But their work was far from over. If he had been forestalled from pitting one bidder against another, Sheikh Ahmad proved to be a very tough negotiator, and was extremely well informed about political developments and concession terms in Iraq, Persia, and Saudi Arabia. Moreover, the Sheikh was not at all pleased with the political agreement about British dominance on which London had insisted. But finally, on December 23, 1934, Sheikh Ahmad, having obtained what he wanted, affixed his signature to the agreement, which granted a seventy-five-year concession to the Kuwait Oil Company. The Sheikh received an upfront payment of £35,700—$179,000. Until oil was found in commercial quantities, he would receive a minimum of £7,150 ($36,000) a year. Once oil was found, he would receive an annual minimum of £18,800 ($94,000), or more—depending on volume. And as his representative to the Kuwait Oil Company in London, the Sheikh appointed his old friend Frank Holmes, who was to hold the position until his death in 1947.[26]

The "Sure Shot"?

The Kuwait concession was signed a year and a half after the Saudi one. By then Standard Oil of California was already busily at work in Saudi Arabia. It had set up Casoc—the California-Arabian Standard Oil Company—to hold the concession, and administrative headquarters had been established in Jidda, in a tall building with multiple balconies and its own electric generating plant. The landlord was none other than H. St. John B. Philby. On the other side of the country, in September of 1933, the first two American geologists had arrived in the town of Jubail on a motor launch from Bahrain. In order to downplay their strangeness to the local population, they had grown beards and had donned Arab headdresses and outer robes. They docked early in the morning, and by evening had made their first excursion into the desert. A few days later, they came to a hilly area they had already spied from Bahrain and identified a promising geological structure, the Dammam Dome. It was a desolate expanse of sand and naked rock, only twenty-five miles from a similar structure on Bahrain where Socal had found oil. It was a "sure shot," they were convinced. Drilling commenced in the summer of 1934. Every item that the geologists, engineers, and construction workers needed—be it equipment or food—had to be brought in over a supply line that stretched back to the port of San Pedro, near Los Angeles. The early optimism notwithstanding, the Dammam Dome was not a sure shot. The first several wells were all failures—either dry or, at best, giving some small shows of oil and gas, but nothing remotely commercial.

Over the next few years, more American geologists arrived; they fanned out through the desert, often traveling by camel, with an escort of ten guards and assorted guides. Conditions were extreme; the daytime temperatures would get up to 115 degrees, while the nights became bitterly cold. They would set out in September from Jubail and not return until the following June. Their guides measured distances for them not in miles or kilometers, but in "camel days." When they got deep into the desert, three weeks away from Jubail, they were beyond the range of the shuttle of supply camels and hunted their own gazelle and Qatar birds, or bought a sheep for five riyals (about $1.35) from a passing Bedouin. But they also put the new techniques of seismography to good use, and they made aerial surveys of the country, using a one-engine Fairchild 71, with a hole cut out of the bottom through which to take photographs with film especially manufactured by Kodak to withstand desert heat. The plane flew straight parallel courses, six miles apart, while geologists sat by the windows, drawing everything they could see for three miles in each direction. There were hints of oil, but only hints.

Socal's management back in San Francisco was becoming increasingly anxious about the project. The mood about the Saudi concession was such, one executive was later to remember, that "sometimes there was an open question whether the venture should be abandoned and the approximately $10 million spent written off as a total loss." Yet there was another alarming possibility—that Socal would find oil in a part of the world where it had no distribution facilities, and at a time when global oil markets, like the rest of the world economy, were depressed and suffering from acute oversupply. In other words, what would Socal do if it actually discovered any petroleum in the desert of Arabia?[27]

The Blue Line Agreement

In fact, Socal was already encountering that nasty problem because of its success on Bahrain, where the existing production capacity was 13,000 barrels per day and the potential capacity was estimated at 30,000 barrels per day. In the first half of 1935, Socal choked down its production in Bahrain to just 2,500 barrels per day due to lack of access to markets. It found great difficulty in selling directly to European refineries because most of them were not equipped to handle a crude such as that from Bahrain, with high sulfur content. A proposed marketing deal with Standard Oil of New Jersey, Shell, and Anglo-Persian fell apart. Socal needed something else—something more stable. The answer was a joint venture of its own.

Early in 1936, a discouraged K. R. Kingsbury, president of Socal, arrived in New York City. James Forrestal, head of the investment bank Dillon, Read, brought Kingsbury (known as "the King") together with the top management of Texaco. Forrestal had recognized that Texaco had a problem no less serious in its own way than that facing Socal. It had an extensive marketing network in Africa and Asia, but did not have its own crude in the Eastern Hemisphere to run through the system, and thus was shipping products from the United States. Without a Middle Eastern supply, Texaco faced the prospect of losing markets or losing money in the years ahead. It was obvious to Forrestal that marrying Socal's low-cost potential Middle East crude to Texaco's Eastern Hemisphere distribution system would make great sense for both companies.

But how to do it? Forrestal, assisted by Dillon, Read vice-president Paul Nitze, worked out a scheme that created a major new enterprise. Socal and Texaco would pool all their assets "East of Suez," with each having an equal interest in the new venture. Socal threw in its Bahrain and Saudi oil concessions, as well as a concession in the East Indies. The joint venture also took over Texaco's widespread marketing system in Africa and Asia. The other companies may have had their Red Line; Socal and Texaco delineated their consolidated area by what they called the "Blue Line." The California-Texas company, or Caltex, as their joint venture became known, would provide the vitally needed outlet both for Bahrain production and for any oil that might eventually be found in Saudi Arabia.

The established international companies, which had been quite concerned about the disruptive impact of Bahrain oil competing for markets, were relieved by Socal's hook-up with the Texas Company. While still complaining that Socal's activities on Bahrain were "irksome" and that it would "probably have to try to buy them out," the IPC, along with Shell and Anglo-Persian, told the Foreign Office that the joint venture would cause "a minimum of disturbance of markets which is all to the good from the point of view of British oil interests." A Jersey executive put it a little differently. The merger "would mean a fair degree of stabilization." The establishment of Caltex also meant that any new oil found in Saudi Arabia could be managed and would not necessarily destroy prices. As for neighboring Kuwait, it was already in the reliable hands of Anglo-Persian and Gulf.[28]

Discovery

Exploration in Kuwait had begun in 1935, but only in 1936 was seismic work undertaken. The Burgan field in southeastern Kuwait was suggested as the most promising area. And there petroleum was struck, unexpectedly and with a surprisingly large flow, on February 23, 1938. In order to gauge the size of the discovery, crude oil was allowed to flow unrestrictedly into an adjoining sand reservoir, then ignited. The heat from the flaming oil was so intense that the sand walls of the banked reservoir were transformed into sheets of glass. The directors of Anglo-Persian and Gulf heaved a mighty sigh of relief. Major Frank Holmes was jubilant, while at Dasman Palace, Sheikh Ahmad needed to have no further worry about the economic threat from cultured pearls.[29]

Meanwhile, exploration next door in Saudi Arabia had met with repeated discouragements, and the Socal board grew increasingly restive. In November 1937, Socal's manager of foreign production cabled Arabia with the firm order that no more projects were to be initiated without the submission of a detailed proposal first. Then, in March 1938—a few weeks after the Kuwait discovery—came the stunning news: at 4,727 feet, large quantities of oil were tapped in Well Number 7 in the Arab Zone. Thus, discovery was finally made, almost three years after drilling had first begun on Dammam Number 1. Ibn Saud and Saudi Arabia were on the road to fortune. The kingdom's unity would no longer be dependent upon—or vulnerable to—the fluctuations in the number of the faithful who made the pilgrimage to Mecca.

The discovery of oil in Saudi Arabia set off fevered efforts to obtain concessions there, not only by the Iraq Petroleum Company, but also, more ominously, by German, Japanese, and Italian interests. It appeared to observers that there was a concerted drive by the Axis powers to obtain drilling rights in Saudi Arabia. The Japanese established diplomatic representation in Saudi Arabia and offered what were, in comparison to existing terms, huge sums for a concession within the country and for the King's interest in the Neutral Zone—altogether what one Saudi official called an offer of "astronomical proportions." The Japanese also presented Ibn Saud with a gift of classic samurai military armor, though far too small to fit the large monarch. In order to try to gain a foothold, the Germans accredited their minister in Baghdad to Saudi Arabia and opened a permanent mission; they also pursued an arms deal with the Saudis. Meanwhile, Italy kept up a steady campaign of pressure on the Saudis for a concession. But Casoc had, by the secret annex to the 1933 agreement, preference rights to Saudi territory, which it successfully exercised on May 31, 1939, expanding the total area of its exclusive concession to 440,000 square miles—equal to about one-sixth the size of the continental United States. Of course, there was a price to be paid for such loyalty. As Saudi financial needs mounted, Socal found itself repeatedly making loans, totaling several million dollars, to the kingdom.

But there was good reason to be forthcoming, considering the stakes. The discovery in Well Number 7 in March 1938 had opened a new era. Work sped up on creating the required industrial, administrative, and residential development at Dhahran—which would eventually become an American middle-class suburb, an oasis, in the midst of the desert. Immediately after the strike at Well Number 7, a pipeline was begun, linking the oil field to Ras Tanura, a coastal spot that had been chosen as the site for a marine terminal. In April 1939 a great procession of four hundred cars, carrying the King and a huge retinue, crossed the desert to Dhahran, where they encamped in 350 tents. The occasion was the arrival of the Socal tanker *D. G. Scofield* at Ras Tanura, to pick up the first cargo of oil. With appropriate pomp, King Ibn Saud himself turned the valve through which the first trickle of oil flowed out of Saudi Arabia.[30]

Socal hastened to spread its exploration over the vast desert. A wildcat well, drilled down to ten thousand feet, indicated the possibility of very large deposits of oil. Meanwhile, production in 1940 got as high as twenty thousand barrels per day. The future seemed ever more promising. But then World War II intruded. In October 1940 the Italians bombed Dhahran, though apparently aiming at Bahrain. A small refinery was started up at Ras Tanura a few months later, in January 1941, but it was closed down the following June. In neighboring Kuwait, operations were suspended because of the war. On orders of the Allied governments, all the wells in Kuwait were plugged with cement, putting them out of commission, for fear that they would fall into German hands.

In Saudi Arabia, too, oil operations were mostly shut down, and the majority of the American employees sent home. A skeleton crew kept production of twelve thousand to fifteen thousand barrels per day going, as feedstock for the Bahrain refinery. But further development was postponed, and the entire enterprise was put in a state of suspended animation. Elsewhere, however, as people began to assimilate what Saudi Arabia's oil potential might be, and what it might mean, the country's petroleum reserves would become the object of political power plays more intricate and intense than anything that might have been imagined by the men from Standard of California, King Ibn Saud, or even Philby, who had originally planted the idea of buried treasure in the King's mind.

Jack Philby had prospered in Saudi Arabia through the 1930s and continued his geographical explorations of the country. After the outbreak of World War II, he tried to become a middleman between Ibn Saud and Chaim Weizmann, leader of the Zionist movement, for a partition of Palestine, but it came to nothing. His habitual anti-British sentiments did not abate. He became obstreperously critical of the Allies, and was arrested on a trip to India and sent back to Britain, where he was put in jail for a half year. He spent the rest of the war writing pamphlets, poetry, and unpublishable books, and tinkering in fringe politics. Returning to Saudi Arabia after the war, he again became an adviser to the King, carried out new explorations, wrote more books, and profitably pursued his trading business in the postwar oil boom. With a young woman presented to him by the King, he also became a father again at the age of sixty-five. After the death of Ibn Saud, however, Philby took to criticizing what he saw as the spendthrift ways under King Saud, Ibn Saud's son. He was expelled from Saudi Arabia, but after a few years was allowed to return. On a trip to Beirut in 1960 to visit his son Kim, he became ill and was rushed to the hospital. The man whose life had been so eventful and panoramic, so daring and theatrical, now lay unconscious. He awoke for only a moment and murmured to his son, "I am so bored." And then he expired.

On his gravestone in a Moslem cemetery in Lebanon, Kim arranged for a simple inscription: "Greatest of Arabian Explorers."

And what of Major Frank Holmes—"Abu Naft," the Father of Oil? He was, of course, the one who imagined, conceived, and promoted the entire Arabian oil venture. In the middle 1940s, when the extent of Arabian oil riches was first beginning to be realized, Holmes—by then ensconced as Kuwait's oil representative in London—was asked an obvious question. What made him so certain of Arabian oil prospects and so confident, in the face of the virtually unanimous verdict of the world's leading oil geologists that Arabia would be "oil dry"? To be sure, his early practical experience as a mining engineer had taught him that theoretical advice, however expert or august, could be quite wrong.

But Holmes offered a more simple answer. He tapped his finger to his nose. "This," he said, "was my geologist." [31]

WAR
AND
STRATEGY

CHAPTER 16

Japan's Road to War

ON THE NIGHT of September 18, 1931, soldiers of the Japanese Imperial Army, based in the semi-autonomous Chinese province of Manchuria, carried out a bomb attack against the South Manchurian Railway. The actual evidence of the explosion was scant; only about thirty-one inches of track were affected, and damage was so negligible that a speeding express train passed over the spot a few minutes later with no difficulty. But this was by intention, for the Japanese controlled the railroad line; their aim was to keep damage to a minimum—and blame it on the Chinese. The Japanese Army now had the desired pretext to launch an attack on Chinese forces, which it proceeded to do without delay. The Manchurian Affair had begun, marking the entry into an era of Japanese history they were to call, when it was all over, the Valley of Darkness.

Japan had gained many economic and political prerogatives in Manchuria, including the right to maintain military forces, as a result of its victories over China in 1895 and Russia in 1905, as well as from a treaty with China. By the end of the 1920s, there was strong support in Japan to take complete control of Manchuria—"Japan's life line," as one Prime Minister described the province. It would supply the raw materials and "living space" thought necessary for the crowded home islands and vital, no less, for Japan's military strength. More-over, Manchuria's geographical location made control seem essential for Japan's security; the Japanese Army had grown to fear the dual threats of Soviet communism and Chinese nationalism. For their part, the other great powers concerned with the Pacific were increasingly suspicious of Japan, which, in the space of only a few short decades, had emerged as a formidable military as well as commercial power.[1]

"Shall We Trust Japan?"

In 1923, responding to the temper of the time, Franklin Roosevelt, who had been Assistant Secretary of the Navy during World War I, wrote an article entitled "Shall We Trust Japan?" In introducing the article, the editors observed that one of Roosevelt's "chief duties during a large part of his term of office was to prepare to fight Japan." In the article Roosevelt observed that "long before the events of 1914 centered attention elsewhere, an American-Japanese war was the best bet of prophets. Its imminence began to be taken for granted." A war now, he said in 1923, might well turn into a military deadlock, and then, "economic causes would become the determining factor." Yet Roosevelt answered the question "Shall we trust Japan?" with a ringing affirmative. Japan had changed. It was honoring its international commitments; it had aligned itself with the Anglo-American postwar military order; and in the Pacific "there would seem to be enough commercial room and to spare for both Japan and us well into the indefinite future."[2]

Indeed, through the 1920s, Roosevelt's analysis proved correct. Japan had a functioning parliamentary system. The 1921 Washington Naval Conference defused a potential naval race in the Pacific among Japan, the United States, and Britain, and, thereafter, Japan had based its security on cooperation with the Anglo-American powers. But that cooperation did not survive the decade. The Japanese military, particularly the Army, came to dominate the government, and Japan embarked on its course of imperial expansion in East Asia—seeking, in the process, to exclude the Western powers from what it would call its "Greater East Asia Co-Prosperity Sphere."

This decisive shift was born of several sources. The Great Depression and the collapse of world trade brought great economic hardship to Japan, heightening the sense of vulnerability that came from lack of raw materials and shrinking access to international markets. At the same time, the Army and important segments of society were gripped by a spirit of extreme nationalism, moral distress, arrogance, and a mystical belief in the superiority of Japanese culture and imperial institutions and "The Imperial Way," all of which was amplified by the conviction that the other great powers were deliberately seeking to restrain Japan to a second-rank position and deny it its due in Asia. To be sure, Prime Minister Osachi Hamaguchi, who favored an extension of the naval treaty arrangements with the United States and Britain, won a smashing electoral victory in February 1930. But the strength of the opposition was brought home a few months later, when a youth, enraged at Japan's cooperation with the United States and Britain, shot Hamaguchi at a railway station in Tokyo. He never fully recovered, and died in 1931. With him perished the spirit of cooperation, and instead, a new cult of ultranationalism—bolstered by "government by assassination"—took hold. Japan also organized its new puppet state in Manchuria, which was dubbed Manchukuo, with the deposed Chinese emperor Pu Yi as its figurehead. When the League of Nations condemned Japan for its actions in Manchuria, it stalked out of the League and embarked on its own path—one that would eventually lead to ruin.[3]

The New Order in Asia

Over the next few years, as Tokyo elaborated its claims to a "mission" and "special responsibilities in East Asia," Japanese politics seethed with conspiracies, ideological movements, and secret societies that rejected liberalism, capitalism, and democracy as engines of weakness and decadence. It was thought that there was nothing more noble than to die in battle for the Emperor. Yet some elements in the Japanese military were also, by the mid-1930s, focusing on the more practical question of how to wage modern warfare. Promulgating a doctrine of total war, they sought to establish a "national defense state" in which the industrial and military resources of the country would all be built up and harnessed for that grim eventuality. Those officers who had either closely observed or studied the German failure in World War I attributed that nation's defeat to its economic vulnerability—its relative lack of raw materials and its inability to withstand the Allied naval blockade. Japan, they gloomily recognized, was far less well-endowed than Germany. Indeed, it faced a unique problem of supply. It was almost bare of the resource of oil. While petroleum held a relatively small place in the country's energy mix—accounting for only about 7 percent of total energy consumption—its significance was in its strategic importance. Most was consumed by the military and in shipping. By the late 1930s, Japan produced only about 7 percent of the oil it consumed. The rest was imported—80 percent from the United States, and another 10 percent from the Dutch East Indies. But America was committed to an "open door" policy, political as well as economic, in Asia, which was wholly at odds with Japan's imperial ambitions. With the United States emerging as Japan's most likely antagonist in the Pacific, where, in the event of war, was the necessary oil to be obtained to fuel Japan's ships and planes?

That question had already sparked an acrimonious split between the Japanese Army and Navy, which would be crucial to the evolution and direction of Japanese policies. The Army was focused on Manchuria, North China, Inner Mongolia, and the threat from the Soviet Union. The Navy, under the doctrine of *hokushu nanshin*—"defend in the north, advance to the south"—had its sights set on the Dutch East Indies, Malaya, Indochina, and a number of smaller islands in the Pacific, in order to provide the empire with secure access to natural resources, particularly the prime and absolutely essential resource—oil. Both military services, however, were united in their central objective: to restructure Asia in a "spirit of co-prosperity and co-existence based upon the Imperial Way"—Asia under Japanese control.[4]

In the early 1930s, soon after the Manchurian Affair began, the Japanese government sought to assert domination over the oil industry to serve its own needs. Sixty percent of the internal market was held by two Western companies—Rising Sun, the Japanese affiliate of Royal Dutch/Shell, and Standard-Vacuum, otherwise known as Stanvac, the amalgam of Jersey's and Standard of New York's operations in the Far East—with the rest split among about thirty Japanese companies, which imported their oil from a number of American producers. With the support of Japanese commercial interests, which wanted to improve their market position, the military won passage in 1934 of the Petroleum Industry Law, which gave the government the power to control imports, set market share quotas for specific companies, fix prices, and make compulsory purchases. The foreign companies were required to maintain six months of inventories beyond the normal commercial working levels. The objective in all this was obvious: to build up the Japanese-owned refining industry, to reduce the role of the foreign companies, and to prepare for war. At the same time, Japan was also establishing a petroleum monopoly in its new colony, Manchukuo, with the aim of squeezing out the Western companies.

The foreign companies recognized that they were going to be squeezed. The American and British governments also disapproved of Japan's new, restrictive oil policies. But how to respond? There was talk in Washington and New York and London about an embargo—full or partial—that would, in retaliation, constrain the supply of crude oil to Japan. In August 1934 Henri Deterding and Walter Teagle went to Washington to see both State Department officials and Harold Ickes, the Oil Administrator. The oil men suggested "frightening" Japan into moderation by merely hinting at an embargo. Word would get back to Tokyo, they hoped, and perhaps would lead to changes in Japanese policy. In November 1934 the British Cabinet endorsed the Foreign Office's position that "the stiffest possible resistance should be offered" to the Japanese oil policies, including government support of a privately organized embargo. However, Secretary of State Cordell Hull made clear that the U.S. government would not support such an action, and that was the end of the embargo talk for the time being. Meanwhile, the pressures and tensions between the oil companies and the Japanese government continued to build, right up to the summer of 1937. Then Japan's circumstances abruptly changed.[5]

"Quarantine"

Over the night and morning hours of July 7 and 8, 1937, two obscure clashes took place between Japanese and Chinese troops at the Marco Polo Bridge near Beijing. As hostilities escalated over the next several weeks, the Chinese Nationalists took a defiant stand against further concessions to Japan. "If we allow one more inch of our territory to be lost," the Nationalist leader Chiang Kai-shek declared, "we shall be guilty of an unpardonable crime against our race." The Japanese, for their part, had decided that the Chinese needed to be chastised and its Army dealt a "thoroughgoing blow." A little over a month after the first incidents, on August 14, the Chinese bombed the Japanese naval station at Shanghai. Japan went to war with China.

Japan immediately accelerated efforts to put its economy on a full war footing. It also proceeded quickly to patch up relations with foreign oil companies. The government did not want to risk any disruption of oil supplies. At the same time, a special session of the Diet, convened to approve mobilization legislation, passed the Synthetic Oil Industry Law. It provided for a seven-year plan aimed at producing, by 1943, synthetic fuels—primarily liquid fuel out of coal—in volumes equivalent to half of Japan's entire 1937 consumption level. The goal was not only ambitious, it was also extremely unrealistic.

From the very first, official American policy and public opinion supported China as the victim of aggression in the Sino-Japanese War. But the United States remained very much in the grip of isolationism. Fourteen years had passed since Franklin Roosevelt, then merely a former assistant navy secretary, had written his "Shall We Trust Japan?" article. Now, as President, Roosevelt felt frustrated, both by the political constraints at home and by the ominous developments abroad. In a speech in October 1937, he obliquely broached the idea of establishing a "quarantine" to check the spreading "epidemic of world lawlessness." After a Japanese air attack on four American ships in the Yangtze River, he privately explained to his Cabinet that, by quarantine, he meant "such a thing as using economic sanctions without declaring war." But neutrality legislation and the prevailing isolationist sentiment prevented the President from putting the idea into practice.[6]

As reports of Japanese attacks on Chinese civilians began to mount, however, American sentiment turned more sharply against Japan. In 1938, after newspaper and newsreel pictures of the Japanese bombing of Canton, polls found that a large majority of the American public was opposed to the continued export of military materiel to Japan. But the Roosevelt Administration was fearful both of undermining Japanese moderates by too strong a stand and of interfering with America's ability to respond to what was seen as the more immediate and serious threat of Nazi Germany. So the Administration went no further than to adopt a "moral embargo" on the export of airplanes and aircraft engines to Japan. Lacking legislative authority, the State Department took to writing letters to American manufacturers, asking them not to sell such goods. Washington was also alarmed by the implications of the growing ties between Japan and Germany, both of which had signed the Anti-Comintern Pact in 1936, ostensibly aimed against the Soviet Union. But Japan was resisting German pressure to move closer—chiefly, Tokyo explained to Berlin, because Japan's dependence on the United States and the British empire for indispensable raw materials, and for oil in particular, meant that it "was not yet in a position to come forward as an opposer of the Democracies."

Here was the deadly paradox for Japan. It wanted to reduce its reliance on the United States, especially for most of its oil, much of which went to fuel its fleet and air force. Japan feared that such dependence would cripple it in a war. But Tokyo's vision of security and the steps it took to gain autonomy—its brutal expansion in pursuit of its "co-prosperity sphere"—created exactly the conditions that would point toward war with the United States. Indeed, in the late 1930s, the supply requirements for the war with China actually increased Japan's trade dependence on the United States. To complicate things further, foreign currency constraints made it more difficult for Japan to pay for imports. This forced strict restrictions on supplies for the domestic economy, including the rationing of oil and other fuels, thus weakening efforts to build up a war economy. The fishing fleet, which was one of the main sources of Japan's food, was ordered to give up oil and instead to depend exclusively upon wind power![7]

By 1939, the United States was explicitly opposed to Japanese actions. Still, Roosevelt and Secretary of State Hull hoped to find a middle ground between overly strong American countermeasures on the one hand, which could provoke a serious crisis in the Pacific, and appeasement on the other, which would only encourage further Japanese aggression. The Japanese bombing of Chinese civilian centers, especially the bombings of Chungking, in May 1939—"milestones in the history of aerial terror," in the words of the journalist Theodore H. White, who covered them for *Time*—shocked and further aroused American public opinion. Various groups, such as the American Committee for Non-Participation in Japanese Aggression, campaigned hard to cut off all American exports. "Japan furnishes the pilot," said one pamphlet. "America furnishes the airplane, gasoline, oil, and bombs for the ravaging of undefended Chinese cities." A Gallup Poll in June 1939 reported that 72 percent of the public favored an embargo on the export of war materials to Japan.

Within the Roosevelt Administration, however, there was intense and sharp discussion about how best to respond, including the ever-present question of direct economic sanctions. But Joseph Grew, the American ambassador to Japan, warned of the possible consequences. The Japanese would submit, he reported from Tokyo, to any deprivation rather than see their nation humbled by Western powers—and lose face. On a visit to Washington in the autumn of 1939, Grew met twice with President Roosevelt and later wrote in his diary: "I brought out clearly my view that if we once start sanctions against Japan we must see them through to the end, and the end may conceivably be war. I also said that if we cut off Japanese supplies of oil and if Japan then finds that she cannot obtain sufficient oil from other commercial sources to ensure her national security, she will in all probability send her fleets down to take the Dutch East Indies."

"Then we could easily intercept her fleet," the President replied.

Grew was expressing foreboding, not commenting on policies that were imminent in the autumn of 1939. There was no plan for an embargo on oil. Nor, despite his remarks, was Roosevelt willing to risk a confrontation. But oil was fast emerging as the critical issue between the two countries.[8]

A year earlier, in September 1938, in The Hague, two American businessmen had sat together, close to a radio set, listening somberly to the latest news. One was George Walden, the head of Stanvac, the Far Eastern joint venture of Jersey and Standard of New York. The other was Lloyd "Shorty" Elliott, president of Stanvac's producing arm in the Dutch East Indies. It was the time of the Munich crisis; Europe had seemed to be on the verge of war. But Britain and France had just given way to Hitler's demands on Czechoslovakia in order to guarantee what Prime Minister Neville Chamberlain would call "peace in our time." But to Walden and Elliott listening intently to the radio reports of the speech Hitler had given that day, war seemed inevitable, not only in Europe, but also in Asia. And when war came to Asia, they were sure the Japanese would attack the East Indies—in Elliott's words, "it was just a question of when and how."

That night in The Hague, Walden and Elliott began working out what to do when the Japanese invasion came. The two men wasted little time in implementing their new plans. As a first step, all German, Dutch, and Japanese employees in the Indies who were of doubtful loyalty were dismissed. Plans were prepared for the destruction of Stanvac's refinery and oil wells—but rather openly, as a deterrent to the Japanese. By early 1940, evacuation plans were also well advanced, and Walden indicated to local Stanvac officers in the Indies that if the United States placed an embargo on oil to Japan, the company "would cooperate fully" and "stop all shipments from all properties under its control all over the world," even though much of that property was not under American jurisdiction. "Shipments from the Netherlands East Indies would be stopped," he made clear, "despite the possibility that the Japanese Navy would attempt to take the properties there and despite the further fact that the American Government, due to cries within the United States against 'fighting for Standard Oil,' might not attempt to protect American interests in the Netherlands East Indies."[9]

Japanese Advance and American Restrictions
—The First Round

Increasingly worried about a cut-off of oil and other supplies from the United States, Tokyo instituted a policy to establish industrial self-sufficiency and to try to eliminate economic dependence upon the United States. The Japanese public, even schoolchildren, were bombarded with propaganda about how the "ABCD" powers, as they were called—America, Britain, China, and the Dutch—were engaged in a conspiracy to deny resources and strangle the empire. But Japan's position appeared stronger after the outbreak of war in Europe in September 1939, and even more so after May and June of 1940, when the Germans swept through Belgium, Holland, and France, overriding all resistance. The Japanese continued their advance through China, and suddenly, with the colonial powers overrun, excepting Great Britain, all of the Far East looked truly vulnerable. As if to underline that threat, the Japanese abruptly demanded far larger supplies of oil from the East Indies, now under the sway of the Dutch government-in-exile in London. Fearful that a beleaguered Britain would withdraw its own forces from the Far East, Washington made a fateful decision; it transferred the American fleet from its base in Southern California to Pearl Harbor on the island of Oahu in Hawaii. Since the fleet was, at the time, already on maneuvers near Hawaii, the move was accomplished with a minimum of fanfare. One purpose was to stiffen British resolve. The other was to serve as a deterrent to Tokyo.

The summer of 1940 was a major turning point. In June, Japan started on the road south. It asked the new, collaborationist government of France to approve the dispatch of a military mission to French Indochina; it demanded that the East Indies guarantee war materials; and it threatened Britain with war if it did not both take its troops out of Shanghai and close the Burma supply route into China. That same month, Roosevelt brought Henry Stimson into the Cabinet as Secretary of War. Stimson was a long-time critic of American exports to Japan and what he saw as insufficient resolve in U.S. policy. On July 2, 1940, Roosevelt signed the National Defense Act, passed hurriedly after the Nazi invasion of Western Europe. Section VI gave the President the power to control exports; that would be the lever with which to regulate oil supplies to Japan.

In Tokyo, leaders who wanted to avoid a collision with the Western powers were rapidly losing ground. A section of the secret police organized a plot to kill those seen as favoring efforts at a settlement with Britain and the United States. The targets included the Prime Minister. The plot was aborted in July, but the message was clear. That same month, the Japanese Cabinet was reconstructed under the new Prime Minister, Prince Konoye. The militant general Hideki Tojo—known as *"Kamisori,"* the "Razor"—became War Minister. He had formerly been Chief of Staff of the Kwantung Army in Manchuria, which had fabricated the original provocation on the South Manchurian Railway in 1931.[10]

In the second half of July 1940, virtually simultaneous developments in Tokyo and Washington pointed Japan and the United States more directly on their collision course. Oil was the linchpin. The Japanese strengthened their commitment to drive into Southeast Asia. That, they thought, would help them win the war in China. To assure adequate supplies, Japan would attempt to get additional oil from the Dutch East Indies one way or another. It also sought to import far larger than normal volumes of aviation gasoline from the United States, setting off alarm bells in Washington. Meeting on July 19, 1940, with senior advisers, Roosevelt pointed to a map across the room. He explained that he sat there day after day eyeing that map, and he had finally come "to the conclusion that the only way out of the difficulties of the world" was by cutting off supplies to the aggressor countries, "particularly in regard to their supply of fuel to carry on the war." There was no dissent in the discussion that followed about taking such a step *vis-à-vis* the European aggressors. However, the question of Japan occasioned very sharp words, and no consensus about whether that would make things better or worse.

The next day Roosevelt signed legislation authorizing the building of a two-ocean Navy, so that the United States could meet the Japanese threat in the Pacific without leaving the Atlantic Ocean to Germany. That being the case, some asked, why continue to provide Japan with oil supplies to fuel *its* Navy? Treasury Secretary Henry Morgenthau and War Secretary Stimson tried to promote a proclamation that would have meant a complete embargo on oil exports to Japan. But the State Department, still fearful of provoking a rupture with Japan, succeeded in redrafting the proclamation so that the ban was limited only to aviation gasoline of 87 octane or higher, as well as some kinds of iron ore and steel scrap. That would protect gasoline supplies for the U.S. military, as American planes used 100-octane fuel. But the ban did not hinder the Japanese, as their planes could operate with fuels below 87 octane. And, if need be, the fuel could be raised to a higher octane in Japan simply by "needling" it with a little tetraethyl lead. As it turned out, Japan bought 550 percent *more* 86 octane gasoline from the United States in the five months after the July 1940 proclamation than *before*. Despite appearances, an embargo had not gone into effect, only a licensing system. Still, Tokyo was alerted to what it might expect down the road.[11]

The alignments were now very clear. On September 26, 1940, responding both to Japanese moves in Indochina and to an imminent new Japanese pact with Germany and Italy, Washington banned the export of all iron and steel scrap to Japan—but not oil. The next day, Japan formally signed the Tripartite Pact with Hitler and Mussolini, tying itself much more tightly into the Axis. "The hostilities in Europe, in Africa and in Asia are all parts of a single world conflict," said

Roosevelt. But he believed that the European war, which threatened the very survival of Britain, took precedence, and thus he remained committed to a "Europe first" strategy. That meant husbanding all possible resources for Europe. Roosevelt had an extra reason for caution: The Presidential election was only a month away, he was running for an unprecedented third term, and he did not want to risk doing anything that looked provocative in the intervening weeks. The United States Army and Navy, concerned to avoid a confrontation with Japan while in the midst of their own build-ups, added their voices to those arguing against the imposition of an oil embargo. Meanwhile, the Japanese were trying to buy up all the petroleum supplies they could obtain, as well as drilling equipment, storage tanks in knocked-down form, and other supplies. The British now wanted to find a way to halt the flow of oil. They feared that if Japan did build large stockpiles, it would become relatively immune to any economic sanctions. Still, Roosevelt and Hull resisted cutting the flow.[12]

Quiet Conversations

Was there some way to find a *modus vivendi*, something short of war that would not leave Japan with an iron grip on Asia? What might have been overlooked? So asked Secretary of State Hull again and again. In an effort to find an answer, he began talking privately with the new Japanese ambassador, Admiral Kichisaburo Nomura, a former Foreign Minister. The two would meet at night, with only a couple of aides, usually in Hull's apartment at the Wardman Park Hotel.

Each man epitomized his respective society. Tall and silver-haired, Cordell Hull was a backwoodsman turned statesman. Born in a log cabin in Tennessee, he had become a circuit court judge, a volunteer in the Spanish-American War, and then Congressman and Senator. Cautious, careful, "given to sifting a difference into its smallest particles," and in his own way implacable, he had devoted himself since becoming Secretary of State in 1933 to one central purpose: breaking down trade barriers in order to promote a liberal international economic order that would also serve the cause of world peace. Now, in 1941, he could see all those labors going for naught. But he was not yet ready to give up. He was willing to explore and re-explore, well beyond most people's ideas of patience, every cranny of U.S.–Japanese relations in order to find some alternative to a total breakdown. And he would seek to buy time.

Admiral Nomura shared Hull's desire to avert a conflict. A political moderate, he was widely respected in Japanese political and military circles. At six feet, the solemn admiral stood out among his countrymen. He had lost an eye in a bomb attack by a Korean nationalist in Shanghai in 1932. That attack had also left him with a limp and more than a hundred metal fragments permanently in his body. During World War I, he had served as naval attaché in Washington, where he had gotten to know Assistant Navy Secretary Franklin Roosevelt. When the two met again, in February 1941, upon Nomura's arrival in Washington as ambassador, Roosevelt greeted him as "friend" and insisted upon addressing him as "Admiral," rather than "Ambassador." Nomura felt comfortable in the United States, had many friends in the United States, and most certainly did not want war between the two countries. As he told the U.S. Chief of Naval Operations, "his lips and his heart" were "at variance." But he was a messenger, not a decision maker. Some years later, trying to explain how he had felt during the tense days of 1941, Nomura simply said, "When a big house falls, one pillar can not stop it."

Beginning in March 1941, Hull and Nomura met on many evenings, perhaps forty or fifty times altogether, reviewing proposals, looking for any steps that would prevent a collision, plowing on and on through the discouraging, unpromising soil. To be sure, Hull had a startling advantage during all these talks. Thanks to the code-breaking operation known as "Magic," the United States and Britain had cracked "Purple," the top-secret Japanese diplomatic code. Thus, Hull was able to read, before the meetings with Nomura, Tokyo's instructions to the ambassador and, afterward, Nomura's reports. Hull played his part adroitly, never giving any hint of knowing more than he was supposed to know. In early May 1941 the Germans informed the Japanese that the United States had broken their codes. But Tokyo discounted that piece of intelligence; the Japanese simply did not believe that Americans were capable of such a feat.

Yet, despite "Magic," there were many things that Hull and his colleagues in Washington did not know. Among them was the Japanese Navy's concern that the American fleet in Hawaii, if left unattended in the midst of an invasion of the East Indies and Singapore, could launch a dangerous flank attack. As a result, the Japanese Navy had begun to plan a daunting and high-risk project—a surprise attack on Pearl Harbor.[13]

Yamamoto's Gamble—"Doubtless I Will Die"

As early as the spring of 1940, Admiral Isoroku Yamamoto, Commander in Chief of Japan's Combined Fleets, had started to outline this wild, almost preposterous gamble. He was the most daring, original, and controversial of the Japanese admirals, widely respected for his physical courage and leadership, though resented by some for his bluntness. He was short and broadly constructed; his face and his whole manner reflected willpower and determination. Of all the personnel on active duty in the Combined Fleets on the eve of World War II, he was the only one who had actually experienced combat duty in the Russo-Japanese War, almost four decades earlier, and to prove it, he was missing two fingers on his left hand, which he had lost in Japan's great victory, the Battle of Tsushima in 1905.

Nothing could have been more in keeping with Yamamoto's strategic vision—or his love of gambling—than his plan for an attack on Pearl Harbor. Yet such a proposal was particularly surprising coming from him. He had spent more than four years in the United States in the 1920s, first as a student at Harvard, then as a naval representative and attaché in Washington. He had read four or five biographies of Abraham Lincoln, and he regularly received and perused *Life* magazine. He had traveled through the United States, he knew the country and prided himself on his understanding of Americans, and he recognized that the United States was rich in resources while Japan was poor and that America's productive capacity far outstripped that of his own country.

Indeed, even while developing his Pearl Harbor plan, Yamamoto continued to challenge the whole idea of war with the United States. Ultimately, he thought, such a struggle would be, at best, very risky and, most probably, a losing proposition. He was one of those naval officers who preferred to seek some accommodation with America and Britain. He was acidly critical of Japan's civilian and Army leaders, and thought they were partly responsible for the tensions with the United States. The complaint about "America's economic pressure," he said, in December 1940, "reminds me of the aimless action of a schoolboy which has no more consistent motive than the immediate need or whim of the moment." He scoffed at the ultranationalists and jingoists, with their "armchair arguments about war" and their mystical fantasies, who had so little understanding of the real costs and sacrifices that war would mean.

Moreover, the oil factor weighed heavily in Yamamoto's mind. He had a special grasp of and sensitivity to the Navy's, and Japan's, oil predicament. He had grown up in the Niigata district, one of the regions responsible for Japan's small domestic oil production, and his home town of Nagaoka was populated by hundreds of tiny factories producing oil for lamps. His time in America had convinced Yamamoto that the industrial world was moving from coal to oil and that air power was the future, even for navies. Acutely conscious of Japan's oil vulnerability, as Commander in Chief of the Combined Fleets he insisted on restricting the Navy, the third-largest in the world, to training only in the waters immediately off Japan. The reason—to conserve petroleum. So concerned was he with Japan's oil problem that he even sponsored experiments, to the chagrin of his naval colleagues, by a "scientist" who claimed he could change water into oil.[14]

Yet, whatever his doubts, Yamamoto was a fervent nationalist to his core, devoted to the Emperor and to his country. He believed that the Japanese were a chosen people and that they had a special mission in Asia. He would do his duty. "It's out of the question!" he exclaimed. "To fight the United States is like fighting the whole world. But it has been decided. So I will do my best. Doubtless I will die."

If Japan had to go to war, Yamamoto believed, it should go for the "decisive blow" and seek to knock the United States off balance, incapacitate it, while Japan secured its position in Southeast Asia. Thus a surprise attack on Pearl Harbor. "The lesson which impressed me most deeply when I studied the Russo-Japanese War was the fact that our Navy launched a night assault against Port Arthur at the very beginning," Yamamoto said in early 1941. "This was the most excellent strategical initiative ever envisaged during the war." What was the most "regrettable," he added, was "that we were not thoroughgoing in carrying out the attack." His plan for the attack on Pearl Harbor—"at the outset of the war to give a fatal blow to the enemy's fleet"—was decided upon in late 1940 and early 1941. Yamamoto's objective was not only "to decide the fate of the war on the very first day" by knocking out the U.S. fleet in the Pacific, but also to destroy the morale of the American people.

The requirements of success for "Operation Hawaii," as it was called, were many: secrecy; first-rate intelligence; superb coordination; high technical skills; many technological innovations, including development of new aerial torpedoes and new techniques of refueling at sea; absolute devotion to the cause at hand;

and the cooperation of the weather and the waves. Yet, early in 1941, despite the secrecy, U.S. Ambassador Grew heard from Peru's minister to Tokyo about a rumor that Japan was planning an attack on Pearl Harbor. Grew reported it to Washington, where it was immediately discounted. American officials simply could not believe—then or in the months following—that such an audacious assault was even possible. Moreover, officials in the Navy and State departments were astonished that an ambassador of Grew's caliber could take seriously such an obviously ridiculous story.[15]

Embargo

From April through June 1941, the arguments continued to rage in the U.S. government about whether or not to cut off oil exports to Japan and to freeze Japanese funds in the United States—most of which were used to purchase oil. The Axis powers and America were clearly moving closer to direct confrontation. On May 27, 1941, President Roosevelt declared an "unlimited national emergency." His aim, in the words of one of his advisers, was "to scare the daylights out of everyone," about the true dangers of the Axis drive for world domination. Immediately following that, but acting on his own authority, Harold Ickes, just appointed Petroleum Coordinator, prohibited oil shipments to Japan from the East Coast. Petroleum supplies were getting short in the eastern United States— primarily because of transportation difficulties—and public opposition to exporting oil from the East Coast, especially to Japan, was building rapidly. The order, however, did not pertain to the Gulf or West Coasts. At the same time, Ickes was trying to promote a general embargo on all oil exports to Japan.

An angry President countermanded Ickes's order, which led to a brittle and bitter exchange. "There will never be so good a time to stop the shipment of oil to Japan as we now have," Ickes argued. "Japan is so preoccupied with what is happening in Russia and what may happen in Siberia that she won't venture a hostile move against the Dutch East Indies. To embargo oil to Japan would be as popular a move in all parts of the country as you could make."

"I have yours of June 23rd recommending the immediate stopping of shipments of oil to Japan," Roosevelt replied sarcastically. "Please let me know if this would continue to be your judgment if this were to tip the delicate scales and cause Japan to decide either to attack Russia or to attack the Dutch East Indies." He also delivered a stern little constitutional lesson, telling Ickes that the question of Japanese exports was "a matter not of oil conservation, but of foreign policy, a field peculiarly entrusted to the President and under him to the Secretary of State."

Complaining about "the lack of a friendly tone in letters that have come from you recently," Ickes, as was his wont, proffered his resignation—as Petroleum Coordinator, though not as Interior Secretary. But Roosevelt, as he had done so often in the past, refused to accept it. "There you go again!" wrote the President on July 1, 1941. "There ain't nothing unfriendly about me, and I guess it was the hot weather that made you think there was a lack of a friendly tone!" Then by way of further explanation, Roosevelt said, "the Japs are having a real drag-down and knock-out fight among themselves . . . trying to figure out which way they are going to jump." And he added, "As you know, it is terribly important for the control of the Atlantic for us to help to keep peace in the Pacific. I simply have not got enough Navy to go around and—every little episode in the Pacific means fewer ships in the Atlantic." [16]

The "knock-out fight" to which Roosevelt referred had been precipitated by Germany's surprise attack on the Soviet Union in June 1941, which forced a major strategic choice for Tokyo: whether to continue on its southern course or to take advantage of Hitler's success, join in the attack on Russia from the east, and help itself to part of Siberia. Between June 25 and July 2, 1941, senior officials in Tokyo grappled with and argued fervently over the choices. Finally, they made the fateful decision: They would put off doing anything about the Soviet Union, and instead concentrate on the southern strategy—and, in particular, seek to secure control of all of Indochina, deemed necessary in order to go for the East Indies. They did so with the recognition that the occupation of southern Indochina could well provoke an all-out American oil embargo, which would be "a matter of life or death to the empire," in the words of the Navy General Staff. But Japan, it was also decided, would not be deterred from its objectives by the threat of war with Britain and the United States.

Through the "Magic" intercepts of the Japanese codes, Washington knew of the momentous debate and, at least to some degree, its outcome. "After the occupation of French Indochina-China," said one intercepted message, "next on our schedule is . . . the Netherlands East Indies." At Roosevelt's Cabinet meeting on July 18, it was reported that the Japanese were virtually certain to advance into southern Indochina in the next few days.

"I would like to ask you a question which you may or may not want to answer," Treasury Secretary Morgenthau said to the President. "What are you going to do on the economic front against Japan if she makes this move?"

"If we stopped all oil," Roosevelt replied, "it would simply drive the Japanese down to the Dutch East Indies, and it would mean war in the Pacific."

But he did indicate that, if Japan moved, he would support a different form of economic sanction: the freezing of Japanese financial assets in the United States, which would restrict Japan's ability to buy oil. Even Hull, quite ill and generally discouraged, called in from a health spa where he was resting to advocate stronger export controls—though "always short of being involved in war with Japan."

With its own back against the wall in Europe, Britain registered its concern that a total embargo might lead Japan to accelerate its southward advance, and the British were far from sure that Washington was prepared for the possible consequences, including war. But, in Washington, only the Army and Navy, focused on the Atlantic and Europe and intent on having as much time as possible for their build-up, were still reluctant to impose new curbs.

On July 24, 1941, the radio reported that Japanese warships were off Camranh Bay, and that a dozen troop transports were on their way south from the Japanese-controlled island of Hainan, in order to effect the occupation of southern Indochina. That same afternoon, Roosevelt, receiving Ambassador Nomura, suggested a neutralization of Indochina. He said that he had kept oil exports flowing, despite "bitter criticism," in order not to provide the Japanese with a pretext to attack the East Indies—an attack that would have the eventual result, he indicated, of direct conflict with the United States. He also clearly suggested that, with "this new move by Japan in Indochina," he might not be able any longer to withstand the domestic political pressure to restrict oil exports to Japan.

Such a shift was already at hand. Roosevelt himself did not want to impose a full embargo. He wanted to tighten controls, but to keep them, as he said, "day to day," a flexible tool that could be adjusted to specific circumstances. His aim was to create maximum uncertainty for Japan, but he did not want to push it over the brink. He thought he could use oil as an instrument for diplomacy, not as the trigger for war. He did not want, as he told the British ambassador, to try to fight two wars at once. Undersecretary of State Sumner Welles proposed a program that fit the President's objective; it would hold petroleum exports to the 1935–36 level, but prohibit the export of any grades of oil or oil products that could be manufactured into aviation gasoline. Export licenses would be required for *all* oil exports. On the evening of July 25, the U.S. government ordered all Japanese financial assets in the United States to be frozen. Licenses—that is, government approval—would be required for each use of the frozen funds, including the purchase of oil. On July 28, Japan began its anticipated invasion of southern Indochina, and with that took another step toward war.

The new American policy was not meant to cut off oil entirely, at least explicitly, but a virtually total embargo was the actual result. A key role was played by Dean Acheson, Assistant Secretary of State for Economic Affairs, and one of the few senior State Department officials to favor an out-and-out embargo. He turned the July 25 order into an embargo, in consultation with the Treasury Department, by completely preventing the release of the frozen funds necessary for the Japanese to buy the oil. "Whether or not we had a policy, we had a state of affairs," Acheson later said. "Until further notice it would continue." From the beginning of August, no more oil was exported to Japan from the United States. Two Japanese tankers were left sitting empty in the harbor at San Pedro, near Los Angeles, waiting for oil that had already been contracted for.[17]

"We must act as drastically as the U.S.A.," said British Foreign Secretary Anthony Eden. But both Britain and the Dutch government-in-exile were, understandably, somewhat baffled as to exactly what American policy was. Still, Britain followed with its own freeze and an embargo, cutting off supplies from Borneo, as did the Dutch East Indies.

By the end of July 1941, Japan had secured its occupation of southern Indochina. "Today I knew from the hard looks on their faces that they meant business," Ambassador Nomura reported to the Foreign Ministry in Tokyo on July 31, after meeting with American officials. "Need I point out to you gentlemen that in my opinion it is necessary to take without one moment's hesitation some appeasement measures." The Foreign Ministry scathingly dismissed the ambassador's concerns. With the Japanese thrust into Indochina and the consequent American freeze on Japanese funds—which, in practice, meant an embargo on oil—the countdown had begun. As Nomura was later to say to Hull, "The Japanese move into south Indochina in the latter part of July" had "precipitated" the "freezing measures, which in turn meant a de facto embargo and had reacted in Japan to increase the tension."

But the embargo itself did not create the impending confrontation. It was virtually the only way left for the United States—and the British and the Dutch—to respond to Japanese aggression, short of military action. With the Japanese move into Southeast Asia and the Nazi sweep into the Soviet Union, the United States faced a horrifying prospect—of both Europe and Asia dominated by the Axis, leaving the United States the last island left between two unsafe seas. Thus, the President sought to use the oil lever. For the Japanese, however, it was the final link in their "encirclement" by hostile powers. Tokyo refused to recognize that it was creating a self-fulfilling prophecy. The embargo was the result of four years of Japanese military aggression in Asia. Tokyo had worked itself into a corner: According to its own calculations, the only oil securely available to Japan was what it held in its own inventories. There were no other significant sources that it could tap to make up for the closing off of American and East Indies supplies. If it were to maintain and secure its capability to wage war, then it would inevitably have to risk—or make—war.[18]

"We Cannot Endure It"

The leaders of the Japanese Navy had previously been far more cautious than the Army about a confrontation with the United States. But that was no longer the case in the light of what was taken to be a complete embargo. As a leading Japanese admiral later said: "If there were no supply of oil, battleships and any other warships would be nothing more than scarecrows." Admiral Osami Nagano, the chief of the Naval General Staff, stressed to the Emperor that Japan's petroleum reserves would, without replenishment, last no more than two years.

The new Japanese foreign minister, Teijiro Toyoda, expressed the paranoia in Japanese policy in secret messages to his ambassadors in both Berlin and Washington: "Commercial and economic relations between Japan and third countries, led by England and the United States, are gradually becoming so horribly strained that we cannot endure it much longer," he wrote on July 31, 1941. "Consequently, our Empire, to save its very life, must take measures to secure the raw materials of the South Seas. Our Empire must immediately take steps to break asunder this ever-strengthening chain of encirclement which is being woven under the guidance and with the participation of England and the United States, acting like a cunning dragon seemingly asleep."

How different it all looked to Cordell Hull. Ill and exhausted, Hull had gone to White Sulphur Springs to take a cure. "The Japanese are seeking to dominate militarily practically one-half the world. . . . Nothing will stop them except force," he told Undersecretary of State Welles over the telephone. Still he sought to postpone what now seemed inevitable. "The point is how long we can maneuver the situation until the military matter in Europe is brought to a conclusion."

Ambassador Grew, in Tokyo, saw the situation all too clearly. "The vicious circle of reprisals and counter reprisals is one," he wrote in his diary. "*Facilis descensus Averno est.* Unless radical surprises occur in the world, it is difficult to see how the momentum of this down-grade movement can be averted, or how far it will go. The obvious conclusion is eventual war." By that time, power shovels were already digging shelters around the perimeter of the Imperial Palace in Tokyo.[19]

There were last-minute diplomatic efforts on both sides to stave off confrontation. With some support from the Navy, Prince Konoye, the Prime Minister, raised the possibility of a summit meeting with Roosevelt. Perhaps he could appeal directly to the American President. Konoye was even willing to try to jettison the Axis alliance with Hitler in order to reach an understanding with the Americans. Worried palace officials endorsed Konoye's idea. "The whole problem facing Japan had been reduced to a very simple factor, and that was oil," Koichi Kido, the Lord Privy Seal, told the premier in private, adding, "Japan could not possibly fight a war of certain victory against the United States."

The Emperor himself gave Konoye's idea his blessing. "I am in receipt of intelligence from the Navy pertaining to a general oil embargo against Japan by America," the Emperor told Prince Konoye. "In view of this, the meeting with the President should take place as soon as possible." Konoye suggested that Roosevelt and he meet in, of all places, Honolulu. The President was at first quite interested in the idea—indeed, taken enough to reply that he and Konoye should meet in Juneau, Alaska, instead of Honolulu. But Hull and the State Department strenuously opposed this breach in diplomatic due process. The Americans did not understand that this was Konoye's last gamble to avoid the calamity, and they no longer had any reason to trust Japan. Nor did they think that Konoye had anything new to offer. Moreover, Roosevelt did not want to risk looking like an appeaser; he did not want "Juneau" to join the vocabulary along with "Munich." No good purpose would be served in meeting Konoye without a reasonable agreement more or less settled in advance; Roosevelt was also reading the "Magic" intercepts, which indicated that the Japanese were intent on further conquest. So, for the time being, Roosevelt, with his talent for ambiguity, neither agreed to nor rejected such a meeting.[20]

"Dwindling Day by Day"

In Tokyo, on September 5 and 6, the most senior Japanese officials met with the Emperor and went through the formality of asking permission to assume a war posture, even while the diplomatic alternatives were still being explored. Again, access to oil was their central concern. "At present oil is the weak point of our Empire's national strength and fighting power," their briefing materials said. "As time passes, our capacity to carry on war will decline, and our Empire will become powerless militarily." Time was running out, the military leaders reiterated in front of the Emperor. "Vital military supplies, including oil," said the Navy Chief of Staff, "are dwindling day by day."

How long would hostilities last in the event of a Japanese-American war? the Emperor asked the Army Chief of Staff.

"Operations in the South Pacific could be disposed of in about three months," the Chief of Staff answered.

"The General had been Minister of War at the time of the outbreak of the

China Incident, and . . . had then informed the Throne that the incident would be disposed of in about one month," the Emperor retorted sharply. "Despite the General's assurance, the incident was not yet concluded after four long years of fighting."

The general tried to explain that "the extensive hinterland of China prevented the consummation of operations according to the scheduled plan."

"If the Chinese hinterland was extensive," the Emperor shot back, raising his voice, "the Pacific was boundless." How could the general "be certain of his three month calculation?"

The Chief of Staff hung his head, with no reply.

The Naval Chief of Staff, Admiral Nagano, stepped in to the general's aid. "Japan was like a patient suffering from a serious illness," he said. "A quick decision had to be made one way or the other." The Emperor tried to ascertain whether the senior advisers were in favor of diplomacy, first, or war, first. He could not get a clear answer.

The next day, when the same question was raised again, the Chiefs of the Army and Navy General Staff remained silent. The Emperor expressed his regret that they had not seen fit to answer. He then drew a piece of paper out of his robe and read a poem by his grandfather, the Emperor Meiji:

> Since all are brothers in this world,
> Why is there such constant turmoil?

The hall was silent. "Everyone present was struck with awe." Then Admiral Nagano rose and said that military force would be used only when all else had failed. The meeting adjourned—"in an atmosphere of unprecedented tenseness."

The coming winter weather put an operational boundary on how much time was left. If the military were to make its moves before the spring of 1942, it would have to do so by early December. Still, Prince Konoye kept hoping to find some alternative short of war. After the conference in front of the Throne on September 6, the Cabinet took up the question of whether synthetic oil production could be greatly and speedily increased. It was better to spend vast sums on such a program, said Konoye, than on war, with all its uncertainties. But the head of the Planning Board said that it would be an immense task—requiring up to four years, many billions of yen, and a vast amount of steel, pipes, and machinery. A huge massing of engineering skills and upward of four hundred thousand coal miners would also be needed. Konoye's proposal was put aside. In late September, four men armed with daggers and short swords sprang at Konoye's car, aiming to assassinate him. They were repulsed, but the Prime Minister was badly shaken.

On October 2, the United States officially rejected a meeting between Konoye and Roosevelt. Shortly after, unable to muster a credible alternative to war, Konoye fell from office. He was replaced as Prime Minister on October 18 by Hideki Tojo, the bellicose war minister, who had consistently dismissed diplomacy as useless and had opposed any compromise with the United States. Back in Washington, Ambassador Nomura futilely described himself as "the bones of a dead horse." With diplomacy at a stalemate, Roosevelt himself fell into the grip of the fatalism that had captured those in both Tokyo and Washington. Yet he pleaded with Nomura that, between their two countries, there be "no last words."

The two Japanese tankers had continued to sit in the harbor near Los Angeles since midsummer, waiting to pick up contracted supplies of oil. In the first part of November, they finally weighed anchor and sailed away, with no oil aboard. Now, no one could doubt the absoluteness of the oil embargo. With winter almost at hand in Tokyo, the Japanese authorities retaliated by cutting off all supplies of heating oil to the American and British embassies.

On through October and into November, Japan's military high command and political leaders, often meeting in a small room in the Imperial Palace, continued to debate the final commitment to war. Again and again, the discussion came back to oil. Japanese oil imports had fallen drastically in 1941. Inventories were declining, too. "From the records available it is clear that this time-oil factor hovered over the conference table like a demon," one historian later wrote. "A decision for war was considered the most readily available means of exorcising it."[21]

On November 5, an Imperial Conference of the most senior leaders convened before the Emperor. He himself remained silent through the proceedings as was the custom in most circumstances. The Razor—Prime Minister Tojo—summarized the majority position. "The United States has from the beginning believed that Japan would give up because of economic pressure," he declared, but on this it would prove to be wrong. "If we enter into a protracted war, there will be difficulties," he said. "We have some uneasiness about a protracted war. But how can we let the United States continue to do as she pleases, even though there is some uneasiness? Two years from now we will have no petroleum for military use. Ships will stop moving. When I think about the strengthening of American defenses in the Southwest Pacific, the expansion of the American fleet, the unfinished China Incident, and so on, I see no end to difficulties. . . . I fear that we would become a third-class nation after two or three years if we just sat tight."

The proposal before the conference called for the presentation of stiff last-ditch demands to the United States. If they were rejected, Japan would go to war. "Do you have any other comments?" Tojo asked the group. Hearing no objection, he ruled the proposal approved.

A Japanese diplomat arrived in Washington the third week of November to present the list of demands. To Secretary of State Hull, it read like an ultimatum. There was another arrival of Japanese origin in Washington that week: an intercepted "Magic" message of November 22, informing Nomura that American agreement to Tokyo's latest proposals had to be received by November 29 at the very latest, for "reasons beyond your ability to guess." For, "after that, things are automatically going to happen."

On November 25, Roosevelt warned his senior military advisers that war could come very soon, even within a week. On the next day, Hull presented a note to the Japanese, proposing that Japanese troops be withdrawn from Indochina and China in exchange for a resumption of American trade with Japan.

Tokyo chose to regard this proposal as an American ultimatum. On that same day, November 26, a Japanese naval task force that had gathered in the Kurile Islands was ordered to set sail, under radio silence. Its destination was Hawaii.[22]

While the Americans did not know about that specific fleet, Secretary of War Stimson did bring Roosevelt an intelligence report indicating that a large Japanese expeditionary force was moving south from Shanghai toward Southeast Asia. "He fairly blew up, jumped into the air, so to speak, and said he hadn't seen it," commented Stimson, "and that that changed the whole situation because it was an evidence of bad faith on the part of the Japanese that while they were negotiating for an entire truce—an entire withdrawal—they should be sending this expedition down there." With that, the President arrived at a final answer to the question he had posed in his article almost two decades earlier. Japan could not be trusted. The following day, November 27, Hull told Stimson that he had completely given up on negotiations with Japan. "I have washed my hands of it," said the Secretary of State. It was now in the hands, he added, of the Army and the Navy. That same day, Washington sent off "a final alert" to American commanders in the Pacific, including Admiral Husband Kimmel, the commander of the Pacific Fleet stationed in Hawaii. The message to Kimmel began, "This dispatch is to be considered a war warning."

Up to the very end, there were those in Tokyo who saw nothing but disaster ahead. On November 29, the Senior Statesmen met with the Cabinet and the Emperor to plead that Japan seek some diplomatic solution as a better alternative than taking on the might of America. In reply, Prime Minister Tojo railed that to continue with broken economic relations would mean a progressive weakening of Japan. The Japanese leaders, in all their studies, all their discussions, had recognized that a long war would increasingly favor the United States because of its resources, capabilities, and endurance, but so strongly were the militarists gripped in the trance of their own making that those committed to war simply waved that consideration aside. War was on a speeding track.[23]

Pearl Harbor

On December 1, the special Japanese task force, still undetected, crossed the international dateline. "Everything is decided," a flight commander on one of the Japanese ships wrote in his diary on December 2. "There is neither here nor there, neither sorrow nor rejoicing." Tokyo gave orders to its embassies and consulates to destroy codes. An American military officer, sent to reconnoiter the Japanese embassy in Washington, found that papers were being burned in the backyard.

On Saturday, December 6, Roosevelt decided to send a personal note directly to the Emperor, seeking to dispel "the dark clouds" that had so ominously gathered. The message did not go off until nine o'clock that evening. Shortly after sending it, Roosevelt told some visitors, "This son of man has just sent his final message to the Son of God."

At 12:30 in the afternoon on December 7, Washington time, Roosevelt received the Chinese ambassador. The President said he expected "foul play" in Asia. He had a feeling, he added, that the Japanese might do something "nasty" within forty-eight hours. At 1:00 P.M. Washington time, he was still chatting with the Chinese ambassador. At that very same moment—it was 3:00 A.M., December 8, in Tokyo—Roosevelt's message was finally delivered personally to the Emperor. In the middle of the Pacific, it was the early morning hours of December 7, and the Japanese fleet was coming upon the Hawaiian Islands. Aloft above the flagship was the flag that had flown on a Japanese battleship in 1905, when the fleet had destroyed the Russian Navy in the Tsushima Strait. Planes were leaving the decks of the aircraft carriers. Their crews had been told that they were going to destroy the ability of the United States to cheat Japan out of its deserved place on earth.

The bombs began to fall on the American fleet in Pearl Harbor at 7:55 A.M., Hawaiian time.

An hour after the attack began on Pearl Harbor, Ambassador Nomura, accompanied by another Japanese diplomat, arrived at the State Department. They were kept in a diplomatic waiting room, while Hull took an urgent call from the President.

"There's a report that the Japanese have attacked Pearl Harbor," said Roosevelt in a steady but clipped voice.

"Has the report been confirmed?" Hull asked.

"No," the President replied.

Both men thought it was probably true. Still, there was one chance in a hundred that it was not, Hull thought, and he had the two Japanese diplomats brought to his office. Nomura, who had learned of the attack from the radio news, diffidently handed a long document to the American Secretary of State. Hull made a pretense of reading Tokyo's justification for its actions. He could not control his rage. "In all my fifty years of public service I have never seen such a document more crowded with falsehoods and distortions—infamous falsehoods and distortions on a scale so huge that I never imagined until today that any Government on this planet was capable of uttering them." What use had been his many months of private conversations in his apartment with Nomura? To Hull, the backwoodsman turned statesmen, the two diplomats looked to him "like a pair of sheep-killing dogs."

Neither Japanese offered any further comment. The meeting ended, but no one came forward to open the door for them, for they were now enemies. They opened the door out of Hull's office themselves and rode down in an empty elevator that was waiting for them, and let themselves out to the street.[24]

All that day, the reports flowed into Washington from Pearl Harbor—disjointed, fragmentary, and finally, dismal. "The news coming from Hawaii is very bad," Stimson noted in his diary at the end of that long Sunday. "It has been staggering to see our people there, who have been warned long ago and were standing on the alert, should have been so caught by surprise." How could such a disaster have occurred?

Senior American officials had fully expected a Japanese attack, and imminently. But they expected it to be in Southeast Asia. Virtually no one, whether in Washington or Hawaii, seriously considered, or even comprehended, that Japan could—or would—launch a surprise assault against the American fleet in its home base. They believed, as General Marshall had told President Roosevelt in

May of 1941, that the island of Oahu, where Pearl Harbor was located, was "the strongest fortress in the world." Most of the American officials seemed to have forgotten—or never knew—that Japan's great victory in the Russo-Japanese War had begun with a surprise attack on the Russian fleet at Port Arthur.

At a fundamental level, each side had underestimated the other. Just as the Japanese did not think the Americans were technically capable of cracking their most secret codes, so the Americans could not conceive that the Japanese would be able to mount so technically complex an operation. Indeed, in the immediate aftermath, some of Roosevelt's senior advisers believed that the Germans had orchestrated the assault; they assumed the Japanese could not have done it alone. And each side mistook the other's psychology. The Americans could not believe that the Japanese would do something so daring and even reckless. They were wrong. And the Japanese, for their part, counted on Pearl Harbor to shatter American morale, when, instead, the attack would revivify national morale and swiftly unite the country. That was a much greater error.

After the fact, of course, the Japanese intentions could be clearly discerned in the mass of information that was available to the United States government, including the bountiful treasure of secret communications that came from "Magic," the cracked Japanese codes. But in those tense months leading up to the attack, the clear signals were lost in the "noise"—the maze of complex, confusing, contradictory, competing, and ambiguous pieces of information. After all, there were also many indications that the Japanese were about to attack the Soviet Union. The dissemination of "Magic" itself and its intelligence was sometimes bungled, in critical ways. This was part of a larger failure, the breakdown of critical communication among key actors on the American side that may have been the second most important cause of the tragedy at Pearl Harbor, following only on the failure to believe that such an attack could take place at all.[25]

The One Mistake

The waiting was over. Japan and the United States were now at war. But Pearl Harbor was not the main Japanese target. Hawaii was but one piece of a massive, far-flung military onslaught. In the same hours as the attack on the U.S. Pacific Fleet, the Japanese were bombing and blockading Hong Kong, bombing Singapore, bombing the Philippines, bombarding the islands of Wake and Guam, taking over Thailand, invading Malaya on the way to Singapore—and preparing to invade the East Indies. The operation against Pearl Harbor was meant to protect the flank—to safeguard the Japanese invasion of the Indies and the rest of Southeast Asia by incapacitating the American fleet and, thereafter, to protect the sea lanes, particularly the tanker routes from Sumatra and Borneo to the home islands. The primary target of this huge campaign remained the oil fields of the East Indies.

Thus, Operation Hawaii was essential to Japan's larger vision. And a critical element in its success—luck—had been with the Japanese attackers right up to the last moment. Indeed, the Japanese far exceeded even their own ambitions. The extent of the surprise and the incapacity of American defenses at Pearl Harbor were both much greater than the Japanese had anticipated. In their attack on Pearl Harbor, two waves of Japanese aircraft succeeded in sinking, capsizing, or severely damaging eight battleships, three light cruisers, three destroyers, and four auxiliary craft. Hundreds of American planes were destroyed or damaged. And 2,335 American servicemen and 68 civilians were killed. All this added up to, perhaps, the most devastating shock in American history. The American aircraft carriers survived only because they happened to be out on missions at sea. The Japanese lost a total of only twenty-nine planes. Admiral Yamamoto's gamble had paid off, handsomely.

Yamamoto himself might well have taken one more chance, but he was thousands of miles away, monitoring events from his flagship, off Japan. The commander of the Hawaiian task force, Chuichi Nagumo, was a far more cautious man; indeed, he had actually opposed the entire operation. Now, despite the entreaties of his emboldened officers and much to their chagrin, he did not want to send planes back to Hawaii, for a third wave, to attack the repair facilities and the oil tanks at Pearl. His luck had been so enormous that he did not want to take more risks. And that, along with the sparing of its aircraft carriers, was America's only piece of good fortune on that day of devastation.

In the course of planning the operation, Admiral Yamamoto had observed that the great mistake made in Japan's surprise attack against the Russians at Port Arthur in 1904 was in not being "thoroughgoing" enough. The same mistake was made once again at Pearl Harbor. Oil had been central to Japan's decision to go to war. Yet the Japanese forgot about oil—at least in one crucial dimension—when it came to planning Operation Hawaii. Yamamoto and his colleagues, who had endlessly reviewed America's preponderant position in oil, all failed to grasp the significance of the supplies on the island of Oahu. An assault on those supplies was not included in their plans.

It was a strategic error with momentous reverberations. Every barrel of oil in Hawaii had been transported from the mainland. If the Japanese planes had knocked out the Pacific Fleet's fuel reserves and the tanks in which they were stored at Pearl Harbor, they would have immobilized every ship of the American Pacific Fleet, and not just those they actually destroyed. New petroleum supplies would only have been available from California, thousands of miles away. "All of the oil for the Fleet was in surface tanks at the time of Pearl Harbor," Admiral Chester Nimitz, who became Commander in Chief of the Pacific Fleet, was later to say. "We had about $4\frac{1}{2}$ million barrels of oil out there and all of it was vulnerable to .50 caliber bullets. Had the Japanese destroyed the oil," he added, "it would have prolonged the war another two years." [26]

CHAPTER 17

Germany's Formula for War

ONE AFTERNOON in June of 1932, an open car appeared at a Munich hotel to pick up two officials of I. G. Farben, the huge German chemical combine. The men—one a chemist, the other a public relations man—were driven to the private apartment of Adolf Hitler at *Prinzregentenplatz*. Hitler had not yet come to power as Chancellor of Germany, but he was the leader of the National Socialist party, which held almost 20 percent of the seats in the Reichstag and looked likely to increase significantly its seats in the election due the following month.

The men from I. G. Farben had sought out the would-be Führer in order to try to bring an end to the continuing Nazi press campaign against their company. The Nazis railed against I. G. Farben as an exploitative tool of "international financial lords" and "money-mighty Jews," and attacked the company for the fact that Jews occupied some senior positions. They even caricatured the company as "Isadore G. Farben." The Nazis also criticized it for pursuing its expensive project to manufacture liquid fuels from coal—otherwise known as synthetic fuels—and for the tariff protection it had obtained for the project from the government. And that pointed to a second problem. I. G. Farben had made a very large financial commitment to synthetic fuels, but it appeared by 1932 that the project could never be profitable without continued government tariff protection and other support. The company's main argument was that a synthetic fuels industry would cut Germany's dependence on foreign oil and thus reduce the acute pressure on the country's foreign exchange. The two I. G. Farben representatives hoped to convert Hitler to their point of view.

Hitler himself was late for the meeting, having just returned from an electoral campaign trip. He intended to give the two I. G. Farben officials only a half hour, but he became so engrossed in the discussion that he spent two and a half hours with them. Mesmerized by his own visions, Hitler did much of the talking, lecturing and declaiming on his plans to motorize Germany and build new highways. But he also asked technical questions about synthetic fuels, and he assured the two men that such fuels fit perfectly with his overall plans for a new Germany. "Today," he told them, "an economy without oil is inconceivable in a Germany which wishes to remain politically independent. Therefore, German motor fuel must become a reality, even if this entails sacrifices. Therefore, it is urgently necessary that the hydrogenation of coal be continued." He strongly endorsed the synthetic fuels effort. He also promised to halt the press campaign against I. G. Farben and to keep the tariff protection for synthetic fuels in place once the Nazis came to power. For its part, I. G. Farben—then or later—promised to deliver what the Nazis wanted: campaign contributions. When the I. G. Farben officials reported back on their conversation with Hitler, the chairman of the company said, "Well, this man seems to be more reasonable than I had thought."[1]

Hitler had good cause to appear reasonable. A successful synthetic fuels program, he had quickly grasped, could prove very valuable, perhaps essential, to his overall objectives for a resurgent and dominant Germany. One of the major obstacles to the achievement of that goal, he knew, was Germany's dependence on imported raw materials—and oil in particular. Domestic petroleum production was tiny; imports, correspondingly high. Moreover, much of the imported oil came from the Western Hemisphere.

Germany's remarkable economic growth over the preceding half century had been based largely on its own plentiful energy source—coal. Whereas in the late 1930s coal provided just about half of the United States' total energy, it supplied 90 percent of Germany's energy—while oil accounted for only about 5 percent. But already in 1932, Hitler was planning for the future, and oil would be essential to his ambitions. He became Chancellor in January 1933 and then, over the next year and a half, seized complete power. He wasted no time in launching a motorcar campaign that he would hail as "a turning point in the history of German motor traffic." Autobahns, limited-access highways without speed limits, were to span the country, and in 1934 planning began for a new type of vehicle. It was called "the people's car," the Volkswagen.

But these were but pieces of his grand plan, which was to subordinate all of Europe to the Nazi Reich—and to himself. To that end, he quickly began to regiment the economy, harness big business to the state, and build the Nazi war machine—including bombers and fighter planes, tanks and trucks, all of which required oil. And the synthetic fuels on which I. G. Farben was working were to be of decisive importance.[2]

The Chemical Solution

Pioneering work on the extraction of synthetic fuels from coal had actually begun in Germany before the First World War. The country was then already acknowledged as the world's leader in chemistry. In 1913, the German chemist Friedrich Bergius first succeeded in extracting a liquid from coal in a process that became known as hydrogenation. Large amounts of hydrogen were added to coal under high temperatures and high pressure in the presence of a catalyst. The end product was a high-grade liquid fuel. A competing German process, Fischer-Tropsch, was detailed a decade later, in the mid-1920s. Here, coal molecules were broken down under steam into hydrogen and carbon monoxide, which, in turn, were made to react together, resulting in the production of a synthetic oil. The Bergius hydrogenation process was deemed the better of the two. Among other things, it could produce aviation fuel, which Fischer-Tropsch could not. In addition, I. G. Farben, which acquired the patent rights in 1926 to the Bergius process, was politically more powerful than the sponsors of Fischer-Tropsch.

I. G. Farben became interested in synthetic fuels in the 1920s because of the same predictions of the imminent exhaustion of the world's conventional petroleum supplies that were stimulating the great oil exploration drive around the world. The government provided support because the increasing demand for foreign oil was causing a hemorrhage of vital and scarce foreign exchange. A pilot plant was built at I. G.'s Leuna works, with initial production beginning in 1927. At the same time, I. G. Farben was busy seeking potential partners in other countries. After negotiations with a leading British chemical group fell through, I. G. Farben found a much more important potential partner—Standard Oil of New Jersey.[3]

At that time, Standard was midway in its strategic transformation from refiner to integrated oil company, well-supplied with its own crude, both in the United States and abroad. It had also been exploring alternatives to crude oil as a source of liquid fuels; as early as 1921, it had purchased twenty-two thousand acres in Colorado with the hope of finding a commercially successful method for extracting oil from shale. But Standard had been dissatisfied with the results; the production of one barrel of synthetic oil from shale required a ton of rock, and the economics were extremely unattractive.

Frank Howard, the head of research at Standard, visited I. G.'s Leuna works in 1926. He was so impressed that he immediately fired off a telegram to Standard's president, Walter Teagle, then visiting in Paris. "Based upon observations and discussions today, I think that this matter is the most important which has ever faced the company since the dissolution," wired Howard. "This means absolutely the independence of Europe in the matter of gasoline supply." Teagle himself, alarmed about the possibility of losing European markets to the new synthetic oil, hurried to Leuna. The research and production facilities awed him. "I had not known what research meant until I saw it," he later said. "We were babies compared to the work I saw."

Teagle, Howard, and other Standard executives hurriedly gathered at a hotel room in Heidelberg, ten miles from the I. G. Farben works. They concluded, Howard later recalled, that the hydrogenation process might be "more significant than any technical factor ever introduced into the oil industry up to this time." Here, in the laboratories of I. G. Farben, was a clear threat to Standard's business. "Although hydrogenation of coal probably could never compete on an economic basis with crude oil," said Howard," 'the nationalistic factor' would lead to hydrogenation's being made the foundation of a protected manufacturing industry in many countries willing to pay the price." Thus, markets could be closed to imported crude oil and refined products; Standard could hardly afford not to become involved.

An initial agreement was therefore reached with I. G. Farben, which allowed Standard to build a hydrogenation plant in Louisiana. But by this time, the world oil shortage was beginning to turn into a surplus, and the American company's interest shifted. Hydrogenation could also be used on crude oil, to increase the gasoline yield. Thus, the new plant in Louisiana would experimentally apply the process not to coal, but to oil, in order to squeeze more gasoline out of each barrel of petroleum.

In 1929, the two companies struck a broader agreement. Standard would have the patent rights to hydrogenation outside Germany. In exchange, I. G. Farben received 2 percent of Standard's stock—546,000 shares—valued at $35 million. Each company agreed to stay out of the other's main fields of activity. As a Standard official put it, "The I. G. are going to stay out of the oil business—and we are going to stay out of the chemical business." The next step came in 1930, with the establishment of a joint company to share developments in the "oil-chemical" field. Overall, a good deal of technical knowledge was flowing to Standard.[4]

In 1931, German science and, in particular, hydrogenation received the highest accolade: Bergius, the inventor of the hydrogenation technique, and Carl Bosch, the chairman of I. G. Farben, shared the Nobel Prize in chemistry. Yet, while the project at Leuna was by then producing at a rate of two thousand barrels per day, it was floundering badly and was in deep financial trouble. Development was proving to be more difficult and much more expensive than anticipated. At the same time, the oil surplus, with the new discoveries in East Texas, had turned into an overwhelming global glut. The resulting collapse in world oil prices made the synthetic fuel effort at Leuna decidedly uneconomical, and I. G. Farben feared that the project might never turn a profit. The cost of producing a liter of *Leunabenzin,* as the fuel was called, was up to ten times the price per gallon at which gasoline was put onto tankers in the Gulf of Mexico bound for Germany. Some I. G. Farben executives said the whole project should be abandoned. The sole reason to keep it going, others replied, was that the costs of shutting it down would be greater than the costs of continuing.

The only real hope for keeping the synthetic fuels project alive, in the midst of the Great Depression, was with some kind of state support or bail-out. The tariff protection from the pre-Hitler Brüning government was not enough. The new Nazi regime was willing to go much further and guarantee prices and markets to I. G. Farben—so long as the company promised to increase substantially its production of synthetic fuels. Even that was not enough, for hydrogenation was still an infant technology. It needed both further development and additional political patronage in the Third Reich. I. G. Farben won support from the Air Force, the Luftwaffe, by proving that it could develop a high-quality aviation gasoline. The German Army, the Wehrmacht, also lobbied for an expanded commitment to a domestic synthetic fuels industry, arguing that Germany's own current supplies would be woefully inadequate to the requirements of the new type of warfare that it was planning.[5]

Girding for War

Two further developments demonstrated to Hitler and his entourage both the dangers of depending upon foreign oil and the concurrent need to develop Germany's own supply. The first was by example. In October 1935, Italy invaded the East African country of Ethiopia, then better known as Abyssinia, which shared uneasy and poorly demarcated borders with adjacent colonies of Italy. Benito Mussolini, the Italian dictator, dreamed of creating a great empire to befit his imperial Roman pretensions, and he began with the attack on Ethiopia. Forthwith, the League of Nations condemned the invasion, imposed some economic sanctions, and considered placing an embargo on oil exports to Italy. The Roosevelt Administration gave indications that the United States, though not a League member, might find a way to cooperate with such an embargo. Mussolini well knew that a shut-off of petroleum supplies would paralyze the Italian military. While his armies advanced, throwing poison gas against the hapless Ethiopians, he resorted to every form of bluff and bluster to intimidate the League. Sanctions, he said, could well be regarded as an act of war. A leading proponent of the oil sanctions was the British Minister for the League, Anthony Eden, who dismissed the threat; Mussolini, he said, would not risk a "mad-dog act" and "has never struck me as the kind of person who would commit suicide." But Mussolini found a willing ally in the French Prime Minister, the cunning Pierre Laval, who cleverly subverted the oil sanctions movement at the moment when it was closest to success.

By the spring of 1936, Mussolini's forces had conquered Ethiopia, the King of Italy had added "Emperor of Ethiopia" to his title, and the entire sanctions movement had collapsed. The oil embargo was never tested, because it was never applied. Mussolini himself later confided to Hitler: "If the League of Nations had followed Eden's advice on the Abyssinian dispute, and had extended economic sanctions to oil, I would have had to withdraw from Abyssinia within a week. That would have been an incalculable disaster for me!" Hitler took the lesson of dependence most seriously.

The second lesson struck closer to home. The Nazi regime was committed to "recovering" the domestic German market from Standard Oil, Shell, and the other foreign companies. But even worse, the hated Bolsheviks owned a large chain of gasoline stations through which they sold the oil products they provided Germany. The Nazi government pushed a German gasoline marketeer to acquire the Soviet chain, which it did in 1935. The aim was to clear out "a wasp's nest." For a time, though unhappily, the Soviets continued to supply the amount of oil previously sold through its distribution system. But then, in February 1936 they abruptly stopped deliveries. The reason given was "difficulties with foreign payments." The deliveries did not resume. And that, too, was a warning to Hitler about the dangers of dependence.

Just at this time, in the middle of February 1936—as the League still debated oil sanctions—the annual German motor show in Berlin was opened by Hitler, who, the *New York Times* observed, "is believed to reel off a higher annual motor mileage than any other ruler or head of State." Hitler took the occasion to announce that Germany "had effectively solved the problem of producing synthetic gasoline." This achievement, he pointedly declared, "possessed political significance." The question of foreign supplies and sanctions was very much on Hitler's mind. For he was on the eve of a critical move. The next month, March 1936, he boldly remilitarized the Rhineland, on the border with France, in violation of treaty agreements. It was the first time that he flexed his muscles on the international front, taking what afterward he was to call his gravest risk—the forty-eight hours that were "the most nerve-racking in my life." He waited to be challenged, but the Western powers did nothing to stop him. The gamble had paid off. The pattern was to be repeated.[6]

Later in 1936, Hitler took decisive steps to gird the German state so that it would be prepared for war by the target date of 1940. He inaugurated his Four-Year Plan, which, among other things, aimed to reduce dependence on foreign oil through new technology and chemistry. "German fuel production must now be developed with the utmost speed," he said in establishing the plan. "This task must be handled and executed with the same determination as the waging of a war, since on its solution depends the future conduct of the war." He added that "the production cost of these raw materials" was of "no importance."

The synthetic fuels industry, with a central place in the overall plan, was supposed to expand output almost sixfold. The program received major financial support, and vast amounts of steel and labor were requisitioned to construct the sprawling industrial establishments required for conversion. Each plant was a huge engineering undertaking, spread over acres, that depended on great industrial companies—in full partnership with the Nazi state. I. G. Farben led the way, adapting itself to the Nazi ideology. By 1937–38, it was no longer an independent company, but rather an industrial arm of the German state, and fully Nazified. All Jewish officials had been removed, including the third of the supervisory board who were Jews. The anti-Nazi chairman of the managing board, Carl Bosch, the man who had made the deal with Standard Oil, was pushed aside, while most of the other members of the managing board who did not already belong to the Nazi party fell all over each other in their rush to sign up.

While the ambitious promises of the Four-Year Plan proved far too grandiose, Germany nevertheless did build up a very substantial synthetic fuel industry. By September 1, 1939, when Germany invaded Poland, starting the Second World War in Europe, fourteen hydrogenation plants were in full operation, with six more under construction. By 1940, synthetic fuel output had increased drastically—72,000 barrels per day, accounting for 46 percent of total oil supply. But the synthetic fuels were even more significant when viewed in terms of military needs. Hydrogenation, the Bergius process, provided some 95 percent of Germany's total aviation gasoline. Without those synthetic fuels, the Luftwaffe could not have taken to the air.

Despite the strength of his military machine, and with a growing supply of synthetic fuels at his disposal, oil was never far from Hitler's mind. Indeed, that concern had helped shape his basic strategic approach to war, which was based on the *blitzkrieg*, or "lightning war"—fierce but short battles with concentrated mechanized forces that would lead to decisive victory before petroleum supply problems could develop. Initially, the strategy worked astonishingly well, not only in Poland in 1939, but also in the spring of 1940, as Hitler's forces overran Norway, the Low Countries, and France with surprising ease. The campaign in the West actually improved Germany's oil position, for German troops captured oil stocks considerably in excess of the fuel they had expended in the invasions. Even though Hitler's subsequent attempts to subdue the British Isles through massive aerial bombardment had failed by the fall of 1940, Germany seemed on the verge of dominating Europe. It had also fallen into the habit of thinking that victory was cheap. So when Hitler turned his sights east toward his next objective, he envisioned another easy victory. The target was the Soviet Union.[7]

The Russian Campaign: "My Generals Know Nothing About the Economic Aspects of War"

Many factors shaped Germany's decision to go to war with the Soviet Union: Hitler's deep-seated hatred of Bolshevism (its eradication, he said, was his "life's mission"); his personal enmity for Stalin; his contempt for the Slavs, whom he regarded as "little worms"; his desire to dominate completely the Eurasian land mass; and his drive for glory. In addition, when he looked East, he saw *lebensraum* ("living space") for the Thousand-Year Reich, his new German empire. Moreover, despite Stalin's almost pathetic eagerness to live up to the Nazi-Soviet Pact of August 1939 and avoid provoking Hitler, the German dictator suspected a secret deal between Britain and the Soviet Union. How else to explain England's refusal to capitulate in 1940 when its cause seemed so obviously lost? Amid all else, there was also the issue of oil.

From the very start, the capture of Baku and the other Caucasian oil fields was central to Hitler's concept of his Russian campaign. "In the economic field," one historian has written, "Hitler's obsession was oil." To Hitler, it was *the* vital commodity of the industrial age and for economic power. He read about it, he talked about it, he knew the history of the world's oil fields. If the oil of the Caucasus—along with the "black earth," the farmlands of the Ukraine—could be brought into the German empire, then Hitler's New Order would have within its borders the resources to make it invulnerable. In that conception, there was a striking similarity to the Japanese drive to encapsulate the resources of the East Indies and Southeast Asia within its empire, an ambition also powered by the belief that such a resource base would make it impregnable. Albert Speer, the German Minister for Armaments and War Production, said at his interrogation in May 1945, "the need for oil certainly was a prime motive" in the decision to invade Russia.[8]

Hitler also saw Soviet might as a permanent threat to the Ploesti oil fields of Rumania, Europe's largest source of petroleum production outside the Soviet Union. They had been a major German objective in the First World War. Now, Rumania was a German ally, and Germany had become heavily dependent upon Ploesti, which provided 58 percent of Germany's total imports in 1940. Oil shipments from the Soviet Union had resumed with the signing of the Nazi-Soviet Pact of 1939, and in 1940, they accounted for another third of Germany's oil imports, leading one senior Nazi to describe them as a "substantial prop to the German war economy." In June 1940 the Soviet Union used the terms of the

Nazi-Soviet Pact as justification to seize a significant part of northeastern Rumania, which put Russian troops all too close to the Ploesti oil fields for Hitler's taste. "The life of the Axis depends on those oilfields," he told Mussolini. An attack on Russia would guarantee the security of Ploesti.

The conquest of Russia would, of course, also make available a far grander prize: the oil resources of the Caucasus—Maikop, Grozny, and Baku itself. To support his plans, Hitler propounded his own bizarre calculations: that the number of German casualties in a war with Russia would be no greater than the number of workers tied up in the synthetic fuel industry. So there was no reason *not* to go ahead.

In December 1940 Hitler issued Directive Number 21—Operation Barbarossa—ordering that preparations begin for an invasion of the Soviet Union. The Germans took care to give no public sign of displeasure to their Russian friend and, indeed, went out of their way to engage in an elaborate charade of deception and disinformation to lull Stalin into disbelief that the Germans might be contemplating such a strike. Warnings of the impending invasion came from many sources—Americans, British, other governments, his own spies—but Stalin resolutely refused to believe them. Scarcely hours before the invasion, a dedicated German communist defected from a German Army unit and slipped over to the Soviets with word of what was about to happen. Stalin suspected a trick and ordered the man shot.[9]

In the early morning hours of June 22, 1941, Russian freight trains were lumbering slowly westward along railway tracks in the Soviet Union, carrying oil and other raw materials destined for Germany. Just after 3:00 A.M. the German Army, three million men strong, with 600,000 motor vehicles and 625,000 horses, struck along a wide front. The German onslaught caught the Soviet Union completely off guard and put Stalin into a nervous collapse that lasted several days. The Germans thought the attack would be a repetition of the *blitzkrieg* that had swept so successfully across Poland, the Low Countries, France, Yugoslavia, and most recently Greece. It would all be over in six or eight weeks, or ten at the most.

Hitler's boast about the Russian campaign, that "we'll kick the door in and the house will fall down," seemed amply borne out in the first weeks of the campaign. Initially, the Germans moved even faster than they had expected, driving back the disorganized Soviet forces. Victory appeared almost at hand, save for some mopping up. Yet there were soon some preliminary signs that the Germans were stretching themselves. They had seriously miscalculated their supply needs, including the need for fuel. On the bad Russian roads and difficult terrain, vehicles burned considerably more fuel than anticipated, sometimes twice as much. The larger vehicles, which sank into the unsurfaced roads and could not move, had to be replaced with small Russian wagons drawn by horses. But warnings about looming fuel shortages were ignored in the initial euphoria of early victories.[10]

In August, German generals sought Hitler's permission to make Moscow the prime target. Hitler refused. "The most important aim to be achieved before the onset of winter is not to capture Moscow," said his directive of August 21, "but to seize the Crimea and the industrial and coal region on the Donets, and to cut off the Russian oil supply from the Caucasus area." The Wehrmacht had to reach Baku. As for the Crimea, Hitler described it as "that Soviet aircraft carrier for attacking the Rumanian oil fields." To the arguments of his generals, he responded with what would become one of his favorite maxims—"My generals know nothing about the economic aspects of war." Intoxicated by conquest, Hitler was already dreaming aloud about the vast autobahn he would build from Trondheim, in Norway, to the Crimea, which would then become Germany's Riviera. And, he said, "the Volga will be our Mississippi."

Later on, Hitler changed his mind, and put Moscow back at the top of his objectives. But critical time had been lost. As a result, while the Germans did manage to reach the outskirts of Moscow, just twenty miles from the Kremlin, they did not do so until the end of autumn 1941. There they bogged down in the mud and snow of fast-approaching winter. The shortages of oil and other essential supplies finally caught up with them. "We have reached the end of our resources of personnel and material," the Quartermaster General said on November 27. Then on December 5 and 6, General Yuri Zhukhov launched the first successful Soviet counterattack, thus preventing the Germans from moving any further and tying them down for the winter.

Nor were the German troops able to get to the Caucasus. The six to eight to ten weeks had already turned into months, and the Germans were now stalemated by winter. They had vastly underestimated the space over which their supply lines would have to stretch; they had no less underestimated the reserves of Soviet manpower—and the capacity of Soviet soldiers and citizens alike to tolerate hardship and deprivation. The numbers were beyond comprehension; six to eight million Soviet soldiers were killed or captured in the first year of war, and still new men were thrown into battle. In addition, the Japanese decision to attack Pearl Harbor and move into Southeast Asia, rather than attack the Soviet Union, allowed Stalin to transfer his crack Siberian divisions westward, to the German front.[11]

Operation Blau

In the early months of 1942, Berlin was making plans for another great offensive in Russia, Operation Blau. The oil of the Caucasus was its main objective, and from there, on to the oil fields in Iran and Iraq and then on to India. Hitler's economic experts had told him that Germany could not continue the war effort without access to Russian oil, and Hitler fully agreed. At the same time, he wanted to strike at the heart of the Russian war economy. Deprived of fuel for its military units and for agriculture, Russia would not be able to stay in the war. Hitler was sure that the Soviet Union would expend its last reserves of manpower to defend the oil fields, and then victory would be his. With considerable confidence, Germany assembled a Technical Oil Brigade, eventually fifteen thousand men strong, with the charge of rehabilitating and running the Russian oil industry. The only thing standing in the way of Germany's exploitation of Russian oil was the requirement to capture it.

By late July 1942 the German armies seemed well on the way to that goal, with the conquest of the city of Rostov and the cutting of the oil pipeline from the

Caucasus. On August 9, they reached Maikop, the most westerly of the Caucasian oil centers—but a small one, with an output equivalent under normal circumstances to only a tenth of that of Baku. Moreover, before withdrawing from Maikop, the Russians had so thoroughly destroyed the oil fields and supplies and equipment, right down to the small tools in the workshops, that by January 1943 the Germans were able to eke out no more than seventy barrels per day there.

Still, the Germans drove on, now thousands of miles from their homeland and their supply centers. In mid-August, German mountain troops planted the swastika at the summit of Mount Elbrus, the highest point in the Caucasus and in Europe. But the German war machine was stopped before it could reach its objectives. Its armies were blocked by mountain passes that could be defended, given time, and stymied further by their own shortage of fuel, which provided that time. In order to fight Russia, the German forces required petroleum resources on a grand scale, but they far outran their supply lines and lost their advantages of speed and surprise. The irony of Operation Blau was that the Germans ran short of oil in their quest for oil.[12]

The Germans had captured Russian oil supplies as they had done with French supplies, but this time to no avail, for Russian tanks ran on diesel, which was useless to the German panzer units, which ran on gasoline. Panzer divisions were sometimes at a standstill for several days at a time in the Caucasus, while they waited for fresh supplies. Trucks carrying oil could not catch up because they too had run short of fuel. Finally, in desperation, the Germans took to transporting oil supplies on the backs of camels. By November 1942 the last German effort to break through the mountain passes toward Grozny and Baku had definitively been repulsed.

The city of Stalingrad, to the northwest of the Caucasus, was meant to be a sideshow to the main campaign, a secondary German objective. But, from the very beginning, its name made its fate pregnant with symbolism for both sides. It became the scene of a titanic, decisive struggle fought in the winter of 1942–43. Again and again, the German Army was crippled by supply shortages, of which that of fuel loomed largest. General Heinz Guderian, the legendary panzer commander, wrote that winter from the Stalingrad front to his wife: "The icy cold, the lack of shelter, the shortage of clothing, the heavy losses of men and equipment, the wretched state of our fuel supplies, all this makes the duties of a commander a misery."

After more than eighteen months of unrelenting effort and extraordinary costs in human and material resources, the tide of battle turned, and the Germans were finally on the defensive in Russia. In a midnight phone call, Field Marshal Erich von Manstein begged Hitler to transfer the German forces in the Caucasus to his command in order to help the embattled Sixth Army at Stalingrad.

Hitler refused. "It's a question of the possession of Baku, Field Marshal," the dictator said. "Unless we get the Baku oil, the war is lost." Hitler then proceeded to deliver a lesson on the central importance oil had assumed in warfare. He repeated himself over and over, but he could not stop his own harangue. How much fuel an aircraft needed. How much fuel a tank needed. On and on, the words kept coming. "If I can no longer get you the oil for your operations, Field Marshal, you will be unable to do anything."

The map legend:

The Axis countries
Greatest extent of Axis Control, late 1942
Areas held by Allies
Neutral nations
Allied advances
Axis advances

WAR IN EUROPE
AND NORTH AFRICA

Manstein tried to stop the onslaught, to argue with him about the immediate strategic issue—the survival of the Sixth Army. Hitler would not listen. Instead he described how German armies would meet up in the Middle East. "Then we shall march with our assembled forces to India, where we shall seal our final victory over England. Goodnight, Heil, Field Marshal!"

"Heil, mein Führer!" was all Manstein could say.

Despite Hitler's rapture, the order to retreat was given in January 1943 to German soldiers in the Caucasus. But that was too late to do anything for the Sixth Army at Stalingrad. Surrounded by Soviet forces, it was trapped, unable to risk a breakout. Its tanks only had enough fuel to move twenty miles, but to escape, they would have had to bridge a gap of thirty miles. It simply could not be done. And so at the end of January and beginning of February 1943, the encircled German forces at Stalingrad—ground down and incapacitated, frozen, hungry, and with their essential element of mobility gone—surrendered.

Stalingrad was Germany's first major defeat in Europe, and it sent Hitler into an uncontrollable rage. German soldiers were supposed to die, not surrender. But Germany was no longer on the offensive. The *blitzkrieg* phase was over. Instead of lightning attacks, the critical factors from here on would be military manpower and economic resources—including oil. And on the eastern front, despite some reverses, the Soviets would press relentlessly forward, pushing the Germans out of all captured Russian territory and moving inexorably on the road toward the final goal, Berlin itself.[13]

Rommel and the Revenge of the Quartermaster

It was not only at Stalingrad, in late 1942 and early 1943, that the tide turned against Germany. Another major reversal unfolded on the sands and brownish grit and barren rocks of North Africa, near the border between Libya and Egypt.

North Africa was, in the words of General Erwin Rommel, the one theater of World War II where the military struggle took place almost totally on the new "principle of complete mobility." That mobility was provided by Germany's Panzer Army in North Africa and its most important element, the Afrika Korps, both the creation of Rommel. He was a brilliant, innovative, imaginative master of tank warfare and the mobile campaign, as well as a consummate risk taker in strategy and tactics. Small, silent, and cold, Rommel had established his reputation as a battlefield leader in the First World War. Hitler was impressed by a book he wrote on infantry tactics and in 1938 appointed him, though he was not a Nazi party member, to head the battalion responsible for the personal safety of the Führer. In 1940, he commanded a panzer division that moved with astonishing speed across France. That sweep had seemed to him more like a romp than a war. "We never imagined war in the West would be like this," he wrote to his wife. The campaign, he added lightheartedly, had "turned into a lightning tour of France."

In February 1941 Rommel was dispatched to North Africa to shore up an Italian Army that was on the brink of being defeated by British forces. War in this theater would also provide him with a tour of North Africa; for though only seventy miles in width, the battle zone would stretch a thousand miles in length, from Tripoli in Libya to El Alamein in Egypt. But despite the rapid movements of forces, there was nothing lightning about this struggle.

Rommel was committed to a war of movement and to boldness. He was scathing about the commander who halted a victorious advance at the urging of the quartermaster. "It has become the habit for quartermaster staffs to complain at every difficulty, instead of getting on with the job and using their powers of improvisation, which indeed are frequently nil," he wrote. "When, after a great victory which has brought the destruction of the enemy, the pursuit is abandoned on the quartermaster's advice, history almost invariably finds the decision to be wrong and points to the tremendous chances which have been missed." Rommel had no intention to be so bound.

At first, Rommel won stunning victories in North Africa against British forces—often with only slender resources and captured booty. At one point, 85 percent of his transport was provided by captured British and American vehicles. He also had a considerable talent for improvisation, and not only in terms of tactics. Early in his campaign, Rommel ordered a number of "dummy tanks" built at workshops in Tripoli, which were then mounted on Volkswagens in order to frighten the British into thinking his armored divisions were much larger than in fact they were. But there was one thing even he could not fake. Mobile warfare was absolutely dependent on ample supplies of fuel—supplies that had to keep pace with rapid advances and be delivered along very long lines. And oil proved to be one of Rommel's most persistent problems; it was sometimes, he said, his greatest. As early as June 1941 he wrote: "Unfortunately, our petrol stocks were badly depleted, and it was with some anxiety that we contemplated the coming British attack, for we knew that our moves would be decided more by the petrol gauge than by tactical requirements."[14]

But, with his forces successfully resupplied with fuel in late 1941 and the first part of 1942, Rommel renewed his offensive, and at the end of May 1942 launched a major assault against the British. It went well—very well, indeed. The British fell back, and in one week, his forces were able to travel three hundred miles. Rather than stop at the frontier between Libya and Egypt, as had been planned and as the nature of his supply lines dictated, and as the quartermaster might have advised, Rommel pushed on across the border, until his advance was finally halted, late in June, near a small railway stop called El Alamein. He was now less than sixty miles from Alexandria; Cairo and the Suez Canal were not far beyond.

The Axis powers thought they were on the verge of a famous victory. Mussolini flew over to North Africa, accompanied, in another airplane, by a white charger, on which he planned to make a triumphant entry into Cairo. Rommel's goals were far more expansive: Cairo would only be a way station for a campaign through Palestine, Iraq, and Iran whose final objective would be Baku and its oil fields. Their capture, in concert with the German forces then battling in the Caucasus, would, Rommel predicted, create "the strategic conditions" to "shatter the Russian colossus." Hitler was swept up by the same intoxicating vision. "Destiny," he wrote to Mussolini, "has offered us a chance which will never occur twice in the same theater of war."

Both Rommel and Hitler spoke too soon. While the Soviets held on in the Caucasus, the Allies had succeeded, despite ferocious German attacks, in retaining the Mediterranean island of Malta, off the coast of Libya, which gave them a base from which to attack the Axis shipping that supplied Rommel's forces in North Africa. The Allies were further aided by their intercepts of the German and Italian codes. Moreover, the supply aircraft of the Luftwaffe were themselves beginning to run short of fuel. The Italian supply ships were no longer getting through to North Africa. And Rommel's very success—the incredible distance that the Afrika Korps had traveled—created a dangerous vulnerability. His own supply lines were very long, and the fuel trucks traveling from Tripoli used more gasoline to get to the front and back than they were able to carry. By driving so far and so incessantly fast, Rommel had not only called up the quartermaster's nightmare, but had also put the Panzer Army in a position of considerable risk. Still, he thought victory was shortly to be his. On June 28, 1942, he wrote his wife to plan on a July vacation together in Italy. "Get passports!" he said.

On the other side of the line, panic was building in Cairo. The British were burning their documents, Allied personnel were being squeezed into cattle trains for a hasty evacuation, and Cairo merchants were hurriedly replacing the photographs of Churchill and Roosevelt in their shop windows with those of Hitler and Mussolini. But the British did not give way in late June and July 1942, and Rommel was too short of gasoline to regain his momentum. The two exhausted armies had fought to a stalemate in what came to be called the First Battle of El Alamein. And there in the desert they waited.[15]

In mid-August, Rommel gained a new and formidable adversary—the austere, ascetic, self-righteous, sometimes insubordinate, but invariably patient General Bernard Montgomery. A cousin of H. St. John B. Philby, the sorcerer's apprentice of Saudi oil, Montgomery had been the best man when Philby was married in India. Early on in life, Montgomery had learned to depend on his own resources, and not much else. After the death of his wife, apparently the bizarre result of an insect bite, he seemed to have been left with virtually no emotional attachments, and precious little of any other kind. "Everything I possessed had been destroyed by enemy bombing in Portsmouth in January 1941," he later wrote about his abrupt summons to take command of the British Eighth Army in Egypt. "I was now going to be given the opportunity to get my own back on the Germans." Some thought him strange, even paranoid. Indeed, in his first address to a group of officers of the Eighth Army, at Ruweisat Ridge near El Alamein, he felt the need to say: "I assure you I am quite sane. I understand there are people who often think I am slightly mad; so often that I now regard it as rather a compliment."

If a little strange, Montgomery was also a methodical, pedagogic, and deeply analytical military strategist. He would sometimes spend hours a day by himself—in his own "mental oases," as he called them—thinking through problems, searching for the key principles, laying out his plans. He hung a drawing of Rommel in his desert trailer, to help him think how Rommel would think. Montgomery knew that, in facing Rommel, he was taking on a modern legend, and one who had cast a spell of fear and awe over the Eighth Army. He had one clear aim, to do what many thought was beyond possibility: turn the tables on the master of mobile warfare and decisively defeat Rommel. For, said Montgomery, Rommel "had never been beaten before though he had often had to 'nip back to get more petrol.' " For the actual execution of his battles, Montgomery would afterward be criticized for being too cautious. But, as one German general would later say, "he is the only Field-Marshal in this war who won all his battles."

As Montgomery contemplated his coming struggle with Rommel, he sought to work out a strategy that would use the Eighth Army, now equipped with Sherman tanks, to full advantage as a unified whole, and that would capitalize on the fact that its own supply line was short and Rommel's very long—and highly vulnerable. Still, by the end of August 1942, Rommel's supply position had, once again, somewhat improved. Would he go on the offensive?

Rommel himself was of two minds about what to do next. He was acutely conscious of the fuel shortages and the constraints that would result; moreover, he was ill with a severe intestinal ailment and sheer exhaustion and had just requested medical leave. But he also wanted to get on with the drive to Cairo—and beyond. He was convinced that time was running short, and that the spirit of the Afrika Korps would win the day, whether the supplies were adequate or not. He gave the order to attack. This round, still in the environs of El Alamein, became known as the Battle of Alam Halfa.

Again and again, during that week-long battle, Rommel recorded how the Afrika Korps was hampered by its fuel shortage. On August 31: "Due to the heavy going, the Afrika Korps' petrol stocks were soon badly depleted and at 16:00 hours we called off the attack on Hill 132." On September 1, "the promised petrol had still not arrived in Africa." It never came. Most of the fuel due for delivery by ship had either been sunk or was still waiting to be loaded in Italy. A small railway that might have been used to transport gasoline had been flooded. Rommel's forces were not able to get past the tactically well-placed British artillery. By September 7, 1942, the battle of Alam Halfa was over. Rommel's latest offensive had been stopped dead in its tracks, and a legend of invincibility was toppling.[16]

In the weeks that followed, Rommel begged Hitler's headquarters for more supplies at all costs, including fuel sufficient for two thousand miles per vehicle. On September 23, Rommel left North Africa to see, first, Mussolini in Rome, then Hitler at his headquarters on the Russian front. Again he pleaded for more supplies; instead, he received the baton of Field Marshal, personally bestowed upon him by his Führer. Hitler made generous promises, which he would not keep.

On October 23, after weeks of careful preparation and resupply, Montgomery opened his counterattack, known as the Second Battle of El Alamein, with a powerful artillery barrage. The Germans were stunned. On the first day, General Georg Stumme, Rommel's replacement, fell out of his car when it came under British aerial bombardment and died of a heart attack. Hitler telephoned Rommel, who was in the Austrian Alps on sick leave, and ordered him to return immediately to North Africa. By the evening of October 25, he was back in Egypt—this time commanding the beginning of a long retreat.

German hopes for new supplies were pinned on aircraft and ships that were being methodically destroyed by the Royal Navy and the RAF. When Rommel learned that four ships, carrying desperately needed supplies of petroleum, had been sunk by the RAF in the supposedly safe harbor at Tobruk, he lay awake all night, his eyes wide open, unable to sleep. He knew what the sinkings meant. "In attacking our petrol transport," he wrote, "the British were able to hit us in a part of our machine on whose proper functioning the whole of the rest depended."

During the ensuing weeks, all Rommel could do was fall back. At times, he believed that he could have turned around and delivered devastating blows to his pursuers, but he simply did not have the fuel to dare it. To Hitler he repeatedly described the fuel situation as "catastrophic." An even greater catastrophe loomed when Allied forces invaded Morocco and Algeria—in the path of his retreat. The days of his Afrika Korps were numbered. On Christmas Eve, 1942, Rommel joined his headquarters company's Christmas party. That day, he had shot a gazelle from his car, which he contributed to the dinner table. In turn, he received a present from the assembled men—a couple of pounds of captured coffee, presented in a miniature oil barrel. "Thus proper homage," he said, "was paid to our most serious problem even on that day." Soon all that Rommel's forces had left was a toehold between the forces advancing from both east and west. The legend had fallen, and in March 1943 Rommel, now regarded by Hitler as defeatist, was removed from command of the Afrika Korps. By May, the last German and Italian troops in North Africa had surrendered.[17]

But Rommel was called to serve the Führer again, first in Italy, and then in France, where he was badly wounded shortly after the Normandy invasion, when his car was hit by Allied bombs. Three days later, a group of army officers tried to assassinate Hitler, but failed. Rommel was suspected both of involvement in the conspiracy and of plotting a separate surrender in the West to the Allies. Hitler ordered his death, but it could not be done publicly, for Rommel was a very popular general, and the adverse impact on German morale might have been enormous. So, instead, in October 1944, two SS generals appeared at his home with an ultimatum. Either he commit a suicide that could be disguised as a natural death, or his entire family would be in danger. Rommel, holding the Field Marshal's baton that Hitler had given him two years earlier, drove off with the two SS men. A few hundred yards from the house, the car pulled over into an open space in the woods; the area had been sealed off by the Gestapo. Rommel was given a poison pill; he swallowed it and fell forward in the seat, dropping the Field Marshal's baton. Rommel was dead. The death was attributed to a brain hemorrhage; a state funeral was organized, and Hitler sent his condolences. Rommel's "heart," said the official funeral oration, "belonged to the Führer."

In Rommel's papers, collected after his death, he left a hard-earned epitaph for the role of supply, and in particular of oil, in the age of mobile warfare. "The battle is fought and decided by the Quartermasters before the shooting begins," he wrote, looking back on El Alamein. He himself had derisively rejected such a thought a few years earlier. But he had learned a bitter lesson across the sands of North Africa. "The bravest men can do nothing without guns, the guns nothing without plenty of ammunition, and neither guns nor ammunition are of much use in mobile warfare unless there are vehicles with sufficient petrol to haul them around." But he had also stated that dictum in more personal terms. Two weeks after the Second Battle of El Alamein, as his army was retreating before Montgomery's forces, Rommel wrote to his wife. "Shortage of petrol! It's enough to make one weep."[18]

Autarchy and Catastrophe

By mid-1943, the Axis had been defeated both in Russia and in North Africa, and the dream of German armies converging at Baku or on the oil fields of the Middle East had been relegated to the realm of fantasy. Thus, Germany had to turn back to its own resources. There was no other choice. Synthetic fuels would be at the heart of the frantic effort to sustain the machines of war. And in that effort, Hitler's Reich would display its technological ingenuity—as well as its total moral bankruptcy.

Belatedly, the Nazi regime began to reorganize the German economy to increase the output of synthetic fuels and other essential materials in preparation for a long struggle. The man in charge was Albert Speer, Hitler's personal architect. The intensely ambitious Speer had much earlier established himself as one of Hitler's favorites. He had caught Hitler's attention a decade before with his various plans for a grand panorama of flags, hundred-foot-high eagles and spot-lights for the 1933 Nazi party rally in Nuremberg. Himself a frustrated artist, Hitler was captivated by Speer's conceptions and by his personality and put him in charge of all the monuments of the Reich. Hitler also gave him a personal commission to construct the new Reich Chancellery and to rebuild Berlin. In 1942, the Führer appointed Speer Minister of Armaments and War Production. In early 1943, as the magnitude of the failures in Russia and North Africa became clear, Speer's brief as Minister of Armaments was much expanded; he was given sweeping powers over the entire German economy. He now controlled, or at least influenced, virtually all phases of economic life.

The architect, formerly in charge of the stone monuments to the eternal glory of the Thousand-Year Reich, proved himself remarkably adept in dealing with the Reich's more immediate and urgent problems of industrial mobilization. Speer drove the slack out of the German economy. The two and a half years after his initial appointment would see a more than threefold increase in the production of aircraft, weapons, and ammunition, and a nearly sixfold increase in tanks. And these remarkable production records were being set at the same time that Allied forces were carrying out an extensive if not particularly successful strategic bombing campaign against a variety of German targets, such as the aviation industry and railway depots and ball-bearings factories. German industrial production was still rising; indeed it registered its highest level of the entire war in June 1944. The great potential claimed for strategic bombing was far from being realized. "Oil, which was Germany's weakest point," wrote the British military historian Basil Liddell Hart, "was scarcely touched." Yet both the German military chiefs and Speer worried. Would the Allies make the destruction of the synthetic fuels industry a major objective? For it offered a critical, concentrated, sensitive target in the way that other industrial activities did not, and a campaign against it could well jeopardize the entire German war economy.

The synthetic fuels industry was headed on the same upward trend as the rest of the war economy. By 1942, the industry had, on all fronts, recorded a considerable advance over the 1930s—new production technologies, better catalysts, higher grade output, and a capability to accept a much wider variety of coals as raw material. And production was increasing quickly. Between 1940 and 1943, synthetic fuel production almost doubled, from 72,000 to 124,000 barrels per day. The synthetic fuels plants were the critical links in the fuel system; in the first quarter of 1944, they provided 57 percent of total supply—and 92 percent of aviation gasoline. And the throttle was open; in the first quarter of 1944, production was running, if annualized, at an even higher rate. Altogether, during the Second World War, synthetic fuels would account for half of Germany's total oil production.[19]

It could not have happened without immense effort and all the normal tools and techniques of the Nazi war economy, including slave labor. Hitler had transformed the Viennese streetcorner anti-Semitism of his youth into a monstrous and diabolical ideology, at the center of which was the murder and destruction of the Jews. The concentration camps were the mechanisms for achieving this "Final Solution," which was decided upon in a mere two hours at the Wannsee Conference in January 1942. But, until the "Final Solution" could be completed, those Jews who were fit—along with Slavs and other prisoners—were to be put to work to further the objectives of the Reich that had already pronounced a death sentence upon them. And so a continuing supply of concentration camp prisoners was drafted into I. G. Farben's hydrogenation plants, as well as into its synthetic rubber plants. The company, in fact, was building synthetic fuel and rubber plants adjacent to the Auschwitz concentration camp in Poland. Auschwitz was the largest of the Nazi mass murder factories; upwards of two million people, mostly Jews, were put to death there with gas manufactured by an I. G. Farben subsidiary. The I. G. Farben officials described the Auschwitz site, with its ample supplies of coal and labor, as "very favorably located." The synthetic fuels plant at Auschwitz was under the directorship of the same chemist who had represented the company in the June 1932 meeting with Hitler in Munich.

I. G. Farben used both so-called "free" and slave labor in this enterprise. The chemical company paid a per diem for each slave laborer—three or four marks for an adult, depending upon skills, and half price for children. The money went, of course, not to the workers but into the coffers of the SS, Hitler's elite military force. The slave laborers subsisted, at most, on a thousand calories a day and slept on wooden racks. They would work for a few months, then die from the horrible living conditions or the beatings or be killed in the death camp, and then would be replaced by others taken from newly arrived trains packed like cattle cars with more prisoners.

I. G. adjusted to the requisites of its partnership with the SS. At one point, it asked that the guards cease severely flogging prisoners in front of the "free" Poles and Germans on the site. "The exceedingly unpleasant scenes" were having a "demoralizing effect. . . . We have therefore asked that they should refrain from carrying out this flogging on the construction site and transfer it to the inside of the concentration camp." Several months later, however, the I. G. Farben management came to agree with the methods of the SS: "Our experience so far has shown that only brute force has any effect on these people."

Eventually, I. G. Farben became disenchanted with the quality of slave labor from the main camp at Auschwitz; the daily four-mile hike in each direction tended to debilitate the prisoners, and they became too prone to the diseases in the main camp. To prevent that, the company built its own private enterprise "branch" concentration camp, Monowitz, along the same model as the main camp. The records that survived indicate that three hundred thousand prisoners passed through I. G. Farben's portals at Auschwitz. The plants were so large that they used more electricity than the entire city of Berlin.

One of those was Prisoner Number 174,517, a young Italian named Primo Levi, who managed to survive only because he could recall enough of the organic chemistry he had studied in Turin to be put to work in a lab. "This huge entanglement of iron, concrete, mud and smoke is the negation of beauty," he said of I. G.'s industrial complex. "Within its bounds not a blade of grass grows, and the soil is impregnated with the poisonous saps of coal and petroleum, and the only thing alive are machines and slaves—and the former are more alive than the latter." Monowitz was a death factory. It was also a business, down to the camp staff who made money selling in the nearby market the clothes and shoes of those who died at Monowitz and those who had been stripped naked to be sent to crematoriums of the neighboring camps. The stench of the crematoriums at Auschwitz and Birkenau suffused the air at Monowitz. To Levi, it was "world of death and phantoms. The last trace of civilization had vanished."

By 1944, according to one estimate, a third of the total work force in the German synthetic fuels industry, throughout the Reich, was slave labor. I. G. Farben had become a deeply involved and enthusiastic partner in its joint venture with the SS at Auschwitz. And, naturally enough, the two parties engaged in a good deal of mutual socializing. Just before one Christmas, the resident I. G. Farben managers at Auschwitz joined men from the local SS in a holiday shooting party. The bag totaled 203 rabbits, one fox, and one wildcat. The head of construction at the Farben complex was "proclaimed champion hunter," with a bag of one fox and ten rabbits. "A good time was had by all," according to the record of the hunt. "The result was the best in this district so far this year and will probably only be surpassed by the hunt the concentration camp is holding in the near future."[20]

"The Primary Strategic Aim"

Following the haphazard and ineffectual Allied strategic bombing campaign against Germany, General Carl Spaatz, the commander of the United States Strategic Air Force in Europe, decided that a change had to be made. On March 5, 1944, he proposed to General Dwight Eisenhower, who was in charge of the preparations for the Normandy invasion, that a new priority target be set—the German synthetic fuels industry. Its output, he promised, could be cut by half within six months. He noted an expected added benefit: So critical to the Germans were these plants that such attacks would flush out the Luftwaffe, and also force the diversion of many planes and pilots from France, where the invasion was targeted.

The British opposed Spaatz's plan, insisting instead on the need to target the French railway system. But, finally, Spaatz received a tacit go-ahead from Eisenhower for the synthetic fuel targets. On May 12, 1944, a combat force involving 935 bombers, plus the fighter escorts, bombed a number of synthetic fuels factories, including the giant I. G. Farben plant at Leuna. As soon as Albert Speer realized what had happened, he rushed by plane to Leuna to see the damage for himself. "I shall never forget the date May 12," he later wrote. "On that day the technological war was decided." The results of the attack, and the broken, twisted pipe systems that he now saw as he toured the plant site, made real "what had been a nightmare to us for more than two years." A week after the attack, Speer flew off to report personally to his Führer. "The enemy has struck us at one of our weakest points," he told Hitler. "If they persist at it this time, we will soon no longer have any fuel production worth mentioning. Our one hope is that the other side has an Air Force General Staff as scatterbrained as ours!"[21]

Yet this initial attack was not so troublesome as it first appeared. Just before Allied invaders forced Italy out of the war, the German military had seized its oil stocks, adding substantially to its own reserves. That provided some cushion. And feverish activity at the damaged fuel plants returned synthetic fuel production to its former levels within a couple of weeks. But then, on May 28–29, the Allies hit the oil facilities in Germany again. Other Allied bombers attacked the oil installations at Ploesti in Rumania. On June 6, D-Day, the Allies staged the long-awaited invasion of Western Europe, gaining a precarious foothold on the beaches of Normandy. It was now more important than ever to knock out the Germans' fuel supply, and on June 8, General Spaatz gave the formal directive— the "primary strategic aim of the United States Strategic Air Forces is now to deny oil to enemy armed forces." Regular bombing attacks on the synthetic fuels industry followed.

In response, Speer ordered that synthetic fuels plants and other oil facilities be rebuilt quickly, or dispersed where possible into smaller, better-protected, and hidden sites—some in the rubble of destroyed factories, some in quarries, some underground. Even breweries were converted to making fuel. Substantial increases in synthetic fuel capacity had been planned for 1944, but now the machinery and components scheduled for the increase had to be cannibalized to repair existing facilities. Upward of 350,000 workers—many of them slave laborers—were engaged in this frantic undertaking. At first, the plants were rebuilt quickly, but as time went on and they were subjected to further air attacks, they became more fragile and vulnerable, and more difficult to get back into operation yet again. Production began to plummet quickly. Before the attacks first began in May 1944, synthetic fuel production by hydrogenation was averaging 92,000 barrels per day; by September, output had fallen to 5,000 barrels per day. Aviation gasoline production that month was just 3,000 barrels per day—only 6 percent of the average production in the first four months of 1944. Meanwhile, the Russians had captured the Ploesti oil fields in Rumania, depriving Hitler of his major source of crude oil.

The output of German planes was still at peak level. But they sat on the ground; they were of little use without fuel. Jet fighters, a new German innovation that would have given the Luftwaffe a significant advantage, were being introduced into operational squadrons in the autumn of 1944. But there was no fuel to train the pilots, or indeed even to get the planes into the air. Altogether, the Luftwaffe was operating on just one-tenth of the minimum required gasoline. The German Air Force was now caught in a fatal trap. Without fighter planes to protect the fuel plants, the destructive impact of the Allied raids grew, further reducing the supplies of aviation gasoline available for the Luftwaffe. In-air training for new pilots was cut to only an hour a week. "This was actually the fatal blow for the Luftwaffe!" General Adolph Galland, commander of German fighter forces, was to say after the war. "From September on, the shortage of fuel was unbearable. Air operations were thereby made virtually impossible."

In the autumn of 1944, bad weather temporarily curtailed the attacks, and in November, the Germans managed to increase the output of synthetic fuels. But production fell again in December. "We must realize that the men on the enemy side who are directing the economic air raids know something about German economic life," Speer told an armaments conference. "Fortunately for us the enemy began following this strategy only in the last half or three-quarters of a year. . . . Before that he was, at least from his standpoint, committing absurdities." At last, the strategic bombing campaign, with its attack on the synthetic fuels industry, was paralyzing major parts of the German war machine. But the battle was not yet over.[22]

The Battle of the Bulge: Europe's Biggest Gas Station

By the fall of 1944, the D-Day invasion of Normandy had widened, in slow and costly stages, to drive the Germans out of France. At the same time, Soviet forces were pushing in on Germany from the east. But to Hitler, the war could not be coming to an end. His Reich could not fail. On December 16, he launched a huge counteroffensive in the hilly, wooded Ardennes forest, in the east of Belgium and Luxembourg. Later known as the Battle of the Bulge, it was Germany's last great concentrated attack, the final *blitzkrieg*. The plan was Hitler's own, and everything was thrown into it, including every bit of fuel that could be scrounged from other units inside Germany. The objective was to throw back the Allies, isolate their armies, recapture the initiative, and buy time for the development of new, more devastating weapons to use against the Allies' soldiers and civilians. The Germans caught the unprepared Allies completely by surprise, managed to work great confusion behind their lines, and succeeded in breaking through.

The Germans had the advantage of surprise, but they had undertaken the assault with vastly insufficient resources and with a strength that on paper appeared much greater than in fact it was. The reserves of manpower that might have been decisive in the battle could not reach the front lines. "They could not be moved," a German commander later said. "They were at a standstill for lack of petrol—stranded over a stretch of a hundred miles—just when they were needed."

In their 1940 *blitzkrieg* in this same area, the Germans' fuel deficiency had mattered little; they had captured more gasoline than they had expended. Now, four and a half years later, they had no such luck. But they did come perilously close. For the area around Stavelot in eastern Belgium was the Allies' largest fuel dump and, indeed, the biggest filling station in Europe. There the Allies had stored 2.5 million gallons of fuel for their troops—along with two million road maps of Europe. The roads throughout the area were lined with hundreds of thousands of five-gallon jerry cans. The Allied forces would stop, fill up with whatever fuel they needed, and drive on.

On the morning of December 17, the second day of the German offensive, a panzer unit, led by a butcherous colonel named Jochem Peiper, overran a small fuel dump nearby. Peiper forced fifty captured American soldiers to fill his vehicles and then coldly ordered them killed. A large number of American prisoners were also gunned down, in what became known as the Malmédy Massacre. By that evening, Peiper's forces were within a thousand feet or so of the much grander prize—the forward edge of the Stavelot dump, fifty times larger than the one they had captured earlier that day. The Allied defenses were light and disorganized. Peiper's forces moved north across the bridge over the Ansleve River and into Stavelot. In a desperate effort to improvise, a small group of Allied defenders poured the contents of some of the jerry cans into a round ditch and set it ablaze, creating a wall of fire. Peiper examined his maps carefully, but they were out of date and failed to show the correct location or magnitude of the dump. He did not know about the bounty that lay at hand. Instead of sending his forces through the thin wall of fire, he ordered them to go back across the bridge and head west, leaving the fuel dump secure. Ironically, Peiper's unit soon ran out of fuel; his tanks got only half a mile to the gallon. Resupply efforts by the Luftwaffe failed, and the unit was captured.

Peiper's U-turn was one of those small battle incidents with momentous consequences. The Stavelot fuel supply was equivalent to the requirements of the first ten days of Germany's entire Ardennes offensive; its capture would have given the Germans the fuel to push on to Antwerp and the English Channel at a time when the Allies were still reeling in disorganization and confusion. Even so, it was not until Christmas Day 1944, ten days after the Germans began their offensive, that they were finally stopped and pushed back.[23]

"Twilight of the Gods"

More fuel would have bought the Germans more time. With the failure of the Ardennes offensive, Germany's war effort, from a strategic point of view, was over. In February 1945 German production of aviation gasoline amounted to just a thousand tons—one-half of one percent of the level of the first four months of 1944. None was produced thereafter. But the delusions of victory lived on. Those around Hitler, Speer would recall, "would listen to him in silence when, in the long since hopeless situation, he continued to commit nonexistent divisions or to order units supplied by planes that could no longer fly for lack of fuel."

Still, the months of bloody fighting went on, on both the western and eastern fronts, as Hitler and his immediate entourage retreated ever more into fantasy, with the Führer himself calling for a scorched earth policy and issuing (in the words of one of his generals) "the last crazy orders." Even as the end approached, he remained in the grip of his insane, violent visions, for which at least 35 million people paid with their lives. He listened to Wagner's *Götterdämmerung*—"Twilight of the Gods"—on the gramophone, waiting for some magical deliverance, and avidly read horoscopes that promised a sudden improvement in his fortunes. Only when Russian soldiers were almost directly above his underground bunker, on the doorstep of the now-ruined Chancellery that Speer had designed for him, did Hitler commit suicide. He left orders that his body be doused in gasoline and burned so that it would not fall into the hands of the hated Slavs. There was enough gasoline at hand to carry out that final order.

But to many around Hitler, the impending disaster resulting from the Nazi fantasies and savagery had been apparent for months. On a night journey to the struggling remnants of Germany's Tenth Army in Italy, Albert Speer had seen before him a clear vision of one of the primary reasons why the Reich that was to last for a thousand years had, in fact, only weeks to go. For on that trip, he had encountered 150 German Army trucks. To each were hitched four oxen, which were dragging the trucks forward. It was the only way the vehicles could move. They had no fuel.[24]

CHAPTER 18

Japan's Achilles' Heel

IN THE FIRST WEEK of December 1941, an American naval squadron was on a courtesy call in the huge, splendid harbor at Balikpapan, in Borneo, in the British East Indies. At the turn of the century, at that then unknown spot on the map, Marcus Samuel had ordered his nephew to carve a refinery complex out of the jungle. Over the four decades that followed, what had seemed Samuel's foolish and reckless dream had not only turned into a great oil-refining center for the island's production, but had also become one of the grand jewels of the Royal Dutch/Shell Group and a major landmark in the world oil industry.

Now, in December 1941, the refinery management had just hosted a party for the visiting American sailors, and the Americans were planning to offer a return party onshore, at the local club. The junior officers, along with crates of liquor, were already assembling at the club when a senior officer suddenly appeared and ordered them back to the ship immediately. On board, fueling began at once, and by midnight, the American ships were steaming out of the harbor. That was how the English and Dutch oil men at Balikpapan learned about the attack on Pearl Harbor. The war for which they had been waiting and preparing had, at last, begun.

A year earlier, in 1940, when a Shell manager named H. C. Jansen arrived at Balikpapan, he had found air raid shelters already built and evacuation plans developed. In the subsequent months, the entrance to the harbor was mined, and 120 men practiced destruction exercises. They all knew that Balikpapan and the surrounding oil fields were one of the great prizes for which the Japanese would go to war. The oil men's job would be to deny the Japanese that prize.

In the days right after Pearl Harbor, the wives and children of the oil men were evacuated from Balikpapan. In the nights that followed, Jansen and his colleagues, now all bachelors, would sit in rattan chairs in his garden, looking down in the darkness at the refinery and the ocean beyond it—the moon did not rise from the sea until late—talking about the depressing radio reports of the Japanese advances into Southeast Asia. What would the Americans do? When would the Japanese come to Balikpapan? What would be the future of this great industrial enterprise? They wondered, as well, about what fate held in store for each of them. They talked, more immediately, about how to strengthen the defenses of Balikpapan. During the days, however, they had precious little time to reflect on anything; they worked themselves to exhaustion, seeking to turn out as much refined product as possible, which would be used, they fervently hoped, for the Allied war effort.

In mid-January 1942, with the Japanese closing in, crews in the outlying oil fields began to destroy the wells, as was being done elsewhere in the Indies. They pulled out the tubing, cut it up, and jammed it down again into the wells, along with pumps, rods, any bolts, nuts, and drilling bits they could lay their hands on, and one other thing—a tin of TNT for each well. The wells were blown up. The crews started with the least productive wells, but finally all were destroyed.

Meanwhile, the first steps toward demolishing the Balikpapan refinery complex were underway. The stills and steam boilers were turned on and allowed to boil dry until they collapsed. No one knew how long it would take. But, after thirty hours, the first still began to fall apart, and the others soon followed. Then on January 20, the men at the refinery received the word they had feared: A Japanese fleet was at sea only twenty-four hours' steaming time away. The Japanese sent, via two captured Dutchmen, an ultimatum: Surrender at once, or all would be bayoneted. A military officer assigned to the refinery gave the order to begin the demolition.

Jansen and the other men blew up the mines store first; the blast shattered all the window panes in the area. Next came the wharves, which had been thoroughly doused with either gasoline or a mixture of kerosene and lubricating oil. By noon, the docks were aflame. The oil men noticed, with some technical curiosity, that when the smoke from the wharves that had been set on fire by gasoline converged with the smoke from those set aflame by kerosene and lubricating oil, the interaction ignited bursts of lightning in an otherwise clear midday sky.

From then on, the huge complex was rocked by one explosion after another. The flames leaped 150 feet in the sky, as the salt water station, the tin plant, the refinery installations, the power plant, and more and more buildings joined the conflagration. The men, covered with sweat and soot, ran about amidst the flames, following plans they had rehearsed many times. They moved on to the tank farm, the area in which oil was stored. To each tank, fifteen sticks of TNT had been affixed. But some of it had deteriorated in the humid weather and would not ignite. Moreover, exhaustion was starting to overcome the men. Would they be able to find some way to set the tanks on fire? A fresh batch of volunteers tried to ignite them with rifles, but to no avail. The alternative was to open the valves. But the keys to the valves, they realized, were back in the tank farm office, which had already been destroyed.

Finally, the tanks on the higher ground were opened, and the oil poured down to the tanks below. Electrical ignition would be used to detonate four or five tanks, with the hope that flaming oil would set off fire in the rest. Jansen took cover with the others behind an empty tank as the ignition was sent. Instantly, a great ball of fire rose up, followed by a terrific explosion, and then a huge hurricane of wind. As the flaming sea of oil poured down the hill toward the other tanks, the tank farm became a hellish conflagration.

There was nothing more to do. Jansen and the others ran down the hill to the wireless station; native guards, in full uniform, saluted them. The oil men, their throats parched, exhausted to the bone, climbed into native boats called *proas*. The sea around them was red from the reflection of the great columns of fire, and still there was one explosion after another. Now came the next phase of the plan, the one that had never been practiced—escape.

The men left the bay and entered the mouth of the Riko River, headed upriver toward an evacuation camp. Eventually, the fiery carnage at the refinery site disappeared, lost in the dark jungle foliage and the heavy night, and the explosions faded away into the endless chorus of cicadas. For hours, they sailed on. Occasionally, they could still spot the red glow from Balikpapan high in the sky. They had done their work well; four decades of industrial creation had been destroyed in less than a day. Finally, they came to the evacuation camp deep in the jungle, on a small tributary of the Riko. They spent interminable hours listening intently for the sound of the airplane that was to be sent to their rescue. It did not come.

The next night, Jansen and a small group sailed back down the tributary to its junction with the Riko. They passed that night in the boat, hoping that rescue would arrive, straining their ears to hear a plane or a boat but worrying that whatever they heard might be Japanese. One man, asleep on the hard seat, fell overboard; the others pulled him back in, making a great din to scare the crocodiles away. The only way to keep the mosquitoes at bay was by smoking pipes and cigarettes. To Jansen, the hours seemed unending. Dawn came, and still they waited.

At about one in the afternoon, a company seaplane appeared out of the sky and set down on the river. The pilot was going on to pick up a wounded man at another location and promised to return. He was as good as his word. He took four people. Jansen was not among them. Later, Jansen and the others received a message to go back to the bay at Balikpapan, and they set off downstream again. That night, two seaplanes appeared and evacuated many of them. Jansen was on the second plane, packed in so tightly that he could hardly breathe. But once they were in the air, a breeze filtered in through the cabin, and some even sank to the floor of the plane, asleep.

When the evacuees arrived at Surabaya, on the north coast of the island of Java, they were greeted by the commander of the local airbase. "It is no longer possible to send planes to Balikpapan; the Japs are there," he said. "I have forbidden the Grumman to go back." Seventy-five people were left stranded by the bay at Balikpapan, still awaiting rescue. It was too late; the Japanese had landed on the south side of the bay. A few hours after midnight on January 24, four blacked-out American destroyers slipped up on a dozen Japanese troop transports, starkly silhouetted against the red fires of the still-burning refinery complex. In what became known as the Battle of Balikpapan, the Americans sank four of the transports, plus a patrol boat. But, owing to faulty torpedoes, they did not sink more. It was the first American sea battle with the Japanese, indeed, the first time the United States Navy had been involved in surface action since Admiral Dewey's victory at Manila in 1898.

It hardly slowed the Japanese landing at Balikpapan at all. The stranded oil men had no choice but to retreat into the jungle. They broke up into small groups for what proved to be a desperate effort to find some route of evacuation through the jungle. It was a terrible ordeal. They made their way by foot or in *proas*, beset by hunger, exhaustion, malaria, dysentery, and fear, their parties growing smaller and smaller as the ill and dead dropped away. From natives they encountered, they learned that the Japanese had landed all over Borneo. Trapped in the jungle, they felt like rats in a cage. A few did finally escape from the island. Of the seventy-five people left behind, only thirty-five survived the jungle, the Japanese firing squads, and the Japanese prisons.[1]

"Victory Drunk"

Preemptive destruction of oil facilities, similar to that at Balikpapan, was carried out elsewhere in the East Indies. But this seemed a minor inconvenience to the Japanese wave sweeping across Southeast Asia and the Pacific. By mid-March 1942, Japan's control of the East Indies was complete. Coming in the wake of other conquests, this meant that Japan had, in just three months, won possession of all the rich resources of Southeast Asia—and the oil, in particular, for which it had gone to war. And still the Japanese war machine rolled on. In Tokyo, Premier Tojo bragged that Hong Kong had fallen in eighteen days, Manila in twenty-six, and Singapore in seventy. A "victory fever" gripped the country; the stunning military successes spawned such a runaway stock market boom in the first part of 1942 that the government had to intervene to dampen it down. Some said the country had become "victory drunk." Only a few warned of the inevitable morning after.

The Japanese elation was more than matched by American shock and despair. On Christmas Day 1941, Admiral Chester Nimitz, newly appointed head of the U.S. Pacific Fleet, arrived by seaplane at Pearl Harbor to begin the job of picking up the pieces. As he was ferried across the harbor to the dock, he passed small boats that were searching for bodies; two and a half weeks after the attack, corpses were still floating to the surface. That grim scene in Hawaii was only part of the larger, dismal outlook facing the United States—war in both hemispheres, a truly global conflict. Pearl Harbor was, almost certainly, the worst humiliation in American history. Fear and panic gripped the nation. Yet the war, so long feared and so long half-expected, was finally here, and the country quickly rallied for the long, arduous struggle with Germany and Japan.

Who would be in charge of America's Pacific War, the Army or the Navy? Each service was loath to commit its entire force in the Pacific to an officer from the other. Personal rivalries and animosities compounded the bureaucratic competition. As a result, two commands and two theaters were set up. The contrast between the Army's and Navy's top commanders was enormous. General

Douglas MacArthur, although a strategist of great shrewdness, was also egoistic, bombastic, and imperious. At one meeting during the war, after three hours of listening to MacArthur, Franklin Roosevelt told an aide, "Give me an aspirin. . . . In fact, give me another aspirin to take in the morning. In all my life nobody has ever talked to me the way MacArthur did." For his part, Admiral Chester Nimitz was a soft-spoken, unassuming team player, who, when waiting word on the outcome of a battle, would practice on his pistol range or toss horseshoes right outside his office. "It simply was not in him to make sweeping statements or to give colorful interviews," noted one correspondent.

The divided command did much more, however, than provide a contrast in styles of military leadership; it also led to bitter and wasteful battles over scarce resources and, worse, poor coordination in key military operations in the far-flung battle zones. The distances American forces would have to cover in their ultimate convergence on the Japanese Home Islands were simply enormous. No war had ever been fought on such a scale. America had a great advantage in resources. But how were American forces to be supplied? And how were the Japanese to be denied the abundant resources they had already captured? The answers to these two questions would help shape strategy and would do much to determine the course of the far-flung war in the Pacific. From the beginning, Nimitz had no question in his own mind as to what his strategy would be. He and Admiral Ernest King, Chief of Naval Operations, agreed, in the words of Nimitz's biographer, "that the primary objectives of the Allied armed forces were to safeguard their own supply lines and then drive westward in order to capture bases from which Japan's indispensable 'oil line' might be blocked."[2]

"The Adults' Hour"

While, in the early months of 1942, the Americans belatedly mobilized for the conflict, the Japanese, glorying in their astonishing string of victories, pondered their next moves. So self-confident had they become that the country's military leaders considered thrusting westward through the Indian Ocean to link with German forces in the Middle East or Russia and help sever the Allies' supply of oil from Baku and Iran. Not all the Japanese, to be sure, fell victim to "victory fever." In April 1942, Admiral Isoroku Yamamoto, the architect of the Pearl Harbor attack, wrote to his favorite geisha: "The 'first stage of operations' has been a kind of children's hour, and will soon be over; now comes the adults' hour, so perhaps I'd better stop dozing and bestir myself."

Yamamoto continued to share with other Japanese naval leaders the deep belief in and commitment to the "decisive battle," which would knock the enemy out of the war. A quick victory was essential, he knew from his years in the United States, because of America's oil and other resources and its industrial might. Thus, the Japanese decided to mount a major attack on Midway Island, just eleven hundred miles west of Hawaii. At the very least, the Japanese planned to use an assault on Midway to extend their defense perimeter. And if the American fleet were drawn out, so much the better, for the Japanese could make this the decisive battle and finish the job they had started at Pearl Harbor, obliterating the U.S. Navy in the Pacific.

The Battle of Midway, in early June 1942, turned out to be decisive, but not in the way most Japanese expected. Instead, it was the "adults' hour" that Yamamoto had feared. Having made a remarkable recovery from the devastation at Pearl Harbor, and with the additional advantage of being able to read the enemy's code (which the Japanese had been slow to change because their forces were so dispersed), the U.S. Navy dealt a resounding defeat to the complacent Japanese, sending four of the Imperial Fleet's aircraft carriers to the bottom, while losing just one of its own.

Midway was the true turning point in the Pacific War, the end of the Japanese offensive. Thereafter, the balance would shift, as the relentless weight of American manpower, resources, technology, organizational ability, and sheer determination began to push the Japanese back across the Pacific in one bloody battle after another. The counterattack began two months after Midway, when American troops landed on the island of Guadalcanal, off New Guinea. Six months of brutal fighting followed, but finally the United States captured the island in what was called the first American offensive of the war. The aura of invincibility that surrounded the Japanese Army had been pierced. But it was only one small and costly step in the long struggle to exhaust the resources, if not the will, of an implacable enemy.[3]

The initial attempts to deny Japan the oil of the Indies did not prove much of an impediment. The Japanese had anticipated the demolition—the likelihood had long been signaled—but they found, despite Shell's efforts at Balikpapan and Stanvac's on Sumatra, that the destruction was neither as severe nor as widespread as they had expected. The Japanese immediately set about rehabilitating the oil industry of the Indies. Drilling teams, refinery crews, and equipment were rushed in. In short order, some four thousand oil field workers, 70 percent of the total in the Home Islands, were shipped south.

The result was astonishing. Before the outbreak of war, the Japanese military had planned on getting sufficient oil within two years from the Indies—what was called the Southern Zone—to make up for shortfalls. That goal was exceeded. Oil production in the Southern Zone had been 65.1 million barrels in 1940. In 1942, the Japanese were able to produce just 25.9 million barrels, but by 1943, they had gotten it back to 49.6 million barrels—75 percent of the 1940 level. In the first three months of 1943, Japanese oil imports had risen to 80 percent of the amount imported in the same period of 1941, just before the imposition of the oil embargo in July 1941 by the Americans, British, and Dutch. As they had planned, the Japanese were able to use the captured East Indies to replenish their stocks of petroleum. Moreover, there was no lack of oil in the Southern Zone. The Japanese fleet could refuel locally at will.

The Japanese also took advantage of efforts by Caltex, the partnership in the Eastern Hemisphere of Standard of California and Texaco. Just before the war, Caltex had identified a very promising field, the Minas structure, in central Sumatra, and had moved in a drilling rig and the required equipment. The Japanese took over the work and, using Caltex's rig, drilled a discovery well—their only wildcat in all of World War II. They struck a super giant field, the largest between California and the Middle East. So successful was the overall effort in the

Southern Zone that in 1943, Premier Tojo announced that the oil problem—which had triggered Japan's aggression—was solved. But Tojo had spoken too soon.[4]

The Battle of the Marus: The War of Attrition

In devising their military strategy, the Japanese had assumed that the rich resources—the oil, the other raw materials, the food supplies—of the Southern Zone could be welded securely to the economy and needs of the barren Home Islands, thus giving Japan the staying power to erect and maintain a "Pacific wall." The Japanese could then take on the Americans and the British, wearing down their resolve until they wearied and made peace, leaving Asia and the Pacific to the Japanese empire. That strategy was a gamble, the success of which depended not only on weakening the opponents' resolve, but also, and absolutely, on the integrity of Japan's own shipping system. Japan had entered the war with oil inventories sufficient to last two years—or so Japanese planners thought. Beyond that time, Japan would have to call upon the oil of the East Indies. And this dependence, in the words of the United States Strategic Bombing Survey, "proved to be a fatal weakness." Or, as one history of Japanese military operations put it, "The shortage of liquid fuel was Japan's Achilles' heel."

The specific weakness was the vulnerability of Japanese shipping to submarines. Military planners had given surprisingly little thought to that risk. They underestimated both American submarines and the men who would sail on them. The Japanese thought Americans would be too soft and luxury-loving to stand up to the rigors of undersea living and warfare. In fact, America's submarines were the best in the war; and once equipped with improved torpedoes, they were a deadly weapon that weakened and then ruptured the critical shipping links between the Southern Zone and the Home Islands. The long, drawn-out confrontation became known as the Battle of the Marus, after the term the Japanese used to describe all merchantmen. But only in late 1943 did the Japanese begin to give serious attention to the protection of shipping against submarines, including the establishment of convoys. Their efforts were inadequate and incomplete. "When we requested air cover," one convoy commander said ruefully, "only American planes showed up." The Japanese shipping losses continued to mount.[5]

Moreover, convoys created their own problems, which actually aided the Allies. Assembling and directing the movement of convoys required the generation of a network of radio signals that, among other things, announced exact "noon positions." The interception of these messages by the Americans, who had cracked the Japanese codes, provided vitally useful information for the submariners. Overall, the consequences would be devastating. Of Japan's total wartime steel merchant shipping, some 86 percent was sunk during the conflict and another 9 percent so seriously damaged as to be out of action by the time the war ended. Less than 2 percent of American naval personnel—the submariners—were responsible for 55 percent of the total loss. Submarines from other Allied nations contributed another 5 percent. The success of this campaign—in effect, an ever-tightening blockade, a war of attrition—was later described by a group of Japanese economists as "a death blow to the war economy of Japan."

Oil tankers were among the favorite targets of the submariners, and tanker sinkings rose very sharply from 1943 on. By 1944, sinkings were far outrunning new tanker construction. Oil imports into Japan had reached their peak in the first quarter of fiscal year 1943. A year later, in the comparable quarter of 1944, the imports were less than half the 1943 figure. By the first quarter of 1945, imports had disappeared altogether. "Towards the end the situation was reached," a Japanese captain said, "that we were fairly certain a tanker would be sunk shortly after departing from port. There wasn't much doubt in our minds that a tanker would not get to Japan."

As their oil situation worsened, the Japanese tried many expedients and improvisations. Oil was put into drums of many sizes, and even into fiber containers that were loaded onto the decks of freighters. The Japanese filled large rubber bags with three hundred to five hundred barrels of oil that tugs were to tow to Japan. As ingenious as this was, the concept failed for a number of reasons: The gasoline attacked the rubber, the filling and emptying of the bags was difficult, and the bags reduced the maneuverability of the tugs, making them better targets for air attack. In desperation, the Japanese even tried to transport oil in their own submarines and sought to force German subs to deliver oil in exchange for access to repair facilities in Japan.

At home, as imports dried up, the Japanese squeezed and squeezed. Civilian gasoline consumption in 1944 was down to just 257,000 barrels—a mere 4 percent of the 1940 figure. Those gasoline-driven vehicles deemed essential were reequipped with charcoal or wood burners. Oils for industrial uses were made from soy beans, peanuts, coconuts, and castor beans. Civilian stocks of potatoes, sugar, and rice wine—even bottles of sake from the shelves of retail stores—were requisitioned for conversion to alcohol, to be used as fuel.

In 1937, the Japanese had made an ambitious and determined commitment to synthetic fuels, and in the months before Pearl Harbor, some in Tokyo had championed synthetic fuels as an alternative to war. But the actual wartime effort failed miserably, crippled by shortages of steel and equipment, and by a never-ending series of technical, engineering, mechanical, and personnel problems. In 1943, Japan's synthetic fuels production amounted to a total of one million barrels—only 8 percent of the 14-million-barrel target that had been set for that year—and never did those fuels meet more than 5 percent of the total oil requirements. Moreover, over half of the capacity was in Manchuria, where it was rendered useless in late 1944 and 1945 by blockade. Synthetic fuels were not only a failure, but an expensive failure because of their drain on resources, manpower, and management—to the degree that one analyst would comment, "The synthetic fuel industry in Japan, in terms of its absorption of materials and manpower and its meager product, was more of a liability than an asset during the war."[6]

"No Sense in Saving the Fleet"

The growing shortage of oil increasingly constrained Japanese military capabilities and directly affected the course of many battles. The pinch had been felt as early as June 1942, with the Battle of Midway, in which, as one admiral said, "We used very much fuel at that time, more than we had expected would be necessary; and the effect of that was felt right through afterwards." Following the victory at Midway, Allied forces were on the offensive, island-hopping westward, in a combination of naval and land operations that steadily removed ever closer to Japan—Tarawa and Makin in the Gilbert Islands, Kwajalein and Eniwetok in the Marshall Islands, Saipan and Guam in the Marianas. For both sides, every yard on every beach seemed to be measured in hundreds of lives. But the Americans had put together a devastating package—amphibious warfare, carrier air power, and industrial might. The Japanese could not match this enormous expenditure of resources. The Americans even exacted their revenge for Pearl Harbor in April 1943, when cryptanalysts learned that Admiral Yamamoto, who had planned the deadly attack, was due to make a visit to the island of Bougainville, near New Guinea. American fighters, waiting in ambush, came out of the clouds and sent the admiral down in flames to his death in the jungle below.

Yet it was not until the early months of 1944 that the submarine campaign finally began to make the Imperial Navy feel the fuel shortage "very keenly," as another admiral put it. Vanishing oil inventories also started to influence strategic decisions, with increasingly devastating consequences. In the Marianas campaign in June 1944, the Japanese battle fleet did not join the action because its fuel was low. Moreover, the carrier force approached the Americans directly rather than circuitously, in order to conserve oil. "It would take too much fuel to take the longer route," the Japanese commander was later to say. The direct approach was a costly one, for the result was what became known as the "Great Marianas Turkey Shoot," in which the Japanese lost 273 planes to the Americans' 29. With their victory in the Marianas, the Americans had finally penetrated the inner Japanese defense ring.

In the aftermath of that battle, it would have made good strategic sense for the Japanese to base the two battle groups of the Imperial Fleet in home waters—either at Okinawa or in the Home Islands themselves—ready to strike in any direction. But the sundering of the oil line to the Home Islands and the rapidly disappearing stocks of fuel did not permit such disposition. Thus, part of the fleet, with the carriers, was based in Japan, where it would await new aircraft and pilots, in the process draining the last of the fuel oil inventories. The heavy battleships were stationed close to Singapore, to be near the East Indies supply, but there, once committed to action, they could not refuel and be ready for action again for about a month. The overall consequence of the oil shortage was to divide naval strength when the Japanese needed a truly combined fleet, strong enough to repel Allied advances.

Japanese air operations were also severely curtailed because of the fuel shortage. Pilot training in 1944 was cut to thirty hours, half of what was considered necessary. And as the shortage worsened in 1945, navigation training was

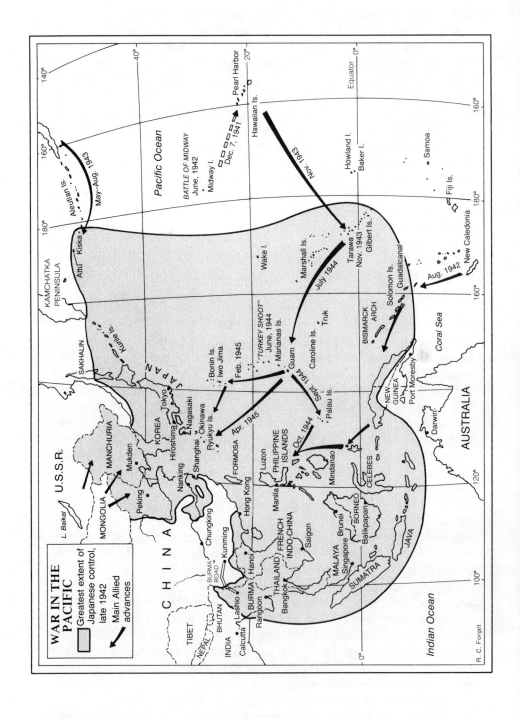

WAR IN THE PACIFIC

Greatest extent of
Japanese control,
late 1942

Main Allied
advances

U.S.S.R.

L. Baikal

MONGOLIA

MANCHURIA

Peking

Mukden

KOREA

Tokyo

Nanking

Shanghai

Nagasaki

Hiroshima

Okinawa

Ryukyu Is.

JAPAN

Kuril Is.

SAKHALIN

KAMCHATKA
PENINSULA

Attu

Kiska

Aleutian Is.

May–Aug. 1943

160°

180°

160°

Pacific Ocean

BATTLE OF MIDWAY
June, 1942

Midway I.

Dec. 7, 1941

Pearl Harbor

Hawaiian Is.

20°

0°

160°

Equator

Howland I.

Baker I.

Samoa

Fiji Is.

180°

New Caledonia

Wake I.

Marshall Is.

Tarawa
Nov. 1943

Gilbert Is.

Solomon Is.

Guadalcanal

Aug. 1942

160°

Coral Sea

Nov. 1943

July 1944

Caroline Is.

Truk

BISMARCK
ARCH.

NEW
GUINEA

Port Moresby

Darwin

AUSTRALIA

120°

"TURKEY SHOOT"
June, 1944

Marianas Is.

Guam

Sept. 1944

Palau Is.

Oct. 1944

Apr. 1945

Bonin Is.

Iwo Jima

Feb. 1945

FORMOSA

Luzon

Manila

PHILIPPINE
ISLANDS

Mindanao

CELEBES

BORNEO

Balikpapan

Hong Kong

FRENCH
INDO-CHINA

Saigon

Brunei

MALAYA

Singapore

SUMATRA

JAVA

100°

Chungking

Kunming

BURMA
ROAD

Lashio

Hanoi

THAILAND

Bangkok

Rangoon

BURMA

C H I N A

TIBET

BHUTAN

NEPAL

INDIA

Calcutta

0°

Indian Ocean

140°

40°

R. C. Forget

eliminated altogether; pilots were simply to follow their leaders to targets. Few were expected to return. Aviation gasoline was made from the only available source, wood turpentine, which was mixed with larger and larger proportions of alcohol. The combination of inferior fuels, poorly trained pilots, and inadequately tested planes was a deadly one. The Japanese lost up to 40 percent of their aircraft on ferrying operations alone.

In order to stretch oil supplies, many Japanese ships burned unrefined Borneo crude, which, as Marcus Samuel had claimed many years earlier, was indeed good for fuel. However, it was also highly flammable, and thus a threat to the ships it powered. Under duress, the Japanese even reversed the historic trend in naval propulsion and rebuilt ships in service to burn coal where available. Ships under construction were converted to coal before launching. That achieved a relative security of supply, but meant a loss in speed and flexibility.[7]

It was the primacy of fuel that finally led the Imperial Navy to throw all its weight into the Battle of Leyte Gulf, off the Philippines, in October 1944. By then the noose was growing very tight. The recapture of Guam in August 1944 brought cities on the Home Islands into the range of the new B-29 bombers. To the south, on September 15, General MacArthur landed on Morotai, in the Moluccas, only three hundred miles from the Philippines. Looking in that direction, he declared, "They are waiting for me there." To the Japanese, there seemed no choice but to put everything they had into trying to prevent the Americans from recapturing the Philippines, which were within air striking distance of the Home Islands and stood astride the sea lanes between Japan and its conquered territories in Southeast Asia. Admiral Soemu Toyoda, Chief of the Naval General Staff, gave the order that would lead to the greatest battle in the history of naval warfare. "Should we lose in the Philippines operations," he later said, "even though the fleet should be left, the shipping lane to the south would be completely cut off so that the fleet, if it should come back to Japanese waters, could not obtain its fuel supply. If it should remain in southern waters, it could not receive supplies of ammunition and arms. There would be no sense in saving the fleet at the expense of the loss of the Philippines. That was my reason for my order."

But the shortage of fuel handicapped the Japanese again and again in the battle for the Philippines. Because of the split in basing, their Navy had to try to concentrate its forces at decisive points from widely separate directions. Two Japanese battleships never even made it to the great battle because of the lack of oil. Instead, they proceeded to Singapore, where they refueled and went home again. Other ships arrived a few critical hours too late because they were slow-speeding to conserve fuel. On October 25, 1944, Admiral Takeo Kurita, commander of the Second Fleet, was in position to enter Leyte Gulf, which would have made it possible for him to annihilate General MacArthur's lightly defended invasion troops and change the course of battle. But, just forty miles away from the invasion beach, Kurita abruptly pulled off and sailed away. After the war, one of the Japanese admirals was asked why. "Because," he replied, "of shortage of fuel."

The three-day battle of Leyte Gulf was a devastating defeat for the Japanese. Their losses included three battleships, all four aircraft carriers, ten cruisers, and thirteen destroyers. It was out of desperation in this battle that the Japanese explicitly introduced a new weapon—the suicide pilots called *kamikazes*. The word meant "divine wind," after a typhoon in the thirteenth century that had shattered the great invasion fleet of Kublai Khan before it could land in Japan. These suicide pilots, who were ordered to crash their planes (including specially designed manned rocket bombs) onto the decks of the American ships, were meant to be the ultimate embodiment of the Japanese spirit, inspiring all their compatriots to total sacrifice. But they also served a very practical purpose for a country extremely short of oil, planes, and skilled pilots. The Japanese had methodically calculated that, whereas eight bombers and sixteen fighters were required to sink an American aircraft carrier or battleship, the same effect could be achieved by one to three suicide planes. Not only was the pilot sure to cause more damage if he crashed his plane, not only would his commitment and willingness to die unnerve an enemy who could not comprehend the mentality of such an act, but—since he was not going to return—his fuel requirement was cut in half.[8]

The End of the Imperial Navy

The Japanese could do little or nothing to interrupt the ever more abundant flow of fuel and other supplies to the American forces in the Pacific, no matter how distant their source. The Americans developed huge floating bases—composed of fuel barges, repair ships, tenders, tugs, floating docks, salvage ships, lighters and store ships—that gave the U.S. Navy "long legs" across the vast expanse of the Pacific. Roving fueling task forces, made up of two or three giant tankers plus destroyer escorts, took up stations at designated areas, large rectangles twenty-five miles wide by seventy-five miles long, where American ships would rendezvous to refuel. When Guam became, in the second half of 1944, the major American base for bombing Japan, 120,000 barrels of aviation gasoline were supplied there daily. At that same time, Japan's entire air forces on all fronts were consuming just 21,000 barrels per day—one-sixth of what was available just at Guam.

The Japanese were being pushed back on almost every side. By early 1945, the Americans had recaptured Manila in the Philippines, as well as Iwo Jima, though at appalling cost—6,800 Americans and 21,000 Japanese dead, with another 20,000 Americans wounded—for an island that measured 4.5 by 2.5 miles. In South Asia, the British had launched their final offensive in Burma. The Japanese abandoned Balikpapan and the other chief oil port in the East Indies, and most of their Home Islands refineries were out of oil. In March 1945, the last Japanese oil tanker convoy left Singapore. It never got to Japan. It was sunk.

At home, oil had virtually disappeared from the domestic economy, in part of a larger pattern of deprivation. Gas, electricity, coal, and charcoal were all in incredibly short supply. It was no longer possible to take private baths, and the public bathhouses became very crowded. People called the experience "washing potatoes in a tub"—with the heat coming from scrap wood gathered in the streets. Many Japanese took to burning their libraries for fuel, figuring that the books would, in any event, be destroyed in the next air raid. Tokyo's fuel distribution for the cold 1944–45 winter season did not commence until May 21, 1945, long after most residents had learned to cook with the charred ruins scavenged from a bombed-out city. Food intake was down to less than 1,800 calories a day, substantially below the minimum requirement of 2,160 calories.[9]

The fuel situation had become so severe for the military that the Navy decided on a dramatic variant of the *kamikaze* raid—to sacrifice the *Yamato*, the world's largest battleship and the pride of the Japanese fleet. It was to be the nucleus of a Special Attack Force that would break through the American ships supporting the invasion of Okinawa, do as much damage as possible, then beach itself and use its huge eighteen-inch guns in defense of the island. "Any large-scale operation requiring heavy supplies of fuel became almost out of the question," said Admiral Toyoda. "Even in getting the squadron together we had a very difficult time getting the necessary 2,500 tons of fuel oil together. But it was felt that, even if there was not a fifty-fifty chance, nothing was to be gained by letting those ships lie idle in home waters, and besides it would have been contrary to the tradition of the Japanese Navy not to have sent them, even though we could not clearly perceive they had a fifty-fifty chance of success. That is how acute the fuel situation was."

It was clearly a suicide mission; the *Yamato* carried only enough fuel for a one-way trip. The monster battleship, with its accompanying vessels, cleared Tokuyama the morning of April 6, deprived of all air cover by the requirements of the *kamikaze* campaign. On April 7, at midday, three hundred American planes came out of the low-hanging overcast and began their barrage. By mid-afternoon, the *Yamato* and most of the other ships had been sunk. To many, the sinking of the *Yamato*, which had been destroyed even before it could commit suicide, marked "the end of the Imperial Navy." The Japanese fleet, which had prided itself on commanding the entire western Pacific, had now been driven even from the immediate seas of the homeland.[10]

A Fight to the Finish?

And still Japan's position worsened. Shortages of fuel were preventing its planes from flying more than two hours a month. Was there no other way to get oil? Desperate for fuel, the Navy launched its fantastic pine root campaign. Guided by the slogan "two hundred pine roots will keep a plane in the air for an hour," people all over the Home Islands began to dig up pine roots. Children were dispatched to the countryside to scour for the roots. The pine roots were to be heated for twelve hours, producing a crude oil substitute. Thirty-four thousand kettles, stills, and small distillation units were put in place, with the aim of each producing three or four gallons of oil per day. The futility of the effort was revealed by the labor requirements. Each gallon produced required 2.5 man-days of work. To meet the official target of twelve thousand barrels per day would have required 1.25 million persons per day!

Some of the results of the pine root campaign were evident to the eye: mountainsides stripped bare of every tree and sapling, huge piles of roots and stumps lining roads. By June 1945, production of the pine root oil had reached seventy thousand barrels per month, but the refining difficulties had not been solved. Indeed, by the time the war ended, only three thousand barrels of gasoline intended for planes had been produced from pine root oil, and there was no evidence that any of it had ever actually been tried in a plane.

The end was nearing for Japan. Under relentless American bombardment, the country's wooden cities flamed into charred ruins, the economy ground ever more slowly, and the military's ability to mount any counterattacks virtually vanished. The Razor—Hideki Tojo—had been forced out as premier the previous July; now, in the spring of 1945, yet another new government came in, with at least some of its members interested in finding a way to end the war short of total annihilation. "Everything had just about come to the bottom level," said one minister. "Looking in whatever direction, we had come to the end of the road." The new government was headed by an eighty-year-old retired admiral, Kantaro Suzuki, who carried some prestige and was seen as a relative moderate. The jockeying became even more intense between those who wanted to carry on the war and those who wanted to find a way to bring it to an end. Those in the latter group were, however, cautious and circumlocutious, deathly afraid of either a coup or their own assassinations.[11]

The Soviet Union renounced its neutrality pact with Japan on April 5, 1945. Under the terms, however, the pact would remain in effect until April 1946. Now senior officers in the Japanese Navy conceived yet another idea, no less fantastic in its own way than the pine root campaign—to approach the Soviet Union directly and ask it to mediate between Tokyo and Washington and London, and to trade Soviet oil for resources from the Southern Zone. Koki Hirota, a former premier and ambassador to Moscow, was charged with opening the dialogue with the Soviet ambassador to Japan. But the Japanese did not know that, at Yalta the preceding February, Stalin had promised Roosevelt and Churchill to bring the Soviet Union into the war against Japan some ninety days after the end of the war in Europe. Moreover, Stalin had worked out a far more attractive exchange than one involving just raw materials. As the price for his entry into the war, the Soviet dictator had obtained rich territorial concessions: the reestablishment of Russian predominance in Manchuria, the recovery of the southern part of the Sakhalin Islands, and the acquisition of the Kurile Islands. Though an ethnic Georgian, Stalin was a Russian nationalist par excellence; with those rewards, he would right the defeat that Czarist Russia had suffered at the hands of the Japanese in 1905. The Soviet ambassador thus dismissed all of Hirota's political proposals when they met in late June. As for exporting oil to Japan, the ambassador added, that would be quite impossible, as the Soviet Union itself was much too short of petroleum.

Premier Suzuki ordered a survey of Japan's fighting capabilities to determine whether they were sufficient to continue the war. The study came back in mid-June 1945 with a picture of a war economy that was almost immobilized because of lack of fuel and the fury of the American aerial attacks. The numbers gave further proof of Japan's desperate situation. Fuel oil inventories had been 29.6 million barrels in April 1937; by July 1, 1945, they were just 0.8 million barrels. Below a million barrels, the Navy could not operate. For all practical purposes, it was out of oil. To some in the Japanese government, "the utter hopelessness of our situation" was clear. But not to all, by any means. The possibility of surrender was far from accepted at the top of the Japanese government, and many bitterly rejected even its mention. The government was still pushing the slogan, "100 million people united and ready to die for the Nation." The Army and elements of the Navy were battling to commit Suzuki's Cabinet to a war to the bitter end.[12]

As if to demonstrate what they had in mind, Japanese resistance to the American invasion of Okinawa in April of 1945 was fierce and fanatic and, in its organized form, did not end until June 21, 1945. The Americans experienced a 35 percent casualty rate in taking that island. Assuming that a similar ratio would hold in an invasion of the Home Islands, the American commanders anticipated a minimum of 268,000 dead and wounded on their side in the first phase of the attack. Altogether, they anticipated up to a million American military casualties—with a similar number for the Japanese, and many, many more millions in civilian deaths.

The bloodiness and stubbornness of the battle on Okinawa contributed strongly to the American decision to use, if necessary, a new weapon that, though unproved, would soon be in the U.S. arsenal—the atomic bomb. The American leaders knew that Japan's fighting capability was crumbling, but they saw no sign that its fighting spirit was fading. And, indeed, it did seem that the whole island nation was being mobilized for a suicidal battle to the death; even young schoolchildren were ordered to start sharpening bamboo shoots with which to kill Americans. The secret messages between Tokyo and Moscow that the Americans intercepted were hardly sufficient to indicate that the Japanese government was prepared to ask for peace—because it was not.

Despite the ever-worsening position, the Japanese government remained ambiguous, indistinct, and tentative in its signals about surrender, reflecting the fact that internally there was no consensus, and that the war party still had the upper hand. Tokyo spitefully rejected the Allies' Potsdam Declaration, which would have enabled Japan to quit the war on reasonable grounds, including the retention of the Emperor. Many Japanese leaders were unwilling to take the steps that would have spared the Japanese people, soldiers and civilians alike, any more of the terrible suffering they had already borne in the name of a fevered nationalistic ideology and relentless militarism. To the Allies, there was little indication of anything from Tokyo save a determination to fight to the finish.[13]

The first atomic bomb was dropped on Hiroshima on August 6, 1945. On August 8, the Soviet Union declared war on Japan and sent its troops pouring into Manchuria, a week earlier than had been planned, in order to ensure that the war did not end before it could get in. On August 9, the second atomic bomb fell on Nagasaki. Yet even as late as the Nagasaki explosion, the Army Chief of Staff insisted on reminding senior officials that Japanese soldiers and sailors were not permitted to surrender under any circumstances; suicide was the only acceptable way out. On August 13, four days after the explosion of the Nagasaki bomb, Vice-Admiral Takijiro Onishi, the creator of the *kamikaze* missions, was still advocating that the government reject surrender. Instead, he said, the Japanese people should fight to the bitter end, and 20 million of them should sacrifice themselves in suicide attacks against the invading troops.

Yet so appalling were Japan's circumstances and so great the shock of the atomic bombs, made worse by the new Soviet threat, that those seeking to end the war finally prevailed over the intense opposition from the military. On the night of August 14, the Emperor made a phonograph recording that bore the surrender message. It was to be broadcast the next day. Even then, that same night, insurgent soldiers assassinated the head of the Imperial Guard and broke into the Imperial Palace, trying to capture the record, to prevent its broadcast, and at the same time kill Premier Suzuki. They were repulsed. The next day, the Japanese people heard over their radios a faint voice, fading in and out because of the uncertain electricity, that most had never heard before. It was the voice of their Emperor, calling on them to surrender. The war in the Pacific was over.

Still, not everyone was willing to heed the call. Earlier that same morning, War Minister Korechika Arami had committed hara-kiri; and the next day, Admiral Onishi did the same. Moreover, there had been notable preparations for Onishi's final *kamikaze* attacks. After the surrender and the establishment of the U.S. Occupation, American authorities found a total of 316,000 barrels of oil, secreted away by the Imperial Army and Navy in remote caves and numerous hiding spots, to be used solely for suicide flights against the invaders. Some stores of the pine root gasoline—one of Japan's last hopes for resistance—were also discovered after the surrender. It was tested in American military jeeps. It turned out to be a terrible fuel, gumming up the engines beyond use.[14]

The Ambulance

From the first moments of the Occupation, the fuel shortage continued to make its impact felt in Japan. On August 30, the Supreme Commander, General Douglas MacArthur, landed in Japan, at the Atsugi airfield. The propellers had been removed from the Japanese planes on the field so they could not be used for *kamikaze* attacks. The general immediately set off in a haphazard motorcade, led by a red fire engine that resembled the Toonerville Trolley. His destination was Yokohama and the battleship *Missouri*, which sat in the harbor, and on which, three days later, the instruments of surrender were to be signed. The motorcade route was lined with Japanese soldiers, their backs turned toward the passing MacArthur—the same sign of obeisance previously reserved for the Emperor. Though the distance was only twenty miles, the motorcade took two hours to make the trip; the battered vehicles, the best the Japanese could provide, were powered not by gasoline, because there was no gasoline, but by charcoal. They repeatedly broke down.

Twelve days later, on September 11, 1945, American officials in Tokyo arrived outside a modest one-story house on the edge of intensively cultivated fields. The house belonged to the Razor—General Hideki Tojo, the wartime premier. Tojo appeared at an open window, to be told that he was under arrest, and that he should immediately get ready to go with the Americans. He agreed and shut the window. A shot rang out. The Americans burst into the house and found Tojo sitting in an oversized chair, bleeding from a self-inflicted bullet wound just below his heart.

Four years earlier, in 1941, as War Minister and then Premier, Tojo had forced Japan's decision to go to war against the United States, arguing that the fate of Imperial Japan hung in the balance because of its shortage of oil. The cost of what Tojo and his collaborators had launched was enormous. The Pacific War in its entirety claimed upwards of 20 million lives, including about two-and-a-half million Japanese. Now, in 1945, Tojo's own life hung in the balance, not because his self-inflicted wound was inevitably fatal, which it was not, but because of the difficulty, first, in locating a suitable doctor, and then in finding an ambulance that had any gasoline in its tank. So widespread was the fuel shortage that it proved easier to find an American doctor than an ambulance with gasoline. But finally a vehicle with sufficient fuel was located, and it arrived at Tojo's house two hours after he had shot himself. Tojo was carried off to the hospital and nursed back to health. The next year he went on trial as a war criminal, was found guilty, and was in due course executed.[15]

CHAPTER 19

The Allies' War

WINSTON CHURCHILL had spent the decade of the 1930s in the political wilderness, his warnings about Nazi intentions and capabilities unheeded. But in September 1939, at the age of sixty-six, he was abruptly recalled to the post he had held more than a quarter century earlier, on the eve of World War I—First Lord of the Admiralty. But this time his charge was not to prepare for a war. It was too late for that. War had already come a few days earlier, with Germany's invasion of Poland on September 1, 1939. Still, there then ensued a lull, a phoney war, which lasted half a year, until the spring of 1940, when Hitler sent his troops smashing across Western Europe. The appeasers fell in London, and Churchill became Prime Minister of Great Britain.

The outlook was grim and disheartening. Norway and Denmark were in German hands, France would surrender the following month, and Britain would stand alone, bearing the brunt of the war. No one was better suited than Churchill to lead his country through its "darkest hour." No one better understood the critical role that oil would play, first in Britain's very survival, and then in the prolonged conflict ahead.

Well before the outbreak of fighting, the British government had initiated a serious evaluation of its oil position in light of what seemed the inevitable conflict with Germany. In late 1937, a special committee examined the possibility of Britain's adopting an "oil from coal" synthetic fuels strategy along the lines of Germany's. After all, Britain possessed very rich coal reserves within its own safe confines, while almost all of its oil had to be imported. Nevertheless, the strategy was rejected. It would have been very costly, while Britain not only had access to substantial supplies of cheaper oil around the world, but was also home to two major international companies, Shell and Anglo-Iranian. Also, it was decided that, appearances notwithstanding, security would not be better served with synthetic oil. A system based on the importation of conventional oil in many ships through many ports would be less vulnerable to air attack than one dependent on a few very large, easily identifiable—and easily bombed— hydrogenation plants.

In its planning for war, the British government anticipated very tight and explicit cooperation with the oil industry of a kind that could not have been easily fashioned in the United States. In the United Kingdom, 85 percent of domestic refining and marketing was in the hands of just three companies—Shell, Anglo-Iranian, and Jersey's British subsidiary. At the time of Munich, in 1938, the government decided that, in the event of war, all the "paraphernalia of competition" would be eliminated, and that the entire British oil industry would be run as one giant combine, under the aegis of the government.

The government also had to cope with a different kind of problem—the future of the Royal Dutch/Shell Group. The current management of the Group was no less concerned and apprehensive. For there was a risk that the Group could pass under the Nazi sway. The heart of the problem was Henri Deterding, the grand master of the company. He had continued to dominate the Group through the 1920s. "Sir Henri's word is law," observed a British official in 1927. "He can bind the Board of the Shell without their knowledge and consent." But by the 1930s, Deterding's grip on the company was slipping, and he was becoming an embarrassment to the management and a source of anxiety to the British government. His behavior was increasingly erratic, disruptive, megalomaniacal.

In the mid-1930s, as he entered his seventies, Deterding had developed two infatuations. One was for his secretary, a young German woman. The other was for Adolf Hitler. The determined Dutchman—who had gravitated to Britain before World War I, had been courted by Admiral Fisher and Winston Churchill, and had become a firm and fervent ally during that war—was now, in his old age, entranced with the Nazis. "His hatred of the Soviets, his admiration for Hitler and his *idée fixe* on the subject of Anglo-German friendship in an anti-Soviet sense are, of course, all well-known," sighed a Foreign Office official. On his own, Deterding initiated discussions in 1935 with the German government about Shell's providing a year's supply of oil—in effect, a military reserve—to Germany on credit. Rumors of these talks so greatly alarmed the Shell management in London that one of the senior directors, Andrew Agnew, asked the government to have the British embassy in Berlin investigate so that Agnew "could take suitable actions with his colleagues on the Board here in good time." Commented one official, "Deterding is getting an old man, but he is a man of strong views, and I am afraid we cannot stop him consorting with political chiefs." He added, "The British members of the Board of the company are keen that the company should not do anything which would be contrary to the views of H.M. Government."

Finally, retiring from Shell at the end of 1936, Deterding acted on both of his new infatuations. He divorced his second wife, married his German secretary, and went to live on an estate in Germany. He also took to urging other European nations to cooperate with the Nazis to stop the Bolshevik hordes, and he himself exchanged visits with the Nazi leaders. By 1937, the Prime Minister of the Netherlands, a former colleague of Deterding's in Royal Dutch, said that he

351

"could not understand how a man, who had made his name and fortune in England and who had received certain assistance from the country of his adoption, should suddenly migrate to Germany and devote himself to furthering the welfare of that country." His activities, the Prime Minister added disparagingly, were "infantile and leaving no doubt as to his feelings." Deterding's last years, not surprisingly, were to undermine what would otherwise have been a considerable reputation as an "international oil man."

Deterding died in Germany in early 1939, six months before the war began. Strange and deeply disturbing rumors immediately reached London. Not only had the Nazis made much of his funeral, but they were also trying to take advantage of the circumstances of his death to gain control of the Royal Dutch/Shell Group. That, of course, would have been a disaster for Great Britain. The company had virtually been Britain's quartermaster general for oil during World War I. Should it now pass under Nazi domination, Britain's entire system of petroleum supply would be undermined. But it was discovered that the key "preference" shares, which embodied control, could only be held by directors, and at his demise, Deterding's shares had been swiftly distributed to the other directors. At best, the Germans could only get their hands on a tiny fraction of the common shares, which would do them no good at all, either before or after the outbreak of war.[1]

As soon as the war began, the British oil companies, including Shell, merged their downstream activities into the Petroleum Board, in effect creating a national monopoly. It was done swiftly and without protest. Petrol pumps were painted dark green and the product was sold under the single brand name "Pool." The industry people continued to operate the business, but they now did so under national control. Britain's oil war was thereafter run out of Shell-Mex House on the Strand in London, just down from the Savoy Hotel. (Shell's own headquarters was moved to a company sports facility on the edge of London.) Overall government direction was eventually lodged in an agency called the Petroleum Department.

The issues facing Britain were global. It had to assume that Germany, as signatory of a new pact with the Soviet Union, would be able to obtain an abundant supply of Russian oil, while British supplies from the Far East would be curtailed if the Japanese invaded Southeast Asia. Closer to home, the rich and convenient resources of Rumania were also available to Germany. A few months after the war began, before France had been overrun by the Germans, the British and French governments, seeking to replicate what had been carried out in World War I, had jointly offered to pay Rumania $60 million to destroy its oil fields, and thus prevent the Germans from taking the output. But the two sides could never agree on a price, the deal was never struck, and Rumanian oil went, as feared, to the Germans. The destruction was left to be done by Allied bombers, much later in the war.

In Britain itself, practical problems of supply had to be quickly dealt with. Rationing was imposed almost at once. The "basic ration" for motorists was first set at eighteen hundred miles a year. It was progressively tightened, as military needs increased and stocks declined, and then it was eliminated altogether. The authorities wanted to see the family car up on blocks in the garage, and not on the road. As a result, there was a great boom in bicycling.

And what was to be done with oil supplies if Britain were invaded—all too real a possibility in the dismal days of 1940, after the Nazi armies had swept across Western Europe and were poised on the French side of the English Channel? The Germans had captured France's oil stockpiles, thus providing the wherewithal to maintain the momentum of their advance. A similar booty of British oil supplies could prove critical to the success or failure of a Nazi cross-Channel attack. As a consequence, plans were made at Shell-Mex House for the immediate destruction of British stocks in the event of an invasion. Meanwhile, the local, friendly, unprotected petrol (or gas) station seemed to be much too convenient for invading Germans, who would simply be able to pull in and fill up. For that reason, some seventeen thousand gasoline-selling establishments in eastern and southeastern England were quickly shut down, and sales and supplies were concentrated in two thousand stations that could be better defended—or, if need be, set ablaze to deny them to the enemy.[2]

The Oil Czar: The Mobilization of American Supply

There was, for the British, the overarching concern—how to supply their war? The outbreak of fighting would mean much larger oil consumption for Britain, and the only place to look was to the United States, which was responsible for almost two-thirds of total world production. To the administrators in Whitehall and the oil men at Shell-Mex House, two questions were paramount: Would the oil be available? And would Britain, already strapped for dollars, be able to pay for it? The answers to both questions would have to be found in Washington.

In December 1940, with his third election safely out of the way, Franklin Roosevelt proclaimed the United States the "arsenal of democracy." In March 1941, Lend Lease was instituted, which removed the problem of finance—or, as Roosevelt put it, "the silly, foolish old dollar sign"—as a constraint on American supply to Britain. Among those things to be "lent," for repayment at some indefinite time in the future, was American oil. The neutrality legislation, which restricted the ability to ship supplies to Britain, was progressively loosened. And, in the spring of 1941, when oil supplies began to plummet in the United Kingdom, fifty American oil tankers were switched from supplying the American East Coast ports to carrying oil for transshipment to England. Thus, by late spring 1941, important steps had been taken to weld together the American and British supply systems and put the United States in position to fuel England's lonely stand. In fact, the United States had a surplus, unused oil production capacity of about one million barrels per day. That was equivalent to about 30 percent of the 3.7 million barrels per day production that year. The extra capacity, a result of the federal-state prorationing system set up in the 1930s, turned out to be an invaluable security margin, a strategic resource of immense significance. Without it, the course of World War II might well have been different.

In May 1941, the day after Roosevelt declared an "unlimited national emergency"—though the United States was not yet in the war—he appointed Interior

Secretary Harold Ickes to the additional post of Petroleum Coordinator for National Defense. That once again made the Old Curmudgeon the nation's top oil man, or as he became known, its Oil Czar. Ickes's first task was to reshape the relationship between the Roosevelt Administration and the oil industry. In 1933, the New Deal had come to the aid of an oil industry drowning in the flood of oil from East Texas. But by the latter 1930s, it had become increasingly critical of "monopoly" in the oil industry, and indeed, in 1940, the Justice Department had launched an antitrust case against the American Petroleum Institute and twenty-two major oil companies—and 345 smaller oil companies—charging them with violations in every aspect of their business. Another shift came with the national emergency and the imminence of war. As Roosevelt was later to explain, "Old Dr. New Deal" had to call in his partner "Dr. Win-the-War." And what Dr. New Deal had found unpalatable and unhealthy about Big Oil—its size and scale, its integrated operations, its self-reliance, its ability to mobilize capital and technology—was exactly what Dr. Win-the-War would prescribe as the urgent medicine for wartime mobilization.[3]

Ickes also had to take the lead in turning around an industry geared to coping with surplus to one that would maximize output and avert shortage, and do it in the face of public skepticism that such a shortage could possibly occur. At the same time, America's oil industry—heterodox, riven by bitter rivalries and suspicions among the large integrated majors, the independent producers, and refiners and marketers—would have to be fused in effect, though not formally, into one giant organization, under government direction and mobilized for war. This had been done swiftly and efficiently in Britain, where even rationing had been accepted with hardly a murmur. The story would be different in America.

Harold Ickes started with a huge liability: He was widely detested by the industry. Though he had come to its assistance in 1933, he had subsequently become a critic, calling for greater regulation by the Federal government of the industry's operations and profitability, even flirting with the idea of nationalization. The companies had a particular and bitter complaint against Ickes. During the Depression, at his behest, the oil companies had established pools to buy up "distress" gasoline. In 1936, after the Supreme Court invalidated the National Industrial Recovery Act, under which authority Ickes had acted, the Justice Department indicted the companies for the pooling. Ickes, thereafter, kept quiet about his promotion of the scheme and conveniently found that he could not get to the trial, held in Wisconsin, to testify about his role. The companies were convicted, and that experience made them leery, to put it mildly, of working with him again. Indeed, upon his appointment as Petroleum Coordinator, the *Oil Weekly* rushed out a special supplement to warn against the imminent "antagonistic management and probably vindictive tampering of a man who has no qualities nor specific traits of ability to handle the job." Ickes would prove otherwise. From the beginning, he demonstrated a willingness to work closely and pragmatically with the industry. He chose as his deputy an experienced oil man, Ralph Davies, a marketing executive from Standard of California. And thereafter, the Oil Czar succeeded in disarming the hostility and cooperating effectively to mobilize this critical industry.[4]

Trial by Sea: The Battle of the Atlantic

The most vulnerable link in the supply chain being forged between America and a beleaguered Britain was the wide expanse of the Atlantic that tankers and freighters had to cross. Here lay the German opportunity to strangle the military capabilities of the British and, later, of the American forces in North Africa and Europe, as well as the Russian war machine, for which American oil would quickly become vital. "The more ruthlessly economic warfare is waged," Admiral Erich Raeder, Commander in Chief of the German Navy, declared, "the earlier it will show results and the sooner the war will end." The weapon was the U-boat, and its ability to disorganize shipping was quickly felt. Early in 1941, the U-boats, now operating in "wolf packs," stepped up the campaign. Oil tankers were their favored targets.

The great success of these attacks terrified the British, and those few Americans to whom the British showed the graphs that depicted the ever-rising tonnage and supplies lost, measured against the dwindling stockpiles in the United Kingdom and the growing needs of war. The battle, expressed so abstractly, frustrated Churchill to his core. "How willingly," he said, "would I have exchanged a full-scale invasion for this shapeless, measureless peril, expressed in charts, curves, and statistics!" In March 1941, he described the onslaughts on shipping as "the blackest cloud which we had to face." He had no doubt how much was at stake in the silent and remote battle in the waters of the Atlantic, and he knew that he could not win it without American help.

"Nothing less than shocking . . . grave in the extreme." That was how Harold Ickes's deputy, Ralph Davies, reported to him on England's perilous oil supply situation in July 1941. There were only five weeks of motor gasoline stocks and only two months of fuel for the Royal Navy, whereas seven months' inventory was considered the rock-bottom minimum for safety. Ickes was convinced that everything that could be done to help the British prosecute the war should be done. And reducing oil consumption on the East Coast would help, by freeing both shipping and supplies that could be redirected to England. Ickes mobilized the nation's railway tank cars to rush supplies to the East Coast. Together with the oil companies, he launched a high-profile voluntary conservation campaign, which included the distribution of windshield stickers that said, "I'm using one-third less gas." He asked service stations to close at 7:00 P.M. and not reopen until 7:00 A.M., and he sought to reinstate the "gasolineless Sundays" of World War I. He even tried to get a car pooling program going in the Department of the Interior to serve as a model for the rest of the country. (Always a reformer at heart, he saw a valuable side benefit: "We may be able to improve parking conditions generally in Washington," he noted in his diary.) But the voluntary conservation program was a flop, and Ickes turned to the companies and pressed them to cut back their deliveries to gasoline stations by 10 to 15 percent.[5]

The one thing that Ickes did not do, and could not do, was explain the real reasons for conservation: the dismal effects of the German U-boat campaign in the Atlantic and the deplorable state of British oil supplies. He feared that if he publicized the severity of the situation, he would be passing on valuable intelligence to the Nazis. Nor did he want unnecessarily to rile the isolationists in the United States. Thus, the whole conservation program aroused a storm of protest—from the politically powerful independent oil producers in Texas, who were deprived of their access to independent tankers, to the independent oil refiners and distributors in the East, who had to absorb the much higher costs of railway transportation. The New Jersey state legislature passed a resolution denouncing the conservation effort because of the threat it posed to the state's fishing and summer resorts. Major elements of the press labeled the situation a "phony shortage," and free-wheeling American motorists rebelled at the thought of even voluntarily cutting back on their driving.

In order to counter the U-boat menace, the United States increased its patrols further out into the Atlantic, and established bases in Newfoundland, Greenland, Iceland and Bermuda. At the same time, the British cracked German naval codes, enabling them to put convoys on evasive routings. That—combined with reduced demand, Lend Lease, and the transfer of the fifty tankers—helped take the pressure off England, at least temporarily. But the danger had been far greater than everyone, except a tiny handful, knew. "It was only by the narrowest of margins," in the words of the official history of British intelligence, that "the U-boat campaign failed to be decisive during 1941."

By the autumn of that same year, the supply situation on the U.S. East Coast had improved considerably, while the British, their oil position temporarily improved, handed back the tankers that had been transferred to them. That, of course, seemed to prove there had been no shortage at all, and Ickes found himself pilloried both in the press and in Congress. A special Congressional investigating committee claimed that the Secretary had invented the shortage. It was nothing more, said the committee, than "a shortage of surplus."

Gasoline stations were now putting up signs advertising that they had no scarcity of fuel and urging motorists to "fill it up," which drivers hastened to do. Ickes felt that he was being made to look like a fool. "I would not again be in favor of putting on restrictions until the people had actually felt the pinch," he angrily complained in private. "It is impossible to carry the American people along with you on a program of caution to forestall a threatening position." Prevention, whether it be an ounce or a pound, was bad politics, he concluded. Thereafter, Ickes resolved never again to go too far out on oil issues.

Yet the threat of shortage again loomed when Germany declared war on the United States on December 11, 1941, four days after Pearl Harbor. U-boats immediately began to operate in American coastal waters, with devastating results. Their priority targets were oil tankers, easily recognizable by their distinctive profile. After a Cabinet meeting in January of 1942, Ickes warned the President that renewed tanker sinkings in the Atlantic would bring new pressures on supply, particularly in the Northeast. Still smarting, however, from the criticism that had rained down on him for his conservation program, he resolutely refused to do anything prophylactic. "In view of the hell that I got last Fall for foreseeing the possibility of such a situation and trying to prevent it, I do not intend to say anything about this publicly until it becomes a certainty. If a shortage actually occurs, I may be able to make the front page with lyrical stories of what I am going to do by way of rationing to take care of that shortage. The shortage itself can be attributed to God while I reap a harvest of praise for my lack of foresight."[6]

Altogether, the number of tankers sunk in the first three months of 1942 was almost four times the number built. The U-boats seemed to operate with impunity along the entire coast. As one U-boat headed home from American waters, with eight sinkings to its credit, the submarine's captain exultantly wrote in his war diary: "It is a pity that there were not twenty U-boats last night, instead of one. I am sure all would have found ample targets."

The dismal toll of sinkings mounted. "The situation is desperate," Ickes wrote to Roosevelt in late April 1942. Yet the initial American response to the U-boat onslaught was feeble. Tankers and other ships were urged to hug the coast; those that could fit took the Cape Cod and the Delaware-Chesapeake canals. The United States had neglected antisubmarine warfare and was unprepared. Onshore, American cities were making the job of sinking cargo ships even easier for the U-boats. They were brightly illuminated at night, thus providing perfect silhouettes of the tanker targets for the stalking submarines. Miami was the worst offender; six miles of its beachfront were lit up by neon lights. Hotel owners and the chamber of commerce insisted that the lights could not be doused; it was still the tourist season. Elsewhere along the coast, such as at Atlantic City, crowds would watch from the shore as the dark horizon at sea was suddenly illuminated. Another tanker had been hit.

Finally, some remedial action was taken. Outdoor lighting was doused on the eastern seaboard, and neighborhood wardens went on patrol to ensure that indoor lights were also out or—at least—the curtains drawn.[7]

Other steps were taken to counter the U-boat menace. Convoys were instituted along the East Coast, which gave tankers a greater degree of protection. But an even better option was to minimize the amount of oil that had to travel by tanker. An alternative to tanker shipment was put forward—the construction of a pipeline on a scale never before attempted, and of an unheard-of length, stretching from Texas to the East Coast. Clearly, the steady movement of oil through a pipeline at five miles per hour would be far safer than using seagoing tankers and far cheaper than using railroad tank cars. Initially turned down in the autumn of 1941 on grounds that it would use too much steel, the pipeline project, dubbed Big Inch, was hastily resuscitated after Pearl Harbor and tanker sinkings in American waters.

Construction finally began in August 1942, and what followed was one of the extraordinary feats of engineering in World War II. Nothing like it had ever been done before. The oil transportation and construction industries were mobilized to build a pipeline that would carry five times as much oil as a conventional one, a conduit that would stretch halfway across the country and require a plethora of newly designed equipment. Within a year and a half, by the end of 1943, Big Inch, 1,254 miles long, was carrying one-half of all the crude oil moving to the East Coast. Meanwhile, Little Inch—an even longer line, 1,475 miles—was built between April 1943 and March 1944 to carry gasoline and other refined products from the Southwest to the East Coast. At the beginning of 1942, just 4 percent of total supplies had arrived on the East Coast by pipeline;

357

by the end of 1944, with Big Inch and Little Inch both completed and in operation, pipelines were carrying 42 percent of all oil.[8]

In the spring of 1942, however, construction had not yet begun on Big Inch; and the other measures taken against German U-boats hardly afforded any security. Moreover, the Allies faced a very determined and wily opponent in the ice-blooded Admiral Karl Doenitz, commander of the German submarine force. Anything was fair game for him. "Rescue no one and take no one with you," he ordered his crews. The main objective of his growing fleet, he said, was "to destroy more enemy tonnage than can be replaced by all Germany's enemies put together." The Germans also gained two other very significant advantages. They changed their code procedures, so that the British lost the ability to read the U-boat signals; and at the same time, they broke the ciphers that governed the movement of the Anglo-American convoys. The resulting destruction of Allied shipping was appalling. Once again looming before the Allies was one of their greatest fears: the choking off of the absolutely essential oil supplies to Britain from the Western Hemisphere.

The Battle of the Atlantic became even more dangerous in the second half of 1942. Improved and larger U-boats joined the German fleet, with a greatly expanded cruising range, greater depth capabilities, improved communications, and access to many of Britain's ciphered convoy signals. In addition, Admiral Doenitz introduced *Milchkuhs* ("milk cows"), large underwater supply ships that could deliver diesel fuel as well as fresh food to the U-boats. The Allied losses at sea mounted. Month by month, Britain's supply picture grew worse. For its part, the United States lost a quarter of its total tanker tonnage in 1942. Petroleum stocks were well below required safety levels in Britain, and London saw sharply rising demand ahead, both because of the requirements in North Africa and because of the prospect of an Allied invasion of Europe. Stalin, too, was making increasingly insistent demands for more oil.[9]

In mid-December, Churchill was told that there remained only about two months' supply of fuel oil for ships, aside from a last-gasp emergency reserve. "This does not look at all good," was his bleak comment. Naval forces were stretched very thin trying to protect the transatlantic shipping. In January, Churchill left England for Casablanca, where he and the British Chiefs of Staff met Roosevelt and the American Chiefs. The main topic of the rancorous discussion was an invasion of the European mainland. But all were agreed on one point, which was summed up by General Alan Brooke, Chief of the Imperial General Staff. "The shortage of shipping," he said, "was a stranglehold on all offensive operations, and unless we could effectively combat the U-boat menace we might not be able to win the war."

Though the defeat of the German U-boats was a top Allied priority in 1943, the actual situation showed no quick sign of improvement. By the spring of that year, British oil stocks were at their lowest levels yet. In March, again operating with virtual impunity, the U-boats sank 108 ships. The Atlantic was so infested with the enemy submarines that evasion seemed impossible. "The Germans," said the British Admiralty, "never came so near to disrupting communication between the New World and the Old as in the first twenty days of March, 1943."

But in the last days of March, the scales tipped dramatically, and just in time. First, there was a decisive shift in the cryptanalytic balance; the Allies thoroughly broke the new U-boat codes, while at the same time they successfully closed off their own convoy ciphers to the Germans. Then the British and Americans added a new coordinated counteroffensive capability to the convoy system that included support groups designed to attack the U-boats. The Allies further improved their radar, and they also introduced newly developed long-range aircraft that could, at last, provide coverage over that part of the Atlantic previously inaccessible to air protection. The tables were abruptly and finally turned. In May 1943 alone, 30 percent of the U-boats at sea were lost. A chastened Admiral Doenitz was forced to report to Hitler: "We are facing the greatest crisis in submarine warfare, since the enemy, by means of new location devices . . . makes fighting impossible, and is causing us heavy losses." On May 24, Doenitz ordered the U-boats to withdraw to safer areas. Though he denied it at the time, he was calling off the submarine campaign in the North Atlantic. Allied convoys—carrying their vital cargoes of oil, other goods, and troops—could now cross the Atlantic in reasonable safety.

A combination of technical innovation, intelligence, organization, new tactics—and persistence—finally assured an abundant flow of oil from America to Britain and on to Europe and the Soviet Union. The path was now clear for a two-front assault on Hitler's Fortress Europe. After forty-five months of deadly warfare and mounting peril, the Battle of the Atlantic was over.[10]

Domestic Push

While the safety of oil transport had been the focal point of the war at sea, Harold Ickes was also strenuously trying to promote higher oil production in the United States. His hand was strengthened when he was promoted from Petroleum Coordinator to Petroleum Administrator for War. Since he was also still Secretary of the Interior, the Old Curmudgeon now wielded unprecedented power. Yet that power was far from absolute. Some forty or so other Federal agencies had a say in one part or another of the oil industry, and Ickes's Petroleum Administration for War (PAW) was in a constant struggle with a number of them—particularly the War Production Board, which allocated steel and other materials; the Office of Price Administration, which set prices; and the War Shipping Administration, which controlled tankers. Ickes made continuing appeals to Roosevelt to muzzle the competing czars at the other wartime agencies and uphold his own authority.

Ickes was further constrained by the unwillingness of the American military to share in any detailed way its projected needs with the PAW. The British, who observed and reported on this reluctance to London, were surprised and puzzled. But the heart of the matter was simple—the American military did not trust civilians to keep secret its projected needs, from which could be inferred its plans. Ickes, caught in the middle of this conflict, looked at the British system with some envy. "On any oil matter, the British Government is a unit— Parliament, the Administration, the oil companies, the press," he explained. "Over here, to the contrary, everyone is in everyone else's hair. There is no unity.

The British know it. They can't help knowing it. Congress is investigating all the time."

Despite all these roadblocks, the PAW gradually molded an effective government-industry combine. It sought an antitrust exemption from the Justice Department, vitally needed so that oil companies could talk to one another and coordinate operations and pool supplies. Justice, still in the midst of its antitrust suit against the major companies, was exceedingly reluctant to grant the exemption, but pressure was exerted by the White House until it finally found forgiveness and understanding in its juridical soul. About three-quarters of the executive and technical management of the PAW actually came out of the industry, which led, not surprisingly, to a good deal of criticism of Ickes. But he insisted that it was necessary to have competent people who understood how the oil business was conducted. The PAW was "flanked" by national and regional committees, organized by function (production, refining, and so forth) and also drawn from oil industry executives and managers. Thus, there was a two-way system of communication through which industry activities were directed and monitored.[11]

Overall, there was widespread support for the mission of the PAW, based upon the growing awareness of how critical oil would be to the war. Even so, the entire supply system frequently veered to the edge of shortage. At one point in the middle of winter, in February of 1944, New York was down to just two days of fuel oil supply. But each time, outright shortage was averted through adept emergency coordination and action by the PAW, and there was never a serious supply crisis in the United States—all the more remarkable considering how hard the system was worked.

Key to the success of the system, of course, was the availability of crude oil. While the United States entered the war with a large shut-in spare production capacity, no one could be sure how high the military demand would rise or how long the war would last. Moreover, concern was growing about America's oil reserve base. There was no room for complacency or even for much confidence. Thus, the PAW strove to increase production and maintain and expand capacity. It used its power to deny or grant allocation of drilling supplies to force oil men to adopt improved petroleum engineering methods. It also fought to assure that explorers could deduct drilling costs as intangible expenses from their taxes, as one way to encourage additional exploration.

But the PAW's biggest battle, on the production side, was to get the Office of Price Administration to raise prices to stimulate exploration and production activity. There the PAW met only limited success, winning increases in prices for California heavy oil, to stimulate production to meet the U.S. Navy's needs in the Pacific, and for stripper wells, which produced less than ten barrels per day. But on the central issue, the Office of Price Administration, fearful of inflation, rebuffed Ickes's efforts to get an across-the-board thirty-five cent increase in all oil prices, above the official $1.19 ceiling. The entire struggle over prices, as could be expected, left the industry with a deep distaste for the Office of Price Administration; one oil-industry spokesman dismissed it as a "commie outfit."

Whatever the complaints, America's overall production record was very good: from 3.7 million barrels per day in 1940 to 4.7 million barrels per day in

1945—a 30 percent increase. With the estimated surplus capacity of 1 million barrels in 1940, the United States essentially called on its full reserve. But that was a more difficult task than it appeared, for when the oil field workers opened the valves all the way on the wells, they often found that actual production capacity was less than had been estimated. Moreover, existing wells naturally declined in output. So the industry had to work very hard to get production up and then to keep it there. And, to stay even, the industry had to maintain a high level of exploration. Altogether, between December 1941 and August 1945, the United States and its allies consumed almost 7 billion barrels of oil, of which 6 billion came from the United States. Its wartime output was equivalent to more than a quarter of all oil produced in the United States from the time of Colonel Drake's well right down to 1941. Even so, if the total Allied call on United States oil had grown larger, there would have been real pressure on the available supplies.[12]

Rationing—Through the Side Door

The other side of the oil equation in the United States was consumption, and here occurred the biggest political battles. Efforts were made to get industrial users to switch from oil to coal. Homeowners who heated with oil were asked to keep their houses at sixty-five degrees during the day and fifty-five at night. President Roosevelt himself took a strong interest in the potential of America's then largely underutilized natural gas resources. "I wish you would get some of your people to look into the possibility of using natural gas," he wrote to Ickes in 1942. "I am told that there are a number of fields in the West and the Southwest where practically no oil has been discovered but where an enormous amount of natural gas is lying idle in the ground because it is too far to pipe to large communities." But gasoline was the focus of contention. There were those who went out of their way to cooperate with the national need to reduce gasoline consumption. Never to be forgotten was the altruism of one Bea Kyle, a daredevil who worked at the Palisades Amusement Park in New Jersey. In 1942, she wrote to Harold Ickes to explain her line of work: "First I cover myself with gasoline and then gasoline is poured on top of the water in my portable tank after which both are ignited and I leap into the flaming tank." She wanted Ickes's honest opinion whether her eighty-foot plunge "worked against defense" and should, therefore, be put on hold until the war was over.

"Without detracting materially from the spectacular character of your diving act," an aide to Ickes wrote back helpfully, "you might be able to use a little less gasoline in your performance, or make a few less dives with the same result, so that it will reduce your consumption in the same proportion as recommended generally." And, the aide added, "Your patriotic interest is appreciated."

There were not many Bea Kyles. Gasoline use had become a national birthright over the preceding three decades, and few were willing to give it up unless they were forced to do so. In the spring of 1942, the first step was taken in that direction: A complete ban was imposed on the use of gasoline for auto racing. Then, in May, rationing was imposed on the East Coast—initially through the use of cards, which, like meal tickets, were punched at the service station.

The cards were succeeded by coupons. Whatever the system, it stirred a great outcry of protest from all sides. The governor of Florida phoned Ickes and pleaded with him to postpone rationing so as not to crimp the tourists. People on the East Coast, who did not understand the problems of logistics and petroleum transportation, "knew for sure" that there were full tanks elsewhere in the country. The Roosevelt Administration was reluctant to blanket the whole country with rationing. In the wide open spaces of the West, there were not that many easy alternatives to the automobile for transportation.[13]

The Administration finally found a way into nationwide rationing through a side door—rubber. The Japanese capture of the East Indies and Malaya had cut off 90 percent of the natural rubber exported to the United States, while the synthetic rubber program had hardly begun to be implemented. As a result, the United States was gripped by a "rubber famine." By rationing gasoline, and thus restricting driving, the civilian demand for tires could be cut, freeing the available rubber supplies for the armed forces. Therefore, in the name of rubber, gasoline could be rationed. But such a step, even in disguise, was something that had to have a very serious imprimatur. Thus, Roosevelt appointed an extremely august committee to sell the idea to Congress and to the public; the two members were the presidents of Harvard and MIT, and as chairman, none other than the venerable and revered Bernard Baruch.

There could not have been a more appropriate choice for this public relations job than Baruch. In Washington, Bernard Baruch was considered very serious business. He had enormous prestige; the Wall Street millionaire had been the great industrial mobilizer of World War I, and was now an adviser to Presidents and the nation's semi-official Elder Statesman. "Public deference was universal," recalled John Kenneth Galbraith, who, in his position as the nation's chief price controller, tangled with Baruch. "At the same time private skepticism was also nearly obligatory." The skepticism extended up to the Oval Office and to the man who had appointed Baruch to head the committee—"that old Poohbah" is the way Roosevelt had once described Baruch.

Nevertheless, Baruch could do the political job. He assured his ivory tower committeemen, the two university presidents, that he could take care of the practical problem—Congress. "Let me handle the Senators and fellows on the Hill, they're mostly good friends of mine," he said. "I'll give a dinner for them some evening." Many of the key Congressmen were not only friends but also regular recipients of what were, by the customs of the times, significant campaign contributions from Baruch, and they, in turn, were totally persuaded of his sagacity. The strategy worked. In September 1942, Baruch's committee vigorously recommended that gasoline should be rationed on a nationwide basis to conserve rubber. The actual plan was, however, not implemented until after the 1942 Congressional elections. As it was, one hundred Congressmen from the West staged a protest against the new system. Presumably, they had not been invited to dinner.[14]

Rationing was supplemented by other measures, including a thirty-five-mile-per-hour speed limit. There was even more outrage, however, when "nonessential driving" was banned in January 1943. But since no one could come up with a good definition of "nonessential driving," that particular prohibition was abandoned a few months later. The rationing system provided for five grades of allocation, depending on need and function of the vehicle and driver. The alphabetical stickers displayed on automobile windshields became a status symbol for the lucky motorists whose driving was considered essential. The most fortunate were those with an X—doctors, clergymen, some repairmen, and government officials—who had unlimited rights to purchase gasoline. A degree of shame was felt by those who fell into categories deemed less important to the war effort. The "basic" A ration, which was what most people got, provided— depending on the availability of supplies and the region—for anywhere between one and a half and four gallons a week. Not surprisingly, the system gave rise to a black market in coupons, real and counterfeit, particularly in the big cities on the East Coast. Still, civilian gasoline use was substantially reduced; the average consumption per passenger vehicle was 30 percent less in 1943 than in 1941. Ickes was right; Americans, who had balked at voluntary conservation, accepted enforced rationing of gasoline, along with restrictions on their consumption of sugar, butter, and meat. After all, as was said so often, "There's a war on."[15]

The organization of production and consumption of oil in the United States was only part of the larger international system, jointly improvised and managed by the United States and Britain. It was a system that took crude oil from the American Southwest, refined it and moved it to the Northeast by ship or tank, or later by pipeline, transported it across the Atlantic, and then saw that it was delivered where it was needed, whether to the storage tanks at the air bases in Britain, or into the five-gallon cans that went to the Allied fighting men at the front, or into the railroad tank cars at Murmansk and Archangel, the Soviet Union's ports on the Barents Sea. No less urgent were the needs of the Pacific theater, which had to be supplied in a similar procession moving westward. The Americans and the British ran this system through a number of formal and informal arrangements. They worked on the principle that, in each theater of war, one or the other would have complete responsibility for supplying the troops and air forces of both countries. Thus, in the United Kingdom and the Middle East, the British filled the Americans' gas tanks; in the Pacific and in North Africa, after the Allied invasion in late 1942, the United States had responsibility for fueling all forces.

The problems of coordination in a global war were immense. Supplies had to be allocated among the hotly contending priorities: Europe, North Africa, the Pacific, and the domestic American economy. Tankers had to serve competing demands of the Atlantic and the Pacific, and of the East Coast of the United States. Moreover, transport and supplies had to be matched, and there were constant and costly mix-ups; tankers would arrive at ports and there would be no ready supplies, or the supplies would be waiting and no tanker would appear. Yet, as difficult and contentious as all these matters were, the system that evolved served the Allies very well.[16]

Innovation

Before World War II, the American military had not anticipated that petroleum supplies might pose any particular problem. The Army did not even have any records of how much oil it used. There was incomplete comprehension of how fundamental would be the difference between the First World War and the Second. The former was a static war; the latter, a war of motion. (It was Stalin who had offered a toast at a banquet in Churchill's honor in the darkest days of the war: "This is a war of engines and octanes. I drink to the American auto industry and the American oil industry.") And thus, it would be a war of much greater petroleum consumption; at the peaks, the American forces in Europe used one hundred times more gasoline in World War II than in World War I. The typical American division in World War I used 4,000 horsepower; in World War II, 187,000 horsepower.

Indeed, it was only the planning for the 1942 invasion of North Africa that opened the Army's eyes to the full significance of the petroleum factor, and thereafter, in response, a centralized, disciplined supply organization came into being. After all, about half the total tonnage shipped from the United States during the war was oil. The Quartermaster Corps calculated that when an American soldier went overseas to fight, to support him in his job he required sixty-seven pounds of supplies and equipment, of which half was petroleum products.

The Army's new oil supply organization pushed a number of innovations to facilitate the flow and use of petroleum. It moved to standardize products—specifically, one all-purpose motor fuel and one all-purpose diesel fuel. It adopted a special narrow portable pipeline system, complete with pumps, developed by Shell, which enabled oil, in a combat area, to be transported efficiently toward the front, rather than be moved by truck. But one of the most significant developments was the eventually ever-present five-gallon gasoline can. The Army discovered that its ten-gallon can was unwieldy and too heavy to be handled by one man. The Germans used a five-gallon can. In the quest for a more manageable receptacle, the Americans, joined by the British, based their design for a five-gallon can on captured German cans. Ironic respect was paid to the German original in the nicknames they adopted; the "blitz can"; and, more commonly, the "jerrycan." But the Americans made an important innovation in the German design. The German troops had to use a funnel, which allowed dirt to enter the engines of their vehicles. The Americans added a built-in pouring spout, which kept the dirt out.

One of the biggest technical disappointments of the war was PLUTO—the acronym for "Pipeline Under the Ocean." This underwater pipeline system was designed to link the British side of the English Channel to the French. The aim was, after the invasion of Western Europe, to supply fully half of the fuel needs for Allied forces on their advance through France and into Germany. The pipeline was built, but serious technical problems were compounded by installation mishaps. As a result, PLUTO's flow during the critical months after the invasion was no more than a trickle. On average, from D-Day in June 1944 to October 1944, just 150 barrels per day moved through PLUTO—a minuscule one-sixth of one percent of what Allied forces in Western Europe consumed during that period.[17]

Perhaps the most daunting challenge of all in the Allied fuel chain was the supply of 100-octane aviation gasoline. Developed in the early and mid-1930s, primarily by Shell researchers in Holland and the United States, 100-octane fuel made possible higher aircraft performance—greater bursts of speed, more power, quicker takeoffs, longer range, greater maneuverability—than the 75- or 87-octane fuels customarily used. Tests demonstrated a 15 to 30 percent increase in power over existing fuels, plus large fuel savings, which would make possible longer-range aircraft. But before the actual outbreak of war, there was no significant market for the much more expensive fuel, and in its absence, some companies, notably Shell, followed by Jersey, took major risks in making large investments in 100-octane research and capacity. Shell put much of the 100-octane gasoline that it did manufacture into storage.

But the eruption of war suddenly meant that there was a market—and an important one. The advantages of 100-octane gasoline were proved in the Battle of Britain in 1940, when the 100-octane-powered British Spitfires outperformed the Messerschmitt 109s, fueled on 87-octane gasoline. Some attributed Britain's critical edge and victory in that life-and-death air battle to the 100-octane gasoline. But special expensive refining facilities were required to make the high-performance fuel, and only small amounts of it were actually available. Production targets were set and then raised again and again. Two aviation petroleum committees—one in Washington and one in London—were established to control the disposition of the limited quantities of 100-octane gasoline among all the military claimants. Despite the chronic shortage, sometimes the allocators found they had to be wasteful. During the U-boat threat, they would dispatch three cargoes to a destination, with the hope that at least one would get through.

Almost all of the Allies' needs for 100-octane fuel had to be met by American production—almost 90 percent of the total by 1944. But production could not keep up with the demand. "From present outlook, the situation will get steadily worse," Undersecretary of War Robert Patterson wrote despairingly to Ickes in April 1943. "I see no relief in sight except the most drastic action." The Americans responded with a huge construction and engineering program, one of the largest and most complex industrial undertakings of the war. Fortunately, in the late 1930s, a new refining technology—catalytic cracking—was under development, principally by a Frenchman named Eugene Houdry and by Sun Oil. The first significant advance over the thermal cracking technique developed by William Burton three decades earlier, catalytic cracking facilitated the production of large amounts of 100-octane gasoline. Without that technology, the United States simply could never have hoped to come anywhere close to meeting the demand for aviation fuel. But when the United States entered the war, limited catalytic cracking operations were only just beginning, and seemed beyond doing on a large scale. The requirements were huge—the units would be up to fifteen stories tall and vastly more expensive than traditional refining installations. Yet thereafter, a multitude of catalytic cracking units were built across the country in double time, with what seemed to be hardly any lapse in going from initial design and pilot plant experimentation to full-scale operations.

Altogether, as part of the 100-octane campaign, dozens and dozens of plants and special facilities were built, and many existing ones were converted to produce 100-octane fuel. The PAW and the oil industry constantly struggled with competing agencies and interests for the steel and other goods necessary to meet their construction goals—which were continually raised, as more and more output was required. On top of this, all the aviation fuel plants had to be melded together, to be run as one giant integrated combine, with the various components shifted around the country, among different companies, in order to maximize output and, in the words of Ickes's agency, "eke out the greatest possible number of barrels of product." Continued improvements in production processes and in the fuel itself were pushed. As a result, pilots gained additional surges of power with which to overtake enemy aircraft, and heavily laden bombers were able to get off the runway.

Though, recurrently, it seemed that the Allies were about to run out of 100-octane fuel, increased production miraculously kept pace with rising need. Demand by 1945 was seven times higher than had been projected at the beginning of the war. Yet that requirement was met. By 1945, the United States—which had total production capability for less than 40,000 barrels per day in 1940—was producing 514,000 barrels per day of 100-octane fuel. As one general explained, government and industry "squeezed it out of a hat."[18]

"The Unforgiving Minute"

"At no time did the Services lack for oil in the proper quantities, in the proper kinds and at the proper places," the Army-Navy Petroleum Board pronounced proudly after the war. "Not a single operation was delayed or impeded because of a lack of petroleum products." Though that judgment was for the most part true, there was an exception—one shuddering moment when the system failed.

By the spring of 1944, the pendulum was clearly swinging in the Allies' favor in the war against Germany. American and British troops had landed in Italy, which proceeded to quit the war. The Russians were driving in on the eastern front. Then on June 6, 1944, D-Day, Allied forces hit the beaches of Normandy, opening the invasion of Western Europe. But immediately, the long and carefully wrought plans of the Allies went awry. The invading armies found themselves, contrary to plans and expectations, pinned in Normandy much longer than they had expected. The Germans, very much taken by surprise, managed to hold the invading forces for a time, even though shortages of fuel greatly constrained their ability to move reinforcements rapidly up to the front. Field Marshal Gerd von Rundstedt, the German commander, was forced to issue the order: "Move your equipment with men and horses—don't use gasoline except in battle." Then, on July 25, 1944, the Allied armies finally burst out of the German ring, and the Germans, disorganized and undersupplied, fell back. Now it was the Allies who were surprised—but this time by the ease and speed with which they could rush forward in hot pursuit of the Germans.

No force moved more hotly in pursuit than the Third Army under the leadership of General George Patton, Jr., who led the breakout. Driven, dynamic, impulsive, and volcanic in his anger (the last perhaps the result of injuries to his head while playing polo), Patton had been barely able to contain himself in the face of what he regarded as timid and overcautious Allied strategy since the June 6 landing. During July 1944, he had written a poem expressing his frustration:

For in war just as in loving you must always keep on shoving
Or you'll never get your just reward . . .
So let us do real fighting, boring in and gouging, biting.
Let's take a chance now that we have the ball.
Let's forget those fine firm bases in the dreary shell-raked spaces,
Let's shoot the works and win! yes win it all.

General Dwight Eisenhower, the Supreme Allied Commander, would publicly describe Patton as "a great leader for exploiting a mobile situation." Privately, while acknowledging his considerable strengths as an operational commander, Eisenhower believed Patton lacked that critical ability required of an overall commander—to see the big picture. Indeed, Eisenhower questioned Patton's ability to function in a team, and even his balance. Patton was too ready to gamble, too prone to "ill-advised action," said Eisenhower. "I am thoroughly weary of your failure to control your tongue," he warned Patton directly, "and have begun to doubt your all-round judgment, so essential in high military position."

Yet, despite his deep reservations, Eisenhower most definitely wanted Patton for the invasion. As he wrote to General Marshall, Patton had fighting qualities that "we cannot afford to lose unless he ruins himself." So long as he was "under a man who is sound and solid, and who has enough sense to use Patton's good qualities without being blinded by his love of showmanship and histrionics," he would do fine. In short, Patton represented a special form of insurance—embodied in "the extraordinary and ruthless driving power that Patton can exert at critical moments." For, Eisenhower added, "There is always the possibility that this war, possibly even this theater, might yet develop a situation where this admittedly unbalanced but nevertheless aggressive fighting man should be rushed into the breach." And, thus, to save the day.[19]

Certainly Patton's force of personality, his determination, his ability to project purpose and confidence, and his "winningness" all made him a superb leader in the field, and if his character did not necessarily win the confidence of his superiors, it did engender fierce loyalty in the troops under his command. He knew the importance of creating a legend about himself—be it the two revolvers, one pearl-handled, that Patton wore on his hips, or the nickname "Old Blood-and-Guts" that he had bestowed on himself in his unsuccessful bid in the 1930s to become Commandant of Cadets at West Point. But beneath the tough and profane exterior and the iron self-discipline, there was the Patton whose stomach knotted up before battle and who had published two volumes of poetry.

Patton was as much a master of mobile warfare as Rommel and he fretted while waiting his chance for glory. "I must get in and do something spectacularly successful if I am to make good," he complained. And he did, proving Eisenhower right about his special talents. Revolvers on his hips, he led the breakout from Normandy at an astonishing speed; he covered a vast territory in a month—almost five hundred miles from Brest to Verdun, liberating most of France north of the Loire. Like Rommel, Patton was contemptuous of the quartermasters. His forces pursued any unconventional methods they could think of to ensure their fuel supplies, which became increasingly short as the Third

Army's lines were extended. Some of Patton's troops impersonated members of different units to get supplies; others hijacked trains and truck convoys or commandeered fuel that supply trucks needed for the return journey. Indeed, Patton even sent spy planes flying back to the rear to spot gasoline supplies that could be seized.

Toward the end of August 1944, however, fuel was becoming a very serious constraint on the Allied advance. There was no physical shortage of gasoline in France. The supplies were simply in the wrong place—back in Normandy, far behind the lines—and there was an immense logistical problem in getting the fuel to the front. In the parlance of supply, the Allies had moved "260 logistical planning days" in just twenty-one days. Railways would have been the efficient way to move fuel forward, but there were no appropriate lines. The endless convoys of fuel trucks, routed in a special one-way system right across France, could not keep up; as the supply lines lengthened, the trucks had to use proportionately more and more of their own fuel supply to get to the front and back. In consequence of their logistical problems, the fast-moving Allied armies simply outran their gasoline supplies. The same thing had happened to Rommel when his forces had raced across North Africa in 1942. Patton fumed about the situation. "At the present time," he wrote to his son on August 28, "my chief difficulty is not the Germans but gasoline. If they would give me enough gas, I could go anywhere I want." The next day, he noted in his diary, "I found that, for unknown reasons, we had not been given our share of gas—140,000 gallons short. This may be an attempt to stop me in a backhand manner, but I doubt it."[20]

Despite Patton's suspicion, the other units were also short of fuel. And, at that moment, Eisenhower, as overall commander of the Allied forces, faced a critical decision: whether to direct the bulk of available supplies to Patton's Third Army or give the fuel to the United States First Army, to the north of the Third Army, in support of the British Twenty-first Army Group, under General Montgomery, which was closest to the coast. Was this the moment, Eisenhower had to ask himself, to forsake his own "broad front" strategy—all flanks protected—and instead go for broke and let Patton and the Third Army try to punch through the Siegfried Line, the Nazi's West Wall, into Germany itself? Or was it more prudent to let Montgomery first capture Antwerp, to assure a first-class supply port and avoid further attenuating the supply lines? There was still a third choice, the one that Montgomery argued vigorously for—a huge forty-division phalanx under his own command that would break through to the Ruhr and grind Germany into defeat.

As Eisenhower struggled with his decision, Patton was raring to go. "We have, at this time, the greatest chance to win the war ever presented," he wrote in his diary. "If they will let me move on . . . we can be in Germany in ten days. . . . It is such a sure thing that I fear these blind moles don't see it." But Eisenhower, attuned to the larger requirements of politics and coalition warfare and in particular to the tense relationship with the brittle Montgomery, decided on a compromise, a splitting of forces, and one that sent the critically needed gasoline to the First Army, in support of Montgomery, and not to Patton's Third Army.

Down to a half-day supply of gasoline, Patton was furious. He appeared, "bellowing like an angry bull," at the headquarters of General Omar Bradley, commander of the American forces. "We'll win your goddam war if you'll keep Third Army going," he roared at Bradley. "Dammit, Brad, just give me 400,000 gallons of gasoline, and I'll put you inside Germany in two days."[21]

Patton would not easily accept the limitation on his supplies. This was the critical moment, the one opportunity to push and shove, to ruthlessly drive on, to end the war quickly, to meet his destiny—and glory. He could barely contain his anger and frustration. "No one realizes the terrible value of the 'unforgiving minute' except me," he wrote in his diary. "We got no gas because, to suit Monty, the First Army must get most of it." He ordered his units to drive on until they ran out of fuel "and then get out and walk." He wrote to his wife: "I have to battle for every yard but it is not the enemy who is trying to stop me, it is 'They'. . . . Look at the map! If I could only steal some gas, I could win this war."

On August 30, the supply to the Third Army was reduced to less than a tenth of its normal level. Nor, it was told, would it get any more until September 3. The next day, August 31, Patton's forces reached the Meuse River. The Third Army could go no further. Its gas tanks were empty. "My men can eat their belts," Patton told Eisenhower, "but my tanks have gotta have gas."

Montgomery's forces captured Antwerp on September 4. "I now deem it important," Eisenhower noted in his own diary the next day, "to get Patton moving once again." And, thereafter, Patton did get more fuel. But the minute passed was to prove most unforgiving; the few days that had elapsed had given the Germans critical time to regroup. At the very beginning of September, Hitler at last modified his "no retreat" order so that German units could fall back, reform, and establish a defensive line. Patton's forces pushed beyond the Meuse, but were stalled at the Moselle River—now no longer by lack of gasoline but by the much stiffened resistance of the Germans. Nine months of bitter and costly warfare would follow. And when the Germans mounted one massive last-ditch counter-attack, the Russians, not the Americans, took Berlin.

In the final months of the war, Patton pushed through Germany and rode on as far as Pilsen, in Czechoslovakia. Yet the "unforgiving minute" had denied him his ultimate moment of glory on the battlefield. In December 1945, eight months after the end of the fighting in Europe, the life of the master of mobile warfare came to an unglorious end when his chauffeured limousine collided with a U.S. Army truck on a German road.[22]

Did the Allies let slip the critical opportunity to bring the war to a swift conclusion? The question was bitterly debated at the time, and long after. Of the million casualties the Allied forces suffered in liberating Western Europe, fully three-quarters occurred after the September check on Patton's advance. Many millions more died as a result of military action and in the German concentration camps in the last eight months of the war. Moreover, if the Allies had broken through into Germany from the West earlier, the postwar map of Europe would have been drawn quite differently, for Soviet power would not have projected so far into the heart of Europe.

For Eisenhower, it had been an immensely difficult decision, one made in fleeting time, with poor information, in the face of high uncertainty and high risks. The cost of acceding to Patton could have been very great, threatening the very foundations of the Allied coalition at a critical moment, leaving the entire

Allied Army in a poorly supplied position, and putting the Third Army into a highly exposed position. There were already reports of a German Army building on Patton's flank. In his memoirs of the war in Europe, Eisenhower replied, with diplomatic tact but nevertheless pointedly, to Patton's accusations that he had made the wrong decision. Patton simply could not see the big picture. To Eisenhower, the overall risks had been enormous, and the likelihood of failure of Patton's plan was too great. "In the late summer days of 1944 it was known to us that the German still had disposable reserves within his own country," he wrote. "Any idea of attempting to thrust forward a small force, bridge the Rhine, and continue on into the heart of Germany was completely fantastic." Even if it had gotten through, such a force would have grown smaller and smaller as it dropped off units to protect its flanks. Eisenhower stood by the judgment he had made in those last days of August 1944: "Such an attempt would have played into the hands of the enemy" and the result, for the Allies, would have been "inescapable defeat."[23]

Others who examined the evidence came to a different conclusion: that the mistake was in splitting the armies, instead of concentrating all Allied forces under Montgomery and relentlessly pushing through to the Ruhr and on to Berlin. Patton and his troops would have been a powerful part of that huge phalanx. If that push had been successful, the result would have been an early end to the carnage in Europe.

One who gave long thought to the entire question was Basil Liddell Hart, the eminent British military strategist and historian. It was his writings after the First World War, with the concept of the "expanding torrent," that gave him claim to being a father of mechanized, mobile warfare and the inspirer, ironically, of the *blitzkrieg*. Shortly before his death in 1970, Liddell Hart delivered his considered judgment about Patton's strategy. He agreed with Patton; those days at the end of August 1944 had been the "unforgiving minute." The Germans were still in shock, unprepared; not one bridge on the Rhine had yet been readied for demolition. A powerful punch by Patton—shooting the works, in the words of Patton's poem—could well have caused the disintegration and defeat of the defending German armies. "The best chance of a quick finish," Liddell Hart concluded, "was probably lost when the 'gas' was turned off from Patton's tanks in the last week of August, when they were 100 miles nearer to the Rhine, and its bridges, than the British."[24]

PART IV

THE HYDROCARBON AGE

CHAPTER 20

The New Center of Gravity

IT WAS KNOWN AFTERWARD, in corporate lore, as the "Time of the Hundred Men." These were the war years, when the number of American oil men working in Saudi Arabia was reduced to a hundred or so, cut off for most of the time from the rest of the world, and the development of Saudi oil was forgotten amid the global clash of arms. In late 1943, the "hundred men" were joined by another— Everette Lee DeGolyer, whose arrival was a sure signal that Saudi Arabia had *not* been forgotten by those who were thinking about the future, when the war would be over.

No man more singularly embodied the American oil industry and its far-flung development in the first half of the twentieth century than DeGolyer. Geologist—the most eminent of his day—entrepreneur, innovator, scholar, he had touched almost every aspect of significance in the industry. Born in a sod hut in Kansas and raised in Oklahoma, DeGolyer had enrolled in a geology course at the University of Oklahoma, to avoid Latin, and thus by accident set the course of his life. While still an undergraduate, he took time off to go to Mexico, where, in 1910, he discovered the fabulous Portrero del Llano 4 well. It blew in at 110,000 barrels per day, inaugurating the Golden Lane and the golden age of Mexican oil. It was the biggest oil well ever discovered, and it laid the basis both for the oil fortune of the Cowdray/Pearson interests and for DeGolyer's unique and everlasting reputation.

That was only the beginning. DeGolyer was more responsible than any other single person for the introduction of geophysics into oil exploration. He pioneered the development of the seismograph, one of the most important innovations in the history of the oil industry, and he championed its use, to such a degree that he was said to be "crazy with dynamite." The chief geologist of Standard Oil of New Jersey said admiringly of DeGolyer that "his interest in oil finding is actively with him day and night." DeGolyer put together a successful independent oil company, Amerada, on behalf of the Cowdray interests, and then, though courted by Standard of New Jersey, he instead went off on his own and in the late 1930s established DeGolyer and McNaughton, which became the world's leading petroleum engineering consulting firm. This, too, was an innovation, for it met the new need for independent appraisal of the value of petroleum reserves to serve as a basis for financing by banks and other investors.

DeGolyer was already a millionaire several times over by his mid-forties, and thereafter his earnings averaged $2 million a year. Eventually, he grew bored with making money and gave a lot of it away. Indeed, DeGolyer's interests were much broader than just oil and money. He was a founder of what became Texas Instruments. He was a considerable historian of chili. He built an extraordinary collection of books. He bailed out the *Saturday Review of Literature*, when it was about to go bust, and became its chairman, though he never did care much for its politics.[1]

For many years, the short, plump, and dynamic DeGolyer, with his leonine head, was a familiar, highly respected figure in the councils of the oil industry, where his words carried much weight. And, of course, as a self-made man, DeGolyer had little use for the New Deal. But when the war broke out, he was called to Washington to be one of the chief deputies to Harold Ickes in the Petroleum Administration for War, and he came, though reluctantly. His charge was to help organize and rationalize production throughout the United States. In 1943, however, he was given a special foreign mission: to appraise the oil potential of Saudi Arabia and the other countries of the Persian Gulf, which by that time had emerged at the center of a critical and very contentious debate.

Three years earlier, in 1940, DeGolyer had made a speech on Middle East oil to a group in Texas. "No such galaxy of fields of the first magnitude over such a wide area has been developed previously in the history of the oil industry," he said. "I will be rash enough to prophesy that the area we have been considering will be the most important oil producing region in the world within the next score of years." Now, in 1943, he would have the chance to see for himself. Even so, he did not look forward to the trip. "A long time ago it seemed pretty important to me for some American to make this trip and size up the situation," he wrote to his wife. But, he added, "it is uncertain—likely to be uncomfortable and a little bit hazardous. I am no Lindbergh."

Getting to the Middle East was no easy thing in wartime. The first stop was Miami, where the plane blew a tire on landing. After waiting around for transport, DeGolyer and the members of the mission finally hitched a ride on military planes, over the Caribbean, to Brazil, to Africa, and finally to the Persian Gulf. Their itinerary would take them to the oil fields of Iraq and Iran, to Kuwait and then to Bahrain, and lastly to Saudi Arabia, to see the existing fields and to visit other structures that had been identified. After one stop, DeGolyer wrote his wife, "We haven't seen anything but a pretty barren land on this whole trip. . . . In fact Texas is a garden in comparison with some places we have been."

DeGolyer mastered the art of eating sheep's eyes when they were ceremoniously offered to him. And he noted many curious sights along his route. But it was the geology—the clues that his experienced eye detected as he toured the desert, the further hints that he extracted from the maps, the well reports, and the seismic work—that captured his imagination. Three structures had already been tapped in Saudi Arabia, with reserves estimated at 750 million barrels. But the identification of similar structures suggested that the reserves might be far larger. The same applied in the other countries along the Gulf. The physical hardships were worth the trouble, many times over. For DeGolyer was an oil man, and to him the barren desert of the Arabian Peninsula was an El Dorado, the stuff of legend. He was overcome by excitement, for he recognized that he was investigating something for which no precedent existed in the history of the oil industry. Even he, who had discovered a well that flowed at 110,000 barrels per day, had never seen anything on so vast a scale in the course of half a century in the business.[2]

When he came back to Washington in early 1944, DeGolyer reported that the proven and probable reserves of the region—Iran, Iraq, Saudi Arabia, Kuwait, Bahrain, and Qatar—amounted to about 25 billion barrels. Of that, Saudi Arabia accounted for about 20 percent—perhaps 5 billion barrels. He was a conservative man, and he applied the same standards for "proven" and "probable" reserves on behalf of the United States government as he would have in appraising the reserves for a bank. In fact, he suspected that the reserves would be much, much larger. And, indeed, estimates that sounded like lunacy—up to 300 billion barrels for the region and 100 billion barrels for Saudi Arabia alone—resulted from his trip. One of the members of his mission told officials in the State Department, "The oil in this region is the greatest single prize in all history."

More important than any specific numbers was DeGolyer's overall judgment of the significance of these huge reserves of oil: "The center of gravity of world oil production is shifting from the Gulf-Caribbean area to the Middle East—to the Persian Gulf area," he said, "and is likely to continue to shift until it is firmly established in that area." That judgment, delivered by a man with roots so deep in the American industry, constituted a eulogy for America's receding place in world oil—the end of its dominion. The United States was to produce almost 90 percent of the oil used by the Allies in World War II, but that was the high-water mark for its role as supplier to the world. Its remaining days as an exporter would soon disappear. Yet DeGolyer's words were more than just a eulogy. They were a prediction about a dramatic reorientation in the oil industry that would have a profound impact on the direction of world politics.[3]

"The Allies Have the Money"

The British government had long been actively involved in politics and oil production in the Middle East. The United States had largely ignored the area. The gingerly approach reflected the fact that, when all was said and done, oil production in the Middle East did not as yet amount to much. In 1940, the area including Iran, Iraq, and the entire Arabian Peninsula produced less than 5 percent of world oil, compared to 63 percent for the United States.

Even then, however, there were those who could see that the "center of gravity" was moving. In the spring of 1941, James Terry Duce, a vice-president of Casoc (as the California-Arabian Standard Oil Company was known), had written to DeGolyer that he was "getting a closer and closer look at the Persian Gulf" and that "fields such as we have in that area are a totally different experience to anything in the United States—even East Texas. The amounts of oil are incredible, and I have to rub my eyes frequently and say like the farmer—'There ain't no such beast.' "

But at that time the Axis powers were still on the offensive in Russia and North Africa, and the Middle East was in jeopardy. As a consequence, the dwindling number of Americans, "the hundred men" in Saudi Arabia, largely devoted themselves not to development, but to the opposite: planning how to protect the wells (by filling them in with cement) in case they were bombed, and how to destroy them in the event that "denial" to advancing German armies became necessary. For the same reason, wells were also plugged in Kuwait and Iran, all of which was carried out in coordination with British and American military and political authorities.

Even so, the American orientation toward Saudi Arabia and the Middle East was about to change. The trigger was much as it had been a decade earlier, at the beginning of the 1930s: another collapse in the pilgrimages to Mecca and a new financial crisis in Saudi Arabia. This time it was not an economic depression but the war that interrupted the flow of pilgrims. Things were made worse by a drought and the resulting crop failure. The traditional industries—sword and knife making, leather working—were hardly going to generate enough to make up for the losses. By 1941, Ibn Saud was once more confronting a stark financial crisis. The King had to face the harsh reality. As he explained to an American in 1942, "The Arabs have the religion, but the Allies have the money."[4]

So Ibn Saud once again had to appeal for help—to the British, in whose political sphere he operated, and to Casoc and its two American parents, Standard of California and Texaco. The oil companies did not want to make any further loans against future production, especially at a time when development itself was constrained, but neither did they want to risk losing the concession. Perhaps Washington would come to the rescue. Perhaps, it was further suggested, some help could be provided under Lend Lease, the wartime aid program. But Congress had authorized Lend Lease to be provided to "democratic allies." Unfortunately, Saudi Arabia was a kingdom and not a democracy, and unlike, say, the King of England, Ibn Saud was not exactly a constitutional monarch. Finally, after laborious discussion, Roosevelt decided against any American assistance at all. "Will you tell the British," he instructed one of his aides in July 1941, "I hope they can take care of the King of Saudi Arabia. This is a little far afield for us."

The British did come forward, providing, among other things, about $2 million worth of newly minted coins, and British subsidies would continue to grow substantially. But the American oil men worked hard to convince King Ibn Saud that this British aid was really American—since Britain, in its turn, was a recipient of American assistance. That meant, explained the oil men, that the aid actually came from the United States. It just came indirectly.[5]

376

"We're Running Out of Oil!"

America's entry into the war would be followed, in 1942 and 1943, by a wholesale redefinition of the importance of the Middle East, based upon a new outlook that completely gripped Washington, if not always the oil companies. Oil was recognized as the critical strategic commodity for the war and was essential for national power and international predominance. If there was a single resource that was shaping the military strategy of the Axis powers, it was oil. If there was a single resource that could defeat them, that, too, was oil. And as the United States almost single-handedly fueled the entire Allied war effort, putting an unprecedented drain on its resource, a fear of shortage began to grow. It was another of those periods of pessimism about the American oil position, similar to that at the end of the First World War, but etched, because of this war, in much greater urgency. What would a pervasive and lasting shortage mean for America's security and for its future?

The late 1920s and early 1930s had seen an explosive growth in the United States in discoveries and in additions to known reserves. But from the middle 1930s onward, while significant "revisions and additions" were made to existing oil fields, the discovery rate of new fields had fallen off very sharply, leading to the view that additions in the future would be more difficult, more expensive, and more limited. The precipitous decline in new discoveries transfixed and frightened those responsible for fueling a global war. "The law of diminishing returns is becoming operative," said the director of reserves for the Petroleum Administration for War in 1943. "As new oil fields are not being formed and as the number is ultimately finite, the time will come sooner or later when the supply is exhausted." For the United States, he added, "the bonanza days of oil discovery, for the most part, belong to history."[6]

Secretary of the Interior Harold Ickes shared that view, and the title of an article he published in December 1943 left no one in doubt where he stood—"We're Running Out of Oil!" In the article, the Old Curmudgeon warned ominously that "if there should be a World War III it would have to be fought with someone else's petroleum, because the United States wouldn't have it . . . America's crown, symbolizing supremacy as the oil empire of the world, is sliding down over one eye."

Such gloomy analyses could lead to only one conclusion. Although oil was then flowing out from American ports to all the war fronts, the United States was destined to become a net importer of oil—a transformation of historic dimensions, and one with potentially grave security implications. The wartime gloom about America's recoverable oil resources gave rise to what became known as the "conservation theory"—that the United States, and particularly the United States government, had to control and develop "extraterritorial" (foreign) oil reserves in order to reduce the drain on domestic supplies, conserve them for the future, and thus guarantee America's security. Even private-enterprise Republicans were calling for direct government involvement in foreign oil concessions. For, declared a prominent Republican Senator, Henry Cabot Lodge, "history does not give us confidence that private interest alone would adequately safeguard the national interest." And where were these foreign reserves to be found?

There was only one answer. "In all surveys of the situation," Herbert Feis, the State Department's Economic Adviser, was to say, "the pencil came to an awed pause at one point and place—the Middle East."[7]

Thus, American policymakers arrived, belatedly, at the same conclusion that had directed British oil policy ever since the end of World War I—the centrality of the Middle East. And there, however trusting their overall wartime collaboration, the two Allies certainly viewed each other with considerable suspicion. The British were fearful that the Americans would try to eject them from the Middle East and deny them even the oil reserves currently under their control. The region was considered central to imperial strategy and to the governance of India. Ibn Saud, as keeper of the holy places of Islam, was a person of very great importance to Britain, which, in India, ruled over the largest number of Moslems of any country in the world. He might also prove to be a most important factor in Britain's efforts to find a way out of its dilemma in Palestine, then a British mandate torn by mounting strife between Jews and Arabs.

Both American oil companies and government officials were deeply worried throughout the war that the British had a nefarious design to somehow preempt the United States when it came to Middle Eastern oil and exclude the American companies, particularly from Saudi Arabia. When the British sent a locust-control mission to Saudi Arabia, Casoc was absolutely sure that it was really a cover for secret oil geology exploration. The overall concern was summed up by Navy Undersecretary William Bullitt, who warned that London would "diddle" the American companies "out of the concession and the British into it."

In truth, the Americans woefully exaggerated British designs on Saudi Arabia—and the capability to carry them out. The British were hardly in a position to dislodge the Americans, on whom they depended so heavily; indeed, on balance, they wanted greater American involvement in the Middle East, for both security and financial reasons, and were actually trying to find ways to reduce their subsidies to Ibn Saud. Still, afflicted as they were by such anxieties and concerns, what could the Americans do? Three alternatives were to emerge. The first was to acquire direct ownership in Middle Eastern oil, on the model of the Anglo-Persian Oil Company. The second was to negotiate some kind of settlement and system with the British. And the third was to leave the whole matter in private hands. But, in the middle of the war with its pervasive uncertainty, even the "private hands" were very nervous about being left to their own devices. They wanted government support, and that again took them to Washington.[8]

The Policy of "Solidification"

Socal and Texaco, the two partners in Casoc, were the only two private companies involved in Arabian oil. They feared the British might obtain a dominant position over King Ibn Saud's finances in order to shoehorn their own way into Saudi oil and to show Socal and Texaco the door. The companies were also worried about something else. Socal and Texaco had made very large investments in and financial commitments to Saudi oil, and much more would be required; they knew that they were sitting on an immensely valuable resource. But, as a unified country, Saudi Arabia was only two decades old. Would Ibn Saud's kingdom—and the oil concession—survive the King himself?

How better to keep Britain in check, bolster their concession and protect this extraordinarily valuable asset against the political risks than with American aid to the Saudi government, and perhaps even with direct U.S. government involvement? It was one thing to throw out private companies—after all, only a few years earlier Mexico had nationalized company concessions with virtual impunity—and another thing to take on the leading power in the world. Direct involvement by the American government in Saudi Arabia was what became known to some as the policy of "solidification."

In mid-February 1943 the presidents of Socal, Texaco, and Casoc came to Washington to see the State Department with hats in hands. They appealed for financial aid from the government to keep the British at bay and assure "the continuation of purely American enterprise there after the war." If Washington were willing to come forward with foreign aid, they would be willing, in exchange, to offer the U.S. government some kind of special access or option on Saudi oil.

Over lunch, on February 16, Harold Ickes, a strong advocate of government involvement, bent President Roosevelt's ear on the subject of Saudi Arabia. It was "probably the greatest and richest oil field in the world," said the Interior Secretary. The British were trying to "edge their way into it" at the expense of Casoc, and the British "never overlooked the opportunity to get in where there was oil." It was the arguments of Ickes and other government officials, and not the petition of the oil company executives, that finally swayed Roosevelt. On February 18, 1943, two days after his lunch with Ickes and a year and a half after he had declared that Saudi Arabia was "a little far afield for us," the President authorized Lend Lease assistance to King Ibn Saud. It was only the beginning. Shortly after, the Army Navy Petroleum Board presented its projections for 1944: A serious shortage of oil was impending, which would threaten military operations. The anxieties of the military men gave a powerful additional push to the United States government in the direction of Saudi Arabia.[9]

Financial assistance to a friendly albeit nondemocratic government, even disguised as Lend Lease, was one thing; seeking to acquire direct ownership of the resources of a foreign country was quite another. But that was exactly what followed, in part carried out in the name of the Petroleum Reserves Corporation, a newly created government entity that the ever-ingenious Ickes appropriated with the aim of acquiring actual ownership of foreign oil reserves. In this, he was strongly backed by the Army and Navy. Only the State Department held back. It was fearful, as Secretary of State Hull told Roosevelt, of creating "intense new disputes." For, the Secretary reminded the President, "In many conferences after the last war the atmosphere and smell of oil was almost stifling."

The target of the Petroleum Reserves Corporation was Saudi Arabia. In June 1943 Ickes met at the White House with War Secretary Henry Stimson, Navy Secretary Frank Knox, and James Byrnes, who was Director of the Office of War Mobilization. They "viewed rapidly dwindling domestic supply with alarm," and agreed on the need for the government to acquire "an interest in the highly important Saudi Arabia fields." In July, Roosevelt confirmed this stunning decision at a meeting in the White House. "The discussion was jovial, brief, and far from thorough," said one participant. "A boyish note of enjoyment was in the President's talk and nod, as usual when it had to do with the lands of the Near East." There was still one crucial point outstanding. How much of Casoc or the concession was to be acquired? In a move that would have done credit to John D. Rockefeller himself, it was decided that the government's "interest" should be nothing less than 100 percent!

In August 1943 the unsuspecting presidents of Texaco and Socal, W. S. S. Rodgers and Harry C. Collier, filed into Ickes's office at the Department of the Interior. They thought they would be discussing aid in exchange for an option on Saudi oil. Ickes came out with his proposition: that the government buy the whole of Casoc from Texaco and Socal. Ickes observed, with some satisfaction, that his astonishing proposal "literally took their breath away." A government-owned oil company operating abroad would be an extraordinary departure for the United States. It would also transform the position of the two private companies concerned. All that Rodgers of Texaco and Collier of Socal could manage to say was that the offer was a "tremendous shock" to them. The companies had wanted assistance, not their assimilation. As one participant in the discussions noted, "They had gone fishing for a cod and had caught a whale." [10]

After more discussions, Ickes whittled down his proposal from 100 percent to a 51 percent share, on the model of the British government's ownership of Anglo-Iranian. He even suggested modeling the name of the venture on Anglo-Iranian—that is, the American-Arabian Oil Company. But some thought that such a name, at least in terms of its order, might be received poorly by Ibn Saud, whose objective was to hold foreign involvement in his kingdom to a minimum.*

While Ickes continued to negotiate with the two companies, he also explored the possibility of making a similar deal with Gulf in Kuwait. But, finally, he struck a bargain with Socal and Texaco. The U.S. government would acquire one-third of Casoc for $40 million; the funds would be used to finance a new refinery at Ras Tanura. Further, the government would have the right to buy 51 percent of Casoc's production in peacetime and 100 percent in wartime.

Thus, the United States was about to go into the oil business. Or so it seemed. But then the rest of the American oil industry rose up in what could only be described as righteous and fiery indignation. None of the other companies wanted the government in the oil business. It would be a formidable competitor; it might well champion foreign oil production over domestic oil; it might be only the first step toward Federal control of the oil industry, or even its nationalization. Strong opposition came not only from the independent companies, but also from Standard of New Jersey and Socony-Vacuum (Mobil), which themselves were interested in Saudi oil and did not want to be preempted.

---

* In 1944, Casoc, the California-Arabian Standard Oil Company jointly owned by Standard of California and Texaco, would indeed change its name, but with the order reversed—to the Arabian-American Oil Company, much more commonly known as Aramco.

Ickes had worked hard to mobilize the oil industry on behalf of America's war effort, and he could not afford to disrupt that effort with a battle over Casoc. So, in late 1943, he abruptly retreated and repudiated the plan, in the process blaming Texaco and Socal for being too greedy and recalcitrant. That was the end of the move by the United States to get into direct ownership of foreign oil.[11]

Still, Ickes would not be stayed; in early 1944, he became champion of another plan: to get the United States government into the foreign pipeline business. Ickes agreed in principle with Socal, Texaco, and Gulf that the U.S. government would, through the Petroleum Reserves Corporation, spend up to $120 million to build a pipeline that would carry Saudi and Kuwaiti oil across the desert to the Mediterranean for transshipment into Europe. As part of the deal, the companies would establish a one-billion-barrel oil reserve for the American military, which could be purchased at a 25 percent discount from market prices.

But many forces mobilized against this new plan in the late winter and spring of 1944. Congressmen were already calling for the abolition of the Petroleum Reserves Corporation. Other oil companies were furious at the notion that they would be, as Herbert Feis put it, "exposed to what they feel is favored competition." Independent oil men denounced it as "a threat to our national security" and as "a move towards fascism." It would promote cutthroat competition in world oil markets, said the Independent Petroleum Association of America, undermining domestic prices and shattering the domestic industry. Liberals opposed the plan because it would help big business and "monopolies"; isolationists opposed it because they did not want the government to, literally, dig itself into the far-off sand of the Middle East. Earlier, the Joint Chiefs of Staff had said that such a pipeline was "a matter of immediate military necessity." But, after D-Day, with the end of the war in Europe coming into sight, the Joint Chiefs would not renew their endorsement. It was a potent coalition of opponents and critics, and eventually, despite the Old Curmudgeon's ire and another of his resignation threats, the government pipeline project was eventually allowed to wither away and then disappeared altogether.[12]

"A Wrangle About Oil"

So the United States government was not going into the oil business in Saudi Arabia after all. But there was still another avenue to be explored: a partnership with Great Britain in managing the world oil market. The two governments had already begun probing each other's views about such an arrangement. While a number of the wells in the Persian Gulf area had been cemented in to keep them out of German hands, those who knew the potential of the region were beginning to worry what postwar production from that area would do to the market. A rising tide of cheap oil from the Persian Gulf after the war could be as destabilizing as the flood of oil from East Texas in the early 1930s. At the same time, many in the United States continued to fear the exhaustion of U.S. reserves and wanted to reduce the call on American oil. As they saw it, the major objective of the United States should be to break the prewar restraints and instead see maximum production from the Middle East and, in particular, from Saudi Arabia. That way, a fundamental transformation in supply arrangements would result: Europe could be principally supplied from the Middle East, and not from Western Hemisphere and especially American reserves, which could instead be conserved for America's own use and security.

The British, for their part, were deeply apprehensive about the disorder that would arise from pell-mell production in the Middle East. They feared a competitive production race by concessionaires seeking to satisfy the rising revenue appetites of the Middle Eastern oil countries. If petroleum matters were not settled before the end of the war, the result would be a devastating glut that could, because of low prices, deprive the oil-producing governments of royalty income and ultimately threaten the stability of concessions. Moreover, despite what many Americans thought, the British still favored further American participation in Middle Eastern oil development. Such involvement would, among other things, said the British Chiefs of Staff, improve "the chances of our obtaining American assistance" in defending the area, especially against "Russian pressure." The British Chiefs added that "American continental resources constitute our most secure supply in war and therefore it is in our interest to take any steps that may assist in their conservation." But how to convince the Americans that joint control, not laissez faire exploitation, would be in the best interests of both nations?[13]

The British campaigned hard to open negotiations with the United States about Middle Eastern oil. In April 1943 Basil Jackson, Anglo-Persian's representative in New York, met with James Terry Duce, on temporary leave from his job as a Casoc executive to be head of the Foreign Division of the Petroleum Administration for War. "It is the first time in history that there have been such enormous quantities of oil overhanging the markets of the world," warned Jackson. But it was not possible, he said, "for the companies themselves to arrive at any agreement regarding the future of Near Eastern oil." The American companies were constrained by the Sherman Antitrust Act. After the war it would be too late to act. Yet without such agreement, Jackson concluded, there would be "a fierce competitive battle."

Duce concurred. Both men recognized the fundamental issue ahead, one that would shape the postwar petroleum order. Royalties on oil were or would soon be the major source of revenues for the countries of the Gulf. As a result, those countries would put continuing pressure—augmented by threats, veiled or otherwise—on the companies to increase production, in order to increase royalty revenues. Some overall system of allocation could help balance that pressure.

The account of Jackson's remarks was widely circulated to American policymakers. Ickes sent it to Roosevelt himself. "We should have available oil in different parts of the world," Ickes noted. "The time to get going is now. I see no reason why we could not come to an understanding with the British with respect to oil." Yet so great were the mutual suspicions that it was no easy matter for the two Allies even to come to an understanding within their respective governments on how to structure the discussions. Any notion of a conference to discuss Middle Eastern oil "should be put in a pigeon hole," Lord Beaverbrook, the newspaper tycoon who was serving as Lord Privy Seal, told Churchill. "Oil is the single greatest post-war asset remaining to us. We should refuse to divide our last asset with the Americans."

But others in the British government persisted in trying to formulate a plan with the Americans. On February 18, 1944, the British ambassador in Washington, Lord Halifax, argued for almost two hours with Undersecretary of State Sumner Welles on oil and how to proceed. Afterward, Halifax telegraphed to London that "the Americans were treating us shockingly." So upset was Halifax with the discussions at the State Department that he immediately requested a personal interview with the President. Roosevelt received him that very evening at the White House. Their discussion focused on the Middle East. Trying to allay Halifax's apprehension and irritation, Roosevelt showed the ambassador a rough sketch he had made of the Middle East. Persian oil, he told the ambassador, is yours. We share the oil of Iraq and Kuwait. As for Saudi Arabian oil, it's ours.[14]

Roosevelt's hand-drawn map was not sufficient to relieve the tension. Indeed, the developments over the previous few weeks led to a brittle exchange between President and Prime Minister. On February 20, 1944, just hours after seeing Halifax's report on his meetings, Churchill sent a message to Roosevelt in which he said that he had been watching "with increasing misgivings" the telegrams about oil. "A wrangle about oil would be a poor prelude for the tremendous joint enterprise and sacrifice to which we have bound ourselves," he declared. "There is apprehension in some quarters here that the United States has a desire to deprive us of our oil assets in the Middle East on which, among other things, the whole supply of our Navy depends." To put it bluntly, he said, some felt "that we are being hustled."

Roosevelt replied tartly that he, in turn, had received reports that Great Britain was "eyeing" and trying to "horn in" on the American companies' concession in Saudi Arabia. In response to another sharp telegram from Churchill, Roosevelt added, "Please do accept my assurances that we are not making sheep's eyes at your oil fields in Iraq or Iran." Churchill cabled back, "Let me reciprocate by giving you fullest assurance that we have no thought of trying to horn in upon your interests or property in Saudi Arabia." But, while Britain did not seek territorial advantage, "she will not be deprived of anything which rightly belongs to her after having given her best services to the good cause—at least not so long as your humble servant is entrusted with the conduct of her affairs."

The acid exchange was a testament to the importance oil had assumed in world politics. But the two men managed to put the wrangle aside, and negotiations ensued in the spring of 1944 in Washington. The issue at hand was stated in the opening remarks of the State Department's Petroleum Adviser at the first meeting: The central objective of the negotiations, he said, "is not a rationing of scarcity, but the orderly development and orderly distribution of abundance." In other words, whatever the prospects for American oil, the problem from a global point of view would be too much oil—and how to control production. The British assessment of the Middle Eastern oil situation had prevailed.[15]

Quotas and Cartels

Lord Beaverbrook, whose suspicion of American economic ambitions was obvious, came to Washington to negotiate the final agreement in July of 1944. "I guess the war is on again," James Terry Duce, now back at Aramco, wrote to Everette DeGolyer in commenting on Beaverbrook's arrival. "The lion wouldn't lay down with lamb—except possibly in the guise of lamb chops."

In Washington, the outspoken Beaverbrook brought up the awkward point on which no one really wanted to focus. In London, he had privately described the agreement that was being shaped as a "monster cartel," and one managed by the Americans to protect their domestic producers at England's expense. In his negotiations with the Americans in Washington, he was more polite, observing that the two sides were really trying to come up with "an agreement of an 'As-Is' character"—not so very different from Achnacarry and the subsequent restrictive agreements among the companies in the late 1920s and 1930s.

The American negotiators were quick to disagree. "The Petroleum Agreement under discussion had been formulated on a basis altogether different from anything associated with the expression 'cartel,' " one of the Americans huffily replied. "This was an intergovernmental commodity agreement predicated upon certain broad principles of orderly development and sound engineering practices. It was directed toward assuring the availability of ample supplies of petroleum to meet market demands."

It was not obvious that Beaverbrook was persuaded to change his mind. Still, a few days later the Anglo-American Petroleum Agreement was completed, and it was signed on August 8, 1944. The objective was to assure "equity" to all parties, including the producing countries. The heart of the agreement was embodied in the establishment of the eight-member International Petroleum Commission. It would prepare estimates for global oil demand. It would then allocate suggested production quotas to various countries on the basis of such factors as "available reserves, sound engineering practices, relevant economic factors, and the interests of producing and consuming countries, and with a view to the full satisfaction of expanding demand." The commission would also report to the two governments on how to promote the development of the world petroleum industry. The governments, in turn, would seek "to give effect to such approved recommendations and, wherever necessary and advisable, to ensure that the activities of their nationals will conform thereto."[16]

Whether seen as a "commodity agreement" aimed at stabilizing an important industry or as a government-run cartel, the Anglo-American Petroleum Agreement was, in fact, a direct link to the market management of the late 1920s and the early 1930s, both to the "As-Is" of Achnacarry and to the Texas Railroad Commission. Its fundamental purpose was the same: to balance discordant supply and demand, to manage surplus, and to bring order and stability to a market laden with oversupply. While the agreement may well have satisfied the Roosevelt Administration and the British, it was immediately and bitterly assailed by the independent American oil men and their Congressional allies. The independents had more political clout than the majors, and if the independents did not like Ickes's Arabian pipeline project, they hated the Petroleum Agreement, fearing that it would open the door to international regulation of domestic oil production. It was one thing to have oil production rates set by the Texas Railroad Commission, whose members were elected in Texas, but quite another to have it done by a commission that was half "limeys" and half appointees of Franklin Roosevelt. More than anything else, the domestic oil companies were motivated in their opposition to the agreement by the specter of vast amounts of cheap Middle Eastern oil taking away their markets in Europe, perhaps even flooding into the United States, weakening prices. The independent oil men feared that the international companies would manipulate the agreement to gain decisive control over the world's reserves and markets and then use such control to put the independents out of business.[17]

The majors were also wary, but for a different reason. They were afraid of legal attack at some point in the future for antitrust violations—price fixing and production rigging—if they cooperated with the International Petroleum Commission. After all, when they had taken action to stabilize markets in the late 1930s in response to what they thought was the government's wishes, and in particular the behest of Harold Ickes, they found themselves hauled into court by the Justice Department on antitrust charges in what had become known as the Madison Case. And the "Mother Hubbard" antitrust suit against them had only been suspended because of Washington's need for cooperation after America's entry into the war. This time the majors did not want to take any chances; they wanted an antitrust exemption before going ahead.

The entire petroleum business, whatever the division between the majors and the independents, now seemed to be arrayed against the agreement. "The oil industry is ganging up on this without any rhyme or reason at all," Ickes complained to Roosevelt. "Some of the industry are seeing ghosts where there are no ghosts." The agreement was submitted to the Senate as a treaty, but it quickly became clear that, as such, it was going to go down to an inglorious defeat. In January 1945 the Roosevelt Administration withdrew it so that the antitrust problem and other issues could be addressed. Shortly after, efforts to revise the agreement were put into abeyance, as Roosevelt and his senior advisers set off for Yalta in the Soviet Crimea for a meeting with Joseph Stalin and Winston Churchill. Their aim was to lay the basis for the postwar international order—and to carve out the shape and size of their spheres of influence in that postwar world.[18]

The "Twins"

Yet the issues of Middle Eastern oil touched even this journey. In mid-February, after the conference at Yalta, the *Sacred Cow*, the President's plane, carried Roosevelt and his advisers back from Russia to the Suez Canal Zone in Egypt, where they boarded the USS *Quincy*, which was anchored in the Great Bitter Lake in the Canal. Another American ship, the USS *Murphy*, pulled up with an honored guest—Ibn Saud.

For the Saudi King, this was perhaps only his second trip outside his kingdom since that day, forty-five years ago, when he had left exile in Kuwait to take his first step—the assault on Riyadh—toward regaining Arabia. He had boarded the *Murphy* a couple of days earlier, in Jidda, with a party of forty-eight. His group was also to include one hundred live sheep, but after some negotiation, the number was reduced to just seven in light of the sixty days' worth of provisions, including frozen meat, on board the American ship. Ibn Saud spurned the offer of the commodore's cabin and slept instead on deck, in an improvised tent made of canvas, stretched over the forecastle, and furnished with Oriental carpets and one of the King's own chairs.

Once Ibn Saud had transferred to the President's ship, the chain-smoking Roosevelt, out of deference to the King's religious precepts, did not light up in his presence. On the way to lunch, however, Roosevelt was taken in his wheelchair into a separate elevator. The President himself pushed the red emergency button, stopping long enough to smoke two cigarettes before meeting again with the King. Altogether, the two men spent more than five very intense hours together. Roosevelt's interests were a Jewish homeland in Palestine, oil, and the postwar configuration of the Middle East. For his part, Ibn Saud wanted to assure continuing American interest in Saudi Arabia after the war, in order to counterbalance what had been for him a chronic threat throughout his reign—British influence in the region. In reply to Roosevelt's call for a Jewish homeland, the bitterly anti-Zionist Ibn Saud suggested that those displaced Jews who had somehow managed to survive the war be given a national homeland in Germany.

Roosevelt and Ibn Saud got along very well. At one point, the King declared that he was the "twin" brother of the President because of their close ages, their responsibilities for their nations' well-being, their interests in farming, and their grave physical infirmities—the President confined by polio to a wheelchair, and the King walking with difficulty and unable to climb stairs because of war wounds in his legs.

"You are luckier than I because you can still walk on your legs and I have to be wheeled wherever I go," said Roosevelt.

"No, my friend, you are more fortunate," the King replied. "Your chair will take you wherever you want to go and you know you will get there. My legs are less reliable and are getting weaker every day."

"If you think so highly of this chair," said Roosevelt "I will give you the twin of this chair, as I have two on board."

The wheelchair went back with Ibn Saud to Riyadh, where thereafter it would remain in the King's private apartment to be shown off by Ibn Saud as a most valued memento, though it was too small to be used by a man of the King's large stature.[19]

The official record was surprisingly silent about what the two men said about oil. One member of the party later reported that the President and the King did have a long talk on that subject. Whatever was or was not said, both men knew that it was central to the emerging relationship between their two countries. The *New York Times* foreign affairs correspondent, C. L. Sulzberger, got right to the point. Immediately after the meeting at the Great Bitter Lake, he wrote, "The immense oil deposits in Saudi Arabia alone make that country more important to American diplomacy than almost any other smaller nation." But Winston Churchill was not exactly thrilled to see the American President conferring with monarchs in an area of traditional British influence—Roosevelt also met King Farouk of Egypt and Haile Selassie of Ethiopia. According to one account, Churchill "burned up the wires to all his diplomats in the area, breathing out threatenings and slaughter unless appointments with him were made with the same potentates after they had seen F.D.R." Churchill rushed to the Middle East, and three days after Roosevelt, the British Prime Minister drove into the Egyptian desert to meet Ibn Saud in a hotel at an oasis.

Once again, the issue of smoking came up, though this time complicated by that of imbibing. Churchill's meeting with the Saudi King was to conclude with a grand banquet. Beforehand, Churchill had been told that, as he later noted, the King "could not allow smoking or drinking alcohol in his presence." Churchill would not be so accommodating as Roosevelt toward Ibn Saud. "I was the host and I said if it was his religion that made him say such things, my religion prescribed as an absolute sacred rite smoking cigars and drinking alcohol before, after, and if need be during, all meals and the intervals between them."

Churchill's high-handed insistence on his own rights and prerogatives may have done little to reassure an already suspicious Ibn Saud about British intentions regarding his kingdom and the region. Churchill faced another problem. He gave Ibn Saud a small case of very choice perfumes—value about one hundred pounds. But Ibn Saud, in turn, presented him and Anthony Eden with diamond-hilted jeweled swords, as well as robes and other presents—including about three thousand pounds' worth of diamonds and pearls for, as Ibn Saud had put it, "your womanfolk." Embarrassed by the disparity in the gifts, Churchill, on the spur of the moment, declared that the perfumes were "but tokens," and promised Ibn Saud "the finest motor car in the world." Churchill realized that he had no personal authorization to make such a gift, but so be it. A Rolls-Royce was delivered to the King, which cost the British Treasury upwards of six thousand pounds. Eventually, all the jewels were sold off, though with the proviso that the sale be kept secret, so as not to offend Ibn Saud.[20]

"What Do We Do Now?"

On Roosevelt's return from his long journey, he found his advisers still battling among themselves over the Petroleum Agreement and the related antitrust issue. Harold Ickes had proposed a meeting with the President and the new Secretary of State, Edward Stettinius. But the President was exhausted after his lengthy trip and was going to rest. "I shall be delighted to have the meeting that Harold proposes just as soon as I get back from Warm Springs," he told Stettinius on March 27, 1945. "Will you remind me of it?"

Stettinius did not have the opportunity. Roosevelt died in Warm Springs on April 12, 1945.

Efforts were made under the new President, Harry Truman, to revise the Petroleum Agreement in order to make it domestically palatable. Ickes, by then its leading sponsor, renegotiated it in London with the British in September of 1945. Whatever teeth had existed in the previous agreement were extracted in London. This time around, the International Petroleum Commission, which had been charged in 1944 with suggesting how to allocate production around the world, was effectively precluded from touching domestic American production—a rather large omission for a global petroleum agreement, as the United States then accounted for two-thirds of total world production. But that was the best that could be done. "There is no prospect of getting any more comprehensive Agreement through the U.S.A. Senate," the British Minister of Fuel and Power told the Chancellor of the Exchequer. "On balance, it is better that we should accept the Agreement rather than reject it."

Meanwhile in America, the gloom about the adequacy of oil reserves was receding. At a Senate hearing in 1945, J. Edgar Pew, vice-president of Sun Oil and chairman on the committee on petroleum reserves of the American Petroleum Institute, lambasted the prospect of an oil shortage as a psychological rather than a geological condition. Expressing the Pew family's traditional disdain for the warnings of depletion, he assured the Senators that domestic production would meet all American needs for two decades or more. "Of that I am as sure as I am that the sun shall rise and set tomorrow," he said. "I am an optimist."

With victory in 1945 over Germany and Japan, there was no longer a crushing demand on American oil reserves, and thus another impetus to securing the agreement with Britain was dissipating. Then, in February 1946, the Anglo-American Petroleum Agreement ran into a new problem. Its chief sponsor, Harold Ickes, got into a bitter scrap with Harry Truman over the President's proposed appointment of Edwin Pauley, a California oil man, as Undersecretary of the Navy. Ickes, as had been his wont under Roosevelt, submitted his resignation. And a long good-bye it was—more than six pages typed, single-spaced. "It was the kind of letter sent by a man who is sure that he can have his way if he threatens to quit," Truman later said.

But Ickes had made a mistake; Truman was not Roosevelt. He accepted Ickes's resignation tersely and with alacrity and delight. Ickes requested six weeks to wind up the many things that only he personally could attend to; Truman gave him two days to clean out his desk. The Old Curmudgeon shot back with one last salvo. Truman, he announced to the nation, "showed a lack of adherence to the strict truth" and was "neither an absolute monarch nor a descendant of a putative Sun Goddess." And with that, the Oil Czar of the New Deal and World War II left office and took up a new career as a newspaper columnist.[21]

Had the Anglo-American Petroleum Agreement any future without its champion, Harold Ickes? Support for the agreement now came from an unlikely source: Navy Secretary James Forrestal. A driving, ambitious and politically conservative former investment banker from Dillon, Read, Forrestal was one of the first senior policymakers to conclude that the United States had to organize itself for a protracted confrontation with the Soviet Union. Oil held a central place in Forrestal's strategy for security in the postwar world. "The Navy," he said, "cannot err on the side of optimism" in its estimates for what supplies might be available. The largest known oil reserves outside the United States were in the area of the Persian Gulf. "The prestige and hence the influence of the United States is in part related to the wealth of the Government and its nationals in terms of oil resources, foreign as well as domestic," he said. "The active expansion of such holdings is very much to be desired." The State Department should work out a program to substitute Middle Eastern oil for American oil, he added, and use its "good offices" to "promote the expansion of United States oil holdings abroad, and to protect such holdings as already exist, i.e., those in the Persian Gulf area."

At Potsdam, the final conference between the Allied powers before the end of the war, Forrestal had lectured the new Secretary of State, James Byrnes, on Saudi Arabia as "a matter of first importance." And now, in early 1946, in the immediate aftermath of Harold Ickes's firing, he saw considerable merit in continuing to fight for the Anglo-American Petroleum Agreement. "As you know, I am not in the forefront of the cheering section of 'Honest Harold,' but I think it might be worth taking a new look at these oil treaty negotiations," he said to Byrnes. "I am of the opinion that he is right about the limitations of American oil reserves—in this I am influenced a good deal by the engineer that I used in private business, E. L. DeGolyer." Forrestal added, "If we ever got into another World War it is quite possible that we would not have access to reserves held in the Middle East but in the meantime the use of those reserves would prevent the depletion of our own, a depletion which may be serious within the next fifteen years."[22]

But Forrestal was a minority. Elsewhere throughout the government, support for the agreement was eroding. Indeed, within days of the Old Curmudgeon's exit, one State Department official, Claire Wilcox, wrote a memo entitled "Oil: What Do We Do Now?" Providing a lengthy list of reasons for killing the agreement, Wilcox declared, "the Agreement is either dangerous or useless. If employed as a cover to set up a cartel to allot quotas and fix minimum prices, it is dangerous. If not so employed, it is useless." Wilcox summed up the issue for the Truman Administration. "Mr. Ickes told the President that he had raised this baby on a bottle. Now the orphan is on our doorstep. Shall we smother it or adopt it?"

The answer was pretty clear. The agreement had no political support. Even local schoolteachers in Texas were mobilized to oppose it. Imported oil, they said, would destroy the economy of Texas. The baby was to be smothered. Events and interests had outrun the political process, and the Anglo-American Petroleum Agreement had become increasingly irrelevant and obsolete. In 1947, the Truman Administration gave up any further effort on behalf of the agreement. It was dead.

But even as the agreement, the last of the major wartime oil initiatives, disappeared from sight, other factors were coming to the fore. Whatever the debate about reserves and discovery rates, the United States was finding that it could not sustain itself on domestic production alone. It was about to become a net importer of oil, and its dependence on foreign sources of petroleum would swell in the years ahead. In short, even without the requirements of a global war, the process of "solidification" had to proceed; American and European interests, public and private alike, were best served by the rapid development of the oil lands of the Middle East.

As for the oil companies, the requirements of markets and competition and the demands from producing countries for revenues—none of these pressures could be stayed. Everything that wartime negotiators had sought to prevent was coming to pass, and the petroleum industry in the postwar years looked to be as competitive, chaotic, and unstable as it had ever been in the past. So, while the unprecedented and controversial possibilities offered by the Anglo-American Petroleum Agreement were fading, the oil companies themselves were moving quickly to work out, in the word used by an executive of Anglo-Iranian, their own "salvation" in the Middle East and for the postwar world.[23]

CHAPTER 21

The Postwar Petroleum Order

GASOLINE RATIONING in the United States was lifted in August 1945, within twenty-four hours of Japan's capitulation. And immediately, the voice of the motorist, silenced for years, was heard throughout the land, rising in a single, deafening crescendo—"Fill 'er up!" The rush was on, as drivers tossed away their rationing books and took to the streets and highways. America was in love once again with the automobile, and now consumers had the means to carry on the romance. In 1945, 26 million cars were in service; by 1950, 40 million. Virtually no one in the oil industry was prepared for the explosion of demand for all oil products. Gasoline sales in the United States were 42 percent higher in 1950 than they had been in 1945, and by 1950, oil was meeting more of America's total energy needs than coal.

While demand was exploding far beyond expectations, the pessimistic predictions about postwar oil supplies were being eviscerated by actual experience. After controls were lifted, price proved to be a powerful stimulus to exploration. New regions were brought into production in the United States as well as in Canada, where, in 1947, Imperial, a Jersey affiliate, drilled a successful discovery well near Edmonton, in the province of Alberta, igniting the first, frenetic oil boom of the postwar years. Despite growing demand and rising production, proven United States reserves were 21 percent higher in 1950 than they had been in 1946. The United States was not, after all, running out of oil in the ground.

There was, however, a shortage of *available* oil in 1947–48. Crude prices rose rapidly, so that by 1948 they were more than double the 1945 level. Politicians declared that the country was in an energy crisis. The major oil companies were accused of deliberately orchestrating a squeeze play to push up prices, and suspicions of oil industry skullduggery and conspiracy launched more than twenty Congressional investigations.

But the reasons for shortage were quite obvious. Consumption rose with unexpected rapidity—"astonishingly," said Shell—while there was the inevitable time lag in adapting to the postwar situation. It took time, money, and materials to redesign refineries to turn out the products that the civilian consumer wanted, such as gasoline and home heating oil, as opposed to 100-octane aviation fuel for fighter planes. In addition, steel was in short supply throughout the world, which slowed the conversion of refineries and the construction of tankers and pipelines, contributing to transportation bottlenecks. The tanker shortage worsened in early 1948, after several such ships broke in half while at sea, and the Coast Guard ordered 288 tankers laid up for emergency reinforcing. For the oil companies, it was a time of enormous pressure on retail supplies, and they became the leading advocates of conservation. Standard of Indiana urged motorists to cut back on their driving, avoid "jackrabbit" starts, and keep their tires properly inflated—all to reduce their consumption. "Helpful Hints" for conserving oil were promoted by Sun in its commercials on the popular daily broadcasts of newscaster Lowell Thomas.[1]

The shortages also drew in larger volumes of oil imports. Up through 1947, American exports of oil exceeded imports. But now the balance shifted; in 1948 imports of crude oil and products together exceeded exports for the first time. No longer could the United States continue its historical role as supplier to the rest of the world. Now it was dependent on other countries for that marginal barrel, and an ominous new phrase was being heard more often in the American vocabulary—"foreign oil."

The Great Oil Deals: Aramco and the "Arabian Risk"

That shift added a new dimension to the vexing question of energy security. The lessons of World War II, the growing economic significance of oil, and the magnitude of Middle Eastern resources all served, in the context of the developing Cold War with the Soviet Union, to define the preservation of access to that oil as a prime element in American and British—and Western European—security. Oil provided the point at which foreign policy, international economic considerations, national security, and corporate interests would all converge. The Middle East would be the focus. There the companies were already rapidly building up production and fashioning new arrangements to secure their positions.

In Saudi Arabia, development was in the hands of Aramco, the Arabian-American Oil Company, the joint venture between Socal and Texaco. But Aramco was troubled. The reason was an embarrassment of riches, the very scale of the Saudi oil fields, which meant an enormous need for capital and for markets. Of the two companies in the joint venture, Socal was the more vulnerable. Texaco, the most important enterprise to have been spawned from the 1901 Spindletop discovery, was a famous American company; it sponsored the Metropolitan Opera on coast-to-coast radio, and Texaco's service station attendant, "the man who wears the star," was one of the familiar icons in the modern pantheon of American advertising. Socal, by contrast, was a regional company, and not very well known. Ever since World War I, it had expended millions of dollars searching for oil around the world. Yet it had nothing to show for its efforts, except for some minor production in the East Indies and Bahrain—and the massive potential of Saudi Arabia.

The Arabian concession was a grand prize that the California company could not have dared to hope for. It presented the company with a splendid opportunity—but it also meant, as Socal's chairman Harry Collier saw it, formidable economic and political risks. By 1946, Standard of California's investment in the Aramco concession totaled $80 million, and tens of millions more would be required. To gain access to European markets, Socal and Texaco wanted to build a pipeline across the desert from the Persian Gulf to the Mediterranean. It was essentially the same pipeline project that Harold Ickes had urged the U.S. government to finance, but now the companies themselves were going to have to come up with $100 million to pay for it. Socal faced another, even more daunting challenge. Once the oil got to Europe, how was it going to be marketed? To buy or build a refining and marketing system in Europe of sufficient size, Collier knew, would be very expensive and would commit Socal and Texaco to a deadly battle for market share against well-established competitors. The risks were further increased by the unstable political conditions. Large communist parties were represented in the coalition governments of both Italy and France, the future of occupied Germany was highly uncertain, and in Britain the Labour government was busying itself with nationalizing the "commanding heights" of the economy.

Yet Socal would have no choice but to go for higher and higher production levels, as the Saudi government recognized the potential of the reserves and would be pushing for higher output and revenues commensurate with the scale of the resources. The concession would always be in jeopardy if Aramco could not satisfy the expectations and demands of Ibn Saud and the royal family. This was the number-one consideration for Socal, and it meant that Aramco would have to move a great deal of oil, one way or another, into Europe. But before it ever got there, the Trans-Arabian Pipeline, called Tapline, would have to traverse several political entities, some of them only just moving toward statehood. A Jewish homeland might soon be established in Palestine, with possible American support, and Ibn Saud was one of the most prominent and adamant opponents of such a state. War could erupt in the region. The area also looked vulnerable to Soviet subversion and penetration in those early days of the Cold War.

Then there was the matter of the King himself, the same concern that had helped bring the chairmen of Socal and Texaco rushing to Washington in 1943. Ibn Saud was now in his mid-sixties, blind in one eye, and failing in health. His personal force and drive had created and held together the kingdom. But what would happen when that force was gone? He had sired upwards of forty-five sons, of whom thirty-seven were thought to be living, but would that be a factor for stability or for conflict and disorder? And, in the event of political problems, what kind of support from the American government could Socal count on? When all the risks were added up, it was clear that Socal would have to pursue its own policy of "solidification" and its drive to assure markets in other ways. The answer to Aramco's many problems was a broader joint venture. Spread the risk. Tie in other oil companies, whose presence would add to the political density, and which could deliver capital, international expertise, and, especially, markets. One other qualification was also essential; Ibn Saud insisted that Aramco had to remain 100 percent American, in which case only two companies qualified: Standard Oil of New Jersey and Socony-Vacuum. In the Eastern Hemisphere, they could offer, recalled Gwin Follis, who handled the matter for Socal, "markets that we could hardly touch."

The logic of wider involvement had been evident for some time, and not only to Collier and other oil men. Various State Department and United States Navy officials had encouraged Aramco to bring in additional partners who would "have sufficient markets to handle the concession" and thus help preserve it. Socal was struck by "the surprising enthusiasm in which the State Department received our notification that such a deal was being contemplated." Whether or not Washington actually acted as an overt marriage broker, it was clear that enlarging the participation would further the fundamental goals of American strategy: to increase Middle Eastern production, thus conserving Western Hemisphere resources, and to enhance the revenues going to Ibn Saud, thus ensuring that the concession remained in American hands. As Navy Secretary James Forrestal put it in 1945, he did not "care which American company or companies developed the Arabian reserves" so long as they were *American*." In the spring of 1946, Socal opened talks with Standard Oil of New Jersey.

To say that Jersey was receptive would be an understatement. The company was facing a shortage of oil, and Europe was its most vulnerable market. How was Jersey going to get the oil it needed? Despite all the *Sturm und Drang* that had gone into setting up the Iraq Petroleum Company in the 1920s, Jersey's share of Iraqi production in 1946 totaled a decidedly insignificant 9,300 barrels per day. Meanwhile, there would be more oil coming from Kuwait, further strengthening competitors, and Jersey greatly feared that Socal and Texaco would thrust into European markets on their own, challenging Jersey's marketing system with limitless quantities of cheap Arabian oil. Socal's overture presented Jersey with an opportunity to be passed up on no account.

While the two sides were haggling about the price of admission, Harry Collier, Socal's chairman, found himself challenged by his own people, who rose up in rebellion against the very thought of inviting Jersey into Aramco. The attack came from Socal's production department in San Francisco, which had been in charge of making the barren desert bloom and did not want to lose control to larger and more powerful partners. For thirteen years, there had been no return to stockholders on the investment in Arabia, and only now, in 1946, was the concession starting to become profitable. Why give it away to Jersey? Even more vociferous were the men on the ground, led by James MacPherson, a Socal engineer who was in charge of field operations for Aramco in Saudi Arabia. The concession, he argued, was a "gold mine." MacPherson was intent on building Aramco into a major independent force in world oil. He would point to a globe and tell his staff, "That is our oil market." Aramco, he proclaimed, was destined to become "the greatest oil company in the world." But now, he scathingly declared, Aramco—and Socal—were to turn themselves into an annex to Jersey's production department.

Harry Collier, by contrast, believed Aramco would be able to sell so much additional oil through access to Jersey's system that Socal would end up with much more "gold" than by going it alone with only Texaco. Moreover, the deal would enable Socal to recoup all of its direct investment. Collier was the boss, a strong-willed man—he was not called one of the "Terrible Tycoons" for nothing. Tying up with Jersey was the safer course as far as he was concerned, and Jersey would be invited in. Aramco was not, after all, destined to become the greatest oil company in the world. Argument ended.[2]

Erasing the Red Line

As the discussions proceeded on how Jersey would enter Aramco, Jersey was also holding side conversations with Socony about its possible participation. But both Jersey and Socony confronted two most formidable obstacles before they could enter Aramco: their own membership in the Iraq Petroleum Company— and Calouste Gulbenkian. The companies had expended six years and many, many thousands of hours of executive frustration in the 1920s to put the IPC arrangement together. One of its key provisions, of course, was the famous Red Line Agreement, which provided that the participants in the IPC could not operate independently anywhere inside the red line that Calouste Gulbenkian said he had drawn on the map in 1928. Saudi Arabia was most definitely inside the red line, and the "self-denying" clause 10 of the IPC agreement effectively prohibited Jersey and Socony from going into Aramco unless they took everyone else along with them—Shell, Anglo-Iranian, the French state company (CFP), and Mr. Gulbenkian himself.

Jersey and Socony had wanted out of the Red Line agreement for some time; it did not do them a lot of good, as it had turned out, to be straitjacketed in the most prolific oil basin in the world for a mere 11.875 percent each of an enterprise they did not control. The United States government had helped them get into the deal in the 1920s, but it was now abundantly clear that Washington was not going to do much to help get them out of it in the 1940s.

But, then, Jersey and Socony found another way to extricate themselves. A Socony executive called it a "bombshell." The device was called the doctrine of "supervening illegality." At the outbreak of World War II, the British government had taken control of the shares of the IPC held by CFP and Gulbenkian, who had packed up and gone off with the collaborationist French government to Vichy, where he had been accredited to the Iranian legation as commercial attaché. London's seizure of the shares had been made on grounds that both CFP, as a company, and Gulbenkian were domiciled in territory under Nazi control and, therefore, were considered "enemy aliens." Under the doctrine of "supervening illegality," the entire IPC agreement was thus "frustrated"—made null and void.

At war's end, their IPC shares reverted to both CFP and Gulbenkian. But then in late 1946, Jersey and Socony took up the concept of "supervening illegality" with what could only be called extreme enthusiasm. In their view, the whole IPC agreement was no longer in effect. A new agreement would have to be negotiated. Representatives of Jersey and Socony hurried to London to see the European members of the IPC in order to break their news: The old agreement was dissolved—Red Line and all. They would be willing to enter into a new understanding, certainly, but without the restrictive Red Line clauses, which "under present world conditions and American Law are inadvisable and illegal." The Americans knew that they would have to persuade four distinct parties to renegotiate—Anglo-Iranian, Shell, CFP, and an enterprise called Participations and Investments (P&I), which was really nothing more than the holding company for their old nemesis, Calouste Gulbenkian.[3]

Anglo-Persian and Shell indicated that they thought the matter could be amiably worked out on the basis of "mutual interest." The French, however, were in no mood to compromise. They rejected, without any qualification, the American contention that an agreement no longer existed. The Iraq Petroleum Company and the Red Line Agreement constituted their sole key to Middle Eastern oil. They were depending on this government-sanctioned allocation and would not relinquish what the French government had struggled so hard to obtain. France's energy position was bad enough already. It was said that General Charles de Gaulle, head of the French government, had exploded in rage when he discovered what small volumes of oil CFP actually produced—though he knew that he could not dispute geology or, as one of his aides put it, "be angry with God."

As for Calouste Gulbenkian, he replied swiftly and defiantly to Jersey and Socony's attempt to abandon the agreement: "We do not acquiesce." The Iraq Petroleum Company, and its predecessor, the Turkish Petroleum Company, had been his life's work, his great personal monument. He had started fashioning it forty years earlier, and he was not going to allow it to be easily dismembered. In 1946, Gulbenkian was in residence in Lisbon; he had moved there from Vichy in the middle of the war. Now, though unwilling to budge from Portugal, he would, through his lawyers and agents, do whatever was necessary to resist efforts to overturn the Red Line Agreement. The American negotiators were of a new generation, and lacking the benefit of Walter Teagle's experience of endless exasperation, they dismissed Gulbenkian's threats. "We have no reason to buy Gulbenkian's signature," Harold Sheets, the chairman of Socony, optimistically said. Confident of their legal position, they decided to go ahead and make their deal with Texaco and Socal, the two Aramco companies.

The danger of litigation over the IPC and the Red Line Agreement was not, however, the only risk with which Jersey and Socony had to wrestle. Would the new quadripartite Aramco combination violate American antitrust laws? That worry stirred the lawyers to dust off the 1911 dissolution decree. After all, three of the four would-be participants in the enlarged joint venture had been spun out of the original Rockefeller Trust. But the lawyers concluded that the proposed combination would violate neither antitrust laws, even under the new interpretations, nor the dissolution decree "because no unreasonable restraints on American commerce would be imposed." After all, Aramco was not going to go into the oil business in the United States. The chief counsel of Socony expressed a larger worry—that seven companies would not be allowed to hold such overwhelming control of crude reserves in the Eastern Hemisphere as well as in the Western Hemisphere "for any great length of time . . . without some sort of regulation." But he added, "This is a political question . . . within the realms of conjecture. Our job seems to be to play the game as best we can under the rules now in force."

And the best way to play was to proceed. By December 1946, the four companies had agreed in general principle to expand Aramco. After an immediate protest from one of Gulbenkian's representatives, a Socony executive in London sought to reassure his chairman in New York: "I have no doubt that P&I and the French may raise quite a song and dance about the matter, but I think they will be careful to wash the linen within the family circle."[4]

The French were not burdened with such modesty. In January 1947 they launched a very public counterattack. Their ambassador in Washington lodged a strong protest with the State Department. Authorities in France started to make Jersey's life, commercially uncomfortable. And in London, CFP's solicitors brought suit, charging breach of contract and asking that any shares that Jersey and Socony acquired in Aramco be held in trust for all the members of the IPC.

The awkward situation with France, a key ally in Western Europe, combined with a continuing antitrust concern, prompted the State Department to promote an alternative to the prospective deal that would both satisfy the French and check the growth of suspiciously close arrangements among giant international oil companies. Advice on petroleum matters in the State Department was largely in the hands of Paul Nitze, head of the Office of International Trade Policy. Nitze proposed that Jersey sell its IPC shares to Socony and then go it alone into Aramco, creating two distinct groups with no overlapping membership. The French could not then charge that their rights under the Red Line Agreement were being challenged, said Nitze. Such a deal, he added, would "arrest the trend toward the multiplication of the interlocking agreements among international oil companies" and "retard the growing consolidation outside of the United States of the interests of the two largest American oil companies, Jersey and Socony." The two companies replied that the proposal was "not a practicable plan." And Undersecretary of State Dean Acheson scotched Nitze's idea.[5]

There was someone else whose voice had yet to be heard—Ibn Saud. He, too, had to be consulted. Aramco executives went to Riyadh to see the King. They explained to him that the "marriage" of the four companies was "a natural" and would mean more royalties for the kingdom. But the King was interested in only one point, and on that he was insistent; he wanted to be positive that neither Jersey nor Socony were "British-controlled." Firmly reassured on the purely American character of the two new companies, the King finally gave his approval to the proposal.

But what would happen if the French won in the courtroom? They could insist upon participating in Aramco. And so, for that matter, could Anglo-Iranian. The King had made absolutely clear that he would not tolerate such a situation. The deal had to be restructured to take into account this contingency. Therefore, the final arrangement provided a clever piece of flexibility just in case the American companies lost in any legal proceedings. Jersey and Socony guaranteed a loan of $102 million, which could be converted into equity valued at $102 million as soon as it was legally safe to do so. In the meantime, Jersey and Socony could start taking oil at once, as though they were already owners. In addition, Jersey and Socony would become partners in Tapline. Socal and Texaco would also get overriding payments on every barrel produced for a number of years. Thus, altogether, Socal and Texaco would receive a total of about $470 million over several years for selling 40 percent of Aramco—getting all of their original investment back and a good deal more. Moreover, as Gwin Follis of Socal later observed, the terms of the sale to Jersey and Socony took "the enormous investment" required for Tapline "off our shoulders."

Initially, Jersey and Socony planned to split the 40 percent evenly. But Socony's president, fretful that Middle Eastern oil "was not absolutely safe" and worried about markets, argued that the company "ought to put more money in Venezuela." After some consideration, Socony decided that it did not need so much oil, and that a lower share would do just fine. Thus, Jersey took 30 percent, the same position to which Socal and Texaco had diluted, while Socony took only 10 percent. It would not be very long before Socony came to regret its parsimony.

There was still last-minute nervousness. Antitrust considerations continued to weigh on the minds of executives from all the companies, until they received a reassurance from the U.S. Attorney General. "Off-hand," the Attorney General said, he saw "no legal objections to the deal. It should be a good thing for the country." But then, in confirmation of Harry Collier's worst fears, political troubles in the eastern Mediterranean, which could have an impact on the entire deal, came to the fore. There was a communist-led insurrection in Greece and a Soviet threat to Turkey, and it was feared that, with Britain pulling back from its traditional commitments in the Middle East, communist power might grow in the region. On March 11, 1947, Socony's directors discussed "the problems affecting the Middle East." But optimism prevailed and they approved the deal. The next day, March 12, 1947, officials of the four American companies met and signed the documents that put the historic transaction into force. The concession in Saudi Arabia had, at last, been "solidified."

March 12 happened to be a historic day for another reason. On that day, President Harry Truman went before a joint session of Congress to deliver what was called the "all-out speech," proposing special aid to Greece and Turkey to enable them to resist communist pressure. The speech, a landmark in the emerging Cold War, initiated what became known as the Truman Doctrine and launched a new era in postwar American foreign policy. While a coincidence, the Truman Doctrine and the sealing of the participation of four giants of the American oil industry in the riches of Saudi Arabia now assured a substantial American presence and interest in a vast area, stretching from the Mediterranean to the Persian Gulf.[6]

Gulbenkian Again

The CFP litigation was still pending. But France had many other things on its political agenda with the United States that it wanted to pursue; and by May of 1947, a deal had been worked out that improved the position of the French in the Iraq Petroleum Company. And, of course, CFP would, in exchange, withdraw its suit.

Gulbenkian, as usual, was another matter. Installed in a first-floor suite in

Lisbon's venerable Hotel Aviz, Gulbenkian kept to his flint-nosed habits. Because it was cheaper, he no longer maintained a car and chauffeur, but hired a driver to take him into the country for his daily walk, carefully checking the odometer on the car to make sure he was not being charged for someone else's trips. "Gulbenkian may be regarded as a man of his word once this has been given," observed a British official. "The difficulty lies in obtaining it. The ability to compromise is not one of his attributes." The official could not help adding that "Gulbenkian's idea of his own financial integrity takes on a peculiar form when it comes to taxation, the avoidance of which constitutes one of his major activities." He escaped income taxes in France and Portugal by maintaining an appointment with the Iranian legation. In order to avoid property tax on his mansion in Paris, he turned a small part of it into a picture gallery. And when he sold the Ritz Hotel in Paris, he insisted on terms that provided that a suite be permanently reserved for him, so that he could always claim that he was "in transit" while in Paris—thus further avoiding French taxation.

Gulbenkian brought this same infuriating attention to detail, along with his reluctance to compromise and his intense powers of concentration, to the struggle over the Red Line Agreement. Though the French had dropped their suit, Gulbenkian was willing to wash every last piece of dirty linen in public if necessary. He filed suit in a British court. Jersey and Socony responded with counter-suits.

The case received wide publicity, which helped Gulbenkian in his counter-attack against Jersey and Socony. After all, it was not he, but the American companies that had to worry about the Justice Department and public opinion. Still, there was a side effect of the notoriety that he very definitely found distasteful. Owing to his short stature, he had ordered a special platform built in the restaurant of the Hotel Aviz, so that he could eat his lunch and survey the scene at the same time. As the publicity from the case increased, "Mr. Gulbenkian at the Hotel Aviz" became one of the tourist "musts" of Lisbon, along with the bull-fights. He objected, but there was nothing much he could do about it.

For well over a year, negotiators shuttled back and forth from New York to London to Lisbon, looking for a compromise. Now the next generation of oil men and lawyers had the opportunity to learn how exasperating it was to deal with Calouste Gulbenkian. "It was my father's practice never to press any claim to a breakdown," said his son Nubar, "but, very able negotiator that he was, to make his demands step by step, so that having obtained satisfaction on one point he would raise another and yet another, thus achieving all he wanted or, at least, much more of what he wanted than he would have obtained if he had started by putting forward all his demands at once."

Negotiations were made even more difficult by Gulbenkian's habitual suspicion, which had deepened into an obsession. Gulbenkian did not attend the various meetings himself. He had four different representatives at the sessions, each of whom had to report to him separately in writing, without collaborating—indeed, without even talking to the others. That way, in addition to analyzing his opponents, he could double check and second guess each of his own negotiators.

But what, in essence, did Gulbenkian want? Some suspected that he actually aimed to get a share of Aramco. That was out of the question. Ibn Saud would never allow it. To a director of Socony, Gulbenkian offered a simple explanation of his objective. He could not respect himself unless he "drove as good a bargain as possible." In other words, he wanted as much as he could get. To another American, not an oil man at all, but one who shared his love of art, Gulbenkian could explain still more. He had made so much money that more money, in itself, did not count for very much. He thought of himself in the same terms he had used to Walter Teagle two decades before—as an architect, even as an artist, creating beautiful structures, balancing interests, harmonizing economic forces. That was what gave him his joy, he said. The artworks he had collected over his lifetime had come to compose the greatest collection ever assembled by a single person in modern times. He called them his "children," and seemed to care more for them than for his actual son. But his masterpiece, the greatest achievement of his life, was the Iraq Petroleum Company. To him, it was as architecturally designed, as faultlessly composed, as Raphael's *The School of Athens*. And if he was Raphael, Gulbenkian made clear, he regarded the executives of Jersey and Socony in much the same company as Giroloma Genga, a third-rate, mediocre, obscure imitator of the masters of the Renaissance.[7]

Under the pressure of the unpleasant arguments soon to begin in a London courtroom, an agreement with Gulbenkian at last seemed to take shape; and the whole "caravan," as it was called, of oil men and their lawyers migrated to Lisbon. Finally, at the beginning of November 1948, on the Sunday before the Monday on which the court arguments were to begin, the new agreement was completed. Nubar, the dutiful and gracious son, had booked a private room in the Hotel Aviz where the signing was to take place, at 7:00 P.M., to be followed by a celebratory dinner.

At five minutes to seven, Gulbenkian announced that he had found one more point that had not been covered in the new agreements. Consternation gripped the room. Telegrams were sent back to directors in London, and replies were awaited. A stunned and depressing silence descended on the Hotel Aviz. Yet the food had been ordered, it would soon be cold, and there was no point in not eating, at least so far as Nubar Gulbenkian could see. He summoned the "caravan" to the table. The dinner that ensued was very somber and funereal; only one bottle of champagne was drunk among twelve people. There was nothing to celebrate.

Around midnight, the telegrams came back from London. Gulbenkian's final demand was acceded to. The agreements were retyped, Gulbenkian signed them at one-thirty in the morning, and they were sent by chartered plane to London. The appropriate officials were informed that the court proceedings scheduled for later that day in London should be called off, and the exhausted group in Lisbon finally adjourned to an all-night cafe to celebrate over sandwiches and cheap wine.

Thus was negotiated the Group Agreement of November 1948, which reconstituted the Iraq Petroleum Company. What Gulbenkian got, in addition to higher overall production and other advantages, was an extra allocation of oil. Mr. Five Percent was no more; he was now something greater. The agreements themselves were "monuments of complexity." An Anglo-Iranian executive (and later a chairman of the company) declared, "We have now succeeded in making the Agreement completely unintelligible to anybody." But there was an advantage to such complexity, for, as one of Gulbenkian's lawyers put it, "No one will ever be able to litigate about these documents because no one will be able to understand them."

Once the granite obduracy of Calouste Gulbenkian had been overcome and the new Group Agreement for the Iraq Petroleum Company had been signed, the Red Line Agreement was no more, and the legal threat to Jersey's and Socony's participation in Aramco was removed. It had been a long, tortured struggle by which the two companies won their entrée to Saudi Arabia. "If you laid all the conversations that went into this deal end to end," said one participant, "they would reach to the moon." In December 1948, two and a half years after the deal had first been discussed, the Jersey and Socony loans could be converted into payments, and the Aramco merger could finally be completed. A new corporate entity, more commensurate with Saudi reserves, had come into existence. With the deal done, Aramco was owned by Jersey and Socony, as well as Socal and Texaco. And it was 100 percent American.

For his part, Gulbenkian had once again succeeded in preserving his exquisite creation, the Iraq Petroleum Company, as well as his position in it, against the combined might of international oil. His last display of artistry was ultimately to earn hundreds of millions of dollars more for the Gulbenkian interests. Gulbenkian himself lived on for another six years in Lisbon, occupying himself by ceaselessly arguing with his IPC partners and by writing and rewriting his will. When seven years later, in 1955, he died at age eighty-five, he left behind three enduring legacies: a vast fortune, a splendid art collection and, most fittingly of all, endless litigation over his will and the terms of his estate.[8]

Kuwait

Another American company, Gulf Oil, faced a quandary in the Middle East. As half owner of the Kuwait Oil Company, Gulf was constrained to some degree from competing with its partner, Anglo-Iranian, particularly in India and the Middle East. Where else could Gulf dispose of its oil? It had a small system in Europe that was hardly adequate for even a fraction of the rapidly rising tide of oil that would be available from Kuwait. Gulf needed outlets, primarily in Europe. So Colonel J. F. Drake, the company's president, went in search of them. The best answer to Gulf's problem soon became apparent: the Royal Dutch/Shell Group. It owned one of the two largest marketing organizations in the Eastern Hemisphere, particularly in Europe. And unlike its competitors, it had access to very little Middle Eastern oil. As Drake explained to the State Department, a deal "between Gulf, which is long on crude oil and short on markets, and Shell, which is long on markets and short on crude oil," made excellent sense.

The two companies developed a unique purchase-and-sale agreement; it was a shadow integration that would allow Gulf's Kuwaiti oil to flow into Shell's refining and marketing system via a long-term contract—initially a ten-year agreement, which was later extended another thirteen years. The total volumes of oil over the life of the contract were estimated to account for fully a quarter of Gulf's proven reserves in Kuwait. In turn, Gulf would be providing Shell with 30 percent of its requirements in the Eastern Hemisphere. No one would be so foolish as to set a fixed price over such a long and uncertain duration. So the two companies came up with an innovative solution—what would become known as "netback pricing." The contract provided for a fifty-fifty split of the profits— profit being defined as "final selling price" minus all costs along the way. The schedules and accounting formulas by which profit would ultimately be calculated were so complicated that they took up over half of the 170 printed pages of the contract.

In truth, Gulf had hardly any alternative to Shell. Kuwaiti production was going up rapidly; the Amir would insist upon such increases, especially when he saw the production curves in his neighboring countries. Very few systems could absorb so much oil. Shell's was about the only one available. Furthermore, there was an aspect to the deal that would certainly win the approval of the State Department. It was, Colonel Drake said, the only option that Gulf could see that would leave its one-half interest in Kuwaiti oil "wholly American owned." In short, first with Aramco, and now with the Gulf-Shell arrangement, American oil interests in the Middle East were being protected. As for Shell, the deal would give it a claim on a substantial part of Kuwait's total output. It was more than merely a long-term buyer. As the Foreign Office put it, "in Her Majesty Government's view," Shell was "to all intents and purposes a partner in the Concession."[9]

Iran

The third of the great postwar oil deals involved Iran. In the course of the first discussions in London on abrogating the Red Line Agreement in the late summer and early fall of 1946, the representatives of Jersey and Socony privately raised the possibility of a long-term contract for Iranian crude with Sir William Fraser, chairman of Anglo-Iranian. "Willie" was certainly receptive. Like Gulf, Anglo-Iranian did not have the wherewithal to build up quickly a large refining and marketing system of its own in Europe, and it feared that it would find itself shut out of Europe by cheap and abundant oil from Aramco.

But political considerations also provided reason for AIOC to tie up long-term relationships with American companies and thus help "solidify" its own position. For Iran was under continuing and considerable pressure from the Soviet Union. In the latter part of World War II, the Soviet Union had demanded an oil concession in Iran, and Soviet troops continued to occupy Azerbaijan in northern Iran after the war. Stalin did not withdraw until the spring of 1946, and then only in response to intense pressure by the United States and Britain. Indeed, what became known as the Iranian Crisis of 1946 was the first major East-West confrontation of the Cold War.

In early April 1946, at the same time that the Soviets were finally beginning to withdraw their troops, the American ambassador in Moscow went to the

Kremlin for a private, late-night meeting with Stalin. "What does the Soviet Union want, and how far is Russia going to go?" the ambassador asked.

"We're not going much farther," was the not particularly reassuring reply of the Soviet dictator, who went on to describe Soviet efforts to extend influence over Iran as a defensive move to protect its own oil position. "The Baku oil fields are our major source of supply," he said. "They are close to the Iranian border and they are very vulnerable." Stalin, who had become a "journeyman of the Revolution" in Baku four decades earlier, added that "saboteurs—even a man with a box of matches—might cause us serious damage. We are not going to risk our oil supply."

In fact, Stalin was interested in Iranian oil. Soviet oil production in 1945 was only 60 percent of that of 1941. The country had desperately mobilized a range of substitutes during the war—from oil imports from the United States to charcoal-burning engines for its trucks. Shortly after the war, Stalin interrogated his petroleum minister, Nikolai Baibakov (who subsequently was to be in charge of the Soviet economy for two decades—until 1985, when Mikhail Gorbachev replaced him). Mispronouncing Baibakov's name, as he always did, Stalin demanded to know what the Soviet Union was going to do in the light of its very bad oil position. Its oil fields were seriously damaged and heavily depleted, with little promise for the future. How could the economy be reconstructed without oil? Efforts, the dictator said, would have to be redoubled.

Toward that end, the Soviet Union made its demands for a joint oil exploration company within Iran. So, certainly, oil was one Soviet objective in Iran, but not the only one, by any means, and not the most important. In 1940, in the context of the Nazi-Soviet Pact, Soviet Foreign Minister Vyacheslav Molotov had declared that "the area south of Batum and Baku in the general direction of the Persian Gulf be recognized as the center of the aspirations of the Soviet Union." That area had a name—Iran. Stalin was seeking to build up his own sphere in bordering countries, and to expand Soviet power and influence wherever he could. In trying to reach into Iran and toward the Persian Gulf, he was also pursuing a traditional objective of Russian foreign policy, one that was almost a century and a half old. The pursuit of that same objective had, at the turn of the century, provided the motivation for the British government to support the original 1901 Iranian concession of William D'Arcy Knox, as one way to blunt the Russian advance.

After Stalin pulled back his soldiers from northern Iran in 1946, the Soviet Union continued to try to gain a favored position in that area and sought to establish a joint Soviet-Iranian oil company. Meanwhile, the communist-led Tudeh party was conducting a campaign of demonstrations and political pressure to gain greater sway over the central government—including a general strike and demonstrations at Anglo-Iranian's Abadan refinery complex, in which a number of people were killed. Iran was unstable, the political institutions in the country were weak, and there was the grave possibility of civil war or even of the disappearance of Iran into the Soviet bloc.

Both the American and British governments were trying to help to preserve the independence and territorial integrity of Iran. And London was categoric:

Anglo-Iranian's oil position in Iran was the company's crown jewels, and it had to be preserved at all costs. In the face of such uncertainty and in light of the high stakes, there was some considerable value in having major American companies take a more direct interest in Iranian oil. Thus, political as well as commercial realities underlay the deal between Anglo-Iranian and the two American companies, Jersey and Socony. In September 1947, the three companies signed a twenty-year contract.[10]

With the completion of the three huge deals—Aramco, Gulf-Shell, and the long-term Iranian contracts—the mechanisms, capital, and marketing systems were in place to move vast quantities of Middle Eastern oil into the European market. In the postwar world, petroleum's "center of gravity"—not only for the oil companies, but also for the nations of the West—was indeed shifting to the Middle East. The consequences would be momentous for all concerned.

Europe's Energy Crisis

The swelling volumes of Middle Eastern oil were crucial to the postwar recovery of a devastated Europe. Destruction and disorganization were everywhere. The workshop in the heart of Europe, Germany, was hardly functioning at all. Throughout Europe, food and raw materials were in desperately short supply, established trading patterns and organizations had broken down, inflation was rampant, and there was a severe shortage of the American dollars that were required to purchase necessary imports. By 1946, Europe was already gripped by a severe energy crisis—a terrible shortage of coal. And then the weather, the longest and coldest winter of the century, brought conditions to a crisis point. In England, the River Thames froze at Windsor. Throughout Britain, coal was in such short supply that power stations had to be shut down, and electricity to industry was either reduced greatly or cut off entirely. Unemployment abruptly increased six times over, and British industrial production was virtually halted for three weeks—something German bombing had never been able to accomplish.

This unexpected shortfall of energy drove home the extent to which Britain had been impoverished by the war. Its imperial role had become an insupportable burden. During those few bleak, freezing, and pivotal weeks of February 1947, the Labour government of Clement Attlee referred its intractable Palestine problem to the United Nations and announced that it would grant independence to India. And on February 21 it told the United States that it could no longer afford to prop up the Greek economy. It asked that the United States take over that responsibility and, by implication, broader responsibilities throughout the Near and Middle East. Still, the situation worsened. Throughout Europe, the economic disarray brought on by the weather and the energy crisis in the winter of 1947 accentuated the shortfall of U.S. dollars, which constrained Europe's ability to import vital goods and paralyzed its economy.[11]

The first step toward averting a massive breakdown was taken in June 1947, in Harvard Yard in Cambridge, Massachusetts. There, at the Harvard commencement, United States Secretary of State George Marshall introduced the concept of a large, broadly based foreign aid program that would help revive and reconstruct the economies of Western Europe in a continental framework and

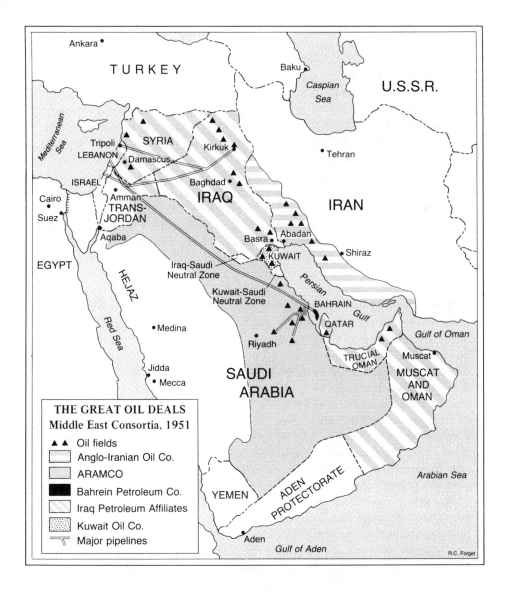

Middle East Consortia, 1951

▲ ▲ Oil fields
Anglo-Iranian Oil Co.
ARAMCO
Bahrein Petroleum Co.
Iraq Petroleum Affiliates
Kuwait Oil Co.
Major pipelines

that would fill the gap created by the shortage of dollars. In addition, the European Recovery Program, or the Marshall Plan, as it was soon known, became a central element in the containment of Soviet power.

Among the first problems to be addressed was Europe's energy crisis. There was not enough coal capacity, productivity was low, and the labor force was disorganized. Moreover, in many countries, communists occupied leading roles in the miners' unions. Oil was part of the solution; it could replace coal in industrial boilers and power plants. Oil was also, obviously, the only source of fuel for Europe's airplanes, motorcars, and trucks. "Without petroleum the Marshall Plan could not have functioned," said a U.S. government report at the time.

Those in Paris running the European Recovery Program did not worry much about the physical availability of oil. They simply counted on the companies to assure that supplies were there. Oil did, however, have to be imported, and that made it not only part of the solution, but also part of the problem. Approximately half of Europe's petroleum came from American companies, which meant that it had to be paid for in dollars. For most of the European countries, oil was the largest single item in their dollar budgets. It was estimated in 1948 that upwards of 20 percent of total Marshall Plan aid over the subsequent four years would go to the imports of oil and oil equipment.[12]

Price became a very contentious matter. The Europeans became particularly outspoken on the subject of the dollar drain from the purchase of oil in 1948, when oil prices, rising quickly, were at their postwar peak. "How unfortunate it was," British Foreign Secretary Ernest Bevin told the American ambassador, that "while the Americans were voting money to help Europe, the rise in oil prices nullified their efforts to a considerable extent." The dollar drain led to a bitter argument as to how much "dollar oil" (from American companies) and how much "sterling oil" (from British companies) would come into the United Kingdom and the rest of Europe. There was also a running battle with the oil companies about cost, particularly for the growing volumes of Middle Eastern oil, and whether those prices were competitively set or could and should be lower. Eventually, after much acrimony, Middle Eastern oil was being pushed down to price levels below what had, until then, been the benchmark U.S. Gulf Coast price. This meant the end of the pricing convention that had been established two decades earlier, at Achnacarry Castle. The last vestige of the prewar "As-Is" system was now gone.[13]

Yet, despite all the controversies, the fundamental fact was that the Marshall Plan made possible and pushed a far-reaching transition in Europe—the change from a coal-based economy toward one based on imported oil. The short supplies of coal, compounded by labor strife and strikes in the mining industry, gave powerful impetus to that change. "It is not a joyful thing, but it is a national necessity to import more oil," Britain's Chancellor of the Exchequer, Hugh Dalton, told Marshall. Government policies also encouraged the conversion of power plants and industry from coal to oil. Because of the great surge of cheap production from the Middle East, oil could effectively compete against coal in price. Moreover, when industrial consumers came to make their choice, they could see a clear distinction between coal, whose tribulations and disruptions were daily fare in the press, and oil, whose supply and distribution were handled efficiently and with little friction.

Where possible, the oil companies moved to capture new markets, both in industry and in homes—in the latter case, with the revolutionary innovation of central heating. In the words of one Shell manager, "The Englishman began to realize that there was no value in being cold, and there was no reason why he should not have the amenities of his American and Canadian cousins." Though Europe remained a coal-based economy, oil became increasingly important, especially to fuel the incremental growth in energy demand. That was where the new production of the Middle East loomed so large. In 1946, 77 percent of Europe's oil supply came from the Western Hemisphere; by 1951, it was expected that a dramatic shift would take place—with 80 percent of supply to come from the Middle East. The synchronization of Europe's needs and the development of Middle Eastern oil meant a powerful and timely combination.[14]

To Market It Goes?

There was still the problem of getting those rapidly growing volumes of oil to market. Aramco and its parent companies, now four, were continuing to battle to build Tapline, which would carry Saudi Arabian oil to the Mediterranean. But several major obstacles stood in its path. Steel, in short supply, was still under United States government controls, and yet a big share of the total steel output in the United States would have to be allocated for the pipes and tubing required for the mammoth undertaking. The independent oil men and their Congressional allies tried to block the allocation in order to prevent the buildup of large volumes of cheap foreign oil that they feared would flow into the American market. But there was significant support for Tapline in the Truman Administration, much of it based on the view that Middle Eastern oil supply was essential to the success of the Marshall Plan. Without the pipeline, warned one State Department official, "the European recovery program will be seriously handicapped."

Another obstacle was the stubbornness of the countries the pipeline had to cross, particularly Syria, all of which were demanding what seemed to be exorbitant transit fees. It was also the time when the partition of Palestine and the establishment of the state of Israel were aggravating American relations with the Arab countries. But the emergence of a Jewish state, along with the American recognition that followed, threatened more than transit rights for the pipeline. Ibn Saud was as outspoken and adamant against Zionism and Israel as any Arab leader. He said that Jews had been the enemies of Arabs since the seventh century. American support of a Jewish state, he told Truman, would be a death blow to American interests in the Arab world, and should a Jewish state come into existence, the Arabs "will lay siege to it until it dies of famine." When Ibn Saud paid a visit to Aramco's Dhahran headquarters in 1947, he praised the oranges he was served but then pointedly asked if they were from Palestine—that is, from a Jewish kibbutz. He was reassured; the oranges were from California. In his opposition to a Jewish state, Ibn Saud held what a British official called a "trump card": He could punish the United States by canceling the Aramco concession. That possibility greatly alarmed not only the interested companies, but also, of course, the U.S. State and Defense departments.

Yet the creation of Israel had its own momentum. In 1947, the United Nations Special Committee on Palestine recommended the partition of Palestine, which was accepted by the General Assembly and by the Jewish Agency, but rejected by the Arabs. An Arab "Liberation Army" seized the Galilee and attacked the Jewish section of Jerusalem. Violence gripped Palestine. In 1948, Britain, at wit's end, gave up its mandate and withdrew its Army and administration, plunging Palestine into anarchy. On May 14, 1948, the Jewish National Council proclaimed the state of Israel. It was recognized almost instantly by the Soviet Union, followed quickly by the United States. The Arab League launched a full-scale attack. The first Arab-Israeli war had begun.

A few days after Israel's proclamation of statehood, James Terry Duce of

Aramco passed word to Secretary of State Marshall that Ibn Saud had indicated that "he may be compelled, in certain circumstances, to apply sanctions against the American oil concessions . . . not because of his desire to do so but because the pressure upon him of Arab public opinion was so great that he could no longer resist it." A hurriedly done State Department study, however, found that, despite the large reserves, the Middle East, excluding Iran, provided only 6 percent of free world oil supplies and that such a cut in consumption of that oil "could be achieved without substantial hardship to any group of consumers."[15]

Ibn Saud could certainly have canceled the concession, but at considerable risk. For Aramco was the sole source of his rapidly rising wealth, and the broader relationship with the United States provided the basic guarantee of Saudi Arabia's territorial integrity and independence. Ever suspicious of the British, the King feared that London might be sponsoring a new coalition to champion the Hashemites, as it had done after World War I, enabling the Hashemites—whom Ibn Saud had driven from Mecca only two decades earlier—to recapture the western part of his country. His apprehension mounted when Abdullah, the Hashemite King of Jordan, was said to have "likened the Saudi regime to the Jewish occupation of Palestine." And the Hashemites constituted, to Ibn Saud, an even greater enemy than the Jews. The Soviet Union and communists, too, were a more dangerous threat, in terms of Soviet pressure to the north and the impact of communist activity within the Arab world.

Indeed, in the face of the Hashemite and communist threats, Ibn Saud pressed the Americans and even the British in late 1948 and 1949 for a tripartite defensive treaty. The British minister to Saudi Arabia observed in his annual review to London, "As Israel became, and was seen by most Arabs to become, a reality which could not be wished away, the Saudi Arabian Government resigned themselves to its existence in practice while maintaining their formal hostility to Zionism." Ibn Saud found he could distinguish between Aramco, a purely commercial enterprise owned by four private companies, and the policy of the U.S. government elsewhere in the region. When other Arab countries declared that Saudi Arabia should cancel the concession to retaliate against the United States and prove its allegiance to the Arab cause, Ibn Saud replied that oil royalties helped to make Saudi Arabia "a stronger and more powerful nation, better to assist her neighboring Arab states in resisting Jewish pretensions."

And so, even while Jew and Arab were at war in Palestine, the feverish oil development continued within Saudi Arabia, and construction proceeded on Tapline. It was finished in September 1950. Two more months were required to fill the line, and in November, the oil began to arrive at Sidon in Lebanon, the terminus on the Mediterranean, where it was picked up by tankers for the last leg of the journey to Europe. Tapline's 1,040 miles would replace 7,200 miles of sea journey from the Persian Gulf through the Suez Canal. Its annual throughput was the equivalent of sixty tankers in continuous operation from the Persian Gulf, via the Suez Canal, to the Mediterranean. The oil it carried would fuel the recovery of Europe.[16]

No Longer "Far Afield": The New Dimension of Security

The overlapping of politics and economics had, in the second half of the 1940s, created a new strategic focus for both the British and the American governments. In the case of the British, even as they withdrew from the far reaches of empire, they could not turn their backs on the Middle East. The Soviets were putting pressure on the "northern tier"—Greece, Turkey, and in particular, Iran. And Iran, plus Kuwait and Iraq, were Britain's major sources of oil. Continued access was required for military security, and the dividends from Anglo-Iranian were a major revenue generator for the Exchequer. "Without the Middle East and its oil," Foreign Secretary Bevin told the Cabinet Defense Committee, he saw "no hope of our being able to achieve the standard of living at which we are aiming in Great Britain."

If the British focus narrowed, the shift in outlook and commitments greatly expanded for the United States. No longer would an American President say, as Franklin Roosevelt had said in 1941, that Saudi Arabia was a little far afield. The United States was becoming an ever more petroleum-based society, which could no longer supply its own needs with domestic production. The world war, just over, had proved how central and critical oil was to national power. American leaders and policymakers were also moving toward a much wider definition of national security—one that reflected the realities of the postwar balance of power, the growing clash with the Soviet Union, and the evident fact that the mantle was passing from Britain to the United States, which was now by far the preeminent power in the world.

Soviet expansionism—as it was, and as it might be—brought the Middle East to center stage. To the United States, the oil resources of the region constituted an interest no less vital, in its own way, than the independence of Western Europe; and the Middle Eastern oil fields had to be preserved and protected on the Western side of the Iron Curtain to assure the economic survival of the entire Western world. Military planners did have considerable doubt about whether the oil fields could actually be defended in an extended "hot war," and they gave at least as much thought to how to destroy them as to how to defend them. But in the Cold War, this oil would be of great value, and everything possible should be done to prevent its loss.

Saudi Arabia became the dominant focus of American policymakers. Here was, said one American official in 1948, "what is probably the richest economic prize in the world in the field of foreign investment." And here the United States and Saudi Arabia were forging a unique new relationship. In October 1950 President Harry Truman wrote a letter to King Ibn Saud. "I wish to renew to Your Majesty the assurances which have been made to you several times in the past, that the United States is interested in the preservation of the independence and territorial integrity of Saudi Arabia. No threat to your Kingdom could occur which would not be a matter of immediate concern to the United States." That sounded very much like a guarantee.

The special relationship that was emerging represented an interweaving of public and private interests, of the commercial and the strategic. It was effected both at the governmental level and through Aramco, which became a mechanism not just for oil development, but also for the overall development of Saudi Arabia—though insulated from the wide range of Arabian society and always within the limits prescribed by the Saudi state. It was an unlikely union—Bedouin Arabs and Texas oil men, a traditional Islamic autocracy allied with modern American capitalism. Yet it was one that was destined to endure.[17]

The End of Energy Independence

Since Middle Eastern oil could not, in the event of war, be very easily protected and was, in the words of the United States Joint Chiefs of Staff, "very susceptible to enemy interference," how could overall security of supply be assured in a future conflict? This question became a major topic of discussion both in Washington and in the oil industry. Some argued for importing more oil in peacetime in order to preserve domestic resources for wartime. Such was the call in *A National Policy for the Oil Industry*, a controversial book by Eugene V. Rostow, a Yale Law School professor. A new Federal agency, the National Security Resources Board, made a similar argument in a major policy review in 1948; importing large amounts of Middle Eastern oil would allow a million barrels per day of Western Hemisphere production to be shut in, in effect creating a military stockpile in the ground—"the ideal storage place for petroleum."

Many advocated that the United States do what Germany had done during the war—build a synthetic fuels industry, extracting liquids not only from coal, but also from the oil shale in the mountains of Colorado and from abundant natural gas. Some were confident that synthetic fuels could soon be a major source of energy. "The United States is on the threshold of a profound chemical revolution," said the *New York Times* in 1948. "The next ten years will see the rise of a massive new industry which will free us from dependence on foreign sources of oil. Gasoline will be produced from coal, air, and water." The Interior Department optimistically declared that gasoline could be made from either coal or shale, for eleven cents a gallon—at a time when the wholesale price of gasoline was twelve cents a gallon!

The more realistic and widespread view in the oil industry was that synthetic fuels were, at best, on the horizon. Still, in late 1947, as the Cold War intensified, the Interior Department called for another Manhattan Project: a huge, crash $10 billion program that would be capable, within four or five years, of producing two million barrels per day of synthetic fuels. As it was, a total of only $85 million was authorized under the Truman Administration for such research. And as time went on, the cost projections became higher and higher, until it was estimated, in 1951, that gasoline from coal would cost three and a half times the market price for conventional gasoline. In the end, it was the ever-growing availability of cheap foreign oil that made synthetic fuels irrelevant and uneconomical. Imported petroleum killed synthetic fuels. And they would remain dead for three decades, until hastily resurrected in response to an interruption in the flow of imported oil.[18]

In the immediate postwar years, technology was opening new domestic frontiers for exploration and development. Much greater depth was attained in drilling, which increased production. And even more innovative was the development of offshore production. As far back as the mid-1890s, operators were drilling off piers near Santa Barbara, but the wells produced no more than one or two barrels per day. In the first decades of the twentieth century, wells were drilled from fixed platforms in lakes in Louisiana and Venezuela. In the 1930s, drillers had walked into the shallow waters immediately off the beaches of Texas and Louisiana, though with little success. They were only wading distance from land. It was quite another thing to go offshore altogether, into deeper waters of the Gulf of Mexico, out of sight of land. That would require the creation of a new industry. Kerr-McGee, an Oklahoma independent, took the gamble. And a very big gamble it was. The technology and know-how did not exist for building a platform, getting it into position, drilling into the ocean floor—or even for servicing the operation. Moreover, essential knowledge about such important matters as weather (including hurricanes), tides, and currents was either rudimentary or almost nonexistent.

Because of its size, the management of Kerr-McGee reasoned that it did not have much chance to win attractive, "real class-one" onshore acreage against the larger companies. But, when it came to offshore sites in the Gulf of Mexico, there was hardly any competition. Indeed, many other companies thought that offshore development was simply impossible. Kerr-McGee put the pieces together, and on a clear Sunday morning in October 1947, on Block 32, ten and a half miles off the Louisiana coast, its drillers struck oil.

The well on Block 32 was a landmark event, and other companies followed Kerr-McGee's lead. Yet the buildup of offshore exploration was not as fast as it might have been, partly because of the expense. An offshore well could cost as much as five times more than a well of similar depth onshore. Development was also slowed by an intense struggle between the Federal government and the states over who actually owned the continental shelf. Of course, what they were really fighting about was who would get the tax revenues, and that matter would not be resolved until 1953.[19]

Given that synthetic fuels would be very expensive and offshore development was only just beginning, was there any other alternative to imported oil? There was. The answer could be seen at night, along the endless highways of Texas, in the bright spears of light that shot up from the flat plains. It was natural gas, considered a useless, inconvenient by-product of oil production and thus burned off—since there was nothing else to do with it. Natural gas was the orphan of the oil industry. Only a fraction of natural gas production was used, mostly in the Southwest. Yet the country appeared to have huge reserves of gas, which could well substitute for oil—or, for that matter, for coal—in residential heating and in industry. But it so lacked in markets that it was sold, on an energy basis, for only a fifth of what oil from the same well would cost.

Natural gas required no complex engineering processes in order to be used. The problem was transmission: how to get it to the markets in the Northeast and the Midwest, where both the large populations and the major industries of the country were to be found. That meant long-distance pipelines, halfway across the country, in an industry for which long distance had heretofore meant 150 miles. But the commercial arguments, compounded by the concerns about national security and dependence on foreign oil, were very compelling. In a judgment that found favor with Defense Secretary Forrestal, the House Armed Services Committee declared that increasing the use of natural gas was "the quickest and cheapest method immediately available to reduce domestic consumption of petroleum" and that, therefore, steel should "be made available for natural gas pipelines in advance of any other use now proposed."

In 1947, both Big Inch and Little Inch—the pipelines hurriedly built in wartime to bring oil from the Southwest to the Northeast—were sold to the Texas Eastern Transmission Company and turned into natural gas pipelines. The same year, in a project championed by Pacific Lighting, the parent of Southern California Gas, Los Angeles was hooked up by large-diameter pipe to the gas fields of New Mexico and West Texas. That pipeline itself, owned by El Paso Natural Gas, was dubbed "Biggest Inch." By 1950, the interstate movement of natural gas reached 2.5 trillion cubic feet—almost two and a half times the level of 1946. Without the additional natural gas use, American oil demand would have been seven hundred thousand barrels per day higher.

By then the new petroleum order had been established, centered in the Middle East, and within it, the oil companies were moving at a feverish pace to satisfy the rapidly rising demand in the marketplace—consumption in the United States jumped 12 percent in 1950 over 1949. Oil would prove to be the favored fuel, not only in the United States, but also in Western Europe and later Japan, providing the energy to power two decades of remarkable economic growth. Forged to meet new political and economic realities, the postwar petroleum order was a great success—in fact, it was, in some ways, far too successful. Already, by 1950, it was clear that the problem now facing the industry was not the immediate postwar anxiety of being unable to keep up with demand. On the contrary, as a Jersey analysis in July of that year described the situation, "It appears that in the future, Mid-East crudes available to Jersey may exceed requirements substantially." What was true for Jersey would be true for the other majors. The Jersey forecast was only a hint of the massive surplus that would confront the industry in the years ahead. In the meantime, even as the new petroleum order was beginning to generate massive profits, bitter battles were already erupting over how these profits were to be divided.[20]

CHAPTER 22

Fifty-Fifty: The New Deal in Oil

REPRESENTATIVES FROM the United States Treasury were conferring in London with British officials in 1950. In the course of their discussions, the Americans mentioned certain developments regarding Saudi Arabian oil policies, the effects of which would surely be felt throughout the Middle East. "The Saudi Arabian government had recently made some startling demands on Aramco," one of the American officials confided. "They covered all possible points ever thought up by a concessionary government." Yet, in one form or another, the demands all came down to one thing: The Saudis wanted more money out of the concession. A good deal more.

Such demands were by no means restricted to Saudi Arabia. In the late 1940s and early 1950s, oil companies and governments grappled continually over the financial terms upon which the postwar petroleum order would rest. The central issue was the division of what has been called "that uneasy and important term in the economics of natural resources"—rents. The character of the struggle varied among countries, but the central objective of those initiating the struggle in each country was the same: to shift revenues from the oil companies and the treasuries of the consuming countries that taxed them to the treasuries of the oil-exporting countries. But money was not the only thing at stake. So was power.

Landlord and Tenant

"Practical men, who believe themselves to be quite exempt from any intellectual influences," John Maynard Keynes once said, "are usually the slaves of some defunct economist." When it came to oil, the "practical men" included not only the businessmen that Keynes had in mind, but also kings, presidents, prime ministers, and dictators—as well as their ministers of oil and finance. Ibn Saud and the other leaders of the time, as well as the various potentates since, were under the thrall of David Ricardo, a fantastically successful stockbroker in late-eighteenth-century and early-nineteenth-century England. (Among other things, he made a killing on Wellington's defeat of Napoleon at Waterloo.) By origin a Jew, Ricardo became a Quaker, then a learned member of the House of Commons, and was one of the founding fathers of modern economics. He and Thomas Malthus, his friend and intellectual rival, constituted between themselves the successor generation to Adam Smith.

Ricardo developed the concept that was to provide the framework for the battle between nation-states and oil companies. It was the notion of "rents" as something different from normal profits. His case study involved grain, but it could also apply to oil. Let there be two landlords, said Ricardo, one with fields much more fertile than the other. They both sell their grain at the same price. But the costs of the one with the more fertile fields are much less than those of the one with the less fertile fields. The latter makes, perhaps, a profit, but the former, the one with the more fertile fields, receives not only a profit, but also something much larger—rents. His rewards—rents—are derived from the particular qualities of his land, which result not from his ingenuity or hard work, but, uniquely, from nature's bountiful legacy.

Oil was another of nature's legacies. Its geological presence had nothing to do with the character or doings of the peoples who happened to reside above it, or of the nature of the particular political regime that held sway over the region in which it was found. This legacy, too, generated rents, which could be defined as the difference between the market price, on one hand, and, on the other, the costs of production plus an allowance for additional costs—transportation, processing, and distribution—and for some return on capital. For example, in the late 1940s, oil was selling for around $2.50 a barrel. Some grizzled stripper-well operator in Texas might only make a 10 cent profit on his oil. But in the Middle East it only cost 25 cents a barrel to produce oil. Deducting 50 cents for other costs, such as transportation, and allowing a "profit" of 10 cents on the $2.50 barrel, that would still leave a very large sum—$1.65 on every barrel of Middle Eastern oil. That sum would constitute rents. Multiply it by whatever the rising production numbers, and the money added up very rapidly. And who—the host country, the producing company, or the consuming country that taxed it—would get how much of those rents? There was no agreement on this elemental issue.

All would have legitimate claims. The host country had sovereignty over the oil beneath its soil. Yet the oil was without value until the foreign company risked its capital and employed its expertise to discover, produce, and market it. The host country was, in essence, the landlord, the company a mere tenant, who would, of course, pay an agreed-upon rent. But, if through the tenant's risk-taking and efforts, a discovery was made and the value of the landlord's property vastly increased, should the tenant continue to pay the same rent as under the original terms, or should it be raised by the landlord? "This is the great divide of the petroleum industry: a rich discovery means a dissatisfied landlord," said the oil economist M. A. Adelman. "He knows that the tenant's profit is far greater than is necessary to keep him producing, and he wants some of the rent. If he gets some, he wants more."[1]

The postwar battle over rents was not exclusively limited to economics. It was also a political struggle. For the landlords, the oil-producing countries, the struggle was interwoven with the themes of sovereignty, nation-building, and the powerful nationalistic assertion against the "foreigners," who were said to be "exploiting" the country, stifling development, denying social prosperity, perhaps corrupting the body politic, and certainly acting as "masters"—in a haughty, arrogant, and "superior" manner. They were seen as the all-too-visible embodiment of colonialism. Nor did their sins end there; they were, in addition, draining off the "irreplaceable heritage" and bounty of the landlord and his future generations. Of course, the oil companies saw it all quite differently. They had taken the risks and held their breath, they chose to put their capital and efforts here and not there, and they signed laboriously negotiated contracts, which gave them certain rights. They had created value where there was none. They needed to be compensated for the risks they had taken—and the dry holes they had drilled. They believed that they were being put upon by greedy, rapacious, and unreliable local powers-that-be. They did not think that they were "exploiting"; their plaintive cry was, "We wuz robbed."

There was yet another political dimension to the struggle. To the consuming countries in the industrial world, access to oil was a strategic prize, not only vital to their economy and their ability to grow, but also a central and essential element in national strategy—and, by the by, also a significant source of tax revenues, both directly in excise taxes and by fueling overall economic activity. To the producing country, oil also meant power, influence, significance, and status—all of which were previously lacking. Thus, it was a struggle in which money signified both power and pride. That was what made the battle often so bitter. The first front in this epic contest was opened in Venezuela.

Venezuela's Ritual Cleansing

The tyrannical dictatorship of Venezuela's General Gómez had come to an end in 1935, through the one sure method when all else fails—the dictator's death. Gómez left a shambles; he had treated Venezuela as a sole proprietorship, a personal hacienda, managed for his own enrichment. Much of the population remained impoverished, while the nation's oil industry had been developed to the point where the country's overall economic destiny depended upon it. Gómez also left behind a wide tableau of opposition. Military men had been humiliated by their treatment at Gómez's hands; they were poorly paid, lacked status, and had to spend part of their time tending the dictator's own multitudinous herds of cattle. No less important was the creation of an opposition on the democratic left, centered in what became known as the "Generation of '28"—students at the Universidad Central in Caracas who had rebelled against Gómez in 1928. They had failed then, of course, and the leaders either went to prison, where sixty-pound leg irons were fastened to their ankles, or went into exile, or were sent by Gómez to work on road crews in disease-ridden jungles in the interior. Many members of the Generation of '28 perished, victims in one way or another of Gómez's terror. Those who did survive became the nucleus of the reformers, liberals and socialists, who worked their way back into Venezuelan political life after Gómez's death. When it finally came to power, Venezuela's Generation of '28 would provide the basis for redefining the relationship between oil companies and producing countries, between tenant and landlord around the world—as well as the methodology for reallocating rents.

With oil already dominating Venezuela's cash economy—accounting for well over 90 percent of total export value in the late 1930s—Gómez's successors set out to reform the chaotic regulation of the industry and effect a wholesale revision in the contractual arrangements between the nation and the companies that produced its oil, including a reallocation of rents. The United States government was catalyst to the process. During World War II, Washington was all too conscious of the continuing strife with Mexico about the nationalization of its oil industry and was intent on protecting access to Venezuela, which was the most important source of oil outside the United States, and one that was relatively secure. Thus, the American government would intervene directly to avoid another Mexico and to safeguard what was, in the midst of wartime, a great strategic prize. For their part, the companies did not want to risk nationalization. Standard Oil of New Jersey and Shell were the dominant producers in Venezuela. They knew they were sitting on some of the most important oil reserves in the world, which they could not afford to lose. Venezuela was the major source of low-cost oil, and Jersey's Creole subsidiary generated half of the company's total worldwide production and half of its total income.[2]

Jersey, however, was sharply divided about what to do in the face of the drive by the Venezuelan government to reallocate the rents. Traditionalists in the company, some of whom had been cronies of the old Gómez regime, wanted to stand firm against any change, whether pushed by Caracas or pushed by Washington. Opposed to them was Wallace Pratt, who had been the company's chief geologist before moving into senior executive positions. Pratt, with long experience in Latin America, thought that the world had changed, and that change and adaptation on the part of the company were not only inevitable but essential to protect its long-term interests. He was also convinced that implacable resistance would probably be not only expensive, but also futile. Better to help create the new order, in Pratt's view, than be a victim of it. The debate came at a time when Jersey itself was the target of traumatic and painful political attacks in Washington because of the controversy over its prewar relationship with I. G. Farben and a new antitrust campaign from the Justice Department. As a result, Jersey altered its attitude and orientation toward public policy and the political environment, and not only in the United States. Moreover, the Roosevelt Administration made it quite clear that, in any dispute with Venezuela that arose from the company's failure to adapt, Jersey could not count on support from Washington.

Jersey simply could not risk losing its position in Venezuela, and Wallace Pratt won. Jersey installed a new top man in Venezuela, Arthur Proudfit, who sympathized with the country's social objectives, and who would show considerable sensitivity to the changing Venezuelan political scene. Proudfit had been part of the migration in the 1920s of American oil men from Mexico to

Venezuela; he brought with him a lasting memory of the disaster of government-company relations and of the bitter labor strife in the oil fields, along with a determination to apply the painful lessons he had learned in Mexico.

All the main players—the Venezuelan and American governments, Jersey and Shell—wanted to work things out. To help facilitate matters, the U.S. Undersecretary of State, Sumner Welles, took the unprecedented step of recommending to the Venezuelan government the names of independent consultants, including Herbert Hoover, Jr., son of the former President and a well-known geologist in his own right, who could help Venezuela to improve its bargaining position vis-à-vis the companies. Welles also pressed the British government to make sure that Royal Dutch/Shell went along. With the assistance of the consultants, a settlement was hammered out based on the new principal of "fifty-fifty." It was a landmark event in the history of the oil industry. According to this concept, the various royalties and taxes would be raised to the point at which the government's take would about equal the companies' net profits in Venezuela. The two sides would, in effect, become equal partners, dividing the rents down the middle. In exchange, the questions about the validity of various concessions would be overlooked—and there were certainly pointed questions about how some had been obtained by Jersey and the companies it had acquired. The title to existing concessions would also be solidified and their life extended, and new exploration opportunities would be made available. These, for the companies, were very desirable gains.

The proposed law was criticized by members of Acción Democrática, the liberal/socialist party that had been formed by the survivors of the Generation of '28. They charged that the law as written would result in a division well below fifty-fifty for Venezuela, and they argued that Venezuela should be compensated for past profits the companies had made. "The total purification of the Venezuelan oil industry, its ritual cleansing, will remain impossible until the companies have paid adequate financial compensation to our country," declared Juan Pablo Pérez Alfonzo, Acción Democrática's spokesman on oil. Despite the abstention of the Acción Democrática deputies, the Venezuelan Congress passed the new petroleum law in March 1943, enshrining the agreement.

The major companies were quite prepared to live under the new system. "What they are after is money," Shell director Frederick Godber said of the Venezuelan government shortly after the law was passed. "Unless they are prompted to do so by our friends across the water, they are not likely yet to reject good money from wherever it comes." But unlike the majors, some of the smaller companies operating in Venezuela were outraged. William F. Buckley, president of the Pantepec Oil Company, telegraphed the Secretary of State to denounce the new law as "burdensome" and to declare that it had been accepted only "under compulsion by the Venezuelan government and our State Department." It was a clear invitation, he added, to further "agitation and attempts to invade the property rights of American oil interests." Buckley's telegram was filed away.

Two years later, in 1945, the interim regime in Venezuela was toppled by a coup by dissatisfied young military officers, acting in collaboration with Acción Democrática. Romulo Betancourt was the first president of the new junta. He had played forward on the university's championship soccer team before emerging as the leader of the Generation of '28, was subsequently twice exiled, became secretary general of the Acción Democrática, and was serving as a first-term member of the Caracas City Council at the time of the coup. The minister of development was Juan Pablo Pérez Alfonzo, who had been the leading Congressional critic of the Petroleum Law of 1943, and who now complained that the promised fifty-fifty split actually worked out to about sixty-forty, in the companies' favor. Pérez Alfonzo instituted significant revisions in the tax laws that were calculated to ensure that the division really was fifty-fifty. Jersey accepted the changes; its resident manager, Arthur Proudfit, told the State Department that "no reasonable objection could be raised to an upward revision of the income tax structure." Altogether, rents were dramatically reallocated between Venezuela and the oil companies by the Petroleum Law of 1943 and Pérez Alfonzo's subsequent adjustments. As the result of those changes, plus the rapid expansion of production, the government's total income in 1948 was six times greater than it had been in 1942.

In another precedent-breaking move, Pérez Alfonzo decided to try to capture income from the downstream segments of the industry; he wanted Venezuela, he said, to "reap the profits of transportation, refining, and marketing." In pursuit of this objective, he insisted on taking some of the royalties due Venezuela not in money, but in kind—that is, in oil. He then turned around and sold the royalty oil directly on the world market. This broke a worldwide "taboo," said President Betancourt. "The name of Venezuela was now known on the world oil market as a country where oil could be bought by direct negotiation. The veil of mystery over the marketing of oil—behind which the Anglo-Saxons had maintained a monopoly of rights and secrets—was removed forever."

In sharp contrast to what had happened in Mexico, the larger oil companies not only adapted to the redivision of rents but also established a successful working relationship with the Acción Democrática during its time in power. Creole moved swiftly to fill its ranks with nationals; in a few short years, its work force became 90 percent Venezuelan. Arthur Proudfit of Creole even lobbied on behalf of the Venezuelan government with the U.S. State Department, at the time when Creole itself was described by *Fortune* as "perhaps the most important outpost of U.S. capital and know-how abroad."

Betancourt may have once called the international companies "imperialist octopi." But he and his colleagues were, essentially, pragmatists; they recognized that they needed the companies and that they could work with them. Oil provided 60 percent of the government's income; the economy was virtually based upon oil. "It would have been a suicidal leap into space to nationalize the industry by decree," Betancourt said afterward. National objectives could be met without nationalization. With some pride, Betancourt noted that, as a result of the tax reforms of the mid-1940s, the Venezuelan government received 7 percent more per barrel than what was paid to the Mexican government by its nationalized industry. And Venezuelan production was six times that of Mexico.

Under Betancourt, the fifty-fifty principle was securely established in Venezuela. But time was running out. A new Acción Democrática government had been elected with over 70 percent of the vote in December 1947. Less than a year later, in November 1948, it was overthrown by members of the same military clique that had been its allies in the 1945 coup.

Some of the oil operators applauded the November 1948 coup. William F. Buckley was delighted for, he said, Betancourt and his allies in Acción Democrática "have used the vast dollar resources of the country to further Russian Communistic interests in the Western Hemisphere, and they have forced American capital to provide the money for this anti-American campaign." That was not, however, the way the major oil companies saw things. Arthur Proudfit found the coup "disheartening and disappointing." It threatened three years of intense efforts to establish a stable relationship with the democratic government.

And Betancourt had demonstrated his pragmatism in many ways. He had even invited a prominent American citizen to establish a new enterprise—the International Basic Economy Corporation, which would fund development projects and new businesses in Venezuela. This particular American owed his considerable fortune to oil. He was the recently resigned Coordinator for Inter-American Affairs in the State Department—Nelson A. Rockefeller, grandson of John D.[3]

The Neutral Zone

Another precedent-shattering redefinition of the relationship between landlord and tenant took place in a remote area of the world that had not one landlord, but two. The Neutral Zone was the two thousand or so square miles of barren desert that had been carved out by the British in 1922 in the course of drawing a border between Kuwait and Saudi Arabia. In order to accommodate the Bedouins, who wandered back and forth between Kuwait and Saudi Arabia and for whom nationality was a hazy concept, it was agreed that the two countries would share sovereignty over the area. If every system has within it the seeds of its own destruction, then it was in the Neutral Zone—and in the way its oil rights were parceled out—that the erosion began that would eventually lead to the end of the postwar petroleum order.

At war's end, the United States government, and specifically the State Department, endorsed and actively supported many of the new oil arrangements in the Middle East, but it had continually worried about one thing: the interlocking relationships among the major oil companies that were emerging from the "great oil deals." It was concerned about the effect on competition and the marketplace. It worried even more about the *perception* both of the dominating role of such a small group of companies and of U.S. government support for them. The whole thing might look too much like a cartel, perfect grist for nationalist and communist mills in and around the region. At the same time, the new system in the Middle East might easily fire up criticism and opposition from diverse groups in the United States, not only trust-busters and liberal critics of big business, but also the independent sector of the domestic oil industry, with its built-in enmity toward "big oil" and now, increasingly, toward "foreign oil."

To forestall these criticisms and perceptions, Washington adopted a surprisingly explicit policy of encouraging "new companies" to participate in Middle

Eastern oil development in order to counterbalance the majors and their consortia. Such a policy would meet two additional political concerns of the State Department. The introduction of more players would stimulate the pace of development of Middle Eastern oil reserves and thus bring higher revenues to the countries in the region, which was an increasingly important objective. At the same time, it was thought that the more sources of oil in the Middle East, the lower the prices to consumers. But there are only so many ways to slice rents, and lower prices to consumers and higher revenues to producing countries were ultimately inconsistent objectives.

In 1947, the State Department, to promote its new policy, circularized American companies, telling them that Kuwait might be putting its rights to the Neutral Zone up for bid and that the United States government would be happy to see them take advantage of this opportunity. Several of the major companies thought it was too risky. They feared that if they got into a bidding match, they might well have to offer considerably better terms than they were paying on their current concessions, which would greatly irritate the countries concerned.

One who was very familiar with the new thrust of American policy, as well as with the opportunities in the Middle East, was Ralph Davies, the former Standard of California marketing executive who had been Harold Ickes's deputy at the Petroleum Administration for War, and then head of the Oil and Gas Division in the Department of Interior, and was now back in private life. In 1947, in order to bid for the Kuwaiti Neutral Zone concession, Davies organized a consortium including such prominent independent companies as Phillips, Ashland, and Sinclair. It was called "Aminoil." What better name could there have been? For Aminoil was short for American Independent Oil Company. Davies warned his partners to expect a rough ride—they were now "bucking the big, big, big time," he said, and the competition with the majors would be very intense.

But Aminoil had a unique entree, derived from Jim Brooks, a Texas oil field welder, who, on his return journey from a working stint in Saudi Arabia, had stopped over at Shepheard's Hotel in Cairo. By coincidence, also staying there was the secretary to the Amir of Kuwait, who had just been instructed to find a Texas oil man unconnected to the majors, in order to bring in new bidders. The welder's cowboy hat had provided reason enough to strike up a conversation, and the welder soon found himself a guest in Dasman Palace in Kuwait City, where he garnered lasting gratitude in the water-starved principality by adjusting the palace's plumbing so that water use was cut 90 percent. When the welder finally got back to the United States, word about his new friendship drifted around the oil patch, though the story was not regarded as particularly credible. But because of his valuable connections, he was recruited for the Aminoil negotiating team—with very positive effects. Aminoil won the Neutral Zone concession from Kuwait with a bid that stunned the industry: $7.5 million in cash, a minimum annual royalty of $625,000, 15 percent of the profits—and a million dollar yacht for the Amir of Kuwait. That settled, there still remained the Saudi Arabian rights in the Neutral Zone, which were also now up for grabs.[4]

"The Best Hotel in Town"

If one aim of United States policy was to spread the wealth by diffusing hold-ings, the fact that the Saudi concession in the Neutral Zone went to an American independent was to have quite the opposite effect. For, within eight years of win-ning the concession, an oil man named Jean Paul Getty, or J. Paul Getty as he called himself, would become the richest man in America. From his earliest days in business, the inward, vain, and insecure Getty had been driven by a powerful need to make money, matched by an extraordinary talent for doing so. "There's always the best hotel in town and the best room in the best hotel in town, and there's always somebody in it," he once said. "And there's the worst hotel, the worst room in the worst hotel, and there's always somebody in that room, too." Clearly, he intended to occupy the best room.

Getty was constantly seeking to win victories, to exert power over people, and then, or so it seemed to some, to betray those who depended upon him or who had put their trust in him. He was certainly no more trusting than Gul-benkian. "One is very nearly always let down by underlings," he explained. "They may be all right for 80 percent of the time, but for 20 percent, they do something quite incredible." There were two things that Getty could not stand: to lose in a contest and to share authority. He had to be in control. "I had a per-fect record with J. Paul Getty," said a business partner. "I had a thousand fights with him, and never won a single one. Getty did not believe in changing his mind. He didn't care what evidence you had. Even if you could show him that a decision was ten to one in his favor, he wouldn't budge—as a matter of princi-ple." Getty was a gambler, but even in his biggest gambles, he was cautious, con-servative, and would do everything he could to bolster his position. As he explained, "If I wanted to gamble on typical gambling games, I'd buy a casino and have the percentages for me, rather than play."

Getty's father was a lawyer for a Minnesota insurance company who had gone to Oklahoma to collect a bad debt and ended up a millionaire oil man. The son began building up his own oil business, alongside his father's, during World War I. The father was a man for whom his word was his bond. The son, in con-trast, engaged in what were called in the oil business "sharp practices," and he did so with such skill and enjoyment that he turned those practices almost into an art. He reveled in his triumphs, business or otherwise. Getty was "well built, pugnacious by nature, and quick," said the boxer Jack Dempsey, who had once sparred with him. "I've never met anybody with such intense concentration and willpower—perhaps more than is good for him. That's the secret."

As a young man, Getty was already launched on a life of wild romance and sexual adventure, with a special predilection for teenage girls. He married five times. But marriage vows were, for him, not even an inconvenience; to engage in some of his more clandestine affairs, he simply operated under a favored and not all that discreet alias, "Mr. Paul." He liked to travel in Europe because it was less noticed that he was in "transit flagrante" with two or three women at a time. Yet the only true love of his life may have been a French woman, the wife of a Russ-ian consul general in Asia Minor, with whom he had a passionate affair in Con-stantinople in 1913. He bade what he hoped was a temporary farewell to her on the dock at Istanbul, but then lost contact with her forever in the turmoil of war and revolution that followed. Even sixty years later, whereas he would discuss his five marriages almost technically, as if they were lawsuits, a mere mention of this lady, Madame Marguerite Tallasou, was enough to bring tears to his eyes.

Getty had other serious pursuits, to be sure. He dabbled at being a man of letters, and wrote at least seven books—including one on how to be rich (for *Playboy*), a history of the oil business, a book on art collecting and a tome entitled *Europe in the Eighteenth Century*. He also launched a significant career as an art collector, creating one of the world's great collections. But while those outside interests, particularly in women, would embroil him in headlines and lawsuits, he never let any of them get in the way of his primary vocation: his single-minded quest for money through oil. "A man is a business failure if he lets his family life interfere with his business record," he once declared. Or, as he more frankly confided to one of his wives, "When I'm thinking about oil, I'm not thinking about girls."

Getty was always looking for a bargain. "He had one idea," said a business associate. "He was obsessed with value. If he thought something had value, he bought it and never sold it." In pursuit of value, he did not hesitate to go against the tide. In the 1920s, he decided that it was cheaper to drill for oil than to buy the overvalued shares of other oil companies. After the 1929 stock market crash, he completely changed tack; he saw that oil shares were selling at a great discount to assets, and he turned to prospecting for oil on the floor of the stock exchange—in the course of which he got into an extended and bitter takeover battle for the Tidewater Oil Company, with Standard of New Jersey as his prime opponent. His pell-mell buying of shares was a great gamble. It was also the right decision. Those purchases were the basis of the rise of his fortune in the 1930s.

Getty always wanted the cheapest price, the best bargain, and he was ruthless in the pursuit. During the Depression, he fired all his employees and then hired them back at lower salaries. In 1938, he picked up the Pierre Hotel on Fifth Avenue in New York for $2.4 million—less than a quarter of its original cost of construction. In that same year, several months after the Nazi seizure of Austria, Getty was in Vienna, where he managed to get himself admitted to the home of Baron Louis de Rothschild. He was there to see not the Baron, who at that moment was being held prisoner by the Nazis, but rather the Baron's valuable furniture, which he understood might soon be available. He liked what he saw and immediately went to Berlin (where he had well-connected girlfriends) and sought to find out what the SS intended to do with the Rothschild furniture. He ended up buying some pieces at a great discount—to his considerable satisfaction. He also, however, lived with his own fears during those years. He told one of his wives that he kept a large yacht in California so that he could make a quick escape in case the communists took power in the United States.

By the end of the 1930s, Getty had become a very rich man. Having made substantial contributions to the Democratic party and to various politicians, he angled for a diplomatic post and then, once America entered the war, for a commission in the United States Navy. His efforts came to nothing, for both the FBI and military intelligence developed suspicions that he had had rather extensive social connections with Nazi leaders—perhaps even sympathy for the Nazis, at least until very late in the day. Some reports went even further; there were bizarre allegations, for instance, that he was staffing the Pierre Hotel with German and Italian spies. His application for a naval commission, according to naval intelligence, "was rejected because of suspected espionage activities." Whatever the truth of the matter, he remained fascinated by dictators for the rest of his life.

During the war, Getty was in Tulsa, managing an airplane factory, a subsidiary of one of his oil companies. By this time, his eccentricities were manifold. He not only ran the operation in Tulsa from a concrete bunker, but also lived in it, in part out of fear of being bombed by the German Luftwaffe. He made a point to chew each mouthful of food thirty-three times, and he had taken to washing his own underwear each night because of his antipathy to commercial detergents. By age fifty-five, he had had his second facelift and was dyeing his hair a funny kind of reddish-brown, all of which gave him a rather wizened, embalmed look.

The end of the war only rekindled his consuming ambition to make much, much more money. He first devoted his efforts to what he was convinced would be the sure route to fabulous wealth as Americans took to the roads and highways in the postwar years: the manufacture of mobile homes. But he gave that up for something he knew much more about—oil. Getty was certain he wanted the Saudi concession for the Neutral Zone even before he had it surveyed. "If one is to be anybody in the world oil business," he declared, "one must have a footing in the Middle East." This was his chance.

The head of exploration in the Rocky Mountain division of Getty's Pacific Western oil company was a young geologist named Paul Walton, a Ph.D. from the Massachusetts Institute of Technology. Walton had worked in Saudi Arabia for Standard of California in the late 1930s and he knew his way around there. Walton would be Getty's point man in making a deal with the Saudis. Getty summoned him to the Pierre Hotel for a few days of discussion and briefings. Getty, Walton remembered ever afterward, had a "half-mad" expression on his face—a sort of angry, disagreeable scowl that he had developed, Walton figured, to keep people at a distance from him and from his money. Walton found Getty overbearing, though a man of considerable intelligence. But their discussions about the Saudi concession went smoothly, and Getty set the boundaries for the deal—at what price to start bidding for the concession and how high Walton could go. He also gave Walton a firm order: When Walton got to Saudi Arabia, he was not to discuss anything with anybody.

Walton left for Jidda and soon found himself face to face with Abdullah Suleiman, the same finance minister who had conducted the negotiations for the original Socal concession almost two decades earlier. Suleiman arranged for Walton to go up in a DC–3 and fly low over the Neutral Zone desert. Walton could barely believe what he saw from the plane: a small mound rising up from the flat expanse. He was elated. It looked almost exactly like the mound in Kuwait's Burgan field, then the largest known oil field in the world.

Though very excited when he came back to Jidda, Walton, remembering Getty's injunction about security, was also very cautious. There were no locks

423

on the rooms in his hotel in Jidda, so he left no pieces of paper around. He did not dare send a message to Getty by wireless, since he was sure it would be intercepted. Instead, he dispatched a handwritten letter by airmail. Judging by that little mound, he told Getty, the odds of a major oil play were fifty-fifty. He would have set the odds higher, but he had been in Saudi Arabia after the original discovery in 1938 and had remembered two seemingly perfect structures that had been drilled, each of which was "as dry as hell." Still, fifty-fifty was a lot more promising than the exploration odds in the Rocky Mountains, which were one in ten or even one in twenty.

Walton opened negotiations with Suleiman, which were mostly conducted on the porch of Suleiman's house in Jidda. Clearly, the deal was going to be expensive. Once again, Saudi Arabia needed money, badly, and as in 1933, Suleiman wanted a large bonus payment up front. As instructed by Getty, Walton opened at $8.5 million. The deal they finally struck was $9.5 million up front, a guaranteed million dollars a year even if no oil was found, and a royalty of fifty-five cents a barrel—far higher than what was being paid anywhere else. Walton also agreed that Getty would establish training programs, build housing, schools, and even a small mosque, and provide free gasoline for the Saudi Army. In addition, Suleiman insisted that Getty pay for a Saudi Army unit to defend the area of the concession against any potential Iranian or Soviet threats. It finally took a telegram signed by Secretary of State Dean Acheson to the Saudi government, explaining that private American companies were legally prohibited from funding another nation's Army, to take that issue off the table.

By the last day of 1948, Suleiman had given assurance to Walton that Getty had won the concession. However, Suleiman also took the precaution of telling Aminoil and a Wall Street firm that, if either would top the Getty offer, the concession would be theirs. But the price tag was too high and the risk too great; neither took it. Of course, Walton, for his part, had played a pretty good game of poker. Suleiman had stopped at $9.5 million. He never found out that, at the Pierre Hotel, Getty had authorized Walton to go up to $10.5 million. Still, Getty's company, Pacific Western, was paying an unprecedentedly high price to wildcat in an unknown desert.[5]

Kuwait and Saudi Arabia each held what was called an "undivided half interest" in the Neutral Zone. That meant that they would split the entire pie. Therefore, their respective concessionaires had to amalgamate operations to a considerable degree. The result was a totally unhappy marriage. Relations between Aminoil and Getty's Pacific Western were terrible; Getty and Ralph Davies, the head of Aminoil, could not stand each other. Pacific Western was a one-man band; Aminoil, an awkward consortium that required approvals from its many members.

Aminoil played the more dominant role in exploring the area. Nothing was easy. It battled hard to keep costs down and to do everything as cheaply as possible. But, no matter what Aminoil did, it was never cheap enough for J. Paul Getty. Exploration took longer, and proved to be more difficult and, thus, much more costly than anticipated. As time passed, anxiety among the American oil men was rising rapidly, and with good reason. By the beginning of 1953, half a decade had elapsed since the concessions had been granted, both groups were looking at expenditures in excess of $30 million, and there was nothing to show for their efforts except five dry holes. Getty sought to allay his anxiety in a variety of ways. He focused on his business interests. He wandered across Europe. He spent several weeks researching Rembrandt's portrait of Marten Looten, which he owned. Like the young John D. Rockefeller a century earlier, the sixtyish Getty relaxed by totting up his income and expenses each evening. One entry from Paris listed sums under "income" that were measured in the thousands and millions, while under "expenses" were such things as "newspaper—10 centimes" and "bus fare—5 centimes." Coming back to the United States, he finally won his twenty-year battle for control of Tidewater Oil, bought a rare Louis XV lacquer table, and enrolled in a $178 course at Arthur Murray's School of Dancing, with a special concentration on the samba and the jitterbug and on improving his ability to lead.

Still, Getty's patience and confidence were wearing thin. Not only was the string of dry holes exasperating, but so was the outflow of expenses, including his million-dollar-a-year payment to Saudi Arabia. Getty made it clear that he was disgusted with the whole approach. The Aminoil team resolutely ignored the little mound that Walton had seen from the airplane. Getty insisted that the sixth hole be drilled at that site. Furthermore, sunk costs were sunk costs; if the sixth hole was dry, he was going to pull out. Such extreme action proved unnecessary. In March 1953 the Aminoil team struck oil where Walton had thought, all along, that oil would be found. To call it a major discovery would prove an understatement. *Fortune* was to describe it as "somewhere between colossal and history-making."

Billionaire

It was only afterward that Getty made his first visits to the region. In anticipation of one trip, he listened to a "Teach Yourself Arabic" course on records, and learned enough to use Arabic in describing the geology of the Neutral Zone at a joint "banquet seminar" he cohosted with Aminoil for the Amir of Kuwait and King Saud, who had succeeded his recently deceased father, Ibn Saud. Getty's rival, Ralph Davies of Aminoil, never made it to the Neutral Zone at all; in the words of one of the other executives of Aminoil, he had "a pathological fear of dust, dirt and germs," which was a good reason to stay close to home.

Getty used his Neutral Zone production, especially the cheap, heavier "garbage oil," to build up vast integrated oil operations in the United States, Western Europe, and Japan. He reorganized all his holdings, putting Getty Oil at the top and making himself the sole commander of a great oil empire. By the end of the 1950s, Getty was the seventh-largest marketeer of gasoline in the United States. *Fortune* magazine announced in 1957 that he was America's richest man and its sole billionaire. He was stoic in the face of that news. "My bankers kept telling me," he said, "that it was so, but I was hoping I wouldn't be found out." He added a sensible admonition. "If you can count your money, you don't have a billion dollars." He achieved further fame as the Billionaire Miser. He spent his final years as squire of Sutton Place, an exquisite, 72-room Tudor manor house in Surrey, and there, amid the splendors of his priceless collection of art and antiques, he installed a pay phone for guests to use.

Paul Walton, the geologist, had come down with amebic dysentery while negotiating in Saudi Arabia in 1948. It took him three years to recover. Getty gave him a $1200 bonus, and Walton returned to Salt Lake City to work as an independent geologist. In the early 1960s, more than a decade after he had first spotted that small mound in the Neutral Zone from the air, Walton was visiting England. He telephoned Getty from London, and the billionaire invited him to Sutton Place. Getty demonstrated the positive effects of his devotion to physical fitness; well into his seventies, he regularly lifted weights, which he kept in his bedroom. The two men reminisced about how furious Getty had become at the refusal of the Aminoil people to drill in the little mound that Walton had spotted from the air. Finally, they had given way, vindicating Walton—and Getty. The Neutral Zone was by far, Getty said, his largest single asset. "He was very favorably impressed with the whole operation," Walton recalled. And well he should have been. It was estimated that his company still had more than a billion barrels of recoverable reserves in place there. The Neutral Zone had made him not only the richest American, but also the richest private citizen in the world. As for Walton, the man who had spotted the site, he continued to put together quite ordinary drilling deals back in Salt Lake City.

When Getty died in 1976, age eighty-three, the eulogy at his funeral was delivered by the Duke of Bedford. "When I think of Paul," said the Duke, "I think of money." For J. Paul Getty, there surely could have been no higher compliment.

The extraordinary deal that Getty made with the Saudis in 1948–49 was exactly what the established companies had feared would result from the arrival of the independents. Still, shock was the general reaction to Getty's Pacific Western bid, whose terms went far beyond what might have been expected. Getty's 55-cent-a-barrel royalty to the Saudis loomed over Aminoil's 35-cent royalty to Kuwait, the roughly 33-cent royalty that Aramco had just been compelled to pay the Saudis—and far overshadowed the 16½ cents that Anglo-Iranian and the Iraq Petroleum Company were paying in Iran and Iraq respectively, as well as the 15-cent royalty that the Kuwait Oil Company was paying. The general manager of Iraq Petroleum pronounced the 55-cent royalty "completely insane, uncalled for, and responsible for the difficulties being encountered in Iran and Iraq." A British diplomat angrily denounced "the notorious Pacific Western" concession.

No one had a greater sense of foreboding about what would follow from the arrival of the independents than that most skillful hand at negotiating Middle Eastern oil concessions, Calouste Gulbenkian. "These new groups lack the experience of developing oil concessions in the Middle East," he wrote to an executive of Standard Oil of New Jersey. "They offer fantastic terms to the local governments who expect similar fantasies from us. The result is trouble all around." Perhaps Gulbenkian harbored personal resentment toward Getty; after all, the American was an *arriviste* in the unsettled vineyards of Middle Eastern oil, which Gulbenkian had so carefully cultivated for a half century. Moreover,

Getty was challenging him in another sphere—with fierce competition for the position of world-class art collector. Yet Gulbenkian did speak out of his long experience and with the perspicacity of the shrewd survivor. "I feel confident that the local governments, although by no means cordially disposed with each other, will come together on this question of petroleum concessions, and do their best to squeeze us," he prophesied. "I fear that the wind of nationalization and other complications . . . may spread to us also." He added a further caution. "I would not be tranquil."[6]

"Retreat Is Inevitable"

The world's demand for Saudi oil, which had been rising rapidly, flattened out in 1949, because of an American recession and economic problems in Britain. With Aramco production down, Saudi revenues were also cut, but the financial commitments of the King and his kingdom were continuing to grow at a rapid rate. It was all too reminiscent of the two previous financial crises of the early 1930s and early 1940s. Soldiers and officials were going unpaid, subsidies to tribes were being withheld, and the government was piling up debt.

Where else to turn, in their current time of need, but to that very profitable concern called Aramco? Finance Minister Abdullah Suleiman had skillfully, with the assistance of "Jack" Philby, negotiated the original Socal concession in 1933. But now, he regularly threatened to close down the entire oil operation unless Saudi Arabia were able to share in what he called "large company profits." Suleiman's demands seemed endless: Aramco should pay for construction projects; Aramco should contribute to a Saudi "welfare fund"; Aramco should advance new loans. "Each time the company agreed to one thing," said Aramco's general counsel, "there was always just one more." But what the Saudis really wanted was a renegotiation of the original concession so that the government's "take," its share of the rents, would be much increased. Aramco was clearly a very profitable company, and they insisted that they get their rightful share. They wanted what the Venezuelans had already gotten.

It was not just word of the deal the Venezuelans had recently struck that had traveled from Caracas. A Venezuelan delegation was promulgating the fifty-fifty concept throughout the Middle East, even going to the trouble of translating its documents into Arabic. The Venezuelans had not taken on these additional expenses sheerly out of altruism. In Caracas, it had become, as Romulo Betancourt observed, "increasingly evident that the competition from low-cost, high-volume production from the Middle East was a grave threat to Venezuela." Best get those costs up, which would be accomplished if the Middle Easterners were to raise their taxes. And so, in the ironic words of a State Department petroleum expert, the Venezuelans "decided to spread the benefits" of the fifty-fifty principle "to the area which was taking business away from them—the Middle East."

The closest the Venezuelan delegation got to Saudi Arabia was Basra in Iraq; the Saudis did not like the way that Venezuela had voted on Israel in the United Nations and would not let the delegation in. Nevertheless, the fifty-fifty concept expeditiously crossed the border, and when the Saudis looked at the numbers for 1949, they could see what a difference it would make. Aramco's profits that year were almost three times Saudi Arabia's own earnings from the concession. What really struck the Saudis, however, was the way in which the United States government's tax take had risen—to the point that, in 1949, taxes paid to the American government by Aramco were $43 million, $4 million higher than Aramco's royalty payments to Riyadh. The Saudis made clear to the Americans that they knew exactly what the company had earned, what it had paid in taxes to the United States, and how that compared to the royalty payments to Saudi Arabia. And they also made clear, as the head of Aramco delicately put it, that they "weren't a darn bit happy about that."

The terms of J. Paul Getty's new Neutral Zone concession certainly demonstrated to them that oil companies could pay much more. Yet the Saudis did not want to squeeze too hard. There was still a very large investment program to be carried out within the concession. Moreover, having just seen Aramco lose market share, the Saudis did not want to hobble the company with additional costs that might make its oil uncompetitive with that produced by other Persian Gulf countries.

Perhaps they could get more money out of Aramco without directly affecting the company's competitive position. The Saudis did their research; they had even, unbeknownst to Aramco, retained their own adviser on American tax law, and to their delight, they learned about a most interesting and intriguing provision in American tax laws that would leave Aramco whole. It was called the "foreign tax credit."

Under legislation dating back to 1918, an American company operating overseas could deduct from its United States income tax what it paid in foreign taxes. The objective was to avoid penalizing American companies doing business abroad. Royalties and other fixed payments—costs of doing business—could not be deducted, only taxes paid on income. That distinction was all important. For it meant that if Saudi Arabia had collected not only $39 million in royalties in 1949, as it had done, but also another $39 million in taxes, then that $39 million tax bill could have been deducted from the $43 million tax bill that Aramco owed to the United States government. As a result, Aramco would have had to pay only $4 million to the United States Treasury—the difference between $43 million and $39 million—not $43 million. For its part, Saudi Arabia would have received not $39 million, but twice that—$78 million. In other words, the overall tax bite on Aramco would have remained the same, but most of it would have been collected in Riyadh, not Washington. And to the Saudis, that was how things should have been, for as far as they were concerned, it was their oil.

Armed with a new weapon, Saudi Arabia kept up the pressure on Aramco, until finally, in August 1950, the company faced reality and authorized negotiations for a fundamental revision in the concession. The company was in continuing contact with the State Department, which was a very strong proponent of meeting Saudi demands. The Korean War had begun in June 1950, and the American government was now even more worried about communist influence and Soviet expansion in the Middle East and about regional stability and secure access to the oil. Anti-Western nationalists had to be kept at bay. Despite the loss to the United States Treasury, the State Department wanted to see more revenues going to Saudi Arabia and other oil-producing countries in the region, in order to maintain pro-Western governments in power and to keep discontent within manageable bounds. In the case of Saudi Arabia, it was particularly urgent to do whatever was necessary to preserve the position of the American companies.

Just twelve years had passed since Mexico had expropriated the American and British oil companies. That stood as the great warning of how badly things could go wrong. "Since company retreat is inevitable," concluded a State Department policy paper, "it would seem useful to make the retreat as beneficial and orderly as possible to all concerned." As George McGhee, Assistant Secretary of State for Near Eastern Affairs, saw it, fifty-fifty had become inescapable. "The Saudis knew the Venezuelans were getting 50/50," he later said. "Why wouldn't they want it too?" At a State Department meeting on September 18, 1950, McGhee told representatives of American oil companies operating in the Middle East that the time was certainly at hand for "rolling with the punch."

There was one last stumbling block—the four Aramco parent companies. Some of them were resolutely opposed to the idea; after all, the original concession terms specifically prohibited an income tax. But at a following meeting, McGhee bluntly told the parent companies that there was no alternative and that long-term contracts created "the practical necessity of horsetrading." Speaking in support of fifty-fifty, an executive vice-president of Aramco said, "From a psychological point of view such a formula sounded fair and would be considered fair in Saudi Arabia." The parent companies were persuaded. On December 30, 1950, after a month of complex negotiations, Aramco and Saudi Arabia signed a new agreement, the heart of which was the Venezuelan fifty-fifty principle.

But if the Saudis were satisfied with their new revenues, there was still the open and very critical question of whether these tax payments would be eligible for an American tax credit. In fact, their eligibility was not confirmed until 1955, when the Internal Revenue Service, in the course of auditing Aramco's 1950 tax return, approved the credit. In 1957, the staff of the Joint Congressional Committee on Internal Revenue Taxation added its approval, based upon the various tax laws, their legislative history, judicial decisions, and IRS rulings with respect to "other similarly situated taxpayers." In later years, some would argue that the United States government, particularly the National Security Council, had bent the tax laws to give Aramco a special dispensation on the matter of the tax credit. But based upon the records available, that was not the case. The Aramco ruling was consistent.

In the meantime, and thereafter, a substantial flow of revenues was diverted from the U.S. Treasury to that of Saudi Arabia. Whereas the Treasury had collected $43 million in taxes from Aramco in 1949, compared to $39 million in royalties paid to Saudi Arabia, by 1951 the division of rents was completely different. In that year, Saudi Arabia collected $110 million from the company, while, after the application of the tax credit, Aramco paid only $6 million to the United States Treasury.[7]

The impact of the Saudi-Aramco deal on neighboring countries was swift. The Kuwaitis insisted on a similar arrangement, and Gulf Oil was fearful about

429

failure to respond. "We might wake up any morning and find that we had lost Kuwait," a worried Colonel Drake, chairman of Gulf, told American officials. Gulf succeeded in overcoming the obdurate objections of Anglo-Iranian's chairman, Sir William Fraser, and got that company, its partner in the Kuwait Oil Company, to agree to fifty-fifty in Kuwait. Britain's Inland Revenue opposed the principle of the tax credit on Anglo-Iranian's share, but pressure came from other parts of the British government until the taxmen too finally saw the light and agreed to an appropriate tax credit mechanism. In neighboring Iraq, a fifty-fifty deal was also in place by early 1952.

Thus, a new foundation had been established for relations between David Ricardo's landlord and tenant. And the tenant oil companies had to grapple with their significance. Inside Jersey, several departments collaborated on a working paper on the fifty-fifty arrangements, to provide internal guidance for the company. The paper noted that Jersey had gone through a considerable process of education since the Mexican expropriation. "We now know that the safety of our position in any country depends not alone on compliance with laws and contracts, or on the rate or amount of our payments to the government, but on whether our whole relationship is accepted at any given moment by the government and public opinion of the country—and by our own government and public opinion—as 'fair.' If it is not so accepted, it will be changed." Unfortunately, " 'fairness' and 'unfairness' are essentially concepts of the emotions rather than fixed and measurable standards." However disconcerting and unpalatable it might be to the engineers, businessmen, and buccaneers who ran international oil companies, that was a fact of life. "Experience already shows that there is something inherently satisfying in the '50/50' concept."

Satisfying or not, it was a necessity. But had the battle over rents ended with a lasting peace treaty or was it only a truce? Did the companies now have a position that they could successfully defend against nationalism, the assertion of sovereignty, and the inevitable thirst of nation-states for more revenues? The paper prepared for Jersey management offered a sharp warning: "If we ever admit in any country that an equal division is less than 'fair,' the ground will be cut out from under our feet in every country." On fifty-fifty, the paper warned, was where Jersey should make its stand: " '50/50' is a good position which needs no defense and is hard to attack; '55/45' or '60/40' would have no such appeal, and could be only rear-guard defense positions in an unlimited retreat."

The Watershed

The Saudi-Aramco fifty-fifty agreement of December 1950 was, with justification, described as a "revolution" by one historian of the decline and fall of the British empire—"an economic and political watershed no less significant for the Middle East than the transfer of power for India and Pakistan." As for the American government, it satisfied the urgent and critical need to increase the income to Saudi Arabia and other governments in order to maintain the postwar petroleum order and to help keep those "friendly" regimes in power. The stakes and risks were enormous. At a time when every dollar of Truman Doctrine and Marshall Plan aid was a battle in Congress, an arrangement that enabled Middle

Eastern governments to tax the profits of the oil companies was more efficient than trying to get additional foreign aid out of the Congress. Moreover, the fifty-fifty principle had the right psychological feel. Both politically and symbolically, it did the job that needed to be done.

Many years later, in 1974, when the international politics of oil had become burningly controversial, George McGhee, who as Assistant Secretary of State had brokered the Saudi-Aramco deal, was quizzed at a Senate hearing about the arrangement that he had helped hammer out in 1950. A Senator asked him if the tax credit wasn't really "a very ingenious way of transferring many millions by executive decision out of the public treasury and into the hands of a foreign government treasury without ever needing any appropriation or authorization from the Congress of the United States?"

McGhee disagreed. It was not a sleight of hand. There was consultation with the Treasury Department at the time, and with Congress. The decision was not secret. The fifty-fifty principle had already been at work in Venezuela for seven years before its adoption in Saudi Arabia. No, McGhee explained, the question missed the point. "The ownership of this oil concession was a valuable asset for our country." The risk of not doing something along these lines was much too great. "In essence," McGhee said, "the threat was the loss of the concession."[8]

And, indeed, Aramco's concession in Saudi Arabia had been preserved. But, within six months of the signing of the Saudi fifty-fifty deal in December 1950, events in neighboring Iran were to prove that the relationship between landlord and tenant had by no means been satisfactorily resolved.

C H A P T E R 2 3

"Old Mossy" and the Struggle for Iran

WHEN WORD REACHED Tehran in 1944 that Reza Pahlavi, the former Shah of Iran, had died in exile in South Africa, his son and successor was devastated. Many years later, he summed up his reaction simply: "My grief was immense." Mohammed Reza Pahlavi had worshiped his father, the resolute and physically towering commander of the Persian Cossack brigade who had seized power and crowned himself Shah in the 1920s. Reza Shah had thereafter brought order to the fractious country, begun modernizing it at a pell-mell rate, and had subjugated the powerful mullahs, whom both father and son regarded as dangerous and deadly enemies from the Middle Ages.

But what made the son's grief and guilt still worse was that, if not the actual usurper of his father's throne, he was part of the agency of his father's downfall. In August 1941, two months after the German invasion of the Soviet Union, the British and the Russians moved their forces into Iran in order to protect the refinery at Abadan and the supply line from the Persian Gulf to the Soviet Union. Alarmed by rapid German advances in Russia and North Africa, the Allies feared a pincer that would converge in Iran. They deposed Reza Shah, who had shown friendliness and sympathy toward the Nazis, and replaced him with his son, only twenty-one at the time.

After Reza Shah's death, Mohammed Pahlavi would be dedicated to, and haunted by, the memory of his father. He would forever try to be worthy of Reza Shah, against whose standard he would be judged—by others and by himself. One day in 1948, the Shah himself even admitted to a visitor, "My sister Ashraf asked me yesterday whether I was a man or a mouse." He had laughed about it, but he obviously didn't think it funny. There was always the implication that he was weak, indecisive, and inadequate compared to his father. There would always be a way, too, in which the Shah was somehow an outsider. At age six, he was entrusted to a French governess; at age twelve, he was sent off to school in Switzerland. His education and experience engendered a certain distance from Iranian society. "It might, of course, be," mused the American ambassador in 1950, "that he is a little too Westernized for an Oriental country." That possibility would dog him for almost four decades.

Yet, whatever the Shah's own anxieties, he had been plunged at a very young age into treacherous circumstances, which would have mightily challenged even the most self-assured and practiced of politicians. The legitimacy of his dynasty was problematic; the role of the monarchy in Iran, a totally unresolved question. He had to contend with chronic intervention by foreign powers, as well as direct Soviet pressure on the country's territorial integrity and a highly visible British economic presence. He was forced to struggle to assert his authority in a political system riven by every kind of division—class, regional, religious, and modern versus traditional. On one side were the Islamic fundamentalists, led by the fiery Ayatollah Seyed Kashani, who were outraged by every intrusion of the modern world—be it the presence of foreign advisers or the fact that Reza Shah had allowed women to dispense with the veil. On the other side were the communists and the Tudeh, a well-organized leftist party tied to Moscow. In between were reformers and nationalists and republicans, all of whom wanted to remake the political system, as well as military officers who were itching to take power for themselves.

The political culture of Iran was itself chaotic and phantasmagoric, given to wild exaggeration and violent emotion. Graft and corruption were a way of life. The British chargé d'affaires summed up the rules of the game as it was played in the Majlis, the parliament in Tehran, with a straightforward maxim: "Deputies expect to be bribed." Out in the countryside lived a multitude of tribes and clans, which hated their subordination to Tehran and the Pahlavis. Virtually no part of the Shah's domain was immune to secessionist drives. And in the late 1940s the nation was gripped by grinding poverty and suffering from economic collapse. A pervasive hopelessness settled over the land.

Only one thing really united the country—hatred of foreigners and, in particular, the British. Never had so much malevolence been attributed to a so rapidly declining power. The English were regarded as almost supernatural devils, controlling and manipulating the entire nation. Every Iranian politician, wherever he might be in the political spectrum, was virtually obliged to accuse his enemies and opponents of being British agents. Even droughts, crop failures, and locust plagues were blamed on the evil designs of those clever Englishmen. But the detestation centered, in particular, on the largest industrial employer in Iran, the major source of the nation's foreign earnings, and the all-too-tangible symbol of the intrusion of the modern foreign world—the Anglo-Iranian Oil Company.

Part of the hatred of Anglo-Iranian was fueled by the battle over the oil rents. Between 1945 and 1950, Anglo-Iranian registered a £250 million profit, compared to Iran's £90 million royalties. The British government received more in taxes from Anglo-Iranian than Iran did in royalties. To aggravate matters still further, a substantial part of the company's dividends went to its majority owner, the British government, and it was rumored that Anglo-Iranian sold oil to the

British Navy at a substantial discount. But, in Iran, far more important than pounds and pence were the emotions and symbols. Those were what drove the politicians and the street mobs to frenzied excitement, and what turned the animosity against Anglo-Iranian into a national obsession. It was very handy to have such a foreign scapegoat when so much was wrong at home.[1]

The Last Chance

Through World War II, Iran was seen by Americans and British alike as London's responsibility, primarily a "British show." Thereafter, however, the development of the Cold War, combined with the growing worry about the security of Persian Gulf oil, shoved Iran toward the fore of American foreign policy concerns. Soviet troops were withdrawn from northern Iran in 1946, but by 1949, the Americans were fearful that Iran was in so advanced a state of economic and political decay that it would be easy prey for the Soviet Union.

Iran's prospects were made still more uncertain, and the political scene more chaotic, by the endemic assassinations and assassination attempts. In February 1949 a Moslem fanatic, posing as a photographer, tried to kill the Shah as he arrived at Tehran University. Though firing a half-dozen shots at point-blank range, the would-be assassin only slightly wounded the Shah, who displayed courage and responded coolly. "The miraculous failure of this assassination attempt," he said afterward, "once again proved to me that my life was protected." It was a turning point in the Shah's view of himself and his vision for his country. He used the incident to impose martial law and to begin a vigorous campaign to assert his personal authority. He ordered the body of his father—to whom he now posthumously gave the title "the Great"—to be exhumed in South Africa, brought back to Iran, and given a state funeral. In due course, huge equestrian statues of Reza Shah would rise up in different parts of his son's domain.

The Shah's drive to extend his political control was paralleled by efforts to readjust the financial ties between Iran and the Anglo-Iranian Oil Company, in parallel to what was happening in the other oil exporting countries. Washington, fearful of Soviet ambitions and with much less to lose than London, pushed the British government and Anglo-Iranian to increase the royalties paid to Iran. The point man for the Americans was George McGhee, the Assistant Secretary of State for Near Eastern and African Affairs who was at the same time brokering the new fifty-fifty deal between Aramco and the government of Saudi Arabia, and who thought the existing division of earnings between Anglo-Iranian and Iran was not reasonable. British officials, not surprisingly, greatly resented intervention and free advice from McGhee and other Americans. They took to calling McGhee, who was just thirty-seven in 1949, "that infant prodigy" and tended to see him, in particular, as a source of their troubles. They thought he was anti-British and anti–Anglo-Iranian. On that they were wrong. As a Rhodes Scholar at Oxford, McGhee had gotten to know the daughters of Sir John Cadman of Anglo-Iranian and had even visited the Cadman country house. In the course of earning a doctorate in geophysics at Oxford, he had shared his seismic research with Anglo-Iranian (in an area of Hampshire where the company was interested in drilling) and then, ironically, had been offered a job as a geophysicist in Iran.

After seriously considering the offer, he turned it down, but only because he was homesick and wanted to get back to America. "I had, however," he later said, "a kindly feeling toward AIOC at that time."

Events showed that he had made the right choice. It was soon after his return from England, at the beginning of World War II, that McGhee discovered a sizeable oil field in Louisiana, which gave him the wealth, independence, and opportunity to devote the rest of his career to public service. He married the daughter of the eminent Everette DeGolyer, and was, until he entered the armed services, a partner in DeGolyer's oil appraisal firm. McGhee was an unabashed Anglophile (and later in life was chairman of the English-Speaking Union). He just thought the British needed to be saved from themselves, especially when it came to their "nineteenth-century" attitude toward oil. McGhee also fairly reflected the view of his colleagues, which was summed up by Secretary of State Dean Acheson when he criticized "the unusual and persistent stupidity of the company and the British Government" on the subject of Iran.[2]

On the other hand, though the Americans never seemed to believe it, the British government was no less at odds with Anglo-Iranian. The British government was a 51 percent owner of Anglo-Iranian, but that did not mean there was any great affection or empathy between the two parties. On the contrary, there was suspicion and rancor, and some of their fiercest fights were with each other, in a classic case of what has been called the "battle between Minister and Manager." Foreign Secretary Ernest Bevin had complained as early as 1946 that Anglo-Iranian "is virtually a private company with state capital and anything it does reacts upon the relationships between the British Government and Persia. As Foreign Secretary, I have no power or influence, in spite of this great holding by the Government, to do anything at all. As far as I know, no other Department has."

To the company, of course, the entire situation looked quite different. It was the third-largest crude oil producer in the world, with most of that oil coming from Iran, and it thought the Iranians had a pretty good deal as things were. Under the 1933 agreement, Iran received not only a royalty, but also 20 percent of the company's worldwide profits—which was better than the terms of any other oil producer. Beyond that, Anglo-Iranian had become one of the major international oil companies. The company was trying to carry on a complex, global business. It operated as a private firm—that had been the original intention in Churchill's share acquisition in 1914—and its senior executives resented and resisted the intrusions and advice of the politicians and civil servants. They thought that the bureaucrats—whom Anglo-Iranian's chairman, Sir William Fraser, dismissingly called "West End gentlemen"—simply did not understand the oil business, or indeed, what it was like to try to do any business at all in Iran. But the pressures were such that, by the summer of 1949, Anglo-Iranian was forced to negotiate with the Iranians a Supplemental Agreement—supplemental to the revised 1933 concession. The new proposal provided for a large hike in royalties, as well as a big lump-sum payment.

Although Anglo-Iranian and the Iranian government had come to agreement, the government, fearful of the Parliament's opposition, held back from submitting the agreement to the Majlis for almost a year—until June of 1950.

435

The oil committee of the Parliament responded by furiously denouncing the new agreement, calling for cancellation of the concession, and demanding the nationalization of Anglo-Iranian. A leading pro-British politician was assassinated, and the fearful Prime Minister, deciding that prudence was the better course, quickly resigned.

The Shah nominated General Ali Razmara, the Army's Chief of Staff, as the new Prime Minister. A lean, young "soldier's soldier," a graduate of the French military academy at St. Cyr, ambitious and cold-blooded, Razmara was known to have done the unheard of—he had once returned a bribe. Razmara sought to distance himself from the Shah and to develop authority of his own. To the Americans and British, he looked like the last chance. Iran appeared ever more vulnerable to both communist subversion and direct Soviet expansion.

That same month, June of 1950, the North Koreans invaded South Korea, turning the Cold War into a hot one. There had already been border clashes between Soviet and Iranian forces, and in the State Department, George McGhee urgently directed the preparation of contingency plans for responding to a Soviet invasion of Iran. Moreover, in the midst of the Korean War, Iranian oil took on a new urgency; it accounted for 40 percent of total Middle Eastern production, and Anglo-Iranian's refinery at Abadan was the major source of aviation fuel in the Eastern Hemisphere.[3]

With the stakes suddenly so much higher, the U.S. government urged the British government even more strongly to pressure Anglo-Iranian to make an offer that the Iranian government could swiftly accept. But Sir William Fraser would not easily budge. With many years' experience dealing with the Iranians, he had little respect for their governmental system and counted on nothing but ingratitude, deception, backbiting, and new demands. He was hardly more kindly disposed toward the Americans. He would bitterly blame the troubles that would befall Anglo-Iranian on America's political meddling in Tehran and on the activities of the American oil companies—Aramco in particular—in the Middle East.

Fraser, very definitely, was the man who determined Anglo-Iranian's position. He was a formidable opponent under any circumstances. With none of John Cadman's diplomatic skill, he was a tough, implacable autocrat who ran Anglo-Iranian one way—his way. Dissent was not tolerated. The chairman of Gulf, which was Anglo-Iranian's partner in Kuwait, observed that Fraser's domination was so total that Anglo-Iranian's other directors "did not dare call their souls their own." It was said of Fraser that he was a "Scotsman to his fingertips." His father had been founder of the leading Scottish shale oil company, and then sold it to Anglo-Iranian—and, as was later said, "Willie came with the shale." Said one who worked with Fraser, "Few, in an industry where tough bargaining is a way of life, were likely to get the better of him."

The same was true when his adversary was the British government. A minister of state in the Foreign Office declared that Fraser appeared "to have all the contempt of a Glasgow accountant for anything which can not be shown on a balance sheet." To another British official who dealt with him, Fraser was an "obstinate, narrow old skinflint." Though many senior government officials felt that he should be removed, and his retirement was often under consideration, they seemed powerless to make it happen. One of Fraser's great strengths against all opponents derived from the huge importance of Anglo-Iranian's earnings to the British Treasury and the overall British economy.[4]

Fraser implacably resisted repeated entreaties from the British government to negotiate further with Iran, and he ignored the American. But then in the autumn of 1950, Fraser had an abrupt and uncharacteristic change of heart. He not only wanted to come up quickly with substantially more money for Iran, but also talked about subsidizing Iranian economic development and supporting Iranian education. What had happened? It was not that Fraser had experienced a sudden conversion to philanthropy. Rather, he had learned about what was known as the "McGhee Bombshell"—the imminent fifty-fifty deal in Saudi Arabia—and knew that he had to do something speedily. But time had already run out. In December, the announcement of the fifty-fifty deal with Aramco forced Premier Razmara to withdraw his support from the Supplemental Agreement, and that was its end.

At last, Anglo-Iranian came forward to offer its own fifty-fifty deal. It was no longer enough. The entire opposition in Iran now focused on the infamous Anglo-Iranian. The leader was an elderly firebrand, Mohammed Mossadegh, who was chairman of the oil committee in the Parliament. "The source of all the misfortunes of this tortured nation is only the oil company," Mossadegh declared. Another deputy thundered that it would be better for the Iranian oil industry to be destroyed by an atomic bomb than remain in the hands of Anglo-Iranian. All were calling for nationalization of the industry and the ouster of Anglo-Iranian. Prime Minister Razmara did not know what to do. Finally, in a speech to Parliament in March 1951, he came out against nationalization. Four days later, as he was about to enter Tehran's central mosque, he was assassinated by a young carpenter, who had been entrusted by Islamic terrorists with the "sacred mission" of killing the "British stooge."

Razmara's murder demoralized the proponents of compromise, weakened the Shah's position, and emboldened the broad opposition. A week and a half later, the minister of education was also assassinated. The Majlis proceeded to pass a resolution nationalizing the oil industry, but it was not implemented immediately. Then, on April 28, 1951, the Majlis chose Mohammed Mossadegh, who was by now the number-one foe of Anglo-Iranian, as the new Prime Minister, with the specific and wildly popular mandate to execute the nationalization law. The Shah signed the law, and it went into effect on May 1. Anglo-Iranian's days seemed finished in Iran, for in the nationalization decree it was designated as the "Former Company." As the British ambassador reported, Anglo-Iranian, though in business around the world, "has been legally abolished" and Tehran "has taken the line it has no further existence."

Mossadegh dispatched the governor of Khuzistan province to Anglo-Iranian's headquarters at Khorramshahr. On arrival, the governor sacrificed a sheep in front of the building, and then announced to a delirious crowd that the concession was voided. Anglo-Iranian's facilities in Iran, as well as the oil they produced, now belonged to the nation of Iran. Mossadegh's son-in-law followed with an emotional speech in which he declared that the days of colonialism were over and those of prosperity at hand. He fainted from the excitement and had to be carried away. Directors of the newly established state oil company, led by Mehdi Bazargan, dean of the Tehran University engineering faculty, appeared at the refinery site at Abadan, carrying with them stationery, rubber stamps, and a large sign, all of which said "Iranian National Oil Company"—the sign to be nailed up on one of the office buildings. Scores more sheep were sacrificed to celebrate the great event, and the huge throng that had gathered to greet the directors went wild with exultation. But, though the sheep had been sacrificed, the deed was not yet done. And for the next five months, the status of Anglo-Iranian's facilities in Iran remained clothed in uncertainty and indecision.[5]

"Old Mossy"

About seventy years old and frail in appearance, with a completely bald head, a very long nose, and bright, buttonlike eyes, Mohammed Mossadegh would completely dominate the drama of the next two years. He would slyly outwit everyone—foreign oil companies, the American and British governments, the Shah, his own domestic rivals. He himself was a man of evident contradictions. Cosmopolitan, educated as a lawyer in France and Switzerland, he was fiercely nationalistic, antiforeigner, and obsessed by his opposition to the British. The son of a high bureaucrat and a great-grandson of a Shah from the preceding dynasty, Mossadegh was an aristocrat with extensive landholdings, including a 150-family village that belonged to him personally. Yet he took on the mantle of reform, republicanism and rabble rousing, appealing to and mobilizing the urban masses. One of the first professors at the Persian School of Political Science, he had become caught up in the constitutional revolution of 1906, which remained the lodestar for the rest of his career. He had gone to the Versailles Peace Conference after World War I, ordered a rubber stamp with the legend "Comité resistance des nations," and sought to plead Persia's case against foreign intervention, particularly that of the British. He had received no hearing, and returned home, feeling that his hopes and idealism had been betrayed by the colonial powers.

In the 1920s, Mossadegh occupied a number of ministerial posts and played a leading role in opposing Reza Shah's moves to turn Persia into a dictatorship and make himself the absolute ruler. For these efforts, Mossadegh was rewarded with various spells in jail and under detention on his estate, where he busied himself with amateur medical studies and the investigation of homeopathic remedies. The expulsion of Reza Shah in 1941 by the British and Americans provided the signal for Mossadegh to burst back upon the political scene. He quickly amassed a following; his long years of dedicated opposition had firmly established him as the "pure" man, devoted to Iran and to cleansing it of foreign domination.

Mossadegh was both unpretentious and eccentric in his personal style; frequently, clothed in pajamas, he received Iranians and important foreigners while stretched out on his bed, where he spent much time because, some said, of dizzy spells. His bodyguards were always nearby; he lived, understandably, in perpetual fear of assassination. Mossadegh would say whatever suited his needs at the moment, no matter how exaggerated or fanciful. But in the next moment, there was no assertion, no matter how strongly expressed, that could not be remolded or changed or, with a joke or only a giggle, repudiated altogether if that was more convenient. All that mattered was that whatever he said contributed to his two overriding objectives: the maintenance of his own political position and the expulsion of the foreigner—the British, in particular. In pursuit of his goals, he proved to be a master at merging theatricality with politics. In public, he would burst into tears, he would moan; it was common for him to faint at the climax of a speech. Once he collapsed on the floor of the Majlis in the midst of an impassioned oration. A Parliamentary deputy, who was also a medical doctor, rushed up, and fearing that the old man might be in his last moments, grabbed for Mossadegh's wrist to check the pulse. As he did so, Mossadegh opened one eye and winked at him.

American and British officials who dealt with Mossadegh took to calling him "Mossy." Anthony Eden remarked that "Old Mossy," with his pajamas and iron bedstead, was the "first real bit of meat to come the way of the cartoonists since the war." Even some of those most exasperated by Mossadegh would ever after also remember how charmed they had been by him. The Americans tended, at first, to see Mossadegh as a rational, nationalistic leader, and one with whom business might be done. He could be a bulwark against the Soviet Union and an agent of reform; the alternative to Mossadegh was communism. And, throughout, Cold War considerations and fears shaped American policies and perceptions more than British. In any event, there were, as far as Washington was concerned, good enough grounds to be opposed to old-fashioned British imperialism. No less an authority than President Harry Truman said that Sir William Fraser of Anglo-Iranian looked like a "typical nineteenth-century colonial exploiter." The Americans understood, better than the British, that Mossadegh's great problem centered on his rivals within Iran; he was always constrained by the need to keep at bay those who were more nationalistic, more extreme, more fundamentalist, and more antiforeigner than he. In the meantime, he would improvise and play off the great powers and never quite compromise. Eventually, the Americans lost patience with him. When it was all over, Dean Acheson delivered a pungent judgment. Mossadegh, he said, "was a great actor and a great gambler."

The British saw things differently, right from the beginning. They thought that the Americans failed to understand how hard it was to negotiate with Mossadegh; some British officials regarded the communist danger as much exaggerated. "Mossadegh was a Moslem, and in 1951 he would not have turned to the Russians," said Peter Ramsbotham, secretary of the British Cabinet's special Persia Committee. The real danger was to an existing interest in Iran and to established political and economic arrangements in the Middle East. Some of the British regarded Mossadegh as a "lunatic." What could be done with such a man? Add to that the fact that Mossadegh, in the words of the British ambassador, Sir Francis Shepherd, had to be watched very carefully because he was "cunning and slippery and completely unscrupulous." In the ambassador's view, the Iranian Prime Minister looked "rather like a cab horse" and diffused a "slight reek of opium." But in what was about to unfold, perhaps nothing galled the British so much as the fact that their national champion, Anglo-Iranian, and Great Britain itself would be outwitted by an old man in pajamas.[6]

Plan Y

In the immediate aftermath of the nationalization of Anglo-Iranian, and facing such a wily and untrustworthy opponent, the British hurriedly reviewed their options. There was strong feeling that something had to be done to save the country's most valuable foreign asset, and its number-one source of petroleum. But what to do? The Cabinet gave consideration to Plan Y, a contingency proposal for military intervention. The inland oil fields were too remote to be easily secured, the Cabinet concluded, but the island of Abadan, site of the world's largest refinery, was quite another thing; it made a much more reasonable target. With the advantage of surprise, Abadan could be taken. Perhaps a quick, strong show of force would be enough to restore a sufficient measure of respect and transform the situation.

But perhaps not. British lives would certainly be lost. Hostages would be taken. The United States government was pressing hard against armed intervention, for fear that such a British action in the south would legitimize a Russian move in the north and that Iran would end up behind the Iron Curtain. There were other obstacles to military action. India had just become independent, and there was no longer an Indian Army upon which to call. Britain could expect to be castigated all over the world for old-fashioned imperialism. Britain's own might was sharply circumscribed; because of its grave balance-of-payments difficulties, it did not have staying power. How could it pay for an extended military involvement?

Yet, were Britain to cave in here, some argued in the Cabinet, its whole position in the Middle East would be undermined. "If Persia were allowed to get away with it, Egypt and other Middle Eastern countries would be encouraged to think they could try things on," declared Defense Minister Emmanuel Shinwell. "The next thing might be an attempt to nationalize the Suez Canal." Outside the Cabinet, the Leader of the Opposition and that venerable defender of empire, Winston Churchill, described himself to Attlee as "rather shocked at the attitude of the United States, who did not seem to appreciate fully the importance of the great area extending from the Caspian to the Persian Gulf: it was more important than Korea." Churchill stressed "the importance of the balance of oil supplies as a factor in deterring the Russians from embarking on aggression." Foreign Secretary Herbert Morrison, denouncing the policy of "scuttle and surrender," discussed the use of force. Parachutists were moved into place in Cyprus in order to be able to protect, and if necessary, evacuate the large number of British workers and families in Abadan. Nevertheless, it looked to some as if Britain might be tempted to implement Plan Y and undertake a military test of its waning imperial power.[7]

Averell in Wonderland

The prospect of armed intervention set off alarm bells in Washington. The British might push Iran right into eager Soviet arms. Dean Acheson hurriedly arranged a meeting with the British ambassador and with Acheson's old friend, Averell Harriman. Sitting on the veranda of Harriman's house, overlooking the Potomac River on a June evening, Acheson made abundantly clear that he wanted to keep the British from doing what, in his eyes, would be silly—or dangerous. He suggested that Harriman might go out to mediate between Britain and Iran. All present thought it a very good idea—all, that is, except Harriman himself, who hardly wanted such an assignment. Nevertheless, he agreed to go.

A tall, austere man, Harriman was a multimillionaire who had given up private business for public service. He had handled many complex and delicate matters: He had been Roosevelt's Special Representative in the early years of World War II, ambassador to Moscow and London, Secretary of Commerce, U.S. representative in Europe for the Marshall Plan. But never was he to be involved in so bizarre a negotiation. He arrived in Tehran in mid-July 1951. Accompanying him were a U.S. Army lieutenant colonel, Vernon Walters, who would serve as interpreter (Mossadegh wanted to conduct business in French), and Walter Levy, who had directed oil affairs in the Marshall Plan and had just set up his own consulting firm.

The British had grudgingly accepted Harriman's efforts to play the honest broker. They were more worried about Levy, known to some American officials as "the real oracle of State" when it came to international petroleum matters. Levy made no secret of his view that Anglo-Iranian's position had so degenerated that it could never go back in its existing form. He expressed what was already becoming a common idea on the American side. If the British wanted to regain their overall oil position, Levy said, they would have to "camouflage" the existence of Anglo-Iranian and "dilute" it within a new operating company, a consortium, that would be controlled by a number of companies, some of them American. The British were outraged at what was called a proposal for the disastrous "mongrelization" of a leading British company. They suspected that the real reason for the consortium proposal was that the American companies were eagerly waiting in the wings, anticipating their opportunity to come into Iran. British suspicions were further stirred when a junketing young Congressman, Representative John F. Kennedy, son of the former American ambassador to London, stopped in Tehran and suggested to the British ambassador that, if no settlement emerged, "it would be a good thing for American concerns to step into the breach."

In Tehran, Harriman and his party were put up in a palace belonging to the Shah. The walls of the great reception room were covered with thousands of tiny mirrors, which gave the impression of shimmering jewels. It all seemed very novel and exotic at first. Harriman and his party hardly knew that they would spend much of the next two months there. After a while, they tired of the setting.

Harriman, accompanied by Walters, went to see Mossadegh at what, in contrast, was his unpretentious private home. They found the Prime Minister lying on a bed, the palms of his hands crossed directly below his neck. Two doors were blocked off with wardrobes to prevent easy access by would-be assassins. Mossadegh fluttered his hands weakly in greeting when Harriman and Walters entered, then wasted no time in telling Harriman exactly how he felt about the British. "You do not know how crafty they are," said the Prime Minister. "You do not know how evil they are. You do not know how they sully everything they touch."

Harriman disagreed. He knew them well; he had been ambassador there. "I assure you they are good and bad and most of all in between," said Harriman.

Mossadegh leaned forward, took Harriman's hand, and merely smiled. It was only in a later talk that Mossadegh happened to mention that his grandson, the apple of his eye, was away at school. "Where?" asked Harriman. "Why, in England, of course," Mossadegh replied. "Where else?"

Soon they established the unusual physical pattern for their discussions: Mossadegh either sitting or stretched out on his bed, hands crossed just below his neck; Colonel Walters sitting at the foot of the bed, perhaps yogi style; and Harriman on a chair almost right up against the bed, midway between the two men. The arrangement helped mitigate Mossadegh's poor hearing. Walter Levy would often join them as well. And here, in this incongruous scene, might hang the fate of the postwar petroleum order and the political orientation of the Middle East. So strong was the sense of constantly moving between reality and fantasy that Walters ordered from Washington a copy of *Alice in Wonderland* to serve as a kind of unofficial guidebook for what might lie ahead.

Day after day, Harriman, assisted by Levy, would try to educate Mossadegh about the realities of the oil business. "In his dream world," Harriman cabled Truman and Acheson, "the simple passage of legislation nationalizing the oil industry creates profitable business and everyone is expected to help Iran on terms that he lays down." Harriman and Levy sought to explain to Mossadegh about the need for marketing outlets, in order to sell oil, but to no avail. That the company was called "Anglo-Iranian," they said, did not mean that all of its oil was produced in Iran. Revenues were also earned from refining and distribution in many countries. At one point, Mossadegh seemed to be demanding more revenue from a barrel of oil than the total selling price realized on all the different products derived from that barrel. "Dr. Mossadegh," said Harriman, "if we are going to talk intelligently about these things, we have to agree on certain principles."

Mossadegh peered at Harriman. "Such as what?"

"Such as: nothing can be larger than the sum of its parts."

Mossadegh stared straight at Harriman and replied, in French, "That is false."

Harriman, though he did not speak French, thought he had caught Mossadegh's drift, but he couldn't believe it. "What do you mean 'false'?" he asked incredulously.

"Well, consider the fox," said Mossadegh. "His tail is often much longer than he is." With that salvo fired, the Prime Minister put the pillow over his head and rolled back and forth in the bed, laughing uproariously.

Yet there were a couple of times when, at the end of the day's discussions, Mossadegh seemed to have agreed on the framework of a settlement. But the fol-

442

lowing morning, the Americans would come back, only to be told by Mossadegh that he could not carry through on the agreement. He would not survive. What mattered to Mossadegh, far more than the oil market or international politics, was how the whole affair would play in domestic politics and how his various rivals on both right and left, as well as the Shah's supporters, would respond. He particularly feared the Moslem extremists, who opposed any truck with the foreign world. After all, it was only a few months since General Razmara had been assassinated by a Moslem fundamentalist.

Harriman, sensing how greatly this fear constrained Mossadegh, went to see the Ayatollah Kashani, the leader of the religious right, who had been imprisoned during World War II for his Axis sympathies. The mullah declared that, though he knew nothing about the British, the one thing he did know was that they were the most evil people in the world. In fact, all foreigners were evil, to be dealt with accordingly. The ayatollah then went on to tell the story of an American who had come to Iran some decades earlier and had involved himself in oil. He had been shot on the street in Tehran, and then rushed to a hospital. A mob, searching for the American, broke into the hospital and found him on the operating table, where they butchered him.

"Do you understand?" asked the ayatollah.

Harriman immediately recognized that he was being threatened. His lips tightening, he struggled to keep his anger under control. "Your Eminence," he replied in a steely voice, "you must understand that I have been in many dangerous situations in my life and I do not frighten easily."

"Well," shrugged the ayatollah, "there was no harm in trying."

In the course of the conversation, Ayatollah Kashani accused Mossadegh of the worst of sins—being pro-British. "If Mossadegh yields," Kashani said, "his blood will flow like Razmara's." There could be no question but that Kashani was an implacable and dangerous opponent. But, as for Mossadegh, Harriman developed a certain affection for the Prime Minister. He was theatrical, entertaining, gracious in a way, and Harriman started calling him "Mossy," though not to his face.[8]

Harriman also thought he saw a glimmer of a solution, a possible *modus vivendi*. He flew back to London, where he recommended that the British send out a special negotiator to follow up. The choice was a socialist millionaire, Richard Stokes, whom Harriman accompanied back to Tehran. Stokes, with some confidence, boldly announced his goal—to put before Mossadegh "a jolly good offer."

On the return trip to Tehran with Stokes was Sir Donald Fergusson, the influential Permanent Undersecretary at the Ministry of Fuel and Power. Fergusson was a consistent critic of Anglo-Iranian and its chairman, Sir William Fraser, whom he regarded as narrow, dictatorial, and insensitive to larger political currents and considerations. But he was also skeptical of the possibility for any settlement, and he feared that a deal would threaten every other British foreign investment with expropriation by rapacious governments, against which there would be no effective sanction otherwise. "It was British enterprise, skill and effort," he declared, "which discovered oil under the soil of Persia, which has got the oil out, which has built the refinery, which has developed markets for Persian oil in 30 or 40 countries, with wharves, storage tanks and pumps, road and rail tanks and other distribution facilities, and also an immense fleet of tankers." For that reason, he thought the call, on moral grounds, for a fifty-fifty split—as the Moslem religious leader the Aga Khan had urged—was "bunk and ought to be shown to be bunk."

In any event, Fergusson recognized that Mossadegh's objective was "not better financial terms but to get rid of this foreign company with its predominating influence out of Persia." And Mossadegh had no intention of allowing Anglo-Iranian back. Moreover, he was now the prisoner of the popular passions he had done so much to stir. Thus, in this second round of negotiations, there was no way to agree on the decisive issue of who, in the event of a settlement, would really run and control the oil industry in Iran. "An evening session of negotiations in the garden of the palace where we stayed was like the last act of *Figaro*," recalled Peter Ramsbotham, who was a senior negotiator on the Stokes mission. "Unknown, dim figures lurked behind the rose bushes. Everybody was spying on everybody else. People were lurking all about. We never knew with whom we were dealing. Neither did Mossadegh." Stokes decided to call it quits. His mission, and the much longer one of Harriman, had failed. "Mossadegh's stock-in-trade was fighting with the British," Harriman was to conclude. "Any settlement of the dispute would end his political power." Still, on the plane out of Tehran, Harriman was forced to make a painful admission. "I am simply not used to failure," he said. But then he had never before tried to do business with anybody like "Old Mossy."[9]

"Stand Firm, You Cads!"—The Farewell to Abadan

Meanwhile, in the oil fields and at the refinery, operations were grinding to a stop. The British managed to mount an embargo by threatening tanker owners with legal action if they picked up "stolen oil." In addition, Britain embargoed goods to Iran, and the Bank of England suspended financial and trade facilities that had been available to Iran. In short, the expropriation was being met with economic warfare.

The Majlis retaliated by passing a law that anybody found guilty of "sabotage or inattention" was liable to the death penalty. A letter was sent to Eric Drake, Anglo-Iranian's general manager in Iran, accusing him of such "sabotage and inattention." On the advice of the British ambassador, Drake hurriedly left the country, slipping out in a small plane. He thereafter ran the Iranian oil operations from an office in Basra in Iraq and then from a ship in the Gulf. After a meeting at Suez with the British chiefs of staff, he was flown back under a pseudonym to Britain, where he was abruptly summoned to a meeting of Attlee's Cabinet. The invitation enraged the autocratic Sir William Fraser, who was not invited and who himself had been much too busy to see Drake; after all, Drake had been nothing more than Anglo-Iranian's man on the spot. Despite Fraser's anger, Drake went to the meeting, entering 10 Downing Street by a secret passage through the back garden in order to evade waiting reporters. Drake told the Cabinet that, if Britain did not do anything about Abadan, it would eventually lose much more, including the Suez Canal. He was then taken to meet the Leader of the Opposition, Winston Churchill, who, after quizzing him about his discussion with the Cabinet, suddenly growled, "You've got a pistol, Drake?" Drake explained that he had turned in his pistol to the Iranian authorities because of a new law that provided the death penalty for unauthorized possession of a gun. "Drake, you can finish a man with a pistol," Churchill admonished him. "I know because I have."

In the aftermath of the failure of Harriman's and Stokes's missions, the British government was once again debating the use of military force to seize the island of Abadan and its refinery. Secret military preparations were, in fact, so far advanced that by September 1951, an operation to seize Abadan Island could have been mounted in less than twelve hours. But what could be accomplished? Would not the whole of Iran then unite against the British? Would they not risk a break with the United States? In any event, the element of surprise had been lost. "It would be humiliating to this country if the remaining British staff at Abadan were expelled," Attlee told his Cabinet. But the British government decided against using armed force to prevent it. In retrospect, some would see the public threat to use force in the first months of the crisis, and then not doing so, as the real beginning of the end of Britain's credibility and position in the Middle East.

On September 25, 1951, Mossadegh gave the last remaining British employees at Abadan exactly one week to clear out. A few days later, Ayatollah Kashani declared a special national holiday—"a day of hatred against the British Government." At the refinery complex at Abadan, the British oil men and the nurses from the infirmary put on an evening of songs and skits for their own entertainment; the revue was entitled "Stand Firm, You Cads!"

On the morning of October 4, the oil men and their families gathered in front of the Gymkhana Club, which had been their social center. They were carrying fishing rods and tennis rackets and golf clubs; a few had their dogs, though most of the pets had been destroyed. The group included not only the refinery men, but the indomitable lady who ran the guest house. Only three days earlier, she had used her umbrella to interdict an Iranian tank commander who had driven across her lawn. The parson joined the others in front of the club, having just locked the little church that housed the history of this island community— "the records of those who had been born, baptized, married or had died, in Abadan."

Waiting for them all was the British cruiser *Mauritius*, which would carry them upriver to safe haven at Basra, in Iraq. The ship's band, in a bizarre display of protocol, played the Iranian national anthem as the launches belonging to the Iranian Navy began to shuttle back and forth between ship and shore. By midday, all were aboard, and the *Mauritius* began to steam slowly upriver toward Basra. The band continued to play, but now it was "Colonel Bogey." The passengers broke into song, a great chorus under the hot sun, singing the unpublished and more obscene versions of that venerable military march. With this musical burst of defiance, Britain bade good-bye to its largest single overseas enterprise, and the world's largest refinery, which now had virtually ceased to operate. It was a particularly humiliating climax to Britain's six postwar years of imperial retreat. The first of the great Middle Eastern oil concessions was also the first to be summarily canceled.[10]

"A Splutter of Musketry"

No oil was flowing out of Iran, owing to the effectiveness of the British embargo and, in particular, Anglo-Iranian's vigilance in bringing legal action against refiners or distributors who took Iranian oil. But the embargo also had the result of removing a substantial amount of petroleum from world commerce at a critical time, during the Korean War. Rationing was put into effect in some parts of Asia; "unnecessary" air flights east of Suez were cut back. The United States Petroleum Administration for Defense came up with the chilling estimate that, without Iran, world oil demand would exceed available supply by the end of 1951.

The machinery was quickly put in place to manage the shortfall. As in World War II, it was based upon Anglo-American cooperation. In the United States, acting under the Defense Production Act of 1950 and with antitrust exemption, nineteen oil companies formed a Voluntary Committee to coordinate and pool supplies and facilities. It worked closely with a similar British committee, moving supplies around the world to eliminate bottlenecks and shortages. The companies themselves also strained to increase production in the United States and in Saudi Arabia, Kuwait, and Iraq. As it turned out, the momentum of the great postwar oil development supported the British embargo, against Iran, and the feared shortage never materialized. By 1952, Iranian production had plummeted to just 20,000 barrels per day, compared to 666,000 in 1950, while total world production had risen from 10.9 million barrels per day in 1950 to 13.0 million in 1952—an increase more than three times greater than Iran's total output in 1950![11]

British policy against Iran had stiffened in October 1951, when the Labour government was replaced by a new Conservative one, headed once again by Winston Churchill, now seventy-seven—and more than half a decade older than Mossadegh. Churchill was showing his age; he complained of " 'my old brain' not working as it used to do." But he had very firm ideas about the Iranian nationalization: The Labour government had been too indecisive and too weak. Had he been in office, he told Truman, "there might have been a splutter of musketry," but Britain "would not have been kicked out of Iran." There was a grand irony in all this. As First Lord of the Admiralty some thirty-seven years earlier, Churchill had purchased the government's stake in Anglo-Persian, as it was then. Now, he had lived and remained in politics long enough to come back and head the government at the time of the company's greatest crisis. He would defend the company to the limits of his ability.

His Foreign Secretary was Sir Anthony Eden, who had a different kind of affinity with the issue. At Oxford, after World War I, Eden's subject had been Oriental languages, and he had been a star student of the Persian language and a devotee of the beauties of Persian literature. Eden kept up his Persian connections. As Undersecretary at the Foreign Office in 1933, he had played a key role in solving the crisis occasioned by Reza Shah's expropriation of Anglo-Persian. Eight years later, in 1941, as Foreign Secretary, Eden had become very anxious about Reza Shah's flirtation with the Nazis and took a major part in the decision to invade Iran and topple him. He was personally fascinated by the country, and had made many trips there. When he returned as Foreign Secretary in 1951, he was still able to call up proverbs in Persian. But a crisis of major proportions—the result of the nationalization and the expulsion from Abadan—awaited him. "Our authority throughout the Middle East had been violently shaken," he said.

The crisis also posed a painful personal dilemma for Eden. A substantial part of his own financial assets were tied up in Anglo-Iranian stock, whose price had plummeted. After much deliberation, he decided that—despite the government's own shareholding and the lack of any rule or convention that would so demand—it was not appropriate for him to own the shares. He sold them at the very bottom of their price. With that went his one chance to build up any substantial financial security, and that action ultimately cost him much, including his country house.

With the Conservatives back in power, the fundamental disagreement that divided London and Washington became more clear-cut. The Americans feared that, if Mossadegh fell, he would be succeeded by the communists, and it was better to try to work with him, as exasperating as it could be, than against him. The British, on the contrary, thought it likely that Mossadegh would be succeeded by, from their point of view, a more reasonable government—and the sooner, the better. Acquiescence in Iran—and Mossadegh's ability to act with impunity—would inevitably and irresistibly tempt countries all over the world, leading to an epidemic of nationalization and expropriation. Britain could not afford risking the rest of its foreign assets. "We ought to tell the U.S.A. at the highest level," said Sir Donald Fergusson of the Ministry of Fuel and Power, "that even supposing their view is right that Dr. Moussadeq has got to be maintained to save Persia from Communism, they have got to choose between saving Persia and ruining this country." There was much frustrating argument within the British government over what to do and whom to blame—and also a great deal of impatience and anger with Anglo-Iranian and what seemed to officials to be its obscurantism. Eden himself complained that the company's chairman, Sir William Fraser, was "in cloud cuckoo land." [12]

In the autumn of 1951, a few weeks after the departure of the British from Abadan, Mossadegh went to the United States, to plead the cause of Iran at the United Nations. He also went to Washington to make his case to Truman and Acheson and to ask for economic aid. The American government wanted stability in Iran, but it was not prepared to bail out Mossadegh to get it. When Mossadegh began to explain to Truman and Acheson that he was "speaking for a very poor country—a country all desert—just sand," Acheson interrupted, "Yes, and with your oil, rather like Texas!" The Prime Minister received only minimal economic assistance.

But Assistant Secretary George McGhee, after some eighty hours of talks with Mossadegh during the visit, believed he had come up, just barely, with an outline for a settlement. It included the acquisition of the Abadan refinery by Royal Dutch/Shell (on the premise that it was a Dutch—that is, non-British—company) and a special oil purchase contract for Anglo-Iranian that would have the effect of a fifty-fifty split. But Mossadegh insisted on an extra condition: that

447

no British oil technicians could work in Iran. Acheson was to personally try out the proposal on Anthony Eden at a luncheon in Paris. When Acheson came over the phone from Paris to McGhee and others waiting eagerly in the State Department, however, he reported that Mossadegh's extra condition had infuriated Eden, who regarded it as humiliating, and that Eden had peremptorily rejected the entire proposal. McGhee, whose hopes were very high, was stunned. His efforts to resolve the Iranian oil crisis had failed. "To me it was almost the end of the world," he said. It was not clear that Mossadegh shared his distress—nor that he ever actually wanted an agreement. "Don't you realize that, returning to Iran empty-handed," Mossadegh told one American the night before leaving the United States, "I return in a much stronger position than if I returned with an agreement which I would have to sell to my fanatics?"

Still, the Truman Administration continued to hope for a settlement with "Mossy." In the State Department, and indeed in the Foreign Office in London, there were proposals for a consortium of companies to assume management of the Iranian oil industry. There was also an ingenious plan under which the World Bank would take over oil operations in Iran as a sort of trustee until a final settlement could be negotiated. All foundered on Iran's lack of interest in a compromise that mitigated the nationalization and its control, or provided a role for Anglo-Iranian.

As the crisis dragged on through the early months of 1952, Mossadegh's government could not sell its oil, it was running out of money, economic conditions were deteriorating. But none of that seemed to count. The important thing was that he was the popular national leader who had achieved the historic objective of throwing the foreigners out and regaining the national heritage. He declared that, as far as he was concerned, the oil could remain in the ground for use by future generations. The U.S. ambassador to Tehran noted Mossadegh's fundamental antipathy to the Shah, which he attributed to the "secret contempt" of a man of an old aristocratic family toward "the weakling son of an upstart tyrannical imposter." But Mossadegh the constitutionalist was turning to extraconstitutional means to run the country, including the use of street mobs to manipulate politics. He was also assuming more dictatorial power. "I always considered this man to be unsuitable for high office," one opposition leader said. "But I never imagined, even in my worst nightmares, that an old man of seventy would turn into a rabble rouser. A man who constantly surrounds the Majlis with thugs is nothing less than a public menace." Mossadegh was also proving himself a notable political innovator; he was the first Middle Eastern leader to use the radio to arouse his followers. When he called, thousands and thousands—sometimes, it seemed, hundreds of thousands—would rush, frenzied, into the streets, chanting, intimidating, destroying the offices of opposition newspapers. The Shah felt powerless in the face of Mossadegh's popularity. "What can I do?" he said to the American ambassador. "I am helpless."[13]

"Luck Be a Lady Tonight"

Around this time, Acheson again met with Eden, who said that "at some stage it might be necessary . . . to impress on the Shah the need" to give Mossadegh a push from power.* But neither the United States nor Britain, by any means, had given up on diplomacy with Mossadegh. Truman implored Churchill to accept the validity of the Iranian nationalization law, "which seems to have become as sacred in Iranian eyes as the Koran. . . . If Iran goes down the communist drain, it will be little satisfaction to any of us that legal positions were defended to the last." Churchill wanted to promote a joint appeal to Mossadegh, of whom he said, "We are dealing with a man at the very edge of bankruptcy, revolution and death but still I think a man. Our combined approach might convince him."

Truman reluctantly agreed to a joint proposal for arbitration to determine compensation for the nationalized properties, but after much dodging and debating, Mossadegh finally rejected the proposal because, he said, it was a "trap" laid by the Anglo-Iranian Oil Company.

By the end of the Truman Administration, both the Americans and the British had about given up on Mossadegh. In late 1952, the British raised with the Americans the possibility of their working together to bring about a change in the Iranian government—in other words, a coup. The American reply was postponed until the Eisenhower Administration had taken over. A go-ahead for such consideration was given, with the support of Secretary of State John Foster Dulles and his brother Allen, the new head of the Central Intelligence Agency.

Still, in the final weeks of the Truman Administration and the first of the Eisenhower Administration, the United States launched yet another diplomatic effort to work out an oil settlement between Iran and Britain. After much frustrating discussion, Mossadegh once again said no. Meanwhile, conditions in Iran had deteriorated even further. Before nationalization, oil exports had generated two-thirds of the country's foreign exchange and half of government revenues. But there had been no oil revenues for two years, inflation was rampant, and the economy was falling apart. The country was much worse off than before nationalization. Law and order were collapsing; the chief of police in Tehran was kidnaped and murdered. Moreover, Mossadegh showed little gift for governing. He conducted his Cabinet meetings from his bed. In the early months of 1953, he tried to bolster his weakening domestic position by taking more power in his own hands—extending martial law, governing by decree, seizing control of military appointments, intimidating and silencing opposition, abolishing the upper house of the parliament and dissolving the lower, and carrying out a Soviet-style plebiscite that gave him a 99 percent victory. Many nationalists and reformers, who had once supported Mossadegh, were antagonized by his efforts to monopolize power and his increasing reliance on "mobocracy" and the Tudeh party. The religious fundamentalists also turned against him as he sought to extend his power. They decided that he was an enemy of Islam. The fact that *Time* magazine had chosen him as "Man of the Year" was, in some eyes, proof that he

* Sometimes Acheson and Eden were mistaken for each other. Eden did not know exactly why. "Acheson," he said, "does not look like a typical citizen of the United States." He thought it might have been because Acheson's mother was Canadian. Once, on a flight from New York to Washington, an American naval officer passed Eden a note: "You are either Dean Acheson or Anthony Eden. Whichever you are, will you autograph my book?"

was an American agent. It also appeared that Mossadegh was setting the stage to eliminate the Shah. And he was moving closer to the Soviet Union. As for the Shah himself, he seemed as helpless as ever.

Mossadegh's tilt toward Moscow became even more ominous when a new Soviet ambassador came to Tehran—the same man who had presided as Soviet ambassador in Prague in 1948, when the communists there staged a coup and took power. Only the naive could believe that the Russians were not organizing to gain political control of Iran through their own agents and the Tudeh party. A long-desired objective of Russians—Romanovs and Bolsheviks alike—appeared, at last, to be at hand; after all, as part of the Nazi-Soviet Pact, the Kremlin had staked out Iran as the center of Soviet "aspirations." The chicken was only waiting to be plucked.

In Washington at a somber meeting of the National Security Council, Secretary of State Dulles predicted that Iran would soon be a dictatorship under Mossadegh, which would be followed by a communist takeover. "Not only would the free world be deprived of the enormous assets represented by Iranian oil production and reserves," Dulles said, "but the Russians would secure these assets and thus henceforth be free of any anxiety about their petroleum resources. Worse still . . . if Iran succumbed to the Communists there was little doubt that in short order the other areas of the Middle East, with some 60 percent of the world's oil reserves, would fall into Communist control."

"Was there any feasible course of action to save the situation?" asked President Eisenhower. There was.

On the British side, Foreign Secretary Eden had temporized. But he was ill, and in July 1953 was still convalescing. Churchill, taking direct charge of the Foreign Office, approved a plan to overthrow Mossadegh. So did the Americans. In Allen Dulles's words, the operation went "active." General Fazlollah Zahedi, loyal to the Shah, would take the lead in challenging Mossadegh. The two Western countries believed they were not supporting a coup—that was what Mossadegh was carrying out—but rather a countercoup by the Shah and Zahedi.[14]

Field control of what was called "Operation Ajax" was vested in the CIA's Kermit Roosevelt, grandson of Theodore Roosevelt. Support was provided by MI6, British intelligence. In mid-July 1953, "Kim" Roosevelt entered Iran by motorcar from Iraq. But before Operation Ajax could begin, the suspicious Shah still had to be convinced that the plan was real and had a chance of success. He knew full well that the U.S. government had been trying to court Mossadegh. He also suspected that Mossadegh was a British agent, though perhaps a slightly errant one. In order to meet clandestinely with the Shah and assuage his doubts, Roosevelt slipped into the palace grounds late one night, hidden under a blanket on the floor of a car. He succeeded in persuading the Shah.

Operation Ajax unfolded over the middle of August 1953 with great suspense and high drama. There were code names for all the main actors. The Shah was the "Boy Scout"; Mossadegh, "the old bugger." One of Roosevelt's code names, owing to a border guard who misread his passport, was "Mr. Scar on Right Forehead." Waiting nervously over several days in the home of one of his operatives in Tehran, Roosevelt took to playing and replaying the song "Luck Be a Lady Tonight," from the musical *Guys and Dolls*, which was then a great hit on Broadway. It became the theme song for the operation.

But it looked like bad luck at the beginning. The operation was scheduled to start when the Shah issued an order dismissing Mossadegh, but delivery of the order was delayed three days, by which time Mossadegh had been tipped off, either by one of his supporters or by Soviet intelligence. He had the officer who carried the order arrested and launched his own effort to topple the Shah. General Zahedi went into hiding. Mossadegh's supporters and the Tudeh party held the streets. They smashed and tore down the statues of the Shah's father in the public squares of Tehran. The Shah himself took flight, first to Baghdad. As far as he was concerned, the countercoup had failed and he had little hope of ever returning to Tehran. He told the American ambassador in Baghdad there that "he would be looking for work shortly as he had a large family and very small means outside of Iran."

The Shah's next stop was Rome, where he and his wife took up residence in a borrowed suite in the Excelsior Hotel. They had virtually no clothes, no retainers, and no money. The Queen wandered through the shops without cash to buy anything. The Royal Couple was reduced to eating their meals in the hotel's public dining rooms and getting their news secondhand from the encamped reporters. Altogether, it was an excruciatingly anxious and nervous time at the Excelsior.

On August 18, Undersecretary of State Walter Bedell Smith explained to Eisenhower that Operation Ajax had failed, adding gloomily, "We now have to take a whole new look at the Iranian situation and probably have to snuggle up to Mossadeq if we're going to save anything there. I dare say this means a little added difficulty with the British." But the next morning the tide turned in Tehran. General Zahedi held a press conference at which he handed out photostats of the Shah's order dismissing Mossadegh. A small pro-Shah demonstration grew into a vast, shouting crowd, led by tumblers doing handsprings and wrestlers showing off their biceps and giant weightlifters twirling iron bars. Growing even larger, it swarmed out of the bazaars to the center of the city to proclaim its hatred of Mossadegh and its support for the Shah. Pictures of the Shah suddenly appeared to be plastered everywhere. Cars turned on their headlights to show support for the Shah. Though street fights broke out, the momentum was clearly with the pro-Shah forces. The Shah's dismissal of Mossadegh and appointment of Zahedi as successor had become known. Key elements of the military rallied to the Shah, and soldiers and police dispatched to quell pro-Shah demonstrators instead joined them. Mossadegh fled over the back wall of his garden, and Tehran now belonged to the Shah's supporters.

At the Excelsior in Rome, a wire service reporter rushed up to the Shah with the bulletin: "Teheran: Mossadeq overthrown. Imperial troops control Teheran." The Queen burst into tears. The Shah turned white, then he spoke. "I knew that they loved me." He returned in triumph to Tehran. The coup—or countercoup—was a very close thing, but it had worked. By the end of August 1953 the Shah was back on his throne, his new Prime Minister was in power, and Mossadegh was under arrest. And the statues of the Shah's father toppled by Mossadegh supporters were being re-erected.[15]

In the years following, there would be much argument about the real significance of the American-British operation. Did it cost less than one hundred thousand dollars, or was it priced in the millions? Did the two Western powers create
the countercoup, or did they only lubricate it? Mossadegh's time was certainly
running out; his base of support had greatly narrowed, and he would fall either to
the left or to the right. What the CIA and MI6 did was to play the facilitating role,
providing the financial and logistical aid, emboldening the opposition and making the critical connections, during uncertain and fluid times. Operation Ajax
succeeded because it accorded with increasing popular support for the Shah and
the existing regime and growing disillusionment with Mossadegh, who was trying to change the regime into one in which he, rather than the Shah, held final
power, and which might, ultimately, come under Soviet control. In the words of
one of the operation's planners, Operation Ajax created "a situation and an atmosphere in Tehran that forced the people to choose between an established institution, the monarchy, and the unknown future offered by Mossadeq." Even so,
success had been by no means certain. On his return to Washington, Kermit
Roosevelt reported directly to Eisenhower, who, with admiration, noted in his
diary that Operation Ajax "seemed more like a dime novel than an historical
fact." [16]

"A Group of Companies"

With the Shah again in power, the stage was set to bring Iranian oil back into production and onto the world market. But how was this to be done? Anglo-Iranian,
of course, was hamstrung. For it to take the lead would only reignite the nationalist fires in Iran. As for the British government, it was, in the words of an official
at the Ministry of Fuel and Power, "completely stumped."

Clearly, Washington would have to lead the way to an oil settlement. The
State Department retained Herbert Hoover, Jr., as the special representative of
Secretary of State Dulles to see if a new consortium of companies could be created to take up Anglo-Iranian's interests. In addition to being the former President's son, Hoover was an oil consultant of some note, who had helped devise
the original fifty-fifty arrangement with Venezuela. He also gave every sign of
disliking the British. The solution that Hoover would pursue had become the
common American prescription and one the British government had considered:
a consortium in which Anglo-Iranian was camouflaged in the midst of a number
of companies, several of them American.

However, the American oil companies, at least the majors, were far from enthusiastic about involving themselves in Iran. Their production elsewhere in the
Middle East was building up rapidly. The Arab producers, who were enjoying
higher revenues, were not particularly keen to have their output or revenues reduced to make room for Iranian oil, and they might take out any displeasure on
the oil companies. The four Aramco partners had more than enough petroleum
in Saudi Arabia to meet their requirements for the foreseeable future; they were
making large capital investments there. Why invest in Iran for oil they did not
need?

Moreover, who needed the aggravation of dealing with the Iranians and their unstable domestic political situation? "There was no assurance whatever that we wouldn't lose the whole thing within a few months again," a Jersey official recalled. "It was touch and go as to whether or not," he explained, "the country would last." The political risks did not end with the nationalists and the religious fundamentalists. The continuing threat of Russian pressure on Iran created what a representative of Standard of California called an "ouchy" situation.

For their part, American officials did not find oil company executives the easiest people to deal with. In mid-1953, Richard Funkhouser, a senior oil strategist in the State Department, advised his colleagues, in a long analysis on Middle Eastern oil, that "it is of critical importance to the success of any approach that oil men be handled most carefully and diplomatically. Oil officials seem overly sensitive to any indication that the industry isn't perfect. . . . Emotion, pride, loyalty, suspicion make it difficult to penetrate to reason."

Thus, considerable diplomacy went into the effort to get the American oil men to do what they did not particularly want to do: go into Iran and help repair the situation. So did a great deal of brute persuasion from Washington, backed up by London. "If the U.S. and British governments hadn't really beat us on the head, we wouldn't have gone back," Howard Page, Jersey's Middle East coordinator, later said. In particular, the State Department harped on one theme: If Iranian oil did not move, the country would collapse economically, and it would fall, one way or another, into the Soviet camp. That would, in turn, threaten the rest of the Middle East—specifically, Saudi Arabia, Kuwait, and Iraq—and the concessions therein. It could also mean serious trouble in strictly commercial terms: The Russians might dump Iranian oil onto the world market. The communist threat to Iran warranted participation in the country and there would be one notable benefit. Participation would give the American companies an important vantage point, if not out-and-out control, over Iranian production rates, which would have to be balanced in any event against those in Kuwait and Saudi Arabia.

Herbert Hoover, Jr., stopped in London and told Sir William Fraser that there was no alternative and that Anglo-Iranian would have to take the initiative. "He who pays the fiddler gets to invite the guests to the ball," said Hoover. And so, in December 1953, Fraser wrote to the chairmen of each of the American majors, inviting them to London to talk about setting up a consortium. To extend such an invitation was a humiliating admission of defeat for Fraser. Nor were the American companies thrilled to accept it. Writing to Secretary of State Dulles, a Jersey vice-president said, "From the strictly commercial viewpoint, our Company has no particular interest in entering such a group but we are very conscious of the large national security interests involved. We, therefore, are prepared to make all reasonable efforts."[17]

But, before Jersey and the other companies would make such efforts, and indeed before anything further could be done, there was another obstacle to be surmounted. It was a most awkward matter: the U.S. government was bringing a giant antitrust case against the major oil companies—the very same companies that it was seeking to push into the new consortium to bolster Iran. The Justice Department once again was busily gearing up a criminal suit against those companies for belonging to an "international petroleum cartel" and for engaging in exactly the sort of business relationships that the State Department was now promoting for Iran. It was an altogether confusing state of affairs and not one likely to fill the companies with enthusiasm for joining a consortium.

The Oil Cartel Case

Two contradictory, even schizophrenic, strands of public policy toward the major oil companies have appeared and reappeared in the United States. On occasion, Washington would champion the companies and their expansion in order to promote America's political and economic interests, protect its strategic objectives, and enhance the nation's well-being. At other times, these same companies were subjected to populist assaults against "big oil" for their allegedly greedy, monopolistic ways and indeed for being arrogant and secretive. Never before, however, had those two policies come together in such sharp and potentially paralyzing collision, with an outcome that could have momentous economic and political consequences.

Antitrust lawyers in the Justice Department were very suspicious of any cooperation among the major oil companies. The system that had emerged under Harold Ickes to assure sufficient petroleum supplies during World War II, they said, was merely an "official rubber stamping" of a prewar cartel. They looked with abhorrence on Aramco and the other great oil deals of the late 1940s. They searched for the hidden hand of the Rockefellers and ignored other explanations for the Aramco joint venture: the economic and political risks, the very large capital requirements of developing the concession, constructing a pipeline, and building a refining and distribution system—and the less-than-gentle nudging from the U.S. government itself. Nor were the antitrust lawyers the only people in Washington whose suspicions were aroused. In 1949, the Federal Trade Commission used its power of subpoena to obtain company documents, and in due course it produced the most extensive, detailed historical analysis of the international relationships among the companies that had ever seen the light of day. It was a landmark study, used by students of the industry down to the present day.

It also had a decidedly pronounced point of view, as evidenced by the title— *The International Petroleum Cartel*. One of the top oil specialists in the State Department commented at the time on "its highly slanted and non-objective approach." It generally interpreted complex events in such a way as to support its overriding thesis that international oil was indeed a cartel. In particular, it showed a fundamental blind spot; in the world of *The International Petroleum Cartel*, oil companies did not have to adapt and bend to the will and requirements of governments—the cartel-minded governments of the 1930s, dictatorships around the world, the British and French governments throughout the years, and the governments of oil-producing countries, which wanted higher revenues and could always cancel a concession.

Those concerned with foreign policy—in such agencies as the State Department, the Department of Defense, and the CIA—were appalled by the FTC report. It would provide immense fodder, they believed, to those trying to undermine the Western position in the Middle East and elsewhere. It would, in the words of the White House's Intelligence Advisory Committee, "greatly assist Soviet propaganda" and "would further the achievement of Soviet objectives throughout the world." It could hardly have come at a worse time; the United States was engaged in the Korean War and was also trying to bring about a solution to the Iranian crisis, and the same companies that were the targets of the FTC had been harnessed to assure sufficient petroleum for the war and to make up for the big gap in supplies that had resulted from the Iranian shutdown.

Fearing unfortunate repercussions, the Truman Administration had classified the report secret. But as word leaked out, political pressure mounted to release the report, especially as the 1952 Presidential election came into sight. Truman finally allowed a Senate subcommittee to publish it, though with some deletions. The report's effect was far-reaching. It found careful readers from Riyadh to Caracas, and, not surprisingly, it became a subject of commentary even on Radio Baku's broadcasts to the Middle East.

Months before its official publication, the FTC report had finally persuaded senior officials in the Justice Department to move ahead with a criminal antitrust suit against what it called the "As-Is conspiracy." Justice's own history of the "oil cartel" contained many errors and odd insinuations. For instance, the "spot market" was said to be "at highest prices in history." Obviously, the report implied, that was the result only of the machinations of the "cartel" and had nothing to do with either the Iranian shutdown and loss of supplies or the Korean War and the economic boom. In the Justice Department's version, there were no foreign governments making demands on the oil companies; there was not even the Texas Railroad Commission.

To the State Department, the FTC report was bad enough; a long, drawn-out criminal investigation would be far more dangerous. The very fact of a grand jury would seem to brand the companies as wrongdoers, and a Justice Department campaign would not only invite all other governments—particularly those in the Middle East—to go after the oil companies, but would sanctify such an assault. More specifically, such a prosecution would make it impossible to settle the Iranian crisis by bringing in the American companies. Nevertheless, in June 1952 President Truman authorized the Department of Justice to begin a criminal investigation, impanel a grand jury, and subpoena documents.

The Justice Department wanted to go after foreign companies, as well—Shell, Anglo-Iranian, and CFP, all of them members of the Iraq Petroleum Company. They were also served with subpoenas and ordered to produce documents. The British government was outraged; the Justice Department's actions represented a violation of sovereignty, it believed, and constituted an unacceptable assertion of extraterritoriality. The case itself was plain stupid in London's view, as it would not only further complicate any settlement of the Iranian crisis, but also disrupt the wide range of relations with the oil-producing countries and threaten basic Western strategic, political, and economic interests. At a Cabinet meeting in September 1952, Foreign Secretary Eden described the report as "stale bread," and the work of "witch-hunters." Even so, he added, FTC disclosures "could be extremely prejudicial to the national interest." The British government very firmly ordered Anglo-Iranian and Shell not to cooperate in any way. The Netherlands government was recruited to give similar instructions to the Royal Dutch side of the Royal Dutch/Shell Group. Both governments, along with the French, protested vigorously to the State Department.

The Justice Department was proceeding under a new and expanded definition of antitrust. Even if the companies had indubitably engaged in cartel behavior outside the United States, that, in itself, did not constitute a violation of the Sherman Antitrust Act. But practices of American businesses outside the United States could be a violation of American antitrust laws, under new interpretations, if those practices had "effects" on domestic prices or other aspects of American commerce.[18]

Not surprisingly, the Justice Department's assault made the oil companies more than a little shy in cooperating with the State Department's efforts to pry them into the oil business in Iran. After all, Washington had blessed and even encouraged the "great oil deals"—the Aramco consortium, the Kuwait Oil Company, the reconstruction of the Iraq Petroleum Company, and long-term contracts involving Jersey, Socony, and Anglo-Iranian—on grounds that they would, in the words of a 1947 State Department memo, "serve the US national interest." Now, the Justice Department was preparing to charge the companies with a criminal conspiracy for those very actions at exactly the same time the State Department was trying to persuade them to go into the Iranian consortium—no doubt risking the further fury of the Justice Department at some later date!

Dean Acheson, fearful of the consequences both in Iran and for overall American foreign policy objectives, threw as much muscle as he could into calling off Justice. Flanked by Defense Secretary Robert Lovett and General Omar Bradley, Chairman of the Joint Chiefs of Staff, he sought to persuade Attorney General James McGranery to back off. But to no avail. The decision whether to launch the suit would be President Truman's. Time was very short. Eisenhower had been elected in November 1952, and the last weeks of the Truman Administration were rapidly dwindling away.

What would the President decide? Harry Truman knew something about oil. As a young man, he had been a partner in a company that was going to wildcat in several states. Back then, Truman had been motivated by the customary dream of a bonanza and great wealth, but it did not work out, and he lost his money. Paradoxically, a group that bought some of Truman's leases soon discovered a very large field, and later in life, Truman would sometimes muse on what might have happened had he and his partners discovered the oil. Perhaps he would have ended up an oil millionaire, instead of President. Truman had remained skeptical and critical of Big Oil; he had chaired the Senate committee that had pilloried Jersey in 1942 for its prewar relations with I. G. Farben. But, whatever Truman's populist bent, whatever his sense of what was right and wrong and what made good domestic politics, the risks, as he saw them, were far too large. Iran was a country he worried about. Once, in the middle of a discussion about the Korean War, Truman put his finger on Iran on a globe. "Here is where they will start trouble if we aren't careful," he told an aide. "If we just stand by, they'll move into Iran and they'll take over the whole Middle East." And "they" were the Soviets.

On January 12, 1953, less than two weeks before the end of his Administration, Truman announced his decision. The criminal investigation was called off. It would be replaced by a civil action. The Eisenhower Administration filed the civil case in April 1953, charging the five American companies with participating "in an unlawful combination and conspiracy to restrain interstate and foreign commerce of the United States in petroleum and products." The only reason that the Justice Department had not gone ahead with the criminal case was, as L. J. Emmerglick, a senior antitrust attorney, put it, "the considered judgment of two Presidents, two Secretaries of State or their principal representatives, two Secretaries of Defense, and in addition, the Chairman of the Joint Chiefs of Staff, the Central Intelligence Agency, and a number of present and former Cabinet members."

To implement the new administration's decision, the National Security Council issued a directive to the Attorney General that "the enforcement of the Antitrust laws of the United States against the Western oil companies operating in the Near East may be deemed secondary to the national security interest." But it was absolutely clear that the oil companies would not go into the Iranian consortium unless they had a specific guarantee of nonprosecution that would survive from administration to administration. In January 1954 both the Attorney General and the National Security Council provided the explicit guarantee. The plan for the Iranian consortium, said Eisenhower's Attorney General, Herbert Brownell, "would not violate the anti-trust laws of the United States."[19]

Building the Consortium

Now the real effort began to create a new consortium of Western companies to operate in Iran. It would become a marvelous case of multidimensional diplomacy. In addition to Anglo-Iranian, the participating companies included the four Aramco partners—Jersey, Socony, Texaco, and Standard of California—plus Gulf, which was Anglo-Iranian's partner in Kuwait; Shell, which was tied to Gulf in Kuwait; and the French company, CFP. The American and British governments were also intimately involved. What qualified those particular seven companies for membership was that they were members of the joint ventures that produced oil elsewhere in the Middle East, and along with Anglo-Iranian, were responsible for most of the oil production throughout the area. During the years that Iran had been out of the world oil market, production in neighboring countries had increased dramatically, and it was evident to all concerned that the surging output in the region would have to be reined in to make room for a resumption of Iranian exports. The only way to assure the acquiescence of all seven of the companies would be to give each of them a stake in the new consortium.

Even before the situation in Iran could be addressed, the other oil-producing countries had to be placated. The Aramco partners went to see the aged King Ibn Saud just prior to his death with the delicate mission of explaining why they would be taking Iranian oil, and thus reducing the growth in Saudi output. They were going into the Iranian consortium, they said, "solely on the basis that there might be chaos out in the area if we didn't." They were doing this not because they wanted more petroleum, because they didn't, but "as a political matter at the request of our government." Ibn Saud understood. The geopolitical considerations were clear: Iran might otherwise go communist, with all the dangers that would entail for Saudi Arabia. The Aramco partners should go ahead, the King said. But he had an important warning for them. "In no case should you lift more than you are obligated to lift to satisfy the requirements of doing that job."

The companies sent a small team to Tehran to negotiate with the Iranians. Once again, it became one of those interminable Persian negotiations, in which the issues and definitions and objectives were forever shifting. Though Mossadegh was gone from power and Iranian officials were very keen to get oil exports going again, they could not allow themselves to be seen as compromising Iran's sovereignty or its capture of economic rents. Moreover, the Shah and his negotiators had an altogether rational fear of another uprising and of being expelled from the country—or much worse. As a result, they were tough and insistent.

At one point, the companies' negotiating team, discouraged and losing hope, prepared to return to London, though they left some members behind in Tehran as, they joked, "hostages." Howard Page of Jersey led the negotiators back to Tehran in June 1954 for another try. Finally, on September 17, 1954, Page, who played a central role for the companies, and the Iranian Finance Minister initialed an agreement between the consortium and the National Iranian Oil Company. The Shah signed it on October 29, 1954. The day after—and three years after the British had been forced to make their ignominious retreat from the refinery at Abadan to the tune of "Colonel Bogey"—there was a different kind of ceremony at Abadan. While Page and the Iranian Finance Minister delivered speeches of celebration, oil began to flow to waiting tankers. The first one to pull away from the dock was the *British Advocate*, owned by Anglo-Iranian. Iran was back in the oil business.

The establishment of the consortium marked one of the great turning points for the oil industry. The concept of the concession owned by foreigners was for the first time replaced by negotiation and mutual agreement. The Mexican experience had been a dictated expropriation. But now, in Iran, all parties acknowledged, again for the first time, that the oil assets belonged, in principle, to Iran. Under this new deal, Iran's National Iranian Oil Company would own the country's oil resources and facilities. But, in practice, it could not tell the consortium what to do. The consortium would, as a contract agent, manage the Iranian industry and buy all the output, with each company in the consortium disposing of its share of the oil through its own independent marketing system. Though humbled, Anglo-Iranian was still the dominant partner, with 40 percent of the consortium. Shell had 14 percent, each of the five American majors had 8 percent initially, and CFP had 6 percent.

After a few months, the consortium took on a slightly different composition. By prearrangement with the American government, each of the American companies surrendered 1 percent to a new entity called Iricon—a sort of "little consortium" inside the larger one. It was composed of nine independent American oil companies, among them Phillips, Richfield, Standard of Ohio, and Ashland. Their entry had been insisted upon by the U.S. government for political and an-

458

titrust reasons. Without their participation, the consortium might not have survived domestic American politics. As Howard Page later joked, there was a feeling that "because people were always yacking about it, we had better put some independents in there." The British were furious with the whole idea. "We didn't know who the independents were," recalled a British official who was central to the negotiations. "We didn't think they were reputable marketers. We thought they would upset the apple cart around the Middle East in various ways, that they weren't the sort of people to do business with." But the British had no choice but to accede to the American insistence.

The "little consortium" was open to any American independent whose financial capability had been examined and approved by the accounting firm of Price Waterhouse. But, seeking to placate the angry British government, the State Department assured London that State itself had taken on "the responsibility for managing the introduction of the independents" and earnestly promised that "only good and reliable independents" would be let in.

With the establishment of the Iranian consortium, the United States was now *the* major player in the oil, and the volatile politics, of the Middle East. Even though the supply disruption resulting from the Iranian imbroglio had been far more easily solved than had been anticipated, there were still a few who worried about the implications of ever-growing dependence on Middle Eastern oil. Some months after the fall of Mossadegh and the return of the Shah, Loy Henderson, the American ambassador to Tehran and formerly the Assistant Secretary responsible for the Middle East, tried to gather his thoughts. He could not be confident that the mere disappearance of Mossadegh meant that longer-term risks were any less, in particular with regard to the security of oil supplies. "It seems almost inevitable that at some time in the future . . . the Middle Eastern countries . . . will come together and decide upon unified policies which might have disastrous effects upon the operations of the companies," he predicted in 1953. "Continuation and enhancement of the dependence which the West now places upon Middle Eastern oil might eventually place European consumers at their mercy."

The antitrust case against the oil companies continued to grind on. The Attorney General's approval of the Iranian consortium had the effect of insulating other joint production arrangements in the upstream, such as Aramco, from antitrust attack. Thus, the case was narrowed to downstream—marketing and distribution facilities—and resulted in the early 1960s in the breakup of Stanvac, Jersey's and Socony's joint company in the Far East. Caltex's downstream system in Europe, jointly owned by Socal and Texaco, was dissolved for commercial reasons. Even though more and more independents and national oil companies entered the world oil market, it was not until 1968 that the American government finally folded its tent on the case. By then, the consortium had been in business for almost a decade and a half in Iran.

For its part, the Anglo-Iranian Oil Company came out surprisingly well from the Iranian turmoil. All along, it had insisted that it had to be compensated for its nationalized properties, as part of any consortium arrangement. The ungiving and unrepentant Sir William Fraser was tenacious in his pursuit of compensation, to the point where he infuriated all the other participants, corporate and governmental, in the various negotiations. But Fraser was not willing to retreat—much. By sheer persistence, he finally won his compensation, though it came not from Iran, which even under the Shah insisted that it owed nothing, but from the other companies who joined the consortium. They paid Anglo-Iranian about $90 million up front for the 60 percent rights that the company was said to be giving up. In addition, Anglo-Iranian would receive a ten cent a barrel royalty on the entire production controlled by the consortium until another $500 million or so had been paid. Thus, despite the official acceptance of the nationalization and of the fact that Iran did indeed own its oil resources and industry, the other companies were paying Anglo-Iranian, and not the Iranian government, for rights to the oil. "It was a wonderful deal for Fraser, the best deal Willie Fraser ever made," said John Loudon, the senior managing director of Royal Dutch/Shell. "After all, Anglo-Iranian actually had nothing to sell. It had already been nationalized."[20]

The other irascible old man who played a major role in the Iranian crisis did not fare nearly as well. Mohammed Mossadegh was put on trial by the reinstated Shah, delivered impassioned speeches in his own defense, and spent three years in prison. He lived out the rest of his days under house arrest on his estate, continuing his experiments with homeopathic medicines, much as he had done, three decades earlier, when the Shah's father put him under house arrest. Meanwhile, swelling oil revenues were transforming the insecurities of the young Shah, now firmly ensconced on the Peacock Throne of Iran, into the pretensions of an assertive monarch with global aspirations.

CHAPTER 24

The Suez Crisis

THE SUEZ CANAL—a narrow waterway a hundred miles long, dug through the Egyptian desert to link the Red Sea to the Mediterranean—was one of the grandest achievements of the nineteenth century. It was the handiwork of Ferdinand de Lesseps, a Frenchman ever after celebrated as "the Great Engineer." In fact, he was no engineer at all, though he was a man of other considerable accomplishments—as a diplomat, entrepreneur, and promoter. And his talents did not end there. At the age of sixty-four, he married a woman of twenty, and then, forthwith, proceeded to father twelve children.

Though long discussed, such a waterway was thought to be impossible until de Lesseps floated a private concern, the Suez Canal Company, which won a concession from Egypt to build a canal and began actual construction in 1859. A decade later, in 1869, the canal was finally completed. The British were quick to recognize a good thing when they saw it, especially when it substantially reduced the travel time to India, the jewel of the empire, and they rued their lack of a direct stake in "our highway to India," as the Prince of Wales dubbed the waterway. Fortunately, in 1875, Egypt's 44 percent ownership in the canal came on the market, owing to the insolvency of the Khedive, the country's ruler. With lightning speed and the timely financial assistance of the English branch of the Rothschilds, Britain's Prime Minister, Benjamin Disraeli, engineered the acquisition of those shares. The Suez Canal Company became an Anglo-French concern, and Disraeli capped his efforts with a pithy and immortal note to Queen Victoria: "You have it, Madam."[1]

A wonderful boon to travelers and businessmen, the canal cut the time required for the journey to India in half. But the canal's greatest significance was strategic; it was indeed the main highway, the lifeline, of the British Empire, linking England to India and the Far East. The "defense of the communications with India" became the fundamental rationale for Britain's security strategy. British forces were stationed permanently in the Canal Zone. The canal's military importance was made starkly clear in World War II when the British mounted their stand at El Alamein to defend the canal against the advancing Rommel.

But, in 1948, the canal abruptly lost its traditional rationale. For in that year India became independent, and control over the canal could no longer be preserved on grounds that it was critical to the defense either of India or of an empire that was being liquidated. And yet, at exactly the same moment, the canal was gaining a new role—as the highway not of empire, but of oil. The Suez Canal was the way most of the swelling volumes of Persian Gulf oil got to Europe, cutting the 11,000-mile journey around the Cape of Good Hope to Southampton down to 6,500 miles. By 1955, petroleum accounted for two-thirds of all the canal's traffic, and in turn two-thirds of Europe's oil passed through it. Flanked to the north by Tapline and the IPC pipelines, the canal was the critical link in the postwar structure of the international oil industry. And it was a waterway of unique importance to the Western powers that were becoming so heavily dependent on petroleum from the Middle East.[2]

The Nationalist: The Role Finds Its Hero

Britain had exercised control over Egypt and hence the Suez Canal for three-quarters of a century, first by outright invasion and military occupation, then by political and economic dominance over a succession of client regimes. But there had long been a restive current of Egyptian nationalism, which grew stronger in the early postwar years. In 1952, a group of military officers successfully carried out a coup and dispatched the sybaritic King Farouk to exile on the Riviera, where, though unmourned, he acquired new renown both for his many girlfriends and for being enormously fat. By 1954, Colonel Gamal Abdel Nasser had toppled General Mohammed Naguib, the titular leader of the 1952 coup, and had emerged as the undisputed dictator of Egypt.

Son of a post office clerk and a born plotter, Nasser had begun his original anti-British maneuverings a decade earlier, during World War II, and he relished the game of subterranean intrigue thereafter. A secret CIA profile of Nasser concluded, "He gets boyish pleasure out of conspiratorial doings." Even when head of state, he would tell visitors and associates that he still felt he was a conspirator. He also had the capability to capture and direct the new spirit of nationalism in the Arab world. A gifted student of Mohammed Mossadegh, he mastered the use of rhetoric and the radio to inflame and mobilize the masses, stirring tens or hundreds of thousands of demonstrators to take to the streets in feverish passion. In turn, he was to become the model for the military officer turned fiery nationalist leader in the emerging nations of the Third World.

Nasser was indeed a nationalist, dedicated to Egypt's restoration and independence. But he also wanted to reach far beyond Egypt's borders, from one end of the Arab-speaking world to the other, from the western edge of North Africa to the shores of the Persian Gulf. "The Voice of the Arabs" was the name of his powerful radio station, and it was beamed across the Middle East, carrying his impassioned speeches over the air waves, calling for the rejection of the West and threatening other Arab regimes in the region. His program included pan-Arabism, the creation of a new Arab world, led by Gamal Abdel Nasser, the elimination of the Israeli wedge dividing the Arab world, and the righting of what he called—the "greatest international crime" in history—the creation of Israel.

The Suez Canal—with ships guided through it under the hot sun by mostly foreign, mostly French and British pilots garbed in impeccable knee socks, shorts, crisp white shirts, and captain's hats—was an all-too-evident and embarrassing symbol of the old nineteenth-century colonialism right in the middle of what was to be Nasser's new Egypt. Symbols, however, were not the only consideration. As with the oil concession in Iran before Mossadegh, most of the canal company's earnings, derived from tolls, were going to European shareholders, including the largest shareholder of all, the British government. If Egypt could secure complete control over the canal, the tolls would open a major new source of income for a desperately poor country, whose new military leaders were far more experienced at nationalist rhetoric than at economic management.

Under any circumstances, the concession's days were numbered. By treaty, it was due to expire in 1968, and British influence was already in retreat. Britain still maintained a military base and a large supply center in the Canal Zone, under the terms of the 1936 Anglo-Egyptian treaty; but Egyptians, impatient for their withdrawal, were conducting a campaign of harassment against them, including terrorist raids, murders, and kidnaping. What was the point of maintaining a base to protect the Middle East when it was under attack from one of the core territories it was supposed to defend? In 1954, Anthony Eden, as Foreign Secretary, directed the negotiation of an agreement whereby the last British troops stationed in the Canal Zone would be withdrawn within twenty months. The next year, just two months before he succeeded Churchill as Prime Minister, Eden made a stop in Cairo, where he startled Nasser by speaking—and telling Arabic proverbs—in Arabic.

There was the hope, of course, that the British government could remain on reasonable terms with Egypt, but this hope faded when Nasser attempted to incorporate the separate country of the Sudan into his Greater Egypt.[3] Nasser was viewed more tolerantly in Washington, where the Administration and many in Congress tended to adopt an attitude of moral superiority toward the European colonial powers, combined with a desire to see them divest themselves of their empires more quickly. The Americans believed that the relics of colonialism were an enormous handicap for the West in its struggle with communism and the Soviet Union. The Suez Canal Company, despite the economic and strategic significance of the waterway, was one of the most visible of these relics. The chairman of the canal company would later observe bitterly that, to Americans, "the company had a certain musty, nineteenth century odor derived from that lamentable colonial period."

Yet alarm about Nasser began to mount not only in London but also in Washington in the autumn of 1955, when it was learned the Egyptian dictator had turned to the Soviet bloc for weapons. Did that mean the expansion of Soviet influence? Might the Suez Canal be closed to Western oil and naval traffic? As early as February 1956 the State Department raised with the oil companies the question of revising the Voluntary Agreement of 1950—used originally to manage the loss of Iranian oil supplies—which would allow them to cooperate with one another and the government in the event the canal was closed, whatever the reason, to oil tanker traffic. To the companies, however, the Administration's proposal for joint action looked unworkable because of the threat of antitrust prosecution. That threat could hardly be regarded as idle; after all, the Justice Department was still pursuing its antitrust case against the major oil companies. Yet the companies themselves were worried about the possibility of supply disruption. In April 1956 Standard Oil of New Jersey commissioned its own study of how to move oil westward from the Persian Gulf if the canal was indeed shut.

Around this same time, the British Foreign Secretary, Selwyn Lloyd, visited Nasser in Egypt. As far as Britain was concerned, Lloyd made clear, the canal was "an integral part of the Middle East oil complex, which was vital to Britain." To this, Nasser replied that the oil-producing countries received 50 percent of the profits from their oil—but Egypt did not get 50 percent of the profits from the canal. If the Suez Canal was an integral part of the oil complex, he declared, then Egypt should have the same fifty-fifty terms as the oil producers. Nothing, however, was done to revise the existing arrangement.

At the end of 1955, in an effort to placate Nasser and strengthen the Egyptian economy, the Americans and British, in conjunction with the World Bank, had begun considering a loan to Egypt to build a huge dam at Aswan on the Nile. That project appeared to be going forward. And Nasser was further gratified when, on June 13, 1956, the last British troops were withdrawn from the Canal Zone in accordance with the agreement that Eden had negotiated two years earlier. But Nasser's arms deals with the Soviet bloc had already alarmed and alienated Washington. It was thought that the Egyptians would mortgage their limited resources to pay for Soviet arms instead of making their contribution to the dam. Moreover, the expected economic difficulties and hardships arising from the massive project could lead to antagonism and blame being directed toward the financing countries, and perhaps it was better to let the Soviets be stuck with that long-term cost. In any event, opposition was mounting in the United States. American senators from the South were hostile to the dam project because they feared it would lead to a much larger Egyptian cotton crop, which would compete with American exports on the world market. Congressmen friendly to Israel were not exactly keen to provide foreign aid to a government relentlessly antagonistic to Israel's existence. Nasser had recognized "Red China," as it was then called, further alarming both the Administration and many Congressmen. But the coup de grace came when Senate Republicans told Dulles that foreign aid could be approved for only one of the two "neutralist" leaders slated for assistance: Tito of Yugoslavia, or Nasser of Egypt. But not both. Dulles chose Tito. Eisenhower confirmed the decision. The British were in accord. On July 19, 1956, Dulles canceled the proposed Aswan Dam loan, taking Nasser and the World Bank by surprise.[4]

Code Word "de Lesseps": Nasser Moves

Nasser was angry, humiliated, and eager for revenge. The tolls from the canal, he thought, could be used to finance the Aswan Dam; the hated symbol of colonialism in his midst would be exorcised. On July 26, he gave a speech in the same square in Alexandria where he had, as a boy, for the first time joined a demonstration against the British. Now, as leader of Egypt, he repeatedly heaped calumny on the name of de Lesseps, the builder of the canal. It was no mere history lesson. "De Lesseps" was the code word to the Egyptian military to move; by the time the speech was completed, the army had seized control of the Canal Zone. The Suez Canal had been expropriated.

It was a resounding and daring act. Tension rose swiftly and dramatically in the immediate aftermath of the seizure. In England, the Chancellor of the Exchequer, Harold Macmillan, echoing the foreshadowing of his beloved Victorian novels, wrote apprehensively in his diary, "There was last night and through today, the most violent gale I ever remember." In Cairo, Nasser, deciding that he needed to escape from the mounting tension, slipped off to the Metro movie theater to see Cyd Charisse in *Meet Me in Las Vegas*.

Now followed three months of diplomatic circus and futile efforts to work out a compromise. In mid-September, the British and French pilots who had continued to guide ships through the canal were withdrawn on the instructions of the Suez Canal Company. The job was considered to be at the pinnacle of the merchant marine, and senior officials in London and Paris assumed that the Egyptians would not be able to run the canal by themselves. And indeed, considerable skill was required to pilot a ship through the canal, owing to the shallowness of the waterway, as well as the fierce crosswinds from the Sinai. But the Egyptian government had insisted for several years that Egyptians be trained to be pilots, and by the time of the nationalization, a considerable cadre of capable Egyptians was ready to take the helm, assisted by hurriedly dispatched ship pilots from the Soviet bloc. So the nationalized canal, under Nasser, continued to function more or less normally.[5]

Right at the beginning and throughout the mounting crisis, the British and French governments made one thing clear: they did not want to do anything to interrupt traffic and especially the passage of oil through the canal. But just where did the U.S. government stand? The American position during these months seemed confusing, not only to the British and French, but even to some American officials. To make matters worse, personal pique and a clash of styles irritated relations between Eden and Dulles. After one discordant meeting between the two men, Eden's principal private secretary wrote to a friend: "Foster talks so slowly that Master [Eden] does not want to hear what he has to say, while our man talks in so roundabout and elusive a style that the other, being a lawyer, goes away having failed to make the right guesses." Eisenhower himself had pinpointed in his diary what seemed to be part of the problem. Dulles, he wrote, "is not particularly persuasive in presentation and, at times, seems to have a curious lack of understanding as to how his words and manner may affect another personality." For his part, Dulles, along with other Americans, found Eden both arrogant and languid. But their discord went beyond style; there were specific grievances as well. Eden and Dulles had already clashed over the French Indo-Chinese war two years earlier. Eden had promoted diplomacy, but Dulles was not interested in that kind of peaceful resolution. Now, over Suez, they would trade roles.

Yet in August 1956, a few days after the nationalization, Dulles reassured the British and French foreign ministers that "a way had to be found to make Nasser disgorge" the canal. That expression was to ring as comfort in Eden's ear for the next couple of months. But the Americans came up with a number of diplomatic stratagems that seemed unrealistic to the British—or, if looked upon more cynically, seemed aimed at postponing more direct action on the part of the British and French.

In point of fact, U.S. policy was determined not by Dulles but by Eisenhower, and the President had no doubt from the beginning as to what the American position should be. Force was neither warranted nor justified in his view, and the essence of policy was to prevent the British and French from intervening militarily. The President believed that the two European countries simply would not be able to establish a pliable government in Egypt that would survive. Meanwhile, any such attempt would arouse not only the Arabs but the entire developing world against the West and would play into the hands of the Soviets, allowing them, in Ike's words, to claim the "mantle of world leadership." Moreover, he said to Eden, "Nasser thrives on drama," and the best thing to do was to let the drama seep out of the situation. To his own advisers, Eisenhower complained that British thinking was "out of date," while Nasser embodied the demands of people in the region for "slapping the white man down." A military assault against Egypt would surely turn Nasser into a hero throughout the developing world and would undermine friendly Arab leaders, jeopardizing Middle Eastern oil. Eisenhower repeatedly and sternly advised London against the use of force; and, to him and his advisers, American policy was crystal clear. Events would, however, prove that U.S. policy was not by any means so crystal clear to those to whom it was directed, the British and the French.

But it was of paramount importance to Eisenhower that the United States not appear associated, even indirectly, with sponsoring what seemed a return to the era of colonial domination. On the contrary, the situation in Egypt might prove to afford an opportunity to win support among the developing nations—even if it entailed alienation from America's traditional allies, Britain and France. After hearing the report of one statement by Eisenhower, Nasser joked to an aide, "Which side is he on?"[6]

There was another factor. Eisenhower was up for reelection in November 1956; he had ended the fighting in Korea at the beginning of his Administration, he was running as a man of peace, and the last thing he wanted now was a military crisis that might frighten the electorate and threaten his campaign. In an appalling blunder, the British and French never really factored the calendar of the American Presidential election into their calculations. While the public diplomatic show went on, they were also working secretly on a second track. They were making plans for military intervention in the Canal Zone, even though neither was well prepared to take such an action. The British found they had to requisition ocean liners at the height of the tourist season and even had to call on a private moving company, Pickford Removals, to cart around tank units.[7]

"We Had No Intention of Being Strangled to Death"

Both London and Paris were strongly motivated toward military intervention. The French saw Nasser as a threat to their position in North Africa. The Egyptian leader was not only egging on rebels in Algeria, who had begun a war for independence two years earlier; he was also training and supplying them. The French were determined to humble Nasser and reclaim the canal that de Lesseps had built with French financing. They had already opened a military dialogue with the Israelis, who had their own reasons for striking at Nasser. The Egyptian President was building up his armaments in apparent preparation for a war against Israel. He was also sponsoring guerilla raids into Israel and had instituted a blockade of Israel's southern port of Eilat, which was, after all, a belligerent act.

But why was the canal so important to the British? Oil was a key part of the answer. The canal was the jugular. Just a few months before the expropriation of the canal, in April of 1956, the traveling team of "Mr. B" and "Mr. K"—as the post-Stalin Soviet leaders, Nikolai Bulganin and Nikita Khrushchev, were known—had come to London. Before their meeting, Eden had thoroughly reviewed with Eisenhower what he planned to say to the Soviets, and Eisenhower was in full accord. "We should not be acquiescent in any measure," advised the President, "which would give the Bear's claws a grip on production or transportation of oil which is so vital to the defense and economy of the Western World." In the course of his discussions with the Soviet leaders, Eden warned them against meddling in the Middle East. "I must be absolutely blunt about the oil," he said, "because we would fight for it." To drive home the point, he added, "We could not live without oil and . . . we had no intention of being strangled to death."[8]

Nasser's seizure of the canal made that prospect all too real. Britain's international finances were precarious; its balance of payments, fragile. It had gone from being the world's greatest creditor to being the world's greatest debtor. Its gold and dollar reserves were sufficient to cover only three months of imports. Britain's oil holdings in the Middle East contributed mightily to its total foreign earnings. Their loss would be devastating economically. And a Nasser victory in Egypt might have the same kind of repercussions as a Mossadegh victory in Iran would have had. British prestige would be in tatters, and prestige mattered greatly when the British already felt they were losing ground everywhere. A triumphant Nasser would go on to subvert and topple regimes friendly to Britain and undermine the British—and American—oil position throughout the Middle East. The moment could come, Eden warned Eisenhower, when "Nasser can deny oil to Western Europe and we shall all be at his mercy."

Eden's anxieties were not only about oil and economics, but also about the possibility of a wholesale influx of Soviet power in a Middle East vacuum. "Eden was very worried about Soviet expansion in the Middle East," recalled a

Foreign Office official who reported directly to Eden on oil matters. "The Americans were not ready to take over the Middle East from the British, so the British were left with the job of keeping the Russians out."

Harold Macmillan, the Chancellor of the Exchequer, thought exactly the same way as Eden about the threat to oil supplies and the dangerous implications thereof. He, too, was convinced that Britain was in a very dangerous and exposed position. To be sure, he did not give off the same outward signs of anxiety as Eden. Indeed, in the first two weeks of the crisis, he managed, midst all his duties, to read through thousands of pages of nineteenth-century novels and still more thousands of pages of other works—Jane Austen's *Northanger Abbey* and *Persuasion*, Dickens's *Our Mutual Friend*, George Eliot's *Scenes from a Clerical Life, Middlemarch*, and *Adam Bede*, before moving on over the next few weeks to Thackeray's *Vanity Fair*, Churchill's *History of the English Speaking Peoples*, various lives of Machiavelli and Savonarola, and a new novel by C. P. Snow. If he had not done all this reading, Macmillan later said, "I'd have gone barmy!" But he was as forceful as anyone in supporting Eden's gloomy prognosis and the need for action. "The truth is that we are caught in a terrible dilemma," he wrote in his diary. "If we take strong action against Egypt, and as a result the Canal is closed, the pipelines to the Levant are cut, the Persian Gulf revolts and oil production is stopped—then U.K. and Western Europe have 'had it.' " Yet "if we suffer a diplomatic defeat; if Nasser 'gets away with it'—and the Middle East countries, in a ferment, 'nationalize oil' . . . we have equally 'had it.' What then are we to do? It seems clear to me that we should take the only chance we have—to take strong action, and hope that thereby our friends in the Middle East will stand, our enemies fall, and the oil will be saved, but it is a tremendous decision."[9]

At the "Rhineland" Again—Twenty Years Later

As they confronted the crisis, Eden, Macmillan, and those around them, along with Premier Guy Mollet of France and his colleagues, were haunted by powerful historical memories. To all of them, Nasser was a resurrected Mussolini, even a nascent Hitler. Just a decade after the Axis defeat, they thought, another conspirator turned demagogue dictator had emerged to strut on the world stage and inflame the masses, promoting violence and war in pursuit of vast ambitions. The central experience of the Western leaders had been the two world wars. For Eden, the failure of diplomacy to head off tragedy had begun in 1914. "We are all marked to some extent by the stamp of our generation, mine is that of the assassination in Sarajevo and all that flowed from it," he later wrote. Looking back on the diplomacy and policies of the Entente in those critical weeks in 1914, he said, "It is impossible to read the record now and not feel that we had a responsibility for always being a lap behind. . . . Always a lap behind, the fatal lap."

The failure of governments to respond in time was etched even more strongly in memory by the events of the 1930s. The year 1956 marked the twentieth anniversary of Hitler's remilitarization of the Rhineland, in violation of treaty obligations. The British and French could have stopped the German dictator in 1936. Hitler would have lost his momentum and his prestige, he might even have been toppled, and then tens of millions of people would not have died. But the Western powers did not act. Again in 1938, the Western nations failed to back Czechoslovakia, and instead appeased Hitler at Munich. There, too, Hitler might have been stopped and the terrible carnage of the second great war averted.

Eden had courageously resigned as Foreign Secretary in 1938 in protest of the appeasement policies toward Mussolini and Hitler. Now, in the summer and early autumn of 1956, it seemed to him that Nasser was embarked on an all-too-familiar program of aggrandizement. To Eden, Nasser's *Philosophy of Revolution* read like Hitler's *Mein Kampf*. Nasser, too, wished to carve out a great empire, and in his book, he emphasized that the Arab world should use the power that came with the control over petroleum—"the vital nerve of civilization"—in its struggle against "imperialism." Without petroleum, Nasser proclaimed, all the machines and tools of the industrial world are "mere pieces of iron, rusty, motionless, and lifeless." Eden had already attempted compromise. He had invested great personal prestige in the 1954 settlement with Egypt over the withdrawal of British forces from the Canal Zone and had been subjected to considerable attack from a segment of his own Tory party because of it. Now, he felt, he had been personally betrayed by Nasser. As with Hitler, Nasser's promises were not worth the paper they were written on. Was the seizure of the canal, in violation of international agreements, another Rhineland? Would further attempts to accommodate and appease Nasser merely be another Munich? Eden did not want to go through all that again. Two of his brothers had died in World War I; his eldest son had been killed in World War II. He personally owed it to them and to all the other millions who had died because the Western countries had been too laggard in 1914 to halt the crisis and too irresolute in the 1930s to stop Hitler. If force had to be used against Nasser, better now than later.

Premier Mollet had been imprisoned in the German concentration camp at Buchenwald, and he thought as Eden did. So, too, the Belgian Foreign Minister, Paul Henri Spaak, who wrote to the British Foreign Secretary during the crisis, "I do not wish to hide from you that I am haunted by the memory of the mistakes which were committed at the outset of the Hitler period, mistakes which have cost us dear." [10]

The analogies were much less compelling in Washington than in Western Europe. But if there was no consensus on how to deal with Nasser, the Western countries were at least making contingency plans for an oil crisis that might result from a duel over Suez. Eisenhower authorized the creation of a Middle East Emergency Committee, which was charged with figuring out how to supply Western Europe if the canal was blocked. The Justice Department granted limited antitrust immunity to companies participating in the plan, but not enough for them to work together to allocate supplies and to exchange information about oil demand, tankers, and all the other logistical data that would have been necessary to mount a joint supply operation. Nevertheless, the committee established close communication with the British Oil Supply Advisory Committee and the Organization for European Economic Cooperation on plans for crisis management.

Overall, the oil companies believed that the bulk of Western Europe's requirements could be met by increased production in the Western Hemisphere, which would call on the large amount of surplus capacity in the United States and Venezuela. On the last day of July, the Executive Committee of Standard Oil of New Jersey finally received its report on alternatives to the Suez Canal that it had commissioned in April. Instead of building larger tankers, the study recommended the construction of a large-diameter pipeline from the head of the Persian Gulf through Iraq and Turkey to the Mediterranean. The estimated cost for the pipeline was half a billion dollars. There was, however, a slight hitch with timing; it would take four years to build. Moreover, the danger of excessive reliance on pipelines was demonstrated a few days later when Syria, as a warning to the West, stopped the oil flowing through Tapline for twenty-four hours.

In September, Eisenhower insisted, in a message to Eden, that there was a danger in "making of Nasser a much more important figure than he is." To this, Sir Ivone Kirkpatrick, the Permanent Undersecretary of the Foreign Office, provided a sharp rejoinder: "I wish the President were right. But I am convinced that he is wrong. . . . If we sit back while Nasser consolidates his position and gradually acquires control of the oil-bearing countries, he can, and is, according to our information, resolved to wreck us. If Middle East oil is denied to us for a year or two our gold reserves will disappear. If our gold reserves disappear the sterling area disintegrates. If the sterling area disintegrates and we have no reserves we shall not be able to maintain a force in Germany or, indeed, anywhere else. I doubt whether we shall be able to pay for the bare minimum necessary for our defense. And a country that cannot provide for its defense is finished."

That same month, with the Suez crisis still brewing, Robert Anderson, a wealthy Texas oil man who was much admired by Eisenhower, made a secret trip to Saudi Arabia as the President's personal emissary. The objective was to get the Saudis to apply pressure on Nasser to compromise. In Riyadh, Anderson warned King Saud and Prince Faisal, the Foreign Minister, that the United States had made great technical advances that would lead to sources of energy much cheaper and more efficient than oil, potentially rendering Saudi and all Middle Eastern petroleum reserves worthless. The United States might feel constrained to make this technology available to the Europeans if the canal were to be a tool of blackmail.

And what might this substitute be, asked King Saud.

"Nuclear energy," replied Anderson.

Neither King Saud nor Prince Faisal, who had done some reading on nuclear power, seemed impressed, nor did they show any worry about the ability of Saudi oil to compete in world energy markets. They dismissed Anderson's warning.

Meanwhile, the key British and French decision makers had become very skeptical about the prospects for a diplomatic settlement of the crisis, which were now centered at the United Nations. Only military force, they concluded, could work with Nasser and stop him at his "Rhineland."[11]

Force Applied

On October 24, 1956, senior British and French diplomatic and military officials, including the respective foreign ministers, met secretly at a villa in Sèvres, outside Paris, with a delegation of top Israelis, including David Ben-Gurion, Moshe Dayan, and Shimon Peres. The three nations came to an understanding: Israel, responding to Egyptian threats and military pressure, would launch a military strike across the virtually uninhabited Sinai Peninsula toward the Suez Canal. Britain and France would issue an ultimatum about protecting the canal, and then, if the fighting continued—as surely it would—they would invade the Canal Zone to protect the international waterway. The ultimate objective for the British and French would be to effect a canal settlement and, if possible, to topple Nasser in the process.

There was much closer understanding between Israel and France than between Israel and Britain, in whose official circles was to be found considerable distaste for Israel and Jews. It was ironic that Eden himself, who had a fondness for Arabs and Arab culture and had said to his private secretary during World War II, "Let me murmur in your ear that I prefer Arabs to Jews," was now preparing to pit himself against the self-appointed leader of the Arab world. By contrast, the Chancellor of the Exchequer, Harold Macmillan, thought that Jews "had character." But at Sèvres, the British Foreign Secretary, Selwyn Lloyd, and his lieutenants treated the Israelis with what seemed like disdain. Indeed, over the preceding several weeks, Britain had also been giving thought to coming to Jordan's aid if war broke out between Israel and Jordan, and had so warned the Israelis. One reason that the French took the lead at Sèvres in bringing the Israelis into the Anglo-French plans was to ensure that Britain and Israel did not end up fighting each other over Jordan in the midst of a confrontation with Egypt.[12]

The day before the secret accord at Sèvres, Egypt and Syria set up a joint military command under Egyptian control. The day after, Jordan acceded to the joint military command. The die was cast. Yet now came one of those strange bunchings of political and personal dramas that would further complicate the Suez crisis. On October 24, the same day as the meeting at Sèvres, Red Army troops entered Budapest to put down a revolution that had erupted in Hungary against Soviet control.

Then there was the medical condition of Anthony Eden. In 1953, in the course of a gall bladder operation, a careless surgeon had damaged his bile duct, which had only been partly repaired by subsequent operations, leaving Eden, as he once said, "with a largely artificial inside," along with a tendency to become ill and suffer pain under pressure. Some later said that such a condition could, literally, slowly poison the mind. To make matters worse, Eden thereafter was on drugs for his stomach pain, as well as on stimulants (apparently amphetamines) to counter the effects of the painkillers. The interaction and side effects of these various medications were not then well known. Eden struck others as very agitated. The dosage of both sets of drugs had to be increased considerably after Nasser's seizure of the canal. In early October, Eden collapsed and was rushed to a hospital with a 106 degree fever. Though he was back in command for much of October, he continued to show signs of ill health and was put on a regimen of ever-heavier medication. A certain character change was evident to some. A British intelligence official confided to an American opposite number, "Chums in Downing Street tell me our old boy is feeling queer and is all nerves." Sometimes, Eden would seek solace from the strain and his own ill health by sitting in his wife's drawing room at 10 Downing and gazing at a Degas bronze of a girl in a bath that the film producer Sir Alexander Korda had given him.[13]

Eden was not the only one who was feeling the effects of illness. Eisenhower had suffered a heart attack in 1955, and then, in June of 1956, had been stricken with ileitis, which had necessitated surgery. So in the impending confrontation, two of the principal actors on both sides of the Atlantic were in poor health. A third would soon join them.

After months of indecision and delay, events began to move swiftly. On October 29, Israel launched its attack into the Sinai, putting the Sèvres accord into action. On October 30, London and Paris issued their ultimatum and announced their intention to occupy the Canal Zone. The same day, Russian troops were withdrawn from Budapest with a promise of nonintervention. The next day, October 31, the British bombed Egyptian air fields, and the Egyptian Army began its hasty retreat across the Sinai.

The entire Suez operation came as a surprise to the Americans. Eisenhower first learned about the Israeli attack while on a campaign swing through the South. He was absolutely furious. Eden had betrayed him; his allies had deliberately deceived him. They might inadvertently set off a far wider international crisis, involving a direct clash with the Soviet Union. And they had done so while America was preoccupied with the Presidential election, only a week away. Eisenhower was so angry that he called 10 Downing Street and personally gave Eden "unshirted hell." Or at least that is what Eisenhower thought he was doing. In fact, he was in such a rage that he mistook one of Eden's aides, who took the call, for the Prime Minister himself, and without waiting for the answerer to identify himself, the President dumped the diatribe on the hapless aide and then hung up before Eden could be summoned to the phone.

On November 3, it was Dulles's turn to be rushed to the hospital; acute stomach cancer was diagnosed, and a substantial part of his stomach was removed. So now three of the key players were sick. With Dulles out from November 3 onward, day-to-day control of U.S. foreign policy rested in the hands of Undersecretary Herbert Hoover, Jr., who had put together the Iranian consortium and who was regarded in London as antipathetic to the British.

For several reasons—logistical problems, poor planning, and Eden's vacillation—there was a gap of several days before the British and French troops could follow up their ultimatum and carry out their invasion of the Canal Zone. Meanwhile, Nasser had been quick to act exactly where the most damage could be done. He scuttled dozens of ships filled with rocks and cement and old beer bottles, effectively blocking the waterway, and thus choking off the supply of oil, the security of which had been the immediate reason for the attack. Syrian engineers, on instructions from Nasser, sabotaged the pumping stations along the pipeline of the Iraq Petroleum Company, further reducing supplies.[14]

During the months of joint planning to avert oil shortages if Nasser closed the canal, the British had always assumed that the United States would step into any breach with American supplies. That assumption proved to be a large and decisive mistake, no less a miscalculation than their inattention to the date of the Presidential election. Eisenhower refused to permit any of the emergency supply arrangements to be put into action. "I'm inclined to think," he told aides, "that those who began this operation should be left to work out their own oil problems—to boil in their own oil, so to speak." Petroleum would provide the way for Washington to punish and pressure its allies in Western Europe. Instead of providing supplies to America's allies, Eisenhower would impose sanctions.

By November 5, the Israelis had consolidated control over the Sinai and the Gaza Strip and had secured the Strait of Tiran. On that same day, British and French forces began their airborne assault on the Canal Zone. "I remember the phone call from Eden," recalled a British diplomat at the United Nations, "and hearing in that clipped First World War aristocratic accent, 'the paraboys are being dropped.' It was an unreal world, as though he were ringing up from Mars." The day before, Soviet troops had reentered Budapest and brutally proceeded to crush the rebellion in Hungary. The coincidence of Suez precluded any effective common Western response to the Hungarian uprising and Soviet intervention. Indeed, without even a hint of embarrassment or self-consciousness, Moscow vituperatively denounced the British, French, and Israelis as "aggressors." The Soviets also threatened military intervention, perhaps even nuclear attacks on Paris and London. Any such assaults would result in devastating counterattacks on the Soviet Union, Eisenhower made clear—"as surely as night follows day."

Purgatory

Despite Ike's reply, the anger of the U.S. government toward Britain, France, and Israel continued unabated. The message from Washington remained the same: It did not approve of the military action, and the British and French would have to stop. On November 6, Eisenhower won a landslide victory over Adlai Stevenson. That same day, the British and French agreed to a cease-fire in place; they had by then achieved only a foothold along the canal. For them, the war had lasted barely a day, and its objective, unrestricted use of the canal, had already been lost. But Washington made clear that a cease-fire was not enough. They would have to withdraw. And so would Israel, or it would face economic reprisals from Washington. Eisenhower told his own advisers that it was imperative not to "get the Arabs sore at all of us," because they might embargo oil shipments from the entire Middle East.

Without American aid, all of Western Europe would soon be short of oil. Winter was coming, and the stock levels were only enough for several weeks. The normal route for three-quarters of Western Europe's oil was now interrupted by the combined loss of transit through the canal and the Middle Eastern pipelines. In addition, Saudi Arabia instituted an embargo against Britain and France. In Kuwait, acts of sabotage shut down that country's supply system. When word was brought to the British Cabinet's Egypt Committee that the United States was considering oil sanctions against Britain and France, Harold

Macmillan threw his arms up in the air. "Oil sanctions!" he said. "That finishes it." On November 7, the British government announced that consumption would have to be cut by 10 percent. Upon entering the House of Commons, Eden was greeted with catcalls from the Labour opposition, which had flipflopped from its initial vigorous endorsement of a strong policy against Nasser. Critics in Parliament now declared that, if ration coupons were issued, they should bear a portrait of Sir Anthony Eden.

On November 9, Eisenhower met with the National Security Council to begin considering help for the Europeans. He talked about getting the oil companies to cooperate on a major supply program. "Despite my stiff-necked Attorney General," he said with a smile, he would provide the companies with a certificate that they were operating in the interests of national security, thus protecting them from antitrust action. But what would happen if the heads of the oil companies landed in jail anyway for participating in such a program? Why, said the President, laughing, he would pardon them. But he also made emphatically clear that all this was only contingency planning. Absolutely no emergency oil supply program would be put into effect until the British and French had actually begun to withdraw from Egypt. Europeans bitterly complained that the United States was going to punish Britain and France by "keeping them in purgatory." The international oil companies, seeing shortages develop, begged the Eisenhower Administration to activate the Middle East Emergency Committee. But, as one oil company executive put it, "The Administration simply refused."

Britain was economically vulnerable in another way. Its international finances were very shaky, and once the military assault began at Suez, an immense run on the pound started. The British strongly believed that the run was carried out with the acquiescence, perhaps with the support and even instigation, of the Eisenhower Administration. The International Monetary Fund, under American prodding, refused London's pleas for emergency financial aid. The economics minister at the British embassy in Washington reported to London that he was meeting a "brick wall at every turn" in Washington in seeking the urgently required financial support. The Americans, he added, "seem determined to treat us as naughty boys who have got to be taught that they cannot go off and act on their own without asking Nanny's permission first." [15]

By the middle of November, United Nations "peacekeeping" troops had begun to arrive in Egypt. But the Eisenhower Administration indicated that purgatory was not over: The Middle East Emergency Committee would not be activated until all the British and French troops were out of Egypt. An oil shortage appeared imminent. Eisenhower wrote to his wartime compatriot, the British general Lord Ismay, now head of NATO, about "the sadness in which the free world has become involved." He was "far from being indifferent to the fuel and financial plight of Western Europe," but reiterated his desire not to "antagonize the Arab world." That last consideration was, he said, "an extremely delicate matter" that "cannot be publicly talked about." Ismay sent back his appreciation for the message, but confidentially warned Eisenhower that "next spring might well find NATO forces practically immobilized by lack of oil." Finally, at the end of November, London and Paris pledged a swift withdrawal of their forces from Suez. Only then did Eisenhower authorize the activation of the Middle East Emergency Committee. The Americans had carried the day. They also added to the burden of defeat and humiliation the British and French had already suffered at Nasser's hands. In the whole messy business, Nasser was the only clear winner.

Yet in mid-November, while British and French troops were still in Egypt, Foreign Secretary Selwyn Lloyd had made a visit to John Foster Dulles in his room at the Walter Reed Hospital. A most puzzling conversation had there ensued, at least according to Lloyd's recollection.

"Selwyn, why did you stop?" Dulles asked. "Why didn't you go through with it and get Nasser down?"

Lloyd was flabbergasted. After all, this was the same Secretary of State who had, seemingly, done everything he could to head off Anglo-French action, and whose government had effectively terminated action once it began.

"Well, Foster," Lloyd replied, "if you had so much as winked at us we might have gone on."

He couldn't have done that, Dulles replied.[16]

The "Oil Lift" and the "Sugar Bowl": Surmounting the Crisis

At the beginning of December, a month after the canal's closing, with Britain and France thwarted and all of Western Europe on the brink of an energy crisis, the emergency supply program finally swung into operation. The Oil Lift, as it was called, was a cooperative venture of governments and oil companies in both Europe and the United States.

For the most part, oil production in the Middle East had not been interrupted. The problem was, first of all, one of transportation. The solution was to tap into other supplies. Owing to the shorter distances and shorter travel time, any given tanker could transport twice as much oil from the Western Hemisphere to Europe as from the Persian Gulf, around the Cape of Good Hope, to Europe. Therefore, the main focus of the emergency committees was the wholesale redeployment of tankers so that the Western Hemisphere would once again be the main supply source for Europe, as it had been up to the end of the 1940s. Tankers were rerouted, they were shared among companies, supplies were swapped—all with the objective of moving oil in the quickest, most efficient way possible.

In Europe, extensive efforts were made to ensure that the emergency supplies—which became known as "the sugar bowl"—were equitably distributed among the various countries. The Organization for European Economic Cooperation (subsequently the OECD) created a Petroleum Emergency Group, which actually made the allocations, based on a formula that reflected pre-Suez oil use, stock levels, and local energy supplies. The Oil Lift was complemented by rationing and other demand-restraint measures. Belgium banned private driving on Sundays. France limited sales by oil companies to 70 percent of pre-Suez levels. Britain slapped new taxes on oil, which resulted in higher gasoline and fuel oil prices and led to a hike in London taxi fares that became immortalized as the "Suez sixpence." Power plants were encouraged to switch from oil to coal; by the end of December, Britain was rationing gasoline.

Though the disposition of tankers was the number-one problem, oil supply itself turned out to be right behind. It was estimated that Western Hemisphere production would have to be increased substantially to meet Europe's needs—much of that extra supply to be drawn from the United States, where there was a great deal of shut-in capacity. The international companies aggressively scoured U.S. crude markets for all the extra supplies they could obtain for the Oil Lift. But neither the companies nor the concerned governments had reckoned with the Texas Railroad Commission, which, to almost everyone else's consternation, barely allowed any increase in production in the critical winter months of 1957, and basically kept the shut-in production shut in. Here was a new arena for the old battle between independent producers and major companies. As an internal memorandum for Jersey's board delicately put it, the railroad commission reflected the independent producers in Texas, "whose interests normally are wholly domestic." The commission feared that a buildup of crude inventories inland, as well as gasoline inventories for which there was no extra call from Europe, would depress prices. And, of course, if anything, it wanted higher, not lower prices.

The commission's refusal to allow significantly higher production set off a storm of disapproval. Eric Drake of British Petroleum called it "nothing less than a calamity for Europe." One of Jersey's European representatives said it was "disastrous" and could lead to a 50 percent drop in the company's supplies to Europe. Both Eden and Macmillan personally protested the policies of the Texas Railroad Commission, and the British press denounced this unknown and mysterious agency deep in the heart of Texas. The *Daily Express* declared: "No Extra Oil—Say Texas Wise Men." The venerable Texas Railroad commissioner, Colonel E. O. Thompson, did not hesitate to give as good as he got. To the complaints from Great Britain, he replied, "We have already shipped her many barrels of crude, but we only get criticism for not going all out at her bidding. England apparently still looks on us as a province or dominion."

The mood among Texas producers improved markedly when they got what they wanted out of the majors and the marketplace. Humble, the Jersey affiliate in Texas, fearing a shortage of supply, announced that it would raise the price at which it bought oil in the fields by thirty-five cents a barrel. Other companies followed suit; additional crude became available from Texas producers, and supply for the Oil Lift surged. But soon there was a new storm of denunciation—this time of the oil companies, which were accused of collusion to raise prices. In the face of a shortage, higher prices sent out a necessary double signal: to increase supply and to reduce demand, both of which were not only welcome and constructive in the midst of the Suez oil crisis, but probably necessary to make the Oil Lift work. Yet, oil and politics being what they were, the price increases resulted in much controversy and led to highly charged Congressional hearings, which filled more than 2,800 pages, and to a new antitrust prosecution of twenty-nine oil companies by the Justice Department. The case would eventually be thrown out in 1960, when a Federal judge ruled that there was "economic justification" for the price increases and that the government's evidence "does not rise above the level of suspicion."

The Oil Lift involved a tremendous job of coordination and logistical dexterity. It drew upon the experience and personnel that had been involved in the Allies' Atlantic oil supply system during World War II. Yet considerable bureaucratic and administrative complexity had to be overcome. A host of governments, companies, and supply committees had to establish guidelines, marshal information and communicate it, and make sure the programs were properly implemented. There was much room for confusion, but the Oil Lift worked so well that it looked almost effortless. It was not. Afterward, one oil company executive tried to explain how misleading it was to believe that, during a crisis, "all you had to do is push a button and everything is all right." It was an admonition that should have been kept in mind for future crises.[17]

By the spring of 1957, the oil crisis was finally coming to an end, owing mainly to the unanticipated effectiveness of the Oil Lift. Almost 90 percent of the lost supplies had been compensated for. In Europe, the immediate conservation measures, abetted by warm weather, made up for much of the rest of the lost supplies, so that the actual shortage was quite small. Overall, the European economy was not as vulnerable to oil disruptions as it would later become. In 1956, oil was only responsible for about 20 percent of total energy consumption. Though shifting, Europe was still primarily a coal economy. That, in the years to come, would change.

In March 1957 the Iraq Petroleum pipelines were partly reopened, and by April, the Suez Canal had been sufficiently cleared for tankers to resume passage. Nasser had won; the canal now unequivocally belonged to, and was operated by, Egypt. While the Egyptian Suez Canal pilots did not dress as crisply as their British and French predecessors, they were quite adequate to the navigational task. The Persian Gulf producers were keen to get supplies moving again; Kuwait had seen its production drop by half owing to the inability to transport oil. In April, the American government suspended the emergency Oil Lift program. In mid-May, the British government terminated gasoline rationing and then took the final grudging step of directing "British shipping to use the Suez Canal." With that, the Suez crisis was really over.[18]

The Exit of "Sir Eden"

One of the American participants was later to remember the crisis as "a curious time, those Suez months. There was high comedy, low conspiracy, and deep tragedy, but mostly tragedy—for individuals and for nations." It constituted a huge personal tragedy for Prime Minister Anthony Eden—whom Nasser called "Sir Eden"—who had until then amassed an extraordinary record of prescience, courage, and diplomatic skill, but whose reputation was to be ignominiously scuttled with the ships that Nasser sent to the bottom of the canal. Eden, who had prepared so long for the prime ministership, was under constant emotional pressure throughout the crisis. In November, while the crisis still raged, his ill health forced him to take a long vacation in Jamaica, in a house loaned to him by Ian Fleming, the creator of James Bond. On his return, Eden's doctors told him that his health would not allow him to function as Prime Minister. He spent the days between Christmas and New Year's quietly at Chequers, thinking about his future. In a letter to a friend at that time, he described himself as "so unrepentant . . . I find it strange that so few, if any, have compared these events to 1936—yet it is so like." In January 1957 he resigned.

One of the first to be told was Harold Macmillan, who was next door at Number 11 Downing Street, when Eden summoned him into the little drawing room at the front of 10 Downing Street. "I can see him now on that sad winter afternoon," Macmillan wrote in his diary, "still looking so youthful, so gay, so debonair—the representative of all that was best of the youth that had served in the 1914–1918 War. . . . The survivors of that terrible holocaust had often felt under a special obligation, like men under a vow of duty. It was in this spirit that he and I had entered politics. Now, after these long years of service, at the peak of his authority, he had been struck down by a mysterious but inescapable fate." Stunned, Macmillan walked sadly back through the connecting passage to the Chancellor's residence at Number 11 Downing Street. The next morning, he sat under a portrait of Gladstone, at Number 11, quietly reading *Pride and Prejudice* to soothe his soul, until interrupted by a phone call asking him to come to the Palace to be invested as Prime Minister.

Suez was a watershed for Britain. It was to cause as severe a rupture in British culture as in that nation's politics and its international position. Yet Suez did not presage Britain's decline; rather, it made obvious what had already come to pass. Britain no longer belonged to the top echelon of world powers. The bleeding of two world wars and the divisions at home had heavily drained not only its exchequer, but also its confidence and political will. Eden had no doubt that he had done the right thing at Suez. Years later, the *Times* of London said of Anthony Eden, "He was the last prime minister to believe Britain was a great power and the first to confront a crisis which proved she was not." It was as much an epitaph for an empire and a state of mind as for a man.[19]

The Future of Security: Pipelines Versus Tankers

The Suez crisis gave the international oil industry considerable food for thought. Despite the return to operation of the canal, the oil companies were no longer confident that they could rely on it. A great deal of discussion followed among companies and governments about building more pipelines. But Syria's interdiction of the Iraq Petroleum Company's line had shown how vulnerable pipelines were to interruption. Clearly, they were not the sole answer to the overriding question of transit safety. The risks were too obvious.

In all the agitated discussion in 1956 about the Suez Canal as the jugular, one point had not been given much attention: If the canal and the Middle Eastern pipelines were vulnerable, there was a safer alternative—the route around the Cape of Good Hope. To be economical and practical, however, supplying Western Europe by that route would require much larger tankers, capable of carrying a great deal more oil. The general view in the industry, however, was that, physically, such tankers could not be constructed. But Japanese shipyards, taking advantage of advances in diesel engines and better steels, were soon proving otherwise. "In 1956, the tanker people were saying larger ships would be too expensive, their fuel costs too high," recalled John Loudon, senior managing director of Shell. "The amazing thing to me was how quickly these tankers were being manufactured by the Japanese." Not only would they prove to be eminently economical, they would also provide the requisite security. Thus, supertankers, along with the decline of British influence and prestige and the ascendancy of Gamal Abdel Nasser, were among the consequences of the Suez Crisis. As one British official put it, "Tankers clearly were less subject to political risk."[20]

To End the Suez Schism

In the wake of Suez, much bitterness toward the Americans remained on the part of the British and French. Eisenhower, the British ambassador in Washington acidly commented in early 1957, "has the American boy scout views about colonialism, the United Nations and the effectiveness of phrases as act of policy. . . . The need to save his health, added to natural inclination, have made him one of the idlest (if the most revered) Presidents known to American history."

During the crisis, the United States had focused on trying to bolster its position with Arab oil producers. Eisenhower himself put great emphasis on "building up King Saud as a major figure in the Middle Eastern area" as an alternative to Nasser and on making clear to Arab oil producers that the United States intended to work to "restore Middle East oil markets in Western Europe." Beyond that consideration, there was also the thrust to support stable pro-Western governments in the Middle East as a bulwark against Soviet expansionism. Britain and France certainly shared both those strategic objectives. Their differences were over means, not ends.[21]

Yet on both sides of the Atlantic there was recognition of the need to heal the Suez schism. The process would be much aided by the fact that the new Prime Minister was Harold Macmillan, who was to become known for his "unflappability," though he would later admit that, internally, he often "suffered from agonies of nervous apprehension." He and Eisenhower had worked together during World War II and had maintained friendship and high regard for each other. When Macmillan was mentioned as a possible successor to Eden, Eisenhower described him as "a straight, fine man." Somehow it did not hurt at all that Macmillan's mother came from a small town in Indiana. Macmillan was also a realist. After the hard lesson of Suez, he said, it was the "rulers in Washington, in whose hands all our destinies largely lay." That was simply a fact. To Eisenhower's good wishes, Macmillan replied, "I have no illusions about the headaches in store for me, but thirty-three years of parliamentary life have left me pretty tough, without, I hope, atrophying my sense of humor."

The Middle East and oil were, of course, among his most troublesome headaches, along with the rift in the American alliance. The formal process of healing was to take place at the Bermuda Conference between Eisenhower and Macmillan, which was held in March 1957 at the Mid-Ocean Golf Club. Oil was much on Macmillan's mind as he prepared for the meeting. He requested a map showing the various oil companies' positions in the Middle East, as well as a "family tree" of the companies themselves. The intertwined subjects of oil and Middle Eastern security formed one of the major subjects at the meeting itself. As Eisenhower later said, there was "some very plain talk" about oil, including the possibility of promoting the construction of supertankers. Suez had taught all the Western powers about the volatility of the Middle East; and now at Bermuda, the British emphasized the importance of maintaining the independence of Kuwait and the other states along the Gulf, all of whose rulers seemed very vulnerable to Nasserite coups d'état. Both sides agreed on the need for Britain to do everything it could to ensure the security of the Gulf. Calling Middle Eastern oil "the biggest prize in the world," Macmillan urged cooperation between the two governments to achieve long-term peace and prosperity in the area—the sort of "common approach," he said, that they had during the war.

The Bermuda Conference did help to close the schism between Britain and America. Eisenhower and Macmillan promised to write privately and "freely" to each other as often as once a week. After all, the two countries did indeed have common objectives in the Middle East. But as Suez had so dramatically demonstrated, in the years ahead it would be America, not Britain, whose power would predominate.

In 1970, fourteen years after the Suez Crisis, the Conservatives won the general election in Britain and Edward Heath became Prime Minister. He arranged a dinner at 10 Downing Street for Lord Avon, as Anthony Eden had now become, and whose chief whip Heath had been in 1956 during the Suez Crisis. For Eden, his own return as an honored guest to Number 10 was a wonderfully sentimental evening. Heath made a witty and charming speech, and Eden rose to respond impromptu. He offered a special prayer for the British people—that they might discover "a lake of oil" under the North Sea. That was exactly what they were just beginning to do in 1970, though they would not be able to benefit from it in time to prevent Edward Heath from being toppled by another energy crisis. How different things might have been in 1956 had the British then known of, or even suspected, the existence of such a lake.[22]

CHAPTER 25

The Elephants

IN THE PARLANCE of the oil industry, a giant oil field is called an "elephant." And by the early 1950s, the tally of elephants discovered in the Middle East was lengthening rapidly. In 1953, the geologist Everette DeGolyer wrote to a friend, the chief geologist of the Iraq Petroleum Company, F. E. Wellings, whose company had just found three elephants in three successive months. "The Middle East," said DeGolyer, "is rapidly getting into a condition which has been almost chronic in the United States since the earliest days of the industry, that is, the problem becomes market rather than production." He added that his firm of De-Golyer and McNaughton, the leading petroleum engineers of the day, was just finishing a secret study for the Saudi Arabian government of its reserves. Much had been learned about that country's oil resources since 1943, when DeGolyer had made his first trip to Saudi Arabia at the behest of Harold Ickes. DeGolyer knew that the reported reserves in the new study would far dwarf the preliminary estimate he had made a decade earlier. Though the results would not be "altogether too astronomic figures," he told Wellings, "they are so great that adding a billion barrels or so makes no real difference in the oil available."[1]

The oil industry had clearly entered a new era when a billion barrels or so no longer made much difference. From the early 1950s to the end of the 1960s, the world oil market was dominated by extraordinarily rapid growth, a tremendous surge that, like a powerful and somewhat frightening undertow, swept everyone in the industry forward with its seemingly irresistible force. Consumption grew at a pace that simply would not have been conceivable at the beginning of the postwar era. Yet, as rapidly as it grew, the availability of supplies grew even more swiftly.

The increase in free world crude oil production was gargantuan: from 8.7 million barrels per day in 1948 to 42 million barrels per day in 1972. While U.S.

production had grown 5.5 to 9.5 million barrels per day, America's share of total world production had slipped from 64 percent to 22 percent. The reason for the percentage decline was the extraordinary shift to the Middle East, where production had grown from 1.1 million barrels per day to 18.2 million barrels per day—a 1,500 percent increase!

Even more dramatic was the shift in proven oil reserves—that is, oil in a particular reservoir that, with some confidence, people knew about and that could be produced economically. Proven world oil reserves in the noncommunist world increased from 62 billion barrels in 1948 to 534 billion barrels in 1972, almost a ninefold rise. American reserves had grown from 21 billion barrels in 1948 to 38 billion barrels by 1972. But their statistical significance had shrunk, from 34 percent of total world reserves to a mere 7 percent. While major growth was being registered in Africa, by far the most staggering part of the increase was in the new center of gravity, the Middle East, whose reserves had increased from 28 billion barrels to 367 billion barrels. Out of every ten barrels added to free world oil reserves between 1948 and 1972, more than seven were found in the Middle East. What these enormous numbers meant was that, as fast as the world was using oil, it was even more quickly adding new reserves. In 1950, the industry estimated that, with current proven reserves at current rates of production, the world had enough oil to last nineteen years. By 1972, after all the years of rapid growth, of the intoxicating increase in consumption and the pell-mell pace of production, the estimated reserve life was thirty-five years.[2]

The sheer abundance of Middle Eastern "elephants" led inevitably both to vigorous efforts by new players to get into the game and to fierce competition for markets, a nonstop battle in which price cuts were the most potent weapon. To the companies, such price cuts were necessary business decisions. But they proved to be highly flammable tinder thrown onto the pyre of nationalism in the oil-producing countries—already ablaze, in the Middle East, because of Nasser's victory at Suez.

The postwar petroleum order rested on two foundations. One was composed of the great oil deals of the 1940s, which had established the basic relationships among companies operating in the Middle East. These arrangements mobilized the resources required for the rapid development of the petroleum reserves, tied production to the refining and marketing systems required by the scale of reserves, and developed and secured the much larger demand that was required. The second foundation was composed of the concessional and contractual relationships between the companies and the governments of the oil-producing countries, at the heart of which was the hard-won fifty-fifty profit-sharing arrangement. On those two bases, it was hoped, relative stability would be built.

Fearing what might otherwise follow, neither major oil companies nor the governments of the consuming nations wanted to budge from the fifty-fifty principle. As the Middle East Oil Committee of the British Cabinet Office said in late 1954, "A reasonable basis has now been reached for a partnership between the oil companies and the Middle East Governments. . . . Any further encroachment by the Middle East Governments . . . would seriously damage our oil supply system." But for the governments of the producing countries, it was quite a different matter. Why not increase their revenues if they could, so long as they did not irrevocably alienate the companies and Washington and London?

So, certainly, thought the Shah of Iran. By the mid-1950s, the days were past when he worried aloud whether he was a mouse or a man. He had already taken to declaring in private, "It was Iran's destiny to become a great power." To satisfy his ambition and appetite, he wanted more oil income. And he wanted to pursue a policy of greater independence in oil, thus reducing and restricting the power of the consortium of companies that had been one of the results of his humiliating struggle with Mossadegh. But he could not afford to upset the basic foreign relations and security of Iran. He needed an interlocutor. But such a person could not be from the ranks of the majors, nor from among the prominent American independents, almost all of which had enrolled in the consortium. Who else?

A European, an Italian who had his own oil agenda—Enrico Mattei.[3]

A New Napoleon

At a time when the major companies had become large bureaucracies, too large and complex and well established to reflect the image of any single man, Enrico Mattei was determined to create a new major, Italy's state-owned AGIP, that would be very much cast in his own image. He was a bold swashbuckler, a condottiere, a throwback to the earlier days of Napoleonic entrepreneurs. The stocky hawk-featured Mattei had the look of a fervent but worldly Jesuit of the sixteenth century. His dark, somber eyes were set under high, arching eyebrows; his thin hair was combed straight back. He was willful, ingenious, manipulative, and suspicious. He had a talent for improvisation, and a propensity for gambling and taking risks, combined with a steely commitment to his basic objective, which was to obtain for Italy and AGIP, and Enrico Mattei, a place in the sun.

The unruly son of a policeman from northern Italy, Mattei had left school at age fourteen to go to work in a furniture factory. By his early thirties, he was running his own chemical plant in Milan, and there during World War II he became a partisan leader (of the Christian Democrats). At the end of the war, his managerial and political skills led to his being entrusted with running the remnants of AGIP in northern Italy. AGIP—Azienda Generali Italiana Petroli—was, at that point, slightly less than two decades old. Italy, following the example of France in the 1920s, had set out to create a state-owned refining company, a national champion, to compete with the international companies. By the mid-1930s, AGIP had attained a market share in Italy roughly commensurate with Esso and Shell. But outside Italy it counted not at all. With a great burst of energy and a masterly grasp of politics, Italian-style, Mattei set about to turn AGIP into a much greater enterprise. Yet this could not be done without cash, and postwar Italy was a cash-starved country. The requisite money was found in the Po Valley, in northern Italy, with the discovery and development of significant reserves of natural gas, which generated the high earnings to fund both AGIP's expansion in Italy and its ambitions overseas.

In 1953, Mattei took a major step toward the realization of his ambitions when the various state hydrocarbon companies were gathered into a new entity,

ENI—Ente Nazionale Idrocarburi—a sprawling conglomerate, with a total of thirty-six subsidiaries ranging from crude oil and tankers and gasoline stations to real estate, hotels, toll highways, and soaps. Though there was meant to be government influence over ENI, the various operating companies (AGIP for oil, the pipeline company SNAM, and a number of others) were to be run somewhat autonomously, as commercial entities. Still, the president of ENI and, coincidentally, the presidents or managing directors of AGIP and all the other operating companies were one and the same, Enrico Mattei. "For the first time in the economic history of Italy," the United States embassy in Rome reported with some wonder in 1954, a government-owned entity in Italy "has found itself in the unique position of being financially solvent, capably led, and responsible to no one other than its leader." ENI's future would be profoundly shaped, the report continued, by "the limitless ambition evidenced in the person of Enrico Mattei."

Mattei himself became a popular hero, the most visible man in the country. He embodied great visions for postwar Italy: antifascism, the resurrection and rebuilding of the nation, and the emergence of the "new man" who had made it himself, without the old boy network. He also promised Italians their own secure supply of oil. Italy was a resource-poor country that was not only very conscious of its shortages but also blamed many of its woes, including its military reverses, on them. Now, with Mattei, these problems, at least in the energy realm, were to be solved. He appealed to national pride and knew how to capture the imagination of the public. Along the roads and autostradas of Italy, AGIP built new gas stations that were larger, more attractive, and more commodious than those of the international competitors. They even had restaurants.

Mattei soon became, it was said, the most powerful man in Italy. ENI owned the newspaper *Il Giorno*, subsidized several others from the right to the far left, and financed the Christian Democrats, as well as politicians of other parties. Mattei did not particularly like politicians, but he would use them to whatever extent was necessary. "To deal with a government," he would complain, "is like sucking needles." Mattei spoke a clipped, prosaic, rough-hewn Italian that compared poorly to the eloquence and rhetoric of Italian politics. Still, he was magnetic, potently charming, and persuasive, with a combination of intense emotion and sincerity, and all of it backed up by a driving, volcanic, irresistible energy. Many years later, one of his aides would recall, "Anybody who worked with him would go into the fire for him, although you couldn't really explain why."

As ENI grew, so it appeared, so did Mattei's sense of himself, which sometimes worked against him. Once Mattei came to London for a lunch date with John Loudon, the senior managing director of Royal Dutch/Shell. Here was the old and the new, the establishment and the parvenu. Loudon's father, Hugo, had been one of the founders of Royal Dutch/Shell; and by midcentury, his tall, aristocratic son was not only the outstanding corporate leader of international oil, but its leading diplomat as well. He was also a shrewd judge of character. At this point, Mattei badly wanted something that Shell was not particularly willing to give. That was the reason for the lunch date. "Mattei was a very difficult man," Loudon recalled. "He was also intensely vain." At least that was how he seemed to Loudon and his colleagues at Shell. So Loudon, at the start of the meal, asked

Mattei with apparent innocence how he had gotten into the oil business. Mattei, no doubt flattered to be taken so seriously at the heart of a major, thereupon talked nonstop through virtually the entire lunch, with no further prompting, telling the whole story of his life. "Finally, when we went to dessert, he asked us for something," said Loudon. "We couldn't do it, and that was the end of the conversation." It was not, however, the last he would hear from Enrico Mattei.[4]

Mattei's Greatest Battle

Mattei's overriding objective was to ensure that ENI—and Italy—had its own international petroleum supply, independent of the "Anglo-Saxon" companies. He wanted to share in the rents of Middle Eastern crude. He loudly and continually attacked the "cartel," as he called the major companies, and he was credited with coining the term "Sette Sorrelle," the Seven Sisters, in derisive reference to their close association and multiple joint ventures. The "Seven Sisters" included the four Aramco partners—Jersey (Exxon), Socony-Vacuum (Mobil), Standard of California (Chevron), and Texaco—plus Gulf, Royal Dutch/Shell, and British Petroleum, which were tied together in Kuwait. (In 1954, Anglo-Iranian, taking up the name of the subsidiary it had acquired in World War I, had rechristened itself British Petroleum). Actually there was an eighth sister, the French national champion, CFP, which was in both the Iranian consortium, with the Seven Sisters, and the Iraq Petroleum Company, along with Jersey, Socony, British Petroleum, and Royal Dutch/Shell. But since CFP did not fit under the rubric of "Anglo-Saxon," Mattei conveniently dropped it. His real complaint against this exclusive club of major companies was not its existence, but rather that he was not in it.

Mattei certainly tried to win membership. He believed that, because of his scrupulous cooperation with the embargo instituted by the majors against Iranian oil after Mossadegh's nationalization, he had earned a place in the Iranian consortium that the companies and the British and American governments fashioned after the fall of Mossadegh. Because of their membership in the Iraq Production Company, the French were invited into this new Iranian consortium. Because of American antitrust concerns, the nine independent American companies were shoehorned in too, though they had for the most part no foreign interests and no real need to produce in Iran. But Italy, which had hardly any resources and was so dependent upon the Middle East, was excluded. Mattei was furious. He would look for his opportunity—and for his revenge.

He found both when the 1956 Suez Crisis put the established oil companies on the defensive and made clear the degree to which British power and influence were in retreat in the Middle East. That meant a vacuum, which Mattei would step forward to fill. And in his own anticolonial rhetoric and his attacks on "imperialism," he was a fair match for the nationalistic fervor of the exporting countries.[5]

Mattei began talking seriously to Iran and to the Shah. If the majors had become expert at corporate intermarriages through their joint ventures, Mattei would go them one better: Thinking dynastically in his quest for Italian access to Iranian oil, he floated the idea of marrying off an Italian princess to the Shah, who was in urgent need of a male heir. The Shah was also in urgent need of a larger share of oil revenues than he obtained from the consortium. One of the legacies of Mossadegh, nationalization, gave the Shah comparative flexibility. In the other oil-producing countries, the concessionaires—the foreign companies—still owned the reserves in the ground. By contrast, the government owned all the oil resources in Iran, and the Shah was no less committed than Mossadegh to control of the country's petroleum resources.

Mattei took advantage of that situation, and over the spring and summer of 1957 he moved to hammer out a wholly unprecedented arrangement with Iran, one that took into account both Iran's new position and the Shah's ambitions. The Shah personally championed and pushed the deal through his government, under the terms of which the National Iranian Oil Company would be ENI's partner as well as its landlord. That meant, in practice, that Iran would get 75 percent of the profits to ENI's 25 percent—breaking the treasured fifty-fifty arrangement. As J. Paul Getty and others had already found, it cost more to get into the game when you were a latecomer.

As the proposed terms of the new deal between the Shah and Mattei leaked out, they greatly perturbed much of the rest of the oil world. The companies already established in Iran and the Middle East were appalled, as were the American and British governments. What did Mattei want? Why was he doing this? Some wondered if the new agreement was merely "a form of blackmail destined to secure Italian admission to the Consortium." Certainly, Mattei was not embarrassed to suggest that he was willing to be bought off. Just small pieces, he murmured, say 5 percent of the Iranian consortium and 10 percent of Aramco. The companies were shocked by the boldness of his demands. Enrico Mattei did not come cheap.[6]

Thought was given to trying to work with Mattei. "The Italians are determined somehow or other to muscle in on Middle East oil," said one British official in March of 1957. "My personal view, which I am sure would not be kindly regarded by the oil companies, is that B.P. and Shell and the Americans would be wise to consider whether it might not be a lesser evil to find room for the Italians than to give them cause to run amok in the Middle East." That was, however, a decidedly minority opinion, and one roundly condemned. "Signor Mattei is an unreliable person," said another official. "I doubt if we want to add to his megalomania by hinting that we come to terms with him." Indeed, the general view was that Mattei could not be admitted into the consortium, because if he were, then the Belgian company, Petrofina, would soon be knocking at the door, as would various German oil companies, and who knew who else? And, fundamentally, it would be impossible to work with Mattei. All conceivable persuasion had to be applied to try to stop the seventy-five-twenty-five deal.

The Americans and British protested to the Iranian government and to the Shah, warning that upsetting the fifty-fifty principle would "seriously prejudice the stability of the Middle East," and threaten the security of Europe's oil supplies. The Secretary General of the Italian Foreign Ministry, resentful of Mattei's power and independence, advised the British in great confidence, indeed in such secrecy that his remarks were not reported through normal channels, to take a very tough line with Mattei. Any hint of a desire to reach an accommodation with him, even any politeness, said the Secretary General, would be seen by Mattei "as a sign of weakness."[7]

All the objections were to no avail. By August 1957 Mattei's deal had been pretty much done, and there was reason to suspect that he was actually in Tehran. "The Italian Embassy were at first keeping him under close wraps," reported the British ambassador from Iran. "But we were pretty sure that he was here, so I took on a chance on Saturday evening . . . and rode my horse over from Gulhak to the Italian summer Embassy at Farmanieh." Sure enough, the ambassador came upon none other than Enrico Mattei himself, sitting under a tree, relaxing with a whiskey and soda, happily celebrating his victory—for that very day he had signed the agreement with Iran. He was affable and talked very freely. "No mysteries about the AGIP agreement," Mattei genially said. "Anyway it is all public property now." He then proceeded to offer a dissertation "on the thesis that the Middle East should now be industrial Europe's Middle West." Afterward, the ambassador reflected, with some understatement, that "Mattei certainly uses bold strokes and a big brush on a wide canvas."

To his own inner circle, Mattei expressed perplexity at the reaction of the majors. "They've given us two tiny places in Iran, and everybody makes a great noise." Of course, he knew why. Yet the partnership between ENI and Iran did not work out very well, not because of the deal itself, but because of the geology. No significant amounts of commercial oil were found in the areas of partnership. So entry into Iran did not fulfill Mattei's dream of obtaining Italy's own secure supply. But he did achieve another part of his ambition; the fifty-fifty principle had been pierced, much weakening the foundation on which, he believed, the power of the "Seven Sisters" had rested. "By various verbal acrobatics the Shah and his ministers are trying to maintain a facade of innocence about this and to pretend that it is still intact," the British embassy in Tehran reported with some resignation in August 1957. "But in fact we all know that 50/50 is about as dead as the four minute mile. And as inevitably."[8]

Japan Enters the Middle East

Italy was not the only industrial country seeking its own seat at the Middle Eastern oil table. Japan was acutely oil-sensitive because of both its history and position at the time of almost total dependence on imports as it was beginning its extraordinary economic climb. The Suez Crisis made Japan even more nervous. It, too, wanted its own secure supplies, and several key policy committees, both public and private, were coming to the conclusion that—whatever the diligent efforts to protect the domestic coal industry—imported oil was going to become Japan's most important fuel. But the flow of oil into Japan was for the most part controlled by the major American and British companies, through their own Japanese subsidiaries, through joint ventures, or by means of long-term contracts with independent Japanese refiners, who had been allowed to start up business again a few years earlier.

In the spring of 1957, just at the time the Suez Crisis was coming to its end and while Mattei was fashioning his new partnership with Iran, it became known that a consortium of Japanese companies was pursuing a concession from the

Saudis and Kuwaitis to explore the area offshore from the Neutral Zone. It was a bold maneuver; after all, a powerful group—Shell, British Petroleum, Gulf, and Jersey—was expressing interest in the same area.

The whole idea had emerged in the course of a train ride in Italy, when an employee of the Japan Development Bank happened to run into another Japanese businessman, who mentioned that he had contacts with people who knew about Middle Eastern oil. Back in Japan, the banker reported the conversation to his father, Taro Yamashita, an entrepreneur who had made a fortune before World War II constructing rental homes in Manchuria for the employees of the South Manchurian Railway. After the war, in addition to his business interests in Japan, he also became enormously well-connected politically. Yamashita seized on the idea, put together the consortium—which became known as the Arabian Oil Company—organized the financing, and won the blessing and support of the Japanese government. The entire thing had to be improvised; none of the participating companies had any significant experience in the oil industry.

That lack of experience, however, was not what worried the established companies and the Western governments. Rather, it was that the Japanese, in their eagerness to break in, would commit that cardinal sin—what the British Foreign Office called "a real breach of the 50/50." To be sure, Mattei's deal had maintained a fig leaf over the fifty-fifty with rhetoric about "partnership." If the fifty-fifty could not be kept sacrosanct, at least in principle, what foundation could there be for stable relationships between companies and governments? Yet how else, except by flouting the principle, could a newcomer like the Japanese, lacking the financial strength of the established players, gain entrance to the Middle East?[9]

The Japanese began their negotiations first with the Saudis, who insisted on various large down payments. But Japan was a very capital-short country, and the Japanese group did not have the wherewithal for such payments. The Saudis then proposed to reduce the required up-front payments if the Japanese would go below 50 percent in their take. After much back and forth, the Japanese agreed to take only 44 percent, leaving the Saudis with 56 percent. In addition, the Saudis would have the right to acquire an equity stake in the company should it strike oil.

When word of the terms spread among the American and British companies, the alarm bells went off. The whole structure of Middle Eastern relations might be threatened. But what could be done? Should London and Washington protest to the Japanese? "The feeling in the Foreign Office is that there is not much to be gained from approaching the Japanese direct," one official said. "They would be very likely to take such an approach as a hint that they had been rather clever, and the upshot might be that they would conclude their non-50/50 deal amid a great deal of diplomatic apology signifying nothing."

Could the Japanese government be persuaded to withdraw its support of the project? On the contrary, the Japanese cabinet reconfirmed its endorsement. As for the Saudis, they too were pleased with the arrangement. "An agreement has been reached in principle between us and the company," King Saud telegraphed to the Amir of Kuwait in early October 1957, adding that the Japanese were awaiting an invitation to Kuwait. "Both of us undoubtedly are keen to protect the interests of both our countries," the Amir replied, "and hope, God willing, that we will succeed in our endeavor to get in touch with a good firm." Shortly after, Kuwait also signed with the Arabian Oil Company. Its caution in letting the Saudis go first paid immediate dividends; whereas the Saudis had won 56 percent of the income, the Kuwaitis were able to go one point higher, and won 57. In due course, the Saudis corrected this disparity.

Arabian Oil started drilling offshore in July 1959 and made its first discovery in January 1960, whereupon the Saudi and Kuwaiti governments took up a 10 percent equity share in the company. Since Arabian Oil did not have any outlets of its own, Japan's Ministry of International Trade and Industry, dubbing it a "national project," made the Japanese refiners take the oil on a proportional basis. Arabian Oil stood, for some time, as an exception for Japan, which continued to be a country very short of capital as well as of upstream expertise. For the most part, it remained dependent on the system developed in the postwar years, based on supply coming through the hands of the majors. But Arabian Oil did give Japan an independent source of oil and by the mid-1960s was providing almost 15 percent of Japan's total supply.[10]

Even the Americans . . .

Whatever their nationalities, others wanting to enter the Middle Eastern fray would from now on have to pay a higher price and observe the new precedents— even American companies. Standard Oil of Indiana had long regretted that, at the bottom of the Great Depression in 1932, it had sold its Venezuelan production to Jersey. Now, in the late 1950s, Indiana decided that it, too, would have to become part of the great expansionist move of American companies and once again go overseas to seek, as the stockholders were told, "opportunities for profitable operations wherever they may exist." Staying at home was too risky.

An agreement in principle was swiftly made with Iran in 1958, along the lines of Mattei's seventy-five-twenty-five joint venture—except that Indiana also had to pay a very large up-front bonus. The Shah had recently divorced his wife for failing to provide him with what he described, in the course of a conversation with one visitor, as "continuity," which meant a male heir. To the visitor, the Shah seemed, in the aftermath of his divorce, "a man at an emotional crossroads. . . . We have a man in a sensitive, delicate mood, a lonely man with scarcely one really intimate friend and few relationships, plunging himself still more deeply into his work." This was a convenient time for the Standard of Indiana deal to be turned into another landmark in the Shah's quest for status and his struggle against the consortium and the major oil companies. After all, Indiana was no Italian newcomer; it was a well-established, well-respected American company, one of the most prominent, distinguished, and technologically advanced successors to Rockefeller's Standard Oil. In order to underline the significance of the deal, the Shah personally insisted that Frank Prior, the chairman of Indiana, fly to Tehran for the signing.

The Shah opened their first meeting with a surprising lecture that caught

Prior off-balance. "You know—we are not Arabs," the Shah said. "We are Aryans, and we are the same race as you are, and we have a great history. We have great pride."

"Oh, yes," replied Indiana's chairman, "we know, Your Majesty."

With the Shah's pride mollified, the rest of the discussions went very well, and the deal was forthwith signed, to the further outrage of the other oil companies. Unlike ENI, Indiana found a good deal of oil, beginning with a great off-shore field south of Kharg Island in the Persian Gulf. To flatter the Shah, it was named for Darius, the ancient Persian King. Soon thereafter, the Shah remarried and his new wife gave birth to a male heir. His "continuity" now seemed assured.[11]

Nasser Ascendant

The Shah was hardly alone in his campaign of national assertion against the established position of the major oil companies. Throughout the Middle East, nationalism was building to a crescendo, and Nasser was its driving force. Suez had been a great victory for him, proving that a Middle Eastern country could triumph not only against "imperialistic" companies but also against the might of Western governments. He had extirpated the ignominy of Mossadegh's failure. And now a notable technological innovation, the cheap transistor radio, was carrying his rousing voice to the poor masses throughout the Arab world, making him a hero everywhere.

In 1958, further adding to Nasser's laurels, Egypt finally bamboozled a reluctant and skeptical Soviet Union into providing the funding to build the Aswan Dam. In the same year, in a great symbol of Nasser's appeal, Syria joined Egypt to form the United Arab Republic, seemingly the first step in the realization of his dream of pan-Arabism. The apparent merger, ominously, brought together the two countries which—with the Suez Canal in Egypt and the Saudi and Iraqi pipelines passing through Syria—dominated the transit routes for Middle Eastern petroleum. Nasser was, at least theoretically, in a position to threaten single-handedly or actually even to choke off most of that supply. In order to counter what the British ambassador to Iraq called Nasser's "stranglehold," discussions ensued about quickly building Iraqi pipelines to the Persian Gulf as well as an export terminal at Fao, on the Gulf. But then the situation in the region, and in Iraq itself, went from bad to what seemed a total disaster.

For three years, Nasser had been conducting a virulent propaganda war against Iraq and the Hashemites, the British-backed Royal Family that had been installed by Great Britain on a newly created throne in Baghdad after World War I. In July 1958, officers plotting a coup told their troops the far-fetched story that they had been ordered to march to Israel and surrender their weapons. That was sufficient to get the soldiers to support a rebellion. The coup that followed set off an explosion of violence and savagery. Crowds surged through the streets, holding aloft huge photographs of Nasser, along with live squirming dogs, which represented the Iraqi Royal Family. King Faisal II himself was beheaded by troops that stormed the palace. The Crown Prince was shot, and his hands and feet were hacked off and carried on spikes through the city. His mutilated body, along with those of a number of other officials, was dragged through the streets, and then hung from a balcony at the Ministry of Defense. The pro-Western Prime Minister, Nuri es-Said, was recognized as he tried to flee the city, apparently disguised as a woman, and was lynched on the spot by a mob. His body, too, was hauled through the streets, and then a car was driven back and forth over it until it was flattened almost beyond recognition.

The new government in Baghdad immediately demanded vast revisions in the far-flung concession of the Iraq Petroleum Company. The grisly coup in Baghdad sent shudders through almost all the governments in the region; Nasserism seemed destined to reign supreme in the Middle East.[12]

Oil was a central focus of the rising Arab nationalism. Since the early 1950s, a number of meetings and contacts among what were semi-officially called "Arab Oil Experts" had been taking place in the Middle East. Initially, the dominant subject had been economic warfare against Israel: the establishment of an oil blockade against the new state and its enforcement upon the international companies through the threat of blacklist, harassment, and expropriation. Over time, however, the agenda broadened. Though Egypt was not an oil exporter, Nasser used such meetings to involve himself directly in petroleum politics. He sought to arouse and shape public opinion, hammering on the issues of sovereignty and the struggle against "colonialism," and to assert his influence over the oil, and the countries, of the Gulf. It was a case of a "have not" seeking to better his position with the wherewithal of the "haves." At a conclave of Arab Oil Experts in Egypt, in the spring of 1957, the delegates proposed the building up of domestic refining capacity and the establishment of an Arab tanker fleet and an Arab pipeline to the Mediterranean. They also discussed creating an Arab "international body" or "international consortium" that would manage Middle Eastern oil production, increase revenues, and counterbalance the power of the petroleum companies. The group emphasized the need to build up Arab expertise and technical skills in order to challenge their mystique.

The targets of the strong spirit of nationalism and confrontation at that meeting extended beyond the majors to the Western nations themselves. Oil, declared Abdullah Tariki of Saudi Arabia, is "the strongest of weapons the Arabs wield." And as if to celebrate the growing awareness of their power, the delegates took time out to observe and celebrate the passage of the first tanker through the Suez Canal after its reopening under unchallenged Egyptian control. The ship carried oil from the Neutral Zone for J. Paul Getty.[13]

Yet the talk among the delegates about a consortium or organization of oil-exporting states remained inchoate, still too much concerned with the Arab world alone. To become a reality it would require participation by other major producers, particularly Venezuela and Iran. And it would further require the catalytic role of one man, Juan Pablo Pérez Alfonzo.

Juan Pablo Pérez Alfonzo

In 1948, shortly after codifying the fifty-fifty principle, the new democratic government in Venezuela had been overthrown by a military coup, and control passed to the brutal, corrupt dictatorship of Colonel Marcos Pérez Jiménez.

Under his regime, oil production advanced at a very rapid pace, doubling by 1957. Support for Pérez Jiménez drained away, and his regime collapsed in January 1958, making way for a return to democracy in Venezuela. Many of the leaders of the new government had been prominent figures in the democratic government of the 1940s and, subsequently, veterans of exile and Pérez Jiménez's jails. The new President was Romulo Betancourt, who had been head of the 1945 revolutionary junta. During his years in exile, he had been not only an eloquent opponent of Pérez Jiménez, but also a passionate critic of the international oil companies, whose "close identification with the dictatorship," he said, had turned Venezuela into a "petroleum factory" that represented a throwback to the dark days of the Gómez dictatorship.

Yet Betancourt and his colleagues had learned the lessons of the coup of 1948: the need to maintain coalitions and unity across the democratic political spectrum and not to alienate other parties and interests. In the first years, the new government had to contend with assaults from both the right and the left, including communist guerillas. There was much anti-American sentiment in the country because of the friendliness shown to Pérez Jiménez by the Eisenhower Administration. Indeed, visiting the country in 1958, Vice-President Richard Nixon came close to being hurt or even killed when an enraged mob attacked his motorcade as it made its way into Caracas from the airport. In 1960, Betancourt himself was badly burned when his car was bombed in an assassination attempt. With the coup of 1948 all too vivid in his mind, Betancourt proceeded with caution. However much he may have denounced the oil companies, he needed them. As he said, he and his colleagues were "not impractical romantics." The man to whom Betancourt turned, when it came to oil, was Juan Pablo Pérez Alfonzo. Though Pérez Alfonzo, too, was a realist, and indeed a careful and pragmatic analyst, he was also an austere, self-sufficient moralist, with the fervor not of the politician, but of the intellectual. "He was a man of iron determination," said one Venezuelan who worked with him, "and yet soft-spoken and almost monastic."

Born to a well-to-do Caracas family, Pérez Alfonzo had studied medicine at Johns Hopkins University in Baltimore and then came back to Caracas, where he studied law. But then the family lost its money, and immense burdens fell on Pérez Alfonzo, for as, the oldest son, he found himself responsible for his 10 brothers. The whole experience profoundly shook him; and a commitment to conservation and planning thereafter became part of his character. Already a nonconformist of stern judgments when he married in 1932, he refused to have the ceremony performed by a certain judge in Caracas whom he regarded as incompetent and corrupt. Instead, Pérez Alfonzo and his wife went out to the countryside and found a local judge to marry them. After the end of the Gómez regime, Pérez Alfonzo, working in concert with Betancourt, emerged as the opposition's expert on the oil industry in the Chamber of Deputies. From 1945 on, first in the revolutionary junta and then in the democratic government, he was Minister of Development. As such, he set about repairing what he saw as the inadequacies of the 1943 law, to assure that Venezuela really got 50 percent of the profits, as well as greater control over the industry.

In November 1948 Pérez Alfonzo received a phone call from the United States ambassador in Caracas. A coup was in motion, said the ambassador, and he wanted to offer Pérez Alfonzo the hospitality of the embassy. Pérez Alfonzo thought it over, said no, he would take his chances, and went home for lunch, to wait. He was arrested and, regarded as a gray eminence of the democratic government, was thrown into jail. He would later joke to his family that he had been working too hard as minister, and that jail was like a holiday, an opportunity for him to rest. But, in fact, it was no joke; he was treated harshly and spent some of the time in solitary confinement.

Finally, he was allowed to go into exile, and he left the country, disgusted with day-to-day politics, and promising his family that he would never go back into active public life. He found refuge first in the Wesley Heights section of Washington, D.C., where he and his family scraped by on the rents they received from their house in Caracas. He wrote articles for exile newspapers and took up woodworking, but more than anything else, he devoted himself to the study of the oil industry. He was a regular reader at the Library of Congress. He carefully perused the gamut of American magazines to which he subscribed, from *Forbes* and *Fortune* to *The Nation* and *Oil and Gas Journal*. And he devoted considerable time to the study of an institution that particularly fascinated him—the Texas Railroad Commission, the agency that had begun regulating oil production in Texas, and thus in the nation, in the early 1930s, during the darkest days of ten-cent-a-barrel oil. After several years of exile in Washington, Pérez Alfonzo ran short of money, and he and his family moved to Mexico City. Another reason for the move was his worry that his children would become too Americanized to return comfortably to Venezuela—assuming that day would ever come.

The day came in 1958, when the dictatorship toppled. Pérez Alfonzo's wife begged him not to go back into government. But Betancourt insisted that he return to Caracas to take up the position of Minister of Mines and Hydrocarbons, which Pérez Alfonzo did. He was struck by how much more affluent the Caracas of 1958, buoyed by oil earnings, looked than the Caracas he had fled a decade earlier. His response was not altogether favorable. Oil wealth, he believed, was the gift of nature and politics, and not of hard work, and he soon found a perfect symbol for what he saw as its pernicious consequences. While still in exile in Mexico, the family had managed to get the money together to buy a 1950 Singer, a British motorcar that looked rather like an MG. Pérez Alfonzo treasured the car; it was one of his very few indulgences. When he came back to Venezuela, he arranged to have the Singer shipped after him. The car arrived at the docks, where it sat for two months, rusting away; no one bothered to tell Pérez Alfonzo that it was there. At last, hearing of its arrival, Pérez Alfonzo sent a mechanic down to the harbor to drive the car up to Caracas. Along the way, it broke down. The mechanic had forgotten to check the oil, and it turned out that there had been no oil in the engine. The car could no longer be driven at all; the engine was completely burned out. A truck had to be sent for it. Eventually, it was delivered to his villa in the suburbs. But the corrosion had eaten through the car. Pérez Alfonzo viewed it all like a sign from heaven; he had the car installed near a ping-pong table in his garden, as a corroded, overgrown shrine and symbol of what he saw as the dangers of oil wealth for a nation—laziness, the spirit of not caring, the commitment to buying and consuming and wasting.

Pérez Alfonzo had vowed never to allow himself to be seduced by the trappings of power, and once back in office, he kept to a simple, disciplined, and parsimonious life. He carried his own sardine sandwiches to the office for lunch. He also brought to his new office a sophisticated understanding of the structure of the oil industry as well as his own clearly defined objectives. He wanted not only to increase the government's share of the rents, but also to effect a transfer to the government, and away from the oil companies, of power and authority over production and marketing. To sell oil too cheaply, he argued, was bad for consumers, as the result would be the premature exhaustion of a nonrenewable resource and the discouragement of new development. For the producing countries, oil was a national heritage, the benefits of which belonged to future generations as well as to the present. Neither the resource nor the wealth that flowed from it should be wasted. Instead, the earnings should be used to develop the country more widely. Sovereign governments, rather than foreign corporations, should make the basic decisions about the production and disposition of their petroleum. Human nature should not be allowed to squander the potential of this precious resource.[14]

Yet Pérez Alfonzo was also motivated by shrewd commercial judgment. He knew that while Venezuela had a kinship with the oil-producing nations of the Middle East, those nations were also dangerous competitors. Venezuela was a relatively high-cost producer, at about eighty cents a barrel, according to one estimate, compared to twenty cents among the Persian Gulf producers. So Venezuela would inevitably be at a disadvantage in an out-and-out production race. It would lose market share. Venezuela, thus, had a very good reason to seek to persuade the Middle Eastern producers to raise their taxes on the companies and thus the cost of their oil.

The twist that Pérez Alfonzo introduced to improve Venezuela's position was based, in effect, on the Texas Railroad Commission, to which he had devoted so much study while in exile. He went so far as to contact the commission and to retain one of its consultants to explicate the mysteries and wonders of prorationing and how to apply it in Venezuela. He also saw that the way to get beyond mere talk with the Middle Eastern producers was by establishing a global alliance modeled on the Texas Railroad Commission. Venezuela could protect its market share not only by helping to raise costs in the Middle East, but also by getting the lower-cost producers to agree to a system of international prorationing and allocation along the lines that had become an art in Texas. The establishment of such a common front would, by regulating production, prevent Venezuela from having its oil industry, its main revenue source, swamped by millions and millions of barrels of cheap Middle Eastern oil.

The Eisenhower Administration's reluctant decision in early 1959 to put quotas on foreign oil in order to protect domestic producers hit Venezuela harder than any other country, since the United States was the destination of 40 percent of its total exports. Then the United States took an additional step. To placate its immediate neighbors, it provided exceptions to the quota for oil shipped overland—that is, from Canada and Mexico—on grounds of national security. With the World War II "Battle of the Atlantic" in mind, the Eisenhower Administration said that oil shipped overland was more secure because it could not be interdicted by enemy submarines. To the Venezuelans, that was simply a convenient fiction, employed to reduce friction with Canada and Mexico, and they were outraged. "The Americans are throwing us the bones," Pérez Alfonzo acidly said to one of his aides. Venezuela protested vigorously. After all, it had been the major, reliable supplier during World War II, and it would be a strategic resource in the future, as well. And it was Mexico, not Venezuela, that had nationalized the American oil companies. Why was Venezuela being punished?

Bitterly complaining, Pérez Alfonzo flew to Washington, now no longer a political exile trying to carve out a precarious life, but the Minister of Mines and Hydrocarbons of one of the world's oil powers. He came with a proposal to create a Western Hemisphere oil system, but one that would be run by the governments, not the oil companies. Under it, Venezuela would, as a nation, be given a quota—a guaranteed share of the U.S. market. No longer would it be the prerogative of the companies to decide from which producing country to bring in petroleum. What Pérez Alfonzo was asking was not so bizarre; after all, he could point out, it was exactly the way the American sugar quota system worked—each country had its share. But, then, oil was not sugar.

The United States government was not interested in Pérez Alfonzo's proposal; indeed, it did not really ever respond. The new democratic government in Caracas was insulted. And Pérez Alfonzo would seek a more attentive hearing elsewhere—in Cairo.[15]

The "Red Sheikh"

Abdullah Tariki, a Saudi Arabian, was the son of a camel owner who organized caravans between towns in Saudi Arabia and Kuwait. His father wanted him to follow in the same tracks. But Tariki's intelligence was noted early, and he was sent off to school in Kuwait. He later spent a dozen years studying in Cairo, where he imbibed the nationalist springs that nourished Nasserism. A scholarship carried him to the University of Texas, where he studied both chemistry and geology, and then he took a job as a trainee geologist with Texaco. His view of America was shaped in Texas, where on several occasions, it was said, he was excluded from bars and other facilities because he was taken for a Mexican. In 1948, he returned to Saudi Arabia, virtually the first of the American-educated Saudi technocrats, and certainly the first Saudi trained in both geology and chemistry. He also had an American wife. In 1955, at age thirty-five, Tariki was appointed to head a newly created Directorate of Oil and Mining Affairs. And, from the beginning, he intended to do more than merely assemble petroleum statistics from Aramco and pass them on to the Royal Family. He created a team of experts, including both an American lawyer and a young Saudi technocrat, Hisham Nazer, and prepared to challenge not only the basis of the Aramco concession but the Western oil companies themselves.

Tariki was an odd combination—not only an ardent supporter of Nasser but also a fervent Arab nationalist critical of the family that had created modern Saudi Arabia, while, at the same time, a servant of that same family in perhaps the single most important economic position in the kingdom. That Tariki, known to some as the "Red Sheikh," occupied that position, despite his views, resulted from the fact that an internal power struggle was being waged within the Royal Family between King Saud and his younger brother, Faisal. As had been feared in the last years of old Ibn Saud's life, the erratic King Saud, the eldest son, was getting the country into foreign policy trouble, was proving to be weak and indecisive, and was all too obviously a spendthrift. Faisal, by contrast, was the shrewd, coldly calculating son, the one to whom his father had entrusted the most important diplomatic and political business, beginning with Faisal's official visit to England at age fourteen. Faisal insisted that the wastrel spending had to be reined in. In contrast to Saud, who preferred to dally with Nasser, Faisal wanted to align with more traditional regimes, and with the United States and the West. With so much attention and energy focused on this power struggle, and in the absence of a single dominating personality, Tariki was able to shape policy with considerable autonomy in an absolutely critical area, the one that happened to generate the kingdom's entire wealth.

At first, Tariki concentrated on trying to gain control over refining and marketing assets as a means to raise Saudi oil revenues. He wanted to create a Saudi oil company integrated "right down to the service station" in the consuming countries. He would even raise an idea calculated to send a chill down the back of the American majors: the outright nationalization of Aramco. But then, in early 1959, his entire strategy abruptly changed. Control over prices and production, he suddenly decided, was more important than nationalization and integration. The reason for his change of mind was a sudden cut in the price of oil.[16]

Competitive Pressures

While world oil demand was growing during the 1950s, production capacity was growing still more rapidly. Always in search of higher revenues, the exporting countries, for the most part, sought to gain them by increasing the volume sold, rather than by raising prices. More oil was in search of markets than there were markets for oil. As a result, the companies were forced to offer bigger and bigger discounts on the prices at which they in turn sold their Middle Eastern oil.

Discounting was leading to a crucial divergence in the world oil industry between the "posted" or official price, which was held constant, and the actual market price at which crude was sold, which was dropping. It was the former, the posted price, against which the producing country's "take"—taxes and royalties—was computed. The posted price was supposed to more or less match the market price, and originally, it had. But, as discounting spread, a gap between the two had appeared and expanded. The posted price could not easily be lowered because of its importance to the revenues of the producing countries. That meant that they continued to take 50 percent of the profits based upon the posted price. But by the late 1950s, that was a fictional price, existing only as a basis for calculating revenues. In fact, the producing countries were taking a higher percentage—perhaps 60 or 70 percent—of the profit realized from the actual price. In other words, the Middle Eastern governments were being kept whole, while the companies absorbed all the effects of the price cuts. The problem of discounting became more acute from 1958 onward. The imposition of the import quotas in the United States closed off, to a considerable degree, the largest oil market in the world to the rapidly rising production outside the United States. As a result, those additional barrels had to fight their way into a less-than-global market.

But there was an even more important reason for the increasing prevalence of discounting: the arrival in the world market of a major new entrant, or rather a reentrant—the Soviet Union. Hardly a dozen years had passed since Stalin had ruminated bitterly on the weaknesses and inadequacies of the Soviet oil industry. But enormous investment and effort had paid off in a rebound that carried the Russian industry far beyond its previous levels of output. The new Volga-Urals area proved to be a bonanza. Between 1955 and 1960, Soviet oil output actually doubled, and by the end of the 1950s, the Soviet Union had displaced Venezuela as the second-largest oil producer in the world, after the United States. Indeed, Soviet output was equal to about three-fifths of the total production in the Middle East.

At first, most of the Soviet production had been consumed within the Soviet bloc. But by 1955, Russia had resumed oil exports to the West on a commercial scale. From 1958 onward, the exports surged and became a major factor in the world market—"a force to be reckoned with in the international petroleum field," said the Central Intelligence Agency. The Soviet Union was ready to reassume Russia's nineteenth-century role as a significant supplier to the West. It wanted whatever buyers it could obtain, and it cut prices to get them as part of what became known in Washington as the "Soviet Economic Offensive." At a Cabinet meeting in 1958, Allen Dulles, director of the CIA, warned, "The free world faces a quite dangerous situation in the Soviet capacity to dislocate established markets."[17]

For the oil companies, the one way to meet the challenge and keep the Russians at bay, barring major restrictions by Western governments on the importation of Soviet oil, was the competitive response—price cuts. But the companies faced a dilemma. If only the market price were reduced, then the companies alone would be absorbing the entire cut. Could they dare to cut the posted price as well, so that the producing countries would share the burdens of competing with the Russians?

They did so in early 1959. British Petroleum made the first cut—eighteen cents a barrel, about a 10 percent reduction. Its action instantly set off a torrent of denunciation from the oil exporters. Juan Pablo Pérez Alfonzo was outraged. Abdullah Tariki was furious. With a stroke of the pen, a major oil company had unilaterally slashed the national revenues of the oil producers. The exporters were galvanized into action.

The Arab Oil Congress

For some time, an Arab Oil Congress had been scheduled to open in Cairo in April of 1959. That the meeting was being held there was symbolic of Nasser's ascendancy over the Arab world. Four hundred people attended the conference, including, of course, Tariki. Juan Pablo Pérez Alfonzo—angry about both BP's price cut and the restrictions on Venezuelan oil under the new American quotas, and still smarting over the recent summary rejection in Washington of his Western Hemisphere oil scheme—came as an "observer," accompanied by a Venezuelan delegation that carried texts of the country's tax laws and other oil legislation translated into Arabic. The one notable absence was that of Iraq. Despite the sway of Nasserite ideology in the Arab world, the new rulers in Baghdad were not disposed to subordinate themselves to Nasser, and very shortly after the bloody coup, Iraq was almost completely at odds with Egypt. As a result, Iraq officially boycotted the Arab Oil Congress because it was being held in Cairo, and because it threatened to give Nasser a decisive say over oil matters.

The attendees sat through a large number of papers that had been planned well in advance, most of them technical. But BP's price cut on the eve of the conference had transformed the mood, driving the key participants, seething with anger, to seek some common front against such a practice. Worried that there might be talk of nationalization, the major oil companies sent their own observers to the Cairo meeting. But what the representatives saw and heard eased their minds. "Conference can be considered successful insofar as political issues did not get the upper hand," Michael Hubbard, a British Petroleum representative, assured that company's chairman. Informal discussions between Arab and Western delegates were conducted, he added, "in an atmosphere which was remarkably friendly. Ignorance of what to western minds are the elementary facts about the oil industry was the main feature of the Congress." Another BP representative said that the conclave "can be marked off as a 'plus' for the oil industry's future relations with the Arab host countries."

BP did try some of its own private diplomacy at the meeting. Hubbard reported to the chairman that, through "Miss Wanda Jablonski of *Petroleum Week*"—who "was active behind the scenes"—he was able to arrange a meeting with Abdullah Tariki. Jablonski assured him "from personal knowledge" that "it was possible to discuss economic facts" with the Saudi Arabian. "Unfortunately," said Hubbard, "this proved not to be the case and we were subjected to a diatribe on the inequity of oil production in Kuwait with its population of a few hundred thousand growing more rapidly than production in Saudi Arabia with its many millions of under-privileged." Hubbard added, "It proved quite impossible to establish any point of contact." (Aramco officials later complained that when the Western oil men began talking with Tariki along the "when you have been in the oil business as long as I have, my boy" line, "they had already done more harm than they possibly could do good.")[18]

"Regards to All, Wanda"

But Wanda Jablonski was even busier in Cairo than Hubbard knew. As correspondent of *Petroleum Week* and, later, editor of *Petroleum Intelligence Weekly*, she was the most influential oil journalist of her time. Blonde and stylish, she carried the European *savoir faire* required to get her through all sorts of situations. While she had the resoluteness and independence of Ida Tarbell, she was not a critic of the industry, but rather provided a channel for communication and intelligence in its great years of global expansion. Wisecracking and tough, a solo woman navigating her way through a super-masculine world of engineers and nationalists, she intuitively grasped just how far to go in jousting with and needling her contacts, though always in an engaging way, until she got the story she wanted. She knew virtually everybody of significance in the oil industry. Periodically, over the years, she would infuriate one or another company or country with her scoops; sometimes, companies would cut off their subscriptions en masse, until she shamed them into resubscribing. In the final analysis, no one in a position of power or responsibility in the oil industry could easily do without her journal.

Born in Czechoslovakia, Jablonski was the daughter of a prominent botanist turned geologist who joined a Polish company that eventually became part of Socony-Vacuum, later Mobil. His job was to travel around the world, investigating the geological likelihood that competitive local oil might be discovered in countries where Socony planned to market. As it turned out, Jablonski learned more about plants than about oil from her father; she would be given a penny for every plant she could identify and once earned over a hundred dollars doing so on an auto trip across America. She trailed after her father as he worked around the globe, though often with long separations, and by the time she entered Cornell University, she had already been to school in New Zealand, Egypt, England, Morocco, Germany, Austria, and Texas, and had spent almost a month traveling by camel from Cairo to Jerusalem (and afterward had to be deloused). "I have a different attitude towards the world," she once said. "I can't fit in any one spot, except New York."

In 1956, just after the Suez Crisis, Jablonski made a memorable reporting trip through twelve countries of the Middle East, even wangling an invitation to interview King Saud in Riyadh. "Guess where I spent yesterday evening?" she wrote back to her colleagues in New York. "In the harem of the King of Saudi Arabia! Before you jump to any conclusions, let me hasten to add that I was there . . . drinking tea (with rose water), eating dinner, and having a perfectly gay 'hen party.' . . . Forget what you've seen in the movies, or read in the 'Arabian Nights.' None of that fancy, filmy stuff. Just plain, ordinary, warm home and family atmosphere—just like our own, though admittedly on a considerably larger family scale! Regards to all, Wanda." She did not mention the eunuchs guarding the King's harem, who looked right through her.

Jablonski met not only King Saud, but also Abdullah Tariki, whom she described as "the No. 1 man to watch in the Middle East—as far as oil concession policies are concerned . . . he is a young man with a mission." She quoted at length Tariki's virulent denunciation of the American oil companies operating in Saudi Arabia. During a second meeting a couple of years later, during which Tariki was no less truculent in his criticism, she also passed on an important piece of information. "There's another guy who's just as nuts as you," she told Tariki. She meant Juan Pablo Pérez Alfonzo, and she promised to bring them together.

In 1959 at the Arab Oil Congress in Cairo, she kept her word and invited Pérez Alfonzo up to her room at the Cairo Hilton for a Coke. There she introduced Abdullah Tariki. "You're the one I've been hearing so many things about," said Pérez Alfonzo. Now the real business for which Pérez Alfonzo had come to the conference could begin. The two men agreed that they should talk secretly with representatives from the other major exporters. But where? There was a yacht club in Maadi, a suburb of Cairo; it was off-season and the club was virtually deserted. They could reconvene there, unobserved.

The ensuing discussions in Maadi were conducted in such great secrecy and with such extreme precautions that, afterward, the Iranian participant would say, "We met in a James Bond atmosphere." Those involved, in addition to Pérez Alfonzo and Tariki, included a Kuwaiti; the Iranian, who kept insisting that he was present only as an observer and that he had no mandate to represent his government; and an Iraqi, who, since his country was boycotting the conference, was there in his role as an official of the Arab League. Given all these considerations, they could not make an official accord. But Pérez Alfonzo knew how to sidestep that obstacle; they would make a "Gentlemen's Agreement," which would merely contain recommendations to their governments. All signed the agreement without hesitation, with the exception of the Iranian. He was so frightened about acting without authorization from the Shah that he disappeared, and the others had to call upon the Cairo police to find him so that he, too, could affix his signature.

The recommendations in the Gentlemen's Agreement reflected ideas that Pérez Alfonzo had had in mind before leaving Caracas: that their governments establish an Oil Consultative Commission, that they defend the price structure, and that they establish national oil companies. The governments were also urged to jettison officially the much-treasured fifty-fifty principle—much-treasured, that is, in the West—and move to at least a sixty-forty split in their favor. In addition, they should build up their domestic refining capacity, move downstream, and become more integrated in order to "assure stable markets" for themselves, and thus better protect government revenues. In all its dimensions, the Gentlemen's Agreement, though secret, was a milestone in the changing dynamics of the petroleum industry. It marked the first real steps toward creating a common front against the oil companies. As for Wanda Jablonski, she was as usual near the center of the action; she had just been the matchmaker for an alliance that would develop into the Organization of Petroleum Exporting Countries—OPEC.[19]

C H A P T E R 2 6

OPEC and the Surge Pot

STILL, THE SURPLUS OF oil continued to mount. Further discounts off the posted price followed, in large measure as a result of the aggressive marketing by the Soviet Union, which stepped up its drive to sell oil in the West, slashing prices and making barter deals. In these Cold War years, many in the West believed that the intensifying Soviet petroleum campaign represented not only a commercial venture, but also a political assault, the purpose of which was to create dependence in Western Europe, weaken the unity of NATO, and subvert the Western oil position in the Middle East. "Economic warfare is especially well adapted to their aims of worldwide conquest," Senator Kenneth Keating would say of the Russians. And of the blustering leader of the Soviet Union, he said, "Khrushchev has threatened to bury us on more than one occasion. It is now becoming increasingly evident that he would also like to drown us in a sea of oil if we let him get away with it."

Certainly, the Soviet Union was proving to be a very tough competitor. The Soviets needed dollars and other Western hard currencies to buy industrial equipment and agricultural products. Oil exports then, as now, were one of the few things that they had to sell to the West. On sheer economic terms, Soviet prices could not easily be resisted. At one point, Russian oil could be picked up in Black Sea ports at about half the posted price of Middle Eastern oil. The companies feared significant losses of sales to the Russians in Western Europe, which was also the primary market for Middle Eastern oil. The agitation among the Western companies was further increased when they observed that the most prominent buyer of Russian oil was none other than their number-one bête noire, the Italian Enrico Mattei.[1]

Once again, as in 1959, the only way open to the companies to cope with the general oversupply and, in particular, to counter the Soviet threat (barring gov-

ernmental restrictions on the importation of Soviet oil), was the competitive response, price cuts. But which price? If only the market price were reduced, then
the oil companies alone would be absorbing the entire loss. Yet could they risk
making another cut in the posted price? The first one, in February of 1959, had
inflamed the Arab Oil Congress and led to the Gentlemen's Agreement. What
would happen if they did it again?

T Square Versus Slide Rule

In July of 1960, fifteen months after the Arab Oil Congress in Cairo, the board of
Standard Oil of New Jersey met in New York to consider the vexing question of
the posted price. The meeting was contentious. The company had a new chairman, the no-nonsense Monroe Rathbone, known as "Jack." Rathbone's life was
practically a textbook of the American oil industry. Both his father and his uncle
had been Jersey refiners in West Virginia. Rathbone himself had studied chemical engineering, and then had gone to work right after World War I in Jersey's
huge refinery at Baton Rouge. He was the first member of the new wave that, as
a Jersey man once said, took refining from being "a combination of guesswork
and art" and turned it into a science.

By the age of thirty-one, Rathbone was the general manager of the Baton
Rouge refinery. There he developed considerable political skills fending off the
predatory attacks of Huey Long, the demagogic political boss of Louisiana, who
"customarily ran for office against Standard Oil." (As part of his personal war
against Standard, Long had once offered the by-then-elderly Ida Tarbell one
hundred dollars for an out-of-print copy of her Standard Oil history.) Rathbone
soared through the Jersey organization to the top position. As the boss, he was
self-confident, decisive, unemotional, and uninterested in small talk. A colleague described him as "an engineer with a T Square." The great drawback was
that his entire career had been spent in positions in the United States, so that he
did not intuitively grasp the changing mentality of foreign oil producers. Fending off the populist Huey Long was not as good a preparation as Rathbone might
have thought for dealing with the nationalistic leaders of the oil-exporting countries. He simply did not realize how another reduction in the posted price would
be received. It did not seem to him necessary even to consult with the producers,
for whom he had a certain impatience. "Money is heady wine for some of these
poor countries, and some of these poor people," he once said.

Jersey was then managed by what seemed such an endless number of committees that it was known in-house as the "Standard Committee Company of
New Jersey." This system was intended to forestall precipitous decisions, to
make sure a problem was carefully analyzed and looked at from all sides. But
Rathbone had, as an associate once said, "the kind of determination that takes an
awful lot of evidence to break down." And, at this moment, Rathbone, preoccupied by the strategic problem of gaining markets midst the glut, was determined
to overwhelm the committee system and force a cut in the posted price.[2]

Howard Page, Jersey's expert Middle East negotiator, and the man who had
put together the Iranian consortium, vigorously disagreed with Rathbone. He
and others on the Jersey board thought that Rathbone did not fully comprehend

the problem or the likely reactions. He had been at odds with Rathbone for some time on this issue. Page had broad international experience; he had helped organize oil supplies between the United States and Britain during the war under Harold Ickes; afterward, he became Jersey's Middle East coordinator. "He was a very tough man," said one who negotiated against him. "He always had a slide rule on his lap, so that he could calculate down to the last half cent on a barrel. But he was also a man of some vision, and was very well able to understand other people's vision." Page grasped the explosive force of nationalism in the Middle East, and he feared that his colleagues at Jersey, and Rathbone in particular, did not.

In an effort to educate his fellow directors, Page arranged for the intrepid journalist Wanda Jablonski, just back from the Middle East, to meet with the board of Jersey. She told them, according to a report by a British diplomat who talked to Jablonski afterward, that there was "almost universal adulation for Nasser among all classes, and a hostility towards the west that had deepened markedly. In the field of oil this took the form of a growing outcry against absentee landlordism. She had listened to many a bitter diatribe against those international oil companies who, from foreign capitals, were draining off the wealth of the Arab countries! It was intolerable that, from their remote vastness in London, New York, Pittsburgh etc., top executives of oil companies should control the economic destinies of Middle East oil producing states." Jablonski even told the Jersey board that the existing structure of the Iraq Petroleum Company and Aramco might prove to be "short-lived," which was about the last thing in the world they wanted to hear.

At a separate meeting with Jablonski, Rathbone vehemently disagreed with her disquisition on the force of nationalism. He dismissed her concerns. He had just come back from the Middle East, and he told her that she was being excessively pessimistic.

"You never got beneath the surface," Jablonski replied tartly. "Jack, do yourself a favor. You got the red carpet treatment, you were there only a few days. You'd be wiser if you didn't make those statements."

Now, as the Jersey board debated cutting the posted price, Page argued against it. Jersey would be cutting the countries' national revenues. Consult with the governments, he said, compromise with them, but don't do anything unilateral. Page made a motion that a cut be made, but that it be effected only after some discussion and agreement with the governments. The other directors backed the motion. Jack Rathbone did not, and he was the chairman. He privately dismissed Page as "a know-it-all." He decided that Jersey would go ahead and cut the price, and that the company would do it the way he wanted it done, which was without first consulting any governments or anyone else. That was that.

On August 9, 1960, with no direct warning to the exporters, Jersey announced cuts of up to fourteen cents a barrel in the posted prices of Middle Eastern crudes—about a 7 percent reduction. The other companies followed suit, though without any enthusiasm and, in some cases, with a good deal of alarm. To John Loudon of Shell, it was "the fatal move. You can't just be guided by market forces in an industry so essential to various governments. You had to take these other things into consideration. You had to be so terribly careful." BP, which had learned its lesson when it cut the posted price in 1959, complained that it "heard the news with regret."

The reactions on the part of oil-producing countries went far beyond "regret." Standard Oil of New Jersey had suddenly made a significant slice in their national revenues. Moreover, that decision, so critical to their fiscal position and their national identity, had been carried out unilaterally, without any consultation. They were outraged. "All hell broke loose," recalled Howard Page. Another Jersey executive, who had also opposed the cut, was in Baghdad when it was announced. He was, he later said, "glad to get out alive."[3]

"We've Done It!"

The exporters were furious and they wasted no time. Within hours of Standard Oil's announcement of the cut in the posted price in August 1960, Abdullah Tariki telegraphed Juan Pablo Pérez Alfonzo, and then hurriedly departed to Beirut for a twenty-four-hour visit. What would happen, he was asked by journalists? "Just wait," he replied. Tariki and Pérez Alfonzo wanted to bring the other signatories of the Cairo Gentlemen's Agreement together again as quickly as possible. In the maelstrom of anger and outrage, the Iraqis recognized a political opportunity. The revolutionary government of Abdul Karim Kassem had no wish to subordinate itself to a Nasserite order in the Middle East, and it was deeply opposed to the influence that Nasser could wield on oil policy through his domination of the Arab League and the various Arab oil conferences. The Iraqis now saw that, by using the price cut as the catalyst to establish a new organization composed exclusively of oil exporters (including two non-Arab countries, Iran and Venezuela), they could isolate oil policy from Nasser. The Iraqis also hoped that such a grouping would bolster them in their confrontation with the Iraq Petroleum Company—and provide the additional revenues they desperately needed. And so, jumping at the chance to bring the other exporters together under Iraqi auspices, they quickly dispatched invitations to meet in Baghdad.

When the telegram from the Iraqi government was brought into Pérez Alfonzo's office in Caracas, he was exultant. Here were the makings of the international "Texas association" that he so ardently advocated. "We've done it!" he declared excitedly to his aides as he held the telegram aloft. "We've achieved it!"

The oil companies quickly realized that the unilateral price cut was a dreadful error. On September 8, 1960, Shell offered an olive branch; it raised its posted prices by two to four cents. The gesture came too late. By September 10, representatives of the major exporting countries—Saudi Arabia, Venezuela, Kuwait, Iraq, Iran—had arrived in Baghdad. Qatar attended as an observer. The omens for the meeting were not particularly good. Pérez Alfonzo had to delay his departure from Caracas owing to an attempted coup against the new democratic government. Baghdad itself was filled with tanks and armed soldiers; the new revolutionary regime was on alert against an anticipated coup. Armed guards were posted behind each delegate during the discussions.

Yet, by September 14, the group had completed its work. A new entity had been established with which to confront the international oil companies. It was called the Organization of Petroleum Exporting Countries, and it made its intention clear: to defend the price of oil—more precisely, to restore it to its precut level. From here on, the member countries would insist that the companies consult them on the pricing matters that so centrally affected their national revenues. They also called for a system of "regulation of production," Tariki's and Pérez Alfonzo's dream of a worldwide Texas Railroad Commission. And they committed themselves to solidarity in case the companies sought to impose "sanctions" on one of them.

The creation of OPEC gave the companies good reason for second thoughts, creative backstepping, and outright apologies. "If you disapprove of what we have done, we regret it," a representative of Standard Oil abjectly told an Arab oil conference a few weeks later. "Whenever in any manner, large or small, you disagree with something that we have done, we are sorry that this is so. Whether what we have done is in fact right or wrong, the fact that you feel it is wrong or that you do not understand why we have done it, is a failure on our part."

To apologize was prudent, for OPEC's five founding members were the source of over 80 percent of the world's crude oil exports. Moreover, OPEC's creation represented "the first collective act of sovereignty on the part of the oil exporters," in the words of Fadhil al-Chalabi, later an OPEC Deputy Secretary General, as well as "the first turning point," as he put it, "in the international economic relations towards the states' control over natural resources."

Yet, despite all the motion and rhetoric, the newly created OPEC did not seem very threatening or imposing. And, whatever their initial apologies, the companies certainly did not take the organization all that seriously. "We attached little importance to it," said Howard Page of Standard Oil, "because we believed it would not work." Fuad Rouhani, the Iranian delegate to the founding conference in Baghdad and OPEC's first Secretary General, observed that the companies initially pretended that "OPEC did not exist." Western governments did not pay much attention either. In a secret forty-three-page report on "Middle East Oil," in November 1960, two months after OPEC's founding, the CIA devoted a mere four lines to the new organization.[4]

OPEC in the 1960s

Indeed, OPEC could claim only two achievements in its early years. It ensured that the oil companies would be cautious about taking any major step unilaterally, without consultation. And they would not dare cut the posted price again. Beyond that, there were many reasons why OPEC had so little to show for its first decade. In all the member countries, with the exception of Iran, the oil reserves in the ground actually belonged by contract to the concessionaires, the companies, thus limiting the countries' control. Furthermore, the world oil market was overwhelmed with surplus, and the exporting countries were competitors; they had to worry about holding on to markets in order to maintain revenues. Thus they could not afford to alienate the companies on which they depended for access to those markets.

The 1960s saw a continuing process of decolonization and the rise of "Third World" questions and controversies. Yet the issues of sovereignty in the oil world, which was so stark and central in OPEC's formation in 1960, subsided over the next few years, as the companies sought to meet the exporters' demands for higher revenues by pushing up production. There were broader political factors as well. In Saudi Arabia, King Faisal was now firmly in charge, and he, in contrast to his brother Saud, was oriented toward the West. Indeed, a political competition soon developed between Saudi Arabia and Egypt, which culminated in their proxy war in Yemen. Outside the Middle East, Venezuela was interested in pursuing a stable relationship with the United States and became a key country in the Alliance for Progress of the Kennedy and Johnson Administrations. Overall, the circumstances of international politics, including the dominance of the United States and its importance to the security of several of the producing nations, prevented them from challenging too directly the United States and other Western industrial countries.

And if the OPEC member countries had a common economic goal—to increase their revenues—the political rivalries among them were considerable. In 1961, when Kuwait became completely independent of Britain, Iraq not only claimed ownership of the small country, but also threatened invasion. Iraq backed off only after Britain dispatched a small military squadron to help defend Kuwait. But Iraq did temporarily suspend its membership in OPEC in protest. The two major producers, Iran and Saudi Arabia, looked upon each other with apprehension and envy, even as the ascent of Nasser and nationalism in Egypt and throughout the Middle East posed a threat to their dynasties as well as their political leadership in the region. The Shah wanted to increase his revenues as speedily as possible, and he believed that could only be achieved by selling more oil, not by holding back production and raising the price. And he wanted to be sure that Iran regained and held on to a position of preeminence that befitted his own ambitions. "Iran must be restored to number one producer," he said. "International oil prorationing is nice in theory but unrealistic in practice."[5]

Abdullah Tariki, the Saudi proponent of prorationing, had aligned himself with King Saud. It was an unwise choice, as Faisal won the power struggle. In 1962, Tariki was fired and replaced as oil minister by the young legal adviser to the Cabinet, Ahmed Zaki Yamani, who had no particular attachment to the notion of creating an international Texas Railroad Commission. Thus was Tariki severed from OPEC. He spent the next decade and a half in itinerant exile, as a consultant, advising other petroleum countries, and as a journalist and polemicist, denouncing the oil companies and urging the Arabs to seize full control of their resources.

OPEC's other father, Pérez Alfonzo, grew disillusioned not only with politics, but also with OPEC. The physical strain of being minister and of all his travels also took its toll, and he finally resigned in 1963. He said that his mission had been to get the oil producers together; he had done that, and there was nothing more for him to do. A few weeks after his resignation, he let loose a blast at OPEC for its ineffectiveness and for failing to produce any benefits for Venezuela. He then retreated to his villa to read and write and study philosophy, managing to preserve the house and its gardens as an enclave for contemplation and criticism in a city bursting with growth, din, and automobiles. But Pérez Alfonzo no longer talked about "sowing the oil"; instead, he took to calling petroleum "the excrement of the devil." He kept the rusted Singer automobile in the garden as a monument to what he saw as the waste of oil wealth. His concerns, in his last years, continued to focus on the need to husband, not dissipate, resources, and on the pollution created by industrial society. "I am an ecologist first of all," he said shortly before his death in 1979. "I have always been an ecologist first of all. Now I am not interested in oil any more. I live for my flowers. OPEC, as an ecological group, has really disappeared."

The oil companies strenuously sought to avoid direct negotiations with OPEC during much of the 1960s. "Our position was that we owned the concessions, and we would deal with the countries in which the concessions were located," recalled an executive of one of the majors. OPEC itself continued through the 1960s to be, as another executive called it, a sideshow: "The reality of the oil world was U.S. import quotas, Russian oil exports, and competition. This was what filled the columns of the trade press, the minds of oil executives and the memos of government policy makers. These were the important underlying preoccupations of the oil industry." What loomed over everything else was the dizzying growth in demand and the even more dizzying growth in available supply. It seemed that OPEC's moment to mount an effective challenge to the power of the major oil companies had passed—or would never come.[6]

"The New Frontier"—and More Elephants

Almost as soon as OPEC was established, its member countries lost what had been their almost total grip on world oil exports. Wholly new oil provinces were found and opened up in the 1960s, adding to the supplies that were swamping the market. And while most of the producing countries would eventually become members of OPEC, they entered the world market first as competitors, capturing the market share of the more established exporters.

Africa was considered the "new frontier" for world oil in those years. France took a lead in exploring it, drawing on the policies that had been enunciated after World War I, when Clemenceau had said that oil was the "blood of the earth"—and had decided that he could no longer depend upon his "grocer," supplied by foreign companies, for so critical a commodity. If France were to remain a great power, it had to have its own petroleum resources. Within a few months of the end of World War II, Charles de Gaulle ordered a maximum drive to develop oil supplies within the French empire. The objective was to have French oil production around the world at least equivalent in volume to France's own consumption, thus helping the balance of payments and promoting security. Yet France's national champion, CFP, was preoccupied with sorting out the Iraq Petroleum Company and its position in the Middle East, and so the government charged a new group of state companies, under the Bureau de Recherches Pétroliers (BRP), to explore for oil elsewhere in the empire. After several years, oil was discovered in Gabon in West Africa.

In North Africa, France's High Commissioner for Morocco had been promoting the potential of the Sahara, though in the face of much skepticism. The most prominent professor of geology at the Sorbonne announced that he was so sure that there was no oil in the Sahara that he would happily drink any drops of oil that happened to be found there. Nevertheless, the territory was large, there was very little competition for permits, and another state company, Régie Autonome des Pétroles (RAP), began exploration. And, in 1956, RAP discovered oil in Algeria.

The Algerian find, in the Sahara, fired excitement in France. Here, for the first time, France would have control of oil resources that were outside the Middle East and beyond the reach of the "Anglo-Saxons" (though Shell was a partner in the Algerian venture). The Suez Crisis, later in that same year, only reinforced the significance of the "Sahara" to France, again demonstrating the danger of dependence upon the unreliable "Anglo-Saxons"—in this case the Americans—for oil and indeed for political support. The French felt that they had been betrayed by their American ally. Moreover, the crisis had been a severe blow both to French pride and to economic stability. The government's Economic Council called for a stepped-up international exploration campaign, especially in Africa. "The diversification of sources of supply," declared the council, "is for our country an essential condition of security."

All this made the new Algerian oil discoveries and their rapid development even more crucial. The "Sahara" became a magical word in France. The "Sahara" would free France from dependence upon foreigners and from the acute pain of foreign exchange crises; the "Sahara" would make possible the revitalization of French industry; the "Sahara" would be France's answer to Germany's Ruhr, where the German postwar economic miracle was taking place. De Gaulle himself made a private visit to the Saharan oil fields in 1957, a year before his return to power. "Here is the great opportunity for our country that you have brought into the world," he told the oil men at the desert camp. "In our destiny, this can change everything."

Getting the oil out was very difficult. The fields were deep in the desert; even the simplest things, like water, had to be trucked hundreds of miles across roadless wastelands. Yet by 1958, within two years of discovery, the first oil began to flow out of the desert, for export to France. There was, however, one great inconvenience with Saharan oil. Algeria was caught up in a bloody war for independence, which had begun in 1954, and the Algerian rebels regarded the Sahara as integrally part of Algeria, French protestations to the contrary. The future of Saharan oil production could not exactly be regarded as secure. Indeed, in some French circles, it was believed that the "Anglo-Saxons," as well as Signor Mattei of Italy, were colluding with the rebels to gain preferential access to the Saharan petroleum in a postindependence Algeria.

Yet the thrust of French policy worked. Whatever the risks, by 1961 companies primarily belonging to and controlled by the French state were producing oil around the world equivalent in volume to 94 percent of French demand. The next year, Algeria formally won its independence. But the Evian Agreement that de Gaulle negotiated with the Algerians guaranteed retention of France's position in Saharan oil.

Nevertheless, there was no telling how long the deal with Algeria would hold. To strengthen the overall French oil position and to compete more effectively against the established majors, RAP was merged in 1965 with the BRP group of state companies, which among other things had discovered and developed a significant gas field in France. "We have chosen to adapt in a realistic manner to the international situation," explained André Giraud, the Director of Fuels. The combined company was named Enterprise de Recherches et d'Activ-ités Pétrolières, or more commonly Elf-ERAP. It eventually became better known simply as Elf, which was one of its gasoline brand names. Building upon its Algerian base, Elf launched a global exploration campaign and became not only a new major, one of the largest oil companies, but also one of the biggest in-dustrial groups in the world.

Production was beginning to build up in other countries as well, promoted by eager independents hoping to strike a bonanza. The majors also moved in ex-peditiously. Despite their vast holdings in the Middle East, they wanted to diver-sify their sources in order not to find themselves hostage to what might happen in the countries around the Persian Gulf. As a Shell managing director put it in 1957, they wanted to be in a position "which is more commercially defensible than having all our eggs in one basket." A joint venture between Shell and BP, which had begun exploration in Nigeria in 1937, finally in 1956 hit the first signs of oil in the swampy delta of the Niger River. But nothing anywhere else in the world would compare to the extraordinary phenomenon that unfolded in the desolate desert kingdom of Libya. It transformed the world oil industry, and would ultimately transform world politics.[7]

The Libyan Jack-Pot

Thousands of tanks had rolled back and forth over the gravelly rock of Libya during World War II, in the titanic desert struggle between the Germans and the British. And it was there that Rommel's forces, chronically short of fuel, had ul-timately been overwhelmed. Neither side knew that, even as their fuel gauges were falling, they were fighting at times only a hundred miles or so from one of the world's great reserves of oil.

In the decade after World War II, Libya was seen as having some moderate significance from a military point of view. It was the site of Wheelus Air Force Base, one of the main American bomber bases in the Eastern Hemisphere. Be-yond that, it did not count for much in international terms. Three distinct "provinces" had been rather arbitrarily joined to create the thinly populated country. At the top of its rickety political system sat old King Idris, who did not really like being king. He actually wrote out a resignation letter once, but desert tribesmen heard about it and prevented him from abdicating. Libya was a very poor country, plagued by droughts and locusts. Its economic prospects could hardly be called promising; in the years after World War II, its leading exports were two: esparto, a type of grass used to make paper for currency bills, and scrap metal scavenged from the rusting tanks and trucks and weaponry that had been left behind by the Axis and Allied Armies.

But by the middle 1950s, there was growing suspicion among geologists that the country might produce oil. To encourage exploration and development, the Libyan Petroleum Law of 1955 provided for a host of much smaller conces-sions, instead of the very large concession areas characteristic of the Persian Gulf countries. "I did not want Libya to begin as Iraq or as Saudi Arabia or as

Kuwait," explained the Libyan petroleum minister who oversaw the law. "I didn't want my country to be in the hands of one oil company." Libya would give many of the concessions to independent companies, which did not have oil production and concessions to protect in other countries in the Eastern Hemisphere, and thus would have no reason to hold back from exploring and producing as much, and as rapidly, as they could in Libya. The law offered another incentive. The government's take would be pegged to the actual market price for its oil, which was lower than the increasingly fictional posted price. This meant that Libyan oil would be more profitable than oil from other countries, which was an excellent reason for any company to maximize its Libyan output. The central objective of the arrangement was summed up by the Libyan petroleum minister: "We wanted to discover oil quickly."[8]

The strategy of diffusion worked: In the first round of negotiations, in 1957, seventeen companies successfully bid for a total of eighty-four concessions. The Libyan juggernaut was starting to roll. Working conditions, however, were hardly convenient. The country was very backward. There were no telephones to the outside world. Those wanting to make an overseas call to the United States had to fly to Rome to do it. Progress by the geologists in the field was impeded by obstacles they had never encountered before: an estimated three million land mines left over from World War II. Geologists and oil field workers were not infrequently injured or killed by undetected mines. The companies formed mine detection and clearance squads, and in due course, some of the Germans who had laid the mines for General Rommel were recruited to remove them.

The early exploration results were disappointing, and discouragement soon settled in. BP was already beginning to dispose of its warehouse supplies, its leases, and its villas in preparation to exit. Then, in April 1959, at a spot called Zelten, about a hundred miles south of the Mediterranean coast, Standard Oil of New Jersey made a big strike. The State Department summed it up for the British Foreign Office: "Libya has hit the jack-pot." Ironically, Jersey had come close to deciding to give Libya a pass. After all, it owned 30 percent of the Aramco concession, which seemed capable of providing endless oil; it was also a member both of the Iraq Petroleum Company and of the Iranian consortium, and the largest producer in Venezuela. Yet, though the risks looked very large, there would also be a very important advantage to having oil in Libya. "One of our purposes in going into Libya was to try to find oil that would compete with the Middle East," said M. A. Wright, who had been Jersey's worldwide production coordinator. "We would have an improved position with the Saudis by having another source of crude." Moreover, Jersey, like other companies, tended to believe that political risk in Libya was much lower than the risk in the Persian Gulf countries or in Venezuela.

With the discovery at Zelten, the rush was on. By 1961, ten good fields had been discovered, and Libya was exporting oil. It was a very high-quality, "sweet" (i.e., low-sulfur) crude. In contrast to the heavier Persian Gulf crudes, which provided a large proportion of fuel oil, Libyan crudes could be refined into a much higher proportion of gasoline and other light, "clean" products, perfect for the growing automobile fleets of Europe and excellently suited to the dawning age of environmentalism. Moreover, Libyan production could hardly have been better located; it was not in the Middle East nor did it require transit either through the Suez Canal or around the Horn of Africa. From Libya, it was a quick, secure jaunt across the Mediterranean to the refineries in Italy and on the southern coast of France. By 1965, Libya was the world's sixth-largest exporter of oil, responsible for 10 percent of all petroleum exports. By the end of the 1960s, it was producing over three million barrels per day, and in 1969 its output actually exceeded that of Saudi Arabia. It was an incredible accomplishment in a country in which, a decade earlier, no petroleum reserves had yet been discovered.[9]

But with such quick and unexpected prosperity, the Libyan business environment became redolent with corruption. Everyone seemed to have his hand out. One oil executive complained that his company was being "nickeled and dimed" to death. Most of the takers, however, were looking for much more than mere nickels and dimes. "If you were using any local contractors, there would be a shakedown," remembered Bud Reid, a geologist with Occidental Petroleum, a small American independent that had obtained significant Libyan concessions. "The pressures came from all kinds of places. If the brother-in-law was an official in the customs department, then suddenly some piece of equipment you needed to get through customs wasn't coming through fast enough. If you wanted to ensure your equipment got in, then you did business with a certain trucking or a certain contracting firm." The family running the royal household was particularly well known for its interest in extremely large gratuities. The death of the senior member of that family in an auto accident set off a crisis of sorts in the country; for his demise, explained an American oil man, "created real uncertainty about who to bribe."

The vast surge of Libyan oil dramatically affected world oil prices, giving further force to the descent that had begun after Suez. The flood of Libyan oil picked up where Soviet oil left off. In Libya, more than half the production was in the hands of independent oil companies, many of which, unlike the majors, did not have their own outlets. Nor did they have any reason for restraint, since they had no other supply sources to protect. Moreover, they were shut out of the American market by quotas that protected and encouraged high-cost domestic oil. So politics, as well as economics and geography, forced the independents operating in Libya to crowd into a single market in Europe and to seek aggressively to sell their oil at whatever cost. And not only in Europe, but throughout the world, there was more petroleum looking for markets than there was demand. The result was cutthroat competition. Between 1960 and 1969, the market price for oil fell by 36 cents a barrel, a drop of 22 percent. Correcting for inflation, the fall was even steeper—a 40 percent decline. "Oil was available for anybody, anytime, any place and always at a price as low as you were charging for it," recalled Howard Page of Jersey. "I mean, I have never seen such a competitive market. The market was just falling on its face."[10]

Mattei's Last Flight

And what of the man who had set in motion this challenge to the power of the majors and the very structure of the industry, Enrico Mattei? As he turned ENI and its oil subsidiary AGIP into a world force, Mattei went from battle to battle, finally taking on not only the established oil companies, but also the United States government and the North Atlantic Treaty Organization, both of which were alarmed by his bid to become a massive buyer of cheap Soviet oil. He intended to link his Mediterranean-based pipeline system to the westward-marching Soviet system and, in the process, to barter Italian pipe for Russian oil. But he was also working toward a compromise in his bitter fights with Standard Oil of New Jersey and the other majors and was preparing for a trip to the United States to meet its new President, John Kennedy. The American government supported oil company efforts to work out a détente with Mattei and, in the words of the U.S. ambassador to Italy in April 1962, "assuage his damaged ego sufficiently to minimize future polemics."

On October 27, 1962, Mattei took off from Sicily in his private jet. The only other passenger aboard was *Time* magazine's Rome bureau chief, who was researching a cover story on the Italian tycoon in anticipation of his upcoming trip to America. Their destination was Milan. They never made it. The plane crashed in a terrible thunderstorm, about seven miles short of the runway at Linate Airport in Milan.

Because it was Italy, because it was Mattei, and because he was so controversial, there was much speculation about the cause of the crash. Some said that the Western intelligence services had sabotaged his jet because of his oil deals with the Soviet Union. Some said that the French Secret Army Organization, the diehards who were fighting against Algerian independence, had sabotaged the plane because of Mattei's criticism of colonialism and the French role in Algeria and in revenge for his flirting with the Algerian rebels, which was aimed at positioning AGIP for Algerian independence. But it is more likely that his death was an accident, with the weather and character as fate. Mattei was always in a rush, and his impatient, driving personality would not permit a storm to deter him from landing when he had important things to do on the ground. He had often pushed his reluctant pilot to battle through the Milanese weather, and always with impunity. This time he had simply pushed too hard.

At the time of his death, Mattei was fifty-six, and at the height of his empire building. He had seemed invincible and invulnerable. The foreign affairs columnist of the *New York Times* called him "the most important individual in Italy," more important than either the premier in Rome or the Pope in the Vatican. It was said that he was more responsible than any other man for the sustained postwar boom known as "the Italian Miracle." Afterward, the location of ENI's headquarters in Rome was named "Piazza Enrico Mattei," and ENI and AGIP continued their growth and quest for expansion. But without Mattei, ENI's buccaneer days as the world's number-one maverick oil company were over.[11]

The New Competitors

Though Mattei was gone, he had instigated a revolution that would eventually overthrow the majors' global dominance. To be sure, and contrary to the customary image, the industry's structure was constantly changing. The history of international oil in the twentieth century was one in which "newcomers" continually broke in on the established order. But, for the most part, up until the 1950s, there always seemed to be a way that they could be accommodated; they, too, could, more or less, become part of the establishment. That possibility ended in 1957 when Mattei did his deal with Iran, and the Japanese followed suit offshore of the Neutral Zone. The frenetic activity in Libya in the 1960s carried on the revolution Mattei had begun and dramatized how much had changed. Now, there were many participants in the international oil game—with strikingly divergent interests and far too many for the clubby collaboration of the era of the Seven Sisters.

The reasons for the explosion in the number of players were several. The advance and diffusion of technology reduced the geological risk and made exploration and production expertise readily available. Governments in producing and would-be producing countries adopted concessionary policies that favored the entry of independents and new players. Improvements in travel, communications, and information made Latin America, the Middle East, and Africa all less remote and more accessible. The high rate of return on international oil investment, at least up until the middle 1950s, provided a great appeal. The United States tax code made foreign investment less risky and more attractive. Prorationing in the United States also encouraged companies to go abroad to seek capacity that they could produce at full rate. And petroleum demand among the industrialized nations was climbing to new highs, while the governments of both the consuming and the producing countries increasingly looked to oil as the motor for economic growth and as a tangible symbol of security, pride, and power.

There was one other factor at work: the preeminence of the United States in the Western alliance system and the world economy. Despite crises generated by nationalism and communism, American influence was pervasive, supplanting that of the old colonial empires. America's military might was widely respected, and its economic success was an object of admiration and envy. The dollar ruled supreme, and the United States was at the center of an economic order that encouraged, among other things, the outflow of American capital, technology, and managerial expertise in oil, as in other industries. And the United States was in a position to shape a political order in which risks and threats were manageable. Private enterprise responded.

The proliferation of players in the oil game was remarkable, especially in the Middle East. In 1946, 9 oil companies operated in the region; by 1956, 19; and by 1970, the number reached 81. Yet even this was only part of a larger expansion. Between 1953 and 1972, by one estimate, more than 350 companies either entered the foreign (that is, non-U.S.) oil industry or significantly expanded their participation. Among these "new internationals" were 15 large American oil companies; 20 medium-sized American oil companies; 10 large American natural gas, chemical, and steel companies; and 25 non-American firms. How different this was from the situation at the beginning of the postwar period, when only six American firms, in addition to the five acknowledged American majors, had any active exploration interests anywhere overseas at all. In 1953, no private oil company anywhere in the world, other than the seven largest, had as much as 200 million barrels of proven foreign reserves; by 1972, at least thirteen of the "new internationals" each owned more than 2 billion barrels of foreign reserves. Altogether, the new entrants owned 112 billion barrels of proven reserves—a quarter of the free world total. By 1972, the "new internationals" had a total daily output among them of 5.2 million barrels per day.

One of the most obvious results of such a crowded arena was a decline in profitability. The industry had earned high rates of return on its foreign investment until the mid-1950s—the rewards, some would say, for the risks taken in distant, inaccessible regions in the turbulent postwar days, or, others would say, the result of an oligopoly, an industry dominated by a handful of major players. The series of crises—Mossadegh and Iran, the Korean War, and Suez—continued to buoy the profit rate above 20 percent. But with the reopening of the Suez Canal in 1957, the intense competition to sell supplies started to force both prices and profits down. Thereafter, and continuing through the sixties, investment in foreign oil yielded 11 to 13 percent returns, which were pretty much the same as for manufacturing industries. While the exporting countries were counting more money than they had ever seen before, the oil industry itself was no longer being as well rewarded as in the past.[12]

Walking the Tightrope—Iran Versus Saudi Arabia

The global battle of production intensified the long-standing rivalry between the two key oil countries of the Middle East—Iran and Saudi Arabia. The swell of output around the world put the major companies in a political quandary. They had to try to balance supply against demand, even while production was coming on from the newcomers, and that meant restraining output in the world's greatest reserve, the Persian Gulf region. And while Persian Gulf production would grow quickly, it would not increase as rapidly as its reserves would have allowed, or as the governments in the region wanted. In the United States, production was managed and restricted by the Texas Railroad Commission and similar agencies in other states. In the far-more-bountiful oil provinces around the Persian Gulf, output was reined in to what the major companies estimated was necessary to fill the gap between projected demand and available production from the rest of the world. Thus, the Persian Gulf became the stabilizer, the control mechanism for balancing supply and demand. It was the "swing area" or, as some oil men liked to call it, the "surge pot." But allocating growth, particularly between Iran and Saudi Arabia, was hardly easy. It took considerable ingenuity and application to try, even half successfully, to satisfy an Iran whose Shah was already swelling with grand ambition, and a Saudi Arabia that had no intention of acknowledging Iranian leadership in petroleum production, or indeed in anything else.

There were many points of conflict between the two nations: one was Arab and one was not; one was Sunni Moslem and the other Shia Moslem. Each wanted to be the leader, both in the region and among oil producers, and each had unfulfilled territorial ambitions. Their competition over levels of oil production underscored the fundamental jealousy and suspicion between the two countries. For output translated into wealth; and wealth in turn meant power, influence, and respect.

The rivalry between Iran and Saudi Arabia created enormous problems for the major companies. It was like "walking a tightrope," said J. Kenneth Jamieson, later an Exxon chairman. The stakes were very large. The companies did not want to lose their position in either country. A single point stood out to the four Aramco companies—Jersey, Mobil (as Socony-Vacuum had become), Standard of California, and Texaco. Nothing should be done to jeopardize the Saudi concession. The challenge, said Howard Page, Jersey's director responsible for the Middle East, was to keep the Saudis sufficiently happy to maintain Aramco's position "because this was the most important concession in the entire world and we didn't want to take any chances of losing it." Appearing in Saudi eyes to be tilting toward Iran when it came to output could threaten the concession.

But Iran was potentially the dominant power in the region, and the Shah did have to be placated, if not always satisfied. "Nobody could have lifted enough crude to satisfy all the governments in the Persian Gulf during this period," said George Parkhurst, who was Standard of California's Middle East coordinator. The potential to supply, assuming that the appropriate investment was made, simply would outrun demand at any given point. Somehow the available growth in requirements had to be allocated in such a way that neither government would feel that the other was getting a better deal. Saudi Arabia's gain would be Iran's loss, and vice versa. "This is like a balloon," said Page of Jersey. "Push it in one place, it comes out in another, and so if we acceded to all those demands, we would get it in the neck."

To make matters more complicated, the major companies were dealing as partners in the various countries, and they had divergent, competing interests. Some had more crude than they needed, and some were short of crude. "What you have to do is in effect negotiate with your partners all the time, day and night," Page said. "They're always fighting." Then, to make the fighting worse, there were the American independents who had been shoehorned into the Iranian consortium. They did not have other sources of crude or other major concessions to protect, and they were less concerned about the overall world situation than about getting as much oil out of Iran as possible and marketing it as aggressively as they could. They were continually pushing for higher Iranian production, and the majors suspected that they were egging the Shah on. But if Iranian production went up, that would mean the independents would have more oil, in Page's phrase, to "peddle" against the majors, while the majors would have to hold back Saudi production and explain it all to an irate Ahmed Zaki Yamani and perhaps to King Faisal himself.

The question of how to allocate production between Saudi Arabia and Iran was not, strictly speaking, a matter of economics. The cost differential between the two countries was usually only a penny or two—"peanuts," said Page. Rather, it was a strategic and political decision, and on many occasions, the responsibility fell to Howard Page, on behalf of the four Aramco partners, to explain and justify the companies' actions. Zaki Yamani, the Saudi oil minister, was a formidable foe. He knew that Page liked Iranians personally, and he did not hesitate to express bluntly his suspicion that Page was showing favoritism toward Iran, at the expense of Saudi oil production.[13]

Dealing with the Iranians was no less difficult. The 1954 consortium agreement had pledged that Iran's output would grow at least as fast as the average annual growth rate for the entire region, but the Shah was convinced that he was being hoodwinked by the oil companies. At a luncheon at the White House in 1964, he told Lyndon Johnson of his fear that oil companies would give preferential treatment to Arab oil producers. OPEC, the Shah added, had become an "instrument of Arab imperialism." Fired by his own imperial vision and intent on regaining the number-one Middle Eastern export role for his country, the Shah tried any number of tactics and approaches to bring the companies around, even attempting to get the American State Department and the British Foreign Office to pressure the companies on geopolitical grounds.

The Shah made clear exactly where he stood in a meeting with his old friend, Kim Roosevelt, who had helped orchestrate the countercoup that put the Shah back in power a decade earlier. He "was tired of being treated like a schoolboy" by the United States, the Shah now told Roosevelt. He listed all the ways he was helping Western interests, including "Iran's stand-up fight against the incursions of Nasser." But "indifference" and "maltreatment" were all he got in return. "America does better by its enemies than it does by its friends," he added. The special relationship between Iran and America, he warned, "is coming to an end." To drive home his point, he patched up relations with the Russians, made a gas deal with Moscow, and threatened to reorient Iranian imports away from the West and toward the Soviet Union.

The Shah's tactics worked. Both the American and British governments urged the oil companies to "do their best" to meet Iranian demands. The Iranians also kept constant pressure directly on the companies to increase output. All sorts of remedies were tried to keep the Shah happy. The companies even switched from a Western to an Iranian calendar in order to push more production into that particular year. In negotiations, no one would dare tell the Shah when he had made an error, even in simple arithmetic, and he did make such errors. The additional pressure that he brought to bear in the mid-1960s achieved the desired effect. For most years between 1957 and 1970, Iranian production grew at a faster rate than Saudi output: Altogether Iran's production over those years grew by 387 percent, compared to Saudi Arabia's 258 percent. But because Saudi Arabia had started with a larger base, the two countries' respective outputs, in absolute terms, were within 5 percent of each other in 1970. The highwire balancing act had, despite the running controversies, succeeded.

For this achievement, however, the companies, as well as the Saudis and Iranians, owed a considerable debt to still another party, radical Iraq, though the service provided by that country was quite inadvertent. At the beginning of the 1960s, Iraq revoked 99.5 percent of the concession held by the Iraq Petroleum Company, the company originally created by Calouste Gulbenkian, leaving it only the region where it was actually producing oil. The IPC in turn ceased investing in new exploration and production in that area. The result was that Iraqi output, which could have surged along with Iran's and Saudi Arabia's, creating an impossible problem of allocation, only edged up gradually through the 1960s.

At one point during those years, Oman, at the southeast corner of the Arabian peninsula, emerged as a very interesting oil play. Standard Oil of New Jersey, as might be expected, had the chance to get in. But when the issue came up in the company's executive committee, Howard Page recommended against it. He had spent so much time negotiating with the Saudis and the Iranians that it required little effort on his part to conceive of how furious they would be. He could well imagine, in particular, what Yamani would say to him if Jersey and Aramco sought to restrain Saudi output to make room for production from a new concession in a neighboring country. That would surely contradict Jersey's principle number one, which was not to do anything that "would endanger our Aramco concession."

But the members of Jersey's production department disagreed with Page. After all, they were geologists, and as far as they were concerned, discovering and developing new reserves were what the game was all about. Their ambition was to find new elephants, and they were very excited about Oman. "I am sure there is a 10 billion barrel oil field there," a geologist who had just returned from Oman told the executive committee.

"Well, then," replied Page, "I am absolutely sure we don't want to go into it, and that settles it. I might put some money in if I was sure we weren't going to get some oil, but not if we are going to get oil because we are liable to lose the Aramco concession." With that logic, Jersey stayed out of Oman. The geologists, however, were right. Oman did become a significant oil producer, with Shell in the lead.[14]

"Us Independent Oil Suckers"

Consumers around the world welcomed cheap oil from Venezuela and the Middle East. So, after some hesitation, did the governments of the industrial countries. There was one exception—in the United States. No longer was the growing abundance of cheap foreign oil something to be encouraged and applauded as a way to relieve the pressure on U.S. reserves. Rather, the rising flood of imported oil was seen, at least among the independent American producers, as a dangerous threat that was undercutting domestic prices and undermining the domestic industry itself. As early as 1949, an irate geologist from Dallas named "Tex" Willis had written to his Senator, Lyndon B. Johnson, to ask if he was "going to be able to do anything for us independent oil suckers on that foreign oil that has destroyed the market for two billion dollars worth of Texas independent oil this year?" Tex Willis wanted to be sure that Johnson understood how he and his fellow oil men felt. There was, he said, "no sense in bankrupting every independent oil man in Texas for a few Arabian princes and because . . . Standard Oil of New Jersey claims they need the money."

Johnson and the other members of the Congressional delegations from the oil states clearly heard Tex Willis and his compatriots and pressed hard to give the domestic oil industry some protection against Venezuelan and Middle Eastern oil. At one point, Johnson sent his aide, John Connally, to the State Department along with a number of Texas congressmen, all of whom wanted to impress on rather unsympathetic officialdom that their "re-election might turn upon whether or not they could provide a satisfactory answer to their constituents." The oil state representatives sought to raise the tariff on imported oil tenfold, from 10.5 cents to $1.05 per barrel, and to limit imports to 5 percent of domestic consumption. Such efforts found no favor with President Harry Truman, who told one Congressman, "Something must be radically wrong with the reasoning of the people who would like to cut off our foreign trade for the benefit of the oil crowd."

After the end of the Korean War and the return of Iranian oil to the market with the fall of Mossadegh, petroleum imports made even further inroads on domestic petroleum and coal. As a result, coal-producing and oil-producing states formed an unlikely coalition to seek to limit such imports. But one of the last things the new Eisenhower Administration wanted to do was put tariffs or quotas on imported oil. It wanted to encourage freer trade, to broaden economic relations with developing countries, and to keep them in the Western orbit. Yet Congress insisted on giving the President the power to restrict oil imports through a "National Security Amendment" to the 1955 trade act, which would enable him to control their level when he concluded that the nation's security or its economic well-being was threatened.

Eisenhower was loath to use this new power. Instead of mandatory restrictions on foreign oil, his administration called for "voluntary" restrictions on the part of importers. It launched a vigorous campaign of letter writing and moral suasion directed at the importing companies, but the campaign proved rather ineffective in the face of the continuing buildup of Middle Eastern supply capacity and the price advantage of imported oil.

The Suez crisis of 1956 highlighted concerns about national security. The price fall that followed the crisis further increased the clamor among independents for protection in the form of tariffs or quotas. The majors, with their foreign production, did not join in the clamor. Eisenhower, himself still opposed to protectionism, came up with an alternative. If access to oil in an emergency was required for national security, he asked, then why not have the government stockpile oil in abundance? At one Cabinet meeting, he reminded his colleagues of what he called "an old suggestion"—that the government purchase low-cost foreign oil and store it in exhausted wells. Perhaps he remembered what had happened in 1944, when General Patton had run out of gas, and he had faced the thankless job in the "unforgiving minute" of allocating supplies between the furious Patton and the inflexible Montgomery. Storage might not improve the health of the domestic oil industry, but it would reconcile national security concerns with the administration's free trade economic policies. But Eisenhower could not get any support for his idea. Indeed, the special committee he appointed to report on the entire oil import and security question rejected this particular alternative as simply too impractical.[15]

National Security and "a Nice Balance"

The independent oil men wanted mandatory controls—and soon. They intensified their campaign for a tariff as imports continued to rise, from the equivalent of 15 percent of domestic production in 1954 to over 19 percent in 1957. Meeting with three pro-restriction Senators in June of that year, a reluctant Eisenhower outlined the large number of considerations that he was trying to juggle:

"the health of the domestic industry, national defense, the tax income of the various states, overall depletion of U.S. reserves, and the encouragement of exploration without causing the marketing of too much domestic oil and thereby unduly reducing our domestic reserves." In short, said the President, "a nice balance should be obtained." In an attempt to achieve that balance, the Administration adopted, in 1957, a system of more-explicit voluntary controls. The government was now in the business of informally allocating import rights.

No one particularly liked the "voluntary" allocation machinery. Still, it would work if *everybody* cooperated. But several companies were decidedly uncooperative. One reason was obvious: they were disproportionately disadvantaged because they had made large commitments to foreign oil. This applied not only to majors. J. Paul Getty had embarked on a $600 million expansion program to build tankers, gas stations, and a big new refinery, all of it to be based upon his new production from Kuwait's Neutral Zone. Getty adopted the simple expedient of ignoring the voluntary quota system. After all, it was only voluntary. Sun Oil was much worried about the antitrust implications of cooperating with a "voluntary" program, the effect of which was to maintain prices. At that very moment, the Justice Department was suing the majors under the Sherman Antitrust Act for actions they had taken during the Suez Crisis in response to encouragement from other branches of the Federal government, which had been worried about shortages. Robert Dunlop, the president of Sun, also remembered the "Madison Case" in the 1930s, when the Justice Department had successfully prosecuted the oil industry on antitrust for going along with a market stabilization scheme promoted by Harold Ickes and the Interior Department. What assurance could the government now give that Sun, along with other companies, would not later be hauled up once again on antitrust charges for cooperating with the so-called voluntary system—which did look rather like a government-sponsored scheme to bolster prices!

The recession of 1958 did in the voluntary program. While oil demand dropped substantially, imports increased further, and the political pressure for mandatory controls was becoming irresistible. Clarence Randall, the chairman of the Council on Foreign Economic Policy, exasperatedly told Secretary of State Dulles that those who were invoking "national security" to restrict imports were all mixed up. If national security was the concern, then the best thing to do was to encourage imports in order to preserve domestic reserves. "Our policy should be to conserve that which we have," he said, "rather than to take measures which would cause our supplies to be exhausted more rapidly."

Still, the Eisenhower Administration resisted mandatory quotas. "This business about the national security is a good deal of window dressing," Dulles complained in a phone conversation with Attorney General Herbert Brownell. "What they are doing," Dulles continued, referring to the Texans who were calling for mandatory controls, "is to try to put the price of oil up and put more of the Texas wells into production and accelerate new drilling which will only happen if the price goes up." Between seniority and adroit politicking, the oil states and the interests of the independents were powerfully represented in the Congress. The Speaker of the House, Sam Rayburn, was from Texas, and for him, as his biographer wrote, "oil and Texas were inseparable." The Senate Majority Leader,

Lyndon Johnson, was from Texas, and was no less sensitive to his constituents. He had already by 1940 made himself the key link in fundraising for Democratic politicians among wealthy Texas oil men. One of the most powerful Senators was Robert Kerr, a millionaire oil man from Oklahoma. Eisenhower could see what was coming. "Unless the Executive takes some action, Congress will," he finally told Dulles, and he doubted that a Presidential veto would be upheld.

The President, unhappy about the position in which he found himself, unloaded his anger at a Cabinet meeting, criticizing the "tendencies of special interests in the United States to press almost irresistibly for special programs like this" that were "in conflict with the basic requirement on the United States to promote increased trade in the world." Nevertheless, four days later, on March 10, 1959, Eisenhower announced the imposition of mandatory quotas on oil imports into the United States. A full decade after the battle had begun, the United States finally adopted formal controls. They may well have constituted the single most important and influential American energy policy in the postwar years. The independent oil "suckers" were jubilant; the majors, disappointed.[16]

"A Very Healthy Domestic Industry"

The quotas lasted for fourteen years. Under Eisenhower, imported oil could not exceed 9 percent of total consumption. The Kennedy Administration tightened the quotas a bit in 1962. Later, in the second half of the 1960s, the Johnson Administration did make some effort to relax the quotas as a way to bring down oil prices and thus help to counteract the inflation that was beginning to build with the Vietnam War. But, essentially, the quota system remained intact.

Oil import quotas sounded simple, straightforward. These were not. As time went on, their management became more and more Byzantine. Indeed, under the Mandatory Oil Import Program, as it was known, there were continuing fights over allocations, struggles over interpretations, searches for loopholes, and the ever-more-intense hunt for exceptions and exemptions. Over the years, the program became increasingly barnacled and distorted. A brisk market developed, not in oil itself, but in oil import "tickets" or rights to bring in oil. Some parts of the refining industry ended up, in effect, subsidizing others.

But there was nothing to compare with what became known variously as the "Mexican Merry-Go-Round" or the "Brownsville U-Turn." Since memories were fresh of World War II and the U-boat attacks on tankers, and since "national security" was what the quotas were supposed to be all about, oil that came "overland" to the United States from Mexico or Canada was deemed more secure than oil shipped in tankers, and was given certain preference and exemption, which also happened to help political relations with Mexico and Canada. But here was the catch: There were no oil pipelines from Mexico, and oil was certainly not going to be trucked several hundred miles from Mexico's production centers. Therefore, Mexican oil was shipped by tanker to the bordertown of Brownsville, Texas, put into trucks, driven across the bridge into Mexico, around a traffic circle, and then back across the bridge into Brownsville, where it was reloaded into tankers for shipment to the northeast. Thus shipped "overland," it perfectly legally qualified for an exemption.

By the time of the Johnson Administration, one official was calling the entire quotas program "an administrative nightmare." It also had far-reaching effects. It led, as was intended, to higher levels of investment in domestic oil exploration, relative to exploration outside the United States, than would have otherwise been the case. It tilted foreign investment by U.S. companies toward Canada, on account of that country's preferential access to the American market. It resulted in the building of substantial refining capacity in the American Virgin Islands and Puerto Rico because of special exemptions to the quotas that were granted to refineries there on economic development grounds. And, finally, the program gave an important impetus to the global oil trade. If companies could not bring foreign oil into their own systems in the United States, which was the objective of integration, then they would have to find and develop markets elsewhere in the world.

As a further result of the program, prices were higher in the United States than they would have been without the protection. Moreover, the quotas put the prorationing systems in Texas and the other states back into a position where they could stabilize domestic prices. Indeed, domestically, the ten-year period following the introduction of the mandatory quotas was reminiscent of the price stability that followed the full implementation of prorationing in the 1930s. The average price of oil at the wellhead in the United States in 1959 was $2.90 a barrel; a decade later, in 1968, it was $2.94—stable, certainly, and also 60 to 70 percent above Middle Eastern crude in East Coast markets. In contrast, by closing off the American market, the mandatory controls resulted in lower prices outside the United States.

Despite all the exemptions, complications, and administrative nightmares, the import quotas did achieve their fundamental goal: They provided ample protection for domestic oil production against lower-cost foreign oil. By 1968, United States crude oil output was 29 percent higher than it had been in 1959, the year the mandatory quotas were introduced. Without that protection, American production would surely have plateaued or declined. Companies, large and small, adapted to the mandatory quotas. The majors, despite their initial vociferous criticism of the quotas, eventually came to see merit in a program that protected the profitability of their own domestic operations, albeit at the expense of their foreign ones. Their adjustment was facilitated by the fact that demand elsewhere was growing with sufficient rapidity to absorb their foreign production.

The mandatory program also taught the international companies a lesson. They may have had the financial resources, they may have had the scale and the know-how, but the independents had the political clout, and it was to them that the senators and congressmen from the oil patch responded. Sometimes the point had to be made explicit. In the mid-1960s, Senator Russell Long of Louisiana felt constrained to deliver a little homily to a group of executives from the larger oil companies. Congressmen from the oil states, he explained, "are especially interested in the domestic phases of the industry, because that is the part that gives employment to our people and means revenue to our state governments, and it is essential to our economy." Long wanted the oil executives to ponder the implications. "We would like you fellows that produce oil overseas to realize this, that when problems come up with regard to your tax credit overseas or even your depletion allowance overseas, or the special tax treatment to your employees that you have overseas, the fellows you are going to rely upon to protect your activities in that respect are the same people who are interested in the domestic production of oil." To summarize his message, Long added, "It is very much to your advantage to have a very healthy domestic industry and do everything within your power to cooperate to that end." [17]

The international companies grudgingly absorbed the lesson.

CHAPTER 27

Hydrocarbon Man

WHATEVER THE TWISTS AND TURNS in global politics, whatever the ebb of imperial power and the flow of national pride, one trend in the decades following World War II progressed in a straight and rapidly ascending line—the consumption of oil. If it can be said, in the abstract, that the sun energized the planet, it was oil that now powered its human population, both in its familiar forms as fuel and in the proliferation of new petrochemical products. Oil emerged triumphant, the undisputed King, a monarch garbed in a dazzling array of plastics. He was generous to his loyal subjects, sharing his wealth to, and even beyond, the point of waste. His reign was a time of confidence, of growth, of expansion, of astonishing economic performance. His largesse transformed his kingdom, ushering in a new drive-in civilization. It was the Age of Hydrocarbon Man.

The Explosion

Total world energy consumption more than tripled between 1949 and 1972. Yet that growth paled beside the rise in oil demand, which in the same years increased more than five and a half times over. Everywhere, growth in the demand for oil was strong. Between 1948 and 1972, consumption tripled in the United States, from 5.8 to 16.4 million barrels per day—unprecedented except when measured against what was happening elsewhere. In the same years, demand for oil in Western Europe increased fifteen times over, from 970,000 to 14.1 million barrels per day. In Japan, the change was nothing less than spectacular; consumption increased 137 times over, from 32,000 to 4.4 million barrels per day.

What drove this worldwide surge in oil use? First and foremost was the rapid and intense economic growth and the rising incomes that went with it. By the end of the 1960s, the populations of all the industrial nations were enjoying a standard of living that would have seemed far beyond their reach just twenty years before. People had money to spend, and they spent it buying houses, as well as the electrical appliances to go inside those houses and the central heating systems to warm them and the air conditioning to cool them. Families bought one car and then a second. The number of motor vehicles in the United States increased from 45 million in 1949 to 119 million in 1972. Outside the United States, the increase was even more monumental, from 18.9 million vehicles to 161 million. To produce the cars and appliances and package goods, to satisfy directly and indirectly the needs and wants of consumers, factories had to turn out ever-increasing supplies, and those factories were increasingly fueled by oil. The new petrochemical industry transformed oil and natural gas into plastics and a host of chemicals, and in every kind of application, plastics began to replace traditional materials. In a memorable scene in the 1967 motion picture *The Graduate*, an older man confided the true secret of success to a young man who was undecided about his future: "Plastics." But, by then, the secret was already everywhere evident.

During the 1950s and 1960s, the price of oil fell until it became very cheap, which also contributed mightily to the swelling of consumption. Many governments encouraged its use to power economic growth and industrial modernization, as well as to meet social and environmental objectives. There was one final reason that the market for oil grew so rapidly. Each oil-exporting country wanted higher volumes of *its* oil sold in order to gain higher revenues. Using various mixtures of incentives and threats, many of these countries put continuing pressure on their concessionaires to produce more, and that, in turn, gave the companies powerful impetus to push oil aggressively into whatever new markets they could find.

The numbers—oil production, reserves, consumption—all pointed to one thing: Bigger and bigger scale. In every aspect, the oil industry became elephantine. For all the growth in production and consumption could not have been accomplished without infrastructure. Multitudes of new refineries were built—larger and larger in size, as they were designed to serve rapidly growing markets and to go for economies of scale. New technologies enabled some refiners to increase the yield of high-value products—gasoline, diesel and jet fuel, and heating oil—from less than 50 percent to 90 percent of a barrel of crude. The result was a sweeping conversion to jet planes, diesel locomotives and trucks, and oil heat in homes. Tanker fleets multiplied, and tankers of conventional size gave way to the huge, seagoing machines called supertankers. Gasoline stations, more and more elaborate, popped up at intersections and along highways throughout the industrial world. Bigger is better—that was the dominant theme in the oil industry. "Bigger is better" also enthralled the consumers of oil. Powered by huge engines and bedecked with chrome and extravagant tail fins, American automobiles grew longer and wider. They got all of eight miles to a gallon of gas.[1]

Old King Coal Deposed

In the buoyant decades following World War II, a new war was being fought, though not of the sort that was reported in communiqués on the front page, but rather one buried in the pages of the day-to-day trade press. An astute student of oil affairs, Paul Frankel, called it "a war of movement." It was also a war that reflected a great historical transformation for modern industrial societies. It had enormous economic and political consequences and profound impact on international relations and on the organization and patterns of daily life. It was the battle between coal and oil for the hearts and minds, and pocketbooks, of the consumer.

Coal had powered the Industrial Revolution of the eighteenth and nineteenth centuries. Cheap and available, it was truly King. Coal, wrote the nineteenth-century economist W. S. Jevons, "stands not beside but entirely above all other commodities. It is the material energy of the country, the universal aid, the factor in everything we do. With coal almost any feat is possible or easy; without it we are thrown back into the laborious poverty of early times." King Coal held on to his throne through the first half of the twentieth century. Yet he could not resist, he could not stand unmoved, in the face of the great tidal wave of petroleum that surged out of Venezuela and the Middle East and flowed around the planet after World War II. Oil was abundant. It was environmentally more attractive and easier and more convenient to handle. And oil became cheaper than coal, which proved the most desirable and decisive characteristic of all. Its use provided a competitive advantage for energy-intensive industries. It also gave a competitive advantage to countries that shifted to it.

The wave broke first across the United States. Despite the motorcar, even the United States had remained primarily a coal economy until mid-twentieth century. By then, however, the coal industry's own cost structure made it a sitting duck. With repeated price cuts, oil was becoming cheaper than coal in terms of energy delivered per dollar. There was yet another compelling reason to switch to oil: labor strife in America's coal fields. Strikes by coal miners, led by John L. Lewis, the combative president of the United Mine Workers, were virtually an annual ritual. Lewis's bushy eyebrows became a familiar totem among the nation's editorial cartoonists, while his bellicose pronouncements shook the confidence of coal's traditional consumers. Interrupt the production of coal, he boasted, and you can stop "every part of our economy." To any manufacturer worried about the continuity of his production line, to a utility manager anxious about his ability to meet electricity requirements in the dead of winter, Lewis's fiery rhetoric and the militancy of his United Mine Workers constituted a powerful invitation to find a substitute for coal. That meant oil, to which there was no such obvious threat, and in particular, fuel oil, a very high proportion of which was imported from Venezuela. "We ought," a Venezuelan oil man once mused, "to take up a public subscription throughout Venezuela to erect a statue of John L. Lewis in the central square in Caracas—to honor him as one of the greatest benefactors and heroes to the Venezuelan oil industry."[2]

The Conversion of Europe

The decline of King Coal followed a somewhat different course in Europe, spurred chiefly by the cheap and readily available oil of the Middle East. The first postwar energy crisis, in 1947, was Europe's severe shortfall of coal. Its legacy for Britain was a specter of shortage. Fearing that coal supplies would be inadequate, the government began to encourage power plants to switch from coal to oil as a stopgap. But oil was hardly a stopgap. It was a relentless competitor. The 1956 Suez crisis did create a large question mark for Britain and other European countries about the security of Middle Eastern oil supplies. In the immediate aftermath of Suez, Britain decided to push ahead with its first major nuclear energy program to reduce dependence on imported oil. Plans were discussed among the industrial countries to maintain inventories beyond commercial needs—that is, emergency stocks—as a form of insurance against future disruptions. But the security concerns dissipated with surprising quickness, and Europe's move away from coal continued unabated.

Part of the reason for oil's victory over coal was environmental, especially in Britain. London had long suffered from "Killer Fogs" as the result of pollution from coal burning, particularly the open fires in houses. So thick were those fogs that confused motorists literally could not find their way home to their own streets and instead would drive their cars onto lawns blocks away from their own houses. Whenever the fogs descended, London's hospitals would fill with people suffering from acute respiratory ailments. In response, "smokeless zones" were established where the burning of coal for home heating was banned, and in 1957 Parliament passed the Clean Air Act, which favored oil. Still, the biggest force promoting the switch was cost; oil prices were going down, and coal prices were not. From 1958 onward, oil was a cheaper industrial fuel than coal. Homeowners switched to oil (as well as to electricity and, later, to natural gas). The coal industry responded with a vigorous advertising campaign based upon the theme of the "Living Fire." Despite the rhetoric, when it came to heating homes, coal was a dying ember.

Trying to balance the economic advantages of lower-priced oil against the costs and dislocations and job loss of an embattled coal industry, the British government struggled with policies that would give domestic coal some protection against cheap imported oil. But by the middle 1960s, the government had pretty much concluded that Britain's international trade position required rapid growth in oil use. Otherwise, British manufacturers would be disadvantaged competing against foreign firms that used cheap petroleum. A government official summed up the transformation: "Oil has become the lifeblood of the economy, as of all other industrialized countries, and it affects every part of it."

The pattern was indeed being repeated right across Western Europe. By 1960, the French government had officially committed itself to the rationalization and contraction of the domestic coal industry, and to a wholesale switch to oil. The use of oil, it emphasized, provided a way to promote the modernization of its industrial establishment. John Maynard Keynes had once said that "the German empire was built more truly on coal and iron, than on blood and iron." But Germany, too, converted, as oil became cheaper than coal. The full extent of the conversion was dramatic. In 1955, coal provided 75 percent of total energy use in Western Europe, and petroleum just 23 percent. By 1972, coal's share had shrunk to 22 percent, while oil's had risen to 60 percent—almost a complete flip-flop.[3]

Adolf Hitler made oil central to his plans for conquest in World War II. His ill-conceived invasion of the Soviet Union was halted just short of the rich oil resources of the Caucasus.

51

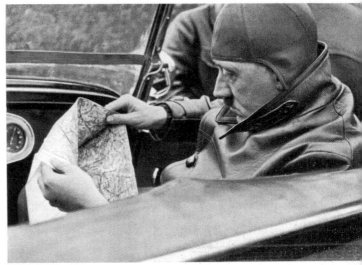

52

General Erwin Rommel, the master of mobile warfare, swept across North Africa, planning to join forces with the German invaders of the Caucasus. Later, as the battle turned against him, he wrote to his wife, "Shortage of petrol! It's enough to make one weep."

The opening of the Magdeburg synthetic fuels plant in 1937 (left). Such fuels provided more than half of Germany's total oil supply during the war. The same plant (right) after three thousand bombs were dropped on it by Allied bombers.

53

54

Oil was also central to Japanese strategy in the war in the Far East. Admiral Isoroku Yamamoto planned the attack on Pearl Harbor to protect Japan's flank as it went for the oil of the East Indies.

55

Almost a year of preparation went into the assault, including the construction of a miniature Pearl Harbor.

56

The American strategy in the Pacific was elemental: cut off Japan's supply of oil. Here a Japanese tanker sinks after being hit by torpedoes from an American sub.

57

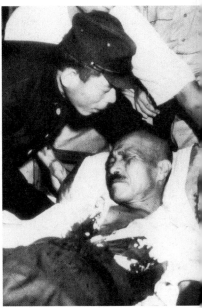

In 1941, the militarist general Hideki Tojo used the American oil embargo as the reason to attack Pearl Harbor. In 1945, with Japan devastated and defeated—and completely out of oil—he tried to commit suicide, unsuccessfully.

58

59

With victory in sight, Churchill and Roosevelt met in Quebec in September 1944. America's abundance of oil had been decisive in the battles in both Europe and the Pacific.

"The Old Curmudgeon," Roosevelt's Interior Secretary Harold Ickes, dominated America's oil policy for more than a decade and became "oil czar" in World War II.

60

61

General George Patton with Supreme Commander Eisenhower. When Patton's tanks ran out of fuel as he pursued the retreating Germans across France, he said bitterly, "If only I could steal some gas, I could win this war."

"Pump girls" in London, where, as in America, women took over at gasoline stations and other vital jobs when the men went to war.

62

King Ibn Saud met with President Franklin Roosevelt in 1945 aboard an American ship in the Suez Canal. The immensity of the oil riches of the Middle East was just beginning to be recognized.

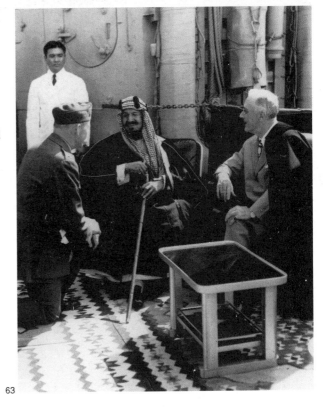

63

Happy days are here again. Victory meant the end of gasoline rationing in the United States.

64

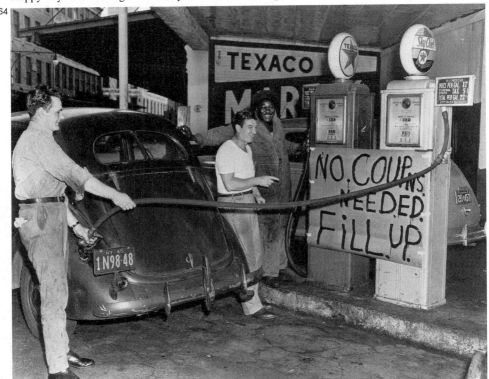

65

As late as 1940, J. Edgar Hoover, director of the FBI, warned that "motels" were centers of vice and the "hot pillow trade," even though some advertised their moral probity. In postwar America, motels became welcome—and respectable—havens as millions of American families took to the roads.

66

A milkshake maker salesman with vision opened this, the first real McDonald's, in Des Plaines, outside Chicago, in 1955.

The drive-in became a central part of American life in the 1950s, for families—and for teenagers. Forty percent of marriages were proposed in automobiles.

68

Oil and petrochemicals became the bricks and mortar of the Hydrocarbon Age. "Shell girls" demonstrated plastic hula hoops atop Shell's headquarters in London in 1957.

69

Customers were dazzled by the auto industry's introduction of tail fins in the 1950s, no matter that the engines were "gas guzzlers."

Iranian Prime Minister Mohammed Mossadegh nationalized British Petroleum in 1951, setting off the first postwar oil crisis and unleashing political forces he could not control.

J. Paul Getty, taking along "Teach Yourself Arabic" tapes, met with King Saud in the desert shortly after the discovery that was to make Getty a billionaire.

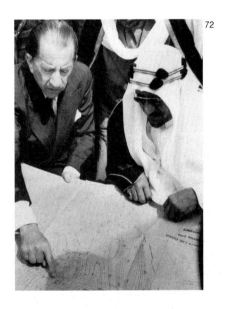

George McGhee, the State Department's "infant prodigy," promoted the "fifty-fifty" profit split with Middle Eastern producers in 1950 to help save the oil companies from outright nationalization.

Gamal Abdel Nasser and Anthony Eden before Nasser nationalized the Suez Canal in 1956, triggering the second postwar oil crisis.

74

75

President Dwight Eisenhower, with Secretary of State John Foster Dulles, was so furious with the British-French invasion of Suez that he cut off emergency oil supplies to them and angrily declared that they could "boil in their own oil."

In 1957, General Charles DeGaulle, visiting France's new oil field in Algeria, declared, "In our destiny, this can change everything." France had long sought its own independent oil supply.

76

77

The Italian tycoon Enrico Mattei (left) risked the combined wrath of the major international oil companies when he made his deal with the Shah of Iran (right) in 1957.

The intrepid oil journalist Wanda Jablonski, here with Sheikh Shakbut of Abu Dhabi in 1956.

78

Jersey's chairman, Monroe Rathbone, holding a tiger's tail, unilaterally cut the price of oil in 1960 to counter a surge of Soviet oil sales. The result was OPEC.

79

80

OPEC's two founding fathers: Abdullah Tariki, the "Red Sheikh" and first Saudi oil minister, and the Venezuelan oil minister, Juan Pablo Pérez Alfonzo.

King Faisal of Saudi Arabia and his protégé, oil minister Ahmed Zaki Yamani, during a negotiation with American oil men.

82

Dr. Armand Hammer built a puny, bankrupt oil company into the giant Occidental on the basis of the newly discovered oil riches of Libya.

The Yom Kippur War in 1973 sent gasoline prices soaring and Americans into gas lines. The same thing happened again in 1979, when the Shah of Iran fell from power.

Senator Henry Jackson put chagrined oil executives under oath at 1974 hearings, then accused them of making "obscene profits."

85

Japanese Prime Minister Masayoshi Ohira modeled the "energy conservation look" after the 1973 embargo. Though the fashion never caught on in Japan, energy conservation certainly did.

Richard Nixon's last hurrah. In June 1974, in the aftermath of the Yom Kippur War, he was cheered in the streets of Cairo. Two months later, he was forced to resign as President because of the Watergate scandal, which had crippled America's response to the 1973 oil crisis.

87

President Jimmy Carter and Energy Secretary James Schlesinger on an oil platform in the Gulf of Mexico, shortly after Carter described the energy challenge facing the United States as the "moral equivalent of war."

88

The Alaska pipeline, which took five years to win approval to build.

When Jimmy Carter greeted the Shah of Iran at the White House in 1978, tears came to their eyes—not from emotion, but from the tear gas used to break up demonstrations.

With the fall of the Shah, the Ayatollah Ruhollah Khomeini was deliriously greeted in Tehran when he returned from exile in February 1979.

The frenzied trading floor at the New York Mercantile Exchange, which became the center for establishing global oil prices after 1983.

92

Mesa's Boone Pickens (left) shakes hands with Gulf chairman Jimmy Lee in December 1983, in the midst of the fierce struggle for Gulf, one of the biggest of the big merger battles of the 1980s.

93

94

Just after midnight, on Good Friday, March 24, 1989, the supertanker *Exxon Valdez* went aground in Alaska's Prince William Sound, spilling 240,000 barrels of oil—and giving a great boost to the environmental movement.

96

Oil man George Bush, then president of Zapata Off-Shore, with his son George, dedicat-
ing a new drilling rig in 1956. Thirty years
later, in 1986, as Vice President, he met with
Saudi Arabia's King Fahd, and in 1990, as
President, he squared off against Saddam Hussein over Iraq's invasion of Kuwait and the
struggle for Middle East oil.

95

97

Saddam Hussein, the Iraqi
strongman who triggered
the first post–Cold War oil
crisis.

Japan: No Longer Poor

Japan was somewhat slower in initiating the shift to oil. Coal was its traditional basic energy source. Before and during World War II, oil had primarily fueled the military, with a small consuming sector in civilian transportation and a continuing use of kerosene for lighting. Refineries and the rest of the oil infrastructure were devastated in the finale of World War II. Not until 1949 did the American Occupation even permit the reestablishment of oil refining in Japan, and then only under the tutelage of Western companies—Jersey, Socony-Vacuum, Shell, and Gulf. With the end of the Occupation, the regaining of political independence, and the Korean War, Japan embarked on its remarkable process of economic growth.

So successful was the first phase, based on the rapid development of heavy industry and chemicals, that by 1956, the government was already able to make an epochal declaration: "We are no longer living in the days of postwar reconstruction." Japan would not be poor forever, and coal, it was expected, would fuel the continuing growth. At the beginning of the 1950s, coal provided more than half of Japan's total energy, and oil only 7 percent—less than firewood! But oil prices kept falling. By the beginning of the 1960s, there could be no question that the government and Japanese industry would bet on oil. As elsewhere, it would free the economy from the threat of labor unrest among coal miners, and once again, oil was much cheaper than coal.

As oil itself became more important to the Japanese economy, the government, as a matter of high policy, sought to reduce the foreign influence in its oil industry. The Ministry of International Trade and Industry, MITI, restructured the Japanese petroleum industry so that independent Japanese refiners would gain a substantial market share in competition with those companies linked directly to the international majors. The independents were seen as more reliable, more committed exclusively to Japanese economic objectives, more assuredly tied in to the Japanese economic and political system. A new oil law in 1962 gave MITI the authority to grant permits to import oil and to allocate sales. It used this power to bolster the independent refiners and to promote competition that would help keep the cost of oil as low as possible. Price wars resulted, as refiners battled strenuously for markets. And as though making up for lost time, Japan completed the conversion to an oil economy at a phenomenally rapid rate. In the second half of the 1960s, while the Japanese economy itself was growing at an altogether extraordinary 11 percent a year, oil demand was growing at an even more extraordinary rate, an annual 18 percent. By the end of the 1960s, oil was providing 70 percent of the total energy consumed in Japan, compared to 7 percent at the beginning of the 1950s!

Much of the increase in the demand for oil reflected the dynamism of Japanese industry. But another force was also at work—the Japanese automotive revolution. In 1955, the Japanese industry produced only 69,000 cars; just thirteen years later, in 1968, that same industry produced 4.1 million cars, of which 85 percent were bought and used within Japan, and only 15 percent were exported. That meant a tremendous rise in domestic gasoline consumption. The great auto export boom, which would help establish Japan as a formidable global economic power, had yet to begin.

The two *wunderkinder* of the postwar world were Japan and Germany, each of which not merely recovered from defeat but set envied and astonishing standards for economic performance. Looking back on their achievements, the economic historian Alfred Chandler succinctly summed up the recipe for their success: "The German and Japanese miracles were based on improved institutional arrangements and cheap oil." Not all their allies and competitors had, by any means, the same access to "improved institutional arrangements," but all had the benefits of abundant petroleum. As a result, in the boom years of the 1950s and the 1960s, economic growth throughout the industrial world was powered by cheap oil. In a mere two decades, a massive change in the underpinnings of industrial society had taken place. On a global basis, coal had provided two-thirds of world energy in 1949. By 1971, oil, along with natural gas, was providing two-thirds of world energy. What the economist Jevons had said in the nineteenth century about coal was now, a century later, true not for coal, but for oil. It stood above all other commodities; it was the universal aid, the factor in almost everything we did.[4]

The Struggle for Europe

Because of rapid economic growth and industrial expansion, combined with the switch from coal to oil and the advent of the popularly priced automobile, Europe was the most competitive market in the world in the 1950s and 1960s. With protectionist quotas now limiting the amount of oil that could be imported into the United States, all the American companies that had gone overseas to discover oil had to find other markets, and that meant Europe more than anywhere else. At the same time, the producing nations kept up the pressure on the companies to increase volume. "Each year our people made an annual pilgrimage to Kuwait City," said Gulf executive William King. "It was universally the same. It would be a difficult meeting, with lots of threats and blandishments on both sides. The Kuwaitis would tell us how much they wanted us to increase our liftings there, and we would tell them that it was too much, there weren't markets for it." The Kuwaitis would point out that the Iranians had already won an increase. "Finally, both sides would agree on a number—a five or six percent increase."

Where could all that additional oil be sold? There were some opportunities in the developing world. Gulf constructed a fertilizer plant in South Korea to help it obtain the right to build a refinery and distribution system in that country. It loaned money to such Japanese companies as Idemitsu and Nippon Mining so they could build refineries, with the collateral being a long-term crude contract. But Europe was by far the most significant market. Entry and expansion took not only economic ability, but also political skill, as there was much more direct and indirect government regulation and control than in the United States. For instance, companies could not simply go out and buy a plot of land and put up a gasoline station; governments exerted tight controls in allocating locations, with the result that there was tremendous jockeying for spots. "Competition was horrendously intense in Europe because the amounts of money involved were so great," King said. "The people from the different companies would speak politely to each other and act friendly, and then we'd all go out and try to steal each other's markets."

Shell was the leading European marketer, which meant that it was on the defensive, and had to learn to be much more competitive. In West Germany, for instance, Deutsche Shell announced proudly that its 220 young salesmen were trained in "American-style aggressive selling." Jersey had to be even more aggressive because it was trying to build its relative position. In Britain, a single gasoline station would often have pumps representing several companies, sometimes selling as many as six different brands. That was heresy to Jersey. It wanted stations that sold its Esso gasoline and only Esso, and it pretty much attained that objective. In order to win the affection of farmers across the Continent who were mechanizing, it sponsored a World Plowing Match in Europe. Calling on the great American tradition, its stations in Europe began to offer road maps and local tour information without charge, to win patronage both from Europeans and from the increasing number of American tourists who had grown up expecting the maps as a constitutional right, gratis.

Among the Goliaths striding across Europe, there were also a number of agile Davids that had developed production and were scrambling for markets, and in so doing, further stimulating the thirst for oil. Among them the most notable was the Continental Oil Company, later Conoco. Continental had begun life in 1929 as the result of a merger between a Rocky Mountain marketing company, originally part of the Standard Oil empire, and an Oklahoma-based crude producer and refiner. The new enterprise was a tightly defined American regional company. Then, in 1947, the board brought in a new president, Leonard McCollum, who had been Standard Oil of New Jersey's worldwide production coordinator. McCollum wanted to focus on building up the company's North American production. But he soon found that Continental was at a competitive disadvantage. Lower-cost foreign oil was pouring into the United States in the late 1940s, winning the incremental demand, while Continental's domestic production was being restricted by prorationing in Texas, Oklahoma, and elsewhere. Continental, McCollum decided, would have to go overseas. The company spent a good deal of money drilling dry holes in Egypt and elsewhere in Africa over the next decade. Yet, despite the headaches and disappointments, McCollum was convinced that it was better, when it came to crude oil, to be a "have" than to be a "have-not" company. "If you set out to be a 'have,' " he said, "you must have the audacity to acquire as much acreage as possible—to take a big bite. Though a small piece may look like a sure thing, you better take as much as you can so you don't miss."

In the mid-1950s, Continental took a considerable bite in Libya, in a partnership with Marathon and Amerada that was called the Oasis Group. At the end of the 1950s, Oasis began to strike it very big in Libya. But, just at that moment, the rules were being drastically changed in Washington, completely undercutting McCollum's original strategic rationale. The new import quotas pretty much precluded Continental at the time from bringing its cheaper Libyan oil into the U.S. market, as it had planned. That meant the oil had to go elsewhere, and "elsewhere," of course, meant Western Europe, the most competitive oil market in the world.

At first, Continental sold its surging Libyan output to the established majors and independent refiners in Europe. "We were brand new, and we had to go out and beat the bushes," recalled one Continental executive. But the company had little flexibility, and it also had to offer considerable price concessions to its buyers. Thus, it faced the classic dilemma—dependence on others. At the turn of the twentieth century, William Mellon had turned Gulf into an integrated company, with its own refining and distribution, so that he would not have to say "by your leave" to Standard Oil or anyone else. Now, sixty years later, McCollum would do the same.

So, in three years, beginning in 1960, the company established its own downstream refining and distribution system in Western Europe and Britain, acquiring where it could, starting from scratch where it could not. Its higher-quality Libyan oil, which was particularly suited to making gasoline, pushed Continental to develop its own networks of gasoline stations. In addition, Continental negotiated long-term contracts with strategically placed independent refiners. It built a very efficient refinery in Britain, where it sold low-cost gasoline under the "Jet" name. By 1964, sixteen years after McCollum had initiated the foreign oil search, Continental was producing more overseas than in the United States. It had become a significant integrated international oil company, which had never been in McCollum's original plan.

The multiplication of such companies, each organized as more-or-less autonomous chains, increased the competitive pressures in the marketplace and gave further push to the falling oil price. Their success would also stir up nationalist sentiment in the countries that supplied their oil. In short, the companies were most vulnerable at the extreme ends of the production chain, the wellhead and the pump.[5]

Courting the Consumer

The consumer, particularly the gasoline consumer in his motorized status symbol, was riding high, wide, and handsome in the 1950s and 1960s in America. The deprivation and rationing of the war years were but a distant memory. Huge investments in refinery construction and upgrading, combined with the growing volume of available oil, provided a perfect recipe for hard competition among the gasoline suppliers, driving down the price.

That suited American motorists just fine, especially when they were the beneficiaries of the frequent "price wars." With gasoline stations sprouting on every street corner, their operators would rush hand-lettered signs out to the very edge of the property, proclaiming that their price was half a cent less than that of the station across the street. The first shot in the price wars was often fired by independent stations, unaffiliated with the larger companies, that picked up cheap surplus gasoline from the secondary market. The majors did not particularly like price wars—they could find themselves vulnerable to charges of predatory pricing—and often adopted a "we were forced" attitude. But despite the protests, the majors would sometimes initiate price wars when aggressively trying to break into new markets.

Competition took other forms as well. Never had motorists been better served. Tires and oil were checked, windows were washed, drinking glasses and sweepstakes entry forms were handed out—and all for free—in order to win and hold the affection of motorists. Credit cards were introduced in the early 1950s to tie the customer to a particular company. Television provided a whole new medium to advertise national brands and lure consumer loyalty. Texaco reached beyond its Metropolitan Opera radio fans to a far broader audience on television, with the *Texaco Star Theater* and Milton Berle, urging the millions and millions of loyal viewers to "Trust Your Car to the Man Who Wears the Star." Texaco also proudly assured its patrons that, for their benefit, it had gone so far as to "register" all its restrooms throughout the forty-eight states.

Then there was the great hullabaloo over gasoline additives. Their whole purpose was to carve out brand identification for a product, gasoline, that was, after all, a commodity that was more-or-less the same, whatever its brand name. In a period of a year and a half, in the mid-1950s, thirteen of the top fourteen marketers began to sell new "premium" gasolines, racing hard to outdo one another in extravagant claims. In the years that saw the first tests of a hydrogen bomb, Richfield proclaimed that its gasoline "uses hydrogen for peace," a bold but rather unexceptional statement, since all hydrocarbons, including gasoline, are composed of molecules that contain hydrogen. Shell claimed that its TCP (tricresyl phosphate), meant to counteract spark-plug fouling, was the "greatest gasoline development in thirty-one years." Sinclair's Power-X contained additives that were supposed to inhibit engine rust. Cities Service, not to be left behind, had concluded that, if one additive was grand, five would be fantastic, and introduced "5-D Premium." As the list went on, the one common claim by all the companies, whatever the additive, was that it was the result of "years of research."

Shell, pepped up with TCP, increased its sales by 30 percent in one year. This outrage could not be allowed to go unchallenged. Socony-Vacuum rushed out a "confidential" memorandum to its Mobilgas distributors warning that TCP was of little value and might actually damage automobile engines. "No other gasoline like it!" Socony proclaimed of its own "double powered" Mobil gas. Standard Oil of New Jersey went even further, declaring that TCP was a marketing hoax, a cure for a nonexistent problem, as spark-plug fouling did not really ever occur anymore. Instead, Jersey raised octane levels and introduced its own new "Total Power." With the proliferation of "real" gasoline from so many companies, consumers now had a choice between so-called "regular" or "standard" gasolines and a variety of "high-octane" or "high-test" premiums. In due course, Mobil provided yet another variant, "high energy gasoline," explaining that in its exclusive refining process, "light, low-energy atoms are replaced by huskier, high-energy atoms." Naturally enough, a driver of a "high performance" automobile felt compelled to purchase a "high-performance" fuel—for several cents more a gallon—if only for the pleasure, real or imagined, of leaving some other hapless motorist eating his dust back at the stoplight.

Additives were but one way to win the hearts of consumers. In Britain in 1964, as part of its effort to develop a "new look" in gasoline marketing, Jersey came up with the first version of the Esso tiger and the slogan "Put a tiger in your tank." The tiger made its way throughout Esso's European marketing system, helping to provide a comprehensive brand identification. Its first appearance in the United States, however, was not all that successful. "This is not a very nice looking tiger," was the sour judgment of one Jersey executive. Half a decade or so after the tiger's original appearance, he was redrawn by a young artist who had once worked for Walt Disney Productions. This new, improved tiger, also the result of "years of research," was friendly looking, cheerful, good-humored, helpful—and one hell of a salesman. A tiger in your tank, it seemed, could be even more effective in selling gasoline than any of the new additives. Irritated by the popularity of the Esso tiger and its advance into increasing numbers of motorists' tanks in the United States, managers at Shell Oil began to refer to their star additive, TCP, more colloquially as "tom cat piss."[6]

The New Way of Life: "Six Sidewalks to the Moon"

The inexorable flow of oil transformed anything in its path. Nowhere was that transformation more dramatic than in the American landscape. The abundance of oil begat the proliferation of the automobile, which begat a completely new way of life. This was indeed the era of Hydrocarbon Man. The bands of public transportation, primarily rail, which had bound Americans to the relatively high-density central city, snapped in the face of the automotive onslaught, and a great wave of suburbanization spread across the land.

While the move to the suburbs had begun in the 1920s, its development had been halted for a decade and a half, first by the Depression and then by World War II. It began anew immediately after the war. Indeed, the starting point may well have been 1946, when a family of builders named Levitt acquired what eventually added up to four thousand acres of potato farms in the town of Hempstead on Long Island, twenty-five miles east of New York City. Soon bulldozers were leveling the land, and building materials were being dropped off by trucks at exactly sixty-foot intervals. Saplings—apple, cherry, and evergreen trees—were planted on each lot. This first Levittown, its houses priced between $7,990 and $9,500, would eventually encompass 17,400 houses and become home to 82,000 people. And Levittown would in due course become the prototype of the postwar suburb, the embodiment of one version of the American dream and an affirmation of American values in an uncertain world. As William Levitt explained, "No man who owns his own house and lot can be a Communist. He has too much to do."

Suburbanization quickly gathered amazing speed. The number of single-family new housing starts rose from 114,000 in 1944 to 1.7 million in 1950. With the developer's magic, every kind of terrain—broccoli and spinach and dairy farms, apple orchards, avocado and orange groves, plum and fig groves, old estates, racetracks and garbage dumps, hillside scrub, and just plain desert—gave way to subdivisions. Between 1945 and 1954, 9 million people moved to the suburbs. Millions more followed thereafter. Altogether, between 1950 and

1976, the number of Americans living in central cities grew by 10 million, the number in suburbs by 85 million. By 1976, more Americans lived in suburbs than in either central cities or rural areas. In time it became intellectually fashionable to criticize suburbs for everything from their architecture to their values; but to the millions and millions who made their homes there, the suburbs provided better housing in which to nurture the baby boom as well as privacy, autonomy, space, yards for children to play in, better schools, and greater safety—and a haven for optimism and hope in post-Depression and postwar America.

Suburbanization made the car a virtual necessity, and formerly rural landscapes were reshaped by the pervasive automobile. The skyline of this new America was low, and new institutions emerged to serve the needs of suburban homeowners. Shopping centers, with acres of free parking, became meccas for the consumer and the strategic focus for retailers. There were only eight shopping centers in all of the United States as late as 1946. The first major planned retail shopping center was built in Raleigh, North Carolina, in 1949. By the early 1980s, there were twenty thousand major shopping centers, and they rang up almost two-thirds of all retail sales. The first all-enclosed, climate-controlled mall made its appearance near Minneapolis in 1956.

The word "mo-tel," it is thought, was coined as early as 1926 in San Luis Obispo, California, and was applied to the clusters of cabins that grew up, often near gasoline stations, along the nation's highways. But the reputation of this particular creation of the gasoline era was not, at first, anything to brag about. As late as 1940, J. Edgar Hoover, the director of the FBI, put the nation on alert that motels were "camps of crime" and "dens of vice and corruption." Their main purpose, said the nation's top G-man, was either as a rendezvous for illicit sex or a hideout for criminals. Warning the nation about the imminent dangers of the "hot pillow trade," Hoover revealed that the cabins in some motels were rented out for sex as often as sixteen times a night. But respectability emerged out of necessity, as the American family took to the road in the postwar years. It was in 1952 that two entrepreneurs opened a "Holiday Inn" in Memphis. Thereafter, motels popped up, like mushrooms, everywhere. To parents at the end of their tether because of tired, cranky, quarrelsome children in the back seat, the green Holiday Inn sign, coming into view far down the road at dusk after a long day's driving, was a desperately craved and infinitely welcome beacon of respite, relief, and even salvation. And throughout America, the whole family could find plenty of room at the motor inn, respectably equipped with televisions, individually wrapped bars of soap, "magic fingers" vibrator beds—and, out in the corridor, ice and soda pop machines.

People had to be fed, too, whether they were simply out for a drive in their own suburb or making a long-distance trip, so the nature of eateries changed, too. The nation's first drive-in restaurant, Royce Hailey's Pig Stand, had opened in Dallas in 1921. But it was not until 1948 that two brothers named McDonald fired the carhops at their restaurant in San Bernardino, California, sharply reduced the offerings on their menu, and introduced assembly-line-like food production. The new era of fast food, however, really began in earnest in 1954, when a milkshake machine salesman named Ray Kroc teamed up with the two

McDonald brothers. The following year, they opened the first of their new outlets, called McDonald's, in a suburb of Chicago, and yes, the rest was history.

America had become a drive-in society. In Orange County, California, it was possible to attend religious services sitting in your car at the "world's largest drive-in church." In Texas, you could sign up for your courses at a community college at the drive-in registration window. Movies flickered on the huge screens of drive-in theaters, dubbed "passion pits" by their teenage patrons. The annual model change in automobile showrooms in early autumn was a time of national celebration, when the entire populace seemed to pause reverentially to "ooh" and "aah" at Detroit's latest innovations—wrap-around bumpers, more chrome, or longer tail fins—the tail fins being proffered on the grounds that they were needed to contain the complex system of lights that were now to be found at the back of a car. Sometimes, the additional argument was made that "these elaborate fins had a stabilizing effect on the vehicle's motion." As one scholar noted, that "might well have been so if the car was airborne." But it was not. Still, it burned rubber on the ground, whizzing along at ever greater speeds, carrying its passengers to and from work and even serving as a mobile office—a boon to traveling salesmen. Ninety percent of American families took their vacations by car, and in 1964, some lucky motorist shoved into his glove compartment the five billionth road map provided free by an American gasoline station. Obtaining a learner's permit and then a driver's license became the major rites of passage for teenagers, their own "wheels" the most important symbol of their maturity and independence. The automobile was also absolutely central to dating, going steady, the acquisition of carnal knowledge, and the rituals of courtship. One survey in the late 1960s found that almost 40 percent of all marriages in America were proposed in a car.

The arteries and veins of this new way of life were the roads and highways. And here, as in so many other ways, public policy supported the requisite development. With a hike in the California gasoline tax in 1947, the building of the Los Angeles freeways began in earnest—including the all-critical downtown "interchange," which tied individual freeways into a single grand system. In the same year, across the country in New Jersey, Governor Alfred E. Driscoll revealed in his inaugural address a vision for the great society in his state—a turnpike stretching from one end of the Garden State to the other, which would bring an end to the congestion and permanent traffic jam that was threatening New Jersey in the postwar years, and would save cross-state motorists an hour and ten minutes. Nothing, Driscoll believed, was more important for New Jersey's future than a turnpike.

Construction started in 1949, exciting enormous enthusiasm in the state, where it was hailed as the "miracle turnpike" and "tomorrow's highway built today." There were no environmental impact studies in those days, no anti-development litigation, only the sense that in America you could get important things done quickly, and the whole job, from the first planning to the last toll booth, was accomplished in less than two years. The opening was celebrated with a breakfast to remember, every detail of which was personally overseen by none other than the grand maestro of America's highway menu, the restaurateur Howard Johnson himself.

The New Jersey Turnpike soon became the busiest toll road in the United States, and probably the world. Its landmarks were the rest areas—the Walt Whitman near exit 3, the Thomas Edison, the Dolley Madison, the Vince Lombardi, and the rest—and the orange-tiled roofs of Howard Johnson's restaurants. At the turnpike's opening, Governor Driscoll announced: "The Turnpike has permitted New Jersey to emerge from behind the billboards, the hot dog stands, and the junkyards. Motorists can now see the beauty of the real New Jersey." Few motorists would agree with that description. Other turnpikes were far more beautiful: the Merritt Parkway in Connecticut, the Taconic Parkway in New York. "It's difficult to obscure major features of the landscape altogether," one critic wrote, "but the [New Jersey] Turnpike manages it." It was built, however, for speed and convenience, not beauty, a functional monument to Hydrocarbon Man's urgent need to get expeditiously from one place to another. And it was only one short pathway in an ever-lengthening maze.

In 1919, Major Dwight D. Eisenhower had led his motorized military expedition across the United States with barely a road system to follow. It got him thinking about the motorways of the future. Thirty-seven years later, in 1956, President Dwight D. Eisenhower signed the Interstate Highway Bill, which provided for a 41,000-mile superhighway system (later raised to 42,500 miles) that would crisscross the nation. The Federal government would pay 90 percent of the cost, with most of the money coming from a specially designated, non-divertible highway trust fund accumulated out of gasoline taxes. The program was actively advocated and promoted by a broad coalition of interests that became known as the "highway lobby"—automobile makers, state governments, truckers, car dealers, oil companies, rubber companies, trade unions, real estate developers. There was even the American Parking Association; after all, no matter how great the distance covered, drivers would eventually have to come to the end of their trips—and park their cars.

Eisenhower himself advocated the interstate highway program on several grounds: safety, congestion, the many billions of dollars wasted because of inefficient road transport, and, evoking the darkest fears of the Cold War, the requirements of civil defense. "In case of atomic attack on our cities," he said, "the road net must permit quick evacuation of target areas." The resulting program was massive, and Eisenhower took great pride in the scale of the construction, using wondrous and mesmerizing comparisons. "The total pavement of the system would make a parking lot big enough to hold two third of all the automobiles in the United States," he said. "The amount of concrete poured to form these roadways would build eighty Hoover Dams or six sidewalks to the moon. To build them, bulldozers and shovels would move enough dirt and rock to bury all of Connecticut two feet deep. More than any single action by the government since the end of the war, this would change the face of America." His words were, if anything, an understatement. Meanwhile, public transport and the railroads would be the losers as Americans and American goods began to move in ever-larger streams along an endless ribbon of roads. If, in those expansive years, bigger was better, so was longer and wider.

Even in their living rooms, oil became part of the lives of Americans. Upwards of 60 million of them were entertained each week by a situation comedy

535

called *The Beverly Hillbillies*, which became an instantaneous hit when it took to the airwaves in 1962 and was the number-one-rated show for a couple of years. Many millions more watched it elsewhere in the world. It was the story of the Clampetts, an Ozark hillbilly family that struck it rich when an oil well hit in their front yard and that forthwith left Hooterville for a mansion in Beverly Hills. The joke was their naiveté and innocence of "big city ways." Viewers not only adored the show and the lovable oil multimillionaires but found that they couldn't get the theme song out of their heads:

> Come and listen to a story 'bout a man named Jed,
> A poor mountaineer, barely kept his family fed,
> Then one day he was shootin' at some food,
> When up through the ground come 'a bub-a-lin' crude,
> —oil that is, black gold, Texas tea.

The Beverly Hillbillies was a celebration of sorts. For, in truth, oil was not only "black gold" for the lucky Clampetts, but was also the "black gold" for consumers, enriching the industrial world by what it made possible. And yet there was a haunting question: How reliable was the flow of petroleum on which Hydrocarbon Man had come to depend? What were the risks?[7]

Crisis Again: "A Recurring Bad Dream"

Though Egypt's Gamal Abdel Nasser had no oil with which to assert his will, he had military force. He was intent on shoring up his prestige, which had declined in the Arab world in the 1960s. He wanted to avenge Israel's battlefield successes in 1956, and he reiterated his calls for the "liquidation" of Israel. His ultimate victory in 1956 made him overly confident of his luck. He was also being dragged along by Syria, which was sponsoring terrorist attacks on Israel, and he could not allow himself to be seen as insufficiently militant. In May 1967, Nasser ordered the United Nations observers, who had been on duty since the conclusion of the 1956 Suez Crisis, out of Egypt. He instituted a blockade against Israeli shipping in the Gulf of Aqaba, cutting off its southern port of Eilat and threatening to interrupt its ability to import petroleum, and he sent Egyptian troops marching back into the Sinai. King Hussein of Jordan put his armed forces under Egyptian command in case of conflict. Egypt began airlifting soldiers and military materiel into Jordan; and other Arab states were already sending, or planning to send, their own troops to Egypt. On June 4, Iraq adhered to the new Jordanian-Egyptian military agreement. To the Israelis, watching the mobilization of Arab military might all around them, the noose seemed to be growing very tight.

The next morning, June 5, at about eight o'clock, they responded by preempting and going on the offensive. The Third Arab-Israeli War, the Six-Day War, had begun. Gambling everything, Israel succeeded in the very first hours in catching on the ground the entire air forces of Egypt and the other belligerent states and quickly obliterated them. With mastery of the air thus assured, Israeli forces threw back the Arab armies. Indeed, insofar as Egypt and Jordan were concerned, the outcome of the Six-Day War was decided within three days. The Egyptian forces in the Sinai collapsed. By June 8, the Israeli Army had completely traversed the Sinai, destroying in the process 80 percent of all Egyptian materiel, according to Nasser himself, and had reached the eastern bank of the Suez Canal. Over the next few days, cease-fires were hastily put into place. But they left Israel in command of the Sinai, all of Jerusalem and the West Bank, and the Golan Heights.[8]

Among the Arabs, there had been talk for more than a decade about wielding the "oil weapon." Now was their chance. On June 6, the day after fighting began, Arab oil ministers formally called for an oil embargo against countries friendly to Israel. Saudi Arabia, Kuwait, Iraq, Libya, and Algeria thereupon banned shipments to the United States, Britain, and, to a lesser degree, West Germany. "In compliance with the decision of the Council of Ministers taken in the session held last night," Ahmed Zaki Yamani informed the Aramco companies on June 7, "you are requested hereby not to ship oil to the United States of America or the United Kingdom. You should see that this is strictly implemented and your company will be gravely responsible if any drop of our oil reaches the land of the said two states."

Why would the oil-exporting countries deliberately cut off their own major source of revenues? For some, the decision was influenced by disturbances within their own borders—strikes by oil field workers, riots, sabotage—and by their fear of the ability of even a politically crippled Nasser to fire up the masses and street mobs over transistor radios. The worst disturbances were in Libya, where foreign oil company offices and personnel were assaulted by mobs; and a huge evacuation program, with planes leaving Wheelus every half hour, was quickly instituted for the Western oil workers and their families. Production was also interrupted by strikes and sabotage in Saudi Arabia and Kuwait.

By June 8, the flow of Arab oil had been reduced by 60 percent. Saudi Arabia and Libya were completely shut down. The huge Iranian refinery at Abadan was closed because Iraqi ship pilots were refusing to work in the Shatt-al-Arab waterway. The overall initial loss of Middle Eastern oil was six million barrels per day. Moreover, logistics were in total chaos not only because of the interruptions but also because, as in 1956, the Suez Canal and the pipelines from Iraq and Saudi Arabia to the Mediterranean were closed. "The crisis is more serious than at the time of the Suez blockage in 1956–57," said a United States Assistant Secretary of the Interior on June 27. "At that time no major producer except northern Iraq was closed and the problem was exclusively one of transportation. Now . . . three-quarters of [Western Europe's oil] comes from the Arab regions of the Middle East and North Africa, one-half of which is now out of production. Europe is, therefore, facing an immediate petroleum shortage of critical proportions."[9]

The situation grew more threatening in late June and early July when, coincidentally, civil war broke out in Nigeria. That country's Eastern Region, in which newly developed significant oil production was concentrated, wanted a bigger share of government oil revenues. The Nigerian government said no. Beneath the struggle over oil revenues were deep-seated ethnic and religious conflicts. The Eastern Region, calling itself Biafra, seceded, and the Nigerian government instituted a blockade against oil exports. The resulting conflict removed another 500,000 barrels per day from the world market at a critical moment.

Partly because of the overwhelming concentration on Vietnam, American policymaking on the Six-Day War took on so ad hoc a quality that it became known to participants as "the floating crap game." In an effort to coordinate policy better, President Johnson established a special Ex-Com, chaired by McGeorge Bundy and modeled on the Ex-Com that John Kennedy had used during the Cuban Missile Crisis—and known thereafter as the "Unknown Ex-Com." Bundy's committee devoted much of its time to considering the implications of the closure of the Suez Canal. Meanwhile, the oil companies were compelled to take hasty and drastic action. The Interior Department in Washington, reverting once again to authorization dating from the Korean War, activated a Foreign Petroleum Supply Committee, composed of about two dozen American oil companies. If necessary, the antitrust laws could have been suspended so that the companies could jointly manage logistics and institute another oil lift to Europe. This was the same committee that had been called into action during the Iranian nationalization crisis of 1951–53 and again during the Suez Crisis in 1956–57. Said a lawyer who became, as he had in the previous crises, an adviser to the committee, "It's like a recurring bad dream."

The working assumption was that the oil committee of the Organization for Economic Cooperation and Development, representing the industrial countries, would, in the event of a crisis, declare an emergency and implement a "Suez system," as in 1956, and coordinate the overall allocation among the Western countries. Yet when the United States requested such a step, many OECD countries, confident that they would be able to make their own supply arrangements, resisted. American officials were shocked. Without an OECD resolution to the effect that an emergency was at hand, the Justice Department would not grant the antitrust waiver necessary for American companies to cooperate with each other. Only when the United States warned that, without an OECD statement, American companies would not share information (and, by implication, oil) with foreign companies did the OECD unanimously, though with abstentions by France, Germany, and Turkey, approve a motion that the "threat of an emergency" existed, so allowing both American and international coordination measures to be put into effect.

The major problem once again turned out to be tankers and logistics. The normal flow of oil had to be massively reorganized. Petroleum from non-Arab sources was diverted to the embargoed countries (or, in the case of the United States, transported from the Gulf Coast to the East Coast), while the Arab oil originally destined for the United States, Britain, and Germany was sent elsewhere. The closure of the Suez Canal and the Mediterranean pipelines meant, as in 1956, much longer journeys around the Cape of Good Hope and thus resulted in a mad scramble for tankers. BP found the job of reorganizing transportation so complex that it gave up using computers—it could not write the programs quickly enough—and went back to pencil and paper. Yet the requirements of the much-longer voyages could be more easily met than was expected owing to the development of "supertankers," an innovation spurred by the 1956 Suez crisis.

By 1967, a mere eleven years after that crisis, supertankers five times larger than the tankers of 1956 were available. And six Japanese-built supertankers, each 300,000 deadweight-tons, seven times larger than the standard 1956 tanker, were poised to go into service, shuttling between the Persian Gulf and Europe.

Despite the high anxiety and uncertainty, the problems proved less severe than might have been expected. The domestic scene in the Arab countries calmed down, and once the Arab exporters got their production back into operation, the maximum loss was about 1.5 million barrels per day—the amount of Arab oil that normally went to the three embargoed countries, the United States, Britain, and Germany. The missing 1.5 million barrels could be made up in the very short term by drawing on the high stock levels and, over time, by additional production elsewhere. Seven years earlier, in 1960, the U.S. National Security Council had described American shut-in production as "Europe's principal safety factor in the event of denial of Middle East oil." That hypothesis was borne out in 1967. The national security arguments by some in the American government—and by the Texas independents—in favor of prorationing were vindicated; America had a large reserve capacity of shut-in oil that could be quickly called into production (though the reserve may not have been as large as was publicly claimed). With dispensation from the Texas Railroad Commission and the corresponding agencies in other states, American output surged by almost a million barrels per day. Venezuela's output increased by over 400,000 barrels per day, and Iran's by about 200,000. Indonesia also stepped up its production.[10]

By July 1967, a mere month after the Six-Day War, it was clear that the "Arab oil weapon" and the "selective embargo" were a failure; supplies were being redistributed to where they were needed. The Foreign Petroleum Supply Committee stuck to an informational and advisory role; the formal emergency machinery for joint operations and antitrust exemptions never needed to be implemented. The international companies themselves, working individually, had managed to handle the situation.

The biggest losers turned out to be the countries that instituted the embargoes. They were giving up substantial revenues to no obvious effect. Moreover, they were being called upon to pick up the bill and provide large, continuing subsidies to Egypt and the other "front-line" Arab states. Zaki Yamani began to question publicly the value of the embargo under such circumstances. Not everyone agreed. Iraq called for a complete three-month embargo on all shipments of oil to all customers, to teach the West a lesson. But Iraq found no takers among its Arab brethren. At an Arab summit meeting in Khartoum in late August 1967, Nasser, who had left 150 senior officers under arrest in Cairo to forestall a coup, admitted that his country was totally broke and desperately needed money. The assembled leaders concluded that pumping oil, and earning oil revenues, was the right thing to do; such represented an assertion of "positive" Arab strategy. By the beginning of September, the embargo on exports to the United States, Britain, and Germany had been lifted.

At this point, the risk of any shortage had disappeared. Even in August, while still observing the selective embargo, Arab oil producers had boosted their overall output to make up for lost volume and to hold on to overall market share.

As a result, total Arab production was actually 8 percent higher in August than it had been in May, before the Six-Day War! The increases in Arab output alone were double what had been lost because of the Nigerian civil war.

Though this latest disruption had been dealt with rather easily, it could have been far more severe and difficult had overall production in the various exporting countries continued to be interrupted, whether by decision or by political unrest, or had it taken place under different market conditions. The U.S. Department of the Interior, in its report on the management of the crisis, drew two lessons: the importance of diversifying sources of supply and of maintaining "a large, flexible tanker fleet." In the aftermath of the crisis, the Shah, always eager for higher production, came up with an ingenious notion that he thought would appeal to policymakers in Washington and win their backing in his continuing struggle with the oil companies. Iran, he said, should obtain a special American import quota for oil that would be stockpiled as a strategic reserve in old salt mines. This would give the United States greater security and supply flexibility, and would also give him a new market outlet. But it would take another oil crisis before the sensible idea of a stockpile was acted upon.

It was clear by the autumn of 1967 that available supplies would, at least in the short term, actually exceed demand as a result of the worldwide surge in production following the Six-Day War. In October, a lead story in the *Wall Street Journal* was headlined, "Shortage Fears Raised by Mideast War Yield to Threat of New Glut." *Oil and Gas Journal* was already warning about a new crisis—"over-supply." Executives in the industry no longer worried about availability of supply but instead recalled how the response to the 1956 Suez crisis had intensified the glut in the late 1950s, resulting in the imposition of U.S. import quotas, cuts in the posted price—and the birth of OPEC. Once again the pendulum appeared to be swinging in the all-too-familiar progression from shortage to glut.[11]

The Cassandra at the Coal Board

The outcome of the Six-Day War seemed to confirm how secure the supply of oil was. And Hydrocarbon Man continued to take his petroleum for granted. It defined and motivated his life, but because it was so pervasive, and so readily available, he hardly thought about it. After all, the oil was there, it was endlessly abundant, and it was cheap. It flowed like water. The surplus had lasted for almost twenty years, and the general view was that it would continue indefinitely, a permanent condition. That was certainly the way things looked to most people within the oil industry. "The 'over-hang' of surplus crude avails is very large," said a study by Standard Oil of California (Chevron) in late 1968. "Pressures will exist to continue to produce in many areas in excess of market requirements." If consumers gave the matter any consideration at all, they too would have expected cheap oil to continue as virtually a birthright, rather than the product of certain circumstances that could change; and their main concern would have been nothing more than whether to drive a couple of extra blocks to save two cents a gallon during a price war.

There were mavericks, skeptics, who questioned and said unfashionable things, but they were few. One was a German-born economist, E. F. Schu-macher, who had studied at Oxford as a German Rhodes Scholar, and then at Co-lumbia University, and then emigrated permanently to England in the late 1930s. A sometime writer for the *Economist* and the *Times* of London, he became in 1950 the economic adviser to the National Coal Board, which controlled the in-dustry that Britain had nationalized after the war. It was a position he would hold in virtual anonymity for two decades. But Fritz Schumacher had a fertile, broad-ranging mind. He became fascinated with Buddhism, and he investigated what he called "intermediate technologies" for developing countries as an alternative to the high-cost, showcase industrial projects that were copied from the West.

As economic adviser to the Coal Board, Schumacher also had a specific agenda to defend. He was charged with providing the intellectual fodder for the coal industry in its great struggle against oil for market share. He was one of the strongest minds on what turned out to be—inevitably, it seemed—the losing side of that battle, and it was with bitterness and regret that he observed coal so unceremoniously deposed as "the universal aid." Later, he would be much cele-brated by environmentalists, yet he was defending coal, a dirtier fuel, against oil. But his focus was on the problem of depletion, not on the effects of combustion, which would much concern his followers two decades later.

"There is no substitute for energy," Schumacher said in 1964, echoing Jevons, the nineteenth-century economist and celebrator of coal. "The whole ed-ifice of modern life is built upon it. Although energy can be bought and sold like any other commodity, it is not 'just another commodity,' but the precondition of all commodities, a basic factor equally with air, water and earth." Schumacher argued vigorously for the use of coal to supply the world's energy needs. Oil, he believed, was a finite resource that should not be used wantonly. He also thought that it would not always be cheap, as reserves dwindled and exporters sought to capture a larger and larger share of the rents. More specifically, he warned about dependence on Middle Eastern oil. "The richest and cheapest reserves are lo-cated in some of the world's most unstable countries," he wrote. "Faced with such uncertainty, it is tempting to abandon the quest for a long-term view and simply to hope for the best."

In an age of optimism, Schumacher's own long-term view was glum. He ex-pressed the risks in economic terms. With the fast consumption growth rates and low prices, he warned, "the world's oil supply will not be ensured for the next twenty years, certainly not at current prices." Once he was even tempted to put his warning in more metaphysical terms. Citing an eminent Oxford economics professor, he declared that "the twilight of the fuel gods will be upon us in the not very distant future."

But his was a voice in the wilderness. There still was a huge oil surplus, and Schumacher continued to issue his jeremiads to an indifferent and uninterested public. In 1970, discouraged and figuring that he had done all he could in his bat-tle against oil, he retired from the Coal Board. He had scored little with his argu-ments; indeed, his years at the Coal Board coincided almost exactly with the two decades in which petroleum had mercilessly dethroned Old King Coal and as-sumed suzerainty over industrial society. "The chickens are about to come home to roost," Schumacher took to saying with resignation. At that point, he seemed an irritable, crankish killjoy, who had refused to enjoy the party. But he would soon publish a book that would challenge the precepts of the Hydrocarbon Age and the very foundations of the enthrallment of "Bigger is Better." And before very long, events would make him look less like a killjoy and more like a prophet.[12]

PART V

THE BATTLE FOR WORLD MASTERY

CHAPTER 28

The Hinge Years: Countries Versus Companies

PERSEPOLIS, THE CAPITAL of the ancient Persian empire, was sacked by Alexander the Great in 330 B.C. and left for over two millennia in deserted ruins. In October 1971 it was brought back to dazzling life. Three huge tents and fifty-nine smaller ones were erected on the desolate site. The occasion was what *Time* magazine called "one of the biggest bashes in all history," put on by the Shah of Iran to celebrate the founding of the Persian empire twenty-five hundred years before. The assembled dignitaries included the President of the Soviet Union, the Vice-President of the United States, Marshal Tito of Yugoslavia, twenty kings and sheikhs, five queens, twenty-one princes and princesses, fourteen other presidents, three other vice-presidents, three premiers, and two foreign ministers. In the course of the ceremonies, the Shah publicly communed with the ghost of the king Cyrus the Great, the empire's founder, and promised to continue the tradition and works of that monarch, now dead for some twenty-five centuries. And the jeweled and medal-bedecked guests were taken in buses up to the hill above Persepolis for a stunning *son et lumière* under the stars that dramatized, strangely enough, Alexander's destruction of Persepolis.

In anticipation of the Persepolis celebration, the Iranian government had urgently sought top-secret advice from Britain on a most critical matter of high diplomacy: how to handle the seating plan with so many VIPs in attendance. Likelihood of offense to various potentates was great. The protocol section in the Foreign Office in London came up with an innovative plan based on a specially constructed table with undulating curves so that no one would be too far from a prominent member of the Pahlavi family.

As signal proof of his grandeur, the Shah had invited Queen Elizabeth II to attend his party. But Her Majesty's ambassador in Tehran had the unhappy task of explaining that the Queen was already committed to a state visit elsewhere.

The "elsewhere," however, happened to be neighboring Turkey, which could not but aggravate the Shah. He then asked for Prince Charles. Sorry, Charles was not available; he was on naval duty on a frigate in the North Sea. Never mind that Persepolis was not just another party, but a once-in-a-twenty-five-hundred-year celebration—and that the Shah was, among other things, in the process of ordering several hundred British-built Chieftain tanks, which happened to be critically important to Britain's balance of payments. London offered him Prince Philip and Princess Anne. The Shah accepted, but he was not exactly placated.

The event was catered by Maxim's of Paris. The menus, created and carried by 165 chefs, bakers, and waiters, all flown in from Paris, were wondrous. To accompany the meals, twenty-five thousand bottles of wine were also flown in from France. (Since the whole affair had so strong a French accent, France's President, Georges Pompidou, was notable by his absence. "If I did go," he explained privately just beforehand, "they would probably make me the headwaiter.") The cost of the pageantry and celebrations was estimated at somewhere between $100 and $200 million. When some questioned such extravagance, the Shah could not contain his irritation. "So what are people complaining about?" he asked. "That we are giving a couple of banquets for some 50 Heads of State? We can hardly offer them bread and radishes, can we? Thank heavens, the Imperial Court of Iran can still afford to pay for Maxim's services."

After Persepolis, the British, in an effort to mollify the Shah and reduce various tensions between the two countries, invited him to spend Royal Ascot Weekend with the Royal Family at Windsor Castle. The visit proved to be a great success. The only hitch emerged when the Shah was going to go riding *à deux* with the Queen. Some hours before they were to mount up, it was realized with horror that the Shah, as an Iranian male, would not ride a mare or gelding, only a stallion. But no stallion was available. Then, just as despair was settling over the British side, the Queen remembered that Princess Anne had a stallion. But it was noted, with new horror, that the horse was named Cossack. The Shah, of course, was the son of an officer of the Cossack Brigade, who had taken power in the 1920s. Given the Shah's sensitivity about his father, and Britain's role in deposing him, as well as his general suspicion of the British, he might well take the proffering of such a steed as a new and blatant insult and, indeed, a gross putdown. The Shah did ride Cossack, though the horse's name was successfully kept from him, and the ride and the rest of the weekend went swimmingly. Queen Elizabeth and Prince Philip and the Shah and the Empress rode around the racecourse at Ascot in an open coach, and thereafter the Shah was to write to the Queen as "My Dear Sovereign Cousin." Britain was back in favor.

The overall aim of the Shah's grand celebration at Persepolis was to establish himself firmly in the line of Cyrus the Great, the Appointed One of God. His visit to the Queen enhanced his stature as every bit her royal equal. He was no longer a puppet, a pawn, a mere appointee to his throne. He was now a man of enormous wealth, power—and pride—who was stepping into pivotal new roles in the Middle East and on the international stage.[1]

The Anglo-American Retreat

The postwar petroleum order in the Middle East had been developed and sustained under American-British ascendancy. By the latter half of the 1960s, the power of both nations was in political recession, and that meant the political basis for the petroleum order was also weakening. The United States had been mired in Vietnam for several years in a costly, unpopular, and ultimately unsuccessful war. At the same time, anti-Americanism had become a great fashion and a great industry throughout much of the world, organized around denunciations of imperialism, neocolonialism, and economic exploitation. Americans themselves were deeply divided not only by the Vietnam War but also by arguments over the "lessons of Vietnam," which had to do with the extent and character of America's global role. For some in the developing world, however, the lessons of Vietnam were quite different: that the dangers and costs of challenging the United States were less than they had been in the past, certainly nowhere near as high as they had been for Mossadegh, while the gains could be considerable.

The United States was a newcomer to the Middle East compared to Britain, which had been involved in the Persian Gulf since the early 1800s, when it first began to put down the pirates plying those waters and established truces in the chronic maritime wars among the sheikhs along the Arabian side of the Gulf. In exchange, the British took on the responsibility of keeping the peace with agreements that evolved into guarantees to protect the independence and integrity of these various "trucial" states. In the late nineteenth and early twentieth centuries, similar treaties and understandings were extended to Bahrain, Kuwait, and Qatar. But Britain in the 1960s was a country preoccupied with its economic decline, which had combined with politics, both domestic and international, to make the liquidation of empire the central drama for Britain in the postwar world. Great Britain bowed out of the port city of Aden, at the southern tip of the Arabian peninsula. Entirely a British creation, Aden was strategically located on the oil routes from the Persian Gulf and was one of the busiest ports in the world. Now it was in anarchy. As the British governor departed, a military band played "Fings ain't wot they used to be." And, indeed, they were not. With the British withdrawal, Aden disappeared into the harsh Marxist-Leninist state of South Yemen. Then, at the beginning of January 1968, in response to a balance of payments crisis, Prime Minister Harold Wilson announced that Britain would end its defense commitments east of Suez. It would completely withdraw its military presence from the Persian Gulf by 1971, thus eliminating the last major remnant of the great Pax Britannica of the nineteenth century and of the British Raj.

The sheikhs and other rulers in the Persian Gulf were dumbfounded by the decision of the Wilson government; only three months earlier they had all been reassured by the Foreign Office that Britain had no intention of leaving the Gulf. The sheikhs begged the British to maintain their presence. "Who asked them to leave?" asked the ruler of Dubai. The Amir of Bahrain was blunter. "Britain could do with another Winston Churchill," he said. "Britain is weak now where she was once so strong. You know we and everybody in the Gulf would have welcomed her staying."

The British position in the Gulf actually involved only about six thousand ground troops, plus air support units, at a non-sterling cost of 12 million pounds a year. That might have been seen as a rather small amount, an insurance premium, given the vast investment by British oil companies in the region, generating both corporate earnings that had a very positive impact on the British balance of payments and very large revenues for the government's exchequer. Some of the sheikhs said they would be glad to put up the 12 million pounds themselves in order to keep British forces in the area. The offer was angrily rejected. Defense Secretary Denis Healey derided the notion of the British becoming "mercenaries for people who like to have British troops around." But, as some pointed out, such offset payments were accepted for garrisoning British troops in West Germany or Hong Kong. But it was not only economic necessity that motivated Healey; the growth of nationalism had persuaded him that it would be "politically unwise" to maintain a military presence in the Middle East.

The British did help set up a federation, the United Arab Emirates, to bind together a number of the small states, which, it was hoped, would afford them some measure of protection. With that accomplished, the British packed up and left the Gulf in November 1971. Their departure marked the most fundamental change in the Gulf since World War II and meant the end of the security system that had operated in the area for over a century. It left behind a dangerous power vacuum in a region that supplied 32 percent of the free world's petroleum and that, at the time, held 58 percent of the proven oil reserves.

The Shah of Iran, as he had demonstrated the previous month with the great celebration at Persepolis, was eager to fill the vacuum. "The safety of the Persian Gulf had," he said, "to be guaranteed, and who but Iran could fulfill this function?" The Americans were not happy to see the British go. But if not the British, there was the Shah. This was, after all, the era of the Nixon Doctrine, which attempted to deal with the new political and economic constraints on American power by depending upon strong, friendly local powers as regional policemen. No one seemed better fitted to play that role than the Shah. Nixon himself had developed a high regard for the Shah, whom he first met in 1953, a few months after the Shah had regained his throne. "The Shah is beginning to have more guts," he had told President Eisenhower then. "If the Shah would lead, things would be better." When Nixon lost the California gubernatorial election in 1962, he set out on a round-the-world trip. The Shah had been one of the few heads of state to receive him cordially. Nixon never forgot that show of respect when he was down. Now, in the early 1970s, the Shah was intent on leading, not only in Iran but throughout the region, and the Nixon Administration supported him. Though the fact often went unrecognized, there was no other obvious choice. Soviet arms were flowing in large volumes into neighboring Iraq, which had its own long-held ambitions for hegemony over the Gulf and its oil. From here on, a very different security system would reign over the Gulf.[2]

The End of the Twenty-Year Surplus: To a Seller's Market

The 1970s also saw a dramatic shift in world oil. Demand was catching up with available supply, and the twenty-year surplus was over. As a result, the world was rapidly becoming more dependent on the Middle East and North Africa for its petroleum. The late 1960s and early 1970s were, for the most part, years of high economic growth for the industrial world and, in some years, outright boom. This growth was fueled by oil. Free world petroleum demand rose from almost 19 million barrels per day in 1960 to more than 44 million barrels per day in 1972. Oil consumption surged beyond expectation around the world, as ever-greater amounts of petroleum products were burned in factories, power plants, homes, and cars. In America, gasoline use increased not only because people were driving more miles but also because cars were getting heavier and were carrying more "extras," such as air-conditioning. The cheap price of oil in the 1960s and early 1970s meant that there was no incentive for fuel-efficient automobiles.

The late 1960s and early 1970s were the watershed years for the domestic U.S. oil industry. The United States ran out of surplus capacity. For decades, going back to Dad Joiner, the East Texas field, and Harold Ickes, production had been regulated by the Texas Railroad Commission, the Oklahoma Corporations Commission, the Louisiana Conservation Commission, and similar bodies in other states. They had prorationed output, keeping actual production well below capacity in order to promote conservation, and maintained prices in a situation of chronic potential oversupply. The inadvertent result of their work had been to provide the United States and the entire Western world with a security reserve, a surge capacity, that could be called upon in time of crisis—whether of the extended magnitude of World War II or the much more limited crises of 1951, 1956, and 1967.

But the need to restrain production was erased by rising demand, low investment because of low prices, and relatively low discovery rates, along with the import quotas. Now there was an eager buyer for every barrel of oil that could be produced in the United States. In the period 1957 to 1963, surplus capacity in the United States had totaled about 4 million barrels per day. By 1970, only a million barrels per day remained, and even that number may have been overstated. That was the year, too, that American oil production reached 11.3 million barrels per day. That was the peak, the highest level it would ever reach. From then on, it began its decline. In March 1971, for the first time in a quarter century, the Texas Railroad Commission allowed all-out production at 100 percent of capacity. "We feel this to be an historic occasion," declared the chairman of the commission. "Damned historic, and a sad one. Texas oil fields have been like a reliable old warrior that could rise to the task when needed. That old warrior can't rise anymore." With consumption continuing to rise, the United States had to turn to the world oil market to satisfy the demand. The quotas, originally established by Eisenhower, were eased, and net imports rose rapidly from 2.2 million barrels per day in 1967 to six million barrels per day by 1973. Imports as a share of total oil consumption over the same years rose from 19 percent to 36 percent.

The disappearance of surplus capacity in the United States would have major implications, for it meant that the "security margin" upon which the Western world had depended was gone. In November 1968, the State Department had told the European governments at an OECD meeting in Paris that American production would soon reach the limits of capacity. In the event of an emergency, there would then be no security cushion; the United States would not be able to provide stand-by supply. The other participants at the meeting were taken by surprise. This was only one year after the 1967 embargo effort by OPEC, and the Middle East was manifestly no more secure.

Indeed, the razor's edge was the ever-increasing reliance upon the oil of the Mideast. New production had come on from Indonesia and Nigeria (in the latter case, after the end of its civil war in early 1970), but that output was dwarfed by the growth in Middle Eastern production. Between 1960 and 1970, free world oil demand had grown by 21 million barrels per day. During that same period, production in the Middle East (including North Africa) had grown by 13 million barrels per day. In other words, two-thirds of the huge increase in oil consumption was being satisfied by the wells in the Middle East.[3]

Environmental Impact

Another significant shift was taking place in the industrial countries. Man's view of the environment and his relationship to it was also changing, with the paradoxical effect of both increasing the demand for oil and regulating its use. Beginning in the mid-1960s, environmental issues began to compete successfully for their place in the political process, in the United States and elsewhere. Air pollution prompted utilities around the world to shift from coal to less-polluting oil, adding another major stimulus to demand. In 1965, New York's mayor pledged to banish coal from the city. An air pollution crisis hit New York City on Thanksgiving Day, 1966; smog gripped the city, and coal burning was restricted. Within two years, Consolidated Edison, the utility serving New York City, switched to oil. In 1967, a clean air bill passed the United States Senate by a vote of eighty-eight to three. In 1970, Federal legislation was enacted that established what became known as environmental impact statements: The possible consequences for the environment of major new projects had to be projected and taken into account before a go-ahead would be given. That same year, one hundred thousand people paraded down Fifth Avenue in New York City to mark Earth Day.

Nothing else so much reflected the new environmental consciousness as the extraordinarily wide and intense public response to *The Limits to Growth: A Report for the Club of Rome's Project on "The Predicament of Mankind."* Published in 1972, the book argued that if several basic world trends—in population, industrialization, pollution, food production, energy consumption, and resource depletion (including oil and natural gas)—continued unabated, they would make contemporary industrial civilization unsustainable and "the limits to growth on this planet will be reached sometime within the next hundred years." The study warned not only of resource depletion, but also of the environmental consequences of hydrocarbon burning, the buildup of carbon dioxide in the atmosphere and a new concern about global warming. It was a general caution: The timing of future crises was highly uncertain.

The study itself was released at a critical juncture. A worldwide economic boom, with high inflation and an even higher growth in resource use, was taking place at the same time that American oil reserves were declining and both American imports and worldwide energy use were rising dramatically. Moreover, the new environmental consciousness was beginning to reconfigure public policy in the industrial world and to force changes in corporate strategies. It meant, in the words of a Sun Oil executive, a "new game" for energy companies. *Limits to Growth* became a lodestar in the debates over energy and the environment. Its arguments were a potent element in the fear and pessimism about impending shortages and resource constraints that became so pervasive in the 1970s, shaping the policies and responses of both oil-importing and oil-exporting countries.

The impact of environmentalism on the energy balance was manifold. The retreat from coal was accelerated, and reliance on cleaner-burning oil grew. Nuclear power was bruited as an environmental improvement over the combustion of hydrocarbons. Efforts were accelerated to search for new sources of oil, and toward the end of the 1960s, hopes grew for major production offshore of California. There, after all, before the end of the nineteenth century, the very first drilling in water had taken place from piers near Santa Barbara. Now, more than seventy years later, rigs were being positioned along the scenic Southern California coastline. But then, in January 1969, the drilling of an offshore well in the Santa Barbara channel encountered an unexpected geological anomaly, and as a result, an estimated six thousand barrels of oil seeped out of an uncharted fissure and bubbled to the surface. A gooey slick of heavy crude oil flowed unchecked into the coastal waters and washed up on thirty miles of beaches. The public outcry was nationwide and reached right across the political spectrum. The Nixon Administration imposed a moratorium on California offshore development, in effect shutting it down. However great the need for oil, the leak increased opposition to energy development in other environmentally sensitive areas, including the most promising area in all of North America, the one most likely to stem the decline in American production and counterbalance rising dependence on the Middle East—Alaska.[4]

The Alaskan Elephant

As early as 1923, President Warren Harding had created a naval petroleum reserve on the Arctic coast of Alaska, and wildcatters poked around the region in the years thereafter. Following the 1956 Suez Crisis, Shell and Standard Oil of New Jersey began exploring in Alaska, but in 1959, after drilling what proved to be the most expensive dry hole up to that time, they suspended operations.

Another interested company was British Petroleum. In the aftermath of Mossadegh in Iran, and then the Suez Crisis, BP was dead keen on reducing its virtually total dependence on the Middle East. In 1957, the year after Suez, it made the strategic decision to seek diversified sources of production, particularly in the Western Hemisphere. In this it was strongly supported by the British government. "The British oil companies are well aware of the insecurity of their hold on oil supplies from the Middle East, on which they mainly rely to sustain their business in Western Europe and indeed throughout the Eastern Hemisphere," Prime Minister Harold Macmillan wrote privately to the Australian Prime Minister, Robert Menzies, in 1958. "They also know that the United

Kingdom Government for political and economic reasons, would welcome any action they can take to reduce their dependence on the Middle East. The British Petroleum Company in particular has its own commercial reasons to broaden the base of its supply: it was worse hit by the Suez crisis than any of the other major international oil supplying companies and within the resources it commands it is trying to cut down its vulnerability to a stoppage of supplies from the Middle East."

Sinclair Oil offered BP a nostrum to relieve its dependence on the Middle East—joint exploration in Alaska. But, after drilling six expensive dry holes in succession on the North Slope in the frigid far north of Alaska, the two companies suspended the effort. Gulf Oil also showed some interest in Alaska. Some of its explorationists argued valiantly that, despite the dry holes, the geology was promising and the company ought to try its hand at exploration on the North Slope. The top management absolutely refused even to consider the request. "It would cost $5 a barrel," one senior executive flatly declared, "and oil will never get to $5 a barrel in our lifetime."

Yet another company was still investigating Alaska—Richfield, a California-based independent. It was particularly attracted by the thick marine sediments in the virtually inaccessible North Slope. In 1964, Jersey decided to reenter Alaska, and for payments and commitments totaling a little over $5 million, its Humble subsidiary became Richfield's partner. In 1965, this new joint venture won about two-thirds of the exploration leases on the North Slope's Prudhoe Bay structure. The BP-Sinclair combination was the other main winner.

That same year, Richfield merged with Atlantic Refining, forming Atlantic Richfield, which later became ARCO. The combined company was headed by Robert O. Anderson. Though Anderson would often strike other people as surprisingly relaxed, almost casual, perhaps even a little absent-minded, he had the determination and concentration that befit a man who was one of the last of the great wildcatters and oil tycoons of the twentieth century. Anderson was the son of a Chicago banker who had made a speciality during the 1930s of lending prudently to Texas and Oklahoma independent oil men at a time when others wouldn't lend at all. The young Anderson grew up around the University of Chicago, attended it during the heyday of the Great Books curriculum, and gave some thought to becoming a philosophy professor. But the oil men who were his father's clients captured his imagination much more than the academics he saw on campus, and in 1942 he went down to New Mexico to take over a fifteen-hundred-barrel-per-day refinery. He soon moved over to exploration, and became one of the better-known independents in the business. He had the same gift for quick mental arithmetic as Rockefeller and Deterding. In the early days, he could beat a slide rule and, later on, a pocket calculator, and he had a habit in meetings of correcting people on their decimal points. "I was never particularly conscious of the ability," he explained. "The biggest thing it does is to help you discard a lot of things and move on. In negotiations, you can casually allow something that the other guy doesn't understand the importance of. You stay ahead of the curve."

Over the years, Anderson would prove to be a man of wide and diverse interests, consistently an intellectual maverick in the oil industry. Drawn to ideas, comfortable with social science professors, curious about such things as values and governance and social change, he liked seminars where businessmen discussed such varied subjects as technology and humanism, the environment, and Aristotle. In short, despite many successes, he never exactly fit the mode of the typical oil tycoon. He believed in many things that were quite anomalous among his peers. Yet, at heart, he was the quintessential wildcatter, and in nothing else did he believe so fervently as crude oil and reserves in the ground—"the absolute heart of the industry," he would say. "Over and over, the lesson in this business is that if you can't take disappointment, you ought not to be in this business, since 90 percent of what you drill are failures. You really have to take defeat regularly." Still, the other 10 percent would prove very good to Anderson, not only making him very wealthy but also, among other things, enabling him to end up the largest individual landowner in the United States.

But in the winter of 1966 it looked as if Alaska might well end up in the 90 percent failure column. ARCO, with Humble's participation, drilled a costly well sixty miles south of Alaska's north coast. It was dry. One more wildcat was on the schedule, at Prudhoe Bay on the North Slope. There was considerable doubt about whether to continue. It was up to Anderson. It would be his decision. He believed in exploration, he believed in crude oil. But ARCO's own dry hole had come on top of the six dry holes of BP and Sinclair, and he wasn't in the oil business to lose money. He gave the okay, though without a great deal of conviction. It was just that the drilling rig was already in Alaska, and it only had to be moved sixty miles. "It was more a decision not to cancel a well already scheduled than to go ahead," he later said.

In the spring of 1967, the ARCO-Humble venture began drilling what would certainly be the last wildcat if it failed. The well was dubbed Prudhoe Bay State Number 1. On December 26, 1967, a loud, vibrating sound drew a crowd of about forty men to the well. They were wrapped in heavy clothes—it was thirty degrees below zero—and they had to struggle to hold their places in the thirty-knot wind. The noise grew louder and louder—the roar of natural gas. To one geologist it sounded like four jumbo jets flying directly overhead. A natural gas flare from a pipe shot defiantly thirty feet straight up in the strong wind. They had struck oil. In mid-1968, a "step-out well," drilled seven miles from the discovery well, confirmed that this was a great structure, a world-class oil field. A true elephant. The petroleum engineering firm DeGolyer and McNaughton estimated that Prudhoe Bay might hold as much as 10 billion barrels of recoverable reserves. However grudgingly Anderson had given the go-ahead, it was the most important decision he would ever make as an oil man. Prudhoe Bay was the largest oil field ever discovered in North America, one-and-a-half times larger than Dad Joiner's East Texas field, which had destroyed the price of oil in the early 1930s.

In the tightening world oil market, Prudhoe Bay was not going to destroy any price structures, but it did have the potential to slow the growth of American oil imports greatly and to reduce dramatically the tautness in the global oil balance. Estimates suggested that total output would quickly be upwards of two million barrels per day, making it the third-largest producing field in the world, after Saudi Arabia's Ghawar and Kuwait's Burgan. Initially, ARCO and Jersey, along with Jersey's Humble subsidiary, thought that the field would be in operation within three years. Its development seemed likely to be expedited when the management structure on the North Slope was simplified; ARCO acquired Sinclair, snatching it just in time from the jaws of the conglomerate Gulf & Western, in what was the largest merger to that date in the United States. Now the Big Three on the North Slope were ARCO, Jersey, and BP. The merger also made ARCO the seventh-largest oil company in the United States.

A great obstacle to development was the physical environment in the isolated north: inaccessible, extreme in weather, harsh, and fiercely hostile—"a mean, nasty, unforgiving place to work," said one geologist. It was a place unlike any other from which oil had yet been recovered. The technology did not exist for production in such an environment. The tundra, a few feet in thickness, froze concrete-hard in winter as temperatures fell to as much as sixty-five degrees below zero. It then thawed into a spongy prairie land in summer. There were no roads across the tundra, and beneath was the permafrost, that part of the soil that was permanently frozen, extending sometimes a thousand feet in depth. Normal steel piling would crumble like soda straws when driven into the permafrost.

If that obstacle could be overcome, there would still be the daunting challenge of getting the oil to market under very difficult conditions. Icebreaker tankers that would travel through the frozen Arctic seas to the Atlantic were seriously considered. Other suggestions included a monorail or a fleet of trucks in permanent circulation on an eight-lane highway across Alaska (until it was calculated that it would require most of the trucks in America). A prominent nuclear physicist recommended a fleet of nuclear-powered submarine tankers that would travel under the polar ice cap to a deep-water port in Greenland—the port to be created, in turn, by a nuclear explosion. Boeing and Lockheed explored the idea of jumbo jet oil tankers.

Finally, it was decided to build a pipeline. But in which direction? One proposal was for an eight-hundred-mile pipeline south from the oil fields to the port of Valdez, where the oil would be picked up by tankers for shipment through environmentally sensitive Prince William Sound and on to market. The other was for an entirely overland pipeline, east through Alaska to Canada, then south into the United States, eventually terminating perhaps in Chicago. Opponents of a trans-Alaskan pipeline argued that it could result in "probable major discharges from tanker accidents," while the Canadian route was environmentally sounder overall and would reduce the cost of an otherwise expensive pipeline for Alaskan natural gas. The Alaskan route, however, had the advantage of being an "all-American route," thus supposedly more secure, and it had the added plus of flexibility: Alaskan oil could go either to the United States or to Japan. And the oil men would only have to deal with two governments—one state and one Federal, and both American, and not, in addition, Canada's Federal government in Ottawa and three or four provincial and territorial jurisdictions, all with their own fiscal systems, as well as Canadian environmentalists and a couple more American states. Moreover, the Canadian government seemed to be turning against a trans-Canada pipeline. Given all these considerations, one argument

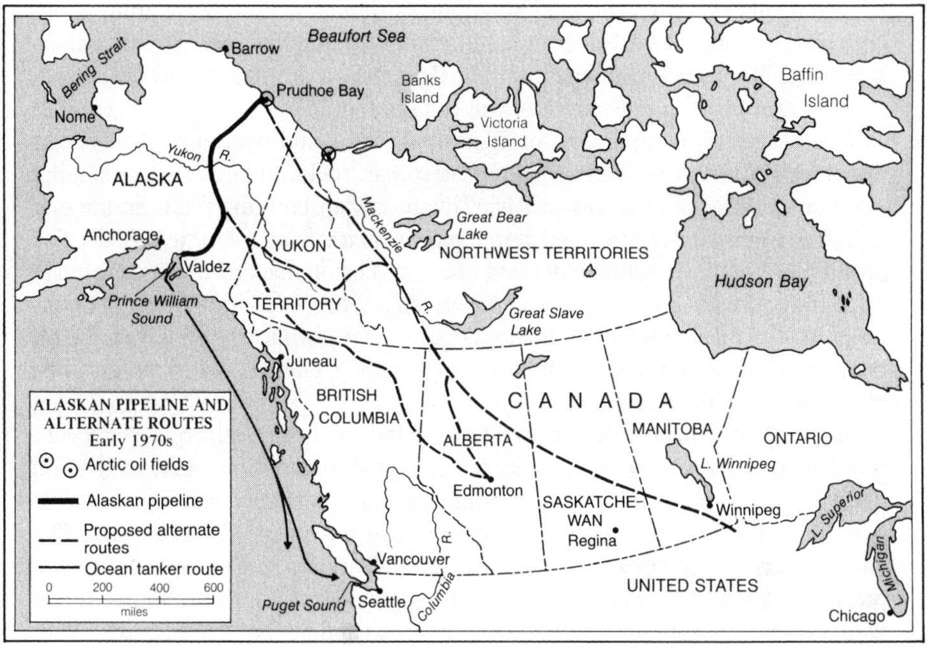
Beaufort Sea
Barrow
Bering Strait
Nome
Prudhoe Bay
Banks Island
Victoria Island
Baffin Island
Yukon R.
ALASKA
Mackenzie R.
Great Bear Lake
Anchorage
YUKON TERRITORY
NORTHWEST TERRITORIES
Valdez
Hudson Bay
Prince William Sound
Great Slave Lake
Juneau
BRITISH COLUMBIA
C A N A D A
MANITOBA
ONTARIO
ALBERTA
L. Winnipeg
Edmonton
SASKATCHEWAN
Regina
Winnipeg
L. Superior
L. Michigan
Vancouver
Columbia R.
UNITED STATES
Puget Sound
Seattle
Chicago
ALASKAN PIPELINE AND ALTERNATE ROUTES
Early 1970s
Arctic oil fields
Alaskan pipeline
Proposed alternate routes
Ocean tanker route
0 200 400 600
miles

stood out: The trans-Alaskan pipeline was likely to be built much more quickly than one taking the Canadian route. The trans-Alaskan route was chosen.

Construction of the pipeline posed a host of engineering challenges that would require a great deal of innovation and ingenuity. For instance, the oil came out of the ground at 160 degrees; yet it would then have to enter a pipeline running through permafrost many degrees below zero. If it ran beneath the surface through areas where the permafrost had a high water content, it could turn the area into a mushland; and the pipeline, deprived of support, might snap. But no matter what construction problems it might encounter, the group organized to build the Trans-Alaskan Pipeline—composed of ARCO, Jersey, and BP, plus companies with much smaller positions on the North Slope—rushed out and hurriedly bought 500,000 tons of forty-eight-inch pipe from a Japanese company; they did not think there was time to wait for American manufacturers to gear up. They were wrong. The pipeline was to come to a dead halt before it even started.

It was slowed by claims by Eskimos and other Alaskan natives, and by wrangling among the partners. But it was completely stopped by something entirely different: a Federal court injunction won by environmentalists in 1970. Energized by the 1969 Santa Barbara oil spill, the newly formed but diverse environmental movement converged on blocking the Alaska pipeline. Some of the environmentalists said that the companies were trying to move too fast, without sufficient study, understanding, skills, or care and that the proposed pipeline was poorly planned. The consequences of an accident would be environmentally catastrophic. The Canadian route was better, posing less environmental risk. Before going ahead, they said, the United States should institute a program of energy conservation. Other environmentalists argued that irreplaceable natural assets and a unique environment would be damaged or destroyed, and the project should never go ahead. Alaskan oil was not needed.

The eager oil companies, confident that they could overcome the opposition, brought in $75 million worth of Caterpillar trucks and trailers to a staging point on the banks of the Yukon River, ready to start building roads and laying pipe. The trucks and tractors and stored pipe were not to move for five years. The prohibitions on the pipeline remained in effect. The oil that had been expected in 1972 from Alaska did not flow, and American imports of foreign oil went up instead. As for those trucks and tractors on the banks of the Yukon River, the oil companies spent millions of dollars keeping their engines tuned up and just plain warm, waiting for the day.

Just when it became clear that Alaska and offshore California were very problematic as new sources, another promising alternative appeared with an oil discovery in the North Sea. But development in the North Sea was highly uncertain. The effort would be both enormous in scale and very, very expensive. The environment was harsh and treacherous. Like Alaska's North Slope, North Sea production would require an entirely new generation of technology. And it would take time, a great deal of time. Yet Alaska and the North Sea had another common bond: Though their reserves were in very difficult places, physically, they were not in unstable places, politically. Even so, neither could provide any imminent relief for the global supply-demand balance, which was drawing ever more taut. That meant there was still only one place to turn for the additional supplies required to satisfy the world's almost-insatiable appetite for oil—the Middle East.[5]

The Doctor

It was just before dawn, one day toward the end of August 1970, when a chartered French Falcon jet entered Libyan air space. Soon it was on the ground at the airport at Tripoli. The door of the jet opened, and a short, stocky man, just turned seventy-two, stepped out into the early light. He was a very worried man, so worried that he had flown virtually nonstop from Los Angeles, landing in Turin only long enough to change jets. He feared that he was about to lose what he called his "shining star," his company's richly prolific oil concession in Libya. But, as always, he tried to be confident. His life had been devoted to making deals, and he believed firmly, as an article of faith, that, as he once said, "There's nothing worse than a deal that wasn't closed."

The man was Dr. Armand Hammer, chairman of Occidental Petroleum.

When it came to deal-making, there have been few in the entire twentieth century to rival Armand Hammer. He was born in 1898 in New York City to an immigrant Jewish family with roots in Odessa, on the Black Sea. Among other things, a rich uncle had held the local Ford distributorship there. In the nineteenth century, Odessa was a great trading emporium where Western industrialists met Middle Eastern merchants, and in some ways, Odessa always remained in the blood of Armand Hammer. His father, Dr. Julius Hammer, was not only a practicing physician and drug manufacturer, but also a partisan of the left; he had met Lenin in Europe in 1907 and was one of the founders of the American Communist party. Armand did not share his father's socialist tendencies; he was interested in making money and doing deals—in short, a capitalist.

In 1921, having just graduated from medical school, Hammer set off for Russia. He carried relief medical supplies for the strife-torn country, and aimed, in the process, to collect $150,000 that the Soviets owed the family drug business. Through his father's connections, he became known to Lenin, who was allowing some competition in the ruined Russian economy and encouraging trade with the bourgeois West. Lenin applied a special benediction to Hammer, commending him to Stalin as "a little path leading to the American 'business' world and we should use it in every possible way." So Hammer, assisted by his brother Victor, stayed in Russia to do business under Lenin's New Economic Policy— an asbestos concession, an agency for Ford tractors and other products, the national concession for pencils. He even ran his own fur stations in Siberia, with his own personal trappers. But when Stalin came to power at the end of the decade, Hammer correctly read the portents and packed his bags. He and Victor carried out with them a load of Russian art, which they sold through department stores in the United States. Hammer then went on to make millions in a variety of other businesses, from beer barrels and bourbon whiskey to bull semen.

He came to Los Angeles in 1956, at fifty-eight, a wealthy man—like so many others, intent on retiring. He was now a prominent art gallery owner and collector. Seeking a tax shelter, he made some investments in oil, and then, almost as a sport, bought a small, nearly bankrupt company called Occidental. He knew nothing about the petroleum business. Yet by 1961, Occidental had made its first significant discovery in California. Hammer, the inveterate dealmaker, acquired a number of other companies, and by 1966, Occidental's annual sales were almost $700 million.

Through clever deal making and good timing, Hammer was eventually to build Occidental into one of the world's great energy companies. Not for him the normal chains of command. Phoning anywhere in the world at almost any time, he ran things himself, out of his hat, like a modern Marcus Samuel. His political connections were unparalleled; his ability to get into places unstoppable; and his personal fortune vast. In his never-ending negotiations, Hammer could be, as one opponent once said, "fatherly and very loving," always breaking the tension with an anecdote. But he was also deadly serious in seeking what he wanted. In advancing his interests, he had a great talent for letting people hear what they wanted to hear. "The Doctor is one of the greatest actors in the world," was the acid comment of one of the many men who had mistakenly thought himself Hammer's heir apparent.

Hammer renewed his contacts with the Soviet Union under Nikita Khrushchev and ended up as a go-between for five Soviet General Secretaries and seven U.S. Presidents. His access to the Kremlin was unique. He was virtually the only person who could tell Mikhail Gorbachev firsthand about Lenin, who had died a decade before Gorbachev's birth. As late as 1990, at age ninety-two, Hammer was still the active chairman of Occidental, and loyal stockholders continued to sing his praises. He was indeed in the line of the great buccaneer-creators of oil: Rockefeller, Samuel and Deterding, Gulbenkian, Getty and Mattei. He was also an anachronism, a privateer from the past, in spirit a "merchant from Odessa" circling the globe in his corporate jet in search of the next great deal. But it was a deal in Libya that had made possible his global tycoonery.[6]

The mad Libyan oil rush was already well along when, in 1965, Occidental won concessions there in the second round of bidding. Oxy's thick bid stood out midst the 119 others because it had been done up, under Hammer's personal supervision, on sheepskin manuscripts and was wrapped in red, black, and green ribbons—the colors of the Libyan flag. As one sweetener, Oxy promised to establish an agricultural experimental farm at a desert oasis that had been the childhood home of King Idris and the burial spot of his father. Hammer gave the King a gold chess set. The company also paid the expected overrides and special commissions to those who could help it obtain the concessions.

The blocks that Occidental won, Numbers 102 and 103, covered almost two thousand square miles of bleak, gravelly, scorched desert in the Sirte Basin, more than a hundred miles from the Mediterranean coast. "The hardest thing to live with is dry holes," Hammer once said, and the first few holes drilled on the site were very dry. They were also costly. Occidental's board of directors began to grumble loudly about "Hammer's Folly." Libya was the place for the big boys. But Hammer was persistent.

His persistence paid off. In the autumn of 1966, Occidental struck oil on Number 102. But this paled when compared to what happened on Number 103, forty miles to the west, subsequently called the Idris field. Occidental drilled right under the site of the former base camp of Mobil Oil, which had previously held but then relinquished the concession. The first well came in at forty-three thousand barrels per day; then another at a phenomenal seventy-five thousand barrels per day. Occidental had struck one of the most prolific deposits of oil in the world. It was the use of newly developed seismic technology that had enabled this puny California producer to find what giant Mobil had missed. With the discovery, said Hammer, "All hell broke loose. We became one of the Big Boys."

By another stroke of luck in 1967, the Six-Day War left the Suez Canal shut, and so Libyan oil took on an even greater value. The Libyan oil boom turned into a frenzy. DeGolyer and McNaughton, the petroleum engineering firm, estimated that, on the basis of the discoveries to date, Occidental alone was in possession of three billion barrels of recoverable reserves, almost a third of the reserves discovered at the same time on Alaska's North Slope! But what could not be done in Alaska—build a pipeline—certainly could be done in Libya. Normally, it would have taken three years to construct a 130-mile pipeline across the desert; but, force-paced, the line was built in less than a year, and Occidental was actually shipping oil to Europe less than two years after receiving its concessions. It was soon producing upwards of eight hundred thousand barrels per day in Libya. From nothing Occidental Petroleum had become the sixth-largest oil-producing company in the world, and it had fought its way into the competitive European market both by contract and by buying its own downstream system.

Yet this sudden giant was very vulnerable, owing its very success to its lopsided dependence on Libya. Elderly King Idris could not last all that long. To diversify, Hammer sought to acquire Island Creek Coal, a major United States coal producer. But before agreeing, William Bellano, the president of Island Creek, decided that he had better investigate the outlook for political stability in Libya. Bellano talked to people at such places as the State Department, the Chase Manhattan Bank, and Citibank. The general answer was the same: One could well expect political stability in Libya for five or six years, and "it was easy to anticipate the orderly transfer of power when the King died." The merger went ahead. That was 1968. The experts were all dead wrong.[7]

The Libyan Squeeze

On the night of August 31–September 1, 1969, a senior Libyan military officer, finding himself unexpectedly awakened in his own bedroom by a junior officer, told the insistent young man that he was too early; the coup was scheduled for a few days later. Alas, for the senior officer, this was a different coup. For months, the whole Libyan military had been seething with conspiracies, as various groups of officers and politicians prepared to topple the ailing regime of King Idris. A group of radical young officers, led by the charismatic Muammar al-Qaddafi, beat all the others to the punch, including their military superiors, who had scheduled their own coup for just three or four days later. Indeed, many military men participated in the September 1 coup without knowing who was in charge or even which coup it was.

Qaddafi and his associates had begun their conspiring a full decade earlier, while teenagers in secondary school, inspired by Gamal Abdel Nasser, his book, *Philosophy of Revolution,* and his radio station, the Voice of the Arabs. They decided to model their lives and their cause on Nasser. They also decided that the road to power was not directly through party politics, but the same route Nasser had taken, through the military academy. In Qaddafi's mind, as one acute observer put it, the revolutionary doctrines of Nasser were amalgamated "to the ideas of Islam at the time of Mohammed." The group was, indeed, in thrall to Nasser and the Nasserite vision of Arab unity. In due course, Qaddafi would seek to pick up the mantle of Nasser. A born conspirator like Nasser, but also eccentric and erratic, with the wide mood swings of a manic-depressive, he would attempt to make himself not only the leader of, but indeed the very embodiment of the Arab world. In doing so, he would plot and campaign endlessly against Israel, Zionism, other Arab states, and the West, and—equipped with huge oil revenues—he would become banker, sponsor, and paymaster for many terrorist groups around the world.

Among the first acts of Qaddafi's new Revolutionary Command Council after the successful September coup were the shutting down of the British and American military bases in Libya and the expelling of the large Italian population. Qaddafi also closed all the Catholic churches in the country and ordered their crosses removed and their contents auctioned off. Then, in December 1969, a countercoup was aborted and Qaddafi's consolidation of power was complete. He was now ready to deal with the oil industry. In January 1970, officers of the Revolutionary Command Council launched their offensive with a call for an increase in the posted price. Qaddafi warned the heads of the twenty-one oil companies operating in Libya that he would shut down production, if necessary, to get what he wanted. "People who have lived without oil for 5,000 years," he said, "can live without it again for a few years in order to attain their legitimate rights."

Initially, much pressure was put on Esso-Libya. The military government asked for a 43-cent-a-barrel increase in the posted price: "43 cents in those days," recalled the head of Esso-Libya. "Good God! That was out of the world." Esso offered five cents. The companies were not about to budge. Stymied by Jersey and the other major companies, most of which had alternative sources, the Libyans turned to the one company that did not have any alternatives, Occidental. They understood its vulnerability. As one Libyan explained, "It had all its eggs in one basket." In late spring of 1970, Occidental was ordered to cut its production, the lifeblood of the company, from eight hundred thousand barrels per day to around five hundred thousand. Just in case Occidental was missing the point, Libyan policemen began to stop, search, and harass the company's executives. Though cutbacks and harassments were also applied to other companies, Occidental was especially favored with such attention.

The Qaddafi regime had chosen a very propitious time to initiate its campaign. Libya was supplying 30 percent of Europe's oil. The Suez Canal was still closed, maintaining pressure on transportation. Then, in May 1970, a tractor ruptured the Tapline at a point inside Syria, which prevented the export of five hundred thousand barrels per day of Saudi oil through that pipeline to the Mediterranean. Tanker rates immediately tripled. There was no shortage of oil, but there was a shortage of transportation, and this put Libya and Qaddafi, sitting directly across the Mediterranean from European markets, in a central position. The Libyans were not loath to exploit their advantage. Their own cutbacks added dramatically to the tightness and tensions in the market; between the Tapline closure and the Libyan cutbacks, a total of 1.3 million barrels per day had been abruptly taken out of the market. The young Libyan military officers, moreover, were not exactly operating in the dark when it came to oil economics and strategy; Abdullah Tariki, the radical and anti-Western oil nationalist who had been fired eight years earlier as Saudi oil minister, was now in Tripoli, advising the revolutionary government.

As the pressure mounted, Hammer became agitated. He went to Egypt to ask Qaddafi's hero, President Nasser, to intercede with his "disciple." Concerned that a shutdown of oil production in Libya would threaten Libya's subsidies to the Egyptian Army, Nasser advised Qaddafi to go easy. He also told the Libyan leader not to repeat his own mistakes—Egypt had paid dearly for his policy of nationalization and the expulsion of key foreign technicians. The advice was not heeded.

Hammer tried to find other companies that would provide replacement oil at cost to Occidental if it did not cave in to Libyan demands but stood firm against Qaddafi and was then nationalized. He was not successful. Even a visit to Kenneth Jamieson, chairman of Exxon, did not produce the oil that Hammer wanted, at least not on the terms that he wanted. Hammer was very bitter. But perhaps

Jamieson simply could not take Hammer seriously. It was "perfectly understandable that Jamieson turned Hammer down," one of Hammer's chief advisers was to say privately. "Here the President of Exxon, the world's largest corporation, was confronted with an art dealer, a non-fraternity type, coming out of the cold with a worldwide scheme."

Desperate to find an alternative source of oil, Hammer came up with yet another global scheme. Over dinner at Lyndon Johnson's ranch in Texas, he tried to put together a barter deal whereby he would be the middleman in trading McDonnell Douglas warplanes for Iranian oil. That effort failed as well. He had just about run out of alternatives when, at the end of August 1970, he received an urgent phone call from George Williamson, his manager in Libya, warning him that the Libyans might nationalize Occidental's operations. And it was that warning that sent him streaking through the night skies to Tripoli.

On the Libyan side, the negotiations were led by Deputy Prime Minister Abdel Salaam Ahmed Jalloud, considered more fun-loving than the puritanical Qaddafi, but still a relentless negotiator. Once, to show his displeasure during a discussion with representatives of Texaco and Standard of California, he rolled their proposal up into a paper ball and threw it back into their faces. On another occasion, he charged into a room full of oil executives with a submachine gun slung over his shoulder. At his first session with Hammer, Jalloud, after offering the Doctor hot rolls and coffee, unbuckled his belt and set down his .45 revolver on the table directly in front of Hammer. Hammer smiled. But he was disconcerted. He had never before negotiated across a gun barrel.

Each day Hammer worked through the arduous, draining negotiations. Each night, he flew back to Paris, so that, from his suite at the Ritz Hotel, he could telephone with some security back to his board of directors in Los Angeles. There was another reason for this daily commute. Despite Jalloud's offer of the hospitality of a palace that had belonged to the deposed King Idris, Hammer worried that he might be "detained" for an extended stay. Yet he did relax his guard a little. On that first day, he had used a chartered French jet to reach Tripoli out of caution that the Libyans might seize his private plane. Thereafter, a bit reassured, he switched back, on each morning's commute from Paris, to his own more familiar Gulfstream II, with its cork-lined bedroom. He would arrive back in Paris at 2:00 A.M., and would be off again by 6:00 A.M. Throughout his life, he had had a remarkable capacity to nap under all kinds of conditions, and he put that ability to very good use on the flights.

As the intense discussions dragged on, crowds outside, preparing to celebrate the first anniversary of the coup, were chanting for death to opponents of the regime. Yet, finally, there came a point in the negotiations when Hammer and Jalloud went into a corner and shook hands. They had an agreement in principle, and the deal seemed set for signing when a new hitch, regarding the form of the contract, suddenly appeared. Suspicious, Hammer decided to depart the country immediately, leaving it to George Williamson to complete the deal. The next day, ensconced at the Ritz in Paris, Hammer learned that the final bargain had been signed. The Libyans got a 20 percent increase in royalties and taxes. Occidental would be able to stay. As for the other companies, they wavered, but by the end of September, virtually all of them had given in, though with immense reluctance. The Libyans solemnly promised that they would stick to these new agreements for five years.

But what had happened was of far greater significance than a thirty-cent increase in the posted price and a hike in Libya's share of profit from 50 to 55 percent. The Libyan agreements decisively changed the balance of power between the governments of the producing countries and the oil companies. For the oil-exporting countries, the Libyan victory was emboldening; it not only abruptly reversed the decline in the real price of oil, but also reopened the exporters' campaign for sovereignty and control over their oil resources, which had begun a decade earlier with the foundation of OPEC, but then had stalled. For the companies, it was the beginning of a retreat. The Jersey director who had responsibility for Libya succinctly summarized the significance of the new agreements when he said, "The oil industry as we had known it would not exist much longer." George Williamson of Occidental had a premonition of how great the changes would be. As he prepared to affix his signature to the final documents, he observed to another Occidental manager, "Everybody who drives a tractor, truck, or car in the Western world will be affected by this." They signed the documents, and then Williamson and his associates sat with the Libyans, sipping orange soda, the best that could be found in the alcohol-less land, silently contemplating the uncertain future.[8]

Leapfrogging Prices

The Shah of Iran was certainly not going to let himself be outdone by some upstart young military officers in Libya. In November 1970, he broke through the old fifty-fifty profit-sharing barrier and won 55 percent of the profits from the consortium companies. The companies then decided that they had no choice but to offer 55 percent to the other Gulf countries. And with that, a game of leapfrog began. Venezuela introduced legislation that would raise its share of profits to 60 percent and would also provide for oil prices to be raised unilaterally, without reference to or negotiation with the companies. An OPEC conference endorsed 55 percent as the minimum country share and threatened a cutoff of supplies to the companies if demands were not met. It also insisted that the oil companies negotiate with regional groupings of exporters, rather than with OPEC as an entity. Then, at the beginning of 1971, Libya jumped over Iran and made new demands. Obviously the game could be endless unless the companies established a common front and succeeded in sticking with it.

David Barran, chairman of Shell Transport and Trading, became the foremost advocate of the common front. "Our Shell view was that the avalanche had begun," said Barran. Without a united front, the companies would "be picked off one by one." Out of Barran's urging came the makings of a joint approach, by which the companies would as a single unit negotiate with OPEC as a group, rather than with individual countries. That way, they hoped, the avalanche of competing demands could be stayed. Winning an antitrust waiver from the U.S. Justice Department, the oil companies set about to create a "Front Uni," harking back to the one they had formed to confront the Soviet Union in the 1920s. But now it was a much more complicated world, with many more players. This modern Front Uni eventually comprised two dozen companies—American and non-American—representing about four-fifths of free world oil production. The companies also created the Libyan Safety Net, a secret understanding that if any company had its production cut back because it had stood up to the Qaddafi government, the other companies would provide replacement oil. The agreement institutionalized the kind of deal that Hammer had tried, and failed, to get out of Exxon six months earlier. It also represented, in the words of James Placke, the American petroleum attaché in Libya, a "truce" between the majors and the independents.

On January 15, 1971, the companies hurriedly despatched a "Letter to OPEC," which called for a global, "all-embracing" settlement with the oil exporters. The aim was to maintain a united front and to negotiate with OPEC as a whole, rather than with individual exporters or subgroups, as OPEC wanted. Otherwise, the companies would be endlessly vulnerable to leapfrogging.

But the Shah resolutely opposed the companies' plan for an "all-embracing" settlement because, he said, the "moderates" would not be able to restrain the "radicals"—Libya and Venezuela. Yet if the companies would only be reasonable and deal with the Gulf countries separately, the Shah promised a stable agreement that would be kept for five years. "If the companies tried any tricks," he added, "the entire Gulf would be shut down and no oil would flow."

The negotiating process opened in Tehran, with the Front Uni represented by George Piercy, the Exxon director responsible for the Middle East, and Lord Strathalmond, a director of BP and a lawyer by profession. The latter was a friendly, genial man, who liked to kid the Kuwaiti oil minister by calling him, owing to his appearance, "Groucho." He was also the son of William Fraser, who had been chairman of BP at the time of the Mossadegh affair, and who had remained a most unpopular man in Iran—so much so that Lord Strathalmond felt constrained to tell some confused Iranians, "I am not my father."

The companies thought that they had the support of the U.S. government in their struggle with the Shah; but Piercy and Strathalmond discovered, on arriving in Tehran, that Washington had acquiesced to the Shah's position. The two oil men were flabbergasted and furious. "It made the whole exercise look silly as hell," Piercy said.

On January 19, Piercy and Lord Strathalmond met with the OPEC Gulf committee: Iranian Finance Minister Jamshid Amouzegar (educated at Cornell and the University of Washington), Saudi Arabian Oil Minister Zaki Yamani (educated in part at New York University and Harvard Law School), and the Iraqi Oil Minister, Saadoun Hammadi (Ph.D. in agricultural economics from the University of Wisconsin). The three ministers were unbending. They would discuss oil pricing for the Gulf countries, and only for the Gulf countries, and not the rest of OPEC, and that was that. The Shah himself denounced the companies and brandished the possibility of an embargo if the companies did not agree to his point of view. He even summoned up the ghost of Mossadegh. "The conditions of the year 1951 do not exist anymore," he sternly warned. "No one in Iran is cuddled up under a blanket or has shut himself off in a barricaded room." The effort to get a single, "all-embracing" negotiation, he said, was either "a joke or was an intent to waste time."

No conclusion emerged from this first phase of the Tehran negotiation. In a private meeting, Yamani warned Piercy that yes, it was true what Piercy had heard, that there was talk among the exporting countries about an embargo to strengthen their hand against the companies. Moreover, Yamani admitted, the Saudis and the other Gulf producers supported the idea. Piercy was shocked. The Saudis had never embargoed oil except in wartime. Did this, he asked, have the support of King Faisal? Yes, said Yamani, and also that of the Shah. Piercy urged Yamani not to take such a step.

"I don't think you realize the problem in OPEC," said Yamani. "I must go along."[9]

The companies concluded, however reluctantly, that they would have to give up their insistence on the all-embracing approach—there was no longer a choice. They agreed to have the negotiations split. Otherwise, there would be no negotiated settlement at all; the exporters would simply decide prices on their own. At all costs, the companies needed to maintain the pretense, even if it was only a pretense, that the exporters negotiated matters with them, rather than simply deciding things themselves.

So now there were to be two sets of negotiations: one in Tehran and one in Tripoli. On February 14, 1971, the companies capitulated in Tehran. The new agreement interred the fifty-fifty principle. The hallowed fifty-fifty had done its job, it had lasted for two decades, but now its time was past. The new accord established 55 percent as the minimum government take and raised the price of a barrel of oil by thirty-five cents, with a commitment to further annual hikes. The exporters gave their solemn promise: no increases for the next five years beyond what had already been agreed.

The Tehran Agreement marked a watershed; initiative had passed from the companies to the exporting countries. "It was the real turning point for OPEC," said one OPEC official. "After the Tehran Agreement, OPEC got muscles." In the immediate aftermath, the Shah, from the ski slopes at St. Moritz in Switzerland, offered his blessings to the undertaking. "Whatever happens," he pledged, "there will be no leapfrogging." David Barran, the chairman of Shell, was a better prophet. "There is no doubt," he said, "that the buyer's market for oil is over."

Now it was time for the other half of the negotiation, over the price of OPEC oil in the Mediterranean. The Mediterranean committee included Libya and Algeria, as well as Saudi Arabia and Iraq, since part of their production came to the Mediterranean via pipelines. A few days after the Tehran Agreement, discussions began in Tripoli, with Libya—and Major Jalloud—very definitely in charge of the negotiations on the Arab side. Jalloud resorted to all his now-familiar tactics—bluster, intimidation, revolutionary sermons, and threats of embargo and nationalization. On April 2, 1971, agreement was announced. The posted price was raised by ninety cents—well beyond what would have been implied by the Tehran Agreement. The Libyan government had increased its oil revenues by almost 50 percent.

The Shah was beside himself with rage. Once again, he had been leapfrogged.[10]

Participation: "Indissoluble, like a Catholic Marriage"

The pledge of five years' stability in the Tehran and Tripoli agreements proved to be illusionary. A fresh battle soon ensued when OPEC sought increases in the posted price to compensate for the devaluation of the dollar in the early 1970s. But that fight was overshadowed by a more momentous conflict, which would dramatically alter the relationship of the companies and countries. The battle was over the issue of "participation": the acquisition by the exporting countries of partial ownership of the oil resources within their borders. If the exporters won that battle, it would mean a radical restructuring of the industry and a fundamental change in the roles of all players.

For the most part, oil operations outside the United States had been based on the concession system, the history of which reached back to William Knox D'Arcy and his bold and blind venture into Persia in 1901. Under the concession system, the oil company contractually obtained rights from a sovereign to explore for, own, and produce oil in a given territory, be it as large as D'Arcy's original 480,000 square miles in Persia or Occidental's 2,000 square miles in Libya. But now, as far as the oil exporters were concerned, concessions were already a thing of the past, holdovers from the defunct age of colonialism and imperialism, wholly inappropriate to the new age of decolonization, self-determination, and nationalism. Those countries did not want to be mere tax collectors. It was not only a question of garnering more of the rents. For the exporters, the greater question was sovereignty over their own natural resources. Everything else would be measured against that objective.

Outright nationalization had been an obvious resort for some exporting regimes—in Russia after the Bolshevik Revolution, in Mexico, in Iran. The concept of "participation," partial ownership achieved by negotiation, was consciously devised as an alternative to nationalization and full ownership because it satisfied the interests of some of the main oil exporters. Oil was not only a symbol of national pride and power; it was also a business. Outright nationalization would disrupt relations with the international companies and put the oil-producing country directly into the business of selling oil. The country would thus face the same obstacle that bedeviled the independents who had built up large reserves of oil in the Middle East—the problem of disposal. That would lead to a battle royal with other exporters for markets. The oil companies not only would be free to shop around for the cheapest barrel, but also would have strong incentive to do so, since they would now be making their profits on sales in the consuming markets, rather than in production.

"We in the producing countries—having become the operators and sellers of our oil—would find ourselves in a competitive production race," said Sheikh Yamani in 1969, warning against outright nationalization. The result would be "a dramatic collapse of the price structure, with each of the producing countries trying to maintain its budgeted income requirements in the face of the declining prices by moving larger volumes of oil to the market." The costs, and risks, would not only be economic. "Financial instability would inevitably lead to political instability." Yamani insisted that participation—joint ownership with the major companies rather than their ejection—was the way to meet the exporters' objectives and yet maintain the system that held up the price. It would, he said, create a bond that "would be indissoluble, like a Catholic marriage."

Participation fit Saudi Arabia's situation; it meant gradual change, rather than a radical overturning of the oil order. But for other exporters, gradual participation was insufficient. Algeria, with no pretense of negotiation, took 51 percent ownership in the French oil operations there, which had been preserved a decade earlier when Algeria won its independence. Venezuela passed legislation under which all concessions would revert to the government automatically upon their expiration in the early 1980s.

OPEC itself called for immediate implementation of participation with the threat of "concerted action"—cutbacks—if it received no satisfaction. Yamani was put in charge on the OPEC side. The pressure on the companies to go along was rising. With the British withdrawal from the Gulf at the end of 1971, Iran had seized some tiny islands near the Strait of Hormuz. For the militants among the Arabs, it was a great affront, the seizure of Arab territory by non-Arabs. To punish the British for "collusion" in this dastardly act, Libya, twenty-four hundred miles away, nationalized BP's holdings in that country. Iraq nationalized the Iraq Petroleum Company's last remnant in the country, the Kirkuk concession, the great producing field that had been discovered in the 1920s and had been both the focus of all Gulbenkian's wrangling with the major companies and the basis of much of Iraq's production. "There is a worldwide trend toward nationalization and Saudis cannot stand against it alone," Yamani warned the companies. "The industry should realize this and come to terms so that they can save as much as possible under the circumstances."

Yet before any agreements could be made, several fundamental issues had to be thrashed out, including the basic question of valuation. For instance, depending on the accounting formula that was chosen, 25 percent of the Kuwait Oil Company could be worth anywhere from sixty million to one billion dollars. Finally, in that case, the two sides came together by inventing a new accounting concept, "updated book value," which included inflation and large fudge factors. And in October 1972 a "participation agreement" was finally reached between the Gulf states and the companies. It provided for an immediate 25 percent participation share, rising to 51 percent by 1983. But, despite all the OPEC endorsements, the application of the agreement was less popular in the rest of OPEC than Yamani hoped. Algeria, Libya, and Iran all stood outside it. Kuwait's oil minister approved it, but the Kuwaiti parliament rejected it, and so Kuwait was also out.

The Aramco companies had finally agreed to participation with Saudi Arabia because the alternative was worse—outright nationalization. The chairman of Exxon said hopefully that he expected "more stable future relationships" to result from the agreement, which "maintained the essential intermediary role of the private international oil companies." Others were not so sure. At a meeting in New York of oil company executives, chaired by John McCloy, Aramco announced its initial decision to agree to participation. At the end of the fractious discussion, McCloy asked the opinion of Ed Guinn, an executive of the independent Bunker Hunt oil company, which operated in Libya. Guinn was upset. Any concession made in the Persian Gulf, he believed, would only goad Libya into even greater demands. The Aramco plan he had just heard, he added, reminded him of the story about two skeletons hanging in a closet, which he proceeded to tell. One skeleton said to the other, "How did we get here?" The other replied, "I don't know, but if we had any guts we'd get out."

"Meeting adjourned!" McCloy immediately shouted, and everyone left.

After Yamani's deal with Aramco, Libya took over 50 percent of the operations of ENI, the Italian state oil company, then proceeded to expropriate Bunker Hunt's holdings altogether. Standing with Uganda's brutal dictator, Idi Amin Dada, by his side, Qaddafi proudly announced that, by taking over Bunker Hunt, he had given the United States "a big hard blow" on "its cold insolent face." He then went on to nationalize 51 percent of the other companies that were operating in Libya, including Hammer's Occidental Petroleum.

The Shah was intent on assuring that he ended up with a better deal than Saudi Arabia. But for Iran, participation was irrelevant. Owing to the 1951 nationalization, Iran already owned the oil and facilities; but the consortium set up in 1954, and not the National Iranian Oil Company, actually operated the industry. So the Shah insisted on gaining not only higher production and financial equivalency with the agreement that Yamani had hammered out, but also much greater control. And he got what he wanted. NIOC was to become not only owner but also operator; the 1954 consortium companies would set up a new company to act as a service contractor to NIOC, replacing the old consortium. Having the National Iranian Oil Company officially recognized as operator was a first for a state oil company, a victory of considerable symbolism in the Shah's quest to turn NIOC into the world's premier oil company. And it was a victory for himself. He was now moving into his grandest phase. "Finally I won out," the Shah declared. "Seventy-two years of foreign control of the operations of our industry was ended." [11]

The Hinge Years

In gaining greater control over the oil companies, whether by participation or outright nationalization, the exporting countries also gained greater control over prices. Where they had only recently sought to increase their income by emphasizing volume, competing to put more and more barrels into the marketplace, which just seemed to push the price down, they would now seek higher prices. Their new approach was supported by the tight supply-demand balance. The result was the new system that was forged in Tehran and Tripoli, under which prices were the subject of negotiation between companies and countries, with the countries taking the lead in pushing up the posted price. The companies had been unable to put up a successful new Front Uni. Nor had their governments. In fact, the governments of the consuming countries did not particularly want to support or bolster the companies in their confrontations with the exporters. They were distracted by other matters. Oil prices did not seem a terribly high priority, and some thought that the price increases were, in any event, justified and would be helpful in stimulating conservation and new energy development.

But there was that further element to the response of the two key Western governments. Both Great Britain and the United States had powerful incentives to seek to cooperate with, rather than confront, both Iran and Saudi Arabia, and in passing not to gainsay them larger incomes. By the early 1970s, Iran and Saudi Arabia had heeded the Sultan of Oman's appeal for help to put down a radical rebellion and were taking up their roles as regional policemen. Their arms purchases were increasing rapidly, one calibration of the interaction of rising oil prices with the new security calculus for the Gulf.

Yet putting aside politics and personalities, the supply-demand balance that emerged at the beginning of the 1970s was sending a most important message: Cheap oil had been a tremendous boon to economic growth, but it could not be sustained. Demand could not continue growing at the rate it was; new supplies needed to be developed. That was what the disappearance of spare capacity meant. Something had to give, and that something was price. But how, and when? Those were the all-critical questions. Some thought the decisive year would be 1976, when the Tehran and Tripoli agreements were due to expire. But the supply-demand balance was already very taut.

While, of course, the recoverable reserves in the Middle East were huge, available production capacity was much more closely attuned to actual demand. As late as 1970, there were still about 3 million barrels per day of excess capacity in the world outside the United States, most of it concentrated in the Middle East. By 1973, the additional capacity, in pure physical terms, had been cut in half; it was down to about 1.5 million barrels of daily capacity—roughly 3 percent of total demand. In the meantime, some of the Middle Eastern countries, led by Kuwait and Libya, had been instituting cutbacks in their output. By 1973, the surplus production capacity that could be considered actually "available" added up to only 500,000 barrels per day. That was just one percent of free world consumption.

Not only in oil, but in almost any industry, even in the absence of politics, a 99 percent utilization rate and a one percent security margin would be considered an extraordinarily precarious supply-demand balance. Politics was adding to the dangers.

What might all this mean for the future? One who watched with a growing sense of foreboding was James Placke, an American diplomat who had been an economic officer at the U.S. embassy in Baghdad a decade earlier, when OPEC was formed, and was now the petroleum officer in the U.S. embassy in Tripoli. In late November 1970, he sat down to collect his thoughts on paper for a dispatch to the State Department. Fifteen months had passed since an unknown group of officers had staged their military coup in Tripoli, and almost three months had elapsed since those same young officers had carried out a coup in oil pricing. Placke had been reporting on a daily basis all through the period as the Libyans had battled with Occidental, Esso, Shell, and the other companies, but now there was time to stand back. The weather had cooled and become somewhat stormy, with squalls coming in off the Mediterranean and the tangy smell of salt and sea in the air. A permanent sense of unease and even fear had settled over the western community in Libya. There were constant rumors about who had been harassed, detained, or deported, and both company officials and Western diplomats found themselves followed by security men, usually recognizable in the rearview mirror because of their white Volkswagen beetles.

It took Placke several weeks to work out what he wanted to say to Washington. He did not want to so overstate matters that his cable would be disregarded. As he wrote, he could look out through the one window in his broom closet of an office across a narrow alley to an Occidental office, where engineers bent over draughting tables as though it were business as usual and as if nothing had changed. But as Placke saw it, everything had changed. The old game in oil was over, even if no one in Washington or London quite grasped it. The international petroleum order had been irrevocably changed. In the report he finally sent to Washington in December, he argued that what had happened in Libya made it much more likely that the producing countries "will be able to overcome their divisions to cooperate in controlling production and raising prices."

But it was not only a question of price, but of power. "The extent of dependence by western industrial countries upon oil as a source of energy has been exposed, and the practicality of controlling supply as means of exerting pressure for raising the price of oil has been dramatically demonstrated." As he saw it, the United States and its allies, along with the oil industry, were simply unprepared intellectually and politically to "deal with the changed balance of power in the petroleum supply situation." The stakes were high. Among other things, though the "oil weapon" had not worked in 1967, the "rationale of those who call for the use of Arab oil as a weapon in Middle East conflict has also been strengthened in current circumstances."

He added one over-arching point: "Control of the flow of resources has been of strategic concern throughout history. Asserting control over a vital source of energy would permit Middle Eastern states to regain the power position vis-a-vis the West, which this area lost long ago." Placke emphasized that he was not pleading for maintaining the status quo. That was impossible. But what was important was to understand how the world was changing and to prepare for it. The greatest sin was inattention.

The U.S. ambassador was so impressed by Placke's paper that, in order to give it added weight, he had it sent out over his own name. But to Placke's knowledge, no one in Washington paid any serious attention to the message. He certainly never heard a word back on it.[12]

CHAPTER 29

The Oil Weapon

JUST MOMENTS BEFORE 2:00 P.M. on October 6, 1973—on what, by that year's calendar, was Yom Kippur, the holiest of Jewish holidays—222 Egyptian jets roared into the sky. Their targets were Israeli command posts and positions on the eastern bank of the Suez Canal and in the Sinai. Minutes later, more than 3,000 field guns opened fire along the entire front. At almost exactly the same time, Syrian aircraft launched an attack on Israel's northern border, followed immediately by a barrage from 700 pieces of artillery. Thus began the October War, the fourth of the Arab-Israeli wars—the most destructive and intense of all of them, and the one with the most far-reaching consequences. The armaments on both sides of the conflict had been supplied by the superpowers, the United States and the Soviet Union. But one of the most potent weapons was unique to the Middle East. It was the oil weapon, wielded in the form of an embargo—production cutbacks and restrictions on exports—that, in the words of Henry Kissinger, "altered irrevocably the world as it had grown up in the postwar period."

The embargo, like the war itself, came as a surprise and a shock. Yet the pathway to both in retrospect seemed in some ways unmistakable. By 1973, oil had become the lifeblood of the world's industrial economies, and it was being pumped and circulated with very little to spare. Never before in the entire postwar period had the supply-demand equation been so tight, while the relationships between the oil-exporting countries and the oil companies continued to unravel. It was a situation in which any additional pressure could precipitate a crisis—in this case, one of global proportions.

The United States Joins the World Market

In 1969, as the new Administration of Richard Nixon settled into Washington, oil and energy were beginning to rise on the American political agenda. The number-one concern was the rapid increase in oil imports. The Mandatory Oil Import Program, reluctantly established by President Eisenhower a decade earlier, was laboring under mounting strain, creating controversies and gross disparities among companies and regions. Its loopholes and exceptions were very lucrative to those who had figured out how to capitalize on them, and all too visible. Nixon established a Cabinet Task Force on Oil Import Control, headed by Labor Secretary George Shultz, to review the quota program and recommend changes.

Politicians from oil-consuming states and oil users, such as utilities and petrochemical companies, were eager to see the restrictions loosened so they could get cheaper supplies. The independent oil men, however, were resolute in defense of quotas, which assured them prices higher than the world market. As for the majors that had fought the quotas a decade earlier, they had by now generally reconciled and adjusted themselves to the system, and were content with it. Prices were protected for their domestic production, and the companies had devised distribution systems outside the United States to dispose of their foreign oil. Many of them, therefore, were alarmed at the prospect of change and argued against it.

As it turned out, George Shultz's committee recommended that the quotas be scrapped altogether and replaced with a tariff, thus removing the necessity for allocation by administrative fiat, leaving that task to the market instead. The political response to the Shultz study was not only vigorous but also overwhelmingly negative. The American oil and gas industry was already in a deep trough; the number of drilling rigs had declined steadily since 1955, hitting its lowest levels in 1970–71—little more than a third the level of the mid-1950s. A hundred Congressmen, fearing that the proposal would mean still more oil imports, signed a letter denouncing the Shultz report as a threat to the domestic industry. Nixon, the shrewd politician, shelved the report and kept the quotas.

That, of course, disappointed those who wanted to see the quota system dismantled, a group that was not limited to oil-consuming interests in the United States. The Shah of Iran wrote to Nixon, arguing that Iran's security and economic development required it to surmount the quota barriers and sell larger volumes of oil directly into the United States. The Nixon Administration was sympathetic to Iran's quest for higher production and thus higher revenues, owing, in the words of one White House adviser, to the "power vacuum in the Persian Gulf" consequent on the British withdrawal. But the Administration was not about to tear down the restrictions on imports, not even to please the Shah. "Your disappointment that it has not been possible to find a way to provide for increased sales of Iranian oil in the United States is understandable," Nixon wrote to the Shah. "Our lack of success is due to the great complexity of our oil import policy problems." Though apologetic, Nixon did promise to send a copy of the Cabinet Task Force report on oil import policy to the Shah for his own personal edification.[1]

By this time, however, there were already clear and politically worrisome signs of stress throughout the U.S. energy supply system. During the 1969–70 winter, the coldest in thirty years, both oil and natural gas were in short supply. The demand for low-sulfur oil, which had to be imported from such countries as Libya and Nigeria, grew dramatically in those months as electric utilities switched from coal to oil. The following summer, capacity constraints in the electric utility industry led to brownouts up and down the Atlantic Coast. Meanwhile, surplus oil production capacity in the United States had disappeared, as the industry pumped every single barrel that it could to meet the swelling demand.

With supply problems becoming chronic in the early 1970s, the phrase "energy crisis" began to emerge as part of the American political vocabulary, and in limited circles, there was agreement that the United States faced a major problem. The central reason was the rapid growth in demand for all forms of energy. Price controls on oil, imposed by Nixon in 1971 as part of his overall anti-inflation program, were discouraging domestic oil production while stimulating consumption. Natural gas supplies were becoming tighter, primarily because of a regulatory system that controlled prices and could not keep up with changes in markets. The artificially low prices provided little incentive either for new exploration or for conservation. Electricity generating plants were operating near capacity in many parts of the country, continuing to threaten brownouts or even blackouts. Utilities were hurriedly ordering new nuclear power plants as the solution to a host of problems, which included the growing demand for electricity, the prospect of rising oil prices, and new environmental restrictions on coal burning.

As oil demand continued to surge in the first months of 1973, independent refiners were having trouble acquiring supplies, and a gasoline shortage was looming for the summer driving season. In April, Nixon delivered the first ever Presidential address on energy, in which he made a far-reaching announcement: He was abolishing the quota system. Domestic production, even with the protection of quotas, could no longer keep up with America's voracious appetite. The Nixon Administration, responding to political pressure from Capitol Hill, immediately followed up on its abolition of quotas with the introduction of a "voluntary" allocation system, meant to assure supplies to independent refiners and marketers. Those two acts, coming one on top of the other, perfectly symbolized how circumstances had changed: Quotas were meant to manage and limit supplies in a world of surplus, while allocations were aimed at distributing whatever supplies were available in a world of shortage.

"The Wolf Is Here"

With energy matters rising on the political agenda, James Akins, a tall, somber Foreign Service officer who was one of the State Department's chief oil experts, had been detailed to the White House to work on those issues. He had recently, at State, directed a secret oil study, in which he had concluded that the world petroleum industry was passing through "the last gasp in the buyers market." He went on to add, "By 1975, and possibly earlier, we will have entered a permanent sellers' market, with any one of several major suppliers being able to create a supply crisis by cutting off oil supplies." It was time, he had said, for "an end to the 'interminable' studies on energy problems." Instead the United States should act to reduce the growth rate of consumption, raise domestic production, and strive to import from "secure sources." Such actions, he said, "will be as unpopular as they will be costly." Neither the unpopularity nor the cost was ever tested, for none of those steps was taken. Indeed, with the rapid growth in imports, quite the opposite was happening.

In April 1973, the same month as Nixon's abolition of quotas, Akins, now from his White House post, tried again. He prepared a secret report filled with proposals to counter the growing energy threat, among which were expanded coal use, development of synthetic fuels, stepped-up conservation efforts (including a stiff gasoline tax), and much-increased research and development spending in order to get beyond hydrocarbons. His ideas were met with incredulity. "Conservation is not the Republican ethic," John Ehrlichman, Nixon's chief domestic adviser, told him flatly. That same month, Akins went public with his concerns. He published an article in *Foreign Affairs* whose title captured the economic and political trends: "The Oil Crisis: This Time the Wolf Is Here." It was very widely read. But it was also very controversial, and Akins's arguments were far from widely endorsed or even accepted. For instance, at the same time, *Foreign Policy* magazine, the brash upstart rival to *Foreign Affairs*, published an essay entitled "Is the Oil Shortage Real?" That article emphatically said it was not. Announcing that "the world 'energy crisis' or 'energy shortage' is a fiction," it seemed to suggest that Akins himself was part of a State Department/oil exporter/company cabal. So the warning flags were up, but there was no particular response, nor, it must be said, was there the requisite consensus either in the United States or among the industrial countries as a group that would have been needed for more concerted prophylactic action.[2]

Yet, without the import barriers, the United States was now a full-fledged, and very thirsty, member of the world oil market. It joined with the other consuming countries in the clamorous call on the Middle East. There was hardly any choice but to remove the quotas; but their abolition meant a major new demand on an already-fevered market. Companies were buying any oil they could find. "In spite of all the crude we had," recalled the president of Gulf Oil's supply and trading arm, "I thought we had to go out and buy oil. We needed diversification." By the summer of 1973, United States imports were 6.2 million barrels per day, compared to 3.2 million barrels per day in 1970 and 4.5 in 1972. Independent refiners also rushed out to the world markets, joining a frenetic group of purchasers in bidding up the price for such supplies as were available. The trade journal *Petroleum Intelligence Weekly* reported in August 1973 that "near-panic buying by U.S. and European independents as well as the Japanese" was sending "oil prices sky-rocketing."

As demand worldwide bobbed up against the limit of available supply, market prices exceeded the official posted prices. It was a decisive change, truly underlining the end of the twenty-year surplus. For so long, reflecting the chronic condition of oversupply, market prices had been below posted prices, irritating relations between companies and governments. But now the situation was reversed, and the exporting countries certainly did not want to be left behind; they did not want to see the growing gap between the posted price and the market price go to the companies.

Wasting little time, the exporters sought to revise their participation and buy-back arrangements so that they would be able to obtain a larger share of the rising prices. Libya was the most aggressive. On September 1, 1973, the fourth anniversary of Qaddafi's coup, it nationalized 51 percent of those company operations it had not already taken over. President Nixon himself issued a warning in response: "Oil without a market, as Mr. Mossadegh learned many, many years ago, does not do a country much good." But the stern admonition fell flat. It was not just twenty years that separated Mossadegh and Qaddafi; so did a dramatic difference in market conditions. When Mossadegh nationalized Anglo-Iranian, a great deal of new production capacity was being developed elsewhere in the Middle East. But now, in 1973, there was no spare capacity, the market was certainly there, and it was hungry. Libya had no trouble selling its environmentally desirable low-sulfur oil.

The radicals in OPEC—Iraq, Algeria, and Libya—began pushing for revision in the two supposedly sacred texts, the Tehran and Tripoli agreements. By the late spring and summer of 1973, the other exporters, observing the upward bound of prices on the open market, came around to that same point of view. They cited rising inflation and the dollar's devaluation, but more than anything else, they cited what was happening to price. Between 1970 and 1973, the market price for crude oil doubled. The exporters' per-barrel revenues were going up, but the companies' part of revenues was also increasing in the buoyant market, which was in sharp contradiction to the objectives and ideology of the exporters. The companies' part of the pie was supposed to decline, not grow. The price system, based on the 1971 Tehran Agreement, was "now out of whack," Yamani told the president of Aramco in July 1973. By September, Yamani was ready to deliver a eulogy on the Tehran Agreement. It was, he said, "either dead or dying." If the companies did not cooperate in framing a new price agreement, he added, the exporters would "exercise our rights on our own." Even as the economics of oil were changing, so were the politics that surrounded it—and dramatically so.[3]

The Secret: Sadat's Gamble

Upon coming to power after Nasser's death in 1970, Anwar Sadat had been regarded by many as a nonentity and was judged likely to last only a few months, or even weeks. The new President of Egypt was much underestimated. "The legacy Nasser left me was in a pitiable condition," he was later to say. He inherited a country that, in his view, had been politically and ethically bankrupted under the lofty rhetoric of pan-Arabism. The vaulting ambitions and self-confidence derived from Egyptian success in the 1956 Suez crisis had turned to dust, particularly in the aftermath of the 1967 defeat, and the country was in economic ruin. Sadat did not have the ambition to lead a united Arab nation stretching from the Atlantic to the Persian Gulf; as an Egyptian nationalist, he wanted to concentrate not on pan-Arabist visions but on the restoration of Egypt.

Egypt was pouring over 20 percent of its gross national product into military expenditures. (Israel was close behind, at 18 percent.) How could Egypt ever make economic progress in such circumstances? Sadat wanted to break out of the cycle of conflict with Israel, and out of stalemated diplomacy. He wanted some kind of stabilization or settlement, but after a couple of years of fruitless negotiation and discussion, he concluded that this could not occur while Israel sat on the eastern bank of the Suez Canal. Israel would have little incentive to negotiate, and he could not negotiate from such a position of weakness and humiliation, certainly not while the entire Sinai was in Israeli hands. He would have to do something. He moved first to consolidate his own domestic position and to assure himself a free hand internationally. He purged the pro-Soviet Egyptians; and then, in July 1972, he threw out the arrogant Soviet military advisers, as many as twenty thousand of them, though he continued to receive Soviet military supplies. Still, Sadat did not get the response he expected from the West, and in particular from the United States.

In late 1972 and early 1973, Sadat came to his fateful decision. He would go to war; it was the only way to obtain his political objectives. "What literally no one understood beforehand was the mind of the man," Henry Kissinger was later to say. "Sadat aimed not so much for territorial gain but for a crisis that would alter the attitudes into which the parties were then frozen—and thereby open the way to negotiations. The shock would enable both sides, including Egypt, to show a flexibility that was impossible while Israel considered itself militarily supreme and Egypt was paralyzed by humiliation. His purpose, in short, was psychological and diplomatic, much more than military."

Sadat's decision was calculated; he was operating on Clausewitz's dictum that war was the continuation of politics by other means. Yet, at the same time, he made his decision with a profound sense of fatalism; he knew he was gambling. While the possibility of a war was hinted and even talked about in a general way, it was not taken very seriously, especially by those who would be its object—the Israelis. Yet by April of 1973 Sadat had begun formulating with Syria's President Hafez al-Assad strategic plans for a joint Egyptian-Syrian attack. Sadat's secret—the specifics and the reality of his preparations for war—was tightly kept. One of the few people outside the high commands of Egypt and Syria with whom he shared it was King Faisal of Saudi Arabia. And that meant oil would be central to the coming conflict.[4]

The Oil Weapon Unsheathed: Faisal Changes His Mind

Ever since the 1950s, members of the Arab world had been talking about using the hazily defined "oil weapon" to achieve their various objectives regarding Israel, which ranged from its total annihilation to forcing it to give up territory. Yet the weapon had always been deflected by the fact that Arab oil, while it seemed endlessly abundant, was not the supply of last resort. Texas, Louisiana, Oklahoma—those states could always and quickly put additional oil into the world market. But once the United States hit 100 percent in terms of production rates, that old warrior, American production, could not rise up again to defend against the oil weapon.

In the early 1970s, as the market tightened, various elements in the Arab world became more vocal in calling for use of the oil weapon to achieve their economic and political objectives. King Faisal of Saudi Arabia was not among them. He hated Israel and Zionism as much as any Arab leader. He was sure there was a Zionist-communist plot to take over the Middle East; he told both Gamal Abdel Nasser and Richard Nixon that the Israelis were the real paymasters for radical Palestinian terrorists. Yet Faisal had gone out of his way to reject use of the oil weapon. In the summer of 1972, when Sadat called for the manipulation of oil supplies for political purposes, Faisal was quick to disagree strongly. It was not only useless, he said, but "dangerous even to think of that." Politics and oil should not be mixed. So Saudi Arabia had discovered for itself during the 1967 war, when it had cut its exports to no avail, except in terms of its own loss of revenues and markets. Faisal believed that the United States was unlikely to be affected by any cutoffs because it would not need Arab Gulf oil before 1985. "Therefore my opinion is that this proposal should be ruled out," he said emphatically. "I see no profit in discussing it at this time."

There were political as well as economic reasons for Faisal's caution. One Marxist state was already established on the Arabian Peninsula in South Yemen, where only recently the British flag had flown over the port of Aden, and Marxist guerillas were fighting elsewhere on the peninsula. In 1969, the same year that military cabals overthrew both the monarchy in Libya and the civilian government in the Sudan, a plot was uncovered among some air force officers in Saudi Arabia. Faisal feared the spread through the Arab world of radicalism, which challenged the legitimacy of kingship, and he rejected its agenda. He knew that his country was tied very closely to the United States in an economic and strategic relationship that was fundamental to his kingdom, not only for prosperity, but also for security. It was hardly desirable to take belligerent action against a government that was so important to one's own survival. Yet, by early 1973, Faisal was changing his mind. Why?

Part of the answer lay in the marketplace. Much sooner than expected, Middle Eastern oil, not American, had become the supply of last resort. In particular, Saudi Arabia had become the marginal supplier for everybody, including the United States; American dependence on the Gulf had come not by the widely predicted 1985, but by 1973. Saudi Arabia had, at last, graduated to the position once held by Texas; the desert kingdom was now the swing producer for the entire world. The United States would no longer be able to increase production to supply its allies in the event of a crisis, and the United States itself was now, finally, vulnerable. The supply-demand balance was working to make Saudi Arabia even more powerful. Its share of world exports had risen rapidly, from 13 percent in 1970 to 21 percent in 1973, and was continuing to rise. Its average output in July 1973, 8.4 million barrels per day, was 62 percent higher than the July 1972 level of 5.4 million barrels per day, and it was still going up. Aramco was producing at capacity; indeed, it had pushed production up very quickly to meet the unexpected rush of demand, and some alleged that, whatever happened, Saudi Arabia would have to cut back production to prevent damage to the fields and to permit the development of more capacity.

In addition, there was a growing view within Saudi Arabia that it was now earning revenues in excess of what it could spend. Two devaluations of the American dollar had abruptly cut the worth of the financial holdings of countries with large dollar reserves, including Saudi Arabia. Libya and Kuwait had imposed production restraints. "What is the point of producing more oil and selling it for an unguaranteed paper currency?" the Kuwaiti oil minister had rhetorically asked. "Why produce the oil which is my bread and butter and strength and exchange it for a sum of money whose value will fall next year by such-and-such a percent?" Perhaps, some Saudis argued, their own country should do the same and substantially cut back on its output.

These changing conditions in the marketplace, which with each passing day made the Arab oil weapon more potent, coincided with significant political developments. Faisal had been estranged, for the most part, from Nasser, whom he saw as a radical pan-Arabist who wanted to topple traditional regimes. Anwar Sadat, Nasser's successor, was cut from a different cloth; he was an Egyptian nationalist who was seeking to dismantle much of Nasser's legacy. Sadat had become close to the Saudis through the Islamic Conference, and Faisal was sympathetic to Sadat for trying to escape from the suffocating bearhug of the alliance that Nasser had made with the Soviet Union. Without Saudi support, Sadat might be forced to revert to the Soviet connection, and the Russians would thereupon use every opportunity to expand their influence throughout the region, which was exactly counter to Saudi Arabia's interests. By the spring of 1973, Sadat was strongly pressing Faisal to consider using the oil weapon to support Egypt in a confrontation with Israel and, perhaps, the West. King Faisal also felt growing pressure from many elements within his kingdom and throughout the Arab world. He could not afford to be seen as anything other than forthright in his support both for the "frontline" Arab states and the Palestinians. Otherwise, Saudi assets, beginning with oil installations, would be at risk from guerilla activity. Underlining that vulnerability, armed men attacked the Tapline terminal at Sidon in the spring of 1973, destroying one tank and damaging others. A few days later, the pipeline itself was hit. There were a number of other incidents, including an assault that ruptured the line inside Saudi Arabia.

Thus, politics and economics had come together to change Faisal's mind. Thereupon the Saudis began a campaign to make their views known, warning they would not increase their oil production capacity to meet rising demand, and that the Arab oil weapon would be used, in some fashion, unless the United States moved closer to the Arab viewpoint and away from Israel. In early May 1973, the King himself met with Aramco executives. Yes, he was a staunch friend of the United States, he said. But it was "absolutely mandatory" that the United States "do something to change the direction that events were taking in the Middle East today."

"He barely touched on the usual conspiracy idea but emphasized that Zionism and along with it the Communists were on the verge of having American interests thrown out of the area," the president of Aramco reported afterward. "He mentioned that except for Saudi Arabia today it was most unsafe for American interests" in the Middle East. "He stated that it was up to those Americans and American enterprises who were friends of the Arabs and who had interests in the area to urgently do something to change the posture" of the United States government. "A simple disavowal of Israeli policies and actions" would "go a long way toward overcoming the current anti-American feeling," the Aramco president said, adding there was "extreme urgency" to the King's remarks.

To the relief of the apprehensive Aramco executives, the subject of oil itself was never specifically mentioned at that meeting. It did, however, come up explicitly a few weeks later, when executives from Aramco's parent companies were meeting with Yamani at the Geneva Intercontinental Hotel. Would they like, asked Yamani, to make a courtesy visit to the King, who happened to be resting in Geneva after a trip to Paris and Cairo? The oil men were, of course, pleased by the invitation. In passing, Yamani happened to mention that the King had just had a "bad time" in Cairo; Sadat had put intense pressure on him for greater political support. "Time is running out with respect to U.S. interest in Middle East," the King said to the oil men when they met. "Saudi Arabia is in danger of being isolated among its Arab friends, because of failure of the U.S. Government to give Saudi Arabia positive support." Faisal was emphatic; he would not let such isolation occur. "You will lose everything," he told the oil men.

They had no doubt what he meant. "Concession is clearly at risk," said one of the executives afterward. They blamed the American media and indicated that they themselves were not immune to the conspiracy theories. The agenda, as these oil executives saw it, was clear: "Things we must do (1) inform U.S. public of their true interests in the area (they are now being misled by controlled news media) and (2) inform Government leaders—and promptly."

A week later, the company executives were in Washington, at the White House and the State and Defense departments. They summarized Faisal's warnings. "Action must be taken urgently; otherwise, everything will be lost." They found an attentive audience, but receptive only up to a point. There was a problem, certainly, acknowledged the government officials. But they also expressed, according to the company representatives, "a large degree of disbelief that any drastic action was imminent or that any measures other than those already underway were needed to prevent such from happening." The Saudis, the government officials said, had faced much greater pressure from Nasser in the past. They "had handled such successfully then and should be equally successful now." In any event, there was little the United States could do in the short term, the oil men were told in Washington. "Some believe" that the King "is calling wolf when no wolf exists except in his imagination." One of the U.S. senior officials opined that the King's remarks at the Geneva meeting were for "home consumption"—to which one of the executives replied sharply that no one from "home" was there.

Three of the companies—Texaco, Chevron, and Mobil—all publicly called for a change in American Middle East policy. So did Howard Page, the retired Exxon director for the Middle East. King Faisal suddenly made himself available to the American press, which, despite its supposedly being "controlled," responded eagerly. In short order, he was interviewed by the *Washington Post*, the *Christian Science Monitor, Newsweek,* and NBC Television. His message was the same to each. "We have no wish to restrict our oil exports to the United States in any way," he told American television viewers, but "America's complete support for Zionism and against the Arabs makes it extremely difficult for us to continue to supply the United States with oil, or even to remain friends with the United States."[5]

Nervous Leaders

In June 1973, Richard Nixon was hosting Soviet General Secretary Leonid Brezhnev at a summit meeting at Nixon's estate in San Clemente, California. After the two leaders had retired for the night on the last evening of the meeting, something unexpected occurred. An agitated Brezhnev, troubled and unable to sleep, suddenly demanded an unscheduled meeting with Nixon. Despite the obvious breach of diplomatic protocol, Nixon was awakened by the Secret Service. Though suspicious, the President received Brezhnev in his small study, overlooking the dark Pacific, in the middle of the night. For as much as three hours, in front of a little fire, Brezhnev in rough and emotional terms argued that the Middle East was explosive, that a war might soon start there. The only way to avoid war, insisted the Soviet leader, was with a new diplomatic initiative. Brezhnev was communicating that the Soviets knew something about Sadat's and Assad's intent, in general terms, if not the specifics—after all, they were supplying the weapons—and that the consequences would threaten the new Soviet-American "détente." But Nixon and National Security Assistant Henry Kissinger thought Brezhnev's strange démarche was a heavy-handed ploy to force a Mideast settlement on Soviet terms, rather than a genuine warning, and dismissed it.

On August 23, 1973, Sadat made an unannounced trip to Riyadh to see King Faisal. The Egyptian had news of great moment. He told the King that he was considering going to war against Israel. It would begin with a surprise attack and he wanted Saudi Arabia's support and cooperation. He got it. Faisal allegedly went so far as to promise half a billion dollars for Sadat's war chest. And, the King pledged, he would not fail to use the oil weapon. "But give us time," the King was reported to have added. "We don't want to use our oil as a weapon in a battle which only goes on for two or three days, and then stops. We want to see a battle which goes on for long enough time for world opinion to be mobilized."

The effect of Sadat's plan on Faisal was quite evident. Less than a week later, on August 27, Yamani told an Aramco executive that the King had suddenly taken to asking for detailed and regular reports on Aramco's production, its expansion plans, and the consequences of curtailment in its production on consuming countries—in particular, on the United States. At one point, the King asked what would be the effect if Aramco's production were reduced by two million barrels per day. "This is a completely new phenomenon," explained Yamani. "The King never bothered with such details."

Yamani sounded a warning. There were elements in the United States led by Kissinger, he said, that "are misleading Nixon as to the seriousness" of Saudi Arabia's intentions. "For that reason, the King had been giving interviews and making public statements designed to eliminate any doubt that might exist" in the United States. "Anyone who knows our regime and how it works realizes that the decision to limit production is made only by one man, i.e., the King, and

579

that he makes that decision without asking for anybody's concurrence." The King, continued Yamani, is "one hundred percent determined to effect a change in U.S. policy and to use oil for that purpose. The King feels a personal obligation to do something and knows that oil is now an effective weapon." Yamani continued, "He is additionally under constant pressure from Arab public opinion and Arab leaders, particularly Sadat. He is losing patience." Yamani added one other detail. The King was now often quite nervous.

September 1973: "Pressure All Around"

By September 1973, talk about both the security of supply and an impending energy crisis was becoming widespread. The *Middle East Economic Survey* headlined: "The Oil Scene: Pressure All Around." That same month, the major oil companies and the Nixon Administration were discussing their common worry that Libya might shut down all production by the majors. The administration, after a good deal of debate, decided to impose mandatory allocations for some oil products that were in tight supply domestically.

King Faisal had told the oil company executives that a "simple disavowal" of Israeli policies by the United States would help deflect the use of the oil weapon. And, to a certain degree, there were now such disavowals. "While our interests in many respects are parallel to the interests of Israel," Joseph Sisco, the American Assistant Secretary of State, told Israeli television, "they are not synonymous with the state of Israel. The interests of the U.S. go beyond any one nation in the area. . . . There is increasing concern in our country, for example, over the energy question, and I think it is foolhardy to believe that this is not a factor in the situation." He was then asked whether the Arab oil producers might use petroleum as a political weapon against the United States sometime in the future—say, in the 1980s. "I'm in no position to be clairvoyant and predict it," Sisco replied. But "there are obvious voices in the Arab world who are pressing for a linking of oil and politics."

The American "disavowals" also came from even higher levels. At a press conference, President Nixon was asked whether the Arabs would "use oil as a club to force a change in the Middle East policy." It was "a subject of major concern," he replied. All the consuming countries, including the United States, could be affected. "We are all in the same bag when you really come down to it," Nixon said, and then he went on to blame both sides, including Israel, for the impasse. "Israel simply can't wait for the dust to settle and the Arabs can't wait for the dust to settle in the Mid-East. Both sides are at fault. Both sides need to start negotiating. That is our position. . . . One of the dividends of having a successful negotiation will be to reduce the oil pressure."

That pressure was being felt by all the major consumers. In Germany, in September, the Bonn government finally unveiled its first energy program, which had some focus on security of supply. The leading proponent of the program was State Secretary Ulf Lantzke, whose own anxiety had been sparked in 1968 by the American warning to the OECD that its spare capacity was evaporating. "For me," Lantzke later said, "that was a triggering point. From that point onwards, I was trying to turn around energy policies in Germany. The issue was no longer how to resolve our coal problems, but how do we get security of supplies into our policies. It was very, very cumbersome. It took me five years to prepare the ground and convince people, so deep-seated was the political belief that energy supplies were no problem."

In Japan, in that same anxious month of September, a newly created Agency for Resources and Energy in the Ministry of International Trade and Industry (MITI) released a White Paper on energy that addressed the overall insecurity of oil supply and emphasized the need to institute measures to cope with an emergency. It was the result of concerns that had risen over the previous year or so about what Japan's rampant growth in oil demand meant in terms of dependence and vulnerability. Most of the nation's oil was supplied directly or indirectly by the international companies, and both government officials and businessmen could discern the rapid shift of power away from the companies toward the oil-exporting countries. "The oil-supply management system, until now run by the international oil companies, has crumbled," the September 1973 White Paper bluntly observed. And for Japan, that meant "the passive international response of the 1960s can no longer be permitted."

By that time, in accord with the changing situation, a new strand had emerged in Japanese foreign policy, which had heretofore been firmly anchored in the United States–Japan alliance. It was called "resource diplomacy," and it aimed to reorient Japan's foreign policy in such a way as to try to guarantee access to oil. Its most prominent champion was MITI Minister Yasuhiro Nakasone (later Prime Minister), who argued, "It is inevitable that Japan will competitively follow her own independent direction. The era of blindly following has come to an end." It was the United States that should no longer be so followed. In June 1973, Nakasone had called for a new resource policy "standing on the side of the oil producing countries." By then, concern about an energy crisis was already commonplace in some circles in Japan. The previous winter had seen shortages of kerosene and gasoline, and now, in the summer of 1973, there were signs, as in the United States, of electricity brownouts. It seemed, at least to one visitor, that almost every Japanese policymaker concerned with energy had read James Akins's article, "The Oil Crisis: This Time the Wolf Is Here," and was persuaded. The only question was "When?" On September 26, Prime Minister Kakuei Tanaka told a television interviewer, "Regarding the energy crisis, an oil crisis ten years from now is clearly seen."

It was more like ten days. For, at that very moment, Anwar Sadat was beginning his countdown to war.[6]

Nothing Further to Negotiate

At a meeting in Vienna in mid-September 1973, the OPEC countries had called for a new deal with the oil companies. The Tehran and Tripoli agreements were dead. The members of OPEC were determined to capture for themselves what they said were the "windfall profits" that the companies were earning from the increase in market prices. Oil company representatives were summoned to meet in Vienna on October 8 with a team headed by Yamani.

In order to negotiate as a group at the meeting, the oil companies would once again have to obtain a business review letter from the Justice Department, giving them assurance that they were not in violation of antitrust laws. The companies' joint lawyer, the venerable John J. McCloy, requested such permission from Washington on September 21, initiating strenuous diplomacy not only between the companies and the Justice Department, but between a skeptical Justice and a worried State Department. At one heated meeting with Justice, McCloy invoked the names of former attorney generals, going back to Robert Kennedy, who had permitted the companies to work out joint strategies on difficult matters involving foreign affairs. "If Justice failed to give clearance," he said, "Justice would be responsible for the companies being picked off one by one." Kenneth Jamieson, chairman of Exxon, argued, "The industry was indispensable in holding the line against the volatile Arab world." Justice Department attorneys, citing a book by an MIT professor that bore little relevance to the political crisis at hand, insisted that oil prices were going up because of the machinations of large integrated oil companies, not because of market conditions and OPEC's move to capitalize on them. Jamieson stared in disbelief. But, finally, on October 5, three days before the Vienna meeting was scheduled to begin, the Antitrust Division reluctantly gave McCloy's clients the clearance they needed to negotiate jointly.

Though the previous spring there had been some anxiety in Washington about the possibility of military conflict in the Middle East, it had dissipated over the summer, and for months the American intelligence community had, for the most part, been dismissing the likelihood of war. It did not make sense: The Israelis had no reason to open hostilities, nor could they dare launch a preemptive attack as they had in 1967. Since the Israelis were thought to hold military superiority, it appeared irrational for the Arabs to consider starting a war in which they would be badly beaten. The Israelis, whose survival was at stake, also consistently dismissed the prospect of war, which greatly influenced the American reading of the situation.

There was an exception to the consensus. In late September, the National Security Agency reported that the sudden intensification of military signals suggested that war might be imminent in the Middle East. The warning was passed over. On October 5, the Soviets suddenly airlifted dependents out of Syria and Egypt. The obvious significance of that move was also disregarded. A CIA analysis for the White House on that day reported: "The military preparations that have occurred do not indicate that any party intends to initiate hostilities." At 5:30 P.M. on October 5, the latest Israeli estimate was given to the White House: "We consider the opening of military operations against Israel by the two armies [of Egypt and Syria] as of low probability." The Watch Committee, representing the entire American intelligence community, reviewed developments and prospects. War, it said, was unlikely.

On that same day, while it was still afternoon in Washington, it was already nighttime in the Middle East, and in Israel, the country was coming to a halt, as the somber and most sacred Jewish holiday of Yom Kippur began. In Riyadh, the members of the Saudi delegation to OPEC boarded their jet for Vienna. They spent the time on the plane focused on their technical dossiers—such matters as prices, inflation, company profits, and gravity differentials. Only when the delegates arrived in Vienna on October 6 did they learn the dramatic news—that Egypt and Syria had launched their surprise attacks against Israel. And that morning, U.S. East Coast time, senior American officials and oil executives awoke to find the Middle East at war.

The outbreak of hostilities created great commotion among OPEC delegates in Vienna. As the oil company officials arrived for the discussions, they found that the Arab delegates were excitedly passing around newspaper articles and photographs. There could be no question that the Arab members of OPEC, at least, were taking sustenance and gaining confidence from what appeared to be an Arab victory on the battlefield. The oil men, for their part, could not have been anything but nervous. Not only were they on the defensive when it came to price, but at any moment the oil weapon in some form could be called into play. The Iranian petroleum minister noted that the oil men were "a little panicky." He sensed something else more profound, as well: "They were losing their strength."

At the negotiating table, even as war raged in the Middle East, the companies offered a 15 percent increase in the posted price, about forty-five cents more per barrel. To the oil exporters, that was wholly and laughably inadequate. They wanted a 100 percent increase—another three dollars. The gap was huge. The companies' negotiating team, led by George Piercy of Exxon and André Bénard of Shell, could not reply without consulting their principals in Europe and the United States. Could they negotiate further? What kind of new offer could they next put on the table? When the crucial replies came from London and New York, the answer was essentially "no offer," at least for the time being. The difference was so great that the companies did not dare begin the dangerous effort to bridge it without, in turn, first consulting the governments of the major industrial countries. What would be the effect on the economies of the Western world? Could the large increases be passed on to consumers? Moreover, the companies had been criticized for giving way too easily in the past before OPEC, and the decision now was too momentous, too political, to be theirs alone. So the various corporate headquarters told Piercy and Bénard not to negotiate further, but to ask for a delay while the consultations with the Western governments could be carried out. Between October 9 and 11, the United States, Japan, and a half-dozen Western European governments were canvassed. The reply was virtually unanimous: The increase sought by the exporters was much too big, and the companies should definitely not improve their offer to the point where OPEC might actually accept it.

It was after midnight, in the very early hours of October 12, six days since the war had begun, when Piercy and Bénard went to see Yamani in his suite at the Intercontinental Hotel. They could make no further offer at that time, they explained, and they asked for two weeks in which to frame a reply. Yamani did not say anything. Then he ordered a Coke for Piercy, cut a lime, and squeezed it into the drink. He was waiting. He wanted to keep things going. He gave Piercy the Coke, but neither Piercy nor Bénard had anything in return to give him.

"They won't like this," Yamani said at last. He put through a call to Baghdad, talked vigorously in Arabic, and then told the two oil men, "They're mad at you."

Yamani dialed one of the rooms of the delegation of Kuwait, whose members were also staying in the Intercontinental. The Kuwaiti oil minister soon appeared, in his pajamas. There were further excited conversations. Yamani began looking up airline timetables. Still there was nothing further to negotiate. Finally, in the dark hours of the morning, the impromptu session broke up. In leaving, George Piercy asked what would happen next.

"Listen to the radio," replied Yamani.[7]

Sadat's Surprise

The choice of Yom Kippur as the day to launch the latest Arab attack was designed to catch Israel when it was least prepared. Its entire defense strategy depended on rapid and total mobilization and deployment of its ready reserves. On no other day would such a response be so difficult; the country was shut down for meditation, introspection, soul searching, and prayer. Moreover, Sadat was intent on strategic surprise, and to that end had put considerable effort into deception. At least twice before he had feinted, appearing to prepare for war. Both times, Israel had mobilized at great cost and budgetary stress but to no purpose, and that experience did what Sadat had hoped—made them skeptical and complacent. Indeed, the Israeli Chief of Staff had found himself publicly criticized for a costly and unnecessary mobilization in May 1973. Assad had joined in the deception. A terrorist organization with links to Syria kidnapped some Soviet émigrés traveling to Vienna from Moscow, and Israeli Prime Minister Golda Meir went to Austria to deal with the crisis, which took up the attention of the Israeli leadership until October 3.

There were, however, genuine warning signs of an impending attack. The Israelis had disregarded them, as had the Americans. A few weeks before the attack, a Syrian source provided the United States with astonishingly exact information, including Syria's order of battle, but that intelligence was only identified afterward, lost as it had been among hundreds of other pieces of information, some quite contradictory. Assad, in Syria, had ordered the preparation of large graveyards, another ominous sign. On October 3, a member of the U.S. National Security Council asked a CIA official about the large Egyptian troop movements. "The British, when they were still in Egypt, used to hold their fall maneuvers at this time of year," the CIA man replied. "The Egyptians are just carrying on." Some American officials noted reports that Egyptian hospital beds were suddenly being emptied, but these reports were waved away as merely another element in the Egyptian military exercises and without significance. On October 1 and again on October 3, a young Israeli lieutenant submitted reports to his superiors on the movement of Egyptian forces that pointed to an imminent war. They, too, were ignored. The Israeli military, and its intelligence especially, were in thrall to the "conception," a particular view of the necessary preconditions for war, which, by definition, precluded an Egyptian attack under the current circumstances. Yet in the first days of October, a key Israeli source in Egypt had sent an urgent signal; he was hurriedly extricated and rushed to Europe, where he was debriefed. There could be no doubt about what he said. But, inexplicably, there was a delay of a day in transmitting his warning to Tel Aviv. By then, it was too late.

The Americans, as much as the Israelis, made the fundamental error of not thinking the way Sadat thought, of not putting themselves into his shoes, and of not taking him, and what he said, seriously enough. Attitudes and ideas prevented key intelligence warnings from being identified, analyzed, and correctly interpreted. Until October 1973, Kissinger later admitted, he thought of Sadat as more actor than statesman. Sadat's gamble paid off, and the enormity of the surprise of the Arab attack would be for the Israelis what Pearl Harbor had been thirty-two years earlier for Americans. Afterward, the Israelis would ask themselves how they could have been caught so completely off guard. The signals were all so clear. But those signals were not so easily extracted from the noise of contradictory information mixed with deliberate deception, especially when complacency and overconfidence had taken hold.

When finally, nine and a half hours before the attack, the Israelis got what they took as confirmation of imminent hostilities, they were still hamstrung. This was not 1967. They could not go first, they could not take preemptive action. Also, in a fatal piece of misinformation, they thought the war would start four hours later than it actually did. They were not, in any manner, ready, and in the initial few days of the attack, the Israelis fell back, disordered, before the onslaught, while the Egyptians and Syrians scored massive victories.[8]

"The Third Temple Is Going Under"

Once war broke out, America's number-one objective quickly became to arrange a truce, whereby the belligerents would pull back to their prehostilities lines, to be followed by an intensified search for a diplomatic solution. The United States wanted, as a top priority, to keep out of direct involvement; it did not want to become too obviously engaged in supplying the Israelis against the Soviet-supplied Arabs, but this was thought to be unlikely owing to the purported Israeli superiority. While U.S. policy would not countenance an Israeli defeat, it regarded the best outcome, in the words of one senior official, as one in which "Israel won, but had its nose bloodied in the process," thus making it more amenable to negotiation.

Something far worse, however, than a bloody nose suddenly appeared to be at hand, owing to the second big miscalculation on the part of Israel (the first being that there would be no war at all). Israel had assumed that it had enough supplies to last for three weeks of war, a premise that was derived from the experience of the 1967 Six-Day War. But the 1967 war had been much easier from Israel's point of view, for then it had had both the upper hand militarily and the advantage of surprise. Now, immediately thrown on the defensive by an Egypt and a Syria both richly equipped with Soviet weapons, the Israelis found themselves devouring materiel at an alarming pace, one far greater than anything they had anticipated. That miscalculation of requirements would prove a grave one for Israel; it would also lead directly to a staggering change in world oil.

On Monday, October 8, two days after the surprise attack, Washington told the Israelis that they could pick up some supplies in the United States in an un-marked El Al plane. That, it was thought, would be sufficient. But Israel was still reeling from the initial attack. A distraught Moshe Dayan, Israel's Defense Minister, told Premier Golda Meir that "the Third Temple is going under," and Meir herself prepared a secret letter to Richard Nixon, warning that Israel was being overwhelmed and might soon be destroyed. On October 9, the United States realized that the Israeli forces were in deep trouble, and were becoming desperately short of supplies. On October 10, the Soviet Union began a massive resupply to Syria, whose forces had started to retreat, and then to Egypt. The Soviets also put airborne troops on alert and began encouraging other Arab states to join in the battle. The United States then started discussions about having more unmarked El Al planes come to the United States for additional supplies. At the same time, the State Department began pressing American commercial carriers to provide charters to ferry materiel to Israel. Kissinger thought such an approach could be relatively low profile and would avoid out-and-out United States identification with Israel. "We were conscious of the need to preserve Arab self-respect," Kissinger later said. But the huge scale of the Soviet resupply soon became evident. And on Thursday, October 11, the Americans realized that Israel could lose the war without resupply. In Kissinger's and even more so in Nixon's formulation, the United States could not allow an American ally to be defeated by Soviet arms. Moreover, who could know the consequences of a fight to the death?

On Friday, October 12, two private letters were sent to Nixon. One was from the chairmen of the four Aramco companies—Exxon, Mobil, Texaco, and Standard of California—hurriedly sent, via John McCloy. They said that the 100 percent increase in the posted price of oil that the OPEC delegation in Vienna was demanding would be "unacceptable." But some kind of price increase was warranted since "the oil industry in the Free World is now operating 'wide open,' with essentially no spare capacity." Yet they had something even more urgent they wanted to communicate. If the United States increased its military support for Israel, there could be a "snowballing effect" in terms of retaliation "that would produce a major petroleum supply crisis." There was a further warning. "The whole position of the United States in the Middle East is on the way to being seriously impaired, with Japanese, European, and perhaps Russian interests largely supplanting United States presence in the area, to the detriment of both our economy and our security."

The second letter was a desperate message from Israel's Premier, Golda Meir. Her nation's survival and the lives of its people, she wrote, now hung in the most precarious balance. Her warning was confirmed around midnight on that Friday, when Kissinger learned that Israel might well run out of critical munitions over the next several days. He also learned from Secretary of Defense James Schlesinger that all efforts to arrange commercial charters had failed. The American airlines did not dare risk either an Arab embargo or terrorist attacks, and they certainly did not care to send their planes into a war zone. In order for the United States government to draft them into service, they said, the President should declare a national emergency. "If you want supplies there," Schlesinger told Kissinger, "we are going to have to use U.S. airlift all the way.

There's no alternative. There are not going to be new supplies without a U.S. airlift."

Kissinger had to agree. But he asked Schlesinger to get the Israelis' word that United States Air Force planes could land under cover of darkness, be unloaded, and be back off the ground by daybreak. If they were not seen, the resupply could be kept as inconspicuous as possible. Before daylight Saturday morning, October 13, Schlesinger had the Israeli promise, and the Military Airlift Command began to move supplies from bases in Rocky Mountain and Midwestern states to an airfield in Delaware. But the American planes would need a refueling stop on the way to Israel. On Saturday morning, the United States asked Portugal for landing rights in the Azores. It took direct and blunt pressure from President Nixon himself to get the required permission.

Still, Washington hoped to keep the low profile, but the presumption of secrecy did not take into account an unexpected act of nature. There were powerful crosswinds at Lajes airfield in the Azores, which would have put the huge C-5A transports at risk, so they were held back in Delaware, their bellies crammed full of supplies. The crosswinds did not diminish until late afternoon, which meant a half-day's delay. As a result, the C-5As did not arrive in Israel in the darkness of Saturday night. Instead, they came lumbering out of the sky on Sunday during the day, October 14, their immense white stars visible for all to see. The United States, instead of keeping to its position of honest broker, was now portrayed as an active ally of Israel. The aid had been extended to counterbalance the huge Soviet resupply to the Arabs, but that did not matter. Not knowing of the strenuous efforts to keep American aid in the background, Arab leaders assumed that it was meant to be a dramatic and highly visible sign of support.

The Israelis had succeeded in stopping the Egyptian offensive before it could break through the critical mountain passes in the Sinai, and on October 15, they launched the first of a series of successful counteroffensives against the Egyptians. Meanwhile, in Vienna on October 14, OPEC had announced the failure of its negotiations with the companies, and the Gulf OPEC countries scheduled a meeting in Kuwait City to resume the oil price issue on their own. But most of the delegates had remained in Vienna, since the breakdown of the talks with the companies, and now they found themselves stranded. They were frantically trying to book airplane seats, but because of the war, the airlines had canceled virtually all flights into the Middle East. It appeared that the delegates would not be able to leave at all, meaning that the scheduled meeting in Kuwait would not take place. Then, at last, it was discovered that one flight was still operating—an Air India jet through Geneva that made an intermediate stop at Kuwait City—and, on the evening of October 15, many delegations rushed to the airport and hastily boarded the plane.

On October 16, the delegates of the Gulf states—five Arabs and an Iranian—met in Kuwait City to pick up where the discussions had been left off a few days earlier in Yamani's suite in Vienna. They would not wait any longer for the companies to reply. They acted. They announced their decision to raise the posted price of oil by 70 percent, to $5.11 a barrel, which brought it into line with prices on the panicky spot market.

The significance of their action was twofold—in the price increase itself, and in the unilateral way in which it was imposed. The pretense that the exporters would negotiate with the companies was now past. They had taken complete and total charge of setting the price of oil. The transition was now complete from the days when the companies had unilaterally set the price, to the days when the exporters had at least obtained a veto, to the jointly negotiated prices, to this new assumption of sole suzerainty by the exporters. When the decision was taken, Yamani told one of the other delegates in Kuwait City, "This is a moment for which I have been waiting a long time. The moment has come. We are masters of our own commodity."

The exporters were ready for the expected anguished complaints about the size of the increase. They announced that consuming governments were taking 66 percent of the retail price of oil in taxes, while they received the equivalent of only 9 percent. The Iranian oil minister, Jamshid Amouzegar, said that the exporters were merely keying prices to the forces of the market, and they would set prices in the future on the basis of what consumers were willing to pay. It was for the momentous October 16 decision on price that Yamani had advised George Piercy of Exxon to listen to the radio. As events turned out, however, Piercy learned about it from the newspapers.

If the OPEC exporters could unilaterally raise the price of oil, what might they do next? And what would happen on the battlefield? At the White House, the next day, October 17, Richard Nixon expressed his concern to his senior advisers on national security. "No one is more keenly aware of the stakes: oil and our strategic position." Historic meaning was being given to that statement the same day, halfway around the world, again in Kuwait City. The Iranian oil minister had left the meeting, and the rest of the Arab oil ministers arrived for an exclusively Arab conclave. Their subject was the oil weapon. It was on everybody's mind. The Kuwaiti oil minister declared, "Now the atmosphere is more propitious than in 1967."[9]

Embargo

Yet there remained the question of what exactly Saudi Arabia would do. Despite Sadat's importuning, King Faisal was reluctant to take any action against the United States without more contact with Washington. He sent a letter to Nixon, warning that if American support for Israel continued, Saudi-American relations would become only "lukewarm." That was on October 16.

On October 17, at the time that the oil ministers were meeting in Kuwait City, first Kissinger and then Kissinger and Nixon together received four Arab foreign ministers. They were led by the Saudi, Omar Saqqaf, whom Kissinger would characterize as "gentle and wise." The discussions were cordial, and there seemed to be some common ground. Nixon had pledged to strive for a ceasefire that would make it possible to "work within the framework of Resolution 242," the United Nations resolution that would return Israel to its 1967 borders. The Saudi Minister of State seemed to affirm that Israel had a right to exist, so long as it was within its 1967 borders. Kissinger explained that the American resupply should not be taken as anti-Arab, but rather was "between the U.S. and the

USSR." The United States had to react to the Russian supply. He added that the status quo ante in the region was untenable, and that, after the war was over, the United States would undertake an active diplomatic role and work for a positive peace settlement.

To Saqqaf, Nixon made the ultimate promise: the services of Henry Kissinger as negotiator, which seemed to be Nixon's idea of a sure-fire guarantee of success. Nixon also assured Saqqaf and the other foreign ministers that, despite his Jewish origins, Kissinger "was not subject to domestic, that is to say Jewish, pressures." He went on to add, "I can see that you are concerned about the fact that Henry Kissinger is a Jewish-American. A Jewish-American can be a good American, and Kissinger is a good American. He will work with you." Kissinger was writhing with embarrassment and anger as the President made his gratuitous remark, but Saqqaf was nonplussed. "We are all Semites together," Saqqaf deftly replied. And then the Minister of State made his way to the White House Rose Garden, where he told reporters that the talks had been constructive and friendly, and where, according to the press, it was all smiles, graciousness, and mutual compliments. After the meetings with Saqqaf and the other Arab foreign ministers, Kissinger told his staff he was surprised that there had been no mention of oil, and that it was unlikely that the Arabs would use the oil weapon against the United States.

That, however, was exactly what the Arab oil ministers in Kuwait City were contemplating. Early in 1973, in one of his "thinking aloud" speeches about Egypt's options, Sadat had discussed the oil weapon. And around that time, at his urging, experts from Egypt and other Arab countries had begun drawing up a plan for the use of the oil weapon, taking into account the growing energy crisis in the United States. The Arab delegations in Kuwait City were familiar at least with the concept before the October 17 meeting. But at the meeting itself, radical Iraq had a different notion. The chief Iraqi delegate called on the Arab states to target their ire on the United States—to nationalize all American business in the Arab world, to withdraw all Arab funds from American banks, and to institute a total oil embargo against the United States and other countries friendly to Israel. The chairman of the meeting, the Algerian minister, dismissed such a proposal as impractical and unacceptable. Yamani, on instructions of his King, also resisted what would have been a declaration of all-out economic warfare against the United States, the consequences of which would have been, to say the least, very uncertain for all parties concerned. The angry Iraqi delegates withdrew from the meeting and from the whole embargo plan.

Instead, the Arab oil ministers agreed to an embargo, cutting production 5 percent from the September level, and to keep cutting by 5 percent in each succeeding month until their objectives were met. Oil supplies at previous levels would be maintained to "friendly states." The nine ministers present also adopted a secret resolution recommending "that the United States be subjected to the most severe cuts" with the aim that "this progressive reduction lead to the total halt of oil supplies to the United States from every individual country party to the resolution." Several of the countries immediately announced that they would start with 10 percent, rather than 5 percent, cutbacks. Whatever their size, the production cuts would be more effective than a ban on exports against a single country, because oil could always be moved around, as had been done in the 1956 and 1967 crises. The cutbacks would assure that the absolute level of available supplies went down. The overall plan was very shrewd; the prospect of monthly cutbacks, plus the differentiation among consuming countries, would maximize uncertainty, tension, and rivalry within and among the importing countries. One clear objective of the plan was to split the industrial countries right from the start.

The two meetings in Kuwait City—October 16 and October 17—were not formally connected. The price increase and the OPEC seizure of sole price-setting authority were a logical continuation of what had long been in motion. The decision to use the oil weapon moved on a separate track. "Suffice it to say, however," the *Middle East Economic Survey* commented, "that the new Arab-Israeli war probably stiffened the resolve of the Arab price negotiators." And then, in what proved to be a momentous understatement, it added, "Probably, also, the cuts in output will incidentally serve to push up oil prices still further."

Events moved rapidly after the Kuwait City meetings. On October 18, Nixon met with his Cabinet. "When it became clear that the fighting might be prolonged and the Soviets began a massive resupply effort, we had to act to prevent the Soviets from tilting the military balance against Israel," he told the Cabinet officers. "This past weekend, therefore, we began a program of resupply to Israel." Recalling his discussions the day before with Saqqaf and the others, he continued, "In meeting with the Arab foreign ministers yesterday, I made the point that we favor a cease fire and movement towards a peace settlement based on U.N. Resolution 242. Arab reaction thus far to any resupply effort has been restrained and we hope to continue in a manner which avoids confrontation with them." He was being optimistic.

The next day, October 19, Nixon publicly proposed a $2.2 billion military aid package for Israel. It had been decided on a day or two earlier, and word of it was conveyed to several Arab countries in advance so that they would not be surprised by the announcement. The strategy was to try to assure that neither Egypt nor Israel ended up in a position of ascendancy, with the result that both would have reason to go to the negotiating table. That same day, Libya announced that it was embargoing all oil shipments to the United States.

At two o'clock Saturday morning, October 20, Kissinger departed for Moscow to try to devise a cease-fire formula. On board the plane, he learned further stunning news. In retaliation for the Israeli aid proposal, Saudi Arabia had gone beyond the rolling cutbacks; it would now cut off all shipments of oil, every last barrel, to the United States. The other Arab states had done or were doing the same. The oil weapon was now fully in battle—a weapon, in Kissinger's words, "of political blackmail." The three-decade-old postwar petroleum order had died its final death.

The embargo came as an almost complete surprise. "The possibility of an embargo didn't even enter my mind," said a senior executive of one of the Aramco companies. "I thought that if there was an outbreak of war and if the United States was on the side of Israel, there was no way that the U.S. companies in Arab countries would not be nationalized." Nor was much thought given in the United States government to the prospect of an embargo, despite the evidence at hand: almost two decades of discussion in the Arab world about the "oil weapon," the failed attempt in 1967, the threats of an embargo made in 1971 at the time of the Tehran negotiations, Sadat's public discussion of the "oil option" in early 1973, and the exceedingly tight oil market of 1973. To be sure, whatever the nature of Faisal's discussions with Sadat, and whatever promises Sadat may have heard, Faisal and the other conservative Arab leaders were reluctant to directly challenge the United States, a country on which they depended for their security. Moreover, they might have been surprised, even shocked in a way, had the United States not provided supplies to Israel. What transformed the situation and finally galvanized the production cuts and the embargo against the United States was the very public nature of the resupply—the result of the crosswinds at the Lajes airfield in the Azores—and then the $2.2 billion aid package. Not to have acted, some Arab leaders thought, could have put certain regimes at the mercy of street mobs. Yet the public show of support for Israel also provided them with a sufficient pretext to take on the United States, as others clearly wanted to do.

Even the embargo itself was not the end of stunning events on October 20. It was only in Moscow on Sunday morning that Kissinger learned what had transpired in Washington the night before. In what became known as the Saturday Night Massacre, a critical point in his Presidency, Nixon fired the special prosecutor, Archibald Cox, who had been appointed to investigate the Watergate scandal, and who had subpoenaed the President's secret office tape recordings. Access to those tapes had become the centerpiece of the struggle between the President and the Senate—to ascertain how much Nixon himself had been directly involved in a maze of illegalities. Immediately after the firing, Attorney General Elliot Richardson and his chief deputy, William Ruckelshaus, resigned in protest. "And now," White House Chief of Staff Alexander Haig told Kissinger over the telephone, "all hell has broken loose." [10]

The Third-Rate Burglary

Throughout the clash of arms in the Middle East and the weeks of crisis over oil, one key actor was otherwise preoccupied. Richard Nixon was thoroughly entangled in the series of events that escalated from what he called a "third-rate burglary" into the unprecedented series of Watergate scandals, at the center of which was the President himself. The United States had seen nothing remotely like it since Teapot Dome. The unfolding of the Watergate saga during the October War; the country's obsession with it; its effects on the war and the embargo, on American capabilities, and on perceptions within the United States—all interacted to create a strange, surrealistic dimension to the central drama on the world stage. For instance, on October 9, the day that a desperate Golda Meir signaled that she wanted to fly to Washington to plead personally for aid, Nixon was working out the resignation of Vice-President Spiro Agnew, who asked Nixon to help him find work as a consultant and complained that the Internal Revenue Service was trying to find out how much he had paid for his neckties. On October 12, the day that senior American officials realized that Israel might lose the war and were grappling with how to resupply, they were summoned to the White House for what Kissinger described as an "eerie ceremony," in which Nixon introduced Gerald Ford as his choice for his new Vice-President.

In the weeks that followed, though Nixon would temporarily depart his own personal crisis and weave in and then out again of the world crisis, effective control over American policy was lodged in the hands of Henry Kissinger, who, in addition to being Special Assistant for National Security, had also just been appointed Secretary of State. Kissinger's original base had been twofold—Harvard's Center for International Affairs, housed in space borrowed from the Harvard Semitic Museum, and his service to Nixon's great rival, Nelson Rockefeller. This former professor, who had fled to the United States as a boy, a Jewish refugee from Nazi Germany, and whose ambition once had been no loftier than to become a certified public accountant, now came, through the strange twists of Watergate and the crumbling of Presidential authority, to be the very embodiment of the legitimacy of the American government. Kissinger's public personality expanded to oversized dimensions to fill the vacuum created by a discredited Presidency. He emerged—for Washington, for the media, for the capitals around the world—as the desperately needed figure of authority and continuity at a time when confidence in America was being severely tested.

Too much seemed to be happening. The media and the public mind were overloaded. But Watergate, and the President's predicament, had direct and major consequences for the Middle East and for oil. Sadat might, at least arguably, never have gone to war had a strong President been able, after the 1972 election, to use his influence to open a dialogue between Egypt and Israel. An undistracted President might also have been able to address the energy issue with greater focus. And once the war began, Nixon was so preoccupied, his credibility so diminished, that he could not provide the Presidential leadership required for dealing with the belligerents, the oil exporters and the explicit economic warfare against the United States—and the Russians. For their part, foreign leaders could not comprehend this strange Watergate process, part ritual, part circus, part tragedy, part thriller, that had gripped American politics and the American Presidency.

Watergate gave, as well, a lasting cast to the energy problems of the 1970s. The accident of coincidence—the embargo and the Saturday Night Massacre, Watergate and the October War—seemed to imply logical connections. Things meshed in hazy, mysterious ways, and this impression left deep and abiding suspicions that fed conspiracy theories and obstructed more rational responses to the energy problems at hand. Some argued that Kissinger had masterminded the oil crisis to improve the economic position of the United States vis-à-vis Europe and Japan. Some believed that Nixon had deliberately started the war and actually encouraged the embargo to distract attention from Watergate. The oil embargo and illegal campaign contributions by some oil companies, which were part of the illegal loot extracted from corporate America by the Committee to Reelect the President, flowed together in the public's mind, greatly expanding the traditional distrust of the oil industry and leading many to think that the October War, the embargo, and the energy crisis had all been created and masterfully manipulated by the oil companies in the name of greed. Such various perceptions were to last much longer than the October War or the Nixon Presidency itself.

Alert

In Riyadh, on the afternoon of October 21, the day following the Saturday Night Massacre, Sheikh Yamani met with Frank Jungers, the president of Aramco. Using computer data about exports and destinations that the Saudis had requested from Aramco a few days earlier, Yamani laid out the ground rules for the cutbacks and the embargo the Saudis were about to impose. He acknowledged that the administration of the system would be very complicated. But the Saudis were "looking to Aramco to police it," he said. "Any deviations by the Aramco offtakers from the ground rules," he added, "would be harshly dealt with." At one point, Yamani departed from the operational details to ask Jungers a more philosophical question. Was he surprised by what had just happened? No, Jungers replied, "except that this cutback was greater than we had anticipated."

Yamani then pointedly asked if Jungers would be "surprised at the next move if this one didn't produce some results."

"No," said Jungers, "I would not be surprised."

Jungers's own guess, based upon his previous conversations with Yamani and other information, was that the subsequent move would be "complete nationalization of American interests if not a break in diplomatic relations." This was suggested by Yamani in his final ominous comment to Jungers: "The next step would not just be more of the same."

In Moscow, meanwhile, Kissinger and the Russians completed a cease-fire plan. But in its implementation over the next few days, it ran into serious snags. Neither the Israelis nor the Egyptians seemed to be observing the cease-fire, and there was the imminent possibility that Egypt's Third Army would be captured or annihilated. Then came a blunt and provocative letter from Leonid Brezhnev to Nixon. The Soviet Union would not allow the Third Army to be destroyed. If that happened, Soviet credibility in the Middle East would also be destroyed and Brezhnev, in Kissinger's words, would "look like an idiot." Brezhnev demanded that a joint American-Soviet force move in to separate the two sides. If the United States would not cooperate, the Soviets would intervene unilaterally. "I shall say it straight," Brezhnev menacingly wrote. His threat was taken very seriously. It was known that Soviet airborne troops were on alert, and Soviet ships in the Mediterranean seemed to be proceeding in a belligerent fashion. Most worrying, neutron emissions from what might be nuclear weapons had been detected on a Soviet freighter passing through the Dardanelles into the Mediterranean. Was Egypt the destination?

A half dozen of the most senior American national security officials were summoned to a hurriedly called late-night emergency meeting in the White House Situation Room. Nixon himself was not awakened for the meeting on the advice of Alexander Haig, who told Kissinger that the President was "too distraught" to join them. Some of the participants were surprised to find that the President was not there. The officials grimly reviewed the Brezhnev message.

Direct Soviet military intervention could not be tolerated; it could upset the entire international order. Brezhnev could not be allowed to assume that the Soviet Union could take advantage of a Watergate-weakened Presidency. There was further reason for alarm. Over the previous few hours, United States intelligence had "lost" the Soviet air transport, which it had been tracking as the planes ferried arms to Egypt and Syria. No one knew where the planes now were. Could they be on their way back to Soviet bases to pick up the airborne troops, already on alert, and carry them into the Sinai?

The officials in the White House Situation Room concluded that the risks had suddenly escalated; the United States would have to respond resolutely to Brezhnev's challenge. Force would have to be prepared to meet force. The readiness state of American forces was raised to DefCon 3 and in some cases even higher, which meant that, in the early morning of October 25, the American military went on a nuclear alert around the world. The message was clear. The United States and the Soviet Union were squaring off directly against each other, something that had not happened since the Cuban Missile Crisis. Miscalculation could lead to a nuclear confrontation. The next hours were very tense.

But the following day, the fighting in the Middle East stopped, Egypt's Third Army was resupplied, and the cease-fire went into effect. It was just in time. The superpowers pulled back from their alerts. Two days later, Egyptian and Israeli military representatives met for direct talks for the first time in a quarter of a century. Egypt and the United States, meanwhile, opened a new dialogue. Both had been objectives of Sadat when he first conceived his gamble a year earlier. The nuclear weapons were sheathed. But the Arabs continued to wield the oil weapon. The oil embargo remained in place, with consequences that would extend far beyond the October War.[11]

C H A P T E R 3 0

"Bidding for Our Life"

THE EMBARGO SIGNALED a new era for world oil. As war was too important to be left to the generals, so oil was now clearly too important to be left to the oil men. Petroleum had become the province of presidents and premiers, of foreign and finance and energy ministers, of congressmen and parliamentarians, of regulators and "czars," of activists and pundits, and in particular, of Henry Kissinger, who, by his own proud admission, had before 1973 known nothing about oil, and precious little about international economics. Politics and grand strategy were what he relished. In the months after the embargo, he would tell aides, "Don't talk to me about barrels of oil. They might as well be bottles of Coca Cola. I don't understand!" Yet once the oil weapon had been brought into play, this diplomatic acrobat would do more than anybody else to get the sword back into its scabbard.

"The Loss"

What was called the "Arab oil embargo" had two elements. The broader one was composed of the rolling production restraints that affected the entire market— the initial cutbacks, then the additional 5 percent each month. The second element was the total ban on export of oil, which was initially imposed on only two countries, the United States and the Netherlands, though subsequently extended to Portugal, South Africa, and Rhodesia. In a bizarre twist, the embargo also extended to United States military forces in the Eastern Hemisphere, including the Sixth Fleet, among whose de facto responsibilities was the protection of some of those states that were imposing the embargo. The oil companies may have thought that they carried out that particular cutoff with a "wink," implying that the missing oil would be made up from other sources; but, if so, the wink was not seen in the Pentagon, where, in the midst of a major military crisis, which might involve American forces, the reaction was one of fury. Nor was it seen in the Congress, where an amendment was hurriedly passed making it a criminal act to discriminate against the Department of Defense. Meanwhile, supplies were resumed to the U.S. forces.

At the beginning of November 1973, only two weeks after the initial decision to use the oil weapon, the Arab ministers decided to increase the size of the across-the-board cuts. But how much oil was really lost? Available Arab oil in the first part of October had totaled 20.8 million barrels per day. In December, at the most severe point in the embargo, it was 15.8 million barrels per day, a gross loss of 5 million barrels per day of supply from the market. This time, however, there was no spare capacity in the United States. Its disappearance represented a major change in the underlying dynamics of politics and oil as they had existed as recently as six years earlier, during the 1967 Six-Day War. America's spare capacity had proved to be the single most important element in the energy security margin of the Western world, not only in every postwar energy crisis but also in World War II. And now that margin was gone. Without it the United States had lost its critical ability to influence the world oil market. Other producers, led by Iran, were able to increase their output by a total of 600,000 barrels per day. Iraq, its proposal for total economic warfare against the United States rejected by the other Arab producers, not only sulked, it actually increased production and thus its revenues. Seeking to explain his country's policies, Iraq's Saddam Hussein blasted the governments of Saudi Arabia and Kuwait as "reactionary ruling circles well known for their links with America and American monopolistic interests" and attacked the cutbacks on supplies to the Europeans and Japanese as likely to throw them back into the arms of the reviled Americans.

The production increases elsewhere meant that the net loss of supplies in December was 4.4 million barrels per day, or about 9 percent of the total 50.8 million barrels per day that had been available in the "free world" two months earlier—not, at first blush, a particularly large loss, proportionately. But it amounted to 14 percent of internationally traded oil. And its effects were made even more severe because of the rapid rate at which world oil consumption had been growing—7.5 percent a year.

Yet all the knowledge about the dimensions of the loss and its limits was gained only after the fact. In the midst of the cutbacks, there was great uncertainty about how much oil was available, combined with an inevitable tendency to exaggerate the loss. The confusion resulted from the contradictory and fragmentary nature of information and from the massive disruption of established supply channels, all overlaid by rabid and violent emotions. Unanswered questions made the fear and disarray even worse. Would further cutbacks be imposed each month? Would new countries be embargoed? Would countries be shifted from the "neutral" to the "preferential" or even to "most favored" list—meaning that the Arabs would reward them for good behavior by giving them more oil? Would other countries find themselves punished more severely?

There was one other very large uncertainty. The oil exporters, in the final analysis, thought in terms of revenues. In 1967, they had pulled back from an embargo because they had found that their total earnings fell. Having learned that lesson, King Faisal had been reluctant, at least through 1972, to resort to the oil weapon. But now, with the price per barrel skyrocketing, the exporters could cut back on volumes and still increase their total income. They could sell less and still earn more. Looking at their earnings, they might decide to make the cuts permanent and never bring the missing barrels back to the market, which could mean a chronic shortage, permanent fear—and even higher prices.[1]

Panic at the Pump

What better recipe could there have been for panic prices than the oil supply situation in the memorable final months of 1973? The ingredients included war and violence, cutbacks in supply, embargoes, shortages, desperate consumers, the specter of further cutbacks, and the possibility that the Arabs would never restore production. Fear and uncertainty were pervasive and had a self-fulfilling effect: both oil companies and consumers frantically sought additional supplies not only for current use but also for storage against future shortages and the unknown. Panic buying meant extra demand in the market. Indeed, buyers were scrambling desperately to get any oil they could find. "We weren't bidding just for oil," said one independent refiner who did not have a secure source of supply. "We were bidding for our life."

The bidding propelled prices even further upward. The posted price for Iranian oil, in accord with the October 16 agreement, was $5.40 a barrel. In November, some Nigerian oil was sold for over $16. In mid-December, Iran decided to hold a major auction to test the market. The bids were dramatic—over $17 a barrel, 600 percent greater than the pre-October 16 price. Then, at a rumor-laden and skillfully managed Nigerian auction, a Japanese trading company—inexperienced in buying oil, under pressure to help assure Japan's supply, and competing with eighty or so other companies—bid $22.60 a barrel. As events turned out, the trading company could never find buyers at that price, and the deal was not consummated, but no one could have known that at the time. There were reports of even higher bids.

The embargo and its consequences sent shock radiating through the social fabric of the industrial nations. The Club of Rome's pessimistic outlook seemed to have been substantiated. E. F. Schumacher, it appeared, had been a prophet after all. His jeremiads on the dangers of the stupendous growth rates in oil demand and on the risks of dependence upon the Middle East had been proved right. With the timely publication of *Small is Beautiful* in 1973, he had found, after decades of anonymity, his metier as spokesman for those who challenged the tenets of unbridled growth and the "bigger is better" philosophy that had dominated the 1950s and 1960s. Now, in his later years, the erstwhile coal champion and energy Cassandra had become a man for his time. The title of his book and its interpretation—"less is more"—became catch phrases of the environmental movement following the embargo, and Schumacher was lionized around the world. Queen Elizabeth awarded him a CBE and invited him to lunch at Buckingham Palace, and he had a private dinner with Prince Philip. "The party is over," Schumacher declared to the world. But then he asked, "Whose party was it anyway?"

The age of shortage was at hand. The prospect, at best, was gloomy: lost economic growth, recession, and inflation. The international monetary system could be subject to extreme dislocation. Most of the developing world would surely suffer a significant setback, and there was good reason to wonder darkly about the political effects on the industrial democracies from the loss of sustained economic growth, which had provided social glue in the postwar years. Would protracted economic problems mean a rebirth of the domestic conflict of the interwar years, which had had such terrible consequences? Moreover, the United States, the world's foremost superpower and the underwriter of the international order, had now been thrown on the defensive, humiliated, by a handful of small nations. Would the international system now unravel? Would the decline of the West mean the inevitable rise of world disorder? As for individual consumers, they began to worry about higher prices, their pocketbooks, and the disruption of their way of life. They feared that the end of an era was at hand.

The effects of the embargo on the psyches of the Western Europeans and the Japanese were dramatic. The disruption instantaneously transported them back to the bitter postwar years of deprivation and shortages. Suddenly their economic achievements of the 1950s and 1960s seemed very precarious. In West Germany, the Ministry of Economics took on the task of allocating supply and found itself buried almost immediately under a sea of telexes from desperately worried industries. The first was from the sugar beet industry, whose season was in full swing. If it ran out of fuel oil for only twenty-four hours, its entire operations would come to a stop, and the sugar would crystallize in the tubes. The specter of Germany's sugar beet industry immobilized, its supplies lost forever to the market, was such that its sugar refineries were swiftly given a sufficient allotment of fuel oil.

In Japan, the embargo came as an even more devastating shock. The confidence that had been built up with strong economic growth was suddenly shattered; all of the old fears about vulnerability rushed back. Did this mean, the Japanese asked themselves, that, despite all their exertions, they would be poor again? The fears aroused by the embargo ignited a series of commodity panics that recalled the violent "rice riots" that had shaken Japanese governments in the late nineteenth and early twentieth centuries. Taxicab drivers staged angry demonstrations, and housewives rushed out to buy and hoard laundry detergent and toilet paper—in some cases, upwards of a two-year supply. The price of toilet paper would have quadrupled, like that of oil, had it not been for government controls. As a result, in Japan, the oil shortage was accompanied by a toilet paper shortage.

In the United States, the shortfall struck at fundamental beliefs in the endless abundance of resources, convictions so deeply rooted in the American character and experience that a large part of the public did not even know, up until October 1973, that the United States imported any oil at all. But, inexorably, in a matter of months, American motorists saw retail gasoline prices climb by 40 percent—and for reasons that they did not understand. No other price change had such visible, immediate, and visceral effects as that of gasoline. Not only did motorists have to shell out more money to fill their tanks, they also passed stations that upped the price of a gallon of gas as often as once a day. But the shortage became even more vivid with the emergence of what John Sawhill, of the Federal Energy Office, called "one-time supply curtailment measures"—better known as "gas lines."

These gas lines became the most visible symbol of the embargo and America's most direct experience of it. An allocation system had been introduced in the United States just before the embargo because of the growing tightness of the market. It was meant to distribute supplies evenly around the country. Now it assured, perversely, that gasoline could not be shifted from an area already well-supplied to one where it was needed. As reports and rumors proliferated, Americans became prey to their own commodity panic—not for laundry detergent or toilet paper, but for gasoline itself. Motorists who had been content to drive until the gauge was virtually on "empty" now hastened to "top off" their tanks, even if it was only a one-dollar purchase, thus contributing to the lengthening gas lines. It was prudent; no one wanted to take a chance on there not being any gasoline tomorrow. At some stations, purchases were allocated by the day of the week and whether the motorist's license plate ended with an odd or even number. Motorists waiting in line an hour or two, with their engines running and their tempers rising, sometimes seemed to burn more gas than they were able to purchase. In many parts of the country, gasoline stations sprouted "Sorry, No Gas Today" signs—very different from the "gas war" signs advertising discounts, which had been so common in the preceding decade of surplus. The embargo and the shortage it caused were an abrupt break with America's past, and the experience would severely undermine Americans' confidence in the future.[2]

"Beef Prices"

Richard Nixon would seek to restore that confidence. In early November, in a Cabinet meeting devoted to energy, one Cabinet officer suggested turning off lights on public buildings. "But you have to hire more police," the practical law-and-order President pointed out. He had something much more significant and far-reaching in mind. On November 7, 1973, he made a major Presidential address on energy to an alarmed and fearful nation. He offered a wide range of proposals. Citizens should turn down their thermostats and car pool. He would seek authority to relax environmental standards, halt the switching of utilities from coal to oil, and establish an Energy Research and Development Administration. He called for a grand new national undertaking, Project Independence. "Let us set as our national goal," he said, "in the spirit of Apollo, with the determination of the Manhattan Project, that by the end of this decade we will have developed the potential to meet our own energy needs without depending on any foreign energy source." To call his plan ambitious was a considerable understatement; it would require many technological advances, vast amounts of money, and a sharp swerve away from the new road of environmentalism. His staff had told him that the goal of energy independence by 1980 was impossible and suggested that it was thus silly to proclaim. Nixon overruled his staff. For energy was now both a crisis and high politics.

The President fired his old energy czar, John Love, a holdover from the pre-embargo days, and replaced him with Deputy Treasury Secretary William

Simon. Telling the Cabinet about Simon's new post, Nixon likened it to Albert Speer's position as armaments overlord in the Third Reich. Had Speer not been given the power to override the German bureaucracy, Nixon explained, Germany would have been defeated far earlier. Simon was somewhat discomfited by the comparison. Nixon further said that Simon would have "absolute authority." But that was one thing he surely did not have in fragmented, contentious Washington. The new energy czar found himself caught up in a never-ending series of hearings by Congressional committees and subcommittees that seemed to run almost around the clock. Once, rushing out of a meeting to answer a summons to one such hearing, Simon was rapidly walking backward so that he could finish a conversation with two lieutenant governors. As he backed into his car, he cracked his head, cutting his scalp. Though Simon needed stitches, the committee chairman would not put off the hearing, and thus the energy czar sat through five hours of interrogation, with blood oozing from the gash in his head. So intense were emotions during these months of gasoline lines that Simon's wife gave up using any charge cards that carried her husband's name.

The Administration resisted the continual clamor for gasoline rationing. Finally, as the din grew ever louder, Nixon ordered rationing stamps printed and held in reserve. "Maybe that will shut them up," he said. Though his Administration continued to generate policies and programs, Nixon himself wanted to be deliberate in reacting to the crisis. One of his aides, Roy Ash, wrote a memo to him advising great caution. "I urge that we not allow pressures of the next month or two, based on a real and immediate shortage, seriously compounded by trendiness and news-magazine hysteria, to result in unnecessary and even counter-productive energy policies," said Ash. "In a few months, I suspect, we will look back on the energy crisis somewhat like we now view beef prices—a continuing and routine governmental problem—but not a Presidential crisis." On this note, Nixon penned two hand-written comments: "absolutely right" and "makes great sense." But to the public, what had happened to gasoline prices far transcended the price of beef. Their birthright as Americans seemed to be in jeopardy.

Who was to blame? The oil industry was held responsible by many for the embargo, the shortages, and the price increases. After the oil companies themselves, the Nixon Administration became the other major target of animosity. In early December, public opinion analyst Daniel Yankelovich forwarded to General Alexander Haig, for the President's attention, a memorandum he had prepared at the behest of Treasury Secretary George Shultz on the "incipient signs of panic" among the public. People are "growing fearful that the country has run out of energy," Yankelovich explained. "A combination of circumstances has shaped an unstable public mood compounded of misinformation, mistrust, confusion and fear." Those circumstances included Watergate, mistrust of the petroleum industry (which was thought to be using the gasoline shortage as an excuse for unfairly raising prices), a general decline of confidence in business, and the belief that the Nixon Administration was "too close to big business." Watergate, Yankelovich added, "has bred a widespread feeling of gloom about the state of the nation," and, as a direct result, public confidence that "things are going well in the country" had plummeted from 62 percent in May 1973 to a mere 27 percent in late November 1973.

Clearly, Watergate increased the need for the weakened Administration to do "something" positive; yet the scandal dogged the Nixon White House at every step of its efforts to respond to the oil crisis, and continually diverted the attention not only of the public but of the top policymakers. "Because of Watergate, there was a general sense of paralysis," recalled Steven Bosworth, who was director of the State Department's Office of Fuels and Energy. "Congress was mesmerized by Watergate, the executive branch was mired in it, and the White House was circling the wagons. It was difficult to get a political decision on anything on an inter-agency basis. There was no real decision-making apparatus in Washington—other than Henry Kissinger."

Watergate was, as Kissinger himself would say, a "hydra-headed monster." He was the only one who seemed able to surmount it. He strove to keep foreign policy, including that involving oil, insulated from Watergate, but domestic energy policy had no such luck. In November 1973, one White House official argued with Haig about a planned announcement of Administration energy actions. "I fully recognize the desire for a heavy newsday on Monday with an eye toward burying the tapes to Sirica issue," he said, referring to the transfer of Nixon's secret Oval Office tape recordings to a federal judge. "However, we are deluding ourselves if we think that any action will bury that issue." Several weeks later, White House adviser Roy Ash added that news on any Presidential action on energy, no matter what day it was released, would not get good press coverage. "Perhaps nothing could win against Watergate," he said. It seemed to those around Nixon that the President was always searching for some political "spectacular" involving oil and the Middle East to try to divert the country from its obsession with Watergate and each new revelation in the scandal. If that was his strategy, it failed.[3]

"Equal Misery"

In such circumstances of global agitation, anger, and suspicion, how was the shortfall in oil supplies to be allocated among nations—by governments or by companies? Several of the companies had warned about the instability of oil supplies. For the American companies, particularly the Aramco partners, the major problem was the Arab-Israeli dispute. If only the United States would jettison Israel, or at least greatly restrain its support, then all could return to normal. In this formulation, the Israelis were generally intransigent; the Arabs were not. To the European companies, the problem was different: The tight supply-demand balance had become inherently unstable and unsafe. The industrial world had become overdependent on a volatile part of the world, and thus was vulnerable to it. The real answer was to slow the growth in oil demand and, in the meantime, put some formal energy security measures in place. Royal Dutch/Shell had been distributing to government leaders a confidential "Pink Book," warning that the supply situation was dangerously out of whack and that an "oil scramble" might ensue. Unlike the American companies, Shell had been campaigning for an intergovernmental agreement to share supplies in a crisis and, indeed, had already begun outlining, in its planning group, how such a system might work.

There had also been discussions about a sharing plan among the Western governments before October 1973, as had been put in place in 1956 and 1967. Each government, however, insisted on a system that would fit its own needs and position. Moreover, before the actual crisis, the issues had been too complex, there was too little agreement on the stakes and risks, the impetus was insufficient, and such coordination would have been too controversial for American politics. Thus, there was hardly any preparation. In June 1973, the industrial nations had agreed to establish "an informal working group to develop and evaluate the various options." And that was as far as they got before the crisis.

In the midst of the crisis—with all its uncertainty, with American-European relations so brittle, and with the Arabs shrewdly intent on splitting the Western allies—no such machinery could quickly be established. Agreement for emergency sharing did exist among the members of the European Economic Community, but it was never invoked. After all, the main target of the cutback was the United States. Moreover, by putting the various European countries into different categories, from embargoed to "most favored," the Arab exporters had gone a long way toward frustrating the ability of the Europeans to unite and implement a sharing agreement.

The U.S. government might have invoked the Defense Production Act of 1950, which provided antitrust exemption so that companies could pool supplies and information during a designated crisis, and which had been used to varying degrees in crises going back to the Korean War and the Iranian nationalization of 1951–53. Its use this time, however, might have added tinder to the fire, hindering the ability of the oil companies to finesse their way through the crisis and making the conflicts with the Arabs and the Western allies more explicit and more difficult. Moreover, its invocation in the midst of Watergate would have provoked suspicion and vocal criticism of collusion between the administration and the oil companies; Nixon simply was not in a position to appeal credibly to national interests.

That left only one choice for coping with the crisis—the companies themselves, primarily the majors. Heretofore, the companies had expressed pride in their buffer role between the producing and consuming countries—"the thin lubricating film," David Barran of Shell had called it. But now they found how painfully grating it was to be that buffer during a time of acute stress, when the lubrication of oil was abruptly withdrawn.

On the one side there was the intense, deadly serious pressure from the Arab governments. The threat was explicit; the companies could lose their entire position in the Middle East. When the Saudis ordered the first 10 percent cutback on October 18, Aramco responded immediately, and it cut just a little extra for good measure. Here would unfold the anomalous, unpalatable sight of an American company—the most important jewel, some thought, of all American investments abroad—in a position that it had worked to avoid at all costs: actually implementing an embargo against the United States. But what choice did it have? Was it not better to cooperate and move as much oil onto the world market as possible than to be nationalized and thrown out? "The only alternative was not to ship the oil at all," Chevron's George Keller, an Aramco director, said afterward. "Obviously it was in the best interests of the United States to move 5, 6, 7 million barrels per day to our friends around the world rather than to have that cut off."

Yet on the other side were the consuming governments, all of which sought the oil their publics needed. The most powerful of them happened to be the United States, not only the home government of five of the seven majors, but also the primary target for the whole exercise. Any action the companies took, they knew, would be subject to the closest scrutiny and after-the-fact evaluation. They did not want to lose markets, nor find themselves shut out, nor invite investigation and retaliation by consumers and their governments.

In such circumstances, the only logical response was one of "equal suffering" and "equal misery." That is, the companies would try to allocate the same percentage of cutbacks from total supplies to all countries by moving both Arab and non-Arab oil around the world. They had already gained some experience in how to organize a sharing system during the embargo that accompanied the 1967 war. But the scale and the risks in 1973 were much, much larger. As the basis for the prorated cutbacks, they used either actual consumption in the first nine months of 1973 or what had been their projections for the immediate period ahead. Equal suffering "was the only defensible course if governments were not collectively to agree on any alternative preferred system," said a senior executive of Shell. For the companies, he added, "this seemed the only way to avoid inviting their own destruction." Anything else, for international companies, would be suicide. A further factor reinforced the equal-suffering principle—the existence of the "internal market" within the international oil companies. The head of a major's Far East operations, for instance, who had to explain things subsequently to Japan and other governments in his part of the world, would certainly have raised holy hell if he thought his opposite number for Europe were getting a proportionately bigger share.

Though the companies were by long experience expert at juggling supplies under normal circumstances, they now had to improvise frenetically. "It was a very great torment," recalled the head of oil supply for Gulf. "We were working around the clock. There were teams of people in the office all night long, allocating countries, figures, and supply plans, responding to anguished cries. We had to be cutting back on our worldwide commitments. We prorated worldwide. That meant cutting back our own refineries as well as our customers. I had to fight for our third party customers. Gulf and the other companies were being bombarded daily. 'Why are you selling to the Koreans and Japanese when you could bring it into the United States? You're an American company.' We were being attacked in the press every day. The pressures were so great to bring another cargo in for a U.S. refinery. I had to remind our board that we had sold our long-term contracts to customers on the basis that we would treat them like ourselves. We had to contact field people, tell old friends that we were cutting them back, and go around the world explaining the supply/demand balance and, therefore, the prorationing. It was very difficult to make all that happen."

The massive reallocation posed a very considerable logistical problem.

Even under calm and relatively predictable circumstances, managing an integrated oil system was a highly complex matter. Supplies of varying qualities from various sources had to be linked into the transportation system and then moved to refineries that had been designed to handle those specific oils. Free will was not an option when it came to assigning crude oils. The "wrong" crudes could do considerable damage to the innards of a refinery, as well as reducing efficiency and profitability. And once the crude supplies were run through the refinery and turned into a number of products, they then had to be moved into a distribution system and linked with a "market demand" that wanted that particular balance of products—this amount of gasoline, that amount of jet fuel and heating oil.

And to make matters even more difficult, the companies still had to figure out what their oil supplies actually cost, so that they did not sell at a loss or invite attack for excessive profit margins. The costs for oil royalties, extent of government participation, buyback prices, volumes—all these were changing week by week, and were further complicated by leapfrogs and retroactive increases by the various exporting governments. "It was impossible to know whether a calculation made on the basis of all the known facts on one day would not be overturned by a rewriting of those facts a month later," said an executive from Shell. Indeed, the only sure thing was that the price of oil was going up and up, and up again.

The scale of operations was massive; the decision points, almost innumerable. Normally, the complex calculations that guided the movement of oil through an integrated system were masterminded by computers on the basis of economic and technical criteria. Now, the political criteria were at least as important—not crossing the Arabs and their restrictions and yet also satisfying, to the greatest degree possible, the importing nations. Much sleight of hand was required to meet both objectives. Yet to a considerable degree, the companies made it all work.

Different governments responded differently to the companies' prorationing cutbacks. Washington gave little direct guidance. John Sawhill, the head of the new Federal Energy Office, urged the companies "to bring as much as possible" into the United States, but at the same time to recognize "the interest in all of the countries of the world in having some kind of equitable share of the world's supplies." In a meeting with oil executives, Henry Kissinger did make it a point to ask them "to take care of Holland," which had been made a particular target by the Arabs because of its traditional friendship for Israel.

Japan felt a special vulnerability. It had little indigenous energy, and imported oil fueled its formidable economic growth. Not only was its public in a panic, but Japan was highly dependent on the majors, most of which were American. At one meeting, a senior MITI bureaucrat warned representatives of the majors not to divert non-Arab oil from Japan to the United States. The company representatives replied that they were allocating oil as fairly as possible, and that they would be more than happy to hand the whole thankless job over to the governments, Japan included, if they wanted it. The Japanese government backed off; thereafter, it seemed to be satisfied with what was being done, though it continued to monitor the operation very closely.[4]

604

The most turbulent reaction was that of the British government. The United Kingdom had been placed among the "friendly nations" by the Arab governments, and thus was supposed to receive 100 percent of its September 1973 supplies, notwithstanding the cutbacks. The Secretary of Trade and Industry confidently told the House of Commons about "assurances from Arab states." He himself went to Saudi Arabia to negotiate a government-to-government oil deal. The British government also owned half the stock in BP, but, by agreement dating back to Churchill's acquisition of the shares in 1914, was not supposed to intervene in commercial matters. But was this a case of commerce or of security? Moreover, a confrontation was already brewing between the coal miners and the Conservative government of Prime Minister Edward Heath, threatening a major strike that could cut coal supplies at the same time that oil supplies were down. An oil shortfall would greatly strengthen the miners' hand. Heath wanted as much oil as he could get in order to be prepared for a battle with the miners.

Heath summoned Sir Eric Drake, chairman of BP, and Sir Frank McFadzean, chairman of Shell Transport and Trading, to his country house, Chequers. The Prime Minister was accompanied by several of his Cabinet colleagues; it was clear that, if he could not cajole the oil companies into accepting his point of view, he would try to bully them into doing so. Britain must be given preferential treatment, Heath told them. The two companies must not cut their supplies to the United Kingdom; the flow should be maintained at 100 percent of normal requirements.

Both chairmen pointed out that the oil companies were in a position that was not of their choosing; they had been sucked into a vaccuum that had resulted, as McFadzean later expressed it, "from the failure of governments to plan on how to handle an oil shortage." Every company had a web of legal and moral obligations in the many countries in which they did business; in the event of their being left to handle the shortage, the only possible policy they could pursue was equality of sacrifice, though they recognized even that principle would become progressively more difficult to hold to. McFadzean added another dimension. He was terribly sorry, but the Royal Dutch/Shell Group was 60 percent owned by the Dutch and only 40 percent by the British. So, even if he were to agree with Heath, which he "most decidedly did not," it would not be possible to ride roughshod over the Dutch interests.

Rebuffed and irritated, Heath more insistently bore in on Drake in his quest for special treatment. Since the government owned 51 percent of BP, said Heath bluntly, Drake would do as the Prime Minister had ordered. But Drake was nothing if not direct, and he certainly was not accustomed to giving ground. As BP's general manager in Iran in 1951, he had stood up to Mossadegh, for which he had been threatened with death, and thereafter he had stood up no less resolutely to BP's chairman, the autocratic William Fraser, for which he had been threatened with exile to a BP refinery in Australia. He certainly was not going to give way now before Edward Heath and allow the Prime Minister, as he later said, "to destroy my company." Having lived through one nationalization in Iran, Drake did not intend to be party to any more nationalizations, which he was sure would be the fate of BP facilities in other countries if he acquiesced to the Prime Minister.

So, when Heath pressed him, Drake parried with a question of his own. "Are you asking me to do this as a shareholder or as a government? If you're asking me as a shareholder to give the U.K. 100 percent of its normal supplies, you should know that we could get nationalized in retaliation in France, Germany, Holland, and the rest. That would mean a great loss to the minority shareholders." Drake then delivered to Heath an acid-edged lecture on company law, which prohibited giving one shareholder better treatment than another. All directors had a fiduciary responsibility to look after the interests of the company as a whole and not of particular shareholders. So, not only would the company be risking retaliation in countries that had had to suffer deeper cuts, the British government would leave itself open to minority suits for abuse of dominant power. "If you're telling me to do it as a government," Drake continued, "then I must tell you that I must have it in writing. Then we can plead *force majeure* to the other governments because I'm under government instructions. Perhaps, just perhaps, we can avoid nationalization."

At this point, Heath lost his temper. "You know perfectly well that I can't put it in writing," he boomed. After all, he was the great champion of British entry into the European Community and of cooperation with the Europeans.

"Then I won't do it," Drake replied with absolute firmness.

Of course, Heath could always have asked Parliament to pass a law forcing BP to give special preference, but after a few days of reflection—including, no doubt, meditation on the effects such a move would have on Britain's relations with its European allies—Heath's temper cooled, and he gave up on his insistence on special treatment.

The civil servants in Whitehall had a better grasp of the overall picture than the politicians. They recognized the merits of the "fair share" principle and were more deft in seeking to bend it. They brought pressure to bear on the international companies, including reminding them that it was the British government that decided who would get exploration licenses in the North Sea. That way, they could assure that Britain received what *they* chose to interpret as "fair share"— and a bit more.

Diversion was the essence of the application of the principles of equal suffering and fair share. Non-Arab oil was diverted to countries that were embargoed or on the neutral list, while Arab oil was directed to the countries on the preferred lists. Altogether, the five American major companies ended up so diverting about a third of their oil. Overall, the principle of equal suffering was applied relatively effectively. Adjusting the data for the wide differences in energy and oil growth rates, the loss to Japan over the embargo period was 17 percent; to the United States, 18 percent; and to Western Europe, 16 percent. The Federal Energy Administration subsequently prepared, for the Senate Multinational Subcommittee, a retrospective analysis of how the informal allocation system had worked. When all things were considered, the report said, "it is difficult to imagine that any other allocation plan would have achieved a more equitable allocation of reduced supplies." The companies "were called upon to make difficult and potentially volatile political decisions during the embargo—decisions beyond the realm of normal corporate concerns." And that, the report noted, was something they absolutely never wanted to have to do again.[5]

A New World of Prices

It was amid the feverish bidding up of prices on spot markets that the OPEC oil ministers met in Tehran in late December 1973 to discuss the official price. The proposed range extended from that of the Economic Commission of OPEC, which recommended prices as high as $23 a barrel, down to Saudi Arabia's $8. Saudi Arabia was fearful that a suddenly skyrocketing price could result in a depression, which would affect Saudi Arabia along with everybody else. "If you went down," Yamani said, referring to the industrial world, "we would go down." He argued that the recent huge auction prices were not indicative of real market conditions, but rather reflected the fact that the auctioning took place in the midst of politically imposed embargoes and cutbacks. Also, King Faisal wanted to maintain the "political character" of the embargo; he did not want it to look like a cloak for a money grab. Still, the prospect that income from one's only major cash crop was about to be increased many times over could certainly mute such discomfort among the exporters.

By far the most aggressive and outspoken country was Iran. Here was the opportunity, finally, for the Shah to gain the kind of revenues that he deemed necessary to finance his grand ambitions. Iran argued for $11.65 as the new posted price, which would mean a government take of $7. The Iranians had a rationale for their price. It was based not on supply and demand but on what the Shah called a "new concept," the cost of alternative energy sources—liquids and gas made from coal and shale oil. It was the minimum price necessary if these other processes were to be economical, or so the Shah said. In private, he proudly cited a study done for Iran by Arthur D. Little on this subject. The Arthur D. Little premises, in turn, were shared in many oil companies. The study had the ring of hard analysis, but in fact was at best very conjectural, since of all those alternative energy processes, only one was commercially operating, and that was a single, limited coal liquefaction project in South Africa. Shell's chief Middle East adviser summed the matter up thus: "The alternative source existed, in the volume required, only in economic theory, not in reality." As in previous oil shortages, the miracle of shale oil was really a chimera.

After much hard discussion, the Shah's position was accepted by the oil ministers gathered in Tehran. The new price would be $11.65. It was a price rise pregnant with history and significance. The posted price had been raised from $1.80 in 1970 to $2.18 in 1971 to $2.90 in mid-1973 to $5.12 in October 1973— and now to $11.65. Thus, the two price increases since the war had begun—in October and now in December—constituted a fourfold increase. As a "marker," the new posted price was applied to a particular Saudi crude, Arabian Light. The prices of all other OPEC crudes were to be keyed to that, with the "differentials" in prices to be based upon quality (low- or high-sulfur), gravity, and transportation costs to major markets. Pointing out that the new price was considerably lower than the $17.04 that had been tendered in the recent Iranian auction, the Shah graciously said that it was a price fixed out of "kindness and generosity."

At the end of December, Richard Nixon wrote a very strong private letter to the Shah. Outlining "the destabilizing impact" of the price increase and the "catastrophic problems" it could create for the world economy, he asked that it be reconsidered and withdrawn. "This drastic price increase is particularly unreasonable coming as it does when oil supplies are artificially restrained," said the President. The Shah's reply was brief and totally ungiving. "We are conscious of the importance of this source of Energy to the prosperity and stability of the international economy," he said, "but we also know that for us this source of wealth might be finished in thirty years."

The Shah had found a new role; he was now to be the moralist for world oil. "Really oil is almost a noble product," he declared. "We were almost careless to use oil for heating houses or even generating electricity, when this could so easily be done by coal. Why finish this noble product in, say, 30 years' time when thousands of billions of tons of coals remain in the ground." It was also the Shah's inclination to become the moralist for world civilization. He had words of advice for the industrial nations. "They will have to realize that the era of their terrific progress and even more terrific income and wealth based on cheap oil is finished. They will have to find new sources of energy. Eventually they will have to tighten their belts; eventually all those children of well-to-do families who have plenty to eat at every meal, who have their cars, and who act almost as terrorists and throw bombs here and there, they will have to rethink all these aspects of the advanced industrial world. And they will have to work harder. . . . Your young boys and young girls who receive so much money from their fathers will also have to think that they must earn their living somehow." His haughty stance, in the midst of shortages and huge price increases, would cost him dearly only a few years later, when he desperately needed friends.[6]

Alliance Strained

The embargo was a political act that took advantage of economic circumstances, and it had to be met with political action on three interrelated fronts: between Israel and its Arab neighbors; between America and its allies; and between the industrial countries, particularly the United States, and the Arab oil exporters.

On the first front, Kissinger made himself the center of a vortex of activity. He would seek to take advantage of the new reality created by the war: Israel had lost a good deal of confidence, while the Arabs, particularly Egypt, had regained some of theirs. "Shuttle diplomacy" became the hallmark of his tireless, dramatic, and virtuoso performance. There were several landmarks along the way, including an Egyptian-Israeli disengagement agreement in mid-January 1974, and finally a Syrian-Israeli disengagement at the end of May. And while the negotiations were arduous, uncertain, and contingent, they would lay the basis for broader agreement four years later. Throughout, Kissinger had a particular partner, Anwar Sadat, who knew his own objectives. Sadat had resorted to war, in the first place, to effect political changes. In the aftermath of the war, he had a better chance of achieving those changes in collaboration with the Americans. For, as he publicly declared, "The United States holds 99 percent of the cards in this game." Of course, Sadat was a politician, conscious of his audiences, and in private he once admitted, "The United States actually only holds 60 percent of the cards, but it sounds better if I say '99 percent.' " And even 60 percent, in pursuit of his objectives, was certainly more than adequate reason to lean toward the

United States. At a meeting in Cairo less than a month after the war, Sadat left no doubt in Kissinger's mind that, having achieved his shock, he was now ready to begin a peace process—and, at enormous risk, to seek to transform the psychology of the Middle East.

The oil embargo precipitated one of the gravest splits in the Western alliance since its foundation in the days after World War II, and certainly the worst since Suez in 1956. Relations had already been under some strain before the October War. Once the embargo began, the European allies, led by France, hastened to disassociate themselves from the United States as fast as they could and to assume positions more agreeable to the Arabs, a process that was speeded by a tour of European capitals that Yamani made in the company of his Algerian counterpart. At each stop, the two ministers pressed the Europeans to oppose the United States and its Middle East policy, and to support the Arabs. Yamani applied his special touch to the effort. "We are extremely sorry for the inconvenience caused in Europe by the Arab oil cut," he said apologetically. But he left no doubt what he expected of them.

As the Europeans compliantly shifted their policies, seeking to separate themselves from the United States and promote "dialogue" and "cooperation" with the Arab countries and OPEC, senior American officials began to talk caustically about the Europeans' being soft on OPEC and falling all over themselves in their haste to accommodate and appease. For their part, the Europeans insisted that the United States was too confrontational, too belligerent, in its posture toward the oil exporters. To be sure, there were significant variations among countries in Europe. The French and the British were the most keen to distance themselves from the United States and to court the producers; the Germans, less so; and the Dutch, by contrast, resolute in their commitment to traditional alliances. Some of the Europeans emphasized that they had large and immediate interests to protect. "You only rely on the Arabs for about a tenth of your consumption," French President Georges Pompidou bluntly told Kissinger. "We are entirely dependent upon them."

There were elements of both resentment and poetic justice in the European position. The French had long felt that the "Anglo-Americans" had unfairly excluded them from most Middle Eastern oil, especially in Saudi Arabia with the postwar renunciation of the Red Line agreement, and that the Americans had undermined them in the struggle over Algeria. Then there was the 1956 Suez crisis. Seventeen years had passed since the Americans had blatantly undercut France and Britain in the confrontation over the canal with Nasser, hastening the retreat of the two nations from their global roles and giving a great boost to Arab nationalism. But, as Prime Minister Edward Heath now privately and pointedly told the Americans, "I don't want to raise the issue of Suez but it's there for many people." It was vividly "there" for Heath himself; he had been Anthony Eden's loyal whip during those pained Suez days. In mid-November 1973 the European Community passed a resolution supportive of the Arab position in the Arab-Israeli conflict. Some Arab officials were still dissatisfied. One described the statement as "a kiss blown from afar—which is all very nice but we would prefer something warmer and closer." The resolution, however, clearly represented the kind of accommodation that the Arabs were trying to force, and it was enough to win the Europeans a suspension of the 5 percent December cut. But, just to make sure the Europeans kept on good behavior, the Arab oil ministers warned that the cuts would be imposed after all if the Europeans did not keep "putting pressure on the United States or Israel."

There was one very awkward matter for the European Community. While many of its members were being put on the "friendly" list, one of them, the Netherlands, was still embargoed. If the other members decided to embargo transshipments to the Netherlands, they would be violating one of the basic precepts of the community, namely, the free flow of commodities. Nevertheless, they were inclined to do exactly that, until the Netherlands reminded them, forcefully, that it was the major source of natural gas for Europe, including 40 percent of France's total supply and most of the gas used for heating and cooking in Paris. A quiet compromise was worked out, involving an unspecified "common position" among the European Community members and the expediting of non-Arab supplies from the international companies.

The Japanese, who thought themselves quite removed from the Middle East crisis, were distressed to find they had been placed on the "unfriendly" list. Forty-four percent of Japan's oil came from the Arab Gulf states, and of all the industrial countries, it was the one most dependent on oil as a source of total energy—77 percent compared to 46 percent for the United States. It had taken oil for granted as the essential and reliable fuel for economic growth. No longer. Point blank, Yamani spelled out the new Arab export policy to the Japanese: "If you are hostile to us you get no oil. If you are neutral you get oil but not as much as before. If you are friendly you get the same as before."

Before the oil embargo, the "resource faction" in Japanese government and business circles had already begun to call for a reorientation in Japanese policy toward the Middle East. They had not made much progress because, in the words of Vice-Minister for Foreign Affairs Fumihiko Togo, "before 1973, we could always buy the oil if only we had the money," and because Japan bought the oil primarily from the international companies, and not directly from Middle Eastern countries. After the crisis began, the resource faction mightily intensified its efforts. On November 14, the same day that Kissinger was in Tokyo trying to persuade Japan's Foreign Minister not to break with the United States, worried business leaders met privately with Prime Minister Kakuei Tanaka to make a "direct request" for a major change of policy. A few days later, the Arab exporters exempted from further cutbacks the European countries that had issued the pro-Arab statement. Here was tangible proof of the rewards of changing policy. Meanwhile, unofficial Japanese emissaries, who had hurriedly been dispatched on secret trips to the Middle East, were reporting back that the Arabs regarded "neutrality" as insufficient and indeed, as opposition to their cause. On November 22, Tokyo issued its own statement endorsing the Arab position.

That declaration represented Japan's first major split on foreign policy with the United States in the postwar era. Such an action was hardly to be undertaken lightly, as the U.S.-Japan alliance was the basis of Japanese foreign policy—or had been. Four days after the statement's issuance, Japan got its own reward; it was exempted from the December cutbacks by the Arab exporters. As part of its new resource diplomacy, Tokyo sent a host of high-level representatives to the

Middle East to conduct business with a decidedly political cast—economic assistance, loans, new projects, joint ventures, bilateral deals, the works. By the very fact of the allocations and cutbacks, the majors had, in effect, said they could not supply Japan as they had in the past. Therefore, Japan could not rely on them and would have to make its own deals in quest of security of supply. However, the Japanese did refuse, despite continuing Arab insistence, to break all diplomatic and economic relations with Israel. The Japanese who pushed to abrogate those ties, argued Vice-Foreign Minister Togo, were suffering from a familiar epidemic—"oil on the brain."[7]

Even as its traditional allies gave way to Arab demands, the United States tried to promote a coordinated response among the industrial countries. Washington feared that a resort to bilateralism—state-to-state barter deals—would result in a much more rigid, permanently politicized oil market. Already the rush was on. "Bilateral Deals: Everybody's Doing It," headlined the *Middle East Economic Survey* in January 1974. The oil industry looked with skepticism at the scramble among politicians to acquire national supplies. One oil man, Frank McFadzean of Shell, could only marvel, albeit with a certain cynicism, at "the spectacle of two senior Cabinet Ministers flying off with the melodrama normally associated with the relief of a beleaguered fortress, to sign a barter agreement involving a quantity of crude oil equal to less than four weeks of the country's requirements. Delegations and emissaries, politicians and friends of politicians, most of them with little knowledge of the oil business, descended on the Middle East like a latter day plague of near Biblical proportions." For his part, Henry Kissinger feared these bilateral arrangements would undermine his efforts to negotiate a settlement to the Arab-Israeli war. If the industrial countries kept to the helter-skelter approaches already taken—competitive bidding based on panic and lack of information, mercantilism, *sauve qui peut* ("every man for himself")—all were likely to end up worse off.

The United States called for an energy conference to convene in Washington in February 1974. It wanted to assuage fears about competition over supplies, heal the deep rifts in the alliance, and ensure that oil did not become a lasting source of division in the Western alliance. The British had come to the conclusion that being on the "friendly" list did not buy them very much; they still had to face the same price increases, and were most interested to participate. In fact, the political situation in the United Kingdom had changed dramatically. For Britain, the oil shortage was magnified several times over by the coal miners' confrontation with Prime Minister Heath, which turned not only into a strike but into full-scale economic war. There were not sufficient oil supplies to substitute quickly for coal in power stations; the country's economy was paralyzed as it had not been since the coal shortage of 1947. Electricity supplies were interrupted, and industry went on a three-day workweek. Even heating for domestic hot water was in short supply, and clergymen solemnly debated on the BBC the morality of family members' sharing their bath water, and thus keeping demand down to one tub of hot water, as a contribution to the national weal. In what would turn out to be the last weeks of the Heath government, Britain was a supportive participant in the Washington Energy Conference.

So were the Japanese. They believed that a coordinated response by the industrial nations was required. They were also keen to find a framework in which to mute what they regarded, in the words of one official, as the "tendency of U.S. policy to become very confrontational." The Germans were also eager to talk on a multilateral basis. Not so the French. They remained unreconstructed. Though reluctantly making an appearance at the Washington gathering, they were vocal in their antagonism. Foreign Minister Michel Jobert, who took Gaullism to an extreme, opened a caucus in Washington of the European Community partici-pants with the bitter salutation: *"Bonjour les traitres"* ("Hello traitors").

For their part, American officials did hint rather broadly that America's overall security relationship, including the maintenance of American troops in Europe, might be jeopardized by discord on energy issues. Most of the partici-pants in the Washington Energy Conference agreed on the merit of developing consensus and some common policies on international energy matters, and the conference led to the establishment of an emergency sharing program for the "next" crisis and the creation of the International Energy Agency, which would be charged with managing the program and, more broadly, with harmonizing and making parallel the energy policies of the Western countries. The IEA would help to deflect the drive to bilateralism and establish the framework for common response, both politically and technically. Before 1974 was out, the IEA would be set up in a leafy section of the sixteenth arrondissement in Paris, in an annex to the Organization for Economic Cooperation and Development. Yet one major industrial country refused to join—France. The IEA, said the un-reconstructed Foreign Minister Jobert, was a *"machine de guerre,"* an instru-ment of war.[8]

Sheathing the Oil Weapon

When, and how, would the embargo itself end? No one really knew, not even the Arabs. In the last days of December 1973, the Arab producers did relax the em-bargo a little as some initial progress was made on the Arab-Israeli dispute. The embargo had become, in Kissinger's sarcastic formulation in early January, "in-creasingly less appropriate." He himself went twice to Saudi Arabia to see King Faisal. On the first trip, as he walked through a tremendous hall where distin-guished men of the kingdom in black robes and white headdresses sat against the walls, Kissinger, a Jewish emigrant to America, found himself reflecting "in some wonder what strange twists of fate had caused a refugee from Nazi perse-cution to wind up in Arabia as the representative of American democracy." He also found the form of the discussions outside his experience. "The King always spoke in a gentle voice even when making strong points. He loved elliptical comments capable of many interpretations." Kissinger sat at his right in the cen-ter of the room. "In talking to me he would look straight ahead, occasionally peeking from around his headdress to make sure I had understood the drift of some particular conundrum." That would be the case, whether Faisal was talking about the conspiracy of Jews and communists to take over the Middle East or the practical political questions that stood in the way of ending the embargo. The King was apologetic but firm. He did not have a free hand to end the embargo; the invocation of the oil weapon had been a joint decision of all the Arabs. Its withdrawal would also have to be a joint decision. "What I need," said the King, "is the wherewithal to go to my colleagues and urge this." He would also insist that a fundamental condition for him was that Jerusalem become an Arab Islamic city. What about the Wailing Wall, he was asked? Another wall, he replied, could be built somewhere else, against which the Jews could wail.

With little sign that the embargo would be lifted, Washington now turned to a new ally, Anwar Sadat. The chief proponent and beneficiary of the embargo had now become the chief advocate of its cancellation. Like the war itself, he said, the embargo had served its purpose and should be ended. And, indeed, he recognized that the continuation of the embargo would now work against Egypt's interests. The United States itself could go only so far on the path toward Middle Eastern peace under the gun of the embargo. Moreover, the perpetuation of the embargo, which after all was an act of economic warfare, could do long-term damage to the whole range of America's relations with nations like Saudi Arabia and Kuwait, to the disadvantage of those countries. For finally, Watergate or no Watergate, a superpower like the United States could ill afford to be in such a position for any length of time.

But the Arab exporters, having successfully played their trump card, did not want to move quickly to take it off the table. Nor did they want to be seen as giving in too quickly to American blandishments. Still, as the embargo dragged on, more and more oil was seeping back into the market, and the cutoffs were becoming less and less effective. The Saudis indicated to the Americans that the embargo could not be brought to an end without some kind of movement on the Syrian front and at least the tacit acquiescence of Syria's Hafez al-Assad, who was furious with Sadat because of the diplomatic progress being made on the Egyptian front. In effect, Assad had a veto on whether the embargo would be lifted. To help end it, the Saudis opened the door to American-Syrian negotiations concerning disengagement on the Golan Heights. In mid-February 1974, Faisal met in Algiers with Sadat, Assad, and the president of Algeria. Sadat made clear he thought the embargo had outlived its usefulness and might well work against Arab interests. He said that the Americans were leading the way toward a new political reality. Faisal agreed to ending the embargo, so long as there was "constructive effort," promoted by the United States, toward a Syrian-Israeli disengagement. Over the next few weeks, however, Assad maintained a particularly hard-line position, which prevented the others from publicly endorsing an end to the embargo. But they took seriously the warning that the United States peace effort could not proceed without the lifting of the embargo. On March 18, the Arab oil ministers agreed to its end. Syria and Libya dissented.

After two decades of talk and several failed attempts, the oil weapon had finally been successfully used, with an impact not merely convincing, but overwhelming, and far greater than even its proponents might have dared to expect. It had recast the alignments and geopolitics of both the Middle East and the entire world. It had transformed world oil and the relations between producers and consumers, and it had remade the international economy. Now it could be resheathed. But the threat would remain.

In May, Henry Kissinger was able to secure the Syrian-Israeli disengagement, and a peace process seemed to have begun. In June, Richard Nixon made a visit to Israel, Egypt, Syria, and Saudi Arabia. The embargo was now history, albeit very recent history, at least insofar as the United States was concerned. (It was still in place against the Netherlands.) The United States could rightly claim the beginnings of some considerable results in Middle Eastern diplomacy. Watergate, however, was an ever-present reality, and Nixon's behavior on the trip struck some as stark. Sitting with the Israeli Cabinet in Tel Aviv, he suddenly announced that he knew the best way to deal with terrorists. He sprang to his feet and, with an imaginary submachine gun in his hand, pretended to gun down the entire Cabinet, Chicago style, making a "brrrrr" sound as he did so. The Israelis were perplexed and a little concerned. In Damascus, Nixon told Syria's Assad that the Israelis should be pushed back until they fell off the cliff, and to underscore the point, he made strange, chopping gestures. Thereafter, in subsequent meetings with other Americans, Assad would insist on recalling Nixon's exposition.

But it was in Egypt that Nixon had his great moment. His days there could only have been called triumphal. Millions of enthusiastic, delirious Egyptians turned out to cheer. It was truly his last hurrah. There was much there for savorers of irony. After all, this was the land of Gamal Abdel Nasser, who had been a master at turning out massive crowds to denounce Western imperialism and, in particular, the United States. But now the country in which Richard Nixon was much more likely to be greeted by hostile crowds was not Egypt but the United States. The bitter antagonism to him at home in what would prove to be the last months of his Presidency stood in marked contrast to the clamor and excitement that greeted him in the streets of Cairo. For the Egyptians, it was a celebration of Sadat's restoration of Egypt and of its prestige, which had been greatly tarnished in Nasser's final years. For Nixon, it could no less have been a celebration both of the end of the embargo and of the diplomatic effectiveness of his administration. But he was hardly in a mood to enjoy it. He was in poor health during the visit, his leg was swollen with phlebitis, and he spent much of his free time on the trip listening to his own incriminating Oval Office tapes, which would finally compel his resignation.[9]

C H A P T E R 3 1

OPEC's Imperium

TIMES CHANGE, empires rise and fall, and the modern office building on Vienna's Karl Lueger Ring, with the small bookstore on the ground floor, had customarily been called "the Texaco Building," in honor of its lead tenant. But by the middle 1970s, owing to the presence of another tenant, it had abruptly become known as "the OPEC Building." The change symbolized a profound process of global upheaval: the suddenness with which the oil-exporting countries had assumed the position formerly held by the international companies.

As it was, OPEC ended up in Vienna only by accident. In its early days, it had established itself in Geneva; but the Swiss doubted its serious intent and even its significance, and refused to grant the diplomatic status appropriate to an international organization. The Austrians, however, were eager to get what they could in terms of international prestige and were willing to be accommodating, and so in 1965, despite Austria's inferior international air links, OPEC moved to Vienna. OPEC's lodgement in Vienna, in the Texaco Building, clearly showed how little account had been taken, early on, of this rather mysterious and peculiar organization, which, despite the initial clamor at its founding, had fallen well short of its central political objective—the assertion of "sovereignty" by the oil exporters over their resources.

But now, in the middle 1970s, all that had changed. The international order had been turned upside down. OPEC's members were courted, flattered, railed against, and denounced. There was good reason. Oil prices were at the heart of world commerce, and those who seemed to control oil prices were regarded as the new masters of the global economy. OPEC's membership in the mid-1970s was virtually synonymous with all the world's petroleum exporters, excepting the Soviet Union. And OPEC's members would determine if there was to be inflation or recession. They would be the world's new bankers. They would seek to ordain a new international economic order, which would go beyond the redistribution of rents from consumers to producers, to one that established a wholesale redistribution of both economic and political power. They would set an example for the rest of the developing world. The member countries of OPEC would have a significant say over the foreign policies and even the autonomy of some of the most powerful countries in the world. It was not surprising, therefore, that a former Secretary General of OPEC would one day look back on those years, 1974 through 1978, as "OPEC's Golden Age."

Yet nostalgia certainly clothed his recollection. To be sure, the OPEC countries did in the mid-1970s complete the acquisition of control over their own resources. There would no longer be any question about who owned their oil. But those years were dominated by a bitter battle not only with consumers but also within OPEC about the price of that valuable resource. And that one question alone would dominate the economic policies and international politics of the entire decade.

Oil and the World Economy

The quadrupling of prices triggered by the Arab oil embargo and the exporters' assumption of complete control in setting those prices brought massive changes to every corner of the world economy. The combined petroleum earnings of the oil exporters rose from $23 billion in 1972 to $140 billion by 1977. The exporters built up very large financial surpluses, and the fear that they could not spend all the money created grave concern for the world's bankers and economic policymakers: The unspent tens of billions, sitting idle in bank accounts, could spell serious contraction and dislocation in the world economy.

They need not have worried. The exporters, suddenly wealthy and certainly far richer than they might have dreamed, embarked on a dizzying program of spending: industrialization, infrastructure, subsidies, services, necessities, luxuries, weapons, waste, and corruption. With the avalanche of expenditures, ports were clogged far beyond their capacity, and ships waited weeks for their turn to unload. Vendors and salesmen for all sorts of goods and services rushed from the industrial countries to the oil-exporting nations, scrambled for rooms in overbooked hotels, and elbowed their way into the waiting rooms of government ministries. Everything was for sale to the oil producers, and now they had the money to buy.

Transactions in armaments became a huge business. To the Western industrial nations, the 1973 disruption of supplies and their high degree of dependence on the Middle East made the security of access to oil a strategic concern of the first order. Weapons sales, aggressively pursued, were a way to enhance that security and maintain or gain influence. The countries in the region were just as eager to buy. The events of 1973 had demonstrated the volatility of the area; not only were the regional and national rivalries deep and the ambitions large, but the two superpowers had squared off in a nuclear alert over the Middle East.

But weapons were only part of the contents of the vast, post-1973 cornucopia, which included everything from consumer goods to entire telephone systems. The proliferation of Datsun pickup trucks in Saudi Arabia was a sign of the times. "It is very expensive to maintain camels," said one executive of Nissan, "it is cheaper to keep a Datsun." To be sure, a Datsun cost $3,100 in Saudi Arabia in the mid-1970s, while the sticker price for a camel was only $760. But at twelve cents a gallon for gasoline compared to the going price for camel feed, it was much cheaper to fuel a Datsun than feed a camel. Almost overnight, Nissan became the number one purveyor of vehicles to Saudi Arabia, and the Datsun pickup was the favorite among the sheep-tending Bedouins whose fathers and grandfathers had been the camel-riding backbone of Ibn Saud's armies. In total, the massive spending of the exporting countries, combined with the galloping inflation of their overheated economies, ensured that their financial surpluses would soon disappear. And disappear they did, completely—the bankers' initial fears notwithstanding. In 1974, OPEC had a $67 billion surplus in its balance of payments on goods, services, and such "invisibles" as investment income. By 1978, the surplus had turned into a $2 billion deficit.

For the developed countries of the industrial West, the sudden hike in oil prices brought profound dislocations. The oil rents flooding into the treasuries of the exporters added up to a huge withdrawal of their purchasing power—what became known as the "OPEC tax." The imposition of this "tax" sent the industrial countries into deep recession. The U.S. gross national product plunged 6 percent between 1973 and 1975, while unemployment doubled to 9 percent. Japan's GNP declined in 1974 for the first time since the end of World War II. As the Japanese worried that their economic miracle might be over, sobered students in Tokyo stopped chanting "Goddamn GNP" at demonstrations and instead found new virtue in hard work and the promise of lifetime employment. At the same time, the price increases delivered a powerful inflationary shock to economies in which inflationary forces had already taken hold. While economic growth resumed in 1976 in the industrial world, inflation had become so embedded in the fabric of the West that it came to be seen as the intractable problem of the modern age.

The group that suffered the most from the price increases were those developing countries that were not fortunate in having been blessed with oil. The price shock was the most devastating blow to economic development in the 1970s. Not only were those developing nations hit by the same recessionary and inflationary shocks, but the price increases also crippled their balance of payments, constraining their ability to grow, or preventing growth altogether. They suffered further from the restrictions on world trade and investment. The way out for some was to borrow, and therefore, a goodly number of those OPEC surplus dollars were "recycled" through the banking system to these developing countries. Thus, they coped with the oil shock by the expedient of going into debt. But a new category also had to be invented—the "fourth world"—to cover the lower tier of developing countries, which were knocked flat on their backs and whose poverty was reinforced.

The new and very difficult problems of the developing countries put the oil exporters into an awkward, even embarrassing situation. After all, they, too, were developing countries, and they now proclaimed themselves as the vanguard of the "South," the developing world, in its efforts to end the "exploitation" by the North, the industrial world. Their objective, they said, was to force a global redistribution of wealth from North to South. And, initially, other developing countries, thinking of their own commodity exports and overall prospects, loudly cheered OPEC's victory and proclaimed their solidarity. And this was the time when "the new international order" was much discussed. But OPEC's new prices constituted a huge setback for the rest of the developing world. Some oil exporters instituted their own lending and oil supply programs to aid other developing countries. But the main response of the exporters was to champion a broad "North-South dialogue" between developed and developing countries, and to insist on tying the oil prices to other development issues, with the stated aim of promoting that global redistribution of wealth.

The Conference on International Economic Cooperation, meant to embody the North-South dialogue, convened in Paris in 1977. Some of the industrial nations hoped to secure access to oil as a result of their participation. The French, still smarting with indignation at Kissinger's leadership during the oil embargo and long envious of America's position in Middle Eastern oil, promoted the dialogue as an alternative to American policies. More quietly, other countries saw it as a way to mute the confrontation between importers and exporters and to provide a counterweight to higher oil prices. Though the dialogue proceeded for two years, absorbing much effort, there was little to show for it in the end. The participants could not even agree on a communique. What came to matter most for the rest of the developing world, in practical terms, was not the high-minded rhetoric in Paris, but the reality of the depressed markets in the industrial world for their own commodities and manufactures.[1]

The Saudis Versus the Shah

OPEC itself became an international spectacle of the first order in the mid-1970s. The eyes of the world fastened on its meetings, with their drama, pomp, and commotion. Ears eager for any clues about what would happen to the world economy, strained to hear the quick response of a minister to a shouted question as he swept through a hotel lobby. Following in OPEC's wake, oil talk—"differentials," "seasonal swings," "inventory build"—now became the parlance of government policymakers, journalists, and financial speculators. Though OPEC was usually described as a "cartel" during this period, in fact it was not. "You can call OPEC a club or an association but not, properly speaking, a cartel," Howard Page, the former Middle East coordinator for Exxon, observed in 1975. To prove his point, he reached for a Funk & Wagnall's dictionary, which defined a cartel as "a combination of producers to regulate the prices and the output of a commodity." OPEC was certainly trying to set price, but not output—not yet. There were no quotas or assigned production levels. The market was really dominated, according to one formulation, not by a cartel but by a "somewhat unruly oligopoly." During this period, most of the exporters were producing virtually at capacity. The exception was Saudi Arabia, which was setting its production in order to try to achieve its price objectives.

In response to the criticism of the oil price increases, the exporters generally replied by pointing out that, if one broke down the prices that consumers in industrial countries were paying for oil products on a per barrel basis, the Western governments were taking more in terms of tax than the OPEC countries were receiving in their sales price. This was the case in Western Europe, where there was a long history of a large gasoline tax. In 1975, for instance, about 45 percent of what the Western European consumer paid for oil products went to his government, while about 35 percent was accounted for by the OPEC price. The other 20 percent went for shipping, refining, dealer's margins, and so forth. The argument had less validity for the United States, where the tax component was only 18 percent, while the share going to the OPEC exporter was more on the order of 50 percent. In Japan, the government took 28 percent, with 45 percent going to OPEC. Whatever the split, the consumer governments responded to the OPEC claim by saying that what they did within their borders and how they taxed their citizens was their own business, and that the macro-economic consequences of their sales taxes were strikingly different from that of the "OPEC tax."

But the real issue was what would happen in the future. The central concern of the consuming countries in the years 1974 through 1978 came down to a simple question: Would the price of oil continue to go up, or would it be held more-or-less steady and thus eroded by inflation? On the answer to that question would depend, among other things, economic growth or collapse, employment, inflation, and the direction of the flow of tens of billions of dollars around the world. Though OPEC was commonly said to be split between "radicals" and "moderates," this same question was also the focus of a running battle between the two largest producers in the Middle East, Saudi Arabia and Iran. It was not a new rivalry. In the 1960s, the two countries had competed over which would produce the most oil. Now the two nations struggled over price, and primacy.

For the Shah, the December 1973 price increase had been his great victory, and very much a personal one. From then on, he saw his moment and opportunity—the prospect of seemingly endless revenues, provided as if by divine intervention, to fulfill his ambitions to create what he called Iran's Great Civilization, and, by the way, solve Iran's mounting domestic economic problems. "One of the only things my husband likes in life," said the Empress in the mid-1970s, "is flying, driving, driving boats—speed!" The Shah applied his passion for speed to his entire country in an attempt to hurtle Iran into the twenty-first century. In so doing, he would ignore the agitation and disorientation that such rapidity caused, as well as the resentment and unhappiness among the many who did not share his obsession with modernity. Iran, the Shah proclaimed, would become the world's fifth largest industrial power; it would be a new West Germany, a second Japan. "Iran will be one of the *serious* countries in the world," he boasted. "Everything you can dream of can be achieved here."

Cut off from reality by the huge infusion of oil money, the Shah became consumed by his ambitions and dreams. He also began to believe all the imperial trappings. Who would dare disagree with the Shah, to counsel caution, to be the messenger of any bad news? As to criticisms of the price increases, the Shah was sarcastic and haughtily dismissive. Inflation in the West, he said, justified the drive for still higher prices, and he discounted the notion that higher oil prices themselves could possibly stoke inflation. "The day has passed when the big industrial countries can get away with political and economic pressure tactics," he told the United States ambassador. "I want you to know that the Shah will not yield to foreign pressure on oil prices." Moreover, Iran's more limited oil reserves, limited at least when compared to its neighbors, argued for going for higher prices sooner rather than later. For, when "later" arrived, Iran's oil reserves might be exhausted. And, finally, there was the Shah's pride. All the humiliations of the past could now be laid to rest, all the gibes turned around. "There are some people who thought—and perhaps some who still think—that I am a toy in the Americans' hands," he said in 1975. "Why would I accept to be a toy? There are reasons for our power which will make us stronger, so why would we be content to be someone else's catspaw?"

But as he pushed for further price increases, the Shah collided with his neighbors across the Gulf. The Saudis had never approved of the scale of the December 1973 price increase. They thought it was too large, and too dangerous to their own position. They feared the economic consequences. And they had been alarmed to find themselves losing control over OPEC and over the basic decisions about oil, which was so central to the kingdom's existence and future. It was not in their interest to perpetuate the cycles of recession and inflation that would be stimulated by further increases in oil prices. Owing to the size of their oil reserves, the Saudis had a decisive stake in the long-term markets for oil, in contrast to Iran. The Saudis feared that higher prices, and the expectation thereof, could set off a move away from oil to conservation and to other fuel sources that would change and contract that long-term market for oil and thus diminish the value of their reserves.

From those considerations flowed other concerns. Saudi Arabia was a country large in territory but small in population, not much bigger in terms of numbers than, for instance, the geographically minute Hong Kong. The rapid buildup of oil revenues could create social and political tensions, as well as dangerous expectations, weakening the ties that held the kingdom together. Nor did the Saudis want higher prices to interfere with, complicate, or undermine their objectives in the Arab-Israeli conflict. And they worried about the effects of higher prices on the political stability of the industrial and developing world, because such instability could in due course come to threaten them. Economic difficulties in Europe in the middle 1970s seemed to be opening the door of government to communists, particularly in Italy, and the prospect of communists in power on Europe's Mediterranean coast was deeply unsettling to a Saudi government already fearful of Soviet designs to encircle the Middle East.

There was still another concern in Riyadh—Iran. The Shah, they were sure, was too short-sighted in his drive for higher prices, too fired by his own ambitions. Further oil price hikes would only provide Iran with still more money and power, enabling it to buy even more weapons, thus shifting the strategic balance and encouraging the Shah to claim hegemony over the Gulf. Why, the Saudis asked, were the Americans so obsessed with the Shah? In August 1975, the U.S. ambassador to Riyadh reported to Washington that Zaki Yamani had said that "the talk of eternal friendship between Iran and the United States was nauseating to him and other Saudis. They knew the Shah was a megalomaniac, that he was highly unstable mentally, and that if we didn't recognize this there must be something wrong with our powers of observation." Yamani sounded a warning.

"If the Shah departs from the stage, we could also have a violent, anti-American regime in Teheran."

For their many and various reasons, political and economic, the Saudis purposefully and forcefully pursued their line against further price increases at meeting after meeting of OPEC. Their firmness at one point even forced OPEC to accept two different prices: a lower one for the Saudis and its ally, the United Arab Emirates, and a higher price for the eleven other members. When the other exporters were looking for justification to raise prices, the Saudis in opposition would push up their production to try to weaken the market. But in so doing, they made a disconcerting discovery. Their sustainable production capacity was not as high as had been assumed.[2]

Yamani

In all these Saudi maneuvers, the spotlight fell on one man—Ahmed Zaki Yamani. To the global oil industry, to politicians and senior civil servants, to journalists, and to the world at large, Yamani became the representative, and indeed the symbol, of the new age of oil. His visage, with his large, limpid, seemingly unblinking brown eyes and his clipped, slightly curved Van Dyke beard, became familiar the planet over. But in the quest for simplification and personalities, and also in response to the opaque political structure of Saudi Arabia, the world sometimes confused his role and ascribed greater power to him than he had. He was, in the final analysis, the representative of Saudi Arabia, albeit an enormously important one. He could not dictate or solely determine Saudi policy, but he could shape it. His style of diplomacy, his mastery of analysis and negotiation, and his skill with the press all gave him decisive influence. His power was augmented by simple longevity, the fact that he ended up being "there" longer than anyone else.

While Yamani was widely known as "Sheikh," the title was in his case an honorific, assumed by prominent commoners, which he was. By background, Yamani was a Hijazi, an urban man from the more worldly and commercial Red Sea coast of Saudi Arabia, rather than a Nejdi, from the more isolated desert principalities that had provided the original base of Ibn Saud's support and that looked to Riyadh as their center. Yamani was born in Mecca in 1930, the year that St. John Philby convinced King Ibn Saud that the only way out of the kingdom's desperate financial situation was to allow prospecting for oil and minerals. During Yamani's childhood, camels still thronged the streets of Mecca, and Yamani read at night either by oil lamp—"electreeks"—or went to the mosque, where electricity had been hooked up.

Both his grandfather and his father were religious teachers and Islamic lawyers; his father had been grand mufti in the Dutch East Indies and Malaya. This combination of learning and piety shaped Yamani's outlook and intellectual development. After his father's return to Saudi Arabia, the family house in Mecca became a meeting place for his students. "Many of them were famous jurists and they would discuss the law with my father and argue cases," Yamani later said. "I started to join them and often after they left, my father and I would stay up for hours and he would teach me and he would criticize my arguments."

Yamani's own considerable intelligence was recognized early in Saudi schools. He went off to university in Cairo, and then to New York University Law School, followed by a year at the Harvard Law School, where he studied international law. He also developed an intuitive grasp of the West and of the United States in particular, and of how to communicate and be comfortable with Americans. When he returned to Saudi Arabia, he set up the first law office in the country. He worked as an adviser to various government ministries, and he wrote the contract for the 1957 concession with Arabian Oil, the Japanese consortium, that broke in on the majors in the Middle East.

Yamani also wrote commentaries on legal issues for various newspapers. That last is what attracted the attention of a most valuable patron, Prince Faisal, second son to Ibn Saud. Faisal invited Yamani to become his legal counselor, and in 1962, when Faisal emerged triumphant from the power struggle with his brother Saud, one of his first acts was to fire the nationalist oil minister Abdullah Tariki. He appointed as his successor the thirty-two-year-old Yamani, whose initial task, in turn, was to end Tariki's confrontation with Aramco, and to begin with more subtlety and skill—and effectiveness—to drive toward the same eventual objectives. "Give me the rantings and ravings of Tariki any day," an executive of one of the Aramco companies complained. "Yamani drives you to the wall with sweet reasonableness."

By the time of the 1973 embargo, Yamani had already been oil minister for eleven years and had developed considerable experience and skill, and superb negotiating talents. His voice was soft, forcing adversaries to strain and to be silent to hear what he said. He almost never lost his temper; the angrier he got, the more quiet he became. Flamboyant rhetoric was not his style. He went logically from point to point, dwelling on each long enough to draw out the essence, the connections, the imperatives, and the consequences. It was all so simple and persuasive and so overwhelmingly obvious and irrefutable that only a maniac or a simpleton could disagree. It was a manner of presentation that was mesmerizingly irresistible to many, and absolutely infuriating to others.

Yamani carefully crafted his mystique; he was the master of patience and of the unblinking stare. When required, he would just look at his interlocutor, without saying a word, fingering his ever-present worry beads, until the subject was changed. He was always playing chess, carefully pondering his opponent's position and how to get where he wanted. Though an adept tactician, expert at maneuvering as required by short-term needs inside Saudi Arabia and out, he also tried always to think long term, as befitted the representative of a country with a small population and one-third of the world's oil resources. "In my public life, in my personal life, in everything I do I think long term," he once said. "Once you start thinking short term, you are in trouble because short-term thinking is only a tactic for immediate benefit." The Western world, he believed, was afflicted by the curse of short-term thinking, the inevitable result of democracy. By nature, Yamani was also cautious and calculating. "I can't bear gambling," he said in 1975, when he was at his apogee. "Yes, I hate it. It rots the soul. I've never been a gambler. Never." In oil politics, he insisted, he never gambled. "It's always a calculated risk. Oh, I calculate my risks well. And when I take them, it means

I've taken all necessary precautions to reduce them to the minimum possible. Almost to zero."

Yamani generated strong reactions. Many thought him brilliant, a diplomat of high order, with a broad, superb grasp of oil, economics, and politics. "He was a consummate strategist," said one who dealt with him for twenty-five years. "He never went directly to his goal, but he never lost sight of where he wanted to go." In the West, he became the embodiment of the OPEC Imperium and the ascendancy of oil power. To many Western leaders, he was the one reasonable and influential interlocutor, and the most knowledgeable. To many in the public, he was the most visible and, therefore, the most criticized and derided of the exporters' representatives. Some in OPEC itself and in the Arab world hated him, either resenting his prominence, or regarding him as too close to the West, or simply thinking he was given too much credit. Jealous rivals and skeptics said that he was "overrated." One Aramco official who frequently dealt with him was struck, more than anything else, by what he described as Yamani's capacity for being "ostentatiously calm."

Henry Kissinger, who also had many dealings with Yamani, was backhanded and almost deliberately lighthearted in commending him: "I found him extraordinarily intelligent and well read; he could speak penetratingly on many subjects, including sociology and psychology. His watchful eyes and little Van Dyke beard made him look like a priggish young don playing at oil policy but not really meaning the apocalyptic message he was bringing, especially as it was put forward with a gentle voice and a self-deprecatory smile at variance with the implications of his actions. . . . In his country at that time, barred by birth from the political leadership reserved for princes and by talent from an ordinary existence, he emerged in a position as essential as it was peripheral to the exercise of real political power within the Kingdom. He became the technician par excellence."

Yamani was very much a Faisal man, devoted to the King who had chosen him. The King, in turn, regarded Yamani as a favored protégé and rewarded him with extensive grants of real estate, which skyrocketed in value during the oil boom and were a basis for Yamani's personal fortune. Yamani's close, intense relation with the King gave him carte blanche in making oil policy, though always under the final control of Faisal, and always within lines defined by the Royal Family, whose most prominent member when it came to oil policy after the King himself was his half brother, Prince Fahd.

In March 1975, Yamani accompanied the visiting Kuwaiti oil minister to an audience with King Faisal. A nephew of Faisal followed the party into the little reception room, and as the Kuwaiti knelt before the King, the nephew stepped forward and fired several bullets into Faisal's head, killing him almost instantaneously. Afterward, some said that the murder was the revenge for the nephew's brother, who had been slain a decade earlier in the course of leading a fundamentalist attack on a television station to protest the first broadcast of a woman's voice in the kingdom. Others said the young man had been caught up in the miasma of the extreme left. Still others simply said he was deranged, mentally off-balance, and, noting that he had been prosecuted while a student in Colorado for selling LSD, suffering from the effects of drugs.

Then in December of that year, the international terrorist known as "Carlos," a fanatic Marxist from Venezuela, led five other terrorists in an attack on a ministerial meeting in the OPEC Building on Karl Lueger Ring in Vienna. Three people were killed in the first few minutes. The terrorists took the oil ministers and their aides hostage, and eventually embarked on a harrowing air journey, flying first to Algiers, then to Tripoli, and then back to Algiers, threatening all the way to kill the ministers. Again and again, they said that two people had already and absolutely been sentenced to death: Jamshid Amouzegar, the Iranian oil minister, and Yamani, who was their number-one quarry. On the tense flights, Yamani spent the time playing with his worry beads and reciting to himself verses from the Koran, convinced that he would soon be a dead man. Forty-four hours after the initial assault in Vienna, the ordeal finally came to an end in Algiers, with the suspension of the "death sentences" and the release of everyone, including Yamani. Some thought that a faction in one of the Arab governments had provided assistance to the terrorists and may even have promised a large reward.

After 1975, Yamani became, understandably, obsessive on the subject of security. And after Faisal's assassination, he never had the independence on oil that had previously been his. Faisal's successor was his half-brother Khalid, who already suffered from a heart condition and did not project himself as a strong leader. Fahd became Crown Prince and Deputy Prime Minister. He was also the chief policymaker when it came to oil, and the person to whom Yamani now reported. To the outside world, Yamani still was the number-one figure, but within Saudi Arabia, it was the careful, cautious Prince Fahd who had the final say on policy. On those occasions when he spoke for the record, Fahd made clear that the opposition to higher oil prices was not Yamani's position alone, but Saudi policy. Hiking the price further, Fahd declared, would spell "economic disaster." Indeed, at a private meeting with President Jimmy Carter in Washington in 1977, Fahd went so far as to vigorously urge Carter to put pressure on two other OPEC countries, Iran and Venezuela, to prevent further price increases.

At times, the Saudi policies infuriated the other exporters enough to bring down a rain of vituperation, often carefully directed toward Yamani and not the Royal Family. "When you listen to Iranian radio or read Iranian newspapers, you will learn that I am a devil," Yamani complained. One of the leading newspapers in Tehran castigated Yamani as a "stooge of capitalist circles, and a traitor not only to his own King and country but also to the Arab world and the Third World as a whole." The Iraqi oil minister declared that Yamani was acting "in the service of imperialism and Zionism." To such rhetoric, the unflappable Yamani reacted with his enigmatic smile and unblinking stare.[3]

America's Strategy

Whatever the internal rivalries within OPEC, there certainly was a general meeting of minds between Riyadh and Washington when it came to oil prices. The United States government, through the Nixon, Ford, and Carter Administrations, consistently opposed higher prices because of the further damage that such increases might do to the world economy. But Washington did not want to aggressively force prices down. "The only chance to bring oil prices down immediately would be massive political warfare against countries like Saudi Arabia and Iran to make them risk their political stability and maybe their security if they did not cooperate," Kissinger, Ford's Secretary of State, explained in 1975. "That is too high a price to pay, even for an immediate reduction in oil prices. If you bring about an overthrow of the existing system in Saudi Arabia and a Qaddafi takes over, or if you break Iran's image of being capable of resisting outside pressures, you're going to open up political trends that could defeat your economic objectives." Indeed, there was some concern that the oil exporters might themselves suddenly drop the price substantially and thus undermine expensive new developments, such as those in the North Sea. As a result, there were discussions among the members of the International Energy Agency about establishing a "minimum safeguard price" to provide a floor to protect higher-cost energy investment in the Western world against an abrupt, perhaps politically motivated, slash in world prices.

Washington's central objective was stability, and it campaigned hard against further price rises for fear that they would stoke inflation, cripple the international payments and trade system, and retard growth. Before every OPEC meeting, the United States would dispatch a host of emissaries to the interested parties around the world. Backed up by a flood of cables that were filled with statistics about inflation and energy use, the American officials would strenuously argue against any further increases. To be sure, contradictory messages would sometimes emerge from the large, contentious bureaucracies that shaped foreign and economic policies in the United States. At times, the Saudis would even suspect that the United States was playing a trick on them, that it was in secret collusion with the Shah to raise prices. Nixon, Ford, and Kissinger were, in fact, reluctant to push too hard on the Shah, given other strategic considerations. Moreover, on the domestic American scene, there was no consensus, but on the contrary a series of battles that made energy the number-one political issue of the mid-1970s. Internationally, however, the consistent central objective of U.S. policy was to build stability back into the price and let inflation wear it down. All the rhetorical tools, from cajolery and flattery to prophecies of doom and implicit threats, were used by Washington in the pursuit of that stability.

Other, less visible approaches were tried as well. In an effort to help cap prices and ensure additional supplies for the United States, Washington flirted with the idea of going into the oil business itself in partnership with nothing less than the Soviet Union. Kissinger pursued a "barrels-for-bushels" deal under which the United States would import Soviet oil in exchange for American wheat. A preliminary letter of understanding was signed in Moscow in October 1975. Shortly after, senior Soviet officials came to Washington for what proved to be intense negotiations. Here was the chance for Kissinger to score a "victory" for his American-Soviet détente, which, under mounting domestic criticism, could use some victories. It could also mean a "defeat" for OPEC, with the savory irony of using Soviet oil to "break" OPEC's grip.

After a few days of lengthy discussions, the Russians found themselves on a weekend in Washington with nothing to do. For a little light relief, they were whisked off by Gulf Oil, which traded oil with the Soviets, in a company jet to

Disney World. On the trip down to Florida, the head of the Soviet delegation explained to his hosts why the negotiations were so difficult: Kissinger was insisting on maximum publicity to embarrass OPEC. The Russians would love to sell their oil, they would love not to have to spend hard currency on American wheat, but the transaction had to be kept, if not secret, at least very low profile. They could not allow themselves to be seen as undercutting OPEC and Third World nationalism. There was also a problem of terms; Kissinger was insisting that the American wheat be valued at world wheat prices, while the Russian oil should be valued at 12 percent or so below world oil prices. When the Russians asked why the disparity, the Americans explained that American wheat was being sold into an established market, whereas a new market was being opened up for Russian oil, and therefore the Soviets would have to discount to get into it. Ultimately, the deal fell through. But the Soviet officials did have a fabulous time at Disney World.[4]

The American commitment to stable oil prices put it on a collision course with Iran; after all, the Shah was the most vocal and influential of the price hawks, and the United States was frequently urging him otherwise. When President Ford criticized higher prices, the Shah quickly shot back, "No one can dictate to us. No one can wave a finger at us, because we will wave a finger back." To be sure, Iran, no less than Saudi Arabia, was tied politically and economically to the United States. Yet as the government ministers and businessmen and weapons merchants trouped to Tehran, and as the Shah continued to lecture and hector Western society on its weaknesses and ills, some in Washington questioned exactly who was whose client.

At the beginning of the 1970s, Nixon and Kissinger had established a "blank check" policy, giving the Shah a free hand to buy as many American weapons systems as he wanted, even the most technologically advanced, so long as they were not nuclear. The policy was part of the "twin pillars strategy," established for regional security in the wake of the British withdrawal from the Gulf. Iran and Saudi Arabia were together to be the pillars, but of the two, Iran was clearly, as one American official put it, the "Big Pillar," and by the mid-1970s, Iran was responsible for fully half of total American arms sales abroad. The Defense Department was alarmed by this blank check; in its view, Iran needed a strong conventional army, but not ultramodern weapons systems that it would have trouble mastering and that might fall into the hands of the Russians. Defense Secretary James Schlesinger personally cautioned the Shah that Iran lacked the technical resources to assimilate so many new and complex weapons systems. "He fell in love with the F-15," said Schlesinger. While the Shah normally brushed aside such warnings, in the case of the F-15, he heeded the advice and did not purchase that particular aircraft.

Pungent criticism came from Treasury Secretary William Simon. "The Shah," he said, "is a nut." Not surprisingly, the Shah took strong exception to that characterization and Simon quickly apologized: He had been quoted out of context. He had meant, he explained ingeniously, that the Shah was a "nut" on oil prices in the way one might say somebody was a "nut about tennis or golf." The American ambassador was away from Tehran when this incident blew up, and the unhappy task of explaining the remark fell to the chargé. He repeated

Simon's excuse to the Minister of Court, who replied, "Simon may be a good bond salesman, but he does not know a whole lot about oil." The Shah himself was reported to have commented that he knew English as well as Simon and that he understood "exactly what Mr. Simon meant."

Yet despite the carpings and criticism, a consensus held sway through the Nixon and Ford administrations. Iran was an essential ally with a major security role in the Middle East, and nothing should be done to undercut the Shah's prestige and influence. Nixon, Ford, and Kissinger had strategic and personal predilections for the Shah; he had not embargoed oil to the United States in 1973, and Iran could play a key role in geopolitical strategies. The Saudis, Kissinger would tell colleagues, were "pussy cats." But with the Shah he could talk geopolitics; Iran, after all, shared a border with the Soviet Union.

The Shah had good reason to worry in 1977 about the new American President, Jimmy Carter. In the words of the British ambassador in Tehran, "The calculating opportunism of Nixon and Kissinger was far more to the Shah's taste." Two of the most important policies of the Carter Administration, human rights and restriction on arms sales, were directly threatening to the Shah. Despite those policies, the new Administration maintained the pro-Shah orientation of its predecessors. As Gary Sick, a Middle Eastern affairs officer on the National Security Council during the Carter Administration, later wrote, "The United States had no visible strategic alternative to a close relationship with Iran."

Relations were facilitated by the Shah's shift on the subject of oil prices. By the time Carter moved into the White House, the Shah was having second thoughts about the value of pushing for higher prices. The freneticism and euphoria, the onrush of petrodollars, and the oil boom itself were wrecking the fabric of Iranian economy and society. The results were already evident: chaos, waste, inflation, temptation, corruption, and deepening political and social tensions that were broadening opposition to the regime. Increasing numbers of his subjects wanted no part of the Shah's Great Civilization.

At the end of 1976, the Shah himself ruefully summed up the problem: "We acquired money we could not spend." Money, he was now forced to acknowledge, was not the remedy, but the cause of many of his nation's ailments. Higher oil prices would not help him, so why go to the trouble of challenging the United States on this point when he needed, with Carter's arrival, to bolster relations with America more than ever? Early on, the Carter Administration decided to launch and maintain a "price freeze offensive" as a central U.S. policy. And after Secretary of State Cyrus Vance, on a visit to Tehran in May of 1977, reassured the Shah about continued American support, the Iranian government began to surprise the other oil exporters, and even its own officials, by calling for moderation on oil prices. The Shah privately went so far as to tell Treasury Secretary Michael Blumenthal that Iran "does not want to be known as a price hawk." Had the Shah undergone a marketplace conversion? Had the number-one price hawk become a dove?

In November 1977, the Shah came to Washington to meet President Carter. At the very moment the Shah arrived at the White House, a battle erupted on the nearby Ellipse among anti- and pro-Shah demonstrators, primarily Iranian students in the United States. The police broke it up with tear gas. The fumes wafted over the South Lawn of the White House, where President Carter was greeting the Shah. Carter began blinking and rubbing his eyes, while the Shah wiped away his own tears with a handkerchief. The news footage was broadcast not only on American television, but also in Iran, owing to the new liberalization—giving the Iranian population a less-than-grand view of their monarch, a perspective of the sort they had never before been permitted. That scene, along with the fact of the demonstrations themselves, convinced some Iranians that the United States was about to disengage from Mohammed Pahlavi. Why else, they thought, not understanding the American system, would Carter have "allowed" such demonstrations?

In their private meetings, Carter lobbied both for human rights and for stability in oil prices. As the Shah saw it, he was being asked by Carter to make a trade: to join Saudi Arabia in moderation on oil prices in exchange for the continued flow of arms from the United States and relief from the pressure on human rights. Carter stressed "the punishing impact of increased oil prices on the industrial economies." Contradicting much of what he had said since the end of 1973, the Shah agreed with Carter, and he promised to urge the other OPEC countries to "give Western nations a break."

Iran had now joined Saudi Arabia on the side of moderation. With those two countries representing 48 percent of OPEC production, they could dictate to the other members, and oil prices would be held in check. Thus the battle between the Shah and the Saudis was ended. The Shah had been won over. Through the half-decade from 1974 to 1978, there were only two rather small OPEC-wide price increases: from the $10.84 set at Tehran in December 1973 to $11.46 in 1975 and to $12.70 at the end of 1977. But inflation was increasing at a more rapid rate, and as had been anticipated, eroding the real price. By 1978, the oil price, when adjusted for inflation, was about 10 percent below what it had been in 1974, immediately after the embargo. In short, by restricting the increases to those two relatively small ones, the real oil price was in fact lowered somewhat. Oil was no longer cheap by any means, but neither had the price, as many feared it would, gone through the roof.[5]

Kuwait and "Our Friends"

If the oil exporters no longer had to negotiate over price with anybody except one another, there were still the oil concessions, reminders of the times when the companies held sway, relics of the days when the exporters were poor. The very existence of concessions, the oil nations now said, was degrading. The concession in Iran, of course, had been wiped out by Mossadegh's nationalization in 1951, and Iraq had completed its nationalization of the IPC concession in 1972. While some concessions would remain in the aftermath of the 1973 price shock, the termination of the last great ones—in Kuwait, Venezuela, and Saudi Arabia— would mark the final demise of the twentieth century concessional arrangements that had begun with William Knox D'Arcy's bold and risky commitment to Persia in 1901.

The Kuwaiti concession was the first on the block. The Kuwait Oil Company had been established in 1934 by BP and Gulf to bring an end to their acrimonious competition, which was stoked by the irrepressible Major Frank Holmes and sharpened by the determination of Ambassador Andrew Mellon. Forty years later, at the beginning of 1974, Kuwait acquired 60 percent participation in the Kuwait Oil Company, leaving BP and Gulf with 40 percent. Then in early March 1975, Kuwait announced that it was going to take over that last 40 percent and not maintain any special links to BP and Gulf. They would simply be treated like other buyers. And what would happen if BP and Gulf did not agree to Kuwait's terms? "We will just say thank you very much and good bye," said the Kuwaiti oil minister Abdel Mattaleb Kazemi. The objective, he added, was "to gain full control over the country's oil resources." He went to the essence: "Oil is everything in Kuwait."

James Lee of Gulf and John Sutcliffe of BP were quickly summoned to Kuwait City. Sutcliffe told the oil minister, "There should be consideration for the old relationship." The Kuwaiti reply was emphatic. "No compensation was due." Meeting with the Prime Minister, Sutcliffe and Lee offered a brief history of how, as a result of the battle over rents, the profit split had shifted over the years, "from the 50/50 split concept in the early 1960s to the present split of about 98 percent for the Government and 2 percent for the companies." They hoped now to work out a satisfactory arrangement. But they were told, quite firmly, that Kuwait intended to take over 100 percent, that it was a matter of sovereignty, and that the question was not open to debate.

For a number of months Kuwait struggled with the two companies, which continued to try to hold on to some kind of preferential access. At one point, a senior BP negotiator, P. I. Walters, half-jokingly suggested to the Kuwaitis that they would do far better to invest some of their new oil wealth in BP shares rather than to acquire the physical assets of the Kuwait Oil Company. The Kuwaitis were not interested, at least not at that time. Finally, in December 1975, the two sides came to an understanding—on Kuwait's terms. Gulf and BP had asked $2 billion in compensation. At this, the Kuwaitis laughed. The companies got a tiny fraction of that amount, $50 million.

Once the deal was done, the two international companies still assumed that they would retain preferential access. That assumption was much in the mind of Herbert Goodman, the president of Gulf Oil Trading Company, when he was dispatched with a small team to Kuwait City to put the finishing touches on the new relationship, or so he thought. Goodman quickly found out how much had really changed. Not that he could ever have been accused of being naive. Goodman was one of the most experienced oil supply men and traders in the world; indeed, his career embodied the extraordinary development and expansion of the international oil companies in the decade of the 1960s. A former U.S. Foreign Service officer who had joined Gulf in 1959, Goodman had earned his place in any oil hall of fame; for during four years in Tokyo, he had the distinction of selling over a billion barrels of oil in a series of long-term contracts with Japanese and Korean buyers. The 1960s were the glory years, both for an oil man and for an American abroad. "There was tremendous cachet to being an American businessman then, enormous entree everywhere," Goodman was to recall. "You learned to take it as your due. People paid attention. There was a respect for your credibility, clout, power. Why? It was trade following the flag—

the enormous credibility and respect enjoyed by the United States. The American passport was truly a laissez-passer—a safeguard. Then that began to fade. I could feel it everywhere. It was the ebbing of American power—the Romans retreating from Hadrian's Wall. I tell you, I could feel it everywhere." Then came the oil embargo, the price increase, Nixon's humiliation and resignation, and the abrupt American withdrawal from Vietnam. And now Goodman found himself, in 1975, sitting in Kuwait City where the Kuwaitis were also insisting that an era had ended.

Still, Goodman expected, as did the other executives in his party, that Gulf would get some kind of special price or preference, reflecting a relationship almost a half century long, the training of the many young Kuwaitis who had come to Pittsburgh and stayed with Gulf families, all the hospitality, personal relations, and connections. But no, Goodman was told to his surprise that Gulf would be treated just like any other customer. Furthermore, the Kuwaitis said, Gulf would only get enough oil for its own refineries, and not for its third-party customers in Japan and Korea. But those were the markets, Goodman replied, that Gulf had sweated blood to develop. He knew; it was his blood that had been sweated. No, said the Kuwaitis. Those were their own markets, based upon their oil, and they would sell their oil in those markets.

The Gulf men could not help but notice how differently they were being treated from past days. "We would go from our hotel to the ministry, day after day, and wait," said Goodman. "Sometimes, a junior person would come. Sometimes not." At one point in the discussions, Goodman tried to remind a Kuwaiti official of the history, at least as he understood it, as Gulf understood it, of all that Gulf had done for Kuwait. The Kuwaiti became very angry. "Whatever you did, you got paid for," he said. "You never did us any favors." Then he walked out of the meeting.

Ultimately, Gulf got a very small discount on oil going into its own system, but none at all on any oil that it might sell to anyone else. "For the Kuwaitis, it was the overthrow of the colonial power," Goodman reflected afterward. "There was this misunderstanding. Here was the conceit of the Americans that we were loved because we had done so much for these people. This was the American naiveté. We thought we had good relations. They saw it from a different point of view. They had always felt patronized. They remembered it. In all these relationships, there's this love-hate thing.

"Yet," he added, "it was transitory. It was just that they were just about to get very rich."[6]

Venezuela: The Kitty Cat Died

The great concessions in Venezuela were also being swept away. Already, at the beginning of the 1970s, there was no doubt of what was to come. After all, this was the country of Juan Pablo Pérez Alfonzo, oil nationalist and cofounder of OPEC. In 1971, Venezuela passed a "law of reversion," which said that all the oil companies' concessions and other assets in the country would revert to Venezuela when the concession terms ended, with restricted compensation. The first concessions would start expiring in 1983. The economic effect of the reversion law, plus Venezuela's policy of "no new concessions," was inevitable: The companies slowed their investment, which meant that Venezuela's production capacity was declining. The decline, in turn, just as inevitably fueled the nationalistic antipathy to the companies. "It was chicken and egg," recalled Robert Dolph, the president of Creole, the Exxon subsidiary in Venezuela. "The policy was that there were no new areas to explore. So we were not feeding the kitty cat, and then they were complaining that the kitty cat is dying."

By 1972, the government had passed a number of laws and decrees that gave it effective administrative control over every phase of the industry, from exploration to marketing. It also raised the effective tax rate to 96 percent. So it had achieved many of the objectives of nationalization without yet nationalizing. But nationalization was still only a matter of time. The 1973 price increase and OPEC's apparent victories quickly strengthened the spirit of nationalism and self-confidence and speeded up the last act. In the new era, 1983 was too long to wait. Foreign ownership was simply not acceptable anymore, and nationalization would have to come as soon as possible. On this, virtually all political factions seemed to agree.

Two sets of negotiations—not one—would ensue. The first was with the international companies, Exxon and Shell, followed by Gulf and a number of others. The second was only among Venezuelans themselves. The first set did not go smoothly. "As 1974 ended, the country was still in the midst of a feverish debate on the oil nationalization issue," said one participant. "The field had clearly split between those who advocated a violent confrontation with the foreign oil companies and those who preferred a nonviolent, negotiated settlement." Juan Pablo Pérez Alfonzo weighed in from the garden of his house on the side of the confrontationists; he declared that not only the oil industry but all foreign investment in Venezuela should be nationalized immediately.

Yet the process of settlement proceeded with less rancor than might have been expected, partly because of the realism of the companies. Some might have called it fatalism. Venezuela had been the source of a substantial part of their profits in earlier years, at one point half of Exxon's entire global income. It had also been the place to be if you had any hope of getting to the top of Shell, if not necessarily Exxon. But, in the new era, there was no way they could resist. For them, the critical thing was to retain access to oil. "We couldn't win," said Dolph of Creole. "Prices were strong; circumstances in the market were emboldening all the countries, which assumed that what was happening would go on forever. The actual nationalization gave us very little room for maneuver."

Venezuela would have two requirements after nationalization. One was to maintain the flow of technology and skills from the outside world to keep the industry as efficient and up-to-date as possible. The companies negotiated service contracts with Venezuela under which, in exchange for a continuing transfer of technical skills and personnel, the former concessionaires were paid fourteen or fifteen cents a barrel. The second need was access to markets; the nationalized industry would be producing a vast amount of oil. It did not have its own marketing system outside the country, and it would need to be able to sell the oil. Meanwhile, the former concessionaires still needed petroleum for their downstream systems, so they made long-term contracts with Venezuela that would get the oil to market. The first year after nationalization, Exxon signed with Venezuela what was considered to be the largest single oil supply contract ever made up to that date—900,000 barrels per day.

Far more difficult and emotional was the second negotiation—between the Venezuelan politicians and Venezuelan oil men. Two generations of Venezuelans had grown up within the oil industry; by this time, 95 percent of all the positions, right up to the most senior levels, were staffed by Venezuelans, many of whom had been partly trained abroad and had gained international experience within the multinational companies, and they generally thought they had been treated fairly. The question now came down to this: Was the Venezuelan oil industry, on which the government's revenues depended, to be primarily a political entity, with its agenda set by politicians and the interplay of domestic politics, or would it be a government-owned entity that was run as a business, with a longer time horizon and with its agenda set by oil men? Behind that question, of course, was a struggle for power and primacy in post-nationalization Venezuela, as well as a battle over the future of the nation's economy.

Certain inescapable considerations shaped the outcome. The oil industry and its health were central to the overall economic well-being of Venezuela. In Caracas, there was a widespread fear that "another Pemex" might be created, that is, an extraordinarily powerful national company like Petróleos Mexicanos, which was an impenetrable state-within-the-state. Or the result, it was feared, might be a weakened, politicized, corrupted oil industry, with a devastating effect on the Venezuelan economy. The outcome was also affected by the fact that there was a broad, accomplished, technically sophisticated cohort of oil people not only throughout the Venezuelan subsidiaries, but also at the very top. If the industry were politicized, they might just pack up and leave.

In those circumstances, President Carlos Andrés Pérez, who had recently won a landslide victory as the candidate of Acción Democrática, opted for a "moderate" and pragmatic solution, and one in which the oil industry itself was able to participate. A state holding company—Petróleos de Venezuela, known as PDVSA—was established to play a central financial, planning, and coordinating role, and to be a buffer between the politicians and the oil men. A number of operating companies were also created, based upon the pre-nationalization organizations, and eventually consolidated to four and then three. Each was a fully integrated oil company, down to its own gasoline stations. Such quasi-competition, it was hoped, would assure efficiency and prevent the growth of another bloated, bureaucratic state company. Also, this structure would help to maintain the various aspects of corporate culture, tradition, efficiency, and esprit de corps that would improve operations. On the first day of 1976, the nationalization took effect. President Pérez called it "an act of faith." The country's new nationalized oil company was destined quickly to become a major force in its own right in the new world oil industry.[7]

Saudi Arabia: The Concession Surrendered

What remained was the greatest concession of them all—Aramco's in Saudi Arabia. From the bleak years of the early 1930s, when the impoverished King

Ibn Saud had been more desirous of discovering water than oil, Aramco had grown into a vast economic enterprise. In June of 1974, Saudi Arabia, operating on Yamani's principle of participation, took a 60 percent share in Aramco. But, by the end of the year, the Saudis told the American companies in Aramco— Exxon, Mobil, Texaco, and Chevron—that 60 percent was simply not enough. It wanted 100 percent. Anything less, in the new era of oil nationalism, was humiliating. The companies dug in their heels. After all, their number-one dictum was "never to give up the concession." It was the most valuable in all the world. Even if that rule could not stand up to the political pressures of the mid-1970s, the companies would, at least, try to make the best deal they could. The Saudis, for their part, were no less insistent on getting what they wanted and exerted economic pressure when necessary. In due course, the companies were persuaded, and they agreed to the Saudi demand—in principle.

To transform principle into practice, however, took another year and a half, as the two sides argued over the crucial operational and financial questions. These negotiations to determine ownership of fully one-third of the free world's oil reserves were arduous and difficult. They were also nomadic. For a month in 1975, the representatives of the Aramco companies encamped with Yamani in Beit Meri, a hill town above Beirut. Each morning, the oil men would walk down the little street from their hotel to an old monastery that Yamani had converted into one of his homes. There they would debate how to value an extraordinary resource and how to maintain access. Then word got to them that a terrorist group might be planning to attack them or kidnap them, and suddenly the little street looked not quaint but dangerous. They promptly cleared out, and thereafter, the negotiators trailed along with Yamani on his global peregrinations.

Finally, late one night in the spring of 1976, they came to agreement in Yamani's suite at the Al-Yamama Hotel in Riyadh. Forty-three years earlier in Riyadh, after Standard of California had reluctantly agreed to make an up-front payment of $175,000 for the right to wildcat in the trackless desert, Ibn Saud had ordered the original concession document to be signed. By 1976, the proven reserves in that desert were estimated at 149 billion barrels—more than a quarter of total free world reserves. And now the concession was to be disbanded once and for all. "It was truly the end of the era," said one of the Americans who was there at the Al-Yamama Hotel that night.

But the agreement did not by any means provide for a severing of links. The two sides needed each other too much. It was the same old issue that had tied the Aramco partners together in the first place: Saudi Arabia had enough oil to last several lifetimes, while the four companies had the huge marketing systems required to move large volumes of that oil. So, under the new arrangement, Saudi Arabia would take over ownership of all of Aramco's assets and rights within the country. Aramco could continue to be the operator and provide services to Saudi Arabia, for which it would receive twenty-one cents a barrel. In turn, it would market 80 percent of Saudi production. In 1980, Saudi Arabia paid compensation, based on net book value, for Aramco's holdings within the kingdom. With that, the sun finally set on the great concessions. The oil producers had achieved their grand objective; they controlled their own oil. These nation-states had become synonymous with petroleum.

There was just one odd thing about the agreement between Saudi Arabia and the four Aramco companies. The Saudis did not sign it, not until 1990, fourteen years after it had been agreed upon. "It was very practical," said one of the company negotiators. "They got what they wanted—full control—but they didn't want to disrupt Aramco." As a result, about 33 billion barrels of oil were produced and marketed and over $700 billion of business was conducted for fourteen years and all of it in a condition, in the words of an Aramco director, "in limbo."

While, initially, the oil companies were still linked by supply contracts to their former concessions in Saudi Arabia, Venezuela, and Kuwait, those connections would weaken over time due to diversification policies of both countries and governments, and because of the opportunities and alternative ties that existed in the market. Moreover, at the same time that the "great concessions" were being terminated, a new relationship was emerging between various petroleum exporting countries and international oil companies. Instead of being "concessionaires," with ownership rights to the oil in the ground, the companies were now becoming mere "contractors," with "production sharing" contracts that gave them rights to part of any stream of oil they discovered. This new type of relationship was pioneered by Indonesia and Caltex in the late 1960s. The "services" happened to be the familiar ones of exploring for, producing, and marketing oil. But the shift in terminology reflected an all-important political change: The sovereignty of the country was recognized by both parties in a way that was acceptable in the domestic politics of the countries. The lingering aura of a colonial past was banished; after all, the companies were there merely as hired hands. By the mid-1970s, such production sharing contracts were becoming common in many parts of the world.

Meanwhile, the amount of oil sold directly by the exporters themselves into the market, without benefit of the companies in their traditional role as middlemen, was increasing dramatically—quintupling from 8 percent of total OPEC output in 1973 to 42 percent by 1979. In other words, the state-owned companies of the oil-producing countries were moving downstream, beyond production, into the international oil business outside their own borders. Thus, in many ways, the global oil industry had in little more than half a decade taken on a completely new form under the OPEC Imperium. Even more dramatic changes were ahead.[8]

CHAPTER 32

The Adjustment

DID THE END of cheap oil mean the end of the road for Hydrocarbon Man? Would he be able to afford the petroleum that powered his machines and provided the material things he had come to cherish in his daily life? In the 1950s and 1960s, cheap and easy oil had fueled economic growth and thus, indirectly, promoted social peace. Now, it seemed, expensive and insecure oil was going to constrain, stunt, or even eradicate economic growth. Who knew what the social and political consequences would be? Yet the risks loomed large, for one of the great lessons of the miserable decades between the two world wars was how central was economic growth to the vitality of democratic institutions. The oil exporters had, heretofore, complained that their sovereignty was being impaired because of oil. After 1973, it was the industrial nations that found their sovereignty diminished and under assault, their security threatened, and their foreign policies constrained. The very substance of power in international politics seemed to have been transmuted by its oleaginous reaction with petroleum. No wonder that the decade of the 1970s was, for Hydrocarbon Man and for the industrial world as a whole, a time of rancor, tension, unease, and gritty pessimism.

Yet Hydrocarbon Man would not so easily surrender his postwar legacy, and a process of massive adjustment to the new realities now began. The International Energy Agency turned out to be not the agent of confrontation that had been predicted by the French, but rather an instrument for coordination among the Western countries, and a way to bring their energy policies into parallel. The procedures for an energy emergency sharing program were worked out, as were targets for government-controlled strategic reserves of oil, which could be called upon to make up a shortfall in the event of a disruption. The IEA also provided a forum for evaluation of national policies and for research on both conventional and new energy sources.

The central goal for the Western world in the middle of the 1970s came down to what Kissinger described as changing those "objective conditions" in the marketplace from which oil power derived—those conditions being the supply-demand balance and the overall reliance of the industrial economies on oil. Virtually all of the industrial countries, responding to both price and security concerns, embarked on energy policies aimed at reducing dependence on imported oil. In trying to alter those "objective conditions," each of the major consuming countries proceeded in its own characteristic way, reflecting its political culture and idiosyncrasies—the Japanese with a public-private consensus; the French with their tradition of *dirigisme*, state direction; and the United States with its usual fractious political debate. The mix may have been different, but the elements required to roll back the new oil power were the same: the use of alternative fuels, the search for diversified sources of oil, and conservation.

Nations Respond

After the initial panic and shock of the Arab oil embargo, Japan began to marshal its responses. The Ministry of International Trade and Industry made a personal "statement" of sorts; it curtailed elevator service in its own headquarters building. In order to reduce the need for electric air conditioning in the summer months in Japan, a major effort was also made to promote an ingenious innovation in men's fashions: the *shoene rukku* or "energy conservation look"— business suits with short-sleeve jackets. While the elevator service remained curtailed, the new suits, though advocated by Prime Minister Masayoshi Ohira himself, never caught on.

Quite a vigorous struggle for leadership in energy decision making broke out in Japan. Nevertheless, in every dimension, the country was committed to altering drastically its energy position, which since the early 1960s had been based on access to cheap and secure Middle Eastern oil. It was no longer cheap or secure, and Japan's naked vulnerability was, once again, all too evident. The lines of response and change were widely accepted and acted upon. They included the conversion of electric generation and industrial production from oil to other fuel sources, the acceleration of nuclear power development, the expansion of imports of coal and liquefied natural gas, and the diversification of oil imports away from the Middle East and toward the Pacific Rim. "Resource diplomacy" came to the fore in Japan's international relations, as the country arduously sought to woo oil producers and energy suppliers, both in the Middle East and in the Pacific Rim.

Yet no other effort was more concentrated nor more immediately significant than the concerted government-business drive to promote energy conservation in industry and, in particular, to reduce oil use. The campaign's success far exceeded what was anticipated, and was of key importance to the renewed international competitiveness of Japanese business. The response, in fact, set the standard for the rest of the industrial world. "Both workers and business leaders were very apprehensive after 1973," recalled Naohiro Amaya, then a Vice-

Minister of MITI. "They feared for the survival of their companies, and so everybody worked together." In 1971, MITI had conducted a study on the need to move from "energy-intensive" to "knowledge-intensive" industry, based upon the premise that Japanese oil demand was growing so fast that it would put undue pressure on the world oil market. The heavy industries had not particularly liked the study, which implied that they were lame ducks; and the report, developed at a time when prices were still low, was much criticized. But the 1973 crisis provided the force to implement the new strategy at breakneck speed. "Instead of using the resources in the ground, we would use the resources in our head," said Amaya. "The Japanese people are accustomed to crises like earthquakes and typhoons. The energy shock was a kind of earthquake, and so even though it was a great shock, we were prepared to adjust.

"In a way," he added, "it was a kind of blessing, because it forced the rapid change of Japanese industry."

In France, the most senior energy official was Jean Blancard, an engineer and member of the elite Corps des Mines with long experience in the oil industry. As General Delegate for Energy in the Ministry of Industry, he coordinated government policies and those of the state-owned energy companies. In early 1974, even as Paris was trying to pursue its conciliatory bilateral policies toward the oil producers, Blancard was arguing to President Georges Pompidou, "The period from here on will be quite different—a transformation, not a crisis. . . . It is not reasonable for such a country as ours to be hanging on the Arabs' decisions. We must pursue a policy of diversification of energy and try to decrease the need for oil—or at least not allow it to increase."

Blancard found a very receptive audience for his ideas in Pompidou, who, in early 1974, convened a meeting of his senior advisers. Seriously ill, Pompidou had become swollen from the effects of treatment, and was obviously in great pain during the long meeting. Nevertheless, that discussion confirmed the three basic planks of French energy policy: rapid development of nuclear power, a return to coal, and a heavy emphasis on energy conservation—all designed to restore France's autonomy. Pompidou died less than a month after the meeting, but his successor, Valery Giscard d'Estaing, pushed ahead on all three programs. With a governmental system much more impervious than other Western countries to such outside interveners as environmentalists, France within a few years would outstrip all the others in its commitment to nuclear power. But nuclear power was proceeding elsewhere as well; and electricity generation would, by the early 1980s, become one of the major markets in the West lost to oil, as indeed was the intention, though nowhere else to so great a degree as in France.

France also developed the most aggressive government policy on energy conservation. Inspectors would swoop down on banks, department stores, and offices and do "le check up"—take the inside temperature with special thermometers. If the temperature exceeded the officially approved twenty-degree-centigrade level, fines would be levied on the building management. But perhaps the most striking aspect of France's overall energy conservation program, and an altogether French initiative, was the ban on any advertising that "encouraged" energy consumption. A manufacturer could advertise that his portable electric heater was more efficient than comparable heaters, but he could not say that electric heating was the best form of heating, because that encouraged energy use. Officials of the French Energy Conservation Agency were known to hear a radio advertisement on the way to work and, judging it as encouraging consumption, have it pulled from the radio by lunchtime.

The advertising ban created particular perplexity for the oil companies. They were accustomed to waging aggressive campaigns to win even a 1 percent gasoline market share away from their competitors. No more. Now about the best they could do was trumpet the gasoline-saving properties of various additives. Exxon's tiger was tamed in France; no longer in the tank, he was judiciously advising motorists to check their tires and tune their engines to save gasoline. The companies could not give away the kind of trinkets and premiums that gasoline stations habitually offered around the world—mugs, glasses, spoons, and decals. After all, such gifts would encourage consumption. Instead, about the only thing they were permitted to hand out were cheap tool kits, but only so long as they contained a brush for cleaning spark plugs to promote higher efficiency.

One of the two French national oil companies, Total, searched desperately for some way to keep its name in front of the public. At last, it had a brilliant idea. It started putting up billboards, picturing a beautiful piece of green French countryside, with a simple legend announcing "This is France" and signed "Total." The ad was banned. A stunned Total asked why. "It is easy," said Jean Syrota, the director of the Energy Conservation Agency. "Consumers look at this ad and say, 'Oil companies are wasting a great deal of money on such ads, therefore the companies must be rich, therefore there must not be any energy problem, therefore it is all right to waste energy.' "[1]

"Obscene Profits"

The laugh was something that the playwright Eugene O'Neill would never have anticipated, and he might well have been perplexed. In a popular Broadway revival of his play *A Moon for the Misbegotten,* one of the characters bellowed at the beginning of the second act, "Down with all tyrants! Goddamn Standard Oil!" Night after night, the audience broke into laughter, and sometimes into applause. It was early 1974, three decades after the play had been written, but the line resonated with another drama that was being simultaneously acted out in the halls of Congress, as Senators and Congressmen held hearings on the energy crisis and the role of the oil companies. Of all the hearings, the most dramatic were those held by the Senate Permanent Subcommittee on Investigations, chaired by Senator Henry Jackson, who, as a boy, had been nicknamed "Scoop" by his sister owing to his resemblance to a cartoon character, and who was still known as Scoop even as the powerful chairman of the Senate Interior Committee. He saw himself as a hard-headed Truman kind of Democrat, a realist, who, as he liked to say, had his head "screwed on straight." Nixon would fume in private at what he called "Scoop Jackson's demagoguery." But a White House aide tried to explain to the irate Nixon that "our troops in the Interior Committee have a major inferiority complex when it comes to Jackson, because, frankly, he Scoops the hell out of them."

Now, at the hearings, the playing to populism was irresistible, and Jackson would score one of the greatest political scoops of his long career. Senior executives of the seven largest oil companies were lined up at one table and were made to testify under oath. Then, facing Jackson and his colleagues in a crowded hearing room that was flooded by television lights, they were subjected to withering questioning about the operations and profits of their companies. Those executives, whatever their skills at geology or chemical engineering or general management, were no match for Jackson and the other Senators when it came to political theater. They came across as inept, insulated, self-satisfied, and out of touch.

The timing of the hearings was exquisite: The oil companies were reporting huge increases in profits while the Arab oil embargo was still in effect. In an atmosphere that reeked with distrust and hostility, Jackson announced that his subcommittee was going to find out if there really was an oil shortage. "The American people," he declared, "want to know if this so-called energy crisis is only a pretext; a cover to eliminate the major source of price competition—the independents, to raise prices, to repeal environmental laws, and to force adoption of new tax subsidies. . . . Gentlemen, I am hopeful that we will receive the answers to these and other questions before we leave here today." He added menacingly, "If not, I can assure you we will get the answers one way or another in the days ahead."

Jackson and the other Senators then laid into the company executives, who tried to defend themselves. "The contrivance theory is absolute nonsense," the president of Gulf U.S. lamely protested, though adding, "I know people are a bit mystified by the rapid turn of events in the United States." A senior vice-president of Texaco haplessly declared, "We have not cheated or misled anyone and if any member of the subcommittee has proof of any such acts by Texaco, we would like to be presented with such proof." When a senior vice-president of Exxon was unable to recall the amount of his company's 1973 dividend, Jackson scathingly told him that he was being "childish."

The oil men were humiliated, sobered, and infuriated, particularly by Jackson, who found a line that, though perhaps not quite up to Eugene O'Neill, nevertheless won thunderous applause across the country, especially from those who were still spending their time in gas lines in the winter of 1974. The companies, said Jackson, were guilty of making "obscene profits." The oil men, accustomed to some deference, were hardly prepared for the onslaught. "We didn't have a chance," complained Gulf's bloodied president after the hearings. But Jackson knew that he was speaking for many Americans because he felt the way they felt. The two gasoline stations near his house were now always closed by the time he headed home. "We have to send one of the boys in the office out in the middle of the day to try to find an open station," he said with some frustration after the hearings. He was outraged by what he saw as the arrogance and greed of the oil companies, and he proposed that they be chartered by the Federal government. Jackson succeeded in turning "obscene profits" into a national catch phrase, the yardstick for the times. When Exxon had the misfortune coincidentally to release its 1973 earnings, up 59 percent over 1972, on the third day of the hearings, Kenneth Jamieson, the company's chairman, felt constrained to declare, "I am not embarrassed." Many thought otherwise.

The Standard Oil cursed in O'Neill's play had been broken up in 1911, but the reference still seemed appropriate. For John D. Rockefeller was once again casting his dark shadow over the land, with all the sinister intimations about collusion, manipulation, and secret deals. The oil companies were now among the most unpopular institutions in all of America. The same was true in other industrial countries. Some Japanese publications, for instance, ran articles about how the American oil companies had planned the crisis in order to raise their profits. Indeed, such was the public outcry and the demands for accountability and control that the confidential main planning document that went to the board of one of the largest oil companies warned in 1976, "The future for private oil companies is much less certain. The trend for upstream operations to pass into government hands will continue, with companies filling a contractor role, either formally or de facto. Greater government involvement, direct or indirect, is also to be expected downstream" in the consuming countries. The next year, 1977, a senior executive of Shell in London went so far as to opine, "Paradoxically, the threat to the viability of an oil company may today come more from the importing than the exporting governments."

He had a good point. After all, the worst had already happened in the oil-producing nations: The companies had been nationalized, they no longer owned the oil, they no longer set prices or production rates. As far as the oil exporters were concerned, the companies were contractors, hired hands. Was it now, oil executives asked themselves, the turn of the consuming governments to hammer their companies? Some of the industrial countries launched antitrust investigations of oil company practices. Political risk, at least if it were to be measured in terms of senior management time, shifted to the industrial countries, particularly in the United States. The hallowed depletion allowance, which reduced the tax on oil production, was sharply curtailed; to a lesser degree, so was the foreign tax credit, the "golden gimmick" that had been put to such good use after World War II to facilitate oil development in Venezuela and the Middle East and protect the American position in both places. There were continuing efforts in Congress to roll back oil prices and even stronger political pressure to keep down natural gas prices. No less a threat was the movement for "divestiture," by which was meant the breaking up of the integrated companies into totally separate firms for each segment of the business: crude oil and natural gas production, transportation, refining and marketing. At one point, forty-five out of one hundred Senators voted in favor of divestiture. The opinion of the oil industry on this particular movement was summed up by the term it preferred to use—"dismemberment."

And then there were the constant attacks on those "obscene profits." What were the facts underlying this matter of great contention and anger? The profits of the largest oil companies had been almost perfectly flat for the five years through 1972, despite the explosive growth in demand. Profits then rose from $6.9 billion in 1972 to $11.7 in 1973, and shot up to $16.4 billion in the banner year of 1974. The reasons were several. Much of the immediate increase was derived from foreign operations. As the exporting countries pushed prices up, the companies got a free ride in terms of the increases on the non-U.S. equity oil they still owned. The value and market prices of their American oil reserves also went up. Moreover, they had bought oil at lower prices, say $2.90, before the increases, and had it in inventory, and then made money when they finally sold that same oil at $11.65. Their chemical operations had also done well, helped by a weak dollar. But then profits dropped back to $11.5 billion in 1975, lower than 1973. Again, the reasons were several. Overall oil demand was down because of a recession. The exporting countries noticed the profits that the companies were making on equity oil, and hurriedly proceeded to raise taxes and royalties to ensure that rents went into their treasuries and not to the companies. That was the year, as well, that some tax advantages were curtailed. Over the next few years, profits rose again, reaching $15 billion in 1978, which in real terms meant that they were just about keeping up with inflation. The companies' profits were huge in absolute terms, but their rates of return were, except for 1974, somewhat below the average rate for all American industry.

One other feature of the profitability picture was significant. Profits were concentrated in the upstream part of the business—crude oil and natural gas production. The value of the reserves the companies had in places like the United States and the North Sea had increased with the price of oil. The downstream—refineries, tankers, gasoline stations, and so forth—had been built before 1973 on expectations of 7 to 8 percent annual growth in oil demand. Actual demand was much lower, and thus capacity in the downstream was bloated far beyond requirements. A third of the total tanker fleet was in surplus. This overcapacity, combined with the loss of equity crude in the Middle East, made the international oil companies start to question the rationale and value of the large downstream systems they had built up in Europe in the 1950s and 1960s for the disposal of Middle Eastern oil—the oil that had now been taken away from them.[2]

The United States Energy Policy: "Chinese Water Torture"

Despite the surprisingly strong consensus and continuity of the Nixon, Ford, and Carter Administrations on the lines of international energy policy, no similar agreement existed in domestic politics. On the contrary, the domestic side of the energy equation continued to be marked by a divisive, angry, bitter, confused debate about price controls, and about company practices and policies. Nixon had resigned from office in August 1974, but the Watergate debacle had left a crippling crisis of confidence in government and a pervasive suspicion about the energy crisis itself.

Oil and energy were already well on the way to becoming the hottest cauldron in national politics, made all the more difficult by the "threat" to the American way of life and the high stakes in terms of power and money. Back in August 1971, in an effort to stamp out inflation (then running at almost 5 percent, considered unacceptably high), Nixon had imposed price controls on the entire economy. Most controls were allowed to expire by 1974, but not those on oil. Instead, the politics and intense pressures of the time gave rise to an awesome Rube Goldberg system of price controls, entitlements, and allocations that made the mandatory oil import program of the 1960s appear, by comparison, to have the simplicity of a haiku.

The public wanted Washington to do "something"—and the something was to return prices to the good old days but at the same time, assure adequate supplies. Markets were confused and distorted, with unforeseen consequences continually arising from each decision. "For every problem you solve, it seems you create two more," moaned one government regulator. Those who figured out how to work the system could do very well. For instance, acquiring entitlements to crude oil supplies became a big business, and the result was the bringing out of mothballs any piece of "refining junk" that could be found—leading to the return of hopelessly inefficient "tea kettle" refineries of the kind that had not been seen since the flood of oil in the East Texas field in the early 1930s. The various programs gave rise to much wasted motion, endless Congressional hearings, and so much work for attorneys as to comprise one of the great "lawyer's relief" programs of the century. As one scholar has written, "For the oil industry, the Federal Register became more important than the geologist's report." Whatever the short-term gains in terms of equity, the costs were enormous in terms of the inefficiencies, the confusion in the market, diversion of efforts, and misallocation of resources and time. Just the standard reporting requirements for what became the Federal Energy Administration involved some two hundred thousand respondents from industry, committing an estimated five million man-hours annually. The direct costs of the regulatory system—measured simply in terms of expenditures by government agencies and by industry on regulatory matters—added up to several billion dollars in the mid-1970s. The entire regulatory campaign did less to boost the national weal than to cause a chronic migraine of immense proportions in the nation's politics. But such was the temper of the times.

Meanwhile, something big did have to be done. In January 1975 President Gerald Ford, picking up on Nixon's Project Independence theme, proposed a grand ten-year plan to build 200 nuclear power plants, 250 major coal mines, 150 major coal-fired power plants, 30 major oil refineries, and 20 major synthetic fuel plants. Not long after, Vice-President Nelson Rockefeller, grandson of the man who personified the monolith of oil, championed an even grander $100 billion program to subsidize synthetic fuels and other high-cost energy projects that commercial markets would not support. But opponents challenged the cost of these projects, and the Rockefeller initiatives came to naught. There were, however, two highly significant achievements during the Nixon-Ford years. In the immediate aftermath of the embargo, Congress gave a green light to the Alaskan oil pipeline. The project ended up costing ten billion dollars. Environmentalists said that the delays and the reconsiderations had led to a safer and more environmentally sound pipeline. As it was, TAPS—the Trans-Alaskan Pipeline—made possible what proved to be the single most important new contribution to American energy supply since Dad Joiner's discovery of the East Texas field in the 1930s.

The other landmark was the setting in 1975 of fuel efficiency standards for the automobile industry. According to the new standards, the average fuel efficiency of a new car would have to double over a ten-year period, from its then-current 13 miles per gallon to 27.5 miles per gallon. Since one out of every seven barrels of oil used in the world every day at that time was burned as motor fuel on America's roads and highways, such a change would have a major impact not only on America's but also on the world's oil balance. The legislation that included the fuel efficiency standards also established a strategic petroleum reserve: the same idea that Eisenhower had proposed after the 1956 Suez crisis, and that the Shah had tried to sell to the United States in 1969. The plan was excellent; such a reserve would provide the surge capacity to compensate for any interruption of supply. In practice, however, the rate at which the reserve was built up turned out to be fatally slow.[3]

In 1977, Jimmy Carter became President, having campaigned as an outsider who would bring moral renewal to the fallen, Watergate-besmirched politics of America. Energy was a subject that had engaged his attention many years earlier. He had been a submariner in the United States Navy, and he always remembered a warning that Admiral Hyman Rickover, the father of the nuclear submarine, had once delivered about how humanity was exhausting nature's stock of oil supplies. During the campaign, Carter had promised a national energy policy within ninety days of inauguration day, and he was dedicated to making good on his word.

He gave the job to James Schlesinger, a Ph.D. economist, who had originally made his name as a specialist on the economics of national security. Schlesinger combined a powerful analytic intelligence and a strong sense of duty with what has been described as "intellectual zeal and moral fervor." He held clear views about what was right when it came to policy and to governance, and he did not hesitate or beat around the bush when it came to expressing them. He had little patience himself for easygoing give-and-take, and he could certainly try his opponents' patience. He would lay out his thinking in a slow, spare, emphatic manner that sometimes seemed to suggest that his auditors, be they generals or senators or even presidents, were first-year graduate students who had failed to understand the most self-evident theorem.

Richard Nixon had plucked Schlesinger from the Rand Corporation for the Bureau of the Budget, then to be chairman of the Atomic Energy Commission, then made him director of the Central Intelligence Agency, but soon after switched him to Secretary of Defense. On a fine Saturday or Sunday morning, however, he could be found in the countryside around Washington, binoculars in hand. He was not out in his professional capacity, looking for Russians, but pursuing his hobby of bird watching, about which he was passionate. His tenure at the Defense Department came to an end under Gerald Ford, when Schlesinger took exception to Kissinger's détente policy and to the American posture regarding South Vietnam's last agony leading up to the fall of Saigon—and made his feelings abundantly clear in Cabinet meetings. After the Democratic National Convention in 1976, Jimmy Carter phoned Schlesinger and invited him to the Carter home in Plains, Georgia, to talk politics and policy. Schlesinger was a close friend of Senator Henry Jackson, who was after all the single most important Senator when it came to energy, and had been Carter's rival for the nomination. After the election, Jackson pressed Carter to make Schlesinger the energy champion for the new Administration. Carter was more than willing. Not only was he impressed by Schlesinger, but as Schlesinger himself later commented, "It was kind of convenient if the chairman of the Senate Energy Committee was a pal of your proto-energy secretary."

During the first weeks of the Carter Administration, "energy" was cast as its number-one issue. Carter read a CIA report, prepared in late 1976, predicting future oil shortages; he found it compelling and persuasive, and it was important in motivating him to proceed in the way he did. Schlesinger, like Carter, was convinced that hydrocarbons would be under growing pressure, which posed major economic and political dangers for the United States. To be sure, Schlesinger, an economist, did not believe in absolute depletion, but rather that prices would inevitably rise, balancing the market. Both men shared a deep concern about the foreign policy implications of a tight oil market. As Carter wrote in his memoirs, many Americans, obviously including Jimmy Carter and James Schlesinger, "deeply resented that the greatest nation on earth was being jerked around by a few desert states."

In 1972, well before the crisis and while he was still chairman of the Atomic Energy Commission, Schlesinger had expressed a then-heretical idea: that the United States should promote energy conservation on grounds of national security, foreign economic policy, and environmental improvement. "We can do somewhat better than automobiles that move at 10 miles to the gallon and badly insulated buildings that are simultaneously heated and cooled," he had said then. Indeed, he had advised environmentalists that the "heart" of their case should be "challenging the presupposition" that "demand for energy grows more or less automatically." Now, in 1977, he was more convinced than ever that conservation should be central to any energy policy. Unfortunately, that premise was considerably more obvious to him than to many others.

The new Administration remained committed to unveiling its all-encompassing energy program within the first ninety days. Such haste did not leave enough time to build the requisite consensus and working relationships not just with committee chairmen in the Congress, but also with a broader base of interested Congressmen—or, indeed, even within the Administration. The development of the programs themselves was kept as secret as possible. Moreover, Schlesinger had to devote a third of those first ninety days to moving emergency natural gas legislation, in order to help relieve the shortages of 1976–77, and additional time on the legislation setting up the Department of Energy. With so much else happening, Schlesinger asked Carter to consider relenting on the ninety-day commitment. "I said 90 days," Carter firmly replied. "I made the pledge and I intend to keep it."

Yet Carter himself was not altogether happy with the emerging energy plan. "Our basic & most difficult question is how to raise the price of scarce energy with minimum disruption of our economic system and greater equity in bearing the financial burden," he wrote in a note to Schlesinger. "I am not satisfied with your approach. It is extremely complicated." To drive home his complaint, Carter plaintively added, "I can't understand it."

The plan was due to be unveiled in early April in a major Presidential address. The Sunday before, Schlesinger appeared on a television interview show, in which, trying to find a metaphor to capture the magnitude of the energy challenge, he recalled a quotation from William James—"the moral equivalent of war." It turned out that the viewers that Sunday included Jimmy Carter, who was impressed by the phrase and put it into his speech. Thus, Carter, appearing in a cardigan for a fireside chat to the nation in April 1977, introduced his energy program as the "moral equivalent of war"; and by that sobriquet it would often thereafter be known. Its detractors preferred to use the acronym—"Meow."

The Carter program included a host of initiatives aimed at reshaping America's energy position, introducing economic rationality into pricing, and reducing the need for imported oil. In Schlesinger's mind, the number-one priority was to find a way to let domestic oil, which was under price controls, rise to world market prices so that consumers could react to correct price signals. The current system blended the price-controlled domestic oil and the higher-priced imported oil into the final price that consumers paid, which really meant that the United States was subsidizing imported oil. Thus, the Carter program promulgated a procedure to end price controls on domestically produced oil through a "crude oil equalization tax." There was some irony here, as it had been the Republican Administration of Richard Nixon that had originally imposed price controls in August 1971, and it was now a Democratic Administration that was trying to lift them. Carter and Schlesinger also sought an ingenious, if very complex, method for extracting the country from the straitjacket of natural gas price controls. The Administration put much greater emphasis than its predecessors on conservation and on the use of coal. It sought to introduce some competition into the electricity sector and to encourage the development of alternative and renewable energy sources, including solar power.

The Administration was proceeding as though there was a crisis that would rally the nation; the public, however, did not think there was a crisis. And, in the course of pushing his program, Carter received a firsthand education in how special interests operate in the American system, including liberals, conservatives, oil producers, consumer groups, automobile companies, pro- and antinuclear activists, coal producers, utility companies and environmentalists—all with conflicting agendas. To Schlesinger, however, the issue was absolutely clear. The United States faced "a substantial long-run national problem." He did not think that the world was about to run out of oil, but rather that the high growth rates in consumption that had supported economic development in the 1950s and 1960s could no longer be sustained. "We had to stop depending on crude oil for economic growth," he would later explain. "We had to wean ourselves away." Confident in his rigorous analysis of the issue, he was unprepared for the storm of debate and the bitterness of the ensuing battles. Sitting through one Congressional hearing after another, he came to recall a word of advice that an old veteran at the Atomic Energy Commission had given him when he was its chairman: "There are three kinds of lies—lies, damn lies, and energy lies." Later, Schlesinger would say, "I have a sort of World War II mentality. If the President says something is in the national interest, I had assumed he would get a more supportive response than we found. But there had been a change in the nation. As Secretary of Defense, everyone not against you is with you. Here, on energy, you had interest groups against interest groups. You couldn't put a consensus together. It was distressing."

Of all the energy issues, natural gas proved to be the most contentious and intractable. For the Carter Administration had walked right into the middle of a decades-old political and almost theological struggle over natural gas pricing, and whether that price should be controlled by government or set by the market. So bitter was the argument that Schlesinger was moved to observe during the Senate-House conference meetings on natural gas, "I understand now what Hell is. Hell is endless and eternal sessions of the natural gas conference." Yet, somehow, a compromise, a very intricate one, was arranged. Natural gas prices would be allowed to increase in limited increments. Some gas, currently controlled, would be decontrolled, while some gas that was decontrolled would be recontrolled for a time and then decontrolled again. A number of different categories for pricing purposes were created for a commodity that, for the most part, was composed of the same standard molecules of one carbon atom and four hydrogen atoms.

Despite all the bloody political battles, and the consequent exhaustion of much of its political capital, the Carter Administration could claim a series of important accomplishments on the energy front. "Passage of the National Energy Act represents a watershed in that it starts the adjustment of our demands to the means available," Schlesinger told a London audience. "The turn in the road is forced upon us—all of us—by the limits, physical and political, on prospective oil supplies." But, as he looked back on the almost two years of struggle that had followed Carter's initial call to action, Schlesinger could not help but ruefully observe, "The response was less close to William James' moral equivalent of war than to the political equivalent of the Chinese water torture."[4]

Boom Times

By the end of 1978, the post-embargo policies elsewhere, as in the United States, were just beginning to make their influence felt. There was, however, one reaction to the embargo that was virtually instantaneous. The price hikes, the expectation of future increases, much-expanded cash flows, and the eagerness of investors—all combined to ignite a frenetic and inflationary global hunt for oil. When asked to characterize the worldwide craze, Exxon's deputy exploration manager summed it up simply: "It's just wild." What had been a depressed exploration business up through 1972 was now running at capacity, and the cost of everything, be it a semisubmersible drilling rig or a dynamically positioned drilling ship or just an old-fashioned land crew in Oklahoma, was bid up to double what it had been in 1973.

Moreover, the flow of investment was redirected in a very substantial way. The number-one commandment was to avoid, at all costs, nationalism in the Third World. In any event, exploration in most of the OPEC countries was foreclosed because of nationalization, and there was a rather strong presupposition that, if a company had success in other developing countries, the fruits would be seized before they could be ingested, leaving only small pieces and bits for the company. So the companies redirected their exploration spending, to the degree possible, to the industrial countries of the Western world: to the United States, despite a growing pessimism about its oil potential, to Canada, and to the British and Norwegian sectors of the North Sea. In 1975, Gulf undertook a complete review of its global budget. Every investment dollar that was not nailed down and committed was quietly yanked out of the Third World and brought back to North America and the North Sea. By 1976, Royal Dutch/Shell was concentrating 80 percent of its worldwide, non-U.S. production expenditures in the North Sea. "After 1973 and nationalization, you had to go hunt your rabbits in a different field," recalled an Exxon executive, "and we went to places where we could still obtain equity interests, ownership, in oil."

The oil companies also began to diversify into totally different businesses. This was somewhat difficult to justify at the same moment that the companies were calling for the lifting of price controls on the basis that they needed all the money they could get to invest in energy and, indeed, undermined their case. But the diversification reflected the view that the business and political environment for oil companies might grow more and more narrow, more and more harsh, more and more constrained by government intervention and regulation. There was something else, as well, a more-than-nagging fear that the days of oil companies, and oil itself, might be numbered by geological depletion. Between 1970 and 1976, proven American oil reserves dropped by 27 percent, and gas reserves by 24 percent. It seemed that the United States was about to fall off the oil mountain. Though the actual investments outside the energy business were small when compared to the overall financial wherewithal of the companies, they were still large dollar commitments. Mobil bought the Montgomery Ward department store chain, Exxon went into office automation, and ARCO into copper. But nothing aroused so much mirth and derision as Gulf's bid for the Ringling Bros, and Barnum & Bailey Circus. That, more than anything else, did seem to prove that the clamorous new era—of OPEC's imperium and high oil prices, of confusion, bitter debates, and energy wars in Washington—really was a circus.[5]

New Supplies: Alaska and Mexico

OPEC continued to dominate the world oil market throughout the 1970s. It accounted for 65 percent of total "free world" oil production in 1973 and 62 percent in 1978. But, though not yet too obviously, its grip was beginning to loosen. The incentive of price and the motive of security were stimulating oil development outside OPEC, and in a matter of years, these new sources would transform the world's oil supply system. While the activity was spread out all over the world, three new oil provinces would have overwhelming influence: Alaska, Mexico, and the North Sea. Ironically, each had been identified before the 1973 price hike but had not yet been developed because of various mixtures of politics, economics, environmental opposition, technical obstacles, and the simple factor of time, the long leads required by major energy projects.

With the emergency go-ahead on the Alaskan pipeline in the weeks after the embargo, the work there could at last begin. The steel pipe and tractors, so optimistically purchased in 1968, conveniently still sat waiting on the frozen banks of the Yukon River, and the engines in the tractors had been dutifully turned over on schedule for the last five years. Now they could be used, and work proceeded rapidly. By 1977, the eight-hundred-mile pipeline had been completed, some parts of it suspended on stilts above the tundra, and the first oil had made the journey from the North Slope to the shipping port at Valdez on Alaska's southern coast. By 1978 over a million barrels per day were flowing through the line. Within a few years, the flow would be two million barrels per day, a quarter of America's total crude oil production.

In Mexico, after the fiery nationalization battle of the late 1930s, the oil industry had turned inward. No longer did Mexico strive to be one of the world's great exporters. Instead, Pemex, the national oil company and the embodiment of Mexican nationalism, devoted itself to supplying the domestic market. Pemex was also subject to a struggle for control between the government and the powerful oil workers' union, which happened, and not by coincidence, to be in the unusual position of being one of the leading contractors to Pemex. For decades, Pemex was a company under pressure. Its income was limited by the low domestic prices. Its development program was directed by cautious engineers, who were guided by a conservationist ethic based on the conviction that resources should be husbanded for future generations. Pemex did relatively little to expand its reserve base. While production rose, it could not keep up with the demands of the rapid growth of the "Mexican economic miracle." As a result, Mexico not only ceased being an exporter but actually became a minor oil importer, though, to save face, a good deal of effort went into disguising that fact when, for instance, it hurriedly had to buy a cargo of Venezuelan crude from Shell.

In search of more oil, Pemex launched a deep-drilling exploration program in the rolling savannahs in the southern state of Tabasco. In 1972, oil was struck in what proved to be an extraordinary structure called Reforma. The highly productive wells in the Reforma fields led to the region's being dubbed "Little Kuwait," and discoveries there were soon followed by further very large discoveries in the adjacent offshore area, the Bay of Campeche.

It was becoming clear that Mexico possessed world-class oil reserves. In 1974, the country started, in a very small way, to sell petroleum abroad again, though the export of oil was criticized by some as running against the tenets of Mexican nationalism. While production was rising, the engineers in Pemex continued to be very cautious in their estimates of reserves through the last years of the Presidency of the radical, nationalistic Luis Echeverría Álvarez. But matters changed with the election of a new President, José López Portillo, in 1976. López Portillo, who had been Echeverría's Minister of Finance, inherited an economic crisis that was Mexico's worst since the Great Depression. The Mexican economic miracle had run out of gas, the economy had stalled, the value of the peso had collapsed, and the country was regarded as a poor risk by international lenders. To make matters worse, the population was growing faster than the economy—one out of every two Mexicans was under the age of fifteen—and 40 percent of the workforce was either unemployed or underemployed. In the months before López Portillo actually took over, conditions were so bad that there were even rumors of a possible military coup.

The new oil was a godsend. So was the 1973 price shock, which made that oil much more valuable. López Portillo decided to put the new discoveries to work as the central element in a new economic strategy. He appointed an old friend, Jorge Díaz Serrano, as the head of Pemex. Unlike his predecessor, an engineer whose specialty had been bridge construction, Díaz Serrano knew the oil industry. He had become a millionaire supplying it with services, and he grasped the potential that was now at hand. Oil would provide Mexico with its desperately needed foreign earnings, if would remove the constraint of the balance of payments on economic growth, it would provide the collateral for new international borrowing, and it would put Mexico at the center of the new oil-based international economy. In short, it would be the engine of renewed growth.

President López Portillo, however, offered a word of caution: "The capacity for monetary digestion is like that of a human body. You can't eat more than you can digest or you become ill. It's the same way with the economy." But López Portillo's actions spoke considerably louder than his words, and with quite a different accent. Investment, much of it borrowed from abroad, was poured into the industry. The proving and expanding of reserves was pushed at a rapid rate; officially sanctioned rumors were floated of greater and greater and still greater oil potential. Production proceeded at a breakneck pace, even ahead of plan. Daily output rose from 500,000 barrels in 1972 to 830,000 barrels in 1976 to 1.9 million barrels in 1980—almost a fourfold increase in less than a decade.

Whereas Mexico had been a country to be avoided by international lenders through 1976, it now became one of the most active borrowers in the world. "Why the Bankers Suddenly Love Mexico" was the title of an article in *Fortune*. The reason, of course, was oil. "Every Tom, Dick, and Harry in banking is knocking on their door," said the vice-chairman of Manufacturers Hanover Trust. One Mexican official was even chosen "borrower of the year" in 1978 by a New York financial newsletter. The title might well have been won by the whole country. There seemed to be no restraint: The Mexican government was borrowing from abroad, Pemex was borrowing, other state companies were borrowing, private firms were borrowing, everyone was borrowing from abroad. How much was being borrowed altogether? No one knew. But it didn't seem to matter. Mexico's credit, lodged in oil, was good. Or so the bankers and their Mexican counterparts thought. But one thing was certain: Mexico had become a major new force in the world oil market, as it had not been since the 1920s, and it would provide yet another significant alternative source of supply, undermining the OPEC Imperium.[6]

The North Sea: The Biggest Play of All

For many centuries, fishermen had the North Sea to themselves, to catch the herring that was Northern Europe's biggest business in the Middle Ages, and then more recently the haddock and cod. But by the middle 1970s, a new breed of seafarer could be seen from a helicopter on the waters below: floating drilling rigs, supply boats, platforms, pipe-laying barges—singly at first, and then in such profusion, at times, as almost to crowd the sea. Here on the waters of the North Sea, between Norway and Britain, was the biggest new play for the world oil industry, and its single greatest concentration of capital investment and effort. No major company dared to be left out, and many new players were drawn in, ranging from industrial companies to staid Edinburgh investment trusts to Lord Thomson, the newspaper tycoon, who owned the *Times* of London. He was a partner of Armand Hammer.

Ever since 1920, thousands of onshore wells had been drilled by the hopeful in Western Europe. The results had been distinctly disappointing; total production in the area never exceeded 250,000 barrels per day. The Suez crisis of 1956 gave a new boost to the search for secure oil and gas resources in Europe; and in 1959, at Groningen in Holland, Shell and Esso had discovered a vast gas field, the largest then known outside the USSR. Realizing that the geology of the North Sea matched that of Holland, the oil companies began to explore in adjacent waters. In 1965, the same year that Britain and Norway formally agreed on how to divide the North Sea right down the middle between them, insofar as mineral rights were concerned, major natural gas deposits were found in the relatively shallow southern reaches of that sea; and platforms comparatively primitive by future standards were built to exploit the gas. Some companies continued exploring for oil, with, at best, what might be described as mild interest, but hardly with eagerness.

Phillips Petroleum from Bartlesville, Oklahoma, was among them. Its interest had been piqued in 1962, when the company's vice-chairman, while on a vacation in the Netherlands, had noticed a drilling derrick near Groningen. Two years later, after the company's senior executives spent an afternoon crawling around and examining three hundred feet of seismic data unrolled on the company's basketball court in Bartlesville, Phillips decided to pursue an exploration program. But half a decade later, in 1969, after a string of dry holes, the company was ready to call it quits. Including Phillips's own efforts, some thirty-two wells had been drilled on the Norwegian continental shelf, and not one of them was commercial. And hole for hole, the North Sea was much more expensive and much more difficult than anything else the company had ever tried. The message from Bartlesville to Phillips's managers in Norway was clear: "Don't drill any more wells."

But in the grand tradition that stretched back to Colonel Drake's well in Pennsylvania in 1859 and the first discovery in Persia in 1908, Phillips reluctantly decided to give it one more go—but only because it had already paid for the use of a rig, the *Ocean Viking*, and could not find anyone else who wanted to sublease it. Phillips would have to pay the daily charges whether the rig was drilling or not. The weather was getting bad, and the seas were rough. At one point, the rig broke off from its anchors and began to drift away from the bore hole. On another night, it became so stormy that a capsize was likely and an emergency evacuation of the rig was initiated at the first daylight. But the *Ocean Viking* did the job; in November 1969, it made a major find on Block 2/4 in the Ekofisk field, on the Norwegian side of the median line. That happened to be a great moment for technology; American astronauts had just landed on the moon. As the drilling superintendent on the *Ocean Viking* examined an oil sample brought up from a depth of 10,000 feet beneath the seabed, he was quite amazed by its appearance, which bespoke a very high quality. "What the astronauts have done is great," he said to the rig's geologist, "but how about this?" He held up the oil; it had a golden sheen, almost transparent, but definitely almost like gold.

Phillips's discovery set all the companies to re-evaluating their seismic data and stepping up their activity. No longer would any orphan drilling rigs go begging for employment in the North Sea. Some months later, a senior Phillips executive was excitedly asked at a technical meeting in London what methods Phillips had used to diagnose the geology of the field.

"Luck," he replied.

Toward the end of 1970, British Petroleum announced the discovery of oil in the Forties field, on the British side, one hundred miles northwest of Ekofisk. It was a huge reservoir. A series of major strikes followed in 1971, including Shell and Exxon's discovery of the huge Brent field. The North Sea oil rush was on. The 1973 oil crisis turned the rush into a roar.

Fortunately, a new generation of technology was either available or under development that would allow production to proceed in the North Sea, a province of the sort that the industry had never before attempted. The whole venture was risky and dangerous—physically and economically. Drilling rigs had to be able to work through water depths much greater than anything tried heretofore, and then still drill another four miles under the seabed. And all the equipment and workers had to cope with a nasty and vicious sea and some of the worst weather in the world. "There's nothing quite as vile as the North Sea when she's in a temper," was the lament of one skipper. Not only was the weather bad, but it could change three or four times a day; a sudden storm could brew up in hours; waves of fifty feet and winds of seventy miles an hour were not uncommon. The permanent platforms through which the oil was pumped—really small industrial cities set on man-made islands—not only had to stand on mud, quicksand, clay, and sand wave bottoms, but also had to be built to withstand the fury of a "100-year wave," ninety feet high, as well as winds of 130 miles per hour.

Altogether, the development of the North Sea was one of the greatest investment projects in the world, made all the more expensive by rapidly inflating costs. It was also a technological marvel of the first order. And it was carried out in an amazingly expeditious manner. On June 18, 1975, the British Secretary of State for Energy, Anthony Wedgwood Benn, turned a valve at a ceremony on an oil tanker in the estuary of the River Thames. The first North Sea oil, from the Argyll field, flowed ashore to a refinery. Publicly, Benn enthusiastically declared that June 18 should from then on be a day of national celebration. Personally, however, he did not enjoy the inaugural event at all. Benn was a leader of the left wing of the Labour party with a passion for nationalization and an inbred detestation of capitalism, especially as it was represented by the oil industry, and was extremely distrusting by nature. He sourly noted in his diary that he had been forced to participate in the ceremony in the company of "a complete cross-section of the international capitalist and British Tory establishment." And when he turned the valve, the oil "allegedly went on shore," he added with great suspicion.

Benn found a much greater outlet for his animosity toward oil companies, for he was playing a leading role in Britain in replicating the traditional battle between governments and oil companies. The North Sea reserves had been proved and the risks greatly reduced; whereupon the British government had decided that it, too, like many other governments, wanted a larger share of the rents, along with more control over its "destiny," and perhaps outright national-ization. "Oil companies can jump over national boundaries to save on taxes more easily than a kangaroo pursued by a dingo can jump over a fence," complained Lord Balogh, Benn's Minister of State. The result of the struggle was a special tax on oil revenues and the establishment of a new state oil company, British National Oil Corporation. It now held title to the government's participation oil, reflecting the right to buy 51 percent of North Sea production, and was meant to be the national champion. The British government's push for more revenues and more control of North Sea oil led the head of one company finally to explode, "I don't see any difference any more between those OPEC countries and Britain."

In some ways, that same thought was on the mind of Harold Wilson, Britain's Prime Minister. He was sitting in a second-floor study at 10 Downing Street, puffing on his pipe, in the summer of 1975, a few weeks after the celebration over the first barrels of North Sea oil. Wilson had already had one of the longest-running tenures as Prime Minister. He had also made a major contribution to political theory with a line that deserved to be engraved on the wall of every parliament and congress around the world: "In politics, a week is a long time." Wilson had first come to power in 1964 with a promise to lead stagnant Britain into the "white heat of the technological revolution," but now, a decade later, it was the technology of oil development, not computers and aerospace, that seemed Britain's best economic bet. That particular summer day, Wilson was ruminating on how Britain's oil production could grow from a trickle to perhaps two and a half million barrels per day, transforming Britain's economic prospects and certainly affecting the balance of oil power in the world. He was already thinking like the prime minister of an oil country. At the time the Ford Administration was campaigning against higher oil prices. "We do have an interest in seeing that the price of oil doesn't fall too much," Wilson said. "If America wanted to really bring down the price, not many people here would necessarily agree."

There was a great irony in all this. Wilson was sitting in a room that had been used by Anthony Eden two decades earlier, at the time when Eden struggled over what to do about Suez, Nasser, nationalism, and the threats to Britain's oil supply. So grave was the perceived threat in 1956 that Eden had decided to use military force in the form of the aborted attack on the Canal Zone, which did so much to end the historic European role in the Middle East—and, certainly, Eden's career. Wilson faced no such fate. In fact, he confessed an ambition that might have shocked Eden. As leader of a newly emerging major oil power, Wilson genially said, he hoped he might be chairman of OPEC by 1980.[7]

"The Crunch"

One peculiar result of the price shock of 1973 was the rise of a new line of work—oil price forecasting. Before 1973, it had not really been necessary. Price changes had been measured in cents, not dollars, and for many years prices were more-or-less flat. After 1973, however, forecasting blossomed. After all, oil price movements were now decisive not only for the energy industries, but also for consumers, for a multitude of businesses from airlines to banks to agricultural cooperatives, for national governments, and for the international economy. Everybody now seemed to be in the forecasting business. Oil companies did it, governments did it, central banks did it, international organizations did it, brokerage houses and banks did it. Indeed, one might have been reminded of the Cole Porter refrain: "Birds do it, bees do it, even educated fleas do it."

This particular kind of forecasting, like all economic forecasting, was as much art as science. Judgments and assumptions governed the predictions. Moreover, such forecasting was much affected by the "community" in which it was done; thus, it was also a psychological and sociological phenomenon, reflecting the influences of peers and the way individuals and groups groped for certainty and mutual comfort in an uncertain world. The end result was often a strong tendency toward consensus, even if the consensus completely changed its tune every couple of years.

Certainly, by 1978, such a consensus could be observed throughout the community of oil forecasters and among those who made decisions based on those forecasts: While Alaska, Mexico, and the North Sea would together add six to seven million barrels per day to world markets by the early or mid-1980s, those new sources were expected to serve only as a supplement and a sort of modern-day Fabius, holding off and postponing, but not decisively banishing the inevitable day of shortage and reckoning. For, most forecasters agreed, another oil crisis was highly probable a decade or so hence, in the second half of the 1980s, when demand would once again be at the very edge of available supply. The result, in popular parlance, was likely to be an "energy gap," a shortage. In economic terms, any such imbalance would be resolved by another major price increase, a second oil shock, as had happened in the early 1970s. Though variations were to be found among the forecasts, there was considerable unanimity on the central themes, whether the source was the major oil companies, the CIA, Western governments, international agencies, distinguished independent experts, or OPEC itself. Not only were the forecasters convinced, so were the decision makers who relied on the forecasts to make their policies and investments and choose their course of action.

The single most important assumption underlying this common view was the belief in the "Iron Law"—that is, that there was an inevitably and inescapably close relationship between economic growth rates and the growth rates for energy and oil use. If the economy grew at 3 or 4 percent a year, as was generally presumed, oil demand would also grow by 3 or 4 percent a year. Another way of putting it was that income was the main determinant of energy and oil consumption. And the facts, as measured in 1976, 1977, and 1978, seemed to bear out this assessment. Economic growth in the industrial world had rebounded from the deep recession and averaged 4.2 percent in those three years; oil demand had grown at an average rate of about 4 percent. The picture of the future world that thus emerged was a projection of the then-current circumstances: Growing economies would continue to call upon growing volumes of oil. Economic progress in developing countries would add to the demand. The future effects of conservation were discounted. The stage would be set for a repetition of 1973.

Ahmed Zaki Yamani, the leading proponent of a Long-Term Strategy for OPEC, began to depart from his customary advocacy of price stability and instead argued for regular, small increases in the price that would encourage conservation and development of alternative sources. This, he said, was much preferable to and less destabilizing than the wrenching increase in price that had become the common expectation. "From our own studies and from all the reliable studies I have read," he said in June of 1978, "there are very strong indications that there will be a shortage of supply of oil sometime around the mid-1980s if not before. . . . No matter what we do, that date is coming."

Yamani was expressing what had become the general informed outlook in both the oil-importing and the oil-exporting countries. Even in Washington, some, looking at the falling real price of oil and the rising demand, had begun to think that modest price increases sooner might spare much agony later. The crunch would surely come in a decade, give or take a year or two. But it was also generally agreed that conditions did not point to any major price increases in the near term. That was a view based upon looking at the economics. Politics, of course, was something else; politics was never easy to fit into models that dealt with economic growth rates and elasticities of demand. Yet it could not be disregarded. And politics was not about to allow anyone the luxury of a long-term strategy.

On the last day of 1977, President Jimmy Carter, en route from Warsaw to New Delhi in the course of a hectic three-continent trip, arrived in Tehran. He said he had asked Mrs. Carter where she wanted to spend New Year's Eve, and she had said with the Shah and his wife, so delightful a time had the Carters had when the royal couple visited Washington six weeks earlier. Yet there were reasons of realpolitik as much as sentiment behind their choice. Carter had been impressed by the Shah. The Shah, for his part, was taking significant steps toward liberalization and was talking about human rights. With a new understanding between the two men, Carter was now in a position to better appreciate the strategic role of Iran and its leader than when he had first come into office. Iran was a fulcrum country, essential to stability in the region. It was a critical element in counterbalancing Soviet power and ambitions in the area, as well as those of radical and anti-Western forces. It was central to the security of the world's oil supplies, both as one of the world's two major exporters and as a regional power.

Carter also wanted to show his gratitude to the Shah for his progress on human rights and his switch of position on oil prices, which was seen as a major concession on the part of the monarch. Moreover, the President was regretful and embarrassed over the rioting and tear gas that had greeted the Shah's arrival on the South Lawn of the White House, and he wanted to clear up any misunderstandings, within Iran and outside the country, and firmly underline American support. So, at a New Year's Eve banquet, he rose to offer a memorable toast. "Iran, because of the great leadership of the Shah, is an island of stability in one of the more troubled areas of the world," he said. "This is a great tribute to you, Your Majesty, and to your leadership, and to the respect and the admiration and the love which your people give you." On that strong and hopeful note, the President and the Shah welcomed the momentous New Year of 1978.

654

Not everyone saw the same island of stability that the President had described. Shortly after Carter's visit, the president of one of the independent American oil companies active in Iran came back from a trip to Tehran. He had a confidential message he urgently wanted to share with one of his directors. "The Shah," he said, "is in big trouble."[8]

The Second Shock:
The Great Panic

A WEEK AFTER Jimmy Carter's departure from Iran, a Tehran newspaper published a savage attack on an implacable opponent of the Shah's, an elderly Shiite cleric named Ayatollah Ruhollah Khomeini, who was then living in exile in Iraq. The article, though anonymous, appeared to be the work of an official of the Shah's regime. Perhaps the Carter visit had bolstered flagging confidence. Certainly the article was already in the works, for exasperation was increasing with Khomeini's own harsh attacks on the Shah's government, which were being circulated clandestinely in cassettes throughout Iran.

Animosity between the royal house of Iran and the fundamentalists of the dominant Shia Islamic sect dated back to Reza Shah's fierce battle for power with the Shiite clergy in the 1920s and 1930s and was part of a much larger struggle between secular and religious forces. But that newspaper article of January 7, 1978, triggered a wholly new stage in the struggle.

Disillusion and Opposition

It had become evident in the mid-1970s that Iran simply could not absorb the vast increase in oil revenues that was flooding into the country. The petrodollars, megalomaniacally misspent on extravagant modernization programs or lost to waste and corruption, were generating economic chaos and social and political tension throughout the nation. The rural populace was pouring from the villages into the already-overcrowded towns and cities; agricultural output was declining, while food imports were going up. Inflation had seized control of the country, breeding all the inevitable discontents. A middle manager or a civil servant in Tehran spent up to 70 percent of his salary on rent. Iran's infrastructure could not cope with the pressure suddenly thrust upon it; the backward railway system was overwhelmed; Tehran's streets were jammed with traffic. The national electricity grid could not meet demand, and it broke down. Parts of Tehran and other cities were regularly blacked out, sometimes for four or five hours a day, a disaster for industrial production and domestic life and a further source of anger and discontent.

Iranians from every sector of national life were losing patience with the Shah's regime and the pell-mell rush to modernization. Grasping for some certitude in the melee, they increasingly heeded the call of traditional Islam and of an ever more fervent fundamentalism. The beneficiary was the Ayatollah Khomeini, whose religious rectitude and unyielding resistance made him the embodiment of opposition to the Shah and his regime and indeed to the very character and times of Iran in the mid-1970s. Born around 1900 in a small town 180 miles from Tehran, Khomeini came from a family of religious teachers. His father had died a few months after his birth, killed on the way to a pilgrimage by a government official, it was said by some. His mother died when he was in his teens. Khomeini turned to religious studies and, by the 1930s and 1940s, was a popular lecturer on Islamic philosophy and law, promulgating the concept of an Islamic Republic under the stern control of the clergy.

For many years, Khomeini had regarded the Pahlavi regime as both corrupt and illegitimate. But he did not become politically active until about the age of sixty, when he emerged as a leading figure in the opposition to the "White Revolution," as the Shah's reform program was called. In 1962, Khomeini expressed outrage at the proposal that places in local assemblies no longer be restricted exclusively to male Moslems. When, under the rubric of the White Revolution, the government redistributed large estates, including the vast holdings of the Shiite clergy, Khomeini came forward as one of the most unyielding opponents, landing in jail more than once and eventually ending up in exile in Iraq. His hatred of the Shah was matched only by his detestation of the United States, which he regarded as the main prop of the Pahlavi regime. His denunciations from exile in Iraq were cast in the rhetoric of blood and vengeance; he seemed to be driven by an unadulterated anger of extraordinary intensity, and he himself became the rallying point for the growing discontent. The words of other, more moderate ayatollahs were overwhelmed by the exile's harsh and uncompromising voice.

Another dimension of opposition had emerged. With Jimmy Carter's capture of the Democratic nomination and then the Presidency in 1976, human rights became a major issue in United States foreign policy, and the human rights record of the Shah was not good. It was also typical of much of the Third World, and better than that of some other countries in the region. A member of the International Commission of Jurists, who was a leading critic of the Shah and who investigated human rights conditions in Iran in 1976, concluded that the Shah was "way down the list of tyrants. He would not even make the A-list." Still, Savak, the Iranian secret police, was brutal, quick, and particularly nasty in its torture; it was also callous, stupid, intrusive, pervasive, and arbitrary. None of this fit the image of the Great Civilization, the Iran that was pursuing its ambition to be a world power—and whose Shah was lecturing the industrial world on its own character flaws. Thus, Iran's human rights record became more visible and much more reported upon than the abuses in other developing countries, contributing further to the growing hostility, both inside Iran and out, to the Shah and his regime. The Shah himself felt intense pressure on the human rights question from the United States, and ironically, even as the criticism mounted, he had determined to move on a course of political liberalization.[1]

"Doing the 40–40"

Khomeini's words took on a new fury in late 1977, when his eldest son was murdered under mysterious circumstances. The murder was attributed to Savak. Then came that newspaper article of January 7, 1978. It ridiculed Khomeini, questioned his religious credentials and concerns, challenged his Iranian nationality, and luridly accused him of various acts of immorality, including the authorship of risqué love sonnets as a young man. This journalistic assault on Khomeini set off riots in the holy city of Qom, which remained his spiritual home. Troops were called in and demonstrators were killed. The disturbance in Qom ignited a new confrontation between the Moslem religious leadership and the government, which took a very specific form. The Shiite branch of Islam provided for a forty-day mourning period. By plan, the end of the forty-day period for the slain at Qom became the occasion for new demonstrations, more deaths, more mourning, and then, after forty days, more demonstrations—and still more deaths. One leader of this relentless cycle of protests was later to call it "doing the 40–40." The riots and demonstrations spread across the country, with further dramatic clashes, more people killed, and more martyrs.

Attacks by the police and army on critics of the regime only served to swell and broaden the ranks of those antagonistic to the Shah. The withdrawal of subsidies to the Shia religious establishment alienated and further angered the clergy. Indeed, overt opposition was becoming part of the fabric of national life. Yet all through the first half of 1978, its significance was discounted. Yes, the Shah told the British ambassador, the situation was serious, but he was determined to press ahead with liberalization. His most implacable enemies, and the most powerful, were the mullahs, with their hold on the minds of the masses. "There could be no compromise with them," he said. "It was a straightforward confrontation and one side had to lose." The Shah made it clear that he could not imagine being on the losing side.

In the U.S. government, too, hardly anyone could imagine that the Shah might fail. For Washington, any alternative was virtually unthinkable. After all, Iran's powerful monarch had sat on his throne for thirty-seven years. He was courted throughout the world. He was modernizing his country. Iran was one of the world's two great oil powers, with wealth far beyond anything it had known only a few years earlier. The Shah was a critical ally, a regional policeman in a crucial area, the "Big Pillar." How could he possibly be toppled?

American intelligence on Iran was constrained. As the United States became more dependent on the Shah, there was less willingness to risk his ire by trying to find out what was happening among the opposition that he despised. In Washington, there were surprisingly few people with the requisite analytic skills on Iran. And until late in the day, there did not seem to be great demand among the "consumers" of intelligence, as senior American national security officials are sometimes called, for analyses of the stability of the Shah's regime, either because they thought it unnecessary or because they feared, at some level, that the conclusions might be too unpalatable. "You couldn't *give* away intelligence on Iran," was the comment of one frustrated intelligence analyst.

The American intelligence community struggled throughout 1978 to assemble a National Intelligence Estimate on Iran, but could never get it together. There was plenty of daily reporting, but great difficulty in assessing how all the disparate forces of discontent and opposition would interact and play out. The State Department's *Morning Summary* did, in mid-August, suggest that the Shah was losing his grip and that Iran's social fabric was unraveling. But as late as September 28, 1978, the Defense Intelligence Agency's prognosis was that the Shah "is expected to remain actively in power over the next ten years." After all, it was reasoned, he had weathered other crises in the past.

And yet there were at that very moment various signs, some particularly grisly, of the fury of the forces that were rising against the Shah. Over a period of two weeks in August 1978, half a dozen movie theaters around the country were set afire by fundamentalists opposed to "sinful" movies. In mid-August, in Abadan, the home of the great refinery, about five hundred people were crowded into a theater when some group locked the doors and incinerated the trapped moviegoers. Though uncertainty remained, it was thought that the perpetrators were fundamentalists. In early September, bloody demonstrations took place in Tehran itself. That was the turning point. From then on, the Shah's government began to collapse as an effective ruling force. Still, the Shah pushed on with his liberalization, including talk of free elections in June 1979.

To those with access to the monarch, something seemed to be wrong with the Shah himself. He appeared distant and more isolated. Rumors had circulated for years about his health. Did he have cancer? Or an incurable venereal disease? On September 16, the British ambassador went to see the Shah again. "I was worried by the change in his appearance and manner. He looked shrunken; his face was yellow and he moved slowly. He seemed exhausted and drained of spirit." The fact of the matter was that the Shah did have cancer, specifically a form of leukemia, which French doctors had first diagnosed in 1974. But the seriousness of the illness was kept for several years from both the Shah and his wife. As it was, he insisted upon the greatest secrecy for his treatment. Later, some in Washington suspected that elements in the French government would nonetheless have had to know. The British government and most certainly the American did not know. Had they been informed of the fact and nature of his ailment, the calculations on many accounts might have been different. As time went on, the Shah began to feel the effects of the illness more and more, and to fear its consequences, which might help explain the indecisiveness, strange detachment, even malaise and fatalism that seemed to take hold of him.[2]

"Like Snow in Water"

As the political situation in his country deteriorated, the Shah vacillated. He would not wage total war against the growing rebellion; "world public opinion" was watching too closely. And these were his people. But neither would he concede. He was befuddled by the contradictory advice emanating from the U.S. government. He felt betrayed by one and all. Again and again, he expressed his suspicion that the American CIA, British intelligence—and the BBC, the hotline for his opponents' communications—were conspiring against him, though for reasons that were never very clear.

As the weeks passed, more and more of the country went on strike, including oil industry technicians. In early October 1978, at Iran's urging, Ayatollah Khomeini was expelled from Iraq; after all, the Ba'thist regime in Baghdad had to worry about its own Shiite population. Denied refuge in Kuwait, Khomeini went to France and established himself and his entourage in a suburb of Paris. The Iranian government may have thought, out of sight, out of mind, but it was mistaken. France provided Khomeini and his followers with access to the direct-dial international phone service that the Shah had installed in Tehran, greatly facilitating communication. The elderly, irate cleric, who knew so little of the Western world and held it in such scorn, nevertheless proved himself a master of propaganda in front of the media that camped at his door.

Still, the Shah proceeded with his liberalization program. Academic freedom, freedom of the press, freedom of assembly—these were being proffered, but such Western-style rights were of little interest to a population that was rising up against the monarch and his dynasty and the whole process of modernization. At the end of October, the Shah could only say, "We are melting away daily like snow in water." Strikes immobilized the economy and the government, students were out of control, and demonstrations and riots went on unchecked.

The Iranian oil industry was in a state of escalating chaos. The main production area was known as "The Fields." Located in the southeast, it included Masjid-i-Suleiman, where Anglo-Persian had made its original discovery in 1908. Now, seventy years later, operations in The Fields were in the hands of the Oil Service Company of Iran, Osco, which was the descendant of the consortium that had been established in 1954, after the fall of Mossadegh and the Shah's return. Staffed by expatriate oil men, mostly from the member companies, Osco's headquarters was in Ahwaz, about eighty miles north of Abadan. In October, some of the striking Iranian laborers from The Fields moved into Osco's main headquarters building in Ahwaz. No one tried to evict them. By November, a couple of hundred of them were living in the corridors, eating and sleeping there, in a tactic aimed at increasing the pressure on Osco and the National Iranian Oil Company. The Western oil men went about their jobs, carefully trying to avoid stepping on the workers. Meanwhile, in the courtyard outside, impromptu prayer meetings had begun. At first, no more than half a dozen participated. But soon, the oil men could see from their windows that the numbers of the chanting faithful at each meeting had swollen to several hundred.

The impact of the strikes was felt immediately. Iran was the second-largest exporter of oil after Saudi Arabia. Of the upwards of 5.5 million barrels produced daily in Iran, about 4.5 million were exported; the rest were consumed internally. By early November, exports had been reduced to less than a million barrels per day, and thirty tankers were waiting in line at the loading facilities at Kharg Island for oil that was not there at a time when, in the international mar-

660

ket, the winter demand surge was beginning. Petroleum companies, responding to the general softness in the market, had been letting their inventories fall. Would there be a shortage on the world market? Moreover, the stability of Iran itself depended upon oil revenues; they were the basis of the country's entire economy. The head of the National Iranian Oil Company went south to The Fields to seek a dialogue with the striking oil workers—or so he thought. When he got there, he was mobbed by angry strikers. He immediately decided to forgo negotiations and instead fled the country. There seemed to be no way to end the strike.

Trying to contain the growing chaos, the Shah took a critical step that he had always wanted to avoid; he installed a military government. This was his last chance, but he put a weak general in charge. The general immediately suffered a heart attack and never asserted authority. The new government was able, at least temporarily, to restore some order in the oil industry and get production going again. Soldiers now also moved into Osco's headquarters at Ahwaz, where they coexisted uneasily with the striking workers, who continued to camp in the corridors.

As events tumbled toward their conclusion, the policy of the United States, Iran's most important ally, was in confusion, disarray, and shock. During most of 1978 senior officials of the Carter Administration had been distracted and preoccupied by other momentous and demanding developments: the Camp David peace accords with Egypt and Israel, strategic arms negotiations with the Soviets, normalization of relations with China. American policy had been based on the premise that Iran was a reliable ally and would be the Big Pillar in the region. Out of deference to the Shah and because of the desire not to anger him, American officials had kept their distance from the various opponents of his regime, which meant that they lacked channels of communication to the emerging opposition. There was not even any reporting to Washington on what the Ayatollah was actually saying on those by-now-famous tapes. Some in Washington insisted that the unrest in Iran was a secret, Soviet-orchestrated plot. And, as always, there was the same question: What could the United States government do, whatever the case? Only a few American officials thought that the Iranian military could withstand the persistence of nationwide strikes and the defection of religiously minded soldiers. Indeed, the last few months of 1978 saw a fierce bureaucratic battle over policy waged in Washington. How to bolster the Shah or assure continuity to a friendly successor regime? How to support the Shah without being so committed as to assure an antagonistic relationship with his successors, should he fall? How to disengage, if disengagement were required, without undermining the Shah, in case he could survive politically? Indecision and vacillation in Washington resulted in contradictory signals to Iran: The Shah should hang tough, the Shah should abdicate, military force should be used, human rights must be observed, the military should stage a coup, the military should stand aside, a regency should be established. "The United States never sent a clear, consistent signal," one senior official recalled. "Instead of oscillating back and forth between one course of action and the other and never deciding, we would have done better to have flipped a coin and then stuck to a policy." The cacophony from the United States certainly confused the Shah and his senior officials, undermined their calculations, and drastically weakened their resolve. And no one in Washington knew how sick the Shah was.

The efforts to construct hastily some new American position were complicated by the fact that the Shah was an object of dislike and criticism in the media in the United States and elsewhere, which resulted in a familiar pattern— moralistic criticism of U.S. policy combined with the projection by some of a romantic and unrealistic view of the Ayatollah Khomeini and his objectives. A prominent professor wrote in the *New York Times* of Khomeini's tolerance, of how "his entourage of close advisers is uniformly composed of moderate, progressive individuals," and of how Khomeini would provide "a desperately-needed model of humane governance for a third-world country." The American ambassador to the United Nations, Andrew Young, went even further; Khomeini, he said, would eventually be hailed as "a saint." An embarrassed President Carter immediately felt the need to make clear "that the United States is not in the canonization business."

So great was the lack of coherence that one senior official, who had been involved in every Middle Eastern crisis since the early 1960s, noted the "extraordinary" fact that the "first systematic meeting" at a high level on Iran was not convened until early November—very late in the day. On November 9, William Sullivan, the American ambassador in Tehran, finally confronted the unpleasant realities in a dramatic message to Washington entitled "Thinking the Unthinkable." Perhaps the Shah would not be able to survive after all, he said; the United States should begin to consider contingencies and alternatives. But in Washington, where the bureaucratic battles continued to rage, there was no meaningful reaction, save that President Carter sent hand-written notes to his Secretary of State, National Security Adviser, Secretary of Defense, and Director of Central Intelligence to ask why he had not been previously informed of the situation inside Iran. Ambassador Sullivan, meanwhile, came to the conclusion that the United States faced the situation in Iran "with no policy whatsoever."[3]

"Torrents of Blood"

December 1978 was a month for mourning, processions, and self-flagellation in the Shiite creed. The high point was the holiday of Ashura, marking the martyrdom of the iman Hussein and symbolizing unremitting resistance to a tyrant without legitimacy. Khomeini promised that it would be a month of vengeance and "torrents of blood." He called for new martyrs. "Let them kill five thousand, ten thousand, twenty thousand," he declared. "We will prove that blood is more powerful than the sword." Huge demonstrations were held across the country, some truly awe-inspiring in size. All of the opposition seemed to have united, and the Army was crumbling away. The Shah was running out of options. "A dictator may survive by slaughtering his people, a king cannot act in such a way," he said privately. But what should he do? And to add to all his other indignities and humiliations, there had been that telephone prank. The Shah had been told that Senator Edward Kennedy was phoning from Washington. Preparing himself, no doubt, for conversation with one of America's leading liberals and human rights champions, the Shah picked up the phone—only to hear a quiet voice repeating a simple incantation over and over: "Mohammed, abdicate. Mohammed, abdicate."

A task force from the Oil Service Company had already begun preparing, in a low-key way, an evacuation plan for the twelve hundred expatriate oil men and their families in The Fields. The group collected maps, looking for desert airstrips that could be used to land planes if airports were closed. But the effort was not taken very seriously. Then, one afternoon, George Link, an Exxon man who was the general manager for Osco, was being driven back to work after lunch. When his driver stopped the car and got out to open a gate, an assailant leaped from the side of the road and tossed something into the car. Link reflexively threw open his door and jumped out. Moments later, the car exploded. Thereafter, the evacuation planning took on a new seriousness.

Strikes once more gripped The Fields, and Iranian production again fell away rapidly. Tension was very high. Osco's assistant general manager for operations was Paul Grimm, on loan from Texaco. His position put him in direct confrontation with the workers. The big and outspoken Grimm bluntly warned some blue-collar expatriates, who were joining the strikes out of fear and confusion, that they would be dismissed if they did not come back to work, and he, in turn, was singled out as the man who was trying to break the walkout. In the middle of December, as Grimm was driving to work, a shot was fired at him from a car that was following his own, and he died instantly from a bullet in the back of his head. Evacuation of dependents now hurriedly began.

By December 25, Christmas Day, Iranian petroleum exports had ceased altogether. That would prove to be a pivotal event in the world oil market. Spot prices in Europe surged 10 to 20 percent above official prices. The cutbacks in oil production also deprived Iran of domestic oil supplies. Long lines formed in Tehran to get whatever small rations were available of gasoline as well as kerosene, which was the standard fuel for cooking. Soldiers maintained order by firing shots into the air. The oil workers refused to provide any petroleum products to the military, helping to immobilize it. Finally, in an ironic reversal of roles, an American tanker was diverted to Iran to supply badly needed fuel. Over the next critical weeks, the tanker remained in the vicinity, sometimes anchored offshore, sometimes sailing upriver toward Abadan, but it was never able to deliver the cargo, as sufficiently secure arrangements to unload could not be made.

"I Am Feeling Tired"

At the end of December, there was reluctant agreement in the ruling circles that a coalition government would be formed, and that the Shah would leave Iran, ostensibly for medical treatment abroad. But there could be little doubt of what was really happening. The Pahlavi dynasty appeared to be finished. So, virtually, at least for the time being, was petroleum production in The Fields. In the week after Christmas, Osco decided to evacuate all its Western employees. Hardly privy to what was going on either in Tehran in the immediate vicinity of the Peacock Throne or in Washington, the expatriates assumed that their exit was only temporary, a matter of weeks or months at the most, until order was restored.

Thus, they were strictly limited to only two suitcases. They left their houses intact, with pretty much everything in place, for their return. They faced a quandary similar to that of the oil men who had been forced by Mossadegh to leave Abadan in 1951—what to do with their dogs, which they could not bring with them. Since they did not know how long they would be gone, they did what their predecessors had done: took their dogs out back of their houses and either shot them or clubbed them to death.

They gathered at the airport at Ahwaz. Their eventual destination was Athens, where they were supposed to pass the time in sight-seeing, waiting for the all-clear so they could go back. Once again, the heirs of William D'Arcy Knox and George Reynolds were ignominiously leaving Iran. But unlike the "farewell to Abadan" in 1951, there were no honor guards, no salutes, no bands, no singing of verses to "Colonel Bogey's March." Ahwaz had once been a very busy airport, with innumerable domestic flights, plus a constant flow of small planes and whirring helicopters, shuttling back and forth to the various production sites. But now there was no domestic air service, the oil industry was shut down, and the sky over the deserted airport at Ahwaz was empty and silent and foreboding.

On January 8, the British ambassador went to say farewell to the Shah. The monarchy that had survived all sorts of vicissitudes for almost half a century was at its end. The pomp of the celebration at Persepolis of the twenty-five hundredth anniversary of the Persian monarchy was gone. So was the power. Alexander the Great had captured Persepolis in 330 B.C. and burned the royal palace; now the Ayatollah Khomeini was making a mockery of the self-proclaimed heir to Persepolis. Like the Wizard of Oz, Mohammed Pahlavi was revealed to be, after all, but a mere mortal. The show was over.

Talking to the ambassador, the Shah was calm and detached. He spoke about events as though they had no relevance to him personally. That made it all the more emotional for the ambassador, who, despite all his years of steeling himself to be a self-disciplined professional, found himself crying. Awkwardly trying to comfort him, the Shah said, "Never mind, I know how you feel." Considering their relative circumstances, it was a very strange remark. The Shah talked about the conflicting advice he was continuing to receive. Then, in an odd gesture, he looked at his watch. "If it was up to me, I would leave—in ten minutes." The show was indeed over.

At midday on January 16, the Shah appeared at Tehran airport. "I am feeling tired and need a rest," he said to a small group that had gathered, maintaining the pathetic fiction that he was only going on a vacation. Then he boarded his plane and left Tehran for the last time, carrying with his luggage a casket of Iranian soil. His first stop would be Egypt.

With the Shah's departure, all of Tehran erupted into the kind of jubilation that had not been seen since the Shah's own return in triumph in 1953. Car horns blared, headlights were flashed, windshield wipers decorated with pictures of Khomeini swished back and forth, crowds shouted and cheered and danced in the streets, and newspapers were handed out rapid-fire with the unforgettable banner headline "The Shah Is Gone." In Tehran and throughout the country, the great equestrian statues of his father, and of himself, were pulled from their pedestals by wild crowds, and the Pahlavi dynasty and its era crumbled into dust.

And who would rule? A coalition government had been left behind in Tehran, headed by a long-time opponent of the Shah. But on February 1, Khomeini arrived back in Tehran in a chartered Air France 747. Seats on the plane had been sold to Western reporters to finance the flight, while Khomeini himself spent the trip resting on a carpet on the floor of the first-class cabin. He brought with him a second government, a revolutionary council headed by Mehdi Bazargan, whose own credentials as an opponent of the Shah were impeccable. Indeed, in 1951, twenty-eight years earlier, Bazargan had been chosen by Mohammed Mossadegh to be head of the nationalized oil industry, and it was he who had then immediately gone to the oil fields in person with the stamps and wooden sign that said "National Iranian Oil Company." Subsequently, he had served his time in jail under the Shah. And now, despite Khomeini's lasting hatred of Mossadegh as a secularist nationalist, Bazargan was, given the conjunction of political forces, the Ayatollah's candidate to lead the new Iran. So, for a brief time, there were two rival governments in Tehran. But, of course, there could only be one government. In the second week of February, fighting broke out at an air force base in the suburbs of Tehran between non-commissioned officers called "homafars," who were sympathetic to the revolution, and troops of the Imperial Guard. The military support for the coalition government collapsed, and Mehdi Bazargan was in. The American defense attaché provided a succinct summary of the situation in a message to Washington: "Army surrenders; Khomeini wins. Destroying all classified."[4]

The Last Man Out

Not quite all the oil men had departed from The Fields. Twenty or so were asked to stay on in order to maintain the fiction of Osco's legal presence, should there be any argument with the government later on. Among the group was Jeremy Gilbert, an Irish mathematician turned petroleum engineer who had been assigned by BP to Osco and was now Osco's manager for capital planning. They remained only a few days before they, too, decided to leave in the face of the further deteriorating situation. But because he was in the hospital, suddenly stricken with a severe case of hepatitis, Gilbert was not allowed on the evacuation plane. He stayed in the hospital in a feverish haze throughout the tumultuous days of January. At night, from his hospital bed, he could hear the chanting and gunfire and, on the day the Shah left, the clangorous sounds of a giant celebration. His only contact with the world outside Abadan, apart from the BBC, was a huge arrangement of flowers, courtesy of Osco.

Very weak and hardly able to move around the ward, Gilbert was mistaken by the Iranians in the hospital for an American. A group of nurses took to gathering outside his window to chant "Death to Americans." Another patient without warning began beating Gilbert on the head with his crutches, ranting all the while against Americans. Gilbert's actual nationality posed another problem. The only way out of Iran was through Iraq, but because Irish troops on a peacekeeping mission in Lebanon had recently gotten into a shooting match with Iraqi soldiers, Gilbert was denied a visa for Iraq. To obtain it, he literally had to go down on his knees before an agent at the local Iraqi consulate and apologize for all the sins of the Irish.

At the end of January, he finally felt strong enough to make the trip out of the country. At the dusty border crossing, the Iranian officials waved him across with hardly a glance. But the Iraqi guards, suspecting he was a spy, detained, searched, and interrogated him for several hours. Meanwhile, the only available transportation to Basra, a single taxi, had left. When he was finally released, Gilbert asked, "How do I get to Basra?"

"You walk," one of the guards replied.

There was no other choice. Tired, weak, and carrying two bags, he dragged himself off along the dirt road toward Basra. After a couple of hours, a van over-took him and stopped. The driver agreed to take him to Basra for a fee, but laughed uproariously when Gilbert produced Iranian money. It was worthless, the driver chortled. Gilbert used his last dollars to pay the driver to take him to the Basra airport. But now he was broke. How could he get anywhere else? Then he remembered he had just received an American Express card, which he had tucked in his wallet but had not yet used. Thanking heaven that he had not left home without it, he got himself on a flight to Baghdad. He arrived late at night in the Iraqi capital, and after several tries, found a hotel. He called his family. They were appalled; they thought that he was comfortably still in the hospital in Abadan.

Gilbert did not move from the hotel room for three days. When he thought he had enough strength to risk the next leg of the journey, he took another flight from Baghdad to London. Arriving late Friday at Heathrow, he phoned the per-sonnel department at British Petroleum from the airport to announce that he had finally made it. The last Western oil man from The Fields, the great Iranian pe-troleum complex, was finally out. But the personnel officer who took the call, distracted by a conversation about his own weekend plans, misheard Gilbert and thought the caller was phoning with some news of the missing engineer. "Jerry Gilbert," he said, "we've been wondering where he was. Have you been in con-tact?"

That was the final indignity. From the open pay phone at Heathrow, sum-moning up whatever fragments of strength he had left, Gilbert loudly and abu-sively cursed not only the hapless personnel man, but everyone connected with the world oil industry.[5]

Panic Begins

The old regime was gone in Iran, and the new one was in power, though most un-easily; there were already bitter struggles for control. And from Iran, as if it had been shaken by a violent earthquake, a giant tidal wave surged around the world. All were swept up in it; nothing and no one escaped. When the wave finally spent its fury two years later, the survivors would look around and find them-selves beached on a totally new terrain. Everything was different; relations among all of them were altered. The wave would generate the Second Oil Shock, carrying prices from thirteen to thirty-four dollars a barrel, and bringing massive

666

changes not only in the international petroleum industry but also, for the second time in less than a decade, in the world economy and global politics.

The new oil shock passed through several stages. The first stretched from the end of December 1978, when Iranian oil exports ceased, to the autumn of 1979. The loss of Iranian production was partly offset by increases elsewhere. Saudi Arabia pushed up production from its self-imposed ceiling of 8.5 million barrels per day to 10.5 by the end of 1978. It lowered its output to 10.1 million barrels per day in the first quarter of 1979, but that was still well above its 8.5 million "ceiling." Other OPEC countries also boosted output. When all of that was figured in, free world oil production in the critical first quarter of 1979 was about 2 million barrels per day below the last quarter of 1978.

There was, then, an actual shortage, which was not surprising. After all, Iran was the second-largest exporter in the world. Yet when measured against world demand of 50 million barrels per day, the shortage was no more than 4 to 5 percent. Why should a 4 or 5 percent loss of supplies have resulted in a 150 percent increase in price? The answer was the panic, which was triggered by five circumstances. The first was the apparent growth of oil consumption and the signal that it gave to the market. Demand had risen smartly from 1976 onward; the impact of conservation and non-OPEC oil was not yet clear, and it was assumed by virtually all that demand was going to continue to rise.

The second factor was the disruption of contractual arrangements within the oil industry, resulting from the revolution in Iran. Despite major upheavals, world oil had remained an integrated industry. However, the ties were no longer the formal ones of ownership, but rather the looser ties of long-term contracts. The Iranian interruption hit companies unevenly, depending upon their dependence on Iran, and led to disruptions of the contractual flow of supplies. That rupture sent hosts of new buyers hurtling into the marketplace, scrambling to secure the same number of barrels that they had lost. All would do anything they could to avoid being caught short. Here was the real end of the classic integrated oil industry. The links between upstream and downstream were, at last, severed. What had been the fringe, the spot market, became the center. And what had been a somewhat disreputable activity, trading, would now become a central preoccupation.

A third factor was the contradictory and conflicting policies of consumer governments. The international energy-security system, which had been promoted by Kissinger at the 1974 Washington Energy Conference, was still in development, with many aspects yet untested. Actions taken by governments for domestic reasons would be read as major international policies, adding to the stress and tension in the marketplace. While governments were pledging to cooperate to dampen prices, companies from those nations would feverishly be bidding up the price.

Fourth, the upheaval presented the oil exporters with the opportunity to capture additional rents, enormously large rents. Once again, they could assert their power and influence on the world stage. Most, though not all of them, kept pushing the price up at every opportunity, and some manipulated supplies to further agitate the market and gain additional revenues.

Finally, there was the sheer power of emotion. Uncertainty, anxiety, confusion, fear, pessimism—those were the sentiments that fueled and governed actions during the panic. After the fact, when all the numbers were sorted out, when the supply and demand balance was retrospectively dissected, such emotions seemed irrational; they didn't make sense. Yet at the time they were indubitably real. The whole international oil system seemed to have broken down; it was not out of control. And what gave the emotions additional force was the conviction that a prophecy had been fulfilled. The oil crisis expected for the mid-1980s had arrived in 1979, the second phase of the turmoil unleashed in 1973–74. This was not a temporary disruption, but the early arrival of a deeper oil crisis, which would mean permanently high prices. And there was the unanswered question of how far the Iranian Revolution would advance. The French Revolution had reached across all of Europe to the very gates of Moscow before it spent its force. Would the Iranian Revolution reach into nearby Kuwait, to Riyadh, and on to Cairo and beyond? Religious fundamentalism wed to feverish nationalism had caught the Western world by surprise. Though it was still incomprehensible and unfathomable, one of its driving forces was obvious: a rejection of the West and of the modern world. That recognition led to an icy, pervasive fear.

It was the buyers, stunned by the unfolding spectacle, fearing a repetition of 1973, gripped by panic, who inadvertently made the shortage worse by building up inventories—as they had done in 1973. The world oil industry maintains billions of barrels of oil in inventories—supplies in storage—on any given day. Under normal circumstances, they were the stocks necessary for the smooth operations of the highly capital-intensive "machine" that extended from the oil field through the refinery to the gasoline station. A particular barrel of oil could take ninety days to travel from a wellhead in the Gulf through the refining and marketing system into underground storage at a gasoline station. To run short at any point in that system was costly in itself and could also disrupt other parts of the system. Thus, inventories were central to the constant effort to match supply and demand and to keep everything running smoothly. On top of that base requirement, the industry held a sort of insurance cushion: additional stocks to protect against any unexpected shifts in supply or demand—say a sudden surge of oil use in winter because of a January cold snap or a two-week delay in the arrival of a tanker because storms had disrupted loading facilities in the Gulf. Necessary supplies could then be drawn from inventories.

Of course, it was expensive to hold inventories. The oil had to be bought, facilities maintained, money tied up. So companies did not want to hold more inventories than their normal experience suggested they needed. If they thought that prices were going to go down because consumption was sluggish, they reduced inventories, and as quickly as they could, with the idea of buying later when the price would be lower. That was exactly what the industry was doing during the soft market conditions through most of 1978. By contrast, if companies thought that prices were going to go up, they bought more of today's cheaper barrels so that they would have to buy less of tomorrow's more expensive oil. And that was what happened, with extraordinary vengeance and fury, in the panic of 1979 and 1980. In fact, the companies bought well in excess of anticipated consumption, not only because of price, but also because they were not sure they would be able to get any oil later on. And that extra buying beyond the real requirements of consumption, combined with hoarding, dizzily drove up the price, which was exactly what companies and customers were struggling to avoid in the first place. In short, the panic of 1979–80 saw self-fulfilling, and ultimately self-defeating, prophecy on a truly colossal scale. The oil companies were not alone in panic buying. Down the consumption chain, industrial users and utilities also furiously built inventories as insurance against rising prices and possible shortages. So did the motorist. Before 1979, the typical motorist in the Western world drove around with his tank only one-quarter full. Suddenly worried about gasoline shortages, he too started building inventories, which is another way of saying that he now kept his gas tank three-quarters full. And suddenly, almost overnight, upwards of a billion gallons of motor fuel were sucked out of gasoline station tanks by America's frightened motorists.

The rush to build inventories by oil companies, reinforced by consumers, resulted in an additional three million barrels per day of "demand" above actual consumption. When added to the two million barrels per day of net lost supplies, the outcome was a total shortfall of five million barrels per day, which was equivalent to about 10 percent of consumption. In sum, the panic buying to build inventories more than doubled the actual shortage and further fueled the panic. That was the mechanism that drove the price from thirteen to thirty-four dollars a barrel.

Force Majeure

The panic might have been contained if the shortfall had been distributed evenly. But it was not. British Petroleum, as a result of its historical position, was far more dependent on Iran than any of the other companies. Fully 40 percent of its supplies came from Iran, and thus it was hardest hit by the interruption. In the argot of the industry, BP was "crude long"; that is, its crude supplies far outstripped the requirements of its own refining and marketing system. And so it was a "wholesaler," selling much of its oil through long-term contracts to "third parties"—either to the other majors, like Exxon, or to independent refiners, in particular in Japan. But now, having lost Iranian supply, BP invoked *force majeure* (act of God) provisions in its contracts and cut back on its buyers. It canceled altogether its supply contract with Exxon, while at the same time scrambling to buy oil elsewhere. Neither BP nor Shell were members of Aramco, so they had no direct access to the increased Saudi production, which went to the four American Aramco companies.

Thus, the dominoes began to fall. Other worried companies, deprived of oil either directly by the Iranian disruption or indirectly by BP's cutbacks, also invoked *force majeure* to reduce shipments to customers or to cancel contracts altogether. In March, Exxon, faced with an April 1 renewal date for its contracts with Japanese buyers, let it be known that it was going to phase out many of its third-party contracts as they came up for renewal. Exxon had been warning its buyers since 1974 to diversify supply and "not to count on Exxon." In the words of Clifton Garvin, Exxon's chairman, "The handwriting had been on the wall. Venezuela had been taken away from us. We no longer had the concession in

Saudi Arabia. We couldn't see a role in being the middleman between the Saudis and the Japanese consumer. Exxon's decision was not arrived at lightly. The world was just changing." So Exxon had already begun to wind down its third-party contracts. But, in the context of the crisis mood, its March 1979 message took on unexpected significance.

The chain reaction hit Japan hard. After the first oil crisis, it had consistently tried to carve out a niche in Iran, and had succeeded. As a result, it had become relatively more dependent than other industrial countries on Iran, which by 1978 supplied almost 20 percent of Japan's total oil needs. Moreover, the majors now clearly could no longer be counted upon. Japanese refiners were not going to let their refineries go idle through lack of supply. The government was once again face-to-face with Japan's blatant lack of natural resources; Japan's economic miracle was threatened at the jugular, by the fact that its industrial base was largely fueled by oil. Panic was more pervasive in Japan than anywhere else because twenty years of hard-earned economic growth appeared about to come apart. The government ordered the bright electric lights in the Ginza to be dimmed as an energy-saving measure. More important, it urged Japanese buyers to go directly into the world marketplace, something they had not much done before. The formidable Japanese trading companies took the lead, scouring the world for supplies. It often required considerable ingenuity to develop the access they had never before needed. One trading house discovered that an excellent way to get in to see the right officials in ministries and state oil companies was by giving gloves as presents to the secretaries. In order to cultivate the Iraqi oil minister, this same trading house provided him with the services of a world-class acupuncturist.

Other independent refiners from many nations joined established companies and the Japanese in the frantic quest for oil. So did state oil companies, like that of India, which also had been dependent upon Iran. Suddenly, where there had been relatively few buyers, now there were many—a highly desirable situation from the point of view of the sellers, of whom there were still only few. And, suddenly, all the action was in the spot markets, which had, until then, been a kind of sideshow, comprising, in terms of both crude and products, no more than 8 percent of total supplies. It had been a balancing mechanism, a place where buyers would go to get discounted oil, such as refinery overruns, instead of more expensive supplies guaranteed by contracts. But it was the marginal market, and as buyers rushed into it, prices were bid up—and up and up. By late February 1979, spot prices were double official prices. People would call it the "Rotterdam Market," after Europe's huge oil port, but, in fact, it was a global market, connected by a fevered network of telephones and telexes.[6]

Leapfrog and Scramble

Here was the perfect opportunity for the exporters, and they responded in two ways. They began adding premiums to their official prices, with new monthly terms clattering out from telexes around the world. Then the exporters began shifting as many supplies as possible, and as rapidly as possible, from long-term contracts to the much-more-lucrative spot markets. "I would be a fool to give up

$10 extra a barrel in a spot sale," said one OPEC oil minister in private, "when I know that if we don't sell at this price someone else will." The exporters insisted that their long-term buyers take higher-priced spot oil along with the contractual officially priced oil. Also invoking *force majeure,* they canceled contracts altogether. One morning, Shell received a telex from an exporting country announcing that, on grounds of *force majeure*, a contracted supply was no longer available. That same afternoon, Shell received another telex from the same country, informing it that crude oil was available on a spot basis. Miraculously, the volume available was exactly the same amount as that which had been cut off, by act of God, just hours earlier. The only difference? The price was 50 percent higher. Circumstances being what they were, Shell took the offer.

Yet, at the beginning of March 1979, far more rapidly than expected, Iranian exports began to return to the world market, albeit at much lower levels than before the fall of the Shah. Reflecting the apparent easing of supply, spot prices began to fall back toward official selling prices. This was the moment at which some kind of order, well short of disaster, might have been reestablished. At the beginning of March, the member countries of the International Energy Agency pledged to cut demand by 5 percent to help stabilize the market. But the panic and the feverish competition in the marketplace now had a momentum of their own. Who could be confident that Iranian oil would maintain its new availability? Though Khomeini had asserted control over the petroleum industry, The Fields in Iran, as far as the outside world could determine at the time, were controlled by a radical leftist group—a "Committee of 60," composed primarily of militant white-collar workers—that was functioning as a virtual government of its own, jailing oil administrators and other officials at will. Moreover, ominously, other OPEC countries had begun to announce cutbacks in their own output. With prices rising, it would be more valuable to keep the oil in the ground and sell it in the future.[7]

At the end of March, OPEC met. Spot crude prices had risen by 30 percent; products, by as much as 60 percent. OPEC decided that its members could add to their official prices whatever surcharges and premiums "they deem justifiable in light of their own circumstances." What that really meant, Yamani bluntly allowed, was a "free-for-all." The exporters were abandoning any notion of an official price structure. They would charge whatever the market would bear. And now there would be two games in the world oil market. One was "leapfrog": the producers vying with each other to raise prices. The other was "scramble": a bruising competition for supply among purchasers. Anxious buyers—companies that had been cut off, refiners, governments, a new breed of traders, and of course the majors—trampled one another in their rush to court the various exporters. None of this fevered, bruising activity brought forth any new supplies; all it did was intensify the competition for the existing supply, driving up the price. "Nobody controlled anything," said Shell's supply coordinator. "You just fought for it. At every level, you felt you had to buy now; whatever the price, it was good compared to what it would cost you tomorrow. You had to say 'yes' or you would lose out. That was the psychology of the buyer. Horrible as the terms might be from your point of view, they would be worse tomorrow."

Only one exporter stood out clearly in opposition to the surcharges, premiums, and other manifestations of rapid price increases—Saudi Arabia. Having fought further price increases ever since the 1973 quadrupling, it now objected to the leapfrog because it feared that short-term winnings, however great, would be followed by large and perhaps ruinous losses for the exporters. Oil would price itself out of its competitiveness in energy markets; Middle Eastern producers would become, once again, the residual, to be shunned on grounds of energy security. Their importance to the industrial world, and their clout, would decline.

The Saudis issued what became known as the "Yamani Edict," which stated that Saudi Arabia would keep to official prices, no surcharges. In addition, Saudi Arabia insisted that the four Aramco companies sell at those official prices both to their own affiliates and to third-party buyers. If Saudi Arabia found that they were adding premiums to their prices, there would be hell to pay; they would run the risk of having their access to Saudi oil cut off at a time when every company ached over its short supplies. Saudi Arabia was virtually alone among the exporters in taking such a position, both at the OPEC meeting in March and all through the following months. Though its only OPEC ally was the United Arab Emirates, there was much behind-the-scenes pressure and outright imploring from Western nations. One senior official after another traveled from Washington—and Bonn, Paris, and Tokyo—to Riyadh to ask the Saudis for price moderation and to applaud every action they took in that direction.

Yet in the second quarter of 1979, the Saudis cut production, bringing it back to the precrisis "ceiling" of 8.5 million barrels per day. Despite the Saudi insistence on keeping to official prices, that cutback helped to send spot prices soaring. A variety of reasons were proffered. Were the Saudis trying to send a conciliatory and neighborly signal to the new Islamic regime of Ayatollah Khomeini by making room in the market for the returning Iranian production and thus avoiding a regional confrontation? Or were they seeking to express their dissatisfaction with the Camp David peace accords between Israel and Egypt, which had been signed on March 26? Or were they focused on their own financial position? The Saudis were debating among themselves about conservation of oil reserves and "the whole question of production in excess of actual revenue needs," especially at a time when they saw that American oil imports were actually rising. Or did the Saudis, observing the return of Iranian supplies to the market, simply assume that the crisis was easing and would soon be over? Whatever the reason, the blunt fact was that only Saudi Arabia had the kind of spare capacity that the United States had once had, the sort that could, if brought into production, quell the panic. So, even as the Western emissaries extolled the Saudi price moderation, they also urgently and repeatedly asked the Saudis to increase production again and put more supplies into the market to dampen the panic.

"Living Dangerously"

In one of those coincidental tricks of history, several hours after that latest OPEC meeting had adjourned, in the early morning hours of March 28, a pump failed and then so did a valve at the Three Mile Island nuclear plant, near Harrisburg, Pennsylvania. As a result, hundreds of thousands of gallons of radioactive water poured into the building housing the reactor. Days of near-panic passed before the extent of the accident could be evaluated. Some insisted that it was not an "accident," only an "incident." Whatever it was called, the unthinkable and supposedly impossible had happened to a nuclear plant; something had gone seriously wrong.

In itself, the event at Three Mile Island raised large doubts about the future development of nuclear power. It also threatened the assumption throughout the Western world that nuclear power would be one of the major lines of response to the 1973 oil crisis. Did Three Mile Island, by limiting the nuclear option, mean that the industrial world would find itself more dependent on oil than it had expected? Altogether, the accident contributed to the gloom, pessimism, and even fatalism that was now gripping the Western world. "A situation which we had imagined around the middle 1980s where there would be a real scramble for oil is already here," said the commissioner in charge of energy for the European Community. "All the choices are difficult, and most are very costly," declared David Howell, the British Secretary of State for Energy. "Friends, we are living dangerously."

The efforts of Western governments to mobilize cutbacks in demand, to blunt the upward price spiral, were proving insufficient. Yet they were loath to invoke the newly devised emergency oil sharing system of the International Energy Agency for fear that it would introduce more rigidity into the market. And, in any case, it was unclear whether the official trigger point for the system, a 7 percent shortfall, had actually been reached. Governments were torn between two fundamental objectives: obtaining relatively low-priced oil and guaranteeing secure supplies at any price. Once they had been able to do both. But now they found that these two objectives were contradictory. Governments talked the first but, when domestic pressures began to be felt, pursued the second.

The top priority was to keep domestic consumers, who happened to be voters, supplied. Energy questions had become, explained a European energy minister, "short, short, short-run politics." The various Western governments became promoters and champions of aggressive worldwide acquisition hunts, either indirectly through companies or directly through state-to-state deals. The result was suspicion, accusation, finger pointing, and anger among those supposedly allied nations. For the consuming countries as well as for the oil companies, it appeared to be every man for himself. Prices continued to climb.

To the American public the reemergence of gas lines, which snaked for blocks around gasoline stations, became the embodiment of the panic. The nightmare of 1973 had returned. Owing to the disruption of Iranian supplies, there was, in fact, a shortage of gasoline. Refineries that had been geared to Iranian light and similar crudes could not produce as much gasoline and other lighter products from the heavier crude oils to which they were forced to turn as substitutes. Inventories of gasoline were low in California, and after news reports and rumors of spot shortages, all 12 million vehicles in the state seemed to show up at once at gasoline stations to fill up. Emergency regulations around the country made matters worse. Some states, in an effort to avoid running out of supplies, prohibited motorists from buying more than five dollars worth of gasoline at any one time. The results were exactly the opposite of what was intended, for it meant that motorists had to come back to gas stations that much more frequently. Meanwhile, price controls limited the conservation response; and indeed, if gasoline prices had been decontrolled, the gas lines might have disappeared rather quickly. At the same time, the federal government's own allocation system froze distribution patterns on a historical basis and denied the market the flexibility to move supplies around in response to demand. As a result, gasoline was in short supply in major urban areas, but there were more than abundant supplies in rural and vacation areas, where the only shortage was of tourists. In sum, the nation, through its own political immobilism, was rationing gasoline through the mechanism of gas lines. And, to make matters worse, gas lines themselves helped beget gas lines. A typical car used seven-tenths of a gallon an hour idling in a gas line. One estimate suggested that America's motorists in the spring and summer of 1979 may have wasted 150,000 barrels of oil a day waiting in line to fill their tanks!

As gas lines spread across the country, the oil companies were once again public enemy number one. The charges flew thick and fast: The companies were withholding oil, tankers were being held offshore to drive up prices, the industry was deliberately hoarding oil and creating shortages to increase prices. Clifton Garvin, the chairman of Exxon, decided to "go public" personally to try to refute the accusations. Garvin was a cool, measured man, who liked to weigh things very carefully. A chemical engineer by training, he had worked in every facet of the oil business. He was also a passionate bird watcher, as had been his father, an activity he took some ribbing about from his peers. (Later, he was on the board of the National Audubon Society.) Now he made himself available to the media, was interviewed on television, and appeared on the *Phil Donahue Show*, surely a first for the chief executive of the world's largest oil company. But it seemed that whenever Garvin started explaining about the basic matter of inventories and the complex logistics of the business, the interviewers, eyes glazing over, would cut him off and change the subject.

Garvin had no trouble when it came to reading the public mood. "The American is a funny person," he recalled. "He worships the result of things that are big, economies of scale, mass production, but he hates anything that is big and powerful, and the oil industry is seen as the biggest and most powerful industry." It was an impersonal hatred, but Garvin was not about to take any chances. One day, he found himself sitting way at the back of a gas line at his local Exxon station, on Post Road in the heart of Greenwich, Connecticut. The dealer, recognizing the chairman of Exxon, came over to him and offered to have him drive around the back of the station to get to the front of the line.

"How are you going to explain that to everyone else in the line?" asked Garvin.

"Why, I'll tell them who you are," replied the dealer, helpfully.

"I'm sitting right here," Garvin said firmly.[8]

Petroleum and the President

The gas lines marked the beginning of the end of the Presidency of Jimmy Carter. He was one more victim of the revolution in Iran and the upheaval in the oil market. Carter had come to Washington two years earlier, in 1977, with a paradoxical persona that reflected the two sides of his experience: a naval officer turned peanut farmer, and a born-again Christian. He was the Preacher, seeking a moral rehabilitation of post-Watergate America with his unembroidered, down-to-earth Presidency. He was also the Engineer, trying to micromanage the intricacies of the American political machine and to show his command over both big issues and little details.

Carter would have seemed particularly well-suited to leadership in the midst of the 1979 panic; after all, his agenda and interests as both Preacher and Engineer had converged on energy and oil, making them the number-one domestic focus of his Administration. And now he confronted the crisis he had been warning against. But there would be no reward nor credit given to the prophet, only blame. By mid-March of 1979, two months into the crisis, Eliot Cutler, his chief White House energy adviser, was already warning of the "darts and arrows coming at us from all directions—from people who want to get rid of the regulatory structure, from people concerned about inflation, from people who want a sexy and affirmative program, from people who don't want the oil companies to profiteer, and generally from people who want to make life politically miserable for us." Shortly after, there came the accident at Three Mile Island, and the anxious nation saw photographs of the nuclear engineer, Jimmy Carter, wearing little yellow safety booties, tramping through and personally inspecting the control room of the damaged plant.

In April, Carter delivered a major speech on energy policy that merely intensified the barrage. He announced a decontrol of oil prices, which was certain to infuriate liberals, who tended to blame almost everything bad on the oil companies. And he coupled decontrol with a "windfall profits tax" on "excess" oil company earnings, which was no less sure to anger conservatives, who blamed the panic on government meddling, controls, and overweening regulation.

A special Presidential task force on energy met repeatedly in secrecy to try to figure out some solution to the gasoline shortage. The only quick way to fight the global oil disruption, and end the gas lines before they ended the Carter Presidency, was to get the Saudis to increase their output again. In June, the American ambassador to Riyadh delivered an official letter from President Carter as well as a more personal handwritten note. Both implored the Saudis to increase output. The ambassador also met for several hours with Prince Fahd, the head of the Supreme Petroleum Council, seeking a commitment to boost production and try to hold down the price. That same month, Carter went to Vienna to complete negotiation of the SALT II arms control agreement with Soviet President Leonid Brezhnev. The signing of SALT II, in negotiation for seven years through three administrations, might have been celebrated as a landmark achievement. But not then. It simply did not count. The only thing that mattered was the gas lines— and they were Carter's fault.

"The Worst of Times"

Much of the nation now seemed to be in the grip of the gasoline shortage. A survey by the American Automobile Association of 6,286 service stations nationwide showed that 58 percent were closed Saturday, June 23, and 70 percent were closed on Sunday, June 24, leaving Americans with very little gasoline on the first weekend of the summer. Independent truckers were conducting a rowdy, violent nationwide strike, now three weeks old, to protest fuel shortages and rising prices. A hundred truckers snarled rush-hour traffic for thirty miles along the Long Island Expressway, infuriating tens of thousands of motorists. Soaring gas prices were not the only problem. Inflation was also moving up to an unprecedented level.

As had happened before, in times of short supply if not outright panic, support was growing in Washington for a huge "synthetic fuels" program to reduce American dependence on imported oil. In the view of many, Three Mile Island had closed the door on nuclear power. The alternative was a program to produce several million barrels per day of synthetic fuels, primarily oil-like liquids and gases, through chemistry and engineering. The main methods would be hydrogenation of coal—a process similar to that the Germans had used during World War II—and the pulverizing and heating of shale rocks in the Rockies to temperatures up to nine hundred degrees Fahrenheit. Such a program would, to be sure, cost tens of billions of dollars at a minimum, it would take years to implement, it would raise major environmental issues—and it was not at all certain that it would actually work, at least on the scale proposed. Politically, however, the concept seemed increasingly irresistible.

Even as the growing support for "synfuels" added further to the pressures on the beleaguered Administration, Carter himself departed for his next foreign trip, to Tokyo to meet the leaders of the other major Western countries for an economic summit. Fearing the impact of the oil shortage on the overall health of the international economy, the seven leaders of the Western world turned Tokyo into an all-energy summit. It was also a very nasty one. Tempers were badly frayed. "This is the first day of the economic summit, and one of the worst days of my diplomatic life," Carter wrote in his diary. The conference discussions were harsh and acrimonious. Even lunch, Carter noted, "was very bitter and unpleasant." German Chancellor Helmut Schmidt "got personally abusive toward me. . . . He alleged that American interference in the Middle East trying to work for a peace treaty was what had caused the problems with oil all over the world." As for Britain's Prime Minister Margaret Thatcher, Carter found her "a tough lady, highly opinionated, strong-willed, cannot admit that she doesn't know something."

Carter's next stop was supposed to be a vacation in Hawaii. But Stuart Eizenstat, the chief White House domestic policy adviser, feared that a holiday now would be a political disaster of the first order. He thought that the Presidential party, which had been traveling abroad for most of a month, did not comprehend the mood of the country. On the way to the White House one morning, Eizenstat had sat for forty-five minutes in a gas line at his local Amoco station, on Connecticut Avenue, and he had found himself seized by the same almost uncontrollable rage that was afflicting his fellow citizens from one end of the country to the other. And the target of the national fury was not merely hapless station operators and the oil companies, but the Administration itself. "It was a black, dark period," Eizenstat later said. "All the problems, inflation and energy, were coalescing together. There was a sense of siege and an inability to get on top of the issue." The President, preoccupied with foreign affairs, needed to understand what was happening at home.

So, on the last day of the Tokyo Summit, Eizenstat dispatched a grim, depressing memorandum to Carter about the continuing gas shortage: "Nothing else has so frustrated, confused, angered the American people—or so targeted their distress at you personally." He added, "In many respects, this would appear to be the worst of times. But I honestly believe that we can change this to a time of opportunity." The exhausted Carter canceled Hawaii and, returning from Tokyo, found that his approval rating in the polls had plummeted to 25 percent, matched only by Nixon's in the final days before his resignation. He retreated to Camp David, in the Maryland mountains, where, equipped with a 107-page dissection of the national mind by Patrick Caddell, his favorite pollster, he aimed to meditate on the nation's future. He also met with a cross-section of American leaders and embraced a new book that found "narcissism" at the heart of America's problems.

In July, the Saudis pushed up their output from 8.5 to 9.5 million barrels per day. They had heeded the implorings from the United States and had responded to their assessment of their own security interests. The Saudi boost would help ease the shortage over the next few months, but it was not a longterm solution, nor, as events had suggested over the previous couple of months, something on which to base America's and the Western world's entire well-being. Nor could the extra supplies do much immediately to cool the hot temper of the American public.

As a result, Carter was compelled to do something, and to be seen doing something—something big, something positive, something that seemed to offer that long-term solution. He embraced the concept of a vast synthetic fuels plan, essentially based on the hundred-billion-dollar program put forward by Nelson Rockefeller in 1975. This would be the "sexy and affirmative program" that was desperately needed, and his staff worked feverishly to turn it into a specific proposal. Some voices were raised in doubt. The *New York Times* reported in a front-page story on July 12 that a new study from a group of researchers at the Harvard Business School argued that the United States could reduce its oil imports much more cheaply and quickly through a program of energy conservation than through synthetic fuels. Others warned that a synthetic fuel program would have enormous adverse environmental consequences. But in the speech he delivered in July to a distraught nation on America's "crisis of confidence," Carter announced his own plan to make 2.5 million barrels per day of synthetic fuels by 1990, primarily out of coal and shale oil. He had originally wanted to propose 5 million barrels per day, but had been talked out of it. Though he did not use the word, his address became known as Carter's "malaise" speech.

Carter also wanted to make changes in his own Cabinet, and in particular, to force the resignations of two of its members, Treasury Secretary Michael Blumenthal and Health Secretary Joseph Califano. His political advisers, Hamilton Jordan and Jody Powell, had convinced him that the two Cabinet officers were disloyal. Stuart Eizenstat argued otherwise to the President, saying that he had worked with the two men every day and that they were committed to the Admin-

677

istration. Eizenstat urged the President more strongly than he had ever urged him on any other matter not to fire Califano, who had strong political support, and Blumenthal, who was the Administration's chief inflation fighter. But Carter had already made up his mind. They would have to go. But how? Just before a Cabinet meeting, Carter told a few of his senior staff that he had decided to have all his Cabinet members submit their resignations and then he would keep only those he still wanted. Some of the staff members fervently tried to dissuade him. Such an action could create a panic. No, the President insisted, it would be seen as positive by a crisis-weary nation, the turning over of a welcome new leaf.

Carter immediately went into a tense Cabinet meeting dominated by the grim situation that the Administration found itself in. As worked out beforehand, Secretary of State Cyrus Vance proposed that all Cabinet secretaries submit their resignations so that Carter could begin afresh. The President concurred. A few minutes later, Robert Strauss, the Middle East peace negotiator, walked in and, not knowing what had just transpired and why the room was so somber, jokingly said that everybody ought to resign. His remark was greeted with silence. Finally, one of the other Cabinet secretaries leaned over and whispered, "Bob, shut up." They had all just resigned.

Altogether, five people left the Cabinet, some fired and some resigning. The aim was to bolster Presidential leadership. It had quite the opposite effect. The sudden news of the departures sent tremors of uncertainty throughout the country and the Western world. Over lunch that day, the national editor of the *Washington Post* muttered darkly that America's central government had just collapsed.[9]

The Cat-and-Mouse Dialectic

Spot prices in the world oil market eased off in the summer of 1979, but only slightly. Some OPEC countries continued to reduce their output. Iraq announced that it was extending its embargo to preclude shipments to Egypt—the champion of Arab nationalism and the oil weapon in 1973—to punish Anwar Sadat for the sin of having signed the 1978 Camp David peace accords with Israel. Nigeria, in the most dramatic use of the "oil weapon" since 1973, nationalized BP's extensive holdings in that country in retaliation for the British company's alleged indirect sales to South Africa, and then turned around and auctioned off its newly nationalized supply at higher prices.

All the while, buyers continued their quest for oil to build up inventories and fill storage tanks to the very brim in the face of uncertainty and fear for the future. The assumption was that demand was continuing to grow. It was a fatal miscalculation. In fact, a decline had already set in, reflecting the first effects of conservation as well as an economic downturn, but the fall was almost imperceptible at first. And buying remained frantic. As Shell's supply coordinator observed, "Every negotiation with a producer government was a finger-biting exercise: there was one dominant thought in the mind of company presidents and negotiators alike—to hold on to term oil and limit the need for spot acquisition. The suppliers of course sensed this, and a cat-and-mouse dialectic began. . . . Both the terms of contracts and the prices that had to be paid became continuously worse."

As in most panics, information—or rather the lack of information—was the key. If there had existed timely, credible, reliable, widely accepted data, companies would have recognized earlier that they were increasing stocks to an unnecessarily high level and that the underlying demand was weakening. But there was not much in the way of such statistics, and what early indicators did exist were not given much attention. So the stock build continued more or less unabated—as did rising prices.

The increases were not limited, by any means, to the OPEC countries. Britain's new state oil company, BNOC, raised prices on its desirable and secure North Sea crudes, and, for a moment, was even leading the market. "If BNOC, and by implication the British Government, were behaving like OPEC, who could expect OPEC to put an end to the oil price spiral?" asked one observer of the oil market. With the exception of Saudi Arabia, the OPEC countries lost no time in catching up. The market was further churned by the traders, for whom the volatility, disarray, and confusion made this a heyday. Some were from established commodity trading houses, some had entered the business after 1973, and some were johnny-come-latelies who rushed into the fray, their only capital requirement being what was needed to get a telephone and a telex installed. They were everywhere, it seemed, in every transaction, vying with the traditional oil companies for ownership, as cargoes still on the high seas were sold and resold—one cargo, fifty-six times over. The traders' only interest was the quick sale. Enormous sums were at stake; a single cargo on a supertanker could be worth $50 million.

The raison d'être for the traders was the breakdown of the integrated systems of the majors. In the old days, oil had stayed within a company's integrated channels or was swapped among companies. But now state oil companies accounted for larger and larger shares of total production, they had no downstreams of their own, and they sold their oil to a wide spectrum of buyers: major oil companies, independent refiners, and traders. The traders did best when they could take advantage of the enormous arbitrage between the lower prices in the term-contract market and the higher, more-volatile prices in the spot market. "The trader could be in a superb position," observed a senior executive from one of the majors. "All he had to do was manage to get himself a term contract of some sort." He could then turn around and sell it for eight dollars a barrel more on the spot market, making an enormous fortune for himself on a single cargo. And how did the trader obtain his fantastically lucrative term contract? "What he had to do to get his contract was to pay a ridiculously small commission to the appropriate parties. And sometimes the required brown envelopes would be passed." By comparison to what the trader in turn could make, this was hardly more than a gratuity.

Thus, in the summer and early fall of 1979, the world oil market was in a state of anarchy whose global effects far exceeded those of the early 1930s, in the wake of Dad Joiner's discovery in East Texas, and those of the very earliest days of the industry in western Pennsylvania. And while the pockets of the producers and traders swelled with money, consumers were forced to dip ever deeper into their own pockets to pay the price of panic. For many of the exultant exporters, it was another great victory for oil power. There was no limit, they thought, to what the market would bear and what they would earn. Some in the Western world gloomily began to fear that what was at stake was not only the price of the world's most important commodity, not only economic growth and the integrity of the world economy, but perhaps even the international order and world society as they knew it.

"The World Crisis"

Among those leaving Jimmy Carter's Cabinet in that summer of 1979 was James Schlesinger. Depressed by the unfolding situation not only in energy markets but in international politics—and by the position of the United States— Schlesinger decided to give vent to his feelings in a farewell speech in Washington, just as he had done four years earlier, when Gerald Ford forced him out as Secretary of Defense. Unusually somber, even for him, Schlesinger intended his speech this time to be hortatory and a warning. He began by invoking as his text *The World Crisis,* Winston Churchill's history of World War I. It was in the pages of that book that Churchill wrote of his efforts to convert the British Navy from coal to oil, despite the risk that would come from depending upon oil from Iran. Now, six decades later, eerily and uncannily, that risk had become reality.

"Today we face a world crisis of vaster dimensions than Churchill described half a century ago—made more ominous by the problems of oil," said Schlesinger. "There is little, if any, relief in prospect. Any major interruption— stemming from political decision, political instability, terrorist acts or major technical problems—would entail severe disruptions. . . . The energy future is bleak and is likely to grow bleaker in the decade ahead." But as Schlesinger would later say of himself, "I'm not a brooder," and in this instance there was nothing more that he could do about the situation. And so with that foreboding valedictory, yet also with some sense of relief, he departed public life. Very shortly after, the pessimism reflected in his remarks, and the dark anxieties about what he saw as growing Western vulnerabilities and what others were now calling the decline of the West, were to take on yet more bizarre and devastating significance.[10]

CHAPTER 34

"We're Going Down"

SHORTLY AFTER 3:00 A.M. Washington time, on November 4, 1979, Elizabeth Ann Swift, the political officer in the United States embassy in Tehran, got through by phone to the Operations Center, the communications nerve point on the seventh floor of the State Department in Washington, D.C. Her words jolted the officials at the Washington end of the line out of the slumberous quiet. It was already midmorning in Tehran, and Swift reported that a large mob of young Iranians had broken into the embassy compound, surrounded the chancery building, and were forcing their way into other buildings. An hour and a half later, Swift was back on the line to say that the attackers had set fire to part of the embassy. Still another half hour later, she reported that some of the invaders had threatened to kill two unarmed Americans just outside the room, that the table and sofa blocking the door had been pulled aside, and that the Iranians had burst into the office, even as the embassy officials desperately continued to try to make contact by phone with someone of authority in the Iranian government. The hands of the Americans were now being tied, Swift continued in a professional, almost matter-of-fact way to the shocked listeners at the other end. "We're going down" were her last words, before one of the young Iranians, a picture of Khomeini pinned to his shirt, grabbed the phone out of her hand. And then Swift, along with the other Americans, all now blindfolded, was led into captivity. The line remained open for a long time, though no one was there. Then it went dead.

Some sixty-three Americans—the skeleton staff that had remained at the embassy after the personnel had been scaled down from the fourteen hundred officials of the Shah's time—were taken hostage by a large, rowdy, violent band of zealots who were thereafter known to the world as "students." Some of the Americans were soon released, leaving fifty in captivity. The Iranian Hostage

Crisis had begun, and the Second Oil Shock entered a new phase with an even more dire geopolitical cast.

The specific grievance of the hostage takers focused on Mohammed Pahlavi and America's relation to him. His father, Reza Shah, had found a place of exile in South Africa. Not so the son, who, in his own exile, turned into a modern version of the Flying Dutchman. He could find refuge in no port and seemed cursed to wander forever. He went to Egypt, to Morocco, to the Bahamas, to Mexico. But no one wanted him to stay; he was a reject, a pariah, a figure to whom very little world sympathy attached, and virtually no government wanted to risk the ire of the unfathomable new Iran. All the courting of a few years earlier, all the flattery, the ingratiation, the respectful premiers and supplicating cabinet ministers from the industrial nations, the bowing and scraping of the powerful around the world—it was as if none of that had ever happened. To make matters worse, cancer and related illnesses were ravaging the Shah's body. Remarkably, it was only at the end of September 1979, more than eight months *after* he was forced to leave Iran, that senior American officials first learned that the Shah was seriously ill, and only on October 18 did they discover that it was cancer. Carter had adamantly refused to allow the Shah to enter the United States for medical treatment. But, at last, after months of controversy and acrimony at the highest levels of his Administration, compounded by a vigorous campaign by Henry Kissinger, John McCloy, David Rockefeller, and others, the Shah was admitted. He arrived in New York City on October 23. Though he was checked into the New York Hospital, Cornell Medical Center, under a pseudonym, which just happened to be the real name of U.S. Undersecretary of State David Newsom, to the latter's discomfort, his presence was immediately known and widely reported.

A few days later, while the Shah was under treatment in New York, Carter's national security adviser, Zbigniew Brzezinski, was attending the celebration of the twenty-fifth anniversary of the Algerian revolution in Algiers. There he met with the new Iranian Prime Minister, Mehdi Bazargan, and his foreign and defense ministers. The subject of discussion was how the United States might relate to the reborn Iran. The United States, insisted Brzezinski, would not engage in nor support any conspiracies against Iran. Bazargan and his ministers protested the admission of the Shah to the United States. They insisted that Iranian doctors be allowed to examine him, to determine if he was really ill, or if it was only a ruse to disguise a plot.

The news reports of the Algiers meeting, coming on top of the Shah's arrival in the United States, alarmed Bazargan's theocratic and more radical rivals, as well as young militant fundamentalists. The Shah was the enemy and the archvillain. His presence in the United States stoked memories of 1953, of Mossadegh's fall, of the Shah's flight to Rome and his triumphant return to the throne, and it aroused fear that the United States was about to stage another coup and again restore the Shah. After all, the Great Satan—the United States—was capable of carrying out the worst abominations. And here was Bazargan trucking with Zbigniew Brzezinski, one of the chief agents of the Great Satan, and a mere week and a half after the Shah's arrival in New York. For what purpose?

682

"Death to America"

Thus was provided the impetus, and pretext, for the invasion of the embassy. Perhaps it was only intended, originally, as a sit-in, but it soon turned into an occupation and a mass kidnaping, as well as a bizarre circus, complete with vendors in front of the embassy selling revolutionary tape cassettes, shoes, sweatshirts, hats, and boiled sugar beets. The occupiers even took to answering the embassy phone, "Nest of spies." It appeared that the Ayatollah Khomeini and his immediate circle had some idea of the planned assault and encouraged it. That they took advantage of it, seizing on the event for their own purposes, was quite clear. They would use the ensuing crisis to dispose of Bazargan and all others with Western and secular taints, to consolidate their own power, to eliminate their opponents, including what Khomeini called "American-loving rotten brains," and to put in place the elements of the theocratic regime. Until all of that was done, the hostage crisis would grind on for almost fifteen months—444 days, to be precise. Every day, Americans read about "America in Captivity." Each night, Americans were subjected to the televised spectacle of "America Held Hostage," including the repetitive chorus of the zealots chanting "Death to America." Ironically, with its late-night programs on the hostage crisis, ABC finally found a way to compete successfully against Johnny Carson and the *Tonight Show*.

The hostage crisis transmitted a powerful message: that the shift of power in the world oil market in the 1970s was only part of a larger drama that was taking place in global politics. The United States and the West, it seemed to say, were truly in decline, on the defensive, and, it appeared, unable to do anything to protect their interests, whether economic or political. As Carter succinctly summed matters up two days after the hostage seizure, "They have us by the balls." Iran was not the only scene of unrest. The hapless United States was under attack by a variety of opponents in the Middle East who wanted to eject the United States from the area. Later in November 1979, a few weeks after the hostage taking, some seven hundred armed fundamentalists, bitterly opposed to the Saudi government and its links to the West, seized the Great Mosque in Mecca, in what was supposed to be the first stage of an uprising. They were dislodged only with difficulty. The larger Saudi uprising never materialized, but the assault did send shock waves through the Islamic world. In early December, there was a Shia protest in al-Hasa, in the heart of the oil region in the eastern part of Saudi Arabia. Then came another dramatic and much larger shock a few weeks later in December. The Soviet Union invaded Afghanistan, Iran's neighbor to the east, rocking both the Gulf States and the West. Russia, it now seemed to many, was still intent on fulfilling its century-and-a-half-old ambitions to drive toward the Gulf, and was taking advantage of the disarray of the West to position itself to capture as many of the spoils of the Middle East as it could. The bear was also becoming bolder; it was the first large-scale use of Soviet military forces beyond the communist bloc since World War II.

President Carter responded in January 1980 by enunciating what became known as the Carter Doctrine: "Let our position be absolutely clear. An attempt by any outside force to gain control of the Persian Gulf region will be regarded as an assault on the vital interests of the United States of America, and such an assault will be repelled by any means necessary, including military force." The Carter Doctrine made more explicit what American presidents had been saying as far back as Harry Truman's pledge to Ibn Saud in 1950. With even more historical resonance, it also bore striking similarities to the Lansdowne Declaration of 1903, by which the British Foreign Secretary of the day had warned off Russia and Germany from the Persian Gulf.

Carter had earned great respect in the oil world in 1977, his first year in the Presidency, as the man who had forced the Shah to bend, to recant his commitment to higher prices. Carter had been the magician who had tamed the Shah and transformed him from a price hawk into a compliant dove. He had engineered the Camp David accords between Israel and Egypt. Now all those achievements were overwhelmed. The Shah was an outcast, the Iranian Revolution had sparked the oil panic of 1979, and Carter's Presidency continued to be cursed by events in Iran, with Carter himself held hostage, in political terms, by the gang of "student" militants in Tehran.

After the hostages were taken, the dying Shah and his entourage would quickly and apologetically leave the United States, spending their final hours before departure in pathetic isolation in a psychiatric ward, complete with barred windows, on an American Air Force base. They went next to Panama and then back to Egypt, where the wasted Shah finally died in July of 1980, a year and a half after his flight from Tehran. No one really cared. By that point, Mohammed Pahlavi, son of an officer from the Cossack brigade, had become irrelevant to the outcome of the hostage crisis, to the panic in the oil market, and to the international game of nations in which he had once played such a prominent role.[1]

In the immediate aftermath of the hostage taking, Carter responded by placing an embargo on the importing of Iranian oil into the United States and by freezing Iranian assets. The Iranians counterattacked by prohibiting the exporting of Iranian oil to any American firm. The import ban and the asset freeze were virtually the only tools that Carter had easily at hand. The freezing of assets hurt Iran; the ban on oil imports did not. But it did necessitate a redistribution of supplies around the world, further disrupting supply channels and sending into the spot market more frantic buyers, who helped to bid up the price to new heights. Some cargoes went for forty-five dollars a barrel; the Iranians quoted fifty dollars a barrel for their oil to worried Japanese trading companies. The dislocation amplified the overall nervousness and anxiety in the market that followed the hostage taking, contributing further to the seemingly endless cycles of panic buying and price increases. As an executive from one of the majors remarked rather dryly, four days after the hostage taking, "In this environment companies feel a need for higher stocks than previously considered normal." The inventory build was, in industry language, "supply protection"—in other words, insurance.

The hostage crisis had even wider ramifications. It served to demonstrate the apparent weakness, even nakedness, of the consuming countries—in particular, of the United States, whose power was the basis of the postwar political and economic order. And it seemed to establish that world mastery really did lie in the hands of the oil exporters. At least that was the appearance. But there were forces at work in the oil market even more powerful than governments. And now it was the turn of the exporters to make fatal miscalculations of their own.

The Bazaar

Rising oil prices had become the object of constant attention by presidents and prime ministers, as well as the fodder of front pages for months. They were also a subject of intense dismay for the leaders of Saudi Arabia. Once again, they were alarmed both by their own loss of control over the market and by the fact that control seemed to have passed into the hands of such militant and uncompromising rivals as Libya and Iran. They thought that the wildly rising prices threatened the world economy with recession, depression, or even ruin, and thus threatened their own well-being. The days in which Saudi Arabia's economic future was determined by the number of pilgrims who came to Mecca were long since past; now it was the "rates" that mattered to Riyadh—world interest rates, exchange rates, inflation rates, growth rates. The Saudis also feared that their own position would be damaged in another way: that the price increases would destroy consumers' confidence in oil and, thus, stimulate long-lasting competition to OPEC oil, as well as large-scale development of alternative fuels. That would be particularly threatening to a country with huge oil reserves, the life of which would extend far into the twenty-first century.

The Saudis responded to the predicament with jawboning: Yamani became more of a hawk on the need for conservation to blunt price increases than almost any Western leader. The Saudis tried to hold down their own official prices, even if that meant leaving money on the table, at least compared to the prices that the other exporters were exacting. They also sought to counteract rising prices by continuing to push their own production up. Their objective was straightforward: to use the growing weight of excess supply to force prices down. But the effects were not quickly felt. "We're losing control over everything," Yamani had plaintively declared in mid-October 1979, after further Libyan and Iranian price hikes. "We feel so unhappy. We don't like to see it happening like this." Then, a few weeks later, came the beginning of the hostage drama. In a further dislocated and agitated market, prices spiked again and again, despite the Saudi countermeasures. Was some kind of stabilization possible? Eyes focused on the fifty-fifth OPEC meeting, scheduled for Caracas in late December 1979.

When Juan Pablo Pérez Alfonzo first became oil minister of Venezuela in the 1940s, a hillside on the southern side of Caracas had been a sugarcane field. Now it was the site of the Tamanaco, a sprawling international hotel that, with its older wings and new additions and grand outdoor swimming pool, was a monument to the development of the Venezuelan oil industry. It was the place to stay when doing oil business in Caracas, and it was there the OPEC ministers congregated. The issue before them was to try to reunify the OPEC pricing structure, which was in total disarray. Saudi Arabia's official price was $18 a barrel; others were as high as $28, while spot prices were in the $40–$50 range. Before the meeting, the Saudis announced that they would raise their price by $6, to $24, with the idea that others would bring their prices down and into line. It did not seem likely to work; the Iranians immediately leapfrogged their own price up by another $5. Once again, as it had been ever since the 1950s, the sharpest fissure was between Saudi Arabia and Iran.

For much of the year the Saudis had been steadily producing extra oil in order to counter the price increases. In 1979, OPEC production was back up to 31 million barrels per day, even with the Iranian shutdown—3 million barrels per day higher than 1978. Where was the extra oil going? Not into actual consumption, Yamani was sure, but into the inventories of companies fearful that future supplies would be further interrupted. At some point, the extra oil would tumble out of inventories and into the market, depressing prices. "Political decisions cannot permanently negate the divine laws of supply and demand," Yamani would later explain. "Prices go up, and demand goes down, it's simple, it's ABC."

At the Tamanaco, Yamani moved into the top-floor Presidential suite, vacated at his request by the Venezuelan oil minister, and began to campaign for his point of view. The oil ministers met privately in the Saudi's suite for what turned into a marathon. Yamani warned them of the danger as he saw it: that they were damaging their own interests, that demand was already showing signs of weakening, that the continuing price leapfrogging would bring a "catastrophe in the world economy." A few of the other ministers agreed with him; the majority did not. When Yamani said that demand for OPEC oil would dramatically plummet and that they would have to cut production to protect prices and then that prices would collapse in any event, they scoffed. One oil minister said Yamani must be joking; another, that he was obviously on drugs. For fully eleven hours, the ministers argued back and forth in Yamani's suite, but there was no agreement whatsoever. Indeed, there was no official price at all. OPEC and the oil market, Yamani said dismissively, had turned into a bazaar. He also had a warning for the other producers, and a promise to consumers. "There will be a glut in the market. It is coming." Prices would fall.

The other exporters, however, ignored the admonition; they believed their own rhetoric. "In the name of almighty God, there will be no surplus and prices will not fall," intoned the Iranian oil minister. Most of the exporters assumed demand was so inflexible that they could dictate whatever prices they wanted to consumers. Their self-confidence was demonstrated immediately after the meeting, when Libya, Algeria, and Nigeria proceeded to raise prices once again. Others followed suit.

Caracas, in those last dying days of the tumultuous year of 1979, was the moment when the exporters lost touch with the reality of the market. Demand was indeed weakening, new supplies were being developed, panic buying was receding, inventories were being built up, spot prices were declining. And the Saudis continued their steady overproduction. Other producers, however, continued to push up their prices, while some cut back on production, which helped buoy prices. There was now talk of a "miniglut," but that was more than offset by what became known as the new "minipanic." In the face of the hostage crisis, Washington was seeking to promote a general embargo or sanctions against Iran in cooperation with Western Europe and Japan, an effort that accentuated nervousness in the market.[2]

Then in April 1980, frustrated to the extreme by the impasse over the hostages, the Carter Administration mounted a military rescue operation in Iran. Eight helicopters were dispatched from the carrier *Nimitz* to a desolate, isolated spot in Iran that had been dubbed Desert One. There, in the darkness, they were to rendezvous with six C-130s. The large transport planes would refuel the helicopters; they would also carry assault teams, which would switch to the helicopters and continue to Tehran. These squads were to retake the American embassy, free the hostages, and get them to an airfield near Tehran, which was to be secured by other American airborne forces.

But things quickly went awry. One of the helicopters dropped out en route because of navigational problems; another because of mechanical malfunction. Then, in the middle of the night, three Iranian vehicles, including a bus filled with forty-four people, passed the American aircraft close enough to observe them. In a blinding sandstorm, one of the remaining helicopters collided with a C-130 and burst into flames, killing several of the American servicemen. Now there were only five available helicopters; the mission required a minimum of six. It was aborted on direct orders of President Carter. The failure was immediately made public and emblazoned in media around the world. The Iranians wasted no time in dispersing the hostages throughout Tehran in case the United States tried again. The very fact of the hostage rescue mission and its ignominious failure greatly increased the tension in the market. In addition, Iranian output had fallen once more, all of which set off a new round of panic buying. Companies remained focused on their vulnerability and the possibility of new troubles and continued to build their inventories as "insurance."

The general outlook was bleak. The "mini-glut," according to the consensus market view, would be over by the spring of 1981. OPEC's Long Range Strategy Committee came out with its plan for a 10 to 15 percent annual increase in oil prices, starting at the current base, which meant sixty dollars per barrel within five years. There seemed little reason, in the gloom of the moment, to doubt that they could get it. The director of the Central Intelligence Agency, testifying before a Senate committee five days after the hostage rescue attempt turned to disaster in the Iranian desert, said, "Politically, the cardinal issue is how vicious the struggle for energy supplies will become." The bleak mood of the times was summed up in the title of an article in *Foreign Affairs* in the summer of 1980: "Oil and the Decline of the West."

OPEC met again in Algiers in June 1980. The Saudis, now joined by the Kuwaitis, tried yet again to put an end to the bazaar in the oil market and to stabilize prices—and again to no avail. The average oil price was thirty-two dollars per barrel, almost three times what it had been a year and a half earlier. It was at this meeting, sitting in the coffee shop of the hotel in Algiers, being shunned by many of the other delegates and still thinking of those "divine laws of supply and demand," that Yamani unburdened his private feelings to a friend. "They're too greedy, they're too greedy," he said. "They'll pay for it."

In fact, the oil market was again beginning to sag under the distress that Yamani had been predicting, and judging by market trends over the summer of 1980, his prophecy in Algiers seemed likely to come true rather quickly. Inventories were very high; a pronounced economic recession was already emerging; in the consuming countries both product prices and demand were falling; and the inventory surplus was continuing to swell. Companies were even beginning to store oil in supertankers, costly as that was, rather than sell it at a loss in the market. Now it was the buyers' turn to walk away from contracts, and the demand for OPEC oil was coming down. Indeed, in mid-September, a number of OPEC countries agreed to voluntarily cut back production by 10 percent in an effort to firm prices.

Meanwhile, the twentieth anniversary of OPEC was close at hand. Over two decades, the organization had risen from a nonentity to a colossus in the world economy, and a grand celebration was being planned for an OPEC summit in November. In anticipation, a special committee had been working on a long-term strategy. An official history was commissioned, as was a film, and fifteen hundred journalists were to be invited to the great event, which would be held in Baghdad, where OPEC had been founded in 1960. On the morning of September 22, 1980, the oil, finance, and foreign ministers of the OPEC nations self-confidently assembled in the Hapsburg Palace in Vienna to continue planning for the Baghdad celebration. But within minutes of its opening, the meeting erupted in a babble of confusion, anger, and chaos, and the general conference was hurriedly turned into a closed session.

For something else had already been planned in Baghdad.[3]

The Second Battle of Qadisiyah: Iraq versus Iran

On that same day, just as the ministers were preparing to sit down in Vienna, squadrons of Iraqi warplanes attacked, without warning, a dozen targets in Iran, and Iraqi troops began pushing into Iran along a broad front, pounding cities and key installations with heavy artillery. The outbreak of war shook the Persian Gulf yet again and threw the oil supply system into jeopardy, threatening a third oil shock.

There had been incidents along the border between Iraq and Iran for weeks before September 22, and indeed, a war had grown increasingly probable ever since the preceding April. Hostility between Iraq and Iran was long-standing; some would find the present struggle merely a latter-day manifestation of the conflicts that went back almost five thousand years, to the very beginning of recorded civilization in the Fertile Crescent, when the soldiers from Mesopotamia, in what is now in modern Iraq, and Elam, in what is now in modern Iran, had habitually slaughtered one another. An ancient poem lamented the scene after the great and proud city of Ur, where walls had been built "as high as a shining mountain," was ravaged, sacked, and destroyed by the soldiers of Elam four thousand years ago.

> Dead men, not potsherds,
> Covered the approaches,
> The walls were gaping,
> The high gates, the roads,
> Were piled with dead.
> In the side streets, where feasting crowds would gather,

Scattered they lay . . .
Bodies dissolved—like fat in the sun.

The scene would be much the same when, four millennia later, the distant heirs to Mesopotamia, on one side, and to Elam, on the other, savaged each other over the same marshes and the same scorched and scalding deserts.[4]

The war had been sparked by a host of rivalries: ethnic and religious, political and economic, ideological and personal; by a struggle for primacy in the Gulf; by the insecurities of national cohesion; and by the arbitrary way in which "nations" had been created and borders in the Middle East overlaid on the map of the defunct Ottoman Empire. Indeed, geography was decidedly at the heart of the conflict.

The Shah had been at loggerheads with the secular Ba'thist regime in Baghdad since it first came to power in 1968. One of the most important issues between the two countries was the Shatt-al-Arab, a meandering river and delta created by the confluence of the two Iraqi rivers, the Tigris and Euphrates, with several rivers from Iran. The Shatt-al-Arab served as the boundary for 120 miles between the two countries. It was crucial to Iran as its most important avenue to the Gulf—the Abadan oil refinery was built on a mud flat in the river delta—but certainly not its only one. The Shatt-al-Arab, however, was critical to Iraq as its only avenue to the high seas. Iraq's entire coastline was only twenty-six miles or so in length, compared to Iran's fourteen hundred miles of coast. Basra, Iraq's principal port city, was actually almost fifty miles up the Shatt-al-Arab, which had to be frequently dredged owing to its shallowness and the buildup of silt. Sovereignty over the Shatt-al-Arab had, thus, assumed great symbolic significance. To make matters worse, a considerable part of both countries' oil infrastructure—fields, pumping stations, refineries, pipelines, loading facilities, storage tanks—was concentrated around and depended, in one way or the other, on the Shatt-al-Arab. The Shah had prudently built pipelines as an alternative to the river traffic, as well as an offshore terminal at Kharg Island, where supertankers could berth. Iraq, however, exported a good part of its oil through the narrow funnel of the Shatt-al-Arab and its immediate vicinity, though pipelines through Syria and Turkey were in place.

The Shah and the militant Ba'thists had sorted out their various claims in an agreement finalized in Algiers in 1975 and signed on behalf of Iraq by Saddam Hussein. From the viewpoint of sovereignty, Iran came out the better. The Iraqis gave up their insistence, generally accepted for forty years, that the boundary between the two countries was the eastern, that is, Iranian, side of the river, and accepted the Iranian insistence on the midpoint of the navigational channel. But, in turn, the Shah gave the Iraqis something that they needed very badly. He cynically agreed to cut off the considerable aid he was providing to the Kurds, a distinct ethnic group that composed upwards of 20 percent of the total Iraqi population and was then fiercely battling the Ba'thists for autonomy and national identity, in a region where much of Iraq's oil was to be found. The Shah's jettisoning of the Kurds was a major quid pro quo, perhaps a requisite for survival of the Ba'thist regime. Baghdad wasted no time; just six hours after the communiqué with Iran was issued in Algiers, it launched a decisive offensive against the Kurds. Three years later, in 1978, Iraq returned the favor, in what seemed a relatively minor way. At the Shah's request, Iraq expelled the Ayatollah Khomeini, who had been living in exile in Iraq for fourteen years. In light of what later happened, that hardly turned out to be a favor at all.

Khomeini himself was filled with a hatred of the Iraqi regime and a burning desire for revenge for his treatment at its hands. His ire was focused on the President, Saddam Hussein. Certainly, Hussein had proved himself a champion conspirator in the considerable history of Ba'thist conspiracies. The Ba'thist movement itself grew out of an Arab Students Union that two Syrian intellectuals had formed while studying at university in Paris in the early 1930s. A decade later, they launched the Ba'th—or "renaissance"—party back in Damascus. It was militantly pan-Arabic, aiming to create a single Arab nation and fervent in its denunciations of the West and imperialism. All-embracing in its ideology and demands, it was contemptuous and unreservedly hostile to its opponents and those who were outside its grouping. It celebrated violence and absolutism in pursuit of its objectives. The party split into two branches, one of which eventually took power in Syria and the other, in Iraq. Despite their common origins, the two wings became seemingly irreconcilable rivals for preeminence.

Saddam Hussein's father had died just before his birth in 1937, and as he grew up, he found his identity in extreme nationalism and the violent, conspiratorial world of Ba'thism. The decisive influence on him was his uncle Khayr Allah Talfah, who raised him and became his guardian. A fervent nationalist from the Sunni Arab minority, Talfah hated and despised the European culture. For both uncle and nephew, the lodestar event was the pro-Nazi nationalist Rashid Ali coup of 1941, in the course of which German planes attacked British forces in Iraq. When Iraqi troops threatened to fire on a plane evacuating British women and children, British soldiers attacked them and the coup collapsed. Talfah participated in the coup, and when it failed, he was imprisoned for five years, leaving him with a lasting sense of bitterness, resentment, and hatred that he communicated to his fatherless nephew. The Rashid Ali coup became a central myth for the Ba'thist movement. Saddam Hussein was also shaped by the culture of his hometown of Tikrit, which was remote from the national life of Iraq and oriented instead to the harsh desert. Tikrit's values of desert survival— suspicion, stealth, surprise, and the unalloyed use of force to achieve one's objectives—were the ones that Saddam Hussein absorbed.

It was during the tumult and enthusiasm that accompanied Nasser's victory at Suez in 1956 that Saddam Hussein, while still a teenager, was recruited into the Ba'th party. The Nasserite anti-imperialistic rhetoric of the 1950s remained with him ever after. Shortly after joining the party, it is said, he carried out his first assassination—of a local political figure in Tikrit. His commitment to Ba'thism was sealed and the foundations of his reputation established. In 1959, he had been one of the assailants in the assassination attempt, on Baghdad's main street, on Iraq's ruler, Abdul Karim Kassem. The attack failed, and Hussein, wounded in the gunfire and under a death sentence, fled to Egypt. He did not return to Iraq until 1963. Thereafter, he took charge of organizing the Ba'thist party's underground militia. Though already the strongman for several years in the Ba'thist regime that took power in 1968, he assumed the Presidency—replacing Ahmad Hasan al-Bakr, cousin of his uncle—only in 1979 in the course of a purge in which many members of the Ba'thist party were executed. In order to assure that the imprisoned Ba'thists provided the appropriate confessions before their executions, Saddam Hussein took some of their families hostage. By 1979, he had already long since come to be seen as a *shaqawah,* an implacable tough, a man to be feared. He was ruthless and emotionless when it came to those he considered as enemies, threats, obstacles to his objectives, or simply useful or convenient to run over.

The new Iraqi regime—particularly the party, military, and security services—was dominated by Tikritis, many of them related in some way to Hussein. So obvious was their grip that in the mid-1970s the government banned the use of names that indicated clan, tribe, or locality of origin. At the top sat members of Hussein's Talfah family and two other immediately related families, the only people he could trust—to the degree that he could trust anybody. He had already married his cousin, the daughter of his uncle Kahyr Allah Talfah. Now Adnan Khayr Alah Talfah—son of his uncle, brother of his wife, his own cousin—was Minister of Defense (until his death in a mysterious helicopter crash in 1989). Hussein Kamil al-Majid, who happened to be both Hussein's cousin and son-in-law, became chief weapons buyer, and responsible for the development of nuclear and chemical weapons and missiles. And the influence of Khayr Allah Talfah himself continued to be felt. In 1981, the government printing house distributed a pamphlet by Talfah. Its title gave some idea of the thrust of his political thought: *Three Whom God Should Not Have Invented: Persians, Jews, and Flies.*

Though the Ayatollah Khomeini was expelled from Iraq in 1978, before Hussein's complete acquisition of power, the Ayatollah held Hussein personally responsible for his troubles and ranked him among his preeminent opponents. Once asked to list his enemies, Khomeini replied: "First, the Shah, then the American Satan, then Saddam Hussein and his infidel Ba'th Party." Khomeini and his circle saw the secular, socialist Ba'thists as implacable enemies of their own creed and attacked Ba'thism as "the racist ideology of Arabism." As if all that was not bad enough, Khomeini had even worse to say; he denounced Hussein as a "dwarf Pharaoh."

Saddam Hussein had good reason to fear Khomeini's diatribes. Around half of Iraq's population was estimated to be Shia, while the Ba'thist regime was secular and based on the minority Arab Sunnis. Iraq was also the site of the holiest shrines of the Shia faith; and agitation among the Shias, fed from Iran, was growing. After an assassination attempt against his Deputy Prime Minister in April 1980, Hussein ordered the execution of the most prominent Shia ayatollah in Iraq and, for good measure, the ayatollah's sister, and took to denouncing Iran's religious leader as "Khomeini the rotten" and as "a Shah dressed in religious garb."

As incidents and recriminations mounted between the two countries, Iraq thought it saw its opportunity. Iran appeared disorganized and chaotic; in Baghdad, the common saying was "There is a government on every street corner in Iran." Iran's army was demoralized and in disarray and in the grip of a bloody purge. Iraq could strike hard at Iran, topple Khomeini, put an end to the Shiite revolutionary threat to Iraq, and assert sovereignty over the Shatt-al-Arab waterway, protecting Iraq's oil position. There were even more enticing plums to be had. Hussein could appeal to the ethnic Arabs in Iran's Khuzistan (though less than half the population in that southwestern region of Iran was of Arab descent) as their "liberator" and perhaps incorporate that territory, which the Iraqis called Arabistan, into Iraq, or at least cast it under Iraq's sway. The prize was not merely fraternal reunification; 90 percent of Iran's oil reserves happened to be in Khuzistan. On top of all this, the Ba'thists could eradicate the wound of pride; it had been humiliating in 1975 to have to give way to the Iranians on sovereignty over the Shatt-al-Arab. There was more than one vacuum to fill. The Shah, the regional policeman for the Gulf, was gone; Hussein could assert Iraq's primacy, and his own, in this area of great international significance. Moreover, with Egypt alienated from the rest of the Arab world because of the Camp David accord, Iraq could also emerge as the new leader and militant champion of the Arab world, and the one that had crushed the threat from the East. And it could become one of the dominant oil powers. Altogether, the opportunities were irresistible.

From the beginning, Hussein clothed himself in the mantle of Arab leadership, which accorded with the pan-Arabic Ba'thist ideology. If Khomeini was going to base the symbols of his legitimacy on events that had taken place in the seventh century, so would Hussein. He billed the new war as the "Second Battle of Qadisiyah"—the first one having occurred when the Arabs defeated the Persians in A.D. 636/637 near Najaf in what is now south central Iraq. That in turn led to a further triumph over the Persians in 642 that thereafter was celebrated by the Arabs as the "Victory of Victories." It sealed the doom of the Sassanid Persian Empire, whose King fled to the east, where he was eventually assassinated by a local satrap. A century later, Baghdad was founded, and it would be preeminent in the region for several hundred years thereafter. Now in 1980, it would be Baghdad's turn again. Or so it was thought.

Hussein targeted his attack on the heart of the Iranian oil industry, including Abadan and Ahwaz, the same city that had provided the gateway for the final death blow to the Persian empire thirteen centuries earlier. Hussein thought he could achieve all his objectives with an Iraqi *blitzkrieg,* in a hard, sharp series of blows. That view was not, by any means, restricted to Baghdad. In Vienna, where the triministerial OPEC meeting had been disrupted by news of the attack, the virtually universal assumption was that the war would be over in one or at most two weeks. But the Iraqi strategy proved to be based upon grave miscalculation, for the Iranians withstood the first blow and struck back immediately, and no less hard, at Iraqi targets. The assault enabled Ayatollah Khomeini to further consolidate his power, silence his critics, dispose of the nonclerics in his government, and proceed in the fashioning of the Islamic Republic—meanwhile mobilizing the population to resist. Iranians from virtually all political camps rushed to the common defense. The Arabs of Khuzistan showed no desire to be liberated by Iraq and did not welcome the Iraqis as "brothers," but rather saw them as invaders. The Iraqis were unprepared for the "human wave" assaults they encountered on the battlefield. Hundreds of thousands of young people, drawn by the Shiite vision of martyrdom, and with little thought for their own lives, advanced on Iraqi positions in front of regular Iranian troops. Some of the young people arrived at the front carrying their own coffins, exhorted as they had been by Khomeini that "the purest joy in Islam is to kill and be killed for God." They were given plastic keys to heaven to wear around their necks. Children were even used to clear minefields for the far more valuable and much rarer tanks, and thousands of them died.[5]

The End of the Road

The outbreak of the war shook the oil market. On September 23, 1980, the second day of the war, Iraqi warplanes began a sustained assault against the Iranian refinery at Abadan, the world's largest, and proceeded over the next month to damage it severely. They also carried their attack to every Iranian oil port and oil city. The Iranians counterattacked against Iraqi facilities, completely choking off Iraqi oil exports through the Gulf. Moreover, Iran in time persuaded Syria, ruled by a rival Ba'thist party, to cut off Iraqi pipeline exports through Syria, leaving Iraq with only one limited pipeline through Turkey. Iranian oil exports were reduced by the war, but Iraq's almost ceased, something on which Hussein had not counted.

In its initial stages, the Iran-Iraq war abruptly removed almost 4 million daily barrels of oil from the world market—15 percent of total OPEC output and 8 percent of free world demand. Spot prices quickly jumped again. Arab light reached its highest price ever—forty-two dollars a barrel. Fear was once again driving the market. Was this the Third Shock, the next stage in the collapse of the Middle East and its oil industry into chaos and militancy? Would Iraq be eliminated from the world oil balance? Would Iran once again disappear as a supplier? Would the battle between Sunni and Shia, between Arab and Persian, destabilize the entire Gulf? Or, perhaps even worse, would Iran, with three times the population of Iraq, prevail and carry its fundamentalist, anti-Western revolution deeper and deeper into the heart of the Middle East? In pondering those questions, one could read the underlying economic indicators in two ways. One reading certainly did point to a new shock; the other in the opposite direction. Which would prove to be correct?

Oil demand was certainly weakening. Still, one could not yet know whether it was the result of recession, which would mean a temporary downturn, or of conservation, which would have more lasting effects. Economic contraction had already begun, the result of the price increases compounded by a new resolve on the part of the Western nations to fight inflation at all costs, even if it meant deep recession. Whatever the reason, it was clear that demand was going down.

Meanwhile, governments, working through the framework of the International Energy Agency, had learned the lessons of 1979 and concerted their efforts to persuade companies not to buy in panic, not to scramble for supplies, not to drive up prices—but, rather, to draw down their inventories. The message that the IEA put out was meant to reassure: Things were manageable, this was not 1979 all over again, go easy, avoid "undesirable purchases" (which meant oil whose price had been bid up). The message made sense, as the supply positions of the companies were very different this time. Since early 1979, the companies had spent a great deal of money, buying up supplies at all and any cost—including many additional barrels far in excess of demand. Those extra barrels of supply had gone not into the engines of cars, nor into the furnaces of factories, nor into electric power plants, but into storage. The Great Panic had, by its own logic, turned into the Great Inventory Build, and when the war broke out, storage tanks all around the world were brimming over, and oil companies were chartering supertankers to use as additional floating storage. It was expensive to keep oil in storage. In a continuing period of calm, given a choice between buying additional oil and drawing on existing stocks, a company would likely choose to draw from storage.[6]

But now the Iran-Iraq War brutally upset the returning calm, reigniting panic buying, and not all that many companies were inclined, at least initially, to heed the message of the IEA and eschew those "undesirable purchases." "No matter what restraint we show," complained one refiner in November 1980, "there is still someone else out there who is willing to buy at higher prices and thereby pushes up the market." The all-important question was how companies would manage their inventories in this new crisis. In time of anxiety and uncertainty, the inevitable tendency was to hold on and hoard and see what happened. High costs were preferable to no supply, especially if the costs the following day would be even higher. Thus, many players were once again scouring the world for supplies. Among them were the Japanese trading and oil companies, reflecting fears in Tokyo that an extended supply cutoff might be at hand. But the Japanese were hardly alone. An executive of an American company summed up the issue when he said that drawing down stocks "might leave us in deep trouble later on." He explained: "Commercial firms can't afford such moves. Suggestions that we draw down inventories imply some knowledge of when the crisis will end. If I knew Iraq and Iran output would be back to prewar levels by July, sure I would draw down stocks." But he couldn't know.

In December 1980, the OPEC oil ministers met in Bali, yet again to discuss price. There was, however, a most awkward matter that had to be dealt with before any discussions could proceed. In November, the Iranian oil minister had gone to tour the battlefield near Abadan. Unfortunately, no one bothered to tell him that the area had been captured by the Iraqis, and he himself was seized and carted off into captivity. OPEC or no OPEC, the Iraqis would not free him. The Iranians were so angry that they threatened to boycott all OPEC meetings. Could the Bali meeting go ahead? It fell to the artful diplomat, Indonesia's oil minister, Dr. Subroto, to work out a satisfactory compromise. Seating was normally alphabetical, and thus Iran and Iraq were consigned by fate, in the form of the alphabet, to sit next to each other, which could have been most unpleasant. Subroto broke precedent and inserted Indonesia between Iran and Iraq. That led some—thinking of the waterway between the two countries that was at the center of their contention—to say that Indonesia was now occupying the Shatt-al-Arab. Though one problem was now solved, another emerged. The Iranian delegation entered the conference room carrying a large portrait of their captured oil minister, who, they insisted, was still the head of their delegation. They would merely be carrying out his wishes. Dr. Subroto allowed them to put the photograph in the chair designated for the missing minister, so that he could even in his absence continue to inspire, if not quite lead, his delegation. So, with further awkwardness avoided, the conference could begin. It was to conclude with another rise in OPEC prices, to a thirty-six-dollar marker for all except the Saudis. Yes, it seemed, a Third Shock was at hand.

At almost exactly the same moment, but halfway around the world, the energy ministers of the industrial nations were gathered in Paris for their own meeting. Ulf Lantzke, the director of the IEA, would customarily hold an informal get-together in his office, after a ministerial dinner, for a relaxed discussion and exchange of views before the formal sessions the next morning. At this particular post-dinner discussion the mood was gloomy; the IEA's efforts to encourage inventory use, instead of panic buying, were not meeting with great success. As a MITI official had observed, the phrase "undesirable purchases" was "imprecise and could mean different things to different people." The frenetic purchasing by some Japanese trading companies, in particular, had emerged as a sore subject and one that, amid the whiskey and cigars that evening in Lantzke's office, stirred much debate.

Finally, as the hour got close to midnight, the imposing Count Etienne Davignon of Belgium, a prominent and forceful Commissioner of the European Community, lost his patience. He focused on the Japanese representative. "If you don't get your trading companies under control," Davignon bluntly said, "you can forget about any more Toyotas and Sonys coming into Europe."

The room went silent. The Japanese official sat there for a moment, taking in the remark and weighing his response. "You are a great international civil servant," he said at last. And he said no more.

But MITI reinforced its "administrative guidance" to companies to go easy. Those companies got the message and did become more restrained in their buying, as did the American and British oil companies. The market players were, however, responding to more than government policies. For, by the end of 1980, the picture was becoming clearer. While inventories remained very high, demand was continuing its precipitous fall, and market prices were weakening. It was a combination that made it more and more uneconomic to hold on to stocks, so there was a growing incentive to use those inventories, as the IEA wanted, instead of buying additional oil.

And not only was consumption actually declining, but supply from other sources was making up for the lost output from Iran and Iraq. For most of the time since the end of 1978, the Saudis had been cranking out barrel after extra barrel in an effort to stifle the continuing surge in oil prices and to put a cap over their brethren in OPEC. "We engineered the glut," Yamani once said, "and we want to see it in order to stabilize the price." The Saudis were not going to let something so inconvenient as the outbreak of the Iran-Iraq War defeat their strategy, and within days of the first battles, they announced that they were pushing their production up another 900,000 barrels per day, to the very limit of sustainable capacity. In itself, that increase was the equivalent of almost a quarter of the lost production from the two belligerents. Other OPEC producers also upped their output, and even some oil from Iraq and Iran was coming back into the market. At the same time, the production by Mexico, Britain, Norway, and other non-OPEC countries, as well as in Alaska, was continuing to increase, as well. It was no longer just a "miniglut." Under these circumstances, any reluctance to use inventories disappeared; indeed, their use, as opposed to buying oil, became irresistible. Buyers were now beginning to revolt against high prices. Non-OPEC producers, anxious to increase market share, were making significant cuts in their official prices. Their gain was OPEC's loss, and the demand for OPEC oil fell. As a result, OPEC's output in 1981 was 27 percent lower than the 1979 output, and in fact was the lowest it had been since 1970. Yamani's prophecy was at last coming true.

OPEC was nearing the end of the road, though neither the OPEC exporters, nor the industry, nor the Western consuming countries had any idea of what lay ahead. The Carter presidency had also come to an end. In a final humiliation for Jimmy Carter at the hands of the Iranians, the hostages taken at the American embassy in Tehran were not released until the very day he left office, succeeded by Ronald Reagan, whose buoyant confidence in himself and in America had proven much more palatable to the electorate than Carter's "malaise."

Meanwhile, the oil market was responding to the phenomenal rise in prices over the 1970s and consumers' fears for the future. Yet the exporters were still unwilling to face up to the fact that the "objective conditions" of the marketplace were truly shifting. They would not contemplate a price cut. Prices were still in disarray, but finally, in October 1981, they came to a new agreement. Saudi Arabia would raise its price from thirty-two to thirty-four dollars a barrel, while the others agreed to bring their price down from thirty-six to thirty-four dollars. So prices would be reunified. When all the changes were factored in, the average price of oil on the world market would still, because of the Saudi increase, go up a dollar or two. For the other producers, the compromise did, of course, mean a price cut. Yet there were consolations. Saudi Arabia had agreed, at last, as part of the deal to go down to its old 8.5-million-barrel-per-day ceiling.

Iraq and Iran remained locked in bitter battle. Yet even a war between two of the most important exporters could only retard but not cancel out the powerful forces that had been set in motion by the two oil shocks. October 1981 represented the last time that the OPEC price would go up, at least for a decade. The "divine laws of supply and demand" were already in motion to drive prices down, though not yet with the thunderous vengeance that was still to come. It was, as Yamani had said, as simple as ABC.[7]

CHAPTER 35

Just Another Commodity?

NONE OF THE PREVIOUS booms, in an industry characterized by booms, could begin to rival the magnitude and madness of the fever that came at the end of the 1970s with the Second Oil Shock. It was the greatest boom of them all. With the leap in price to thirty-four dollars a barrel, sums of money were involved that dwarfed anything that had ever before been earned or spent in the business. Oil companies plowed their earnings back into new development. Some borrowed from banks, raised more money from eager investors, and leveraged themselves to the hilt so they could play in the wild game. It was the golden age of the independent oil men. They slapped backs, they wheeled and dealed, hired more drilling rigs and explored at greater depths, and they spent and spent. To celebrate it all, at the very end of the 1970s, the television show *Dallas* went on the air, introducing the rapacious J. R. Ewing—in place of the lovable Clampetts of *The Beverly Hillbillies*—to viewers in the United States and around the world, and fashioning for many of them the image of the independent American oil man for years to come.

In the United States, the industry surged to a dizzy and unprecedented level of activity. The frenetic pace meant, inevitably, that costs went out of control. The price of everything connected to oil shot up. Acreage—land on which to drill—skyrocketed. So did real estate in the oil cities—Houston, Dallas, and Denver. The cost of a drilling crew multiplied many times over. Young graduates in geology were wined and dined and courted, and paid fifty thousand dollars a year for their first job out of school. Geologists with twenty years of experience were quitting the majors to put together deals and keep a piece for themselves, dreaming that they might become a new H. L. Hunt or the next J. Paul Getty. These were the years that the doctors and dentists of America put their money into drilling funds. If they did not have oil in their portfolios, they were told, their savings could be devastated by inflation and rising oil prices.

The industry, it was thought, was standing at the perilous edge of what some called the "oil mountain." As from a cliff, supplies would begin precipitously and rapidly to fall away. And depletion, combined with OPEC militancy, would guarantee high and rising prices for an increasingly scarce commodity. As a result, among other things, technology and engineering would have to create alternatives to oil, which in turn would set a ceiling price for oil. What this meant was that at last, after seven decades, the shale oil locked in the rock formations of the Western Slope of the Rockies in Colorado and Utah would be liberated and brought to market, as had been promised each time world oil appeared to be in perilously short supply. It was exactly what President Carter had proposed in 1979 as the solution to the country's energy problems.

Some companies, like Occidental and Unocal, were already working on shale oil technology. In 1980, Exxon, the world's largest oil company, looking ahead to what seemed the inevitable shortage, hastily bought its way into the Colony Shale Oil Project on the Western Slope. Sixty years before, in another period of shortage, the company had acquired acreage in the same area to develop shale oil as a fuel. Nothing had come of it then. Now Exxon became, by far, the leader, spending fully a billion dollars on shale oil development, getting ready for the "new era" of energy. "Exxon had had a love affair with shale oil for a long time," recalled Clifton Garvin, the company's chairman. "It was a huge challenge, technically, and certainly economically." Nevertheless, the country seemed committed to developing secure sources of liquid fuels. And the technology seemed available.

But over the subsequent two years, the economic outlook changed quickly and drastically. In real terms, the oil price was going down; so was demand. So were forecasts for both. Surplus production capacity was building in the oil-exporting countries. And, all along, the cost estimates for the Colony Project kept going up. "We were looking at $6 or $8 billion for 50,000 barrels per day," recalled Garvin. "And there was no expectation that that was the end of it. One night I said to myself, 'I can't spend shareholders' money that way.' " The next day, Garvin assembled a senior management team and asked what would be the consequences of stopping. "It was a tough decision. I rode that decision."

On May 2, 1982, Exxon announced tersely that it was terminating the Colony Project. Nothing that the company now saw in the economic outlook could make the shale oil project viable.

The boom on Colorado's Western Slope ended literally in hours, as work came to an instant stop. The towns of Rifle, Battlement Mesa, and Parachute fell prey to the great tradition established at Pithole, in western Pennsylvania, which in just two years, 1865 and 1866, went from dense forest, to boom town of fifteen thousand people, to eerie ghost town, whose deserted shops and homes were raided for wood to build elsewhere in the Oil Regions. Now, in the three towns in Colorado, newly built homes were empty; weeds quickly covered landscaped lots; half the apartments went unrented; construction workers from the Midwest packed up and headed home; traffic evaporated from the roads; and teenagers with nothing else to do took to vandalizing the partly built homes and office buildings. "My business just died," said the owner of an office supply shop in Rifle. And so did the town. The boom to end all booms could not last.

The Fundamentals

What had happened to the world oil market and to oil prices themselves? Virulent inflation was threatening not only economic performance but the whole social fabric of the Western world. The United States Federal Reserve responded by instituting an exceedingly restrictive monetary policy that resulted in a sharp rise in interest rates, with prime, at one point, reaching the sky-high level of 21.5 percent. Tight money came on top of the drain of spending power from the industrial world because of the oil price increases. The combined consequence was the deepest recession since the Great Depression, with two bottoms, the first in 1980 and the second, and more severe, in 1982. The stalling of economic activity substantially reduced the demand for oil in the industrial nations. The developing world was supposed to be a major source of new oil demand, which would buoy prices. Instead, many developing countries—debt-laden, the markets for their raw materials hit by recession in the industrial world—went into steep economic decline, choking off their oil demand.

Moreover, fundamental changes were taking place in the energy economy itself. Earlier fears of shortage, at the beginning of the 1920s, in the mid-1940s, had ended in surplus and glut because rising prices had stimulated new technology and the development of new areas. The pattern was to be repeated with thirty-four-dollar-a-barrel oil and the expectation of still-higher prices. Huge new developments were taking place outside OPEC. The major buildup of production in Mexico, Alaska, and the North Sea coincided with the turmoil of the Second Oil Shock. Egypt was also becoming a significant exporter. So were Malaysia, Angola, and China. Many other countries became producers and exporters, minor league in themselves, but significant in the aggregate. Major innovations were also improving exploration, production, and transportation technologies. The initial capacity of the Alaskan pipeline was 1.7 million barrels per day. With the addition of a chemical nicknamed "slickem," which reduced drag within the pipeline and thus improved the ease of flow, the pipeline's capacity was boosted to 2.1 million barrels per day. Many things in exploration and production could be done at thirty-four dollars a barrel that were not economical at thirteen dollars a barrel, and output in the "Lower Forty-eight" of the United States continued at higher levels than anticipated. That, along with the stepped-up flow out of Alaska, meant that American oil production actually increased in the first half of the 1980s.

Significant changes were also taking place in demand. The massive twentieth-century march toward higher and higher dependence on oil within the total energy mix was reversed by higher prices, security considerations, and government policies. Coal staged a massive comeback in electricity generation and industry. Nuclear power also made a rapid entry into electricity generation. In Japan, liquefied natural gas increased its share in the energy economy and in electricity generation. All this meant, around the world, that oil was being ejected from some of its most important markets and was rapidly losing ground.

Its share of the market for total energy in the industrial countries declined from 53 percent in 1978 to 43 percent by 1985.

Not only was petroleum experiencing a declining share of the energy pie, but the pie itself was shrinking, reflecting the profound impact of increased energy efficiency, otherwise known as conservation. Though often dismissed or even ridiculed, conservation had turned out to have massive impact. Energy conservation in modern industrial society meant, for the most part, not deprivation, not "small is beautiful," but greater efficiency and technological innovation. The 1975 legislation that mandated a doubling of the average fuel efficiency of new automobile fleets to 27.5 miles per gallon by 1985 would reduce United States oil consumption by 2 million barrels per day from what it would otherwise have been—just about equivalent to the 2 million barrels per day of additional oil production provided by Alaska. Altogether, by 1985, the United States was 25 percent more energy efficient and 32 percent more oil efficient than it had been in 1973. If the United States had stayed at the 1973 levels of efficiency, it would have used the equivalent of 13 million barrels of oil more than it actually did in 1985. The savings were huge. Other countries made their own dramatic savings. Japan over the same period became 31 percent more energy efficient and 51 percent more oil efficient.

By 1983, the first year of economic recovery, the impact of conservation and fuel switching was clear. Oil consumption in the noncommunist world was 45.7 million barrels per day, about 6 million barrels less than the 51.6-million-barrel-per-day level of 1979, which had been the high point. So, while demand had fallen 6 million barrels per day between 1979 and 1983, non-OPEC production had increased by 4 million barrels per day. On top of that, the oil companies eagerly sought to dispose of the tremendous inventories they had built up in anticipation of a demand level that never materialized.

Those three trends—the collapse in demand, the relentless buildup of non-OPEC supply, and the Great Inventory Dump—reduced the call on OPEC by something like 13 million barrels per day, a fall of 43 percent from the levels of 1979! The Iranian Revolution and then the Iran-Iraq War had crippled the exporting capacity of those two countries. Yet suddenly, instead of the feared shortage, there was a large surplus of production capacity over market demand—in short, the makings of a massive glut.[1]

Finally—the Cartel

For OPEC, the day of reckoning was close at hand. As late as 1977, OPEC had produced two-thirds of total free world crude oil. In 1982, for the first time, non-OPEC overtook OPEC production, and indeed was a million barrels per day higher, and still rising. Even Soviet oil exports to the West were substantially increasing, as the USSR sought to take advantage of rising prices to augment its hard currency earnings from the West.

Much of the new oil, particularly that from the North Sea, was sold on spot markets, which made it very responsive to overall market conditions. Only a year or two earlier, spot prices had soared above official prices; now they dropped far below. Many companies paying official prices were losing large amounts of money on their refining and marketing. Spot prices for a particular quality oil could be as much as eight dollars a barrel lower than term contract prices. That gap, as the chief executive of Mobil's German affiliate put it, was the difference between having "a profit margin" and experiencing "formidable losses." In such circumstances, any buyer who could do the simplest arithmetic was going to "go spot" and shop around, seeking the cheapest barrel. The new non-OPEC producers who were trying to enter the market as sellers would have to offer the most "market-responsive," that is, the cheapest, prices in order to win market share.

OPEC was in trouble. The market confronted it with an unpalatable choice: cut prices to regain markets, or cut production to maintain price. But the OPEC countries did not want to reduce prices, for fear that they would undermine their whole pricing structure, lose their great economic and political gains, and so diminish their newly acquired power and influence. Moreover, if they did reduce their prices, they feared, industrial countries would take it as an opportunity to raise excise and gasoline taxes, and transfer the oil rents from OPEC's treasuries back to their own, which was where their battle over rents had begun more than three decades earlier.

But reality had to be faced. If OPEC was not going to cut price in order to defend its production level, then it would have to cut production levels in order to defend price. In March 1982 OPEC, which had produced 31 million barrels per day in 1979, just three years earlier, set an output limit for the group of 18 million barrels per day, with individual quotas for each country, except for Saudi Arabia, which would adjust its production to support the system. OPEC had finally done what it had talked about doing at various times in its history. It had taken on the part played in earlier years by the Texas Railroad Commission in managing production to try to preserve price. In the blunt words of one of the leading analysts from the oil-exporting world, it had turned itself into a cartel, managing and allocating production, as well as setting a price.

In the months after the establishment of the quota, new factors added to the uncertainties of the oil market. Iran was gaining the upper hand in the war with Iraq and becoming more belligerent in its attitude and rhetoric toward Saudi Arabia and the other conservative Gulf countries. That was not the only war in the Middle East. In June 1982 Israel intervened directly in Lebanon. At one meeting of the Organization of Arab Petroleum Exporting Countries, there was some discussion of instituting another embargo against the United States as "punishment." But the distressed condition of the oil market, combined with the immediate geopolitical risks for the Gulf exporters from Iran, was such as to make it an incredible proposal, and one that was quickly scotched as irrelevant, dangerous, and likely to be highly damaging to the interests of the exporters. Meanwhile, in June 1982, King Khalid of Saudi Arabia, an interim figure who had suffered from chronic heart disease, died. He was succeeded by Prince Fahd, who had already been the country's chief administrator and who, among other things, was the Royal Family's oil expert.

The new quotas were meant to be a temporary expedient. But by the autumn of 1982, several things were clear: Demand was not recovering, non-OPEC production was still on the rise, and spot prices were plunging again. Even with production quotas, OPEC oil was still overabundant and overpriced.[2]

"Our Price Is Too High . . ."

In 1983, competition continued to mount rapidly in the oil market. The British sector of the North Sea alone, which had not even started producing until 1975, was now producing more than Algeria, Libya, and Nigeria combined, and still more North Sea oil would be coming on stream. To counter the competition, unofficial discounting and price cutting became the norm among the OPEC countries. Again, the one exception was Saudi Arabia; it maintained the thirty-four-dollar marker that many others were observing only in the breach. Buyers were soon forsaking Saudi Arabia in favor of discounted oil, even including the Aramco partners. They could not easily force more expensive oil on affiliates and customers that were trying to compete with other companies with access to cheaper oil. Saudi production fell to its lowest level since 1970.

Early in 1983, Yamani offered a philosophical disquisition on the origins of what was now, very evidently, an OPEC crisis. "Please excuse the comparison," he said, "but the history of the crisis is similar to that of a pregnant wife. . . . The crisis started just like a normal pregnancy—with passion and joy. At this moment other members wanted us to raise the price of oil even higher despite our warnings of the negative consequences. Moreover, everyone was getting massive financial revenues and rushing into development projects as if this financial revenue would continue to rise forever. . . . We were consumed with our moments of pleasure." But now the consequences had to be faced. "Our price is too high in relation to the world market," Yamani said.

By the end of February 1983 a complete collapse seemed at hand. The British National Oil Company cut the price on North Sea oil by three dollars, to thirty dollars a barrel. That was devastating to Nigeria, a member of OPEC and a country of 100 million people whose economy had become dangerously overdependent on oil. Nigerian oil competed directly in quality with North Sea crudes, and Nigeria's normal buyers, now able to get cheaper oil from the North Sea, deserted the African country. Almost devoid of customers, Nigeria virtually stopped exporting oil. The internal politics of the country, recently returned to civilian rule, shuddered. Nigeria made clear that it would reply in kind. "We are ready for a price war," Yahaya Dikko, the Nigerian oil minister, firmly said.

In early March 1983 the oil ministers and their retinues hurriedly convened, ironically in London, the home court of their leading non-OPEC competitor, Great Britain. They met at the Intercontinental Hotel at Hyde Park Corner, for what turned out to be twelve interminable, frustrating days—an experience that would leave some of them with an allergic reaction whenever, in future years, they set foot inside the hotel. But whatever the ideological and symbolic opposition to a price cut, whatever the ire and frustration, the reality could no longer be resisted. OPEC slashed its prices by about 15 percent—from thirty-four to twenty-nine dollars a barrel. It was the first time in OPEC's history that it had ever done such a thing. The exporters also agreed on a 17.5-million-barrel-per-day quota for the whole group.

But who was going to get what share of the quotas? Billions of dollars were at stake in the division of the quota. Country by country, they wrangled their way through to the allocations. The twelve-day oil marathon in London had averted a price collapse, at least for the time being. OPEC had reset its price to meet the market price, not a rising market price as in the past, but a falling one. It had also established new quotas, without the tentativeness of the preceding year.

One country, to be sure, had no official quota at all. That was Saudi Arabia. If it had been given a quota, Yamani insisted, the number would have been well below the six million barrels per day that he had been instructed was the minimum that was acceptable to Riyadh. Instead, in the words of the communiqué, Saudi Arabia would "act as the swing producer to supply the balancing quantities to meet market requirements." For the first time, Saudi Arabia, with one-third of total Free World reserves, was explicitly charged with the responsibility of raising and lowering its output to balance the market and maintain the price. But OPEC's new system of price administration depended on the eschewing of cheating on the part of twelve members and on the willingness and ability of the thirteenth, Saudi Arabia, to play that pivotal role of swing producer.[3]

The Commodity Market

Behind the evident drama of OPEC's marathon session and its mutation into a true cartel was a far-reaching transformation of the oil industry itself. No longer would it be dominated by large, highly integrated oil companies. Instead, it would turn into a free-for-all, a clanging world of multitudinous buyers and sellers. As was said, sometimes with approval and sometimes with horror, oil was becoming "just another commodity."

Oil, of course, had always been a commodity, from its earliest commercial days in the 1860s and 1870s, when prices fluctuated wildly in western Pennsylvania. But one result of the constant thrust toward integration was to internalize the volatilities of price within the workings of a company tied together from the wellhead to the gasoline pump. Moreover, oil was seen as different from other commodities. "It should be remembered that oil is not an ordinary commodity like tea or coffee," intoned Yamani. "Oil is a strategic commodity. Oil is too important a commodity to be left to the vagaries of the spot or the futures markets, or any other type of speculative endeavor." But that was exactly what began to occur. One reason was the buildup of a huge surplus in the world market. In a complete reversal of the 1970s, producers now had to worry about their access to markets, rather than consumers about their access to supplies. Buyers expected discounts; they would never think of paying security premiums of the sort they had been forced to pay in the late 1970s and early 1980s—premiums that, as one oil man would remark, were "often light on security and heavy on premia!" Security was hardly an issue anymore. What mattered was to be competitive in a glutted market.

The second reason was the changing structure of the industry itself. On grounds of nationalism and in search of rents, the governments of the exporting countries had assumed ownership over the oil resources in their countries and, increasingly, over the international marketing of their oil as well. By so doing, they had broken the links that had tied their reserves to particular companies, refineries, and markets overseas. Shorn of direct access to supplies in many parts of the world, the companies sought to develop new resources elsewhere. But also, it was clear, they would have to find a new identity or they would perish, because they would become obsolete. If they could not be integrated companies anymore, they would become buyers and traders. Thus, their focus shifted from long-term contracts to the spot market. Until the end of the 1970s, no more than 10 percent of internationally traded oil was to be found in spot markets, which were little more than a fringe activity of the main businesses, a way of sopping up excess output of refineries. By the end of 1982, after the upheaval of the Second Oil Shock, more than half of internationally traded crude oil was either in the spot market or sold at prices that were keyed to the spot market.

By no choice of its own, BP led the way. As a result of the upheaval in Iran and nationalization in Nigeria, it lost 40 percent of its supply—on top of nationalizations in Kuwait, Iraq, and Libya. Desperately exposed and acting in its own defense, it went out into spot markets and began buying oil and trading it on a larger and larger scale. With the emergence of the short-term spot markets, the virtues of "old-style" integration were no longer so evident. The new BP could shop around for the cheapest crude; it could push efficiency throughout its operating units; it could beat the competition; it could be more entrepreneurial. The company became much more decentralized, with individual units responsible for their own profitability. The corporate culture changed from that of the 1970s, dominated by the supply planner, to one dominated by traders and commercial people. The company, once seen as a kind of quasigovernmental bureaucracy, adapted what one executive called a "nimble trading-oriented approach." But what about the historic virtues of integration? "It is nice to have some integration, obviously, but it is not something we would pay a premium for," BP's new chairman, P. I. Walters, said at one point. "We see ourselves as being much more opportunistic."

Walters himself led the charge. He had long before concluded that the traditional integration, increasingly governed by computer models, did not make sense. The revelation had come to him when he was mowing the lawn at his home in Highgate in North London one Saturday morning in June 1967, a couple of days after the outbreak of the Six-Day War. He was called into the house to take an urgent call from BP's head of chartering, who told him that tanker tycoon Aristotle Onassis had abruptly canceled all existing chartering arrangements and was offering BP his entire fleet, but at rates double what they had been the day before. BP had until noon to give an answer, and it was up to Walters, who had just been put in charge of BP's worldwide logistics, to give it. Tens of millions of dollars were riding on the decision. With a sudden sinking feeling, he realized that no computer program could help him now, only commercial judgment. He phoned back. Yes, take the offer, he said, and returned to mowing his lawn. Events quickly proved the decision right; by Monday, tanker rates were four times what they had been on the previous Friday.

From that day on, Walters became a campaigner for deintegrating BP's operations. "That set me to thinking about the whole way we did business," he said. "I realized that the proponents of further integration were moving in the wrong direction. They were handing over to machines what should be management judgments." At one point, it seemed that Walters's evangelism might cost him his job, but he held on, and by 1981 had become chairman of BP, at a time when its entire business was in disruption. "So many firm hypotheses about the business were laid to ruin," said Walters. The Iranians had partly deintegrated BP; he would finish the job. "For me, there is no strategy that is divorced from profitability," he explained. Walters became famous for telling managers that "there are no sacred cows in BP" and "you tell us which things make economic sense and which do not, and I'll tell you which we'll keep and which we won't." Necessity had indeed become a virtue.

The other companies were being pushed in the same direction by the same forces. In virtually every company, the result was a struggle between those accustomed to and conditioned by the integrated oil industry of the 1950s and 1960s and those who thought that a new trading world had arrived. Not only established modes of operation, but fundamental, deeply held beliefs were being challenged. "The concept I was taught was that you moved your own crude through your own refining and downstream system," said George Keller, the chairman of Chevron. "It was so obvious that it was a truism." The move to the commodity style of trading would be resisted in many companies by traditionalists who saw this direction as an uncouth, immoral, and inappropriate way to conduct the oil business—almost against the laws of nature. They took a lot of persuading, but in due course they were persuaded. What it came down to in most of the companies was the establishment of trading as a separate profit center, a way to make money on its own terms, and not merely as a method for assuring that supply and demand were balanced within the parent company's own operations. If there had been no loyalty on the part of exporters toward companies when supplies were tight, then, in time of surplus, there would be no loyalty on the part of companies toward the exporters. Buyers would shop for the cheapest barrel anywhere in the world, whether to use it themselves or to turn around and trade it again—all in order to be as competitive as possible.

The four Aramco partners—Exxon, Mobil, Texaco, and Chevron—though somewhat cutting back, had continued to take large volumes of oil from Saudi Arabia even as they found themselves buying oil at "official prices" and thus at costs much higher than competitive crudes. The fundamental precept had always been to preserve access to Saudi oil, and the companies resisted sundering those links. But in 1983 and 1984, they had to acknowledge reluctantly that the price for access was too high. "Those of us in Chevron always viewed Aramco as our operation," said George Keller. "It's something we started, developed, and in which we played a key role. So it was more of a problem. But we couldn't keep pouring money down a hole. We had to back away, and ultimately we had to tell Yamani that we couldn't continue to do this." Though the Aramco links were not broken, they were significantly reduced. Saudi Arabia was no longer the special provider. The alteration in the commercial relationship between the four companies and Saudi Arabia was one of the great symbols of how the oil industry was being transformed.

The shift toward the commodity market was facilitated by a major structural change in the industry. With the decontrol of oil prices and the elimination of various other controls, the United States was no longer insulated from the world oil market. Indeed, it now became tightly tied into the rest of the market. The United States was not only the largest single consuming country; as a result of the fall in production worldwide, American output accounted for almost a quarter of total free world production. It was also a very market-oriented production, which could make its influence felt on the rest of the world. And one particular stream of American crude would even become the new bellwether for the global industry.[4]

From Eggs to Oil

The emergence of this crude stream, West Texas Intermediate, reflected yet another momentous innovation in the operations of the oil industry. It also took place in the turning-point year of 1983, though not in Vienna or Riyadh or Houston, but in Lower Manhattan on the eighth floor of the World Trade Center where the New York Mercantile Exchange, known as the Nymex, introduced a futures contract in crude oil.

When a commodity is largely sold in spot markets, with prices that are very volatile and uncertain, buyers and sellers tend to try to find a mechanism to minimize their risk. That is what gave rise to futures markets, which allow a buyer to acquire the right to buy the commodity at some month in the future at a specific, known price. He is able to lock in his purchase price; he knows his risk. Similarly, a producer can sell his production forward, even before it is produced or, in the case of agricultural products, harvested. He, too, locks in his price. Both buyer and seller are hedgers. Their objective is to minimize their risk and reduce their exposure to volatility. "Liquidity" is provided by speculators, who hope to make a profit by getting themselves on the right side of swings in supply and demand—and market psychology. A number of different commodities, such as grain and pork bellies, have been traded for years on each of the several futures exchanges in the United States. As the world economy became more volatile in the 1970s and regulations fell away, futures emerged for gold, interest rates, currencies, and, finally, oil.

As exchanges go, the New York Mercantile Exchange had not exactly enjoyed the most distinguished career. It had been founded in 1872, the same year that John D. Rockefeller launched "our plan" to take over the American oil industry and squeeze out the competition. The exchange had more modest ambitions, reflecting the interests of sixty-two merchants in New York City who were looking for a place to trade dairy products. Its original name was the Butter and Cheese Exchange. Eggs were soon added to the menu, and in 1880, it became the Butter, Cheese, and Egg Exchange. Two years later, it changed its name again, to the New York Mercantile Exchange. By the 1920s, egg futures had been introduced and were being traded, in addition to the eggs themselves.

Then in 1941, a new commodity entered the portals of the exchange—the Maine potato. Later, futures were added for yellow globe onions, apples (McIntosh and Golden Delicious), Idaho potatoes, plywood, and platinum. But the

Maine potato was the mainstay of the Mercantile Exchange until, unbeknownst to most of the world, the American supply-demand balance for potatoes began to change dramatically. Maine potatoes were losing market share to potatoes produced elsewhere in the country; moreover, the absolute volume of Maine potatoes produced each year was also dropping. As a result, the Maine potato futures contract was running into trouble. In 1976, and again in 1979, scandals hit the potato contract, including the mortifying failure of delivered stocks of potatoes to pass inspection in New York City. The exchange, under pressure, abruptly terminated trading in Maine potatoes and was itself threatened with extinction.

Just in time, however, the Nymex had introduced a new product, a home heating oil contract, which local heating oil distributors found useful. Then in 1981, it started trading futures in gasoline. But the major innovation came on March 30, 1983. On that day, the exchange introduced futures in crude oil, just two weeks after OPEC concluded its marathon meeting at the Intercontinental Hotel in London. The juxtaposition was ironic, for the crude oil futures contract would resolutely undermine OPEC's price-setting powers. And the rights to a single barrel of oil could now be bought and sold many times over, with the profits, sometimes immense, going to the traders and speculators.

Floor traders took enthusiastically to crude futures in New York. Pushing and elbowing themselves into the seething crowd on the floor of the Nymex, they shouted and furiously waved their arms to register their orders for contracts. The traders were also pushing and elbowing their way into the oil industry, which hardly took kindly to them. The initial reaction to the futures market on the part of the established oil companies was one of skepticism and outright hostility. What did these shouting, wildly gesticulating young people, for whom the long term was perhaps two hours, have to do with an industry in which the engineering and logistics were enormously complex, in which carefully cultivated relationships were supposed to be the basis of everything, and in which investment decisions were made today that would not begin to pay off until a decade hence? A senior executive of one of the majors dismissed oil futures "as a way for dentists to lose money." But the practice—of futures, not dentistry—moved quickly in terms of acceptability and respectability. Within a few years, most of the major oil companies and some of the exporting countries, as well as many other players, including large financial houses, were participating in crude futures on the Nymex. Price risk being what it was, none of them could afford to stay out. As the volume of transactions built up astronomically, Maine potatoes became a distant, quaint, and embarrassing memory on the fourth floor of the World Trade Center.

Once it had been Standard Oil that had set the price. Then it had been the Texas Railroad Commission system in the United States and the majors in the rest of the world. Then it was OPEC. Now price was being established, every day, instantaneously, on the open market, in the interaction of the floor traders on the Nymex with buyers and sellers glued to computer screens all over the world. It was like the late-nineteenth-century oil exchanges of western Pennsylvania, but reborn with modern technology. All players got the same information at the same moment, and all could act on it in the next. The "divine laws of supply and demand" still prevailed, but now they were revealed differently and far more widely, and with no delay. The benchmark price in all the transactions was that of WTI, West Texas Intermediate, a plentiful crude stream that could be easily traded, and was thus a good proxy for the world oil price, which had heretofore been embodied in Arab light. Two decades earlier, Arab light had supplanted Texas Gulf Coast oil as the world's marker crude. Now, in almost full circle, it was back to Texas. And with the rapid rise of oil futures, the price of WTI joined the gold price, interest rates, and the Dow Jones Industrial Average among the most vital and carefully monitored measures of the daily beat of the world economy.[5]

New Oil Wars: The Shootout at Value Gap

With the major restructuring of global markets, the oil industry itself also went through a wholesale corporate reorganization from which no major company was immune. Deregulation of an industry removes protection and increases competitive pressure and, thus, typically results in consolidations, spinoffs, takeovers, and a variety of other corporate changes. Oil, completely deregulated in the United States by 1981, was no exception. Overcapacity and weakening prices also encouraged consolidation and shrinking, which would mean greater efficiency—and greater profit. At the same time, institutional investors—the pension, mutual fund, and money managers who typically controlled three-quarters of the stock of major American corporations—were becoming more aggressive and were insisting on higher returns for their investments. Under pressure to show good quarterly performance, they were not willing to wait around for the long term. And, in their eyes, the oil industry was losing its luster in the aftermath of the boom.

At its heart, however, the restructuring of the oil industry was based on what was called the "value gap," the term used when the value of a company's shares did not fully reflect what its oil and gas reserves would fetch in the marketplace. Those companies with the greatest gap between stock price and asset value were the most vulnerable. In such cases, the obvious implication was that a new management might be able to increase the price of the stock and so enhance that noble cause, "shareholders' value," in a way that the old management had failed to do. There was a further twist: It could cost two or three times more to add a barrel of oil by exploration than by buying the assets of an existing operation. To the management of companies, the obvious implication was that it was cheaper to "explore for oil on the floor of the New York Stock Exchange"—that is, buy undervalued companies—than to explore under the topsoil of West Texas or in the seabed of the Gulf of Mexico. Here, again, shareholders' value was a driving force. Many companies had taken the huge cash flows that poured out of the two oil shocks and put them right back into exploration in the United States, seeking secure alternatives to OPEC. The results were very disappointing; reserves were still declining. The expenditure of so much money had proved to be inefficient and wasteful. Rather than continue spending at so helter-skelter a rate, why not give more of the money back to the shareholders through higher dividends or stock buy-backs, and let them decide how to invest it? Or, perhaps even better,

708

why not acquire or merge with other companies of known value and so get reserves on the cheap?

Thus, the value gap, like a geological fault, facilitated a great upheaval throughout the oil industry. The result was a series of great corporate battles, pitting company against company, with a variety of Wall Street warriors mixed in and sometimes in command. It was an entirely new kind of oil war.

The Trigger

Though the industry was ripe for change on the back end of the Second Oil Shock, a trigger would be required. It was to be found in Amarillo, a town of 150,000 on the high, flat, dry plateau of the Panhandle of northwestern Texas— an isolated, arid, wind-swirled region closer to Denver than to Houston. Oil and gas was the biggest business in Amarillo, but it was run mostly by small independents. Cattle was another major business in and around Amarillo. So was nuclear weaponry. Amarillo was the nation's only center for the final fabrication of nuclear bombs, producing by one estimate four warheads a day. It was also the home of an independent oil man with the name of T. Boone Pickens, who, more than anyone else, detonated the explosions that remade the corporate landscape of oil, obliterating in the process some of the best-known landmarks.

Boone Pickens became a celebrity of sorts, expert at turning aside reporters with a dry laugh when they solemnly asked if he were the "real life" J. R. Ewing of the *Dallas* television series. In the financial community, Pickens was widely applauded among investors; he made things happen, he enhanced shareholders' value. In the oil industry, however, while he was admired by some, he was loathed by others. Placing himself strategically at the junction of the oil industry and Wall Street, he said that he was pushing the oil industry back to basics, fighting its self-indulgent waste, saving it from its own excesses and illusions and arrogance, and serving the often-ignored interests of the heretofore-disenfranchised shareholders. His adversaries said that he was merely a clever opportunist, with a gift for salesmanship, who wrapped old-fashioned greed in the mantle of shareholders' rights. One thing was clear all around: Pickens saw the vulnerabilities and weaknesses of the oil industry on the backside of the Second Oil Shock earlier and with more clarity than most. And he not only figured out something to do about it, but even came up with an ideology to explain it. On one level, his campaign, for that was what it was, represented the revenge of the independent oil man on the hated majors.

Born in 1928, Pickens grew up in the oil patch, not far from Seminole, site of one of the greatest Oklahoma discoveries of the 1920s. His father was a landman, who acquired leases from farmers and packaged and sold them to oil companies. His mother was in charge of gasoline rationing for three counties during World War II. He was an only child, who turned into a brash, self-confident, independent-minded, sharp-tongued, and outspoken young man. He did not readily accept the established order but rather would make things happen his way. He was also intensely competitive. He hated to lose.

When his father's luck turned sour, the family moved to Amarillo, where the

709

senior Pickens took a job working for Phillips. Young Boone, after studying geology in college, also went to work for Phillips. He could not stand it. He did not like bureaucracy or hierarchy. And he surely did not like it when one of his bosses told him, "If you're ever going to make it big with this company, you've got to learn to keep your mouth shut." In 1954, after three and a half years, he quit Phillips to set himself up as an independent oil man, consulting, putting together deals to sell to moneyed folk around Amarillo. He traveled the Southwest, getting accustomed to the hot winds and the constant dirt that worked its way into his mouth and nose, and living the gritty, itinerant side of the American dream. He did his shaving in the restrooms of roadside gas stations that carried the names of the big oil companies for which he had already formed a considerable dislike. That was in the dog-eared days of the mid-1950s, during one of the cyclical downturns in the industry. Pickens was one of thousands driving around the oil states, using public phone booths as their offices, hustling, looking at deals, selling them, getting a crew together and a well drilled and, if lucky, hitting oil or gas, dreaming all the while of making it big, really big.

Pickens got farther than most. He was smart and shrewd, with an ability to analyze and think through a problem, step by step. In due course, he went to New York to raise money and later launched a successful operation in Canada. By 1964, he had rolled up his various drilling deals into a single company, Mesa Petroleum. After Mesa went public, he became fascinated by the gap between stock value and the value of the underlying oil and gas assets. Pickens fastened his eyes on Hugoton Production, a sleepy but considerably larger company that had extensive gas reserves in the Hugoton, in southwestern Kansas, then the nation's largest gas field. Its stock price was much lower than its gas reserves would have fetched if sold off. Shareholders could be won over by promising a more generous return, based upon raising the stock price and managing the company differently. Here was the simple concept that would have such vast impact a decade and a half later. In 1969, he completed a hostile takeover of Hugoton and merged the much-larger enterprise into Mesa, creating a significant independent oil company.

Caught up along with almost everybody else in the post-1973 oil fever, Pickens hired as many drilling rigs as he could in the United States, and he went abroad to look for oil, to the North Sea and to Australia. He was still an inveterate trader, and he was a veteran of futures long before almost anybody else in the oil industry had even heard of them. His early speciality had been cattle futures. At one point he even took Mesa into the cattle-feeding business, making the smallish oil company into, as a sideline, the second-largest cattle feeder in the nation. That venture ended poorly, and he turned around and took the company out of the feedlots. Yet even at the height of the oil wars of the mid-1980s, with billions of dollars at stake, Pickens would look out the window as his plane came down over the Texas range and start counting cattle, to see if the herds were big or not, to help him decide whether to go long or short on cattle futures. It was sport.

Pickens had been an intense basketball then racketball player, which meant speed, fast breaks, unexpected moves, quick reflexes, and constant improvisation. That was the way he did business, too. "We used to crowd into Boone's office every Saturday morning, some of us sitting on the floor," said one of his managers, recollecting the 1970s, "and Boone would ask us how we were going to make money the next week." Pickens was proud to be known as the only oil man in Amarillo who still worked Saturdays. His style—game-planned, attentive to detail, but also highly improvisational—would make him a tough rival for the big bureaucratic companies he took on. And he would not shirk from a fight. When his staff brought him word that a competitor or a natural gas pipeline had done something he did not like, Pickens would come back with his standard retort: "Tell them to go kiss a fat man's ass."

By the early 1980s, Pickens was seeing the weaknesses in the oil business. The United States was a declining producer, with increasingly poor prospects and a continually disappointing discovery record. Meanwhile, the stock prices of oil companies did not reflect the sale value of their proven oil and gas reserves. Here was a way for Mesa to make money. It was like Hugoton Production all over.

His initial target, in 1982, was Cities Service, the progeny of Henry Doherty, the oil and utilities tycoon who in the 1920s had first preached the virtues of conservation in oil and gas production to a hostile industry. Cities Service was the nineteenth largest oil company in America; the thirty-eighth largest industrial corporation in the Fortune 500. And it was three times the size of Mesa. But its stock was selling for only a third of the appraised value of its oil and gas reserves, which was not exactly a great service to shareholders. Mesa acquired a block of stock in the larger company. While Mesa was considering acquisition plans, Cities Service tendered for Mesa, which in turn counter-tendered. Gulf intervened with an offer for Cities almost double what its stock had been selling for before the brouhaha, but then backed out. Finally, Armand Hammer's Occidental acquired all of Cities' stock. Mesa made a $30 million profit on its shares. That was the first move.

By this point, restructuring and megamergers were already spreading throughout the oil industry. The starting point was actually 1979, when Shell acquired Belridge, a California heavy oil producer. In the early 1920s, Shell had made a run at Belridge, offering on the order of $8 million, but then pulled back. Now, in 1979, it paid just a tad more, a total of $3.6 billion, in what was the largest corporate acquisition up to that point. In 1981, Conoco escaped a takeover attempt from Canada's Dome Petroleum by falling into the arms of DuPont for $7.8 billion. Mobil took a run at Marathon, a former Standard Oil production company and part owner of the Yates field, one of the nation's great oil fields in the Permian Basin of Texas. Looking for an alternative to Mobil, Marathon sold itself for $5.9 billion to U.S. Steel, which for its part was seeking a way to diversify out of the disaster of the American steel industry. Mesa made a bid for General American, a large crude producer, but Phillips picked it up for $1.1 billion. Deprived of that quarry, Pickens bided his time. Another target would appear.[6]

The Mexican Weekend

All the while, the global oil boom was turning sour. Exploration in the United States fell away. The number of refinancings and bankruptcies among smaller companies jumped. The major companies started the first round of belt-tightening—cutbacks, hiring freezes, and early retirements. Investors, no longer worrying about inflation, began to forsake the oil patch in favor of the stock market; mutual funds and red-hot money managers were becoming a more interesting topic of dinner discussion than oil, drilling programs, and geologists.

As the downturn proceeded, it demonstrated how interdependent oil had become with the global financial system. And nowhere was that more clearly established than in Mexico, which by 1982 had run up a huge international debt, in excess of $84 billion, on the basis of its sudden emergence as a world oil power. That year, Jesús Silva Herzog became Mexico's Finance Minister. His father, who bore the same name, had been head of the national commission that in 1937 had found the oil companies operating in Mexico guilty of making enormous profits and that had provided the rationale for their nationalization by President Cárdenas. Thereafter, the father had headed part of Pemex, the national oil company, until he quit over a confrontation with the oil workers union about wages. The son had taken the path of Mexico's modern technocrats, including graduate education in economics in the United States (at Yale), then rose through the government bureaucracy until April 1982, when President López Portillo appointed him Finance Minister.

To his shock, Silva Herzog realized that the country was on the edge of a grave economic crisis. It was the result of the weakening oil price, high interest rates, a severely overvalued peso, unrestrained government spending, and the drying up of markets for Mexico's non-oil exports because of the recession in the United States. On top of everything else, there was immense capital flight. Silva Herzog recognized that Mexico was unable to service its enormous debt. It could not pay the interest, let alone repay any of the principal. But President López Portillo, who was being told by those around him that he was the most wonderful president in Mexico's history, did not want to listen. "It was," Silva Herzog later said, "a horrible experience."

Silva Herzog began making secret trips to Washington, D.C., leaving Mexico City on Thursday night in order to see Paul Volcker, Chairman of the U.S. Federal Reserve System, on Friday. He would then fly back to Mexico City by Friday night so that he could appear at social events and no one would guess that he had been out of town. He arranged a $900 million emergency loan from the Federal Reserve, but it was dissipated in a week because of the capital flight. On August 12, 1982, Silva Herzog came to the conclusion that improvisation would not work; there was no way that Mexico could pay the interest it owed. It could, of course, default. But that might cause the international financial system to collapse. The loan exposure to Mexico of the nine largest American money center banks was equivalent to 44 percent of their total capital. How many of America's and the world's banks would fall in the first wave, and how many more would be pulled down by them in the second? And how could Mexico function in the world economy?

On August 13, Silva Herzog again flew to Washington, D.C. Those few days would be remembered afterward as "the Mexican Weekend." At his first meeting with Treasury Secretary Donald Regan, Silva Herzog explained that he had run out of foreign exchange. "We have to organize something," he said, "or otherwise there will be very serious international consequences."

At the end of that discussion, Regan said, "You really have a problem."

"No, Mr. Secretary," replied Silva Herzog, "*we* have a problem."

The Mexicans and Americans started working on Friday afternoon and continued virtually nonstop until the early hours of Sunday morning. They put together a multi-billion-dollar package of loans and credits, as well as advance purchases of Mexican oil for the United States Strategic Petroleum Reserve. But then, at around 3:00 A.M. Sunday, it seemed that the negotiations were about to break down. Silva Herzog had discovered a $100 million service fee buried in the agreement and was told by one of the Americans, "Well, when someone is in very serious difficulties, and you lend to them, they have to pay a fee." Silva Herzog was furious. "This is not a business transaction," he snapped. "Sorry, I cannot accept this." He called López Portillo, who angrily said to terminate the discussions and return to Mexico City immediately.

Later in the day, Silva Herzog was glumly eating a hamburger at the Mexican embassy, preparing to leave, when a call came from the United States Treasury saying that the $100 million fee had been rescinded. The Americans could not risk a collapse. Who knew what the effects would be on Monday? And, with that, the Mexican Weekend concluded, with the first part of the emergency package now in place.

Silva Herzog flew back to find Mexico City in an uproar. He went on television for forty-five minutes with a blackboard but no written speech to explain what was happening, and then came up to New York City the following Friday to meet with the Federal Reserve and representatives of terrified banks to work out a restructuring of Mexico's debt. What had been devised was a debt moratorium. But nobody wanted to call it that; instead, they called it a "rollover." It was a polite way to say that, at least in part, Mexico had defaulted.

An exhausted Silva Herzog flew back yet again to Mexico. As soon as he landed, he headed for a small village in the mountains beyond Mexico City. "I needed to separate from all we had gone through. I thought of my father and the role he played in the oil expropriation. I was three years old at the time. In the years after, Father would often talk to me about it. It was one of his favorite themes. And now here I was in Mexico's worst crisis since the one of 1938, and it was also a crisis involving oil. We had just committed terrible mistakes on the basis of oil. But there had been this great mood of victory in Mexico. We had been in the largest boom in Mexican history. And for the first time in our history, in those years 1978 through 1981, we were being courted by the most important people in the world. We thought we were rich. We had oil."

World financial markets teetered on panic in August 1982, but the hasty improvisation of the Mexican Weekend and the days that followed managed to stabilize the global financial system. The Mexican debt drama brought home, however, the reality that the global oil boom was over, and the fact that "oil power" was less powerful than assumed. Oil could mean not only wealth but also weakness for a nation. Moreover, a transition was at hand. The world oil crisis was now giving way to the international debt crisis and many of the world-class international debtors would turn out to be oil nations, which had borrowed heavily on the premise that there would always be markets for their oil, and at a high price.

At the same time that Mexico was balancing on the brink of bankruptcy, a tiny bank with the grand name of Penn Square, located in a nondescript shopping center in Oklahoma City, was also on the edge of insolvency. It had been a go-go energy lender, whose standards of prudence were suggested by one habit favored by its senior energy lending officer—he liked to drink Amaretto and soda out of his Gucci loafers. Penn Square became the focus of intensive deliberation among the Federal Reserve and other regulatory agencies. Why was so much attention being given to a bank in a suburban shopping center, just at the time when Mexico was poised to go under? The reason was that Penn Square had generated a huge volume of oil and gas loans, many of them highly questionable, and then sold them, some two billion dollars worth, "upstream," to major money center banks, like Continental Illinois, Bank of America, and Chase Manhattan. The loan portfolio that Penn Square had retained was worthless, the bank was insolvent, and the regulators closed it down. But that was not the end of the story.

Nationally, the most aggressive major bank when it came to energy lending had been Continental Illinois, largest bank in the Midwest, and seventh largest in the nation. Overall, it was the fastest-growing lender in the country, it was winning awards for good management, and its chairman had been chosen "Banker of the Year." As an energy lender, Continental Illinois was, as a competitor put it, "eating our lunch." It was rapidly increasing its market share in oil and gas loans, as well as in other sectors. The *Wall Street Journal* tagged Continental Illinois as "the bank to beat."

When oil prices started to weaken, it became clear that Continental Illinois, with its huge portfolio of energy loans from Penn Square and other sources, was walking on nothing more solid than thin air. The result, in 1984, was the largest bank run in the history of the world. All around the globe, other banks and companies yanked their money out. Continental Illinois' credit was no good. The integrity of the entire interconnected banking system was now in jeopardy. The Federal government intervened, with a huge bail-out—$5.5 billion of new capital, $8 billion in emergency loans, and, of course, new management. Though the word was hardly ever used in the United States, Continental Illinois had, at least temporarily, been nationalized. The dangers of not responding on such a scale were, however, too frightening to risk.

With the collapse of Continental Illinois, energy lending instantaneously went out of fashion. Any banks still willing or able to lend to energy companies rewrote their guidelines so restrictively that getting an oil and gas loan was now not much easier than passing through the proverbial eye of the needle. And without capital, there was no fuel for exploration and development, let alone a boom.[7]

Dr. Drill

Another drama with lasting repercussions for the oil industry was being played out in remote waters off Alaska. Half of the undiscovered oil and gas in the United States was thought to be in Alaska itself or in adjacent waters, and eyes focused on one place—Mukluk, named for the Eskimo word for sealskin boot. This Mukluk was a vast underground structure, fourteen miles off the north coast of Alaska, where the Beaufort Sea gives way to the Arctic Ocean—some sixty-five miles to the northwest of the prolific North Slope reserves at Prudhoe Bay. Mukluk aroused enormous excitement throughout the oil industry. The many companies that joined to co-venture a wildcat well, led by the BP affiliate, Sohio, and Diamond Shamrock, were hoping for another elephant, another East Texas, another Prudhoe Bay, perhaps even a discovery in the class of one of the preeminent Saudi fields. It was billed as the most exciting prospect to come around for a generation. "It's what you dream about," said the president of Diamond Shamrock's exploration company. BP geologists said it was one of the lowest-risk wildcat wells the company had ever participated in—the odds were one in three instead of the normal one in eight. However, the effort to pierce the riches of Mukluk would prove expensive—over $2 billion. In the daunting physical environment, the companies had to build their own gravel island from which to drill into the frigid waters. That work could only be carried out in the brief summer, before the ocean froze. In the winter, the temperature could fall to eighty degrees below zero.

As the actual drilling proceeded through the summer and fall of 1983, the Mukluk wildcat captured the imagination of the oil industry and the financial community. Stock prices of the companies involved bounded up. If successful, Mukluk would change everything: the position of the companies, the perspective on United States prospects, the world oil balance, even the relation of the industrial world to the oil-exporting countries. But as the great nineteenth-century wildcatter John Galey had said, only Dr. Drill knows for sure. And in the first week of December 1983, Dr. Drill spoke, and word was flashed around the world. At eight thousand feet below the seabed, where the pay sands were supposed to begin, the drill had struck salt water. Mukluk was dry of oil.

There was clear evidence that oil had once been trapped in Mukluk. But either the structure was breached and the oil leaked to the surface—an oil spill of gigantic proportions, though with no environmental gauging of it—or perhaps regional tilting caused the oil to migrate, in one of the jokes of nature, into the Prudhoe Bay structure. "We drilled in the right place," said Richard Bray, the president of Sohio's production company. "We were simply 30 million years too late."

The wildcat at Mukluk was not only the most expensive dry hole in history, it was also the turning point for exploration in the United States. The dry hole seemed to announce that the United States was, after all, a poor prospect. Betting so heavily on exploration was too risky and too expensive. Managements would, in the future, be made to pay a penalty if they continued risking, and losing, money on such a scale. In the minds of many senior oil company executives, Mukluk sent a bracing message; they ought to shift from exploring for oil to acquiring proven reserves, in the form of either individual properties or entire companies. After Mukluk, they were much more in a buying mood.[8]

Family Matters

Not only economics and geology drove the restructuring of the oil industry. So did the hatreds, resentments, and feuds that fester inside families. A war among the heirs to the Keck family fortune resulted in Mobil's acquiring Superior Oil, the nation's largest independent, for $5.7 billion. But the most prominent family troubles were those that fastened on Getty Oil, the great and rich integrated company that J. Paul Getty had begun building in the 1930s and then turned into a world company in the 1950s with the discoveries in the Neutral Zone between Saudi Arabia and Kuwait. Getty, the firm believer in value, had died in 1976. Now, in the 1980s, Getty Oil was not replacing its reserves, and its stock was selling at a very low price relative to the value of the company's holdings in the ground. One of J. Paul Getty's sons, Gordon, was more interested in making music than in searching for oil—he had just composed a song cycle based on the poems of Emily Dickinson—but he wondered where all the value had gone. That put him at odds with the professional managers running Getty. They may have thought they controlled the levers of power, but Gordon Getty and his allies controlled the stock. J. Paul Getty had treated all his sons poorly, including Gordon, and the younger Getty had no great reason to be loyal to the memory, or the handiwork, of his father. When opportunity came knocking, he was ready to open the door.

As events turned out, however, there were two knocks, and Gordon Getty, unfortunately in terms of later complications, apparently answered both. The first was from Pennzoil, a large independent run by a tycoon named Hugh Liedtke, an early partner of George H. W. Bush in the oil industry and a friend of Boone Pickens. In some manner, Getty assented to the Pennzoil offer, though in exactly what manner would become the center of very considerable and critical dispute. The second knock came from Texaco, whose chairman appeared at the Pierre Hotel, once owned by the elder Getty, late one evening to make a counteroffer to young Getty, who definitely accepted that offer. So Texaco got Getty Oil for $10.2 billion. It also got sued by Pennzoil.[9]

The Death of a Major

Early on in the Texaco-Pennzoil-Getty saga, Boone Pickens made a cameo appearance, giving Gordon Getty a personal tutorial on how to assess value in the oil industry. More than once, Pickens also bought Texaco stock. But he had his sights fixed elsewhere. Mesa was suffering from a large problem, one that afflicted almost the entire oil industry. With boom turning into bust, Mesa was committed to spending $300 million on an exploration program. It had fifty-one drilling rigs working, including five very expensive ones in the Gulf of Mexico, supported by a veritable navy and army of workers, boats, and helicopters, all of which were devouring money at an incredible rate. "Boys," Pickens announced at a board meeting in Amarillo in July 1983, "this is it. We've got to figure out a way to make $300 million, and we've got to make it fast. We've lost too much money in the Gulf of Mexico. We can't drill our way out of this one. A field goal won't do it—we need a touchdown."

The place to make that kind of money, quickly, was in the major oil companies, whose stocks sold at only a fraction of the value of their assets. And Pickens' eyes fastened onto the quarry—Gulf Oil, one of the Seven Sisters. It had been built by the Mellon family on the basis of Guffey and Galey's discovery at Spindletop in 1901, and had grown into a major American institution and a global company. It had firmly planted the American flag in Kuwait. The Mellons had long since stepped aside from active management; the family had fragmented and had sold off much of its holdings. As Pickens saw it, Gulf was the most vulnerable of the major oil companies—its stock was selling at little more than a third of its appraised value.

Pickens had already had a close-in view of the Gulf management during the struggle over Cities Service, and he had decided that it was ineffective and indecisive, and that Gulf's heavy bureaucratic structure would make it slow in responding. The company had suffered from ten years of internal problems and deep splits at the top. Illegal political contributions in the United States, some tied to Watergate, and controversial foreign payments had led to an internal upheaval, including a purge of some of the most senior management and their replacement by managers for whom the appearance of probity was among their most important qualifications.

The chairman who took over in the second half of the 1970s was variously nicknamed "Mr. Clean" and "the Boy Scout." Gulf was certainly the only major oil company that placed a nun on its board of directors. "These problems had meant six years of no decision," recalled one Gulf executive. "That was during a crucial time in the oil industry, during the OPEC turmoil, and at a time when everything was up in the air in the Far East, and we were losing our ass in Europe."

Gulf's list of woes was long. In 1975, its concession in Kuwait, responsible for a substantial part of the company's earnings, had been nationalized. The company lost a costly antitrust case related to uranium marketing. Despite the vast sums it had spent since the middle 1970s searching for politically secure sources of oil in the United States and elsewhere, Gulf had little to show in terms of new reserves. Its domestic reserve base was falling away rapidly, declining by 40 percent between 1978 and 1982 alone. It had to distort its exploration spending to try to find hundreds of millions of dollars worth of natural gas in order to meet a disastrous contract it had signed years earlier. With the loss of Kuwait in 1975, Gulf was not well-positioned to be highly competitive; it also lost a great deal of its old raison d'être as a global company undertaking a massive international exploration and production effort. It had not yet found a new one.

The current management was just beginning to tackle the question of making the company leaner, more competitive, and more efficient. The new chairman, who had come in as the successor to "Mr. Clean," was Jimmy Lee. As much as Boone Pickens embodied the independent, so Jimmy Lee's career reflected the evolution of the major oil companies. He had been working in Gulf's refinery in Philadelphia in the late 1940s, when the first shipments of Kuwaiti oil arrived. Thereafter, in the great age of industry expansion, he made his career abroad. He built refineries and marketing systems in the Philippines and Korea; he headed the entire Far Eastern operation; he was Gulf's man in Kuwait at a time when the Middle Eastern producers battled among themselves and pressured the companies to increase their respective outputs. Eventually, he had run, from London, Gulf's entire Eastern Hemisphere operations, which meant that he was in charge of everything from winning the allegiance of European motorists to getting drilling rigs into Angola. But now he was back in Pittsburgh to rebuild the battered company. But he wouldn't have much time.

In August 1983, Mesa started to accumulate Gulf stock through numbered bank accounts scattered around the country, with transfer codes known only to one or two people. In October, Mesa formed a Gulf Investors Group, GIG, providing it with the partners and, thus, the necessary financial clout with which to move on the offensive. Later in the month, the Mesa group turned up the heat. Its objective, it said, was to push Gulf into transferring half of its U.S. oil and gas reserves into a royalty trust, which would be owned directly by the stockholders, giving them the cash flow and eliminating the double taxation on dividends.

Gulf initiated a counterattack. The target was its four hundred thousand shareholders, who would vote either with management or with Pickens. But Gulf had a big problem; its senior management was sharply divided on how to proceed, which hindered Gulf's response to Pickens and made it seem indecisive and ineffective, just as he had predicted. In contrast, Pickens himself was quick, flexible, continually improvising, and imaginative. He knew how to court the institutional shareholders who held the large blocks of Gulf stock. He knew how to play to the public. And he was much more effective with the press than the engineers who ran Gulf. He presented himself as the champion of shareholders, an authentic oil man, a populist, not a faceless bureaucrat from "the good ol' boys club" of "Big Oil."

"I'd never expected to have a proxy fight in my career," said Lee. "I'd never prepared for that." But Gulf did fight back, and hard. Lee and his colleagues wooed the institutional investors, and Gulf managed to squeak to victory in the proxy vote in December 1983, by a bare 52 percent to 48 percent. It was only a reprieve. Pickens continued to move up and down the court. He submitted a written proposal to the Gulf board to spin off oil and gas reserves directly to shareholders. The board turned him down flat. Pickens then went to see the junk bond king, Michael Milken, at Drexel Burnham in Beverly Hills to explore raising the additional money through such bonds to make an outright takeover bid.

Jimmy Lee knew that time was short. He had to get the stock price up. He looked at spinning off refining and marketing and the chemical operations into separate companies. There was one piece of good news; Gulf replaced 95 percent of its reserves in 1983. Still, the company was vulnerable. Very vulnerable. In late January 1984 Lee took a phone call from Robert O. Anderson, chairman of ARCO, who said he wanted to talk about things of "material interest." They met for dinner in Denver, in a private dining room at the Brown Palace Hotel, each accompanied by a single colleague. Anderson knew exactly what he wanted—Gulf's foreign production. He had no interest in its service stations or refineries. He believed that the future of the major oil companies lay in overseas reserves, and that their overall success or failure would depend on the extent to which they got into what he called "the international circuit." As he saw it, a company would have a very hard time developing a major position in international oil—unless it was one of the Seven Sisters and already there. Gulf would provide the shortcut that ARCO needed. "When Gulf lost Kuwait, they lost a great deal," Anderson later said, "but they still had the critical mass." At the dinner, Anderson said he was willing to pay $62 a share for Gulf, a big increase over the $41 it had been selling at a half a year earlier. Lee replied by suggesting that their two companies merge their United States oil operations, which would have given Gulf half of ARCO's immensely valuable North Slope reserves. It hardly took Anderson any time at all to say "no thank you."

Subsequently, Lee received a second phone call from Anderson. "I think I should tell you that I had dinner last night in Denver with Boone Pickens," said Anderson. "I told him that we were prepared to pay $62 a share for Gulf."

"Thanks for telling me this," said Lee, restraining his sarcasm. Anderson's objective in meeting with Pickens was to understand what "Pickens was trying to do and to ascertain to our satisfaction that he would not block any deal." But that certainly was not how Lee saw it. As soon as he hung up on Anderson's call, he summoned his crisis team. "Well," he said, "Bob Anderson has just cut our legs out from under us. For all practical purposes, we're in play."

Anderson's second call ended whatever hopes Lee had that Gulf could remain independent. "The game was up," he later said. There had been a long-observed aversion among major oil companies against making hostile offers for one another. But Anderson's proposition, following on Mobil's recent run at Marathon, made clear that the rule was no longer valid, and the majors had the huge financial resources to go after one another. And now there was a price on Gulf's head, the word would soon be out, and it was just a matter of time before someone bought the company. The only question was who. That being the case, Lee decided to go for the best price. He telephoned the CEOs of the other large companies. It was a most unpleasant task, but the changed circumstances created by Anderson left him no choice. He had the same message for each CEO: "We're vulnerable, and I have indications that someone's going to make a run at us. If you're interested in taking a look, start doing your financials."

Pickens played the next card, topping ARCO's $62 offer with his own $65 tender offer. "I knew that $65 was a low ball," said Lee. "If someone's going to take your company, you might as well get the most you can." He once again went back to the CEOs of the other majors. He was blunt this time. Gulf was for sale.

Among those he called was George Keller, the chairman of Chevron, who had already become interested in Gulf. Descended from the Standard Oil Trust's western operations, Chevron maintained its headquarters in San Francisco, far from the oil patch and an unlikely setting for a major oil company. The company did have an admirable record in taking risks and finding oil, including, of course, in Saudi Arabia in the 1930s. Keller had earlier denounced oil industry takeovers, at least when they were hostile. Companies, he had said, could better spend their money searching for new reserves. But like other top executives in the industry, Keller had been shaken by the enormity of the failure at Mukluk.

"After that," he said, "almost everybody decided to put more money into better bets."

On New Year's Eve of 1984, the chairman of Getty Oil had called Keller and asked him if Chevron would like to take a look at Getty, then in the midst of its own takeover struggle. As soon as he got back to San Francisco, Keller put an analytic team to work, determining how Getty stacked up against other companies: Superior, Unocal, Sun—and Gulf. Getty was soon gone, acquired by Texaco, but Chevron was still looking closely at Gulf. After Jimmy Lee's second phone call, Keller put the Chevron staff to work furiously on the problem, using both published work and material provided by Gulf, under a hastily signed confidentiality agreement. With hardly a week to figure out how much one of the world's largest companies was worth, Chevron frantically set about the business of trying to value Gulf. On February 29, Chevron had reached one valuation; on March 2, another; on March 3 at 4:00 P.M., still another. At the most pessimistic end, Gulf was worth $62 a share; at the most optimistic, $105—that is, anywhere from $10.2 to $17.3 billion. "It was a hell of a range," said Keller. Chevron's board initially accepted management's recommendation and authorized Keller to go up to $78 a share in making an offer, recognizing that the actual proposal might depend on the bidding ground rules. One board member suggested that the board impose no cap at all, but leave the price to Keller's discretion. "For God's sake, put a ceiling on it," begged Keller, unnerved at the thought of sole responsibility. "Every dollar higher on the stock means another $135 million."

On March 5, the Gulf board met at its Pittsburgh headquarters, an ornate structure built in the Depression. The building was virtually deserted; most of Gulfs operations were conducted in Houston, and the Chevron group was given its own floor. The Gulf board was certainly not going to surrender to Pickens's junk bond bid. But there were three other offers on the table. One was Chevron's. Some of the senior executives had come up with an alternative offer, a leveraged buyout by management using junk bonds, to be arranged by the firm of Kohlberg, Kravis, and Roberts. ARCO would also make an offer. So the Gulf board had three serious suitors to consider.

Before the meeting, Lee laid down the ground rules to the bidders: "You have only one chance—no second chances. Make your pitch all at one time." ARCO's president, William Kieschnick, went first, presenting a $72-a-share offer. Kohlberg, Kravis went next, making an offer to the board that was worth $87.50 a share. It was 56 percent in cash—$48.75. The rest—$38.75—was to be in newly issued securities.

Waiting his turn, Keller of Chevron had an offer letter with him, with just one blank spot—the price. He knew there were two great risks: that crude prices would go down and that interest rates would go up. But he didn't think it likely that both would occur at the same time. The Chevron board had insisted that the final offer be at his discretion. Keller wrestled with the problem, knowing so well that each dollar per share of his offer would add another $135 million. But he didn't want to lose Gulf; an opportunity like that would not easily reappear. He picked up his pen and filled in the blank—$80 a share. The offer added up to $13.2 billion, all cash. He presented the letter to the Gulf board and made the case for his offer to the best of his ability. In his four decades with Chevron he had never before been in such a position. The reception seemed chilly.

With no clear clue to the outcome, Keller returned to the Chevron floor, to await the Gulf board's decision. All that he knew for sure was that he had just made the largest cash offer in history. ARCO's Kieschnick waited as well. Robert O. Anderson had convened an ARCO board meeting in Dallas, and he and the other directors conducted the regular business but anxiously waited too, with a phone line open to Pittsburgh. Sometimes they talked to Kieschnick.

Altogether, the Gulf board was in session seven hours that day. It debated the three offers. ARCO's could be summarily dismissed; it was too low. The Kohlberg, Kravis offer could not. It was more money, in theory, but it was more risky, since half would be in the form of securities, and Gulfs financial advisers, Merrill Lynch and Salomon Brothers, could not figure out how much the "paper" in the KKR offer would really be worth. It had the great advantage that the current management would remain in place, but some of the outside directors worried that acceptance of the offer would, for that reason, appear self-serving. Moreover, KKR had not yet secured its financing. "If it did not get the financing together," said Lee, "Boone had a valid tender offer, and he'd be looking down our throats with more shares than he needed" to renew his takeover attempt.

The hours stretched on. Keller was still waiting, reflecting on the risks of his offer, when the phone rang. It was Jimmy Lee. He tried to sound nonchalant. "Hello, George," he said. He paused. "You just bought yourself an oil company." All that Keller could think was that he felt like someone who had, for the first time in his life, bid on a house, and discovered, to his shock, that he now owned it. It was a $13.2 billion house. The Gulf board had decided that the prudent course was to accept Chevron's all-cash offer. The shareholders would be better off. And that was the end of Gulf Oil. Spindletop, Guffey and Galey, the Mellons, Kuwait and Major Holmes—it was all over. It was history.

Anderson was philosophical about ARCO's losing. He had simply never believed that Chevron would go to $80. His absolute limit was $75. "We thought we'd be hanging in there together. But, at least if you lose a merger, you like to lose by a wide margin rather than by a narrow one. You hate to lose by a dollar a share."

As far as Pickens was concerned, it was a great victory for the stockholders; because of his efforts, an ineffective management had been prevented from continuing to waste money in vain pursuit of glory. During the months since he had launched his campaign, Gulf's stock price had gone from about $41 a share to $80 a share, and its total market capitalization had been raised from $6.8 billion to $13.2 billion, giving the Gulf stockholders a profit of $6.5 billion. "That was $6.5 billion that would have never been made if Mesa and GIG hadn't come on the scene," said Pickens. Shareholder rights had been vindicated. Whether Pickens was after a quick profit or actually hoped along the way to become CEO of a major international oil company, his Gulf Investors Group made a $760 million profit, of which about $500 million went to Mesa. After tax, that came out just about to the $300 million that Mesa had so desperately craved in the summer of 1983. As Pickens had said, Mesa needed the money, badly.

Jimmy Lee's first reaction was relief. It was over, and the board had come out with a unified vote, which greatly lessened the likelihood of shareholders' suits. He immediately set out to talk to employees around the country, to try to reassure them about the future. Over the next several days, exhaustion took over. So did sadness, and he sometimes broke into tears. "I never had any intention but that Gulf would be here for ever and ever," he said. "It had been my whole life, my whole career. To think that it wasn't going to be there any more really got to me."

Gulf was fully merged into Chevron, and George Keller never had any reason to regret the $80 offer he had written down at the last moment. Chevron had not overvalued the company. "It was a good buy," he said half a decade later. "We were able to acquire assets on a scale that would never be available otherwise." Then why was Gulf in trouble? "It had ignored its solid existing position," Keller said. "It had decided it had to have one great big elephant. It was as though, instead of risking its future in the town in which it lived, it decided to go to Las Vegas. It missed everywhere." That, of course, could have happened to any of the major oil companies in the feverish climate that followed the oil shocks of the 1970s. But Gulf paid the ultimate price.[10]

Shareholders' Value

Pickens was not yet through. In rapid fire, he made bids for both Phillips, in Bartlesville, Oklahoma, and Unocal, in Los Angeles. On Phillips, he was trailed by an aggressive Wall Street financier, Carl Icahn, who had already bagged Trans World Airlines. Both companies, however, successfully fought off the takeover attempts through the courts and by assuming a great deal of debt, which enabled them to buy back stock at a much higher price than had been the case before the attacks, thus increasing the payout to shareholders. In both cases, however, Mesa made significant profits. Yet the clamor of "shareholders' value" seemed to be losing its populist appeal. After Unocal emerged intact from the assault, Fred Hartley, the company's chairman, received a call from Armand Hammer of Occidental, who told him that he deserved a Nobel Prize for his valor. Another of the large integrated companies, ARCO itself, saw that it, too, might well be vulnerable to a Pickens, or to *the* Pickens, in the financial environment of the mid-1980s. "We were a sitting duck," said Robert O. Anderson, "unless we got our share values up more closely to the values of our company." Thus, ARCO carried out a sort of self-acquisition, leveraging to buy back its stock at a higher price and, at the same time, sharply consolidating its own activities and employment.

The restructuring of even the giants of the industry through mergers and acquisitions continued over the next several years. Royal Dutch/Shell paid $5.7 billion for the 31 percent of Shell Oil U.S.A. that it did not already own. To the senior executives of Royal Dutch/Shell in The Hague and in London, that looked like the best bet among all the investment opportunities available to them. BP had teamed up with Standard Oil of Ohio—John D. Rockefeller's original company, and the basis of the Standard Oil Trust—to assure itself downstream outlets in the United States for Alaskan oil. As part of the Alaska transaction, BP came to own 53 percent of Sohio, while Sohio became BP's American arm. But disillusioned with Sohio's management as a result of its dismal but very expensive exploration program, including the Mukluk fiasco, BP forked over $7.6 billion to Sohio's other shareholders so that it would own the company completely and be able to control directly the huge cash flow from Alaska.

One company had, at least until the beginning of the 1990s, stayed away from the high-visibility arena of mergers and acquisitions. This was Exxon, which had been badly stung by its poor acquisitions record; when *Fortune* celebrated the five worst acquisitions of the 1970s, two of them belonged to Exxon. The billion dollars spent and lost in two years on the Colorado shale oil project was also sobering. Exxon came to the conclusion that there was no way it could adequately spend all of its cash flow on exploration and acquisition or in new businesses. Moreover, the senior management of Exxon believed, as a political judgment and almost as an article of faith, that it could not take over other large oil companies. Exxon had, in the words of CEO Clifton Garvin, "a phobia about acquisition."

All this sharply curtailed the company's options. "We had a huge cash flow and not that many good investments to put it into," Garvin elaborated. Better to take the money it could not spend efficiently and return it to the shareholders, letting them do whatever they wanted with it. This Exxon did through a share buy-back, on which it spent $16 billion between 1983 and mid-1990, guaranteeing the stockholders a rising stock price and a good yield, and in the process ensuring that neither Boone Pickens nor anyone else could claim that shareholders in Exxon were getting the short end of the stick. The $16 billion was much more than Texaco paid for Getty, or even than Chevron paid for Gulf. Exxon did spend a good deal of money on acquisitions, perhaps a billion dollars or so a year, but it was interested in specific properties, not in entire companies, and it went about its work quietly, as far from the headlines as possible. It also cut the number of its employees by 40 percent. As a result, it was a smaller company, both in absolute terms and in relative terms, as measured in reserves and revenues, against its historic competitor and archrival, Royal Dutch/Shell. Marcus Samuel and Henri Deterding would have been proud.

Restructuring meant, overall, a smaller and more consolidated petroleum industry. Beginning geologists were no longer being hired at fifty thousand dollars a year; indeed, they were not being hired at all. Others, supposedly at the height of their careers, suddenly found themselves forced into early retirement. The biggest losers were the people whose jobs were wiped out. "I thought I was working for a great social institution," said one executive whose job disappeared in Chevron's acquisition of Gulf. "I didn't think I was giving 25 years of my life, with all the costs for my family, for some pieces of paper." The great beneficiaries of the industry's restructurings were the shareholders. All the activity—the major mergers and acquisitions, the recapitalizations, and the stock buy-backs—propelled well over $100 billion into the pockets of institutional and individual investors, pension funds, arbitrageurs, and the rest. The shareholders did, eventually, win out.

When management was a shareholder, it won, too. Gulf's chairman, Jimmy Lee, lost his job but made about $11 million on his stock options. But Boone

Pickens was not to be outdone. In 1985, Mesa's board in Amarillo gratefully voted Pickens an $18.6 million deferred bonus for the takeover maneuver with Gulf that had netted Mesa some $300 million. That year Pickens was the highest-paid corporate executive in America.[11]

The New Security

In May of 1985, the leaders of the seven major Western powers met for their annual economic summit, this one in Bonn. The themes were free market politics, deregulation, and privatization. Promising a "new morning" in America, Ronald Reagan had recently been reelected by an enormous margin. His Administration had seen the passing of the defeatism and pessimism that had been so characteristic of the 1970s and that had, to a considerable degree, been the direct and indirect effect of the oil crisis. Instead of the malaise of inflation and recession, the United States was now enjoying a booming economy and bull market. Margaret Thatcher was well embarked on her reconstruction of British society; commerce, hard work, and breakfast meetings had become positive values in Thatcherite England. Even François Mitterrand, the socialist President of France and the most extraordinary survivor in world politics, had jettisoned nationalization and classic French *étatisme* in favor of free markets. The Western world was in the third year of relatively vibrant economic growth. But this economic recovery was fundamentally different from previous periods of postwar growth; it was not being fueled by a rise in oil demand. The economies of the industrial nations had adapted quickly to high oil prices, and oil use was flat.

The only serious question about energy that the leaders had had to grapple with during the previous few years was a divisive battle in the early 1980s over plans by the Western Europeans to increase substantially their purchases of Soviet gas. The Europeans wanted to use the gas as part of their energy diversification strategies and to reduce their dependence on oil. They also hoped to stimulate employment in the engineering and steel industries. The Reagan Administration opposed the plan because they feared that the expanded imports would give the Soviets political leverage over Europe and they did not want to see the Russians gaining additional hard currency earnings, which would strengthen the Soviet economy and military machine. As the controversy mounted, Washington banned the export of American equipment for the project and then sought to prohibit the export of European equipment that contained American technology.

This debatable application of extraterritoriality set off an uproar. The result was the most severe European-American conflict since the October War and the embargo in 1973. Two different views of security were at stake: the European emphasis on jobs and domestic economic stability versus the American focus on the Soviet military threat. The American ban threatened employment in a number of European industries, and it was such a severe setback for the big British engineering firm John Brown that Margaret Thatcher herself called Reagan about it. "John Brown is going under, Ron," she said firmly. And, to drive home her point, she flew to Scotland to be present when John Brown began exporting some of the equipment for the gas deal in explicit defiance of the American ban.

After many angry statements and accusations, a compromise had finally been worked out: The Europeans would restrict their imports from the Soviet Union to 30 percent of total gas, and the development of Norway's huge Troll field would be promoted as an alternative gas source safely within the NATO alliance. With that, the gas pipeline controversy had been brought to an end, and thereafter questions of energy security could be put aside by the Western leaders.

The issues on the agenda at the 1985 Bonn economic summit thus revealed how the world had changed; they primarily concerned trade relations among the industrial countries—protectionism, the dollar, accommodating Japan's economic challenge. They were "West-West" issues. Oil and energy, the preeminent "North-South" issue, was not on the table at all. As in the 1960s, oil and energy were now available in abundance and, thus, they were not a constraint on economic growth. Supplies were safe again. Excess oil capacity around the world exceeded demand by 10 million barrels per day, equivalent to 20 percent of the free world's consumption. In addition, the United States, Germany, and Japan were putting significant amounts of oil into their strategic petroleum reserves. The "security margin" that had been absent during the 1970s was being restored.

Meanwhile, in the Middle East, Iran and Iraq were breaking all the supposed "taboos" in their continuing conflict; they were attacking not only each other's cities, but each other's refineries, oil fields, and tankers, as well as tankers bearing the flags of many other nations. The bombing of a tanker would earlier have been enough to send the price of oil shooting up. Now, however, if a ship were hit, the price of oil could just as easily go down as up on the spot and futures markets. In short, the Western leaders no longer needed to include energy among the limited number of major questions on which they, as heads of state, could focus at any given moment. Oil had often been the dominating, and most acrimonious, issue at previous summits. But now, in 1985, for the first time since those summits had been instituted a decade earlier, the leaders issued a communiqué in which there was nothing about oil and energy. Not a single word.

The omission itself was a powerful statement about the degree to which the world economy had adjusted and accommodated itself to the extraordinary economic and political upheavals of the 1970s that had been related to oil. Now oil did not seem to need any special tending; it was indeed just another commodity. Half of the equation that had contributed to ebullient economic growth in the 1960s—secure oil supplies—appeared to be coming back into place. The other half, however, was not there. Oil was still not cheap—not yet.[12]

CHAPTER 36

The Good Sweating: How Low Can It Go?

THE PRICE OF OIL was balanced precariously in the mid-1980s. So much was at stake that all eyes fastened on its every twitch and twinge. As the president of Esso Europe said in 1984, "Today, the price of oil is the chief variable in our equation and the greatest single source of uncertainty about the future."

Would the price begin to go up again, would it languish, or would it plummet? As the months marched on, "How low can it go?" became a refrain heard more and more around the world, not only in energy companies but also in financial institutions and the corridors of government, and everywhere for good reason. The answer would have a profound impact on the oil companies, of course. But, beyond that, it would determine the future vigor of "oil power" in world politics and would mightily affect global economic prospects and the shifting balance of world economic and political strength. High prices would favor the oil exporters, from Saudi Arabia to Libya to Mexico to the Soviet Union. The USSR depended on oil, along with natural gas, for most of the hard currency that it used to purchase the Western technology it desperately needed for economic modernization. Low prices would favor the oil-importing countries, including the two economic powerhouses, Germany and Japan. In between and uncertain was the United States. It had interests on both sides of the divide. It was the world's largest importer and consumer of oil, and yet it was also the world's second-largest producer of oil, and a good part of its financial system had cast its fate with high oil prices. If push came to shove, on which side would the United States line up?

Despite its becoming even more restrictive in 1984, OPEC's new quota system was not working. Non-OPEC production continued to grow; coal, nuclear power, and natural gas were still taking markets away from oil; and conservation was still shrinking demand. Inevitably, as the various OPEC exporters watched their revenues dwindle, quota cheating among them became more obvious. If they could not attain their revenue ambition on price, then they would discount and try to make it on volume. In an act of self-exasperation, OPEC retained an international accounting firm to police the quota. The accountants were promised access to every invoice, every account, every bill of lading. They did not get such access; in fact, they had great difficulty even in gaining entrée to some OPEC countries and were completely denied entrance to key facilities. Meanwhile, various of the exporters, to get around their quotas and the languishing oil trade, turned to barter and countertrade—exchanging oil directly for weapons, planes, and industrial goods—which had the effect of increasing the oversupply of petroleum on the world market.

High or Low?

The force of the market could not easily be resisted. In establishing the state-owned British National Oil Company in the 1970s, Britain's Labour government not only made it the repository for the government's share of oil and gas reserves in the North Sea, but also gave it a specific trading function; it would buy up to 1.3 million barrels per day of output from North Sea producers and then turn around and sell it to refiners. Thus, BNOC had an important price-setting role in the world oil market, for it announced the prices at which it would buy and sell oil. But, with the weakening oil price, BNOC found itself in the uncomfortable position of buying over a million barrels per day of oil at one price from North Sea operators and selling those same barrels at a lower price! The result was a significant loss for BNOC, and for the British Treasury. As one of the mandarins from Whitehall soulfully explained: "It is obviously very painful to the Treasury to have a body in the public sector buying oil at $28.65 and selling at a lower price; it gives us very, very great pain, be assured of that!" No one was more critical than Margaret Thatcher herself. On principle, she did not like state companies—if anything, she was even more attached to "free markets" and against government intervention than Ronald Reagan; and privatization of state enterprises was a basic plank in her political platform. She saw no security reason to maintain BNOC, and in the spring of 1985 she simply abolished it. With that action, the British government exited from direct participation in the oil business. The elimination of BNOC removed one more important prop from the OPEC price; it was another victory for the market.

In the oil industry, the general view was that, while the price might sink a few dollars, it would recover and start to go up again toward the end of the 1980s or the beginning of the 1990s. Yet weak demand, plus growing supply capacity, plus the shift to the commodity market all pointed much more strongly in one direction, and that was down. But how far?

OPEC's Deepening Dilemma

By the middle 1980s, OPEC was faced with a critical choice. It could cut price; but where would a falling price stop? Or it could continue to prop up the price. But if it did that, it would be holding up an umbrella under which non-OPEC oil, competing energy sources, and conservation would all flourish, guaranteeing itself a shrinking market share. To make matters worse, more oil would be coming out of the OPEC countries themselves. Even as the Iran-Iraq War dragged on, exports from both belligerents were recovering. Nigeria, too, upped its output and, hungry for revenues, adopted for a time a "Nigeria first" policy, aimed at maximizing exports.

As was often the case, much depended on the Saudis. In 1983 Saudi Arabia had explicitly taken on the job of swing producer, varying its output to support the OPEC price. But by 1985, the costs, relative to those of other producers, were becoming disproportionately high. Defending price meant a huge drop in output and a vast loss of market share and a huge fall in revenues. The high point of earnings for Saudi Arabia was $119 billion in 1981. By 1984, its revenues had fallen to $36 billion, and they would fall further, to $26 billion, in 1985. Meanwhile, like the other exporters, Saudi Arabia had embarked on a great spending and development program, which now had to be cut back dramatically. The country started to run a large budget deficit, and foreign reserves were being drawn down. So unsettling was the situation that the promulgation of the national budget was indefinitely postponed.

The loss of market had another consequence; it was reducing Saudi Arabia to a more marginal role on the world stage. The rapid fall-away in political influence and significance, and the likelihood of further erosion, ran counter to the fundamental precepts of the kingdom's security policy, at a time when the Iran-Iraq War was threatening the region and the Ayatollah Khomeini continued to pursue his vendetta against Saudi Arabia. The dramatic market loss also reduced Saudi influence on Middle Eastern politics and the Arab-Israeli dispute, and on the Western industrial countries. Oil power was losing its meaning. "In principle, we must draw a line between economics and politics," Yamani said on Saudi television. "In other words political decisions should not affect economic facts and laws. But crude oil is a political power and no one can deny that Arab political power in 1973 was based on oil and that its influence reached its peak in the Western world in 1979 because of oil. At present we are suffering because of the weakness of Arab political power based on oil. These are elementary facts which are known even to the man in the street."

The Saudis sent out warning after warning to the other OPEC countries and to the non-OPEC producers. It would not continue to accept the loss of market share; it would not indefinitely tolerate and underwrite quota violations by other OPEC countries and increased production by the non-OPEC nations; it could not be counted on to be the swing producer. If need be, Saudi Arabia would flood the market. Did these warnings add up to a serious threat, a clear indication of intention? Or were they only a bluff, intended to frighten? Yet, if the Saudis did not make some change, then, logically, they might expect their production to fall to a million barrels per day or less, as export markets disappeared altogether. Under such circumstances and insofar as oil in a fundamental sense defined the country's identity and influence, Saudi Arabia would on the world stage almost cease being Saudi Arabia.[1]

Market Share

In the first days of June 1985, OPEC ministers congregated at Taif in Saudi Arabia. Yamani read them a letter from King Fahd, who sharply criticized the cheating and discounting by other OPEC countries that had led "to a loss of markets for Saudi Arabia." Saudi Arabia would not abide such a situation forever. "If Member countries feel they have a free hand to act," said the King, "then all should enjoy this situation and Saudi Arabia would certainly secure its own interests."

At the conclusion of the King's message, the Nigerian oil minister said that he hoped "this wise message had sunk in." But in the following weeks there was no particular evidence that it had. And Saudi oil production sagged to as low as 2.2 million barrels per day, half of its quota level and little more than a fifth of what that nation had been producing half a decade earlier. Exports to the United States, which had been as high as 1.4 million barrels per day in 1979, slumped to a mere 26,000 barrels per day in June 1985, which was essentially nothing.

At times, in the summer of 1985, Saudi oil output fell below that of the British sector of the North Sea. Here was the last outrage. To the Saudis, it meant that they were propping up the price so that the British could produce more, while Prime Minister Thatcher continued to celebrate her attachment to free markets and to advertise her indifference to whether oil prices were high or low. An even greater threat was closer to home. The Iraqis were reconstructing their export capacity, expanding and adding new pipelines, some of them passing through Saudi Arabia. Whatever else happened, large additional volumes of Iraqi oil would soon be pushing their way into the already congested market. The situation was untenable. Something had to give, and again, as in the 1970s, it would be price, but in the opposite direction. Still, how low would it go?

A shade from the past was rising again—John D. Rockefeller and the prospect of an all-out price war. In the late nineteenth and early twentieth centuries, Rockefeller and his colleagues had often instituted a "good sweating" against their competitors by flooding the market and cutting the price. Competitors were forced to make a truce according to the rules of Standard Oil, or, lacking the staying power of Standard Oil, they would be driven out of business or taken over. Circumstances were, of course, wholly different in the mid-1980s; yet they were not so different, after all. Once again, a "good sweating" was at hand.

The Saudis moved from a defense of price to a defense of volume—their own desired level of output—and chose an ingenious weapon: netback deals with the Aramco partners and with other oil companies strategically located in key markets. Under such deals, Saudi Arabia would not charge a fixed price to the refiner. Rather, it would be paid on the basis of what the refined products earned in the marketplace. The refiner, however, would be guaranteed a predetermined profit off the top—say $2 a barrel. No matter whether the final selling price of the products was $29 or $19 or $9, he would get his $2 and the Saudis would get the rest (minus various costs). The refiner's profit would be locked in. So he would not have any particular compulsion to strive for a higher rather than a lower price at the point of sale; he would simply want to move as much product as he could. He knew that each additional barrel meant an additional $2 of profit, no matter what the price. But increasing volume and reduced concern about the selling price would add up to a perfect recipe for falling prices. The Saudis, for their part, hoped that what they lost because of lower prices they would make up, and then some, with the higher volumes. But the Saudis were also careful not to be too confrontational; their aim was to regain their quota level, and no more, and they put a cap on the volume that would be covered by their new netback deals. In so doing, they were aiming their new policy as much at other OPEC members, which had been cheating and thus taking away their market share, as at the non-OPEC countries.

In the summer of 1985, a senior executive of one of the Aramco partners received a phone call from Yamani. The oil minister observed that the executive had earlier said that he would be interested in increasing purchases from Saudi Arabia if the pricing was competitive, and, explained Yamani, that was now the case. The executive flew to London in August to discuss the netback terms. "It sounds like it's competitive," the executive said, and he signed forthwith. A number of other companies, Aramco and non-Aramco alike, agreed to similar contracts.

The netback contracts obviously meant that there was no longer an official Saudi price. The price would be whatever oil fetched in the market. And that meant there would no longer be an OPEC oil price. As word circulated around the global marketplace in late September and early October of 1985 about the Saudi netback deals, nervousness and anxiety rose. Yet once the Saudis had committed themselves to a market share strategy, other exporters started to follow suit out of pure competitive self-defense. Netback deals began to proliferate. For the long-suffering downstream part of the oil industry, it was a godsend, an opportunity at last to make money in refining, which had seemed almost beyond human ingenuity since the early 1970s.

Was the price now posed for a great fall? Most of the exporters thought so, but they expected no more than a drop to $18 or $20 a barrel, below which, they thought, production in the North Sea would not be economical. On that, they were mistaken. The tax rates on the North Sea were so high that, for instance, in one field, called Ninian, a drop in the oil price from $20 to $10 would cost the companies only 85 cents. The big loser would be the British Treasury, which was taking most of the rents. Actual operating costs in Ninian—the cash costs to extract oil—were only $6 per barrel, so there would be no reason to shut-in production at any price above that. Furthermore, it was so costly and complicated to close down operations temporarily that there would be reluctance to do so even if the price dropped below $6. As George Keller, chairman of Chevron, said at the time, "There is not much of a price floor." But then few thought matters would be pressed anywhere near that far. It wouldn't be rational.

From the beginning of November 1985, with the approach of winter, the price for bellwether West Texas Intermediate on the futures market continued to climb, reaching what was until then the highest point ever recorded on the Nymex—$31.75—on November 20, 1985, belying the threat of a price collapse. Surely, many thought, the Saudis really did not mean what they were saying. It was merely an elaborate warning designed to scare the other OPEC countries and restore discipline.

A week and a half after the November high, OPEC met yet again. By its actions, Saudi Arabia had already, in effect, declared a war for market share against the other OPEC countries. Now OPEC as a group, including Saudi Arabia, announced its intention to battle non-OPEC to regain lost markets. The communiqué from the conference contained the new formula: OPEC was no longer protecting price; now its objective was to "secure and defend for OPEC a fair share in the world oil market consistent with the necessary income for member countries' development."

Yet how significant, really, were these words? When, on December 9, the text of the communiqué was carried into a room where senior planners of an OPEC country were meeting to talk about the future, one of them said dismissively, "Oh, it's just another OPEC winter communiqué."

Then the price began to collapse.[2]

The Third Oil Shock

What ensued was no less turbulent and dramatic than the crises of 1973–74 and 1979–81. West Texas Intermediate plummeted by 70 percent over the next few months—from its peak of $31.75 a barrel at the end of November 1985, to $10. Some Persian Gulf cargoes sold for around $6 a barrel. In the two previous shocks, marginal losses and disruptions of supply had been enough to send prices shooting up. Here, too, the actual variation in volume was also marginal. OPEC's output in the first four months of 1986 averaged about 17.8 million barrels per day—only about 9 percent higher than 1985 production, and in fact, at about the same level as the 1983 quota. Overall, the additional production meant not much more than a 3 percent increase in total free world oil supply! Yet that, combined with the commitment to market share, was enough to drive prices down to levels so low as to have been virtually unimaginable only months earlier.

It was, indeed, the Third Oil Shock, but all the consequences ran in the opposite direction. Now, the exporters were scrambling for markets, rather than buyers for supplies. And buyers rather than sellers were playing leapfrog, each jumping over the other in pursuit of the lowest price. This unfamiliar situation once again threw up the question of security, but in new dimensions. One was security of demand for the oil exporters—that is, guaranteed access to markets. That concern may have seemed new. But in truth it was the same issue that had made the exporting countries such intense competitors in the 1950s and 1960s, and had led Juan Pablo Pérez Alfonzo to seek a guaranteed market in the United States, before he left for Cairo and the first step on the road to OPEC. For consumers, it seemed that all the concerns of the 1970s about security of supply were being made irrelevant in the battle for market share. But what of the future? Would cheap imported oil undermine the energy security so laboriously rebuilt over the preceding thirteen years?

It was not merely that prices were collapsing; they were also out of control.

731

For the first time in memory, there was no price-setting structure. There was not even an official OPEC price. The market was victorious, at least for the time being. Price would be set not through an arduous negotiation among OPEC countries, but through thousands and thousands of individual transactions. Netback deals, spot deals, "spotback deals," barter deals, processing deals, "topping off" deals, this deal and that deal—there seemed to be no end to the variations and twists adopted as exporters fought to hold on to and regain markets. Not only was OPEC struggling with non-OPEC, but, the December 1985 communiqué notwithstanding, the individual OPEC countries were battling with each other over customers. And, in the fiercely competitive environment, the matter came down to offering discount after discount to assure markets. "Everybody gets tired of endless negotiations for each cargo or each quarter," said the head of the Iraq State Marketing Commission in mid-1986. "In the end, the lazy negotiator for the crude exporter simply gives across-the-board discounts in order to ensure he undercuts all other OPEC barrels." It was not the specific kind of deal—netbacks or whatever—that caused the price collapse, but rather the fundamental facts that more oil was seeking markets than there were markets for oil, and that the hand of a regulator, in this case OPEC and in particular Saudi Arabia, was removed.

Shock was the universal reaction throughout the oil world. Would OPEC do anything? Could it? The organization was bitterly divided. Iran, Algeria, and Libya wanted OPEC to adopt a new and much lower quota and thus to restore price to $29 a barrel. The high-volume countries, principally Saudi Arabia and Kuwait, remained committed to regaining market share, though, almost plaintively, Yamani felt compelled to blame things on the buyers, telling an executive of one of the major companies, "I never sold a barrel someone didn't want." Meanwhile, Iran and Iraq, two of the key OPEC members, were still locked in a death struggle, and Iran's hostility to the Arab exporters was unmitigated.

The non-OPEC countries were suffering no less in their lost revenues. Belatedly, they took OPEC's warnings more seriously and started a "dialogue." Mexico, Egypt, Oman, Malaysia, and Angola attended an OPEC meeting as observers in the spring of 1986. Norway's conservative government initially declared that it was a member of the West, and it would not negotiate with OPEC. However, oil supplied about 20 percent of government revenues, and the government could not meet its budget. The ruling party fell and was replaced by its labor opposition. The new prime minister immediately announced that Norway would take steps to help stabilize the oil price. The oil minister of the new government boarded Zaki Yamani's yacht in Venice for a cruise and a discussion of oil prices. Yet in total, the dialogue between OPEC and non-OPEC did not yield much of substance. Thus, with little meeting of minds both within OPEC and between OPEC and non-OPEC, the "good sweating" continued through the spring of 1986.

"A Little Action"

Many oil companies were unprepared for this latest crisis, their executives having been convinced that "they"—OPEC—would not do something so silly as to eradicate a large part of their own revenues. A few had thought otherwise. Planners at Shell in London, reading the fundamentals carefully, had geared up an "OCS"—an Oil Collapse Scenario. The company had insisted that its senior managers take it seriously even if they thought it was improbable, discuss what their responses would be, and start taking prophylactic action. Thus, when the collapse struck, in contrast to the shock observed in many other oil companies, there was an eerie calm and orderliness at Shell Centre on the south bank of the Thames. Managers there, as well as in the field, went about their jobs as though carrying out a civil defense emergency operation for which they had already practiced.

In general, once the industry had assimilated the reality of the shock, it responded with quick and massive cuts in expenditures. Particularly hard hit was exploration and production in the United States. The country was one of the highest-cost oil provinces and had proved to be one of the most disappointing. Who could forget Mukluk, the $2 billion dry hole off Alaska? And companies had the most flexibility to respond in the United States; they did not have to worry about jeopardizing long-negotiated arrangements with national governments, as they did throughout the developing world.

Consumers were, of course, jubilant. All their fears about a permanent oil shortage were now laid to rest. Their standard of living and lifestyles were no longer at risk. After the years of huffing and puffing, oil was cheap again. The prophecies of doom had been a mirage, it seemed, and oil power was a harmless and empty threat. The "gas wars" at the corner filling stations, which had supposedly disappeared with the 1950s and 1960s, were now back, but writ large as a global oil war. And how low could prices really go? The irreducible minimum was indubitably established at Billy Jack Mason's Exxon station on the north side of Austin, Texas, on a one-day promotion in early April 1986, sponsored by a local country music station. Billy Jack's price for unleaded that day was zero cents a gallon. Free. It was a deal that could not be beat, and the result was a stampede. By nine o'clock in the morning, the line of cars waiting for their fill-up stretched for six miles; some people had driven from as far away as Waco. "The thing you gotta do is create a little action," explained Billy Jack. And when his opinion as an oil expert was sought on the future of prices, Billy Jack declared, "That's overseas. Nothing we can do about that until the Arabs get the prices right."

Another Texan, an adopted one, agreed with Billy Jack Mason that it was up to the Arabs, at least to a large degree. He was the Vice-President of the United States, George H. W. Bush, and while Billy Jack was selling his gasoline for zero cents a gallon, Bush was preparing to leave on an overseas mission to the Middle East to discuss oil, among other things. A visit to Saudi Arabia and the Gulf states had been on his schedule for several months, predating the collapse. But now he was going at a time when the domestic American oil and gas industry, the oil exporters, consumers, America's allies—all were asking the same question. Was the United States government going to do anything about the price collapse? By timing and position, and by his own history, Bush became the point man for the dilemmas of the Reagan Administration and U.S. policy at this very delicate moment in international relations.[3]

George H. W. Bush

A few years later, on the eve of his own inauguration as President in 1989, Bush would say, "I put it this way. They got a President of the United States that came out of the oil and gas industry, that knows it and knows it well." He knew, in particular, the risk-taking, deal-making world of the independent oil men, who were the backbone of exploration in the United States and who were the ones knocked flat on their backs by the price collapse. That had been the world in which he had spent his formative years. On graduating from Yale in 1948, Bush had passed up the obvious jobs on Wall Street for someone of his background; after all, his father had been a partner in Brown Brothers, Harriman, before becoming Senator from Connecticut. Then, having failed to be called back after a job interview with Procter and Gamble, he packed up his red 1947 Studebaker and set off for Texas, first Odessa, then its neighbor Midland, which would soon be calling itself the "oil capital of West Texas." He began at the bottom, as a trainee charged with painting pumping equipment, and then graduated to itinerant salesman, driving from rig to rig, inquiring of the customer what size drill bit he needed and what kind of rock he was drilling through, and then asking for the order.

Bush was an Easterner, with what some would have called a patrician background, but he was not entirely atypical. There was a noble tradition of Easterners coming to seek their fortunes in Texas oil, beginning with the Mellons and the Pews at Spindletop, and continuing through what *Fortune* magazine once called the "swarm of young Ivy Leaguers" who, Bush among them, in the post–World War II years had "descended on an isolated west Texas oil town"—Midland—"and created a most unlikely outpost of the working rich" as well as "a union between the cactus and the Ivy." It was not coincidental that the best men's store in Midland, Albert S. Kelley's, dressed its customers almost exactly the way that Brooks Brothers did.

Soon enough in this little world, as Bush later said, he "caught the fever" and formed an independent oil company in partnership with other ambitious young men no less eager to make money. "Somebody had a rig, knew of a deal, and we were all looking for funds," one of his partners said. "Oil was the thing in Midland." They wanted a name that would be memorable; another partner suggested that it should start either with an A or Z so it would be first or last in the phone book—but not get lost in the middle. The film *Viva Zapata!*, with Marlon Brando in the role of the Mexican revolutionary, was playing in Midland, so they called their company Zapata.

Bush quickly mastered the skills of the independent oil man, flying off to North Dakota in atrocious weather to try to buy royalty interests from suspicious farmers, combing courthouse records to find out who owned the mineral rights adjacent to new discoveries, arranging for a good drilling rig crew as quickly and as cheaply as possible—and, of course, making the pilgrimage back East to round up money from investors. On a brisk morning in the mid-1950s, near Union Station in Washington, D.C., he even closed a deal with Eugene Meyer, the august publisher of the *Washington Post*, in the back seat of Meyer's limousine. For good measure, Meyer also committed his son-in-law to the deal. Meyer remained one of Bush's investors over the years. And did the name Zapata help Bush and his partners in their new venture? "It rubbed both ways," said Hugh Liedtke, a Bush partner. "The shareholders that got in early and had a profitable investment, well, they thought Zapata was a patriot. But those that got in at one of the top swings in the market, and then the market fell, they thought Zapata was a bandit."

Eventually, the partners amicably split Zapata in two, and Bush took the offshore oil services side of the business, making it one of the pioneers and leaders in the dynamic development of offshore drilling and production in the Gulf of Mexico and around the world. Even today, crusty institutional stockbrokers in New York still recall that, when they phoned Zapata's offices in Houston to find out how the next quarter's results might turn out, they got not the Texas drawl of some good ol' boy at the other end of the line, but rather the modified Yankee twang of George Bush. For in addition to CEO, he was part-time investor relations officer. He lived through the erratic cycles of the postwar domestic oil industry. He could see how sensitive industry activity was to the price of oil, as well as how vulnerable it could be to unbridled foreign competition in those years of the great buildup in Middle Eastern oil—at least until Eisenhower imposed quotas in 1959. He also did well. The Bush family was about the first to put in a swimming pool in their neighborhood in Midland.

By the mid-1960s, Bush had decided he had made sufficient money; his father had been a Senator for ten years, and he would head in the same direction. He gave up the oil business for politics. The Republican party was just then getting started in Texas. But the Democrats' perennial lock on the state was not the only political problem to be faced. The would-be reconstituted Republican party was under assault from the right, and at one point Bush had to defend himself against the charge from the John Birch Society that his father-in-law was a communist—on grounds that the gentleman happened to be publisher of the unfortunately named woman's magazine *Redbook*.

Bush went from county chairman to Congress. In contrast to Calouste Gulbenkian, he did not find oil friendships slippery; his associates from his Midland days remained among his closest friends thereafter. As a Congressman from Houston, he was supposed to stand up for the oil industry, which he resolutely did. In 1969, when Richard Nixon was considering ending the quota system that restricted petroleum imports, Bush arranged for Treasury Secretary David Kennedy to meet with a group of oil men at Bush's home in Houston. Afterward, he wrote to Kennedy to thank him for taking time for the discussion. "I was also appreciative of your telling them how I bled and died for the oil industry," Bush said. "That might kill me off in the *Washington Post* but it darn sure helps in Houston." But oil was hardly uppermost on his political agenda once Bush moved on to other jobs—from ambassador to the United Nations and then chairman of the Republican National Committee during Watergate, to United States envoy to the People's Republic of China, to head of the CIA, and then to four years of campaigning unsuccessfully for the Republican nomination for the Presidency. In 1980, the man who beat him, Ronald Reagan, chose him as running mate, which led him to the Vice-Presidency.

Unlike Jimmy Carter, who made energy the centerpiece of his Administration, Ronald Reagan was determined to make it a footnote. The energy crisis resulted mainly, he maintained, from regulation and the misguided policies of the United States government. The solution was to get the government out of energy and return to "free markets." Anyway, Reagan had said during the campaign, there was more oil in Alaska than in Saudi Arabia. One of the very first acts of the Reagan Administration was to speed up the decontrol of oil prices that the Carter Administration had started. In its shift to a policy of "benign neglect" toward energy, the new Administration, to be sure, was helped by what was happening to the world oil market. Jimmy Carter's misfortune, rising oil prices, had been transmuted into Ronald Reagan's good fortune, for just around the time that he moved into the White House in 1981, half a decade before the price collapse, the inflation-adjusted price of oil actually started its long slide in the face of rising non-OPEC supply and falling demand. Not only did the decline in the real price remove energy as a dominant issue, it also served as a major stimulus to renewed economic growth and to the decline in inflation, two key features of the Reagan boom. Of course, the "free market" approach rested upon a contradiction; after all, a cartel, OPEC, was preventing a big fall in the price of oil, thus providing the incentives for conservation and energy development in the United States and elsewhere. But this contradiction remained latent and untroubling until the price collapse of 1986.

What was unleashed that year, in the words of the Acting Secretary General of OPEC Fadhil al-Chalabi, was nothing less than "absolute competition." And the results were devastating to the American oil industry. Pink slips were flying out at an awesome rate; drilling rigs were stacking up throughout the oil patch; and the financial infrastructure of the Southwest was quaking, as the region headed into an economic depression. Moreover, if prices stayed down, United States oil demand would shoot up, domestic production would plummet, and imports would start flooding in again, as they had in the 1970s. Perhaps, when it came to "market forces," there could be too much of a good thing. Still, there was not much that the United States government could do, even if so moved, in the face of those powerful forces of supply and demand. One possibility was to slap on a tariff, and so protect domestic energy production and continue to provide an incentive for conservation. While there were many calls in 1986 for a tariff, none of them came from the Reagan Administration. Another option was to try to jawbone OPEC into getting its act together again. Thus did George Bush's long official inattention to oil abruptly come to an end. Who else in the Reagan Administration could better talk, out of his own long experience, to the Saudis about oil?[4]

"I Know I'm Correct"

In its early planning, the main objective of Bush's trip to the Persian Gulf, amid the apparently interminable Iran-Iraq War, had been to underline United States support for the moderate Arab states in the region. But one could hardly be expected to go to Saudi Arabia and not discuss oil, especially when its price had fallen below ten dollars a barrel. Had the tables turned? In the 1970s, senior American officials had trooped to Riyadh to ask the Saudis to help keep prices down. Now, in 1986, would the Vice-President of the United States go to Saudi Arabia to ask them to push the price up?

Certainly, Bush felt that enough was enough. Conditions were as bad or worse in Texas and in the industry than any he had seen during his own days as an oil man. Moreover, the clamor and criticism from his political base in the Southwest, particularly Texas, was suddenly very intense. Nor was Bush alone with his concerns in the Reagan Administration; Energy Secretary John Herrington was warning that the fall in oil prices had reached the point of threatening national security. But the two belonged to a minority in the Administration.

In early April 1986, on the eve of his trip, Bush said that he would "be selling very hard" to persuade the Saudis "of our own domestic interest and thus the interest of national security. . . . I think it is essential that we talk about stability and that we not just have a continued free fall like a parachutist jumping out without a parachute." He ritualistically acknowledged the central free market precept of the Reagan Administration. "Our answer is market, market—let the market forces work," he incanted. But, he added, "I happen to believe, and always have, that a strong domestic U.S. industry is in the national security interests, vital interests of this country." Bush was clearly saying that market forces had gone too far. And he was quickly and embarrassingly disavowed by the Reagan White House, whose spokesman declared, "The way to address price stability is to let the free market work." The White House pointedly said that Bush would stress to King Fahd that market forces, not politicians, should determine price levels.

Bush's first stop was in Riyadh, where he dedicated the new United States embassy building. At a dinner with several ministers, including Yamani, the talk, of course, dealt in part with oil, and Bush commented that, if prices remained too low, pressure would build in the United States Congress for a tariff, and it would become increasingly difficult to resist that pressure. The Saudis took his remark very seriously. The Vice-President stopped next in Dhahran, in the eastern provinces, where the King was temporarily in residence. The American party was entertained at a banquet in the King's Eastern Palace, served by waiters who wore swords and pistols on their belts and had cartridges strapped across their chests. Their rifles, to the relief of the American Secret Service, were parked along the wall.

A private meeting with the King had been scheduled for the following day, but the Americans were told after the banquet that it had been moved up in the aftermath of an Iranian attack on a Saudi tanker. Bush was summoned into a late-night session with the King, which went on until well after 2:00 A.M. and, altogether, lasted more than two and a half hours. The Saudis were jittery about Iran's military advances and threats, and the prime subject of the meeting, as of the entire Bush mission, was Persian Gulf security and United States weapons supplies. Oil was mentioned only glancingly, but according to American officials, Fahd did express his hope for "stability in the market." The officials added that the King "felt Saudi Arabia was, in non-royal language, being given a bum rap in stories about its role in the oil market."

Though courting criticism at home, the Vice-President stuck to his position on oil prices. "I know I'm correct," he said after his visit with the King. "Some things you're sure of. This I'm absolutely sure of"—that low prices would cripple the domestic American energy industries, with serious consequences for the nation. At a breakfast with American businessmen in Dhahran a day later, Bush declared, "There is some point at which the national security interest of the United States says, 'Hey, we must have a strong, viable domestic industry.' I've felt that way all my political life and I'm not going to start changing that at this juncture. I feel it, and I know the President of the United States feels it."

Bush prided himself on loyalty, and he had proved over the previous five years to be a very loyal Vice-President. Never before had he departed from the White House line. But now he apparently had, and the adverse reaction became more explicit. "Poor George," was the way a senior White House official disparagingly talked about him, adding that Bush's position was "not Administration policy." But Bush refused to retreat—much. "I don't know that I'm defending the [U.S. oil] industry. What I'm doing is defending a position that I feel very, very strongly . . . Whether that's a help politically or whether it proves a detriment politically I couldn't care less."

The general view was that Bush was making not merely a blooper, but an enormous blunder, one that could in itself damage his political ambition and that pointed to a tendency to self-destruct. His gleeful opponents for the Republican nomination for the Presidency could not wait to use clips of the Bush statements in the critical primary state—and resolutely non-oil state—of New Hampshire. Columnists denounced him for cuddling up to OPEC and solemnly intoned that this could be the fatal act of suicide for his upcoming Presidential bid. Of course, within the oil states he was much commended for what he said. But outside the oil patch, it seemed that just about the only voice that had anything good to say about Bush's position was none other than the editorial page of the *Washington Post*, the newspaper he had once feared would kill him off for expressing pro–oil industry sentiments. On the contrary, the *Post* now said that the Vice-President was on to a very important point in his warning of how low prices would undermine the domestic energy industry, even if no one wanted to admit it. "Mr. Bush is struggling with a real question," the *Post* commented. "A steadily increasing dependence on imported oil is, as the man suggests, not a happy prospect." In short, said the *Post*, Bush was right.

But what, in fact, had Bush really said about an oil tariff to the Saudi ministers? Were they merely remarks made in passing, or something stronger than that? Whatever was said or heard—and there is often a big difference between the two in diplomacy—some Saudis thereafter insisted that Bush had explicitly issued a warning that the United States would impose a tariff if prices remained down, even if a tariff was absolutely contrary to the Reagan Administration line. And the Japanese had indicated that if the United States slapped a tariff on imported oil, they would follow suit, in order to protect their own program of energy diversification and to pick up some extra revenues for the Ministry of Finance. Few things could more quickly arouse the exporters to outrage than the prospect of a tariff in the oil-importing countries, for such a levy would transfer revenues from their own treasuries back to the treasuries of the consumers.

But the tariff was only part of a larger consideration. The Saudis, along with the other exporters, were worried about the huge financial losses from a continuing price collapse. Moreover, they were most unhappy with all the external criticism and political pressures that were converging on them because of the collapse. The Bush trip came as an additional incentive to restore some stability to prices. Some of the Vice-President's own advisers may have thought that his remarks about oil were intended only as a placebo for the American oil patch, but that was not how they were interpreted by the Saudis. What they heard was the Vice-President of the United States of America saying that the price collapse was destabilizing and threatened the security of the United States: American petroleum imports would rise substantially, and the United States would be weakened militarily and strategically vis-à-vis the Soviet Union. The Saudis looked to the United States for their own security; surely, they thought in the aftermath of the Bush visit, they would have to be attentive to the security needs of the United States. They had thought about security concerns in 1979, when they had pushed up production. And they thought about security again in the spring of 1986. They were feeling pressure from many countries, including Egypt and battle-strained Iraq. They were very worried about the Iran-Iraq War and its possible outcome. The Bush mission, on top of the turmoil and the other difficulties, provided reason for the Saudis to reconsider the fierce battle for market share that had sent prices into a tailspin—and to look for a way out. Moreover, the other exporters had finally learned that cheating had a cost.[5]

"Hara-Kiri" and $18 a Barrel

Yet no one really had any idea of how to behave in a competitive environment, nor indeed any experience. One OPEC veteran, Alirio Parra, a senior official of Petróleos de Venezuela, struggled to find some historical context. He had begun his career as an assistant to Juan Pablo Pérez Alfonzo during the formation of OPEC and, in fact, had been sitting with Pérez Alfonzo when the invitation to the founding meeting had arrived in 1960. Now the dissolution of OPEC seemed at hand. Searching his mind for some starting point, Parra recollected a book he had read many years earlier, *The United States Oil Policy*, published in 1926 by John Ise, a professor of economics at the University of Kansas. Parra finally found a battered copy in Caracas and brought it with him to London where he read it carefully.

"The unfortunate features of the oil history of Pennsylvania have been repeated in the later history of almost every other producing region," Ise had written then. "There has been the same instability in the industry, the same recurrent or chronic over-production, the same wide fluctuations in prices, with consequent curtailment agreements, the same waste of oil, capital, and of energy." Ise described one episode in the 1920s as a "spectacle of a vast overproduction of this limited natural resource, growing stocks, overflowing tanks, and declining prices, frantic efforts to stimulate more low and unimportant uses, or to sell for next to nothing. . . . It was a case of 'being choked, and strangled, and gagged, by the very thing people most wanted—oil.' " Ise added, "Oil producers were committing 'hara-kiri' by producing so much oil. All saw the remedy, but would not adopt it. The remedy was, of course, a reduction in the production." Although Ise had written the book sixty years ago, the language and the diagnosis sounded all too familiar to Parra. He made notes.

Thereafter, Parra was one of the handful from the exporting countries who began trying to work out a new pricing system that took into account the fact that oil and energy markets were, after all, competitive. Consumers had choices. That led him and the others to focus on a new price range of $17 to $19—and, more specifically, $18 a barrel—$11 less than what had been the official price of $29 a few months earlier. Somehow, this seemed to be the "right" price. Parra and a couple of others spent a week in May holed up in the Kuwaiti embassy in Vienna discussing the rationale for the new price. Correcting for inflation, it took oil prices back to where they had been in the mid-1970s, on the eve of the Second Oil Shock. Now, $18 seemed to be the point at which oil was again competitive with other energy sources and with conservation. It appeared to be the highest level that the exporters could attain and still achieve their goals of stimulating economic growth in the rest of the world and thus energy demand. It would reignite demand for oil and cap or perhaps even reverse the seemingly unstoppable non-OPEC production. Eighteen dollars "is inconvenient for my country," one senior OPEC official said to a friend, "but don't you think it's the best we can do?"

In the last week of May 1986, six oil ministers met in Taif in Saudi Arabia. One of the ministers commented that some of the others were predicting that oil prices would fall as low as $5 a barrel. "None of those present wants to give oil away to the consumer or to give the consumer a present," noted the Kuwaiti oil minister. But, he added, the old $29 a barrel had done OPEC "more harm than good."

Yamani categorically stated Saudi Arabia's position: "We want to see a correction in the trends in the market. Once we regain control of the market by increasing our share, we will be able to act accordingly. We want to regain our market power."

The ministers in attendance all confirmed their support for $17 to $19 a barrel and agreed on the need for a new quota system to go with it. What would have seemed heresy a few months earlier was now becoming the accepted wisdom. For, amid all the confusion and disarray of this latest oil crisis, a new consensus in favor of $18 a barrel was very definitely emerging out of the wreckage of the old. "It was a process of osmosis," said Alirio Parra. And not only producers, but consumers liked it, too. The Japanese, as importers of more than 99 percent of their oil, might have been expected to prefer the lowest possible price. That was not the case. Two problems would result if prices were too low. First, this would undercut the large and expensive commitment they had made to alternative energy sources, leading, they were sure, back to higher oil dependence and eventually to renewed vulnerability, and setting the stage for another crisis. Second, since oil constituted a substantial part of Japan's imports, very low prices would enormously swell Japan's already-huge trade surplus, further accentuating the severe conflicts with American and Western European trading partners. Thus, one found throughout the Japanese oil industry and government a belief in "reasonable prices," which happened to come out to about $18 a barrel.

The new consensus was evident in the United States, as well—in the government, on Wall Street, in banks, among economic forecasters. The gains from falling oil prices (higher growth and lower inflation) would outweigh the losses (the problems of the energy industries and the Southwest). But that was true only up to a point, at least according to the new view. At some level, the pain and the dislocation in the financial system, along with the discomfort to politicians, would start canceling the benefits, and that point, by general consent, fell somewhere between $15 and $18. The Reagan Administration was rooting for the efforts to reestablish the price around $18. Such a price would give a strong boost to economic growth, while helping to restrain inflation, but it was also a price at which the domestic oil industry could scrape by and would thus greatly reduce the pressure for a tariff. As a result, the Administration could maintain its commitment to "free markets" and would not have to take any action. When all things were considered in these circumstances, the most desirable thing to do was nothing.

But consensus was one thing. Putting together a new deal was quite another. And the efforts in that direction were failing, even though the loss of revenue was becoming excruciating for many oil exporters. The Arab Gulf states, which greatly increased the volumes they sold, were the least hurt. Kuwait's revenues were down only 4 percent, Saudi Arabia's 11 percent. The price hawks, which happened to be the countries most belligerent and hostile toward their customers in the West, were the ones hit hardest. Iran's and Libya's oil earnings in the first half of 1986 were down 42 percent from the same period in 1985. Algeria's were down even more. For more than economic reasoning, Iran was the country most disadvantaged. Even as its revenues were plummeting, it was having to finance the war with Iraq, which had entered a new, more intense phase. The Iraqi air war against tankers and facilities was taking a rising toll on Iranian export capabilities. How could Iran continue to fight the Ayatollah Khomeini's holy war against Iraq and Saddam Hussein without money?

Something would have to be done soon. Saudi Arabia had been maintaining its output at its old quota level, but now indicated that it would start pushing its production to higher levels. Even more oil would be coming onto the market. In July 1986, Persian Gulf crudes were going for $7 a barrel or less. Enough was enough, and the leaders of Saudi Arabia and Kuwait were anxious to bring the "good sweating" to an end. They, too, were worrying about prospects for revenues. Moreover, the volatility and uncertainty were too unsettling, too likely to increase broader political risks around the world. Virtually all OPEC decision makers had concluded that the market share strategy was, at least in the short term, a failure. But how to get out of it without sinking back into the bind that had precipitated it in the first place? The only way out was new quotas. But who would get what? Some of the exporters insisted that Saudi Arabia resume its swing role, to which Yamani replied, "Not on your life. We all swing together or not at all. On this point, I am as stubborn as Mrs. Thatcher."

By July, OPEC experts had worked out on paper a detailed rationale for the new pricing: a range of $17 to $19 a barrel would lead to an improved world economic outlook, stimulating oil demand: "It could be an effective instrument for slowing and arresting the pace of fuel substitution" and "will definitely discourage future developments of high cost oil." But, if prices were any lower, the exporters would run a grave risk: "strong protectionist measures in the major consuming countries of the industrialized world," including "the imposition of an oil imports tariff in both the U.S. and Japan." They remembered Eisenhower's import restrictions much better than most Americans.

But still there was the matter of quotas, which would require renewed cooperation among the fractious OPEC exporters, and there appeared to be little hope that anything could be worked out when OPEC next met in Geneva, at the end of July and the beginning of August 1986. Iran, in particular, had signaled its opposition to new quotas. But in the course of the meeting, the Iranian oil minister, Gholam Reza Aghazadeh, appeared in Yamani's suite for a private discussion. He spoke through an interpreter. Yamani was so startled by the message that he insisted that the interpreter translate it again. It was reiterated. Iran, said the minister, was now willing to accept the temporary, voluntary quotas pushed by Yamani and others. Iran had, in fact, backed down. Its oil policy was more pragmatic than its foreign policy.

The market share strategy was over. But in announcing the restoration of quotas, OPEC insisted that it would not carry the burden itself; non-OPEC would have to cooperate. And agreements were subsequently worked out in which various non-OPEC countries indicated that they would do their part. Mexico would cut its output. Norway promised not to cut back but rather to reduce the growth in its output. At least that was something. The Soviet Union had stood aside from most of the discussions. In May 1986 a senior Soviet energy official had derided the notion that the Soviet Union would ever formally cooperate with OPEC. The Soviet Union was not a Third World country, he insisted. "We are not a producer of bananas." It was true in a way; one could not find bananas in Moscow. But, bananas or not, Soviet officials could read their balance of trade accounts, and the loss in terms of hard currency earnings from oil and gas, if continued, could be devastating for the plans to reform and revive the stagnant Soviet economy that were just beginning to be formulated under Mikhail Gorbachev. The Soviet Union promised to contribute a 100,000-barrel-per-day cutback to OPEC's efforts. The pledge was vague enough and the job of tracking Soviet exports sufficiently difficult that the OPEC countries could never be sure that the Russians were really as good as their word. But in the immediate turmoil, the symbolism was important. The next step to cool off the good sweating was for OPEC to formalize quotas and do something about price. But there was an interlude.[6]

Playing It by Ear

In September 1986 Harvard University was celebrating its 350th anniversary. Planning for this grand event had been going on for several years; it would demonstrate Harvard's place in American life and its contribution to global learning. Nothing had been spared for the "350th," from the corralling of illustrious Nobel Prize–winning names to the minting of specially designed commemorative chocolates. To cap the celebration, Harvard had chosen, from among the planet's five billion citizens, two people to give major speeches. One was Prince

Charles, heir to the British throne; after all, it was from England that John Harvard had emigrated to Massachusetts, where eventually in 1636 he had bequeathed his personal collection of three hundred books to the small college that was later renamed in his honor. The other speaker was the Saudi oil minister, Ahmed Zaki Yamani, who had studied for a year at the Harvard Law School and was now a large donor to the university's Islamic collection. A delegation from Harvard even flew to Geneva to personally extend the invitation, which Yamani accepted.

The debonair Prince Charles delivered a lively and entertaining speech, delighting all in attendance. For his part, however, Yamani chose to deliver a very substantive, dense discourse, filled with numbers that went out to the second decimal place. The text was passed out in advance as the members of the audience seated themselves in the crowded ARCO Forum at the Kennedy School of Government. That way they could read along. It was meant to be a speech suitable to the occasion, putting into perspective the tumultuous, globe-shaking events of 1986, which had changed every economic indicator. It was also an explication and a justification. Reading in a soft murmur, only occasionally allowing himself a little half smile or small deviation from the text, Yamani recalled his battles over price with the oil companies in the early 1970s and with his OPEC brethren in the late 1970s and early 1980s. He called for stability and the recognition that oil was a "special commodity." And he held out the hope of a return to that kind of stability: oil priced at fifteen dollars per barrel with a gradual rise in both price and OPEC output. It was a vision of a very orderly world. Did he really believe it?

At the end of his discourse, Yamani agreed to accept questions. For the last question, a tall, thoughtful professor stood up to observe how hard and contentious it was to make energy policy in the United States: the Congress fought with the President, the Senate with the House, various agencies with one another, everybody fought with everybody else. Was it any less contentious in Saudi Arabia? Would Yamani, he asked, describe the process by which oil policy was made inside Saudi Arabia?

Smoothly and without even a moment's hesitation, the oil minister replied, "We play it by ear."

The audience roared with laughter. It was an amusing answer, which captured the truth of improvisation in policymaking, whatever the government. Still, it was a little odd, coming from the self-proclaimed disciple of long-term thinking, who had been at the center of world oil decision making for a quarter of a century. Unbeknownst at the time to those in the audience, those words would be among the last of Yamani's official utterances.

A month later, in October, Yamani was at a meeting in Geneva for the next stage in the reconstruction of OPEC. He followed his role as he was instructed; the kingdom wanted not only to protect its quota and assure its volumes, but also to receive a higher price—the consensus eighteen dollars, which was quite different from the fifteen dollars that Yamani had talked about at Harvard. Yamani went so far as to suggest, semipublicly, that it was contradictory to try to pursue higher volume and higher price at the same time, which seemed to go explicitly against the policy the King had enunciated. Nevertheless, Yamani would do the best he could, and further progress was, in fact, made in reconstructing the quota system. Then, one evening, a week after the meeting, Yamani was back in Riyadh at dinner with friends when he received a phone call advising him to turn on the television news. An item at the end of the broadcast reported tersely and without any adornment that Ahmed Zaki Yamani had been "relieved" of his post as oil minister. That was the way he learned he had been fired. Yamani had been in the job twenty-four years, a good, long run in any position anywhere. Still, it was an abrupt, embarrassing, and disconcerting end to a quarter-century career.

The reasons for his firing, and the way it was carried out, became a subject of intense discussion in Saudi Arabia and throughout the world. And the explanations proffered, as might have been expected, were many and somewhat contradictory: He had embarrassed the Royal Family not only by failing to pursue vigorously his instructions in Geneva but also by criticizing the thrust of those instructions; he had made powerful enemies by opposing barter deals; his firing reflected the jettisoning of policies with which he had publicly been associated. It was also said that there was resentment against him in Riyadh because of what was described by some as his arrogance, his patronizing manner, his high profile, and his celebrity status outside the country. Yamani had been King Faisal's man, but Faisal had been dead for almost a dozen years. Now Fahd was King, and he was the maker of oil policy. By 1986, Yamani had precious few allies, while many of the other ministers and advisers believed that he had usurped their own authority. And, when it came down to it, some said, King Fahd just plain didn't like Yamani.

Perhaps in the long run, it was the decline and then the precipitous collapse in the price of oil that led to Yamani's own downfall. But there was the specific matter of the Harvard speech. Before that event, some in Riyadh had the impression that Yamani was merely going to say a few general words, more or less impromptu, not make a major policy statement. But a seventeen-page speech was not everyone's idea of an impromptu talk. Moreover, the policy it espoused was not exactly identical with the official policy of Saudi Arabia. And the "play it by ear" remark, not a familiar idiom to everyone, was interpreted in Riyadh as a tart criticism of the Saudi government. So Yamani was returned to private life, to manage his fortune, to set up a research institute in London, to try to acquire a Swiss watchmaker, to preside over his perfume factory in Taif, to teach part-time at the Harvard Law School, and not surprisingly, to comment from time to time on world oil.[7]

Price Restored

The OPEC countries finally brought the "good sweating" to an end with a meeting in Geneva in December 1986. It was the first major OPEC meeting at which the new Saudi oil minister, Hisham Nazer, appeared. He, like Yamani, was among the initial generation of Saudi technocrats. Just two years younger than Yamani, he had been educated at UCLA and had been a deputy to Abdullah Tariki, the Saudis' first oil minister. Nazer had then served for many years as Minister of Planning, which made him particularly alert to the links between oil and the national economy and to the overall revenue question that was troubling Riyadh. And he carried no responsibility for or commitment to the now-repudiated market share strategy.

The restoration of revenues was the central consideration at Geneva. The exporters agreed to a "reference price" of eighteen dollars, based upon a composite price of several different crude oils. They also agreed to a quota that, they hoped, would support the price. There was just one loophole. No agreement was possible between Iran and Iraq as to what Iraq's quota should be, in light of the continuing war and Iraq's expanding exports. So the quota only applied to twelve countries; Iraq was outside it, free to do what it could. It had, once again, as it had done at different times going back to 1961, temporarily seceded from OPEC. Still, a "notional" quota was assigned to Iraq, 1.5 million barrels per day, which brought the total up to 17.3 million barrels per day.

To the surprise of many, the framework of agreement managed, though with considerable reconstruction, to hold up through 1987, 1988, and 1989, though in the face of recurrent and sometimes intense pressure in the marketplace. To be sure, the OPEC price was not eighteen dollars, but rather, for the most part, in a range between fifteen and eighteen dollars. Prices were volatile, and at times appeared ready to plunge again. More than once, the quota system seemed about to fall apart. But the producers, faced with the alternative, rallied. After all, the OPEC countries themselves had felt the full brunt of the "good sweating," and they had had enough of it.

The new oil prices, reconstituted in a lower range, completely wiped out the increases of the Second Oil Shock of 1979–81. The economic benefits to consumers were enormous. If the two oil price shocks of the 1970s constituted the "OPEC tax," an immense transfer of wealth from consumers to producers, then the price collapse was the "OPEC tax cut," a transfer of $50 billion in 1986 alone back to the consuming countries. This tax cut served to stimulate and prolong the economic growth in the industrial world that had begun four years earlier, while at the same time driving down inflation. In economic terms, the long crisis was certainly over.

Iran Versus Iraq: The Tide Turns

Politically and strategically, however, there still remained a great threat—the seemingly endless Iran-Iraq War, which could escalate into a wider conflict that would threaten oil production and supplies throughout the region and the security of the oil states themselves. In its seventh year, in 1987, the war broke through the barriers that had largely restricted it to the two belligerents and for the first time became internationalized, drawing in both the other Arab states of the Gulf and the two superpowers. A year earlier, Iran had captured the Fao Peninsula, the southernmost extreme of Iraq, bordering Kuwait. It looked as though Fao could be the gateway to the conquest of the Iraqi city of Basra, making it the potential key to the dismemberment and disappearance of the unified Iraqi state that Britain had created after World War I. But though the Iranians got to Fao, they could get no further. They bogged down in the marshy sands, blocked by a reinvigorated Iraqi Army. Thereafter, the war turned against them. Iraqi successes in the air and its missile attacks on Iranian shipping in the Gulf—

the "tanker war"—led to stepped-up Iranian attacks on third-country tankers. Iran zeroed in on Kuwait, which was assisting Iraq. Khomeini's forces not only hit shipping going to and from Kuwait, but also launched at least five missile attacks directly on Kuwait itself.

Like the other Arab states, Kuwait had taken seriously the United States campaign against the selling of arms to revolutionary Iran. Thus, it was extremely disconcerted by disclosures that the United States had secretly sold weapons to Iran in an attempt to win freedom for American hostages held in Lebanon and to start, somehow, a dialogue with "moderates" in Tehran, whoever they might be. The disclosures greatly increased the small country's inherent sense of insecurity. But it was Iran's attacks that propelled Kuwait, in November of 1986, to ask the United States to protect its shipping (though the American ambassador to Kuwait later insisted that he had relayed such a request in the summer of 1986). Washington was jolted to learn that the Kuwaitis had taken the additional precaution of asking the Russians for protection, and when that information reached the most senior levels in the Reagan Administration, the Kuwaiti request, in the words of one official, "didn't linger." The potential significance of the approach to Moscow provided reason for a quick response. For Russian involvement would have expanded Russian influence in the Gulf—something the Americans had sought to prevent for more than four decades, and the British, for no less than 165 years. But, apart from the East-West rivalries, it was deemed imperative to protect the flow of Middle Eastern oil.

President Reagan himself spoke of the need for self-defense in the Gulf but also restated his guarantee that the United States would safeguard the flow of oil. And in March 1987, the Reagan Administration, intent on excluding the Russians, told the Kuwaitis that the United States would take on the whole job of reflagging or nothing at all. It would not go "halvsies" with the Russians. Thus, eleven Kuwaiti tankers were reflagged with the Stars and Stripes, qualifying the ships for American naval escorts. A few months later, U.S. naval vessels were patrolling the Gulf. All that was left to the Russians was to charter some of their own tankers on the run to Kuwait. British and French naval units, along with ships from Italy, Belgium, and the Netherlands, also entered the Gulf to help protect freedom of navigation. The Japanese, forbidden by their constitution from sending ships but highly dependent on oil from the Gulf, chipped in by increasing the funds that they provided to offset the cost of maintaining American forces in Japan and by investing in a precision locator system in the Strait of Hormuz. West Germany shifted some of its naval vessels from the North Sea to the Mediterranean, freeing, it said, United States ships for duty in and around the Gulf. But, with the United States taking the lead, there was now the possibility of a major military confrontation between the U.S. and Iran.

By the spring of 1988, Iraq, making use of chemical weapons, was manifestly winning. And Iran's ability and will to carry on the war were fading fast. Its economy was in shambles. The defeats were draining support for the Khomeini regime. Volunteers, fervent or otherwise, were no longer forthcoming. War weariness gripped the country; in one month, 140 Iraqi missiles fell on Tehran alone.

Among those maneuvering for position in post-Khomeini Iran—for the

Ayatollah was old and known to be seriously ill—was Ali Akbar Hashemi Rafsanjani, the speaker of the Iranian Parliament and deputy commander of the Army. He was a member of a wealthy family of pistachio growers whose fortune had been augmented by Tehran real estate in the 1970s under the Shah. He himself was a cleric and a Khomeini student and disciple who had begun his opposition to the Shah in 1962. Though deeply involved in the "arms-for-hostages" negotiations with the United States, he had turned aside criticism, and as he navigated his way through the theocratic tangle of Iranian politics, he won himself the nickname of "Kuseh"—the Shark. He was, after Khomeini himself, the top decision maker in the Islamic Republic. And he concluded that it was time to seek an end to the war. Iran no longer had any chance of winning. The costs of the war were enormous, and they had no obvious limit. The Ayatollah's regime, and his own prospects, could be threatened by continuing losses. Moreover, Iran was diplomatically and politically isolated in the world, while Iraq seemed to be growing stronger.

Then, the American naval presence in the Gulf did, in fact, lead to a major confrontation with Iran, but of an unexpected and tragic kind. In early July 1988, the United States destroyer *Vincennes*, engaged in an exchange with Iranian warships, mistook an Iranian Airbus, carrying 290 passengers, for a hostile aircraft and shot it down. It was a horrid mistake. To some in the Iranian leadership, however, it was not a mistake, but a sign that the United States was taking off its gloves and preparing to bring its great power to bear in direct military confrontation with Iran in order to destroy the regime in Tehran. Iran, in its weakened state, might not be able to resist. It could no longer afford to go against the United States. Moreover, in trying unsuccessfully in the aftermath of the accident to marshal diplomatic support, Iran discovered just how isolated politically it had become. All of these factors added to the urgency to reconsider Iran's relentless commitment to the war.

Yet Rafsanjani still had to cope with an implacable force: Ayatollah Khomeini, for whom vengeance, including Saddam Hussein's head, was the price of peace. But the facts of Iran's position were evident to the others around Khomeini, and Rafsanjani finally prevailed. On July 17, Iran informed the United Nations of its willingness to countenance a cease-fire. "Taking this decision was more deadly than taking poison," Khomeini declared. "I submitted myself to God's will and drank this drink for his satisfaction." But vengeance was still his ambition. "God willing," he said, "we will empty our hearts' anguish at the appropriate time by taking revenge on the Al Saud and America," he added. The Ayatollah would not live to see such a day; within a year, he would be dead.

After the Iranian message to the United Nations, another four weeks and much negotiation passed before Iraq, too, would accept a cease-fire. It finally went into effect on August 20, 1988, and Iraq immediately began symbolic oil shipments from its Gulf ports, something it had been unable to do for eight years. Iran announced its intention to rebuild the great refinery at Abadan, which had been the starting point for the oil industry in the Middle East at the beginning of the century and had been almost completely destroyed in 1980, in the first days of the war. One month short of eight years after it began, the Iran-Iraq War ended in a stalemate, though one that favored Iraq. As far as Baghdad was concerned, it had won the war, and it now intended to be the dominant political power in the Gulf, and one of the world's major oil powers. But the significance of the end of the Iran-Iraq War was much more far-reaching. It appeared that the threat to the free flow of Middle Eastern oil had at last been removed; and with the silencing of the guns along the shores of the Persian Gulf, the era of continuing crisis in the world of oil that had begun with the October War fifteen years earlier along the banks of another waterway, the Suez Canal, finally seemed to be at an end.

It was not only the end of the war that pointed to a new era. So did the changing relationship between oil exporting and the consuming countries. The great contentious question of sovereignty had been resolved; the exporters owned the oil. What came to matter for them over the 1980s was sure access to markets. When the producing countries discovered that consumers had more flexibility and wider choices than had been imagined, they came to see that "security of demand" was no less important to them than was "security of supply" to the consumers. Most of the exporters now wanted to establish that they were reliable suppliers and that oil was a secure fuel. With the sovereignty issue settled, with socialism in bad repute, and with the North-South confrontation a fading memory, the exporters could act more on economic than political concerns. In a quest for capital, some were reopening the doors to exploration by private companies within their borders—doors that had been slammed shut in the 1970s.

Others went further, as the logic of integration—so powerful a theme in the history of the industry—reasserted itself, seeking to rejoin the reserves to the markets. The state-owned companies of some of the exporters, following in the historic wake of private companies, went downstream to acquire outlets. Petróleos de Venezuela built up a large refining and marketing system in the United States and Western Europe. Kuwait turned itself into an integrated oil company, with refineries in Western Europe and thousands of gasoline stations in Europe operating under the brand name "Q-8." Kuwait did not stop there. In 1987, Margaret Thatcher reversed Winston Churchill's historic decision of 1914 and sold off the government's 51 percent stake in British Petroleum. In her view, it no longer served any national purpose, and the government would be happy to have the cash. Kuwait thereupon snapped up 22 percent of BP—the very company that, along with Gulf, had developed and owned Kuwait's oil until 1975. The British government was infuriated and forced Kuwait to reduce its holdings to 10 percent.

At almost exactly the same moment that the Iran-Iraq War ended, Saudi Arabia and Texaco, one of the original Aramco partners, announced a new joint venture. Texaco's management was preoccupied not only with the company's immediate problem—a $10 billion judgment that Pennzoil had won against it in a Texas courtroom for the Getty takeover—but also with how to enhance its long-term prospects in a dramatically different world oil industry. Saudi Arabia wanted to ensure that it had access to markets. Under the terms of their new deal, Saudi Arabia acquired a half interest in Texaco's refineries and gasoline stations in 33 states in the eastern and southern United States. The transaction guaranteed the Saudis, if they wanted it, 600,000 barrels per day of sales in the United States, compared to the trickle of 26,000 barrels per day to which they had fallen

748

in 1985, on the eve of the price collapse. Such "reintegration" represented one effort to put greater long-term stability back into the industry and to manage the risks faced both by producers and consumers.

A few months after the Iran-Iraq cease fire, George Bush, the former oil man, succeeded Ronald Reagan as President of the United States. And with the astonishing disintegration of the barriers, both symbolic and actual, that had long divided the countries of the Soviet bloc from the Western democracies, unprecedented prospects for global peace appeared at hand as the decade of the 1980s gave way to the 1990s. The competition among nations in the years ahead, some predicted, would no longer be ideological but instead primarily economic—a battle to sell their goods and services and manage their capital in a truly international marketplace. If that would, indeed, be the case, oil as a fuel would certainly remain a vital commodity in the economies of both the industrialized and the developing nations of the world. As a bargaining chip among the producers and consumers of oil, it would also remain of paramount importance in the politics of world power.

Yet out of the tumult of the 1970s and 1980s, important lessons had emerged. Consumers had learned that they could not regard oil, the fundament of their lives, so easily as a given. Producers had learned that they could not take their markets and customers for granted. The result was a priority of economics over politics, an emphasis on cooperation over confrontation, or at least so it appeared. But would those critical lessons be recalled as the years passed and those who had been part of the great dramas retired from the scene and new players took their place? After all, the temptation to grasp for great wealth and power has been endemic to human society ever since its beginnings. At a discussion in New York City in the late spring of 1989, the oil minister from one of the major exporting countries, a man who had been at the center of all the battles of the 1970s and 1980s, spoke at length about the new realism of producers and consumers and the lessons learned by both. Afterwards, he was asked how long such lessons would be remembered.

The question took him a bit by surprise, and he thought for a moment. "About three years, without reminding," he said.

Within a year of that exchange, he himself was no longer minister. And a month later, his country was invaded.[8]

CHAPTER 37

Crisis in the Gulf

DURING THE SUMMER of 1990, the world was still in the euphoria over the end of the Cold War and the new, more peaceful world that it portended. For 1989 had certainly been the *annus mirabilis*—the miracle year—in which the international order had been remade. The East-West confrontation was over. The communist regimes in Eastern Europe had collapsed, along with the Berlin Wall itself, the great symbol of the Cold War. The Soviet Union was in the midst of a profound transformation arising not only from political and economic change but also from the eruption of long-repressed ethnic nationalisms. Democracy appeared to be taking hold in many countries where, until shortly before, such a possibility would have been dismissed as totally unrealistic. German reunification was no longer an abstract subject for rhetoric and discussion, but an imminent reality; and the reunified Germany would be the predominant power in Europe. Japan was now seen as the global financial powerhouse; and the confrontations of the future would surely become global competitions for money and markets, and for economic growth—a prospect that seemed so down-to-earth that some said what was at hand was not only the end of the Cold War, but also "the end of history."

Oil remained high on the agenda of environmental concerns, but otherwise it appeared to have become rather unimportant, indeed just another commodity. Consumers were happy, for oil prices were low. In real terms, American motorists were actually paying less for gasoline than they had at any time since World War II. There did not seem to be any long-term problem of supply; after all, proven world oil reserves had been mightily increased—from 670 billion barrels in 1984 to 1.0 trillion barrels in 1990.

Yet, amidst the complacency, there were reasons for caution. While vast additions had been made to world oil reserves, they were all concentrated in the five major oil producers of the Persian Gulf, plus Venezuela. There was no large inventory of diversified, non-OPEC oil waiting to come into the system, as had been the case with Alaska, Mexico, and the North Sea at the time of the 1973 crisis. The Persian Gulf's share of world oil reserves had now actually increased to two-thirds of the total.

In economic terms, the oil picture looked a lot less like the early 1980s than the early 1970s, which had set the scene for the 1973 oil shock. The world petroleum market was tightening. Demand was growing with some vigor. American production was plummeting—between its 1986 high and 1990, by two million barrels per day, a volume greater than the individual output of 10 out of 13 OPEC countries in 1989. United States oil imports were at the highest level ever, and still going up. The world was moving back to high dependence on the Persian Gulf. The "security margin"—the gap between demand and production capacity—was shrinking, which would make the market more susceptible to conflict and accidents. That margin had been large enough in the early and mid-1980s to absorb the Iran-Iraq War with all its disruption and loss of output, but no longer.

How high would oil prices go? That was dependent on how rapidly new production capacity was added around the world. With low prices and the renewed confidence about the security of supplies, conservation had run out of steam. The effort to develop alternative sources had even become anemic. In addition, an overall immobility had taken hold in many countries, reflecting the inability to resolve the conflict between energy and environmental concerns. Still, energy crises did seem a thing of the past. At a U.S. Senate hearing in the spring of 1990, it was argued that the probability of a major disruption was minimal, at least for the next several years. And some futurists and analysts announced in the spring of 1990 that there could not possibly be any oil crises in the decade of the 1990s.

Iraq Moves

At 2:00 A.M., on August 2, 1990, those expectations were ripped away. A hundred thousand Iraqi troops began their invasion of Kuwait. Meeting little resistance, Iraqi tanks were soon rolling down the six-lane highway toward Kuwait City. And so the first post-Cold War crisis turned out to be a geopolitical oil crisis.

Over the preceding several years, most of the petroleum exporters had sought to rebuild the links to the consuming countries that had been broken in the 1970s. Owing to the huge additions to reserves, these producers no longer worried that they were rapidly wasting a depleting resource. Instead, they wanted to demonstrate that they were reliable, long-term suppliers, that they could safely be regarded as the energy reserve for the industrial world—and that petroleum could be counted on. Oil needed markets, and markets needed oil; that calculation of mutual self-interest would be the basis of a stable, constructive, non-confrontational relationship that would extend into the twenty-first century.

Iraq was one of the exceptions. It did not hide its hostility to its main customers, the industrial world. In June 1990, Iraq's dictator, Saddam Hussein, warned the West that the oil weapon could well be applied again. Though he claimed to be at the vanguard, Saddam Hussein was a strangely anachronistic figure, a kind of throwback. He asserted himself with the nationalistic rhetoric and anger of the 1950s and 1960s. He would say that Joseph Stalin was one of his models, at the very time that Eastern Europe and the Soviet Union were trying to extricate themselves from the legacy of Stalin's terror and hypocrisy. Saddam Hussein created his own massive cult of personality.

Not only were huge portraits and photographs of him to be seen everywhere in the country, but so omnipresent was his dominance that he would proudly declare in 1990, a month before the invasion of Kuwait, that "Saddam Hussein . . . is to be found in every quantity of milk provided to children and in every clean, new jacket worn by an Iraqi."[1] He also had his own considerable reputation for personal brutality. Videocassettes circulated in the Middle East of the meeting in which Hussein had purged his rivals, and of the bodies of executed military officers displayed on meathooks. Hussein's military forces had used poison gas both against the Iranians and against Kurdish women and children in his own country. When one Western visitor bluntly asked him, in late June of 1990, about his reputation for ruthlessness he blandly replied: "Weakness doesn't assure achieving the objectives required by a leader."[2]

Since 1985, Iraq had been the world's largest purchaser of arms, and it was engaged in a strenuous weapons development and acquisition campaign, supported by an intricate and secret international procurement network. The Israelis had destroyed Hussein's nuclear weapons fuel facility in 1981, but he had resumed the drive for nuclear weapons and had publicly boasted of building up an arsenal of chemical weapons. Iraq was a closed police state, but Saddam Hussein's objectives seemed clear: to dominate the Arab world, to gain hegemony over the Persian Gulf, to make Iraq into the predominant oil power—and ultimately to turn Greater Iraq into a global military power. But Iraq was also suffering considerable financial weakness. The Iran-Iraq War, which Saddam Hussein had launched, had cost the country half a million deaths and serious casualties and had ended in a stalemate. Yet a nation of eighteen million was continuing to support a million-man army. Hussein wanted higher oil prices and very soon; Iraq was devoting about 30 percent of its total gross national product to Saddam's war machine, and even as Iraq scoured the world for new, deadly, and sometimes bizarre weapons, it was not paying its international bills.

In July 1990, Iraq moved 100,000 troops to its border with Kuwait, which was identified with a low oil price strategy. The troops were seen as pieces in a war of nerves, as tools in Saddam Hussein's new role as "enforcer," ensuring that countries like Kuwait and the United Arab Emirates observed their quotas, and forcing up OPEC's prices. After some hesitation, the Kuwaitis responded. The Amir abruptly replaced the Kuwaiti oil minister, the focus of Iraqi criticism; and along with the United Arab Emirates, Kuwait reined in its production and began to observe its OPEC quota. By mid-July 1990, only one country was cheating, exceeding its OPEC quota, and that was OPEC's self-appointed enforcer—Iraq. The Iraqi soldiers were also being used, it was thought, to threaten Kuwait into giving way on a border dispute that involved a large oil field and into handing over two islands to Iraq. Yet Baghdad, all along, had something much more in mind—invasion and annexation of the whole country. It was the ultimate in strategic surprise. The troops were there, they were evident for all to see, they were meant to be threatening, the satellites clicked away, and yet hardly anyone thought that they would be used in the way they were. Last minute urgent warnings from intelligence analysts were discounted in the face of Hussein's personal promises to several other leaders, including President Mubarak of Egypt and King Hussein of Jordan, that he was not planning any hostile action. With the invasion, the Kuwaiti royal family fled, and the small country was in Iraqi hands. The Kuwaitis had survived over two centuries by being smart and knowing how to play neighbors and larger powers off against each other; and even when the Iraqi troops massed on their borders, they thought they could outsmart the Iraqis as they had done for so long. This time, however, they were taken by surprise.

Miscalculation

To justify his actions, Hussein offered a plethora of rationales. He claimed that Kuwait rightfully belonged to Iraq and that the Western imperialists had snatched it away. Actually, Kuwait's origins went back to 1756, two decades before the United States declared its independence, and certainly much before the beginnings of modern Iraq, which was knitted together in 1920 out of three provinces that had been part of the Ottoman Empire for four centuries and, for several centuries before that, had been outlying provinces of various other empires. The Iraqis said that the British had drawn their border with Kuwait to deny Iraq its due—and its oil. In fact, the border adopted at a 1922 conference (which deprived Kuwait of two-thirds of its territory) was a simple copy of the border the Turks had agreed to in 1913, before the First World War. Moreover, in 1922, expert opinion held that there was no oil in Kuwait.

In 1980, in launching the war with Iran, Saddam Hussein had made a grave miscalculation, one that had almost cost him his position: He had assumed that it would take only a few weeks to knock off Iran. He was wrong, and Iraq came close to being defeated. A decade later, in 1990, he assumed that he could swiftly absorb Kuwait and confront the world with a *fait accompli*, which would arouse some complaints but little else. In the meantime, he would have solved his financial problems overnight and would have acquired the wherewithal to finance his grandiose military and political ambitions. He would be the hero of the Arab world; Iraq would be the number one oil power; and, like it or not, the Western countries would have to bow before him.

Once again he had miscalculated. And that was the second surprise. The opposition to his move developed an unprecedented unanimity in the international community and in much of the Arab world. "This will not stand, this aggression against Kuwait," George H. W. Bush announced a few days after the invasion. And he meant it. The United States, making use of personal connections with other leaders that Bush had developed over twenty years, took the lead in marshaling and coordinating the opposition. Led by Secretary of State James Baker, it proved a far more successful and stunning diplomatic achievement than Saddam Hussein—or indeed many others—could possibly have expected. The

Iraqis had failed to recognize how drastically the interests and position of the Soviet Union, until recently an ally, had shifted. The United Nations did what the League of Nations had failed to do in the 1930s—imposed an embargo to frustrate aggression. But Kuwait was not the end of the matter. The disposition of Iraqi forces and the way they were being resupplied suggested that they might plunge on toward the lightly defended Saudi oil fields. Fearful that Saudi Arabia might well be next on Hussein's list, many countries hurriedly sent military forces into the region. American forces were by far the largest component, reflecting guarantees that went back to Harry Truman's letter to Ibn Saud in 1950.

The possible repercussions of the crisis were enormous for the 1990s and the twenty-first century. If successful in holding on to Kuwait, Saddam Hussein would directly control 20 percent of OPEC production and 20 percent of world oil reserves and would be in a position to intimidate neighboring countries, including other major oil exporters. He would be the dominant power in the Persian Gulf, well equipped to resume his war with Iran. He would have the economic freedom to take even larger steps.

The collapse of communism and the agonies of the Soviet Union had left only one superpower in the world—the United States. The absorption of Kuwait could start Iraq on the path to becoming a new superpower. Eleven years earlier, four out of five of the major producers of the Persian Gulf had been pro-Western. With Kuwait absorbed into Iraq, there would be only two friendly producers. George H. W. Bush summed up the dangers as he saw them: "Our jobs, our way of life, our own freedom and the freedom of friendly countries around the world would all suffer if control of the world's great oil reserves fell into the hands of Saddam Hussein."[3]

The public debate in the West was marked by a continuing search for a single factor to explain the response of the Bush Administration. But, as is usually the case in great events, there was not the luxury of a single explanation. Aggression, sovereignty, and the shape of the post-Cold War order were all central considerations. Different people in the United States reached for different analogies. Some warned of another Vietnam and the dangers of a quagmire. While George Bush himself was determined to avoid "another Vietnam," he was a product of his generation and his experience, and he thought back to the late 1930s, Adolf Hitler, and the origins of the Second World War. Fifty million lives had been lost in that conflict. Had Hitler been stopped at the Rhineland in 1936 or in Czechoslovakia in 1938—when Czechoslovakia had more tanks than Germany—those lives might well have been spared. Once again, here was a dictator who blatantly lied and dissembled, who was totalitarian in the way he ran his country, who was obsessed with weapons and power and seemed to have no scruples, and whose ambitions appeared to be unlimited. The Ba'th doctrines provided the rationale to reach far beyond Iraq's current borders. A greater Iraq, which had succeeded in absorbing Kuwait, would be well on its way to turning itself into a formidable nuclear weapons state.

This was the real significance of the "oil factor," the way that oil would be translated into money and power: political, economic—and military. Saddam Hussein knew what it would mean to acquire an additional 10 percent of world oil reserves, especially as not much in the way of population came with them. If he kept the grip on Kuwait, Iraq would be the planet's dominant oil power, and the other petroleum producers would bend to his *diktats*, just as they had begun to do in the summer of 1990, before the invasion. He would gain a decisive say over the world economy, and he would be courted by economic and political leaders. His aggressive arms procurement effort would be much enhanced, and suppliers of every kind of technology, eager to tap into the biggest market for weaponry and related technology, would in due course be at his door offering him the very latest. It would not take Saddam Hussein, equipped with nuclear and chemical weapons know-how, long to turn Iraq into a regional power and perhaps, as Hussein extended his reach, into a global superpower. At a certain point, it would become too costly and too dangerous to try to check him. And the post-Cold War order would turn out to be different and much less benign than was generally imagined—and hoped—at the beginning of 1990. In short, oil was fundamental to the crisis, not "cheap oil," but rather oil as a critical element in the global balance of power, as it had been ever since the First World War. Such is one of the great lessons of the last hundred years.

The New Oil Crisis

Owing to the disruption and the embargo, four million barrels of oil were abruptly removed from the world oil market—on the same scale as the 1973 and 1979 crises. The uncertainty was very high, and, as in previous crises, unsure companies and consumers built inventories. Oil prices skyrocketed, and financial markets plummeted. A new oil shock was at hand—the sixth post-war oil crisis.

OPEC was thrown into its worst crisis ever by the Iraqi invasion. It was now sovereignty and national survival and not merely the price of oil that was at stake, and most of the members explicitly stepped forward to increase production to compensate for the lost output from Kuwait and Iraq, further isolating Iraq and, in effect, underlining their commitment to a new alignment with their customers.

The sharp price rise was driven not only by the supply loss itself, but also by anxiety, fear, and anticipation of conflict. When, in late September 1990, Hussein threatened to destroy the Saudi petroleum supply system, prices on the futures markets leaped toward $40 per barrel, more than double what they had been before the crisis. The high prices reinforced recessionary trends in the U.S. economy. As crude prices rose, so did gasoline prices, accompanied by criticism and investigations. This time, however, in contrast to 1973 and 1979, there were no allocations or controls in the United States to hinder market responses, and neither gas lines nor any significant supply distortions resulted.

The global supply system responded both to higher prices and to urgent appeals for increased production. By December 1990, the lost production had been completely compensated for with "relief" oil produced from other sources. Saudi Arabia alone brought three million barrels per day of shut-in oil back into production, making up for three-quarters of the lost supply. Other major increments of additional supply came from Venezuela and the United Arab Emirates. But any country that could increase its production by 25,000 or 50,000 barrels per day also hastened to do so.

At the same time, demand was weakening, as the United States and other countries headed into economic recession, which in turn meant a reduction in the need for oil. While the International Energy Agency did not formally activate its emergency energy security program, it did take a leading role in informal coordination.

From an oil perspective, one big question stood out. Would the United States use its Strategic Petroleum Reserve, created in the middle 1970s and now holding about 600 million barrels of oil, in the event of further disruption? For a couple of months, there was a spirited debate as to "original intent." Was the SPR to be used only in the event of a "physical shortage," or was it also to be used to head off a major price spike that would seriously damage the economy? Some pointed out that a physical shortage might exist at $20 a barrel but would be eliminated at $40—though, in the meantime, a doubling of the oil price would deal a heavy blow to the economy. By November 1990, the debate was resolved. In the event of conflict, the principle of "early release," previously promoted by the Reagan Administration, would be applied, and the SPR might well be used to flood the market with oil, preventing sharp price hikes driven by a panic build-up of inventories, as had happened in 1973 and 1979.

Altogether, then, by late autumn, the supply-demand picture was improving day by day, and prices began to decline. Still, as the crisis dragged on toward the winter, the fundamental question remained: What would happen if military conflict actually started?

And, as irrational as that seemed, the prospect looked increasingly likely. Despite various diplomatic ploys and the manipulation of Western hostages, Iraq gave little indication that it would pull out from Kuwait. On the surface, it would have seemed, Saddam Hussein was running an enormous risk, but he did not necessarily think so. He was playing for time, which he was convinced was on his side. Iraq was moving quickly to absorb Kuwait, brutally and with terror, and to drive out the Kuwaiti population. At the same time, Hussein was convinced that he could wear down and outlast the unwieldy coalition arrayed against him. He had been 19 at the time of the 1956 Suez Crisis, and he had observed how Nasser had succeeded in splitting the Western alliance. Surely, he would find opportunities to do so with this much larger and ungainly coalition of nations. Somewhere, there would be a chance to play the "Israeli card," and so force the Arab countries out of the coalition. Or he could make overtures to some of the Western countries and sow dissension. Or he could draw off the Soviet Union. With time, he could find ways around the sanctions, or they would simply erode. With Vietnam in mind, as well as the swift U.S. withdrawal from Lebanon after the deaths of several hundred U.S. Marines in 1983, he fundamentally doubted American resolve.

The Bush Administration, too, recognized the factor of time, and that time would work against the thirty-three-nation coalition. How long could the united front be held together? How long would sanctions hold? And how long before Saddam Hussein had effectively dismantled and destroyed Kuwait and

"Iraqized" it? The Soviet Union was in a new and highly unstable situation, with its own political system under great stress. The Soviet military had long had very close ties to Iraq. Would the USSR shift positions away from the coalition and back toward Iraq? And how long would the American public go along with the commitments that had been unfolding since August 2?

The Bush Administration ruefully came to the same conclusion as Saddam Hussein: The longer the crisis went on, the greater the chances that Hussein would be able to claim a "victory." At the end of October and in the first days of November, the U.S. government decided that the coalition had to be prepared to do more than defend Saudi Arabia. It had to put in place an offensive capability. On November 8, Bush announced a large increase in U.S. forces in the Gulf "to insure that the coalition has an adequate offensive military option." That meant a doubling of U.S. forces in the region.

Still, there was propaganda but no movement from Baghdad. Saddam Hussein had imposed 500,000 casualties on his countrymen in the Iran-Iraq War—the equivalent, if adjusted to population, to 7.5 million casualties for a country the size of the United States. This was the result of his own blunder, but he hardly showed any remorse. Indeed, in the huge monument Saddam built to Iraq's "victory" in the Iran-Iraq War, the giant hands holding the two swords were models of his own hands. While Saddam could force further large casualties on Iraq, he doubted that the United States would be willing to absorb even a tiny fraction of the casualties. He looked down on the United States and, like Hitler before WWII, thought its people soft and flabby, without staying power. He had signalled his thinking at a meeting with the U.S. ambassador to Iraq in late July, eight days before the invasion, when he had disdainfully declared that America "is a country which cannot accept 10,000 deaths in one battle." Now, to drive home his point, he flaunted the use of chemical weapons. But Saddam still was not taking seriously the clear signals from the capitals of the coalition members, and he showed continuing indication of underestimating George Bush. Was Hussein miscalculating again?

Now, as had happened before in the twentieth century, the clock began to tick unmistakably. On November 29, the United Nations Security Council passed Resolution 678, giving Iraq "a pause of goodwill"—until January 15, 1991—to comply with Resolution 600 and withdraw from Kuwait. Otherwise, "all necessary means" would be employed to ensure compliance. A variety of public figures—ranging from ex-prime ministers of various coalition partners to a would-be Democratic Presidential candidate to a retired boxer—were trooping to Baghdad to promote peace plans and to help free hostages. In December, Saddam released several hundred foreign hostages, thinking that such an act would undercut the coalition's resolve. But it did not work as he had anticipated, especially as knowledge of Iraqi atrocities in Kuwait seeped out to the rest of the world.

The waiting continued. At newspapers and television stations around the world, newsrooms were being reconfigured to provide for "war desks," and editors and producers were figuring out how they would deploy their staffs in the event of military conflict. And yet many did not believe, fundamentally, that their plans would have to be acted upon. Rationality would prevail; Saddam

Hussein would surely find a way to extricate himself, potentially leaving the leaders of the coalition highly exposed, perhaps even looking quite foolish.

On January 9, 1991, Secretary of State James Baker met in Geneva with Iraqi Foreign Minister Tariq Aziz. Could the deadlock be broken? After more than six hours of talks, a grim Baker emerged to report that he had found no flexibility in the Iraqi position and that Baghdad was about to make "yet another tragic miscalculation." Baker had sought to give Aziz a personal letter from President Bush to carry to Hussein, but Aziz refused to accept it.[4]

On Saturday, January 12, the U.S. Congress concluded a three-day debate by giving the President the authority to go to war—by a slim vote of 52 to 47 in the Senate and, with somewhat more of a margin, by 250 to 183 in the House of Representatives. Many of those who had voted in favor of the resolutions had done so tepidly, and there continued to be a recurrent call to let sanctions work. Protests were held across the United States, and large demonstrations against the coalition were held throughout Western Europe. George H. W. Bush looked lonely and isolated.

The deadline, January 15, came and went. There was no last minute maneuvering—but rather, a grim silence. The "pause of good will" had passed. Would the coalition make good on its authority? It was up to George H. W. Bush. Perhaps the President would let several weeks or even a month go by and allow sanctions to work further. On January 16, Bush talked with two clergymen. He had already publicly warned that if Iraq did not begin withdrawing, the coalition's response to the Iraqi aggressions against Kuwait would be massive—and that it would be swift. And indeed it was. In the early morning hours of January 17, Gulf time, 700 coalition aircraft launched a huge assault on Iraq.

"The Mother of All Battles"

The Gulf Crisis had turned into a war after all, although as some pungently pointed out, the war had actually begun when Iraq had invaded Kuwait on August 2. The air war continued for a month, with systematic attacks on Iraqi command and control centers and a broad range of military and strategic targets. The biggest surprise, at least for the U.S. Air Force, was not that the coalition aircraft and missiles thoroughly knocked out Iraq's air defense capability, but that they did it so easily and so quickly, with so minimal a loss of aircraft.

The scale and impact of the first night's air assault was the key to the reaction in the oil market. At first word, the oil price did as might have been expected—spiked $10 a barrel, from $30 to $40. Within hours, however, it plunged $20—back toward $20 a barrel, below what it had been even prior to the invasion. The supply situation had continued to improve; there was no doubt that the Strategic Petroleum Reserve would be used if needed; demand was falling with the passing of the winter peak. Now the initial air attack appeared to have destroyed Iraq's ability to impose any serious injury on the Saudi supply system. Fear was, thus, removed from the oil price, and the realities of physical supply and demand drove the price down. As a result, the oil price was simply taken off the table at the start of the war, something that would have been impossible two or three months earlier.

The Iraqis responded to the air war with an air war of their own—launching their modified Soviet Scud missiles against Israel and Saudi Arabia. The Iraqis may well have hoped that an attack on Israel would incite Israeli entrance into the war, which they were sure would split the Arab members from the coalition and create in particular an untenable situation for Saudi Arabia. Or perhaps, they could trigger an early initiation of the ground war by an ill-prepared coalition. But the Israelis, under great pressure, held their fire. The Scud attacks did inspire intense fear because of the thought that they would be carrying chemical weapons. But, as events turned out, they never did, and the actual devastation from the Scuds was limited.

As the air war continued, Saddam Hussein was promising that "the mother of all battles" would commence with the opening of hostilities on the ground. But when the ground war began five weeks after the start of the air battle, the predicted "mother of all battles" turned into a rout. The Iraqi soldiers were demoralized, ground down by the air war, restricted by doctrine, and not at all keen to sacrifice themselves to the glory of Saddam Hussein if they could avoid it.

Moreover, the allies had expertly applied deception. The U.S. commander in the Gulf, General Norman Schwarzkopf, kept a copy of a book by the German General Erwin Rommel on the table beside his bed. Rommel was not only the master of mobile warfare, but an expert on fighting in the desert—and he had learned firsthand, in North Africa, the strategic significance of oil. Schwarzkopf imbibed Rommel's strategic lessons, and he had no intention of being drawn into a direct assault on Iraqi positions. "A war in the desert," Schwarzkopf had observed, "is a war of mobility and lethality." Schwarzkopf directed an extensive campaign, including training exercises, to convince the Iraqis that the allies would be attacking head on directly and with a huge amphibious assault. At the same time, substantial forces were being secretly moved far out into the Saudi desert, and, when the ground war began, they swung in a huge arc from the west, coming in behind the entrenched Iraqi positions and cutting them off. The ground war took no more than a hundred hours, and it ended with Iraqi forces in full retreat.[5]

But Saddam's forces had already released what was by far the largest oil spill in history. And they now withdrew from Kuwait with vengeance and vindictiveness. If Hussein could not have Kuwait, he would try to destroy it. In contrast to Hitler's troops who disobeyed the Fuehrer's order in 1944 to set Paris afire as they departed, Hussein's soldiers left Kuwait burning. Upwards of 600 oil wells were set aflame, creating a hellish mixture of fire and darkness and choking smoke and gross environmental damage. As much as six million barrels of oil a day were going up in flames—considerably more than Japan's daily oil imports.

A cease-fire went into effect on February 28, 1991. Meanwhile, rebellions broke out among the Shias, in the southeast of Iraq, and in the north among the Kurds (whose homeland, owing to oil prospects, had been made part of Iraq when the country was invented in 1920). The two uprisings were now taking place in the regions where Iraq's current oil production was concentrated. The Iraqis brutally put down the uprisings, although not before millions of people became refugees. The coalition's forces had stopped short of Baghdad. The coalition partners had expected embittered military officers to quickly topple Saddam Hussein in a coup, but they had underestimated his grip on the country and his fanatical devotion to his own security. Despite all the devastation he had brought on his own country, Saddam Hussein clung to power in the aftermath of the Gulf War. But he no longer had an offensive military machine.

But was the Gulf Crisis really over?

The Lessons of Security

After the 1973 oil shock, it was clear that the oil companies could not and would not manage future crises by themselves, and that it was up to governments to take on that role. In the years after, the industrial countries developed an energy security system built around the International Energy Agency and the strategic stockpiles, such as the U.S. Strategic Petroleum Reserve and similar reserves in Germany and Japan and other countries, which can be brought into play to avert a shortfall and counteract a panic. The IEA provides a framework for coordinated response and for the exchange of timely, accurate information among nations—an absolute requirement to head off any such panic or prevent scrambles for supply. The years of oil crises demonstrated that, given time, markets will adjust and allocate. Those years also provided evidence that governments do well to resist the immediate temptation to control and micromanage the market. Of course, it is hard for governments to resist action when uncertainty is high, panic is building, and accusations are mounting.

Yet the course of the six major disruptions from the early 1950s through 1991 has revealed that the logistical and supply system can adapt to such an extent that the shortages ended up being less dire than had been expected. Indeed, the real problem in the 1970s turned out not to be an absolute shortage, but the disruption of the supply system and the confusion over ownership of oil, with the consequent rush to reorder the system under conditions of high uncertainty. And in 1990 and 1991, the lessons of previous crises, along with the mechanisms developed since the 1970s and improved information, made the impact of the disruption that came with the Gulf Crisis less serious than it might otherwise have been.

Even if experience points the way to better-managed reactions, there are other important questions. During the oil crises in the 1970s, the United States political system was paralyzed in the face of one of the biggest and most costly disruptions of the postwar era. Anger, finger-pointing, scapegoating—all became a substitute for the development of a rational reaction to a very serious problem. Watergate, of course, was part of the explanation. Still, the spectacle of that fragmented, contentious response, provided reason to ponder how, even after the Gulf Crisis, the United States would respond over the long haul to future energy needs and crises.

The Third Environmental Wave

Even as the world continued to move on oil and the economy to live on oil, a new challenge to Hydrocarbon Society emerged, this time from within, portending a great confrontation that would probably affect the oil industry and, indeed, our way of life in the years ahead. The industrial world was now facing a resurgent wave of the environmental movement. The first, in the late 1960s and early 1970s, focused on clean air and water and had a prominent "made-in-America" label on it. It had major energy implications, for it provided the impetus for the rapid switch from coal to fuel oil, which was one of the main forces that tightened the world oil market so quickly, setting the stage for the 1973 crisis. During the 1970s, as security came to the fore and hard economic times led to renewed emphasis on jobs and economic performance, the environmental movement lost some of its momentum. In a second wave, it was more narrowly focused, with much concentration on slowing or stopping the development of nuclear power. It succeeded in so doing in most of the major industrial countries, decisively altering what had been assumed to be the major path of response to the oil crisis.

A powerful third wave began to rise in the 1980s. It engendered wide support, cutting across traditional ideological, demographic, and partisan differences. It was an international phenomenon, born as much in Europe as in North America, whose concerns include every environmental hazard from the depletion of the tropical rain forests to the disposal of waste products and—increasingly—climate change.

Perhaps the single decisive event that catalyzed the new wave of environmentalism had its beginnings in April 1986, when operators of a nuclear reactor in Chernobyl, in Ukraine, in the Soviet Union, lost control. The reactor itself was consumed in a partial nuclear meltdown, and clouds of radioactive emissions spewed forth and were carried by winds across vast stretches of the European continent. The initial reaction of the Soviet government was denial, denouncing the reports of a nuclear disaster as a creation of malevolent Western media. As the days went by, however, rumors reached Moscow of riots at the train station in Kiev, of mass evacuations, of deaths and disaster. International criticism mounted. Still, a blanket of silence was maintained, fueling speculation about horrible disasters. Finally, more than two weeks after the accident, Mikhail Gorbachev went on television. His speech was wholly uncharacteristic of Soviet leadership and represented a sharp break with the way the Kremlin had traditionally communicated with its own people and the rest of the world. There was no propaganda, no denial, in the speech; instead, it was a serious, somber admission that a grave accident had indeed occurred, but steps were now under way to control it. Only then did the Soviet people and the rest of the world realize how incredibly dangerous the first few days had been. Some in the Soviet leadership would afterwards say that Chernobyl had been a major political turning point within the USSR for glasnost and perestroika. Those in Western Europe who had blamed all environmental ills on Western capitalism were forced to rethink their ideology. In both Eastern Europe and the Soviet Union, environmentalism became one of the most important rallying points for opposition to communism, and with good reason. For with the parting of the Iron Curtain, it was revealed that among the leading legacies of cynical communist rule was a frightening pattern of environmental degradation and disasters, some of them perhaps irreversible. Environmental considerations are proving to be among the top issues for the new democratic governments of Eastern Europe.

Chernobyl, with its threat of invisible but deadly danger and its warning of technology out of control, provided a great thrust to the new wave of environmentalism. It helped trigger the "green" movement in Europe. In the United States, the Clean Air Act of 1990 was a landmark in terms of controlling air pollution and smog. But already, a broader—indeed, a global—environmental agenda was beginning to emerge around climate change and global warming.

Yet the 1990s had begun not with another environmental drama, but rather with a struggle over the oil resources of the Persian Gulf, on which the world was again becoming heavily dependent. The Gulf Crisis thrust energy security back onto the political agenda, spurring governments to focus anew on ensuring supplies. But that quest for security would increasingly coexist—and sometimes seem to be at odds—with the third wave of environmentalism.

The Age of Oil

The cry that echoed in August of 1859 through the narrow valleys of western Pennsylvania—that the crazy Yankee, Colonel Drake, had struck oil—set off a great oil rush that has never ceased in the years since. And thereafter, in war and peace, oil would achieve the capacity to make or break nations, and would be decisive in the great political and economic struggles of the twentieth century. But again and again, through the never-ending quest, the great ironies of oil have been made apparent. Its power comes with a price.

Over its entire history as an industry, oil has brought out both the best and worst of our civilization. It has been both boon and burden. Energy is the basis of industrial society. And of all energy sources, oil has loomed the largest and the most problematic because of its central role, its strategic character, its geographic distribution, the recurrent pattern of crisis in its supply —and the inevitable and irresistible temptation to grasp for its rewards. This history has been a panorama of triumphs and a litany of tragic and costly mistakes. It has been a theater for the noble and the base in the human character. Creativity, dedication, entrepreneurship, ingenuity, and technical innovation have coexisted with avarice, corruption, blind political ambition, and brute force. Oil has helped to make possible mastery over the physical world. It has given us our daily life and, literally, through agricultural chemicals and transportation, our daily bread. It has also fueled the global struggles for political and economic primacy. The fierce and sometimes violent quest for oil—and for the riches and power it conveys—will surely continue so long as oil holds a central place. For ours is a civilization that has been transformed by the modern and mesmerizing alchemy of petroleum. This is what has made it the age of oil.

Epilogue
The New Age of Oil

HARDLY A DAY GOES BY in which oil—whether in terms of its price, its impact on the economy, its role in international relations and in the environment—is not in a major newspaper story or in the television news or a hot topic on the blogs.

The questions are many. How does oil change international politics and the strategies and positions of nations? What are the political and economic risks that come with oil, and how to manage them? Is the world going to run out of oil? Or is demand going to change? How, within a ten-year period, could oil be as low as $10 a barrel and as high as $147.27, and then drop by more than a hundred dollars before rebounding, and what is the prospect for prices? There is the whole question of climate change. What is the future for Hydrocarbon Man?

And yet none of these questions, in their essence, is new. In one form or another, they play out again and again across the pages of *The Prize*. Indeed, it is hard to make sense of these questions today without understanding where they come from and how oil has come to have such a defining role in the modern world, in everything from daily life to the game of nations. From these pages readers can draw many lessons and insights that are relevant to sound energy policy, to energy security, and—it is hoped—to clear thinking about energy.

The competition for oil and the struggle for energy security seem to never end. And yet, with the swift victory to the Gulf War in February 1991, the strategic struggle over oil did appear to be over. The threat that a hostile power would dominate the Persian Gulf was no more. That, it now seemed, was part of a larger transformation. For the year that began with Operation Desert Storm in Iraq ended in December 1991 with Mikhail Gorbachev, president of the Soviet Union, going on Russian television to deliver a twelve-minute speech in which he announced what would have seemed almost impossible a few years earlier: the dissolution of the Soviet Union. The communist empire had collapsed, the

Soviet Union had disintegrated, and the Cold War had ended. The threat of nuclear war that had hung over the planet for four decades was lifted, and a new era of peace was at hand.

Although the Soviet Union had been a significant oil exporter, its industry had been isolated behind the Iron Curtain. No longer. With the breakup of the Soviet Union, the petroleum industry of the Russian Federation and the newly independent states, notably Kazakhstan and Azerbaijan, would be integrated with the global industry. Eventually, after years of wrangling, the Baku-Tbilisi-Ceyhan pipeline would link historic Baku, on the Caspian Sea, to a Turkish port on the Mediterranean—in part, a twenty-first-century parallel to the route pioneered by the Nobels, Rothschilds, and Samuels in the late nineteenth century. This pipeline, by providing an alternative to shipping oil through the Russian pipeline system, would help to underwrite the position of those newly independent states of the former Soviet Union.

Off the Agenda?

As it was, in the early 1990s, the outcome of the Gulf War and the collapse of the Soviet Union transformed the international system. Some spoke optimistically of a new world order. The focus of the international community shifted from security to economics and growth and to what was coming to be known as globalization. In the years that followed, there was a vast expansion of international trade, as globalization led to a more open and interconnected world economy and to rising incomes in what had heretofore seemed permanently poor countries.

For most of the 1990s, oil receded as a grand strategic issue. Petroleum supplies were abundant, and prices were low. Much attention was given to the "East Asian economic miracle" and to what was beginning to appear behind it, the emerging role of China in the world economy.[1] But, in 1997–98, Asia's economic miracle, stoked by currency flows and real estate speculation, overheated and then, beginning in Thailand, blew up. The result was a lethal contagion—an epidemic of financial panic, bankruptcies, and defaults and a deep economic downturn that spread across much of Asia (though not China and India) and then engulfed other emerging markets, including Russia and Brazil.

The collapse in GDP led to a drop in oil demand even as oil supplies were increasing. As a result, inventory tanks filled to overflowing until there was no place to put the additional oil. Once again, as in 1986, the price of oil collapsed toward $10 a barrel and, in some cases, even lower. Oil exporters were once more thrown into economic disarray, as in 1986. The collapse in oil prices sent Russia, only in its seventh year as an independent country, into default and bankruptcy and into what turned out to be an agonizing reappraisal of its relationship with the rest of the world.

But for the oil-importing nations—both for developed countries such as the United States, Japan and those of Europe, and for many developing countries— the fall in oil prices was like a giant tax cut, a stimulus package that fired up economic growth. It put a lid on inflation, permitting faster growth. At the gasoline pump in the United States, it pushed prices down, in inflation-adjusted terms, to the lowest level they had ever been. This ignited a great new romance—a passion for fuel-inefficient SUVs and other light trucks, which would soon comprise half of the new vehicles sold in the United States.

Restructuring

Low prices put great pressure on the structure of the industry. As revenues fell away, company managements struggled to find survival strategies. Budgets had to be immediately cut, and projects were either postponed or cancelled altogether. There was another way to survive as well. That was by getting bigger, gaining greater scale. The objective: to bring down costs and increase efficiency. The need for corporate scale in this environment was made more urgent because of the bigger and more complex oil and gas projects that lay ahead and the much greater financial resources that would be required to make them happen. In the 1990s, mega-projects, many of them in offshore waters, might have been defined in hundred of millions of dollars, perhaps even a billion. But the term "mega-project" would need redefining, as the industry was beginning to plan for projects in the twenty-first century that would cost $5 or even $10 billion.

All of this created the imperative for what became known as restructuring. That meant reshaping not only individual companies but the industry itself. The oil majors that the Italian tycoon Enrico Mattei had dubbed the Seven Sisters (minus Gulf, which was already gone) would be remade. The majors combined to become supermajors. BP merged with Amoco to become BPAmoco, and then merged with ARCO, and emerged as a much bigger BP. Exxon and Mobil—once Standard Oil of New Jersey and Standard Oil of New York—became ExxonMobil. Chevron and Texaco came together as Chevron. Conoco combined with Phillips to be ConocoPhillips. In Europe, what had once been the two separate French national champions, Total and Elf Aquitaine, plus the Belgian company Petrofina, combined to emerge as Total. Only Royal Dutch Shell, already of supermajor status on its own, remained as it was. Or, rather, it went through a self-merger. It finally did away with the complex system of two separate holding companies, Royal Dutch and Shell, run from the Hague and London, that Henri Deterding and Marcus Samuel had forged in 1907 as their grand bargain. Instead, it became a unitary company in order, among other things, to improve the efficiency of its operations and speed up decision making. With all these mergers, the landscape of the international oil industry changed.

Overall, in the late 1990s, in the minds of the wider public and many policy makers, oil faded away. So did concerns about energy security. People assumed, if they thought about it at all, that petroleum would be cheap and readily available for years to come. Instead, there were new things and "new new" things about which to get excited. Specifically, that meant the Internet, which brought the New Economy and a revolution in communications. The world would be interconnected twenty-four hours a day and distance would disappear. Information technology, start-ups, Silicon Valley, cyberspace—those were the places to be. Few things seemed as old economy as the petroleum industry, and its relevance seemed to decline. Fewer young people were interested in pursuing jobs in the industry, which was just as well, as there were fewer jobs to pursue.

The Return of Oil

But three things in the first half of this decade were to change the picture.

The first was September 11, 2001. What had been unthinkable, and yet had been forewarned in a passing paragraph in a Presidential Daily Brief in August 2001, took place with the crashing of two hijacked airliners into the World Trade Center and a third into the Pentagon and the planned attack with a fourth on the U.S. Capitol that was aborted over Pennsylvania. For the first time since the assault on Pearl Harbor on December 7, 1941, which had taken the United States into World War II, the United States had been attacked, and with great loss of life. The enemy was the jihadist Al-Qaeda movement.

International relations were transformed. In the autumn of 2001—responding to the September 11 attacks on New York and Washington—the United States and its allies counterattacked in what became known as the "war on terror." They carried the war back to Afghanistan, the base from which Al-Qaeda operated. They quickly drove the ruling Taliban, Al-Qaeda's ally, from power, and achieved what at the time seemed to be a swift victory.

Attention turned back to Iraq. The victory in the Gulf War had also been swift. But it had not been complete. Fearful of a quagmire and the risks of an occupation, the American-led coalition had stopped short of Baghdad in 1991. Saddam Hussein had remained in power, though contained by economic sanctions, isolation, inspections, and no-fly zones in the north, over Kurdistan, and in the predominantly Shia south. But now, after 9/11, proponents of striking at Iraq argued that Saddam had links to Al-Qaeda and was still secretly developing weapons of mass destruction. President George W. Bush determined to launch this war with the advice of some of his father's former top advisers, now in his own administration—but also against the strong warnings of other of his father's advisers. "An attack on Iraq at this time would seriously jeopardize, if not destroy, the global counterterrorist campaign we have undertaken" cautioned his father's national security adviser, Brent Scowcroft, in August 2002. "It would not be a cakewalk. . . . If we are to achieve our strategic objectives in Iraq, a military campaign in Iraq would likely have to be followed by a large-scale, long-term military occupation."[2] But momentum toward war was very strong.

On March 20, 2003, some twelve years and twenty-one days after the end of the earlier Gulf War, the Iraq War began. Historians may well come to call this the Second Gulf War. This time the coalition was much smaller in terms of the number of participating countries—"the coalition of the willing." Britain was the most important partner. Other of America's key allies, specifically France and Germany, opposed war and did not join the coalition. They thought that the U.S. administration was too optimistic and underestimated the risks and the difficulties that woud be waiting in postwar Iraq.

The clear assumption among the war's proponents was that it would be quick—a "lightning victory."[3] The actual war went pretty much as planned and pretty quickly. Already by April 9, 2003, Iraqi civilians and U.S. marines were joining together to pull down the giant statue of Saddarn Hussein in the center of Baghdad. But virtually nothing that followed went as planned. Saddam disappeared into hiding. No weapons of mass destruction were ever found. Multiple insurgencies developed across the country, as out-and-out civil war developed between Sunnis and Shias. More than half a decade after the war began, U.S. troops were still in Iraq; Iraqi politicians were still arguing over responsibility for the oil resources between the central government and the regions; and the Iraqi oil industry, short of technology and skills and security, was still struggling to regain the levels of production that had preceded the war.

Yet, while violence continued to dominate Iraq, striking changes were happening elsewhere in the Persian Gulf and the wider Middle East. In a dramatic reversal, Libya renounced nuclear weapons in December 2003 and rejoined the international community. With the buildup of oil and natural gas revenues, the emirates of Abu Dhabi, Qatar, and Dubai emerged as key players and new centers of the global economy in the twenty-first century. When a tumultuous credit and banking crisis swept the United States and Europe in 2007 and 2008, some of these emirates helped to bail out Western financial institutions.

But the mission the Bush administration had assigned itself—to bring democracy to the region—proved short-lived. Instead, there was widespread apprehension that the ultimate victor from the Iraq War could be Iran, whose Islamic revolution had set off the second oil shock in 1978 and which now saw itself as the regional power in the Gulf—the very role that first the Shah of Iran and then, after him, Saddam Hussein had sought to capture. But it was also clear that neither Saudi Arabia nor the other Gulf countries had any intention of being under such a sway.

The second key feature of this era has been globalization.[4] Between 1990 and 2009, the world economy almost tripled in size. And by 2009, a significant share of the world's GDP was being generated in the developing world, rather than in the traditional grouping of North America, Europe, and Japan.

The New Economy and the Internet notwithstanding, globalization made oil more important again. Of key significance was the period 2003–2007, which saw the best global economic growth in a generation. High economic growth and rising incomes in China, India, the Mideast and other emerging countries meant strong growth in oil demand—to power industry, to generate electricity, and to fuel the rapidly growing fleets of motor cars and trucks.

This surge in oil demand—the third feature—caught by surprise not only consumers but also the global oil industry itself. The preceding decades of slow growth in demand had translated into relatively low levels of investment in new oil and gas supplies. In the late 1990s and first years of the next decade, Wall Street had demanded that the industry be "disciplined"—very cautious and even restrictive in its investment—or face retribution in terms of a lower stock price. Now the industry had to play catch-up in terms of investing in new capacity to produce oil. That effort could not be mustered overnight, or even in a few years. The balance between demand and available supply narrowed dramatically. Geopolitics of one kind or another further constrained supply. At the end of 2002 and in early 2003, strikes and political conflict in Venezuela temporarily shut down its oil production, the first step on the staircase of rising prices. Beginning in 2003, attacks by militias and criminal gangs disrupted output in Nigeria, one of the world's leading suppliers—sometimes cutting production by as much as

40 percent. Over the following years, production capacity declined in both Venezuela, where President Hugo Chávez had placed tight political controls on the national industry, and Mexico, where domestic politics restricted needed investment.

Russia's oil output had plummeted in the 1990s after the collapse of the Soviet Union. But it had begun to recover in the late 1990s and then grew by 50 percent in the first half of the following decade. At times it has been the world's largest producer of oil, ahead of Saudi Arabia. But then the growth rate of its production slowed and then flattened out altogether.

A New Oil Shock

A surge in demand, a slow response in supply, and a narrow balance between the two—this mixture would have led to higher prices in any event. But it was amplified by Iran, which was having, as it recurrently had had over four decades, its own sharp impact on world oil. Iran had relaunched an aggressive program of nuclear development, which included the fuel enrichment technology that would enable it to move easily to nuclear weapons. While trumpeting its program, Iran insisted that it was only seeking to develop civilian nuclear power. The European Union and the United States observed that Iran possesses the second-largest natural gas reserves in the world. They had little doubt that Iran's real objective was a nuclear weapons capability. Certainly, the prospect of a nuclear-armed Iran inevitably caused deep apprehension in Israel when the Iranian president repeatedly threatened to "eliminate this disgraceful stain from the Islamic world" and declared that Israel "must be wiped off the map."[5] Moreover, especially among EU countries, Iran's nuclear ambitions were seen as a major risk for proliferation, as a nuclear Iran might trigger a nuclear race in the Middle East. In these circumstances, an "Iranian premium"—concern over whether stalemate and confrontation over Iran's nuclear program would lead to conflict and threaten oil flows through the Strait of Hormuz—became an additional element in a higher oil price.

Two further factors drove prices to unprecedented levels. One was a dramatic increase in the costs of developing new oil and gas fields—more than doubling between 2004 and 2008. This arose because of shortages—of skilled people, equipment, and engineering capabilities—combined with a rapid rise in the price of other commodities, such as steel, that are required to build offshore oil platforms and other equipment.[6]

The other was the growing involvement by financial investors in oil and other commodities. Oil came to be seen as an asset class that provided an alternative to stocks, bonds, and real estate for pension funds, university endowments, and other investors seeking higher returns. At the same time, traditional commodity investors, speculators, and traders also put more money on the table. The complex role of financial players in the oil market became a very contentious question as people argued over the role of investors and the impact of speculation in the oil price. Continuing weakness of the dollar against the euro and Japanese yen further drove up the price of oil and other commodities as investors sought to hedge against the dollar's decline.[7] A strengthening dollar would be accompanied by a reverse in the oil prices.

Expectations became important as oil prices steadily rose from 2003. There was a widespread apprehension, especially within financial markets, that demand from China and India would go through the roof and that an oil shortage was inevitable in the next several years. All these factors—supply and demand, geopolitics, costs, financial markets, expectations, and speculation—came together to carry oil prices from $30 at the beginning of the Iraq War through $100 and $120 and then over $145 a barrel in 2008. By that point, expectations had created a bubble in which the price was increasingly divorced from the fundamentals. For, as prices went up, demand had—inevitably—begun to weaken.

The oil shocks of the 1970s were precipitated by specific events—the 1973 Yom Kippur War and the Arab oil embargo, and the 1978–79 Islamic revolution in Iran. This time was different in that there was no single dramatic event. Yet there was no question but that the extraordinary rise in prices constituted another oil shock of its own. In the United States, the painful economic impact of the oil shock was made much, much worse by the credit crisis that erupted in mortgage and banking sectors. Moreover, the shock was increasingly felt around the world. But it was only when global oil demand declined—in response to high prices, a financial crisis worse than any since the Great Depression, and a world recession—that the price collapsed.

The preceding half-decade rise in oil prices had brought significant changes in the global economy and dramatic shifts in income. Trillions of dollars flowed from oil-importing countries to the exporters—one of the greatest transfers in income in the history of the world. The accumulation of oil wealth in the savings accounts of the exporters—their sovereign wealth funds—has made many of them powerful forces in the world economy.

The economic shifts also brought political consequences. One of the major themes of *The Prize* is the continuing struggle between consumers and producers over the money and power that accrue from petroleum resources. This is a balance that is always shifting. In the era of high prices, what is called resource nationalism had again come to the fore, although in many different forms. Oil wealth enabled Venezuela's President Chávez to expand his influence over Latin America and pursue his agenda of "socialism for the twenty-first century" across the world stage. In 1998 Russia had been bankrupt. A decade later, bolstered with almost $800 billion dollars of foreign reserves and savings in its sovereign wealth funds, Russia was projecting its power and influence around the world. Its role as an oil exporter and as the key natural gas exporter to Europe—and seven years of strong economic growth—had put it in a new position of primacy—though one that eroded with the financial collapse. In other countries, the government decision-making that is required for new petroleum development slowed down and stalled, and thus the development of new resources also slowed.

"NOCs"—National Oil Companies

It turned out that the restructuring of the world oil industry that had started with the emergence of the supermajors at the end of the 1990s was only the beginning. One more merger—of Norway's Statoil and Norsk Hydro—created Stat-oilHydro, a new supermajor, although partly state-owned. But the balance be-tween companies and governments has shifted dramatically. Altogether, all the oil that the supermajors produce for their own account is less than 15 percent of total world supplies. Over 80 percent of world reserves are controlled by gov-ernments and their national oil companies. Of the world's twenty largest oil companies, fifteen are state-owned.[8] Thus, much of what happens to oil is the re-sult of decisions of one kind or another made by governments. And overall, the government-owned national oil companies have assumed a preeminent role in the world oil industry.

In these circumstances, the cast of characters in the world has become more complex, with a host of companies joining the ones whose names run through this book. Some of these companies have already appeared in these pages; some are new. Saudi Aramco—the successor to Aramco, now state-owned—remains by far the largest upstream oil company in the world, single-handedly producing about 10 percent or more of the world's entire oil with a massive deployment of technology and coordination. The major Persian Gulf producers control for the most part their production, as do the traditional state companies in Venezuela, Mexico, Algeria, and many other countries. The Chinese compa-nies—partly state-owned, partly owned by shareholders around the world—continue to produce the majority of oil in China but have also become increasingly active and visible in the international arena. So have Indian compa-nies. The Russian industry is led by state-controlled giants Gazprom and Ros-neft and by privately held companies, such as Lukoil and TNK-BP, that are majors in their own right.

Petrobras, the Brazilian national oil company, is 68 percent owned by investors and 32 percent by the Brazilian government, though the gov-ernment retains the majority of the voting shares. Petrobras had already estab-lished itself at the forefront in terms of capabilities in exploring for and developing oil in the challenging deep waters offshore. Beginning with the Tupi find in 2006, potentially very large discoveries are being made in what had heretofore been inaccessible resources in Brazil's deep waters, below salt deposits. These discoveries could make Petrobras—and Brazil—into a new powerhouse of world oil. Malaysia's Petronas had turned itself into a significant international company, operating in 32 countries outside Malaysia. State companies in other countries in the former Soviet Union—KazMunayGas in Kazakhstan and SOCAR in Azerbaijan—have also emerged as important players. While Qatar is an oil exporter, its massive natural gas reserves put it at the forefront of the liq-uefied natural gas industry (LNG) and, along with Algeria's Sonatrach and other exporters, at the center of growing global trade in natural gas.

China's companies are increasingly prominent in the global industry. For a few years, there was fear that a battle for oil resources between the United States and China was almost inevitable. Despite disagreements between the two coun-tries over specific issues, the overall fear seemed to fade as it became clear that both countries share common interests as oil importers and consumers and that energy is part of a larger framework of economic integration and overlapping in-terests between the two countries.

"Running Out?"

The rise in oil prices between 2003 and 2008 fueled the fear that the world is running out of oil. This anxiety has taken on a name—"peak oil." Yet such fears are also part of a long tradition in the world. For the current language and apprehensions bear striking resemblance to earlier periods. As described in these pages, there was a widespread conviction in the 1880s that when the wells ran dry in western Pennsylvania, the days of oil would be over. Similar fears were registered in the years after World War I. Such concerns reappeared in the years after World War II, with memories fresh of the strategic role of oil in the war, and as the locus of world production shifted, in accord with Everette DeGolyer's 1944 prophecy, from the Gulf of Mexico to the Persian Gulf. That same conviction about shortage underlay the panic that gripped the oil industry and the world community during the oil crises of the 1970s. In each case, new territories, new horizons, and new technologies banished the fears within a few years and shortage gave way to surplus. A new surplus developed with the recession in 2008 and 2009.

But is this time different? That is a contentious question, which arouses strong passions. It is also a question that appropriately requires thoughtful, careful analysis, for the stakes are very high. Field-by-field analysis suggests that there are ample resources below ground to meet world demand for several years.

But there are three important qualifications. The first is that the aboveground risks—geopolitics, costs, government decision making, complexity, restrictions on access and investments—can hamper development and lead to tight supplies and high prices. It is striking to see the shift toward a focus on these aboveground risks as the key factors. The second is that an increasing share of liquid supplies will be nontraditional oil—whether from such challenging environments as the ultra-deep offshore waters and the Arctic, or Canadian oil sands, or from the associated liquids that are produced with the growing volumes of natural gas. Many of these nontraditionals are more complex and difficult—and expensive—to bring on. The third is the recognition of the sheer scale in the years ahead of demand growth in the new giants, China and India, and other developing economies and the enormous challenge of meeting it.

Climate Change

One further factor critical to the future of oil emerged as a decisive factor only after the turn of the century: climate change. Initially, representatives of eighty-four countries signed the 1997 Kyoto Protocol aimed at reducing CO_2 emissions. The European countries later adopted the treaty and made climate change a cornerstone of their policies. But the U.S. Senate effectively rejected the Kyoto treaty by a vote of 95 to 0. There were three main concerns. The first was the impact of CO_2 restrictions on the overall economy and economic growth. The second specifically concerned restrictions on coal, from which half of the nation's electricity was generated. And the third was that the treaty would require cutbacks from the industrial countries, but not developing countries.

A decade later, attitudes have changed dramatically in the United States. The climate change issue has been embraced across almost the entire U.S. political spectrum, and it is generally expected that a national climate change regime will be enacted. Yet complex questions are still to be addressed. There is a debate as to the costs of moving to a lower-carbon-emissions society as well as the choice between a cap-and-trade system and a carbon tax. Half of U.S. electricity is still generated from coal; and how to move toward a much lower carbon impact for coal has yet to be demonstrated at scale. In the meantime, developments in the world economy have emphasized that third concern cited above—the need to bring the major developing economies into a climate change framework. At the end of 2007, China overtook the United States as the world's number one emitter of CO_2. Carbon management is likely to be a contentious focus for international diplomacy in the years ahead.

Energy Security

Oil imports have been a political and strategic concern since the United States moved from being on oil exporter to an oil importer in the late 1940s. Today those concerns have been amplified both because of the outflow of money and because of turbulence and extremism in parts of the Middle East, the recruiting base for Al-Qaeda. But those import numbers do require some clarification. It became common to assert that the U.S. imports 70 percent of its oil. In fact, in 2008, on a net basis, it was importing about 56 percent of its oil—still a very substantial amount. There was also a widespread belief that most or all of U.S. imports came from the Middle East. That is actually not the case. Some 22 percent of imports come from Canada, part of the overall trade flows with the country that is the United States' largest trading partner, and 12 percent from Mexico. Supplies from the Middle East (including Iraq) constitute 22 percent of total imports and about 12 percent of total U.S. oil consumption. Altogether, petroleum—both domestically produced and imported—provides about 40 percent of the total energy on which the United States' $14 trillion economy operates. Nevertheless, a confluence of concerns turned "ending the addiction to oil" into a common phrase of political discourse in the United States, even if the definition of addiction still needed some clarification.

The need for new supplies—conventional, renewables, and alternatives—plus price and security and climate concerns has unleashed a wave of innovation and research all across energy industries. But how fast will change come? Certainly energy is a major policy focus. President Barack Obama described energy as "priority number one."[9] Technology and markets will provide the answer over time. Many renewables, such as wind and photovoltaics, provide electricity and will do little to supplant oil imports, as only 2 percent of U.S. electricity is generated from oil—unless there is a big growth in electricity-powered transportation. Indeed, transportation is critical. Whatever the innovations—and much is in the works—the auto fleet will not change overnight. It can take five or six years, and a billion dollars to bring new models to market. Only about 8 percent of the auto fleet turns over each year, and so it will take years for the impact to be felt.

But in five or ten years the auto fleet will almost certainly change and will look different from today's fleet, in terms of its energy sources and perhaps its engines. Certainly cars will be more efficient. New automobile fuel efficiency standards, passed at the end of December 2007, represent the first mandated increase in thirty-two years. The original fuel efficiency standards in 1975 were one of the two most important energy policy decisions in the 1970s (the other being the approval of the trans-Alaskan oil pipeline). These new standards will likely have similar impact. But what will be the changes? What kind of breakthroughs lie ahead for biofuels beyond corn-based ethanol? One way or the other, there will be more electricity in auto transportation. Hybrids will gain market share. Will plug-in hybrids mean that the electric power industry will be fueling some part of the auto fleet? Will natural gas become a significant motor fuel?

The dramatic changes in world oil are inevitably leading to a renewed focus on the perennial question of energy security. World War I and World War II had so starkly demonstrated the strategic importance of energy, particularly oil. But, as described in these pages, the present international system for energy security emerged only in the 1970s around the International Energy Agency and has evolved in the decades since. But much has changed in recent years. The major new consumers, China and India, need to be brought into the international energy security system, and that will require greater confidence and communication between them and the traditional importing countries. At the same time, there is an urgent need to address the physical security of energy infrastructure—pipelines, power plants, and transmission lines—and the supply chains that carry oil and natural gas from wellheads in the Persian Gulf, West Africa, Central Asia, and other parts of the world, to consumers. The integration of China and India and the focus on infrastructure are both essential for energy security in the twenty-first century.

Greater efficiency in the use of oil and other energy sources is emerging as a major and common policy objective in countries around the world. The industrial world is twice as energy efficient as it was in the 1970s. The potential for future efficiency is still very large. And yet it seems likely that a growing world economy, with rising incomes and increasing population, will require more oil—perhaps 40 percent or more over the next quarter century, at least according to some estimates. Perhaps innovation will lower that number. The answers depend upon policy and markets and on technology and the scale and character of research and development.

And yet for several decades to come—whether the price is high or low or somewhere in between—oil will be a central factor in world politics and the global economy, in the global calculus of power, and in how people live their lives. And that is why this story provides a framework for the issues we face today and why, hopefully, it helps shed much-needed light on the critical choices we face and on the opportunities, risks—and surely the surprises—that lie ahead. As such, *The Prize* is not only a history of the last 150 years. It is also a starting point for understanding how energy will shape the world of tomorrow.

Chronology

1928	World oil glut leads to meeting at Achnacarry Castle and "As-Is" agreement.	1947	Marshall Plan for Western Europe. Construction begins on Tapline for Saudi oil.
	French petroleum law.	1948	Standard of New Jersey (Exxon) and Socony-Vacuum (Mobil) join Standard of California (Chevron) and Texaco in Aramco.
1929	Stock market collapse heralds Great Depression.		
1930	Dad Joiner's discovery in East Texas.		Israel declares independence.
1931	Japan invades Manchuria.	1948–49	Neutral Zone concessions to Aminoil and J. Paul Getty.
1932	Discovery of oil in Bahrain.		
1932–33	Shah Reza Pahlavi cancels the Anglo-Iranian concession; Anglo-Iranian wins it back.	1950	Fifty-fifty deal between Aramco and Saudi Arabia.
		1950–53	Korean War.
1933	Franklin Roosevelt becomes President of the United States.	1951	Mossadegh nationalizes Anglo-Iranian in Iran (first postwar oil crisis).
	Adolf Hitler becomes Chancellor of Germany.		New Jersey Turnpike opens.
	Standard of California wins concession in Saudi Arabia.	1952	First Holiday Inn opens.
1934	Gulf and Anglo-Iranian gain joint concession in Kuwait.	1953	Mossadegh falls; Shah returns.
		1954	Iranian Consortium established.
1935	Mussolini invades Ethiopia; League of Nations fails to impose oil embargo.	1955	Soviet oil export campaign begins.
			First McDonald's opens in suburban Chicago.
1936	Hitler remilitarizes Rhineland and begins preparations for war, including a major synthetic fuels program.	1956	Suez Crisis (second postwar oil crisis).
			Oil discovered in Algeria and Nigeria.
1937	Japan begins war in China.	1957	European Economic Community established.
1938	Oil discovered in Kuwait and Saudi Arabia.		Enrico Mattei's deal with the Shah.
	Mexico nationalizes foreign oil companies.		Japan's Arabian Oil Company wins Neutral Zone offshore concession.
1939	World War II begins with German invasion of Poland.	1958	Iraqi revolution.
1940	Germany overruns Western Europe.	1959	Eisenhower imposes import quotas.
	United States puts limits on gasoline exports to Japan.		Arab Petroleum Congress in Cairo.
1941	Germany invades Soviet Union (June).		Groningen natural gas field discovered in Netherlands.
			Zelten field discovered in Libya.
	Japanese takeover of Southern Indochina leads United States, Britain and Netherlands to embargo oil to Japan (July).	1960	OPEC founded in Baghdad.
		1961	Iraqi attempt to swallow Kuwait frustrated by British troops.
	Japan attacks Pearl Harbor (Dec.).	1965	Vietnam War buildup.
1942	Battle of Midway (July).	1967	Six Day War; Suez Canal closed (third postwar oil crisis).
	Battle of El Alamein (September).		
	Battle of Stalingrad (begins November).	1968	Oil discovered on Alaska's North Slope.
1943	The first "fifty-fifty" deal in Venezuela.		Ba'thists seize power in Iraq.
		1969	Qaddafi seizes power in Libya.
	Allies win Battle of the Atlantic.		Oil discovered in the North Sea.
1944	Normandy landing (June).		Santa Barbara oil spill.
	Patton runs out of gas (August).	1970	Libya "squeezes" oil companies.
	Battle of Leyte Gulf, Philippines (October).		Earth Day.
		1971	Tehran Agreement.
1945	World War II ends with defeat of Germany and Japan.		Shah's Persepolis celebration.
			Britain withdraws military force from Gulf.

1972	Club of Rome study.		Collapse of Soviet Union
1973	Yom Kippur War; Arab Oil embargo (fourth postwar oil crisis).		Maastricht Treaty provides for Single European currency
	Oil price rises from $2.90 per barrel (September) to $11.65 (December).	1993	U.S. Congress approves North America Free Trade Agreement
	Alaskan pipeline approved.	1994	E-commerce begins
	Watergate scandal widens.	1995	Internet users reach 16 million
1974	Arab Embargo ends.	1997	Asian Financial Crisis Begins
	Nixon resigns.		Kyoto Protocol on Climate Change
	International Energy Agency (IEA) founded.	1998	Oil collapses to $10 a barrel range
1975	Automobile fuel efficiency standards established in the United States.		Financial crisis spreads—Russia defaults
	First oil comes ashore from North Sea.	1998–2002	Consolidation among oil majors
	South Vietnam falls to communists.	2000	Vladimir Putin elected Russia's president
	Saudi, Kuwaiti, and Venezuelan concessions come to an end.		Dot-com boom begins collapse
1977	North Slope Alaskan oil comes to market.		PetroChina first of Chinese oil company IPOs
	Buildup of Mexican production.		Prius hybrid goes on sale in U.S.
	Anwar Sadat goes to Israel.	2001	9/11—Al-Qaeda attacks World Trade Center and Pentagon
1978	Anti-Shah demonstrations, strikes by oil workers in Iran.		War in Afghanistan begins
1979	Shah goes into exile; Ayatollah Khomeini takes power.	2002	Strikes and political conflict disrupt Venezuela oil output
	Three Mile Island nuclear plant accident.	2003	War in Iraq begins—Iraqi oil disrupted
	Iran takes hostages at U.S. Embassy.	2004	World oil demand jumps on strong global economic growth, tightening market
1979–81	Panic sends oil from $13 to $34 a barrel (fifth postwar oil crisis).		National oil companies (NOCs) move to the fore
1980	Iraq launches war against Iran.	2005	U.S. mandates ethanol in gasoline
1982	OPEC's first quotas.		Baku-Tbilisi-Ceyhan pipeline begins operations, linking Caspian and Mediterranean Seas
1983	OPEC cuts price to $29.		
	Nymex launches the crude oil futures contract.	2006	President Bush calls for end to "addiction to oil"
1985	Mikhail Gorbachev becomes leader of Soviet Union.		"Tupi"—first major discovery in new Brazilian offshore oil province
1986	Oil price collapse.		First UN sanctions aimed at Iranian nuclear program
	Chernobyl nuclear accident in USSR.	2007	Credit crisis begins in the U.S.
1988	Ceasefire in Iran-Iraq War.		Chinese auto sales exceed 7 million
1989	*Exxon Valdez* tanker accident off Alaska.	2008	Oil hits $147.27—U.S. gasoline over $4 a gallon
	Berlin Wall falls; communism collapses in Eastern Europe.		"Speculation" and oil prices become major political issue
1990	Iraq invades Kuwait.		"Worst financial crisis since the Great Depression"
	UN imposes embargo on Iraq; multinational force dispatched to Middle East (sixth postwar oil crisis).		U.S. and Europe launch massive bank bailouts
1991	Gulf War		World recession, demand weakens, oil falls to $32.40
	Kuwaiti oil fields set afire		

OIL PRICES AND PRODUCTION

Nominal Crude Oil Prices

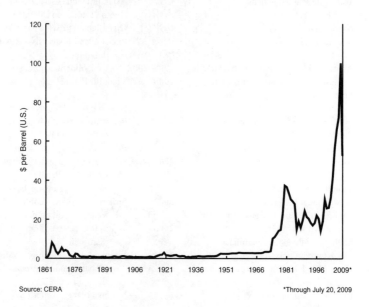

Source: CERA

*Through July 20, 2009

Real Crude Oil Prices
(2007 Base Year)

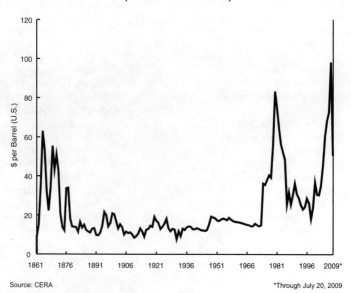

Source: CERA

*Through July 20, 2009

World Liquids Production*
1946–2009

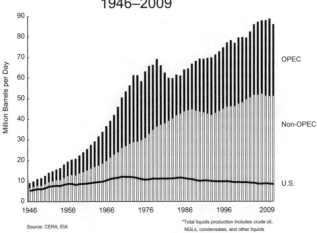

Source: CERA, EIA

*Total liquids production includes crude oil, NGLs, condensates, and other liquids

Nominal U.S. Gasoline Prices*

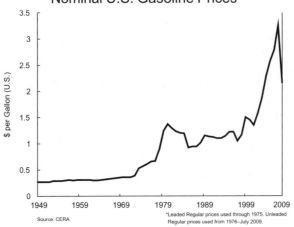

Source: CERA

*Leaded Regular prices used through 1975. Unleaded Regular prices used from 1976–July 2009.

Real U.S. Gasoline Prices*
(2007 Base Year)

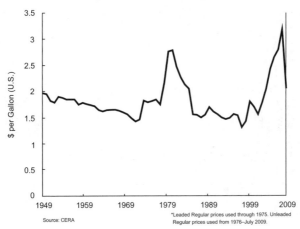

Source: CERA

*Leaded Regular prices used through 1975. Unleaded Regular prices used from 1976–July 2009.

Notes

Upstream, Downstream, All Around the Stream

All of the oil world is divided into three. The "upstream" comprises exploration and production. The "midstream" are the tankers and pipelines that carry crude oil to refineries. The "downstream" includes refining, marketing, and distribution, right down to the corner gasoline station or convenience store. A company that includes together significant upstream and downstream activities is said to be "integrated."

By generally accepted theory, crude oil is the residue of organic waste—primarily microscopic plankton floating in seas, and also land plants—that accumulated at the bottom of oceans, lakes, and coastal areas. Over millions of years, this organic matter, rich in carbon and hydrogen atoms, was collected beneath succeeding levels of sediments. Pressure and underground heat "cooked" the plant matter, converting it into hydrocarbons—oil and natural gas. The tiny droplets of oil liquid migrated through small pores and fractures in the rocks until they were trapped in permeable rocks, sealed by shale rocks on top and heavier salt water at the bottom. Typically, in such a reservoir, the lightest gas fills the pores of the reservoir rock as a "gas cap" above the oil. When the drill bit penetrates the reservoir, the lower pressure inside the bit allows the oil fluid to flow into the well bore and then to the surface as a flowing well. "Gushers"—or "oil fountains" as they were called in Russia—resulted from failure (or, at the time, inability) to manage the pressure of the rising oil. As production continues over time, the underground pressure runs down, and the wells need help to keep going, either from surface pumps or from gas reinjected back into the well, known as "gas lift." What comes to the surface is hot crude oil, sometimes accompanied by natural gas.

But as it flows from a well, crude oil itself is a commodity with very few direct uses. Virtually all crude is processed in a refinery to turn it into useful products like gasoline, jet fuel, home heating oil, and industrial fuel oil. In the early years of the industry, a refinery was little more than a still where the crude was boiled and then the different products were condensed out at various temperatures. The skills required were not all that different from making moonshine, which is why whiskey makers went into oil refining in the nineteenth century. Today, a refinery is often a large, complex, sophisticated, and expensive manufacturing facility.

Crude oil is a mixture of petroleum liquids and gases in various combinations. Each of these compounds has some value, but only as they are isolated in the refining process. So, the first step in refining is to separate the crude into constituent parts. This is accomplished by thermal distillation—heating. The various components vaporize at different temperatures and then can be condensed back into pure "streams." Some streams can be sold as they are. Others are put through further processes to obtain higher-value products. In simple refineries, these processes are primarily for the removal of unwanted impurities and to make minor changes in chemical properties. In more complex refineries, major restructuring of the molecules is carried out through chemical processes that are known as "cracking" or "conversion." The result is an increase in the quantity of higher-quality products, such as gasoline, and a decrease in the output of such lower-value products as fuel oil and asphalt.

Crude oil and refined products alike are today moved by tankers, pipelines, barges, and trucks. In Europe, oil is often officially measured in metric tons; in Japan, in kiloliters. But in the United States and Canada, and colloquially throughout the world, the basic unit remains the "barrel," though there is hardly an oil man today who has seen an old-fashioned crude oil barrel, except in a museum. When oil first started flowing out of the wells in western Pennsylvania in the 1860s, desperate oil men ransacked farmhouses, barns, cellars, stores, and trashyards for any kind of barrel—molasses, beer, whiskey, cider, turpentine, salt, fish, and whatever else was handy. But as coopers began to make barrels specially for the oil trade, one standard size emerged, and that size continues to be the norm to the present. It is 42 gallons. The number was borrowed from England, where a statute in 1482 under King Edward IV established 42 gallons as the standard size barrel for herring in order to end skulduggery and "divers deceits" in the packing of fish. At the time, herring fishing was the biggest business in the North Sea. By 1866, seven years after Colonel Drake drilled his well, Pennsylvania producers confirmed the 42-gallon barrel as their standard, as opposed to, say, the 31½-gallon wine barrel or the 32-gallon London ale barrel or the 36-gallon London beer barrel. And that, in a roundabout way, brings us right back to the present day. For the 42-gallon barrel is still used as the standard measurement, even if not as a physical receptacle, in the biggest business in the North Sea—which today of course is not herring, but oil.

Prologue

1. Randolph S. Churchill, *Winston Churchill*, vol. 2, *Young Statesman, 1901–1914* (London: Heinemann, 1968), p. 529 ("bully"); Winston S. Churchill, *The World Crisis*, vol. 1 (New York: Scribners, 1928), pp. 130–36.
2. Interview with Robert O. Anderson.

Chapter 1

1. "George Bissell: Compiled by his Grandson, Pelham St. George Bissell," Dartmouth College Library; Paul H. Giddens, *The Birth of the Oil Industry* (New York: Macmillan, 1938), p. 52, chap. 3; Harold F. Williamson and Arnold R. Daum, *The American Petroleum Industry*, vol. 1, *The Age of Illumination, 1859–1899* (Evanston: Northwestern University Press, 1959), pp. 23–24. Giddens and Williamson and Daum are basic sources. Paul H. Giddens, *Pennsylvania Petroleum, 1750–1872: A Documentary History* (Titusville: Pennsylvania Historical and Museum Commission, 1947), p. 54 ("Seneca oil"); J. T. Henry, *The Early and Later History of Petroleum* (Philadelphia: Jas. B. Rodgers Co., 1873), pp. 82–83; Henry H. Townsend, *New Haven and the First Oil Well* (New Haven, 1934), pp. 1–3 ("curative powers" and poem).
2. Gerald T. White, *Scientists in Conflict: The Beginnings of the Oil Industry in California* (San Marino: Huntington Library, 1968), pp. 38–45 (on Silliman); *Petroleum Gazette*, April 8, 1897, p. 8; Paul H. Giddens, *The Beginnings of the Oil Industry: Sources and Bibliography* (Harrisburg: Pennsylvania Historical Commission, 1941), pp. 23 ("I can promise"), 62 ("unexpected success"); Giddens, *Beginnings of the Oil Industry: Sources*, pp. 33–35, 40 ("hardest times"), 38, 8 ("turning point"); B. Silliman, Jr., *Report on the Rock Oil, or Petroleum, from Venango Co., Pennsylvania* (New Haven: J. H. Benham's, 1855), pp. 9–10, 20.
3. Abraham Gesner, *A Practical Treatise on Coal, Petroleum, and Other Distilled Oils*, ed. George W. Gesner, 2d ed. (New York: Baillière Bros., 1865), chap. 1; Henry, *Early and Later History of Petroleum*, p. 53; Kendall Beaton, "Dr. Gesner's Kerosene: The Start of American Oil Refining," *Business History Review* 29 (March 1955), pp. 35–41 ("new liquid hydrocarbon"); Gregory Patrick Nowell, "Realpolitik vs. Transnational Rent-Seeking: French Mercantilism and the Development of the World Oil Cartel, 1860–1939" (Ph.D., Massachusetts Institute of Technology 1988), pp. 104–8; *Business History Review*, ed., *Oil's First Century* (Boston: Harvard Business School, 1960), pp. 8 ("coal oils"), 19 ("impetuous energy").
4. R. J. Forbes, *Bitumen and Petroleum in Antiquity* (Leiden: E. J. Brill, 1936), pp. 11–21, 57 ("incredible miracles"), 92 ("eyelashes"), 95–99; R. J. Forbes, *Studies in Early Petroleum History* (Leiden:

E. J. Brill, 1958), pp. 150–53; R. J. Forbes, *More Studies in Early Petroleum History* (Leiden: E. J. Brill, 1959), pp. 20 ("unwearied fire"), 71 ("pitch and tow").

5. S. J. M. Eaton, *Petroleum: A History of the Oil Region of Venango County, Pennsylvania* (Philadelphia: J. B. Skelly & Co., 1865), pp. 211–13; Beaton, "Dr. Gesner's Kerosene," pp. 44–45.

6. "Brief Development of the Petroleum Industry in Penn. Prepared at the Request of and Under the Supervision of James M. Townsend," D-14, Drake Well Museum ("Oh Townsend").

7. E. L. Drake manuscript, D-96, Drake Well Museum, p. 4 ("I had made up my mind"); Herbert Asbury, *The Golden Flood: An Informal History of America's First Oil Field* (New York: Knopf, 1942), pp. 52–53 (Drake to Townsend); Giddens, *Birth of the Oil Industry,* pp. 30–31, 59–61 ("Yankee").

8. Forbes, *More Studies in Early Petroleum History,* p. 141 ("light of the age"); Giddens, *Beginnings of the Oil Industry: Sources,* pp. 81–83 (Bissell to wife), 59 ("I claim"); Leon Burr Richardson, "Brief Biographies of Buildings—Bissell Hall," *Dartmouth Alumni Magazine,* February 1943, pp. 18–19; Henry, *Early and Later History of Petroleum,* p. 349 ("name and fame"); Townsend, "Brief Development," D-14, Drake Well Museum ("whole plan"); Giddens, *Pennsylvania Petroleum,* p. 189 ("milk of human kindness").

9. Giddens, *Birth of the Oil Industry,* pp. 71 ("hive of bees"), 169, 95 ("mine is ruined").

10. Paul H. Giddens, *The American Petroleum Industry: Its Beginnings in Pennsylvania!* (New York: Newcomen Society, 1959), p. 28; Giddens, *Birth of the Oil Industry,* pp. 87, 123–24 ("profits of petroleum" and "assailed Congress"), chap. 9.

11. Giddens, *Birth of the Oil Industry,* p. 137 ("smells"); William C. Darrah, *Pithole: The Vanished City* (Gettysburg, Pa., 1972), pp. 34–35 ("liquor and leases" and "vile liquor"), 230–31; Giddens, *American Petroleum Industry,* p. 21 (song titles); Paul H. Giddens, *Early Days of Oil: A Pictorial History* (Princeton: Princeton University Press, 1948), p. 17 ("Oil on the brain").

12. Williamson and Daum, *Age of Illumination,* pp. 375–77, 759 ("hidden veins"), app. E; August W. Giebelhaus, *Business and Government in the Oil Industry: A Case Study of Sun Oil, 1876–1945* (Greenwich: JAI Press, 1980), p. 2.

13. Andrew Cone and Walter R. Johns, *Petrolia: A Brief History of the Pennsylvania Petroleum Region* (New York: D. Appleton, 1870), pp. 99–100 ("Oil Creek mud"); Henry, *Early and Later History of Petroleum,* p. 286; Giddens, *Birth of the Oil Industry,* pp. 125–26 ("oil and land excitement"); Samuel W. Tait, Jr., *The Wildcatters: An Informal History of Oil-Hunting in America* (Princeton: Princeton University Press, 1946), pp. 26–31.

14. John J. McLaurin, *Sketches in Crude Oil,* 3rd ed. (Franklin, Penn., 1902), 3d ed., pp. 316–21; Giddens, *Birth of the Oil Industry,* pp. 182–83 ("favorite speculative commodity"); John H. Barbour, "Sketch of the Pittsburgh Oil Exchange," *Western Pennsylvania Historical Magazine* 11 (July 1928), pp. 127–43.

Chapter 2

1. John D. Rockefeller, *Random Reminiscences of Men and Events* (New York: Doubleday, Page & Co., 1909), p. 81 ("I'll go no higher"); Allan Nevins, *Study in Power: John D. Rockefeller, Industrialist and Philanthropist* (New York: Scribners, 1953), vol. 1, pp. 35–36 ("I ever point"). Nevins remains the standard biographical source.

2. David Freeman Hawke, *John D.: The Founding Father of the Rockefellers* (New York: Harper & Row, 1980), pp. 2–6, 27; Grace Goulder, *John D. Rockefeller: The Cleveland Years* (Cleveland: Western Reserve Historical Society, 1972), p. 10 ("trade with the boys"); John K. Winkler, *John D.: A Portrait in Oils* (New York: Vanguard Press, 1929), p. 14; Nevins, *Study in Power,* vol. 1, pp. 10–14 ("something big" and "methodical"); Rockefeller, *Random Reminiscences,* p. 46 ("intimate conversations").

3. Nevins, *Study in Power,* vol. 1, p. 19 ("Great Game"); Rockefeller, *Random Reminiscences,* pp. 81 ("All sorts"), 21 ("bookkeeper"); John Ise, *The United States Oil Policy* (New Haven: Yale University Press, 1928), pp. 48–49.

4. Edward N. Akin, *Flagler: Rockefeller Partner and Florida Baron* (Kent, Ohio: Kent State University Press, 1988), pp. 3–18, 19 ("competition" and "Keep your head"), 27 ("A friendship"); Rockefeller, *Random Reminiscences,* pp. 11 ("vim and push"), 13 ("walks"), 19; John T. Flynn, *God's Gold: The Story of Rockefeller and His Times* (London: George Harrap & Co., 1933), p. 172 ("bold, unscrupulous"); John W. Martin, *Henry M. Flagler (1830–1913): Florida's East Coast Is His Monument!* (New York: Newcomen Society, 1956), pp. 8–11 ("American Riviera").

5. John G. McLean and Robert W. Haigh, *The Growth of Integrated Oil Companies* (Boston: Harvard

Business School, 1954), pp. 59–63; W. Trevor Halliday, *John D. Rockefeller (1839–1937): Industrial Pioneer and Man* (New York: Newcomen Society, 1948), p. 14 ("standard quality"); Nevins, *Study in Power,* vol. 1, pp. 80–83 ("Who would ever"), 97 ("independently rich"), 99–100 ("idea was mine"); Hawke, *John D.,* pp. 44–46, 54 ("independence of *woman*"); Dictation by Mr. Rockefeller, June 7, 1904, Rockefeller family, JDR, Jr., Business Interviews, Box 118, "S.O. Company—Misc." folder, Rockefeller archives ("It was desirable").

6. Nevins, *Study in Power,* vol. 1, pp. 107 ("cruelest"), 117 ("Monster" and "Forty Thieves"), 128, 114–15 ("newspaper articles" and "private contracts"), 104 ("try our plan"), 172 ("mining camp"); Chester McArthur Destler, *Roger Sherman and the Independent Oil Men* (Ithaca: Cornell University Press, 1967), pp. 28, 34 ("but one buyer"), 37 ("dry up Titusville").

7. David Freeman Hawke, ed., *John D. Rockefeller Interview, 1917–1920: Conducted by William O. Inglis* (Westport, Conn.: Meckler Publishing, 1984), pp. 4 ("cut-throat"), 6 ("safe and profitable"); Hawke, *John D.,* pp. 79 ("war or peace"), 106 ("good sweating"), 170 ("brass band"); Nevins, *Study in Power,* vol. 1, pp. 216 ("feel sick"), 224 ("barrel famine"), 223 ("Morose"); Akin, *Flagler,* p. 67 ("blankets"); McLean and Haigh, *Integrated Oil,* p. 63.

8. Archbold to Rockefeller, September 2, 1884, Box 51, Archbold folder (1.51.379), Business Interests, 1879–1894, RG 1.2, Rockefeller archives; Jerome Thomas Bentley, "The Effects of Standard Oil's Vertical Integration into Transportation on the Structure and Performance of the American Petroleum Industry, 1872–1884" (Ph.D., University of Pittsburgh, 1976), p. 27.

9. Archbold to Rockefeller, August 15, 1888, Box 51, Archbold folder (1.51.378), Business Interests, 1879–1894, RG 1.2, Rockefeller archives; Destler, *Roger Sherman,* pp. 85 ("overweening"), 95 ("Autocrat"), 132 ("gang of thieves"); Nevins, *Study in Power,* vol. 1, p. 337 ("Rockefeller will get you").

10. Interview with Mr. Rogers, 1903, T-003, Tarbell papers ("every foot" and inheritance); Nevins, *Study in Power,* vol. 1, pp. 132–34 ("pleasant" and "clamorer"); C. T. White folder (87.1.59), Box 134, Business Interests, John D. Rockefeller, Jr., papers, Rockefeller archives (stockholding); Ralph W. Hidy and Muriel E. Hidy, *History of Standard Oil Company (New Jersey)* vol. 1, *Pioneering in Big Business, 1882–1911* (New York: Harper & Brothers, 1955), p. 6 ("You gentlemen").

11. Flynn, *God's Gold,* p. 131 ("everything count"); Standard Oil—Rachel Crothers Group, T-014, Tarbell papers (espionage); Halliday, *Rockefeller,* p. 20; Hawke, *John D.,* p. 50 ("Hope if"); Rockefeller, *Random Reminiscences,* pp. 6 ("not . . . easiest of tasks"), 10 ("just how fast"); Nevins, *Study in Power,* vol. 1, p. 324 ("smarter than I").

12. Goulder, *Rockefeller,* p. 223 ("wise old owl"); Nevins, *Study in Power,* vol. 1, pp. 331, 326 ("expose as little"), 157 ("wonder how old"), 337 ("anxiety"), 328 ("ten letters"); vol. 2, p. 427 ("unemotional man"); Ida M. Tarbell, *The History of the Standard Oil Company* (New York: McClure, Phillips & Company, 1904), vol. 1, pp. 105–06.

13. Vinnie Crandall Hicks to Ida Tarbell, June 29, 1905, T-020 and Marshall Bond to Ida Tarbell, July 3, 1905, T-021, Tarbell papers ("Sunday school" and "Buzz"); Rockefeller, *Random Reminiscences,* pp. 25–26; Nevins, *Study in Power,* vol. 2, pp. 84 ("dentist's chair"), 91–95 ("poulets" and "life principle"), 193–94 ("best investment" and "spare change"); William Manchester, *A Rockefeller Family Portrait, from John D. to Nelson* (Boston: Little, Brown, 1959), pp. 25–26; Flynn, *God's Gold,* pp. 232–35, 280.

14. Rockefeller, *Random Reminiscences,* p. 58 ("volume"); Williamson and Daum, *Age of Illumination,* p. 320 ("length of life"); Catherine Beecher and Harriet Beecher Stowe, *The American Women's Home or Principles of Domestic Science* (New York: J. B. Ford, 1869), pp. 362–63 ("explosions").

15. Williamson and Daum, *Age of Illumination,* pp. 526 ("gas bill"), 678, 249 ("sewing circles"); Gerald Carson, *The Old Country Store* (New York: Oxford University Press, 1954), p. 188 ("lively country store").

16. Hidy and Hidy, *Standard Oil,* vol. 1, pp. 177–78 ("Our business" and "drink every gallon"), 8; Paul H. Giddens, *Standard Oil Company (Indiana): Oil Pioneer of the Middle West* (New York: Appleton-Century-Crofts, 1955), p. 2 ("vanishing phenomena"); S. Cornifort to Archbold, June 27, 1885, Box 51, Archbold folder (1.5.379), Business Interests, 1879–1894, R.G. 1.2, Rockefeller archives ("one hundred to one"); Nevins, *Study in Power,* vol. 2, p. 3; Edgar Wesley Owen, *Trek of the Oil Finders: A History of Exploration for Oil* (Tulsa: American Association of Petroleum Geologists, 1975), pp. 124–26.

17. Giddens, *Standard Oil Company (Indiana),* pp. 2–7 ("skunk juice"); Rockefeller, *Random Reminiscences,* pp. 7–9; Hawke, *John D.,* pp. 182–83 ("conservative brethren"), 185; Nevins, *Study in Power,* vol. 2, pp. 3, 101 ("Buy").

18. Giddens, *Standard Oil Company (Indiana)*, p. 19 ("entirely ignorant"); Hidy and Hidy, *Standard Oil*, vol. 1, pp. 279 (Seep), 87; Gilbert Montagu, *The Rise and Progress of the Standard Oil Company* (New York: Harper & Brothers, 1903), p. 132 ("best possible consensus").

19. Rockefeller, *Random Reminiscences*, pp. 60 ("large scale"), 29; Halliday, *Rockefeller*, pp. 10 ("instinctively realized"), 16 ("conceived the idea"); Hidy and Hidy, *Standard Oil*, vol. 1, pp. 120–21, 38–39 (*Mineral Resources*); Destler, *Roger Sherman*, pp. 47 ("body and soul"), 192; Nevins, *Study in Power*, vol. 2, pp. 54, 78, 129 ("success unparalleled"); J. W. Fawcett, T-082, Tarbell papers.

20. Lockhart interview, p. 3, T-003 (with Rogers interview), Tarbell papers ("Give the poor man"); Nevins, *Study in Power*, vol. 1, p. 402 ("day of combination"); vol. 2, pp. 379–87; Mark Twain with Charles Dudley Warner, *The Gilded Age: A Tale of Today* (New York: Trident Press, 1964), pp. 271 ("giant schemes"), 1; Flynn, *God's Gold*, pp. 4–5; Tarbell, *History of Standard Oil*, vol. 2, p. 31 ("cut to kill").

Chapter 3

1. Giddens, *The Birth of the Oil Industry*, pp. 96–98 ("Yankee invention"); Williamson and Daum, *Age of Illumination*, pp. 488–89 ("drill"); J. D. Henry, *Thirty-five Years of Oil Transport: Evolution of the Tank Steamer* (London: Bradbury, Agnew & Co., 1907), pp. 5, 172–74; Hidy and Hidy, *Standard Oil*, vol. 1, pp. 122–23 ("forced its way").

2. Giddens, *Birth of the Oil Industry*, p. 99 ("safe to calculate"); Robert W. Tolf, *The Russian Rockefellers: The Saga of the Nobel Family and the Russian Oil Industry* (Stanford: Hoover Institution Press, 1976), chaps. 1 and 2, pp. 41–46 ("pillars" and "walnut money"); Boverton Redwood, *Petroleum: A Treatise*, 4th ed. (London: Charles Griffen & Co., 1922), vol. 1, pp. 3–9 (Marco Polo), 36–46; Forbes, *Studies in Early Petroleum History*, pp. 154–62; John P. McKay, "Entrepreneurship and the Emergence of the Russian Petroleum Industry, 1813–1883," *Research in Economic History* 8 (1982), pp. 63–64.

3. Owen, *Trek of the Oil Finders*, pp. 4, 150; Tolf, *Russian Rockefellers*, pp. 108 ("Oil King"), 149 ("Nobelites"); J. D. Henry, *Baku: An Eventful History* (London: Archibald, Constable & Co., 1905), pp. 51–52; Williamson and Daum, *Age of Illumination*, pp. 637–41 ("difficulty"), 517; W. J. Kelly and Tsureo Kano, "Crude Oil Production in the Russian Empire, 1818–1919," *Journal of European Economic History* 6 (Fall 1977), pp. 309–10; McKay, "Entrepreneurship," pp. 48–55, 87 ("greatest triumphs").

4. Charles Marvin, *The Region of Eternal Fire: An Account of a Journey to the Petroleum Region of the Caspian in 1883*, new ed. (London: W. H. Allen, 1891), pp. 234–35 ("chimney-pot"); Sidney Pollard and Colin Holmes, *Industrial Power and National Rivalry, 1870–1914*, vol. 2 of *Documents of European Economic History* (London: Edward Arnold, 1972), pp. 108–10 ("American kerosene"); C. E. Stewart, "Petroleum Field of South Eastern Russia," 1886, Russia File, Oil, Box C-8, Pearson papers; Tolf, *Russian Rockefellers*, pp. 80–86 ("main point" and "speculation"); Williamson and Daum, *Age of Illumination*, p. 519 ("2000 miles"); Bertrand Gille, "Capitaux Français et Pétroles Russes (1884–94)," *Histoire des Enterprises* 12 (November 1963), p. 19; Virginia Cowles, *The Rothschilds: A Family of Fortune* (London: Weidenfeld and Nicolson, 1973), chaps. 7–8; Henry, *Baku*, pp. 74, 79.

5. Archbold to Rockefeller, August 19, 1884, and July 6, 1886, Archbold folder (1.5.381), Box 51, Business Interests, 1878–1894, R.G. 1.2, Rockefeller archives. Tolf, *Russian Rockefellers*, pp. 47–48 ("fountains"); Nevins, *Study in Power*, vol. 2, p. 116; Hidy and Hidy, *Standard Oil*, vol. 1, pp. 138–39 ("Russian competition").

6. Archbold to Rockefeller, July 6, 1886, Archbold folder (1.5.381), Box 51, Business Interests, 1879–1894, R.G. 1.2, Rockefeller archives; Hidy and Hidy, *Standard Oil*, vol. 1, pp. 147–53 (poem and "competitive commerce"); Henry, *Baku*, p. 116; Tolf, *Russian Rockefellers*, pp. 96–97, 107–09; Nicholas Halasz, *Nobel: A Biography of Alfred Nobel* (New York: Orion Press, 1959), pp. 3–5 ("dynamite king"), 211–13.

7. Robert Henriques, *Marcus Samuel: First Viscount Bearsted and Founder of the 'Shell' Transport and Trading Company, 1853–1927* (London: Barrie and Rockliff, 1960), pp. 74–75 ("go-between"), 44 ("lovely day"). Henriques is not only a biography of Samuel but also the most complete work on the origins of Shell. Geoffrey Jones, *The State and the Emergence of the British Oil Industry* (London: Macmillan, 1981), pp. 19–20 ("Shady Lane").

8. Henriques, *Marcus Samuel*, pp. 80 ("powerful company"), 96, 83, 112 ("Hebrew influence"), 108 ("to block"); Henry, *Thirty-five Years of Oil Transport*, pp. 41–47.

Notes for pages 37–51

9. "Petroleum in Bulk and the Suez Canal," *Economist,* January 9, 1892, pp. 36–38; Henriques, *Marcus Samuel,* pp. 109–11 ("got cheaper"), 138–40 ("wire handles"); Henry, *Thirty-five Years of Oil Transport,* p. 50; R. J. Forbes and D. R. O'Beirne, *The Technical Development of the Royal Dutch/Shell, 1890–1940* (Leiden: E. J. Brill, 1957), pp. 529–30.

10. Henriques, *Marcus Samuel,* pp. 52–54 ("two brothers").

11. Archbold to Rockefeller, December 15, 1891, Frank Rockefeller folder, Box 64; Archbold to Rockefeller, July 13 ("quite confident"), July 22, 1892, Archbold folder (1.51.381), Box 51, Business Interests, 1878–1894, R.G. 1.2, Rockefeller archives. Gille, "Capitaux Français et Pétroles Russes," pp. 43–48 ("crisis"); Tolf, *Russian Rockefellers,* pp. 116–17 ("on behalf"); F. C. Gerretson, *History of the Royal Dutch,* vol. 2 (Leiden: E. J. Brill, 1955), p. 35. Gerretson's 4-volume work extensively details the rise of Royal Dutch.

12. Gerretson, *Royal Dutch,* vol. 1, pp. 22 ("earth oil"), 89–90 ("won't bend"), 129–34 ("do not feel" and "mighty storm"), 163–65 ("Half-heartedness" and "stagnate"), 171 ("things go wrong"), 224 ("object of terror"), 174 ("pretend to be poor").

13. Hidy and Hidy, *Standard Oil,* vol. 1, pp. 261–67 (Standard reps in East Indies, "Every day," "Dutch obstacles" and "sentimental barrier"); Gerretson, *Royal Dutch,* vol. 1, pp. 282–84 ("into its power"); vol. 2, p. 48 ("pity"); Henriques, *Marcus Samuel,* pp. 181 ("Dutchman"), 184 ("still open").

Chapter 4

1. Gerald T. White, *Formative Years in the Far West: A History of Standard Oil of California and Its Predecessors Through 1919* (New York: Appleton-Century-Crofts, 1962), pp. 199, 267, 269.

2. Harold G. Passer, *The Electrical Manufacturers, 1875–1900* (Cambridge: Harvard University Press, 1953), pp. 180–81 ("fuzz on a bee"); Arthur A. Bright, Jr., *The Electric Lamp Industry: Technological Change and Economic Development from 1800 to 1947* (New York: Macmillan, 1949), pp. 68–69; Thomas P. Hughes, *Networks of Power: Electrification in Western Society, 1880–1930* (Baltimore: Johns Hopkins University Press, 1983), pp. 55, 73, 176, 227 ("Londoners"); Leslie Hannah, *Electricity Before Nationalization* (London: Macmillan, 1979), chap. 1.

3. James J. Flink, *America Adopts the Automobile, 1895–1910* (Cambridge: MIT Press, 1970), pp. 42–50 ("Get a horse," "skeptical" and "theme for jokers"), 64 ("automobile is the idol"); John B. Rae, *American Automobile Manufacturers: The First Forty Years* (Philadelphia: Chilton Company, 1959), pp. 33 ("Horseless Carriage fever"), 31; George S. May, *A Most Unique Machine: The Michigan Origins of the American Automobile Industry* (Grand Rapids, Mich.: Eerdmans Publishing, 1975), pp. 56–57; Allan Nevins, *Ford: The Times, the Man, the Company,* vol. 1 (New York: Scribners, 1954), pp. 133, 168, 237, 442–57.

4. Williamson and Daum, *Age of Illumination,* pp. 569–81; Arthur M. Johnson, *The Development of American Petroleum Pipelines: A Study in Private Enterprise and Public Policy, 1862–1906* (Ithaca: Cornell University Press, 1956), pp. 173–83 ("gloved hand"); Austin Leigh Moore, *John D. Archbold and the Early Development of Standard Oil* (New York: Macmillan, [1930]), pp. 197–202 ("champions of independence").

5. White, *Standard Oil of California,* pp. 8–13 ("fabulous wealth" and "without limit").

6. Patillo Higgins Oral History, II, pp. 7–9; Carl Coke Rister, *Oil! Titan of the Southwest* (Norman: University of Oklahoma Press, 1949), pp. 3–5, 34, 56–59; James A. Clark and Michael T. Halbouty, *Spindletop* (New York: Random House, 1952), pp. 4–5, 22, 27, 38–42 ("Tell that Captain"); John O. King, *Joseph Stephen Cullinan: A Study of Leadership in the Texas Petroleum Industry, 1897–1937* (Nashville: Vanderbilt University Press, 1970), pp. 12–21, 17 ("Dash and push"). F. Lucas to E. DeGolyer, May 6, 1920, 1074 ("visions"); John Galey to E. DeGolyer, August 22, 1941, 535, DeGolyer papers. Mody C. Boatwright and William A. Owen, *Tales from the Derrick Floor* (Garden City, N.Y.: Doubleday, 1970), p. 14 ("Dr. Drill"); W. L. Mellon and Boyden Sparkes, *Judge Mellon's Sons* (Pittsburgh, 1948), pp. 148–50 ("bewitched"); Robert Henriques, *Marcus Samuel,* p. 346 ("example")

7. Allen Hamill Oral History, I, pp. 20–21 ("Al!"), 34; James Kinnear Oral History, I, pp. 15–19, II, p. 16; T. A. Rickard, "Anthony F. Lucas and the Beaumont Gusher," *Mining and Scientific Press,* December 22, 1917, pp. 887–94; Rister, *Oil!,* pp. 60–67; Clark and Halbouty, *Spindletop,* pp. 88–89 ("X-ray eyes"); Burt Hull, "Founding of the Texas Company: Some of Its Early History," pp. 8–9, Collection 6850, Continental Oil, University of Wyoming.

8. Henriques, *Marcus Samuel,* pp. 353 ("pioneers"), 341–45 ("magnitude" and "opponent"), 349, 350 ("failure of supplies"); Harold F. Williamson, Ralph L. Andreano, Arnold R. Daum, and Gilbert C.

Klose, *The American Petroleum Industry,* vol. 2, *The Age of Energy, 1899–1959* (Evanston: Northwestern University Press, 1963), pp. 16, 22; Clark and Halbouty, *Spindletop,* pp. 100–1.

9. Mellon, *Judge Mellon's Sons,* pp. 153–162 ("epic card game" and "real way"), 269 ("We're out"), 276–78 ("just about as bad" and "good management"), 274–75 ("main problem"); Henriques, *Marcus Samuel,* pp. 462–66 (Samuel's diary).

10. Mellon, *Judge Mellon's Sons,* pp. 272–73 ("Standard made the price," "at the mercy" and "by your leave"), 282 ("marketable"), 284 ("hitch onto"); John G. McLean and Robert Haigh, *The Growth of Integrated Oil Companies,* pp. 78–79; King, *Cullinan,* p. 179 ("throwed me out"). On the fate of the pioneers: Rickard, "Anthony F. Lucas," p. 892; *Oil Investors Journal,* March 1, 1904, p. 3 ("Owing" and "milked too hard"); Clark and Halbouty, *Spindletop,* pp. 123–27 ("whole honor"); Thomas Galey, "Guffey and Galey and the Genesis of the Gulf Oil Corporation," January 1951, P448 (Gulf Oil), Petroleum Collection, University of Wyoming ("Difficult times" and "lost track"); Al Hamill to Thomas W. Galey, February 21, 1951, P448 (Gulf Oil), Petroleum Collection, University of Wyoming ("dribble").

11. August W. Giebelhaus, *Sun Oil, 1876–1945,* pp. 42–43 ("five cents").

12. Curt Hamill Oral History, II, p. 29 ("Hogg's my name"); Robert C. Cotner, *James Stephen Hogg* (Austin: University of Texas Press, 1959), pp. 437–39 ("Northern men"); King, *Cullinan,* pp. 107 ("Tammany"), 180–82 ("time will come"), 186 ("butt into everything"), 190–94 ("Texas deals" and "boarding-house brawl").

13. With the development of Gulf Coast and California output, Standard's control of domestic crude production fell from 90 percent in 1880 to between 60 and 65 percent in 1911. *Business History Review,* ed., *Oil's First Century* (Boston: Harvard Business School, 1960), pp. 73–82; Hidy and Hidy, *Standard Oil,* vol. 1, pp. 416, 473, 462; Joseph A. Pratt, "The Petroleum Industry in Transition: Antitrust and the Decline of Monopoly Control in Oil," *Journal of Economic History* 40 (December 1980), pp. 815–37; Ida Tarbell, *All in the Day's Work* (New York: Macmillan, 1939), p. 215 ("no end of the oil").

Chapter 5

1. Hidy and Hidy, *Standard Oil,* vol. 1, pp. 213–14 ("craze" and "Our friends"); Bruce Bringhurst, *Antitrust and the Oil Monopoly: The Standard Oil Cases, 1890–1911* (Westport, Conn.: Greenwood Press, 1979), pp. 25 ("Clam"), 52–58 ("Democratic Leader"), 63, 90 (Republic Oil ads); Pratt, "Petroleum Industry in Transition," p. 832 ("blind tigers").

2. Nevins, *Study in Power,* vol. 2, pp. 276–78; Hidy and Hidy, *Standard Oil,* vol. 1, pp. 231–32 ("gentlemen"); Peter Collier and David Horowitz, *The Rockefellers: An American Dynasty* (New York: Holt, Rinehart and Winston, 1976), pp. 45–46, 645.

3. E. V. Cary to J. D. Rockefeller, November 8, 1907, 1907–1912 folder, Box 114, John D. Rockefeller, Jr., Business Interests, Rockefeller archives; Moore, *Archbold,* pp. 48–49 ("go ahead" and "hard job"), 17 ("God is willing"), 53 ("oil enthusiasm"), 119 ("not . . . entirely philanthropic"), 109 ("one flash"); Nevins, *Study in Power,* vol. 1, pp. 117–18 ("$4 a barrel"); vol. 2, pp. 285–86 ("really a bank"), 293–94 (three simple rules), 457, n. 8 ("We told him"); Hidy and Hidy, *Standard Oil,* vol. 1, p. 67 ("unfortunate failing").

4. Edward C. Kirkland, *Industry Comes of Age: Business, Labor, and Public Policy, 1860–1897* (New York: Holt, Rinehart and Winston, 1961), p. 312 ("great moral . . . battle"); Lewis L. Gould, *Reform and Regulation: American Politics, 1900–1916* (New York: John Wiley, 1978), pp. 17, 23 ("trust question"); Richard Hofstadter, *The Age of Reform: From Bryan to FDR* (New York: Vintage, 1955), pp. 169, 185–86 ("critical achievement"); Alfred D. Chandler, *The Visible Hand: The Managerial Revolution in American Business* (Cambridge: Harvard University Press, 1977); Naomi R. Lamoreaux, *The Great Merger Movement in American Business, 1895–1904* (Cambridge: Cambridge University Press, 1985), chap. 7; Kathleen Brady, *Ida Tarbell: Portrait of a Muckraker* (New York: Seaview/Putnam, 1984), pp. 120–23 ("great feature" and "new plan of attacking"). H. H. Rogers complained to Ida Tarbell that he could not understand how *Harper's* could have published William Demarest Lloyd's *Wealth Against Commonwealth,* as he "had known Harry Harper socially very well." Tarbell's own theory was that "it was the very desire to keep the Standard Oil people out of society that had something to do with the Harpers publishing that book." Interview with H. H. Rogers, T-004, Tarbell papers.

5. Brady, *Ida Tarbell,* pp. 115 ("holding people off"), 110 ("playing cards"), 123 ("Well, I'm sorry"); Tarbell, *All in the Day's Work,* pp. 19, 204 ("Pithole"), 207 ("Don't do it").

Notes for pages 71–86

6. Joseph Siddell to Ida Tarbell, T-084 ("most interesting figure"); Standard Oil—Rachel Crothers Group, T-014, p. 3 ("confession of failure"), Interviews with H. H. Rogers, T-004 ("ask us to contribute"), T-003 ("made right"), T-001, T-002, Tarbell papers. Albert Bigelow Paine, *Mark Twain: A Biography* (New York: Harper & Brothers, 1912), pp. 971–73 ("stop walking" and "affairs of a friend"), 1658–59 ("best friend"); Justin Kaplan, *Mr. Clemens and Mark Twain* (New York: Simon and Schuster, 1966), pp. 320–23 ("out for the dollars"); Tarbell, *All in the Day's Work,* pp. 217–20 ("born gambler" and "we were prospered"), 211–15 ("by all odds"), 10 ("as fine a pirate"), 227–28; Albert Bigelow Paine, ed., *Mark Twain's Letters* (New York: Harper & Brothers, 1917), pp. 612–13 ("only man I care for"); Hidy and Hidy, *Standard Oil,* vol. 1, p. 662; Brady, *Ida Tarbell,* pp. 125–29 ("straightforward narrative"); "Would Miss Tarbell See Mr. Rogers," *Harper's Magazine,* January 1939, p. 141.

7. Standard Oil—Rachel Crothers Group, T-014, p. 13, Tarbell papers ("turned my stomach"); Brady, *Ida Tarbell,* pp. 137–57 ("very interesting to note," "most remarkable," McClure's comments, "guilty of baldness," "lady friend" and Rockefeller's response); Tarbell, *History of Standard Oil,* vol. 1, p. 158; vol. 2, pp. 207, 60, 230, 288 ("loaded dice"), 24; Hidy and Hidy, *Standard Oil,* vol. 1, pp. 652 ("more widely purchased"), 663; Tarbell, *All in the Day's Work,* p. 230 ("never had an animus"); Hawke, *Rockefeller Interview,* p. 5 ("Miss Tar Barrel").

8. Gould, *Reform and Regulation,* pp. 25–26 ("steamroller," "meteor" and "wring the personality"), 48 ($100,000 donation); Tarbell, *All in the Day's Work,* pp. 241–42 ("muckraker" and "vile and debasing"); George Mowry, *The Era of Theodore Roosevelt, 1900–1912* (New York: Harper & Brothers, 1958), pp. 131–32 ("levees"), 124; Henry F. Pringle, *Theodore Roosevelt* (New York: Harcourt, Brace, and Company, 1931), pp. 350–51 ("read every book" and "Darkest Abyssinia"); United States Congress, Senate, Subcommittee of the Committee on Privileges and Elections, *Campaign Contributions,* 62d Congress, 3d Session (Washington, D.C.: GPO, 1913), vol. 1, p. 133; vol. 2, pp. 1574, 1580; Moore, *Archbold,* p. 260 (1906 visit to TR).

9. Bringhurst, *Antitrust and the Oil Monopoly,* pp. 133, 140 ("Every measure"), 136 ("biggest criminals"). Starr J. Murphy to J. D. Rockefeller, September 7, 1907 ("Administration has started"); Telegram, W. P. Cowan to J. D. Rockefeller, August 3, 1907, 1907–1912 folder, Box 114; Starr Murphy to J. D. Rockefeller, July 9, 1907, Standard Oil Company—Misc. folder, Box 118, J.D.R., Jr., Business Interests, Rockefeller archives. White, *Standard Oil of California,* p. 373 ("inordinately voluminous"); Moore, *Archbold,* pp. 295 ("forty-four years"), 220 ("Federal authorities"); Goulder, *Rockefeller,* pp. 84 ("insolence" and "inadequacy"), 204–5 (Rockefeller on golf course); John K. Winkler, *John D.: A Portrait in Oils* (New York: Vanguard, 1929), p. 147.

10. David Bryn-Jones, *Frank B. Kellogg: A Biography* (New York: Putnam, 1937), p. 66 ("signal triumphs"); Bringhurst, *Antitrust and the Oil Monopoly,* pp. 150, 156–57 ("I have also"); White, *Standard Oil of California,* p. 377 ("No disinterested mind"); *New York Times,* May 16, 1911; Moore, *Archbold,* p. 278 ("one damn thing").

11. Giddens, *Standard of Indiana,* pp. 123–35 ("office boys"); Nevins, *Study in Power,* vol. 2, pp. 380–81 ("young fellows"); Hidy and Hidy, *Standard Oil,* vol. 1, pp. 416, 528, 713–14; White, *Standard Oil of California,* pp. 378–84.

12. Giddens, *Standard of Indiana,* pp. 141–63 (Burton).

13. Moore, *Archbold,* p. 281; Nevins, *Study in Power,* vol. 2, pp. 383 (Roosevelt), 404–5.

Chapter 6

1. Robert Henriques, *Marcus Samuel,* pp. 158 ("Mr. Abrahams"), 272 ("mere production"), 163 ("great disadvantage"), 165 ("berserk").

2. Henriques, *Marcus Samuel,* pp. 186–212 (correspondence), 267 ("tremendous role"), 272; Williamson and Daum, *Age of Illumination,* pp. 336–37.

3. Henriques, *Marcus Samuel,* pp. 300–23.

4. Henriques, *Marcus Samuel,* pp. 319–35, 176–79, 223, 234, 298–99; Gerretson, *Royal Dutch,* vol. 1, pp. 121, 126, 177, 238–39; vol. 2, pp. 324–27, 89, 92–146; Forbes and O'Beirne, *Royal Dutch/Shell,* p. 65.

5. Interview with John Loudon; Henriques, *Marcus Samuel,* pp. 330–31 ("nervous condition"), 333; Henri Deterding, *An International Oilman* (as told to Stanley Naylor), (London and New York: Harper & Brothers, 1934), pp. 28–30 ("lynx-eye" and "go a long way"), 37 ("sniftering"), 9–10 ("Simplicity rules"); Gerretson, *Royal Dutch,* vol. 1, pp. 199–202 ("first-rate businessman"); vol 2,

pp. 173–74 ("not aiming" and "heart and soul"); Robert Henriques, *Sir Robert Waley Cohen, 1877–1952* (London: Secker & Warburg, 1966), p. 98 ("charm"); Lane to Aron, January 11, 1912, Rothschild papers ("terrible sort").

6. Gerretson, *Royal Dutch,* vol. 2, pp. 191–94 ("battledore" and "joint management"); Archbold to Rockefeller, October 15, 1901, GDR to JDR, October 15, 1901, 1877–1906 folder, Box 114, Business Interests, J.D.R., Jr., Rockefeller archives ("There is here").

7. Gerretson, *Royal Dutch,* vol. 2, pp. 195–201 ("no solution" and "cordially"), 234–38 ("Neither of us" and "Delay dangerous"); Henriques, *Marcus Samuel,* pp. 400–3 ("sincere congratulation").

8. Gerretson, *Royal Dutch,* vol. 2, pp. 187–88 ("not . . . worth a white tie"), 244–45 ("rightly and fairly"); Henriques, *Marcus Samuel,* pp. 436–41 (Lane's critique), 446–52 ("rage," "ten Lord Mayors" and "Twenty-one years"), 470.

9. Gerretson, *Royal Dutch,* vol. 2, pp. 298–301 ("seize one's opportunities"), 345–46 (Deterding and Samuel); Henriques, *Marcus Samuel,* pp. 495 ("disappointed man"), 509 ("genius"); Mira Wilkins, *The Emergence of Multinational Enterprise: American Business Abroad from the Colonial Era to 1914* (Cambridge: Harvard University Press, 1970), p. 83; Henriques, *Waley Cohen,* pp. 129–48, chaps. 8–10; Deterding, *International Oilman,* p. 114 ("our chairman").

10. Gerretson, *Royal Dutch,* vol. 3, pp. 303 ("wipe us out"), 297–98 ("I am sorry"), 307 ("To America!"); Kendall Beaton, *Enterprise in Oil: A History of Shell in the United States* (New York: Appleton-Century-Crofts, 1957), pp. 123 ("Oil Capital"), 126 ("we *are* in America!").

11. Geoffrey Jones and Clive Trebilcock, "Russian Industry and British Business, 1916–1930: Oil and Armaments," *Journal of European Economic History* 11 (Spring 1982), pp. 68–69 ("too hurried development"); Serge Witte, *The Memoirs of Count Witte,* trans. and ed. Abraham Yarmolinsky (Garden City: Doubleday, Page & Co., 1921), pp. 27–29, 125, 198 ("imported mediums"), 183 (" 'Byzantine' habits"), 247 ("tangle"), 279; Theodore Von Laue, *Sergei Witte and the Industrialization of Russia* (New York: Atheneum, 1974), pp. 255, 122–23, 250; A. A. Fursenko, *Neftyanye Tresty i Mirovaia Politika* (Moscow: Nauka, 1965), pp. 42–43. On Baku unrest, see Richard Hare, *Portraits of Russian Personalities Between Reform and Revolution* (London: Oxford University Press, 1959), p. 305; Tolf, *Russian Rockefellers,* pp. 151–55 ("revolutionary hotbed"); Adam B. Ulam, *Stalin: The Man and His Era* (New York: Viking, 1973), pp. 37, 59–60; Isaac Deutscher, *Stalin: A Political Biography* (New York: Oxford University Press, 1966), p. 47; Ronald G. Suny, "A Journeyman for the Revolution: Stalin and the Labour Movement in Baku," *Soviet Studies* 23 (January 1972), p. 393.

12. Witte, *Memoirs,* pp. 189 ("monkeys"), 250 ("Russia's internal situation"); Deutscher, *Stalin,* p. 66 ("hour of revenge"); Solomon M. Schwarz, *The Russian Revolution of 1905: The Workers' Movement and the Formation of Bolshevism and Menshevikism,* trans. Gertrude Vaka (Chicago: University of Chicago Press, 1966), pp. 301–14; Adam B. Ulam, *The Bolsheviks & the Intellectual* (New York: Collier Books, 1965), pp. 219, 227; J. D. Henry, *Baku,* pp. 157–59 (Adamoff), 183–84 ("flames"); K. H. Kennedy, *Mining Tsar: The Life and Times of Leslie Urquhart* (Boston: Allen & Unwin, 1986), chaps. 2 and 3; Gerretson, *Royal Dutch,* vol. 3, p. 138; Hidy and Hidy, *Standard Oil,* p. 511; Ulam, *Stalin,* pp. 89–98; Suny, "Stalin," pp. 394, 386 ("unlimited distrust").

13. A. Beeby Thompson, *The Oil Fields of Russia* (London: Crosby Lockwood and Son, 1908), pp. 195–97, 213; Maurice Pearton, *Oil and the Romanian State* (Oxford: Oxford University Press, 1971), pp. 1–45; Tolf, *Russian Rockefellers,* pp. 183–85; Lane to Aron, December 21, 1911 ("I can assure you"), December 13, 1911 ("his intention"), Rothschild papers; V. I. Bovykin, "Rossiyskaya Neft'i Rotshil'dy'," *Voprosy Istorii* 4 (1978), pp. 27–41; Suny, "Stalin," p. 373 ("journeyman for the revolution").

Chapter 7

1. Henry Drummond Woolf, *Rambling Recollections,* vol. 2 (London: Macmillan, 1908), p. 329 ("well versed"); Charles Issawi, ed., *The Economic History of Iran, 1800–1914* (Chicago: University of Chicago Press, 1971), p. 20 (Persian finances); R. W. Ferrier, *The History of the British Petroleum Company,* vol. 1, *The Developing Years, 1901–1932* (Cambridge: Cambridge University Press, 1982), p. 28 ("Shah's prodigality"); T. A. B. Corley, *A History of the Burmah Oil Company, 1886–1927* (London: Heinemann, 1983); Geoffrey Jones, *The State and the Emergence of the British Oil Industry* (London: Macmillan, 1981). The books by Ferrier, Corley, and Jones—all making extensive use of corporate and government archives—are the best works on their respective subjects.

2. Ferrier, *British Petroleum,* pp. 29 ("capitalist"), 31 ("riches"), 35–36 ("morning coffee"). On D'Arcy,

see ibid., pp. 30–32; Corley, *Burmah Oil,* pp. 96–97; Henry Longhurst, *Adventure in Oil: The Story of British Petroleum* (London: Sidgwick and Jackson, 1959), pp. 18–19, 25; David J. Jeremy and Christine Shaw, eds., *Dictionary of Business Biography* (London: Butterworths, 1984), vol. 2, pp. 12–14. On the de Reuter concessions, see Firuz Kazemzadeh, *Russia and Britain in Persia, 1864–1914* (New Haven: Yale University Press, 1968), pp. 100–34, 210–14.

3. Kazemzadeh, *Russia and Britain in Persia,* pp. 3 ("chessboard"), 8, 22 ("Insurance"), 325–28 ("ragamuffins"); Arthur H. Hardinge, *A Diplomatist in the East* (London: Jonathan Cape, 1928), pp. 280 ("elderly child"), 268 ("vassalage"), 328 ("detestable"); Ferrier, *British Petroleum,* pp. 39 ("ready money"), 43 ("no umbrage"); Hardinge to Lansdowne, January 29, 1902, FO 60/660, PRO ("Cossacks"); Briton Cooper Busch, *Britain and the Persian Gulf* (Berkeley: University of California Press, 1967), chap. 4 and pp. 235–42.

4. Issawi, *Economic History of Iran,* p. 41 ("far-reaching effects" and "soil of Persia"); Jones, *State and British Oil,* pp. 131–32; Ferrier, *British Petroleum,* pp. 43 ("wild-catting"), 107.

5. Hardinge, *Diplomatist,* pp. 281, 273–74 ("Shiahs"), 306–11; Ferrier, *British Petroleum,* pp. 57 ("expedite"), 65 ("heat," "Mohamedan Kitchen" and "Mullahs").

6. Ferrier, *British Petroleum,* pp. 59–62 ("Every purse" and "keep the bank quiet"); Jones, *State and British Oil,* pp. 97–99 ("*éminence grise*"), 133; Corley, *Burmah Oil,* pp. 98–103 ("Glorious news").

7. Kazemzadeh, *Russia and Britain in Persia,* pp. 442–44 ("menace" and "Monroe Doctrine"). Lansdowne to Curzon, December 7, 1903, FO 60/731 ("danger"); Cargill to Redwood, October 6, 1904, ADM 116/3807, PRO. Corley, *Burmah Oil,* pp. 99–102 ("imperial," "patriots" and "coincided exactly"); Jones, *State and British Oil,* pp. 133–34 ("British hands").

8. A. R. C. Cooper, "A Visit to the Anglo-Persian Oil-Fields," *Journal of the Central Asian Society,* 13 (1926), pp. 154–56 ("thousand pities"); Kazemzadeh, *Russia and Britain in Persia,* pp. 444–45; Ferrier, *British Petroleum,* pp. 67, 86 ("beer and skittles"), 79 ("dung" and "teeth"); Arnold Wilson, *S. W. Persia: A Political Officer's Diary, 1907–14* (London: Oxford University Press, 1941), p. 112.

9. Wilson, *S. W. Persia,* p. 27 ("dignified" and "solid British oak"); Ferrier, *British Petroleum,* pp. 79 ("reasonable" and "beasts"), 96 ("type machine"), 73; Corley, *Burmah Oil,* p. 110 ("amuse me").

10. Ervand Abrahamian, *Iran Between Two Revolutions* (Princeton: Princeton University Press, 1982), pp. 80–85 ("luxury of Monarchs"); Gene R. Garthwaite, "The Bakhtiar Khans, the Government of Iran, and the British, 1846–1915," *International Journal of Middle East Studies* 3 (1972), pp. 21–44; Ferrier, *British Petroleum,* p. 83 ("nightingale" and "Baksheesh"), 85 ("importance attached"). Harold Nicolson, *Portrait of a Diplomatist* (Boston: Houghton Mifflin, 1930), p. 171 ("spontaneous infiltration"); Spring-Rice to Grey, April 11, 1907, FO 416/32, PRO ("great impetus"); Kazemzadeh, *Russia and Britain in Persia,* pp. 475–500.

11. Ferrier, *British Petroleum,* pp. 86–88 ("last throw," "cannot find" and "Psalm 104"), 96 ("stupid action"); Corley, *Burmah Oil,* pp. 128–39 ("go smash," "abandon operations," "telling no one" and "may be modified"); Wilson, *S. W. Persia,* pp. 41–42 ("endure heat").

12. Ferrier, *British Petroleum,* pp. 105–6 ("making public," "corns" and "immense benefit"), 98 ("great mistake"), 103 ("signing away"), 113 ("just as keen"). While Ferrier places the value of D'Arcy's shares at £895,000, Corley puts them at £650,000—still a healthy return after all. Ferrier, *British Petroleum,* p. 112 and Corley, *Burmah Oil,* p. 142. On Anglo-Persian's operations after the stock issue, see Wilson, *S. W. Persia,* pp. 84, 103 ("spent a fortnight"), 211–12; Ferrier, *British Petroleum,* pp. 152–53 ("one chapter"); Jones, *State and British Oil,* pp. 142, 144 ("serious menace"), 147; Corley, *Burmah Oil,* p. 189 ("hell of a mess").

Chapter 8

1. Ferrier, *British Petroleum,* p. 59; John Arbuthnot Fisher, *Memories* (London: Hodder and Stoughton, 1919), pp. 156–57; Henriques, *Marcus Samuel,* pp. 399–402; John Arbuthnot Fisher, *Fear God and Dread Nought: The Correspondence of Admiral of the Fleet Lord Fisher of Kilverstone,* vol. 1, ed. Arthur J. Marder (Cambridge: Harvard University Press, 1952), pp. 45 ("oil maniac"), 275 ("goldmine" and "bought the south half").

2. Fisher, *Memories,* p. 116 ("God-father of Oil"); Arthur J. Marder, *From the Dreadnought to Scapa Flow: The Royal Navy in the Fisher Era, 1904–1919,* vol. 1, *The Road to War, 1904–1914* (London: Oxford University Press, 1961), pp. 14 ("mixture"), 205 ("tornado"), 19 (Edward VII), 45; Fisher, *Fear God,* vol. 1, pp. 102 ("Full Speed"), 185 ("Wake up"); Ruddock F. Mackay, *Fisher of Kilver-*

stone (Oxford: Clarendon Press, 1973), p. 268 ("Golden rule"); R. H. Bacon, *The Life of Lord Fisher* (Garden City: Doubleday, 1929), vol. 2., pp. 157–59.

3. Paul M. Kennedy, *The Rise of the Anglo-German Antagonism* (London: George Allen & Unwin, 1982), pp. 416 ("naval question"), 417 ("freedom"), 457 ("strident"), 221–29 ("world domination," "mailed fist" and "weary Titan"); Zara S. Steiner, *Britain and the Origins of the First World War* (New York: St. Martin's Press, 1977), pp. 40–57, 127; Samuel Williamson, *The Politics of Grand Strategy: Britain and France Prepare for War, 1904–1914* (Cambridge: Harvard University Press, 1969), pp. 16, 18.

4. William H. McNeil, *The Pursuit of Power: Technology, Armed Force and Society Since A.D.* (Chicago: University of Chicago Press, 1982), p. 277 ("technological revolution"); Marder, *Dreadnought to Scapa Flow,* vol. 1, pp. 71, vii, 139 ("pensions"); Williamson, *Politics of Grand Strategy,* pp. 236, 238. For domestic German politics, see Volker Berghahn, "Naval Armaments and the Social Crisis: Germany Before 1914," in Geoffrey Best and Andrew Wheatcraft, eds., *War, Economy, and the Military Mind* (London: Croom Held, 1976), pp. 61–88. Randolph S. Churchill, *Winston S. Churchill,* vol. 1, *Youth, 1874–1900* (London: Heinemann, 1966), pp. 1888–89.

5. Randolph S. Churchill, *Winston S. Churchill,* vol. 2, *Young Statesman, 1901–1917* (Boston: Houghton Mifflin, 1967), pp. 494 ("nonsense"), 518–19 ("Indeed").

6. Churchill, *Young Statesman,* pp. 545–47 ("whole fortunes"); Churchill, *World Crisis,* vol. 1, pp. 71–78 ("intended to prepare," "important steps" and "veritable volcano"); Fisher, *Memories,* pp. 200–1 ("precipice"); Henriques, *Marcus Samuel,* p. 283; Randolph Churchill, *Winston S. Churchill,* vol. 2, *Companion Volume,* part 3 (Boston: Houghton Mifflin, 1969), p. 1926 ("How right").

7. Churchill, *Churchill,* vol. 2, *Companion Volume,* part 3, pp. 1926–27.

8. Fisher, *Fear God,* vol. 2, p. 404 ("Sea fighting"); Churchill, *World Crisis,* vol. 1, pp. 130–36 (on his decision).

9. Ferrier, *British Petroleum,* p. 158; Jones, *State and British Oil,* p. 170; Corley, *Burmah Oil Company,* p. 186; Fisher, *Fear God,* vol. 2, pp. 451 ("betrayed"), 467 ("no one else"); Mackay, *Fisher,* pp. 437–38; Churchill, *Young Statesman,* pp. 567–68; Churchill, *Churchill,* vol. 2, *Companion Volume,* part 3, p. 1929 ("My dear Fisher").

10. Fisher, *Memories,* pp. 218–20 ("d—d fool"); Lord Fisher, *Records* (London: Hodder and Stoughton, 1919), p. 196; Mackay, *Fisher,* p. 439 ("overwhelming advantages"); Fisher, *Fear God,* vol. 2, p. 438 ("don't grow").

11. Ferrier, *British Petroleum,* p. 94 ("Champagne Charlie" and "decorous"); Jeremy and Shaw, *Dictionary of Business Biography,* vol. 2, pp. 639–41; Corley, *Burmah Oil,* pp. 184, 205; Jones, *State and British Oil,* pp. 96 ("Old Spats"), 151–52 ("Jewishness," "Dutchness," "under the control" and "moderate return").

12. Bacon, *Fisher,* vol. 2, p. 158 ("do our d—st"); Jones, *State and British Oil,* pp. 164 ("embracing as it did" and "pecuniary assistance"), 151 ("Shell menace"); Ferrier, *British Petroleum,* pp. 170–73 ("commercial predominance" and "Evidently").

13. Jones, *State and British Oil,* pp. 166–67 ("speculative risk"); Marian Kent, *Oil and Empire: British Policy and Mesopotamian Oil, 1900–1920* (London: Macmillan, 1976), pp. 47–48 ("keeping alive"); Churchill, *Churchill,* vol.2, *Companion Volume,* part 3, pp. 1932–48; Corley, *Burmah Oil,* p. 191; Asquith to George V, July 12, 1913, CAB 41/34, PRO ("controlling interest"); Ferrier, *British Petroleum,* pp. 181–82.

14. *Parliamentary Debates,* Commons, July 17, 1913, pp. 1474–77 (Churchill statement); Corley, *Burmah Oil,* pp. 187, 191–95 ("scrap heap"); Ferrier, *British Petroleum,* pp. 195–96 ("thoroughly sound," "perfectly safe" and "national disaster").

15. Ferrier, *British Petroleum,* p. 185; Corley, *Burmah Oil,* pp. 195–97; Churchill, *Churchill,* vol. 2, *Companion Volume,* part 3, p. 1964.

16. *Parliamentary Debates,* Commons, June 17, 1914, pp. 1131–53, 1219–32; Bradbury to Anglo-Persian Oil Company, May 20, 1914, POWE 33/242, PRO; Ferrier, *British Petroleum,* p. 199 (Greenway's question).

17. Henriques, *Marcus Samuel,* p. 574; Churchill, *Churchill,* vol. 2, *Companion Volume,* part 3, pp. 1951 ("Napoleon and Cromwell"), 1965 ("*Good Old Deterding*"); Gerretson, *Royal Dutch,* vol. 4, p. 293.

18. Gerretson, *Royal Dutch,* vol. 4, p. 185; Jones, *State and British Oil,* pp. 144, 12 ("*premier cru*"); Ferrier, *British Petroleum,* p. 196; Churchill, *World Crisis,* p. 137; Churchill, *Churchill,* vol. 2, *Companion Volume,* part 3, p. 1999 (war order).

Chapter 9

1. William Langer, "The Well-Spring of Our Discontents," *Journal of Contemporary History* 3 (1968), pp. 3–17; McNeill, *Pursuit of Power,* pp. 334–35; Martin Van Creveld, *Supplying War: Logistics from Wallenstein to Patton* (Cambridge: Cambridge University Press, 1977), pp. 110–111, 124–25 (German general); W. G. Jensen, "The Importance of Energy in the First and Second World Wars," *Historical Journal* 11 (1968), pp. 538–45. Llewellyn Woodward, *Great Britain and the War of 1914–1918* (London: Metheun, 1967), pp. 38–39.

2. Basil Liddell Hart, *A History of the World War, 1914–1918* (London: Faber and Faber, 1934), chap. 4, especially pp. 86–87, 115–22 ("No British officer," "*coups de téléphone,*" "not commonplace" and "forerunner"); Henri Carré, *La Véritable Histoire des Taxis de La Marne* (Paris: Libraire Chapelot, 1921), pp. 11–39 ("How will we be paid?"); Robert B. Asprey, *The First Battle of the Marne* (Westport, Conn.: Greenwood Press, 1977), pp. 127 ("Today destiny"), 153 ("going badly").

3. Woodward, *Great Britain and the War of 1914–1918,* pp. 38–39 ("This isn't war"); Liddell Hart, *The World War,* pp. 332–43 ("antidote," "eyewitness," "black day" and "primacy"); Erich Ludendorff, *My War Memories, 1914–1918* (London: Hutchinson, [1945]), p. 679; J. F. C. Fuller, *Tanks in the Great War, 1914–1918* (London: John Murray, 1920), p. 19 ("present war"); Churchill, *World Crisis,* vol. 2, (New York: Scribners, 1923) pp. 71–91 ("caterpillar" . . . "tank"); A. J. P. Taylor, *English History, 1914–1945* (New York: Oxford University Press, 1965), p. 122; Francis Delaisi, *Oil: Its Influence on Politics,* trans. C. Leonard Leese (London: Labour Publishing and George Allen and Unwin, 1922), p. 29 (truck over the locomotive).

4. Liddell Hart, *The World War,* pp. 457–60 ("good sport"), 554–59; Harald Penrose, *British Aviation: The Great War and Armistice, 1915–1919* (London: Putnam, 1969), pp. 9–12 ("Since war broke out"), 586 ("necessities of war"); Bernadotte E. Schmitt and Harold C. Vedeler, *The World in the Crucible, 1914–1919* (New York: Harper & Row, 1984), pp. 301–4 ("Battle of Britain"); Jensen, "Energy in the First and Second World Wars," pp. 544–45; Richard Hough, *The Great War at Sea, 1914–1918* (New York: Oxford University Press, 1983), pp. 296–97.

5. F. J. Moberly, *History of the Great War Based on Official Documents: The Campaign in Mesopotamia, 1914–1918* (London: HMSO, 1923), vol. 1, p. 82 ("little likelihood"); Ferrier, *British Petroleum,* p. 263 ("build up"); Kent, *Oil and Empire,* pp. 125–26; Corley, *Burmah Oil,* pp. 239, 253 ("All-British Company"); Jones, *State and British Oil,* pp. 182–83.

6. Corley, *Burmah Oil,* p. 258, chap. 16; Henriques, *Marcus Samuel,* pp. 593–619; Henriques, *Waley Cohen,* pp. 200–40; P. G. A. Smith, *The Shell That Hit Germany Hardest* (London: Shell Marketing Co., [1921]), pp. 1–11; Jones, *State and British Oil,* pp. 187–202; Ferrier, *British Petroleum,* pp. 250, 218 ("to secure navy supplies"); Slade, "Strategic Importance of the Control of Petroleum," "Petroleum Supplies and Distribution" and "Observations on the Board of Trade Memorandum on Oil," August 24, 1916, CAB 37/154, PRO.

7. Henriques, *Waley Cohen,* pp. 213–20; *Times* (London), January 14, 1916, p. 5; May 26, 1916, p. 5; G. Gareth Jones, "The British Government and the Oil Companies, 1912–24: The Search for an Oil Policy," *Historical Journal* 20 (1977), pp. 654–64; C. Ernest Fayle, *Seaborne Trade,* vol. 3, *The Period of Unrestricted Submarine Warfare* (London: John Murray, 1924), pp. 465, 175–76, 319, 371, 196–97; George Gibb and Evelyn H. Knowlton, *History of Standard Oil Company (New Jersey),* vol. 2, *The Resurgent Years, 1911–1927* (New York: Harper & Brothers, 1956), pp. 221–23; Beaton, p. 100.

8. Jones, "British Government and the Oil Companies," pp. 661, 665; Paul Foley, "Petroleum Problems of the War: Study in Practical Logistics," *United States Naval Institute Proceedings* 50 (November 1927), pp. 1802–3 ("out of action"), 1817–21; Burton J. Hendrick, *The Life and Letters of Walter H. Page* (London: Heinemann, 1930), vol. 2, p. 288 ("Germans are succeeding"); Ferrier, *British Petroleum,* pp. 248–49 (Walter Long); Henry Bérenger, *Le Pétrole et la France* (Paris: Flammarion, 1920), pp. 41–55; Edgar Faure, *La Politique Française du Pétrole* (Paris: Nouvelle Revue Critique, 1938), pp. 66–69; Pierre L'Espagnol de la Tramerye, *The World Struggle for Oil,* trans. Leonard Leese (London: George Allen & Unwin, 1924), chap. 8; Eric D. K. Melby, *Oil and the International System: The Case of France, 1918–1969* (New York: Arno Press, 1981), pp. 8–20 ("as vital as blood").

9. Mark L. Requa, "Report of the Oil Division 1917–19" in H. A. Garfield, *Final Report of the U. S. Fuel Administrator* (Washington, D.C.: GPO, 1921), p. 261; Gerald D. Nash, *United States Oil Policy, 1890–1964* (Pittsburgh: University of Pittsburgh Press, 1968), p. 27. On American oil policy making during World War I, see Dennis J. O'Brien, "The Oil Crisis and the Foreign Policy of the Wilson Ad-

ministration, 1917–1921" (Ph.D.: University of Missouri, 1974), chaps. 1–2 and Robert D. Cuff, *The War Industries Board: Business-Government Relations During World War I* (Baltimore: Johns Hopkins University Press, 1973).

10. Joseph E. Pogue and Isador Lubin, *Prices of Petroleum and Its Products During the War* (Washington, D.C.: GPO, 1919), pp. 13–33, 289; Rister, *Oil!*, pp. 120–34. On the coal shortage, see David Kennedy, *Over Here: The First World War and American Society* (Oxford: Oxford University Press, 1980), pp. 122–24 ("Bedlam") and Seward W. Livermore, *Politics Is Adjourned: Woodrow Wilson and the War Congress, 1916–18* (Middletown: Wesleyan University Press, 1966), pp. 68–69, 86–88. Requa, "Report of the Oil Division," p. 270 ("no justification"); White, *Standard Oil of California,* p. 542. For auto growth, see Beaton, *Shell,* p. 171; White, *Standard of California,* p. 544. H. A. Garfield, *Final Report of the U.S. Fuel Administrator,* p. 8 ("walk to church").

11. Ludendorff, *War Memories,* pp. 287–88 ("As I now saw"), 358–59 ("did materially"); Liddell Hart, *The World War,* pp. 345–50; Schmitt and Vedeler, *World in the Crucible,* pp. 157–60; *Times* (London), December 5, 1916, p. 7; Pearton, *Oil and the Romanian State* pp. 79–85 ("No efforts"); Gibb and Knowlton, *Standard Oil,* vol. 2, pp. 233–35. On Norton-Griffiths, see R. K. Middlemas, *The Master-Builders* (London: Hutchinson, 1963), pp. 270–83 ("dashing," nicknames and "blasted language"); Mrs. Will Gordon, *Roumania Yesterday and Today* (London: John Lane, 1919), chap. 9 ("sledgehammer"); *New York Times,* January 16, 1917, p. 1; February 20, 1917, p. 4. On the effects on Germany, see Fayle, *Seaborne Trade,* vol. 3, pp. 180–81 ("just the difference"). After the war John Norton-Griffiths was recognized as a "world famous engineer" and contractor. In 1930, he was directing his firm's project of raising the height of the Aswan Dam. A conflict developed with the local Egyptian authorities on the type of steel he had ordered and whether he would be liable for a very large penalty—also with possible great injury to his professional reputation. As was his wont, at 7:45 in the morning on September 27, 1930, he took out a surf boat from his hotel at San Stefano, near Alexandria, and paddled out to sea. A little later, an associate looked out from the hotel and saw Norton-Griffiths's boat floating empty. Observers saw a man swimming or floating at a little distance away. Another boat, dispatched to investigate, recovered the body. It was Empire Jack, "the man with a sledgehammer," with a bullet wound through his right temple—a suicide. *Times* (London), September 28, 1930, p. 12; September 29, 1930, p. 14; *New York Times,* September 28, 1930, II, p. 8, September 29, 1930, p. 11.

12. Erich Ludendorff, *The Nation at War,* trans. A. S. Rappoport (London: Hutchinson, 1936), p. 79; Z. A. B. Zeman, ed., *Germany and the Revolution in Russia, 1915–1918* (London: Oxford University Press, 1958), pp. 107, 134–35; Ronald Suny, *The Baku Commune 1917–1918* (Princeton: Princeton University Press, 1972), pp. 284–85 ("we agreed" and "plunderers"), 328–43; Firuz Kazemzadeh, *The Struggle for Transcaucasia, 1917–1921* (New York: Philosophical Library, 1951), pp. 136–46 ("destroy"); Moberly, *Campaign in Mesopotamia,* vol. 4, pp. 182–212; Ludendorff, *War Memories,* pp. 659–60 ("serious blow"); Anastas Mikoyan, *Memoirs of Anastas Mikoyan,* vol. 1, *The Path of Struggle,* ed. Sergo Mikoyan, trans. Katherine T. O'Connor and Diane L. Burgin (Madison, Conn.: Sphinx Press, 1988), pp. 505–9.

13. Ludendorff, *War Memories,* p. 748; Schmitt and Vedeler, *World in the Crucible,* p. 272; Pearton, *Oil and the Romanian State,* p. 93; Fayle, *Seaborne Trade,* vol. 3, pp. 230, 402; Leo Grebler and Wilhelm Winkler, *The Cost of the World War to Germany and to Austria-Hungary* (New Haven: Yale University Press, 1940), p. 85; Henriques, *Marcus Samuel,* p. 624. On the speeches, see *Times* (London), November 22, 1918, p. 6; Delaisi, *Oil,* pp. 86–91 (Curzon); Bérenger, *Le Pétrole et la France,* pp. 175–80.

Chapter 10

1. *Documents on British Foreign Policy, 1919–1939,* First Series, vol. 4, pp. 452–54, 521; *FRUS: Paris Peace Conference, 1919,* vol. 5, pp. 3–4, 760, 763, 804; David Lloyd George, *The Truth About the Peace Treaties,* vol. 2 (London: Victor Gollancz, 1938), pp. 1037–38.

2. Confidential Memorandum of Negotiations with Turkish Petroleum Company, July 15–August 5,1922, pp. 1–3, 800.6363/T84/48, RG 59, NA; Marian Kent, *Oil and Empire,* pp. 12–80; Edward Mead Earle, "The Turkish Petroleum Company: A Study in Oleaginous Diplomacy," *Political Science Quarterly* 39 (June 1924), 267 ("Talleyrand"); V. H. Rothwell, "Mesopotamia in British War Aims," *Historical Journal* 13 (1970), p. 277.

3. Ralph Hewins, *Mr. Five Percent: The Story of Calouste Gulbenkian* (New York: Rinehart and Com-

pany, 1958), pp. 15–16 ("academic nonsense"), 24 ("fine and consistent"), 11 ("hand"), 188 (Kenneth Clark); Financial Times, July 25, 1955 ("granite"); Gibb and Knowlton, Standard Oil, vol. 2, p. 300; Nubar Gulbenkian, Portrait in Oil (New York: Simon and Schuster, 1965), p. 85 ("very close"); "Memoirs of Calouste Sarkis Gulbenkian, with Particular Relation to the Origins and Foundation of the Iraq Petroleum Company, Limited," March 4, 1948, 890.G.6363/3–448, pp. 6–7 ("wild cat"), 11 ("not, in any way"), RG 59, NA.

4. Kent, *Oil and Empire,* pp. 86–93, 170–71 (Foreign Office Agreement); Hewins, *Mr. Five Percent,* p. 81.

5. Kent, *Oil and Empire,* pp. 109, 121–26; David Fromkin, *A Peace to End All Peace: Creating the Modern Middle East, 1914–1922* (New York: Henry Holt, 1989), pp. 188–95; Elie Kedourie, *England and the Middle East: The Destruction of the Ottoman Empire, 1914–1921* (London: Bowes and Bowes, 1956); Jones, *State and British Oil,* p. 198; Helmut Mejcher, *Imperial Quest for Oil: Iraq, 1910–1928* (London: Ithaca Press, 1976), p. 37; Rothwell, "Mesopotamia in British War Aims," pp. 289–90 (Hankey and Balfour); William Stivers, *Supremacy and Oil: Iraq, Turkey and the Anglo-American World Order, 1918–1930* (Ithaca: Cornell University Press, 1982), pp. 71–72 (Lansing); Lloyd George, *Peace Treaties,* pp. 1022–38.

6. Melby, *France,* pp. 17–23 (Clemenceau's grocer); Jukka Nevakivi, *Britain, France and the Arab Middle East, 1914–1920* (London: Athlone Press, 1969), p. 154; Paul Mantoux, *Les Délibérations du Conseil des Quatre (24 Mars–28 Juin 1919),* vol. 2 (Paris: Editions du Centre National de la Recherche Scientifique, 1955), pp. 137–43; Jones, *State and British Oil,* p. 214; C. E. Callwell, *Field-Marshal Sir Henry Wilson: His Life and Diaries,* vol. 2 (London: Cassell, 1927), p. 194 ("dog-fight"); *Documents on British Foreign Policy,* 1919–1939, First Series, vol. 8, pp. 9–10.

7. Melby, *France,* pp. 67 ("entirely French"), 100–4 ("industrial arm"); Richard Kuisel, *Ernest Mercier: French Technocrat* (Berkeley: University of California Press, 1967), pp. 31–32 ("instrument" and "international difficulties"), 25 ("Anglo-Saxon").

8. Kendall Beaton, *Shell in the United States,* pp. 229–32; B. S. McBeth, *British Oil Policy, 1919–1939* (London: Frank Cass, 1985), p. 41. Waley Cohen to Director, Petroleum Dept., May 15, 1923, FO 371/13540; Proposed Combination of Royal Dutch Shell, Burma Oil, and Anglo-Persian Oil Companies, Notes of Meeting, November 2, 1921, W11691, FO 371/7027; Cowdray to Lloyd-Greame, February 14, 1922, POWE 33/92; Watson to Clarke, October 31, 1921, POWE 33/92, PRO. *Parliamentary Debates,* Commons, March 18, 1920, vol. 126, no. 28, cols. 2375/6; Jones, *State and British Oil,* pp. 223–26 ("over-production," "every action" and "Hottentots"); Ferrier, *British Petroleum,* pp. 372–80 ("whole revenue" and "did not go"); Shaul Bakhash, *The Reign of the Ayatollahs: Iran and the Islamic Revolution* (New York: Basic Books, 1984), pp. 20–23.

9. Martin Gilbert, *Winston S. Churchill,* vol. 5, *The Prophet of Truth, 1922–1939* (Boston: Houghton Mifflin, 1977), pp. 8–17 ("shall not starve"); Corley, *Burmah Oil,* pp. 298–307; Martin Gilbert, *Winston S. Churchill,* vol. 5, *Companion Volume,* part 1, (Boston: Houghton Mifflin, 1981), pp. 54–55 (Churchill on Baldwin), 68–69; Ferrier, *British Petroleum,* pp. 382–85 ("His Majesty's Government").

10. Mark Requa, Letter to the Subcommittee on Mineral Raw Materials, Economic Liaison Committee, May 12, 1919, Baker Library, Harvard Business School; John DeNovo, "The Movement for an Aggressive American Oil Policy Abroad, 1918–1920," *American Historical Review* (July 1956), pp. 854–76; O'Brien, "Oil Crises and the Foreign Policy of the Wilson Administration," p. 176 (Wilson); *National Petroleum News,* October 29, 1919, p. 51 ("two to five years"); Guy Elliott Mitchell, "Billions of Barrels Locked Up in Rocks," *National Geographic,* February 1918, pp. 195 ("gasoline famine"), 201 ("no man who owns"); George Otis Smith, "Where the World Gets Oil and Where Will Our Children Get It When American Wells Cease to Flow?" *National Geographic,* February 1920, p. 202 ("moral support"); *Washington Post,* November 18, 1920 (nine years and three months); George Otis Smith, ed., *The Strategy of Minerals: A Study of the Mineral Factor in the World Position of America in War and in Peace* (New York: D. Appleton, 1919), p. 304 ("within a year"). In 1919, David White, chief geologist of the United States Geological Survey, alarmed at "the widening angle between the flattening curve of production and the rising curve of consumption" in the United States, fixed total recoverable reserves at 6.7 billion barrels. David White, "The Unmined Supply of Petroleum in the United States," paper presented at the annual meeting of the Society of Automotive Engineers, February 4–6, 1919. John Rowland and Basil Cadman, *Ambassador for Oil: The Life of John First Baron Cadman* (London: Herbert Jenkins, 1960), pp. 95, 97. Requa to Adee, May 13, 1920, 800.6363/112; Manning to Baker, March 8, 1920, 811.6363/35; Fall to Hughes, July 15, 1921,

800.6363/324; Memorandum for the Secretary, March 29, 1921, 890g.6363/69; Merle-Smith to the Secretary, February 11, 1921, 800.6363/325; Millspaugh Memorandum, April 14, 1921, 890g.6363/T84/9, RG 59, NA. *Scientific American,* May 3, 1919, p. 474; Cadman to Fraser, December 2, 1920, 4247, Cadman papers ("I don't expect"); Cadman, Notes, Meeting at Petroleum Executive, June 16, 1919, GHC/Iraq/D1, Shell archives; Memorandum on the Petroleum Situation, with Dispatch to HM Ambassador, April 21, 1921, POWE 33/228, PRO.

11. United Kingdom, Admiralty, Geographical Section of Naval Intelligence Division, *Geology of Mesopotamia and Its Borderlands* (London: HMSO, 1920), pp. 84–86, insisted on a "cautious estimate" for the oil potential of the region. *FRUS,* vol. 2, pp. 664–73; Jones, *State and British Oil,* pp. 223, 221; De Novo, "Aggressive American Oil Policy," pp. 871–72; Bennett H. Wall and George S. Gibb, *Teagle of Jersey Standard* (New Orleans: Tulane University Press, 1974), p. 130; Michael Hogan, *Informal Entente: The Private Structure of Cooperation in Anglo-American Economic Diplomacy, 1918–1928* (Columbia: University of Missouri Press, 1977), p. 165; Nash, *United States Oil Policy,* p. 53. Heizer to Ravndal, January 31, 1920, 800.6363/134; Millspaugh Memorandum, November 26, 1921, 890g.6363/134; Tyrrell to Gulbenkian, October 10, 1924, with Wiley to Secretary of State, March 13, 1948, 890 g.6363/3–448 ("instrumental"), RG 59, NA.

12. WWC to Dearing, May 12, 1921, and Memorandum for the Secretary on Proposed Combination of American Oil Companies, 811.6363/73; Bedford to Hughes, May 21, 1921, 890.6363/78. NA 890g.6363/T84: Hoover to Hughes, April 17, 1922, 96; Hughes to Teagle, August 22, 1922, 41a; Allen Dulles Memorandum, December 15, 1922, 81, RG 59. Wall and Gibb, *Teagle,* p. 98 ("queer looking"); Joan Hoff Wilson, *American Business and Foreign Policy, 1920–1933* (Boston: Beacon Press, 1971), p. 189.

13. Wall and Gibb, *Teagle,* pp. 168 ("Boss"), 31–32 ("Come home"), 48–49 ("cigar"), 63–66 ("frequently changes"), 71–72 ("shoes" and "not going to drill"), 176–78 ("present policy"). On the Jersey reorganization, see Alfred D. Chandler, Jr., *Strategy and Structure: Chapters in the American Industrial Enterprise* (Cambridge: MIT Press, 1962) chap. 4, p. 173.

14. NA 890g.6363: Confidential Memorandum of Negotiations with Turkish Petroleum Company, July 15–August 5, 1922, T84/48; Wellman to Hughes, July 24, 1922, 126; Piesse to Teagle, December 12, 1922, T84/62, RG 59.

15. Fromkin, *Peace,* pp. 226 ("ripper"), 306; Elizabeth Monroe, *Britain's Moment in the Middle East, 1914–1971* (London: Chatto and Windus, 1981), 2d ed., pp. 61–64 (Lansing), 68 ("vacant lot"); Peter Sluglett, *Britain in Iraq, 1914–1932* (London: Ithaca Press, 1976), pp. 64, 45, 112; Stivers, *Supremacy and Oil,* p. 78 ("supported"); Briton Cooper Busch, *Britain, India, and the Arabs, 1914–1921* (Berkeley: University of California Press, 1971), pp. 467–69; *Review of the Civil Administration of Mesopotamia,* Cmd. 1061, 1920, p. 94, cited in Elie Kedourie, *The Chatham House Version and Other Middle Eastern Studies* (London: Weidenfeld and Nicolson, 1970), p. 437. Wheeler to Secretary of State, February 2, 1922, 890 g.6363/72. NA 890g.6363/T84: Wadsworth Memo, September 18, 1924, 167; Dulles to Millspaugh, February 21, 1922, 31; Randolph to Secretary of State, March 25, 1926, 214; Allen Dulles Memorandum, November 22, 1924, 208 ("cocked hat"), RG 59. Edith Penrose and E. F. Penrose, *Iraq: International Relations and National Development* (London: Ernest Benn, 1978), pp. 56–74; Gibb and Knowlton, *Standard Oil,* vol. 2, pp. 295–97; "Memoirs of Gulbenkian," p. 25 ("eyewash"); J. C. Hurewitz, *Diplomacy in the Near and Middle East,* vol. 2, *A Documentary Record, 1914–1956* (Princeton: Van Nostrand, 1956), pp. 131–42.

16. "Memoirs of Gulbenkian," pp. 15 ("oil friendships"), 16 ("we worked"), 28 ("hook or . . . crook"); Hewins, *Mr. Five Percent,* p. 161 ("persnickety" and "overbearing"); Gulbenkian, *Portrait in Oil,* pp. 130–39 ("children"), 38–39 ("medical advice"), 94; Henriques, *Waley Cohen,* pp. 285–86; *Financial Times,* July 25, 1955; Gibb and Knowlton, *Standard Oil,* vol. 2, pp. 298–301; Kuisel, *Mercier,* p. 34; Wall and Gibb, *Teagle,* p. 216 ("most difficult"). NA 890g.6363/T84: Allen Dulles Memorandum, January 19, 1926, 236; Houghton to Secretary of State, January 27, 1926, 238; Allen Dulles to Secretary of State, November 11, 1924, 176; Wadsworth Memo, September 18, 1924, pp. 8, 167; Swain to Dulles, December 8, 1925, 245 ("How would you like it"); Piesse to Teagle, January 19, 1926, 284; Oliphant to Atherton, January 12, 1926, 239, RG 59, NA. On the Teagle-Gulbenkian luncheon, Wall and Gibb, *Teagle,* p. 215 and Memorandum of Dulles conversation with Teagle, September 18, 1924, 167, pp. 4–5, RG 59, NA.

17. "Memorandum for Submission to the Foreign Office Setting Out Mr. C. S. Gulbenkian's Position," June 1947, pp. 3–4, POWE 33/1965, PRO; Daniel 3:4–6 ("fiery furnace"); *FRUS, 1927,* vol. 2, pp. 816–27. NA 890g.6363/T84: Allen Dulles Memo, December 2, 1925, 244; Wellman to Dulles, Octo-

ber 8, 1925, 224; Wellman to Secretary of State, April 1, April 11, April 28, 1927, 271, 272, 273; Wadsworth Memo, October 3, 1927, 279; Randolph to Secretary of State, October 19, 1927, 281.

18. William Stivers, "A Note on the Red Line Agreement," *Diplomatic History,* 7 (Winter 1983), pp. 24–25; Hewins, *Mr. Five Percent,* p. 141 ("old Ottoman Empire"); Jones, *State and British Oil,* p. 238. NA 890g.6363/T84: Agreement D'Arcy Exploration Company Limited and Others and Turkish Petroleum Company, July 31, 1928, 360; Wellman to Shaw, December 7, 1927, 292, January 31, 1928, 297; Shaw to Wellman, December 27, 1927, 293. The Quai d'Orsay and Foreign Office Maps are with Wellman to Shaw, March 22, 1928, 307, RG 59, NA. Wall and Gibb, *Teagle,* p. 209 ("bad move!"); Gulbenkian, *Portrait in Oil,* pp. 98–100.

Chapter 11

1. Dwight D. Eisenhower, *At Ease: Stories I Tell to Friends* (Garden City, N.Y.: Doubleday, 1967), pp. 155–68, 386–87 ("genuine adventure"); *New York Times,* July 6, 1920, sec. 4, p. 11.

2. Kendall Beaton, *Shell,* p. 171 ("century of travel"); Williamson et al., *Age of Energy,* pp. 443–46; Frederick Lewis Allen, *Only Yesterday: An Informal History of the Nineteen-Twenties* (New York: Blue Ribbon Books, 1931), p. 164 ("Villages"); Jean-Pierre Bardou, Jean-Jacques Chanaron, Patrick Fridenson, James M. Laux, *The Automobile Revolution: The Impact of an Industry* (Chapel Hill: University of North Carolina Press, 1982).

3. Warren C. Platt, "Competition: Invited by the Nature of the Oil Industry," *National Petroleum News,* February 5, 1936, p. 208 ("new way"); McLean and Robert Wm. Haigh, *The Integrated Oil Companies,* pp. 107–8; Giddens, *Standard Oil Company (Indiana),* pp. 318–20, 283; Thomas F. Hogarty, "The Origin and Evolution of Gasoline Marketing," Research Paper No. 022, American Petroleum Institute, October 1, 1981; Walter C. Ristow, "A Half Century of Oil-Company Road Maps," *Surveying and Mapping* 34 (December 1964), pp. 617 ("uniquely American"); Beaton, *Shell,* pp. 267–79 ("careful in their attendance" and Barton on gasoline); Bruce Barton, *The Man Nobody Knows* (Indianapolis: Grosset & Dunlap, 1925), pp. iv, v, 140.

4. Beaton, *Shell,* pp. 286–87; United States Senate, Subcommittee of the Committee on Manufacturers, *High Cost of Gasoline and Other Petroleum Products,* 67th Congress, 2d and 4th sessions (Washington, D.C.: GPO, 1923), p. 28 ("manipulate oil prices"); John H. Maurer, "Fuel and the Battle Fleet: Coal, Oil, and American Naval Strategy, 1898–1925," *Naval War College Review* 34 (November-December 1981), p. 70 ("failure of supply"). So concerned about supply (and price) was Navy Secretary Josephus Daniels that he argued that the United States government should follow Winston Churchill's lead with Anglo-Persian and go directly into the oil business. John De Novo, "Petroleum and the United States Navy Before World War I," *Mississippi Valley Historical Review* 61 (March 1955), pp. 651–52. Burl Noggle, *Teapot Dome: Oil and Politics in the 1920s* (Baton Rouge: Louisiana State University Press, 1962), pp. 16–17 ("supply laid up"), 3–4 ("looked like a President" and "harmony"). On Albert Fall, Bruce Bliven, "Oil Driven Politics," *The New Republic,* February 13, 1924, pp. 302–3 ("Zane Grey hero"); David H. Stratton, "Behind Teapot Dome: Some Personal Insights," *Business History Review* 23 (Winter 1957), p. 386 ("unrestrained disposition"); Noggle, *Teapot Dome,* p. 13 ("not altogether easy"); John Gunther, *Taken at the Flood: The Story of Albert D. Lasker* (New York: Harper & Brothers, 1960), pp. 136–37 ("it *smells*"); J. Leonard Bates, *The Origins of Teapot Dome: Progressives, Parties, and Petroleum, 1909–1921* (Urbana: University of Illinois Press, 1963).

5. On Harry Sinclair, Sinclair Oil, *A Great Name in Oil: Sinclair Through 50 Years* (New York: F. W. Dodge/McGraw-Hill, 1966), pp. 13–20, 45. Noggle, *Teapot Dome,* pp. 30 ("oleaginous nature"), 35, 51–57 ("my . . . friends" and "illness"), 71–72 ("teapot"), 79, 85 ("little black bag"), 201 ("can't convict"); M. R. Werner and John Star, *The Teapot Dome Scandal* (London: Cassell, 1961), p. 146; Edith Bolling Wilson, *My Memoir* (Indianapolis: Bobbs-Merrill, 1939); pp. 298–99 ("Which way"); Bliven, "Oil Driven Politics," pp. 302–3 ("shoulder deep"); Norman Nordhauser, *The Quest for Stability: Domestic Oil Regulation, 1917–1935* (New York: Garland, 1979), p. 20 (oil lamp); William Allen White, *A Puritan in Babylon: The Story of Calvin Coolidge* (New York: Macmillan, 1938), pp. 272–77; J. Leonard Bates, "The Teapot Dome Scandal and the Election of 1924," *American Historical Review* 55 (January 1955), pp. 305–21.

6. Giddens, *Standard of Indiana,* pp. 366–434 (the battle); M. A. & R., "Continental Trading Co. Ltd.," March 10, 1928, J.D.R., Jr., Business Interests, Rockefeller Archives; Brady, *Ida Tarbell,* pp. 210,

232 (Tarbell and Rockefeller, Jr.). On John D. Rockefeller, Jr., see Collier and Horowitz, *Rockefellers,* pp. 79–83, 104–6.

7. Gibb and Knowlton, *Standard Oil,* vol. 2, pp. 485 (Teagle), 429–30; Owen, *Trek of the Oil Finders,* pp. 449–57, 502–20, 460; Institution of Petroleum Technologists, *Petroleum: Twenty Five Years Retrospect, 1910–1935* (London: Institution of Petroleum Technologists, 1935), pp. 33–73; Henrietta M. Larson and Kenneth Wiggins Porter, *History of Humble Oil and Refining Company: A Study in Industrial Growth* (New York: Harper & Brothers, 1959), pp. 139–42, 276; Frank J. Taylor and Earl M. Welty, *Black Bonanza: How an Oil Hunt Grew into the Union Oil Company of California* (New York: Whittlesley House, McGraw-Hill, 1950), p. 201; E. L. DeGolyer, "How Men Find Oil," *Fortune,* August 1949, p. 97; Walker A. Tompkins, *Little Giant of Signal Hill: An Adventure in American Enterprise* (Englewood Cliffs, N.J.: Prentice-Hall, 1964), p. 2; United States Federal Trade Commission, *Foreign Ownership in the Petroleum Industry* (Washington, D.C.: GPO, 1923), p. x ("rapidly depleted").

8. *Literary Digest,* June 2, 1923, pp. 56–58 ("nearest approach"). Doherty to Smith, February 2, 1929 ("worse than Satan"); Doherty to Veasey, August 13, 1927 ("extremely crude"), Doherty papers. Doherty to Roosevelt, August 14, 1937, Oil, Official File 56, Roosevelt papers; Erich W. Zimmermann, *Conservation in the Production of Petroleum: A Study in Industrial Control* (New Haven: Yale University Press, 1957), pp. 97 ("do likewise"), 122–24; Nordhauser, *Quest for Stability,* pp. 9–18; Williamson et al., *Age of Energy,* pp. 317–19; Nash, *United States Oil Policy,* pp. 82–91; Leonard M. Fanning, *The Story of the American Petroleum Institute* (New York: World Petroleum Policies, [1960]), pp. 68,104–9 ("crazy man"); Linda Lear, "Harold L. Ickes and the Oil Crisis of the First Hundred Days," *Mid-America* 63 (January 1981), p. 12 ("barbarian"); Robert E. Hardwicke, *Antitrust Laws, et. al. v. Unit Operations of Oil or Gas Pools* (New York: American Institute of Mining and Metallurgical Engineers, 1948), pp. 179–186 ("If the public").

9. Williamson et al., *Age of Energy,* p. 311 ("supremacy"); Zimmermann, *Conservation,* pp. 126–28 ("commodity"); Larson and Porter, *Humble,* pp. 257–63 ("production methods"); Henrietta Larson, Evelyn H. Knowlton, and Charles H. Popple, *History of Standard Oil Company (New Jersey),* vol. 3, *New Horizons, 1927–50* (New York: Harper & Row, 1971), pp. 63–64, 88; Giebelhaus, *Sun,* p. 118 ("My father").

10. Rister, *Oil!,* pp. 244–46, 255, 293–97; Hartzell Spence, *Portrait in Oil: How the Ohio Oil Company Grew to Become Marathon* (New York: McGraw-Hill, 1962), pp. 118–29; Phillips Petroleum Company, *Phillips: The First 66 Years* (Bartlesville: Phillips Petroleum, 1983), p. 67; United States Federal Trade Commission, *Prices, Profits, and Competition in the Petroleum Industry,* United States Senate Document No. 61, 70th Congress, 1st Session (Washington, D.C.: GPO, 1928), pp. 108–16; McLean and Haigh, *Integrated Oil Companies,* pp. 90–91; Williamson et al., *Age of Energy,* pp. 394–97; Beaton, *Shell,* pp. 259–60.

11. SC7/G-32, Shell papers; Larson and Porter, *Humble,* pp. 307–9 ("industry is powerless"); Roger M. Olien and Diana D. Olien, *Wildcatters: Texas Independent Oilmen* (Austin: Texas Monthly Press, 1984), p. 52 (Tom Slick); Nordhauser, *Quest for Stability,* pp. 55 ("rather foolish"), 58; Nash, *United States Oil Policy,* pp. 102–3.

12. Joseph Stanislaw and Daniel Yergin, Cambridge Energy Research Associates, "The Reintegration Impulse: The Oil Industry of the 1990s," Cambridge Energy Research Associates Report, 1987; Larson and Porter, *Humble,* pp. 72–75; Gibb and Knowlton, *Standard Oil,* vol. 2, pp. 42, 414; Wall and Gibb, *Teagle,* pp. 140–41, 249; Giddens, *Standard of Indiana,* chap. 9, p. 318; McLean and Haigh, *Integrated Oil Companies,* pp. 95–102; Phillips, *First 66 Years,* p. 37 (Phillips); Beaton, *Shell,* pp. 298–330, 353.

13. McLean and Haigh, *Integrated Oil Companies,* p. 105 ("protection"); Ida M. Tarbell, *The New Republic,* November 14, 1923, p. 301 ("crumbling"); FTC, *Prices, Profits and Competition,* pp. 22–23, xvii–xix ("no longer unity").

14. Beaton, *Shell,* pp. 206–7 (Deterding); FTC, *Prices, Profits and Competition,* p. 29; FTC, *Foreign Ownership,* p. 86 ("parties foreign"); Ralph Arnold to Herbert Hoover, September 22, 1921, Millspaugh to Dearing, September 24, 1921, 811.6363/75 ("viciously inimical"), RG 59, NA; Taylor and Welty, *Union Oil,* pp. 176–78; Phillips, *First 66 Years,* p. 31; Giddens, *Standard of Indiana,* pp. 238–40; Wall and Gibb, *Teagle,* pp. 261–65 ("sunkist").

15. Doherty to Veasey, August 6, 1927; Doherty to Smith, January 26, 1929; Doherty to Smith, February 2, 1929, Doherty papers.

Chapter 12

1. Middlemas, *Master Builders,* pp. 169, 178 ("Dame Fortune" and *"autocrat"*), 211 ("move sharply"), 217 ("craven adventurer"); Jonathan C. Brown, "Domestic Politics and Foreign Investment: British Development of Mexican Petroleum 1889–1911," *Business History Review* 61 (Autumn 1987), p. 389 ("Poor Mexico"); Pearson to Body, April 19, 1901, Box C-43, LCO-2313, Pearson papers ("oil craze"); Pan American Petroleum, *Mexican Petroleum,* (New York: Pan American Petroleum, 1922) pp. 13–28, 185–214; J. A. Spender, *Weetman Pearson: First Viscount Cowdray* (London: Cassell, 1930), pp. 149–55 ("entered lightly" and "superficial").

2. Memorandum, October 7, 1918 ("peace of mind"), Cowdray to Cadman, May 8, 1919 ("carry indefinitely"), Royal Dutch/Shell file, Box C44, Pearson papers; Egan to Frost, memo attached, April 23, 1920, p. 4, 811.6363/352, RG 59, NA; Robert Waley Cohen, "Economics of the Oil Industry," in *Proceedings of the Empire Mining and Metallurgical Congress, 1924,* p. 13; Beeby-Thompson, *Oil Pioneer,* p. 373; Wall and Gibb, *Teagle,* p. 186. Some years later, the Pearson deputy who had originally noted the oil seepages in Mexico commented, "Had the Chief not missed the train connection at Laredo, he would have stepped from one railway carriage to another, gone into a drawing-room compartment—and, as usual, opened his bags containing his books, gone on working, with the exception perhaps of a few minutes in looking at the local paper in search of foreign news–and would thus have missed getting into the oil excitement at Laredo and San Antonio. Such was the coincidence which decided our going into Mexican oil." J. B. Body, "How We Went into Oil," Nov. 21, 1928, Box C43-LCO-2312, Pearson papers.

3. Lufkin to Dearing, April 20, 1921, 800.6363/253; Subcommittee on Mineral Raw Materials, Economic Liaison Committee, "The Petroleum Policy of the United States, p. 11, July 11, 1919, 811.6363/45; "The General Petroleum Situation," February 19, 1921, pp. 32–33, 800.6363/325, RG 59, NA. Gibb and Knowlton, *Standard Oil,* vol. 2, pp. 364–65; George Philip, *Oil and Politics in Latin America: Nationalist Movements and State Companies* (Cambridge: Cambridge University Press, 1982), pp. 16–18; N. Stephen Kane, "Corporate Power and Foreign Policy: Efforts of American Oil Companies to Influence United States Relations with Mexico, 1921–28," *Diplomatic History* 1 (Spring 1977), pp. 170–98; Lorenzo Meyer, *Mexico and the United States in the Oil Controversy, 1917–1942,* trans. Muriel Vasconcellos (Austin: University of Texas Press, 1976), pp. 24–99; O'Brien, "Oil Crisis and the Foreign Policy of the Wilson Administration," chaps. 4–6.

4. FTC, *Foreign Ownership,* pp. 11–13 ("fight for new production"); "General Petroleum Situation," February 19, 1921, p. 44, 800.6363/325, RG 59, NA; Stephen G. Rabe, *The Road to OPEC: United States Relations with Venezuela, 1919–1976* (Austin: University of Texas Press, 1982), pp. 4–5, 20 ("scoundrel"), 38 ("Monarch"); Thomas Rourke, *Gomez: Tyrant of the Andes* (Garden City, N.Y.: Halcyon House, 1936), chap. 11.

5. Philip, *Oil and Politics in Latin America,* pp. 13–15; Gibb and Knowlton, *Standard Oil,* vol. 2, pp. 384–90 ("malaria" and "spent millions"); B. S. McBeth, *Juan Vicente Gomez and the Oil Companies in Venezuela, 1908–1935* (Cambridge: Cambridge University Press, 1983), pp. 17–19, 67, 91–108; Gerretson, *Royal Dutch,* vol. 4, p. 280; Owen, *Trek of the Oil Pioneers,* pp. 1059–60 ("mirage"); Edwin Lieuwen, *Petroleum in Venezuela* (Berkeley: University of California Press, 1954), pp. 36–41; Ralph Arnold, George A. Macready and Thomas W. Barrington, *The First Big Oil Hunt: Venezuela, 1911–1916* (New York: Vantage Press, 1960), pp. 19, 343, 54, 164, 285.

6. McBeth, *Gomez and the Oil Companies,* pp. 114, 163–68; Mira Wilkins, *The Maturing of Multinational Enterprise: American Business Abroad from 1914 to 1970* (Cambridge: Harvard University Press, 1974), pp. 115–16 ("not live forever"), 507, n. 51; Giddens, *Standard Oil of Indiana,* pp. 489–93; Gibb and Knowlton, *Standard Oil,* vol. 2, p. 384 ("nonproducing"); Jonathan C. Brown, "Jersey Standard and the Politics of Latin American Oil Production, 1911–1930," in John D. Wirth, ed., *Latin American Oil Companies and the Politics of Energy* (Lincoln: University of Nebraska, 1985), pp. 38–39.

7. Wall and Gibb, *Teagle,* p. 222 ("bargain basement"); Jones, *State and British Oil,* pp. 209–11 ("be cleared"); Minutes of Meeting Held at Britannic House, November 26, 1919, Russian file 2, Box C-8, Pearson papers ("establishment"); Tolf, *The Russian Rockefellers,* pp. 211–17.

8. Gibb and Knowlton, *Standard Oil,* vol. 2, pp. 332–35 ("no other alternative"); Richard H. Ullman, *Anglo-Soviet Relations, 1917–1920,* vol. 3, *The Anglo-Soviet Accord* (Princeton: Princeton University Press, 1972), pp. 93–99 ("every inch" and "Curzon!"), 117 ("swine"); E. H. Carr, *The Bolshevik Revolution, 1917–1923,* vol. 3 (New York: Norton, 1985), pp. 352 ("cannot by our own strength" and

"quarter"), 349 ("best spies"). NA 861.6363: Teagle to Hughes, August 19, 1920, 18; "Double Victory," 49 ("liquid gold"); Bedford to Hughes, May 11, 1922, 59; Bedford memo, 22; Bedford memo, December 1920, 31, RG 59. *Times* (London), December 22, 1920; Jones, *State and British Oil*, pp. 211–12 ("several good seats"). For the nationalization, William A. Otis, *The Petroleum Industry in Russia: Supplement to Commerce Reports* (Washington: Bureau of Foreign and Domestic Commerce, Mineral Division, 1924) and "Baku Consolidated Oilfields Position of British Property in Russia," *Times* (London), December 23, 1920.

9. *FRUS*, 1922, vol. 2, p. 773; *FRUS*, 1923, vol. 2, pp. 802–04; Tolf, *Russian Rockefellers*, pp. 221–24; Gibb and Knowlton, *Standard Oil*, vol. 2, pp. 340–47 ("sick child," "participation" and "look back"); Wall and Gibb, *Teagle*, pp. 222–25 ("old fashioned"), 350–53 ("encourage the thief," "new hopes" and "so glad"). NA 861.6363: Teagle to Bedford, telegram, July 19, 1922, 84; Sussdorf to Hughes, July 27, 1922, 88, September 19, 1922, 104; unsigned memorandum with Poole memo, October 6, 1922, 112; DeVault memo, October 8, 1923, 169; Deterding telegram, February 1926, 262 (Deterding to J.D.R., Jr.), RG 59.

10. Deterding to Riedemann, October 20, 1927 ("neither honor nor" and "enormous events"), 5–5–35 file, case 6, Oil Companies papers; *Financial Times,* January 16, 1928; Gibb and Knowlton, *Standard Oil*, vol. 2, pp. 352–56 ("thinking people" and "buried Russia"). NA 861.6363: Kelley memo, February 8, 1927, 222; Memo of conversation with Sir John Broderick, Feb. 4, 1928, 239 ("hot water" and "lost his head"); Tobin to Secretary of State, June 18, 1928—Standard Oil Company/4 ("suddenly attacked"); Whaley to Kellogg, March 14, 1928, 240, RG 59. Peter G. Filene, *Americans and the Soviet Experiment, 1917–1933* (Cambridge: Harvard University Press, 1967), p. 118 ("more unrighteous"); Joan Hoff Wilson, *Ideology and Economics: U.S. Relations with the Soviet Union, 1918–1933* (Columbia: University of Missouri Press, 1974), app. D.

Chapter 13

1. Olien and Olien, *Texas Independent*, pp. 15–16 (oil promotion pitches), 56–57 ("trendologist"); James A. Clark and Michael T. Halbouty, *The Last Boom* (Texas: Shearer Publishing, 1984), pp. 4–9 ("treasure trove" and "Medicine Show"), 43 ("Every woman"), 31–32 ("I'll drink"), 67 ("not an oil well"), 80 ("fires!"); Owen, *Trek of the Oil Finders*, p. 857; Oral History interview with E. C. Laster, Texas History Center.

2. *Henderson Daily News*, October 4, 1930; Olien and Olien, *Texas Independent*, pp. 57–58 ("teakettles"); Clark and Halbouty, *Last Boom*, pp. 67–72 ("second Moses"); Larson and Porter, *Humble*, pp. 451–54; Nordhauser, *Quest for Stability* p. 72; Harry Hurt III, *Texas Rich: The Hunt Dynasty from the Early Days Through the Silver Crash* (New York: Norton, 1981), chaps. 3, 5; *C.M. Joiner, et al. v. Hunt Production Company, et al.,* No. 9650, "Plaintiff's Original Petition," November 25, 1932; "Deposition of H. L. Hunt," January 16, 1933, pp. 44 ("flying start"), 83 ("had traded"); "Additional statement of C. M. Joiner," January 16, 1933, District Court of Rusk Country, Texas. Dad Joiner's discovery was, thereafter, an exceedingly sore point with professional geologists. "The discovery of the East Texas field," wrote Wallace Pratt, Jersey's chief geologist in 1941, "is popularly credited to chance. The fact that the well was drilled by an itinerant wildcatter, on a location recommended by a pseudo-geologist, would seem to justify the verdict of a chance discovery, without the benefit of geology. But reflect on the further fact that for fifteen years geologically directed exploration had been carried on in the immediate locality . . . persistent geologic exploration had narrowed down the possible territory still to be drilled to a width of not more than about ten miles." Humble, Shell, Atlantic, and other companies had already drilled scores of wells, and Humble had more than 30,000 acres under lease in what turned out to be the East Texas field. "The work had progressed until only a very narrow gap remained untested. Dad Joiner's discovery well just had to be located in this gap to avoid the existing dry holes. Does geology deserve any credit for this accomplishment?" Pratt to DeGolyer, July 10, 1941, 1513, DeGolyer papers.

3. David F. Prindle, *Petroleum Politics and the Texas Railroad Commission* (Austin: University of Texas Press, 1981), p. 24 ("suicide"); Jacqueline Lang Weaver, *Unitization of Oil and Gas Fields in Texas: A Study of Legislative, Administrative, and Judicial Politics* (Washington: Resources for the Future, 1986), pp. 48–50 ("deadly threat"); Lear, "Harold Ickes," pp. 6–7; Nordhauser, *Quest for Stability*, pp. 66–67 ("physical waste"), 85; Frederick Godber, "Notes of Visit to America," May–June 1931, SC 7/G 30/12, Shell papers; Rister, *Oil!*, p. 264 ("one dollar"); Nash, *United States Oil Policy*, pp. 124, 116; Williamson et al., *The Age of Energy*, p. 561.

4. Clark and Halbouty, *Last Boom,* pp. 168–73 ("insurrection," "rebellion," "worms" and "hot enough"); Olien and Olien, *Texas Independent,* p. 55 ("economic waste"); Owen, *Trek of the Oil Finders,* p. 471 ("water drive"); Larson and Porter, *Humble,* pp. 475–76 ("tooth and claw").

5. Graham White and John Maze, *Harold Ickes of the New Deal: His Private Life and Public Career* (Cambridge: Harvard University Press, 1985), pp. 98 ("plump"), 174 ("Resignation"), 48 ("restless"), 31 ("pick losers"), 116 ("slaved away"), 104–7 ("oil-besmeared"); T. H. Watkins, *Righteous Pilgrim: The Life and Times of Harold L. Ickes, 1874–1952* (New York: Henry Holt, 1990), part 6; Harold L. Ickes, *The Secret Diary of Harold L. Ickes,* vol. 1, *The First Thousand Days, 1933–36* (New York: Simon and Schuster, 1953), p. 82 ("ghost of Albert B. Fall").

6. Ickes to Roosevelt, May 1, 1933 ("demoralization" and "ten cents"), Doherty to Roosevelt, May 12, 1933 ("collapse"); Moffett to Roosevelt, May 31, 1933, Oil, Official File 56, Roosevelt papers. Lear, "Harold Ickes," p. 10 ("unprecedented authority"); Ickes, *Secret Diary,* vol. 1, pp. 31–32 ("beyond the control" and "crawling"); Ickes to Hiram Johnson, May 31, 1933, Box 217, Ickes papers; Harold L. Ickes, "After the Oil Deluge, What Price Gasoline?" *Saturday Evening Post,* February 16, 1935, pp. 5–6 ("age of oil").

7. Ickes, "After the Oil Deluge," p. 39 ("cunning"). Roosevelt to Rayburn, May 22, 1934 ("wretched conditions"); Grilling to Pearson, telegram, with Ickes to McIntyre, June 9, 1934 ("hot oil boys"); Personal Assistant to McIntyre, October 19, 1934 ("heaven and earth"); Cummings to Roosevelt, December 30, 1934 ("good progress"), Oil, Official File 56, Roosevelt papers. Ickes, *Secret Diary,* vol. 1, pp. 65 ("broad powers"), 86 ("prepared the allocation"); Hardwicke, *Antitrust Laws,* pp. 51–53; Nordhauser, *Quest for Stability,* p. 124 ("now to doomsday"); James A. Veasey, "Legislative Control of the Business of Producing Oil and Gas," in *Report of the 15th Annual Meeting of the American Bar Association* (Baltimore: Lord Baltimore Press, 1927), pp. 577–630.

8. Thompson to Roosevelt, n.d., 1937 ("this treaty"); Ickes to Roosevelt, May 4, 1935, Oil, Official File 56, Roosevelt papers. Joe S. Bain, *The Economics of the Pacific Coast Petroleum Industry,* pt. I, *Market Structure* (Berkeley: University of California Press, 1944), pp. 60–66; Zimmermann, *Conservation,* p. 207; Wilkins, *Maturing of Multinational Enterprise,* pp. 210–11; Fanning, *American Petroleum Institute,* pp. 133–36; Lieuwen, *Petroleum in Venezuela,* pp. 56–60; United States Department of Commerce, *Minerals Yearbook, 1932–1933* (Washington, D.C.: GPO, 1933), p. 497 (tariff).

9. Thompson to Roosevelt, n.d., 1937, Oil, Official File 56, Roosevelt papers ("cooperation and coordination"); McLean and Haigh, *Integrated Oil Companies,* p. 113; Robert E. Hardwicke, "Market Demand as a Factor in the Conservation of Oil," in *First Annual Institute on Oil and Gas Law* (New York: Matthew Bender, 1949), pp. 176–79; Nordhauser, *Quest for Stability,* p. 127; Williamson et al., *Age of Energy,* pp. 559–60.

Chapter 14

1. "Particulars Regarding Achnacarry Castle, Season 1928," SC7/A24, Shell archives (Malcolm and Hillcart); *Daily Express,* August 13, 1928 ("no warning"); Wall and Gibb, *Teagle,* pp. 259–61 ("hellions").

2. Loxley and Collier minutes, April 4, 1930, N2149/FO 371/14816, PRO; Deterding to Riedemann, Oct. 20, 1927, 5–5–35 file, case 6, Oil Companies papers; Jones, *State and British Oil,* p. 236; Larson, *Standard Oil,* vol. 3, p. 306; Leslie Hannah, *The Rise of the Corporate Economy,* 2d. ed. (London: Methuen, 1976), chaps. 2, 4, 7; Wilkins, *Multinational Enterprise.*

3. Rowland, *Cadman,* p. 55 (Cadman's academic opponent). Cadman discussion with Fisher, Barstow et al., February 1928, T161/284/533045/2; Hopkins to Chancellor of Exchequer, February 10, 1928, T161/284/533048/1 ("alliance"); Committee on Imperial Defense, Proposed Agreement, February 16, 1928, T8/T10, T161/284/33045/2; Treasury and Admiralty, "Anglo-Persian Oil Company: Scheme of Distribution in the Middle East," T161/284/533048/1 ("irritability," "long run" and "similar alliances"); Churchill to Hopkins, February 12, 1928, T161/284/533048 ("singularly inopportune"); Oliphant minute, Feb. 15, 1928, A1270/6, FO 371/12835; Barstow and Packe to the Treasury, March 15, 1928, T161/284/33045/2; Wilson to Waterfield, February 13, 1928, T161/284/533048/1 ("amalgamation"), PRO. Ferrier, *British Petroleum,* pp. 514, 510.

4. Weill to the Baron, March 5, 1929, 132 AQ 1052, Rothschild papers; United States Congress, Senate, Foreign Relations Committee, Subcommittee on Multinational Corporations, *Multinational Corporations and United States Foreign Policy,* part 8 (Washington, D.C.: GPO, 1975), pp. 30–33 ("As-Is"), 35–39 ("problem," "destructive" and "Association")—hereafter *Multinational Hearings;*

Larson, *Standard Oil*, vol. 3, pp. 308–9; U.S. Congress, Senate. Committee on Small Business, Subcommittee on Monopoly, *The International Petroleum Cartel: Staff Report to the Federal Trade Commission* (Washington, D.C.: 1952)—hereafter FTC, *International Petroleum Cartel*, pp. 199–229; Ferrier, *British Petroleum*, p. 513; Jones, *State and British Oil*, p. 236; Tolf, *Russian Rockefellers*, p. 224.

5. Campbell to Cushendun, October 29, 1928, A7452/1270/45, FO 371/12835; Jackson to Broderick, September 26, November 17, 1930, A6632, FO 371/14296, PRO. FTC, *International Petroleum Cartel*, p. 270 ("fringe"); Kessler to Teagle, September 13, 1928, "misc." file, case 9, Oil Companies papers ("figures").

6. Roy Leigh, "Interview with Deterding," February 18, 1930, SC7/G32, Shell archives; *Multinational Hearings*, part 8, pp. 39–51 ("local arrangements" and "local cartels"). Sadler to Harden et al., March 2, 1931, 6–9–18 file, case 1 ("abrogated"); Sadler memo to Teagle, June 15, 1931, "misc." file, case 9 ("great sacrifice" and "price war"), Oil Companies papers. Weill to the Baron, March 14, 1930, 132 AQ 1052; March 23, 1932, 132A AQ 1052, p. 572 ("bad everywhere"), Rothschild papers. Larson, *Standard Oil*, vol. 3, p. 311; John Cadman, "Petroleum and Policy," in American Petroleum Institute, *13th Annual Meeting: Proceedings, 1932*; FTC, *International Petroleum Cartel*, pp. 235–50.

7. Shuckburgh minute, January 15, 1934, F.W.S., December 12, 1933, Petroleum Dept. Memorandum, January 12, 1934, p. 4, W 488, FO 371/18488, PRO; *Multinational Hearings*, part 8, pp. 51–70 (on economies); FTC, *International Petroleum Cartel*, pp. 255, 264 ("standardized"), 266. Teagle to Kessler, August 14, 1931, "various nos." file, case 2; Harden memo, January 19, 1935, 12–1–3 file, case 6; Sadler memo, June 15, 1931, case 9 ("ambition"); Riedemann to Teagle, June 26, 1935, and extract from June 6, 1935, Executive Committee meeting, 4–2–9 file, case 4; to Harper, September 29, 1933, Brown Envelope, case 9; "Gulf, SONJ, others" file, case 1, Oil Companies papers. Deterding to Riedemann, November 4, 1936, SC7/A14/1 ("much needed munitions"); Emmert to Parker, December 21, 1934, SC7/A12 ("unanimously opposed" and "private walls"); Godber to Agnew, December 31, 1934, SC7/A12, Shell archives. Peter F. Cowhey, *The Problems of Plenty: Energy Policy and International Politics* (Berkeley: University of California Press, 1985), pp. 90–93.

8. Wilkins, *Multinational Enterprise*, pp. 234–38 ("defensive manner," "failure to cooperate" and "90 percent political"); Shuckburgh minute, January 15, 1934, F.W.S., December 12, 1933, Petroleum Dept. Memorandum, January 12, 1934, p. 4 ("general tendency"), W 488, FO 371/18488 PRO; Harden memo, January 19, 1935, file 12–1–3, case 6 ("nationalistic policies"), Oil Companies papers.

9. Peter J. Beck, "The Anglo-Persian Oil Dispute of 1932–33," *Journal of Contemporary History* 9 (October 1974), pp. 127–43; Rowland, *Cadman*, pp. 123–33; Ferrier, *British Petroleum*, p. 610 ("suspicion"); Stephen H. Longrigg, *Oil in the Middle East: Its Discovery and Development* (Oxford: Oxford University Press, 1968), 3d. ed., pp. 59–60 ("Persianization").

10. Jonathan C. Brown, "Why Foreign Oil Companies Shifted Their Production from Mexico to Venezuela During the 1920s," *American Historical Review* 90 (April 1985), pp. 362–85; Roosevelt to Daniels, February 15, 1939, Official File 146, Roosevelt papers; Meyer, *Oil Controversy*, pp. 102, 127–54; Philip, *Oil and Politics*, p. 211.

11. O'Malley, "Leading Personalities in Mexico," March 15, 1938, A 1974/26, FO 371, PRO ("obsidian eyes," "chief" and "bugbear"); William Weber Johnson, *Heroic Mexico: The Violent Emergence of a Modern Nation* (Garden City, N.Y.: Doubleday, 1968), pp. 403–22; Meyer, *Oil Controversy*, pp. 152–56 ("conquered territory"); Anita Brenner, *The Wind That Swept Mexico: The History of the Mexican Revolution, 1910–1942* (Austin: University of Texas Press, 1977), p. 91. Body to DeGolyer, March 21, 1935, 128 ("quite Red"); DeGolyer to McCollum, August 23, 1945, 1110 (DeGolyer and Holman), DeGolyer papers. J. B. Body, "Aguila," August 2, 1935, pp. 4, 6, box C44, Pearson papers; Philip, *Oil and Politics*, pp. 206–9 ("incapable" and "half a Bolshevik"); Clayton R. Koppes, "The Good Neighbor Policy and the Nationalization of Mexican Oil: A Reinterpretation," *Journal of American History* 69 (June 1982). Assheton letter, February 21, 1934, A 1947, FO 371; Murray to Foreign Office, September 17, 1935, A8586, FO 371/18708 (manager's fulminations), PRO. Deterding to Riedemann, November 4, 1936, SC7/A14/1, Shell archives. On other Latin America confrontations, see Stephen J. Randall, *United States Foreign Oil Policy, 1919–1948* (Kingston: McGill-Queen's University Press, 1985), pp. 69–77, 91–96 and Herbert S. Klein, "American Oil Companies in Latin America: The Bolivian Experience," *Inter-American Economic Affairs* 18 (Autumn 1964), pp. 47–72.

12. Philip, *Oil and Politics*, p. 218 ("Men without respect"). Gallop to Eden, June 17, 1937,

149/16/31/37, FO 371/20639 ("notorious but sincere"); Memo, "Regarding the Circumstances Attending Expropriation," A 2306/10/26, FO 371/21464; O'Malley to Foreign Office, December 27,
1937, A9313, FO 371/20637; Murray to Foreign Office, February 6, 1937, A1623, FO 371/20639
("advisers and officials" and "completely unanimous"); O'Malley to Foreign Office, March 8, 1938,
A1835, FO 371/21463, PRO. Godber to Starling, May 25, 1938, SC 7/G3/1, Shell archives; Meyer,
Oil Controversy, pp. 158–70; *FRUS, 1938,* pp. 724–27; on Cardenas's program, see *Antología de la
Planeación en México (1917–1985),* vol. 1, *Primeros Intentos de Planeación en Mexico
(1917–1946)* (Mexico City: Ministry of Budget and Planning, 1985), p. 207.

13. Shell archives, SC7/G3: Davidson to Godber, 3; Godber to Starling, October 27, 1938, 4; Legh-Jones
to Coleman, August 25, 1938, 3 ("precedent"); Memorandum of conversation with Mr. Hackworth,
August 24, 1938, 3; Telephone conversation with New York, June 27, 1938, 1; Wilkinson to Godber
with memo, August 30, 1938, 3. Roosevelt to Daniels, February 15, 1939, OF 146, Roosevelt papers
("fair compensation"). Hohler to Halifax, Aug. 28, 1938, A 7045/10/ 26, FO 371/21476; Davidson to
Godber, March 5, 1940, FO 371/24215; Petroleum Department, "The Expropriation by the Mexican
Government of the Properties of the Oil Companies in Mexico," April 8, 1938, FO 371/21469
("doubtful sources" and "Mexican policy"); Committee of Imperial Defense, "Expropriation of the
Properties of the Oil Companies in Mexico," May 1938, 1428–B, A3663, FO 371/29468; "Mexican
Oil Dispute," October 11, 1940, A4486/57/26, FO 371/24216; Note by the Oil Board, May 9, 1938, A
3663/10/26/21469; Memorandum, "The Mexican Oil Question," December 1, 1938, pp. 2, 18, A
8808/10/26, FO 371/21477 ("paramount consideration"), PRO.

14. Meyer, *Oil Controversy,* pp. 219–24 ("Julius Caesar"). Halifax to Cadogan, June 11, 1941, A4467,
FO 371/26063; Cadogan to Halifax, June 12, 1941, FO 371/26063 ("put ideas"), PRO. Philip, *Oil and
Politics,* p. 34; Arthur W. MacMahon and W. R. Dittman, "The Mexican Oil Industry Since Expropri-
ation II," *Political Science Quarterly* 57 (June 1942), pp. 169–78.

Chapter 15

1. Archibald H. T. Chisholm, *The First Kuwait Oil Concession Agreement: A Record of the Negotiations*
(London: Frank Cass, 1975), pp. 5–6, 93–95, 161; Thomas E. Ward, *Negotiations for Oil Concessions in Bahrein, El Hasa (Saudi Arabia), the Neutral Zone, Qatar, and Kuwait* (New York: privately
printed, 1965), pp. 11, 255; H. St. J. B. Philby, *Arabian Oil Ventures* (Washington: Middle East Institute, 1964), p. 98 ("bluff, breezy").

2. Fox to Secretary of State, June 24, 1933, 890F.6363/Standard Oil Co./17, RG 59, NA ("mischief").
Meeting Relating to Oil in the Persian Gulf, April 26, 1933, paragraph 16, POWE 33/241/114869
("rover"); Interview Regarding Koweit Oil Concession, January 4, 1934, P.Z. 145/1934; p. 4, POWE
33/242/114864 (not . . . "particularly satisfactory"), PRO. Longrigg, *Oil in the Middle East,* pp. 42,
98–99; Chisholm, *Kuwait Oil Concession,* p. 161 ("Father of oil"); Ward, *Negotiations,* pp. 23–26.

3. Randolph to Secretary of State, May 19, 1924, 741.90G/30; August 15, 1924, 890G.6363/T84/164;
November 26, 1924, 890G.6363/T84/189, RG 59, NA. Chisholm, *Kuwait Oil Concession,* pp. 127
("little room"), 162; Ferrier, *British Petroleum,* p. 555 ("devoid").

4. Ballantyne to Gibson, December 16, 1938, P.Z. 8299/38, POWE 33/195/114869, PRO; Chisholm,
Kuwait Oil Concession, pp. 106–9, ("not . . . any . . . promise" and "pure gamble"), p. 13; Jerome
Beatty, "Is John Bull's Face Red," *American Magazine,* January 1939 ("worst nuisance").

5. P. T. Cox and R. O. Rhoades, A Report on the Geology and Oil Prospects of Kuwait Territory, June
11, 1935, 638–107–393, Gulf archives; Standard Oil of California, "Report on Bahrein and Saudi
Concessions," December 5, 1940, 3465, DeGolyer papers; Ward, *Negotiations,* pp. 80–81 ("New
York Sheikhs"); Chisholm, *Kuwait Oil Concession,* pp. 13–14 ("greasy substance"); Frederick Lee
Moore, Jr., "Origin of American Oil Concessions in Bahrein, Kuwait, and Saudi Arabia" (Senior the-
sis, Princeton University, 1951), pp. 22–34; Irvine H. Anderson, *Aramco, the United States, and
Saudi Arabia, 1933–1950* (Princeton: Princeton University Press, 1981), pp. 22–23.

6. Stone and Wellman to Piesse, October 5, 1928, Brown to Piesse, November 12, 1928, 5–5–35 file,
Case 6, Oil Companies papers; Longrigg, *Oil in the Middle East,* pp. 26–27 ("clause" and "interests").

7. Bahrein Oil Concession and U.S. Interests, Rendel Memo, May 30, 1929, E 2521/281/91, FO
371/13730/115395, PRO; Standard Oil of California, "Report on Bahrein and Saudi Concessions,"
December 5, 1940, pp. 7–9, 21–22, 3465, DeGolyer papers.

8. Dickson to Political Resident, April 27, 1933, POWE 33/241/114869, PRO ("astute Bin Saud");

H. St. J. B. Philby, *Arabian Jubilee* (London: Robert Hale, 1952), p. 49; Elizabeth Burgoyne, ed., *Gertrude Bell: From Her Personal Papers, 1914–1926* (London: Ernest Benn, 1961), p. 50 ("well-bred Arab").

9. Philby, *Arabian Jubilee,* pp. 5, 75; Karl S. Twitchell, *Saudi Arabia: With an Account of the Development of Its Natural Resources,* 3d ed. (Princeton: Princeton University Press, 1958), pp. 144–54; Jacob Goldberg, *The Foreign Policy of Saudi Arabia: The Formative Years* (Cambridge: Harvard University Press, 1986), chap. 2 (Mubarak), p. 136 ("our advantage"); H. St. J. B. Philby, *Sa'udi Arabia* (London: Ernest Benn, 1955), pp. 261–68 ("thirty thousand"), 280–92; Christine Moss Helms, *The Cohesion of Saudi Arabia: Evolution of Political Identity* (Baltimore: Johns Hopkins University Press, 1981), p. 211 ("neutral zones"); David Holden and Richard Johns, *The House of Saud* (London: Pan Books, 1982), pp. 51, 80.

10. Clive Leatherdale, *Britain and Saudi Arabia, 1925–1939: The Imperial Oasis* (London: Frank Cass, 1983), pp. 114–20.

11. Mohammed Almana, *Arabia Unified: A Portrait of Ibn Saud* (London: Hutchinson Benham, 1980), p. 90.

12. Kim Philby to Monroe, Oct. 27, 1960, file 3, box 23, Philby papers; Philby, *Arabian Jubilee,* p. 54; Kim Philby, *My Secret War* (MacGibbon & Kee, 1968), p. 99; Almana, *Arabia Unified,* pp. 153, ("true replica"), 151; Elizabeth Monroe, *Philby of Arabia* (London: Faber and Faber, 1973), pp. 158–62 ("how nice"); Philby, *Oil Ventures,* p. 126 ("traditional western dominance"); H. St. J. B. Philby, *Arabian Days: An Autobiography* (London: Robert Hale, 1948), pp. 282–83, 253 ("I was surely"); Memo to S. Wilson, Offices of the Cabinet, August 13, 1929, CO 732/41/3 ("Since he retired"), PRO; Leatherdale, *Britain and Saudi Arabia,* p. 194 ("humbug").

13. Diary of Crane visit to Jidda, February 25–March 3, 1931, chap. 9 of Edgar Snow manuscript, Crane papers; Philby, *Arabian Jubilee,* pp. 175–77 ("Oh, Philby"); "Oil Negotiations," file 3, box 29, Philby to Crane, Dec. 29, 1929, file 2, box 16, Philby papers ("one of his eyes").

14. Notes on Sheikh Ahmad's Trip to Raith, Enclosure 2 in No. 44, April 6, 1932, E 2469/27/25, FO 406/69/115218, PRO; H. S. Villard, Memo of Conversation with Twitchell, November 1, 1932, 890 F.6363/10, RG 59, NA.

15. Lombardi to Philby, January 30, 1933, Philby to Hamilton, March 4, 1933, Aramco/Socal files, Philby papers; Loomis to Secretary of State, October 25, 1932, 890 F.6363/Standard Oil of California/1, RG 59, NA; Almana, *Arabia Unified,* pp. 191–99 (Suleiman); Ryan to Warner, March 15, 1933, E 1750/487/25, POWE 33/320/114964, PRO ("stage is set").

16. Hamilton to Philby, February 28, 1932 ("get in touch"); Philby to Lees, Dec. 17, 1932 ("disposed to help"); Philby to Loomis, April 1, 1933, Aramco/Socal files, Philby papers. Twitchell to Murray, March 26, 1933, 890 F.6363/Standard Oil Co./9, RG 59, NA; Philby, *Oil Ventures,* p. 83 ("It is no good"); Wallace Stegner, *Discovery: The Search for Arabian Oil* (Beirut: Middle East Export Press, 1974), p. 19.

17. Philby, *Oil Ventures,* p. 106 ("did not need"); Philby to Hamilton, March 14, 15, 1933, Aramco/Socal files, Philby papers. Ryan to Warner, March 15, 1933, E 1750/487/25, POWE 33/320 ("pig in a poke"); Jedda Report for April 1933, May 9, 1933, E 2839/902/25, FO 4061/71, PRO. Longrigg, *Oil in the Middle East,* pp. 58–60, 73–75; Benjamin Shwadran, *The Middle East, Oil and the Great Powers,* 3d ed. (New York: John Wiley, 1973), pp. 43–47, 238.

18. Twitchell to Philby, March 26, 1933; Philby to Loomis, April 1, 1933; Hamilton to Suleiman, April 21, 1933, Aramco/Socal files, Philby papers. Telegram from Ryan, May 30, 1933, E 2844/ 487/25, POWE 33/320/114964, PRO; Contract between Saudi Arabian Government and Standard Oil Company of California, May 29, 1933, with Loomis to Hull, May 2, 1938, 890F.6363/Standard Oil Co./97, RG 59, NA; Philby, *Oil Ventures,* pp. 100 ("unfortunate impasse"), 119 ("pack up"), 99 ("detente"), 124 ("pleasure"); Wilkins, *Maturing of Multinational Enterprise,* p. 215.

19. Chancery to Department, August 24, 1933, E 5455/487/25, CO 732/60/10/115125, PRO; Philby, *Oil Ventures,* pp. 125 ("thunderstruck"), 46–48; Monroe, *Philby,* pp. 208–9 (Kim Philby).

20. Chisholm, *Kuwait Oil Concession,* pp. 19 ("stab to my heart"), 176 ("flank" and "sphere"). Ryan telegram, June 1, 1933, E 3073/487/25, POWE 33/320/114964; Rendel, Tour in the Persian Gulf and Saudi Arabia, February-March 1937, CO 732/79/17/115218 ("dangerous policy"); Letter from the Political Resident in the Persian Gulf, December 13, 1927, P1341, CO 732/33/10; Warner memo, November 2, 1932, E 5764/121/91, FO 371/16002/115578 ("jackal"); Rendel to Warner, February 3, 1933, POWE 33/241/114869 ("frittering away"), PRO.

21. Biscoe to Foreign Office, October 29, 1931, No. 18, FO 371/15277/115659; Bullard to Halifax,

Chapter I–Arabia, January 10, 1939, E246/246/25, FO 406/77, PRO. R. I. Lawless, *The Gulf in the Early 20th Century: Foreign Institutions and Local Responses* (Durham: Centre for Middle Eastern and Islamic Studies, 1986), pp. 91–92; Chisholm, *Kuwait Oil Concession*, pp. 19, 37; Jacqueline S. Ismael, *Kuwait: Social Change in Historical Perspective* (Syracuse: Syracuse University Press, 1982), pp. 61–71; Fatimah H. Y. al-Abdul Razzak, *Marine Resources of Kuwait: Their Role in the Development of Non-Oil Resources* (London: KPI Limited, 1984), pp. 59–60; Committee for the Study of Culture Pearls, *Report on the Study of the Mikimoto Culture Pearl* (Tokyo: Imperial Association for the Encouragement of Inventions, 1926).

22. Admiralty, Oil Concession in Kuwait, March 15, 1932, FO 371/16001/115578; Rendel memo, Proposed Koweit Oil Concession, January 30, 1932, FO 371/160001/115578 ("protection"); Political Resident to Secretary of State for India, February 7, 1932, FO 371/16001, 115578 ("losing influence"); Oliphant to Vansittart, January 20, 1932, FO 371/16001/115578; Oliphant to Wakely, January 22, 1932, FO 371/16001/115578 ("oil war"); Oil in Koweit re: Cabinet Conclusions, April 6, 1932, E 1733/121/91, FO 371/16002/115578; Simon to Atherton, April 9, 1932, E 1733/121/91, FO 371/16002/115578; Oliphant to Secretary of State, April 11, 1932, FO 371/16002/115578 ("Americans are welcome"), PRO.

23. Dickson to Political Resident, May 1, 1932, POWE 33/241/114869 ("wonderful victory"); Memo, February 20, 1933, p. 2, POWE 33/241/114869, PRO. David E. Koskoff, *The Mellons: The Chronicle of America's Richest Family* (New York: Thomas Y. Crowell, 1978), pp. 271–98 ("precisely the same"); Chisholm, *Kuwait Oil Concession*, p. 160.

24. P. T. Cox, "A Report on the Oil Prospects of Kuwait Territory," May 12, 1932, pp. 26–27, 638–107–393, Gulf Archives; Chisholm, *Kuwait Oil Concession*, pp. 26 ("two bidders"), 160 ("personal benefit"), 141 ("go easy"), 67, 27–30 ("dead body"). Rendel memo, Dec. 23, 1932, with Oil Concession in Kuwait, E 6801/121/91, FO 371/16003/115659 ("so keen a personal interest"); Oliphant to Cadman, December 30, 1932, E 6830/121/191, POWE 331/241/114869, PRO.

25. Fowle to Colonial Office, re: Kuwait Oil, June 27, 1933, POWE 33/241/114869, PRO; Chisholm, *Kuwait Oil Concession*, pp. 27–28, 175–79 ("keep his hands" and Cadman and Sheikh Ahmad); Ward, *Negotiations*, p. 227.

26. Rendel to Laithwaite, December 14, 1933, E 7701/12/91, POWE 33/241/114869 ("blessing" and "British hands"); Koweit Oil: Political Agreement of March 4, 1934, E 2014/19/91, FO 905/17/115218; Oil Concessions in Kuwait, March 8, 1935, pp. 8–11, POWE 33/246/114964, PRO. 1934 Concession Agreement, December 23, 1934, 78–135–043, Gulf archives; Chisholm, *Kuwait Oil Concession*, p. 45 ("heavenly twins"); Ward, *Negotiations*, p. 229 ("pure in heart").

27. Nomland to Knabenshue, June 7, 1935, with Knabenshue to Murray, June 20, 1935, 890F.6363/Standard Oil Co./82 ("sure shot"), RG 59, NA. *Sun and Flare* (Aramco magazine), February 6, 1957; "Persian Gulf Pioneer," [1956] ("camel days"); "Exploration Comes of Age in Saudi Arabia," *Standard Oil Bulletin*, December 1938, pp. 2–10; "A New Oil Field in Saudi Arabia," *Standard Oil Bulletin*, September 1936, pp. 3–16, Chevron files. Wilkins, *Multinational Enterprise*, pp. 215–17 ("total loss").

28. Seidel to Teagle, November 20, 1935, February 10, 1936, 5–5–36 file, Case 6; Walden memo, 7/26/34, various nos. file, case 2; Halman to Sadler, November 15, 1938, various nos. file, case 2, Oil Companies papers. Rendel Memo, Oil in Arabia, July 7, 1937, P.Z. 612/37, POWE 33/ 533/115294 ("irksome" and "buy them out"); Starling to Clauson, July 3, 1936, P.Z. 674/36, FO 371/19965/115659 ("all to the good"), PRO. William Lenahan to Abdulla Suleiman, February 10, 1934, with Loomis to Hull, May 2, 1938, 890 F.6363/Standard Oil Co./97, RG 59, NA; Anderson, *Aramco*, pp. 26–28; FTC, *International Petroleum Cartel*, pp. 73–74, 115; Wilkins, *Multinational Enterprise*, pp. 214–17.

29. P. T. Cox and R. O. Rhoades, "Report on the Geology and Oil Prospects of Kuwait Territory," June 1, 1935, 638–107–393; Memo to Bleecker, Summary Review of Burgan No. 1, 537–149–501; L. W. Gardner, Case History of the Burgan Field, 621–74–107, Gulf archives. Chisholm, *Kuwait Oil Concession*, pp. 81, 250.

30. Murray, "The Struggle for Concessions in Saudi Arabia," August 2, 1939, 890F.6363/Standard Oil Co./118 ("astronomical proportions"), RG 59, NA. Standard Oil of California, "Report on Bahrein and Saudi Concessions," pp. 75–77, December 5, 1940, pp. 75–77, 3465, DeGolyer papers; Hull to Roosevelt, June 30, 1939, OF 3500, Roosevelt papers; *New York Times*, August 8, 1939; Wilkins, *Multinational Enterprise*, p. 217; Uriel Dann, ed., *The Great Powers in the Middle East, 1919–1939* (New York: Holmes & Meier, 1988), chap. 19.

31. Standard Oil of California, "Report on Bahrein and Saudi Concessions," December 5, 1940, p. 80, 3465, DeGolyer papers; Holden and Johns, *House of Saud*, pp. 121–22; Monroe, *Philby*, pp. 295–96 ("so bored" and "Greatest"); Chisholm, *Kuwait Oil Concession*, pp. 93–95 ("my geologist").

Chapter 16

1. Takehiko Yoshihashi, *Conspiracy at Mukden: The Rise of the Japanese Military* (New Haven: Yale University Press, 1963), p. 14 ("life line" and "living space"); Seki Hiroharu, "The Manchurian Incident, 1931," trans. Marius B. Jansen, in *Japan Erupts: The London Naval Conference and the Manchurian Incident, 1928–1932,* ed. James William Morley (New York: Columbia University Press, 1984), pp. 139, 225–30; Sadako N. Ogata, *Defiance in Manchuria: The Making of Japanese Foreign Policy, 1931–32* (Berkeley: University of California Press, 1964), pp. 59–61, 1–16; G. R. Storry, "The Mukden Incident of September 18–19, 1931," in *St. Antony's Papers: Far Eastern Affairs* 2 (1957), pp. 1–12.
2. Franklin D. Roosevelt, "Shall We Trust Japan?" *Asia* 23 (July 1923), pp. 475–78, 526–28.
3. James B. Crowley, *Japan's Quest for Autonomy: National Security and Foreign Policy, 1936–1938* (Princeton: Princeton University Press, 1966), pp. 244–45 ("government by assassination"); Mira Wilkins, "The Role of U.S. Business," in *Pearl Harbor as History: Japanese-American Relations, 1931–1941,* eds. Dorothy Borg and Shumpei Okamoto (New York: Columbia University Press, 1973), pp. 341–45; Stephen E. Pelz, *Race to Pearl Harbor: The Failure of the Second London Naval Conference and the Onset of World War II* (Cambridge: Harvard University Press, 1974), p. 15; Yoshihashi, *Conspiracy at Mukden,* chap. 6; *FRUS: Japan, 1931–1941,* vol. 1, p. 76.
4. *FRUS: Japan, 1931–1941,* vol. 1, pp. 224–25 ("mission" and "special responsibilities"); Crowley, *Japan's Quest,* pp. 86–90 ("national defense state"), 284–86 (*hokushu*), 289–97 ("spirit"); Robert J. C. Butow, *Tojo and the Coming of the War* (Princeton: Princeton University Press, 1961), pp. 23, 55–70; Akira Iriye, *Across the Pacific: An Inner History of American-East Asian Relations* (New York: Harcourt, Brace & World, 1967), pp. 207–8; Jerome B. Cohen, *Japan's Economy in War and Reconstruction* (Minneapolis: University of Minnesota Press, 1949), pp. 133–37; Irvine H. Anderson, *The Standard-Vacuum Oil Company and United States East Asian Policy, 1933–1941* (Princeton: Princeton University Press, 1975), pp. 221–31. Anderson is a key source on the oil side. Michael A. Barnhart, *Japan Prepares for Total War: The Search for Economic Security, 1919–1941* (Ithaca: Cornell University Press, 1987), pp. 28–29.
5. Laura E. Hein, *Fueling Growth: The Energy Revolution and Economic Policy in Postwar Japan* (Cambridge: Harvard University Press, 1990), pp. 46–52; Anderson, *Standard-Vacuum,* pp. 81–90 ("frightening" and "resistance"); Ickes, *Secret Diary,* vol. 1, p. 192.
6. Crowley, *Japan's Quest,* p. 335 ("unpardonable crime"); Herbert Feis, *The Road to Pearl Harbor: The Coming of War Between the United States and Japan* (New York: Atheneum, 1966), pp. 9–10 ("thoroughgoing blow"), 12. Feis remains the classic diplomatic history, to be supplemented by Jonathan G. Utley, *Going to War with Japan, 1937–1941* (Knoxville: University of Tennessee Press, 1985). James William Morley, ed., *The China Quagmire: Japan's Expansion on the Asian Continent, 1933–1941* (New York: Columbia University Press, 1983), pp. 233–86; Michael A. Barnhart, "Japan's Economic Security and the Origins of the Pacific War," *Journal of Strategic Studies* 4 (June 1981), p. 113; Robert Dallek, *Franklin D. Roosevelt and American Foreign Policy, 1932–1945* (Oxford: Oxford University Press), pp. 147–55 ("quarantine" and "without declaring war").
7. Utley, *Going to War,* pp. 36–37 ("moral embargo"); Feis, *Pearl Harbor,* p. 19 ("not yet").
8. Joseph Grew Diary, 1939, pp. 4083–84, Joseph Grew Papers ("intercept her fleet"); Theodore H. White, *In Search of History: A Personal Adventure* (New York: Harper & Row, 1978), pp. 280–83 ("aerial terror"); Utley, *Going to War,* p. 54 ("Japan furnishes").
9. Anderson, *Standard-Vacuum,* pp. 118–21 (Walden and Elliott).
10. *New York Times,* January 11, 1940; Ickes, *Secret Diary,* vol. 3, pp. 96, 132, 274; Edwin P. Hoyt, *Japan's War: The Great Pacific Conflict* (New York: McGraw-Hill, 1986), p. 215 ("ABCD"); Butow, *Tojo,* p. 7 ("Razor"); James William Morley, ed., *The Fateful Choice: Japan's Advance into Southeast Asia, 1939–1941* (New York: Columbia University Press, 1980), pp. 122, 241–86.
11. Henry Stimson Diary, July 18, 19 ("only way out"), 24, 26, 1940, Henry Stimson Papers; Morgenthau Diary, vol. 319, p. 39, October 4, 1940; John Morton Blum, *From the Morgenthau Diaries: Years of Urgency, 1938–1941* (Boston: Houghton Mifflin, 1965), pp. 349–59; Nobutaka Ike, ed., *Japan's Decision for War: Records of the 1941 Policy Conferences* (Stanford: Stanford University Press, 1967), pp.

7, 11; Ickes, *Diaries,* vol. 3, pp. 273, 297–99 ("needling"); Morley, *Fateful Choice,* pp. 142–45, chap. 3; Cohen, *Japan's Economy,* p. 25. See H. P. Willmott, *Empires in the Balance: Japanese and Allied Pacific Strategies to April 1942* (Annapolis: Naval Institute Press, 1982), p. 68: "It was concern about the security of her oil supplies that primarily molded Japanese strategy at the beginning of the war."

12. Roosevelt to Grew, January 21, 1941, Grew Diary, p. 4793 ("single world conflict"); Sir Llewellyn Woodward, *British Foreign Policy in the Second World War,* vol. 2 (London: Her Majesty's Stationery Office, 1971), p. 137; Ickes, *Secret Diary,* vol. 3, p. 339; Ike, *Japan's Decision for War,* p. 39; Anderson, *Standard-Vacuum,* p. 143 ("Europe first").

13. United States Congress, 79th Congress, 1st Session, *Hearings Before the Joint Committee on the Investigation of the Pearl Harbor Attack* (Washington, D.C.: GPO, 1946), part 17, p. 2463; Feis, *Pearl Harbor,* pp. 38–39 ("smallest particles"); Kichisaburo Nomura, "Stepping Stones to War," *United States Naval Institute Proceedings* 77 (September 1951); *FRUS: Japan, 1931–1941,* vol. 2, p. 387 ("friend"); *FRUS, 1941,* vol. 4, p. 836 (lips and heart); Gordon W. Prange, *At Dawn We Slept: The Untold Story of Pearl Harbor,* with Donald M. Goldstein and Katherine V. Dillon (New York: McGraw-Hill, 1981), pp. 6 ("one pillar"), 119; Cordell Hull, *The Memoirs of Cordell Hull* (New York: Macmillan, 1948), vol. 2, p. 987; David Kahn, *The Codebreakers: The Story of Secret Writing* (New York: Macmillan, 1967), pp. 22–27; Roberta Wohlstetter, *Pearl Harbor: Warning and Decision* (Stanford: Stanford University Press, 1962), p. 178.

14. Prange, *At Dawn We Slept,* pp. 10–11 ("schoolboy" and "armchair arguments"); Hiroyuki Agawa, *The Reluctant Admiral: Yamamoto and the Imperial Navy,* trans. John Bester (Tokyo: Kodansha International, 1982) pp. 2–13, 32, 70–91, 141, 148–58 ("scientist"), 173–89.

15. Prange, *At Dawn We Slept,* pp. 28–29 ("lesson" and "regrettable"), 15–16 ("fatal blow" and "first day"); Morley, *Fateful Choice,* p. 274 ("whole world"); Grew to Secretary of State, January 27, 1941, 711.94/1935, PSF 30, Roosevelt papers (Grew's warning).

16. Feis, *Pearl Harbor,* p. 204 ("emergency"); Roosevelt to Ickes, June 18, June 30, Ickes to Roosevelt, June 23, July 1, 1941, Ickes files, PSF 75, Roosevelt papers (Ickes-FDR exchange).

17. Morley, *Fateful Choice,* p. 255, chap. 4; Ike, *Japan's Decision,* pp. 56–90 ("life or death"); United States Congress, Joint Committee on the Investigation of the Pearl Harbor Attack, 79th Congress, 1st Session, *Pearl Harbor: Intercepted Messages Sent by the Japanese Government Between July 1 and December 8, 1941* (Washington, D.C.: GPO, 1945), pp. 1–2 ("next on our schedule"); Morgenthau Presidential Diaries, vol. 4, 09146–47, July 18, 1941 ("question" and "mean war"); "Exports of Petroleum Products, Scrap Iron and Scrap Steel," Office of Secretary of the Treasury, Weekly Reports, PSF 918, Treasury, Roosevelt papers; United States Congress, *Pearl Harbor Hearings,* part 32, p. 560; Feis, *Pearl Harbor,* pp. 228–29 ("always short"); *FRUS: Japan, 1931–1941,* vol. 2, pp. 527–30 ("bitter criticism" and "new move"). For the criticism, see Eliot Janeway, "Japan's Partner," *Harper's Magazine,* June 1938, pp. 1–8; Henry Douglas, "America Finances Japan's New Order," *Amerasia,* July 1940, pp. 221–24; Douglas, "A Bit of History—Successful Embargo Against Japan in 1918," *Amerasia,* August 1940, pp. 258–60. Woodward, *British Foreign Policy,* vol. 2, p. 138; Blum, *Morgenthau: Years of Uncertainty,* p. 378 ("day to day"); Waldo Heinrichs, *Threshold of War: Franklin D. Roosevelt and America's Entry into World War II* (New York: Oxford University Press, 1988), pp. 134, 153, 178, 246–47; Dean Acheson, *Present at the Creation: My Years in the State Department* (New York: New American Library, 1970), pp. 50–52 ("state of affairs"); *FRUS, 1941,* vol. 4, pp. 886–87.

18. Peter Lowe, *Great Britain and the Origins of the Pacific War: A Study of British Policy in East Asia, 1937–1941* (Oxford: Clarendon Press, 1977), pp. 239–40 ("as drastically"); Woodward, *British Foreign Policy,* vol. 2, pp. 138–39; United States Congress, *Intercepted Messages,* pp. 8 ("hard looks"), 11; Iriye, *Across the Pacific,* p. 218; *FRUS: Japan, 1931–1941,* vol. 2, p. 751 ("Japanese move").

19. Grew Diary, July 1941, p. 5332 ("vicious circle"); Feis, *Pearl Harbor,* p. 249 ("cunning dragon"); Akira Iriye, *Power and Culture: The Japanese-American War, 1941–1945* (Cambridge: Harvard University Press, 1981), p. 273, n. 32; Arthur J. Marder, *Old Friends, New Enemies: The Royal Navy and the Imperial Japanese Navy* (Oxford: Oxford University Press, 1981), pp. 166–67 ("scarecrows"); United States Congress, *Intercepted Messages,* p. 9; Blum, *Morgenthau: Years of Urgency,* p. 380 ("except force"); *FRUS, 1941, vol. 4, pp. 342, 359.*

20. Butow, *Tojo,* pp. 236–37 ("whole problem"); Fumimaro Konoye, "Memoirs of Prince Konoye," in United States Congress, *Pearl Harbor Attack,* part 20, pp. 3999–4003 ("receipt of intelligence"); Hull, *Memoirs,* vol. 2, p. 1025; Gordon W. Prange, *Pearl Harbor: The Verdict of History,* with Donald M. Goldstein and Katherine V. Dillon (New York: McGraw-Hill, 1986), p. 186.

21. Ike, *Japan's Decision,* pp. 154 ("weak point"), 139 ("day by day"), 133–57, 188, 201–16; Konoye, "Memoirs," pp. 4003–12 (Emperor); United States Congress, *Intercepted Messages,* pp. 81–82 ("dead horse"), 141; Hull, *Memoirs,* vol. 2, pp. 1069–70 ("no last words"); *FRUS, 1941,* vol. 4, pp. 590–91; Grew Diary, October 1941, p. 5834; Cohen, *Japan's Economy,* p. 135.

22. Grew to Secretary of State, November 3, 1941, 711.94/2406, PSF 30, Roosevelt papers; Stimson Diary, November 25, 1941; United States Congress, *Intercepted Messages,* pp. 92, 101, 165 ("beyond your ability" and "automatically"); Ike, *Japan's Decision,* pp. 238–39 (Tojo's summation); Hull, *Memoirs,* pp. 1063–83; *FRUS: Japan, 1931–1941,* pp. 755–56.

23. Stimson Diary, November 26 ("fairly blew up"), 27 ("washed my hands"), 1941; Prange, *At Dawn We Slept,* p. 406 ("war warning"); Konoye, "Memoir," pp. 4012–13; United States Congress, *Intercepted Messages,* p. 128.

24. Kahn, *Codebreakers,* p. 41; Agawa, *Yamamoto,* p. 245 ("here nor there"); United States Congress, *Intercepted Messages,* p. 215; Dallek, *Roosevelt,* p. 309 ("clouds" and "son of man"); *FRUS: Japan, 1931–1941,* vol. 2, pp. 784–87; Feis, *Pearl Harbor,* pp. 340–42 ("foul play" and "nasty"); Hull, *Memoirs,* pp. 1095–97 ("Japanese have attacked"); Woodward, *British Foreign Policy,* vol. 2, p. 177 ("infamous falsehoods" and "dogs").

25. Stimson Diary, November 28, 30, December 6, 7 ("caught by surprise"); Prange, *At Dawn We Slept,* p. 527, 558; Forrest C. Pogue, *George C. Marshall: Ordeal and Hope, 1939–1942* (New York: Viking, 1966), p. 173 ("fortress"); Wohlstetter, *Pearl Harbor,* pp. 3, 386–95; Prange, *Verdict of History,* p. 624.

26. Hoyt, *Japan's War,* pp. 236, 246; Anderson, *Standard-Vacuum,* p. 192; Prange, *At Dawn We Slept,* pp. 504, 539; Agawa, *Yamamoto,* pp. 261–65; Prange, *Verdict of History,* p. 566 (Nimitz).

Chapter 17

1. Joseph Borkin, *The Crime and Punishment of I. G. Farben* (New York: Free Press, 1978), p. 54 ("financial lords" and "money-mighty"); Nuremberg Military Tribunals, *Trials of War Criminals,* vol. 7 (Washington, D.C.: GPO, 1953), pp. 536–41 ("economy without oil"), 544–54; Peter Hayes, *Industry and Ideology: I. G. Farben in the Nazi Era* (Cambridge: Cambridge University Press, 1987) pp. 64–68. Hayes is the main academic source on I. G. Farben. Henry Ashby Turner, Jr., *German Big Business and the Rise of Hitler* (New York: Oxford University Press, 1987), pp. 246–49 ("this man").

2. United States Strategic Bombing Survey, *The Effects of Strategic Bombing on the German War Economy* (Washington, D.C.: USSBS, 1945), p. 90; Raymond G. Stokes, "The Oil Industry in Nazi Germany, 1936–45," *Business History Review* 59 (Summer 1985), p. 254; Terry Hunt Tooley, "The German Plan for Synthetic Fuel Self Sufficiency, 1933–1942" (Master's thesis, Texas A & M University, 1978), pp. 25–26 ("turning point"); United States Strategic Bombing Survey, *Oil Division Final Report* (Washington, D.C.: USSBS, 1947), p. 14.

3. Arnold Krammer, "Fueling the Third Reich," *Technology and Culture* 19 (June 1978), pp. 397–399; Neal P. Cochran,, "Oil and Gas from Coal," *Scientific American* May 1976, pp. 24–29; U.K. Ministry of Fuel and Power, *Report on the Petroleum and Synthetic Oil Industry of Germany* (London: HMSO, 1947), p. 82; Thomas Parke Hughes, "Technological Momentum in History: Hydrogenation in Germany, 1898–1933," *Past and Present* 44 (August 1969), pp. 114–23.

4. Teagle to Bosch, February 27, 1930, various nos. file, Case 2, Oil Companies papers; Borkin, *I. G. Farben,* pp. 47–51 (Howard's telegram, "We were babies" and "the I. G."); Frank A. Howard, *Buna: The Birth of an Industry* (New York: Van Nostrand, 1947), pp. 15–20 (Howard on hydrogenation); *New York Times,* May 23, 1945, p. 21; W. J. Reader, *Imperial Chemical Industries: A History,* vol. 1, *The Forerunners, 1870–1926* (London: Oxford University Press, 1970), pp. 456–66.

5. Tooley, "Synthetic Fuel," pp. 14, 28 ("fixed in principle"), 72; Edward L. Homze, *Arming the Luftwaffe: The Reich Air Ministry and the German Aircraft Industry, 1919–1939* (Lincoln: University of Nebraska Press, 1976) p. 140; Nuremberg Tribunals, *Trials,* vol. 7, pp. 571–73; Stokes, "Oil Industry in Nazi Germany," p. 261; Berenice A. Carroll, *Design for Total War: Arms and Economics in the Third Reich* (The Hague: Mouton, 1968), pp. 123–30.

6. Anthony Eden, *The Eden Memoirs: Facing the Dictators* (London: Cassell, 1962), pp. 296–306 ("mad-dog" and Laval); Robert Goralski and Russell W. Freeburg, *Oil & War: How the Deadly Struggle for Fuel in WW II Meant Victory or Defeat* (New York: William Morrow, 1987), pp. 23–24 ("incalculable disaster"). Goralski and Freeburg are an important source for this and the following war chapters. John R. Gillingham, *Industry and Politics in the Third Reich: Ruhr Coal, Hitler, and Europe*

(London: Methuen, 1985), pp. 69, 75 ("wasp's nest"); *New York Times,* February 16, 1936, p. 1 ("motor mileage" and "political significance"); Alan Bullock, *Hitler: A Study in Tyranny* (New York: Harper Torch Books, 1964), rev. ed., p. 345 ("nerve-wracking").

7. Nuremberg Tribunals, *Trials,* vol. 7, pp. 793–803 (Hitler's Four Year Plan); Borkin, *I. G. Farben,* p. 72; Hayes, *I. G. Farben,* pp. 196–202, 183. USSBS, *Oil Division Final Report,* pp. 15–27, figures 22, 23; Krammer, "Fueling the Third Reich," pp. 398–403; USSBS, *German War Economy,* p. 75; Anne Skogstad, *Petroleum Industry of Germany During the War* (Santa Monica: Rand Corporation, 1950), p. 34; Homze, *Luftwaffe,* p. 148; War Cabinet, Committee on Enemy Oil Position, December 1, 1941, Appendix 10, POG (L) (41) 11, CAB 77/18, PRO.

8. Norman Stone, *Hitler* (Boston: Little, Brown, 1980), pp. 107–8 ("life's mission"); Alan Clark, *Barbarossa: The Russian-German Conflict, 1941–1945* (London: Macmillan, 1985), p. 25 ("little worms"); Walter Warlimont, *Inside Hitler's Headquarters, 1939–1945,* trans. R. H. Barry (London: Weidenfeld and Nicolson, 1964), pp. 113–14; Paul Carell, *Hitler Moves East, 1941–1943* (Boston: Little, Brown, 1965), pp. 536–37 ("Hitler's obsession"); USSBS, *German War Economy,* p. 17; Robert Cecil, *Hitler's Decision to Invade Russia, 1941* (London: Davis-Poynter, 1975), p. 84; Barry A. Leach, *German Strategy Against Russia, 1939–1941* (London: Oxford University Press, 1973), pp. 146–48; USSBS, *Oil Division Final Report,* pp. 36–39 ("need for oil").

9. Pearton, *Oil and the Romanian State,* pp. 232–33, 249; USSBS, *German War Economy,* pp. 74–75; John Erickson, *The Road to Stalingrad* (London: Panther, 1985), pp. 80–87 ("substantial prop"), chap. 3; W. N. Medlicott, *The Economic Blockade,* vol. 1 (London: HMSO, 1952), pp. 658, 667; B. H. Liddell Hart, *History of the Second World War* (New York: Putnam, 1970), pp. 143–50 ("those oilfields"); Barton Whaley, *Codeword Barbarossa* (Cambridge: MIT Press, 1973); Gerhard L. Weinberg, *Germany and the Soviet Union, 1939–1941* (London: E. J. Brill, 1954), p. 165.

10. Earl F. Ziemke, *Stalingrad to Berlin: The German Defeat in the East* (Washington, D.C.: Office of the Chief of Military History, U.S. Army, 1968), p. 7; USSBS, *German War Economy,* p. 18; Heinz Guderian, *Panzer Leader* (London: Michael Joseph, 1952), p. 151; Stone, *Hitler,* p. 109; Franz Halder, *The Halder Diaries* (Boulder, Colo.: Westview Press, 1976), p. 1000; B. H. Liddell Hart, *The Other Side of the Hill* (London: Cassell, 1973), p. 126.

11. Van Creveld, *Supplying War,* p. 169; H. R. Trevor-Roper, *Hitler's War Directives, 1939–1945* (London: Sidgwick and Jackson, 1964), p. 95 ("seize the Crimea"); Guderian, *Panzer Leader,* p. 200 ("aircraft carrier" and "My generals"); Ronald Lewin, *Hitler's Mistakes* (New York: William Morrow, 1984), pp. 122–23 ("our Mississippi"); Leach, *German Strategy,* p. 224 ("end of our resources"). On destroying the oil fields, Lord Hankey's Committee on Preventing Oil from Reaching Enemy Powers, August 19, September 19, October 30, December 4, 1941, POG (41) 16, CAB 77/12, PRO.

12. Warlimont, *Hitler's Headquarters,* pp. 226, 240; F. H. Hinsley, E. E. Thomas, C. F. G. Ranson, and L. C. Knight, *British Intelligence in the Second World War,* vol. 2 (London: HMSO, 1981), pp. 80–100. An oil analyst, Walter J. Levy, working in the OSS, conducted a study of German railway tariffs. He discovered a new entry covering shipments from Baku. This gave the clue that the prime German effort would be toward the Caucasus. Walter J. Levy, *Oil Strategy and Politics, 1941–1981,* ed. Melvin Conant (Boulder, Colo.: Westview Press, 1982), p. 36. Trevor-Roper, *Hitler's War Directives,* p. 131; Liddell Hart, *Other Side of the Hill,* pp. 301–5; USSBS, *German War Economy,* p. 18; Albert Seaton, *The Russo-German War, 1941–1945* (London: Arthur Barker, 1971), pp. 258, 266; Halder, *Halder Diaries,* p. 1513. Albert Speer, *Inside the Third Reich,* trans. Richard and Clara Winston (New York: Macmillan, 1970), pp. 238–39.

13. USSBS, *Oil Division Final Report,* fig. 23; Ziemke, *Stalingrad,* pp. 19, 355; Guderian, *Panzer Leader,* p. 251 ("icy cold"); Erich von Manstein, *Lost Victories,* trans. Anthony G. Powell (London: Methuen, 1958), p. 339; Felix Gilbert, ed., *Hitler Directs His War* (New York: Octagon, 1982), pp. 17–18; USSBS, *German War Economy,* pp. 19, 24, Alexander Stahlberg, *Bounden Duty: The Memoirs of a German Officer, 1932–1945,* trans. Patricia Crampton (London: Brassey's, 1990), pp. 226–27 (Manstein phone call).

14. B. H. Liddell Hart, ed., *The Rommel Papers,* trans. Paul Findlay (1953; reprint, New York: Da Capo Press, 1985), pp. 198 ("complete mobility"), 58 ("never imagined"), 85 ("lightning tour"), 96 ("quarter master staffs"), 141 ("petrol gauge"), 191; James Lucas, *War in the Desert: The Eighth Army at El Alamein* (New York: Beaufort Books, 1982), pp. 49–51.

15. Liddell Hart, *Rommel Papers,* pp. 514–15 ("conditions" and "colossus"), 235–37 ("Get passports"), 269; Goralski and Freeburg, *Oil & War,* pp. 203–7 ("Destiny"); Carell, *Hitler Moves East,* p. 519; Halder, *Halder Diaries,* p. 885; van Creveld, *Supplying War,* chap. 6.

16. Bernard Montgomery, *The Memoirs of Field Marshal Montgomery* (1958; reprint, New York: Da Capo Press, 1982), pp. 72 ("Everything I possessed"), 126 ("nip back"); Nigel Hamilton, *Monty,* vol. 1, *The Making of a General, 1887–1942* (London: Sceptre, 1984), p. 589 ("slightly mad"); Liddell Hart, *Other Side of the Hill,* p. 247 ("all his battles"); Liddell Hart, *Rommel Papers,* pp. 278–80 ("badly depleted").

17. Liddell Hart, *Rommel Papers,* pp. 359 ("petrol transport"), 380 ("proper homage"), 394 ("two years"); Hinsley, *British Intelligence,* vol. 2, pp. 454–55 ("catastrophic"); Denis Richards and Hilary St. George Saunders, *Royal Air Force, 1939–1945,* vol. 2 (London: HMSO, 1954), pp. 239–41; Hamilton, *Monty,* vol. 1, pp. 795–98. For Rommel's constant refrain about fuel, see *Rommel Papers,* pp. 342–89.

18. Alan Bullock, *Hitler,* p. 751 ("heart"); Liddell Hart, *Rommel Papers,* pp. 328 ("bravest men"), 453 ("weep").

19. Leach, *German Strategy,* p. 151. Speer's own memoir, *Inside the Third Reich,* should be supplemented with Matthias Schmidt, *Albert Speer: The End of a Myth* (New York: Collier Books, 1982); J. K. Galbraith, *Economics, Peace and Laughter* (Boston: Houghton Mifflin, 1971), pp. 288–302; and the report on Galbraith's original interrogation of Speer as part of the 1945 U.S. Strategic Bombing Survey, reprinted in the *Atlantic Monthly,* July 1979, pp. 50–57. USSBS, *German War Economy,* pp. 23–25, 7, 76; Liddell Hart, *Second World War,* p. 599 ("weakest point"); Williamson Murray, *Strategy for Defeat: The Luftwaffe, 1933–1945* (Maxwell: Air University Press, 1983), pp. 272–74; Tooley, "Synthetic Fuel," p. 110; USSBS, *Oil Division Final Report,* pp. 19–20.

20. Lucy S. Dawidowicz, *The War Against the Jews, 1933–1945* (New York: Bantam, 1978), pp. 199–200; Nuremberg Tribunals, *Trials,* vol. 8, pp. 335 ("favorably located"), 386, 375, 393 ("unpleasant scenes"), 405 ("brute force"), 436–37, 455, 491–92 (shooting party); Borkin, *I. G. Farben,* pp. 117–27; Tooley, "Synthetic Fuel," p. 106; Goralski and Freeburg, *Oil & War,* pp. 282–83; Krammer, "Fueling the Third Reich," p. 416 ("not run away"); Primo Levi, *Survival in Auschwitz and the Reawakening: Two Memoirs,* trans. Stuart Woolf (New York: Summit Books, 1985), pp. 72, 85, 171. For the Wannsee Conference, see J. Noakes and G. Pridham, eds., *Nazism 1919–1945: A History in Documents and Eyewitness Accounts,* vol. 2 (New York: Schocken Books, 1990), pp. 1127–36.

21. Speer, *Third Reich,* pp. 553, n. 3, 346–48 ("technological war" and "scatter-brained"); Wesley Frank Craven and James Lea Cate, *The Army Air Forces in World War II,* vol. 3 (Chicago: University of Chicago Press, 1951), pp. 172–79, 287 ("nightmare"); David Eisenhower, *Eisenhower at War, 1943–1945* (New York: Random House, 1986), pp. 154–57, 184–86; USSBS, *German War Economy,* p. 80; Murray, *Luftwaffe,* pp. 272–76. In *The Collapse of the German War Economy, 1944–1945: Allied Air Power and the German National Railway* (Chapel Hill: University of North Carolina Press, 1988), Alfred C. Mierzejewski argues that, in terms of attacking the German war economy, the railway marshaling yards were the number-one target. But he acknowledges that the destruction of the synthetic fuel plant would have immobilized the military, ibid., p. 185.

22. Craven and Cate, *Army Air Forces,* vol. 3, p. 179; USSBS, *German War Economy,* pp. 4–5 ("primary strategic aim"); Borkin, *I. G. Farben,* pp. 129–30; Goralski and Freeburg, *Oil & War,* pp. 247–48 ("fatal blow"); Speer, *Third Reich,* pp. 350–52 ("committing absurdities"); USSBS, *Oil Division Final Report,* pp. 19–29, 87; United Kingdom Ministry of Fuel and Power, *Synthetic Oil Industry,* p. 116; Milward, *War, Economy, and Society,* p. 316; Krammer, "Fueling the Third Reich," p. 418; Paul H. Nitze, *From Hiroshima to Glasnost: At the Center of Decision* (New York: Grove Weidenfeld, 1989), pp. 35–36.

23. Bullock, *Hitler,* pp. 759–61; Liddell Hart, *Other Side of the Hill,* pp. 450–51 ("stand still"), 463; Hugh M. Cole, *The Ardennes: The Battle of the Bulge* (Washington, D.C.: Department of the Army, 1965), pp. 259–69; John S. D. Eisenhower, *The Bitter Woods* (New York: Putnam, 1969), pp. 235–42.

24. USSBS, *German War Economy,* p. 80; Speer, *Third Reich,* pp. 472 ("nonexistent divisions"), 406; Liddell Hart, *Second World War,* p. 679; Bullock, *Hitler,* pp. 772–73, 781; Warlimont, *Hitler's Headquarters,* p. 497 ("last crazy orders"); Stone, *Hitler,* p. 179.

Synthetic fuels, in the German war economy, reached as high as 60 percent of total supply. The fall-off in output toward the end of the war reflects the Allied bombing campaign. Most of the synthetic fuels were produced by hydrogenation and Fischer-Tropsch, but also included alcohol, benzol, and the product of coal tar distillation.

German Oil Supply, 1938–1945
(barrels per day)

Year	Synthetic	Other	Total	Synthetic Share
1939	47,574	121,973	169,547	28.1%
1941	89,007	119,614	208,621	42.7%
1943	124,299	112,865	237,164	52.4%
1944				
Q1	131,666	100,782	232,448	56.6%
Q2	107,120	66,862	173,981	61.6%
Q3	48,473	40,245	88,719	54.6%
Q4	43,240	36,455	79,695	54.3%
1945				
Q1	5,437	17,726	23,163	23.5%

Source: USSBS, *German War Economy,* tables 37, 38 and 41, pp. 75–76, 79.

Chapter 18

1. Johan Fabricus, *East Indies Episode* (London: Shell Petroleum Company, 1949), pp. 1, 41–67, 57 ("no longer possible").

2. S. Woodburn Kirby, *The War Against Japan,* vol. 1, *The Loss of Singapore* (London: HMSO, 1957), p. 449; Butow, *Tojo,* p. 416; Cohen, *Japan's Economy,* pp. 52–53 ("victory fever"). Cohen provides a most useful analysis of the Japanese economy. United States Strategic Bombing Survey (Pacific), *Interrogations of Japanese Officials,* vol. 2 (Toyoda), OPNAU-P-03-100, p. 320 ("victory drunk"); Ronald H. Spector, *Eagle Against the Sun: The American War with Japan* (New York: Vintage, 1985), pp. 418 (FDR), 146 (Nimitz). Spector is an excellent source on the Pacific War. E. B. Potter, *Nimitz* (Annapolis: Naval Institute Press, 1976), p. 48 ("primary objectives").

3. Agawa, *Yamamoto,* p. 299 ("adults' hour").

4. Jiro Horikoshi, *Eagles of Mitsubishi: The Story of the Zero Fighter* (Seattle: University of Washington Press, 1981), p. 130; United States Strategic Bombing Survey, *The Effects of Strategic Bombing on Japan's War Economy* (Washington, D.C.: GPO, 1946), pp. 18, 135; *Pipeline to Progress: The Story of PT Caltex Pacific Indonesia* (Jakarta: 1983), pp. 27–34; Saburo Ienaga, *The Pacific War, 1931–1945* (New York: Pantheon, 1978), p. 176.

5. USSBS, *Japan's War Economy,* p. 46 ("fatal weakness"); Japan, Allied Occupation, *Reports of General MacArthur: Japanese Operations in the Southwest Pacific Area,* vol. 2, part 1 (Washington, D.C.: U.S. Army, 1966), pp. 48 ("Achilles heel"), 45 (originally printed but not published by General MacArthur's headquarters in 1950); Goralski and Freeburg, *Oil & War,* pp. 191–93; Kirby, *War Against Japan,* vol. 3, *The Decisive Battles* (London: HMSO, 1961), p. 98; United States Strategic Bombing Survey, Oil and Chemical Division, *Oil in Japan's War* (Washington, D.C.: USSBS, 1946), p. 55 ("only American planes").

6. Ronald Lewin, *The American Magic: Codes, Ciphers and the Defeat of Japan* (New York: Farrar Straus Giroux, 1982), pp. 223–24 ("noon positions"), 227–28; Cohen, *Japan's Economy,* pp. 104, 58 ("death blow"), 137–46 (Japanese captain and "synthetic fuel"); Clay Blair, Jr., *Silent Victory: The U.S. Submarine War Against Japan* (Philadelphia: J. B. Lippincott, 1975), pp. 361–362, 435–39, 553–54.

7. USSBS, *Interrogations of Japanese Officials* (Toyoda), p. 316 ("much fuel"); Cohen, *Japan's Economy,* pp. 142–45 ("very keenly" and "too much fuel"); Spector, *Eagle Against the Sun,* p. 370 ("Turkey Shoot"); Kirby, *War Against Japan,* vol. 4, *The Reconquest of Burma* (London: HMSO, 1965), p. 87; *Reports of General MacArthur: Japanese Operations,* vol. 2, part 1, p. 305; United States Army, Far East Command, Military Intelligence Section, "Interrogation of Soemu Toyoda," September 1, 1949, DOC 61346, pp. 2–3; USSBS, *Japan's War Economy,* p. 46.

8. Spector, *Eagle Against the Sun,* pp. 294 (MacArthur), 440 ("divine wind"); USSBS, *Interrogations of Japanese Officials* (Toyoda), p. 317; Cohen, *Japan's Economy,* pp. 144–45 ("shortage"); Rikihei Inoguchi, Tadashi Nakajima, and Roger Pineau, *The Divine Wind: Japan's Kamikaze Force in World War II* (Westport, Conn.: Greenwood Press, 1978), pp. 74–75; *Reports of General MacArthur: Japanese Operations,* vol. 2, part 2, p. 398. Toshikaze Kase, *Journey to the Missouri,* ed. David N. Rowe

(New Haven: Yale University Press, 1950), pp. 247–48. Liddell Hart in his *History of the Second World War* offers other reasons for Kurita's swerve, pp. 626–27.

9. Samuel Eliot Morison, *History of United States Naval Operations in World War II*, vol. 7, pp. 107–9; vol. 8, pp. 343–45; James A. Huston, *The Sinews of War: Army Logistics, 1775–1953* (Washington, D.C.: U.S. Army, 1966), p. 546 ("long legs"); Goralski and Freeburg, *Oil & War*, pp. 316, 310 ("potatoes"); USSBS, *Japan's War Economy*, p. 32; Thomas R. H. Havens, *Valley of Darkness: The Japanese People and World War II* (New York: Norton, 1978), pp. 122, 130.

10. USSBS, *Interrogations of Japanese Officials* (Toyoda), p. 316 ("large-scale operation"); Spector, *Eagle Against the Sun*, p. 538 ("the end").

11. *Reports of General MacArthur: Japanese Operations*, vol. 2, part 2, pp. 617–19, 673–74; Cohen, *Japan's Economy*, pp. 146–47; USSBS, *Oil in Japan's War*, p. 88 ("end of the road").

12. Robert J. C. Butow, *Japan's Decision to Surrender* (Stanford: Stanford University Press, 1954), pp. 30, 64, 77, 90–92, 121–22; United States Strategic Bombing Survey, *Japan's Struggle to End the War* (Washington: GPO, 1946), pp. 16–18; Kase, *Journey to the Missouri*, pp. 171–76 ("utter hopelessness" and "ready to die").

13. Lewin, *American Magic*, p. 288; Richard Rhodes, *The Making of the Atomic Bomb* (New York: Touchstone, 1988), pp. 617–99; and Daniel Yergin, *Shattered Peace: The Origins of the Cold War* (New York: Penguin, 1990), pp. 120–22.

14. United States Army, Far East Command, Military Intelligence Section, "Statements by Koichi Kido," May 17, 1949, DOC 61476, pp. 13–15, DOC 61541, pp. 7–8; Butow, *Japan's Decision*, pp. 161, 205–19; Kase, *Journey to the Missouri*, p. 247; Cohen, *Japan's Economy*, pp. 144, 147.

15. D. Clayton Jones, *The Years of MacArthur*, vol. 2 (Boston: Houghton Mifflin, 1975), pp. 785–86; Courtney Whitney, *MacArthur: His Rendezvous with Destiny* (New York: Knopf, 1956), pp. 214–16; Robert L. Eichelberger, *Our Jungle Road to Tokyo* (New York: Viking, 1950), pp. 262–63; John Costello, *The Pacific War, 1941–1945* (New York: Quill, 1982), p. 599; Butow, *Tojo*, pp. 449–54.

Chapter 19

1. D. T. Payton-Smith, *Oil: A Study of War-Time Policy and Administration* (London: HMSO, 1971), pp. 21–23, 44 ("paraphernalia of competition"), 62 ("strategic oil reserve"). "Spanish Petroleum Monopoly," November 18, 1927, W 10770, FO 371/12719 ("Sir Henri's word"); J. V. Perowne, Minute, September 30, 1935, C6788, FO 371/18868 ("hatred of the Soviets"); Faulkner to Vansittart, September 30, 1935, C6788, FO 371/18868 ("suitable actions" and "getting an old man"); Thornton to Montgomery, January 1, 1937, H2/1937, FO 371/2075 with C137/105/2/37 (Dutch prime minister); Draft, Personalities Series, 1938, FO 371/21795, PRO. On the effort to gain control of Shell, see Bland to Halifax, April 27, 1939, no. 228, 233, C6277, C6278, Watkins memo, April 12, 1939, C5474, FO 371/23087, PRO and Anthony Sampson, *The Seven Sisters: The Great Oil Companies and the World They Shaped*, rev. ed. (London: Coronet, 1988), pp. 96–97; In the autumn of 1939, the British and French allocated $60 million to pay the Rumanians to destroy their oil wells, in order to deny Rumanian oil to the Germans. The Rumanians, however, wanted more, and the Rumanian oil went to the Germans. War Cabinet, Meeting Notes, November 22, 1939, POG (S), CAB 77/16, PRO.

2. Payton-Smith, *Oil*, p. 85 ("basic ration"); George P. Kerr, *Time's Forelock: A Record of Shell's Contribution to Aviation in the Second World War* (London: Shell Petroleum Company, 1948), p. 40; Arthur Bryant, *The Turn of the Tide* (Garden City, N.Y.: Doubleday, 1957), p. 203.

3. Payton-Smith, *Oil*, pp. 195–99 ("arsenal"), 210–11; Huston, *Sinews of War*, p. 442 ("dollar sign"); Dallek, *Roosevelt and American Foreign Policy*, p. 443 ("Dr. Win-the-War"). Roosevelt to Ickes, May 28, 1941, OF 4435; FDR to Smith, May 6, 1941, OF 56, Roosevelt papers. Surplus capacity number derived from John W. Frey and H. Chandler Ide, *A History of the Petroleum Administration for War, 1941–1945* (Washington, D.C.: GPO, 1946), p. 444, which is an important source on Allied oil supplies. For the suit, called the "Mother Hubbard" case because the defendants seemed to include almost all of the American oil industry, see United States Tariff Commission, *Petroleum*, Report No. 17, in *War Changes in Industry Series* (Washington, D.C.: GPO, 1946), p. 94.

4. Everett DeGolyer, "Government and Industry in Oil," 813; PAW, "Transportation of Petroleum to Eastern United States," May 15, 1942, 4435, DeGolyer papers. Ickes to Roosevelt, July 18, 1939, OF 56, Roosevelt papers; Nash, *United States Oil Policy*, pp. 152–63; Ickes, *Secret Diary*, vol. 3, p. 530; *Oil Weekly*, June 2, 1941; Harold Ickes, *Fightin' Oil* (New York: Knopf, 1943), p. 71.

5. Goralski and Freeburg, *Oil & War,* p. 109 (Raeder); Martin Gilbert, *Winston S. Churchill,* vol. 6, *Finest Hour, 1939–1941* (Boston: Houghton Mifflin, 1983), pp. 1020–21 ("measureless peril"), 1036 ("blackest cloud"); Davies to Ickes, July 8, 1941, Ickes to Roosevelt, July 9, 1941, PSF 12, Roosevelt papers ("shocking"); Ickes, *Secret Diary,* vol. 3, pp. 561, 543 ("parking conditions"); Williamson et al., *The Age of Energy,* p. 758 (gasless Sundays); Frey and Ide, *Petroleum Administration,* pp. 118–19 ("one-third less").

6. Beaton, *Shell,* p. 604 ("phony shortage"); Hinsley, *British Intelligence,* vol. 2, pp. 169–74 ("narrowest of margins"); Frey and Ide, *Petroleum Administration,* p. 119 ("shortage of surplus"); Ickes, *Secret Diary,* vol. 3, p. 617 ("fill it up"), 630–33 (Ickes's complaints). Wirtz to Ickes, May 15, 1941, Ickes to Roosevelt, May 19, 1941, OF 4435; Uoyd to Ickes, November 24, 1941, OF 4226; Ickes to Roosevelt, January 17, 1942, PSF 75 (Ickes's new strategy), Roosevelt papers.

7. Goralski and Freeburg, *Oil & War,* pp. 108 ("ample targets"), 114–15. Davies to Ickes, March 21, 1942, Ickes to Roosevelt, March 23, 1942, PSF 75; Ickes to Roosevelt, April 21, 1942, PSF 12 ("desperate"), Roosevelt papers. Morison, *Naval Operations,* vol. 1, pp. 254, 200–1, 130; Bryant, *Turn of the Tide,* pp. 295–96.

8. Ickes to Nelson, June 17, 1942, box 209, Hopkins papers; Nash, *U.S. Oil Policy,* pp. 164–65.

9. NA 800.6363: Minutes of Federal Petroleum Council, March 20, 1942, 411; Thorburg to Collado et al., June 25, 1942, 786 RG 59. John Keegan, *The Price of Admiralty: The Evolution of Naval Warfare* (New York: Viking, 1989), p. 229 ("Rescue no one"); Morison, *Naval Operatons, vol. 1,* pp. 157, 198 ("enemy tonnage"); Michael Howard, *Grand Strategy,* vol. 4, *August 1942–September 1943* (London: HMSO, 1972), p. 54; Bryant, *Turn of the Tide,* p. 387; Goralski and Freeburg, *Oil & War,* pp. 113 ("milk cows"), 116; Stanton Hope, *Tanker Fleet: The War Story of the Shell Tankers and the Men Who Manned Them* (London: Anglo-Saxon Petroleum, 1948), chap. 9.

10. Wilkinson to Ickes, December 5, 1942, Ickes to Roosevelt, December 10, 1942, PSF 75, Roosevelt papers; Martin Gilbert, *Winston S. Churchill,* vol. 7, *Road to Victory, 1941–1945* (Boston: Houghton Mifflin, 1986), pp. 265, 289; S. W. Roskill, *The War at Sea, 1939–1945,* vol. 2 (London: HMSO, 1956), pp. 355, 217 ("not look at all good"); Howard, *Grand Strategy,* vol. 4, pp. 244–245 ("stranglehold"), 621; Liddell Hart, *Second World War,* pp. 387–90 ("never came so near" and "heavy losses"); Larson, *Standard Oil,* vol. 3, p. 529.

11. Ickes to Roosevelt, August 4, 1942, August 7, 1942, September 3, 1942, Smith to Roosevelt, October 1, 1942, OF 4435; Roosevelt to Land, November 6, 1941, OF 56; Nelson file memo, May 1, 1942, OF 12, Roosevelt papers. Ickes to Nelson, November 26, 1942, Box 209, Hopkins papers; Board of Petroleum Reserves Corporation, Record, April 25, 1944, pp. 88–91, RG 234, NA ("any oil matter"); Petroleum Administration for War, *Petroleum in War and Peace* (Washington, D.C.: PAW, 1945), pp. 39–44.

12. Pratt to Farish, May 16, 1941, Pratt to DeGolyer, March 17, 1942, 1513; DeGolyer to Hebert, January 16, 1943, 3470, DeGolyer papers. Cole to Roosevelt, October 22, 1942, pp. 20, 22, OF 4435, Roosevelt papers; Ickes to Brown, April 7, June 10, 1943, Davies to Hopkins, July 26, 1943, box 209, Hopkins papers; Frey and Ide, *Petroleum Administration,* p. 5 and statistical tables; John G. Clark, *Energy and the Federal Government: Fossil Fuel Policies, 1900–1946* (Urbana: University of Illinois Press, 1987), p. 327 ("commie outfit"); E. DeGolyer, "Petroleum Exploration and Development in Wartime," *Mining and Metallurgy,* April 1943, pp. 188–90.

13. Roosevelt to Ickes, August 12, 1942, OF 4435, Roosevelt papers ("natural gas"); Clark, *Energy and Federal Government,* p. 316 (Bea Kyle to Ickes); Frey to Kyle, August 1941, Davies papers; Minutes of Federal Petroleum Council, March 20, 1942, 811.6363/411, RG 59, NA ("knew for sure").

14. Hertz to the Undersecretary of War, August 13, 1942, Hertz to Hopkins, August 13, 1942, box 209, Hopkins papers; John Kenneth Galbraith, *A Life in Our Times: Memoirs* (Boston: Houghton Mifflin, 1981), p. 130 ("private skepticism"); James Conant, *My Several Lives: Memoirs of a Social Inventor* (New York: Harper & Row, 1970), p. 314 (Baruch's dinner).

   On the "rubber famine," see United States Congress, Senate, Special Committee Investigating the National Defense Program, *Investigation of the National Defense Program,* part 11, *Rubber,* 77th Congress, 1st Session (Washington, D.C.: GPO, 1942) (hereafter, *Truman Hearings*); Howard, *Buna;* Larson, *Standard Oil,* vol. 3, pp. 405–18, chap. 15.

   In a case brought by Thurman Arnold, the trust-busting assistant Attorney General, and in a series of Congressional hearings, Jersey was charged with collusion and cartel making in its relationship with I. G. Farben involving synthetic rubber. The arrangements between the two companies, the critics said, had deprived the United States of synthetic rubber know-how and production. Natural rubber had, before Pearl Harbor, constituted the single largest import of the United States. The abrupt cessation of supply resulting from Japan's capture of the primary sources in Southeast Asia created a "rubber famine" in the United States, threatening the entire Allied war effort.

Arnold insisted that the source of the rubber famine was what he called the "full marriage" between Jersey and I. G. Farben (*Truman Hearings,* p. 4811). As to the substance of the charges themselves, Arnold pursued his case in a singular way, though sometimes taking events out of context (*Truman Hearings,* pp. 4313, 4427, 4598). Jersey had made its arrangements with I. G. Farben before the Nazis came to power. As a result of the deal, considerable benefits in chemistry and research organization flowed to the American side, including synthetic rubber know-how. After all, Germany and I. G. Farben, not America and Jersey, held the leadership in world chemistry. The Jersey management did show considerable obtuseness and political naiveté from 1937 on in not recognizing the degree to which I. G. Farben had become a captive and tool of the Nazi state. See Hayes, *Industry and Ideology.* But the charge that Jersey prevented the diffusion of synthetic rubber technology before World War II disregarded the economic realities. In a Depression world of low commodity prices and large surpluses, there was no economic incentive or rationale to develop synthetic technologies, unless a country was preparing for war. If the United States had so prepared, then innovation and implementation would have required substantial government subsidies or tariff protection. Though the import price of natural rubber fluctuated substantially in the years before America's entry into World War II, the production costs for synthetic rubber were estimated at five times those of natural rubber. No company could have been expected to make a commitment to a large level of production in the face of those economics. Indeed, from 1939 on, Jersey, along with other companies, had tried to get Washington to support the development of synthetic rubber technology and production, but the effort fizzled in the face of administrative disarray and rivalry in Washington, lack of consensus on the need, and a strong aversion to the commitment of large amounts of government money. The general view was that the supply of natural rubber from Southeast Asia could not be interrupted, and there was also skepticism about the viability of the synthetic substitutes (see *Truman Hearings,* pp. 4285–89, 4407–79, 4805, 4937). The "rubber famine" resulted not from the patent exchanges between Jersey and I. G. Farben, which on the contrary increased American knowledge about synthetic rubber, but from a failure of the government's program of preparedness in the three years before Pearl Harbor. The "rubber famine" arose primarily from the same psychology that excluded the possibility that there could be a Pearl Harbor.

15. Clark, *Energy and the Federal Government,* pp. 337–44 ("nonessential driving").
16. Payton-Smith, *Oil,* pp. 249–53; Standard Oil (New Jersey), *Ships of the Esso Fleet in World War II* (New York: Standard Oil, 1946), pp. 151–54.
17. Ickes, *Fightin' Oil,* p. 6 (Stalin's toast); Erna Risch, *Fuels for Global Conflict* (Washington, D.C.: Office of Quartermaster General, 1945), pp. 1–2, ix–x, 59–60 (gas cans).
18. United States Congress, Senate, Committee on Foreign Relations, Subcommittee on Multinational Corporations, *A Documentary History of the Petroleum Reserves Corporation* (Washington, D.C.: GPO, 1974) (Patterson to Ickes); Agnew to Lloyd, June 15, 1942, POWE 33/768 121286, PRO; "100 Octane Aviation Gasoline: Report to the War Production Board," March 16, 1942, May 29, 1942, pp. 9–10 ("eke out"), October 15, 1942; Ickes to Roosevelt, October 19, 1942, Nelson to Roosevelt, October 28, 1942, Roosevelt to Ickes, November 7, 1942, PSF 12, Roosevelt papers. Beaton, *Shell,* pp. 560–76, 579–87 ("out of a hat"); Charles Sterling Popple, *Standard Oil Company (New Jersey) in World War II* (New York: Standard Oil, 1952), pp. 29–30; War Production Board, *Industrial Mobilization for War: History of the War Production Board and Its Predecessor Agencies, 1940–1945,* vol. 1 (Washington, D.C.: GPO, 1947), pp. 39–41; James Doolittle Oral History (Shell and 100 octane); Giebelhaus, *Sun,* chaps. 7 and 9.
19. Petroleum Administration for War, *Petroleum in War and Peace* (Washington, D.C.: GPO, 1945), p. 204 ("Not a single operation"); van Creveld, *Supplying War,* p. 213; Roland G. Ruppenthal, *Logistical Support of the Armies,* vol. 1 (Washington, D.C.: Department of the Army, 1953), pp. 499–516; Goralski and Freeburg, *Oil & War,* p. 254 ("men and horses"); Martin Blumenson, *The Patton Papers,* vol. 2, *1941–1945* (Boston: Houghton Mifflin, 1974), p. 492 (poem); Dwight Eisenhower, *Crusade in Europe* (Garden City, N.Y.: Doubleday, 1948), p. 275; Alfred D. Chandler, Jr., and Stephen E. Ambrose, *The Papers of Dwight David Eisenhower,* vol. 4, *The War Years* (Baltimore: Johns Hopkins University Press, 1970), p. 2060, n. 4 ("great leader"); Martin Blumenson, *Patton: The Man Behind the Legend, 1885–1945* (New York: William Morrow, 1985), p. 216; Forrest C. Pogue, *George C. Marshall, vol. 3, Organizer of Victory, 1943–1945* (New York: Viking, 1973), pp. 385 ("thoroughly weary" and "into the breach"), 371–72 ("Patton's good qualities").

20. Van Creveld, *Supplying War*, p. 221; Nigel Hamilton, *Monty, vol. 2, Master of the Battlefield, 1942–1944* (London: Sceptre, 1987), p. 754 ("spectacularly successful"); Eisenhower, *Eisenhower at War*, p. 438 ("planning days"); Blumenson, *Patton Papers*, vol. 2, pp. 841, 571, 533, 529–30 ("chief difficulty").

21. Stephen E. Ambrose, *The Supreme Commander: The War Years of General Dwight D. Eisenhower* (Garden City, N.Y.: Doubleday, 1970), p. 515; Blumenson, *Patton Papers*, vol. 2, p. 523 ("blind moles"); Omar N. Bradley, *A Soldier's Story* (New York: Henry Holt, 1951), pp. 402–5 ("angry bull"); Ruppenthal, *Logistical Support*, vol. 1, table 10, p. 503; Hamilton, *Monty, vol. 2*, p. 777.

22. Blumenson, *Patton Papers*, vol. 2, p. 531 ("unforgiving minute"); Liddell Hart, *Second World War*, pp. 562–63 ("eat their belts"); Robert Ferrell, ed., *The Eisenhower Diaries* (New York: Norton, 1981), p. 127 ("get Patton moving").

23. Cole, *Battle of the Bulge*, pp. 13–14; Liddell Hart, *Second World War*, p. 563; Goralski and Freeburg, *Oil & War*, pp. 264–65; Blumenson, *Patton*, chap. 10, p. 216; Eisenhower, *Crusade in Europe*, pp. 292–93 ("late summer . . . inescapable defeat"); Ruppenthal, *Logistical Support*, esp. pp. 515–16; General George Marshall, Army Chief of Staff, shared Eisenhower's point of view. A decade after the war ended, he said: "Of course he [Patton] wanted more gasoline; of course Montgomery wanted more gasoline and a larger freedom of action. That is just natural to commanders under these circumstances. What was going on was the First Army was making very rapid moves in a very positive manner and getting very little public credit for it in this country. The Third Army was getting far more credit because of Patton's dash and showmanship. . . . Patton wanted to get free—with the great temptation of running right up to the Rhine—and there was almost no gasoline. . . . I think Eisenhower's control of the operations at that time was correct. And that all the others were yelling as they naturally would yell. There is nothing remarkable about that except one was the supreme commander of the British forces, which at that time was very small, and the other was a very high-powered dashing commander who had the press at his beck and call—General Patton. . . . In trying to judge what was the correct disposition of the available gasoline, one has to remember a great many subsidiary facts and prospects. For example, take the German operation in the Bulge later on. If it was successful, it was a grand thing. But it wasn't successful. . . . You can sometimes win a great victory by a very dashing action. But often or most frequently the very dashing action exposes you to a very fatal result if it is not successful." Pogue, *Marshall, vol. 3*, pp. 429–30.

24. Hamilton, *Monty, vol. 2*, pp. 776–821; Nigel Hamilton, *Monty, vol. 3, The Field-Marshal* (London: Sceptre, 1987), pp. 3–8; Liddell Hart, *Second World War*, pp. 565–67 ("best chance").

This table of world oil supply shows how the United States continued to dominate world oil production throughout the first 85 years of the industry. The table also demonstrates the importance that Russian and Mexican production gained and then lost, the significance of Venezuela by World War II, and the beginning of the impact of the Middle East on world supply.

World Crude Oil Production, 1860–1945
(thousands of barrels per day)

Year	United States	Mexico	Venezuela	Russia/USSR	Rumania	East Indies	Persia/Iran	All Others	Total
1865	6.8			0.2	0.1			0.3	7
1875	32.8			1.9	0.3			1.1	36
1885	59.9			38.2	0.5			2.2	101
1895	144.9			126.4	1.6	3.3		7.9	284
1905	369.1	0.7		150.6	12.1	21.5		35.3	589
1915	770.1	90.2		187.8	33.0	33.7	9.9	58.9	1,184
1925	2,092.4	316.5	53.9	143.7	45.6	70.4	93.3	112.8	2,929
1935	2,730.4	110.2	406.2	499.7	169.2	144.4	156.9	317.0	4,534
1945	4,694.9	119.3	885.4	408.1	95.3	26.6	357.6	521.6	7,109

East Indies includes Indonesia, Sarawak, and Brunei. Source: American Petroleum Institute, *Petroleum Facts and Figures: Centennial Edition, 1959* (New York: API, 1959), pp. 432–37.

Chapter 20

1. Pratt to Farish, August 3, 1934, 1513, obituaries, DeGolyer papers; Anderson, *Aramco,* p. 111; Philip O. McConnell, *The Hundred Men* (Peterborough: Currier Press, 1985); Lon Tinkle, *Mr. De: A Biography of Everette Lee DeGolyer* (Boston: Little, Brown, 1970), pp. 212, 227, 255; Herbert K. Robertson, "Everette Lee DeGolyer," *Leading Edge,* November 1986, pp. 14–21.

2. E. DeGolyer, "Oil in the Near East," Speech, May 10, 1940, 2288 ("No such galaxy"); notes, 3466; itinerary, 3459; and letters to wife, November 7, 10 ("no Lindbergh"), 14, December 1 ("pretty barren land"), 1943, DeGolyer papers.

3. Leavell to Alling, February 3, 1943 ("single prize"), Summary of Report on Near Eastern Oil, 800.6363/1511–1512, RG 59, NA; E. DeGolyer, "Preliminary Report of the Technical Oil Mission to the Middle East," *Bulletin of the American Association of Petroleum Geologists* 28 (July 1944), pp. 919–23 ("center of gravity").

4. Moffett to Roosevelt, April 16, 1941, PSF 93; Hull to Roosevelt, June 30, 1939, OF 3500, Roosevelt papers. Duce to DeGolyer, April 29, 1941, 360, DeGolyer papers ("closer look"); Conversation with Ibn Saud, May 10, 1942, with Alling memo, June 18, 1942, 890F.7962/45, RG 59, NA ("have the money"); Aaron David Miller, *Search for Security: Saudi Arabian Oil and American Foreign Policy, 1939–1949* (Chapel Hill: University of North Carolina Press, 1980), pp. 29–35.

5. Knox to Roosevelt, May 20, 1941, Hull with memo to Roosevelt, April 25, 1941, Hopkins to Jones, June 14, 1941, PSF 68, Roosevelt papers; Miller, *Search for Security,* pp. 38–39; Michael B. Stoff, *Oil, War, and American Security: The Search for a National Policy on Foreign Oil, 1941–47* (New Haven: Yale University Press, 1980), pp. 52–54. Stoff, along with Anderson in note 1, Miller in note 4, and Painter in note 9, are the major monographs on postwar oil policy.

6. Pratt to Farish, May 16, 1941, 1513; William B. Heroy, "The Supply of Crude Petroleum Within the United States," July 29, 1943, pp. 4–9, 3417 ("diminishing returns" and "bonanza days"), DeGolyer papers; E. DeGolyer, "Petroleum Exploration and Development in Wartime," *Mining and Metallurgy,* April 1943, pp. 189–90; Foreign Office Research Dept., "A Foreign Policy for Oil," United States Memoranda, May 16, 1944, AN 1926, FO 371/38543/125169, PRO; United States Congress, Senate, Special Committee Investigating Petroleum Resources, *Investigation of Petroleum Resources* (Washington, D.C.: GPO, 1946), pp. 276–77; "Wartime Evolution of Postwar Foreign Oil Policy," May 29, 1947, 811.6363/5–2947, RG 59, NA.

7. Harold Ickes, "We're Running Out of Oil!," *American Magazine,* December 1943 ("America's crown"); Campbell to Eden, September 28, 1943, A9193, FO 371/34210/120769, PRO ("private interest"); Herbert Feis, *Seen from E. A.: Three International Episodes* (New York: Knopf, 1947), p. 102 ("one point and place"). Later, in mid-1944, Roosevelt called a halt to efforts by the American ambassador in Mexico City to negotiate the reentry of private American capital and instead suggested that the United States finance oil exploration for the Mexican government. "When a new and adequate dome is found," said Roosevelt, "it could be set aside in toto by the Mexican Government for the purpose of aiding the defense of the Continent" and the United States government would pay an annual holding fee to Mexico. Roosevelt to Ickes, February 28, 1942, Roosevelt to Hull, July 19, 1944, OF 56, Roosevelt papers.

8. Moose to Hull, April 12, 1944, 890F.6363/124; Stimson to Hull, May 1, 1944, 890F.6363/123, RG 59, NA. Kline to Ickes, Summary of Dillon Anderson report, March 4, 1944, 3459, DeGolyer papers; Multinational Subcommittee, *History of the Petroleum Reserves Corporation,* p. 4 ("diddle"); Woodward, *British Foreign Policy,* vol. 4, pp. 402–5, 410; Feis, *Seen from E. A.,* pp. 110–111. Standard Oil of California, "Plans for Foreign Joint Venture," December 7, 1942, 25391–25617 file, case 1, Oil Companies papers.

9. Kline to Ickes, Summary of Dillon Anderson report, March 4, 1944, 3549, DeGolyer papers; Vice Chief of Naval Operations to Joint Chiefs of Staff, May 31, 1943, U69139 (SC) JJT/E6, RG 218, NA; The Position of the Department on the Petroleum Reserves Corporation, p. 1, 800.6363/ 2–644, RG 59, NA. Feis, *Seen from E. A.,* p. 105; United States Congress, Senate, Special Committee Investigating the National Defense Program, *Investigation of the National Defense Program,* Hearings, part 42, pp. 25435, 25386–87; Anderson, *Aramco,* pp. 46–48 ("purely American enterprise"), 51; David Painter, *Oil and the American Century: The Political Economy of U.S. Foreign Oil Policy, 1941–1954* (Baltimore: Johns Hopkins University Press, 1986), p. 37 ("richest oil field"); Stoff, *Oil, War, and American Security,* p. 54 ("far afield").

10. Thornburg to Hull, March 27, 1943, 800.6363/1141–1/2; Feis to Hull, June 10, 1943, 890F.6363/80,

RG 59, NA. Hull to Roosevelt, March 30, 1943, OF 3500, Roosevelt papers; Painter, *Oil and the American Century*, pp. 41 ("intense new disputes" and "smell of oil"), 43 ("breath away"). Notes, June 12, 1943, 3468 ("rapidly dwindling"); Petroleum Reserves Corporation, Record of Negotiations, August 2–3, 1943, 3463 ("tremendous shock"), DeGolyer papers. Feis, *Seen from E. A.*, pp. 122 ("boyish note"), 129–30 ("caught a whale").

11. NA 890F.6363 Feis to Hull, September 16, 1943, 65; September 23, 1943, 70; Merriam, memo of conversation with Paul Bohannon, October 4, 1943, 84, RG 59; Minutes of Special Meeting of Directors of Petroleum Reserves Corporation, November 3, 1943, 3463, DeGolyer papers.

12. Herbert Feis, *Petroleum and American Foreign Policy*, (Stanford: Food Research Institute, 1944), p. 45 ("favored competition"); Ralph Zook, *The Proposed Arabian Pipeline: A Threat to Our National Security* (Tulsa: IPAA, 1944) ("move towards fascism"); Anderson, *Aramco*, p. 101 ("monopolies" and "military necessity"); RGH Jr. to Berle, April 20, 1944, 890F.6363/122–1/2, RG 59, NA; Ickes to Roosevelt, May 29, 1944, Roosevelt to Ickes, May 31, 1944, PSF 68, Roosevelt papers; Kline to DeGolyer, May 22, 1944, 946, DeGolyer papers ("understatement").

13. Chiefs of Staff to War Cabinet, April 5, 1944, WP (44) 187, FO 371/42693/120769 ("American assistance" and "continental resources"); Cabinet Paper, "Oil Policy," MOC (44) 5, CAB 77/ 15/184, PRO. Minutes, Special Committee on Petroleum, September 21, 1943, 3468, DeGolyer papers.

14. Ickes to Roosevelt, August 18, 1943 ("available oil"), with Duce memo on conversation with Jackson, August 13, 1943 (Jackson), PSF 68, Roosevelt papers. Eden to the Prime Minister, February 11, 1944, POWE 33/1495; Beaverbrook to the Prime Minister, February 8, 1944, POWE 33/1495 ("pigeon hole"); Halifax to Foreign Office, February 19, 1944, No. 846, FO 371/42688 (Roosevelt's map), PRO. NA 800.6363: Feis to Ickes, with memo, October 1, 1943, /1330A; Ailing memo, December 3, 1943, /1402; Sappington to Murray, December 13, 1943, /1466, RG 59; Feis, *Seen from E. A.*, p. 126; Woodward, *British Foreign Policy*, vol. 4, pp. 393–94 ("shockingly"). For DeGolyer's comment, memo with DeGolyer to Snodgrass, n.d., 3468, DeGolyer papers.

15. *FRUS, 1944*, vol. 3, pp. 101–05; Francis L. Loewenheim, Harold D. Langley, and Manfred Jonas, eds., *Roosevelt and Churchill: Their Secret Wartime Correspondence* (New York: E. P. Dutton, 1975), pp. 440–41 ("wrangle"), 459 ("assurances"); Painter, *Oil and the American Century*, p. 55 ("horn in"); Stoff, *Oil, War, and American Security*, p. 156 ("rationing of scarcity").

16. Duce to DeGolyer, August 1, 1944, 360, DeGolyer papers ("lamb chops"); Stoff, *Oil, War, and American Security*, p. 167 ("monster cartel"); Minutes of Anglo-American Conversations on Petroleum: Plenary Sessions, August 1, 1944, 800.6363/7–2544, RG 59, NA (" 'As-Is' character" and "Petroleum Agreement"); Anderson, *Aramco*, pp. 218–23 ("reserves" and "give effect").

17. Duce to DeGolyer, September 11, 1944, 360, DeGolyer papers. NA 800.6363: Pew to Connally, August 17, 1944, with Pew to Hull, August 23, 1944, 8–2344, Rayner memo, Meeting with Senate Committee, August 17, 1944, 8–1744, RG 59. Zook to Roosevelt, November 28, 1944, PSF 56, Roosevelt papers.

18. DeGolyer to Duce, November 13, 1944, 360, DeGolyer papers; Ickes to Roosevelt, November 29, 1944, 800.6363/12–344 RG 59, NA ("seeing ghosts").

19. Roosevelt to Ibn Saud, February 13, 1942, OF 3500, Roosevelt papers; William A. Eddy, *F.D.R. Meets Ibn Saud* (New York: American Friends of the Middle East, 1954), pp. 19–35 (FDR and Ibn Saud); *FRUS, 1945*, vol. 8, pp. 1–3, 7–9; Miller, *Search for Security*, pp. xi–xii, 130–31; Robert E. Sherwood, *Roosevelt and Hopkins: An Intimate History* (New York: Harper & Brothers, 1948), pp. 871–72; Charles E. Bohlen, *Witness to History, 1929–1969* (New York: Norton, 1973), p. 203.

20. Miller, *Search for Security*, p. 131 ("immense oil deposits"); William D. Leahy, *I Was There* (New York: Whittlesey House, 1950), pp. 325–27; Martin Gilbert, *Winston S. Churchill*, vol. 7, *Road to Victory, 1941–1945* (Boston: Houghton Mifflin, 1986), pp. 1225–26 ("allow smoking" and "finest motor car"); Laurence Grafftey-Smith, *Bright Levant* (London: John Murray, 1970), pp. 253, 271 (Rolls-Royce). Churchill's irritation is vividly described in the draft of Eddy, *F.D.R. Meets Ibn Saud*, p. 5, with Kidd to DeGolyer, October 22, 1953, 3461, DeGolyer papers.

21. Roosevelt to Stettinius, March 27, 1945, PSF 115 ("remind me"), Roosevelt papers; Shinwell to Chancellor of Exchequer, September 24, 1945, PREM 8/857/122019, PRO; Anderson, *Aramco*, pp. 224–28 (text of Revised Agreement); United States Congress, Senate, *Investigation of Petroleum Resources*, pp. 278–79, 34, 37 ("optimist"); Robert E. Wilson, "Oil for America's Future," *Stanolind Record*, October-November 1945, pp. 1–4; Ickes to Truman, February 12, 1946, Davies papers; Harry S. Truman, *Year of Decisions* (Garden City, N.Y.: Doubleday, 1955), p. 554 ("kind of letter"); Alonzo L. Hamby, *Beyond the New Deal: Harry S. Truman and American Liberalism* (New York:

Columbia University Press, 1973), p. 73 ("lack of adherence"); Margaret Truman, *Harry S. Truman* (New York: William Morrow, 1973), p. 291 ("monarch").

22. Forrestal to Secretary of State, December 11, 1944, 890F.6363/12–1144 ("cannot err"); Forrestal to Byrnes, April 5, 1946, 811.6363/4–546 ("cheering section"); Collado to Clayton, March 27, 1945, 890F.6363/3–2745, RG 59, NA. Walter Millis, ed., *The Forrestal Diaries* (New York: Viking, 1951) p. 81 ("first importance").

23. Wilcox to Clayton, February 19, 1946, 800.6363/2–1946 ("dangerous or useless" and "orphan"), RG 59, NA; Stoff, *Oil, War, and American Security,* p. 97 ("salvation").

Chapter 21

1. NA 811.6363: Sandifer to McCarthy, July 2, 1948, 6–1847; Department of State, Current and Prospective Worldwide Petroleum Situation, February 17, 1948, 2–1748, RG 59. Larson, *Standard Oil,* vol. 3, pp. 667–72; Beaton, *Shell,* pp. 637–42; Shell Transport and Trading, *Annual Report, 1947,* p. 8 ("astonishingly"); Giddens, *Standard Oil of Indiana,* pp. 682–84 ("jackrabbit"); Arthur M. Johnson, *The Challenge of Change: The Sun Oil Company, 1945–1977* (Columbus: Ohio State University Press, 1983), p. 40 ("Helpful Hints").

2. R. Gwin Follis to author, September 18, 1989 ("hardly touch" and "surprising enthusiasm"); Anderson, *Aramco,* p. 120 ("sufficient markets"), 140–45 (Forrestal); Hart to Secretary of State, July 2, 1949, 890F.6363/7–249, RG 59, NA ("our oil market" and "greatest"); Robert A. Pollard, *Economic Security and the Origins of the Cold War, 1945–1950* (New York: Columbia University Press, 1985), p. 213; "The Great Oil Deals," *Fortune,* May 1947, p. 176 (Collier).

3. Sellers to Foster, June 12, 1946, "IPC memos, 1946" file, case 5 ("bombshell"); "IPC Memorandum on Present Legal Position," July 10, 1946, 127274–127448 file, case 2, Oil Companies papers. *Multinational Hearings,* part 8, pp. 111–15 ("inadvisable and illegal"), 124 ("supervening illegality"); Anderson, *Aramco,* pp. 148–51 ("aliens" and "frustrated"). NA 890F.6363: Meloy to Secretary of State, December 12, 1948, 12–1248; Hart to Secretary of State, July 2, 1949, 7–249, August 6, 1949, 8–649; Sappington to Secretary of State, December 5, 1945, 800.6363/12–545, RG 59.

4. *Multinational Hearings,* part 8, pp. 115–19 ("mutual interest," Sheets, "restraints," "political question" and "family circle"); Interview with Pierre Guillaumat ("angry with God"); CFP, "Events Arising from the War," February 27, 1945; Gulbenkian to Near East Development Corporation, January 6, 1947, 4–5–35 file, case 6, Oil Companies papers ("not acquiesce").

5. FTC, *International Petroleum Cartel,* p. 104; Nitze to Clayton, February 21, 1947, 800.6363/2–2147, RG 59, NA ("arrest" and "retard"); Letter from Paul Nitze to author, October 3, 1989. Sellers and Shepard to Harden and Sheets, February 7, 1947, "IPC memos, 1946" file; Earl Neal, "Alternatives to IPC," February 19, 1947, 126898–127063 file, case 2, Oil Companies papers. *Multinational Hearings,* part 8, pp. 160–61 ("practicable plan").

6. Childs to Secretary of State, January 3, 1947, 890F.6363/1–347, RG 59, NA (Aramco and Ibn Saud); R. Gwin Follis to author, September 18, 1989 ("off our shoulders"); Anderson, *Aramco,* pp. 158, 152 (Socony president); *Multinational Hearings,* part 8, pp. 156–66 ("good thing" and "problems"); Daniel Yergin, *Shattered Peace: The Origins of the Cold War* (New York: Penguin, 1990), pp. 282–83 ("all-out").

7. "Notes on Calouste Sarkis Gulbenkian," June 6, 1947, with Berthoud to Butler, June 9, 1947, PE 650, POWE 33/1965, PRO (British official); Gulbenkian, *Portrait in Oil,* pp. 210–15 ("musts"), 251 ("father's practice"); Interview with John Loudon; Anderson, *Aramco,* pp. 155–59 ("drove as good"). Turner to Johnson, September 15, 1948, 127274–127448 file; Dunaway to Grubb, April 4, 1946, "various nos." file, case 2, Oil Companies papers. Gulbenkian and Raphael in John Walker, *Self-Portrait with Donors: Confessions of an Art Collector* (Boston: Atlantic Monthly Press, 1974), pp. 234–37.

8. Harden to Holman, November 3, 1948, Harding to Vacuum, November 3, 1948, 128167–128229 file, case 7, Oil Companies papers; Gulbenkian, *Portrait in Oil,* pp. 225–27 (complexity of agreements and "caravan"); Belgrave memo, "Gulbenkian Foundation," January 13, 1956, POWE 33/2132, PRO; "The Great Oil Deals," *Fortune,* May 1947, p. 176 ("moon").

9. Memo, Meeting, including Clayton and Drake, February 3, 1947, 811.6363/2–347 ("long on crude oil" and "wholly American owned"); Loftus to Vernon, September 5, 1947, FW 811.6363/8–2047, RG 59, NA. Chisholm, *Kuwait Oil Concession,* p. 187. Jennings to Sheets, September 27, 1946, 17–3–4 file, case 5; "Kuwait—Supply," "Kuwait" file, case 1; "Shell Negotiation," "various nos.,"

incl. Gulf & Jersey" file, case 2, Oil Companies papers. "Shell in Kuwait," Middle East Oil Committee, CME (55), May 16, 1955, CAB 134/1086, PRO ("partner").

10. Yergin, *Shattered Peace,* chap. 7, pp. 163 ("What does . . . how far"), 180; Interview with Nikolai Baibakov; *FRUS, 1946,* vol. 6, pp. 732–36 (Stalin's oil fears); Bruce R. Kuniholm, *The Origins of the Cold War in the Near East: Great Power Conflict and Diplomacy in Iran, Turkey, and Greece* (Princeton: Princeton University Press, 1980), p. 138 ("south of Batum"); William Roger Louis, *The British Empire in the Middle East, 1945–1951: Arab Nationalism, the United States, and Postwar Imperialism* (Oxford: Clarendon Press, 1985), pp. 55–62; Arthur Meyerhoff, "Soviet Petroleum," in Robert G. Jensen, Theodore Shabad, and Arthur W. Wright, eds., *Soviet Natural Resources in the World Economy* (Chicago: University of Chicago Press, 1983), pp. 310–42; Owen, *Trek of the Oil Finders,* pp. 1371–73.

11. European Economic Cooperation, London Committee, Drafts for chaps. 1–3, August 2, 1947, UE 7237, FO 371/62564, PRO; Alec Cairncross, *Years of Recovery: British Economic Policy, 1945–51* (London: Methuen, 1985), pp. 367–70; Alan Bullock, *Ernest Bevin: Foreign Secretary* (London: Heinemann, 1984), pp. 361–62.

12. "Anglo-American Responsibility for Petroleum Prices," January 4, 1951, FOA 0453–4351 file, case 1, Oil Companies papers; Painter, *Oil and the American Century,* pp. 155–56; European Recovery Program, *Petroleum and Petroleum Equipment Commodity Study* (Washington, D.C.: Economic Cooperation Administration, 1949), p. 1 ("Without petroleum"); Walter J. Levy, "Oil and the Marshall Plan," paper presented at the American Economic Association, December 28, 1988.

13. Holman to Hoffman, February 23, 1949; Harden to Foster, April 19, 1950; Foster to Harden, August 22, 1950; Suman to Foster, September 1, 1950; Harden to Daniels, December 27, 1950; Foster to Holman, January 18, 1951, FOA 0453–4351–2 file, case 1, Oil Companies papers. David Painter, "Oil and the Marshall Plan," *Business History Review* 58 (Autumn 1984), pp. 382, 376. Cabinet Programme Committee, January 9, 1949, P49, POWE 33/1772; McAlpine to Trend, September 8, 1948, POWE 33/1557 (Bevin); "Oil Prices," to R. W. B. Clarke, February 3, 1947, T2361/2161, PRO. Levy, *Oil Strategy and Politics,* p. 75; W. G. Jensen, *Energy in Europe, 1945–1980* (London: G. T. Foulis, 1967), p. 21.

14. European Economic Co-Operation, London Conversations, August 2, 1947, Drafts for chaps. 1–3, pp. 56–57, 65–66, UE 7237, FO 371/62564, PRO; Miller, *Search for Security,* pp. 177–78; Ethan Kapstein, *The Insecure Alliance: Energy Crisis and Western Politics Since 1949* (New York: Oxford University Press, 1990), p. 61 (Dalton); Interview with T. C. Bailey, GHS/2B/75, Shell archives ("no value").

15. Miller, *Search for Security,* p. 196 ("handicapped"); "Visit of Abdul Aziz to Aramco," January 1947, pp. 36, 45, Aramco papers; Forrest C. Pogue, *George C. Marshall,* vol. 4, *Statesman, 1945–1954* (New York: Viking, 1987), p. 350 ("famine"). Henderson to Marshall, May 26, 1948, 890 F.6363/5–2648 (Duce); Eakens to Martin and "Impact of Loss of Arab Oil Production on World Petroleum Situation," July 8, 1948, 800.6363/7–848 ("hardship"), RG 59, NA.

16. "Remarks made to Colonel Eddy by King Ibn Saud," November 17, 1947, with Merriam memo, November 17, 1947, 890F.6363/11–1347, RG 59, NA; Trott to McNeil, "Annual Review for 1949," February 28, 1950, ES 1011, FO 371/82638, PRO ("formal hostility"); *FRUS, 1949,* vol. 6, pp. 170, 1618, 1621; Louis, *British Empire,* p. 204 ("Jewish pretensions"); James Terry Duce Statement, House Interstate and Foreign Commerce Committee, January 30, 1948, pp. 10–11, 3461, DeGolyer papers.

17. James Terry Duce Statement, House Armed Services Committee, February 2, 1948, 3461, DeGolyer papers; Bullock, *Bevin,* p. 113 ("no hope"); James Forrestal, "Naval Policy," Speech, June 18, 1947, National War College; David A. Rosenberg, "The U.S. Navy and the Problem of Oil in a Future War: The Outline of a Strategic Dilemma, 1945–1950," *Naval War College Review* 29 (Summer 1976), pp. 53–61; Miller, *Search for Security,* p. 203 ("economic prize"); *FRUS, 1950,* vol. 5, pp. 1190–91 (Truman letter to Ibn Saud); "Saudi Arabia: Economic Report," September 24, 1950, POWE 33/323, PRO.

18. "Problem of Procurement of Oil for a Major War," Joint Chiefs of Staff paper 1741, January 29, 1947, pp. 3, 6 ("very susceptible"), RG 218, NA; McGinnis to Daniels, November 26, 1948, CS/A, 800.6363/11–2648, RG 59, NA; Eugene V. Rostow, *A National Policy for the Oil Industry* (New Haven: Yale University Press, 1948), pp. 147–48. National Security Resources Board, "A National Liquid Fuels Policy," August 1948, p. 1, 3526 ("storage place"); API National Oil Policy Committee, Synthetics Subcommittee of Long Range Availability Subcommittee, July 14, 1948, 3508, DeGolyer papers. On synthetic fuels, see Bernard Brodie, "American Security and Foreign Oil," *Foreign Policy*

Reports, March 1, 1948, pp. 297–312. Richard H. K. Vietor, *Energy Policy in America Since 1945: A Study of Business-Government Relations* (Cambridge: Cambridge University Press, 1984), pp. 44 (*New York Times*), 54–59; Crauford D. Goodwin, ed., *Energy Policy in Perspective: Today's Problems, Yesterday's Solutions* (Washington, D.C.: Brookings Institution, 1981), pp. 148–56.

19. Owen, *Trek of the Oil Finders,* p. 801; John S. Ezell, *Innovations in Energy: The Story of Kerr-McGee* (Norman: University of Oklahoma Press, 1979), pp. 152–69 ("real class-one"); William Rintoul, *Spudding In: Recollections of Pioneer Days in the California Oil Fields* (Fresno: California Historical Society, 1978), pp. 207–9.

20. Standard Oil Company (New Jersey), "Natural Gas," August 1945, 3680, DeGolyer papers. Standard Oil of New Jersey, "Cost Considerations in Mid-East Crudes," July 28, 1950, "various nos. 1937–47" file, case 2; Holman to Hoffman, February 23, 1949, case 1, Oil Companies papers ("crudes available"). Deale to Forrestal, May 8, 1948, Office of the Secretary of Defense, RG 218, NA ("pipelines"); Douglass R. Littlefield and Tanis C. Thorne, *The Spirit of Enterprise: The History of Pacific Enterprises from 1886 to 1989* (Los Angeles: Pacific Enterprises, 1990).

Chapter 22

1. Meeting at the Treasury, September 1950, ES 1532/18, FO 371/82691, PRO ("startling demands"); Richard Eden, Michael Posner, Richard Bending, Edmund Crouch, and Joseph Stanislaw, *Energy Economics: Growth, Resources, and Policies* (Cambridge: Cambridge University Press, 1981), p. 264 ("uneasy"); John Maynard Keynes, *The General Theory of Employment, Interest and Money* [1936], volume 7 of *The Collected Writings of John Maynard Keynes* (London: Macmillan, St. Martin's Press for the Royal Economic Society, 1973), p. 383; David Ricardo, *On the Principles of Political Economy and Taxation* [1817], volume 1 of *The Works and Correspondence of David Ricardo,* ed. Piero Sraffa (Cambridge: Cambridge University Press for the Royal Economic Society, 1951), pp. 11–83; M. A. Adelman, *The World Petroleum Market* (Baltimore: Johns Hopkins University Press, 1972), p. 42.

2. Romulo Betancourt, *Venezuela: Oil and Politics,* trans. Everett Bauman (Boston: Houghton Mifflin, 1979), pp. 29, 43, 67; Franklin Tugwell, *The Politics of Oil in Venezuela* (Stanford: Stanford University Press, 1975), p. 182; Rabe, *The Road to OPEC,* pp. 64–73.

3. Rabe, *The Road to OPEC,* pp. 102 ("suicidal leap"), 103 ("tax structure"); Larson, *Standard Oil,* vol. 3, pp. 479–85; Romulo Betancourt, *Venezuela's Oil,* trans. Donald Peck (London: George Allen & Unwin, 1978), p. 162 ("ritual cleansing"); Godber to Starling, April 10, 1943, A786/94/ 47, FO 371/34259, PRO (Godber); Christopher T. Landau, "The Rise and Fall of Petro-Liberalism: United States Relations with Socialist Venezuela, 1945–1948" (Senior Thesis, Harvard University, 1985), pp. 5 ("octopi"), 10 ("vast dollar resources"), 75–76 ("disheartening"); Betancourt, *Venezuela,* pp. 128–36 ("taboo"). Holman to Hoffman, November 1, 1948, "FOA 0453–4357" file; McCulloch to Orton, December 1948, "CT 3028–3293" file; Miller to McCollum, September 3, 1947, "Gulf 6, 9, 18, etc." file ("reap the profits"), case 1, Oil Companies papers. "Creole Petroleum: Business Embassy," *Fortune,* February 1949, pp. 178–79.

4. Loftus and Eakens to McGhee and Nitze, March 4, 1947, 800.6363/3–447; "Saudi Arabia's Offshore Oil," August 6, 1948, 890F.6363/8–1148, RG 59, NA; Interview with Jack Sunderland; *FRUS: Current Economic Developments, 1945–1954,* July 19, 1948, p. 10 ("new companies"); John Loftus, "Oil in United States Foreign Policy," Speech, July 30, 1946; Monsell Davis memo, "Kuwait Neutral Zone Concession," August 16, 1947, POWE 33/478, PRO; Painter, *Oil and the American Century,* p. 165 ("Aminoil"); Duce to DeGolyer, December 16, 1944, 360, DeGolyer papers; Tompkins, *Little Giant of Signal Hill,* pp. 156–63 (Davies).

5. Somerset de Chair, *Getty on Getty* (London: Cassell, 1989), pp. 15–20 ("best hotel" and "a casino"), 143 ("always let down"), 145, 76 and 158 (Madame Tallasou), 70 ("family life"), 156; Interview with Jack Sunderland ("thousand fights" and "value"); Robert Lenzner, *Getty: The Richest Man in the World* (London: Grafton, 1985), pp. 59–60 (Dempsey), 101, 118–34 ("espionage"); Russell Miller, *The House of Getty* (London: Michael Joseph, 1985), p. 207 ("thinking about girls"); "The Fifty-Million Dollar Man," *Fortune,* November 1957, pp. 176–78; Ralph Hewins, *The Richest American: J. Paul Getty* (New York: E. P. Dutton, 1960), pp. 289 ("Middle East"); Interview with Paul Walton.

6. Miller, *House of Getty,* pp. 191–93 ("expenses"), 200–4 ("Teach" and "seminar"); Lenzner, *Getty,* pp. 156–57 ("insane"), 159–60 ("pathological fear"), 182 ("garbage oil"); Hewins, *Richest American,* pp. 309 ("favorably impressed"), 313 ("My bankers"); Munro to Rowe-Dutton, February 22, 1949, T

236/2161; to Furlonge, Foreign Office, November 1, 1950, ES 1532/24, FO 371/82692 ("notorious"), PRO. Proctor to Drake, June 28, 1949, 12–2–4 file, case 4, 644a; Dunaway to Grubb, 44a, April 4, 1946, "various nos." file, case 2 (Gulbenkian), Oil Companies papers. Interviews with Paul Walton and Jack Sunderland; Bernard Berenson, *Sunset and Twilight: From the Diaries of 1947–1958 of Bernard Berenson,* ed. Micky Mariano (New York: Harcourt, Brace & World, 1963), p. 309 (richest man); *Multinational Hearings,* part 8 (Washington, D.C.: GPO, 1975), pp. 282–84.

7. Trott to Bevin, "Saudi Arabia: Annual Review for 1950," March 19, 1951, ES 1011/1, FO 371/91757; Trott to McNeil, "Saudi Arabia: Annual Review for 1949, February 28, 1950, ES 1011/1, FO 371/82638, "Saudi Arabia: Economic Report," to Foreign Office, September 24, 1950, POWE 33/323, "Saudi Arabia: Economic Report," January 28, 1951, POWE 33/324, PRO. Cable with Duce to Wilkins, May 25, 1950, 886A.2553/5–2550 ("large company profits"); "Arabian-American Oil Company's Tax Problems," July 20, 1949, 890F.6363/7–2049, RG 59, NA. *Multinational Hearings,* part 8, pp. 342–50 ("rolling" and "horsetrading"), 357 (IRS); part 7, pp. 168 ("spread the benefits"), 130–35 ("retreat"); Anderson, *Aramco,* pp. 188–96 ("welfare" and "Each time"); Betancourt, *Venezuela,* p. 89 ("grave threat"); Painter, *Oil and American Century,* p. 166 ("darn bit"); Interview with George McGhee ("Saudis knew"); John Blair, *The Control of Oil* (New York: Pantheon, 1976), pp. 196–99 (criticism of tax credit).

8. Gulf to Anglo-Iranian, June 20, 1951, "various nos. inc. Gulf and Jersey" file, case 2; LWE to Larsen, March 11, 1952, Butte to EPL, Jan. 25, 1952 (Jersey working paper), nos. 128253–128255 file, case 3 ("We now know"), Oil Companies papers. "Aramco December 30 Agreement" Memo, January 10, 1951, 886A.2553/1–1051; "Gulf Oil Talks with Anglo-Iranian," March 29, 1951, 886A.2553/3–2951; "Gulf Oil Company Difficulties," June 4,1951,886D.2553/6–451, NA. Louis, *British Empire,* pp. 595, 647 (historian); *Multinational Hearings,* part 4, pp. 86, 89 (McGhee and senator).

Chapter 23

1. Mohammed Reza Pahlavi, *The Shah's Story,* trans. Teresa Waugh (London: Michael Joseph, 1980), pp. 31–47 ("grief"); Barry Rubin, *Paved with Good Intentions: The American Experience in Iran* (New York: Penguin, 1984), p. 383, n. 9 ("mouse"); *FRUS, 1950,* vol. 5, pp. 463 ("Westernized"), 512; Brian Lapping, *End of Empire* (London: Granada, 1985), p. 205 ("bribed").

2. Pahlavi, *Shah's Story,* p. 39 ("miraculous failure"); Ervand Abrahamian, *Iran Between Two Revolutions* (Princeton: Princeton University Press, 1982), pp. 249–50 ("the Great"); Interview with George McGhee; Louis, *British Empire,* pp. 636, 596 ("infant prodigy" and "nineteenth-century"); George McGhee, *Envoy to the Middle World: Adventures in Diplomacy* (New York: Harper & Row, 1983), pp. 320 ("kindly feeling"); Acheson, *Present at the Creation,* p. 646 ("stupidity").

3. Berthoud memo, April 18, 1951, EP 1531/204, FO 371/91527; Bevin to Frank, April 12, 1950, EP 1531/37, FO 371/82395, PRO. "The Iranian Oil Crisis," 3460, DeGolyer papers; Raymond Vernon, "Planning for a Commodity Oil Market," in Daniel Yergin and Barbara Kates-Garnick, eds., *The Reshaping of the Oil Industry: Just Another Commodity?* (Cambridge: Cambridge Energy Research Associates, 1985), pp. 25–33 ("Minister and Manager"); Louis, *British Empire,* p. 56 ("no power or influence"); Francis Williams, *A Prime Minister Remembers: The War and Postwars of Earl Attlee* (London: Heinemann, 1961), pp. 178–79; Robert Stobaugh, "The Evolution of Iranian Oil Policy, 1925–1975," in *Iran Under the Pahlavis,* ed. George Lenczowski (Stanford: Hoover Institution Press, 1978), p. 206; James A. Bill and William Roger Louis, eds., *Mossadiq, Iranian Nationalism, and Oil* (London: I. B. Tauris & Co., 1988), p. 8 ("West End gentlemen").

4. NA 886D.2553 "Gulf Oil Company Difficulties," June 4, 1951, 6–451; "Gulf Oil Talks with Anglo-Iranian," March 29, 1951, 3–2951 ("did not dare"), RG 59, NA. Bill and Louis, *Mossadiq,* p. 247 ("fingertips" and "tough bargaining"); *Time,* August 1, 1949, p. 58 ("came with the shale"); Minutes of Meeting, August 2, 1950, EP 1531/40, FO 371/82375, PRO; Sampson, *Seven Sisters,* p. 134 ("Glasgow accountant"); Interviews with Robert Belgrave ("skinflint") and George McGhee.

5. Louis, *British Empire,* p. 645 (Fraser); Interview with Peter Ramsbotham ("Bombshell"); Rouhollah K. Ramazani, *Iran's Foreign Policy, 1941–1973: A Study of Foreign Policy in Modernizing Nations* (Charlottesville: University of Virginia Press, 1975), pp. 192–96 ("misfortunes"); Abrahamian, *Iran,* p. 266 ("sacred mission" and "stooge"); Norman Kemp, *Abadan: A First-Hand Account of the Persian Oil Crisis* (London: Allan Wingate, 1953), pp. 27–28. Meeting at Foreign Office, January 16, 1951, EP 1531/112, FO 371/91524; Shepherd to Morrison, "Political Situation in Persia," July 9,

1951, EP 1015/269, FO 248/1514 (1951), part IV ("Former Company," "abolished" and "no further"), PRO. On the governor and the sheep, see Lapping, *End of Empire*, pp. 208–9; *Times* (London), June 9, 11, 1951; *New York Times*, June 9, 10, 11, 1951.

6. Roy Mottahedeh, *The Mantle of the Prophet: Religion and Politics in Iran* (London: Penguin, 1987), pp. 122–25 ("pure"). "Biographic Outline, Mohammed Mossadeq," Memorandum for the President, October 22, 1951; CIA, "Probable Developments in Iran Through 1953," NIE-75/1, January 9, 1953, President's Secretary's File, Truman papers. H. W. Brands, *Inside the Cold War: Loy Henderson and the Rise of the American Empire, 1918–1961* (Oxford: Oxford University Press, forthcoming), chap. 18 (fainting spells); Anthony Eden, *Full Circle* (Boston: Houghton Mifflin, 1960), p. 219 ("Old Mossy"); Painter, *Oil and the American Century*, p. 173 ("colonial exploiter"); Acheson, *Present at the Creation*, p. 651 ("great actor"); Interviews with George McGhee and Peter Ramsbotham ("Moslem"); Vernon Walters, *Silent Missions* (Garden City, N.Y.: Doubleday, 1978), p. 262; C. M. Woodhouse, *Something Ventured* (London: Granada, 1982), pp. 113–14; Louis, *British Empire*, pp. 651–53 ("lunatic" and "cunning"); Paul Nitze, *From Hiroshima to Glasnost*, pp. 130–37.

7. Interviews; Louis, *British Empire*, pp. 667–74 ("Suez Canal"); Notes, June 27, 1951, EP 1531/870, FO 371/91555, PRO (Churchill); Alistair Home, *Harold Macmillan*, vol. 1, *1894–1956*, (New York: Viking, 1988), p. 310; H. W. Brands, "The Cairo-Tehran Connection in Anglo-American Rivalry in the Middle East, 1951–1953," *International History Review*, 11 (1989), pp. 438–40 ("scuttle and surrender").

8. Interview with Richard Funkhouser, Multinational Subcommittee Staff Interviews ("oracle"); Interview with Walter Levy. On Levy's proposals, see Logan memo, July 31, 1951, with Minute, July 29, 1951, EP 1531/1290, FO 371/91575 ("camouflage"); Shepherd to Foreign Office, October 10, 1951, EP 1531/1837, FO 371/91599 (John Kennedy); Cabinet Minutes, July 30, 1951, CM (51), CAB 128/20, PRO. Acheson, *Present at the Creation*, p. 655; Louis, *British Empire*, p. 677, n. 5 ("mongrelization" and "dilute"); Walters, *Silent Missions*, pp. 247–56 ("crafty," "Where else?," "certain principles" and Kashani); *FRUS: Iran, 1951–1954*, pp. 145 ("dream world").

9. Louis, *British Empire*, p. 678 ("jolly good"); Fergusson to Stokes, October 3, 1951, with Fergusson to Makins, October 4, 1951, EP 1531/1839, FO 371/91599; Ramsbotham to Logan, August 20, 1951, EP 1531/1391, FO 371/91580, PRO. Interview with Peter Ramsbotham ("last act of *Figaro*"); Peter Ramsbotham to author, July 4, 1990; Painter, *Oil and American Century*, p. 177. John F. Thynne, "British Policy on Oil Resources 1936–1951 with Particular Reference to the Defense of British Controlled Oil in Mexico, Venezuela and Persia" (Ph.D., London School of Economics, 1987), pp. 211–12, 273 ("stock-in-trade"); Walters, *Silent Missions*, p. 259 ("failure").

10. Cabinet, Persia Committee, "Measures to Discourage or Prevent the Disposal of Persian Oil," December 13, 1951, PO (0)(51)26, CAB 134/1145 ("stolen oil"); Cabinet Minutes, September 27, 1951, CM (51), CAB 128/20 ("humiliating"), PRO. Interview with Eric Drake ("sabotage" and "pistol"); Kemp, *Abadan*, pp. 235 ("day of hatred"), 241 ("Stand Firm"); Longhurst, *Adventure in Oil*, pp. 143–44 ("records").

11. "Steps Taken to Make Up the Loss of Persian Production," Appendix D to "Measures to Discourage or Prevent the Disposal of Persian Oil," December 13, 1951, PO (0) (51), CAB 134/1145; "Persian Oil: Future Policy," April 15, 1953, CAB 134/1149; "Persian Ability to Produce and Sell Oil," November 22, 1951, PO (0) (51) 17, CAB 134/1145, PRO. "Plan of Action No. 1 Under Voluntary Agreement," July 1951; Lilley to Longon, April 26, 1951, "Texas Co. 1951" file, Case 9, Oil Companies papers. Shell Transport and Trading, "Survey of Current Activities, 1951," Shell archives ("unnecessary"). C. Stribling Snodgrass and Arthur Kuhl, "U.S. Petroleum's Response to the Iranian Shutdown," *Middle East Journal* 5 (Autumn 1951) pp. 501–4; Lenczowski, *Iran*, p. 212.

12. Robert Rhodes James, *Anthony Eden* (New York: McGraw-Hill, 1987), pp. 355 (" 'old brain' "), 346 ("splutter of musketry"), 60, 347 (Anglo-Iranian stock); Eden, *Full Circle* pp. 212–25 ("shaken"). Eden Minute on Bullard to Foreign Office, May 7, 1941, No. 202, FO 371/27149; P. Dixon, "Informal Conversation about Persia," November 14, 1951, CAB 134/1145; Fergusson to the Minister, "Persian Oil," January 30, 1952 and February 7, 1952 ("tell the U.S.A."), PO (M) (52), POWE 33/1929; Butler to the Secretary, May 22, 1952, POWE 33/1934, PRO. Bill and Louis, *Mossadiq*, pp. 244, 246 ("cloud cuckoo").

13. Interview with George McGhee ("end of the world"); George McGhee to author, July 5, 1990; Peter Ramsbotham to author, July 4, 1990; McGhee, *Envoy*, pp. 401–3; Acheson, *Present at the Creation*, pp. 650 ("like Texas"); Walters, *Silent Missions*, p. 262 ("my fanatics"); Abrahamian, *Iran*, pp. 267–68 ("rabble rouser"); Sepehr Zabih, *The Mossadegh Era: Roots of the Iranian Revolution* (Chi-

cago: Lake View Press, 1982), p. 46; Brands, *Loy Henderson,* chap. 20 ("secret contempt"); *FRUS: Iran, 1951–1954,* pp. 179 (future generations), 186 ("helpless").

14. "Record of Meeting," June 28, 1952, EP 15314/163, CAB 134/1147 ("some stage"); Makins to Foreign Office, May 21, 1953, EP 1943/1, FO 371/104659; Churchill to Makins, June 5, 1953, EP 1943/3G, FO 371/104659; Makins to Foreign Office, June 4, 1953, No. 473, FO 371/104659, PRO. Churchill to Truman, August 16, August 20 ("very edge"), August 22, September 29, 1952 (with Acheson to Truman, October 1, 1952), Truman to Churchill, August 18, 1952 ("communist drain"), Henderson and Middleton to Bruce and Byroade, August 27, 1952 ("trap"), PSF, Truman papers. Acheson, *Present at the Creation,* p. 650; Eden, *Full Circle,* p. 221 ("autograph"); Woodhouse, *Something Ventured,* pp. 110–27; Kermit Roosevelt, *Countercoup: The Struggle for the Control of Iran* (New York: McGraw-Hill, 1979), pp. 114–20; Pahlavi, *Shah's Story,* p. 55; *FRUS: Iran, 1951–1954,* pp. 742 ("mobocracy"), 693 ("Communist control" and "feasible course"), 737–38 ("active"), 878.

15. Makins to Foreign Office, June 4, 1953, FO 371/104659 (Shah's suspicions); Shuckburgh to Strang, August 29, 1953, FO 371/104659; Roe to Foreign Office, August 25, 1953, EP 1914/1, FO 371/104658; Bromley to Salisburg, August 26, 1953, EP 1941/12, FO 371/104658 (Shah in Rome and Baghdad), PRO. *FRUS: Iran, 1951–54,* pp. 748 ("snuggle up"), 780–88 (description of events); William Shawcross, *The Shah's Last Ride* (New York: Simon and Schuster, 1988), pp. 68–70 ("bulletin" and "I knew they loved me"); Roosevelt, *Countercoup,* pp. 156–72, and passim; Mark T. Gasiorowski, "The 1953 Coup d'Etat in Iran," *International Journal of Middle Eastern Studies* 19 (1987), pp. 261–86; Woodhouse, *Something Ventured,* pp. 115–16. Woodhouse was, at the time, Kim Roosevelt's opposite number, in charge of the 1953 coup enterprise from the British side.

16. Robert Belgrave to author, March 16, 1989; Interview with Wanda Jablonski; "Persia: Quarterly Political Report," July-September 1953, November 19, 1953, EP 1015/263, POWE 33/2089, PRO; Donald N. Wilber, *Adventures in the Middle East: Excursions and Incursions* (Princeton, N.J.: Darwin Press, 1986), p. 189; Stephen E. Ambrose, *Eisenhower: The President* (New York: Simon and Schuster, 1984), p. 129 ("dime novel"); Richard and Gladys Harkness, "The Mysterious Doings of CIA," *Saturday Evening Post,* November 6, 1954, pp. 66–68; Brands, *Loy Henderson,* chap. 20.

17. Butler to Secretary, August 24, August 26, 1953, POWE 33/2088 ("stumped"); "Skeleton Memo on Middle East Oil," August 17, 1953, PO (0) (53) 72, CAB 134/1149; "Draft Proposal/Walter Levy," October 20, 1952, POWE 33/1936, PRO. Interview with Wanda Jablonski; Bennett Wall, *Growth in a Changing Environment: The History of Standard Oil (New Jersey), 1950–1972, and the Exxon Company, 1972–1975* (New York: McGraw-Hill, 1988), pp. 487–88; Wilber, *Adventures in the Middle East,* p. 184; Nitze, *From Hiroshima to Glasnost,* pp. 133–37; United States Congress, Senate, Committee on Foreign Relations, Subcommittee on Multinational Corporations, 93rd Congress, 2d Session, *Multinational Corporations and U.S. Foreign Policy* (Washington, D.C.: GPO, 1975), p. 60; *Multinational Hearings,* part 7, p. 301 ("touch and go"); Interviews with George Parkhurst ("ouchy") and Howard Page ("beat us on the head"), Multinational Subcommittee Staff Interviews; Burton I. Kaufman, *The Oil Cartel Case: A Documentary Study of Antitrust Activity in the Cold War Era* (Westport, Conn.: Greenwood Press, 1978), pp. 162–70 (Funkhouser); Wilkins, *Maturing of Multinational Enterprise,* p. 322; Interview with George McGhee ("fiddler"); United States Congress, Senate, Committee on Foreign Relations, Subcommittee on Multinational Corporations, *The International Petroleum Cartel, the Iranian Consortium, and U.S. National Security* (Washington, D.C.: GPO, 1974), pp. 57–58 ("strictly commercial viewpoint").

18. Wall, *Exxon,* pp. 453–55, 947, n. 33; Kaufman, *Oil Cartel Case,* pp. 27 ("rubber stamping"), 163 ("highly slanted"), 30 ("Soviet propaganda"); FTC, *International Petroleum Cartel.* For Justice Department version, see Multinational Subcommittee, *Iranian Consortium,* pp. 5–16 ("spot market"). For the British view, see Eden in "Notes for Secretary of State on U.S. Federal Trade Commission's Report," September 4, 1952, POWE 33/1920 and "International Oil Industry," Memo by the Foreign Secretary, September 30, 1952, C (52) 315, PREM 11/500; Churchill to Foreign Secretary, August 30, 1952, M 463/ 52, PREM 11/500, PRO. Lloyd to Anglo-Iranian, October 2, 1952, brown wrapper, Case 9, Oil Companies papers ("stale bread," "witch-hunters" and "prejudicial"). On antitrust policy, see Raymond Vernon and Debra L. Spar, *Beyond Globalism: Remaking American Foreign Economic Policy* (New York: Free Press, 1989), pp. 113–17 and Kingman Brewster, Jr., *Antitrust and American Business Abroad* (New York: McGraw-Hill, 1958), pp. 8, 72–74, 330–31.

19. *FRUS: Current Economic Developments, 1945–1954,* January 6, 1947 ("national interest"); Multinational Subcommittee, *Iranian Consortium,* pp. 30–36 ("unlawful combination"), 52 ("enforce-

ment"), 77 ("would not violate"); Burton I. Kaufman, "Oil and Antitrust: The Oil Cartel Case and the Cold War," *Business History Review* 51 (Spring 1977), p. 38 ("start trouble"); Truman, *Memoirs*, pp. 126–27 (Truman's oil experience); Wall, *Exxon*, pp. 481–86 ("considered judgment"); John Foster Dulles, "Iranian Oil" memorandum, January 8, 1954, *DDRS*, 1983, doc. 257C.

20. *Multinational Hearings*, part 7, pp. 304 ("political matter" and "no case"), 297 ("yacking"), 248–49; Wall, *Exxon*, pp. 492–96 ("hostages"); Interview with Robert Belgrave ("apple cart"). "Iran—Basis for Settlement with Anglo-Iranian," March 16, 1954, CAB 134/1085; Cabinet, Middle East Oil Committee, "Middle East Oil Policy," April 2, 1954, O.M.E. (54) 21, CAB 134/1085 ("reliable independents"), PRO. *New York Times*, November 1, 1954, p. 1; Henderson to Jernegan, November 12, 1953, 880.2553/11–1253 ("almost inevitable"), RG 59, NA; Interview with Howard Page, Multinational Subcommittee Staff Interviews; Interviews with Pierre Guillaumat, John Loudon ("wonderful deal") and Wanda Jablonski.

Chapter 24

1. Chester L. Cooper, *The Lion's Last Roar* (New York: Harper & Row, 1978), pp. 12 ("Great Engineer"), 16, 18 ("highway"), 20; Robert Blake, *Disraeli* (New York: St. Martin's, 1967), pp. 584–85 (Disraeli).

2. Office of Intelligence Research, Department of State, "Traffic and Capacity of the Suez Canal," p. 10, August 10, 1956, National Security Council records; Harold Lubell, "World Petroleum Production and Shipping: A Post-Mortem on Suez," P-1274 (Rand Corporation, 1958), pp. 17–18.

3. Selwyn Lloyd, *Suez 1956: A Personal Account* (New York: Mayflower Books, 1978), pp. 45, 69, 24, 2–19; Donald Neff, *Warriors at Suez: Eisenhower Takes the United States into the Middle East* (New York: Simon and Schuster, 1981), p. 83 (CIA profile); Anthony Nutting, *Nasser* (New York: E. P. Dutton, 1972), p. 75 ("Voice of the Arabs"); Elizabeth D. Sherwood, *Allies in Crises: Meeting Global Challenges to Western Security* (New Haven: Yale University Press, 1990), chap. 3; Gamal Abdel Nasser, *The Philosophy of the Revolution* (Buffalo: Smith, Keynes, and Marshall, 1959), p. 61; Y. Harkabi, *Arab Attitudes to Israel*, trans. Misha Louvish (Jerusalem: Israel Universities Press, 1974), p. 61 ("crime"); Jacques Georges-Picot, *The Real Suez Crisis: The End of a Great Nineteenth-Century Work*, trans. W. G. Rogers (New York: Harcourt Brace Jovanovich, 1978), pp. 34, 61–62; W. S. C. to Minister of State, August 19, 1952, Prime Minister's Personal Minutes, Egypt (main file), part 3, PREM 11/392, PRO. C. Mott-Radclyffe to Ambassador, May 4, 1954, D7 107–83, Middle East Centre archives. Mohammed H. Heikal, *Cutting Through the Lion's Tale: Suez Through Egyptian Eyes* (London: Andre Deutsch, 1986), pp. 6, 13, 61–62 (Eden's Arabic).

4. Interview with Robert Bowie; Jacques Georges-Picot, *Real Suez Crisis*, p. 68 ("musty . . . odor"); Wall, *Exxon*, pp. 547–51; Mohammed Heikal, *The Cairo Documents* (Garden City, N.Y.: Doubleday, 1973), pp. 84–85 ("oil complex"); Anthony Nutting, *No End of a Lesson: The Story of Suez* (London: Constable, 1967), p. 40; Anthony Moncrieff, ed., *Suez: Ten Years After* (New York: Pantheon, 1966), pp. 40–41 (cotton).

5. Cooper, *Lion's Last Roar*, p. 103 ("De Lesseps"); Alistair Home, *Harold Macmillan*, vol. 1, *1894–1956* (New York: Vintage, 1989), p. 397 (Macmillan); Wm. Roger Louis and Roger Owen, eds., *Suez 1956: The Crisis and its Consequences* (Oxford: Clarendon Press, 1989), p. 110; Interview with John C. Norton (pilots).

6. Evelyn Shuckburgh, *Descent to Suez: Diaries, 1951–1956* (London: Weidenfeld and Nicolson, 1986), p. 23 ("Master"); Neff, *Warriors at Suez*, p. 39 (Ike on Dulles); Interview with Winthrop Aldrich, p. 27, Tape 27, Box 244, Aldrich papers; Eden, *Full Circle*, p. 487 ("disgorge"); Louis and Owen, *Suez 1956*, pp. 198–99 ("out of date" and "white men"), 210 ("mantle"); Dwight D. Eisenhower, *Waging Peace: The White House Years, 1956–1961* (Garden City, N.Y.: Doubleday, 1965), p. 670 ("drama"); Interview with Robert Bowie; Heikal, *Cairo Documents*, p. 103 ("Which side"); Deborah Polster, "The Need for Oil Shapes the American Diplomatic Response to the Invasion of Suez" (Ph.D., Case Western Reserve University, 1985), pp. 65–66.

7. Herman Finer, *Dulles Over Suez: The Theory and Practice of His Diplomacy* (Chicago: Quadrangle Books, 1964), p. 397; Eisenhower to Hoover, October 8, 1956, Dulles papers, White House Memoranda Series, Eisenhower Library; Polster, "The Need for Oil," chap. 4.

8. April 6, 1956, Personal Telegram Serial, T 221/56, PREM 11/1177 ("Bear's claws"); Cabinet, Egypt Committee, August 24, 1956, E.C. (56), CAB 134/1216, PRO. Eden, *Full Circle*, p. 401 ("absolutely blunt").

9. Eden, *Full Circle,* pp. 520 (Eden to Eisenhower), 475; Lloyd, *Suez,* p. 42 ("very worried"); Home, *Macmillan,* vol. 1, p. 411 (Macmillan's reading and diary).

10. Eden, *Full Circle,* pp. 576–78 ("stamp of our generation"); Interview with Robert Belgrave; Nasser, *Philosophy of the Revolution,* pp. 72–73 ("vital nerve"); Lloyd, *Suez,* p. 120 (Spaak).

11. Kenneth Love, *Suez: The Twice-Fought War* (New York: McGraw-Hill, 1969), pp. 367, 403; Wall, *Exxon,* pp. 549–61; Louis and Owen, *Suez 1956,* p. 123 (Kirkpatrick); Wilbur Crane Eveland, *Ropes of Sand: America's Failure in the Middle East* (New York: Norton, 1980), pp. 209–13 (Anderson); Eisenhower to King Saud, August 20, 1956, *DDRS,* 1985, doc. 655. Views about nuclear power similar to Anderson's were expressed in the Joint Intelligence Staff meetings in London. Chester Cooper to author, May 30, 1989.

12. Moshe Dayan, *Story of My Life* (New York: William Morrow, 1976), p. 218. Lloyd's attitude reminded Dayan "of a customer bargaining with extortionate merchants." Hugh Thomas, *The Suez Affair* (London: Weidenfeld and Nicolson, 1986), pp. 95–109, 224. On attitude towards Jews, see Shuckburgh, *Descent to Suez,* passim; for Eden on Jews, see Neff, *Warriors at Suez,* p. 206 and John Harvey, ed., *The War Diaries of Oliver Harvey* (London: Collins, 1978), pp. 191–94, 247. Harold Macmillan, *Riding the Storm, 1956–59* (London: Macmillan, 1971), p. 149; Louis and Owen, *Suez 1956,* p. 160; Stuart A. Cohen, "A Still Stranger Aspect of Suez: British Operational Plans to Attack Israel, 1955–56," *International History Review* 10 (May 1988), pp. 261–81.

13. James, *Eden,* p. 597 ("artificial inside"); Cooper, *Lion's Last Roar,* p. 128 ("Chums"). On Eden's medication and collapse, see James, *Eden,* pp. 523, 597; Thomas, *Suez Affair,* pp. 43–44; Neff, *Warriors at Suez,* p. 182.

14. Ambrose, *Eisenhower,* p. 357; "Memorandum of Conference with the President," October 30, 1956, Dulles papers, White House Memoranda Series (Eisenhower); Cooper, *Lion's Last Roar,* p. 167 ("unshirted hell"); Lloyd, *Suez,* p. 78 (Hoover); Heikal, *Cairo Documents,* pp. 112–13 (Nasser's instructions); Cabinet, Egypt Committee, "Political Directive to the Allied Commander-in-Chief," November 3, 1956, E.O.C. (56) 12, CAB 134/1225, PRO.

15. Cabinet, Egypt Committee Minutes, September 7, 1956, EC (56), CAB 134/1216; Chiefs of Staff, "Review of the Middle Eastern Situation Arising Out of the Anglo-French Occupation of Port Said," November 8, 1956, E.C. (56) 67, CAB 134/1217, PRO. Ambrose, *Eisenhower,* pp. 359 ("boil in"), 371 ("Attorney General"); Interview with Peter Ramsbotham ("paraboys"); Richard K. Betts, *Nuclear Blackmail and Nuclear Balance* (Washington, D.C.: Brookings Institution, 1987), pp. 62–65 ("night follows day"); Polster, "Need for Oil," p. 114 ("get the Arabs sore"); Wall, *Exxon,* p. 557 ("simply refused"); Macmillan, *Riding the Storm,* p. 164 (IMF); Lloyd, *Suez,* pp. 211, 206 (Macmillan on oil sanctions); Louis and Owen, *Suez 1956,* p. 228 ("naughty boys"); United States Congress, Senate, Committee on the Judiciary and Committee on Interior and Insular Affairs, *Emergency Oil Lift Program and Related Oil Problems: Joint Hearings,* 85th Congress, 1st session (Washington, D.C.: GPO, 1957), p. 2401 ("purgatory"). Statistics from *Emergency Oil Lift Program,* pp. 1046–64; Office of Intelligence Research, Department of State, "Economic Consequences of the Closure of the Suez Canal (and IPC Pipelines)," January 7, 1957; Lubell, "World Petroleum Production and Shipping," p. 21; Harold Lubell, *Middle East Oil Crisis and Western Europe's Energy Supplies* (Baltimore: Johns Hopkins University Press, 1963); Peter Hennessy and Mark Laity, "Suez—What the Papers Say," *Contemporary Record* 1 (Spring 1957), p. 8.

16. Eisenhower to Ismay, November 27, 1956, *DDRS,* 1989, doc. 2941 ("sadness" and "delicate"); Dillon to Director (re: Ismay), *DDRS,* 1989, doc. 859 (Ismay); Lloyd, *Suez,* p. 219 (hospital meeting with Dulles). Robert Rhodes James, while accepting Lloyd's recollection, quotes his somewhat more ambiguous report to Eden at the time, in which Dulles criticized "our methods," but "deplored that we had not managed to bring down Nasser." James, *Eden,* p. 577.

17. Cabinet minutes, November 26, 1956, CAB 134/1216; Macmillan to Eden, January 7, 1957, PREM 11/2014, PRO. *Emergency Oil Lift Program,* pp. 2406 ("sugar bowl"), 2353–57 ("whose interests"), 2404 (Drake and Jersey representative), 810; Love, *Twice-Fought War,* p. 655 ("Suez sixpence"); Wall, *Exxon,* pp. 559 ("have already shipped"), 579 ("push a button"); *Financial Times* and *Daily Express,* January 19, 1957 ("No Extra Oil"); Johnson, *Sun,* pp. 84–86 (antitrust case).

18. United States Congress, Senate, Judiciary Committee, Subcommittee on Antitrust and Monopoly, *Petroleum, the Antitrust Laws and Government Policy,* 85th Congress, 1st session (Washington, D.C.: GPO, 1957), pp. 97–98; Office of Intelligence Research, Dept. of State, "Economic Consequences of the Closure of the Suez Canal," January 7, 1957; Wall, *Exxon,* p. 582 ("British shipping"); "Middle East Oil," January 23, 1957, UE S 1171/39, FO 371/127281, PRO.

19. Cooper, *Lion's Last Roar*, p. 281 ("curious time"); Moncrieff, *Suez: Ten Years After*, p. 45 ("Sir Eden"); James, *Eden*, p. 593 ("so unrepentant"); Macmillan, *Riding the Storm*, p. 181 ("see him now"); Neff, *Warriors at Suez*, p. 437 (*Times*); Home, *Macmillan*, p. 460.

20. Interview with John Loudon ("tanker people"); Wall, *Exxon*, p. 582. On pipelines, see "Transport of Middle East Oil," n.d., UE S 1171/228, FO 371/127213; Bridgeman to Ayres, March 11, 1957, POWE 33/1967; Memorandum from Shell, March 11, 1957, POWE 33/1967, PRO. On tankers, see JWH to Secretary of State, October 11, 1956, and Draft Memorandum for the President, Dulles papers, White House Memoranda series; "Bermuda Conference: Long-Term Tanker Prospects," Note by Ministry of Power and Ministry of Transport, March 15, 1957, UE S 1172/5, FO 371/127210; "Middle East Oil," January 18, 1957, to Mr. Beely, FO 371/127200 ("political risk"); "Long-Term Requirements for the Transport of Oil from the Middle East," January 28, 1957, UE S 1141/29, FO 371/127201, PRO.

21. Caccia to Foreign Office, February 12, 1957, AU 1051/A2, FO 371/126684, PRO ("boy scout"); "Memorandum of Conferences with the President," November 21, 1956, 4:00 P.M., 5:30 P.M., Dulles papers, White House Memoranda series (Ike's Middle East policy).

22. Macmillan, *Riding the Storm*, pp. 198 ("agonies"), 133 ("rulers"), 258 (weekly letters); "Memorandum of Conference with the President," November 21, 1956, Dulles papers, White House Memoranda series ("straight, fine man"). Macmillan to "Dear Friend" (letter to Eisenhower), January 16, 1957, PREM 11/2199 ("no illusions"); COIR to Jarrett, March 5, 1957, PREM 11/2010 ("family tree"); Macmillan to C.E., March 6, 1957, PREM 11/2014; "Middle East: General Questions," Bermuda Conference Notes, PREM 11/1838 (Macmillan in Bermuda), PRO. Eisenhower, *Waging Peace*, p. 123 ("plain talk"); James, *Eden*, p. 617 ("lake of oil").

Chapter 25

1. Wellings to DeGolyer, December 10, 1953; DeGolyer to Wellings, December 24, 1953, 1982, De-Golyer papers.

2. American Petroleum Institute, *Basic Petroleum Data Book*, vol. 6, September 1986, IV-1, II-1.

3. Cabinet, Middle East Oil Committee, October 7, 1954, O.M.E. (54) 36, CAB 134/1086 ("reasonable basis"); Russell to Lloyd, October 11, 1957, EP 1013/4, FO 371/127073, PRO.

4. Hohler to Foreign Office, August 20, 1957, UE S 1171/228, FO 371/127211, p. 6; Foreign Office, "Signor Enrico Mattei," FO 371/127210, PRO. "Enrico Mattei and the ENI," with Tasca and Phelan to Department of State, December 16, 1954, 865.2553/12–1654, RG 59, NA ("economic history"); Paul Frankel, *Mattei: Oil and Power Politics* (New York and Washington, D.C.: Praeger, 1966), pp. 122, 41–51 ("sucking needles"); Interviews with Marcello Colitti ("into the fire") and John Loudon ("difficult" and "dessert").

5. Interview with Robert Belgrave; Frankel, *Mattei*, p. 83 (Mattei on the Seven Sisters). Beckett to Jardine, May 22, 1957, FO 371/127208; Falle to Gore-Booth, May 6, 1957, UE S 1171/105/4, FO 371/127205; Cabinet, Committee on the Middle East, March 26, 1957, O.M.E. (57), CAB 134/2338; Record of Conversation between Mr. Hannaford and Signor Mattei, with Cabinet, Committee on the Middle East, OME (57) 35, Eevise, May 24, 1957, CAB 134/2339; Hohler to Lloyd, August 20, 1957, UE S 1171/228, FO 371/127211, PRO.

6. Frankel, *Mattei*, p. 141 (Italian princess). Cabinet, Committee on the Middle East, May 1, 1957, O.M.E. (57) 29, PREM 11/2032; Stevens to Foreign Office, April 12, 1957, no. 47, PREM 11/2032 ("blackmail"); Foreign Office to Rome, March 27, 1957, no. 469, FO 371/127203; Ashley Clarke, "Signor Mattei and Oil," September 25, 1957, FO 371/127212, PRO.

7. Wright to Coulson, with Cabinet, Middle East Committee, March 25, 1957, O.M.E. (57) 24, CAB 134/2339 ("lesser evil"); Foreign Office, "European Interest in Mid East Oil," Cabinet, Committee on the Middle East, OME 57 (35), May 24, 1957, CAB 134/2339; Pridham minute, August 27, 1957, UE S 1171/228, FO 371/127211; Anglo-American Talks, April 15, 1957, FO 371/127206 ("unreliable person"); Joseph Addison to Wright, April 11, 1957, UE S 1171/120 (C), FO 371/127206; Cabinet, Committee on the Middle East, "Italian-Iranian Oil Agreement," March 25, 1957, O.M.E. 57 (24), CAB 134/2339 and Foreign Office to Rome, March 27, 1957, no. 469, FO 371/127203 ("prejudice"); Hohler to Wright, August 20, 1957, UE S 1171/228, FO 371/127211 ("sign of weakness"); Hohler to Foreign Office, August 20, 1957, no. 141 E, FO 371/127211, PRO.

8. Wright to Coulson, with Cabinet, Committee on the Middle East, "Italian-Iranian Oil Agreement," March 25, 1957, OME 57 (24), CAB 134/2339; Lattimer notes on Mattei visit, June 1957, E.G. 9956,

FO 371/127210; Russell to Wright, August 10, 1957, UE S 1171/223, FO 371/127211 (meeting with Mattei in Tehran and "four minute mile"); M. to Macmillan, May 6, 1957, PREM 11/2032; Wright to Russell, August 16, 1957, UE S 1171/223, FO 371/127211, PRO. Interview with Marcello Colitti ("tiny places").

9. "Japanese Interest in the Oil Concession for the Kuwait-Saudi Neutral Zone Sea-Bed," September 27, 1957, FO 371/127170; British Embassy, Tokyo, to Eastern Department, Foreign Office, November 12, 1957, 1532/107/57, FO 371/127171; S. Falle, "Japanese and Middle East Oil Concessions," October 4, 1957, POWE 33/2110 ("real breach"), PRO. Martha Caldwell, "Petroleum Politics in Japan: State and Industry in a Changing Policy Context" (Ph.D., University of Wisconsin, 1981), pp. 84–86; CIA, "Taro Yamashita," *Biographic Register,* August 1964, *DDRS,* doc. 31A.

10. Halford to Foreign Office, December 19, 1957, No. 22, POWE 33/2110; Cabinet, Official Committee on the Middle East, Minutes, November 7, 1957, p. 5, OME (57), CAB 134/2338; Kuwait to Foreign Office, October 21, 1957, S 1534/29, FO 371/127171; P. J. Gore-Booth to J. A. Beckett, November 4, 1957, PD 1146/17, POWE 33/2110 ("feeling"); Kuwait to Foreign Office, October 8, 1957, nos. 363, 364, FO 371/127171 (royal telegrams); J. C. Moberly, "Kuwait-Saudi Neutral Zone Seabed Concession," November 6, 1957, ES 1534/37, FO 371/127171; British Embassy, Tokyo, to Eastern Department, Foreign Office, November 12, 1957, S 1534/41; Shell to BPM, December 11, 1957, FO 371/127171; Halford to Foreign Office, December 22, 1957, No. 477, POWE 33/2110, PRO. Caldwell, "Petroleum Politics," pp. 86–87 ("national project"); Tadahiko Ohashi to author, August 16, 1989 (information from Mr. Sakakibara).

11. Emmett Dedmon, *Challenge and Response: A Modern History of Standard Oil Company (Indiana)* (Chicago: Mobium Press, 1984), pp. 203–5 ("opportunities" and "Aryans"); "Conversation with His Imperial Majesty," FO 371/1330A, PRO (Shah's home life).

12. Wright to Foreign Office, February 11, 1958, VQ 1015/11, FO 371/134197, PRO ("stranglehold"). On the coup in Iraq, see Wright to Foreign Office, July 17, 1958, VQ 1015/100, FO 371/134199; Embassy, Ankara to Rose, July 17, 1958, VQ 1015/71 (c), FO 37/134199; Stout to Middle East Secretariat, August 7, 1958, VQ 1015/195, FO 371/134202; Johnston to Rose, July 28, 1958, VQ 1015/171, FO 371/134/201; Wright to Lloyd, "The Iraqi Revolution of July 19, 1958," EQ 1015/208, PREM 11/2368; Wright memo on Howard Page, September 1, 1958, EQ 1531/15, FO 371/133119, PRO.

13. Report of the Commission of Arab Oil Experts, April 15–25, 1957, to the Secretariat-General of the Arab League, pp. 2, 5, FO 371/127224; Nuttali to Falk, October 4, 1957, with minutes of Second Session of Fifth Meeting of Arab Oil Experts, pp. 2, 9, UE 511717/2, FO 371/127224, PRO. Interview with Fadhil al-Chalabi.

14. Bridgett to Department of State, March 4, 1959, 831.2553/3–459, RG 59, NA; Philip, *Oil and Politics,* p. 83; Betancourt, *Venezuela,* pp. 323–24, 342 ("factory"); Richard M. Nixon, *Six Crises* (Garden City, N.Y.: Doubleday, 1962), pp. 213–27; Rabe, *The Road to OPEC,* p. 157 ("romantics"); Tugwell, *Politics of Oil,* chap. 3; Interviews with Alirio Parra, Alicia Castillo de Pérez Alfonzo, Juan Pablo Pérez Castillo, Oscar Pérez Castillo.

15. NA 831.2553: Leddy to Department of State, January 10, 1951, 1–1051; Davis memo, September 8, 1953, 9–853; Swihart to Department of State, May 9, 1956, 5–956; Chaplin to Department of State, May 25, 1956, 5–2556; Sparks to Secretary of State, December 20, 1958, 12–2058; Anderson-Rubottom memo, June 5, 1959, 6–559; Boonstra memo, June 5, 1959, 6–559; Eisenhower to Betancourt, April 28, 1959, 4–2859, RG 59; Cox to Department of State, September 10, 1959, 631.86B/9–1059, RG 59. Interview with Alirio Parra ("bones").

16. Pierre Terzian, *OPEC: The Inside Story,* trans. Michael Pallis (London: Zed Books, 1985), pp. 85–97; *New York Times,* June 4, 1958, p. 8; Diary of J. B. Slade-Baker, January 20, 1958, Middle East Center; Nadav Safran, *Saudi Arabia: The Ceaseless Quest for Security* (Cambridge: Harvard University Press, 1985), pp. 88–103; *Fortune,* August 1959, pp. 97, 146 ("service station"); *Petroleum Week,* June 20, 1958, p. 41; CIA, "Abdullah Ibn Hamud al-Tariqi, February 26, 1970, *DDRS,* 1984, doc. 788.

17. CIA, "Middle East Oil," NIE 30–60, November 22, 1960, 83–542–9, paper 12–17 ("force"); Cabinet Minutes, July 25, 1958, Whitman Files, 1953–1961, Cabinet Series, Box 11 ("dangerous situation"); Eisenhower Library. NA 861.2553: Sundt to Department of State, January 28, 1954, 1–2854; February 3, 1954, 2–354, RG 59, NA. Wall, *Exxon,* p. 332; J. E. Hartshorn, *Oil Companies & Governments: An Account of the International Oil Industry in Its Political Environment* (London: Faber and Faber, 1962), pp. 211, 215.

18. Terzian, *OPEC*, pp. 23, 26; Bridgett to Department of State, June 4, 1959, 831.2553/6–459, RG 59, NA; Hubbard to BP, April 29, 1959 ("considered successful" and "Miss Wanda Jablonski"); Chisholm to Chairman of BP, April 30, 1959, Deighton File, "Cairo 'Arab Petroleum Congress' " (" 'plus' "); Weir to Walmsley, June 17, 1959, B51532/8, FO 371/140378, PRO ("my boy").

19. Interviews with Wanda Jablonski and Alirio Parra; "Wanda Jablonski Reports on the Middle East," Supplement, *Petroleum Week*, 1957; "Eugene Jablonski Returns to Botany," *Garden Journal*, May-June 1963, pp. 102–3; Terzian, *OPEC*, pp. 26–29, 7 ("Gentlemen's Agreement").

Chapter 26

1. Wall, *Exxon*, pp. 332–33; Angela Stent, *From Embargo to Ostpolitik: The Political Economy of Soviet–West German Relations, 1955–1980* (Cambridge: Cambridge University Press, 1981), p. 99 (Keating); Hartshorn, *Oil Companies & Governments*, pp. 252–53, 218.

2. "Rathbone of Jersey Standard," *Fortune*, May 1954, pp. 118–19; "How Rathbone Runs Jersey Standard," *Fortune*, January 1963, pp. 84–89, 171–79.

3. Wright memo on Howard Page, September 1, 1958, EQ 1531/15, FO 371/133119 ("tough man"); Williams to Stock, June 23, 1958, 58/6/112, POWE 33/2200 (Jablonski at Jersey), PRO. Interviews with Wanda Jablonski and John Loudon; Ian Skeet, *OPEC—Twenty-five Years of Prices and Politics* (Cambridge: Cambridge University Press, 1988), p. 22 ("regret").

4. Terzian, *OPEC*, pp. 33–34 ("Just wait"), 42–46 ("regulation," "sanctions," "disapprove" and Page); Interviews with Alirio Parra ("We've done it") and Fadhil al-Chalabi; Skeet, *OPEC*, p. 23; Fadhil al-Chalabi, *OPEC and the International Oil Industry: A Changing Structure* (Oxford: Oxford University Press, 1980), p. 67; *New York Times*, September 25, 1960; CIA, "Middle East Oil," NIE 30–60, November 22, 1960, 83–542–9, Eisenhower Library; Robert Stobaugh and Daniel Yergin, eds., *Energy Future: Report of the Energy Project at the Harvard Business School* 3d. ed. (New York: Vintage, 1983), p. 24 (Rouhani).

5. Abdul-Reda Assiri, *Kuwait's Foreign Policy: City-States in World Politics* (Boulder, Colo.: Westview, 1990), pp. 19–26 (Iraq and Kuwait); Skeet, *OPEC*, p. 29 ("nice in theory").

6. Interviews with Alicia Castillo de Pérez Alfonzo, Juan Pablo Pérez Castillo, Oscar Pérez Castillo ("sowing" and "devil"), Alfred DeCrane, Jr., and Fadhil al-Chalabi; Terzian, *OPEC*, pp. 80–85 ("ecologist"); Skeet, *OPEC*, p. 32 ("reality of the oil world").

7. Interview with Gilbert Rutman; Melby, *France*, pp. 253 (Economic Council), 302 (Giraud); Alistair Home, *A Savage War of Peace: Algeria, 1954–1962* (London: Penguin, 1979), p. 242 (De Gaulle); Gilbert Burck, "Royal Dutch Shell and Its New Competition," *Fortune*, October 1957, pp. 176–78 ("all our eggs").

8. Ruth First, *Libya: The Elusive Revolution* (London: Penguin, 1974), p. 141; Multinational Subcommittee, *Multinational Oil Corporations*, p. 98 ("one oil company" and "quickly").

9. Commercial Secretariat to African Department, Foreign Office, June 11, 1957, JT 1534/3, FO 371/126063; Washington to Foreign Office, May 21, 1959, PREM 11/2743/1239 ("jack-pot"), PRO. Wall, *Exxon*, pp. 668–72 (Wright); Interviews with Robert Eeds, Ed Guinn and Mohammed Finaish, Multinational Subcommittee Staff Interviews. The gap in government takes between Libya, which based its taxes on market prices, and other producers, which used posted prices, became so great that, in 1965, Libya revised its system to increase its revenues by reverting to a posted price base.

10. Multinational Subcommittee Staff Interviews (corruption in Libya); *Multinational Hearings*, part 7, p. 287 (Page).

11. Interview with Marcello Colitti; Reinhardt to McGhee, April 25, 1962, *DDRS*, 1981, doc. 206B ("damaged ego"); *New York Times*, October 28, 1962, p. 16; October 29, 1962, p. 16; November 5, 1962, p. 30 ("most important individual"); *Times* (London), October 29, 1962, p. 12; *Time*, November 2, 1962, p. 98; January 18, 1963, p. 26.

12. Neil H. Jacoby, *Multinational Oil: A Study in Industrial Dynamics* (New York: Macmillan, 1974), pp. 138–39 ("new internationals"); *Multinational Hearings*, part 7, p. 352.

13. Multinational Subcommittee, *Multinational Oil Corporations*, p. 95 ("surge pot"); Wall, *Exxon*, pp. 616 (Jamieson), 610 ("always fighting"); *Multinational Hearings*, part 7, pp. 287 ("most important concession"), 314, 288 ("balloon"); Interviews with George Parkhurst, Kirchner, Merrill, Shaffer, Howard Page, Multinational Subcommittee Staff Interviews.

14. Department of State to Tehran Embassy, June 8, 1964, CM/Oil, Conference Files ("Arab imperialism"); Carroll to Vice President, July 27, 1966, EX FO 5, 6/30/66-8/31/66, Box 42 (Shah to Kim Roo-

sevelt), Johnson Library. Multinational Subcommittee, *Multinational Oil Corporations,* p. 108 ("do their best"); Interview with Parkhurst, Multinational Subcommittee Staff Interviews; *Multinational Hearings,* part 7, p. 309 (Oman).

15. Vietor, *Energy Policy,* p. 96 ("Tex" Willis); "Effect of Petroleum Imports Upon Oil Industry in Texas," July 15, 1949, 811.6363/7–1549, RG 59, NA ("re-election"); Goodwin, *Energy Policy,* pp. 227–28 ("old suggestion").

16. President's Appointment with Senators, June 3, 1957 ("nice balance"); Cabinet Minutes, July 24, 1957, p. 3; Eisenhower to Anderson, July 30, 1957, Box 25; Eisenhower to Moncrief, May 12, 1958, Box 33; Dulles-Brownell telephone call, July 2, 1957, box 7, Eisenhower diary, Whitman files ("window dressing"); Memorandum of Conversation with the President, November 10, 1958, Box 7, Dulles White House memos ("some action"), Eisenhower Library. Interview with Robert Dunlop; Lenzner, *Getty,* pp. 217–19; Goodwin, *Energy Policy,* pp. 247–51 (Randall and economic advisers); D. B. Hardeman and Donald C. Bacon, *Rayburn: A Biography* (Austin: Texas Monthly Press, 1987), p. 349; Robert Caro, *The Years of Lyndon Johnson: The Path to Power* (New York: Knopf, 1982); Robert Engler, *The Politics of Oil: Private Power and Democratic Directions* (Chicago: University of Chicago Press, 1967), pp. 230–47.

17. Vietor, *Energy Policy,* pp. 119 ("nightmare"), 134 (Russell Long); *Fortune,* June 1969, pp. 106–7.

Chapter 27

1. Jacoby, *Multinational Oil,* pp. 49–55.

2. P. H. Frankel, *Essentials of Petroleum: A Key to Oil Economics,* new ed. (London: Frank Cass, 1969), p. 1. Frankel's book, though written in 1946, remains essential to understanding the oil industry. David Landes, *The Unbound Prometheus: Technological Change and Industrial Development in Western Europe from 1750 to the Present* (Cambridge: Cambridge University Press, 1979), p. 98; Carlo M. Cipolla, *The Economic History of World Population,* 7th ed. (London: Penguin, 1979), p. 56 (Jevons); Senate, *Emergency Oil Lift Program,* pp. 2739, 2749, 2731–42. Joseph C. Goulden, *The Best Years, 1945–50* (New York: Atheneum, 1976), pp. 123–24 (Lewis); Interview (statue).

3. Senate, *Emergency Oil Lift Program,* pp. 2371–78; G. L. Reid, Kevin Allen, and D. J. Harris, *The Nationalized Fuel Industries* (London: Heinemann, 1973), p. 23 ("Killer Fogs," "smokeless zones" and "Living Fire"); U.K. Department of Energy archives; William Ashworth, *The History of the British Coal Industry,* vol. 5, (Oxford: Clarendon Press, 1986), pp. 672–73; Melby, *France,* pp. 227, 236, 303–4; W. O. Henderson, *The Rise of German Industrial Power, 1834–1914* (London: Temple Smith, 1975), p. 235 (Keynes); Raymond Vernon, ed., *The Oil Crisis in Perspective* (New York: Norton, 1976), pp. 94, 92.

4. Chalmers Johnson, *MITI and the Japanese Miracle: The Growth of Industrial Policy, 1925–1975* (Stanford: Stanford University Press, 1982), p. 237 ("no longer living"); Hein, *Fueling Growth,* chaps. 7, 10, 11; Richard J. Samuels, *The Business of the Japanese State: Energy Markets in Comparative Historical Perspective* (Ithaca: Cornell University Press, 1987), pp. 191–92, 196; Michael A. Cusamano, *The Japanese Automobile Industry: Technology and Management at Nissan and Toyota* (Cambridge: Harvard University Press, 1985), pp. 392–94; Alfred D. Chandler, Jr., "Industrial Revolution and Institutional Arrangements," *Bulletin of the American Academy of Arts and Sciences* 33 (May 1980), pp. 47–48.

5. Interviews with William King and James Lee; Conoco, *The First One Hundred Years* (New York: Dell, 1975), pp. 169, 193; *Fortune,* April 1964, p. 115 ("big bite"); Interviews with Kirchen ("beat the bushes"), Shaffer and Merrill, Multinational Subcommittee Staff Interviews.

6. *Fortune,* September 1961, pp. 98, 204–6 (Texaco); September 1953, pp. 134–37, 150–62; September 1954, pp. 34–37, 157–62; Robert O. Anderson, *Fundamentals of the American Petroleum Industry* (Norman: University of Oklahoma Press, 1984), pp. 280–81; Johnson, *Sun,* pp. 82–83; Wall, *Exxon,* pp. 300–1, 308, 132–33 (tiger). For specific oil company ads, see *Life,* July 5, July 12, July 26, 1954; July 17, July 24, 1964.

7. Kenneth T. Jackson, *Crabgrass Frontier: The Suburbanization of the United States* (New York: Oxford University Press, 1987), pp. 231–38 (Levitt), 248–49 (Eisenhower on atomic attack), 254 ("drive-in church"); Robert Fishman, *Bourgeois Utopias: The Rise and Fall of Suburbia* (New York: Basic Books, 1987), p. 182; Warren James Belasco, *Americans on the Road: From Autocamp to Motel, 1910–1945* (Cambridge: MIT Press, 1979), pp. 141, 168; James J. Flink, *The Automobile Age* (Cambridge: MIT Press, 1988), pp. 166, 162; Ristow, "Road Maps," *Surveying and Mapping* 34 (De-

cember 1964), pp. 617, 623; John B. Rae, *The American Automobile: A Brief History* (Chicago: University of Chicago Press, 1965), p. 109 (tail fins). Angus Kress Gillespie and Michael Aaron Rockland, *Looking for America on the New Jersey Turnpike* (New Brunswick: Rutgers University Press, 1989), pp. 23–37. Dwight D. Eisenhower, *White House Years,* vol. 1, *Mandate for Change, 1953–1956* (Garden City, N.Y.: Doubleday, 1963), pp. 501–2, 547–49 ("six sidewalks to the moon").

8. Rusk to American Diplomatic Posts, "Middle Sitrep as of June 7," June 8, 1967, Mideast Crisis Cable, vol. 4, June 1967, NSF Country File, Johnson Library; Nadav Safran, *Israel: The Embattled Ally* (Cambridge: Harvard University Press, 1978), pp. 240–56. When President Havai Boumedienne of Algeria complained in Moscow in 1967 of inadequate Soviet support for the Arab cause during the war, Leonid Brezhnev replied, "What is your opinion of nuclear war?" Betts, *Nuclear Blackmail and Nuclear Balance,* p. 128.

9. Moore to Bryant, June 27, 1967, with Moore to Califano, June 28, 1967, Pricing Files, January-June 1967, Ross-Robson papers, Aides Files, White House Central Files, Johnson Library ("compliance" and "crisis"); Harkabi, *Arab Attitudes,* pp. 2–9 ("liquidation").

10. Interview with Harold Saunders ("floating crap game"); *Oil and Gas Journal,* July 17, 1967, p. 43 ("bad dream"); September 11, 1967, p. 41; "Wright/Summary," June 26, 1967, Pricing Files, January-June 1967, Ross-Robson papers, Aides Files, White House Central Files, Johnson Library; Letter from James Akins to author, July 27, 1989; Kapstein, *Insecure Alliance,* pp. 130 ("Suez system"), 147 ("threat of an emergency"), 136 ("principal safety factor").

11. "World Export Picture," July 27, 1967, October 15, 1967, Pricing Files, Oil, July and August, 1967, Ross-Robson papers, Aides Files, White House Central Files, Johnson Library; Wall, *Exxon,* pp. 624–26 ("tanker fleet"); *Multinational Hearings,* part 8, p. 589 (salt mines); *Wall Street Journal,* October 27, 1967, p. 1; *Oil and Gas Journal,* August 7, pp. 96–98; September 11, p. 45; August 14, 1967, p. 8.

12. *Multinational Hearings,* part 8, p. 764 ("surplus crude"); Geoffrey Kirk, ed., *Schumacher on Energy* (London: Sphere Books, 1983), pp. 1–5, 82, 14; Barbara Wood, *E. F. Schumacher: His Life and Thought* (New York: Harper & Row, 1984), p. 344 ("chickens").

The following table shows the explosive growth in automobile usage in the U.S. and the rest of the world since World War II.

Passenger Car Registration
(millions of cars)

	U.S.	Rest of World	Total
1950	40.3	12.7	53.0
1960	61.7	36.6	98.3
1970	89.2	104.2	193.4
1980	121.6	198.8	320.4
1990*	147.9	292.8	440.7

*Estimate, Cambridge Energy Research Associates
Source: Motor Vehicle Manufacturers Assn. of the U.S., *World Motor Vehicle Data, 1990,* p. 35.

Chapter 28

1. Interview with Peter Ramsbotham; *Time,* October 25, 1971, pp. 32–33 (Pompidou); James A. Bill, *The Eagle and the Lion: The Tragedy of American-Iranian Relations* (New Haven: Yale University Press, 1988), pp. 183–85 (Shah on Maxim's).

2. Denis Healey, *The Time of My Life* (London: Michael Joseph, 1989), pp. 284 (band), 299 ("unwise"); J. B. Kelly, *Arabia, the Gulf & the West* (New York: Basic Books, 1980), pp. 47–53 ("mercenaries"), 80 (Dubai), 92 (Bahrain); Pahlavi, *Shah's Story,* p. 135 ("safety of the Persian Gulf"); *FRUS: Iran, 1951–1954,* pp. 854–57 (Nixon on the Shah); Interview with James Schlesinger.

3. Interview with Ulf Lantzke; Steven A. Schneider, *The Oil Price Revolution,* p. 110 ("old warrior"); Stobaugh and Yergin, *Energy Future,* p. 1 (1968 State Department notice); Wall, *Exxon,* p. 828 (1972 OECD meeting); Vernon, *Oil Crisis,* pp. 31, 18, 23, 28.

4. On New York utilities, interviews with Pierce, Swartz and Doyle, Multinational Subcommittee Staff Interviews and *New York Times,* November 25, 1966, p. 1; November 26, 1966, p. 1; December 18,

1966, p. 24. Donella Meadows, Dennis Meadows, Jorgen Randers and William Behrens III, *The Limits to Growth: A Report for the Club of Rome's Project on the Predicament of Mankind,* 2d ed. (New York: Signet Books, 1974), pp. 29, 85–86, 75; Johnson, *Sun,* p. 217 ("new game"). On Santa Barbara spill, see Cole to Nixon, "Santa Barbara Channel Oil Leases," November 9, 1973, White House Special Files, President's Office Files, Nixon papers; *New York Times,* February 27, 1969, p. 1; William Rintoul, *Drilling Ahead: Tapping California's Richest Oil Fields* (Santa Cruz: Valley Publishers, 1981), chap. 12.

5. Macmillan to Menzies, "Prime Minister: Personal Telegram," T267/58, PREM 11/2441/PRO ("well aware"); Interviews with Robert Belgrave, James Lee ("never get to $5"), Robert O. Anderson and Frank McFadzean; Peter Kann, "Oilmen Battle Elements to Tap Pools Beneath Alaskan Water, Land," *Wall Street Journal,* February 16, 1967, p. 1; Charles S. Jones, *From the Rio Grande to the Arctic: The Story of the Richfield Oil Corporation* (Norman: University of Oklahoma Press, 1972), chap. 45; Kenneth Harris, *The Wildcatter: A Portrait of Robert O. Anderson* (New York: Weidenfeld and Nicolson, 1987), pp. 77–93; Cabinet Task Force on Oil Import Control, *The Oil Import Question: A Report on the U.S. Relationship of Oil Imports to the National Security* (Washington, D.C.: GPO, 1970). For the environmental battle, see David R. Brower, "Who Needs the Alaska Pipeline?" *New York Times,* February 5, 1971, p. 31; Charles J. Cicchetti, "The Wrong Route," *Environment* 15 (June 1973), p. 6 ("probable major discharges"); United States Department of the Interior, *An Analysis of the Economic and Security Aspects of the Trans-Alaskan Pipeline,* vol. 1 (Washington, D.C.: GPO, 1971).

6. Interviews with Armand Hammer, Victor Hammer, James Placke and others.

7. Interviews with Deutsch (gold chess set) and William Bellano ("orderly transfer"), Multinational Subcommittee Staff Interviews; Armand Hammer with Neil Lyndon, *Hammer* (New York: Putnam, 1987), passim, esp. pp. 337 ("Hammer's Folly"), 340; Steve Weinberg, *Armand Hammer: The Untold Story* (Boston: Little, Brown, 1989), chap. 15.

8. Interview with James Placke; First, *Libya,* chap. 7, pp. 103, 265; Mohammed Heikal, *The Road to Ramadan* (London: Collins, 1975), pp. 185 ("ideas of Islam"), 70; Wall, *Exxon,* pp. 704–11 ("5000 years," "Good God!" and Jersey director); Interviews with William Bollano, Charles Lee, Northcutt Ely, Jack Miklos, Thomas Wachtell, Henry Schuler, James Akins, Mohammed Finaish ("eggs"), George Williamson ("perfectly understandable" and "Everybody who drives"), George Parkhurst and Dennis Bonney, Multinational Subcommittee Staff Interviews; Hammer, *Hammer,* p. 383 ("disciple"); *Fortune,* August 1971, p. 116; John Wright, *Libya: A Modern History* (Baltimore: Johns Hopkins University Press, 1982), p. 239; *Multinational Hearings,* vol. 7, pp. 377–78.

9. *Multinational Hearings,* part 8, pp. 771–73 ("picked off"); part 6, pp. 64 ("tricks"), 84–87 ("year 1951"), 70–71 ("must go along"). Interviews with Fadhil al-Chalabi and James Placke; Interviews with George Williamson and John Tigrett (Libyan Safety Net), James Placke ("truce"), Dudley Chapman and John McCloy, Multinational Subcommittee Staff Interviews. *Fortune,* August 1971, pp. 113, 197 ("Groucho"), 190 ("not my father"); Wall, *Exxon,* pp. 774–76 ("silly as hell"); Thomas L. McNaugher, *Arms and Oil: U.S. Military Strategy and the Persian Gulf* (Washington, D.C.: Brookings Institution, 1985), p. 12.

10. Interview with Fadhil al-Chalabi ("OPEC got muscles"); Kelly, *Arabia,* p. 357 ("no leapfrogging"); Wright, *Libya,* p. 244 ("buyer's market . . . is over"); Interviews with Henry Schuler, pp. 10–12, Joseph Palmer II, Henry Moses, George Parkhurst and Dennis Bonney, Multinational Subcommittee Staff Interviews; Multinational Subcommittee, *Multinational Hearings,* part 6, p. 221 (Jalloud).

11. Zuhayr M. Mikdashi, Sherrill Cleland and Ian Seymour, *Continuity and Change in the World Oil Industry* (Beirut: Middle East Research and Publishing Center, 1970), pp. 215–16 (Yamani on participation); Sampson, *Seven Sisters,* p. 245 ("Catholic marriage"); *Multinational Hearings,* part 6, pp. 44–45 ("concerted action"), 50 ("trend toward nationalization"); Schneider, *Oil Price Revolution,* pp. 176 ("updated book value" and "participation agreement"), 179, 182 (Exxon chairman); Interview with Ed Guinn, Multinational Subcommittee Staff Interviews (skeletons); Wall, *Exxon,* pp. 840–42 ("hard blow" and "I won").

12. Sampson, *Seven Sisters,* pp. 240–42; *Multinational Hearings,* vol. 7, pp. 332–37 (surplus capacity). Embassy in Tripoli to Washington, December 5, 1970, 02823; Embassy in Tripoli to Washington, November 23, 1970, A-220, State Department papers.

Chapter 29

1. Anwar el-Sadat, *In Search of Identity: An Autobiography* (New York: Harper & Row, 1978), pp. 248–52; Henry Kissinger, *Years of Upheaval* (Boston: Little, Brown, 1982), p. 854 ("altered irrevocably"). On quotas, see Kissinger to Nixon, November 21, 1969; Jamiesen and Warner to Nixon, November 26, 1969, White House Central Files [EX] CO 1–7; Flanigan to Staff Secretary, November 20, 1969, [CF] TA 4/Oil, White House Special Files, Confidential Files. Flanigan to Kissinger, January 23, 1970, [EX] CO 128 ("power vacuum"); Nixon to Mohammed Reza Pahlavi, April 16, 1970, [EX] CO 68 ("disappointment"), White House Central Files, Nixon archives.

2. Flanigan to Nixon, March 11, 1972, [EX] UT; Charles DiBona to John Ehrlichman and George Shultz, March 19, 1973, Darrell Trent to the President, April 4, 1973, [EX] CM 29, White House Central Files, Nixon archives. Akins's study is Department of State, "The International Oil Industry Through 1980," December 1971, in Muslim Students Following the Line of the Iman, *Documents from the U.S. Espionage Den,* vol. 57 (Tehran: Center for the Publication of the U.S. Espionage Den's Documents, [1986]), pp. 42, i, ii; James Akins interview (Ehrlichman); James Akins, "The Oil Crisis: This Time the Wolf Is Here," *Foreign Affairs,* April 1973, pp. 462–90; M. A. Adelman, "Is the Oil Shortage Real? Oil Companies as OPEC Tax Collectors," *Foreign Policy,* Winter 1972–1973, pp. 73, 102–3.

3. Interview with Herbert Goodman ("In spite"); Schneider, *The Oil Price Revolution,* pp. 195 ("near-panic buying"), 202 (Nixon), 205–6 ("either dead or dying"); Vernon, *Oil Crisis,* p. 47 (market prices); *Multinational Hearings,* part 7, p. 538 ("out of whack").

4. Sadat, *In Search of Identity,* pp. 210 ("legacy"), 237, 239 (Sadat and Faisal); Kissinger, *Years of Upheaval,* pp. 460 ("Sadat aimed"), 297–99; *New York Times,* December 21, 1977, p. A14.

5. For Faisal on Israelis, see Richard Nixon, *RN: The Memoirs of Richard Nixon* (New York: Grosset & Dunlap, 1978), p. 1012 and Heikal, *Road to Ramadan* (London: Collins, 1975), p. 79. Terzian, *OPEC,* pp. 164–65 (Faisal on oil weapon), 167 (Faisal to American press); *MEES,* September 14, 1973, pp. 3–5; June 23, 1973, p. ii (Kuwaiti oil minister); April 20, 1973; September 21, 1973, p. 11; *Multinational Hearings,* part 7, pp. 504–9 (Faisal's meetings with Aramco, Aramco's with Washington); Interview with Alfred DeCrane, Jr.

6. Raymond Garthoff, *Detente and Confrontation: American-Soviet Relations from Nixon to Reagan* (Washington, D.C.: Brookings Institution, 1985), pp. 364–66; Heikal, *Road to Ramadan,* p. 268 ("give us time"); *Multinational Hearings,* part 7, p. 542 ("new phenomenon"); *MEES,* August 3, 1973, p. 8 (Sisco); September 7, 1973, pp. iii–iv (Nixon press conference); September 21, 1973, p. 1; Interviews with William Quandt, Harold Saunders and Ulf Lantzke; Caldwell, "Petroleum Politics in Japan," pp. 182–88 (White Paper, Nakasone and Tanaka), 264 (Akins article).

7. *MEES,* September 21, 1973, p. 2 ("windfall profits"); "Mr. McCloy Comes to Washington: Highlights of John J. McCloy's Recent Oil Diplomacy," Multinational Subcommittee Staff interviews ("picked off" and "indispensable"); Kissinger, *Years of Upheaval,* pp. 465 (CIA analysis), 466 (Israeli estimate); Interview with William Colby (Watch Committee); Multinational Subcommittee, *Multinational Oil Corporations,* p. 149. A dramatic account of the crucial October 12 meeting is in chaps. 1 and 12 of Anthony Sampson's classic history of the international oil industry, *The Seven Sisters: The Great Oil Companies and the World They Shaped,* rev. ed. (London: Coronet, 1988), esp. pp. 262–64 and 32–33.

8. Interviews with William Quandt and Harold Saunders ("fall maneuvers"); Sadat, *In Search of Identity,* pp. 241–42; Kissinger, *Years of Upheaval,* pp. 482, 459–67; Safran, *Israel,* pp. 285–86, 484; Avi Shlaim, "Failures in National Intelligence Estimates: The Case of the Yom Kippur War," *World Politics* 28 (1975), pp. 352–59 ("conception"); Moshe Ma'oz, *Asad: The Sphinx of Damascus* (New York: Grove Weidenfeld, 1988), pp. 91–92.

9. Safran, *Israel,* pp. 482–90 ("Third Temple" and Meir's letters); Kissinger, *Years of Upheaval,* pp. 493–96 ("conscious"), 536 ("stakes"); *Multinational Hearings,* part 7, pp. 546–47 (Aramco letter), 217; Interviews with William Quandt, James Schlesinger, and Fadhil al-Chalabi; Schneider, *Oil Price Revolution,* pp. 225–26 (Kuwaiti oil minister); *MEES,* October 19, 1973, p. 6. For analysis of the Soviet resupply, see William Quandt, "Soviet Policy in the October Middle East War," part II, *International Affairs,* October 1977, pp. 587–603. Owing to the surprise, said General Haim Barlev, there was on the Israeli side "not a single field in which things were handled according to plan." The massive and confused improvisation further strained the supplies and matériel. Louis Williams, ed.,

Military Aspects of the Arab-Israeli Conflict (Tel Aviv: Tel Aviv University Publishing Project, 1975), pp. 264–68.

10. Interviews with William Quandt and Fadhil al-Chalabi; Kissinger, *Years of Upheaval*, pp. 526 ("luke-warm"), 534–36 (Saqqaf meeting), 854 ("political blackmail"), 552 ("All hell"); Terzian, *OPEC*, pp. 170–75 (secret resolution); Heikal, *Road to Ramadan*, pp. 267–70; Sampson, *Seven Sisters*, pp. 300–1; William Quandt, *Decade of Decisions: American Policy Toward the Arab-Israeli Conflict, 1967–1976* (Berkeley: University of California Press, 1977), p. 190; *New York Times*, October 18, 1973, p. 1; *MEES*, October 19, 1973, p. 1 ("Suffice it"); Cabinet Meeting, October 18, 1973, White House Special Files, President's Office File, President's Meetings, Nixon archives ("had to act"); Nixon, *Memoirs*, p. 933.

11. Interviews with William Colby and James Schlesinger; Nixon, *Memoirs*, p. 923 (Agnew); Kissinger, *Years of Upheaval*, pp. 501, 511 ("eerie ceremony"), 576 ("idiot"), 583 ("say it straight"), 585 ("too distraught"); Quandt, *Decade of Decisions*, pp. 194–200; *Multinational Hearings*, part 7, pp. 515–17 (Jungers); Garthoff, *Détente and Confrontation*, pp. 374–85.

Chapter 30

1. Interviews with Steven Bosworth ("Coca Cola") and James Schlesinger; *MEES*, November 2, 1973, pp. 3, 14–16 (Saddam Hussein); Vernon, *Oil Crisis*, pp. 180–81.

2. Stobaugh and Yergin, *Energy Future*, p. 27 ("bidding for our life"); Wood, *Schumacher*, pp. 352–55 ("party is over"); Interview with Ulf Lantzke. On Japan, Letter from Takahiko Ohashi, August 19, 1989; Daniel Yergin and Martin Hillenbrand, eds., *Global Insecurity: A Strategy for Energy and Economic Renewal*, (New York: Penguin, 1983), pp. 134, 174–75; Caldwell, "Petroleum Politics in Japan,": *A Strategy for Energy and Economic Renewal* (New York: Penguin, 1983), pp. 224–91. Laird to Haig, November 5, 1973, CM 29; Sawhill to Rush, June 26, 1974, UT ("measures"), White House Central Files, Nixon archives.

3. Cabinet Meeting Notes, November 6, 1973, White House Special Files, President's Office Files, President's meetings; Ash to Nixon, "Federal Role in Energy Problem," White House Special Files, President's Office Files, President's Handwriting ("I urge"); Yankelovich to Haig, December 6, 1973, with memorandum; Parker to Haig, November 23, 1973 ("heavy newsday"); Ash to Nixon, February 28, 1974 ("nothing could win"), UT, White House Central Files, Nixon archives. D. Goodwin, *Energy Policy*, pp. 447–48 ("national goal"); William E. Simon, *A Time for Truth* (New York: Berkley Books, 1978), pp. 55–66; Wall, *Exxon*, p. 883; Interviews with Steven Bosworth and Charles Di-Bona; Henry Kissinger, *Years of Upheaval*, pp. 805 ("hydra-headed"), 567, 632 ("spectacular").

4. Pierre Wack, "Scenarios: Uncharted Waters Ahead," *Harvard Business Review*, 63 (September-October 1985), pp. 72–89; "Apportionment of Oil Supplies in an Emergency Among the OECD Countries," with Knubel memo, National Security Council, November 8, 1973, [EX] MC, White House Central Files, Nixon archives ("working group"); *Multinational Hearings*, part 5, p. 187; part 7, p. 418 (Keller); part 9, pp. 190 ("only defensible course"), 33–34 ("equitable share"); Interviews with Eric Drake, Herbert Goodman ("torment") and Yoshio Karita; Federal Energy Administration and Senate Multinational Subcommittee, *U.S. Oil Companies and the Arab Oil Embargo: The International Allocation of Constricted Supply* (Washington, D.C.: GPO, 1975), p. 4; Skeet, *OPEC*, p. 106 ("impossible to know"); Vernon, *Oil Crisis*, pp. 179–88 ("Holland"); Geoffrey Chandler, "Some Current Thoughts on the Oil Industry," *Petroleum Review*, January 1973, pp. 6–12; Geoffrey Chandler in "The Changing Shape of the Oil Industry," *Petroleum Review*, June 1974.

5. Vernon, *Oil Crisis*, pp. 189–90 ("assurances"), 197; Interviews with Eric Drake and Frank McFadzean; Letters to the author from Drake, July 2, 1990, and McFadzean, August 23, 1990; Sampson, *Seven Sisters*, pp. 275–77; FEA, *International Allocation*, pp. 9–10 ("difficult to imagine").

6. Interview with James Akins; Sampson, *Seven Sisters*, p. 270 ("If you went down"); Schneider, *Oil Price Revolution*, p. 237; Interview with Shah by Robert Stobaugh ("new concept"); Skeet, *OPEC*, p. 103 ("alternative source"); Kissinger, *Years of Upheaval*, p. 888 (Nixon to Shah); Mohammed Reza Pahlavi to Richard M. Nixon, January 10, 1974, with Department of State to NSC Secretariat, CM 29, White House Central Files, Nixon archives; *MEES*, December 28, 1973, Supplement, pp. 2–5 ("noble product").

7. Sadat, *In Search of Identity*, p. 293 ("99 percent"); Interviews with Steven Bosworth and William Quandt ("60 percent of the cards"); Schneider, *Oil Price Revolution*, p. 233 ("extremely sorry" and "If you are hostile"); Kissinger, *Years of Upheaval*, pp. 897 (Pompidou), 720 (Heath), 638–44, 883

("putting pressure"); *MEES,* November 30, 1973, p. 13; November 11, 1973 ("kiss blown from afar"); Robert J. Lieber, *Oil and the Middle East War: Europe in the Energy Crisis* (Cambridge: Harvard Center for International Affairs, 1976), p. 15; Michael M. Yoshitsu, *Caught in the Middle East: Japan's Diplomacy in Transition* (Washington, D.C.: Heath, 1984), pp. 1–3 ("always buy" and "oil on the brain"); Caldwell, "Petroleum Politics in Japan," pp. 206–7 ("direct request"), 211 ("neutrality"), 217.

8. *MEES,* January 18, 1974; Frank McFadzean, *The Practice of Moral Sentiment,* (London: Shell, n.d.), p. 30 ("spectacle"); Interviews with Ulf Lantzke and Yoshio Karita; Helmut Schmidt, *Men and Power: A Political Perspective* (New York: Random House, 1989), pp. 161–64; Robert J. Lieber, *The Oil Decade: Conflict and Cooperation in the West* (New York: Praeger, 1983), p. 19 (Jobert).

9. Interviews with William Quandt and Harold Saunders; *MEES,* January 4, 1974, p. 11 ("increasingly less appropriate"); *MEES,* March 22, 1974, pp. 4–5 ("constructive effort"); *MEES,* November 30, 1973, p. 11 ("Wailing Wall"); Kissinger, *Years of Upheaval,* pp. 663–64 (Kissinger and Faisal), 659; Quandt, *Decade of Decision,* pp. 231, 245.

Chapter 31

1. Ali M. Jaidah, "Oil Pricing: A Role in Search of an Actor," *PIW,* Special Supplement, September 12, 1988, p. 2 ("Golden Age"); *Business Week,* May 26, 1975, p. 49 (Datsun); Interview with Chief M. O. Feyide.

2. Howard Page, "OPEC Is Not in Control," 1975, Wanda Jablonski papers. Raymond Vernon describes the period 1973 through 1978 as one of "a somewhat unruly oligopoly, composed of a dominant member (Saudi Arabia), a dozen followers barely prepared to acknowledge its leadership, and a large outer circle of producers pricing under the shelter of the oligopoly. It was clear at this stage that the majors had lost control of prices, but not at all clear what organizing force had taken its place." Raymond Vernon, *Two Hungry Giants: The United States and Japan in the Quest for Oil and Ores* (Cambridge: Harvard University Press, 1983), p. 29. Tax shares calculated from OPEC, *Petroleum Product Prices and Their Components in Selected Countries: Statistical Time Series, 1960–1983* (Vienna: OPEC, [1984]). Shawcross, *Shah's Last Ride,* pp. 166–82 ("speed," "*serious*" and Yamani on Shah); Helms to Secretary of State, September 10, 1974, Tehran 07611 ("day has passed"); Yamani-Ingersoll Meeting Transcript, October 1974, State Department Papers. *MEES,* September 5, 1975, p. 49 ("toy").

3. Jeffrey Robinson, *Yamani: The Inside Story* (London: Simon and Schuster, 1988), pp. 41 (Yamani on his father), 153, 204 ("long term"); *New York Times,* October 8, 1972, section 3, p. 7 ("sweet reasonableness"); Oriana Fallaci, "A Sheikh Who Hates to Gamble," *New York Times Magazine,* September 14, 1975, p. 40 ("can't bear gambling"); Interviews ("consummate strategist" and "ostentatiously calm"); Kissinger, *Years of Upheaval,* pp. 876–77 ("technician"); *Time,* January 6, 1975, pp. 9, 27; Pierre Terzian, *OPEC,* chap. 11; *MEES,* April 25, 1977 ("economic disaster"); May 1, 1978; January 10, 1977, p. 10 ("devil"); December 27, 1976, p. iii ("stooge" and "in the service"). On Prince Fahd's meeting with Carter, see William B. Quandt, *Camp David: Peacemaking and Politics* (Washington, D.C.: Brookings Institution, 1986), p. 68; "Secretary's Lunch for Prince Fahd," May 24, 1977, Vance to Crown Prince Fahd, June 18, 1977, State Department Papers. On division of responsibility in Saudi oil policymaking, see Cyrus Vance to the President, Memorandum, "Saudi Arabian Oil Policy," October 1977, Dhahran to Secretary of State, February 3, 1977, Dhahran 00149, State Department Papers.

4. *Business Week,* January 13, 1975, p. 67 ("only chance"); Cyrus Vance, *Hard Choices: Critical Years in America's Foreign Policy* (New York: Simon and Schuster, 1983), pp. 316–20. A review of 1,021 State Department cables and papers obtained under the Freedom of Information Act shows a consistent opposition of the United States government to higher oil prices from 1974 onward and right across the Nixon, Ford, and Carter Administrations. For instance, the outgoing (Ford Administration) Under Secretary of State for Economic Affairs, William D. Rodgers, wrote a long, private letter to his successor (Carter Administration), Richard N. Cooper, outlining the main international economic issues and policies for the United States. "Our oil diplomacy," Rodgers noted, "concentrate[s] on what we can do to head off an oil price rise." Rodgers to Cooper, January 11, 1977, State Department papers. Indeed, in the last days of the Ford Administration, Kissinger met with the Saudi Arabian ambassador to explain that he was "conscience-bound" to argue against price increases on behalf of the incoming Carter Administration! Secretary Kissinger's meeting with Saudi Ambassador Alireza on

OPEC Price Decision, November 9, 1977, State Department papers. Also see Kissinger to Ford, August 27, 1974; Kissinger Meeting with Senators and Congressmen, June 10, 1975; President Ford to King Khalid, December 31, 1976, State 314138, State Department papers. On Soviet deal, interviews with Herbert Goodman and in Moscow; Hormats to Scowcroft, November 14, 1975, TA 4/29, 10/1/75–12/11/75 file, White House Central Files; Russell to Greenspan, October 29, 1975, "Russell (6)" file; Russell to Greenspan, November 4, 1975, "Russell (7)" file, Box 141, CEA papers, Ford Library.

5. United States Department of State, Briefing Paper on Iran, January 3, 1977, in Muslim Students Following the Line of the Iman, *U.S. Interventions in Iran (1)*, vol. 8 of *Documents from the U.S. Espionage Den* (Tehran: Center for the Publication of the U.S. Espionage Den's Documents, [1986]), p. 129; Barry Rubin, *Paved with Good Intentions: The American Experience and Iran* (New York: Penguin, 1984), pp. 140 ("wave a finger"), 172; Interviews with Harold Saunders ("Big Pillar"), James Schlesinger, Steven Bosworth ("pussy cats") and James Akins. On nut: *New York Times,* July 16, 1974, p. 4, July 18, 1974, p. 57, and letter from Jack C. Miklos to author, Sept. 4, 1990. Minister of Court Asadollah Alam's retort was, "Simon may be a good bond salesman, but he does not know a whole lot about oil." Anthony Parsons, *The Pride and the Fall: Iran, 1974–1979* (London: Jonathan Cape, 1984), p. 47 ("calculating opportunism"); Gary Sick, *All Fall Down: America's Tragic Encounter with Iran* (New York: Penguin, 1987), pp. 16, 26 ("no visible"), 32–33; Robert Graham, *Iran: The Illusion of Power* (New York: St. Martin's, 1979), p. 20 ("acquired money"). Richard Cooper to the Secretary, August 12, 1978 ("price freeze offensive"); Blumenthal to the President, October 28, 1977, Dhahran 01261; Cyrus Vance to the President, November 4, 1977 ("price hawk"), State Department papers. Hamilton Jordan, *Crisis: The Last Year of the Carter Presidency* (New York: Putnam, 1982), pp. 88–89; Vance, *Hard Choices,* pp. 321–22 ("punishing impact" and "break"). Real prices derived from International Monetary Fund, *International Financial Statistics Yearbook, 1988,* p. 187.

6. *PIW,* April 14, 1975, p. 10 ("good bye"); *MEES,* March 7, p. 2; July 18, ("Oil is everything"). On oil companies' meetings in Kuwait, "Kuwait: Summary of Situation as of March 15, 1975," March 17, 1975; "Meetings at Ministry of Oil, March 12 and March 15, 1975," March 17, 1975, pp. 1, 3, 4, 8; "Meeting with the Prime Minister, March 29, 1975," April 2, 1975, Goodman papers; Interview with Herbert Goodman.

7. Interviews with Frank Alcock, Alberto Quiros, and Robert Dolph; Gustavo Coronel, *The Nationalization of the Venezuelan Oil Industry: From Technocratic Success to Political Failure* (Lexington, Mass.: Lexington Books, 1983), pp. 66–71 ("feverish debate"); Rabe, *Road to OPEC,* p. 190 ("act of faith").

8. On Saudi Arabia's purchase of Aramco, Schneider, *The Oil Price Revolution,* pp. 407–8; Aramco Annual Reports; Interviews. On direct sales, Vernon, *Two Hungry Giants,* p. 32 and *PIW,* February 25, 1980, p. 3.

Chapter 32

1. *Business Week,* January 13, 1975, p. 67 ("conditions"). On Japan, interviews with Naohiro Amaya and Yoshio Karita; Letter from Tadahiko Ohashi, August 14, 1989; Samuels, *Business of the Japanese State,* chaps. 5–6. On French policy and advertising, interviews with Jean Blancard, Jean Syrota and Charles Mateudi.

2. Interview with Henry Jackson ("screwed on"); Carol J. Loomis, "How to Think About Oil Company Profits," *Fortune,* April 1974, p. 99; Karalogues to Nixon, December 19, 1973, White House Special Files, President's Office Files, Nixon archives ("Scoops the hell"). On Jackson Committee Hearings, see United States Congress, Senate, Committee on Government Operations, Permanent Subcommittee on Investigations, 93d Congress, 1st Session, *Current Energy Shortages, Oversight Series: Conflicting Information on Fuel Shortages* (Washington, D.C.: GPO, 1974), pp. 113–14, 154, 399, 400, 472–73 and *New York Times,* January 22–25, 1974. Yergin and Hillenbrand, *Global Insecurity,* pp. 119–20; "The Eighties: An Update," Company Document, January 1976, p. 22 ("less certain"); Geoffrey Chandler, "The Innocence of Oil Companies," *Foreign Policy,* Summer 1977, p. 67 ("threat"); Chase Manhattan Bank, *Annual Financial Analysis of a Group of Petroleum Companies, 1970–1979.* On inflation, see Energy Information Administration, *Annual Energy Review, 1988* (Washington, D.C.: GPO, 1989).

3. Pietro S. Nivola, *The Politics of Energy Conservation* (Washington, D.C.: Brookings Institution, 1986); Vietor, *Energy Policy,* pp. 253 ("every problem"), 256 ("refining junk"), 238 (Federal Regis-

ter), 258; Cole to the President, Decision on Signing of Alaska Pipeline Legislation, November 13, 1973, White House Special Files, President's Office Files, Nixon archives; Interview with Robert O. Anderson.

4. Goodwin, *Energy Policy,* pp. 554–55 ("zeal"); Interviews with Stuart Eizenstat and James Schlesinger; Stuart E. Eizenstat, "The 1977 Energy Plan: M.E.O.W.," Case note for the Kennedy School of Government, Harvard University; Jimmy Carter, *Keeping Faith: Memoirs of a President* (London: Collins, 1982), pp. 92–106 ("deeply resented" and "most difficult question"); James Schlesinger, "The Energy Dilemma," *Oak Ridge National Laboratory Review,* Summer 1972, p. 13; Stobaugh and Yergin, *Energy Future,* p. 70 ("Hell"); *MEES,* December 11, 1978, p. i ("water torture").

5. *Business Week,* February 3, 1975, p. 38 ("just wild"); Interview with Robert Dolph ("rabbits"); E. C. G. Werner, "Presentation to the Frankfurt Financial Community," Nov. 25, 1976, p. 3.

6. George W. Grayson, *The Politics of Mexican Oil* (Pittsburgh: University of Pittsburgh Press, 1980), pp. 58, 77 ("digestion"); "Why the Bankers Love Mexico," *Fortune,* July 16, 1979, pp. 138, 142.

7. Anthony Benn, *Against the Tide: Diaries, 1972–1976* (London: Hutchinson, 1989), p. 403 ("cross-section"); Interviews with Harold Wilson and Thomas Balogh; Stig S. Kvendseth, *Giant Discovery: A History of Ekofisk Through the First 20 Years* (Tanager, Norway: Phillips Petroleum Norway, 1988), pp. 9–31; Daniel Yergin, "Britain Drills and Prays," *New York Times Magazine,* November 2, 1975, pp. 13, 59.

8. On oil price forecasting, see Arthur Andersen & Co. and Cambridge Energy Research Associates, *The Future of Oil Prices: The Perils of Prophecy* (Houston: 1984). On the extent of the consensus in 1978, see "Threatening Scramble for Oil," *Petroleum Economist,* May 1978, pp. 178–79; Stobaugh and Yergin, *Energy Future,* pp. 351–52, n. 34; Francisco Parra, "World Energy Supplies and the Search for Oil," *MEES,* Supplement, April 12, 1978, pp. 1–6. "By general consensus," Parra commented, "the next energy crisis is programmed for the 1980s, when a shortage of oil will occur that threatens further economic growth because alternative supplies of energy will not be available in the required quantities." *MEES,* June 26, 1978, p. iv ("our own studies"); *Public Papers of the Presidents of the United States: Jimmy Carter, 1977,* book 2 (Washington, D.C.: GPO, 1978), pp. 2220–21 ("island of stability"); Interview ("big trouble").

Chapter 33

1. Parsons, *Pride and Fall,* pp. 10, 8, 50, 54–55; Graham, *Iran,* p. 19; *New York Times,* June 5, 1989, p. A11; Rubin, *Paved with Good Intentions,* p. 176 ("A-list").

2. Bill, *Eagle and Lion,* pp. 235, 51; Sick, *All Fall Down,* p. 40 ("40–40"). Sick is a significant source for the Iranian revolution and American policy. Parsons, *Pride and Fall,* pp. 62–64 ("no compromise"), 71 ("I was worried"); United States Congress, House of Representatives, Permanent Select Committee on Intelligence, Subcommittee on Evaluation, *Iran: Evaluation of U.S. Intelligence Performance Prior to November 1978, Staff Report* (Washington, D.C.: GPO, 1979), pp. 2, 6–7 (intelligence); Shawcross, *Shah's Last Ride,* chap. 14 (Shah's ill health); Interview with Robert Bowie.

3. IEA archives; Sick, *All Fall Down,* pp. 57 ("public opinion"), 123–25 (Soviet plot), 132; Parsons, *Pride and Fall,* pp. 85 ("snow"); Interviews with Jeremy Gilbert ("The Fields"), James Schlesinger and Harold Saunders ("first systematic meeting"); Richard Falk, "Trusting Khomeini," *New York Times,* February 16, 1979, p. A27 ("entourage"); *New York Times,* February 8, 1979, p. A13; February 9, 1979, p. A17 ("saint"); William H. Sullivan, *Mission to Iran* (New York: Norton, 1981), pp. 200–3 ("Thinking the Unthinkable"), 225 ("no policy").

4. Mohamed Heikal, *Iran, The Untold Story: An Insider's Account of America's Iranian Adventure and Its Consequences for the Future* (New York: Pantheon, 1982), pp. 145–46; Sick, *All Fall Down,* pp. 123 ("torrents of blood"), 108 (prank), 182–83 ("Khomeini wins"); Parsons, *Pride and Fall,* pp. 114 ("dictator"), 124–26 ("I would leave"); Interview with Jeremy Gilbert and Jeremy Gilbert to author, Nov. 15, 1989. On the American tanker, Robert E. Huyser, *Mission to Tehran* (New York: Harper & Row, 1986), pp. 96–247. Shawcross, *Shah's Last Ride,* p. 35 ("feeling tired"); Paul Lewis, "On Khomeini's Flight," *New York Times,* Feb. 2, 1979, p. A7.

5. Interview with Jeremy Gilbert.

6. IEA archives; Daniel Badger and Robert Belgrave, *Oil Supply and Price: What Went Right in 1980?* (London: Policy Studies Institute, 1982), pp. 106–7 (motorists); M. S. Robinson, "The Crude Oil Price Spiral of 1978–80," February 1982, pp. 1–2. Katz to Cooper, "U.S. Oil Strategy Toward Saudi

Arabia," January 12, 1979; Richard Cooper to John West, January 15, 1979, State 011064; Vance to Embassy, Saudi Arabia, January 26, 1979; Cooper to the Secretary, February 8, 1979, 7902573; West to Vance, "Oil Matters: Meeting with Crown Prince Fahd," February 15, 1979, State Department papers. *PIW*, March 19, 1979, pp. 1–2 ("not to count on Exxon"); Interview with Clifton Garvin.

7. Interviews with Ulf Lantzke, J. Wallace Hopkins and others; Muslim Students Following the Line of the Iman, *Documents from U.S. Espionage Den,* vol. 40, *U.S. Interventions in the Islamic Countries: Kuwait (2)* (Tehran: Center for the Publication of the U.S. Espionage Den's Documents, [1986]), p. 58 ("fool"); CIA, outgoing message, April 4, 1979, *DDRS,* 1988, doc. 1300.

8. OPEC, "Communique: 53rd Extraordinary Meeting," March 27, 1977; Stobaugh and Yergin, *Energy Future,* 2d ed., pp. 342 ("free-for-all"), 346 ("short-run politics"); Interview with M. S. Robinson ("Nobody controlled"). On the Saudis and "the whole question of production," see Riyadh to Secretary of State, March 25, 1979, Riyadh 00484; Jidda to Secretary of State, April 17, 1979, Jidda 03094; Yamani edict in Daniels to Secretary of State, May 23, 1979, Jidda 03960, State Department papers. IEA archives; *PIW,* May 14, 1979, pp. 1, 9; United States Department of Justice, Antitrust Division, *Report of the Justice Department to the President Concerning the Gasoline Shortage of 1979* (Washington, D.C.: GPO, 1980), pp. 153–65; Interviews with Richard Cooper and Clifton Garvin.

9. Interviews with Stuart Eizenstat, James Schlesinger and Eugene Zuckert; Eliot Cutler to Jim McIntyre and Stuart Eizenstat, "Synthetics and Energy Supply," June 12, 1979; Benjamin Brown and Daniel Yergin, "Synfuels 1979," draft case, Kennedy School, 1981, pp. 15 ("darts and arrows"), 46 (Eizenstat memo); Richard Cooper to John West, June 8, 1979, State 147000, State Department papers; Carter, *Keeping Faith,* pp. 111–13 ("one of the worst days"); *New York Times,* June 27, 1979, p. A1 (Harvard Business School); July 12, 1979, p. A1; July 19, 1979, p. A14; July 20, 1979, p. A1, July 21, 1979, p. A1. The *Washington Post's* national editor was Lawrence Stern.

10. M. S. Robinson, "Crude Oil Price Spiral," pp. 10, 12 ("cat-and-mouse"); Skeet, *OPEC,* p. 159 ("If BNOC"); Interviews with Ulf Lantzke, James Schlesinger, and industry executive; Shell Briefing Service, "Trading Oil," 1984; *PIW,* August 27, 1979, p. 1, Special Supplement (Schlesinger).

Chapter 34

1. Tim Wells, *444 Days: The Hostages Remember* (San Diego: Harcourt Brace Jovanovich, 1985), pp. 67–69; Warren Christopher et al., *American Hostages in Iran: The Conduct of a Crisis* (New Haven: Yale University Press, 1985), pp. 35–41, 57 (Elizabeth Ann Swift), 58–60, 112 (Carter Doctrine); Terence Smith, "Why Carter Admitted the Shah," *New York Times Magazine,* May 17, 1981, pp. 36, 37ff.; On the Algiers meeting, see Zbigniew Brzezinski, *Power and Principle: Memoirs of the National Security Adviser, 1977–1981* (New York: Farrar Straus Giroux, 1985), pp. 475–76. John Kifner, "How a Sit-in Turned into a Siege," *New York Times Magazine,* May 17, 1981, pp. 58, 63 ("Nest of spies"); Sick, *All Fall Down,* pp. 239 ("rotten brains"), 248 ("by the balls"); Steven R. Weisman, "For America, A Painful Reawakening," *New York Times Magazine,* May 17, 1981, pp. 114ff.; Shawcross, *Shah's Last Ride,* pp. 242–52.

2. IEA archives. Mansfield to Secretary of State, December 14, 1979, Tokyo 21956; Mansfield to Secretary of State, January 4, 1980, Tokyo 00125; Vance to Tokyo Embassy, February 5, 1980, State 031032, State Department papers. *MEES,* October 22, p. 6 ("losing control"); December 31, 1979; *New York Times,* December 21, 1979, p. D3 ("catastrophe"); December 20, 1979, p. D5 ("glut"); Terzian, *OPEC,* p. 275 ("almighty God").

3. *PIW,* Supplement, pp. 1, 4 ("cardinal issue"); Walter Levy, "Oil and the Decline of the West," *Foreign Affairs,* Summer 1980, pp. 999–1015; Interviews with Rene Ortiz and others.

4. Joan Oates, *Babylon* (London: Thames and Hudson, 1979), pp. 51–52 (poem); Georges Roux, *Ancient Iraq* (London: Penguin, 1985), p. 168; Ilya Gershevitch, ed., *The Cambridge History of Iran,* vol. 2, *The Medean and Achaemenian Periods* (Cambridge: Cambridge University Press, 1985), pp. 1–25.

5. Phebe Marr, *The Modern History of Iraq* (Boulder, Colo.: Westview Press, 1985), pp. 217–20 (*shaqawah*), 228; Christine Moss Helms, *Iraq: Eastern Flank of the Arab World* (Washington, D.C.: Brookings Institution, 1984), pp. 147–60 ("infidel Ba'th Party"), 165 ("every street corner"); Anthony H. Cordesman, "Lessons of the Iran-Iraq War: The First Round," *Armed Forces Journal International,* 119 (April 1982), p. 34 ("dwarf Pharaoh"); R. K. Ramazani, *Revolutionary Iran: Challenge and Response in the Middle East* (Baltimore: Johns Hopkins University Press, 1986), p. 60 ("Kho-

meini the rotten"); Bakhash, *Reign of the Ayatollahs,* p. 126; Interview with Rene Ortiz; R. M. Grye, ed., *The Cambridge History of Iran,* vol. 4, *The Period from the Arab Invasion to the Saljuqs* (Cambridge: Cambridge University Press, 1975), pp. 9–25 ("Victory of Victories"); David Lamb, *The Arabs: Journeys Beyond the Mirage* (New York: Vintage, 1988), pp. 287–91 (coffins, "purest joy" and minefields); Samir al-Khalil, *Republic of Fear: The Politics of Modern Iraq* (Berkeley and Los Angeles: University of California Press, 1989).

6. IEA archives; M. S. Robinson, "The Great Bear Market in Oil 1980–1983" (Nyborg: Shell, 1983). Ryan to Secretary of State, October 6, 1980, Paris 31399; Sherman to Secretary of State, October 7, 1980, Tokyo 17911; Salzman to Secretary of State, October 22, 1980, Paris 33213; Muskie to Embassies, October 24, 1980, State 283948, State Department papers.

7. *PIW,* November 17, 1980 ("still someone else"); November 24, 1980, p. 2 ("deep trouble"); April 17, 1981, Supplement, p. 1 ("stabilize the price"). Mansfield to Secretary of State, December 23, 1980, Tokyo 22437 (MITI official on "undesirable purchases"); Vance to Tokyo Embassy, October 11, 1980, State 277058, State Department papers. Interviews with Ulf Lantzke, J. Wallace Hopkins, William Martin (D'Avignon) and Alfred DeCrane, Jr.; Schneider, *Oil Price Revolution,* p. 453.

Chapter 35

1. Interview with Clifton Garvin; *New York Times,* May 3, 1982, p. Al; October 10, 1982, p. A33; Andrew Gulliford, *Boomtown Blues: Colorado Oil Shale, 1885–1985* (Niwot, Colo.: University Press of Colorado, 1989), chaps. 4–6. On autos, see Marc Ross, "U.S. Private Vehicles and Petroleum Use," Cambridge Energy Research Associates Report, October 1988.

2. *PIW:* Alirio Parra, "OPEC Move May Lead to 'Structured' Market," Special Supplement, April 12, 1982; "Spot Products Nosedive Spreads Everywhere," Special Supplement, February 22, 1982; Herbert Lewinsky, "Oil Seen Becoming Even More International," Special Supplement, July 12, 1982, p. 3 (Mobil Executive); Robert Mabro, "OPEC's Future Pricing Role May Be at Stake," Special Supplement, April 19, 1982; December 3, 1982, p. 1; June 4, 1982, pp. 1–3 ("punishment"). Skeet, *OPEC,* p. 178 (rejection of embargo proposal).

3. Interviews with Yahaya Dikko and Alberto Quiros; *PIW,* February 14, 1983 (Yamani on pregnancy); March 21, 1983 ("swing producer"); Terzian, *OPEC,* pp. 313–19.

4. *PIW,* April 11, 1983, pp. 8–9 ("strategic commodity"); John G. Buchanan, "How Trading Is Reshaping the Industry," in Yergin and Kates-Garnick, *Reshaping of the Oil Industry,* pp. 41–44 ("light on security," "nimble" and "opportunistic"); Interviews with P. I. Walters, George Keller and M. S. Robinson; Chevron, *Annual Report,* 1983, "Presentation on Downstream Oil Supply Policy," December 1983.

5. See New York Mercantile Exchange, *A History of Commerce at the New York Mercantile Exchange: The Evolution of an International Marketplace, 1872–1988* (New York: New York Mercantile Exchange, 1988).

6. A. G. Mojtabai, *Blessed Assurance: At Home with the Bomb in Amarillo, Texas* (Albuquerque: University of New Mexico Press, 1986), pp. 47, 199; T. Boone Pickens, *Boone* (Boston: Houghton Mifflin, 1987), passim, and pp. 11, 31 ("mouth shut"), 34; Interviews with T. Boone Pickens and Taylor Yoakam ("Saturday morning"); Adam Smith, *The Roaring '80s* (New York: Summit Books, 1988), pp. 193–95; T. Boone Pickens, "The Restructuring of the Domestic Oil and Gas Industry," in Yergin and Kates-Garnick, *Reshaping of the Oil Industry,* pp. 60–61.

7. Interviews with Jesus Silva Herzog and Patrick Connolly ("eating our lunch"); Fausto Alzati, "Oil and Debt: Mexico's Double Challenge," Cambridge Energy Research Associates Report, June 1987; Philip L. Zweig, *Belly Up: The Collapse of the Penn Square Bank* (New York: Fawcett Columbine, 1986), pp. 198–99 (Gucci loafers); William Greider, *Secrets of the Temple: How the Federal Reserve Runs the Country* (New York: Touchstone, 1989), pp. 518–25 ("bank to beat"), 628–31; Mark Singer, *Funny Money* (New York: Knopf, 1985).

8. *Wall Street Journal,* September 15, 1983, p. 1; December 5, 1983, p. 60; April 19, 1984, p. 1; Interviews with Richard Bray and P. I. Walters.

9. Thomas Petzinger, Jr., *Oil & Honor: The Texaco-Pennzoil Wars* (New York: Putnam, 1987); Steve Coll, *The Taking of Getty Oil* (London: Unwin Hyman, 1988); Lenzer, *Getty,* pp. 331–38; Miller, *House of Getty,* pp. 331–46.

10. Interviews with James Lee, George Keller, Robert O. Anderson, Philippe Michelon and M. S. Robinson; Pickens, *Boone,* pp. 182–83 ("need a touchdown"), 216; *Wall Street Journal,* March 7, 1984,

p. 1; John J. McCloy, Nathan W. Pearson and Beverley Matthews, *The Great Oil Spill: The Inside Report—Gulf Oil's Bribery and Political Chicanery* (New York: Chelsea House, 1976).

11. *Time,* June 3, 1985, p. 58 (Armand Hammer); Interviews with Robert O. Anderson and Clifton Garvin; *Time,* March 17, 1985, p. 46 and *Business Week,* May 6, 1985, p. 82.

12. On the Soviet gas pipeline, Interview with William F. Martin; Angela Stent, *Soviet Energy and Western Europe,* Washington paper 90 (New York: Praeger, 1982); Bruce Jentleson, *Pipeline Politics: The Complex Political Economy of East-West Trade* (Ithaca: Cornell University Press, 1986), chap. 6; Anthony Blinken, *Ally Versus Ally: America, Europe, and the Siberian Pipeline Crisis* (New York: Praeger, 1987).

Chapter 36

1. Richard Reid, "Standing the Test of Time," Speech at University of Surrey, March 23, 1984 ("chief variable"); *PIW,* March 18, 1985, p. 8 ("very painful"); Arthur Andersen & Company and Cambridge Energy Research Associates, *Future of Oil Prices,* p. iii; Joseph Stanislaw and Daniel Yergin, "OPEC's Deepening Dilemma: The World Oil Market Through 1987," Cambridge Energy Research Associates Report, October 1984; I. C. Bupp, Joseph Stanislaw and Daniel Yergin, "How Low Can It Go? The Dynamics of Oil Prices," Cambridge Energy Research Associates Report, May 1985; *MEES,* June 2, 1985, p. A6 ("draw a line").

2. "OPEC Ministers, Taif, June 2–3, 1985" (King's letter); Skeet, *OPEC,* p. 195; Interviews with Alfred DeCrane, Jr. and George Keller; *PIW,* December 16, 1985, p. 8 (communiqué).

3. *PIW,* September 29, 1986; August 11, 1986 (Iraqi official); Interview with Alfred DeCrane, Jr.; Arie de Geus, "Planning as Learning," *Harvard Business Review* 66 (March-April, 1988), pp. 70–74; *Washington Post,* April 4, 1986, p. 3 (Billy Jack Mason).

4. *New York Times,* January 13, 1989, p. D16 ("They got a President"); February 21, 1980, p. B10 (Reagan on Alaska); George Bush, with Victor Gold, *Looking Forward: An Autobiography* (New York: Bantam, 1988), pp. 46, 55 (partner), 64–66, 72 ("rubbed both ways"), 78; Seymour Freedgood, "Life in Midland," *Fortune,* April 1962; Bush to Kennedy, November 12, 1969, White House Special Files, Confidential Files, Nixon archives; Fadhil J. al-Chalabi, "The World Oil Price Collapse of 1986: Causes and Implications for the Future of OPEC," Energy Paper no. 15, International Energy Program, School of Advanced International Studies, Johns Hopkins University, p. 6 ("Absolute competition").

5. *New York Times,* April 2, 1986, pp. Al, D5 ("selling very hard" and "Our answer"); April 3, 1986, p. D6 ("way to address"); April 7, 1986, pp. Al, D12 ("stability" and "bum rap"); *Washington Post,* April 10, 1986, p. A 26 ("I'm correct"); April 9, 1986 ("Poor George" and "couldn't care less"); April 8, 1986 (editorial); *Wall Street Journal,* April 7, 1986, p. 3 ("national security interest"); Interviews with Richard Murphy, Walter Cutler and Frederick Khedouri.

6. Interviews with Alirio Parra and Robert Mabro; Ise, *United States Oil Policy,* pp. 123, 109, 113; "Meeting of Group of Five Oil Ministers," May 24–25, 1986 (Taif meeting); *PIW,* September 22, 1986, p. 3 ("reasonable prices"); July 28, 1986, p. 4; Briefing to Press Editors, Brioni, July 1, 1986 ("Not on your life"); "The Impact of the U.S. $17–19/Barrel Price Range on OPEC Oil," July 24, 1986 (OPEC paper); Discussions in Moscow, May 1986 ("bananas").

7. Ahmed Zaki Yamani, "Oil Markets: Past, Present, and Future," Energy and Environmental Policy Castes, Kennedy School, Harvard University, September 1986, pp. 3, 5, 11, 20; *MEES,* May 25, 1987, p. A2: Interviews.

8. Interview with Richard Murphy; Thomas McNaugher, "Walking Tightropes in the Gulf," in Efraim Karsh, ed., *The Iran-Iraq war: Impact and Implications* (London: Macmillan, 1989), pp. 171–99; Anthony M. Cordesman, *The Gulf and the West: Strategic Relations and Military Realities* (Boulder, Col.: Westview Press, 1988), chaps. 10–11; *New York Times,* July 21, 1988, p. Al ("poison"); *MEES,* May 23, 1988, p. A3; *MEES,* May 30, 1988 p. C1; *MEES,* July 25, 1988, p. C1 ("God willing"): *MEES* August 22, 1988, p. Al; *MEES,* August 29, 1988. pp. A3, C1.

Chapter 37

1. Interview of Saddam Hussein by Diane Sawyer, *Foreign Broadcast Information Services,* July 2, 1990, p. 8.

2. Karea Elliott House, "President Sees New Mideast War Unless America Acts," *Wall Street Journal,*

June 28, 1990, p. A10 ("Weakness"); Marr. *Modern History of Iraq,* chap. 8; Samir al-Khall, *Republic of Fear: The Politics of Modern Iraq* (Berkeley and Los Angeles: University of California Press, 1989).

3. H. R. P. Diction, *Kuwait and Her Neighbors* (Kuwait borders); Thomes B. Allon, F. Clinton Berry, and Norman Polmar, *CNN: War in the Gulf* (Atlanta: Turner Publishing Co., 1991 (Bush on aggression): *New York Times,* August 16, 1990, p. A14 (Bush on freedom).

4. Michael L. Sifry and Christopher Cerf, *The Gulf War Reader: History, Documents, Opinions* (Times Books, New York: 1991), p. 229 ("millitary option"), p. 125 ("10,000 deaths"), p. 173 ("tragic miscalculation").

5. Schwarzkopf quoted in Allen, Berry, and Polmar, *CNN: War in the Gulf,* p. 211.

Epilogue

1. World Bank, *The East Asian Economic Miracle: Economic Growth and Public Policy.* New York: Oxford University Press, 1993.

2. Brent Scowcroft, "Don't Attack Saddam." *Wall Street Journal,* August 15, 2002.

3. Michael R. Gordon and Bernard Trainor, *Cobra II: The Inside Story of the Invasion and Occupation of Iraq.* New York: Pantheon, 2006, Chapters 8–9, epilogue ("lightning victory," p. 506).

4. Daniel Yergin and Joseph Stanislaw, *The Commanding Heights: The Battle for the World Economy.* New York: Touchstone, 2002.

5. *New York Times,* October 30, 2005.

6. CERA Special Report, *Capital Costs Analysis Forum—Upstream: Market Review, 2008.*

7. On the dollar, see Stephen P. A. Brown, Raghav Virmani, and Richard Alm, *Economic Letter—Insights from the Federal Reserve Bank of Dallas,* May 2008, p. 6.

8. J. S. Herold, Financial and Operational Data Base.

9. www.cnn.com/2008/POLITICS/10/07/video.transcript/index.html.

Bibliography

Interviews

Many people generously made themselves available for interviews, which were essential to the writing of this book. I would like to express my great appreciation to all of them for their graciousness and consideration. None of them are responsible for the interpretation and judgments in this book. Most of the interviews were expressly conducted for this book. A few of them were originally for projects that preceded the book.

In some cases, the identification of interviewees may not be the one most familiar to readers. However, in the interests of clarity, I have generally indicated the position that seems most apposite.

Frank Alcock, Vice-President, Petróleos de Venezuela, SA
Robert O. Anderson, Chairman, ARCO
Alicia Castillo de Pérez Alfonzo
James Akins, U.S. Ambassador to Saudi Arabia
Naohiro Amaya, Vice-Minister, Ministry of International Trade and Industry, Japan
Nikolai Baibakov, Minister of Oil, Chairman, Gosplan, USSR
Lord Balogh, Minister of State, United Kingdom Department of Energy
Robert Belgrave, Oil and Middle East Desk, U.K. Foreign Office; Policy Adviser to the Board of British Petroleum
André Bénard, Managing Director, Royal Dutch/Shell
Jean Blancard, General Delegate for Energy, French Ministry of Industry; President, Gaz de France
Steven Bosworth, U.S. Deputy Assistant Secretary of State for Energy, Resources, and Food Policy
Robert R. Bowie, Director, Policy Planning Staff, U.S. Department of State; Deputy Director, U.S. Central Intelligence Agency
Richard Bray, President, Standard Oil Production Company
Juan Pablo Pérez Castillo
Oscar Pérez Castillo
Fadhil al-Chalabi, Deputy Secretary General, OPEC
William Colby, Director, U.S. Central Intelligence Agency
Marcello Colitti, Vice-Chairman, ENI
Patrick Connolly, Vice-President and Head of Energy Group, Bank of Boston
Richard Cooper, U.S. Undersecretary of State
Walter Cutler, U.S. Ambassador to Saudi Arabia
Alfred DeCrane, Jr., Chairman, Texaco

Charles DiBona, Deputy Director, White House Energy Policy Office; President, American Petroleum Institute
Yahaya Dikko, Minister of Petroleum Resources, Nigeria
Robert Dolph, President, Exxon International
Sir Eric Drake, Chairman, British Petroleum
Charles Duncan, U.S. Secretary of Energy
Robert Dunlop, Chairman, Sun Oil
Stuart Eizenstat, Director, Domestic Policy Staff, White House
Chief M. O. Feyide, Secretary General, OPEC
Clifton Garvin, Chairman, Exxon
Jeremy Gilbert, Manager, Capital Planning Oil Services Company of Iran
Herbert Goodman, President, Gulf Oil and Trading
Pierre Guillaumat, Chairman, Société Nationale Elf Aquitaine
Armand Hammer, Chairman, Occidental Petroleum
Victor Hammer
Sir Peter Holmes, Chairman, Shell Transport and Trading
J. Wallace Hopkins, Deputy Executive Director, International Energy Agency
Wanda Jablonski, Editor and Publisher, *Petroleum Intelligence Weekly*
Henry Jackson, U.S. Senator
Yoshio Karita, Director, Resources Division, Ministry of Foreign Affairs, Japan
George Keller, Chairman, Chevron Corporation
Frederick Khedouri, Deputy Chief of Staff to the U.S. Vice-President
William King, Vice-President, Gulf Oil
Ulf Lantzke, Executive Director, International Energy Agency
James Lee, Chairman, Gulf Oil
Walter Levy, oil consultant
John Loudon, Senior Managing Director, Royal Dutch/Shell
Robert Mabro, Director, Oxford Institute for Energy Studies
William Martin, U.S. Deputy Secretary of Energy
Charles Mateudi
Lord McFadzean of Kelvinside, Chairman, Shell Transport and Trading
George McGhee, U.S. Assistant Secretary of State for Near East, South Asia and Africa
Philippe Michelon, Director, Strategic Planning, Gulf Oil
Edward Morse, U.S. Deputy Assistant Secretary of State; Publisher, *Petroleum Intelligence Weekly*
Richard Murphy, U.S. Assistant Secretary of State for Near East and South Asia
George N. Nelson, President, BP Exploration Alaska
John Norton, Partner, Arthur Andersen & Company
Tadahiko Ohashi, Director, Corporate Planning Department, Tokyo Gas Co., Ltd.
Rene Ortiz, Secretary General, OPEC
Alirio Parra, President, Petróleos de Venezuela, SA
T. Boone Pickens, Chairman, Mesa Petroleum
James Placke, U.S. Deputy Assistant Secretary of State
William Quandt, Director, Middle East Office, U.S. National Security Council
Alberto Quiros, President, Maraven, Lagoven, Venezuela
Sir Peter Ramsbotham, British Ambassador to Iran
M. S. Robinson, President, Shell International Trading Company
Gilbert Rutman, Vice Chairman, Société Nationale Elf Aquitaine
Harold Saunders, U.S. Assistant Secretary of State for Near East and South Asia
James Schlesinger, U.S. Secretary of Defense; U.S. Secretary of Energy
Ian Seymour, Editor, *Middle East Economic Survey*
Jesús Silva Herzog, Minister of Finance, Mexico
Sir David Steel, Chairman, British Petroleum
Jack Sunderland, Chairman, Aminoil
Jean Syrota, Director, Agency for Energy Conservation, France
Sir Peter Walters, Chairman, British Petroleum
Paul Walton, geologist, Pacific Western
Harold Wilson, Prime Minister, Great Britain

Taylor Yoakam, Mesa Petroleum
Eugene Zuckert

Archives

Amoco archives, Chicago
Chevron archives, San Francisco
Gulf archives, Houston
Shell International archives, London
Public Records Office, Kew Gardens, London (PRO)
    Foreign Office
    Cabinet Office
    War Cabinet
    Cabinet Committees
    Prime Minister's Office
    Treasury
    Admiralty
India Office, London
National Archives, Washington, D.C. (NA)
    RG 59 State Department
    RG 218 Joint Chiefs of Staff
Franklin D. Roosevelt Library, Hyde Park, New York
    Official File
    President's Secretary's File
    Harry Hopkins papers
Dwight D. Eisenhower Library, Abilene, Kansas
    Eisenhower Presidential Papers (Ann
        Whitman File)
    John Foster Dulles papers
Harry S. Truman Library, Independence, Missouri
    President's Secretary's File
John F. Kennedy Library, Boston, Massachusetts
    White House Staff Files
Lyndon B. Johnson Library, Austin, Texas
    White House Central Files
        Subject Files
        Confidential File
        Name File
        Joseph Califano papers
        Robson-Ross papers
Richard M. Nixon papers, National Archives
    White House Central Files
    White House Special Files
        Confidential Files
        President's Office Files
Gerald Ford Library, Ann Arbor, Michigan
    White House Central Files
    Presidential Handwriting File
    White House Staff Files
    Energy Resources Council papers
International Energy Agency, Paris
United Kingdom Department of Energy
Senate Multinational Subcommittee Interviews
U.S. State Department papers (1970–80) (Freedom of Information)
National War College, Washington, D.C.

Air Ministry
Colonial Office
Ministry of Power (including Petroleum
    Department)
Board of Trade
War Office

RG 234 Reconstruction Finance
    Corporation

Henry Morgenthau Diary
Henry Morgenthau Presidential Diary

Christian Herter papers

Ralph K. Davies papers

National Security Files
Drew Pearson papers

Cabinet Task Force on Oil Import Controls

Council of Economic Advisors papers
Arthur F. Burns papers
Frank G. Zarb papers

Other Manuscript Collections

Juan Pablo Pérez Alfonzo papers, Caracas
Aramco papers, Middle East Center, Oxford
BBC Written Archives Centre, Reading
George Bissell collection, Dartmouth College
Sir John Cadman papers, University of Wyoming
Churchill Archives Centre, Cambridge, England
Continental Oil Collection, University of Wyoming
Charles R. Crane papers, Middle East Center, Oxford
Ralph K. Davies papers, University of Wyoming
Everette Lee DeGolyer papers, Southern Methodist University
Henry L. Doherty papers, University of Wyoming
Colonel Drake manuscript, Drake Well Museum, Titusville, Penn.
James Terry Duce papers, University of Wyoming
Herbert Goodman papers
Joseph Grew papers, Harvard University
Harold L. Ickes papers, Library of Congress
Wanda Jablonski papers
Joiner v. *Hunt* case records, Rusk County District Court, Henderson, Texas
R. S. McBeth papers, University of Texas at Austin
Philip C. McConnell papers, Hoover Institution
A. J. Meyer papers, Harvard University
Oil Companies papers (Justice Department antitrust case), Baker Library, Harvard Business School
Pearson Collection, Imperial College, London
H. St. J. B. Philby papers, Middle East Center, Oxford
Mark L. Requa papers, University of Wyoming
Rockefeller Archives, Tarrytown, New York
Collection Banque Rothschild, Archives Nationales, Paris
Rusk County Historical Commission, Henderson, Texas
W. B. Sharp papers, University of Texas at Austin
Slade-Barker papers, Middle East Center, Oxford
George Otis Smith papers, University of Wyoming
Stimson Diary, Yale University
Ida Tarbell papers, Drake Well Museum, Titusville, Penn.
James M. Townsend papers, Drake Well Museum, Titusville, Penn.
Private archives

Oral Histories

Winthrop Aldrich, Baker Library, Harvard Business School
James Doolittle, Columbia University
Alan W. Hamill, University of Texas at Austin
Curt G. Hamill, University of Texas at Austin
Patillo Higgins, University of Texas at Austin
James William Kinnear, University of Texas at Austin
E. C. Laster, University of Texas at Austin
Torkild Rieber, University of Texas at Austin

Other

Middle East Economic Survey (MEES)
Petroleum Intelligence Weekly (PIW)
Grampian Television, *Oil.* 8–part television series, 1986.

Government Documents

Declassified Documents Reference System. Washington, D.C.: Carrollton, 1977–81, and Woodbridge, Conn.: Research Publications, 1982–90.

Documents from the U.S. Espionage Den. Tehran: Center for the Publication of the U.S. Espionage Den's Document, [1986].

International Energy Agency. *Energy Policies and Programmes of IEA Countries.* Paris: IEA/OECD.

———. World Energy Outlook. Paris: IEA/OECD.

Japan. Allied Occupation. *Reports of General MacArthur: Japanese Operations in the Southwest Pacific Area.* 4 vols. Washington, D.C.: U.S. Army, 1966.

Mexico. Secretaria de Progamacíon y Presupuesto. *Antología de la Planeación en México (1917–1985).* Vol. 1, *Primeros Intentos de Planeación en México (1917–1946).* Mexico City: Ministry of Budget and Planning, 1985.

Nuremberg Military Tribunals. *Trials of War Criminals.* Vols. 7–8. Washington, D.C.: GPO, 1952–53.

Pogue, Joseph E., and Isador Lubin. *Prices of Petroleums and Products.* Washington, D.C.: GPO, 1919.

Requa, Mark L. "Report of the Oil Division, 1917–1919." H. A. Garfield. *Final Report of the U.S. Fuel Administrator. 1917–1919.* Washington, D.C.: GPO, 1921.

United Kingdom. Ministry of Fuel and Power. *Report on the Petroleum and Synthetic Oil Industry of Germany.* London: HMSO, 1947.

United Kingdom. Admiralty. Geographical Section of the Naval Intelligence Division. *Geology of Mesopotamia and Its Borderlands.* London: HMSO, 1920.

U.S. Army. Far East Command. Military Intelligence Section. *Intelligence Series and Documentary Appendices.* Washington, D.C.: Library of Congress, 1981. Microfilm.

U.S. Cabinet Task Force on Oil Import Control. *The Oil Import Question: A Report on the Relationship of OH Imports to the National Security.* Washington, D.C.: GPO, 1970.

U.S. Central Intelligence Agency. *CIA Research Reports: Middle East, 1946–1976.* Ed. Paul Kesaris. Frederick, Md.: University Publications of America, 1983. Microfilm.

U.S. Congress. House of Representatives. Permanent Select Committee on Intelligence. Subcommittee on Evaluation. *Iran: Evaluation of U.S. Intelligence Performance Prior to November 1978.* Staff Report. Washington, D.C.: GPO, 1979.

U.S. Congress. Joint Committee on the Investigation of the Pearl Harbor Attack. *Pearl Harbor: Intercepted Messages Sent by the Japanese Government Between July 1 and December 8, 1941.* 79th Cong. 1st sess. Washington, D.C.: GPO, 1945.

———. *Pearl Harbor Attack.* 79th Cong. 2d sess. Washington, D.C.: GPO, 1946.

U.S. Congress. Senate. Committee on Foreign Relations. Subcommittee on Multinational Corporations. *A Documentary History of the Petroleum Reserves Corporation.* 93d Cong. 2d sess. Washington, D.C.: GPO, 1974.

———. *The International Petroleum Cartel, the Iranian Consortium and U.S. National Security.* 93d Cong. 2d sess. Washington, D.C.: GPO, 1974.

———. *Multinational Corporations and United States Foreign Policy.* 93rd Cong. 1st sess. Washington, D.C.: GPO, 1975 (Multinational Hearings).

———. *Multinational Oil Corporations and U.S. Foreign Policy.* 93rd Cong. 2d sess. Washington, D.C.: GPO, 1975.

———. *U.S. Oil Companies and the Arab Oil Embargo: The International Allocation of Constricted Supply.* Committee Print. Washington, D.C.: GPO, 1975.

U.S. Congress. Senate. Committee on Government Operations. Permanent Subcommittee on Investigations. *Current Energy Shortages Oversight Series.* 93rd Cong. 1st Session. Washington, D.C.: GPO, 1974.

U.S. Congress. Senate. Committee on the Judiciary. Subcommittee on Antitrust and Monopoly. *Petroleum, the Antitrust Laws and Government Policies.* 85th Cong. 1st sess. Washington, D.C.: GPO, 1957.

U.S. Congress. Senate. Select Committee on Small Business. Subcommittee on Monopoly. *The International Petroleum Cartel: Staff Report to the Federal Trade Commission.* 82d Cong. 2d sess. Washington, D.C.: GPO, 1952 (FTC, *International Petroleum Cartel*).

U.S. Congress. Senate. Special Committee Investigating Petroleum Resources. *Investigation of Petroleum Resources.* 79th Cong. 1st and 2d sess. Washington, D.C.: GPO, 1946.

U.S. Congress. Senate. Special Committee Investigating the National Defense Program. *Investigation of the National Defense Program.* Part 11, *Rubber.* 77th Cong. 1st sess. Part 41, *Petroleum Arrangements with Saudi Arabia.* 80th Cong. 1st sess. Washington, D.C: GPO, 1948.

U.S. Congress. Senate. Subcommittees of the Committee on the Judiciary and Committee on Interior and Insular Affairs. *Emergency Oil Lift Program and Related Oil Problems.* 85th Cong. 1st sess. Washington, D.C.: GPO, 1957.

U.S. Congress, Senate. Subcommittee of the Committee on Manufactures. *High Cost of Gasoline and Other Petroleum Products.* 67th Cong. 2d and 4th sess. Washington, D.C.: GPO, 1923.

U.S. Department of the Interior. *An Analysis of the Economic and Security Aspects of the Trans-Alaskan Pipeline.* Washington, D.C.: GPO, 1971.

U.S. Department of Justice. Anti-Trust Division. *Report of the Department of Justice to the President Concerning the Gasoline Shortage of 1979.* Washington, D.C.: GPO, 1980.

U.S. Department of State. *Foreign Relations of the United States.* Washington, D.C.: GPO 1948–90 (FRUS).

U.S. Economic Cooperation Administration. European Recovery Program. *Petroleum and Petroleum Equipment Commodity Study.* Washington, D.C: GPO, 1949.

U.S. Federal Trade Commission. *Foreign Ownership in the Petroleum Industry.* Washington, D.C.: GPO, 1923.

———. *Prices, Profits, and Competition in the Petroleum Industry.* Washington, D.C.: GPO, 1928.

U.S. National Response Team. *The Exxon Valdez Oil Spill: A Report to the President from Samuel K. Skinner and William K. Reilly.* May 1989.

U.S. National Security Council. *Documents of the NSC, 1947–77.* Ed. Paul Kesaris. Washington, D.C. and Frederick, Md.: University Publications of America, 1980–87. Microfilm.

———. *Minutes of Meetings of the NSC, with Special Advisory Reports.* Ed. Paul Kesaris. Frederick, Md.: University Publications of America, 1982. Microfilm.

U.S. Office of Strategic Services and Department of State. *O.S.S./State Department Intelligence and Research Reports.* Ed. Paul Kesaris. Washington, D.C: University Publications of America, 1979. Microfilm.

U.S. Petroleum Administration for War. *Petroleum in War and Peace.* Washington, D.C: PAW, 1945.

U.S. President. *Public Papers of the Presidents of the United States: Jimmy Carter, 1977.* Washington, D.C: GPO, 1978.

U.S. Strategic Bombing Survey. Oil and Chemical Division. *Oil in Japan's War.* Washington, D.C: USSBS, 1946.

———. Oil Division. *Final Report.* 2d ed. Washington, D.C: USSBS, 1947.

———. Overall Economic Effects Division. *The Effects of Strategic Bombing on Japan's War Economy.* Washington, D.C: GPO, 1946.

———. Overall Economic Effects Division. *The Effects of Strategic Bombing on the German War Economy.* Washington, D.C: USSBS, 1945.

U.S. Strategic Bombing Survey (Pacific). Naval Analysis Division. *Interrogations of Japanese Officials.* 2 vols. Washington, D.C: USSBS, [*1945*].

U.S. Tariff Commission. *War Changes in Industry.* Report 17, *Petroleum.* Washington, D.C: GPO, 1946.

U.S. War Production Board. *Industrial Mobilization for War: History of the War Production Board and Predecessor Agencies, 1940–1945.* Vol. 1, *Program and Administration.* Washington, D.C: GPO, 1947.

Woodward, E. L., and Rohan Butler. *Documents on British Foreign Policy, 1919–1939.* 3 series. London: HMSO, 1946–86.

Selected Books, Articles, and Dissertations

Abir, Mordechai. *Saudi Arabia in the Oil Era: Regime and Elites; Conflict and Collaboration.* London: Croom Helm, 1988.

Abrahamian, Ervand. *Iran Between Two Revolutions.* Princeton: Princeton University Press, 1982.

Acheson, Dean. *Present at the Creation: My Years in the State Department.* New York: New American Library, 1970.

Adelman, M. A. "Is the Oil Shortage Real? Oil Companies as OPEC Tax Collectors." *Foreign Policy* (Winter 1972–73): 69–108.

————. *The World Petroleum Market*. Baltimore: Johns Hopkins University Press, 1972.

Agawa, Hiroyuki. *The Reluctant Admiral: Yamamoto and the Imperial Navy*. Trans. John Bester. Tokyo: Kodansha International, 1979.

Ajami, Fouad. *The Arab Predicament: Arab Political Thought and Practice Since 1967*. Cambridge: Cambridge University Press, 1981.

Akin, Edward N. *Flagler: Rockefeller Partner and Florida Baron*. Kent, Ohio: Kent State University Press, 1988.

Akins, James E. "The Oil Crisis: This Time the Wolf Is Here." *Foreign Affairs* 51 (April 1973): 462–490.

Alexander, Yonah, and Allan Nanes, eds. *The United States and Iran: A Documentary History*. Frederick, Md.: University Publications of America, 1980.

Alfonzo, Juan Pablo Pérez. *Hundiéndos en el Excremento del Diablo*. Caracas: Colleción Venezuela Contemporánea, 1976.

————. *El Pentágono Petrolero*. Caracas: Ediciones Revista Politica, 1967.

Almana, Mohammed. *Arabia Unified: A Portrait of Ibn Saud*. London: Hutchinson Benham, 1980.

Ambrose, Stephen E. *Eisenhower*. 2 vols. New York: Simon and Schuster, 1983–84.

————. *The Supreme Commander: The War Years of General Dwight D. Eisenhower*. Garden City, N.Y.: Doubleday, 1970.

American Bar Association, Section of Mineral Law. *Legal History of Conservation of Oil and Gas: A Symposium*. Chicago: American Bar Association, 1939.

Anderson, Irvine H. *Aramco, the United States, and Saudi Arabia: A Study of the Dynamics of Foreign Oil Policy, 1933–1950*. Princeton: Princeton University Press, 1981.

————. *The Standard-Vacuum Oil Company and United States East Asian Policy, 1933–1941*. Princeton: Princeton University Press, 1975.

Anderson, Robert O. *Fundamentals of the Petroleum Industry*. Norman: University of Oklahoma Press, 1984.

Arnold, Ralph, George A. Macready, and Thomas W. Barrington. *The First Big Oil Hunt: Venezuela, 1911–1916*. New York: Vantage Press, 1960.

Arthur Andersen & Co. and Cambridge Energy Research Associates. *The Future of Oil Prices: The Perils of Prophecy*. Houston: 1984.

Asbury, Herbert. *The Golden Flood: An Informal History of America's First Oil Field*. New York: Alfred A. Knopf, 1942.

Ashworth, William. *The History of the British Coal Industry*. Vol. 5, *1946–1982: The Nationalized Industry*. Oxford: Clarendon Press, 1986.

Asprey, Robert B. *The First Battle of the Marne*. 1962. Reprint. Westport, Conn.: Greenwood Press, 1979.

Assiri, Abdul-Reda. *Kuwait's Foreign Policy: City-State in World Politics*. Boulder, Colo.: Westview Press, 1990.

Bacon, R. H. *The Life of Lord Fisher of Kilverstone*. 2 vols. Garden City, N.Y.: Doubleday, Doran, 1929.

Badger, Daniel, and Robert Belgrave. *Oil Supply and Price: What Went Right in 1980?* London: Policy Studies Institute, 1982.

Bain, Joe S. *The Economics of the Pacific Coast Petroleum Industry*. 3 parts. Berkeley: University of California Press, 1944–47.

Bakhash, Shaul. *The Reign of the Ayatollahs: Iran and the Islamic Revolution*. New York: Basic Books, 1984.

Bardou, Jean-Pierre, Jean-Jacques Chanaron, Patrick Fridenson, and James M. Laux. *The Automobile Revolution: The Impact of an Industry*. Trans. James M. Laux. Chapel Hill: University of North Carolina Press, 1982.

Barnhart, Michael A. *Japan Prepares for Total War: The Search for Economic Security, 1919–1941*. Ithaca, N.Y.: Cornell University Press, 1987.

————. "Japan's Economic Security and the Origins of the Pacific War." *Journal of Strategic Studies* 4 (June 1981): 105–24.

Bates, J. Leonard. *The Origins of Teapot Dome: Progressives, Parties, and Petroleum, 1909–1921*. Urbana: University of Illinois Press, 1963.

————. "The Teapot Dome Scandal and the Election of 1924." *American Historical Review* 55 (January 1955): 303–22.

Beaton, Kendall. "Dr. Gesner's Kerosene: The Start of American Oil Refining." *Business History Review* 29 (March 1955): 28–53.

————. *Enterprise in Oil: A History of Shell in the United States.* New York: Appleton-Century-Crofts, 1957.

Beck, Peter J. "The Anglo-Persian Oil Dispute of 1932–33." *Journal of Contemporary History* 9 (October 1974): 123–51.

Beeby-Thompson, A. *Oil Field Development and Petroleum Mining.* London: Crosby Lockwood, 1916.

————. *The Oil Fields of Russia and the Russian Petroleum Industry.* 2d ed. London: Crosby Lockwood, 1908.

————. *Oil Pioneer.* London: Sidgwick and Jackson, 1961.

Belasco, Warren James. *Americans on the Road: From Autocamp to Motel, 1910–1945.* Cambridge: MIT Press, 1979.

Benn, Anthony. *Against the Tide: Diaries, 1973–76.* London: Hutchinson, 1989.

Bentley, Jerome Thomas. "The Effects of Standard Oil's Vertical Integration into Transportation on the Structure and Performance of the American Petroleum Industry, 1872–1884." Ph.D. dissertation, University of Pittsburgh, 1976.

Bérenger, Henry. *Le Pétrole et la France.* Paris: Flammarion, 1920.

Bergengren, Erik. *Alfred Nobel: The Man and His Work.* Trans. Alan Blair. London: Thomas Nelson, 1960.

Betancourt, Romulo. *Venezuela: Oil and Politics.* Trans. Everett Bauman. Boston: Houghton Mifflin, 1979.

————. *Venezuela's Oil.* Trans. Donald Peck. London: George Allen & Unwin, 1978.

Betts, Richard K. *Nuclear Blackmail and Nuclear Balance.* Washington, D.C.: Brookings Institution, 1987.

Bill, James A. *The Eagle and the Lion: The Tragedy of American-Iranian Relations.* New Haven: Yale University Press, 1988.

Bill, James A., and William Roger Louis, eds. *Mossadiq, Iranian Nationalism, and Oil.* London: I. B. Tauris, 1988.

Blair, Clay, Jr. *Silent Victory: The U.S. Submarine War Against Japan.* Philadelphia: J. B. Lippincott, 1975.

Blair, John M. *The Control of Oil.* New York: Pantheon, 1976.

Blum, John Morton. *From the Morgenthau Diaries.* 3 vols. Boston: Houghton Mifflin, 1959–67.

Blumenson, Martin. *Patton: The Man Behind the Legend, 1885–1945.* New York: William Morrow, 1985.

————, ed. *The Patton Papers.* 2 vols. Boston: Houghton Mifflin, 1972–74.

Boatwright, Mody C, and William A. Owen. *Tales from the Derrick Floor.* Garden City, N.Y.: Doubleday, 1970.

Bonine, Michael E., and Nikkie R. Keddie, eds. *Continuity and Change in Modern Iran.* Albany, N.Y.: State University of New York Press, 1981.

Borkin, Joseph. *The Crime and Punishment of I. G. Farben.* New York: Free Press, 1978.

Bowie, Robert R. *Suez 1956.* London: Oxford University Press, 1974.

Bradley, Omar N. *A Soldier's Story.* New York: Henry Holt, 1951.

Brady, Kathleen. *Ida Tarbell: Portrait of a Muckraker.* New York: Seaview/Putnam, 1984.

Brands, H. W. "The Cairo-Tehran Connection in Anglo-American Rivalry in the Middle East, 1951–1953." *International History Review* 11 (August 1989): 434–56.

————. Inside the Cold War: *Loy Henderson and the Rise of the American Empire, 1918–1961.* Oxford: Oxford University Press, 2001.

Brenner, Anita. *The Wind That Swept Mexico: The History of the Mexican Revolution, 1910–1942.* 1943. Reprint. Austin: University of Texas Press, 1971.

Brewster, Kingman, Jr. *Antitrust and American Business Abroad.* New York: McGraw-Hill, 1958.

Bright, Arthur A., Jr. *The Electric Lamp Industry: Technological Change and Economic Development from 1800 to 1947.* New York: Macmillan, 1949.

Bringhurst, Bruce. *Antitrust and the Oil Monopoly: The Standard Oil Cases, 1890–1911.* Westport, Conn.: Greenwood Press, 1979.

Brodie, Bernard. "American Security and Foreign Oil." *Foreign Policy Reports* 23 (1948): 297–312.

Brown, Benjamin, and Daniel Yergin. "Synfuels 1979." Draft case, Kennedy School of Government, Harvard University, 1981.

Brown, Jonathan C. "Domestic Politics and Foreign Investment: British Development of Mexican Petroleum, 1889–1911." *Business History Review* 61 (Autumn 1987): 387–416.

————. "Jersey Standard and the Politics of Latin American Oil Production, 1911–1930." *Latin Ameri-*

can Oil Companies and the Politics of Energy, ed. John D. Wirth. Lincoln: University of Nebraska Press, 1985.

————. "Why Foreign Oil Companies Shifted Their Production from Mexico to Venezuela during the 1920s." *American Historical Review* 90 (April 1985): 362–385.

Bryant, Arthur. *The Turn of the Tide: A History of the War Years Based on the Diaries of Field-Marshall Lord Alanbrooke.* Garden City, N.Y.: Doubleday, 1957.

Brzezinski, Zbigniew. *Power and Principle: Memoirs of the National Security Adviser, 1977–1981.* Rev. ed. New York: Farrar Straus Giroux, 1985.

Bullock, Alan. *Ernest Bevin: Foreign Secretary, 1945–1951.* London: Heinemann, 1984.

————. *Hitler: A Study in Tyranny.* Rev. ed. New York: Harper & Row, 1964.

Bupp, I. C, Joseph Stanislaw, and Daniel Yergin. "How Low Can It Go? The Dynamics of Oil Prices." Cambridge Energy Research Associates Report, May 1985.

Busch, Briton Cooper. *Britain and the Persian Gulf.* Berkeley: University of California Press, 1967.

————. *Britain, India, and the Arabs, 1914–1921.* Berkeley: University of California Press, 1971.

Bush, George, with Victor Gold. *Looking Forward: An Autobiography.* New York: Bantam, 1988.

Business History Review, ed. *Oil's First Century.* Boston: Harvard Business School, 1960.

Butow, Robert J. C. *Japan's Decision to Surrender.* Stanford: Stanford University Press, 1954.

————. *Tojo and the Coming of the War.* Stanford: Stanford University Press, 1961.

Caldwell, Martha Ann. "Petroleum Politics in Japan: State and Industry in a Changing Policy Context." Ph.D. dissertation, University of Wisconsin at Madison, 1981.

Cambridge Energy Research Associates. *Energy and the Environment: The New Landscape of Public Opinion.* Cambridge: Cambridge Energy Research Associates, 1990.

Carell, Paul. *Hitler Moves East, 1941–1943.* Trans. Ewald Osers. Boston: Little, Brown, 1965.

Carré, Henri. *La Veritable Histoire des Taxis de La Mame.* Paris: Libraire Chapelot, 1921.

Caro, Robert. *The Years of Lyndon Johnson: The Path to Power.* New York: Alfred A. Knopf, 1982.

Carter, Jimmy. *Keeping Faith: Memoirs of a President.* London: Collins, 1982.

de Chair, Somerset. *Getty on Getty: A Man in a Billion.* London: Cassell, 1989.

al-Chalabi, Fadhil J. *OPEC and the International Oil Industry: A Changing Structure.* Oxford: Oxford University Press, 1980.

————. *OPEC at the Crossroads.* Oxford: Pergamon, 1989.

Chandler, Alfred D., Jr. *Strategy and Structure: Chapters in the History of the American Industrial Enterprise.* Cambridge: MIT Press, 1962.

————. *The Visible Hand: The Managerial Revolution in American Business.* Cambridge: Harvard University Press, 1977.

Chandler, Alfred D., Jr., and Stephen E. Ambrose, eds. *The Papers of Dwight David Eisenhower.* Vol. 4, *The War Years.* Baltimore: Johns Hopkins University Press, 1970.

Chandler, Geoffrey. "The Innocence of Oil Companies." *Foreign Policy* (Summer 1977): 52–70.

de Chazeau, Melvin G., and Alfred E. Kahn. *Integration and Competition in the Petroleum Industry.* New Haven: Yale University Press, 1959.

Chester, Edward W. *United States Oil Policy and Diplomacy: A Twentieth-Century Overview.* Westport, Conn.: Greenwood Press, 1983.

Chisholm, Archibald H. I. *The First Kuwait Oil Concession Agreement: A Record of the Negotiations, 1911–1934.* London: Frank Cass, 1975.

Christopher, Warren, Harold H. Saunders, et al. *American Hostages in Iran: The Conduct of a Crisis.* New Haven: Yale University Press, 1985.

Churchill, Randolph S. *Winston S. Churchill.* Vols. 1–2. 1966–67.

Churchill, Winston S. *The World Crisis.* 4 vols. New York: Charles Scribner's Sons, 1923–29.

Cicchetti, Charles J. *Alaskan Oil: Alternative Routes and Markets.* Baltimore: Resources for the Future, 1972.

Clark, Alan. *Barbarossa: The Russian-German Conflict, 1941–1945.* 1965. Reprint. London: Macmillan, 1985.

Clark, James A., and Michael T. Halbouty. *Spindletop.* New York: Random House, 1952.

————. *The Last Boom.* Fredericksburg, Tex.: Shearer Publishing, 1984.

Clark, John G. *Energy and the Federal Government: Fossil Fuel Policies, 1900–1946.* Urbana: University of Illinois Press, 1987.

Cohen, Jerome B. *Japan's Economy in War and Reconstruction.* Minneapolis: University of Minnesota Press, 1949.

Cohen, Stuart A. "A Still Stranger Aspect of Suez: British Operational Plans to Attack Israel, 1955–56." *International History Review* 10 (May 1988): 261–81.

Cole, Hugh M. *The Ardennes: Battle of the Bulge.* Washington, D.C.: Department of the Army, 1965.

Colitti, Marcello. *Energia e Sviluppo in Italia: La Vicenda de Enrico Mattel.* Bari: De Donata, 1979.

Coll, Steve. *The Taking of Getty Oil.* New York: Atheneum, 1987.

Collier, Peter, and David Horowitz. *The Rockefellers: An American Dynasty.* New York: Holt, Rinehart and Winston, 1976.

Cone, Andrew, and Walter R. Johns. *Petrolia: A Brief History of the Pennsylvania Petroleum Region.* New York: D. Appleton, 1870.

Continental Oil Company. *Conoco: The First One Hundred Years.* New York: Dell, 1975.

Cooper, Chester L. *The Lion's Last Roar: Suez, 1956.* New York: Harper & Row, 1978.

Cordesman, Anthony H. *The Gulf and the West: Strategic Relations and Military Realities.* Boulder, Colo.: Westview Press, 1988.

Corley, T. A. B. *A History of the Burmah Oil Company* Vol. 1, *1886–1924.* Vol. 2, *1924–1966.* London: Heinemann, 1983–88.

Coronel, Gustavo. *The Nationalization of the Venezuelan Oil Industry: From Technocratic Success to Political Failure.* Lexington, Mass.: Lexington Books, 1983.

Costello, John. *The Pacific War.* New York: Quill, 1982.

Cotner, Robert C. *James Stephen Hogg.* Austin: University of Texas Press, 1959.

Cottam, Richard W. *Iran and the United States: A Cold War Case Study.* Pittsburgh: University of Pittsburgh Press, 1988.

———. *Nationalism in Iran.* 2d ed. Pittsburgh: University of Pittsburgh Press, 1979.

Cowhey, Peter F. *The Problems of Plenty: Energy Policy and International Politics.* Berkeley: University of California Press, 1985.

Craven, Wesley Frank, and James Lea Cate. *The Army Air Forces in World War II.* 7 vols. Chicago: University of Chicago Press, 1948–58.

Crowley, James B. *Japan's Quest for Autonomy: National Security and Foreign Policy, 1930–1938.* Princeton: Princeton University Press, 1966.

Cusamano, Michael A. *The Japanese Automobile Industry: Technology and Management at Nissan and Toyota.* Cambridge: Harvard University Press, 1985.

Dallek, Robert. *Franklin D. Roosevelt and American Foreign Policy, 1932–1945.* Oxford: Oxford University Press, 1981.

Dann, Uriel, ed. *The Great Powers in the Middle East, 1919–1939.* New York and London: Holmes & Meier, 1988.

Darrah, William C. *Pithole: The Vanished City.* Gettysburg, Pa., 1972.

Dawidowicz, Lucy S. *The War Against the Jews, 1933–1945.* New York: Bantam, 1976.

Dedmon, Emmett. *Challenge and Response: A Modem History of Standard Oil Company (Indiana).* Chicago: Mobium Press, 1984.

Delaisi, Francis. *Oil: Its Influence on Politics.* Trans. C. Leonard Leese. London: Labour Publishing and George Allen & Unwin, 1922.

Denny, Ludwell. *We Fight for Oil.* 1928. Reprint. Westport, Conn.: Hyperion, 1976.

DeNovo, John. *American Interests and Policies in the Middle East, 1900–1939.* Minneapolis: University of Minnesota Press, 1963.

———. "The Movement for an Aggressive American Oil Policy Abroad, 1918–1920." *American Historical Review* 61 (July 1956): 854–76.

———. "Petroleum and the United States Navy Before World War I." *Mississippi Valley Historical Review* 41 (March 1955): 641–56.

Destler, Chester McArthur. *Roger Sherman and the Independent Oil Men.* Ithaca: Cornell University Press, 1967.

Deterding, Henri. *An International Oilman* (as told to Stanley Naylor). London and New York: Harper and Brothers, 1934.

Deutscher, Isaac. *Stalin: A Political Biography.* 2d ed. New York: Oxford University Press, 1966.

Dickson, H. R. P. *Kuwait and Her Neighbors.* London: George Allen & Unwin, 1956.

Dixon, D. F. "Gasoline Marketing in the United States—The First Fifty Years." *Journal of Industrial Economics* 13 (November 1964): 23–42.

———. "The Growth of Competition Among the Standard Oil Companies in the United States, 1911–1961." *Business History* 9 (January 1967): 1–29.

Dower, John W. *War Without Mercy: Race and Power in the Pacific War.* New York: Pantheon, 1986.

Earle, Edward Mead. "The Turkish Petroleum Company—A Study in Oleaginous Diplomacy." *Political Science Quarterly* 39 (June 1924): 265–79.

Eaton, S. J. M. *Petroleum: A History of the Oil Region of Venango County, Pennsylvania.* Philadelphia: J. P. Skelly & Co., 1866.

Eddy, William A. *F.D.R. Meets Ibn Saud.* New York: American Friends of the Middle East, 1954.

Eden, Anthony. *Memoirs.* 3 vols. London: Cassell, 1960–65.

Eden, Richard, Michael Posner, Richard Bending, Edmund Crouch, and Joseph Stanislaw. *Energy Economics: Growth, Resources, and Policies.* Cambridge: Cambridge University Press, 1981.

Eisenhower, David. *Eisenhower at War, 1943–1945.* New York: Random House, 1986.

Eisenhower, Dwight D. *At Ease: Stories I Tell to Friends.* Garden City, N.Y.: Doubleday, 1967.

———. *The White House Years.* 2 vols. Garden City, N.Y.: Doubleday, 1963–65.

———. *Crusade in Europe.* Garden City, N.Y.: Doubleday, 1948.

Eisenhower, John S. D. *The Bitter Woods.* New York: G. P. Putnam's Sons, 1969.

Eizenstat, Stuart E. "The 1977 Energy Plan: M.E.O.W." Case note for the Kennedy School of Government, Harvard University.

Elwell-Sutton, L. P. *Persian Oil: A Study in Power Politics.* London: Laurence and Wishart, 1955.

Engler, Robert. *The Brotherhood of Oil: Energy Policy and the Public Interest.* Chicago: University of Chicago Press, 1977.

———. *The Politics of Oil: A Study of Private Power and Democratic Directions.* New York: Macmillan, 1961.

Erickson, John. *The Road to Stalingrad.* London: Panther, 1985.

Esser, Robert. "The Capacity Race: The Future of World Oil Supply." Cambridge Energy Research Associates Report, 1990.

Eveland, Wilbur Crane. *Ropes of Sand: America's Failure in the Middle East.* New York: W. W. Norton, 1980.

Ezell, John S. *Innovations in Energy: The Story of Kerr-McGee.* Norman: University of Oklahoma Press, 1979.

Fabricus, Johan. *East Indies Episode.* London: Shell Petroleum Company, 1949.

Fanning, Leonard M. *American Oil Operations Abroad.* New York: McGraw-Hill, 1947.

———. *The Story of the American Petroleum Institute.* New York: World Petroleum Policies, [1960].

Faure, Edgar. *La Potitique Française du Pétrole.* Paris: Nouvelle Revue Critique, 1938.

Fayle, C. Ernest. *Seaborne Trade.* 4 vols. London: John Murray, 1924.

Feis, Herbert. *Petroleum and American Foreign Policy.* Stanford: Food Research Institute, 1944.

———. *The Road to Pearl Harbor: The Coming of War Between the United States and Japan.* Princeton: Princeton University Press, 1950.

———. *Seen from E. A.: Three International Episodes.* New York: Alfred A. Knopf, 1947.

Ferrier, R. W. *The History of the British Petroleum Company.* Vol. 1, *The Developing Years, 1901–1932.* Cambridge: Cambridge University Press, 1982.

Finer, Herman. *Dulles over Suez: The Theory and Practice of His Diplomacy.* Chicago: Quadrangle Books, 1964.

First, Ruth. *Libya: The Elusive Revolution.* London: Penguin Books, 1974.

Fischer, Louis. *Oil Imperialism: The International Struggle for Petroleum.* New York: International Publishers, 1926.

Fisher, John Arbuthnot. *Fear God and Dread Nought: The Correspondence of Admiral of the Fleet Lord Fisher of Kilverstone.* 2 vols. Ed. Arthur J. Marder. Cambridge: Harvard University Press, 1952.

———. *Memories.* London: Hodder and Stoughton, 1919.

———. *Records.* London: Hodder and Stoughton, 1919.

Fishman, Robert. *Bourgeois Utopias: The Rise and Fall of Suburbia.* New York: Basic Books, 1987.

Flink, James J. *America Adopts the Automobile 1895–1910.* Cambridge: MIT Press, 1970.

———. *The Automobile Age.* Cambridge: MIT Press, 1988.

Flynn, John T. *God's Gold: The Story of Rockefeller and His Times.* New York: Harcourt, Brace, 1932.

Foley, Paul. "Petroleum Problems of the World War: Study in Practical Logistics." *United States Naval Institute Proceedings* 50 (November 1924): 1802–32.

Forbes, R. J. *Bitumen and Petroleum in Antiquity.* Leiden: E. J. Brill, 1936.

———. *More Studies in Early Petroleum History, 1860–1880.* Leiden: E. J. Brill, 1959.

———. *Studies in Early Petroleum History.* Leiden: E. J. Brill, 1958.

Forbes, R. J. and D. R. O'Beirne. *The Technical Development of the Royal Dutch/Shell, 1890–1940.* Leiden: E. J. Brill, 1957.

Frankel, Paul. *Common Carrier of Common Sense: A Selection of His Writings, 1946–1980.* Ed. Ian Skeet. Oxford: Oxford University Press, 1989.

———. *The Essentials of Petroleum: A Key to Oil Economics.* New ed. London: Frank Cass, 1969.

———. *Mattei: Oil and Power Politics.* New York and Washington: Praeger, 1966.

———. "Oil Supplies During the Suez Crisis: On Meeting a Political Emergency." *Journal of Industrial Economics* 6 (February 1958): 85–100.

Frey, John W., and H. Chandler Ide. *A History of the Petroleum Administration for War, 1941–1945.* Washington, D.C.: GPO, 1946.

Friedman, Thomas L. *From Beirut to Jerusalem.* New York: Farrar Straus Giroux, 1989.

Fromkin, David. *A Peace to End All Peace: Creating the Modern Middle East, 1914–1922.* New York: Henry Holt & Co., 1989.

Fuller, J. F. C. *Tanks in the Great War, 1914–1918.* London: John Murray, 1920.

Fursenko, A. A. *Neftianye Tresty i Mirovaia Politika: 1880-e gody-1918 god.* Moscow: Nauka, 1965.

Galbraith, John Kenneth. *A Life in Our Times: Memoirs.* Boston: Houghton Mifflin, 1981.

Garthoff, Raymond. *Detente and Confrontation: American-Soviet Relations From Nixon to Reagan.* Washington, D.C.: Brookings Institution, 1985.

Gasiorowski, Mark T. "The 1953 Coup d'Etat in Iran." *International Journal of Middle Eastern Studies* 19 (1987): 261–86.

Georges-Picot, Jacques. *The Real Suez Crisis: The End of a Great Nineteenth Century Work.* Trans. W. G. Rogers. New York: Harcourt Brace Jovanovich, 1978.

Gerretson, F. C. *History of the Royal Dutch.* 4 vols. Leiden: E. J. Brill, 1953–57.

Gesner, Abraham. *A Practical Treatise on Coal, Petroleum, and Other Distilled Oils.* 2d ed. Ed. George W. Gesner. New York: Baillière Bros., 1865.

de Geus, Arie P. "Planning as Learning." *Harvard Business Review* 66 (March–April 1988): 70–74.

Gibb, George Sweet, and Evelyn H. Knowlton. *History of Standard Oil Company (New Jersey).* Vol. 2, *The Resurgent Years 1911–1927.* New York: Harper & Brothers, 1956.

Giddens, Paul H. *The Beginnings of the Petroleum Industry: Sources and Bibliography.* Harrisburg, Pa.: Pennsylvania Historical Commission, 1941.

———. *The Birth of the Oil Industry.* New York: Macmillan, 1938.

———. *Early Days of Oil: A Pictorial History of the Beginnings of the Industry in Pennsylvania.* Princeton: Princeton University Press, 1948.

———. *Pennsylvania Petroleum, 1750–1872: A Documentary History.* Titusville, Pa.: Pennsylvania Historical and Museum Commission, 1947.

———. *Standard Oil Company (Indiana): Oil Pioneer in the Middle West.* New York: Appleton-Century-Crofts, 1955.

Giebelhaus, August W. *Business and Government in the Oil Industry: A Case Study of Sun Oil, 1876–1945.* Greenwich, Conn.: JAI Press, 1980.

Gilbert, Martin. *Winston S. Churchill.* Vols. 5–8. Boston: Houghton Mifflin, 1977–88.

Gille, Betrand. "Capitaux français et pétroles russes (1884–1894)." *Histoire des Entreprises* 12 (November 1963): 9–94.

Gillespie, Angus Kress, and Michael Aaron Rockland. *Looking for America on the New Jersey Turnpike.* New Brunswick: Rutgers University Press, 1989.

Gillingham, John R. *Industry and Politics in the Third Reich: Ruhr Coal, Hitler and Europe.* London: Methuen, 1985.

Goldberg, Jacob. *The Foreign Policy of Saudi Arabia: The Formative Years, 1902–1918.* Cambridge: Harvard University Press, 1986.

Goodwin, Craufurd D., ed. *Energy Policy in Perspective: Today's Problems, Yesterday's Solutions.* Washington, D.C.: Brookings Institution, 1981.

Goralski, Robert, and Russell W. Freeburg. *Oil & War: How the Deadly Struggle for Fuel in WWII Meant Victory or Defeat.* New York: William Morrow, 1987.

Gould, Lewis L. *Reform and Regulation: American Politics, 1900–1916.* New York: John Wiley, 1978.

Goulder, Grace. *John D. Rockefeller: The Cleveland Years.* Cleveland: Western Reserve Historical Society, 1972.

Graham, Robert. *Iran: The Illusion of Power.* New York: St. Martin's Press, 1979.

Grayson, George W. *The Politics of Mexican Oil.* Pittsburgh: University of Pittsburgh Press, 1980.

Greene, William N. *Strategies of the Major Oil Companies*. Ann Arbor, Mich.: UMI Research, 1982.

Greider, William. *Secrets of the Temple: How the Federal Reserve Runs the Country*. New York: Touchstone, 1989.

Gulbenkian, Nubar. *Portrait in Oil*. New York: Simon and Schuster, 1965.

Gulliford, Andrew. *Boomtown Blues: Colorado Oil Shale, 1885–1985*. Niwot, Colo.: University Press of Colorado, 1989.

Gustafson, Thane. *Crisis amid Plenty: The Politics of Soviet Energy Under Brezhnev and Gorbachev*. Princeton: Princeton University Press, 1989.

Halasz, Nicholas. *Nobel: A Biography of Alfred Nobel*. New York: Orion Press, 1959.

Halberstam, David. *The Reckoning*. New York: William Morrow, 1986.

Halder, Franz. *The Halder Diaries*. 2 vols. Boulder, Colo.: Westview Press, 1976.

Halliday, W. Trevor. *John D. Rockefeller, 1839–1937: Industrial Pioneer and Man*. New York: Newcomen Society, 1948.

Hamilton, Adrian. *Oil: The Price of Power*. London: Michael Joseph/Rainbird, 1986.

Hamilton, Nigel. *Monty*. 3 vols. London: Sceptre, 1984–1987.

Hammer, Armand, with Neil Lyndon. *Hammer*. New York: G. P. Putnam's Sons, 1987.

Hannah, Leslie. *Electricity Before Nationalization*. London: Macmillan, 1979.

———. *The Rise of the Corporate Economy*. 2d ed. London: Methuen, 1976.

Hardinge, Arthur H. *A Diplomatist in the East*. London: Jonathan Cape, 1928.

Hardwicke, Robert E. *Antitrust Laws, et al. v. Unit Operation of Oil or Gas Pools*. New York: American Institute of Mining and Metallurgical Engineers, 1948.

———. "Market Demand as a Factor in the Conservation of Oil." Southwestern Law Foundation. *First Annual Institute on Oil and Gas Law*. New York: Matthew Bender, 1949.

———. *The Oil Man's Barrel*. Norman: University of Oklahoma Press, 1958.

Hare, Richard. *Portraits of Russian Personalities Between Reform and Revolution*. London: Oxford University Press, 1959.

Harris, Kenneth. *The Wildcatter: A Portrait of Robert O. Anderson*. New York: Weidenfeld and Nicolson, 1987.

Hartshorn, J. E. *Oil Companies and Governments: An Account of the International Oil Industry in Its Political Environment*. London: Faber and Faber, 1962.

Havens, Thomas R. H. *Valley of Darkness: The Japanese People and World War II*. New York: W. W. Norton, 1978.

Hawke, David Freeman. *John D.: The Founding Father of the Rockefellers*. New York: Harper & Row, 1980.

———. comp. *John D. Rockefeller Interview, 1917–1920: Conducted by William O. Inglis*. Westport, Conn.: Meckler Publishing, 1984.

Hayes, Peter. *Industry and Ideology: I. G. Farben in the Nazi Era*. Cambridge: Cambridge University Press, 1987.

Heikal, Mohamed. *The Cairo Documents*. Garden City, N.Y.: Doubleday, 1973.

———. *Cutting the Lion's Tale: Suez Through Egyptian Eyes*. London: Andre Deutsch, 1986.

———. *Iran, the Untold Story: An Insider's Account of America's Iranian Adventure and Its Consequences for the Future*. New York: Pantheon, 1982.

———. *The Return of the Ayatollah: The Iranian Revolution from Mossadeq to Khomeini*. London: Andre Deutsch, 1981.

———. *The Road to Ramadan*. London: Collins, 1975.

Heilbroner, Robert L. *The Worldly Philosophers: The Lives, Times, and Ideas of the Great Economic Thinkers*. 6th ed. New York: Simon and Schuster, 1986.

Hein, Laura E. *Fueling Growth: The Energy Revolution and Economic Policy in Postwar Japan*. Cambridge: Harvard University Press, 1990.

Heinrichs, Waldo. *Threshold of War: Franklin D. Roosevelt and America's Entry into World War II*. Oxford: Oxford University Press, 1988.

Helms, Christine Moss. *The Cohesion of Saudi Arabia: Evolution of Political Identity*. Baltimore: Johns Hopkins University Press, 1981.

———. *Iraq: Eastern Flank of the Arab World*. Washington, D.C.: Brookings Institution, 1984.

Henriques, Robert. *Marcus Samuel: First Viscount Bearsted and Founder of the 'Shell' Transport and Trading Company, 1853–1927*. London: Barrie and Rockliff, 1960.

———. *Sir Robert Waley Cohen, 1877–1952*. London: Secker & Warburg, 1966.

Henry, J. D. *Baku: An Eventful History.* London: Archibald Constable & Co., 1905.

———. *Thirty-five Years of Oil Transport: The Evolution of the Tank Steamer.* London: Bradbury, Agnew & Co., 1907.

Henry, J. T. *The Early and Later History of Petroleum.* Philadelphia: Jas. B. Rodgers Co., 1873.

Hewins, Ralph. *Mr. Five Percent: The Story of Calouste Gulbenkian.* New York: Rinehart and Company, 1958.

———. *The Richest American: J. Paul Getty.* New York: E. P. Dutton, 1960.

Hidy, Ralph W., and Muriel E. Hidy. *History of Standard Oil Company (New Jersey).* Vol. 1, *Pioneering in Big Business, 1882–1911.* New York: Harper and Brothers, 1955.

Hinsley, F. H., E. E. Thomas, C. F. G. Ranson, and L. C. Knight. *British Intelligence in the Second World War.* Vol. 2. London: HMSO, 1981.

Hiroharu, Seki. "The Manchurian Incident, 1931." Trans. Marius B. Jansen. *Japan Erupts: The London Naval Conference and the Manchurian Incident, 1928–1932,* ed. James William Morley. New York: Columbia University Press, 1984.

Hofstadter, Richard. *The Age of Reform: From Bryan to FDR.* New York: Vintage, 1955.

Hogan, Michael. *Informal Entente: The Private Structure of Cooperation in Anglo-American Economic Diplomacy, 1918–1928.* Columbia, Mo.: University of Missouri Press, 1977.

———. *The Marshall Plan: America, Britain, and the Reconstruction of Europe.* Cambridge: Cambridge University Press, 1987.

Hogarty, Thomas F. "The Origin and Evolution of Gasoline Marketing." Research Study No. 022. American Petroleum Institute. October 1, 1981.

Holden, David, and Richard Johns. *The House of Saud.* London: Pan Books, 1982.

Hope, Stanton. *Tanker Fleet: The War Story of the Shell Tankers and the Men Who Manned Them.* London: Anglo-Saxon Petroleum, 1948.

Horne, Alistair. *Harold Macmillan.* 2 vols. New York: Viking, 1988–1989.

———. *A Savage War of Peace: Algeria, 1954–1962.* London: Penguin Books, 1979.

Hough, Richard. *The Great War at Sea, 1914–1918.* Oxford: Oxford University Press, 1983.

Howard, Frank A. *Buna Rubber: The Birth of an Industry.* New York: D. Van Nostrand, 1947.

Howard, Michael. *Grand Strategy.* Vol. 4, *August 1942–September 1943.* London: HMSO, 1972.

Hughes, Thomas P. *Networks of Power: Electrification in Western Society, 1880–1930.* Baltimore: Johns Hopkins University Press, 1983.

———. "Technological Momentum in History: Hydrogenation in Germany, 1898–1933." *Past and Present* 44 (August 1969): 106–32.

Hull, Cordell. *The Memoirs of Cordell Hull.* 2 vols. New York: Macmillan, 1948.

Hurt, Harry, III. *Texas Rich: The Hunt Dynasty from the Early Oil Days Through the Silver Crash.* New York: W. W. Norton, 1981.

Huston, James A. *The Sinews of War: Army Logistics, 1775–1953.* Washington, D.C.: U.S. Army, 1966.

Ickes, Harold L. *Fightin' Oil.* New York: Alfred A. Knopf, 1943.

———. *The Secret Diary of Harold L. Ickes.* 3 vols. New York: Simon and Schuster, 1953–54.

Ienaga, Saburo. *The Pacific War, 1931–1945: A Critical Perspective on Japan's Role in World War II.* New York: Pantheon, 1978.

Ike, Nobutaka, ed. and trans. *Japan's Decision for War: Records of the 1941 Policy Conferences.* Stanford: Stanford University Press, 1967.

Inoguchi, Rikihei, and Tadashi Nakajima, with Roger Pineau. *The Divine Wind: Japan's Kamikaze Force in World War II.* Westport, Conn.: Greenwood Press, 1978.

Iraq Petroleum Company. *The Construction of the Iraq-Mediterranean Pipe-Line: A Tribute to the Men Who Built It.* London: St. Clements Press, 1934.

Iriye, Akira. *After Imperialism: The Search for a New Order in the Far East, 1921–1931.* Cambridge: Harvard University Press, 1965.

———. *The Origins of the Second World War in Asia and the Pacific.* London: Longman, 1987.

———. *Power and Culture: The Japanese-American War, 1941–1945.* Cambridge: Harvard University Press, 1981.

Ise, John. *The United States Oil Policy.* New Haven: Yale University Press, 1926.

Ismael, Jacqueline S. *Kuwait: Social Change in Historical Perspective.* Syracuse: Syracuse University Press, 1982.

Issawi, Charles, ed. *The Economic History of Iran, 1800–1914.* Chicago: University of Chicago Press, 1971.

854

Issawi, Charles, and Mohammed Yeganeh. *The Economics of Middle Eastern Oil*. London: Faber and Faber, 1962.

Jackson, Kenneth T. *Crabgrass Frontier: The Suburbanization of the United States*. Oxford: Oxford University Press, 1987.

Jacoby, Neil H. *Multinational Oil: A Study in Industrial Dynamics*. New York: Macmillan, 1974.

James, D. Clayton. *The Years of MacArthur*. Vol. 2, *1941–1945*. Boston: Houghton Mifflin, 1975.

James, Marquis. *The Texaco Story: The First Fifty Years, 1902–1952*. New York: Texas Company, 1953.

James, Robert Rhodes. *Anthony Eden*. New York: McGraw-Hill, 1987.

Jensen, Robert G., Theodore Shabad, and Arthur W. Wright, eds. *Soviet Natural Resources in the World Economy*. Chicago: University of Chicago Press, 1983.

Jensen, W. G. *Energy in Europe, 1945–1980*. London: G. T. Foulis, 1967.

———. "The Importance of Energy in the First and Second World Wars." *Historical Journal* 11 (1968): 538–54.

Jentleson, Bruce. *Pipeline Politics: The Complex Political Economy of East-West Energy Trade*. Ithaca: Cornell University Press, 1986.

Johnson, Arthur M. *The Challenge of Change: The Sun Oil Company, 1945–1977*. Columbus: Ohio State University Press, 1983.

———. *The Development of American Petroleum Pipelines: A Study in Private Enterprise and Public Policy, 1862–1906*. Ithaca: Cornell University Press, 1956.

———. *Petroleum Pipelines and Public Policy, 1906–1959*. Cambridge: Harvard University Press, 1967.

Johnson, Chalmers. *MITI and the Japanese Miracle: The Growth of Industrial Policy, 1925–1975*. Stanford: Stanford University Press, 1982.

Johnson, William Weber. *Heroic Mexico: The Violent Emergence of a Modern Nation*. Garden City, N.Y.: Doubleday, 1968.

Jones, Charles S. *From the Rio Grande to the Arctic: The Story of the Richfield Oil Corporation*. Norman: University of Oklahoma Press, 1972.

Jones, Geoffrey. "The British Government and the Oil Companies, 1912–24: The Search For an Oil Policy." *Historical Journal* 20 (1977): 647–72.

———. *The State and the Emergence of the British Oil Industry*. London: Macmillan, 1981.

Jones, Geoffrey, and Clive Trebilcock. "Russian Industry and British Business, 1910–1930: Oil and Armaments." *Journal of European Economic History* 11 (Spring 1982): 61–104.

Jordan, Hamilton. *Crisis: The Last Year of the Carter Presidency*. New York: G. P. Putnam's Sons, 1982.

Kahn, David. *The Codebreakers: The Story of Secret Writing*. New York: Macmillan, 1967.

Kane, N. Stephen. "Corporate Power and Foreign Policy: Efforts of American Oil Companies to Influence United States Relations with Mexico, 1921–28." *Diplomatic History* 1 (Spring 1977): 170–98.

Kaplan, Justin. *Mr. Clemens and Mark Twain*. New York: Simon and Schuster, 1966.

Kapstein, Ethan B. *The Insecure Alliance: Energy Crises and Western Politics Since 1944*. Oxford: Oxford University Press, 1990.

Kase, Toshikaze. *Journey to the Missouri*. Ed. David N. Rowe. New Haven: Yale University Press, 1950.

Kaufman, Burton I. "Oil and Antitrust: The Oil Cartel Case and the Cold War." *Business History Review* 51 (Spring 1977): 35–56.

———. *The Oil Cartel Case: A Documentary Study of Antitrust Activity in the Cold War Era*. Westport, Conn.: Greenwood Press, 1978.

Kazemzadeh, Firuz. *Russia and Britain in Persia, 1864–1914*. New Haven: Yale University Press, 1968.

———. *The Struggle for Transcaucasia, 1917–1921*. New York: Philosophical Library, 1951.

Keddie, Nikki R., ed. *Scholars, Saints, and Sufis: Muslim Religious Institutions Since 1500*. Berkeley: University of California Press, 1972.

Kedourie, Elie. *England and the Middle East: The Destruction of the Ottoman Empire, 1914–1921*. London: Bowes and Bowes, 1956.

Keegan, John. *The Price of Admiralty: The Evolution of Naval Warfare*. New York: Viking Press, 1989.

Kelly, J. B. *Arabia, the Gulf and the West*. New York: Basic Books, 1980.

Kelly, W. J., and Tsureo Kano. "Crude Oil Production in the Russian Empire, 1818–1919." *Journal of European Economic History* 6 (Fall 1977): 307–38.

Kemp, Norman. *Abadan: A First-Hand Account of the Persian Oil Crisis*. London: Allan Wingate, 1953.

Kennedy, K. H. *Mining Tsar: The Life and Times of Leslie Urquhart*. Boston: George Allen & Unwin, 1986.

Kennedy, Paul M. *The Rise of the Anglo-German Antagonism, 1860–1914*. London: George Allen & Unwin, 1982.

———. *Rise and Fall of the Great Powers: Economic Change and Military Conflict from 1500 to 2000*. New York: Random House, 1987.

Kent, Marian. *Oil and Empire: British Policy and Mesopotamian Oil, 1900–1920*. London: Macmillan, 1976.

Kent, Marian, ed. *The Great Powers and the End of the Ottoman Empire*. London: George Allen & Unwin, 1984.

Keohane, Robert O. *After Hegemony: Cooperation and Discord in the World Political Economy*. Princeton: Princeton University Press, 1984.

Kerr, George P. *Time's Forelock: A Record of Shell's Contribution to Aviation in the Second World War*. London: Shell Petroleum Company, 1948.

King, John O. *Joseph Stephen Cullinan: A Study of Leadership in the Texas Petroleum Industry, 1897–1937*. Nashville: Vanderbilt University Press, 1970.

Kirby, S. Woodburn. *The War Against Japan*. 4 vols. London: HMSO, 1957–1965.

Kirk, Geoffrey, ed. *Schumacher on Energy*. London: Sphere Books, 1983.

Kissinger, Henry A. *White House Years*. Boston: Little, Brown, 1979.

———. *Years of Upheaval*. Boston: Little, Brown, 1982.

Klein, Herbert S. "American Oil Companies in Latin America: The Bolivian Experience." *Inter-American Economic Affairs* 18 (Autumn 1964): 47–72.

Knowles, Ruth Sheldon. *The Greatest Gamblers: The Epic of America's Oil Exploration*. 2d ed. Norman: University of Oklahoma Press, 1978.

Koppes, Clayton R. "The Good Neighbor Policy and the Nationalization of Mexican Oil: A Reinterpretation." *Journal of American History* 69 (June 1982): 62–81.

Koskoff, David E. *The Mellons: The Chronicle of America's Richest Family*. New York: Thomas Y. Crowell, 1978.

Krammer, Arnold. "Fueling the Third Reich." *Technology and Culture* 19 (July 1978): 394–422.

Kuisel, Richard. *Ernest Mercier: French Technocrat*. Berkeley: University of California Press, 1967.

Kuniholm, Bruce R. *The Origins of the Cold War in the Near East: Great Power Conflict and Diplomacy in Iran, Turkey, and Greece*. Princeton: Princeton University Press, 1980.

Kvendseth, Stig S. *Giant Discovery: A History of Ekofisk Through the First 20 Years*. Tanager: Phillips Petroleum Norway, 1988.

Lamb, David. *The Arabs: Journeys Beyond the Mirage*. New York: Vintage, 1988.

Landau, Christopher T. "The Rise and Fall of Petro-Liberalism: United States Relations with Socialist Venezuela, 1945–1948." Senior Thesis, Harvard University, 1985.

Landes, David. *The Unbound Prometheus: Technological Change and Industrial Development in Western Europe from 1750 to the Present*. Cambridge: Cambridge University Press, 1969.

Lapping, Brian. *End of Empire*. London: Granada, 1985.

Larson, Henrietta M., Evelyn H. Knowlton, and Charles S. Popple. *History of Standard Oil Company (New Jersey)*. Vol. 3, *New Horizons, 1927–1950*. New York: Harper & Row, 1971.

Larson, Henrietta M., and Kenneth Wiggins Porter. *History of Humble Oil and Refining Company: A Study in Industrial Growth*. New York: Harper & Brothers, 1959.

Leach, Barry A. *German Strategy Against Russia, 1939–1941*. London: Clarendon Press, 1973.

Lear, Linda J. "Harold L. Ickes and the Oil Crisis of the First Hundred Days." *Mid-America* 63 (January 1981): 3–17.

Leatherdale, Clive. *Britain and Saudi Arabia, 1925–1939: The Imperial Oasis*. London: Frank Cass, 1983.

Lebkicher, Roy. *Aramco and World Oil*. New York: Russell F. Moore, [1953].

Lenzner, Robert. *Getty: The Richest Man in the World*. London: Grafton Books, 1985.

L'Espagnol de la Tramerye, Pierre. *The World Struggle for Oil*. Trans. C. Leonard Leese. London: George Allen & Unwin, 1924.

Levi, Primo. *Survival in Auschwitz and the Reawakening: Two Memoirs*. Trans. Stuart Woolf. New York: Summit Books, 1985.

Levy, Walter J. *Oil Strategy and Politics, 1941–1981*. Ed. Melvin A. Conant. Boulder, Colo.: Westview Press, 1982.

Lewin, Ronald. *The American Magic: Codes, Ciphers and the Defeat of Japan*. New York: Farrar Straus Giroux, 1982.

————. *Hitler's Mistakes.* New York: William Morrow, 1984.

Liddell Hart, B. H., ed. *History of the Second World War.* New York: G. P. Putnam's Sons, 1970.

————. *A History of the World War, 1914–1918.* London: Faber and Faber, 1934.

————. *The Other Side of the Hill: Germany's Generals; Their Rise and Fall, with Their Own Account of Military Events, 1939–1945.* 2d ed. London: Cassell, 1973.

————. *The Rommel Papers.* Trans. Paul Findlay. 1953. Reprint. New York: Da Capo Press, 1985.

Lieber, Robert J. *Oil and the Middle East War: Europe in the Energy Crisis.* Cambridge: Harvard Center for International Affairs, 1976.

————. *The Oil Decade: Conflict and Cooperation in the West.* New York: Praeger, 1983.

Lieuwen, Edwin. *Petroleum in Venezuela: A History.* Berkeley: University of California Press, 1954.

Littlefield, Douglas R., and Tanis C. Thorne. *The Spirit of Enterprise: The History of Pacific Enterprises from 1886 to 1989.* Los Angeles: Pacific Enterprises, 1990.

Lloyd, Selwyn. *Suez 1956: A Personal Account.* London: Jonathan Cape, 1978.

Longhurst, Henry. *Adventure in Oil: The Story of British Petroleum.* London: Sidgwick and Jackson, 1959.

Longrigg, Stephen H. *Oil in the Middle East: Its Discovery and Development.* 3d ed. London: Oxford University Press, 1968.

Louis, William Roger. *The British Empire in the Middle East 1945–1951: Arab Nationalism, the United States, and Postwar Imperialism.* Oxford: Clarendon Press, 1985.

Louis, William Roger, and Roger Owen, eds. *Suez 1956: The Crisis and its Consequences.* Oxford: Clarendon Press, 1989.

Love, Kenneth. *Suez: The Twice-Fought War.* New York: McGraw-Hill, 1969.

Lowe, Peter. *Great Britain and the Origins of the Pacific War: A Study of British Policy in East Asia, 1937–1941.* Oxford: Clarendon Press, 1977.

Loewenheim, Francis L., Harold D. Langley, and Manfred Jonas, eds. *Roosevelt and Churchill: Their Secret Wartime Correspondence.* New York: E. P. Dutton, 1975.

Lubell, Harold. *Middle East Oil Crises and Western Europe's Energy Supplies.* Baltimore: Johns Hopkins University Press, 1963.

————. "World Petroleum Production and Shipping: A Post-Mortem on Suez." P-1274. Rand Corporation, January 2, 1958.

Lucas, James. *War in the Desert: The Eighth Army at El Alamein.* New York: Beaufort Books, 1982.

Ludendorff, Erich. *My War Memories, 1914–1918.* London: Hutchinson, [1945].

————. *The Nation at War.* Trans. A. S. Rappaport. London: Hutchinson, 1936.

Mackay, Ruddock F. *Fisher of Kilverstone.* Oxford: Clarendon Press, 1973.

MacMahon, Arthur W., and W. R. Dittman. "The Mexican Oil Industry Since Expropriation." *Political Science Quarterly* 57 (March 1942): 28–50, (June 1942): 161–88.

Macmillan, Harold. *Riding the Storm, 1956–59.* London: Macmillan, 1971.

Manchester, William. *A Rockefeller Family Portrait, From John D. to Nelson.* Boston: Little, Brown, 1959.

von Manstein, Erich. *Lost Victories.* Trans. Anthony G. Powell. London: Methuen, 1958.

Mantoux, Paul. *Paris Peace Conference, 1919: Proceedings of the Council of Four* (March 24–April 18). Trans. John Boardman Whitton. Geneva: Droz, 1964.

Ma'oz, Moshe. *Asad: The Sphinx of Damascus.* New York: Grove Weidenfeld, 1988.

Marder, Arthur J. *From the Dreadnought to Scapa Flow: The Royal Navy in the Fisher Era, 1904–1919.* Vol. 1, *The Road to War, 1904–1914.* London: Oxford University Press, 1961.

Marr, Phebe. *The Modern History of Iraq.* Boulder, Colo.: Westview Press, 1985.

Marvin, Charles. *The Region of Eternal Fire: An Account of a Journey to the Petroleum Region of the Caspian in 1883.* New ed. London: W. H. Allen, 1891.

Maurer, John H. "Fuel and the Battle Fleet: Coal, Oil, and American Naval Strategy, 1898–1925." *Naval War College Review* 34 (November–December 1981): 60–77.

May, George S. *A Most Unique Machine: The Michigan Origins of the American Automobile Industry.* Grand Rapids, Mich.: Eerdmans Publishing, 1975.

McBeth, B. S. *British Oil Policy, 1919–1939.* London: Frank Cass, 1985.

————. *Juan Vicente Gomez and the Oil Companies in Venezuela, 1908–1935.* Cambridge: Cambridge University Press, 1983.

McCloy, John J., Nathan W. Pearson, and Beverly Matthews. *The Great Oil Spill: The Inside Report— Gulf Oil's Bribery and Political Chicanery.* New York: Chelsea House, 1976.

McFadzean, Frank. *The Practice of Moral Sentiment*. London: Shell, n.d.

McGhee, George. *Envoy to the Middle World: Adventures in Diplomacy*. New York: Harper & Row, 1983.

McKay, John P. "Entrepreneurship and the Emergence of the Russian Petroleum Industry, 1813–1883." *Research in Economic History* 8 (1982): 47–91.

McLaurin, John J. *Sketches in Crude Oil*. 3d ed. Franklin, Pa., 1902.

McLean, John G., and Robert Haigh. *The Growth of Integrated Oil Companies*. Boston: Harvard Business School, 1954.

McNaugher, Thomas L. *Arms and Oil: U.S. Military Strategy and the Persian Gulf*. Washington, D.C.: Brookings Institution, 1985.

———. "Walking Tightropes in the Gulf." *The Iran-Iraq War: Impact and Implications*, ed. Efraim Karsh. London: Macmillan, 1989.

McNeill, William H. *The Pursuit of Power: Technology, Armed Force, and Society Since A.D. 1000*. Chicago: University of Chicago Press, 1982.

Meadows, Donella, Dennis Meadows, Jorgen Randers, and William Behrens, III. *The Limits to Growth: A Report for the Club of Rome's Project on the Predicament of Mankind*. 2d ed. New York: Signet Books, 1974.

Mejcher, Helmut. *Imperial Quest for Oil: Iraq, 1910–1928*. London: Ithaca Press, 1976.

Melby, Eric D. K. *Oil and the International System: The Case of France, 1918–1969*. New York: Arno Press, 1981.

Mellon, W. L., and Boyden Sparkes. *Judge Mellon's Sons*. Pittsburgh, 1948.

Meyer, Lorenzo. *Mexico and the United States in the Oil Controversy, 1917–1942*. 2d ed. Trans. Muriel Vasconcellos. Austin: University of Texas Press, 1977.

Middlemas, R. K. *The Master-Builders*. London: Hutchinson, 1963.

Mierzejewski, Alfred C. *The Collapse of the German War Economy, 1944–1945: Allied Air Power and the German National Railway*. Chapel Hill: University of North Carolina Press, 1988.

Mikdashi, Zuhayr M., Sherrill Cleland, and Ian Seymour. *Continuity and Change in the World Oil Industry*. Beirut: Middle East Research and Publishing Center, 1970.

Miller, Aaron David. *Search for Security: Saudi Arabian Oil and American Foreign Policy, 1939–1949*. Chapel Hill: University of North Carolina Press, 1980.

Miller, Russell. *The House of Getty*. London: Michael Joseph, 1985.

Moberly, F. J. *The Campaign in Mesopotamia 1914–1918*. 4 vols. London: HMSO, 1923–1927.

Moncrieff, Anthony, ed. *Suez: Ten Years After*. New York: Pantheon, 1966.

Monroe, Elizabeth. *Britain's Moment in the Middle East, 1914–1971*. 2d ed. London: Chatto and Windus, 1981.

———. *Philby of Arabia*. London: Faber and Faber, 1973.

Montagu, Gilbert. *The Rise and Progress of the Standard Oil Company*. New York: Harper & Row, 1903.

Montgomery, Bernard. *The Memoirs of Field-Marshal the Viscount Montgomery of Alamein*. 1958. Reprint. New York: Da Capo Press, 1982.

Moore, Austin Leigh. *John D. Archbold and the Early Development of Standard Oil*. New York: Macmillan, [1930].

Moore, Frederick Lee, Jr. "Origin of American Oil Concessions in Bahrein, Kuwait, and Saudi Arabia." Senior Thesis, Princeton University, 1948.

Moran, Theodore H. "Managing an Oligopoly of Would-Be Sovereigns: The Dynamics of Joint Control and Self-Control in the International Oil Industry Past, Present, and Future." *International Organization* 41 (Autumn 1987): 576–607.

Morison, Samuel Eliot. *History of United States Naval Operations in World War II*. 8 vols. Boston: Little, Brown, 1947–1953.

Morley, James William, ed. *Japan's Road to the Pacific War*. 4 vols. New York: Columbia University Press, 1976–84.

Mosley, Leonard. *Power Play: Oil in the Middle East*. New York: Random House, 1973.

Mottahedeh, Roy. *The Mantle of the Prophet: Religion and Politics in Iran*. London: Penguin Books, 1987.

Nash, Gerald D. *United States Oil Policy, 1890–1964*. Pittsburgh: University of Pittsburgh Press, 1968.

Nasser, Gamal Abdel. *The Philosophy of the Revolution*. Buffalo, N.Y.: Smith, Keynes, and Marshall, 1959.

Neff, Donald. *Warriors at Suez: Eisenhower Takes the United States into the Middle East*. New York: Simon and Schuster, 1981.

Nevakivi, Jukka. *Britain, France and the Arab Middle East, 1914–1920*. London: Athlone Press, 1969.

Nevins, Allan. *John D. Rockefeller: The Heroic Age of American Enterprise*. 2 vols. New York: Charles Scribner's Sons, 1940.

———. *Study in Power: John D. Rockefeller, Industrialist and Philanthropist*. 2 vols. New York: Charles Scribner's Sons, 1953.

Nevins, Allan, with Frank Ernest Hill. *Ford: The Times, the Man, the Company*. 2 vols. New York: Charles Scribner's Sons, 1954.

New York Mercantile Exchange. *A History of Commerce at the New York Mercantile Exchange: The Evolution of an International Marketplace, 1872–1988*. New York: NYMEX, 1988.

Nicolson, Harold. *Portrait of a Diplomatist*. Boston: Houghton Mifflin, 1930.

Nitze, Paul, with Ann M. Smith and Steven L. Reardon. *From Hiroshima to Glasnost: At the Center of Decision—A Memoir*. New York: Grove Weidenfeld, 1989.

Nivola, Pietro S. *The Politics of Energy Conservation*. Washington, D.C.: Brookings Institution, 1986.

Nixon, Richard M. *RN: The Memoirs of Richard Nixon*. New York: Grosset & Dunlap, 1978.

Noakes, J., and G. Pridham, eds. *Nazism: A History in Documents and Eyewitness Accounts, 1919–1945*. 2 vols. New York: Schocken Books, 1989.

Noggle, Burl. *Teapot Dome: Oil and Politics in the 1920s*. Baton Rouge: Louisiana State University Press, 1962.

Nomura, Kichisaburo. "Stepping Stones to War." *United States Naval Institute Proceedings* 77 (September 1951): 927–31.

Nordhauser, Norman. *The Quest for Stability: Domestic Oil Regulation, 1917–1935*. New York: Garland, 1979.

Nowell, Gregory Patrick. "Realpolitik vs. Transnational Rent-seeking: French Mercantilism and the Development of the World Oil Cartel, 1860 1939." Ph.D. dissertation, Massachusetts Institute of Technology, 1988.

Nutting, Anthony. *Nasser*. New York: E. P. Dutton, 1972.

———. *No End of a Lesson: The Story of Suez*. London: Constable, 1967.

O'Brien, Dennis J. "The Oil Crisis and the Foreign Policy of the Wilson Administration, 1917–1921." Ph.D. dissertation, University of Missouri, 1974.

Odell, Peter R. *Oil and World Power: Background of the Oil Crisis*. 8th ed. New York: Viking Penguin, 1986.

Ogata, Sadako N. *Defiance in Manchuria: The Making of Japanese Foreign Policy, 1931–32*. Berkeley: University of California Press, 1964.

Ohashi, A. Tadahiko. *Enerugi No Seiji Keizai Gaku* (The Political Economy of Energy). Tokyo: Diamond, 1988.

Olien, Roger M., and Diana Davids Olien. *Wildcatters: Texas Independent Oilmen*. Austin: Texas Monthly Press, 1984.

Owen, Edgar Wesley. *Trek of the Oil Finders: A History of Exploration for Oil*. Tulsa: American Association of Petroleum Geologists, 1975.

Pahlavi, Mohammed Reza. *Mission for My Country*. New York: McGraw-Hill, 1961.

———. *The Shah's Story*. Trans. Teresa Waugh. London: Michael Joseph, 1980.

Painter, David S. *Oil and the American Century: The Political Economy of U.S. Foreign Oil Policy, 1941–1954*. Baltimore: Johns Hopkins University Press, 1986.

———. "Oil and the Marshall Plan." *Business History Review* 58 (Autumn 1984): 359–83.

Parsons, Anthony. *The Pride and the Fall: Iran, 1974–1979*. London: Jonathan Cape, 1984.

Passer, Harold G. *The Electrical Manufacturers, 1875–1900*. Cambridge: Harvard University Press, 1953.

Payton-Smith, D. T. *Oil: A Study of War-time Policy and Administration*. London: HMSO, 1971.

Pearce, Joan, ed. *The Third Oil Shock: The Effects of Lower Oil Prices*. London: Royal Institute of International Affairs, 1983.

Pearton, Maurice. *Oil and the Romanian State*. Oxford: Clarendon Press, 1971.

Penrose, Edith T. *The Large International Firm in Developing Countries: The International Petroleum Industry*. London: George Allen & Unwin, 1968.

Penrose, Edith, and E. F. Penrose. *Iraq: International Relations and National Development*. London: Ernest Benn, 1978.

Philby, H. St. J. B. *Arabian Days: An Autobiography*. London: Robert Hale, 1948.

———. *Arabian Jubilee*. London: Robert Hale, 1952.

———. *Arabian Oil Ventures.* Washington, D.C.: Middle East Institute, 1964.

———. *Saudi Arabia.* London: Ernest Benn, 1955.

Philip, George. *Oil and Politics in Latin America: Nationalist Movements and State Companies.* Cambridge: Cambridge University Press, 1982.

Phillips Petroleum Company. *Phillips: The First 66 Years.* Bartlesville, Okla.: Phillips Petroleum, 1983.

Pickens, T. Boone, Jr. *Boone.* Boston: Houghton Mifflin, 1987.

Pogue, Forrest C. *George C. Marshall.* 4 vols. New York: Viking Press, 1963–87.

Polster, Deborah. "The Need for Oil Shapes the American Diplomatic Response to the Invasion of Suez." Ph.D. dissertation, Case Western Reserve University, 1985.

Popple, Charles Sterling. *Standard Oil Company (New Jersey) in World War II.* New York: Standard Oil, 1952.

Potter, E. B. *Nimitz.* Annapolis, Md.: Naval Institute Press, 1976.

Prange, Gordon W., with Donald M. Goldstein and Katherine V. Dillon. *At Dawn We Slept: The Untold Story of Pearl Harbor.* New York: McGraw-Hill, 1981.

———. *Pearl Harbor: The Verdict of History. New York: McGraw-Hill,* 1986.

Pratt, Joseph A. "The Petroleum Industry in Transition: Anti-Trust and the Decline of Monopoly Control in Oil." *Journal of Economic History* 40 (December 1980): 815–37.

Prindle, David F. *Petroleum Politics and the Texas Railroad Commission.* Austin: University of Texas Press, 1981.

Quandt, William B. *Camp David: Peacemaking and Politics.* Washington, D.C.: Brookings Institution, 1986.

———. *Decade of Decisions: American Policy Towards the Arab-Israeli Conflict, 1967–1976.* Berkeley: University of California Press, 1977.

———. "Soviet Policy in the October Middle East War." *International Affairs* 53 (July 1977): 377–389, (October 1977): 587–603.

Rabe, Stephen G. *The Road to OPEC: United States Relations with Venezuela, 1919–1976.* Austin: University of Texas Press, 1982.

Rae, John B. *American Automobile Manufacturers: The First Forty Years.* Philadelphia: Chilton Company, 1959.

———. *The American Automobile: A Brief History.* Chicago: University of Chicago Press, 1965.

———. *The Road and Car in American Life.* Cambridge: MIT Press, 1971.

Ramazani, Rouhallah K. *Iran's Foreign Policy, 1941–1973: A Study of Foreign Policy in Modernizing Nations.* Charlottesville: University of Virginia Press, 1975.

———. *Revolutionary Iran: Challenge and Response in the Middle East.* Baltimore: Johns Hopkins University Press, 1986.

Rand, Christopher. *Making Democracy Safe for Oil: Oil Men and the Islamic Middle East.* Boston: Little, Brown, 1975.

Randall, Stephen J. *United States Foreign Oil Policy, 1919–1948: For Profits and Security.* Kingston: McGill-Queen's University Press, 1985.

Redwood, Boverton. *Petroleum: A Treatise.* 4th ed. 3 vols. London: Charles Griffin & Co., 1922.

Rhodes, Richard. *The Making of the Atomic Bomb.* New York: Touchstone, 1988.

Rintoul, William. *Drilling Ahead: Tapping California's Richest Oil Fields.* Santa Cruz, Calif.: Valley Publishers, 1981.

———. *Spudding In: Recollections of Pioneer Days in the California Oil Fields.* Fresno: California Historical Society, 1976.

Risch, Erna. *Fuels for Global Conflict.* Washington, D.C.: Office of Quartermaster General, 1945.

Rister, Carl Coke. *Oil! Titan of the Southwest.* Norman: University of Oklahoma Press, 1949.

Ristow, Walter. "A Half Century of Oil-Company Road Maps." *Surveying and Mapping* 34 (December 1964): 617–37.

Roberts, Glyn. *The Most Powerful Man in the World: The Life of Sir Henri Deterding.* New York: Covici Friede, 1938.

Robinson, Jeffrey. *Yamani: The Inside Story.* London: Simon and Schuster, 1988.

Robinson, M. S. "The Crude Oil Price Spiral of 1978–80." Shell, 1982.

———. "The Great Bear Market in Oil 1980–1983." Shell, 1983.

Rockefeller, John D. *Random Reminiscences of Men and Events.* New York: Doubleday, Page & Co., 1909.

Rondot, Jean. *La Compagnie Française des Pétroles.* Paris: Plon, 1962.

Roosevelt, Kermit. *Countercoup: The Struggle for the Control of Iran.* New York: McGraw-Hill, 1979.

Rosenberg, David A. "The U.S. Navy and the Problem of Oil in a Future War: The Outline of a Strategic Dilemma, 1945–1950." *Naval War College Review* 29 (Summer 1976): 53–61.

Roskill, S. W. *The War at Sea, 1939–1945.* 3 vols. London: HMSO, 1954–61.

Rostow, Eugene V. *A National Policy for the Oil Industry.* New Haven: Yale University Press, 1948.

Rothwell, V. H. "Mesopotamia in British War Aims." *Historical Journal* 13 (1970): 273–94.

Rouhani, Fuad. *A History of O.P.E.C.* New York: Praeger, 1971.

Rourke, Thomas. *Gomez: Tyrant of the Andes.* Garden City, N.Y.: Halcyon House, 1936.

Roux, Georges. *Ancient Iraq.* 2nd ed. London: Penguin Books, 1985.

Rowland, John, and Basil Cadman. *Ambassador for Oil: The Life of John, First Baron Cadman.* London: Herbert Jenkins, 1960.

Rubin, Barry. *The Great Powers in the Middle East, 1941–1947: The Road to the Cold War.* London: Frank Cass, 1980.

———. *Paved with Good Intentions: The American Experience and Iran.* New York: Penguin Books, 1984.

Ruppenthal, Roland G. *Logistical Support of the Armies.* 2 vols. Washington, D.C.: Department of the Army, 1953–58.

Rustow, Dankwart A. *Oil and Turmoil: America Faces OPEC and the Middle East.* New York: W. W. Norton, 1982.

el-Sadat, Anwar. *In Search of Identity: An Autobiography.* New York: Harper & Row, 1978.

Safran, Nadav. *Israel: The Embattled Ally.* Cambridge: Harvard University Press, 1978.

———. *From War to War: The Arab-Israeli Confrontation, 1948–1967.* New York: Pegasus, 1969.

———. *Saudi Arabia: The Ceaseless Quest for Security.* Cambridge: Harvard University Press, 1985.

Sampson, Anthony. *The Seven Sisters: The Great Oil Companies and the World They Created.* Rev. ed. London: Coronet, 1988.

Samuels, Richard J. *The Business of the Japanese State: Energy Markets in Comparative Historical Perspective.* Ithaca: Cornell University Press, 1987.

Schlesinger, James R. *The Political Economy of National Security: A Study of the Economic Aspects of the Contemporary Power Struggle.* New York: Praeger, 1960.

Schmitt, Bernadotte E., and Harold C. Vedeler. *The World in the Crucible, 1914–1919.* New York: Harper & Row, 1984.

Schneider, Steven A. *The Oil Price Revolution.* Baltimore: Johns Hopkins University Press, 1983.

Schumacher, E. F. *Small Is Beautiful: A Study of Economics As If People Mattered.* London: Blond and Briggs, 1973.

Seaton, Albert. *The Russo-German War, 1941–1945.* London: Arthur Barker, 1971.

Seymour, Ian. *OPEC: Instrument of Change.* London: Macmillan, 1980.

Shawcross, William. *The Shah's Last Ride: The Fate of an Ally.* New York: Simon and Schuster, 1988.

Sherrill, Robert. *The Oil Follies of 1970–1980: How the Petroleum Industry Stole the Show (and Much More Besides).* Garden City, N.Y.: Anchor Press/Doubleday, 1983.

Sherwood, Elizabeth D. *Allies in Crises: Meeting Global Challenges to Western Security.* New Haven: Yale University Press, 1990.

Shlaim, Avi. "Failures in National Intelligence Estimates: The Case of the Yom Kippur War." *World Politics* 28 (April 1976): 348–80.

Shuckburgh, Evelyn. *Descent to Suez: Diaries, 1951–1956.* Ed. John Charmley. London: Weidenfeld and Nicolson, 1986.

Shwadran, Benjamin. *The Middle East, Oil and the Great Powers.* 3d rev. ed. New York: John Wiley, 1973.

Sick, Gary. *All Fall Down: America's Tragic Encounter with Iran.* New York: Viking Penguin, 1986.

Silliman, Jr., B. *Report on the Rock Oil, or Petroleum, from Venango Co., Pennsylvania.* New Haven: J. H. Benham's, 1855.

Simon, William E. *A Time for Truth.* New York: Berkley, 1978.

Sinclair Oil. *A Great Name in Oil: Sinclair Through 50 Years.* New York: F. W. Dodge/McGraw-Hill, 1966.

Singer, Mark. *Funny Money.* New York: Alfred A. Knopf, 1985.

Skeet, Ian. *OPEC—Twenty-five Years of Prices and Politics.* Cambridge: Cambridge University Press, 1988.

Sluglett, Peter. *Britain in Iraq, 1914–1932.* London: Ithaca Press, 1976.

861

Smith, Adam. *Paper Money.* New York: Summit Books, 1981.

———. *The Roaring '80s.* New York: Summit Books, 1988.

Smith, George Otis, ed. *The Strategy of Minerals: A Study of the Mineral Factor in the World Position of America in War and in Peace.* New York: D. Appleton, 1919.

Smith, P. G. A. *The Shell That Hit Germany Hardest.* London: Shell Marketing Co., [1921].

Smith, Robert Freeman. *The United States and Revolutionary Nationalism in Mexico, 1916–1932.* Chicago: The University of Chicago Press, 1972.

Solberg, Carl E. *Oil and Nationalism in Argentina: A History.* Stanford: Stanford University Press, 1979.

Spector, Ronald H. *Eagle Against the Sun: The American War with Japan.* New York: Vintage, 1985.

Speer, Albert. *Inside the Third Reich.* Trans. Richard and Clara Winston. New York: Macmillan, 1970.

Spence, Hartzell. *Portrait in Oil: How the Ohio Oil Company Grew to Become Marathon.* New York: McGraw-Hill, 1962.

Spender, J. A. *Weetman Pearson: First Viscount Cowdray, 1856–1927.* London: Cassell, 1930.

Standard Oil Company (New Jersey). *Ships of the Esso Fleet in World War II.* New York: Standard Oil, 1946.

Stegner, Wallace. *Discovery: The Search for Arabian Oil.* Beirut: Middle East Export Press, 1974.

Steiner, Zara S. *Britain and the Origins of the First World War.* New York: St. Martin's Press, 1977.

Stent, Angela. *From Embargo to Ostpolitik: The Political Economy of Soviet-West German Relations 1955–1980.* Cambridge: Cambridge University Press, 1981.

———. *Soviet Energy and Western Europe.* Washington paper 90. New York: Praeger, 1982.

Stivers, William. *Supremacy and Oil: Iraq, Turkey, and the Anglo-American World Order, 1918–1930.* Ithaca: Cornell University Press, 1982.

Stobaugh, Robert. "The Evolution of Iranian Oil Policy, 1925–1975." *Iran Under the Pahlavis,* ed. George Lenczowski. Stanford, Calif.: Hoover Institution Press, 1978.

Stobaugh, Robert, and Daniel Yergin, eds. *Energy Future: Report of the Energy Project at the Harvard Business School.* 3d ed. New York: Vintage, 1983.

Stocking, George. *Middle East Oil: A Study in Political and Economic Controversy.* Knoxville, Tenn.: Vanderbilt University Press, 1970.

Stoff, Michael B. *Oil, War, and American Security: The Search for a National Policy on Foreign Oil, 1941–47.* New Haven: Yale University Press, 1980.

Stokes, Raymond G. "The Oil Industry in Nazi Germany." *Business History Review* 59 (Summer 1985): 254–77.

Stone, Norman. *Hitler.* Boston: Little, Brown, 1980.

Storry, G. R. "The Mukden Incident of September 18–19, 1931." *St. Antony's Papers: Far Eastern Affairs* 2 (1957): 1–12.

Sullivan, William H. *Mission to Iran.* New York: W. W. Norton, 1981.

Suny, Ronald G. *The Baku Commune, 1917–1918.* Princeton: Princeton University Press, 1972.

———. "A Journeyman for the Revolution: Stalin and the Labour Movement in Baku, June 1907–May 1908." *Soviet Studies* 23 (January 1972): 373–94.

Tait, Samuel W., Jr. *The Wildcatters: An Informal History of Oil-Hunting in America.* Princeton: Princeton University Press, 1946.

Tarbell, Ida M. *All in the Day's Work: An Autobiography.* New York: Macmillan, 1939.

———. *The History of the Standard Oil Company.* 2 vols. New York: McClure, Phillips & Co., 1904.

Taylor, Frank J., and Earl M. Welty. *Black Bonanza: How an Oil Hunt Grew into the Union Oil Company of California.* New York: Whittlesley House, McGraw-Hill, 1950.

Terzian, Philip. *OPEC: The Inside Story.* Trans. Michael Pallis. London: Zed Books, 1985.

Thompson, Craig. *Since Spindletop: A Human Story of Gulfs First Half-Century.* Pittsburgh: Gulf Oil, 1951.

Thynne, John F. "British Policy on Oil Resources, 1936–1951, with Particular Reference to the Defense of British Controlled Oil in Mexico, Venezuela and Persia." Ph.D. dissertation, London School of Economics, 1987.

Tinkle, Lon. *Mr. De: A Biography of Everette Lee DeGolyer.* Boston: Little, Brown, 1970.

Tolf, Robert W. *The Russian Rockefellers: The Saga of the Nobel Family and the Russian Oil Industry.* Stanford, Calif.: Hoover Institution Press, 1976.

Tompkins, Walker A. *Little Giant of Signal Hill: An Adventure in American Enterprise.* Englewood Cliffs, N.J.: Prentice-Hall, 1964.

Tooley, Terry Hunt. "The German Plan for Synthetic Fuel Self-Sufficiency, 1933–1942." Master's thesis, Texas A & M University, 1978.

Townsend, Henry H. *New Haven and the First Oil Well*. New Haven, 1934.

Trevor-Roper, H. R. *Hitler's War Directives, 1939–1945*. London: Sidgwick and Jackson, 1964.

Truman, Harry S. *Memoirs*. 2 vols. Garden City, N.Y.: Doubleday, 1955–56.

Tugendhat, Christopher. *Oil: The Biggest Business*. New York: G. P. Putnam's Sons, 1968.

Tugwell, Franklin. *The Politics of Oil in Venezuela*. Stanford: Stanford University Press, 1975.

Turner, Henry Ashby, Jr. *German Big Business and the Rise of Hitler*. New York: Oxford University Press, 1987.

Turner, Louis. *Oil Companies in the International System*. London: George Allen & Unwin, 1978.

Twitchell, Karl S. *Saudi Arabia: With an Account of the Development of Its Natural Resources*. 3d ed. Princeton: Princeton University Press, 1958.

Ulam, Adam B. *Stalin: The Man and His Era*. New York: Viking Press, 1973.

Ullman, Richard H. *Anglo-Soviet Relations, 1917–1921*. 3 vols. Princeton: Princeton University Press, 1961–72.

Utley, Jonathan G. *Going to War with Japan, 1937–1941*. Knoxville: University of Tennessee Press, 1985.

Vance, Cyrus. *Hard Choices: Critical Years in America's Foreign Policy*. New York: Simon and Schuster, 1983.

van Creveld, Martin. *Supplying War: Logistics from Wallenstein to Patton*. Cambridge: Cambridge University Press, 1977.

Vernon, Raymond, ed. *The Oil Crisis in Perspective*. New York: W. W. Norton, 1976.

———. *Two Hungry Giants: The United States and Japan in the Quest for Oil and Ores*. Cambridge: Harvard University Press, 1983.

Vietor, Richard H. K. *Energy Policy in America Since 1945: A Study of Business-Government Relations*. Cambridge: Cambridge University Press, 1984.

Von Laue, Theodore. *Sergei Witte and the Industrialization of Russia*. New York: Columbia University Press, 1963.

Wack, Pierre. "Scenarios: Uncharted Waters Ahead." *Harvard Business Review* 63 (September-October 1985): 72–89.

Waley Cohen, Robert. "Economics of the Oil Industry." *Proceedings of the Empire Mining and Metallurgical Congress, 1924*.

Wall, Bennett H. *Growth in a Changing Environment: A History of Standard Oil Company (New Jersey), 1950–1972, and Exxon Corporation, 1972–1975*. New York: McGraw-Hill, 1988.

Wall, Bennett H. and George S. Gibb. *Teagle of Jersey Standard*. New Orleans: Tulane University, 1974.

Walters, Vernon A. *Silent Missions*. Garden City, N.Y.: Doubleday, 1978.

Ward, Thomas E. *Negotiations for Oil Concessions in Bahrein, El Hasa (Saudi Arabia), the Neutral Zone, Qatar and Kuwait*. New York: 1965.

Warlimont, Walter. *Inside Hitler's Headquarters, 1939–45*. Trans. R. H. Barry. London: Weidenfeld and Nicolson, 1964.

Watkins, T. H. *Righteous Pilgrim: The Life and Times of Harold L. Ickes, 1874–1952*. New York: Henry Holt, 1990.

Weaver, Jacqueline Lang. *Unitization of Oil and Gas Fields in Texas: A Study of Legislative, Administrative, and Judicial Politics*. Washington, D.C.: Resources for the Future, 1986.

Weinberg, Steve. *Armand Hammer: The Untold Story*. Boston: Little, Brown, 1989.

Wells, Tim. *444 Days: The Hostages Remember*. San Diego: Harcourt Brace Jovanovich, 1985.

Werner, M. R., and John Star. *The Teapot Dome Scandal*. New York: Viking Press, 1959.

Whaley, Barton. *Codeword Barbarossa*. Cambridge: MIT Press, 1973.

White, Gerald T. *Formative Years in the Far West: A History of Standard Oil Company of California and Predecessors Through 1919*. New York: Appleton-Century-Crofts, 1962.

———. *Scientists in Conflict: The Beginnings of the Oil Industry in California*. San Marino, Calif.: Huntington Library, 1968.

White, Graham, and John Maze. *Harold Ickes of the New Deal: His Private Life and Public Career*. Cambridge: Harvard University Press, 1985.

Wilber, Donald N. *Adventures in the Middle East: Excursions and Incursions*. Princeton, N.J.: Darwin, 1986.

Wildavsky, Aaron, and Ellen Tenenbaum. *The Politics of Mistrust: Estimating American Oil and Gas Resources*. Beverly Hills: Sage, 1981.

Wilkins, Mira. *The Emergence of Multinational Enterprise: American Business Abroad from the Colonial Era to 1914*. Cambridge: Harvard University Press, 1970.

———. *The Maturing of Multinational Enterprise: American Business Abroad from 1914 to 1970*. Cambridge: Harvard University Press, 1974.

———. "The Role of U.S. Business." *Pearl Harbor as History: Japanese-American Relations, 1931–1941*, ed. Dorothy Borg and Shumpei Okamoto. New York: Columbia University Press, 1973.

Williams, Louis, ed. *Military Aspects of the Arab-Israeli Conflict*. Tel Aviv: Tel Aviv University Publishing Project, 1975.

Williamson, Harold F., Ralph L. Andreano, Arnold R. Daum, and Gilbert C. Klose. *The American Petroleum Industry*. Vol. 2, *The Age of Energy, 1899–1959*. Evanston: Northwestern University Press, 1963.

Williamson, Harold F., and Arnold R. Daum. *The American Petroleum Industry*. Vol. 1, *The Age of Illumination, 1859–1899*. Evanston: Northwestern University Press, 1959.

Williamson, J. W. *In a Persian Oil Field: A Study in Scientific and Industrial Development*. London: Ernest Benn, 1927.

Williamson, Samuel. *The Politics of Grand Strategy: Britain and France Prepare for War, 1904–1914*. Cambridge: Harvard University Press, 1969.

Willmott, H. P. *Empires in the Balance: Japanese and Allied Pacific Strategies to April 1942*. Annapolis, Md.: Naval Institute Press, 1982.

Wilson, Arnold. *S. W. Persia: Letters and Diary of a Young Political Officer, 1907–1914*. London: Oxford University Press, 1941.

Wilson, Joan Hoff. *American Business and Foreign Policy, 1920–1933*. Boston: Beacon Press, 1971.

Winkler, John K. *John D.: A Portrait in Oils*. New York: Vanguard, 1929.

Wirth, John D., ed. *Latin American Oil Companies and the Politics of Energy*. Lincoln: University of Nebraska Press, 1985.

Witte, Serge. *The Memoirs of Count Witte*. Trans. and ed. Abraham Yarmolinsky. Garden City, N.Y.: Doubleday, Page, 1921.

Wohlstetter, Roberta. *Pearl Harbor: Warning and Decision*. Stanford: Stanford University Press, 1962.

Wood, Barbara. *E. F. Schumacher: His Life and Thought*. New York: Harper & Row, 1984.

Woodhouse, C. M. *Something Ventured*. London: Granada, 1982.

Woodward, Sir Llewellyn. *British Foreign Policy in the Second World War*. 5 vols. London: HMSO, 1970–1975.

Woolf, Henry Drummond. *Rambling Recollections*. 2 vols. London: Macmillan, 1908.

Wright, John. *Libya: A Modern History*. Baltimore: Johns Hopkins University Press, 1982.

Yamani, Ahmed Zaki. "Oil Markets: Past, Present, and Future." Energy and Environmental Policy Center, Kennedy School of Government, Harvard University, September 1986.

Yergin, Daniel. *Shattered Peace: The Origins of the Cold War*. Rev. ed. New York: Penguin Books, 1990.

Yergin, Daniel, and Martin Hillenbrand, eds. *Global Insecurity: A Strategy for Energy and Economic Renewal*. New York: Penguin Books, 1983.

Yergin, Daniel, and Barbara Kates-Garnick, eds. *The Reshaping of the Oil Industry: Just Another Commodity?* Cambridge, Mass.: Cambridge Energy Research Associates, 1985.

Yergin, Daniel, Joseph Stanislaw, and Dennis Eklof. "The U.S. Strategic Petroleum Reserve: Margin of Security." Council on Foreign Relations Paper/Cambridge Energy Research Associates Report, 1990.

Yoshihashi, Takehiko. *Conspiracy at Mukden: The Rise of the Japanese Military*. New Haven: Yale University Press, 1963.

Yoshitsu, Michael M. *Caught in the Middle East: Japan's Diplomacy in Transition*. Lexington, Mass.: D. C. Heath, 1984.

Young, Desmond. *Member for Mexico: A Biography of Weetman Pearson, First Viscount Cowdray*. London: Cassell, 1966.

Zabih, Sepehr. *The Mossadegh Era: Roots of the Iranian Revolution*. Chicago: Lake View Press, 1982.

Ziemke, Earl F. *Stalingrad to Berlin: The German Defeat in the East*. Washington, D.C.: U.S. Army, Center of Military History, 1968.

Zimmermann, Erich W. *Conservation in the Production of Petroleum: A Study in Industrial Control*. New Haven: Yale University Press, 1957.

Zweig, Philip L. *Belly Up: The Collapse of the Penn Square Bank*. New York: Fawcett Columbine, 1985.

Data Sources

American Petroleum Institute. *Basic Petroleum Data Book.*
———. *Petroleum Facts and Figures: Centennial Edition,* 1959. New York; API, 1959.
Arthur Andersen & Co. and Cambridge Energy Research Associates. *World Oil Trends.*
———. *Natural Gas Trends.*
———. *Electric Power Trends.*
British Petroleum. *Statistical Review of the World Oil Industry.* 1955–80.
———. *Statistical Review of World Energy.* 1981–89.
Chase Manhattan Bank. *Annual Financial Analysis of a Group of Petroleum Companies.* 1955–79.
Darmstadter, Joel, with Perry D. Teitelbaum and Jaroslav G. Paloch. *Energy in the World Economy: A Statistical Review of Trends in Output, Trade, and Consumption since 1925.* Baltimore: Resources for the Future, 1971.
DeGolyer & MacNaughton. *Twentieth Century Petroleum Statistics.*
Eurostat (Statistical Office of the European Communities). *Monthly Energy Statistics.*
International Energy Agency. *Energy Balances of OECD Countries.*
International Monetary Fund. *International Financial Statistics Yearbook.*
McGraw-Hill. *Platt's Oil Price Handbook and Oilmanac.*
Motor Vehicle Manufacturers Association of the U.S. *MVMA Motor Vehicle Facts & Figures.*
Organization of Petroleum Exporting Countries. *Annual Statistical Bulletin.*
———. *Petroleum Product Prices and Their Components in Selected Countries: Statistical Time Series, 1960–1983.* OPEC, [1984].
Organization for Economic Co-Operation and Development. *OECD Economic Outlook.*
United Nations. *International Trade Statistics Yearbook.*
U.S. Department of Commerce. Bureau of the Census. *Historical Statistics of the United States, 1789–1945.* Washington, D.C.: GPO, 1949.
U.S. Department of Commerce. Bureau of Mines. *Mineral Resources of the United States.* 1882–1931.
U.S. Department of Energy. Energy Information Administration. *Annual Energy Review.*
———. *Annual Report to Congress.*
———. *International Petroleum Annual.*
———. *Monthly Energy Review.*
U.S. Department of Interior. Bureau of Mines. *Minerals Yearbook.* 1932/33–87.
U.S. Department of Treasury. *Statistical Abstract of the United States.*
World Bank. *World Development Report.*

Acknowledgments

IN THE COURSE of writing a book of this scope, one accumulates many debts, and I want to acknowledge the people who have helped me in so many different ways over the years.

First and foremost, my gratitude goes to my editor at Simon & Schuster, Frederic Hills, who brought great commitment and considerable conceptual talents and insights to this project and who from the beginning focused on the great themes. Burton Beals proved his gifts for language, nuance, and the psychology of authors, saw this as an epic saga, and was dedicated to word and story. My agent Helen Brann has always provided the support her author needed.

Five people were particularly important over a number of years. Sue Lena Thompson's commitment, manifold talents, and mastery have been essential. She truly ensured that this book would happen. Robert Laubacher brought his high standards and his rigor as a historian and sleuth to the venture. Both have given a great deal of time and energy, and I am very grateful to both of them. Geoffrey Lumsden, an intrepid researcher, helped me through the labyrinth of British archives and sources.

I am especially indebted to my colleagues, James Rosenfield and Joseph Stanislaw, for their commitment, untiring support, intellectual vigor, and willingness to take on extra responsibilities to ensure that the book would be done. They really made it possible.

For their considerable advice and careful readings, I am particularly grateful to Nicholas Rizopoulos, who began as my teacher and became my friend and introduced me to diplomatic history when I was still a student; C. Napier Collyns, who insisted that this subject would properly bring together international affairs, oil and energy, and narrative history; and to James Schlesinger, who was extremely gracious with his insights, experience, and time.

I benefited greatly from extensive critical readings and other assistance from John Loudon, George McGhee, Wanda Jablonski, Tadahiko Ohashi, Philip Oxley, James Placke, Ian Skeet, Ronald W. Stent, Ernst Von Metzsch, Bennett Wall, Julian West, and Mira Wilkins. I extend my gratitude to all of them.

I must acknowledge the support from everybody at Cambridge Energy Research Associates. Some people made special contributions. Welton Barker and Mary Alice Sanderson went far beyond what might have been reasonably expected, gave extraordinary care to all they did, and were always forgiving. I thank Kathleen Fitzgerald, Susan Leland, and Leta Sinclair, all of whom worked directly with me and who managed two tracks that only sometimes met. I. C. Bupp, Dennis Eklof, and James Newcomb all gave welcome support. Peter Bogin helped a great deal with archives in Paris, and Michael Williams worked ingeniously with me on the numbers. I would also like to thank, in Cambridge, members of our Production staff—Christine Marchuk, Patricia Ingalls, Roberta Klix, Mary Moineau-Riegel, and Deanna Troust—and also Steven Aldrich, Sam Atkinson, Alice Barsoomian, Jennifer Battersby, Barbara Blodgett, Laurent Hevey, Matthew MacDonald, Kathleen Moineau, Geoffrey Morgan, Jeff Pasley, and Robin Weiss; in Paris, Micheline Manoncourt and James Long; in Oslo, Odd Hassel; and, in London, Michael Clegg. I also want to thank Barbara Kates-Garnick, Gregory Nowell, and Elizabeth Michelon.

For their sagacious perspectives, I am indebted to Raymond Vernon, a distinguished student of international economics and politics from whom I have learned much, and Edward Jordan, who has a unique grasp of business and public policy.

At Simon & Schuster, I am especially appreciative to Daphne Bien for her essential coordination. And for their great care for the manuscript, I offer deep thanks to Leslie Ellen, Ted Landry, Gypsy da Silva, and Sophie Sorkin. And I thank Irving Perkins Associates for their design and Karolina Harris for her text art direction. Also special appreciation to Sydney Cohen, Ron Doucette and crew, Robert Forget, Ursula Obst, and Karen Weitzman.

The opportunity to be a Lecturer and then a Research Associate at the John F. Kennedy School of Government at Harvard University was most constructive, and I want to express my gratitude to William Hogan, Henry Lee, Graham Allison, and Irwin Stelzer. Although many libraries and archives were helpful, one stands out, and appreciation must be expressed—to the wondrous and accessible Harvard University Library system, especially Widener Library, the Center for International Affairs, and the Baker Library at the Harvard Business School—and to the staffs thereof.

I also want to thank Jay Carlson, Herbert Goodman, and Jerome Levinson for their special and kind help.

I interviewed many people for this book. A number of them subsequently read over the relevant chapters. I thank all in the interviews section, but want to add further appreciation here, as well. Their contribution was enormous. In some cases, people preferred to assist anonymously.

In addition, I would like to express my appreciation to the following people for their comments, readings, help, and dialogues over the years: Richard Adkerson, Frank Alcock, Fausto Alzati, Hans Bär, Joseph Barri, Robert R. Bowie,

Benjamin A. Brown, Elizabeth Bumiller, Victor Burk, Scott Campbell, Guy Caruso, James Chace, Fadhil al-Chalabi, Marcello Colitti, Chester K. Cooper, Richard Cooper, Brian Coughlin, Alfred DeCrane, Jr., Richard Fairbanks, Russell Freeburg, Vera de Ladoucette, Charles DiBona, Robert Dunlop, Margaret Goralski, David Gray, Thane Gustafson, Laura Hein, Peter Holmes, J. Wallace Hopkins, Akira Iriye, Kazuhiko Itano, John Jennings, David Jones, Yoshio Karita, Milton Katz, William Kieschnick, Leonard Kujawa, Kenjiro Kumagai, the late Ulf Lantzke, Kenneth Lay, Quincy Lumsden, Robert L. Maby, Jr., Phebe Marr, William F. Martin, Thomas McNaugher, Robert McClements, John Mitchell, N. Nakahara, E. V. Newland, John Newton, John Norton, Michael O'Donnell, H. Okuda, Rene Ortiz, Alirio Parra, David Painter, Wolf Petzell, George Piercy, Maria Rodriguez, William Quandt, Beatrice Rangel, Gilbert Rutman, Peter Schwartz, Gary Sick, Robert Stobaugh, Nadir Sultan, Katsuhiko Suetsugu, Elizabeth P. Thompson, L. Paul Tempest, Robert W. Tucker, Enzo Viscusi, Hillman Walker, Barbara Weisel, Steven R. Weisman, Mason Willrich, and M. Yamao.

And I express my deepest appreciation to my wife, Angela Stent, for contributing to this work in so many different ways and for her judgment, encouragement, and caring through so many seasons. And I happily thank Alexander and Rebecca, who were patient and curious beyond their years and in their own ways helped their father a lot.

It has taken seven years to research and write this book. I am sure that I have left out some people whom I should acknowledge, and I ask their forbearance and express my appreciation. As I look back over the names herein, I can see how many so graciously helped and how multitudinous are the obligations that I have acquired. The work would not have been possible without their contributions and their cooperation and enthusiasm. But I alone in this venture am responsible for any errors and, of course, for all interpretations and judgments.

—Daniel Yergin

Acknowledgments
for the New Edition

FOR THIS NEW EDITION, first and foremost, I want to thank Martha Levin of Free Press for her commitment to this book, its new edition, and its continuing significance. Thanks also to Sharbari Bose for coordinating and expediting the publication. At CERA, my special appreciation to Amy Kipp for so effectively managing everything. I also thank Ellen Perkins for her continuing help in so many ways on this project and beyond. For their dialogue on matters oil, thanks to David Hobbs, James Burkhard, Susan Ruth, James Placke, and Ruchir Kadakia, who also helped me with the graphics. Thanks to Bethany Genier, for her involvement over many years, and to Levi Tillemann-Dick and Samantha Gross. At IHS and CERA, I'd like to acknowledge the important dialogue with Jerre Stead, Ron Mobed, Jeff Tarr, Jeff Sisson, Steve Green, and Jonathan Gear.

Continuing thanks to James Rosenfield for his creative thought and collaboration since the start of this project. And my great appreciation for her insight and perspective over the years to my colleague and friend Vera de LaDoucette. I certainly want to acknowledge Bill Cran, who saw the book as a film and ensured its translation into that medium. My appreciation for their support—Suzanne Gluck and James Wiatt.

My deep gratitude to my family—Angela, Alex, and Rebecca—from whose thinking and support, over many more seasons, I have so benefitted.

The Prize has been a much bigger project than I ever imagined. I continue to be very grateful to Sue Lena Thompson, who was devoted to it from beginning to end, in book and video, and whose spirit was essential to its completion. And finally, lasting appreciation to Fred Hills, a man of uncommon insight and an editor of great distinction, who saw the prize long before I did.

—Daniel Yergin

Photo Credits

1. BP America
2. Drake Well Museum
3. New York Public Library, courtesy American Petroleum Institute (API)
4. Drake Well Museum
5. Western Reserve Historical Society, Cleveland
6. API
7. BP America, courtesy of API
8. Standard Oil (New Jersey) Collection, University of Louisville Photographic Archives
9. Standard Oil (New Jersey) Collection, University of Louisville Photographic Archives
10. Drake Well Museum
11. BP America
12. Texas Energy Museum
13. BP America, courtesy of API
14. Barker Texas History Center, The University of Texas at Austin
15. Keystone, Paris
16. Topham, England
17. L'Illustration/Sygma, Paris
18. Shell Photographic Library, London
19. Shell Photographic Library, London
20. Shell Photographic Library, London
21. Shell Netherlands
22. API
23. API
24. Standard Oil (New Jersey) Collection, University of Louisville Photographic Archives
25. British Petroleum, London
26. British Petroleum, London
27. British Petroleum, London
28. Hulton-Deutsch Collection, London
29. API
30. Imperial War Museum, London
31. William Higgins Collection, University of Louisville Photographic Archives
32. Dwight D. Eisenhower Library
33. Amoco Corporation, courtesy of API
34. Exxon Corporation, courtesy of API
35. Phillips Petroleum
36. Sun Oil Company, courtesy of API
37. Amoco Corporation, courtesy of API
38. Standard Oil (New Jersey) Collection, University of Louisville Photographic Archives
39. Calouste Gulbenkian Foundation, Lisbon
40. British Petroleum, London
41. Standard Oil (New Jersey) Collection, University of Louisville Photographic Archives
42. DeGolyer Library
43. Shell Photographic Library, London
44. Standard Oil (New Jersey) Collection, University of Louisville Photographic Archives
45. Exxon Corporation, courtesy of API
46. The Hunt Family Archives
47. George C. McGhee
48. H. Roger Viollet, Paris
49. Kyodo Photo Service, Tokyo
50. Secretaria de Energia, Minas E Industria Paraestatal, Mexico City
51. Heinrich Hoffman
52. Roger-Viollet, Paris
53. National Archives
54. National Archives
55. U.S. Navy, courtesy Topham Picture Library, England
56. Admiral Nimitz, courtesy of Margaret Goralski and Russell W. Freeburg
57. U.S. Navy
58. National Archives, courtesy of Margaret Goralski and Russell W. Freeburg
59. U.S. Army
60. Franklin D. Roosevelt Library
61. U.S. Army, courtesy of the Patton Museum, Fort Knox, Kentucky
62. Standard Oil (New Jersey) Collection, University of Louisville Photographic Archives
63. Franklin D. Roosevelt Library
64. Standard Oil (New Jersey) Collection, University of Louisville Photographic Archives
65. Standard Oil (New Jersey) Collection, University of Louisville Photographic Archives
66. Standard Oil (New Jersey) Collection, University of Louisville Photographic Archives
67. McDonald's Corporation
68. UPI/Bettmann
69. Shell Photographic Library, London
70. Frederic Lewis/Gates
71. Topham Picture Library, England
72. Texaco
73. George C. McGhee
74. Topham Picture Library, England
75. National Park Service
76. Keystone, Paris
77. ENI, Center of Photographic Documentation
78. Wanda Jablonski
79. Standard Oil (New Jersey) Collection, University of Louisville Photographic Archives
80. Wanda Jablonski
81. Juan Pablo Pérez Castillo
82. Aramco
83. Sygma, Paris
84. Wide World Photos
85. UPI/Bettmann
86. The Mainichi Newspaper, Tokyo
87. The Nixon Project
88. Zapata Corporation, courtesy of API
89. Alyeska Pipeline Service Company, courtesy of API
90. Sygma, New York
91. UPI/Bettmann
92. New York Mercantile Exchange
93. AP/Wide World Photos
94. Sygma, New York
95. Zapata Corporation
96. White House Photo
97. Reuters/Bettmann

873

Index

Australia, 48, 119, 121
Austria, 422
Austro-Hungarian Empire, 130, 151, 164, 166
Automobile Gasoline Company, 192
automobiles:
    change in, 772
    cross-country caravan of, 190–91
    drive-in services for, 533–34
    early use of, 63
    electric power for, 64
    first races of, 63
    fuel efficiency of, 524, 642–43, 700
    "get a horse" epithet and, 63, 64
    growth in use of, 64, 178, 191–92, 391, 524, 767
    hybrid, 773
    Japanese industry of, 527–28
    in Saudi Arabia, 271–72, 616–17
    service stations and, 192–94, 208, 250, 524,
        528–29, 530–31, 638
    social effects of, 191–92, 532–36
    as status symbols, 64
    steam power for, 64
    see also gasoline; internal combustion engines
Avon, Anthony Eden, Lord, see Eden, Sir
    Anthony
Azerbaijan, 402, 764, 770
Azienda Generali Italiana Petroli (AGIP), 483,
    484, 512
Aziz, Tariq, 758

B-29 bombers, 343
Baba Gurgur, 188, 226
Babylon, 7–8
Baghdad, 157, 172, 185
Bahrain, 254, 264, 265–66, 273, 275, 277, 278,
    282, 375, 547
Bahrain Petroleum, 265, 266
Baibakov, Nikolai, 403
Baker, James A., III, 758
Bakhtiari, 129
Bakr, Ahmad Hasan al-, 691
Baku, 39, 68, 69, 121, 126, 403
    cholera epidemic in, 55
    as East-West trade center, 42
    Gulbenkian in, 170
    oil industry of, 41–47, 116–17, 166–67, 170,
        171, 764
    post-Revolutionary activity in, 221, 224
    pre-Revolutionary activity in, 113–15
    in World War I, 166–67
    in World War II, 317, 318, 319, 320, 323, 325
Baku-Tbilisi-Ceyhan pipeline, 764
Baldwin, Stanley, 177
Balfour, Arthur James, Lord, 159, 172, 173
Balikpapan, 333–36, 338, 344
Balogh, Thomas, Lord, 652
Bangkok, 52

banking industry, 714
Bank of America, 714
Bank of Scotland, 131
Banque Worms, 51
Baptist Church, 24, 32, 33
Barran, David, 562, 564, 602
Barton, Bruce, 193–94
Baruch, Bernard, 362
Basra, 157
Bataafsche Petroleum Maatschappij, 110
Ba'th Party, Ba'thism, 689–92, 693, 754
Batum, 44, 45, 47, 50, 52, 56, 113, 114, 225
Bayonne, N.J., 30
Bazargan, Mehdi, 438, 665, 682–83
Beaumont, Tex., 66, 67, 68, 69, 75, 76
Beaverbrook, William M. Aitken, Lord, 382–83,
    384
Bedford, Alfred, 162
Bedford, John Russell, Duke of, 426
Belgium, 295, 331, 475, 746
Bell, Alexander Graham, 86
Bellano, William, 559
Belridge, 210, 711
Bénard, André, 583
Ben-Gurion, David, 471
Benn, Anthony Wedgwood, 651–52
Berenger, Henry G., 160–61, 167, 168
Bergius, Friedrich, 312, 314
Berle, Milton, 531
Berlin, 63, 322, 327
Berlin Wall, 750
Bermuda Conference, 479–80
Betancourt, Romulo, 417–18, 419, 427, 492
Beverly Hillbillies, The, 536, 697
Bevin, Ernest, 406, 409, 435
Biafra, 537–38
Big Hill, see Spindletop
"Bilateral Deals: Everybody's Doing It," 611
bilateralism, 611
Birkenau, 329
Bissell, George, 3–6, 8, 9, 12, 18, 65
bitumen, 7–8
Black Giant oil field, 211, 229–31
Black Spot, 99
Black Town, 43
Blancard, Jean, 637
Blau, Operation, 319–22
"blind tigers," 81
blitzkrieg, 316, 318, 321–22, 331, 370
Blue Line Agreement, 282
Blumenthal, Michael, 627, 677–78
Bnito (Caspian and Black Sea Petroleum), 45, 50,
    52
Boer War, 101, 108
Bolivia, 257
Bolshevik Party, 114, 115, 166, 172, 220–23
Bombay, 48

Central Command, U.S., xvi
Central Intelligence Agency (CIA), 449, 452, 462, 497, 505, 584, 643, 644, 735
certificates of clearance, 238
Chalabi, Fadhil al-, 505, 736
Chamberlain, Austen, 124
Chamberlain, Neville, 294
Chandler, Alfred, 528
charcoal, as fuel, 348, 403
Charles, Prince of Wales, 546, 742–43
Chase Manhattan Bank, 559, 714
Chávez, Hugo, 768, 769
chemical industry, xviii
chemical weapons, 746, 752, 755, 757
Chernobyl, 761–62
Chevron, *see* Standard Oil of California
Chiah Surkh, 122, 124, 126
Chiang Kai-shek, 292
Chicago, University of, 33
Chicago Times Herald, 193
Chicago Tribune, 28, 37
China, Imperial, 9, 48, 54, 101
China, Nationalist, 289–93, 306
China, People's Republic of, 464, 661, 699, 735
    atmospheric carbon dioxide and, 772
    emerging economic role of, xvii, 764, 767, 770
    as energy consumer, 773
    oil demand from, 769, 771
    oil production in, 770
Chisholm, Archibald, 280, 498
Christian Democratic Party (Italy), 482, 483
Christian Science Monitor, 578
Chungking, 294
Churchill, Jennie Jerome, 137
Churchill, Randolph S., 137
Churchill, Sir Winston S., xv–xvi, xix, 151, 246, 254, 323, 346, 351, 382–83, 435, 440, 445, 468, 547, 680
    Faisal installed as King of Iraq by, 184–85, 490
    Ibn Saud's meeting with, 387
    political rise of, xv, 137
    as prime minister in 1950s, 446
    and proposed Anglo-Persian and Royal Dutch/Shell amalgamation, 176–77
    Royal Navy oil conversion and, xv–xvi, 138–40, 143–48, 157, 194
    tanks championed by, 154–55
    as wartime prime minister, 350, 355, 358, 364, 385
Citibank, 559
Cities Service, 255, 531, 711, 717
Civil War, U.S., 14, 19, 20, 21
Clark, Maurice, 19, 20, 21, 181
Clark, Sir Kenneth, 170
Clausewitz, Carl von, 575
Clean Air Act (1990), 762

Clemenceau, Georges, 161, 168, 173, 507
Cleveland, Ohio, 19, 20, 21, 30, 32
climate change, xix, 771–72
Club of Rome, 550, 597
coal, 109, 406, 726
    cannel, 7
    electricity generated from, 772
    environmental effects of, xviii–xix, 526, 572
    Europe's post–World War II reliance on, 404, 477, 526
    Germany's reliance on, 312, 526
    Japanese exports of, 51
    labor troubles and, 525
    reliance on oil vs., xvi, xviii, 71, 100, 525–28, 550, 611–12, 699
    restricted by Kyoto Protocol, 771
    synthetic fuels from, 146, 178, 292–93, 311–14, 316, 326, 340, 350–51, 410, 573, 676–77
    toluol extracted from, 158–59
    town gas distilled from, 7, 34, 63, 76
    as U.S. energy source, 312, 391, 518, 524–25, 573, 642, 645
    World War I shortage of, 162–63
"coal-oils," *see* kerosene
Coast Guard, U.S., 392
Cochrane, Thomas, 7
Cohen, Robert Waley, 110, 111, 133, 176
Cold War, 409, 535
    development of, 392, 393, 410, 434
    economics and, 501
    end of, xvi, xviii, 750, 764
    Iranian Crisis of 1946 and, 402–3
    October War and, 577, 579–80, 586–87, 588, 590–91, 593–94
Collier, Harry C., 380, 393, 394–95, 398
Colony Shale Oil Project, 698
Colorado, 65
combinations, *see* trusts
"Committee of 60," 671
Committee of Netherlands Indian Producers, 107, 108
Committee to Reelect the President (CREEP), 592
communism, 513
    collapse of, 750, 763–64
    in Middle East, 402–3, 408, 433, 436, 439, 576
    in Western Europe, 393, 398, 620
Compagnie Française des Petroles (CFP), 174, 181, 395, 396, 397, 398, 455, 457, 483, 485, 507
Compañía-Venezolana de Petróleo, 219
concentration camps, 328–29
Concession Syndicate, 126, 130
Conference on International Economic Cooperation, 618

dollar equivalency values:
  of 1850s, 5
  of 1901, 106
  of 1912, 97
Dolph, Robert, 631
Dome Petroleum, 711
Draft Memorandum of Principles, 249–50, 251
Drake, Edwin L. "Colonel," 10–13, 18, 65, 86,
    131, 650, 762
Drake, J. F., 401, 402, 430
Drake, Sir Eric, 444–45, 476, 605–6
drawbacks, 23, 25, 28
Drexel, Burnham, 718
drilling:
  offshore, 411, 551
  rotary, 67, 68
Driscoll, Alfred E., 534–35
Dubai, 767
Duce, James Terry, 375–76, 382, 384, 407
Dulles, Allen, 185, 449, 450, 497
Dulles, John Foster, 449, 450, 452, 453, 519–20
  Suez Crisis and, 464, 465–66, 472, 475
Dunlop, Robert, 519
DuPont, 711
Dutch East Indies, 180
  Japan and oil industry of, xvii, 291, 294–95,
    296, 300, 301, 302, 309, 317, 336
  Mineral Leasing Act (1920) and, 179
  Royal Dutch oil and, 56, 58–60, 99, 106, 108
dynamite, 42, 46, 233, 234, 373

Earth Day, 550
Eastern and General Syndicate, 263, 264–65, 277
East Indies, see Dutch East Indies
East Sumatra Tobacco, 57
Echeverría Alvarez, Luis, 648
Economist, 51, 52
Economy, Pa., 72
Eden, Sir Anthony (Lord Avon), xix, 302, 315,
    387, 610, 652
  Iran and, 439, 446–47, 448, 450, 455
  Suez Crisis and, 463, 465–66, 476, 477, 480
Edison, Thomas Alva, 62–63
Edison Illuminating, 64
Edmonton, 391
Edward VII, King of England, 135, 136
Egypt, 112, 171, 172, 322, 560, 678, 699, 732
  nationalism of, 506, 574, 678
  Saudi Arabia and, 506
  Six-Day War and, 536–37
  Soviet relations with, 575, 577, 579, 582, 586,
    593–94
  see also Suez Crisis
Ehrlichman, John, 573
Eisenhower, Dwight D., xix, 548, 735, 742
  on cross-country motor caravan, 190–91, 535
  elections of, 456, 473

final Allied push and, 367–70
  interstate highway program and, 535
  Normandy invasion and, 329
  on oil quotas, 518–20
  Suez Crisis and, 464, 465, 466, 467, 469, 470,
    472, 473–75, 479
Eisenhower Administration, 457, 463, 492
  oil quotas of, 494–95, 518–20, 528, 530, 540,
    549, 571, 735
Eizenstat, Stuart, 676–78
El Alamein, 322, 324, 325, 326, 462
Elam, 688–89
elections, U.S.:
  of 1904, 91
  of 1920, 195
  of 1924, 198, 204
  of 1952, 455, 456
  of 1956, 466, 472, 473
  of 1976, 657
electric automobiles, 64
electricity, 526, 572
  alternate sources for, 699
  brownouts of, 581
  coal-generated, 772
  cost of, 63
  Edison's development of, 63
  oil-generated, 772
Elektropolis, 63
Elf Aquitaine, Petrofina and Total merger with,
    765
Elf-ERAP (Enterprise de Recherches et
    d'Activités Pétrolières), 509, 634
Elizabeth II, Queen of England, 545, 597
Elliott, Lloyd "Shorty," 294
El Paso Natural Gas, 412
embargoes:
  in Iran Hostage Crisis, 684
  in Iran oil nationalization crisis, 444, 446, 485
  against Iraq (1990), xvi, 754
  of Iraq against Egypt (1979), 678
  in Suez Crisis, 473
  in World War II, 296–97, 300–304, 315, 338
  see also Arab oil embargo (1967); Arab oil
    embargo (1973)
Emmerglick, L. J., 457
energy crisis (1973), 571–74
Energy Department, U.S., 644
engines, see internal combustion engines
ENI (Ente Nazionale Idrocarburi), 484, 485, 486,
    487, 490, 511–12, 567
environmentalists, xviii–xix, 511, 541, 599
  impact on energy policy by, 550–51, 599
  resurgent international wave of, 761–62
  synthetic fuels and, 676, 677
  and Trans-Alaskan Pipeline, 554–56, 642
Esso gasoline, 529, 532
Esso-Libya, 560

ethanol, 773
Ethiopia, 315
Euclid Avenue Baptist Church, 24, 32, 33
Europe, 30, 102
  as competitive oil market, 528–30
  earliest use of oil in, 8
  gasoline prices in, 96
  government oil regulation in, 251–52
  growth of electricity in, 63
  kerosene exports from Russia to, 45
  oil consumption in, 523
  oil conversion of, 525–26
  oil exports from U.S. to, 14, 40–41, 73, 162, 241
  post-World War II energy crisis of, 404–7, 526
  Venezuelan oil imports of, 241
European Economic Community, 598, 602, 605–6, 609–10, 612, 768
European Petroleum Union (EPU), 116
European Recovery Program, see Marshall Plan
European Union, see European Economic Community
Europe of the Eighteenth Century (Getty), 422
Evian Agreement, 508
Export Petroleum Association, 248, 249
Exxon (Standard Oil of New Jersey), 89, 96, 206, 207, 209, 210, 265, 282, 365, 374, 380, 412, 422, 430, 464, 470, 485, 516–17, 527, 669–70
  acquisitions of, 723
  in agreements with other companies, 243–51
  in Alaska, 552–56
  Aramco and, 394–401; see also Aramco
  British subsidiary of, 351
  Colony Shale Oil Project of, 698–99
  Creole subsidiary of, 416, 418, 631
  Esso gasoline of, 529, 532
  40 percent of employees cut by, 723
  global strategy of, 183, 186
  as holding company, 82
  I. G. Farben partnership with, 313–14, 316, 416, 456
  in Inter-Allied Petroleum Conference, 161
  as largest Standard Oil successor, 94
  in Libya, 510
  in Mexico, 255, 257, 262
  Mobil Oil merger with, 765
  in National Petroleum War Service Committee, 162, 163
  in Netherlands, 650
  in North Sea, 651
  Open Door policy sought by, 179, 181
  posted price cut by, 502–4, 505
  profits of, 639–40
  share buy-back of, 723
  Soviet Union foray of, 221–26
  symbol of, 193, 638

  tankers sunk, 160
  in Venezuela, 217, 219–20, 416–19, 489, 631–32
  World War I production of, 180
  see also Humble Oil; Stanvac
ExxonMobil, 765

Fahd, King of Saudi Arabia, 623, 675, 701, 729, 737, 743
Faisal, King of Saudi Arabia, 496, 515
  Arab oil embargo (1973) and, 575–80, 588, 591, 596–97, 607, 612–13
  assassination of, 623, 624
  Suez Crisis and, 470
  Yamani and, 506, 564, 578–80, 589, 622–24, 743–44
Faisal I, King of Iraq, 184–85
Faisal II, King of Iraq, 490
Falkland Islands, Battle of, 156
Fall, Albert B., 195–98, 200, 235, 236
Farish, William, 205, 206, 234
Farouk I, King of Egypt, 387, 462
FBI (Federal Bureau of Investigation), 422
Federal Energy Administration, 642
Federal Oil Conservation Board, 204, 205
Federal Register, 642
Federal Reserve, U.S., 699, 712, 713, 714
Federal Trade Commission (FTC), 203, 209, 210, 454–55
Feis, Herbert, 378, 381
Fergusson, Sir Donald, 443–44, 447
fertilizer, xviii
Findlay, Ohio, 36
First Oil Shock, see Arab oil embargo (1973), oil prices affected by
Fischer-Tropsch process, 313
Fisher, John Arbuthnot, 134–35, 137, 151, 351
  background of, 135
  character of, 135
  Churchill advised by, 138, 139–41, 142, 143, 147, 157, 194
  D'Arcy and, 134–35
  as "God-father of oil," 135
  Royal Navy career of, 135
Fiume, 44
Flagler, Henry, 244
  character of, 22
  Florida developed by, 22–23, 88
  Rockefeller's relationship with, 22, 23, 32
  Standard Oil role of, 22, 23, 24, 26, 29, 30, 88, 91
Fleming, Ian, 477
Florida, 23, 88, 362
flush production, 15–16, 205, 206, 232, 239
Foch, Ferdinand, 155, 167
Follis, Gwin, 394, 398
force majeure, 669–70, 671

882

Gulbenkian, Rita, 189
Gulf & Western, 554
Gulf Crisis (1990–91), 753–60, 763
  air war in, 758–59
  anti-Iraq coalition in, 753, 754, 757, 759
  ground war in, 759
  Kuwait invaded in, xvi, 751, 752, 754, 756–57
  oil and, xvi–xvii, 751, 752, 754–56, 757, 759
  UN role in, xvi, 754, 757
  U.S. role in, 753–54, 756–58
Gulf Investors Group (GIG), 718, 721
Gulf Oil, 193, 208, 210, 249, 255, 381, 485, 552,
  631, 716–22
  in Bahrain and Saudi Arabia, 265–66, 277–80,
    282, 380, 429–30, 717
  Chevron's purchase of, 719–22, 723
  circus bid of, 647
  establishment of, 75, 717
  in Iranian consortium, 457
  in Japan, 527, 528
  in Kuwait, 380, 429–30, 628–30, 717
  in North Sea, 647
  Shell's deal with, 401–2, 404
  in South Korea, 528
  Texas Company's near-merger with, 78
  William Mellon on size of, 76
Gulf Refining, 73, 75
Gulf War, *see* Gulf Crisis
gunpowder, 8
gushers (fountains), 14, 45, 46, 68–69, 100, 131,
  170, 188, 229

Haig, Alexander, 591, 593, 600–601
Haile Selassie, Emperor of Ethiopia, 387
Halifax, Edward F. Lindley, Lord, 383
Hall, A. W., 64
Hamaguchi, Osachi, 290
Hamill, Al, 68, 69, 75–76
Hamilton, Lloyd, 272–73
Hammadi, Saadoun, 563
Hammer, Armand, xix, 556–59, 560–62, 569,
  650, 711, 722
Hammer, Julius, 556–57
Hammer, Victor, 557
Hankey, Sir Maurice, 172
Hannibal, HMS, 134
Harden, Orville, 252
Harding, Warren G., 180, 195–97, 551
Harding Administration, 196–97
Hardinge, Sir Arthur, 119, 120, 121, 122, 125, 147
Harriman, Averell, 441–44, 445
Hart, Basil Liddell, 154, 327, 370
Hartley, Fred, 722
Harvard University, 742–43
Hashemites, 408, 490
Heads of Agreement for Distribution, 249
Healey, Denis, 548

Heath, Edward, 480, 605–6, 609, 611
Henderson, Loy, 448, 459
Henderson Daily News, 229
Herrington, John, 737
Higgins, Patillo, 66–67, 68, 70, 75, 78
highway lobby, 535
Hindenburg, Paul von, 155
Hirohito, Emperor of Japan, 304–5, 307–8, 348
Hiroshima, atomic bombing of, 347
Hirota, Koki, 346
History of Standard Oil Company, The (Tarbell),
  89
Hit, bitumen source at, 7
Hitler, Adolf, xvii, xix, 294, 296, 301, 304,
  311–12, 329, 331, 351, 359, 369, 468–69,
  759
  death of, 332
  Four-Year Plan of, 316
  ideology and "Final Solution" of, 327–29
  oil as concern of, 311–12, 315–17, 319–22
  rise to power of, 312
  Rommel and, 322, 323, 325, 326
  Soviet Union invaded by, 317–22
Hogg, James, 77, 232
Holborn Viaduct Station, 63
holding companies, 82, 110
Holiday Inn, 533
Holman, Eugene, 257
Holmes, Frank, 629
  background of, 263–64
  Saudi and Kuwait oil industry role of, 263–65,
    266, 275, 277, 278–80, 283, 285
home heating oil, 392, 406, 524, 707
Homer, 8
Hong Kong, 51, 336
Hoover, Herbert, 180–81, 278, 279
Hoover, Herbert, Jr., 417, 452, 453, 472
Hoover, J. Edgar, 533
hostile takeovers, 710, 719
hot oil, 234, 238, 239, 240
Houdry, Eugene, 365
House, Edward, 163
House Armed Services Committee, 412
House of Representatives, U.S., 758
Houston, Tex., 78
Howard, Frank, 313
Howell, David, 673
Hubbard, Michael, 498
Hughes, Charles Evans, 181
Hughes, Howard, Sr., 78
Hugoton Production, 710, 711
Hull, Cordell, 261, 292, 294, 297–98, 301, 302–4,
  306, 308, 379
Humble Oil, 205, 206, 207, 233, 234, 476,
  552–54
  see also Exxon; Stanvac
Hungary, 471, 472

885

886

896

899

About the Author

DANIEL YERGIN is a leading authority on energy, international politics, and economics. He is chairman of Cambridge Energy Research Associates (CERA) and executive vice president of IHS, the parent company of CERA. He also serves as global energy expert for the CNBC business news network. Dr. Yergin received the Pulitzer Prize for *The Prize: The Epic Quest for Oil, Money & Power.* A number one bestseller, it was also made into an eight-hour PBS/BBC series seen by 20 million people in the United States. The book has been translated into seventeen languages and also received the Eccles Prize for best book on an economic subject for a general audience.

Of Dr. Yergin's subsequent book, *Commanding Heights: The Battle for the World Economy*, the *Wall Street Journal* said: "No one could ask for a better account of the world's political and economic destiny since World War II." It has been translated into thirteen languages. Dr. Yergin led the team that turned it into a six-hour PBS/BBC documentary—the major PBS television series on globalization. The series received three Emmy nominations, a CINE Golden Eagle Award, and the New York Festival's Gold World Medal for best documentary. Dr. Yergin's other books include *Shattered Peace,* an award-winning history of the origins of the Cold War, (coauthored) *Russia 2010 and What It Means for the World* and *Energy Future: The Report of the Energy Project at the Harvard Business School.*

Dr. Yergin received the United States Energy Award for "lifelong achievements in energy and the promotion of international understanding" and chaired the U.S. Department of Energy's Task Force on Strategic Energy Research and Development. He is a director of the United States Energy Association and the

U.S.-Russian Business Council, a member of the U.S. National Petroleum Council, a trustee of the Brookings Institution, and a director of the New America Foundation. Dr. Yergin received his B.A. from Yale University, where he co-founded *The New Journal*, and his Ph.D. from Cambridge University, where he was a Marshall Scholar.